PHYSIOLOGY

PHYSIOLOGY

Edited by

ROBERT M. BERNE, M.D., D.Sc.(Hon.)

Chairman and Charles Slaughter Professor of Physiology,
Department of Physiology,
University of Virginia School of Medicine,
Charlottesville, Virginia

MATTHEW N. LEVY, M.D.

Chief of Investigative Medicine, Mount Sinai Medical Center;
Professor of Physiology, Medicine, and Biomedical Engineering,
Case Western Reserve University,
Cleveland, Ohio

with 896 *illustrations*

The C.V. Mosby Company

St. Louis   Toronto   1983

MOSBY

A TRADITION OF PUBLISHING EXCELLENCE

Editor: Samuel E. Harshberger
Assistant editor: Anne Gunter
Manuscript editors: Marjorie L. Sanson, Connie Povilat, Susan K. Hume
Design: Nancy Steinmeyer
Production: Linda R. Stalnaker, Teresa Breckwoldt, Margaret B. Bridenbaugh,
Judith Bamert

Printed in the United States of America

The C.V. Mosby Company
11830 Westline Industrial Drive, St. Louis, Missouri 63146

Library of Congress Cataloging in Publication Data

Main entry under title:

Physiology.

   Bibliography: p.
   Includes index.
   1. Human physiology.   I.   Berne, Robert M.
II.   Levy, Matthew N.
QP34.5.P496   1983   612   82-22943
ISBN 0-8016-0644-6

C/VH/VH 9 8 7 6 5 4 3 2   01/B/089

Contributors

MURRAY D. ALTOSE, M.D.
Associate Professor of Medicine, Case Western Reserve University; Associate Director, Department of Medicine, and Chief, Pulmonary Division, Cleveland Metropolitan General Hospital, Cleveland, Ohio

ROBERT M. BERNE, M.D., D.Sc.(Hon.)
Chairman and Charles Slaughter Professor of Physiology, Department of Physiology, University of Virginia School of Medicine, Charlottesville, Virginia

NEIL S. CHERNIACK, M.D.
Professor of Medicine, Case Western Reserve University, and Chief, Pulmonary Division, University and Cleveland VA Hospitals, Cleveland, Ohio

DAVID H. COHEN, Ph.D.
Professor and Chairman, Department of Neurobiology and Behavior, State University of New York at Stony Brook, Stony Brook, New York

BRIAN R. DULING, Ph.D.
Professor, Department of Physiology, University of Virginia School of Medicine, Charlottesville, Virginia

SAUL M. GENUTH, M.D.
Professor of Medicine, Case Western Reserve University School of Medicine, Mount Sinai Medical Center, Cleveland, Ohio

STEVEN G. KELSEN, M.D.
Associate Professor of Medicine, Case Western Reserve University, Cleveland, Ohio

HOWARD C. KUTCHAI, Ph.D.
Professor of Physiology and Director of Program in Biophysics, University of Virginia School of Medicine, Charlottesville, Virginia

MATTHEW N. LEVY, M.D.
Chief of Investigative Medicine, Mount Sinai Medical Center; Professor of Physiology, Medicine, and Biomedical Engineering, Case Western Reserve University, Cleveland, Ohio

RICHARD A. MURPHY, Ph.D.
Professor, Department of Physiology, University of Virginia School of Medicine, Charlottesville, Virginia

OSCAR D. RATNOFF, M.D.
Professor of Medicine, Case Western Reserve University; Career Investigator of the American Heart Association; Co-Director, Division of Hematology/Oncology, Department of Medicine, University Hospitals of Cleveland, Cleveland, Ohio

S. MURRAY SHERMAN, Ph.D.
Professor, Department of Neurobiology and Behavior, State University of New York at Stony Brook, Stony Brook, New York

Preface

This textbook has been designed to emphasize broad concepts and to minimize the compilation of isolated facts. At the beginning of the book itself, as well as at the beginning of several of the sections, many of the important physicochemical principles of physiology are covered in considerable detail. When these principles could be represented profitably by equations, the bases of the equations and the major underlying assumptions have been discussed. This approach provides students with a satisfactory understanding of these basic principles, so that a mastery of certain topics will involve a minimum of pure memorization. This coverage of physicochemical principles is most evident in the cellular, cardiovascular, and respiratory sections.

In keeping with this emphasis on broad principles, homologies are grouped wherever possible. For example, in the endocrine section, discussions of the male and female gonads are included in the same chapter to highlight the similarities between the Sertoli cell functions in spermatogenesis and the granulosa cell functions in oogenesis.

Throughout the section on muscle physiology the three types of muscle are not described in sequence but are consistently considered together. It is emphasized that the basic mechanisms of contraction of skeletal, cardiac, and smooth muscles are very similar and that differences lie mainly in the relative importance of certain components of the underlying process.

In the renal physiology section, homologies again influence the presentation of material. The mechanisms whereby the kidneys handle a few important solutes are described in detail. The specific details of the transport of the myriad individual substances that pass through the kidneys are intentionally ignored.

The section on the nervous system reveals a functional neuroanatomical approach while still capturing the spirit of contemporary cellular neurophysiology. A major portion of the section is devoted to sensory and motor systems because of their relevance to clinical problems. The theoretical framework common to all sensory systems is constructed so as to facilitate the learning of the various components.

The section on hematology provides minimal coverage of blood composition. However, its emphasis on a complete and up-to-date picture of blood coagulation and its aberrations establishes a sound foundation for future clinicians.

To provide a clearer understanding of cardiovascular physiology, the entire system is initially dissected into its major components, and the functions of these components are examined in detail. The system is then reconstructed and considered as a whole, thus indicating how the various parts interact in physiological and pathophysiological states.

In brief, the framework of this textbook comprises firmly established facts and principles. Isolated phenomena are generally ignored unless they are considered to be highly significant, and few experimental methods are described unless they are essential for the comprehension of a specific topic. Although theoretical controversies exist in virtually all areas of physiology, such controversies are not described unless they provide a deeper understanding of the subject. Thus each author has described what he believes to be the most likely mechanism responsible for the phenomenon under consideration. We have decided to make this compromise to achieve brevity, clarity, and simplicity, while recognizing that future advances might prove many of our conjectures wrong.

In keeping with our philosophy of presenting broad principles and minimizing controversies and isolated facts, we have not documented most of the assertions made throughout the book. Only a few relevant references are given at the end of each chapter. These references have been selected because they provide a current and comprehensive review of the topic, they include a clear and detailed description of important mechanisms, or they provide a complete and up-to-date bibliography of the subject.

We wish to express our appreciation to our colleagues who generously provided constructive criticism during the preparation of this book. We also want to give special thanks to Frances S. Langley, who skillfully drew the illustrations for the entire volume.

Robert M. Berne
Matthew N. Levy

Contents

ix

SECTION VII

THE GASTROINTESTINAL SYSTEM

Howard C. Kutchai

SECTION VIII

THE KIDNEY

Brian R. Duling

SECTION IX

THE ENDOCRINE SYSTEM

Saul M. Genuth

CELLULAR PHYSIOLOGY

Howard C. Kutchai

chapter I

Cellular membranes and transmembrane transport of solutes and water

Each cell is surrounded by a plasma membrane that separates it from the extracellular milieu. The plasma membrane serves as a permeability barrier that allows the cell to maintain a cytoplasmic composition far different from the composition of the extracellular fluid. The plasma membrane contains enzymes, receptors, and antigens that play central roles in the interaction of the cell with other cells and with hormones and other regulatory agents in the extracellular fluid.

The membranes that enclose the various organelles divide the cell into discrete compartments and allow the localization of particular biochemical processes in specific organelles. Many vital cellular processes take place in or on the membranes of the organelles. Striking examples are the processes of electron transport and oxidative phosphorylation, which occur on, within, and across the mitochondrial inner membrane.

Most biological membranes have certain features in common. However, in keeping with the diversity of membrane functions there are substantial differences in membrane composition and structure from one cell to another and among the membranes of a single cell.

Proteins and phospholipids are the most abundant constituents of cellular membranes. A phospholipid molecule has a polar head group and two extremely nonpolar, hydrophobic fatty acyl chains. In an aqueous environment it is most energetically stable for phospholipids to form structures that allow the fatty acyl chains to be kept from contact with water. One such structure is the *lipid bilayer* (Fig. 1-1). Many phospholipids, when dispersed in water, spontaneously form lipid bilayer structures. Most of the phospholipid molecules in biological membranes have a lipid bilayer structure.

The proteins of biological membranes are associated with the membrane phospholipids in two major ways: (1) by charge interactions between the polar head groups of the phospholipids and acidic or basic amino acid residues of the protein, and (2) by hydrophobic interactions of the phospholipid acyl chains with hydrophobic amino acid residues of the proteins.

Fig. 1-2 depicts the "fluid mosaic" model of membrane structure. This model is consistent with many of the properties of biological membranes. Note the bilayer structure of most of the membrane phospholipids. The membrane proteins can be divided into two major classes: (1) *integral or intrinsic* membrane proteins that are embedded in the phospholipid bilayer, and (2) *peripheral or extrinsic* membrane proteins that are associated with the surface of the phospholipid bilayer. The peripheral membrane proteins interact with membrane lipids predominantly by charge interactions with the phos-

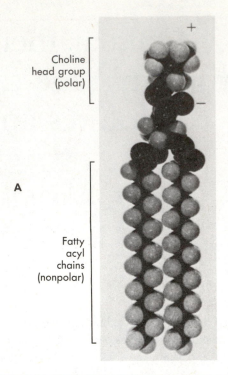

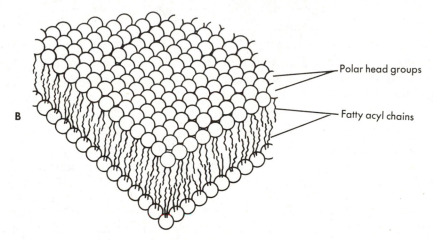

Fig. 1-1 ■ **A,** Structure of a membrane phospholipid molecule, in this case a phosphatidylcholine. **B,** Structure of a phospholipid bilayer. The open circles represent the polar head groups of the phospholipid molecules. The wavy lines represent the fatty acyl chains of the phospholipids.

pholipid polar head groups and thus sometimes may be removed from the membrane by altering the ionic composition of the medium. Integral membrane proteins have important hydrophobic interactions with the interior of the membrane. These hydrophobic interactions can be disrupted only by detergents that solubilize the integral proteins by forming their own hydrophobic interactions with nonpolar amino acid side chains.

In some cases membrane components clearly are *not* free to diffuse in the plane of the membrane. An example of this motional constraint is the sequestration of acetylcholine receptors (integral membrane proteins) at the motor endplate of skeletal muscle. At

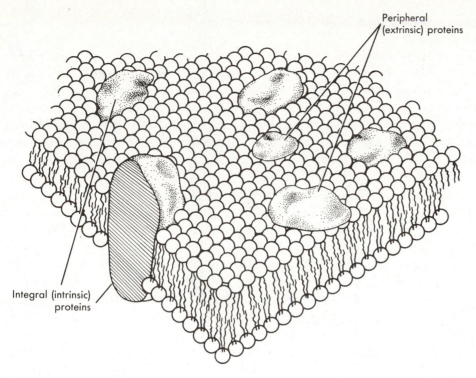

Fig. 1-2 ■ Schematic representation of the fluid mosaic model of membrane structure showing integral proteins embedded in the lipid bilayer matrix of the membrane and peripheral proteins associated with the polar head groups.

present little is known about the ways in which membrane constituents are restrained from lateral diffusion, but there is increasing evidence that the cytoskeleton may play a role in anchoring certain membrane proteins. Recent experiments suggest that specific proteins serve to link certain membrane proteins to the cytoskeleton.

Major phospholipids. In animal cell membranes the most abundant phospholipids are often the choline-containing phospholipids: the lecithins (phosphatidylcholines) and the sphingomyelins. Next in abundance are usually the amino phospholipids: phosphatidylserine and phosphatidylethanolamine. Other important phospholipids that are present in smaller amounts are phosphatidylglycerol, phosphatidylinositol, and cardiolipin.

Cholesterol. Cholesterol is a major constituent of animal cell plasma membranes. The steroid nucleus of cholesterol lies parallel to the fatty acyl chains of membrane phospholipids. Thus cholesterol alters the molecular packing of the membrane phospholipids. In natural membranes cholesterol tends to diminish the lateral mobility of the lipids and proteins of the membrane.

Glycolipids. Glycolipids are present in rather small quantities, but they have important functions. Glycolipids are present mostly in plasma membranes, where their carbohydrate moieties protrude from the external surface of the membrane. The blood group antigens and certain other antigens are the carbohydrate side chains of specific glycolipids or glycoproteins.

Asymmetry of lipid distribution in the bilayer. In many membranes the lipid components are not distributed uniformly across the bilayer. As just mentioned, the glycolipids of the plasma membrane are located exclusively in the outer monolayer and thus show absolute asymmetry. Asymmetry of phospholipids occurs but is not absolute. In the red blood cell membrane, for example, the outer monolayer contains most of the

■ ***Membrane composition***
■ *Lipid composition*

choline-containing phospholipids, whereas the inner monolayer is enriched in the amino phospholipids.

■ *Membrane proteins*

The protein composition of membranes may be simple or complex. The highly specialized membranes of the sarcoplasmic reticulum of skeletal muscle and the disks of the rod outer segment of the retina contain only a few different proteins. Plasma membranes perform many functions and may have more than 100 different protein constituents. Membrane proteins include enzymes (such as adenylate cyclase), transport proteins (such as the Na, K-ATPase), hormone receptors, receptors for neurotransmitters, and antigens.

Glycoproteins. Some membrane proteins are glycoproteins with covalently bound carbohydrate side chains. As with glycolipids, the carbohydrate chains of glycoproteins are located exclusively on the external surfaces of plasma membranes. Cell surface carbohydrate has important functions. The negative surface charge of cells is almost entirely due to the negatively charged sialic acid of glycolipids and glycoproteins. Receptors for viruses may involve surface carbohydrate. Certain surface antigenic determinants reside in carbohydrate moieties on the cell surface. Surface carbohydrate has been implicated in cellular aggregation phenomena and other forms of cell-cell interactions.

Asymmetry of membrane proteins. The absolute asymmetry of glycolipids and glycoproteins is mentioned earlier. The Na, K-ATPase of the plasma membrane and the Ca^{++} pump protein (Ca^{++}-ATPase) of the sarcoplasmic reticulum membrane are other examples of the asymmetrical functions of membrane proteins. In both cases ATP is split on the cytoplasmic face of the membrane, and some of the energy that is liberated is used to pump ions in specific directions across the membrane. In the case of the Na, K-ATPase, K^+ is pumped into the cell, and Na^+ is pumped out, whereas the Ca^{++}-ATPase actively pumps Ca^{++} into the sarcoplasmic reticulum.

It appears that most, if not all, integral membrane proteins are inserted into the membrane lipid bilayer during protein synthesis. The configuration of the protein in the membrane is the result of this specific insertion and the tertiary structure the protein assumes.

■ *Membranes as permeability barriers*

Biological membranes serve as permeability barriers. Most of the molecules present in living systems have high solubility in water and low solubility in nonpolar solvents. Such molecules have low solubility in the nonpolar environment in the interior of the lipid bilayer of biological membranes. As a consequence, biological membranes pose a formidable permeability barrier to most water-soluble molecules. The plasma membrane is a permeability barrier between the cytoplasm and the extracellular fluid. This permeability barrier allows the maintenance of cytoplasmic concentrations of many substances that differ greatly from their concentrations in the extracellular fluid. The localization of various cellular processes in certain organelles depends on the barrier properties of cellular membranes. For example, the inner mitochondrial membrane is impermeable to the enzymes of the tricarboxylic acid cycle, allowing the localization of these enzymes in the mitochondrial matrix. The spatial organization of chemical and physical processes in the cell depends on the barrier functions of cellular membranes.

The passage of important molecules across membranes at controlled rates plays a central role in the life of the cell. Examples are the uptake of nutrient molecules, the discharge of waste products, and the release of secreted molecules.

In some cases molecules move from one side of a membrane to another without actually moving through the membrane itself. Endocytosis and exocytosis are examples of processes that transfer molecules across, but not through, biological membranes. In other cases molecules cross a particular membrane by actually moving *through* the membrane by passing through or between the molecules that make up the membrane.

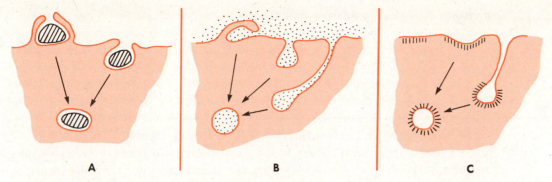

Fig. 1-3 ■ Schematic depiction of endocytotic processes. **A,** Phagocytosis of a solid particle. **B,** Pinocytosis of extracellular fluid. **C,** Receptor-mediated endocytosis by coated pits. (Redrawn from Silverstein, S.C., et al.: Ann. Rev. Biochem. **46:**669, 1977. Reproduced, with permission, from the Annual Review of Biochemistry. © 1977 by Annual Reviews Inc.)

Endocytosis. Endocytosis allows material to enter the cell without passing through the membrane (Fig. 1-3). When particulate material is taken up, the process is termed *phagocytosis* (Fig. 1-3, *A*). When soluble molecules are taken up, it is called *pinocytosis* (Fig. 1-3, *B*). Sometimes special regions of the plasma membrane, whose cytoplasmic surface is covered with bristles made primarily of a protein called *clathrin,* are involved in endocytosis. These bristle-covered regions are called *coated pits,* and their endocytosis gives rise to *coated vesicles* (Fig. 1-3, *C*). The coated pits appear to be involved primarily in *receptor-mediated endocytosis.* Specific proteins to be taken up are recognized and bound by specific membrane receptor proteins in the coated pits. The binding often leads to aggregation of receptor-ligand complexes, and the binding appears to trigger endocytosis in ways that are not yet understood. Endocytosis is an active process that requires metabolic energy. Endocytosis also can occur in regions of the plasma membrane that do not contain coated pits.

Exocytosis. Molecules can be ejected from cells by exocytosis, a process that resembles endocytosis in reverse. The release of neurotransmitters, which is considered in more detail in Chapter 4, takes place by exocytosis. Exocytosis is responsible for the release of secretory proteins by many cells; the release of pancreatic zymogens from the acinar cells of the pancreas is a well-studied example. In such cases the proteins to be secreted are stored in secretory vesicles in the cytoplasm. A stimulus to secrete causes fusion of the secretory vesicles with the plasma membrane and release of the vesicle contents by exocytosis.

Fusion of membrane vesicles. The contents of one type of organelle can be transferred to another organelle by fusion of the membranes of the organelles. In some cells secretory products are transferred from the endoplasmic reticulum to the Golgi apparatus by fusion of endoplasmic reticulum vesicles containing the secretory protein with membranous sacs of the Golgi apparatus. Membrane fusion also occurs between phagocytic vesicles and lysosomes and allows intracellular digestion of phagocytosed material to proceed.

■ *Transport across, but not through, membranes*

The traffic of molecules through biological membranes is vital for most cellular processes. Some molecules move through biological membranes simply by diffusing among the molecules that make up the membrane, whereas the passage of other molecules involves the mediation of *specific transport proteins* in the membrane.

Oxygen, for example, is a small molecule with fair solubility in nonpolar solvents.

■ *Transport of molecules through biological membranes*

It crosses biological membranes by diffusing among membrane lipid molecules. Glucose, on the other hand, is a much larger molecule with low solubility in the membrane lipids. Glucose enters cells via a specific glucose transport protein in the plasma membrane.

First to be considered is diffusion and osmosis through membranes, and next is the passage of molecules through biological membranes via specific transport proteins.

◼ Diffusion

Diffusion is the process whereby atoms or molecules intermingle because of their random thermal (Brownian) motion. Imagine a container divided into two compartments by a removable partition. A much larger number of molecules of a compound is placed on side A than on side B, and then the partition is removed. Every molecule is in random thermal motion. It is equally probable that a molecule which begins on side A will move to side B in a given time as it is that a molecule beginning on side B will end up on side A. Since there are many more molecules present on side A, the total *number* of molecules moving from side A to side B will be greater than the number moving from side B to side A. In this way the number of molecules on side A will decrease, while the number of molecules on side B will increase. This process of *net diffusion* of molecules will continue until the number of molecules on side A equals that on side B. Thereafter the rate of diffusion of molecules from A to B will equal that from B to A, so no further net movement will occur. A state of dynamic *equilibrium* exists when concentrations on side A and side B are equal.

Diffusion leads to a state in which the concentration of the diffusing species is constant in space and time. Thus diffusion across cellular membranes tends to equalize the concentrations on the two sides of the membrane. The rules that govern diffusion processes were written in the last century by Adolf Fick, a German physician, physiologist, and physicist. Fick realized that the diffusion rate across a particular planar surface must be proportional to the area of the plane and to the difference in concentration of the diffusing substance on the two sides of the plane. *Fick's first law of diffusion* is

$$J = -DA \frac{dc}{dx} \tag{1}$$

where

J	= net rate of diffusion in moles or grams per unit of time
A	= area of the plane
dc/dx	= concentration gradient across the plane
D	= constant of proportionality called *diffusion coefficient*

The nature of the concentration gradient dc/dx in Fick's first law requires further explanation. The concentration profile of the diffusing substance may take many forms. In Fig. 1-4 two possible forms of the concentration profile are shown, and the locations of planes across which the rates of diffusion are to be determined are represented by x_1 and x_2. In Fig. 1-4, *A,* the concentration profile is a straight line. In this case the value of the concentration gradient is simply the slope of the line $\Delta c / \Delta x$, and the rate of flux is the same at both planes, that is:

$$J = -DA \frac{dc}{dx} = -DA \frac{\Delta c}{\Delta x} \tag{2}$$

In Fig. 1-4, *B,* however, the concentration profile is nonlinear. At each plane, x_1 and x_2, the value of dc/dx is equal to the slope of the tangent to the curve at that point. Since the slopes of those tangents are different, the rates of diffusion across planes placed at x_1 and x_2 will be different at this time.

Note also that the equation for Fick's first law contains a minus sign. The minus sign indicates the *direction* of diffusion. In Fig. 1-4 the slope of the concentration profile

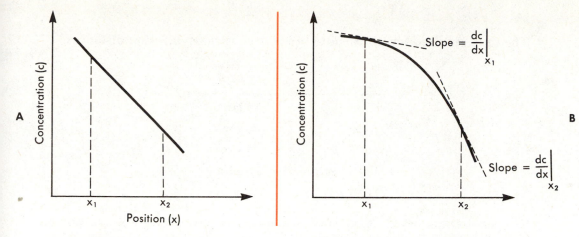

Fig. 1-4 ■ **A,** Linear concentration profile. **B,** Nonlinear concentration profile. The concentration gradient at any point is the slope of the concentration profile at that point.

is *negative,* but the direction of diffusion is in the *positive* x direction. The need for the minus sign in the equation arises because molecules flow down a concentration gradient, that is, from higher to lower concentration.

The diffusion coefficient. The diffusion coefficient, D, has units of square centimeters per second. Solving equation 1 for D and expressing all the variables on the right-hand side of the resulting equation in cgs units, we obtain

$$D = \frac{J}{A \dfrac{dc}{dx}} = \frac{\text{moles/second}}{\text{cm}^2 \dfrac{\text{moles/cm}^3}{\text{cm}}} = \text{cm}^2/\text{second}$$

D can be thought of as proportional to the speed with which the diffusing molecule can move in the surrounding medium. D is smaller the larger the molecule and the more viscous the medium.

For spherical solute molecules that are much larger than the surrounding solvent molecules Albert Einstein obtained the following equation:

$$D = kT/ (6\pi r\eta) \tag{3}$$

where

k = Boltzmann's constant
T = absolute temperature (kT is proportional to the average
    kinetic energy of a solute molecule)
r = molecular radius
η = viscosity of the medium

The equation is called the Stokes-Einstein relation, and the molecular radius defined by this equation is known as the Stokes-Einstein radius.

For large molecules equation 3 predicts that D will be inversely proportional to the radius of the diffusing molecule. Since the molecular weight (MW) is approximately proportional to r^3, D should be inversely proportional to $(\text{MW})^{1/3}$, so that a molecule which is ⅛ the mass of another molecule will have a diffusion coefficient only twice as large as the other molecule. For smaller solutes, with a molecular weight less than about 300, D is inversely proportional to $(\text{MW})^{1/2}$ rather than $(\text{MW})^{1/3}$.

Diffusion is a rapid process when the distance over which it must take place is small. This can be appreciated from another relation derived by Einstein. He considered the random movements of molecules that are originally located at x = 0. Since a given

TABLE 1-1

The time required for diffusion to occur over various diffusion distances*

Diffusion distance (μm)	Time required for diffusion
1	0.5 msec
10	50 msec
100	5 seconds
1000 (1 mm)	8.3 minutes
10,000 (1 cm)	14 hours

*The time required for the ''average'' molecule (with diffusion coefficient taken to be 1×10^{-5} cm^2/second) to diffuse the required distance was computed from the Einstein relation.

molecule is equally likely to diffuse in the $+x$ or $-x$ direction, the *average* displacement of all the molecules that begin at $x = 0$ will be zero. The average displacement squared, $\overline{(\Delta x)^2}$, which is a positive quantity, is represented by

$$\overline{(\Delta x)^2} = 2\,Dt \tag{4}$$

where t is the time elapsed since the molecules started diffusing. The *Einstein relation* (equation 4) tells us how far the average molecule will diffuse in time (t), and it is useful as a rough estimate of the time scale of a particular diffusion process.

Einstein's relation shows us that the time required for a diffusion process increases with the square of the distance over which diffusion occurs. Thus a tenfold increase in the diffusion distance means that the diffusion process will require about 100 times longer to reach a given degree of completion. Table 1-1 shows the results of calculations using Einstein's relation for a typical, small, water-soluble solute. It can be seen that diffusion is an extremely rapid process on a microscopic scale of distance. For macroscopic distances diffusion is a rather slow process. A cell that is 100 μm away from the nearest capillary can receive nutrients from the blood by diffusion with a time lag of only 5 seconds or so. This is sufficiently fast to satisfy the metabolic demands of many cells. However, a nerve axon that is 1 cm long cannot rely on diffusion for the intracellular transport of vital metabolites, since the 14 hours it would take for diffusion over the 1 cm distance is too long on the time scale of cellular metabolism. Some nerve fibers are as long as 1 m; it is no wonder then that intracellular axonal transport systems are involved in transporting important molecules along nerve fibers. Because of the slowness of diffusion over visible distances, it is not surprising that even rather small multicellular organisms have evolved circulatory systems to bring the individual cells of the organism within reasonable diffusion range of nutrients.

■ *Diffusive permeability of cellular membranes*

Permeability of lipid-soluble molecules. The plasma membrane serves as a diffusion barrier enabling the cell to maintain cytoplasmic concentrations of many substances that differ markedly from their extracellular concentrations. As early as the turn of the century, the relative impermeability of the plasma membrane to most water-soluble substances was attributed to its ''lipoid nature.''

The hypothesis that the plasma membrane has a lipoid character is supported by experiments showing that compounds which are soluble in nonpolar solvents (such as ether or olive oil) enter cells more readily than do water-soluble substances of similar molecular weight. Fig. 1-5 shows the relationship between membrane permeability and lipid solvent solubility for a number of different solutes. The ratio of the solubility of

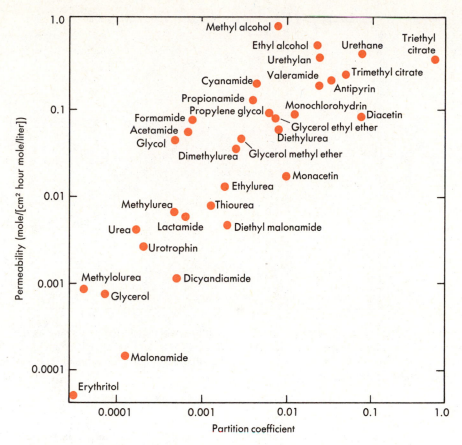

Fig. 1-5 ■ The permeability of the plasma membrane of the alga *Chara ceratophylla* to various nonelectrolytes as a function of the lipid solubility of the solutes. Lipid solubility is represented on the abscissa by the olive oil/water partition coefficient. (Redrawn from Christensen, H.N.: Biological transport, ed. 2, Menlo Park, Calif., 1975, W.A. Benjamin. Data from Collander, R.: Trans. Faraday Soc. **33**:985, 1937.)

the solute in olive oil to its solubility in water is used as a measure of solubility in nonpolar solvents. This ratio is called the olive oil/water partition coefficient. The permeability of the plasma membrane to a particular substance increases with the "lipid solubility" of the substance. For compounds with the same olive oil/water partition coefficient there is decreasing permeability with increasing molecular weight. As described previously, the fluid mosaic model of membrane structure envisions the plasma membrane as a lipid bilayer with proteins embedded in it. The data of Fig. 1-5 support the idea that the lipid bilayer is the principal barrier to substances which permeate the membrane by simple diffusion.

The positive correlation between lipid solvent solubility and membrane permeability suggests that lipid-soluble molecules can dissolve in the plasma membrane and diffuse across it. Consider a substance that dissolves in the lipid bilayer and then diffuses across the plasma membrane to the other side (Fig. 1-6). If the substance equilibrates with the lipid bilayer, its concentration at the outer face of the bilayer, $C_m(o)$, will be βC_o: the partition coefficient (β) times the concentration in the extracellular medium. Its concentration at the inner face of the membrane, $C_m(i)$, will be βC_i: β times the concentration in the cytoplasm. The concentration difference within the membrane itself is not $C_o - C_i$, but $C_m(o) - C_m(i)$. Since $C_m(o) - C_m(i) = \beta [C_o - C_i]$, the concentration gradient relevant to diffusion across the membrane is

$$\frac{\Delta C_m}{\Delta x} = \beta \frac{C_o - C_i}{\Delta x}$$

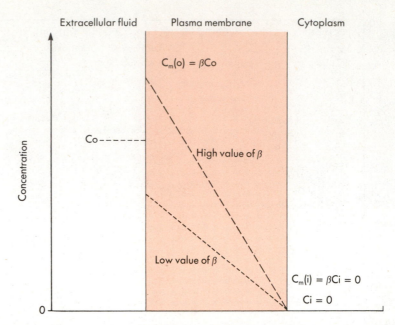

Fig. 1-6 ■ Influence of the membrane lipid/water partition coefficient (β) of a solute on the concentration profile of the solute within the membrane. The cytoplasmic concentrations of the solutes are assumed to be zero for the sake of illustration. The higher the value of β, the steeper the intramembrane diffusion gradient.

The more lipid soluble the substance, the larger is β, and the larger is the effective concentration gradient within the membrane that causes it to diffuse. According to Fick's first law the flux of the substance across the membrane is

$$J = -DA \frac{\Delta C_m}{\Delta x} = -DA\beta \frac{C_o - C_i}{\Delta x} = -\frac{D\beta}{\Delta x} A (C_o - C_i)$$

$^{D\beta}/_{\Delta x}$ is a constant for a particular substance and a particular membrane, and it is called the *permeability coefficient* (k_p). In terms of the permeability coefficient:

$$J = -k_p A(C_o - C_i)$$

Since $k_p = \dfrac{D\beta}{\Delta x}$, β is dimensionless, and the units of D are expressed in square centimeters per second, the units of k_p are centimeters per second.

Permeability of water-soluble molecules. Very small, uncharged, water-soluble molecules pass cell membranes much more rapidly than predicted by their lipid solubility. For example, water permeates the cell membrane much more readily than do larger molecules with a similar olive oil/water partition coefficient. The reason for the unusually high permeability of water is not understood. Some investigators believe that cell membranes contain small pores which allow water and small water-soluble molecules to pass. Others believe that very small water-soluble molecules can pass between adjacent phospholipid molecules without actually *dissolving* in the region occupied by the fatty acid side chains. Some experiments suggest that membrane proteins are responsible for the high membrane permeability of water.

As the size of uncharged, water-soluble molecules increases, their membrane permeability decreases. Most plasma membranes are essentially impermeable to water-soluble molecules whose molecular weights are greater than about 200.

Because of their charge, ions are relatively insoluble in lipid solvents and thus have low permeability to cell membranes. Most ionic diffusion that occurs across membranes is believed to occur through protein "channels" that cross the membrane. Some chan-

*Cellular membranes and
transmembrane transport
of solutes and water* **13**

nels are highly specific with respect to the ions allowed to pass, whereas others allow all ions below a certain size to pass.

Certain water-soluble molecules such as sugars and amino acids, which are essential for cellular survival, do not cross plasma membranes at appreciable rates by simple diffusion. Plasma membranes have specific proteins that allow the transfer of vital metabolites into or out of the cell. The characteristics of membrane protein-mediated transport are discussed later.

Definitions. *Osmosis* is the flow of water across a semipermeable membrane from a compartment in which the *solute* concentration is lower to one in which the *solute* concentration is greater. A *semipermeable membrane* is defined as a membrane permeable to water but impermeable to solutes. Osmosis takes place because the presence of solute results in a decrease of the *chemical potential* of water, and water tends to flow from where its chemical potential is higher to where its chemical potential is lower. Other effects caused by the decrease of the chemical potential of water (because of the presence of solute) include reduced vapor pressure, lower freezing point, and higher boiling point of the solution as compared with pure water. Because these properties, and osmotic pressure as well, depend on the *concentration* of the solute present rather than on its chemical properties, they are called *colligative properties*.

■ *Osmosis*

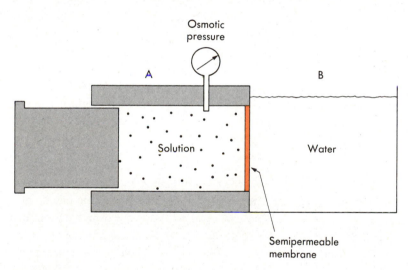

Fig. 1-7 ■ Schematic representation of the definition of osmotic pressure. When the pressure applied to the solution in chamber *A* is equal to the osmotic pressure of that solution, there will be no net water flow across the membrane.

Osmotic pressure. In Fig. 1-7 a semipermeable membrane separates a solution from pure water. Water flow from side B to side A by osmosis occurs because the presence of solute on side A reduces the chemical potential of water in the solution. Pushing on the piston will increase the chemical potential of the water in the solution of side A and slow down the net rate of osmosis. If the force on the piston is increased gradually, a point is reached eventually at which net water flow stops. Application of still more pressure will cause water to flow in the opposite direction, from side A to side B. The pressure on side A that is just sufficient to keep pure water from entering is called the *osmotic pressure* of the solution on side A.

The osmotic pressure of a solution depends on the number of particles in solution. Thus the degree of ionization of the solute must be taken into account. A 1M solution of glucose, a 0.5M solution of NaCl, and a 0.333 . . . M solution of $CaCl_2$ theoretically should have the same osmotic pressure. (Actually their osmotic pressures will differ due to the deviations of real solutions from ideal solution theory.) Important equations that pertain to osmotic pressure and the other colligative properties were derived by the Dutch chemist van't Hoff in the last century. One form of *van't Hoff's law* for calculation of osmotic pressure is

$$\pi = iRTm \tag{5}$$

where

π = osmotic pressure
i = number of ions formed by dissociation of a solute molecule
R = ideal gas constant
T = absolute temperature
m = molal concentration of solute

This equation applies more exactly as the solution becomes more dilute.

In the biological sciences molar concentrations are used more frequently than molal concentrations, and van't Hoff's law is approximated by

$$\pi = iRTc \tag{6}$$

where c is the molar concentration of solute (moles of solute per liter of solution). The discrepancy between theory and reality is a bit larger for this form of van't Hoff's law than for equation 5.

Equations 5 and 6 do not predict precisely the osmotic pressures of real solutions. At the concentrations of many substances in cytoplasm and extracellular fluids the deviations from ideality may be substantial. For example, sodium is the principal cation of the extracellular fluids, and chloride is the main anion. Na^+ is present at about 150 mEq/L and Cl^- at about 120 mEq/L. NaCl solutions in this concentration range differ considerably in their osmotic pressures from the pedictions of van't Hoss's law.

One way of correcting for the deviations of real solutions from the predictions of van't Hoff's law is to use a correction factor called the *osmotic coefficient* (ϕ). Including the osmotic coefficient, equation 6 becomes

$$\pi = \phi iRTc \tag{7}$$

The osmotic coefficient may be greater or less than one. It is less than one for electrolytes of physiological importance, and for all solutes it approaches one as the solution becomes more and more dilute. ϕic can be regarded as the osmotically effective concentration, and ϕic often is referred to as the *osmolar concentration*, with units in osmoles per liter. Values of the osmotic coefficient depend on the concentration of the solute and on its chemical properties. Table 1-2 lists osmotic coefficients for several solutes. These values apply fairly well at the concentrations of these solutes in the extracellular fluids of mammals. Solutions of proteins deviate greatly from van't Hoff's law, and different proteins may deviate to different extents. The deviations from ideality are frequently more concentration dependent for proteins than for smaller solutes.

Sample calculations

1. What is the osmotic pressure (at 0° C) of a 154mM NaCl solution?

$$\pi = \phi iRTc$$

Taking ϕ for NaCl from Table 1-2, we obtain

$$\pi = 0.93 \times 2 \times 22.4 \text{ liter-atm/mole} \times 0.154 \text{ mole/liter} = 6.42 \text{ atm}$$

2. What is the osmolarity of this solution?

$$\text{Osmolarity} = \phi ic = 0.93 \times 2 \times 0.154 = 0.286 \text{ osmole/liter} = 286 \text{ mOsm}$$

Measurement of osmotic pressure. The osmotic pressure of a solution can be obtained by determining the pressure required to prevent water from entering the solution across a semipermeable membrane (Fig. 1-7). However, this method is time consuming

Osmotic coefficients (φ) of certain solutes of physiological interest

Substance	i	Molecular weight	φ
NaCl	2	58.5	0.93
KCl	2	74.6	0.92
HCl	2	36.6	0.95
NH_4Cl	2	53.5	0.92
$NaHCO_3$	2	84.0	0.96
$NaNO_3$	2	85.0	0.90
KSCN	2	97.2	0.91
KH_2PO_4	2	136.0	0.87
$CaCl_2$	3	111.0	0.86
$MgCl_2$	3	95.2	0.89
Na_2SO_4	3	142.0	0.74
K_2SO_4	3	174.0	0.74
$MgSO_4$	2	120.0	0.58
Glucose	1	180.0	1.01
Sucrose	1	342.0	1.02
Maltose	1	342.0	1.01
Lactose	1	342.0	1.01

and technically difficult. Consequently the osmotic pressure more often is estimated from another colligative property, such as depression of the freezing point. The relation that describes the depression of the freezing point of water by a solute is

$$\Delta T_f = 1.86 \ \phi i c \qquad (8)$$

where ΔT_f is the freezing point depression in degrees centigrade. Thus the effective osmotic concentration (in osmoles per liter) is

$$\phi i c = \frac{\Delta T_f}{1.86} \qquad (9)$$

When the freezing point depression of a multicomponent solution is determined, the effective osmolar concentration $\phi i c$ (in osmoles per liter) of the solution as a whole can be obtained.

If the total osmotic pressures of two solutions (as measured by freezing point depression or by the osmotic pressure developed across a true semipermeable membrane) are equal, the solutions are said to be *isosmotic* (or iso-osmotic). If solution A has greater osmotic pressure than solution B, A is said to be *hyperosmotic* with respect to B. If solution A has less total osmotic pressure than solution B, A is said to be *hypoosmotic* to B.

The plasma membranes of most of the cells of the body are relatively impermeable to many of the solutes of the interstitial fluid but are highly permeable to water. There-

■ *Osmotic swelling and shrinking of cells*

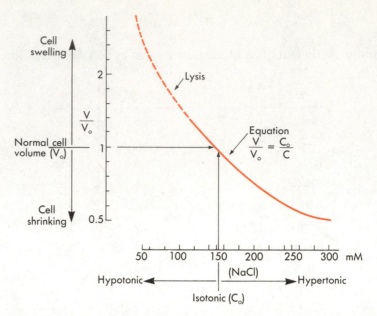

Fig. 1-8 ■ The osmotic behavior of human red blood cells in NaCl solutions. At 154mM NaCl (isotonic) the red cell has its normal volume. It shrinks in more concentrated (hypertonic) solutions and swells in more dilute (hypotonic) solutions.

fore, when the osmotic pressure of the interstitial fluid is increased, water leaves the cells by osmosis, the cells shrink, and cellular solutes become more concentrated until the effective osmotic pressure of the cytoplasm is again equal to that of the interstitial fluid. Conversely, if the osmotic pressure of the extracellular fluid is decreased, water enters the cells, and the cells will swell until the intracellular and extracellular osmotic pressures are equal.

Red blood cells often are used to illustrate the osmotic properties of cells because they are readily obtained and are easily studied. Within a certain range of external solute concentrations the red cell behaves as a "perfect osmometer," since its volume is inversely related to the solute concentration in the extracellular medium. In Fig. 1-8 the red cell volume, as a fraction of its normal volume in plasma, is shown as a function of the concentration of NaCl solution in which the red cells are suspended. At a NaCl concentration of 154mM (308mM particles) the volume of the cells is the same as their volume in plasma; this concentration of NaCl is called *isotonic* to the red cell. A concentration of NaCl greater than 154mM is called *hypertonic* (cells shrink), and a solution less concentrated than 154mM is termed *hypotonic* (cells swell). When red cells have swollen to about 1.4 times their original volume, some cells lyse (burst). At this volume the properties of the red cell membrane abruptly change, and hemoglobin leaks out of the cell. The membrane becomes transiently permeable to other large molecules at this point as well.

The intracellular substances that produce an osmotic pressure which just balances that of the extracellular fluid include hemoglobin, K^+, organic phosphates (such as ATP and 2,3-diphosphoglycerate), and glycolytic intermediates. Regardless of the chemical nature of its contents, the red cell behaves as though it were filled with a solution of *impermeant* molecules with an osmotically effective concentration of 286 milliosmolar (154mM NaCl = 286 milliosmolar).

Osmotic effects of permeant solutes. Permeating solutes eventually equilibrate across the plasma membrane. For this reason permeating solutes exert only a transient effect on cell volume.

Consider a red blood cell placed in a large volume of 0.154M NaCl, containing 0.050M glycerol. Initially, because of the extracellular NaCl and glycerol, the osmotic

TABLE 1-3

Osmotic water flow across a porous dialysis membrane caused by various solutes*

Gradient producing the water flow	Net volume flow (µl/minute)*	Solute radius (Å)	Reflection coefficient (σ)
D$_2$O	0.06	1.9	0.0024
Urea	0.6	2.7	0.024
Glucose	5.1	4.4	0.205
Sucrose	9.2	5.3	0.368
Raffinose	11	6.1	0.440
Inulin	19	12	0.760
Bovine serum albumin	25.5	37	1.02
Hydrostatic pressure	25		

Data from Durbin, R.P.: J. Gen. Physiol. **44**:315, 1960. Reproduced from The Journal of General Physiology by copyright permission of The Rockefeller University Press.

*Flow is expressed as microliters per minute caused by a 1M concentration difference of solute across the membrane. The flows are compared with the flow caused by a theoretically equivalent hydrostatic pressure.

pressure of the extracellular fluid will exceed that of the cell interior, and the cell will shrink. With time, however, glycerol will equilibrate across the plasma membrane of the red cell, and the cell will swell back toward its original volume. The steady-state volume of the cell will be determined only by the *impermeants* in the extracellular fluid. In this case the impermeants (NaCl) have a total concentration that is isotonic, so that the final volume of the cell will be equal to the normal red cell volume. Because the red cell ultimately returns to its normal volume, the solution (0.050M glycerol in 0.154M NaCl) is *isotonic*. Since the red cell initially shrinks when put in this solution, the solution is *hyperosmotic* with respect to the normal red cell. The transient changes in cell volume depend on equilibration of glycerol across the membrane. Had we used urea (a more rapidly permeating substance), the cell would have reached steady-state volume sooner.

The following rules help predict the volume changes a cell will experience when suspended in solutions of permeant and impermeant solutes:

1. The steady-state volume of the cell is determined by the concentration of impermeant particles in the extracellular fluid.
2. Permeant particles cause only transient changes in cell volume.
3. The time course of the transient changes is more rapid the greater the permeability of the permeant molecule.

The magnitudes of osmotic flows caused by permeating solutes. In the preceding example it was explained that permeants, such as glycerol, exert only a transient osmotic effect. It is sometimes important to determine the magnitude of the osmotic effect exerted by a particular permeant.

The flow of water across a membrane is directly proportional to the osmotic pressure difference ($\Delta\pi$) of the solutions on the two sides of the membrane; thus

$$\dot{V}_w = L\Delta\pi \tag{10}$$

where $\dot{V}_w$ is water flow, and L is a proportionality constant called the *hydraulic conductivity*. Equation 10 holds only for osmosis caused by impermeants; permeants cause less osmotic flow. The greater the permeability of a solute, the less the osmotic flow it causes. Table 1-3 shows the osmotic water flows induced across a porous membrane by solutes of different molecular size. The solutions have identical freezing points, so their total osmotic pressures are the same. Note that the larger the solute molecule, and thus the more impermeable it is to the membrane, the greater the osmotic water flow it causes.

Equation 10 can be rewritten to take solute permeability into account by including σ, the *reflection coefficient*.

$$\dot{V}_w = \sigma L \Delta \pi \qquad (11)$$

σ is a dimensionless number that ranges from 1 for completely impermeable solutes down to 0 for extremely permeable solutes. σ is a property of a particular solute and a particular membrane and represents the osmotic flow induced by the solute as a fraction of the theoretical maximum osmotic flow (Table 1-3). In kidney physiology the reflection coefficients of the renal tubular epithelium for certain solutes help explain the effect of these solutes on water movement across the renal tubule.

■ *Protein-mediated membrane transport*

Certain substances enter or leave cells by way of specific carriers or channels that are intrinsic proteins of the plasma membrane. Transport via such protein carriers or channels is called *protein-mediated transport* or simply *mediated transport*. Specific ions or molecules may cross the membranes of mitochondria, endoplasmic reticulum, and other organelles by mediated transport. Mediated transport systems include *active transport* and *facilitated transport* processes. As discussed later, active transport and facilitated transport have a number of properties in common. The principal distinction between these two processes is that active transport is capable of "pumping" a substance against a gradient of concentration (or electrochemical potential), whereas facilitated transport tends to equilibrate the substance across the membrane.

Properties of mediated transport. The basic properties of mediated transport are the following:

1. The rate of transport is more rapid than that of other molecules of similar molecular weight and lipid solubility which cross the membrane by simple diffusion.

2. The transport rate shows saturation kinetics: as the concentration of the transported compound is increased, the rate of transport at first increases, but eventually a concentration is reached after which the transport rate increases no further. At this point the transport system is said to be saturated with the transported compound.

3. The mediating protein has *chemical specificity:* only molecules with the requisite chemical structure are transported. The specificity of most transport systems is not absolute, and, in general, it is broader than the specificity of most enzymes. The transport of glucose into red blood cells occurs by facilitated transport. Glucose is the preferred substrate, but mannose, galactose, xylose, L-arabinose, and certain other sugars also are transported by the same membrane protein. D-Arabinose, however, is a poor substrate for the glucose system, and sorbitol and mannitol do not enter red cells at all. Mediated processes have *stereospecificity* as well. In the red cell sugar transport system D-glucose is transported well, but L-glucose is barely transported at all.

4. Structurally related molecules may compete for transport. Typically the presence of one transport substrate will decrease the transport rate of a second substrate by competing for the transport protein. The competition is analogous to *competitive inhibition* of an enzyme.

5. Transport may be inhibited by compounds that are not structurally related to transport substrates. This may be due to an inhibitor binding to the transport protein in a way that decreases its affinity for the normal transport substrate. The compound phloretin does not resemble a sugar molecule, yet it is a strong inhibitor of red cell sugar transport. Active transport systems, which require some link to metabolism, may be inhibited by metabolic inhibitors. The rate of Na^+ transport out of cells by the Na^+, K^+-ATPase is decreased by substances that interfere with ATP generation.

Facilitated transport, sometimes called facilitated diffusion, occurs via a transport protein that is not linked to metabolic energy. Facilitated transport has the properties discussed previously, except that facilitated transport generally is not depressed by metabolic inhibitors. Because they have no link to energy metabolism, facilitated transport processes cannot move uncharged substances against concentration gradients or ions against electrochemical potential gradients. Facilitated transport systems act to equalize the concentrations (or electrochemical potentials for ions) in the cytoplasm and extracellular fluid of the substances they transport.

Monosaccharides enter muscle cells by a facilitated transport process. Glucose, galactose, arabinose, and 3-*O*-methylglucose compete for the same carrier. The rate of transport shows saturation kinetics. The nonphysiological stereoisomer L-glucose enters the cells very slowly, and nontransported sugars such as mannitol or sorbose enter muscle cells very slowly if at all. Phloretin is an inhibitor of sugar uptake. The sugar transport system in muscle cells is stimulated by insulin. In the absence of insulin, glucose transport is rate limiting for glucose use, so that insulin is a major regulator of muscle glucose metabolism.

The molecular details of protein-mediated transport processes are not well understood at present. Current evidence suggests that most transport proteins span the membrane and are multimeric. Fig. 1-9 depicts a hypothetical model that has been proposed for the monosaccharide transport protein of the membrane of the human red blood cell. The protein is postulated to be a tetramer. Conformational changes of the protein, induced by monosaccharide binding, may allow a sugar molecule to enter and leave the central cavity. It is emphasized that this is a hypothetical model, not an established mechanism. The molecular mechanisms of various membrane transport processes are subjects of current investigations.

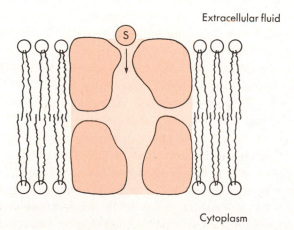

Extracellular fluid

Cytoplasm

Fig. 1-9 ■ Hypothetical model of a transport protein. This is a model of the mechanism of monosaccharide transport by the sugar transport protein of the human red blood cell. The protein is postulated to be a tetramer. Binding of sugar is proposed to cause conformational changes that allow sugar molecules to enter and leave the central cavity of the transport protein. (Redrawn from Jones, M.N., and Nickson, J.K.: Biochim. Biophys. Acta **650:**1, 1981.)

■ *Active transport*

Active transport processes have most of the properties of facilitated transport and, in addition, can concentrate their substrates against electrochemical potential gradients. This requires energy, so active transport processes must be linked to energy metabolism in some way. Active transport systems may use ATP directly, or they may be linked more indirectly to metabolism. Because of their dependence on metabolism, active transport processes may be inhibited by any substance that interferes with energy metabolism.

One possible mechanism for active transport. It is frequently useful to think of mediated transport as analogous to an enzyme-catalyzed chemical reaction. When the transport process is in the steady state across the plasma membrane, the unidirectional flux from the extracellular fluid to the cytoplasm is exactly equal to the unidirectional flux from cytoplasm to extracellular fluid. Since influx equals efflux, there is no net flux across the plasma membrane, and the cellular concentration is constant in time in the steady state.

For a facilitated transport process across the plasma membrane influx and efflux are often symmetrical processes. In that case, in the terms of enzyme kinetics, influx and efflux have the same K_m and V_{max}. Imagine that the K_m of a glucose transport system is 1mM. Then, when the glucose concentrations in cytoplasm and extracellular fluid are 1mM, influx and efflux will be equal (each being equal to $V_{max}/2$). This example illustrates that the steady state of a facilitated transport system is characterized by equilibration of substrate across the membrane.

An active transport system may be one for which the influx and efflux processes are not symmetrical. Consider an imaginary transport system for glycine that has a K_m for influx of 0.5mM, a K_m for efflux of 5mM, and the same V_{max} for both influx and efflux. When the intracellular glycine concentration is 5mM and the extracellular glycine concentration is 0.5mM, influx of glycine will equal glycine efflux (both being equal to $V_{max}/2$), and the steady state will be obtained. Note that the system is in steady state with a 10 times higher glycine concentration inside the cell than outside, so this is an active transport system.

In creating the tenfold concentration difference of glycine the transport system does work. The ultimate source of the energy to do transport work is metabolic energy. How is metabolic energy harnessed to do transport work? One way involves the cyclical phosphorylation and dephosphorylation of the transport protein. In the glycine transport system just described imagine that phosphorylation of the protein at the inner aspect of the membrane (by a protein kinase using ATP) converts the protein to the form where $K_m = 5$mM. If dephosphorylation occurs at the outer face of the plasma membrane, the protein "reverts" to the form where $K_m = 0.5$mM. In this way the K_m for efflux is made greater than that for the influx, allowing a tenfold concentration difference to be created. Since the imaginary glycine active transport system described uses ATP directly, this is termed a *primary active transport system,* as would be any transport system that is rather directly linked to any high-energy metabolic intermediate.

Another way in which active transport is powered involves a less direct link to metabolic energy. The transmembrane concentration difference of a second molecule that itself is transported actively may be used to allosterically modify the affinity of the transport protein for its substrate. Sodium is actively extruded from most cells by the Na, K pump (Na^+, K^+-ATPase) by primary active transport, so that the extracellular concentration of Na^+ is often about tenfold higher than its intracellular concentration. In the imaginary glycine transport system described previously, imagine that at the high Na^+ concentrations present in the extracellular fluid Na^+ binds to the transport carrier at the external face of the membrane and that Na^+ binding converts the protein to the form where $K_m = 0.5$mM. At the inner surface of the membrane the Na^+ concentration is much lower, so much less Na^+ will bind to the transport protein, and it thus will be predominantly in the form where $K_m = 5$mM. In this way the transmembrane gradient of Na^+ (which is created by primary active transport) is harnessed to bring about a transmembrane concentration difference of glycine (in this hypothetical example). In this case active glycine transport is said to occur by *secondary active transport.*

■ *Examples of active transport processes*

Transport powered by phosphorylation of the transport protein. A sodium-potassium pump is present in the plasma membrane of most mammalian cells. It is linked directly to a supply of metabolic energy because ATP itself is involved in conversion of the carrier from one affinity state to the other. The carrier is autophosphorylated by ATP at the inner surface of the membrane, converting it from a form that prefers K^+ to a form that prefers to bind Na^+. At the outer surface of the membrane the binding of K^+ promotes hydrolysis of the phosphoprotein by a mechanism that involves a phosphoprotein phosphatase activity. Cleavage of the phosphate converts the carrier back to the form that preferentially binds K^+. Since the Na^+, K^+ pump splits ATP, the pump also is called the Na^+, K^+-ATPase. The cyclical phosphorylation and dephosphorylation of the pump results in transport of K^+ into the cell and Na^+ out of the cell. The rates of splitting of ATP and of pump activity are stimulated by increasing intracellular $[Na^+]$ and extracellular $[K^+]$. The sodium-potassium pump extrudes 3 Na^+ ions for every 2 K^+ taken up.

Transport powered by the gradient of another species: secondary active transport. The neutral amino acid transport system of the small intestinal epithelial cell is an example of an active transport system that is believed to use the energy present in the gradient of another actively transported species (Na^+). Because of the action of the Na^+, K^+ pump in the epithelial cells of the small intestine, the intracellular concentration of Na^+ is much lower than the Na^+ concentration in the intestinal lumen. It is believed that some of or all the energy required for the active uptake of the neutral amino acids is derived from the energy of the Na^+ gradient.

It appears that in the case of the neutral amino acid transport system the high concentration of Na^+ at the outer surface of the epithelial cell microvilli causes the transport protein to bind Na^+, which causes the transport protein to bind neutral amino acids with greater affinity. At the cytoplasmic membrane surface the concentration of Na^+ is lower, Na^+ dissociates, and the amino acid then is bound less tightly to the transport protein. Thus the apparent K_m for alanine uptake is lower in the presence of high external Na^+ than in the absence of Na^+ (Fig. 1-10, *A*), whereas the apparent V_{max} of the

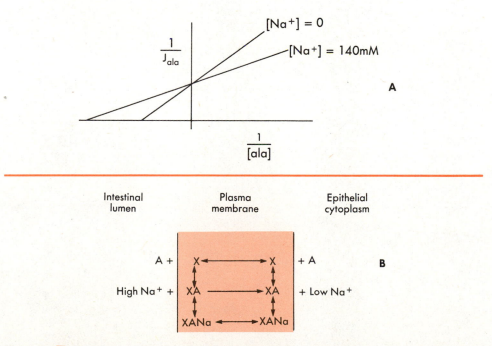

Fig. 1-10 ■ Alanine transport into epithelial cells of rabbit ileum. **A,** Double reciprocal plot showing that the presence of Na^+ in the intestinal lumen causes a lower K_m for transport (higher affinity) but does not alter V_{max} of transport. **B,** A highly simplified model of alanine *(A)* transport. Either alanine or Na^+ may bind first to the transport protein *(X)*. Whichever solute binds first increases the affinity of the protein for the other solute.

transport process does not depend on the external Na^+ concentration. The transport of Na^+ into the intestinal epithelial cell also is enhanced by the presence of neutral amino acids in the lumen. It is believed that Na^+ and amino acids are transported into the cell on the same transport protein. Subsequently amino acids are transported across the serosal surface of the epithelial cell by a facilitated transport mechanism, while Na^+ is extruded from the cell by the Na^+ pump.

The model shown in Fig. 1-10, *B,* has been proposed to account for the active transport of neutral amino acids into intestinal epithelial cells. The binding of Na^+ to the transport protein *(X),* either before or after binding of the amino acid *(A),* results in an increase in the affinity of the transport protein for the amino acid. At the intracellular face of the plasma membrane the concentration of Na^+ is quite low, so the Na^+ tends to dissociate from the XANa complex. This converts the transport protein back to the form with a lower affinity for amino acid and in this way promotes the dissociation of A from X and the release of amino acid to the cytoplasm.

Since the active transport of amino acid depends on the Na^+ gradient but does not depend *directly* on ATP or some other high-energy metabolic intermediate, the active transport of amino acid is *secondary active transport*.

■ *Bibliography*

Christensen, H.N.: Biological transport, ed. 2, Menlo Park, Calif., 1975, W.A. Benjamin, Inc.

Davson, H.: A textbook of general physiology, ed. 4, Baltimore, 1970, Williams & Wilkins Co.

Finean, J.B., Coleman, R., and Michell, R.H.: Membranes and their cellular functions, ed. 2, New York, 1978, John Wiley & Sons, Inc.

Fox, C.F., and Keith, A., editors: Membrane molecular biology, Stamford, Conn., 1972, Sinauer Associates.

Kotyk, A., and Janacek, K.: Cell membrane transport: principles and techniques, ed. 2, New York, 1975, Plenum Publishing Corp.

chapter 2

Ionic equilibria and resting membrane potentials

Most animal cells have an electrical potential difference (voltage) across their plasma membranes. The cytoplasm is usually electrically negative relative to the extracellular fluid. Since the electrical potential difference across the plasma membrane is present even in resting cells, it sometimes is referred to as the *resting membrane potential*. The resting membrane potential plays a central role in the excitability of nerve and muscle cells and in certain other cellular responses.

The major purpose of this chapter is to discuss the ways that electrochemical potential gradients of certain ions across the plasma membrane generate the resting membrane potential. The first part of the chapter deals with some fundamental definitions and concepts that describe the flow of ions across membranes.

■ *Ionic equilibria*

■ *Electrochemical potentials of ions*

A membrane separates aqueous solutions in two chambers (A and B). Na^+ is at a higher concentration on side A than on side B. If there is no electrical potential difference between side A and side B, Na^+ will tend to diffuse from side A to side B, just as if it were an uncharged molecule. If, however, side A is electrically negative with respect to side B, the situation is more complex. The tendency for Na^+ to diffuse from side A to side B because of the concentration difference remains, but now there is also a tendency for Na^+ to move in the opposite direction (from B to A) because of the electrical potential difference across the membrane. The direction of net Na^+ movement depends on whether the effect of the concentration difference or the effect of the electrical potential difference is larger. By quantitatively comparing the two tendencies—concentration and electrical—one can predict the direction of net Na^+ movement.

The quantity that allows us to compare the relative contributions of ionic concentration and electrical potential is called the *electrochemical potential* (μ) of an ion. The electrochemical potential is defined as

$$\mu = \mu^\circ + RT\ln C + zFE \qquad (1)$$

where

μ° = electrochemical potential of the ion at some reference state (say 1M concentration at 0° C with zero electrical potential)
R = gas constant
T = absolute temperature
C = concentration
z = charge number of the ion ($+2$ for Ca^{++}, -1 for Cl^-, etc.)
F = Faraday's number
E = electrical potential

The units of μ and of each term in equation 1 are energy per mole. The electrochemical potential represents the *chemical potential energy* possessed by a mole of ions due to their concentration and the electrical potential.

The net flow of an ion will be from where its electrochemical potential is higher to where its electrochemical potential is lower. Consider a membrane separating two chambers (A and B), each of which contains ion x in solution. The tendency of the ion to move from A to B is proportional to $\mu(x)$ on side A, and the tendency of the ion to move from B to A is proportional to $\mu(x)$ on side B. The *net* tendency for x to flow from A to B is $\mu_A(x) - \mu_B(x)$. This difference is the *electrochemical potential difference* of x across the membrane ($\Delta\mu$).

$$\Delta\mu(x) = \mu_A(x) - \mu_B(x) \tag{2}$$

Substituting into equation 2 for $\mu_A(x)$ and $\mu_B(x)$ from equation 1, we obtain

$$\mu_A(x) = \mu°(x) + RT1n[x]_A + zFE_A$$
$$\mu_B(x) = \mu°(x) + RT1n[x]_B + zFE_B$$

so that

$$\Delta\mu(x) = \mu_A(x) - \mu_B(x) = RT1n\frac{[x]_A}{[x]_B} + zF(E_A - E_B) \tag{3}$$

The first term on the right-hand side of equation 3 is the tendency for the ion x to move from A to B because of the concentration difference, and the second term is the tendency for the ion to move from A to B because of the electrical potential difference. The first term ($RT1n\ [x]_A/[x]_B$) represents the chemical potential energy difference between a mole of x ions on side A and a mole of x ions on side B as a result of the concentration difference. The second term $[zF(E_A - E_B)]$ represents the chemical potential energy difference between a mole of x ions on side A and a mole of x ions on side B caused by the electrical potential difference between A and B. Thus $\Delta\mu$ describes the difference in chemical potential energy between a mole of x ions on side A and a mole of x ions on side B resulting from *both* concentration and electrical potential difference.

The x ions will tend to move from where their electrochemical potential is higher to where it is lower. $\Delta\mu$ is defined as the electrochemical potential of the ion on side A minus that on side B. If $\Delta\mu$ is positive, the ions will tend to move from A to B; if $\Delta\mu$ is zero, there is no net tendency for the ions to move at all; and if $\Delta\mu$ is negative, the ions will tend to move from side B to side A.

■ *Electrochemical equilibrium and the Nernst equation*

$\Delta\mu$ may be thought of as the *net* force on the ion, whereas $RT1n\ [x]_A/[x]_B$ is the force caused by the concentration difference, and $zF(E_A - E_B)$ is the force caused by the electrical potential difference. When the two forces are equal and opposite, $\Delta\mu = 0$, and there is no net force on the ion. When there is no net force on the ion, there will be no net movement of the ion, and the ion is said to be in *electrochemical equilibrium* across the membrane. At equilibrium $\Delta\mu = 0$. Since

$$\Delta\mu = RT1n\frac{[x]_A}{[x]_B} + zF(E_A - E_B) \tag{4}$$

at equilibrium, then

$$RT1n\frac{[x]_A}{[x]_B} + zF(E_A - E_B) = 0$$

Solving for $E_A - E_B$, we obtain

$$E_A - E_B = \frac{-RT}{zF}1n\frac{[x]_A}{[x]_B} = \frac{RT}{zF}1n\frac{[x]_B}{[x]_A} \tag{5}$$

Equation 5 is called the *Nernst equation,* after the nineteenth century physical chemist who derived it. The condition of equilibrium was assumed in its derivation, and *the Nernst equation is valid only for ions at equilibrium.* It allows one to compute the electrical potential difference, $E_A - E_B$, required to produce an electrical force, $zF(E_A - E_B)$, that is equal and opposite to the concentration force, $RT1n ([x]_A/[x]_B)$, tending to move the ion from A to B.

Using the Nernst equation. It is often convenient to convert the Nernst equation to a form involving $\log_{10}$ rather than natural logarithms ($1n\ x = 2.303\ \log_{10} x$). Since biological potentials usually are expressed in millivolts (mV), the units of R are selected so RT comes out in millivolts. At 29.2° C the quantity $2.303\ ^{RT}/_F$ is equal to 60 mV. Since this quantity is proportional to the absolute temperature, it changes roughly by only $\frac{1}{273}$ for each centigrade degree. Thus the value of 60 mV for $2.303\ ^{RT}/_F$ holds approximately for most experimental conditions, and a useful form of the Nernst equation is

$$E_A - E_B = \frac{-60\ mV}{z} \log_{10} \frac{[x]_A}{[x]_B} = \frac{60\ mV}{z} \log_{10} \frac{[x]_B}{[x]_A} \tag{6}$$

Examples of uses of the Nernst equation

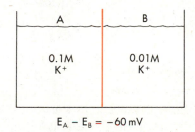

$$E_A - E_B = -60\,mV$$

Fig. 2-1 ■ A membrane separates chambers containing different K^+ concentrations. At an electrical potential difference ($E_A - E_B$) of -60 mV, K^+ is in electrochemical equilibrium across the membrane.

Example 1. In the situation shown in Fig. 2-1 K^+ is 10 times more concentrated in chamber A than in chamber B. Following is a calculation of the electrical potential difference between the chambers that is required for K^+ to be in equilibrium across the membrane.

Since we have specified that K^+ should be *in equilibrium,* the Nernst equation will hold.

$$\begin{aligned} E_A - E_B &= \frac{-60\ mV}{+1} \log_{10} \frac{[K+]_A}{[K^+]_B} \\ &= -60\ mV \log_{10} \frac{0.1}{0.01} = -60\ mV \log(10) \\ &= -60\ mV \end{aligned}$$

The Nernst equation tells us that at equilibrium side A must be 60 mV negative relative to side B. We can see that this polarity is correct, since it will cause K^+ to tend to move from B to A due to the electrical force, which will counteract the tendency for it to move from A to B because of the concentration difference.

This example shows that an electrical potential difference of about 60 mV is required to balance a tenfold concentration difference of a univalent ion. This is a useful rule of thumb.

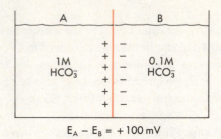

$$E_A - E_B = +100 \, mV$$

Fig. 2-2 ■ A membrane separates chambers containing different HCO_3^- concentrations. $E_A - E_B = +100$ mV. HCO_3^- is not in electrochemical equilibrium. If $E_A - E_B$ were $+60$ mV, HCO_3^- would be in equilibrium. $E_A - E_B$ is stronger than it needs to be to just balance the tendency for HCO_3^- to move from *A* to *B* because of its concentration difference. Thus net movement of HCO_3^- from *B* to *A* will occur.

Example 2. In the situation diagramed in Fig. 2-2 the Nernst equation can help decide whether HCO_3^- is in equilibrium. If HCO_3^- is not in equilibrium, the Nernst equation can help determine the direction of net flow of HCO_3^-.

The Nernst equation tells us the electrical potential difference, $E_A - E_B$, that will *just balance* the concentration difference of HCO_3^- across the membrane.

$$E_A - E_B = \frac{-60 \, mV}{z} \log_{10} \frac{[HCO_3^-]_A}{[HCO_3^-]_B}$$

$$= \frac{-60 \, mV}{-1} \log_{10} \frac{1}{0.1}$$

$$= 60 \, mV \log_{10} (10) = 60 \, mV$$

Thus a potential difference of $+60$ mV between A and B would just balance the tendency of HCO_3^- to move from A to B due to its concentration difference. Since $E_A - E_B$ is actually $+100$ mV, the electrical potential is of the right sign to oppose the concentration force, but it is 40 mV larger than it needs to be to just balance the concentration force. Since the electrical force on HCO_3^- is larger than the concentration force, it will determine the net direction of HCO_3^- movement. Net HCO_3^- flow will occur from B to A.

Summary

1. If the potential difference *measured* across a membrane is equal to the potential difference *calculated* from the Nernst equation for a particular ion, then *that particular ion* is in electrochemical equilibrium across the membrane, and there will be no net flow of that ion across the membrane.

2. If the measured electrical potential is of the same sign as that calculated from the Nernst equation for a particular ion but is larger in magnitude, then the electrical force is larger than the concentration force, and net movement of that particular ion will tend to occur in the direction determined by the electrical force.

3. When the electrical potential difference is of the same sign but is numerically less than that calculated from the Nernst equation for a particular ion, then the concentration force is larger than the electrical force, and net movement of that ion tends to occur in the direction determined by the concentration difference.

4. If the electrical potential difference measured across the membrane is of the opposite sign from that predicted by the Nernst equation for a particular ion, then the electrical and concentration forces are in the same direction. In such a case that ion cannot be in equilibrium, and it will tend to flow in the direction determined by both electrical and concentration forces.

In the discussion on electrical activity of nerve and muscle cells that follows, the concept of what it means for an ion to be at equilibrium and how the balance of electrical and concentration forces determines the direction of net ion movement underlies the discussion of the ionic mechanisms of the resting membrane potential and the action potential.

Cytoplasm typically contains proteins, organic polyphosphates, and other ionized substances that cannot permeate the plasma membrane. Cytoplasm also contains Na^+, K^+, Cl^-, and other ions to which the plasma membrane is somewhat permeable. The steady-state properties of this mixture of permeant and impermeant ions are described by the *Gibbs-Donnan equilibrium.*

Consider a membrane separating a solution of NaCl from a solution of NaY, where Y^- is an anion to which the plasma membrane is completely impermeable (Fig. 2-3). The membrane is permeable to water, Na^+, and Cl^-. Suppose that *initially* on side A there is a 0.1M solution of NaY, and an equal volume of 0.1M NaCl is on side B. Because $[Cl^-]_B$ exceeds $[Cl^-]_A$, there will be a net flow of Cl^- from chamber B to

■ *The Gibbs-Donnan equilibrium*

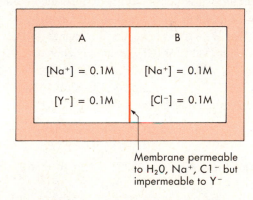

Membrane permeable to H_2O, Na^+, Cl^- but impermeable to Y^-

Fig. 2-3 ■ Prior to establishing a Gibbs-Donnan equilibrium a membrane separates two aqueous compartments. The membrane is permeable to H_2O, Na^+, and Cl^- but impermeable to Y^-.

chamber A. Negatively charged Cl^- ions flowing from side B to side A will create an electrical potential difference, with side A being negative, that will cause Na^+ also to flow from side B to side A. Essentially the same number of Na^+ ions as Cl^- ions will flow from side B to side A to preserve *electroneutrality*. *The principle of electroneutrality* states that any macroscopic region of a solution must have an equal number of positive and negative charges. In reality slight separation of charges does occur in certain situations, but in chemical terms the imbalance between positive and negative charges is negligible.

If enough time passes, those components of the system which can permeate the membrane—Na^+ and Cl^- in this example—will come to equilibrium. At equilibrium both $\Delta\mu_{Na^+}$ and $\Delta\mu_{Cl^-}$ must equal zero. From equation 3

$$\Delta\mu_{Na^+} = RT\ln\frac{[Na^+]_A}{[Na^+]_B} + F(E_A - E_B) = 0$$

$$\Delta\mu_{Cl^-} = RT\ln\frac{[Cl^-]_A}{[Cl^-]_B} - F(E_A - E_B) = 0$$

Adding these two equations and dividing the result by RT gives

$$\ln\frac{[Na^+]_A}{[Na^+]_B} + \ln\frac{[Cl^-]_A}{[Cl^-]_B} = 0$$

This gives

$$\ln \frac{[Na^+]_A}{[Na^+]_B} = -\ln \frac{[Cl^-]_A}{[Cl^-]_B} = \ln \frac{[Cl^-]_B}{[Cl^-]_A}$$

Thus

$$\frac{[Na^+]_A}{[Na^+]_B} = \frac{[Cl^-]_B}{[Cl^-]_A}$$

Cross multiplying gives

$$[Na^+]_A [Cl^-]_A = [Na^+]_B[Cl^-]_B \qquad (7)$$

Equation 7 is called the *Donnan relationship* (or the Gibbs-Donnan equation) and holds for any pair of univalent cation and anion in equilibrium between the two chambers. If other ions that could attain an equilibrium distribution were present, the same reasoning and an equation similar to equation 7 would apply to them as well.

   Example of a Gibbs-Donnan equilibrium. Using the Donnan relationship and the principle of electroneutrality, it is possible to determine the equilibrium concentrations of the components in the problem posed at the beginning of this section. The initial situation is shown in Fig. 2-3. If b represents the change in $[Cl^-]$ when Cl^- moves from B to A, the equilibrium value of $[Cl^-]_B$ can be denoted as $0.1 - b$. By the electroneutrality principle $[Na^+]_B = [Cl^-]_B = 0.1 - b$. If the volumes of A and B are kept the same, at equilibrium, $[Cl^-]_A = b$, and $[Na^+]_A = 0.1 + b$. Substituting these concentrations into the Donnan relationship:

$$[Na^+]_A [Cl^-]_A = [Na^+]_B [Cl^-]_B$$
$$(0.1 + b)\, (b) = (0.1 - b)\, (0.1 - b)$$

Solving this equation for b gives $b = 0.0333 \ldots$, so that at equilibrium we obtain the concentrations shown in Fig. 2-4.

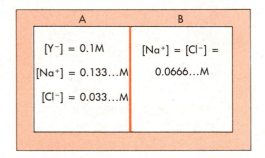

Fig. 2-4 ■ Ion concentrations after Gibbs-Donnan equilibrium has been attained. The initial ion concentrations were as shown in Fig. 2-3.

   In this Gibbs-Donnan equilibrium both Na^+ and Cl^- (but *not* Y^-) are in electrochemical equilibrium. This means that both Na^+ and Cl^- must satisfy the Nernst equation, so that the equilibrium transmembrane potential difference can be computed from the Nernst equation for *either* Na^+ or Cl^-.

$$E_A - E_B = \frac{-60\ mV}{+1} \log_{10} \frac{[Na^+]_A}{[Na^+]_B}$$

$$= -60\ mV \log_{10} \frac{0.1333 \ldots}{0.0666 \ldots}$$

$$= \frac{-60\ mV}{-1} \log_{10} \frac{[Cl^-]_A}{[Cl^-]_B} = 60\ mV \log_{10} \frac{0.0333 \ldots}{0.0666 \ldots}$$

$$= -60\ mV \log_{10} 2 = -60\ mV\,(0.3) = -18\ mV$$

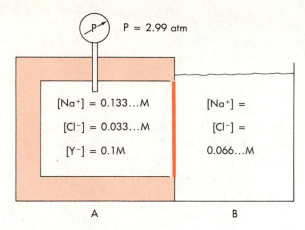

Fig. 2-5 ■ A hydrostatic pressure of 2.99 atm is required to prevent water from flowing from *B* to *A* in the Gibbs-Donnan equilibrium in Fig. 2-4.

Note that only the permeable ions attain equilibrium. The impermeant ion, Y^-, *cannot* reach an equilibrium distribution. What may not yet be evident is that water also will not achieve equilibrium, unless provision is made for that to occur. The total number of Na^+ and Cl^- ions on side A in the preceding example exceeds that on side B. This is a general property of Gibbs-Donnan equilibria. Taking the impermeant Y into account as well, the total concentration of osmotically active ions is considerably greater on side A than on side B. Because Na^+ and Cl^- are present at equilibrium, they will have their full osmotic force even though the membrane is quite permeable to them. Water will tend to flow by osmosis from side B to side A until the total osmotic pressure of the two solutions is equal. But then ions will flow to set up a new Gibbs-Donnan equilibrium, and that requires there be more osmotically active ions on the side with Y^-. All the water from side B will end up on side A unless water is restrained from moving. This can be done by enclosing the solution on side A in a rigid container (Fig. 2-5). Then, as fluid flows from side B to side A, pressure will build up in A that will oppose further osmotic water flow. The pressure in chamber A at equilibrium is equal to the difference between the total osmotic pressures of the solutions in chambers A and B. In this example the approximate hydrostatic pressure (P) in chamber A at equilibrium (at $0°$ C) is

$$
\begin{aligned}
P &= \Delta\pi_{Na^+} + \Delta\pi_{Cl^-} + \Delta\pi_{Y^-} \\
&= RT\,(\Delta[Na^+] + \Delta[Cl^-] + \Delta[Y^-]) \\
&= RT\,(0.06667 - 0.03333 + 0.1) \\
&= (22.4\ \text{atm})\,0.13333) = 2.99\ \text{atm}
\end{aligned}
$$

■ *Regulation of cell volume*

K^+ and Cl^- are nearly in equilibrium across many plasma membranes, and their distribution is influenced by the predominantly negatively charged impermeant ions in the cytoplasm such as proteins and nucleotides. K^+ and Cl^- approximately satisfy the Donnan relationship. This being the case, why does the osmotic imbalance discussed previously not cause the cell to swell and finally burst? One reason is that cells actively pump Na^+ out of the cytoplasm to the extracellular fluid, decreasing the osmotic pressure of the cytoplasm and increasing that of the extracellular fluid. Much of the pumping of Na^+ is done by the Na^+ pump, the Na^+, K^+-ATPase, in the plasma membrane. The Na^+, K^+-ATPase (as discussed earlier) splits an ATP and uses some of the energy released to extrude 3 Na^+ from the cytoplasm and to pump 2 K^+ into the cell. Whereas K^+ is only slightly removed from an equilibrium distribution, Na^+ is pumped out against a large electrochemical potential difference.

When the ATP production of the cell is compromised (in the presence of metabolic inhibitors or low O_2 levels), or when the Na^+, K^+-ATPase is specifically inhibited (by cardiac glycosides), cells swell. Some investigators believe that the irreversible brain damage which occurs after acute oxygen deprivation is due to the injury that results from inhibition of Na^+ pumping and osmotic swelling of brain cells. Consistent with this hypothesis, when brain cells are "preshrunk" with hypertonic solutions, they may have a greater ability to survive periods of low oxygen.

■ Resting membrane potentials

Communication between nerve cells depends on an electrical disturbance, propagated in the plasma membrane, that is called an *action potential*. In striated muscle an action potential propagates rapidly over the entire cell surface, allowing the cell to contract synchronously. The action potential and the ionic mechanisms that account for its properties are discussed in Chapter 3. All cells that are able to produce action potentials have sizable *resting membrane potentials* across their plasma membranes. Most inexcitable cells also have a resting membrane potential.

The resting membrane potential of many cells can be measured using glass microelectrodes with tip diameters of about 0.1 μm that can puncture the plasma membrane of some cells without greatly injuring the cell. The electrical potential difference between the tip of a microelectrode inside a skeletal muscle cell and a reference electrode in the extracellular fluid is about −90 mV. The myoplasm is thus almost 0.1 V electrically negative with respect to the cell's surroundings. The resting membrane potential is necessary for the cell to fire an action potential. If the resting membrane potential is decreased to −50 mV or less, the cell is no longer able to produce an action potential.

Ions that are actively transported are not in electrochemical equilibrium across the plasma membrane. It is shown later that the flow of ions across the plasma membrane, down their electrochemical potential gradients, is directly responsible for generating most of the resting membrane potential. To understand how the electrochemical potential gradient of an ion can give rise to a transmembrane difference in electrical potential, let us first consider a model system known as a concentration cell.

■ The concentration cell

Consider the situation illustrated in Fig. 2-6, *A*. The membrane that separates chambers A and B is permeable to cations but not to anions. Initially there is no electrical

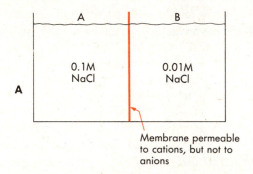

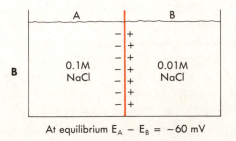

Fig. 2-6 ■ **A,** A concentration cell. The membrane, which is permeable to cations but not to anions, separates NaCl solutions of different concentrations. **B,** The concentration cell after electrochemical equilibrium has been established. The flow of an infinitessimal amount of Na^+ from side *A* to side *B* has generated a transmembrane electrical potential of −60 mV.

potential difference across the membrane. Na^+ will flow from A to B due to the concentration force acting on it. Cl^- has the same force on it but cannot flow from A to B because the membrane is impermeable to anions. The flow of Na^+ from A to B will transfer net positive charge to side B and leave a very slight excess of negative charges behind on side A, causing A to become electrically negative in relation to side B (Fig. 2-6, *B*). This electrical force is oppositely directed to the concentration force on Na^+. The more Na^+ that flows, the larger the opposing electrical force. Net Na^+ flow will stop when the electrical force just balances the concentration force, that is, when the electrical potential difference is equal to the equilibrium (Nernst) potential for Na^+, when

$$E_A - E_B = \frac{-60 \text{ mV}}{+1} \log_{10} \frac{1}{0.1} = -60 \text{ mV} \log(10) = -60 \text{ mV}$$

It should be emphasized that *only a very small amount of Na^+ flows* from A to B before equilibrium is reached. This is because the separation of positive and negative charges requires a large amount of work. The potential difference that builds up to oppose further Na^+ movement is a manifestation of that work.

It is important to appreciate that the Na^+ concentration difference in this example acts much like a *battery*. The natural tendency for any ion that can flow is to seek equilibrium; thus Na^+ tends to flow until its equilibrium potential difference is established. As is explained later, in a system such as a cell membrane with more than one permeable ion *each* ion "strives" to make the transmembrane potential difference equal to the same value as its equilibrium potential. The more permeable the ion, the greater its ability to force the electrical potential difference toward its equilibrium potential.

■ *The distribution of ions across plasma membranes*

In most tissues a number of ions are not in equilibrium between the extracellular fluid and the cytoplasm. Table 2-1 gives the concentrations of Na^+, K^+, and Cl^- in the extracellular fluid and in the cytoplasmic water of frog skeletal muscle and squid giant axon. Intracellular ion concentrations for mammalian muscle are similar to those for frog muscle.

Chloride is close to being in equilibrium across the plasma membrane of both frog muscle and squid axon. This is known because chloride's potential difference for equilibrium, as calculated from the Nernst equation, is about equal to the measured transmembrane potential difference. In both tissues K^+ has a concentration force tending to make it flow out of the cell. The electrical force on K^+ is oppositely directed to the concentration force. If the $E_{in} - E_{out}$ in frog muscle were -105 mV, electrical and

TABLE 2-1

Distribution of Na^+, K^+, and Cl^- across the plasma membranes of frog muscle and squid axon

	Extracellular fluid (mM)	*Cytoplasm (mM)*	*Approximate equilibrium potential (mV)*	*Actual resting potential (mV)*
Frog muscle				
$[Na^+]$	120	9.2	$+67$	
$[K^+]$	2.5	140	-105	
$[Cl^-]$	120	3 to 4	-89 to -96	-90
Squid axon				
$[Na^+]$	460	50	$+58$	
$[K^+]$	10	400	-96	
$[Cl^-]$	540	About 40	About -68	-70

Data from Katz, B.: Nerve, muscle, and synapse, New York, 1966, McGraw-Hill Book Co. Copyright © 1966 by McGraw-Hill Book Co. Used with permission of McGraw-Hill Book Co.

concentration forces on K^+ would exactly balance. Since $E_{in} - E_{out}$ is only -90 mV, the concentration force is greater than the electrical force, and K^+ therefore has a net tendency to flow out of the cell. In muscle and in squid axon *both* the concentration and electrical forces on Na^+ tend to cause it to flow into the cell. Na^+ is the ion furthest from an equilibrium distribution. The larger the difference between the measured membrane potential and the equilibrium potential for an ion, the larger the net force tending to make that ion flow.

■ *Active ion pumping and the resting potential*

The Na^+, K^+-ATPase, located in the plasma membrane, uses the energy of the terminal phosphate ester bond of ATP to actively extrude Na^+ from the cell and to actively take K^+ into the cell. The Na^+, K^+ pump is responsible for the high intracellular K^+ concentration and the low intracellular Na^+ concentration. Since the pump moves a larger number of Na^+ ions out than K^+ ions in (3 Na^+ to 2 K^+), it causes a net transfer of positive charge out of the cell and thus contributes to the resting membrane potential. Because it brings about net movement of charge across the membrane, the pump is termed *electrogenic*. The size of the pump's contribution to the resting potential can be estimated by completely inhibiting the pump with a cardiac glycoside, such as ouabain. Such studies show that in some cells the electrogenic Na^+, K^+ pump is responsible for a large fraction of the resting potential. In most vertebrate nerve and skeletal muscle cells, however, the direct contribution of the pump to the resting potential is small under most circumstances—less than 5 mV. The resting membrane potential in nerve and skeletal muscle results primarily from the diffusion of ions down their electrochemical potential gradients. The ionic gradients are maintained by active ion pumping. In other types of excitable cells electrogenic pumping of ions may make a larger contribution to the resting membrane potential. In certain vertebrate smooth muscle cells, for example, the electrogenic effect of the Na^+, K^+ pump may be responsible for 20 mV or more of the resting membrane potential.

■ *Generation of the resting membrane potential by the ion gradients*

The earlier section on concentration cells shows how an ion gradient can act as a battery. When a number of ions are distributed across a membrane, all being removed from electrochemical equilibrium, *each* ion will tend to force the transmembrane potential toward *its own* equilibrium potential, as calculated from the Nernst equation. The more permeable the membrane to a particular ion, the greater strength that ion will have in forcing the membrane potential toward its equilibrium potential. In frog muscle (Table 2-1) the Na^+ concentration difference can be regarded as a battery that tries to make $E_{in} - E_{out}$ equal to $+67$ mV. The K^+ concentration difference is like a battery that attempts to make $E_{in} - E_{out}$ equal to -105 mV. The Cl^- concentration difference resembles a battery trying to make $E_{in} - E_{out}$ equal to -90 mV.

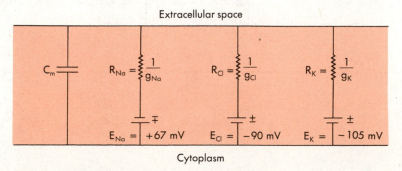

Fig. 2-7 ■ An electrical equivalent circuit model of the plasma membrane of a skeletal muscle cell. The equilibrium potentials of Na^+, K^+, and Cl^- are represented as batteries. The resistances *(R)* and conductances (*g*, reciprocal of resistance) to each ion are shown. C_m represents the membrane capacitance.

We can draw an electrical equivalent circuit for the plasma membrane. In Fig. 2-7 each ion gradient is represented by a battery of the appropriate polarity. The resistor in series with each battery represents the resistance (R) to the passage of that ion through the membrane. The reciprocal of each resistance is the conductance (g) of the membrane for that ion. C_m represents the membrane capacitance that stores the transmembrane potential difference. In this circuit, if the transmembrane resistance to a particular ion is decreased, the total transmembrane electrical potential difference will move toward the battery potential of that ion.

■ *The chord conductance equation*

The way in which the interplay of ion gradients creates the resting membrane potential also is illustrated by a simple mathematical model. If the transmembrane electrical potential difference is equal to the equilibrium potential for a particular ion, there is no net force on that ion, and there will be no net flow of that ion. However, if the membrane potential is *not* equal to the equilibrium potential for a given ion, then the difference between the membrane potential and the ion's equilibrium potential can be regarded as the driving force for that ion. Since ions bear charge, ionic flow is equivalent to electrical current. The net current (I) of an ion across the membrane is equal to the driving force on the ion times the conductance of the membrane for that ion. For Na^+, K^+, and Cl^-

$$I_K = g_K(E_m - E_K)$$
$$I_{Na} = g_{Na}(E_m - E_{Na}) \tag{8}$$
$$I_{Cl} = g_{Cl}(E_m - E_{Cl})$$

where E_K, E_{Na}, and E_{Cl} are the equilibrium (Nernst) potentials for the indicated ion.

In the steady state, when the transmembrane electrical potential difference is constant, the sum of all the ionic currents across the membrane must be zero. This must be so because, if there *were* net current flowing across the membrane, the membrane capacitor would be charging or discharging, and the transmembrane potential difference would be changing. Any net charge transfer across the membrane leads to a change in the degree of charge separation across the membrane and hence in the membrane potential. If K^+, Na^+, and Cl^- are the only important ions, then the algebraic sum of their currents must be zero in the steady state.

$$I_K + I_{Na} + I_{Cl} = 0 \tag{9}$$

Substituting from equation 8 for the currents, we obtain

$$g_K (E_m - E_K) + g_{Na} (E_m - E_{Na}) + g_{Cl} (E_m - E_{Cl}) = 0 \tag{10}$$

For a cell membrane across which chloride is in equilibrium, $E_m = E_{Cl}$, and $E_m - E_{Cl} = 0$, so the third term can be dropped. However, it should be kept in mind that Cl^- is not in equilibrium for all excitable cells. Also, whenever E_m moves away from E_{Cl}, Cl^- will exert a restoring force to bring E_m back toward E_{Cl} just as any ion tries to force E_m toward its equilibrium potential. Without the chloride term equation 10 becomes

$$g_K (E_m - E_K) + g_{Na} (E_m - E_{Na}) = 0 \tag{11}$$

Solving for E_m yields

$$E_m = \frac{g_K}{g_K + g_{Na}} E_K + \frac{g_{Na}}{g_K + g_{Na}} E_{Na} \tag{12}$$

Equation 12 is one form of the *chord conductance equation*. It expresses the transmembrane electrical potential difference as a weighted average of the equilibrium potentials of K^+ and Na^+. The weighting factor for each ion is that fraction of the total ionic conductance caused by that particular ion. To consider more ions, we need only add appropriate terms to the chord conductance equation. In a cell in which Cl^- is not in equilibrium and Ca^{++} plays an important role, the chord conductance equation becomes

$$E_m = \frac{g_K}{g_T} E_K + \frac{g_{Na}}{g_T} E_{Na} + \frac{g_{Cl}}{g_T} E_{Cl} + \frac{g_{Ca}}{g_T} E_{Ca} \tag{13}$$

where

$$g_T = g_K + g_{Na} + g_{Cl} + g_{Ca}$$

For the frog muscle fiber discussed earlier $E_{in} - E_{out} = -90$ mV. The membrane potential is much closer to E_K (-105 mV) than to E_{Na} ($+67$ mV). The chord conductance equation predicts that in resting muscle g_K is 10.5 times larger than g_{Na}. This has been confirmed by ion flux measurements with radioactive tracers. In resting frog sartorius muscle g_K is about 10 times g_{Na}. In resting squid axon ($E_m = -70$ mV) the chord conductance equation predicts that g_K is about five times larger than g_{Na}. In other types of excitable cells the relationship between g_K and g_{Na} is somewhat different. Other ions also may play a role in generating the resting membrane potential. Resting membrane potentials vary from about -7 mV or so in human erythrocytes to -30 mV in some types of smooth muscle and up to -90 mV in vertebrate skeletal muscle and cardiac ventricular cells.

The chord conductance equation shows that, if g_{Na} were suddenly increased, the membrane potential would move toward E_{Na} (toward $+67$ in frog muscle). This is what occurs during an action potential when there is a transient increase in g_{Na}. The ionic mechanism of the action potential is discussed in Chapter 3.

■ Summary of the ionic mechanism of the resting membrane potential

The Na^+, K^+ pump establishes gradients of Na^+ and K^+ across the plasma membranes of cells. Since the amount of Na^+ pumped out is larger than the amount of K^+ pumped in, the pump transfers net charge across the membrane and contributes to the resting membrane potential. The pump therefore is said to be electrogenic. In vertebrate skeletal and cardiac muscle and nerve the electrogenic activity of the pump is directly responsible for only a small fraction of the resting membrane potential. The major portion of the resting membrane potential in these tissues is a result of the diffusion of Na^+ and K^+ down their electrochemical potential gradients, with Na^+ flowing into the cell and K^+ flowing out. The principal role of the pump in these tissues is to maintain the ion gradients it has established. K^+ and Na^+ each tend to force the transmembrane potential toward their own equilibrium potential. The resulting E_m is a weighted average of E_K and E_{Na}, with the weighting factor for each ion being the fraction of the total membrane conductance caused by that ion. The resting membrane potential is *directly* due to the diffusion of K^+ and Na^+ down their respective electrochemical potential gradients. The resting membrane potential is *indirectly* due to the Na^+, K^+ pump, which maintains the gradients. In mammalian smooth muscle cells the electrogenic effect of the active pumping of Na^+ and K^+ may contribute a substantial fraction of the resting membrane potential.

■ Bibliography

Aidley, D.J.: The physiology of excitable cells, ed. 2, Cambridge, 1978, Cambridge University Press.

Davson, H.: A textbook of general physiology, ed. 4, Baltimore, 1970, Williams & Wilkins Co.

Dowben, R.M.: General physiology: a molecular approach, New York, 1969, Harper & Row Publishers, Inc.

Katz, B.: Nerve, muscle, and synapse, New York, 1966, McGraw-Hill Book Co.

Kuffler, S.W., and Nicholls, J.G.: From neuron to brain, Sunderland, Mass., 1976, Sinauer Associates.

chapter 3

Generation and conduction of action potentials

An *action potential* is a rapid change in the membrane potential followed by a return to the resting membrane potential (Fig. 3-1). The size and shape of action potentials differ considerably from one excitable tissue to another. An action potential is propagated with the same shape and size along the whole length of a nerve axon or muscle cell. The action potential is the basis of the signal-carrying ability of nerve cells. It allows all parts of a long muscle cell to contract almost simultaneously. This chapter discusses the ionic currents that generate action potentials and the ways in which action potentials are propagated and conducted.

Our knowledge of the ionic mechanism of an action potential first was obtained from experiments on the squid giant axon. The large diameter (up to 0.5 mm) of the squid giant axon makes it a convenient object for electrophysiological research with intracellular electrodes. The frog sartorius muscle is another extremely useful preparation. The sartorius muscle can be removed from the frog without damage, and the surface muscle cells can be visualized with a microscope and penetrated with one or more microelectrodes.

The following experiment illustrates some basic electrophysiological properties of cells: a sartorius muscle is removed from a frog and put in a dish of fluid with composition similar to the frog's extracellular fluid. Two microelectrodes (tip diameter less than 0.5 μm) are placed in the extracellular fluid. When both microelectrodes are placed in the extracellular fluid, no electrical potential difference between them is observed. One electrode then is moved slowly toward a muscle cell until it penetrates the plasma membrane. At the instant the electrode pops through the membrane, an abrupt change of the potential difference between the two electrodes is observed. The intracellular electrode suddenly becomes about 90 mV negative with respect to the external electrode. This -90 mV potential difference is the *resting membrane potential* of the muscle fiber. A third microelectrode placed in the same cell also will register -90 mV relative to the external solution. At rest there is no net internal current in the muscle cell and thus no potential difference between these two intracellular electrodes.

An intracellular electrode now is placed in the cell for the purpose of passing current across the plasma membrane. When a small pulse of current is passed out of the cell (positive charges move out of the cell), this will decrease the transmembrane electrical potential difference, or *depolarize* the membrane (Fig. 3-2). A pulse of inward current (positive charges move into the cell) will increase the transmembrane electrical potential difference, or *hyperpolarize* the cell (Fig. 3-2).

■ *Experimental observations of membrane potentials*

■ *Subthreshold responses: the local response*

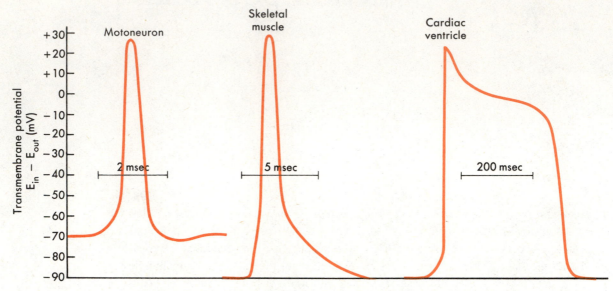

Fig. 3-1 ■ Action potentials (note the different times scales) from three vertebrate cell types. (Redrawn from Flickinger, C.J., et al.: Medical cell biology, Philadelphia, 1979, W.B. Saunders Co.)

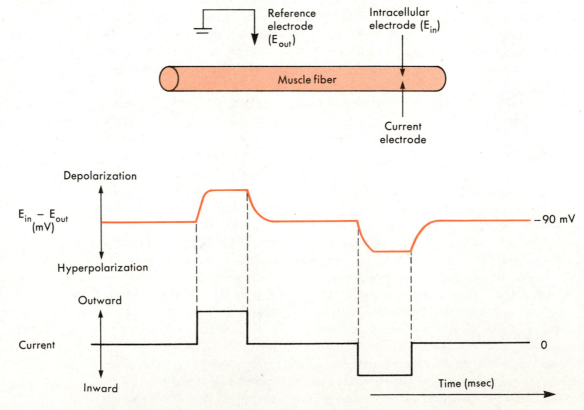

Fig. 3-2 ■ Responses of the transmembrane electrical potential difference to square waves of inward and outward current.

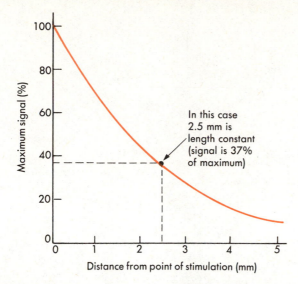

Fig. 3-3 ■ The decrease in magnitude of the local response with distance from the point of stimulation.

The terms *depolarization* and *hyperpolarization* may be confusing at times. A change of the membrane potential from −90 mV to −70 mV is a depolarization, since it is a decrease of the potential difference, or polarization, across the cell membrane. If the membrane potential changes from −90 mV to −100 mV, the polarization across the membrane has increased, so this is hyperpolarization.

When subthreshold current pulses are passed (Fig. 3-2), the size of the potential change observed depends on the distance of the recording microelectrode from the point of current passage. The closer the recording electrode to the site of current passage, the larger the potential change observed. The size of the potential change is found to decrease exponentially with distance from the site of current passage (Fig. 3-3). The distance over which the potential change decreases to 1/e (37%) of its maximum value is called the *length constant* (or *space constant*). A length constant of 2 to 3 mm is typical for mammalian nerve or muscle cells. Because these potential changes are observed primarily near the site of current passage and the changes are not propagated along the length of the cell (as are action potentials), they are called *local responses*.

If progressively larger depolarizing current pulses are applied, a point is reached at which a different sort of response, the *action potential*, occurs (Fig. 3-4). An action potential is triggered when the depolarization is sufficient for the membrane potential to reach a *threshold* value, which is near −60 mV for frog sartorius muscle. The action potential differs from the local depolarizing response in two important ways: (1) it is a much larger response, with the polarity of the membrane potential actually reversing (the cell interior becoming positive with respect to the exterior), and (2) the action potential is *propagated without decrement* down the entire length of the muscle fiber. The size and shape of an action potential remain the same as it travels along the muscle fiber; it does not decrease in size with distance as does the local response. When a stimulus larger than the threshold stimulus is applied, the size and shape of the action potential do not change; there is no increase in the size of the action potential with increased stimulus strength. A stimulus either fails to elicit an action potential (a subthreshold stimulus), or it produces a full-size action potential. For this reason the action potential is an *all-or-none response*.

■ *Action potentials*

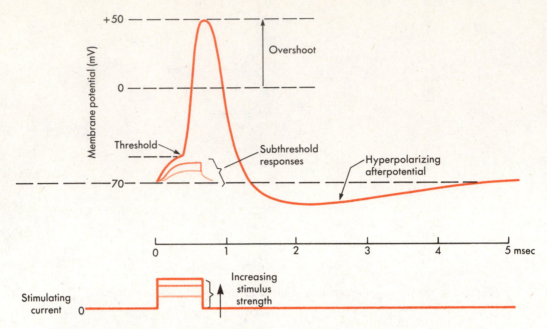

Fig. 3-4 ■ Responses of the membrane potential to increasing pulses of depolarizing current. When the cell is depolarized to threshold, it fires an action potential.

■ *The shape of the action potential*

The form of an action potential of a squid giant axon is shown in Fig. 3-4. Once the membrane is depolarized to the threshold, an explosive depolarization occurs, which completely depolarizes the membrane and even *overshoots* so that the membrane becomes polarized in the reverse direction. The peak of the action potential reaches about +50 mV. The membrane potential then returns toward the resting membrane potential almost as rapidly as it was depolarized. After repolarization a transient hyperpolarization occurs that is known as the *hyperpolarizing afterpotential*. It persists for about 4 msec. The following section discusses the ionic currents that cause the various phases of the action potential.

■ *Ionic mechanisms of the action potential*

In Chapter 2 the resting membrane potential is seen to be a weighted sum of the equilibrium potentials for Na^+, K^+, Cl^-, etc., with the weighting factor for each ion being the fraction it contributes to the total ionic conductance of the membrane (the chord conductance equation). E_K is about -100 mV in squid axon, so an increase in g_K would hyperpolarize the membrane. E_{Cl} is about -70 mV, so an increase in g_{Cl} would stabilize E_m at -70 mV. An increase in g_{Na} of sufficient magnitude would cause depolarization and reversal of the membrane polarity, since E_{Na} is about $+60$ mV in squid axon. A decrease in g_K also would have the effect of depolarizing the membrane.

It has been known for a long time that Na^+ is involved in the action potential. Overton showed in 1902 that Na^+ in the extracellular fluid is required for excitability. Bernstein (also in 1902) proposed that the action potential was caused by a transient breakdown of the barrier properties of the plasma membrane, so that the membrane conductances of all ions increased transiently. Some 50 years later Hodgkin and Huxley showed that the action potential of squid giant axon is caused by successive conductance increases to sodium and potassium ions. Hodgkin and Huxley found that the conductance to Na^+, g_{Na}, increases very rapidly during the early part of the action potential (Fig. 3-5). The sodium conductance reaches a peak about the same time as the peak of the action potential, then it decreases rather rapidly. A delayed increase in the K^+ conductance, g_K, occurs somewhat more slowly, reaches a peak about the middle of the repolarization phase, and then more slowly returns to resting levels.

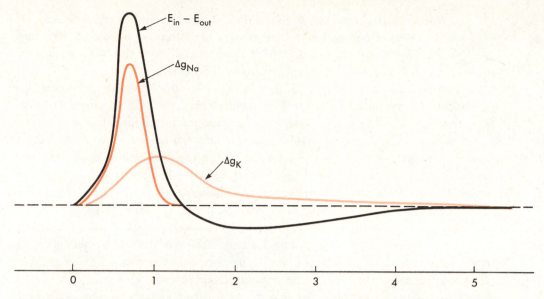

Fig. 3-5 ■ The action potential of squid giant axon with changes in the membrane conductance to sodium and potassium superimposed.

As described in Chapter 2, the chord conductance equation shows that the membrane potential is a result of the opposing tendencies of the K^+ gradient to bring E_m toward the equilibrium potential for K^+ and the Na^+ gradient to bring E_m toward the equilibrium potential for Na^+. Increasing the conductance of either ion will increase its ability to pull E_m toward its equilibrium potential. The rapid increase in g_{Na} during the early part of the action potential causes the membrane potential to move toward the equilibrium potential for Na^+ ($+58$ mV). The peak of the action potential never reaches $+58$ mV but only about $+50$ mV. This is because the conductance of K^+ also increases, albeit more slowly, providing an opposing tendency, and also because g_{Na} tends to turn off after a certain time. The rapid return of the membrane potential toward the resting potential is due to the rapid decrease of g_{Na} and the continued increase in g_K, both of which decrease the size of the Na^+ term in the chord conductance equation and increase the size of the K^+ term. During the hyperpolarizing afterpotential, when the membrane potential is actually more negative than the resting potential (more polarized), g_{Na} has returned to baseline levels, but g_K remains elevated above resting levels so that E_m is pulled closer to the K^+ equilibrium potential (-100 mV) for a short time.

Early in this century it was hypothesized that the action potential was caused by a transient increase of the permeability of the plasma membrane to all ions. Later evidence indicated that the entry of Na^+ into the cell plays a central role in generating the action potential. Removal of Na^+ from the extracellular fluid was found to abolish the action potential. Both the height of the action potential and the maximum rate of depolarization during the action potential depend on the extracellular concentration of Na^+, increasing with increasing external Na^+.

By repetitively stimulating the squid giant axon in a bath containing radioactive $^{22}Na^+$, it was found that 3 to 4 $\times$ 10^{-12} mole of Na^+ enter the axon through each square centimeter of surface area during one action potential. This amount of Na^+ is in good agreement with calculations of the amount of Na^+ that should enter the axon based on electrical considerations. The capacitance (C) of the membrane is defined as dq/dv, the amount of charge that must flow to cause a 1 V change in potential difference. It is known that the capacitance of 1 cm^2 of squid axon membrane is about 1 microfarad. The amount of charge that must flow to discharge the membrane capacitor by 100 mV,

■ *The basis for knowledge of the action potential*

the approximate height of the action potential, is C times V, or 10^{-6} farad $\times$ 10^{-1} V = 10^{-7} coulombs. Dividing this by the amount of charge on 1 mole of Na^+ ions (96,500 coulombs/mole) gives about 10^{-12} mole of Na^+ ions that must enter each square centimeter to depolarize the membrane by 100 mV.

If only Na^+ and K^+ are involved, the amount of K^+ that leaves the cell during the repolarization phase must be equal to the amount of Na^+ that enters during depolarization. The Na^+, K^+-ATPase pumps out the Na^+ that enters and reaccumulates the lost K^+. Because of the extremely small number of ions that cross the membrane during each impulse, the Na^+, K^+ pump is not required in the short run. A squid axon that has been poisoned with cyanide or ouabain so that ion pumping is abolished can fire nearly 100,000 action potentials before failing. Smaller axons, with a much larger ratio of surface to volume, will fail after fewer impulses if the Na^+ pump is poisoned.

■ The voltage clamp technique

Much of the current knowledge of the ionic mechanism of the action potential comes from experiments using the *voltage clamp* technique. The voltage clamp technique uses electronic feedback to rapidly set the transmembrane potential difference to whatever value the experimenter chooses. The voltage clamp circuit holds the membrane potential at this level and measures the net ionic current that flows across the membrane during the clamp.

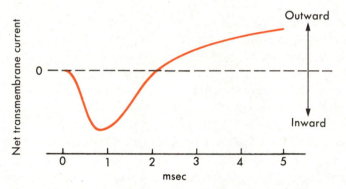

Fig. 3-6 ■ Net transmembrane current flow in a squid giant axon after clamping the membrane potential to zero.

When the membrane of the squid giant axon is depolarized rapidly to some point beyond the threshold, there is an almost instantaneous current that is caused by discharge of the membrane capacitance. This is followed by currents which flow over the same time course as a normal action potential (Fig. 3-6). The inward phase of the current is due to Na^+ entering the axon, and the outward current is due to K^+ leaving. The Na^+ current can be separated from the K^+ current as follows.

1. Enough of the external Na^+ is replaced by choline$^+$ (an impermeant cation), until the Na^+ concentration inside the axon equals that in the extracellular fluid. If the membrane potential then is clamped to zero, there will be no net force causing Na^+ to flow, since there is no concentration difference across the membrane and no electrical potential difference. When this is done, the inward current component disappears (Fig. 3-7, curve *B*), indicating that the inward current is normally caused by Na^+. If E_m then is clamped to a positive value, there is a net electrical force causing Na^+ to leave the axon, so an outward Na^+ current occurs (Fig. 3-7, curve *C*).

2. The involvement of Na^+ in the inward current during the action potential also can be shown without changing the external $[Na^+]$. If E_m is clamped to +58 mV, the Na^+ equilibrium potential, no net flow of Na^+ occurs, and the inward current disappears. Clamping the axon to positive E_m greater than +58 mV causes a net outward Na^+ current. The potential at which the Na^+ current reverses direction (+58 mV) is

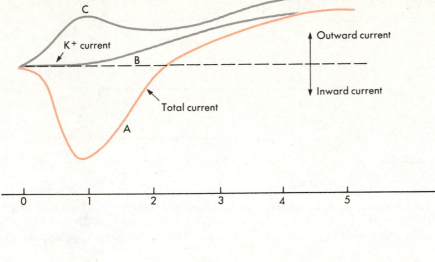

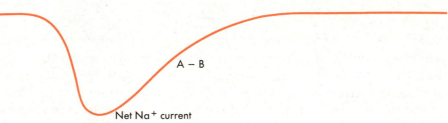

Fig. 3-7 ■ Voltage clamp currents in squid giant axon. *Curve A,* The net membrane current during a clamp to 0 mV in normal extracellular fluid. *Curve B,* Net membrane current during a clamp to 0 mV in a medium in which enough choline has been substituted for sodium so that intracellular and extracellular Na^+ concentrations are equal. This eliminates the Na^+ current. *Curve C,* Membrane current during a clamp to a positive voltage under the conditions of curve B. There is a net tendency for Na^+ to leave the cell during the clamp so that the Na^+ current hump is reversed. *Lower tracing,* The net Na^+ current is computed as curve A (Na^+ and K^+ currents) minus curve B (only K^+ current).

called the *reversal potential* for the Na^+ current. The fact that the reversal potential for the inward current phase of the squid giant axon action potential is equal to the equilibrium potential for Na^+ supports the contention that Na^+ carries most of the current in the inward phase.

In their pioneering studies of action potentials in squid giant axons Hodgkin and Huxley found that rapidly depolarizing the membrane caused a rapid increase in the flow of Na^+ into the axon and a slower increase in the flow of K^+ out of the axon. By separating the Na^+ current from the K^+ current, they were able to determine the conductances, g, for Na^+ and K^+ as a function of time during voltage clamp. [If I and E_m are known, g can be calculated, since $I_{Na} = g_{Na} (E_m - E_{Na})$, and $I_K = g_K (E_m - E_K)$.] When the membrane potential was clamped to zero, the conductances shown in Fig. 3-8 were obtained. The increase in g_{Na} shuts off even though the clamp is maintained. The increase in g_K, however, remains and does not decrease to resting levels until the clamp is released. Smaller depolarizations cause smaller changes in g_K and g_{Na}.

In summary, it can be said that there are three basic events in the action potential: the increase of g_{Na}, the return of g_{Na} to resting levels, and the increase of g_K. The rate and extent of each of these depends on the level of the membrane potential. By doing voltage clamp experiments at different E_m levels, Hodgkin and Hukley were able to reconstruct the changes in g_{Na} and g_K that must occur during the normal action potential.

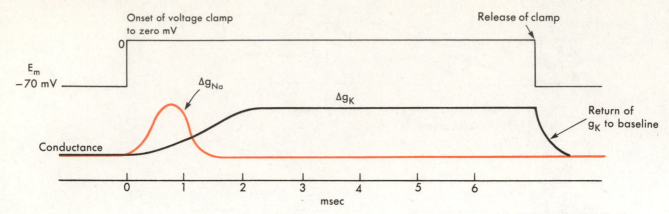

Fig. 3-8 ■ Changes in sodium and potassium conductances in squid giant axon following a clamp to $E_m = 0$. The increase in sodium conductance (g_{Na}) is self-inactivating. The increase in potassium conductance (g_K) is delayed and persists as long as the clamp is maintained.

■ *Ion channels and gates*

Hodgkin and Huxley proposed that ion currents pass through separate Na^+ and K^+ channels, each with distinct characteristics, in the plasma membrane. Recent research supports this interpretation and has determined some of the properties of the channels (Fig. 3-9). The Na^+ channel appears to have both an *activation gate* and an *inactivation gate* that account for the changes in g_{Na} during voltage clamp. Digestion of the interior of the squid axon with the proteolytic enzyme pronase does away with the self-inactivation of g_{Na} by destroying the inactivation gate. To enter the part of the channel known as the selectivity filter, it is believed that K^+ and Na^+ must shed most of their water of hydration. This allows Na^+ and K^+ to form coordination bonds with oxygen atoms in the interior of the selectivity filter. The configuration of the oxygen atoms in the selectivity filter determines the specificity of the Na^+ and K^+ channels. Tetrodotoxin (TTX) is a specific blocker of the Na^+ channel, and tetraethylammonium ion (TEA^+) blocks the K^+ channel. The sites at which these agents are believed to block the channels are shown in Fig. 3-9.

The gates of the channels may be charged and probably are peptide chains. The opening and closing of the gates cause redistributions of charges across the membrane that are measurable as small *gating currents*.

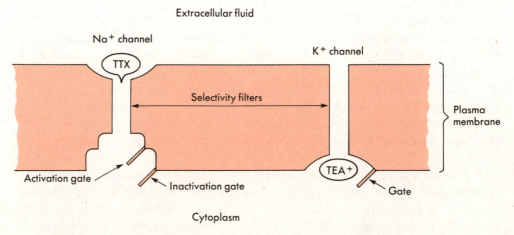

Fig. 3-9 ■ Certain features of the Na^+ and K^+ channels of squid giant axon.

An action potential in a cardiac ventricular cell is schematically shown in Fig. 3-1. The initial rapid depolarization and overshoot is caused by the rapid entry of Na^+ through channels that are very similar to the Na^+ channels of nerve and skeletal muscle. Because of the rapid kinetics of opening and closing of these channels, they are called *fast Na^+ channels*.

After the initial depolarization and overshoot (together called the spike) the cardiac ventricular action potential has a *plateau phase*. The plateau is due to another set of channels that are distinct from the fast Na^+ channels. These channels open and close much more slowly than the fast Na^+ channels and are called *slow channels*. The slow channels conduct both Na^+ and Ca^{++} ions, and the Ca^{++} that enters the cell during the plateau phase plays an important role in initiating contraction of the ventricular cell.

The repolarization of the ventricular cell is brought about by the closing of the slow channels and by a much delayed opening of K^+ channels.

The ionic mechanism of the cardiac action potential is discussed in more detail in Chapter 27.

Action potentials vary considerably among different types of smooth muscle. Characteristically action potentials in smooth muscle have slower rates of depolarization and repolarization and less overshoot than skeletal muscle action potentials. Smooth muscle cells lack fast Na^+ channels. The depolarizing phase of smooth muscle action potentials is primarily caused by channels that resemble the cardiac slow channels in their kinetics and in conducting both Na^+ and Ca^{++}. The Ca^{++} that enters via the slow channels is often vital for excitation-contraction coupling in smooth muscle. Repolarization is caused by the closing of the slow Na^+/Ca^{++} channels and a simultaneous opening of K^+ channels.

If a neuron or skeletal muscle cell is depolarized, for example, by increasing the concentration of K^+ in the extracellular fluid, its action potential has a slower rate of rise and a smaller overshoot. This is a result of two factors: (1) a smaller electrical force driving Na^+ into the depolarized cell, and (2) *voltage inactivation* of the Na^+ channels. The increase in g_{Na} during the action potential is self-inactivating. Once the Na^+ channels are inactivated, the membrane must be repolarized toward the normal resting membrane potential before the channels can be reopened. As the membrane potential is restored toward normal resting levels, more and more of the Na^+ channels again become capable of being activated. Since the action potential mechanism requires a critical density of open Na^+ channels, an action potential may not be generated in response to stimulation when a considerable fraction of Na^+ channels is inactivatable because of depolarization. This is called *voltage inactivation* of the action potential because of voltage inactivation of the Na^+ channels. Voltage inactivation of the Na^+ channels is involved in important properties of excitable cells, such as refractoriness and accommodation.

During much of the action potential the membrane is completely refractory to further stimulation. This means that, no matter how strongly the cell is stimulated, it is unable to fire a second action potential. This is called the *absolute refractory period* (Fig. 3-10). The cell is refractory because a large fraction of its Na^+ channels is voltage inactivated and cannot be reopened until the membrane is repolarized.

During the last part of the action potential the cell is able to fire a second action potential, but a stronger than normal stimulus is required. This is the *relative refractory period*. Early in the relative refractory period, before the membrane potential has returned to the resting potential level, some Na^+ channels are voltage inactivated, so a

■ *Action potentials in cardiac and smooth muscle*
■ *Cardiac muscle*

■ *Smooth muscle*

■ *Properties of action potentials*
■ *Voltage inactivation of the action potential*

■ *The refractory periods*

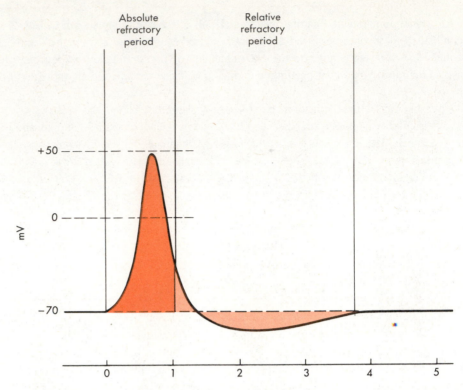

Fig. 3-10 ■ The action potential of nerve showing the absolute and relative refractory periods.

stronger than normal stimulus is needed to open the critical number of Na^+ channels needed to trigger an action potential. Throughout the relative refractory period the conductance to K^+ is elevated, which results in increased opposition to depolarization of the membrane. This also contributes to the refractoriness.

■ *Accommodation to slow depolarization*

When a nerve or muscle cell is depolarized slowly, the threshold may be passed without an action potential being fired. This is called *accommodation*. Na^+ and K^+ channels both are involved in accommodation. During slow depolarization some of the Na^+ channels that are opened by depolarization have enough time to become inactivated before the threshold potential is attained. If depolarization is slow enough, the critical number of open Na^+ channels required to trigger the action potential may never be achieved. In addition, K^+ channels open in response to the depolarization. The increased g_K tends to repolarize the membrane, making it still more refractory to depolarization. Because of accommodation, a very weak stimulus will not elicit an action potential no matter how long it is applied.

■ *The strength-duration curve*

A stimulus depolarizes the cell by causing charge to flow across the plasma membrane. The relevant quantity is the *total amount* of charge that flows across the membrane (current × time). A strong stimulus depolarizes the membrane to threshold quickly. A weaker stimulus must be applied longer for the critical amount of charge to flow across the membrane. This often is depicted in the *strength-duration curve,* which is a plot of the strength of stimuli versus the minimum time they must be applied to cause an action potential (Fig. 3-11).

A very weak stimulus will not cause an action potential even if applied for a very long time (accommodation). The smallest stimulus strength that can elicit an action potential from a particular preparation is called the *rheobase.* The time needed by a stimulus *twice* as strong as the rheobase to elicit an action potential is called the *chron-*

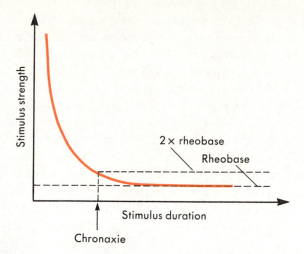

Fig. 3-11 ■ The strength-duration curve. The ordinate shows stimulus strength, and the abscissa shows the minimum time a stimulus of that strength must be applied to produce an action potential.

axie of the preparation. Chronaxie is useful as an index of the excitability of a preparation. The larger the chronaxie, the less excitable the preparation.

A principal function of neurons is to transfer information by the conduction of action potentials. The axons of the motor neurons of the ventral horn of the spinal cord conduct action potentials from the cell body of the neuron in the spinal cord to a number of skeletal muscle fibers. The distance from the motor neuron to one of the muscle fibers it innervates may be as long as 1 m. The mechanism of action potential conduction and the factors that determine the speed of conduction are considered next.

■ *Conduction of the
action potential*

Fig. 3-12, *A,* shows the membrane of an axon or muscle fiber that has been depolarized in a small area. In the depolarized region the external aspect of the membrane is

■ *The local response*

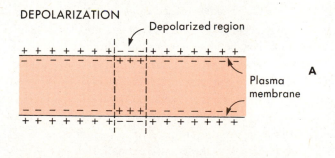

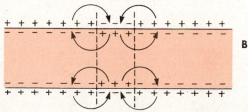

Fig. 3-12 ■ Mechanism of electrotonic spread of depolarization. **A,** The reversal of membrane polarity that occurs with local depolarization. **B,** The local currents that flow to depolarize adjacent areas of the membrane and allow spreading of the depolarization.

negative relative to the adjacent membrane, and the internal face of the depolarized membrane is positively charged relative to neighboring internal areas. The potential differences cause local currents to flow (Fig. 3-12, *B*), which depolarizes the membrane adjacent to the initial site of depolarization. These newly depolarized areas then cause current flows which depolarize other segments of the membrane that are still further removed from the initial site of depolarization. This depolarization spread is called the *local response*. This mechanism of conduction is known as *electrotonic conduction*.

■ *Conduction velocity*

The speed of electrotonic conduction along a nerve or muscle fiber is determined by the membrane capacitance and the electrical resistance to the flow of current. The cell membrane is a capacitor (an insulator separating two conductors). The membrane potential is a measurable manifestation of the charge stored by the membrane capacitor. The amount of charge that must flow to depolarize 1 cm^2 of membrane is proportional to the membrane capacitance (the coulombs of charge stored per volt of potential difference across the membrane) per unit area. A typical value of membrane capacitance (C_m) is about 10^{-6} farad/cm^2 membrane. To depolarize the membrane from -100 mV (-0.1 V) to 0 mV, 10^{-7} coulombs of charge must flow across each square centimeter of membrane:

$$\text{Charge flow/cm}^2 = \text{Capacitance/cm}^2 \times \text{Voltage change}$$
$$= 10^{-6} \text{ farad/cm}^2 \times 0.1 \text{ V}$$
$$= 10^{-7} \text{ coulombs/cm}^2$$

The membrane capacitance thus determines *how much charge* must flow to depolarize the membrane. The larger the membrane capacitance, the greater the amount of charge that must flow, and the slower the rate of electrotonic spread.

The resistance to electrotonic current flow determines *how rapidly* charge can flow. The resistance to electrotonic current flow depends on the resistance to current flow across the membrane (R_m) and the resistance to longitudinal current flow in the cytoplasm (R_{in}). The effective resistance is proportional to the geometric mean of R_m and R_{in} ($\sqrt{R_m R_{in}}$). The larger $\sqrt{R_m R_{in}}$, the slower will electrotonic current flow, and the slower will be the rate of electrotonic conduction.

The product $\sqrt{R_m R_{in}} \, C_m$ has units of time and is proportional to the time required for a given voltage change to occur by electrotonic conduction. The smaller $\sqrt{R_m R_{in}} \, C_m$, the more rapidly electrotonic conduction can occur, and vice versa.

Effect of fiber size on conduction velocity. Consider a cylindrical nerve or muscle cell. The surface area of the cell increases with increasing radius ($A_s = 2\pi rl$). The cross-sectional area of the cell increases with the *square* of the radius ($A_x = \pi r^2$). The membrane capacitance increases in direct proportion to membrane area. The membrane resistance decreases in proportion to an increase in membrane area. The internal resistance is inversely proportional to the cross-sectional area. The effect of doubling the radius of the cell thus will increase C_m by a factor of 2, decrease R_m by a factor of 2, and decrease R_{in} by a factor of 4. Thus C_m will increase twofold, but $\sqrt{R_m R_{in}}$ will decrease by a factor of $\sqrt{2 \cdot 4}$, or $2\sqrt{2}$. The product $\sqrt{R_m R_{in}} \, C_m$ thus will decrease to $1/\sqrt{2}$ of its former value with a doubling of cell radius, and the velocity of electrotonic conduction will increase to $\sqrt{2}$ times the conduction velocity of the smaller fiber. Thus larger fibers have larger conduction velocities.

■ *Electrotonic conduction involves decrement*

Earlier in this chapter it is noted that the local response dies away to almost nothing over the course of several millimeters (Fig. 3-3). A nerve or muscle fiber has some of the properties of an electrical cable. In a perfect cable the insulation surrounding the core conductor prevents all current loss so that a signal is transmitted along the cable with undiminished strength. The plasma membrane of an unmyelinated nerve or muscle fiber serves as the insulation. The membrane has a resistance much higher than the resistance of the cytoplasm, but (partly because of its thinness) the plasma membrane is not a perfect insulator. The higher the ratio of R_m to R_{in}, the better the cell can function as a cable, and the longer the distance that a signal can be transmitted electrotonically without significant decrement. $\sqrt{R_m/R_{in}}$ determines the *length constant* of a cell. The length constant is the distance over which an electrotonically conducted signal falls to 37% (1/e) of its initial strength. A typical length constant for unmyelinated mammalian nerve and muscle fibers is about 2 to 3 mm. Some axons in the human body are about 1 m long, so it is clear that the local response cannot conduct a signal over so great a distance.

■ *The action potential as a self-reinforcing signal*

Many nerve and muscle fibers are much longer than their length constants. The action potential serves to conduct an electrical impulse with undiminished strength along the full length of those fibers. To do this, the action potential *reinforces itself* as it is propagated along the fiber. The propagation of the action potential occurs by the mechanism depicted in Fig. 3-12. When the areas on either side of the depolarized region reach threshold, these areas also fire action potentials, which locally reverses the polarity of the membrane potential. By local current flow the areas of the fiber adjacent to these areas are brought to threshold, and then these areas in turn fire action potentials. There is a cycle of depolarization by local current flow followed by generation of an action potential in a restricted region that then travels along the length of the fiber, with "new" action potentials being generated as they spread. In this way the action potential propagates over long distances, keeping the same size and shape.

Since the shape and size of the action potential are ordinarily invariant, only variations in the frequency of the action potentials can be used in the code for information transmission along axons. The maximum frequency is limited by the duration of the absolute refractory period (about 1 msec) to about 1000 impulses/second in large mammalian nerves.

■ *Effect of myelination on conduction velocity*

A squid giant axon with 500 µm diameter has a conduction velocity of 25 m/second and is unmyelinated. If conduction velocity were directly proportional to fiber radius, a human nerve fiber with a 10 µm diameter would conduct at 0.5 m/second. With this conduction velocity a reflex withdrawal of the foot from a hot coal would take about 4 seconds. Even though our nerve fibers are much smaller in diameter than squid giant axons, our reflexes are much faster than this. The myelin sheath that surrounds certain vertebrate nerve fibers results in a much greater conduction velocity than that of nonmyelinated fibers of similar diameters. A 10 µm *myelinated* fiber has a conduction velocity about 50 m/second, which is two times greater than that of the 500 µm squid giant axon. The high conduction velocity permits reflexes that are fast enough to allow us to avoid dangerous stimuli. Fig. 3-13 shows the large increase in conduction velocity

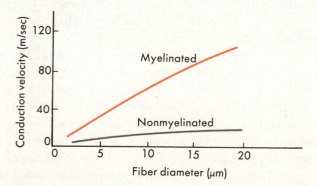

Fig. 3-13 ■ Conduction velocities as functions of fiber diameter for myelinated and nonmyelinated vertebrate axons. (Redrawn from Flickinger, C.J., et al.: Medical cell biology, Philadelphia, 1979, W.B. Saunders Co.)

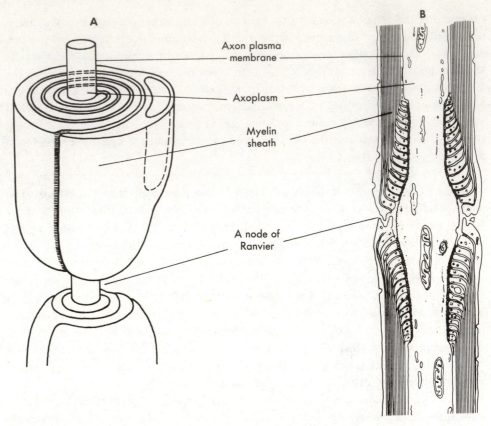

Fig. 3-14 ■ The myelin sheath. **A,** Schematic drawing of Schwann cells wrapping around an axon to form a myelin sheath. **B,** Drawing of a cross section through a myelinated axon near a node of Ranvier. (Redrawn from Elias, H., et al.: Histology and human microanatomy, ed. 4, New York, 1978, John Wiley & Sons, Inc.)

caused by myelination. As discussed later, the myelin sheath increases the velocity of action potential conduction by decreasing the capacitance of the axon and by allowing action potentials to be generated only at the nodes of Ranvier.

The evolution of myelinated fibers in vertebrates is significant. If each of our peripheral nerve fibers had to be as large as a squid giant axon, then peripheral nerve trunks (each containing hundreds of nerve fibers) would be so large that the human form would have to be altered just to accommodate the larger peripheral nerves.

Myelination is caused by the wrapping of Schwann cell plasma membrane around a nerve axon (Fig. 3-14). This results in a myelin sheath consisting of several to 100 or more layers of plasma membrane. Gaps that occur in the sheath every 1 to 2 mm are known as nodes of Ranvier. Nodes of Ranvier are about 1 μm wide and are the lateral spaces between different Schwann cells along the axon.

Myelination greatly alters the electrical properties of the axon. The many wrappings of membrane around the axon greatly increase the effective *membrane resistance,* so that R_m/R_{in} and thus the length constant is much greater. Less of a conducted signal is lost through the electrical insulation of the myelin sheath, so that the amplitude of a conducted signal declines more slowly with distance along the axon.

Each Schwann cell membrane has a capacitance similar to that of the plasma membrane of the axon. The capacitances of the membranes in the myelin sheath act as though they were connected in series. Capacitors in series are added according to the equation $1/C_t = 1/C_1 + 1/C_2 + 1/C_3 + \ldots$ etc., where C_t is the effective overall capacitance, and C_1, C_2, C_3, etc., are the individual capacitances. If there are 50 identical capacitances in series (25 Schwann cell wraps), then $1/C_t = 50/C$, and $C_t = C/50$. A myelin sheath with 50 membranes thus *lowers* the membrane capacitance by a factor of 50. Resistances in series, however, add directly, so the myelination will increase the membrane resistance by fiftyfold. Earlier in this chapter we saw that the time constant $\sqrt{R_m R_{in}} \, C_m$ determines the electrotonic conduction velocity. The smaller this product, the greater the conduction velocity. The 25 Schwann cell wraps should have no effect on R_{in} but will lower C_m by a factor of 50 and increase R_m by a factor of 50. Thus $\sqrt{R_m R_{in}} \, C_m$ will decrease by $50/\sqrt{50}$, resulting in sevenfold increase in electrotonic conduction velocity because of myelination.

Another property of conduction in myelinated fibers that enhances conduction velocity is called *saltatory conduction,* because the impulse "jumps" from one node of Ranvier to the next. Saltatory conduction occurs because the action potential is regenerated only at the nodes (1 to 2 mm apart). The action potential in myelinated fibers is not regenerated continuously as the impulse is propagated. The internodal plasma membrane cannot produce action potentials. This is because depolarization of the internodal membrane is divided among 50 or so membranes of the myelin sheath. The resulting depolarization of the internodal plasma membrane is only 1 or 2 mV—not nearly sufficient to reach threshold.

In summary, conduction in myelinated axons is characterized by rapid electrotonic conduction with little decrement between the nodes of Ranvier. Only at the nodes does the action potential "pause" to be regenerated.

Myelinated axons are also more metabolically efficient than nonmyelinated axons. The sodium-potassium pump is responsible for extruding the sodium that enters and for reaccumulating the potassium that leaves the cell during action potentials. In a myelinated axon, ionic currents are restricted to the small fraction of the membrane surface at the nodes of Ranvier. For this reason fewer Na^+ and K^+ ions traverse a unit area of membrane, and less ion pumping is required to maintain Na^+ and K^+ gradients.

■ *Bibliography*

Aidley, D.J.: The physiology of excitable cells, ed. 2, Cambridge, 1978, Cambridge University Press.

Davson, H.: A textbook of general physiology, ed. 4, Baltimore, 1970, Williams & Williams Co.

Hodgkin, A.L.: The conduction of the nervous impulse, Springfield, Ill., 1964, Charles C Thomas, Publisher.

Kandel, E.R., and Schwartz, J.H.: Principles of neural science, New York, 1981, Elsevier.

chapter 4

Synaptic transmission

A synapse is a site at which an impulse is transmitted from one cell to another. There are two types of synapses: electrical synapses and chemical synapses. At an *electrical synapse* two excitable cells communicate by the direct passage of electrical current between them. This is called *ephaptic* or *electrotonic* transmission. *Gap junctions* link electrotonically coupled cells and are thought to be low-resistance pathways for current flow directly between the cells. There are few well-studied examples of ephaptic transmission in the vertebrate central nervous system.

Information more often is transferred between excitable cells by means of *chemical synapses*. Chemical synapses may be better suited for the complex modulation of synaptic activity and the integration that occurs at synapses in vertebrate central nervous systems. At a chemical synapse an action potential causes a transmitter substance to be released from the presynaptic neuron. The transmitter diffuses across the extracellular synaptic cleft and binds to receptors on the membrane of the postsynaptic cell to cause a change in its electrical properties. Chemical synapses have *synaptic delay*—the time required for these events to occur. The neuromuscular (or myoneural) junction is a particularly well-studied vertebrate chemical synapse.

Although the nature of the presynaptic and postsynaptic cells, the structure of the synapse, and the transmitter substance vary, there are certain characteristics that chemical synapses have in common.

General characteristics of transmission at chemical synapses

Action potential in presynaptic cell

↓

Depolarization of the plasma membrane of the presynaptic axon terminal

↓

Release by the presynaptic terminal of the transmitter

↓

Chemical combination of the transmitter with specific receptors on the plasma membrane of the postsynaptic cell

↓

Transient change in the conductance of the postsynaptic plasma membrane to specific ions

↓

Transient change in the membrane potential of the postsynaptic cell

The synapses between the axons of motor neurons and skeletal muscle fibers are called neuromuscular junctions, myoneural junctions, or motor endplates. The neuromuscular junction was the first vertebrate synapse to be well characterized. The neuromuscular junction serves as a model chemical synapse that provides a basis for understanding more complex synaptic interactions among neurons in the central nervous system. Katz and his associates have made many notable contributions to our understanding of events at the neuromuscular junction.

■ *Neuromuscular junctions*

Near the neuromuscular junction the motor nerve loses its myelin sheath and divides into fine terminal branches (Fig. 4-1). The terminal branches of the axon lie in synaptic troughs on the surfaces of the muscle cells (Fig. 4-1). The plasma membrane of the muscle cell lining the trough is thrown into numerous junctional folds. The axon terminals contain many 400 Å smooth-surfaced synaptic vesicles that contain acetylcholine. The axon terminal and the muscle cell are separated by the synaptic cleft, which contains a carbohydrate-rich amorphous material.

There is evidence which suggests that the acetylcholine receptor molecules are concentrated on the crests of the junctional folds. Acetylcholinesterase appears to be evenly distributed on the external surface of the postsynaptic membrane. The synaptic vesicles in the nerve terminals are concentrated opposite the mouths of the junctional folds.

■ *Structure of the neuromuscular junction*

The action potential is conducted down the motor axon to the presynaptic axon terminals. Depolarization of the plasma membrane of the axon terminal brings about a transient increase in its calcium conductance. Ca^{++} flows down its electrochemical potential gradient into the axon terminal. In ways that are not yet well understood the influx of Ca^{++} causes synaptic vesicles to fuse with the plasma membrane and to empty their acetylcholine into the synaptic cleft by exocytosis. Acetylcholine diffuses across the synaptic cleft and combines with a specific receptor protein on the external surface of the muscle plasma membrane of the motor endplate. The combination of acetylcholine with the receptor protein causes a transient increase in the conductance of the post-

■ *Overview of neuromuscular transmission*

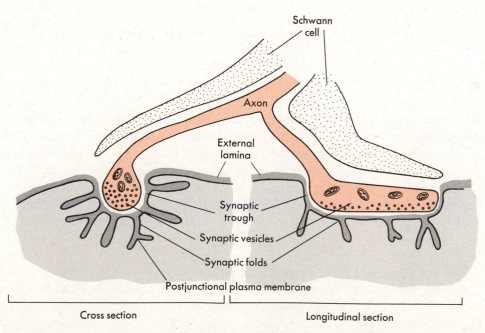

Fig. 4-1 ■ The ultrastructure of the neuromuscular junction in skeletal muscle. (Redrawn from Flickinger, C.J., et al.: Medical cell biology, Philadelphia, 1979, W.B. Saunders Co.)

junctional membrane to Na^+ and K^+. Ionic currents (Na^+ and K^+) result in a transient depolarization of the endplate region. The transient depolarization is called the *endplate potential,* or *EPP*. The EPP is transient because the action of ACh is ended by the hydrolysis of ACh to form choline and acetate. The hydrolysis is catalyzed by the enzyme acetylcholinesterase, which is present in high concentration on the postjunctional membrane. The molecular mechanism whereby binding of acetylcholine to its receptors brings about increased Na^+ and K^+ conductance is not yet understood in detail.

The postjunctional plasma membrane of the neuromuscular junction is not electrically excitable and does not fire action potentials. After it is depolarized, adjacent regions of the muscle cell membrane are depolarized by electrotonic conduction. When those regions reach threshold, action potentials are generated. Action potentials are propagated along the muscle fiber at high velocity and induce the muscle cells to contract. The steps involved in neuromuscular transmission are listed below, and some are considered next in more detail.

Summary of events occurring during neuromuscular transmission

Action potential in presynaptic motor axon terminals

↓

Increase in Ca^{++} permeability and influx of Ca^{++} into axon terminal

↓

Release of ACh from synaptic vesicles into the synaptic cleft

↓

Diffusion of ACh to postjunctional membrane

↓

Combination of ACh with specific receptors on postjunctional membrane

↓

Increase in permeability of postjunctional membrane to Na^+ and K^+ causes EPP

↓

Depolarization of areas of muscle membrane adjacent to endplate and initiation of an action potential

■ Synthesis of acetylcholine

Motor neurons and their axons synthesize acetylcholine. Most other cells are not able to make acetylcholine. The enzyme choline-*O*-acetyltransferase in the motor neuron catalyzes the condensation of acetyl coenzyme A (acetyl CoA) and choline. Acetyl CoA is produced by the neuron (as well as by most cells). Choline cannot be synthesized by the motor neuron and is obtained by active uptake from the extracellular fluid. The plasma membrane of the motor neuron has a transport system that can accumulate choline against a large electrochemical potential gradient. About half the choline that is freed in the synaptic cleft when acetylcholine is hydrolyzed is actively taken back up into the motor neuron to be used in the resynthesis of acetylcholine.

■ Quantal release of transmitter

Even if the motor neuron is not stimulated, small depolarizations of the postjunctional muscle cell occur spontaneously. These small spontaneous depolarizations are known as *miniature endplate potentials,* or *MEPPs*. They occur at random times with a frequency that averages about 1 per second. Each MEPP depolarizes the postjunctional membrane by only about 0.4 mV. The MEPP has the same time course as an EPP that is evoked by an action potential in the nerve terminal. The MEPP is similar to the EPP in its response to most drugs. The EPP and the MEPP both are prolonged to the same extent by drugs that inhibit acetylcholinesterase, and both are similarly depressed by compounds that compete with acetylcholine for binding to the receptor protein. The frequency of MEPPs varies greatly in time, but their amplitude is quite constant. The MEPP is caused by the spontaneous release of a very small number of vesicles (probably only one vesicle) of transmitter into the synaptic cleft.

The quantal nature of transmitter release has been shown in another way. Extracellular Ca^{++} is absolutely required for transmitter release. If the extracellular Ca^{++} is reduced to very low levels, the size of the EPP that is evoked by stimulation of the motor neuron is greatly reduced. Under these conditions spontaneous variations occur in the size of stimulation-evoked EPPs. The size of the EPP does not vary continuously, but in small steps. Statistical analysis shows that the size of the steps corresponds to the size of a single MEPP.

Recently evidence has been presented that acetylcholine release also can occur via a pathway that does not involve membrane-bounded vesicles. The relative roles of transmitter release from synaptic vesicles and via the vesicle-independent pathway remain to be determined.

■ *Action of cholinesterase and re-uptake of choline*

Acetylcholinesterase is concentrated on the external surface of the postjunctional membrane and the basal lamina. The termination of the EPP is brought about by the hydrolysis of acetylcholine. Eserine and edrophonium are inhibitors of the enzyme and thus are called *anticholinesterases*. In the presence of an anticholinesterase the EPP is larger and dramatically prolonged.

Approximately half the choline released by hydrolysis of ACh is taken back up into the presynaptic terminal to be used in resynthesizing the transmitter. Hemicholiniums are drugs that block the choline transport system and inhibit choline uptake. Prolonged stimulation of the motor nerve in the presence of a hemicholinium results in depletion of the store of transmitter and ultimately brings about a decrease in the acetylcholine content of the quanta.

■ *The ionic mechanism of the EPP*

The cation channels that ACh causes to open in the postjunctional membrane differ from the cation channels of the nerve and muscle membranes by being independent of the membrane potential. The postjunctional channels are gated by the action of ACh rather than by the transmembrane potential. The ACh-dependent cation channels of the postjunctional membrane are permeable only to cations, but the channels are not very selective among small cations. Na^+, K^+, Rb^+, and NH_4^+ pass through these channels with roughly equal ease.

In Chapter 2 it is shown that the membrane potential may be determined primarily by the membrane conductances to K^+ and Na^+, as shown by the chord conductance equation (short form).

$$E_m = \frac{g_K}{g_K + g_{Na}} E_K + \frac{g_{Na}}{g_K + g_{Na}} E_{Na}$$

If g_{Na} were to become equal to g_K due to the action of acetycholine, the conductance equation would become

$$E_m = \frac{1}{2}(E_K) + \frac{1}{2}(E_{Na})$$

During an EPP the membrane potential would tend toward the average of the equilibrium potentials of K^+ and Na^+, or about -15 mV. This value for the ''equilibrium potential'' of the EPP was determined by experiments in which EPPs were evoked by stimulation of the motor neuron while the postjunctional resting potential before stimulation was set at various levels by passing current into the postjunctional muscle cell

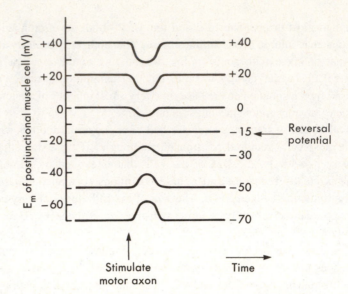

Fig. 4-2 ■ The results of an experiment to determine the reversal potential of the EPP. The resting potential of the postjunctional muscle cell before the nerve impulse is adjusted to various levels by passing current with an intracellular electrode. The motor axon then is stimulated, and the resulting postsynaptic potential change is recorded. (Redrawn from Flickinger, C.J., et al.: Medical cell biology, Philadelphia, 1979, W.B. Saunders Co.)

(Fig. 4-2). When the postjunctional resting potential was more negative than about -15 mV, the EPP caused depolarization of the postjunctional membrane. The depolarizations caused by the EPP became smaller and smaller as the postjunctional E_m was made less negative. For a postjunctional resting potential of about -15 mV the EPP caused no change in the postjunctional membrane potential, showing that this is the "equilibrium" value for the EPP. When the postjunctional E_m before stimulation was set to values less negative than -15 mV, the direction of the EPP reversed, with the EPP tending to return the postjunctional E_m toward -15 mV. Thus the equilibrium value of the postjunctional resting potential is more appropriately called the *reversal potential* of the EPP.

■ *The acetylcholine receptor protein*

Recently the acetylcholine receptor protein has been studied intensively. Development of methods for isolating and purifying hydrophobic membrane proteins and the availability of snake venom neurotoxins that bind very tightly to the ACh receptor have been essential in these studies. α-Bungarotoxin, from the venom of the Formosan krait (a relative of the cobra), is a useful toxin that binds to the ACh receptor almost irreversibly.

Binding of radioactively labeled α-bungarotoxin by neuromuscular junctions suggests that there are 10^7 to 10^8 binding sites per motor endplate. At mouse diaphragm neuromuscular junctions the ACh binding sites concentrated on the crests of the postjunctional folds have a density of about 20,000 per μm^2. This suggests that the receptor molecules are quite tightly packed, since the maximum density possible has been estimated to be about 50,000 per μm^2.

The acetylcholine receptor protein is an integral membrane protein and is deeply embedded in the hydrophobic lipid matrix of the postjunctional membrane. Cholinesterase, on the other hand, is loosely associated with the surface of the postjunctional membrane by hydrophilic interactions.

Acetylcholine receptor protein has been isolated from certain neuromuscular junctions and from the electroplax of electrical fish, where the receptor concentration is especially high, and the receptor protein has been extensively purified by affinity chromatography. Purified receptor protein has been reconstituted into lipid bilayer model membrane. In such systems binding of acetylcholine by the receptor protein leads to an increase in the Na^+ and K^+ conductance of the lipid bilayer. One peptidic fragment of the receptor has been found to increase the cation permeability of lipid bilayers. This suggests that the cation channel which is opened during the EPP may be one part of the receptor protein complex.

Defects in neuromuscular transmission are involved in myasthenia gravis. Experiments involving tritiated bungarotoxin binding suggest that in this disease there is a decreased density of ACh receptors on the postjunctional membrane. Most myasthenic patients have circulating antibodies to ACh receptor protein. Whether these antibodies are the cause of myasthenia gravis or are produced secondarily as the result of endplate degeneration remains to be determined.

■ Synapses between neurons

Chemical synapses between neurons are more prevalent than electrical synapses and are better understood. Chemical transmission between neurons has many of the same properties that characterize the neuromuscular junction. Electrical synapses have been described in the central nervous systems of animals from invertebrates to mammals. The prevalence of electrical synapses in the nervous systems of higher animals and their physiological roles are still poorly defined.

■ Electrical synapses

Electrical synapses do not have the synaptic delay (about 0.5 msec) that is characteristic of chemical synapses. For this reason electrical synapses may be important in reflex pathways where speed of transmission is essential. In general, electrical synapses allow transmission in both directions, and in this respect they differ from chemical synapses, which are obligatorily unidirectional. Electrical synapses that conduct more readily in one direction than in the other have been described. This property is called *rectification*.

Responses of chemical synapses to drugs have been useful in working out the physiological mechanisms of chemical transmission. Because of their simple physical mechanism of conduction, electrical synapses are not readily altered by pharmacological agents. The *coupling ratio* (the ratio of sizes of presynaptic and postsynaptic impulses) does vary from one electrical synapse to another, but it is not known whether the coupling ratio is subject to physiological regulation.

The vertebrate electrical synapses that are best characterized are those on the Mauthner cells of the medullas of teleost fishes. Mauthner cells are very large and have numerous synaptic inputs, both chemical and electrical. Excitatory and inhibitory synapses of both types have been described. The interaction of both chemical and electrical synaptic inputs to the same postsynaptic neuron suggests that complex integrative activity may occur at the level of this postsynaptic neuron. The functions of electrical synapses in mammalian species remain to be elucidated.

■ Chemical synapses

When one neuron makes a chemical synapse with another, the presynaptic nerve terminal characteristically broadens to form a *terminal bouton*. At the synapse itself the presynaptic and postsynaptic membranes come more or less into close apposition and lie parallel to one another. Substantial structures stabilize the synaptic association, so that when nervous tissue is disrupted, the relationship of the presynaptic and postsynaptic membranes at the synapse often is preserved.

Electron micrographs of synapses in the central nervous system often show areas of high electron density subjacent to the plasma membranes in the region of synaptic contact. Synapses where this electron-dense region is of roughly similar extent and intensity on both presynaptic and postsynaptic sides of the synapse are called *symmetrical syn-*

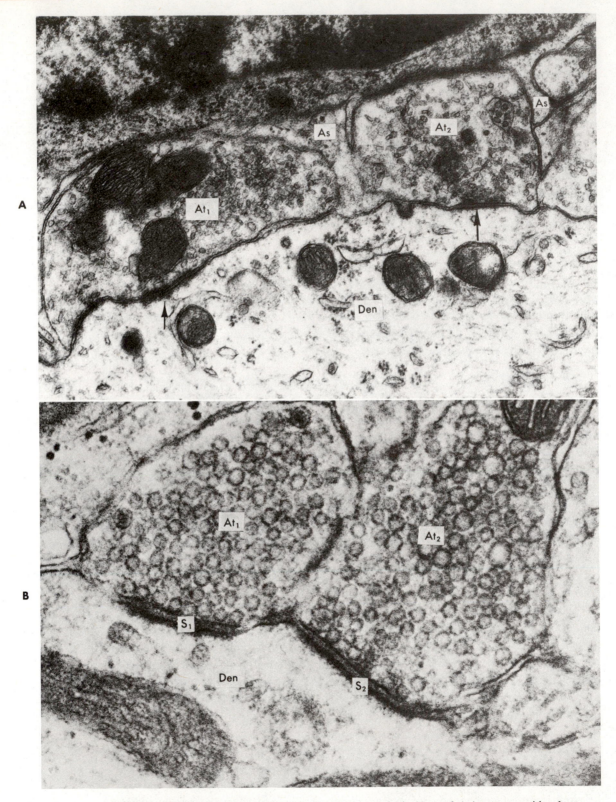

Fig. 4-3 ■ **A,** Symmetrical synapses. Two axon terminals (At_1 and At_2) synapse with a large dendrite *(Den)* in the anterior horn of the spinal column. Note that the presynaptic and postsynaptic electron densities *(arrow)* are similar in extent. Astroglial cells *(As)* invest the axon terminals. **B,** Asymmetrical synapses (S_1 and S_2) in the cerebral cortex. Two axon terminals synapse with dendrite of a stellate cell. Note the greater extents of the electron densities on the postsynaptic (dendrite) side of the two synapses. (From Peters, A., Palay, S.L., and Webster, H. de F.: The fine structure of the nervous system, Philadelphia, 1976, W.B. Saunders Co.)

apses (Fig. 4-3, *A*). Synapses characterized by a postsynaptic electron density that is greater in extent and intensity than the presynaptic electron density are called *asymmetrical synapses* (Fig. 4-3, *B*). These classifications may indicate the two ends of a spectrum of synaptic structure rather than two distinct classes. The presynaptic nerve terminals at asymmetrical synapses often contain spherical synaptic vesicles, whereas symmetrical synapses are characterized by flattened or ellipsoidal vesicles. There is evidence which suggests that the asymmetrical synapses (round vesicles) involve excitatory transmitters, whereas the symmetrical synapses (flattened vesicles) involve inhibitory transmitters. This interpretation is, however, controversial.

Because of the structure and organization of chemical synapses, conduction is necessarily one way. *One-way conduction* of chemical synapses contributes to the organization of central nervous systems of vertebrates. A *synaptic delay* of about 0.5 msec is a property of transmission at chemical synapses. Synaptic delay is primarily caused by the time required for transmitter release. In polysynaptic pathways synaptic delay accounts for a significant fraction of the total conduction time.

At chemical synapses the transmitter released by the presynaptic neurons alters the conductance of the postsynaptic plasma membrane to one or more ions. A change of the conductance of the postsynaptic membrane to an ion that is not in equilibrium across the membrane brings about a flow of that ion which causes a change in the membrane potential of the postsynaptic cell. In most cases that have been studied transmitters produce their effect by increasing the conductance of the postsynaptic membrane to one or more ions, but some invertebrate transmitters (and perhaps vertebrate transmitters as well) act by decreasing the postsynaptic conductances to specific ions.

The part of the membrane of the postsynaptic neuron that forms the synapse is specialized for *chemical sensitivity* rather than electrical sensitivity. Action potentials are not produced at the synapse. The change in membrane potential, be it depolarization or hyperpolarization, that occurs at the synapse is conducted electrotonically over the membrane of the postsynaptic neuron to the *axon hillock–initial segment* region (Fig. 4-4). The axon hillock–initial segment has a lower threshold than the rest of the plasma membrane of the postsynaptic cell, and an action potential will be generated here if the sum of all the inputs to the cell exceeds threshold. Once the action potential has been generated, it is conducted back over the surface of the soma of the postsynaptic cell and is propagated along its axon.

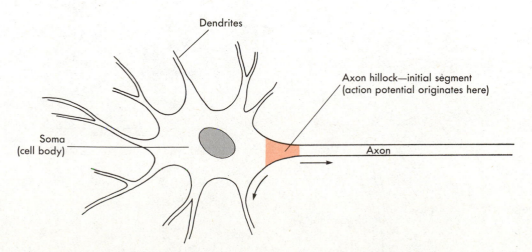

Fig. 4-4 ■ A neuron. The axon hillock–initial segment has the lowest threshold, and, as a result, action potentials tend to originate here.

■ *Input-output relations*

The neuromuscular junction is representative of a particularly simple type of synapse in which one action potential in the presynaptic cell (the input) results in a single action potential in the postsynaptic cell (the output). In other types of synapses the output may differ from the input. Synapses can be classified as one-to-one, one-to-many, or many-to-one, based on the relationship between input and output.

In a *one-to-one synapse,* like the neuromuscular junction, the input and output are the same. A single action potential in the presynaptic cell evokes a single action potential in the postsynaptic cell. Because the output is the same as the input, no integration can occur at this type of synapse.

In a *one-to-many synapse* a single action potential in the presynaptic cell elicits many action potentials in the postsynaptic cell. One-to-many synapses are not common, one example being the synapse of motor neurons on Renshaw cells in the spinal cord. One action potential in the motor neuron induces the Renshaw cell to fire a burst of action potentials.

In a *many-to-one synaptic arrangement* one action potential in the presynaptic cell is not enough to make the postsynaptic cell fire an action potential. The nearly simultaneous arrival of presynaptic action potentials in several input neurons that synapse on the postsynaptic cell is necessary to depolarize the postsynaptic cell to threshold. The spinal motor neuron has this type of synaptic organization. One hundred or more presynaptic axons synapse on each spinal motor neuron (Fig. 4-5). Some of these are excitatory inputs that *depolarize* the postsynaptic cell and bring it closer to its threshold. Other inputs are inhibitory and *hyperpolarize* the motor neuron, taking it farther away from threshold. The changes in postsynaptic potential caused by an action potential in a single input are about 1 to 2 mV. Thus no one excitatory input is capable of bringing the motor neuron to threshold. A transient depolarization of the postsynaptic neuron as the result of an action potential in the presynaptic cell is called an *excitatory postsynaptic potential (EPSP)* (Fig. 4-6, *A*). The transient hyperpolarization caused by an action potential in an inhibitory input is called an *inhibitory postsynaptic potential (IPSP)* (Fig. 4-6, *B*). At any instant the postsynaptic cell *integrates* the various inputs, essentially adding them algebraically. If the momentary sum of the inputs depolarizes the postsynaptic cell to its threshold, it will fire an action potential. This is integration at the level of a single postsynaptic neuron.

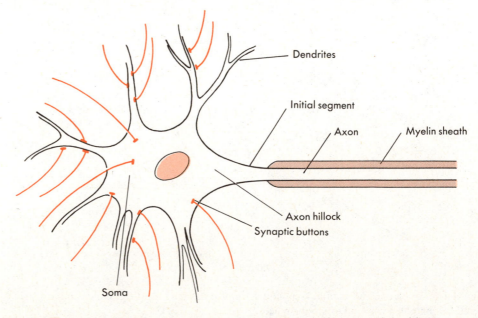

Dendrites

Initial segment

Axon   Myelin sheath

Axon hillock

Synaptic buttons

Soma

Fig. 4-5 ■ A spinal motor neuron with multiple synapses on both soma and dendrites.

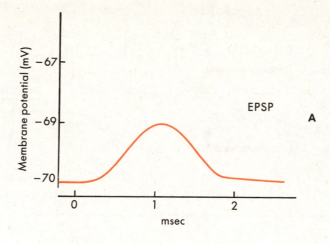

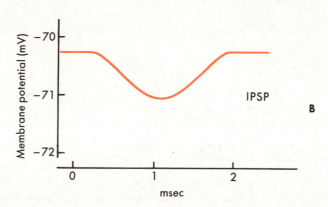

Fig. 4-6 ■ **A,** An excitatory postsynaptic potential *(EPSP)* and **B,** an inhibitory postsynaptic potential *(IPSP)* recorded from spinal motor neurons of cats.

■ *Summation of synaptic inputs*

The summation (or integration) of inputs can occur by either *spatial summation* or *temporal summation* (Fig. 4-7). *Spatial* summation occurs when two separate inputs arrive simultaneously. The two postsynaptic potentials are added so that two simultaneous excitatory inputs will depolarize the postsynaptic cell about twice as much as either input alone. By adding algebraically, one EPSP and one IPSP that occur simultaneously will tend to cancel one another. Even inputs that synapse at opposite ends of the postsynaptic cell will act in this way. The postsynaptic potentials (EPSPs and IPSPs) are conducted rapidly over the entire cell membrane of the postsynaptic cell with almost no decrement. This is because cellular dimensions (about 100 μm) are much smaller than the length constant (about 1 to 2 mm) for electrotonic conduction.

Temporal summation occurs when two or more action potentials in a single presynaptic neuron are fired in rapid succession, so that the resulting postsynaptic potentials overlap in time. A train of impulses in a single presynaptic neuron can cause the potential of the postsynaptic cell to change in a stepwise manner, each step being caused by one of the presynaptic impulses.

Integration at the spinal motor neuron takes place because many positive and negative inputs impinge on a single motor neuron. This allows for fine control of the firing pattern of the spinal motor neuron.

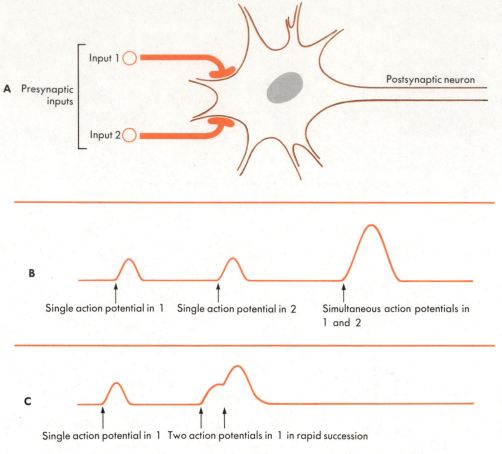

Fig. 4-7 ■ A, Spatial and temporal summation at a postsynaptic neuron with two synaptic inputs (*1* and *2*). **B,** Spatial summation. The postsynaptic potential in response to single action potentials in inputs *1* and *2* occurring separately and simultaneously. **C,** Temporal summation. The postsynaptic response to two impulses in rapid succession in the same input.

■ *Modulation of synaptic activity*

Usually the postsynaptic response evoked by stimulating a particular presynaptic axon is relatively constant in size. Under some conditions, however, the size of the postsynaptic potential is influenced by the immediate history of the synapse.

Facilitation. When a presynaptic axon is stimulated with several consecutive individual stimuli, each stimulus may evoke a larger postsynaptic potential than that evoked by the preceding stimulus. This phenomenon is known as *facilitation* (Fig. 4-8, *A*). It is believed that each succeeding stimulus causes greater release of transmitter, but the mechanism whereby this occurs is not clear.

Synaptic fatigue. If the successive stimuli are continued, eventually the postsynaptic response diminishes. This phenomenon is called synaptic fatigue (Fig. 4-8, *A*). It is called *neuromuscular depression* or Wedensky inhibition at the neuromuscular junction. In many cases synaptic fatigue involves a decrease in the amount of transmitter released, either by a decrease in the number of quanta released or a decrease in the amount of transmitter in each quantum.

Posttetanic potentiation. If a presynaptic axon first is stimulated with a single stimulus, then with a volley of stimuli (say 100 stimuli/second for 2 seconds), and then again with another single stimulus, the second single stimulus may evoke a larger postsynaptic response than did the first single stimulus. Since the larger response follows a tetanic stimulation of the presynaptic nerve, this phenomenon is known as *posttetanic*

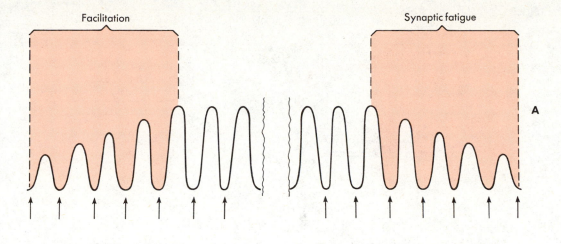

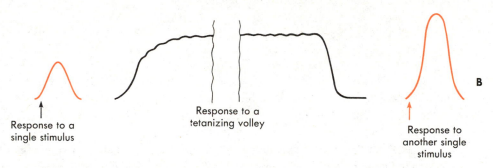

Fig. 4-8 ■ Modulation of synaptic activity. Changes in the membrane potential of a postsynaptic cell are schematically shown. The *arrows* represent single action potentials in the presynaptic neuron. **A,** Facilitation and synaptic fatigue. **B,** Posttetanic potentiation.

potentiation (Fig. 4-8, *B*). As in the case of facilitation, increased transmitter release apparently is involved in posttetanic potentiation.

■ *Ionic mechanisms of postsynaptic potentials in spinal motor neurons*

Much of our current knowledge of synaptic mechanisms in the mammalian central nervous system is derived from studies of cat spinal motor neurons, most notably by Eccles and his associates.

The EPSP. The EPSP (Fig. 4-6, *A*) of the cat spinal motor neuron is caused by a transient increase of the conductance of the postsynaptic membrane to both Na^+ and K^+. One way in which this was demonstrated was by injecting Na^+ and K^+ into the cell to raise the intracellular concentration of that particular ion. Injection of either Na^+ or K^+ results in a smaller EPSP because, when the conductance increase occurs, there is a smaller tendency for Na^+ to flow in and depolarize the cell after Na^+ injection and a greater tendency for K^+ to flow out and oppose depolarization after K^+ injection. If the postsynaptic membrane is progressively depolarized, the EPSP progressively decreases in size because of a decreased tendency for Na^+ to enter and an increased tendency for K^+ to leave the cell. When E_m reaches about zero, the EPSP disappears, and if E_m is made positive, the EPSP changes direction. Thus zero is the *reversal potential* for EPSP. If Na^+ were the only ion flowing during the EPSP, the reversal potential for the EPSP should equal the equilibrium potential for Na^+ (about $+65$ mV). If K^+ were the only ion involved in the EPSP, the reversal potential should equal the K^+ equilibrium potential (about -100 mV). The reversal potential for the EPSP is a

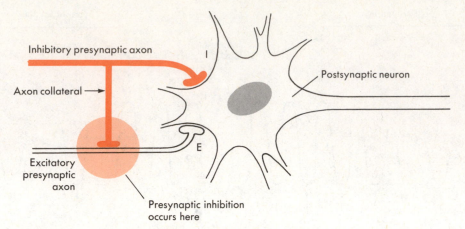

Inhibitory presynaptic axon

Axon collateral →

Postsynaptic neuron

I

E

Excitatory presynaptic axon

Presynaptic inhibition occurs here

Fig. 4-9 ■ Presynaptic inhibition. Axon collaterals of the inhibitory axon *(I)* synapse on the excitatory axon terminal *(E)*. An action potential in the inhibitory axon depolarizes the excitatory axon terminal. The depolarized excitatory axon terminal will release less transmitter in response to an action potential in the excitatory neuron.

weighted average of E_{Na} and E_K. Injection of Cl^- into the cell does not alter the EPSP, so Cl^- apparently is not involved in the EPSP.

The IPSP. The IPSP (Fig. 4-6, *B*) of cat spinal motorneurons is caused by an increased chloride conductance of the postjunctional membrane. The reversal potential for the IPSP is a bit more negative than the normal resting membrane potential of the postsynaptic neuron. The reversal potential is equal to the equilibrium potential for Cl^-. At rest there is a net tendency for Cl^- to enter the cell. The increase in chloride conductance, as the result of transmitter release at the inhibitory synapse, allows Cl^- to enter the postsynaptic cell and hyperpolarize it. Injecting Cl^- into the cell or hyperpolarizing it decreases the net tendency for Cl^- to enter the cell and decreases the size of the IPSP. Injection of Na^+ or K^+ produces no change in the IPSP, suggesting that neither Na^+ nor K^+ is involved in the IPSP.

Presynaptic inhibition. Inhibitory interactions are believed to be vital in stabilizing the central nervous system. Another type of inhibition is called *presynaptic inhibition*. If an inhibitory input to a spinal motor neuron is stimulated tetanically and then an excitatory input is stimulated, the EPSP elicited by stimulating the excitatory input may be reduced in magnitude after the inhibitory volley. This is believed to occur by a mechanism, first described in invertebrates, in which axon collaterals of the inhibitory axons synapse on the excitatory nerve terminals (Fig. 4-9). An action potential in the inhibitory nerve produces a depolarization (rather long lived) of the excitatory nerve terminal. This brings the excitatory nerve terminal closer to threshold. What is important here is that the partly depolarized excitatory terminal *will release less transmitter in response to an action potential*. The smaller release of transmitter results in a smaller EPSP. The phenomenon of decreased transmitter release from a partially depolarized nerve terminal is well known at the neuromuscular junction.

The presumed anatomical arrangement that underlies presynaptic inhibition is diagramed in Fig. 4-9. The mechanisms described for presynaptic inhibition have been worked out in invertebrates. They are believed to apply in mammalian nervous systems, but they have not been fully characterized in mammals. Synaptic physiology is a frontier of modern biology, and our knowledge of the details of synaptic transmission can be expected to increase in the near future.

A number of compounds have been proposed to function as neurotransmitters in the central nervous system. Such compounds are called *candidate neurotransmitters,* or *putative neurotransmitters.* Candidate neurotransmitters usually are concentrated in specific neurons or in specific neuronal pathways. Microapplication of the putative transmitter to particular areas of the central nervous system may evoke specific responses. Correlation of information about the localization of the putative transmitter with knowledge of the location of neurons that respond to the candidate transmitter and the ways in which they respond allow intelligent speculation about the functions of a putative neurotransmitter substance.

It is often difficult to *prove* that a substance is the transmitter at a particular synapse. A putative transmitter (X) must satisfy the following criteria before it is accepted as a proven transmitter at a particular synapse:

1. The presynaptic neurons must contain X and must be able to synthesize it.
2. X must be released by the presynaptic neurons on appropriate stimulation.
3. Microapplication of X to the postsynaptic membrane must mimic the effect of stimulation of the presynaptic neuron.
4. The effects of presynaptic stimulation and of microapplication of X should be altered in the same way by pharmacological agents.

Our knowledge of neurotransmitters has increased greatly in recent years. Some transmitters, like those discussed so far, have rapid and transient effects on the postsynaptic cell. Other transmitters have effects that are much slower in onset and may last for minutes or even hours. Most of the candidate neurotransmitters that have been discovered so far fall into three major chemical classes: amines, amino acids, and oligopeptides.

Acetylcholine. As discussed previously, acetylcholine is the transmitter used by all motor axons that arise from the spinal cord. Acetylcholine plays a central role in the autonomic nervous system, being the transmitter for all preganglionic neurons and also for postganglionic parasympathetic fibers. The Betz cells of the motor cortex use acetylcholine as their transmitter. The basal ganglia, which are involved in the control of movement, contain high levels of acetylcholine, and ACh is believed to be a transmitter in the basal ganglia.

Among the amines that may serve as neurotransmitters are norepinephrine, epinephrine, dopamine, serotonin, and histamine.

Dopamine, norepinephrine, and epinephrine are catecholamines and share a common biosynthetic pathway that starts with the amino acid tyrosine. Tyrosine is converted to L-dopa by tyrosine hydroxylase. L-Dopa is converted to dopamine by a specific decarboxylase. In dopaminergic neurons the pathway stops here. Noradrenergic neurons have another enzyme, dopamine β-hydroxylase, that converts dopamine to norepinephrine. Other cells add a methyl group to norepinephrine to produce epinephrine. *S*-Adenosylmethionine is the methyl donor, and the reaction is catalyzed by phenylethanolamine-*N*-methyltransferase.

As discussed earlier, norepinephrine is the primary transmitter for postganglionic sympathetic neurons. In the brain norepinephrine-containing cell bodies are found in the locus ceruleus. The neurons of the locus ceruleus project to the cortex, hypothalamus, cerebellum, and spinal cord. The noradrenergic neurons in the brain may be involved in arousal, in regulation of mood, and in dreaming.

Neurons that contain high levels of dopamine are prominent in the midbrain regions known as the substantia nigra and the ventral tegmentum. Some of the axons of these neurons travel to the forebrain, where they may play a role in emotional responses.

■ ***Transmitters in the central nervous system***

■ *Identification of transmitter substances*

■ *Transmitters and putative transmitters in the central nervous system*

■ *Biogenic amine transmitters*

Other dopaminergic axons terminate in the corpus striatum, where they are believed to play an important role in control of complex movements. The degeneration of these dopaminergic synapses in the corpus striatum occurs in Parkinson's disease and is believed to be a major cause of the muscular tremors and rigidity that characterize this disease.

Some evidence suggests that hypersecretion of dopamine by neurons in the limbic system and/or increased sensitivity to dopamine may be involved in schizophrenia. Antipsychotic drugs such as chlorpromazine and related compounds are dopamine-receptor antagonists and thus diminish the effects of endogenous dopamine.

Serotonin (5-hydroxytryptamine)-containing neurons are present in high concentration in the raphe nuclei located in the midline of the brainstem. These nuclei project widely to brain and spinal cord. Serotonergic neurons may be involved in regulating temperature, sensory perception, onset of sleep, and control of mood.

Histamine is present in certain neurons in the hypothalamus. The functions of these presumably histaminergic neurons are not yet known.

◼ *Amino acid transmitters*

Glycine, the simplest amino acid, is an inhibitory neurotransmitter released by certain spinal interneurons.

The dicarboxylic amino acids, glutamate and aspartate, have strong excitatory effects on many neurons in the brain. It may be that glutamate and aspartate are the most prevalent excitatory transmitters in the brain.

γ-Aminobutyric acid (GABA) is not incorporated into proteins, nor is it present in all cells (as are the other naturally occurring amino acids). GABA is produced from glutamate by a specific decarboxylase present only in the central nervous system. Among the cells that contain GABA are some cells in the basal ganglia, the cerebellar Purkinje cells, and certain spinal interneurons. In all known cases GABA functions as an inhibitory transmitter. It is the most common transmitter in the brain, and it may be that as many as a third of the synapses in the brain have GABA as their neurotransmitter.

GABA is believed to be important in many different central control pathways. A deficit in the level of GABA and other neurotransmitters in the corpus striatum occurs in patients with Huntington's chorea. The uncontrolled movements that characterize this disease may be partly due to diminished effectiveness of GABA-mediated inhibition in central motor pathways. GABA also may be involved in control of mood and emotion, since some evidence suggests that benzodiazepines (such as diazepam) which are effective antianxiety drugs may act to enhance the effectiveness of GABA as a transmitter.

◼ *Neuroactive peptides*

Relatively recently it has become appreciated that certain neurons contain peptides which act at very low concentrations to excite or inhibit other neurons. To date, about 25 of these so-called neuropeptides, ranging from 2 amino acids to about 40 amino acids long, have been identified. Some of these neuropeptides may serve as true neurotransmitters, whereas others may act primarily to modulate the release of or the response to the actual transmitter. Most of the neuropeptides that have been discovered so far are listed on the opposite page. It is likely that more neuropeptides will be added to this list in the near future.

Like other peptides, neuropeptides are encoded by the cells' DNA, transcribed into messenger RNA, and translated into a peptide chain on ribosomes. In this respect neuropeptides differ from other transmitters, which are synthesized from commonly available precursors by specific enzymatic pathways. The ribosomes are located in the cell bodies of the neurons, so the newly synthesized neuropeptides must travel to the nerve terminals via axonal transport. Nonpeptide transmitters, however, are synthesized in the nerve terminals themselves.

Neuroactive peptides

Gut-brain peptides
Vasoactive intestinal polypeptide (VIP)
Cholecystokinin octapeptide (CCK-8)
Substance P
Neurotensin
Methionine enkephalin
Leucine enkephalin
Insulin
Glucagon
Hypothalamic-releasing hormones
Thyrotropin-releasing hormone (TRH)
Luteinizing hormone–releasing hormone (LHRH)
Somatostatin (growth hormone release–inhibiting factor, or SRIF)
Pituitary peptides
Adrenocorticotropin (ACTH)
β-Endorphin
α-Melanocyte–stimulating hormone (α-MSH)
Others
Angiotensin II
Bradykinin
Vasopressin
Oxytocin
Carnosine
Bombesin

From Snyder, S.H.: Science **209**:976, 1980. Copyright 1980 by American Association for the Advancement of Science.

Substance P is a peptide containing 11 amino acids that is present in certain neurons in the brain and spinal cord and in peripheral sensory fibers. Primary sensory neurons have their cell bodies in the dorsal root ganglia and send their axons to the substantia gelatinosa of the spinal cord. There is considerable evidence that substance P is one of the transmitters at the synapses between primary pain sensory neurons and spinal neurons in the substantia gelatinosa. Substance P also is present in certain neurons in the brain, particularly in the substantia nigra, the basal ganglia, the hypothalamus, and the limbic system. Substance P may be a transmitter in the striatonigral pathway, which participates in control of motor activities.

Opioid peptides. Certain neurons in the brain and spinal cord contain peptides that bind with high affinity to the same receptors which bind opiate drugs (such as morphine and its derivatives). Two pentapeptides called *enkephalins* differ only in their C-terminal amino acids: methionine enkephalin (try-gly-gly-phe-met) and leucine enkephalin (try-gly-gly-phe-leu). Enkephalins are present in certain neurons in the brain and spinal column, and many are involved in the perception and integration of pain and with the emotional responses to pain. Spinal interneurons that contain enkephalins have been proposed to release enkephalins near the nerve terminals of the primary sensory neurons that use substance P as transmitter. Some investigators believe that enkephalins act by suppressing the release of substance P by presynaptic inhibition. Enkephalins apparently impinge directly on some spinothalamic neurons. The enkephalins also inhibit pain pathways at higher levels of the central nervous system by neuronal mechanisms that remain to be elucidated. The methionine enkephalin sequence and even stronger opiate activities are found in longer peptides, the *endorphins,* that have been found in certain cells of the pituitary gland and more recently in certain neurons in the brain. Three endorphins have been discovered thus far: α-, β-, and γ-endorphin. The three endorphins appear to be different peptide fragments produced by cleavage of a large precursor peptide called β-lipotropin that is produced by certain cells in the anterior pituitary gland (and perhaps elsewhere). The functions that may be affected by enkephalins and endorphins include pain perception and its emotional correlates, other emotional effects (euphoria), certain reflexes (respiratory, cardiovascular, and gastrointestinal), pupillary constriction, and secretion of certain hormones.

Behavioral effects of certain neuropeptides. Some neuropeptides apparently are involved in mediating highly complex integrated behaviors. Angiotensin II, in extremely small amounts, elicits drinking behavior in well-hydrated animals. LHRH (luteinizing hormone–releasing hormone) injected into the cerebrospinal fluid of a female rat evokes copulatory behavior. These responses suggest that the neuropeptides which elicit them act as modulators of neuronal function at several levels in integrated neural pathways.

Mechanisms of action of neuropeptides. In several neuronal types the neuron contains both a neuropeptide and a transmitter substance (ACh or a biogenic amine). In these cases it is believed that the neuropeptide functions not as a conventional transmitter but as a modulator of presynaptic release of the transmitter or the response of the postsynaptic cell to the transmitter. In some cases the neuropeptide may act to elicit a reaction complementary to the effect of the neurotransmitter. For example, the vagal neurons that elicit salivary secretion release both acetylcholine and vasoactive intestinal polypeptide (VIP). ACh acts directly on the gland cells to elicit secretion of saliva, whereas VIP acts to cause dilation of neighboring blood vessels. In other cases a neuropeptide apparently functions as a true transmitter; that is, binding of the neuropeptide to its receptor results in a change in an ion conductance and causes a change in the membrane potential of the postsynaptic cell. In those cases where neuropeptides act as modulators, rather than transmitters, a number of mechanisms are possible. Some neuropeptide modulators are believed to use cyclic AMP or cyclic GMP (or another compound) as a second messenger. Binding of the neuropeptide to its receptor brings about a change in the cytoplasmic concentration of cyclic AMP or cyclic GMP (or another second messenger), which in turn alters the responsiveness of the cell to transmitter.

Our knowledge of the functions of neuropeptides and their mechanisms of action can be expected to increase in the near future.

■ Bibliography

Aidley, D.J.: The physiology of excitable cell, ed. 2, Cambridge, 1978, Cambridge University Press.

Davson, H.: A textbook of general physiology, ed. 4, Baltimore, 1970, Williams & Wilkins Co.

Eccles, J.C.: The physiology of synapses, Berlin, 1964, Springer Verlag.

Kandel, E.R., and Schwartz, J.H.: Principles of neural science, New York, 1981, Elsevier Biomedical Press.

Katz, B.: Nerve, muscle, and synapse, New York, 1966, McGraw-Hill Book Co.

Kuffler, S.W., and Nicholls, J.G.: From neuron to brain, Sunderland, Mass., 1976, Sinauer Associates.

THE NERVOUS SYSTEM

David H. Cohen
S. Murray Sherman

chapter 5

The nervous system and its components

The contemporary era of the neurosciences began in the mid-1950s with the introduction of neurophysiological techniques that provided the tools for studying the dynamics of neural activity. Concomitantly there were important developments in neuroanatomical methods that helped specify the detailed connectivity of neural pathways. Together these advances initiated an era in which the focus has been on a rigorous cellular approach to understanding brain function. In the succeeding three decades gains in knowledge of the nervous system have been virtually explosive, and the neurosciences have rapidly evolved into an integrated discipline that has most recently woven molecular approaches into its fabric. It is indeed fair to say that this exciting venture is generating a true cell biology of the nervous system and its ultimate output—behavior.

When the nervous system is discussed in a medical physiology textbook, selection of material is a critical factor. Space constraints prevent comprehensive coverage, and to preserve some historical continuity it is thus necessary to bias the presentation toward the inclusion of traditional material. Selectivity is particularly difficult in a field that is advancing as rapidly as neuroscience. One would like to present the most recent developments and capture the spirit and excitement of this explosive field; however, it is not appropriate to do this at the expense of firmly established findings that are basic to the training of a physician.

For these reasons we have chosen to dedicate the major portion of this section to sensory and motor systems. Disorders of these features of brain function are most prominent and most common in the clinic. These chapters are followed by a description of cortical organization, which incorporates the basic concepts of higher function, by a review of the autonomic nervous system and its central control, and finally by a chapter on the plasticity or modifiability of the nervous system. This final chapter is perhaps the only one that is not ordinarily included in a traditional text; however, we believe it is of sufficient importance and clinical relevance to dictate at least a limited presentation.

Although the subject matter is indeed traditional, we have attempted to treat it in a contemporary manner. It is our firm conviction that the most useful approach is to present the nervous system with a functional neuroanatomical orientation. We have attempted to meet this objective while still capturing the spirit of contemporary cellular approaches. We have not made an effort to establish a neuroanatomical foundation because it would be inappropriate in this forum. It is assumed that the student has been exposed to fundamental neuroanatomy as background for this text.

Throughout we have made an effort to stress approaches to the analysis of neural function. In particular, we have emphasized the importance of analyzing the brain from the sensory or motor peripheries centrally, allowing one to relate central activity to either sensory stimulus patterns or motor behavior. Such an approach relies first on establishing the relevant neural circuitry. This, in turn, provides the substrate for applying contemporary analytical techniques to study at a cellular level how the brain organizes its sensory inputs, motor outputs, and their coupling. Through this approach remarkable achievements have been realized in describing how the brain processes sensory information to establish a basis for perception and how goal-directed motor behavior is organized. Understanding of sensorimotor integration and more complex functions has not advanced as rapidly. In fact, it seems that the level of understanding of any given structure and its function varies inversely with its proximity to the periphery. Nevertheless, we have tried to impart a spirit of optimism regarding future acquisition of knowledge of more complex activities of the nervous system. The techniques and strategies for such advances are at hand, and some of what we present here might be obsolete by the time this is read. This need not arouse anxiety, however, because the core of material contained in this section will remain essential information for some time.

■ Glial cells and neurons

The cellular elements of the nervous system are *glial cells* and *neurons*. Although attention will be focused almost exclusively on neurons, glial cells actually outnumber neurons by an order of magnitude. Roughly speaking, there are 10^{13} glial cells versus 10^{12} neurons. In this section, we shall briefly review some salient features of these cellular elements; details of cellular biology are found elsewhere in this book (Section 1).

■ Glial cells

Glial cells neither conduct action potentials nor form functional synapses with other cells, although they can be passively polarized in response to nearby neural activity. Their functions are complex and not completely understood. Glial cells generally provide support for neurons. They form a mechanical matrix in which neurons are embedded; they probably play metabolic and nutritive roles; they may help to regulate blood flow through the brain; they may act as a sink or source of ions; they can electrically insulate axons and synapses from one another; and they phagocytose neural debris in response to damage.

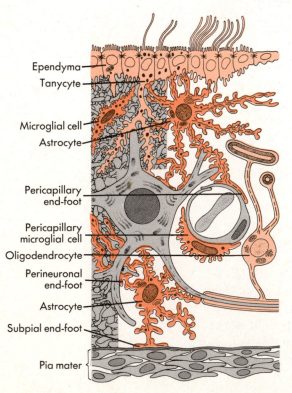

Ependyma
Tanycyte
Microglial cell
Astrocyte
Pericapillary end-foot
Pericapillary microglial cell
Oligodendrocyte
Perineuronal end-foot
Astrocyte
Subpial end-foot
Pia mater

Fig. 5-1 ■ Schematic representation of nonneural elements in the central nervous system. Two astrocytes *(darker color)* are shown ending on a neuron's soma and dendrites. They also contact the pial surface and/or capillaries. An oligodendrocyte *(lighter color)* provides the myelin sheaths for axons. Also shown are microglia *(darker color)* and ependymal cells *(lighter color).* (Redrawn from Williams, P.L., and Warwick, R.: Functional neuroanatomy of man, Edinburgh, 1975, Churchill Livingstone.)

Five broad types of glial cells have been described: *astrocytes (astroglia), oligoden-drocytes (oligodendroglia), ependymal cells, microglia,* and *Schwann cells* (Figs. 5-1 and 5-2). The first four are associated with the central nervous system and the last with peripheral nerves.

Astrocytes, so named because they are star shaped, are found abundantly throughout the brain and spinal cord. Their many processes surround the neurons and axons and often end on the walls of blood vessels. They probably serve metabolic and nutritive functions for the neurons as well as provide a mechanical matrix. Furthermore, many single synapses are surrounded by astrocytic processes that separate and perhaps electrically insulate these synapses from one another.

Oligodendrocytes are found chiefly among myelinated axons in the brain and spinal cord. In fact, processes from these cells wrap many times around an axon to form the myelin sheath (Fig. 5-2). This sheath not only insulates axons from one another, but also because of the discontinuities at *nodes of Ranvier,* the action potential is conducted in a saltatory fashion (Chapter 3). Such conduction is much more rapid than would result if myelin were absent. Schwann cells provide the myelin sheaths, in much the same fashion, for peripheral nerves.

Ependymal cells line the surfaces of the brain's ventricles and the central canal of the spinal cord. Their precise function is unclear.

Finally, microglia are the smallest cells in the central nervous system. Under normal conditions, they have no obvious function. However, if nervous tissue is damaged, these cells enlarge and act as phagocytes to eliminate any debris that forms.

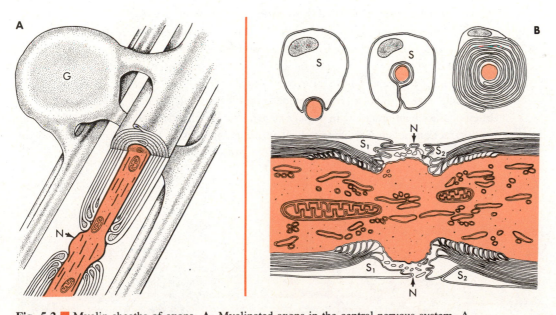

Fig. 5-2 ■ Myelin sheaths of axons. **A,** Myelinated axons in the central nervous system. A single oligodendrocyte *(G)* emits several processes, each of which winds concentrically around an axon to form the myelin sheath. The axon *(color)* is shown in cutaway. The myelin from a single oligodendrocyte ends before the next wrapping from another oligodendrocyte. The bare axon between sheaths is the node of Ranvier *(N)*. Conduction of action potentials is saltatory down the axon, skipping from node to node. **B,** Myelinated axon in the peripheral nervous system. A Schwann cell forms a myelinated sheath for peripheral axons in much the same fashion as oligodendrocytes do for central ones, except that each Schwann cell myelinates a single axon. The top shows a cross-sectional view of progressive stages in myelin sheath formation by a Schwann cell *(S)* around an axon *(color)*. The bottom shows a longitudinal view of a myelinated axon *(color)*. The node of Ranvier *(N)* is shown between adjacent sheaths formed by two Schwann cells *(S₁ and S₂)*.(Redrawn from Patton, H.D., et al.: Introduction to basic neurology, Philadelphia, 1976, W.B. Saunders Co.)

■ *Neurons*

Neurons are the cells responsible for information processing and transfer in the nervous system. With the exception of certain interneurons that lack long axons, neurons possess axons that conduct regenerative action potentials over considerable distances. Neurons communicate with one another via synapses.

Fig. 5-3 illustrates a "typical" neuron, but many variations in morphology will be described in later chapters. The neuron has several distinct regions. The *soma* (or *cell body*) and *dendritic processes* comprise the major input zone of the neuron; that is, axons from other neurons form synapses onto the dendrites and soma. The soma is also the metabolic factory for the neuron. Many dendrites issue from the soma, but only a single *axon* does so. The *myelin sheath,* if present, begins near the soma. Although not shown in Fig. 5-3, many axons branch repeatedly at nodes of Ranvier to provide collateral fibers. Finally, the myelin sheath ends, and the axon exhibits unmyelinated, preterminal branches that form synapses onto other neurons.

The soma, dendrites, and perhaps the preterminal portions of the axon are not electrically excitable; that is, their membranes do not possess voltage-dependent ion channels that can lead to an action potential. Electrical excitability and thus the capacity to propagate an action potential are properties of the axon (Chapter 3). In the normal course of events, excitatory and inhibitory postsynaptic potentials (EPSPs and IPSPs) spread electrotonically from the dendrites and soma to the most proximal portion of electrically excitable axon (the *initial segment*). These potentials sum linearly, and if this sum sufficiently depolarizes the initial segment, an action potential ensues. Although the action potential might cease at the preterminal axon branches, the depolarization it creates propagates electrotonically to the synapses, thus causing release of transmitter.

This is the traditional view of the neuron, but contemporary research has raised some qualifications and notes of caution. In addition to synapses formed from axons onto dendrites and somata (axodendritic and axosomatic synapses), synapses between axons (axoaxonic synapses) are not uncommon. In many parts of the central nervous system, dendrodendritic synapses have also been described; that is, some dendrites form synapses onto others. Finally, recent evidence indicates that, at least in some cells, dendrites possess membranes with voltage-dependent ion channels, and these dendrites can conduct action potentials. Thus not all information transfer between synapses and the axon's initial segment need be conducted electrotonically.

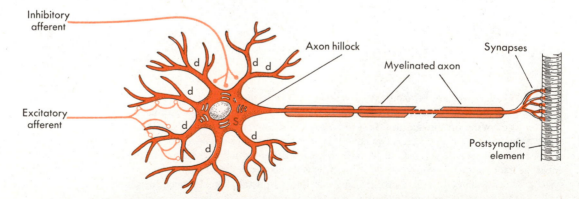

Fig. 5-3 ■ Schematic diagram of an idealized neuron and its major components. Most afferent input from axons of other cells terminates in synapses on the dendrites *(d),* although some may terminate on the soma *(S).* Excitatory terminals tend to terminate more distally on dendrites than do inhibitory ones, which often terminate on the soma. (Redrawn from Williams, P.L., and Warwick, R.: Functional neuroanatomy of man, Edinburgh, 1975, Churchill Livingstone.)

The central nervous system of vertebrates consists of five major components: the *spinal cord*, the *brainstem*, the *cerebellum*, the *diencephalon*, and the *telencephalon* (or *cerebral hemispheres*) (Figs. 5-4 to 5-6). Because these regions will be considered in more detail in later chapters, the purpose here is merely to indicate their gross relationships with one another and to identify certain prominent structures.

The brainstem is further subdivided into the *medulla, pons,* and *midbrain.* The cerebellum is rather prominent in humans and is located between the brainstem and posterior cortex. The *thalamus* and *hypothalamus* are located in the diencephalon. The massive telencephalon includes the *corpus striatum* and *cerebral cortex.* The cortex is divided into the *occipital, temporal, parietal,* and *frontal lobes,* and its convolutions form numerous *sulci* and *gyri* (Fig. 5-6). Joining the cerebral hemispheres is a massive commissure, the *corpus callosum.* Its posterior end is called the *splenium,* and anteriorly it bends downward at the *genu.* The *optic chiasm,* formed by the junction of the optic nerves, is found just anterior to the *hypophysis* and ventral to the hypothalamus.

■ Gross topography of the brain and spinal cord

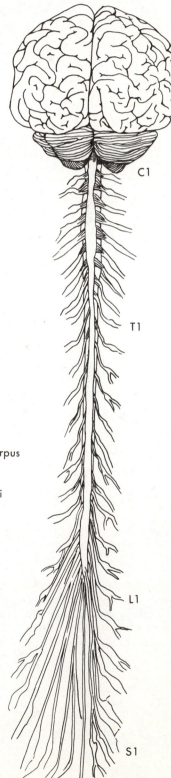

Fig. 5-4 ■ Brain and spinal cord with attached spinal nerves. Note the relative size of various components. *C1, T1, L1,* and *S1,* First cervical, thoracic, lumbar, and sacral segments, respectively. (Redrawn from Williams, P.L., and Warwick, R.: Functional neuroanatomy of man, Edinburgh, 1975, Churchill Livingstone.)

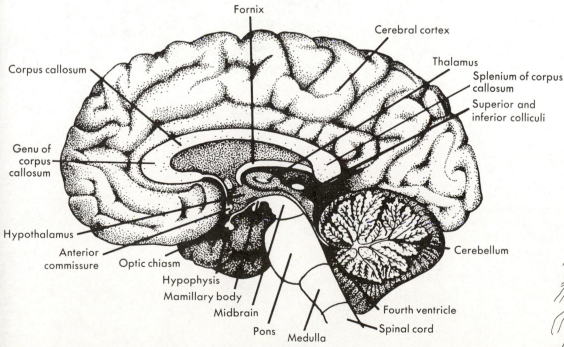

Fig. 5-5 ■ Midsagittal section of the brain. Note the relationships among the cerebral cortex, cerebellum, thalamus, and brainstem plus the location of various commissures. (Redrawn from Kandel, E.R., and Schwartz, J.H.: Principles of neuroscience, New York, 1981, Elsevier North-Holland, Inc.)

Fig. 5-7 shows the *ventricles,* which are continuous with the *central canal* of the spinal cord and contain *cerebrospinal fluid.* The *fourth ventricle* is enclosed by the cerebellum, pons, and anterior portion of the medulla, and the *third ventricle* is located in the diencephalon. Between these ventricles is the *cerebral aqueduct.* The paired *lateral ventricles* communicate with the third ventricle via the *interventricular foramen* on each side. The relationship of the extensive lateral ventricles to the hemispheres is shown in Fig. 5-8.

This information is meant to serve as a rough framework for later chapters. The reader should consult standard neuroanatomical reference books if more detailed anatomical information is required.

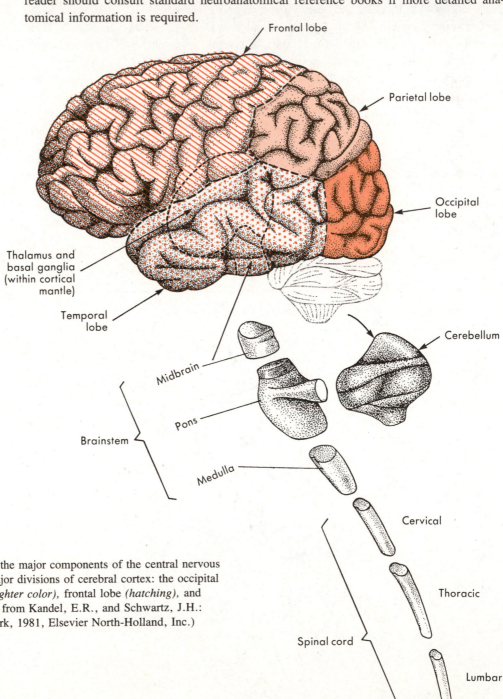

Fig. 5-6 ■ Exploded view showing the major components of the central nervous system. Also shown are the four major divisions of cerebral cortex: the occipital lobe *(darker color),* parietal lobe *(lighter color),* frontal lobe *(hatching),* and temporal lobe *(stippling).* (Redrawn from Kandel, E.R., and Schwartz, J.H.: Principles of neuroscience, New York, 1981, Elsevier North-Holland, Inc.)

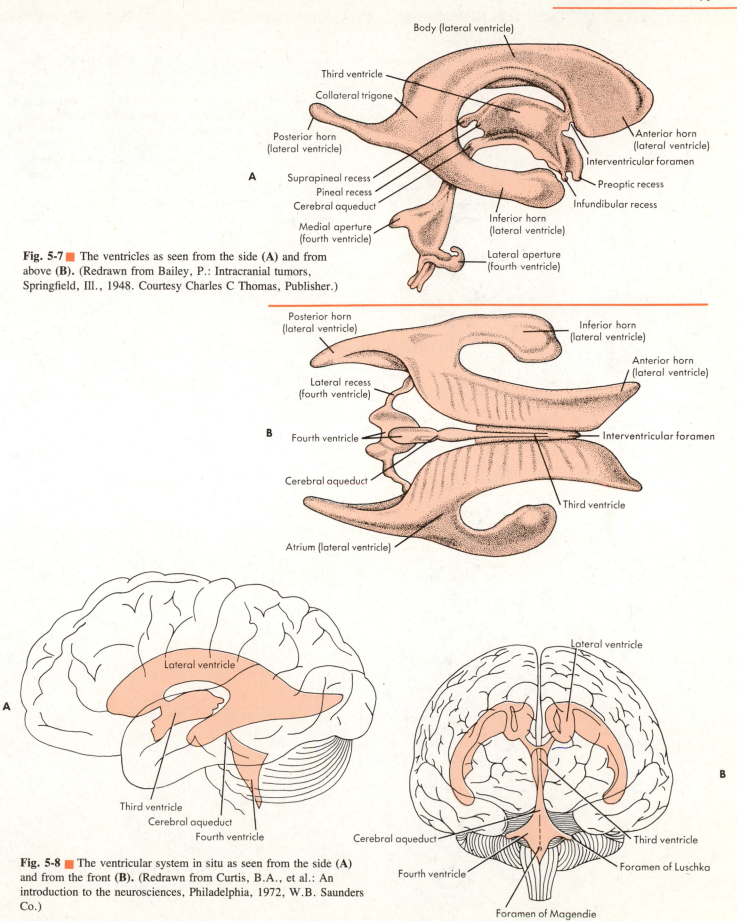

Body (lateral ventricle)

Third ventricle

Collateral trigone

Posterior horn (lateral ventricle)

Anterior horn (lateral ventricle)

Interventricular foramen

Preoptic recess

A

Suprapineal recess

Pineal recess

Cerebral aqueduct

Infundibular recess

Inferior horn (lateral ventricle)

Medial aperture (fourth ventricle)

Lateral aperture (fourth ventricle)

Fig. 5-7 ■ The ventricles as seen from the side (**A**) and from above (**B**). (Redrawn from Bailey, P.: Intracranial tumors, Springfield, Ill., 1948. Courtesy Charles C Thomas, Publisher.)

Posterior horn (lateral ventricle)

Inferior horn (lateral ventricle)

Anterior horn (lateral ventricle)

Lateral recess (fourth ventricle)

B

Fourth ventricle

Interventricular foramen

Cerebral aqueduct

Third ventricle

Atrium (lateral ventricle)

A

Lateral ventricle

Lateral ventricle

B

Third ventricle

Cerebral aqueduct

Fourth ventricle

Cerebral aqueduct

Fourth ventricle

Third ventricle

Foramen of Luschka

Foramen of Magendie

Fig. 5-8 ■ The ventricular system in situ as seen from the side (**A**) and from the front (**B**). (Redrawn from Curtis, B.A., et al.: An introduction to the neurosciences, Philadelphia, 1972, W.B. Saunders Co.)

■ *Bibliography*

Journal articles

Llinas, R., and Sugimori, M.: Calcium conductances in Purkinje cell dendrites: their role in development and integration, Prog. Brain Res. **51:**323, 1979.

Wong, R.K.S., Prince, D.A., and Basbaum, A.I.: Intradendritic recordings from hippocampal neurons, Proc. Natl. Acad. Sci. USA **76:**986, 1979.

Books and monographs

Carpenter, M.B.: Core text of neuroanatomy, Baltimore, 1972, The Williams & Wilkins Co.

Eccles, J.C.: The physiology of nerve cells, Baltimore, 1957, The Johns Hopkins University Press.

Kuffler, S.W., and Nicholls, J.G.: From neuron to brain, Sunderland, Mass., 1976, Sinauer Associates, Inc.

Peters, A., Palay, S.L., and de F. Webster, H.: The fine structure of the nervous system: the neurons and supporting cells, Philadelphia, 1976, W.B. Saunders Co.

Williams, P.L., and Warwick, R.: Functional neuroanatomy of man, Philadelphia, 1975, W.B. Saunders Co.

chapter 6

Peripheral units of the nervous system

The central nervous system can be viewed as an integrator that analyzes signals from the sensory pathways and uses this information to generate command signals along the motor pathways to muscles and other effectors. The nervous system thus senses and analyzes the environment in order to generate behavior appropriate to that environment. All its interactions with the environment must pass through the peripheral units, which are sensory receptors or transducers and motor units, since these are the only interfaces between the nervous system and the environment. Because it is simpler to understand the functional organization of the nervous system by starting at the periphery and proceeding centrally, the peripheral units are considered before detailed descriptions of specific sensory or motor pathways.

■ Sensory receptors or transducers

Sensory systems start with receptors or transducers. (For reasons given later, *receptor* and *transducer* are terms that will be used interchangeably.) These receptors initiate neural signals in response to sensory stimuli, and these signals are transmitted to the central nervous system to be integrated with other neural signals at a number of synaptic sites. Each sensory system has its own unique population of transducers and neural pathways.

■ General considerations

To encode the sensory environment, it is first necessary to transduce or convert the energy of environmental signals (light, heat, touch, etc.) into electrochemical energy that can be used to generate neural signals. This is accomplished by the transducers. Each transducer is able to convert a particular form of stimulus energy (light, heat, etc.) into a graded, slow potential called a *generator potential*. The generator potential is similar in many ways to the excitatory postsynaptic potential seen in the central nervous system and will be considered in more detail later.

Transducers thus exhibit a high degree of modality specificity; that is, they are relatively sensitive to one stimulus modality and not to others. However, this specificity is relative and not absolute. A photoreceptor in the retina has a low generator potential threshold for photic energy but will also exhibit a generator potential to sufficiently intense stimuli of other modalities. Thus pressure against the eyeball, if great enough, excites these photoreceptors. Nonetheless, the sensation perceived is one of flashes or spots of light, not pressure, and the significance of this is considered later in the description of sensory systems. The various classes of transducers found in mammals include mechanoreceptors involved in the somatosensory, auditory, and vestibular systems; photoreceptors for vision; thermoreceptors involved in the somatosensory system; chemoreceptors used for the senses of smell and taste as well as for signaling the chemical composition of the blood; and nociceptors involved in pain sensation (Table 6-1).

TABLE 6-1

TABLE 6-1

Classification of neural transducers

Stimulus energy	*Transducer types*
Mechanical	1. Mechanoreceptors of skin and deep tissues, including free nerve endings and specialized structures 2. Mechanoreceptors of joints 3. Stretch receptors of muscle and tendon 4. Hair cells, including vestibular and cochlear 5. Visceral pressure receptors
Photic	Photoreceptors of retina, including rods and cones
Thermal	Thermoreceptors for "cool" and "warm" stimuli
Chemical	1. Chemoreceptors for taste 2. Olfactory receptors 3. Osmoreceptors 4. Carotid and aortic body receptors
Extremes of mechanical, thermal, or chemical energy	Nociceptors

Sensory receptors are located at or near the periphery, often in specialized tissue such as the retina or the organ of Corti. These receptors can be either specialized portions of the distal endings of primary afferent neurons or specialized nonneural cells that form synapses upon the primary afferents. Each primary afferent is associated with only one or a very few transducers of the same type. The generator potential initiated by the transducer, if sufficiently large, typically leads to action potentials in the primary afferents (an exception noted later is the retina, in which action potentials first occur several synapses central to the photoreceptors). To explain more completely the properties of transducers, specific detailed examples of three kinds of receptors are considered next.

■ *Pacinian corpuscle*

Pacinian corpuscles are mechanoreceptors located in the skin and deep tissues. They are involved in the sensation of pressure or vibration. Fig. 6-1 illustrates their general appearance and functional properties. The pacinian corpuscle consists of a body composed of concentric layers, much like an onion, into which the distal end of a primary afferent fiber penetrates. The soma of this afferent fiber lies in a dorsal root ganglion of the spinal cord, and its axon enters the spinal cord via the dorsal root. Pressure applied to the "onion" leads to a graded generator potential in the axon (Fig. 6-1, *A*). Its size increases monotonically with increasing pressure, and if it is large enough to surpass the axon's threshold, a regenerative action potential occurs. It should be noted that a subthreshold generator potential, which is electrotonically conducted by the axon, decays to an insignificant size long before the spinal cord is reached. Consequently, the only message transmitted to the central nervous system results from the action potentials.

The distal end of the axon, and not the onion, is the actual transducer; that is, when the onion is stripped away, force applied to the nerve ending (Fig. 6-1, *B*), but not more centrally along the axon (Fig. 6-1, *C*), leads to a generator potential. Evidently such a mechanical distortion of the membrane opens up ionic channels, particularly for sodium. The passage of such ions down their electrochemical gradients depolarizes the axon and is represented by the generator potential.

Although not the actual transducer, the onion nonetheless serves an important mechanical function. Fig. 6-2 shows that the pacinian corpuscle is a *phasic* or rapidly adapting transducer; that is, the generator potential rapidly returns to the baseline (Fig. 6-2, *left*). This potential occurs only during rapid pressure changes such as the onset or

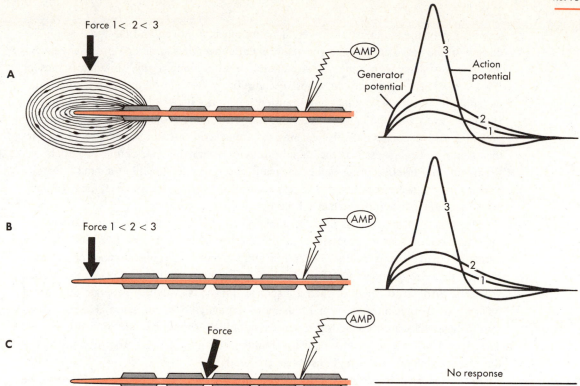

Fig. 6-1 ■ Schematic illustration of response properties of the pacinian corpuscle. A myelinated axon *(color)* penetrates an onionlike structure that forms the pacinian corpuscle; the axon is the distal end of a primary afferent fiber that enters the spinal cord centrally via a dorsal root. In all drawings the electrophysiological responses of the axon (as recorded through an electrode and amplified by appropriate electronic equipment) are shown to three forces applied to the receptor. Force *1* is smaller than force *2,* which is smaller than force *3,* and the responses are numbered accordingly. **A,** Intact corpuscle. Forces applied to the surface of the corpuscle result in electrotonically transmitted generator potentials that demonstrate temporal and spatial summation (not illustrated). The amplitude of each potential monotonically reflects the size of the applied force. If the force and resultant potential are sufficiently large, a regenerative action potential results and travels to the spinal cord. Smaller generator potentials that do not trigger an action potential are functionally irrelevant because they decay to undetectable size long before reaching the spinal cord. **B,** Experiment to demonstrate that the onionlike structure is not necessary for transduction. Forces applied directly to the distal tip of the axon result in responses as in **A. C,** Experiment to demonstrate that the actual transducer is the distal tip of the axon, since forces applied elsewhere evoke no electrophysiological response. (Redrawn from Loewenstein, W.R.: Biological transducers, Scientific American **203:**98, 1960.)

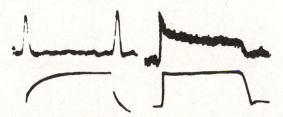

Fig. 6-2 ■ Stimulus-response records of a pacinian corpuscle. The top traces represent the generator potentials recorded in the axon in response to the application of forces 50 msec in duration as shown in the lower traces. On the left is the stimulus-response pattern of an intact pacinian corpuscle. Note the extremely phasic responses that occur only when the stimulus changes, that is, at onset and termination. The right half shows the pattern of stimuli applied directly to the distal tip of the axon after removal of the onionlike structure. Here the response is relatively tonic. (From Lowenstein, W.R., and Mendelsohn, M.: J. Physiol. [Lond.] **177:**377, 1965.)

termination of the stimulus. However, if the onion is removed and the force is applied directly to the nerve ending, the generator potential becomes *tonic* or very slowly adapting; it is sustained with rather slow temporal decay as long as the stimulus is present (Fig. 6-2, *right*). Apparently the onion, because of its viscoelastic properties, transforms the stimulus in such a way that only transient pressure changes reach the nerve ending. However, even with the onion removed so that a stimulus evokes a sustained generator potential, the responses actually reaching the spinal cord (action potentials) reflect further significant adaptation at the impulse-initiation site; that is, action potentials are generated only briefly during rapid changes in the generator potential although the site remains very depolarized. The phasic responses of the pacinian corpuscle, then, are only partially explained by the mechanical properties of the onion.

Phasic responses, such as those exemplified by the pacinian corpuscle, subserve an important role in sensory processing. Phasic activity is the most sensitive indicator of a temporal change in a stimulus. Because a tonic response or generator potential would increase or decrease monotonically with stimulus strength, small changes in the stimulus would be difficult to detect in such a response. Phasic transducers generate a potential only when such changes occur and thus more dramatically signal these changes. However, phasic cells are ill suited to signal quantitatively the magnitude of such changes. They are even ill suited to distinguish stimulus intensity increases from decreases. For this, tonic or slowly adapting transducers are needed.

■ Tonic mechanoreceptors

A tonic receptor is one from which the evoked generator potential is maintained for the duration of the stimulus with only slight adaptation; that is, the generator potential decays very slowly during a maintained stimulus in a fashion similar to that illustrated in Fig. 6-2, *right*. A sustained transducer and afferent fiber, unlike the pacinian corpuscle, exhibits no adaptation at its spike-initiation region, and thus the presence of action potentials closely corresponds to the maintained level of depolarization. If the generator potential exceeds the sensory axon's threshold, action potentials ensue. The nature of a neuron's membrane properties are such that the more the maintained generator potential exceeds this threshold, the sooner threshold can be reached again following an action potential. Thus the greater the generator potential is, the higher the rate of action potentials or firing frequency (Fig. 6-3). From this series of processes, a neural code that signals stimulus strength reaches the central nervous system; that is, firing frequency is

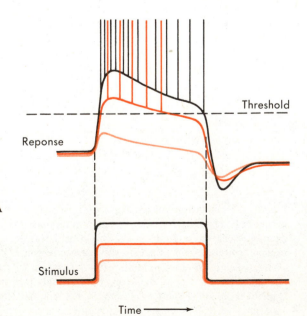

Fig. 6-3 ■ Relationships among stimulus intensity, generator potential amplitude, and frequency of generated action potentials for a tonic transducer. A monotonic relationship exists between stimulus intensity and generator potential amplitude. If this amplitude surpasses some threshold value, action potentials result with a frequency that is monotonically related to the generator potential's amplitude because the greater this amplitude, the sooner the cell's relative refractory period can be overcome. Thus a code is established for transmission to the central nervous system whereby the frequency of action potentials relates to stimulus intensity in a reliable monotonic fashion.

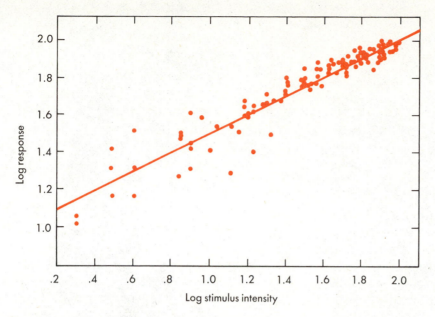

Fig. 6-4 ■ Experimental data from a tonic sensory neuron that verify the relationships shown in Fig. 6-3. Recordings of responses from a single cutaneous fiber are shown as a function of stimulus intensity applied to the skin. The response *(Log R)* monotonically encodes stimulus intensity *(Log S)*. (From Werner, G., and Mountcastle, V.B.: J. Neurophysiol. **28:**359, 1965.)

monotonically related to stimulus intensity (Fig. 6-3). This is an important property of tonic transducers and their associated axons.

Recordings from primary afferent fibers from glabrous skin in the monkey hand have shown that these fibers respond tonically in such a way that the firing frequency accurately encodes stimulus strength. This is illustrated by Fig. 6-4. A phasic transducer, such as an intact pacinian corpuscle, could never encode stimulus strength in this fashion. The actual transducers that provide the generator potentials for the fibers shown in Fig. 6-4 have not been identified. Excellent candidates are specialized cells in the skin that seem to form synaptic contacts with the distal end of primary afferent fibers. The generator potential might be evoked in these cells and synaptically transmitted to the afferent nerve fiber.

The preceding sections illustrate the differences between fast-adapting (phasic) and slowly adapting (tonic) transducers, as well as the functional significance of these differences. It is important to remember, however, that tonic transducers do adapt, albeit slowly. A consideration of photoreceptors in the retina serves to illustrate this point and its significance.

Humans have two broad types of photoreceptors, more sensitive rods for black-and-white vision and less sensitive cones for color vision. At low light levels *(scotopic illumination)*, cones do not respond, and only the rods function; thus colors are not appreciated in very dim light. At higher light levels *(photopic illumination)*, cones function and rods are saturated; color vision is now possible. These photoreceptors transduce photic energy into generator potentials that are synaptically transmitted to retinal interneurons (Chapter 8).

Rods and cones have similar morphological features (Fig. 6-5). An outer segment (cylindrical in rods and conical in cones) is attached to an inner segment and the rest of the cell by a thin stalk. The outer segment is stacked with parallel membranes into which the photosensitive chemicals (or photopigments) are embedded. In rods the photopigment is called *rhodopsin* and consists of a small vitamin A derivative attached to

■ *Photoreceptors: rods and cones*

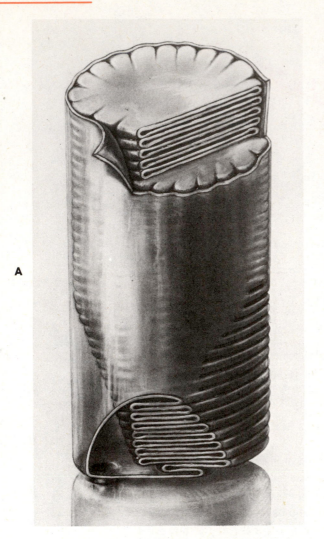
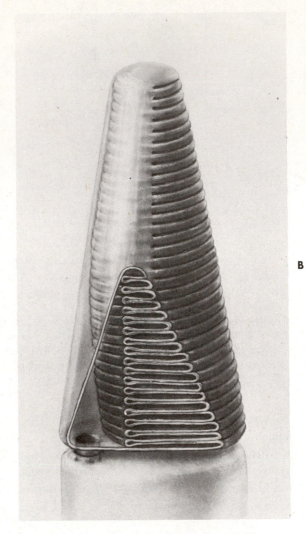

Fig. 6-5 ■ Schematic drawings of the outer segments of a vertebrate rod (**A**) and cone (**B**). Because of its much greater length, only the proximal portion of the rod outer segment is shown. (Courtesy R.W. Young.)

a large protein. The vitamin A derivative is *retinene,* and the protein is known as an *opsin.* The photochemistry and transduction processes of rods are described later; those of cones are thought to be similar.

When a photon strikes rhodopsin, it causes a change in isomerization of the retinene portion of the molecule, and this eventually leads to a breakdown of rhodopsin into retinene or vitamin A and the opsin. This breakdown is known as *bleaching* and is reversible (Fig. 6-6). Bleaching is the first in a series of incompletely understood steps that ultimately lead to the generator potential. A particularly unusual feature of rods and cones is the polarity of their generator potentials: these neurons are hyperpolarized by light. In the dark, sodium ions flow into photoreceptors. This is the "dark current," and it depolarizes the cell. In response to light, a series of incompletely understood steps ultimately leads to a blockage of many sodium channels in the membrane; the dark current is reduced, and the cell becomes hyperpolarized with respect to its resting state in the dark. This hyperpolarization signals the presence of light quite effectively to other retinal neurons. Indeed, in a cell without action potentials, for example, photoreceptors, there is no reason to suppose that the more conventional depolarization is any better a means of signaling stimulus features than is hyperpolarization.

This hyperpolarization, or generator potential, slowly adapts in the presence of con-

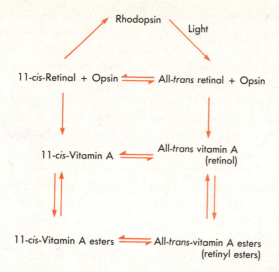

Fig. 6-6 ■ Chemical reactions involved in synthesis and breakdown of rhodopsin. For simplicity enzymes and metabolic prerequisites are not shown. (Redrawn from Wald, G.: Science **162**:230, 1968. © The Nobel Foundation 1968.)

stant light. Only a small portion of this adaptation can be related to the photochemistry of bleaching, and even this is significant only at high levels of illumination. The magnitude of the generator potential is related to the rate of the bleaching reaction. For a given number of photons, the rate of bleaching (and thus the generator potential size) is monotonically related to the concentration of intact rhodopsin still available. When a constant light is first applied to the retina, more intact rhodopsin is present, and the potential is maximal. As rhodopsin bleaches, the reaction slows down, and the potential diminishes. In other words, it slowly adapts. However, both in terms of rate and extent, the amount of photoreceptor adaptation actually measured far exceeds that predicted on the basis of bleaching alone. Other unknown factors contribute more significantly to this process.

This slow adaptation serves a useful purpose at the expense of slightly degrading the stimulus intensity/frequency code described earlier. Each photoreceptor at any one time has a dynamic response range that can encode only about a 5 log unit intensity range. However, the human visual system can operate over a 9 to 10 log unit range. Gradual adaptation is the key to this difference, and in vision virtually all such adaptation has its origin in the transduction process. Each receptor can gradually adjust its limited dynamic range to suit the light level present. For instance, when a person remains in a dark room, the photoreceptors become *dark adapted* for maximum sensitivity. If the person then emerges into a bright, sunlit area, he is "dazzled," and everything seems equally bright. As the retina *light adapts* to become less sensitive, shades of gray are appreciated, and shapes can be perceived. The converse is true when a person goes from a bright to a dark environment. Therefore slow adaptation serves to increase the dynamic range of a sensory system beyond that of the individual transducers. An alternative strategy for such an increase would require transducers to maintain fixed sensitivity ranges that differ from one another. To a very limited extent, the sensitivity difference between rods and cones achieves this, but not nearly enough to account for the extensive dynamic range of human vision given the individual ranges of rods or cones. This alternate strategy of different, fixed ranges for transducers would result in fewer responsive receptors and neural elements to analyze a given stimulus than is the case with slowly adapting cells. Again, this most important feature of sensory processing is a property of the transducers.

■ *Effectors and their control*

In the previous section the peripheral units at the input or sensory end of the nervous system were discussed. We shall now focus on the peripheral units of the output or motor end of the nervous system. These outputs are ultimately expressed by three classes of effector action: contraction of striated or skeletal muscle, contraction of smooth muscle of the viscera and eye, and glandular secretion. The neural control of glandular secretion is discussed elsewhere in this text (Chapter 43), and the neural (autonomic) control of the viscera is treated in Chapter 20. Here we will focus on the peripheral units of movement.

■ *Basic unit of movement*

With respect to the contraction of skeletal muscle, which induces movement, it is most appropriate to consider first the motoneuron. Muscles receive their innervation, and thus commands, from only one source, the motoneurons of the spinal cord and cranial nerve motor nuclei. These motoneurons are polygonal, multipolar cells that vary in size up to 70 μm. They give rise to large myelinated axons, 9 to 20 μm in diameter, that leave the central nervous system to reach their target muscle through either cranial or peripheral nerves.

In the spinal cord the motoneurons are located in the ventral horn. In a transverse section through the spinal cord different groups of motoneurons varying in mediolateral location can clearly be identified. In sagittal or horizontal section it can be seen that these groups are arranged in longitudinal columns (Fig. 6-7). As described in greater detail in Chapter 13, this cytoarchitectural arrangement of the spinal motoneurons reflects a highly topographic organization with respect to target musculature. Indeed, within each column are subsets of spatially contiguous motoneurons that innervate a single muscle, and this complement of motoneurons innervating a given muscle is defined as the *motoneuron pool* of that muscle (Fig. 6-8).

In mammals each muscle fiber is innervated by a single motoneuron. However, any given motoneuron will innervate a group of fibers within a single muscle. A motoneuron and the fibers it innervates are referred to as a *motor unit* (Fig. 6-8), and this can be viewed as the basic unit of movement. The innervation of a given muscle fiber by only

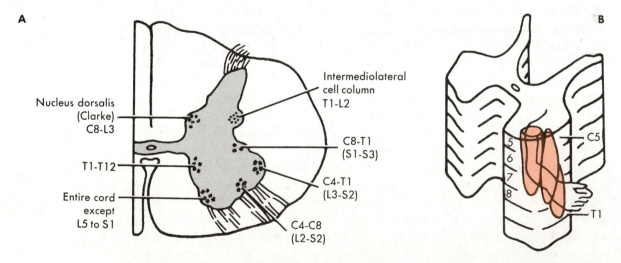

Fig. 6-7 ■ Diagrammatic illustration of the topographical organization of the motoneurons and the muscles they innervate. **A** shows a transverse section of the spinal cord with the positions of various cellular columns and their longitudinal extents. **B** shows three longitudinal columns of motoneurons *(color)* at cervical levels in relation to the parts of the arm they innervate. It should be appreciated that the more proximal muscles of the arm are innervated by more ventromedially situated motoneurons and the more distal muscles by more dorsolaterally situated motoneurons. (Redrawn from Brodal, A.: Neurological anatomy, ed. 3. Copyright © 1981 by Oxford University Press, Inc. Reprinted by permission.)

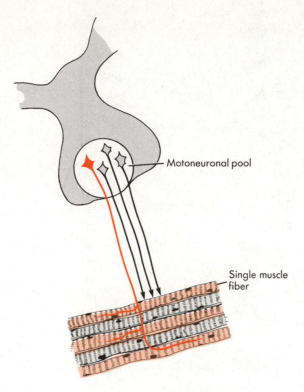

Motoneuronal pool

Single muscle fiber

Fig. 6-8 ■ Schematic illustation of a motoneuron pool, that is, the group of motoneurons innervating a given muscle. The motoneuron in color illustrates the concept of the motor unit, that is, a single motoneuron and all the muscle fibers it innervates.

one motoneuron leads to some important clinical applications. It is possible in humans to record the electrical activity of muscles during contraction *(electromyography)*. Also, the action potentials of a single muscle fiber can be studied with needle electrodes implanted in muscles. This activity directly reflects the discharge of a single motoneuron and is useful in various clinical contexts such as the differential diagnosis of atrophies and dystrophies and the assessment of regeneration after peripheral nerve injury.

The *size of motor units* varies considerably. Some muscles, such as the extrinsic eye muscles and the intrinsic muscles of the digits, are used for highly differentiated movements. Such muscles may have but a few fibers innervated by a single motoneuron. In contrast, the *innervation ratio* (the number of muscle fibers innervated by each motoneuron) may be as high as 2000:1 in large postural muscles. The muscle fibers of a motor unit are rather evenly distributed throughout the muscle, as opposed to being spatially restricted in a contiguous group.

Because a single motoneuron innervates many muscle fibers, it is clear that on reaching a muscle the axon of the motoneuron must ramify extensively. At each ramifying branch the termination forms the *motor endplate,* which is a specialized ending that constitutes the neural component of the neuromuscular junction. The morphology and function of the neuromuscular junction are discussed in Chapter 4. However, it is important to reemphasize that motoneurons produce only an excitatory influence on muscle; that is, activation of a motoneuron results in contraction of the fibers it innervates, and relaxation or lengthening of muscle fibers can only be achieved by a reduction of the motoneuron discharge. The tension generated in a motor unit will thus be a function of the discharge rate of its motoneuron, and the tension generated in an entire muscle will be the sum of the tensions of all the active motor units of that muscle.

Graded muscle contraction is achieved (1) by activating an increasingly larger num-

■ *Control of contraction*

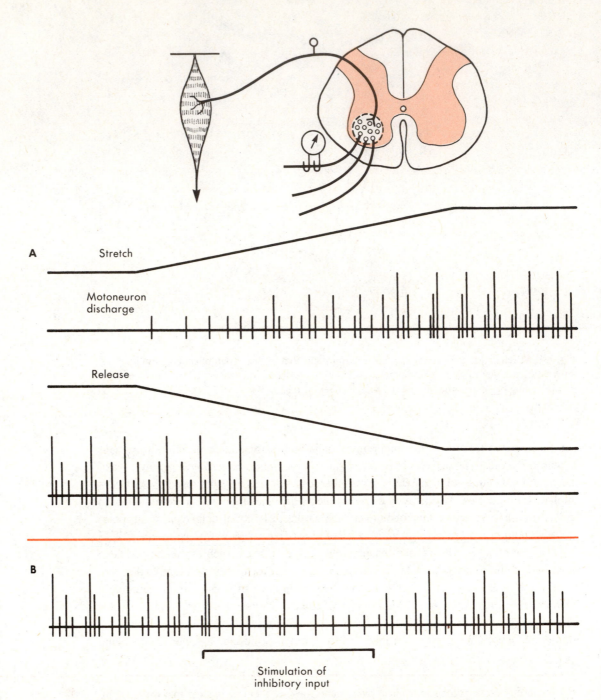

Fig. 6-9 ■ The size principle in the recruitment of motoneurons. The schematic at the top shows a Ia afferent fiber from a muscle spindle and recording electrodes on a dissected ventral root filament arising from an homonymous motoneuron. **A,** Excitation. As a muscle is stretched, the increased activity of the Ia afferent fibers first recruits the smaller motoneurons. As the stretch increases, successively larger motoneurons are recruited. On release from stretch, the larger motoneurons stop discharging first and the smaller motoneurons last. **B,** Inhibition. Activating an inhibitory input to the motoneuron pool first silences the larger motoneurons and then successively smaller motoneurons. (Reproduced with permission from Eyzaguirre, C., and Fidone, S.J.: Physiology of the nervous system, ed. 2. Copyright © 1975 by Year Book Medical Publishers, Inc., Chicago.)

ber of motor units *(recruitment)* and (2) by increasing the frequency of discharge of the motoneurons of each motor unit. With regard to motor unit recruitment, there is evidence that a *size principle* governs the order in which motor units come into play (Fig. 6-9). The excitability threshold of a motoneuron appears to be a function of its size, such that the first motor units recruited are those of the smaller motoneurons. These generate smaller contractile tensions than the larger motor units. As successively larger motor units are recruited, larger and larger increments of contractile tension are added. With regard to discharge frequency, the smaller motoneurons have an initial phasic burst, followed by a maintained discharge at a lower frequency. The larger motoneurons tend to discharge phasically, followed by a highly irregular discharge. Smoothly graded contraction and sustained contraction are thus a function of how motor units are recruited, the rate coding properties of different sized motor units, and a rotation among the motor units of periods of activation. The properties of different types of muscle fibers are also relevant to the characteristics of contraction, and these are discussed in Chapter 23.

■ *The final common path*

It is clear from the previous discussion that considerable central integration is required for even the smooth onset of contraction in a single muscle. The coordinated action of a group of muscles around a joint requires still more integration. Complex sequences of muscle contractions, to produce spatially and temporally organized movement, demand truly impressive neuronal interactions. This is reflected in the large number of pathways involved in the control of movement and the thousands of synaptic contacts distributed over the soma and dendrites of each motoneuron.

It must be appreciated, however, that, despite the complexity of the integration of motor activity that occurs at many levels of the neuraxis, this must ultimately be expressed through the motoneuron if muscle contraction and therefore movement are to be generated. It is for this reason that Sherrington described the motoneuron as the *final common path*. An important clinical implication of this is that any lesion compromising the final common path of a given muscle will preclude the ability of that muscle to contract and thereby result in its paralysis. Such lesions are sometimes referred to as *lower motoneuron lesions,* and they may result from compromise of different segments of the final common path (Fig. 6-10). For example, diseases such as poliomyelitis and amyotrophic lateral sclerosis destroy the motoneuronal cell body, whereas the various peripheral neuropathies, such as the Guillain-Barré syndrome, can destroy the axon of

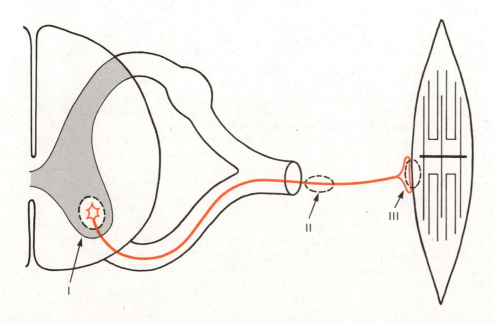

Fig. 6-10 ■ Sites involved in diseases that affect the final common path for skeletomotor activity. *I,* The motoneuron cell body that can be directly affected by such diseases as poliomyelitis and amyotrophic lateral sclerosis. *II,* The axon of the motoneuron that can be affected by injury and various diseases such as the demyelinating Guillain-Barré syndrome. *III,* The neuromuscular junction where transmission can be compromised by a disease such as myasthenia gravis or by a toxin such as botulinus toxin. Compromise of the motoneuron or its axon produces lower motoneuron disease. (Modified from Ruch, T., and Patton, H.D.: Physiology and biophysics, vol. IV, Excitable tissues and reflex control of muscle, Philadelphia, 1982, W.B. Saunders Co.)

the motoneuron. If the lesion produces actual motoneuronal death or interruption of its connection with a muscle, then degeneration of the muscle fibers of the motor unit will occur. This leads to muscle atrophy and reflects the trophic relationship that exists between the motoneuron and muscle fibers of a motor unit (Chapter 23).

Beyond lower motoneuron lesions, the function of the motor unit can be compromised by any disease process affecting transmission at the neuromuscular junction, such as myasthenia gravis. Still more peripherally, the motor unit can be affected by a number of diseases involving the muscle itself—the myopathies, which include diseases such as the muscular dystrophies.

■ Bibliography

Journal articles

Attwell, D., and Wilson, M.: Behavior of the rod network in the tiger salamander retina mediated by membrane properties of individual rods, J. Physiol. (Lond.) **309:**287, 1980.

Bader, C.R., MacLeish, P.R., and Schwartz, E.A.: A voltage-clamp study of the light response in solitary rods of the tiger salamander, J. Physiol. (Lond.) **296:**1, 1979.

Baylor, D.A., Lamb, T.D., and Yau, K.W.: The membrane current of single rod outer segments, J. Physiol. (Lond.) **288:**589, 1979.

Baylor, D.A., Matthews, G., and Yau, K.W.: Two components of electrical dark noise in toad retinal rod outer segments, J. Physiol. (Lond.) **309:**591, 1980.

Buchtal, F.: The electromyogram: its value in the diagnosis of neuromuscular disorders, World Neurol. **3:**16, 1962.

Copenhagen, D.R., and Owen, W.G.: Current-voltage relations in the rod photoreceptor network of the turtle retina, J. Physiol (Lond.) **302:**159, 1980.

Detwiler, P.B., Hodgkin, A.L., and Mc-Naughton, P.: Temporal and spatial characteristics of the voltage response of rods in the retina of the snapping turtle, J. Physiol. **300:**213, 1980.

Fain, G.L., Gold, G.H., and Dowling, J.E.: Receptor coupling in the toad retina, Cold Spring Harbor Symp. Quant. Biol. **40:**547, 1975.

Fain, G.L., and Lisman, J.E.: Membrane conductances of photoreceptors, Prog. Biophys. Mol. Biol. **37:**91, 1981.

Griff, E.R., and Pinto, L.H.: Interactions among rods in the isolated retina of *Bufo marinus,* J. Physiol. (Lond.) **314:**237, 1981.

Hagins, W.A.: The visual process: excitatory mechanisms in the primary receptor cells, Ann. Rev. Biophys. Bioeng. **1:**131, 1972.

Hubbell, W.L., and Bownds, M.D.: Visual transduction in vertebrate photoreceptors, Ann. Rev. Neurosci. **2:**17, 1979.

Iggo, A.: Is the physiology of cutaneous receptors determined by morphology? Prog. Brain Res. **43:**15, 1976.

Lasanskey, A.: Synaptic actions mediating cone responses to annular stimulation in the retina of the larval tiger salamander, J. Physiol. (Lond.) **310:** 205, 1981.

Rushton, W.A.H.: Visual adaptation (Ferrier Lecture), Proc. R. Soc. Lond. (Biol.) **162:**20, 1965.

Sterling, P., and Kuypers, H.G.J.M.: Anatomical organization of the brachial spinal cord of the cat. II. The motoneuron plexus, Brain Res. **4:**16, 1967.

Tomita, T.: Electrical activity of vertebrate photoreceptors. Q. Rev. Biophys. **3:**179, 1970.

Wald, G.: The receptors of human color vision, Science **145:**1007, 1964.

Wald, G.: Molecular basis of visual excitation, Science **162:**230, 1968.

Books and monographs

Brodal, A.: Neurological anatomy, ed. 3, New York, 1981, Oxford University Press, Inc.

Burgess, P.R., and Perl, E.R.: Cutaneous mechanoreceptors and nociceptors. In Iggo, A., editor: Handbook of sensory physiology,vol. II, Somatosensory system, New York, 1973, Springer-Verlag New York, Inc.

Cohen, A.I.: Rods and cones. In Fuortes, M.G.F., editor: Handbook of sensory physiology, vol. VII, Physiology of photoreception organs, New York, 1972, Springer-Verlag New York, Inc.

Gordon, A.M.: Muscle. In Ruch, T., and Patton, H.D., editors: Physiology and biophysics, vol. IV, Philadelphia, 1982, W.B. Saunders Co.

Henneman, E.: Organization of the motoneuron pool. In Mountcastle, V.B., editor: Medical physiology, ed. 14, vol. I, St. Louis, 1980, The C.V. Mosby Co.

Henneman, E.: Skeletal muscle, the servant of the nervous system. In Mountcastle, V.B., editor: Medical physiology, ed. 14, vol. I, St. Louis, 1980, The C.V. Mosby Co.

Kenshalo, D.R., editor: The skin senses, Springfield, Ill., 1968, Charles C Thomas, Publisher.

Lowenstein, W.R.: Mechano-electrical transduction in the pacinian corpuscle. Initiation of sensory impulses in mechanoreceptors. In Lowenstein, W.R., editor: Handbook of sensory physiology, vol. I, Principles of receptor physiology, New York, 1971, Springer-Verlag New York, Inc.

Miller, W.H., editor: Current topics in membranes and transport, vol. 15, Molecular mechanisms of photoreceptor transduction, New York, 1981, Academic Press, Inc.

Ripps, H, and Weale, R.A.: Visual adaptation. In Davson, H., editor: The eye, New York, 1976, Academic Press, Inc.

Rodieck, R.W.: The vertebrate retina: principles of structure and function, San Francisco, 1973, W.H. Freeman Co. Publishers.

Rowland, L.P.: Diseases of the motor unit: the motor neuron, peripheral nerve, and muscle.

In Kandel, E.R., and Schwartz, J.H., editors: Principles of neuroscience, New York, 1981, Elsevier North-Holland, Inc.

Sherrington, C.S.: The integrative action of the nervous system, ed. 2, New Haven, Conn., 1947, Yale University Press.

Teorell, T.: A biophysical analysis of mechano-electrical transduction. In Lowenstein, W.R., editor: Handbook of sensory physiology, vol. I, Principles of receptor physiology, New York, 1971, Springer-Verlag New York, Inc.

Yoshikami, S., and Hagins, W.A.: Control of the dark current in vertebrate rods and cones. In Langer, H., editor: Biochemistry and physiology of visual pigments, New York, 1973, Springer-Verlag New York, Inc.

chapter 7

General principles of sensory systems

Mammalian sensory systems have evolved to provide the organism with information about its environment so that these systems can guide its behavior accordingly. These systems must tell the organism *what* is in the environment, or the precise form of stimulus energy present, that is, photic, mechanical, thermal, or chemical. They must specify *when* changes in the environment occur, or the sudden presence or absence of a stimulus. In addition to these qualitative considerations, mammalian sensory systems have evolved the capability to quantify stimulus parameters in terms of *how much* (intensity) and *where* (location).

A major goal is to describe how the various sensory systems accomplish this. It is important to appreciate that these systems share many basic organizational features, such that once a fundamental framework is understood, it is relatively easy to elaborate the details unique to each specific modality. Consequently, the visual system will be described in relatively great detail because an in-depth understanding of this system establishes the framework common to other sensory systems. Also, more is known about the neural basis of vision than about any other modality.

Mammalian sensory systems are complex combinations of interacting pathways. A great deal of the complexity of these pathways derives from their evolutionary history, and their organization can more readily be grasped with some understanding of this evolution. Although the details of such an evolutionary history are not known, the general processes can be suggested. An imaginary but plausible process by which sensory systems might have evolved is outlined in following sections. Although this should not be taken literally, it should facilitate understanding the enormous complexity of mammalian sensory systems. Finally, the concept of a receptive field will be discussed because this has proven important in analyzing sensory pathways.

■ General evolution of sensory systems

Imagine a simple, multicellular organism swimming in the primordial seas (Fig. 7-1, *A*). Two behaviors, among others, are essential to this organism's survival; it must find food and must avoid serving as another beast's meal. The evolution of these behaviors involves the development of a rudimentary analogue of a nervous system, including various sensory systems.

Sensory receptors or transducers sensitive only to specific nourishing chemicals in the sea might evolve as laterally paired cells on or near the organism's surface. These could be functionally connected to paired contractile units (analogous to muscles) in an ipsilateral fashion. This arrangement would cause the organism to move along an appropriate chemical gradient toward the source of food and thereby provide a means of securing nourishment (Fig. 7-1, *A*). Note that this scheme requires both that the receptors induce contraction only in the presence of the appropriate chemicals and that infor-

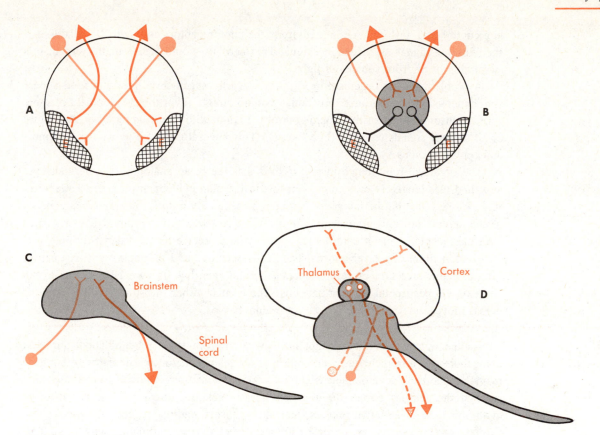

Fig. 7-1 ■ Schema for a plausible but imaginary course of evolution for sensory systems. **A,** A primitive animal with bilateral symmetry swimming in primordial seas must find food and avoid predators in order to survive. For food acquisition, a pair of specialized chemosensory elements *(colored triangles)* have evolved sensitivity to chemical nutrients in the environment; these are analogous to neural transducers. By virtue of their ipsilateral activation of contractile effector *(E)* elements (analogous to muscles), their activation causes the organism to orient toward the food source and swim down its chemical gradient to feed. To avoid predators it might be useful to keep to dark places to avoid visual detection. For this purpose a pair of photosensitive elements *(colored circles)* exists that activates the contralateral effector elements, thereby establishing a negative phototaxic response for the organism. **B,** A problem with the primitive design in **A** is that when food and light appear from the same direction, the effector elements receive contradictory signals from the chemoreceptors and photoreceptors. Thus an integrator *(I)* has evolved to analyze the raw sensory signals and send a coordinated command to the effectors. **C,** The integrator may be viewed as a primitive brain that performs the same basic function as the spinal cord and brainstem as it may have appeared at the beginning of vertebrate evolution. For simplicity only two transducers are shown (analogous to the two pairs in **A** and **B**) that represent an early photoreceptor *(colored circle)* and pressure-sensitive receptor *(colored triangle)*. **D,** Finally, as mammals appear, a great expansion of thalamus and cortex appears (although homologous structures seem to exist in nonmammalian vertebrates). Corresponding to this is the evolution of newer transducers and their central connections through thalamocortical pathways. However, note that the older spinal cord and brainstem pathways do not disappear as the newer portions of the sensory systems evolve. Indeed, these older pathways continue to contribute significantly to sensory processing.

mation from these receptors be conducted along specific lines or channels. Thus activity along these lines can uniquely reflect the presence or absence of appropriate chemicals.

If the same contractile elements are also used in the behavior to avoid predators, however, confusion can result. For instance, the organism might have evolved the strategy of avoiding predators that use vision by avoiding light. This could be accomplished by the evolution of laterally paired photoreceptors connected with the contractile units in a contralateral fashion. Again, activity in the photoreceptors and their "labeled lines" specifies the signal as visual in origin. This activity would result in the organism's swimming away from sources of light (Fig. 7-1, *A*).

A problem for the organism in Fig. 7-1, *A*, arises each time light and food appear simultaneously from the same direction. Evolution might favor the development of some sort of sensory integrator that could control the contractile elements based on the nature of all sensory signals received (Fig. 7-1, *B*). This integrator can be viewed as a primordial central nervous system.

The organism represented in Fig. 7-1, *B*, has the basic analogue of a nervous system, and it is relatively easy to convert this to the plan evident in primitive vertebrates (Fig. 7-1, *C*); that is, the integrator can be represented by a dorsal neuraxis that consists of the spinal cord and brainstem. This neuraxis receives sensory input that originates with the receptors, and it controls the muscles (and other effector organs) through various motoneurons. Each receptor is specifically sensitive to a particular sensory quality, such as light, pressure, or certain chemicals, and each has its own labeled line to the neuraxis via centripetal axon projections. These enter either the spinal cord by way of dorsal roots or the brainstem by way of cranial nerves.

■ *Basic plan of mammalian sensory systems*

As the mammalian brain might have evolved from such a hypothetical ancestral plan, there was a tremendous proliferation of the telencephalon, particularly with regard to thalamocortical development (Fig. 7-1, *D*). New sensory pathways developed that reached the sensory cortex through appropriate thalamocortical pathways. However, evolution is a conservative process, and older sensory pathways generally are not discarded as the newer thalamocortical pathways are evolved. Mammalian sensory systems are consequently a combination of older (subcortical and extrathalamic) and newer (thalamocortical) pathways that cooperate and interact in analyzing the sensory environment. Unfortunately it is never possible in existing mammalian species to identify unambiguously the relative phylogenetic age of these components in any sensory system. Such tentative identifications are inferential rather than factual. Also, the properties and functions of "older" pathways or components may change as "newer" ones evolve. The important lesson here is that older subcortical sensory pathways may be important to perception. Too often in the past these older pathways have been swept under the rug like some evolutionary debris, but recently much deserved attention has been directed toward these subcortical pathways.

As an example of this organizational principle, the visual system will be considered briefly. In nonmammalian vertebrates, the optic tectum is the largest and most important central structure for vision. It is likely that a similar condition held true for vertebrates before the evolution of mammalian geniculocortical pathways, although even the most primitive existing vertebrates, for example, elasmobranchs, exhibit small but distinct thalamic-telencephalic visual pathways. In any case, as these newer pathways evolved, the presumably older retinotectal pathways that involve the superior colliculus (the mammalian homologue of the optic tectum) did not disappear. Indeed, substantial recent evidence, considered in more detail later, emphasizes the importance of the superior colliculus and other subcortical structures to mammalian vision. It is thus insufficient to consider only the newer (retinogeniculocortical) part of the visual system. The same principle applies to other sensory systems such as the somatosensory and auditory systems. The following sections are organized around this concept.

How mammalian sensory systems are able to analyze the nature of stimuli in the environment can now be considered in a general way. Because transducers are relatively sensitive to one specific form of stimulus energy and because they are connected with little or no convergence or divergence to axons projecting toward the central nervous system, these pathways are *labeled lines* and their activity specifies *what* the stimulus is. Thus if a transducer is excited by extreme forms of inappropriate stimulus energy (remember that their sensitivity is relative), the percept is as if the appropriate stimulus energy were present. For instance, extreme pressure applied to the retina is perceived as flashes of light and not as a mechanical stimulus.

Fast-adapting or phasic cells are ideally suited to signal *when* a change occurs in the environment. Many transducers are phasic and thus can serve this function. As described later, central synaptic integration can convert tonic input signals into phasic signals. For example, the visual system, which has only tonic photoreceptors, can transform these within the brain to phasic responses that reliably signal temporal changes.

As indicated in the previous description of sensory receptors, however, tonic receptors and neurons are needed to signal stimulus intensity, or *how much* is present. This is accomplished with monotonic relationships among intensity, the size of the generator potential, and the resultant firing frequency of the central neurons involved. The slow adaptation of these cells sacrifices some precision of the intensity code in favor of an increased dynamic range.

Finally, the ability to determine precisely *where* a stimulus is located, or its spatial distribution, is also largely a function of the transducers. In the various sensory systems, these transducers are spatially arranged in peripheral tissue (retina, skin, etc.), and the spatial pattern of active transducers conveys the basic information required to determine the stimulus location. The central nervous system has mechanisms that then sharpen this spatial image.

■ *Receptive fields*

Receptive field analysis has proven a useful tool in analyzing the functional organization of sensory systems. For any sensory neuron in the central nervous system, a *receptive field* can be described in terms of that area of transducer-containing tissue which, when properly stimulated, causes a change in the neuron's electrochemical properties. This change is usually revealed as a change in the firing rate. The precise region of transducer-containing tissue involved in a receptive field often depends on the nature of the stimulus used. For instance, a visual neuron has a receptive field related to a region of retina. If the retina is explored with small spots of light, flashed on and off, a map of the receptive field can be constructed with separate regions excited by light turned off, excited by light turned on, inhibited by light turned off, and/or inhibited by light turned on. If the same receptive field is plotted by a spot of light moving across the retina, the map might look somewhat different from the map based on flashing spots, and it might also change with changes in the speed or direction of movement. Indeed, some visual cells might exhibit strong responses to some stimulus parameters applied to a proscribed region of retina and none to other parameters.

Receptive fields, then, can be complicated to describe, but they are always related to areas of transducers that have been stimulated. Fig. 7-2 shows several simple examples of different receptive fields for visual and somatosensory cells. The receptive field for a neuron could be readily inferred if one could know the detailed functional connections between transducers, intervening neurons, and the neuron in question. Such detailed information on connectivity is rarely available. The value of a receptive field analysis is the converse. A knowledge of receptive field properties of many neurons at various levels in a sensory system allows inferences regarding the functional interactions among these neurons.

■ ■ ■

Each sensory system is a complex arrangement of interacting pathways that is perhaps best appreciated from an evolutionary perspective; that is, these systems can be roughly and inferentially divided into phylogenetically older and newer portions. It must be remembered that the older portions are present and probably contribute significantly

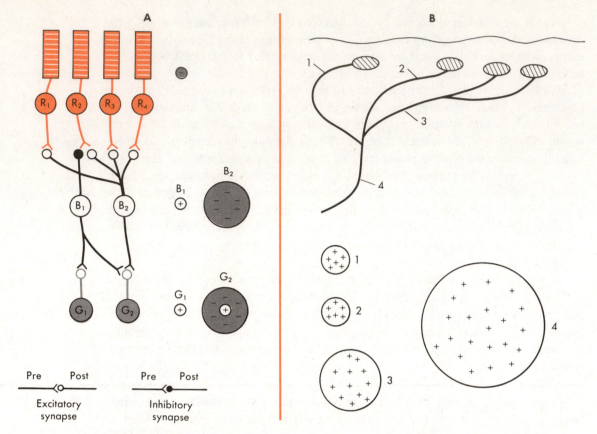

Fig. 7-2 ■ Examples of neuronal receptive fields based on functional connections between sensory transducers and neuronal element under study. **A,** Visual receptive fields. The receptive field of each photoreceptor (R_1, R_2, R_3, or R_4) is equal to the patch of retina occupied by the photoreceptor. Because photoreceptors hyperpolarize to photic stimulation, the photoreceptor receptive field by convention is drawn as inhibitory and indicated with minus signs. For bipolar cell B_1, which receives an inhibitory input from photoreceptor R_2, the receptive field is equal in size but opposite in sign to that of R_2. For bipolar cell B_2, which receives excitatory input from each photoreceptor, the receptive field is equal to the spatial sum of those of R_1, R_2, R_3, and R_4. Likewise, the receptive fields of ganglion cells G_1 and G_2 can be inferred from the circuitry as shown. **B,** Somatosensory receptive fields. For this example assume that the action potential is generated near each distal tip of the axonal branches by the transducer mechanism and that these potentials are not antidromically (retrogradely) conducted toward a distal tip from a branch point. Thus receptive fields of distal branches *1* or *2* are simply equivalent to the skin area innervated by the single transducer attached to each. The receptive field recorded at point *3* is equal to the skin area innervated by the right pair of transducers, and the receptive field recorded more proximally at point *4* is the skin area innervated by all four transducers.

to sensory processing. A consideration of only the newer, or thalamocortical, portions of each system is insufficient.

The basic plan of sensory systems involves transducers with both relative sensitivity to a single form of stimulus energy and also labeled lines to the central nervous system. This allows identification of the stimulus modality. Phasic transducers can best signal temporal changes in the stimulus, and tonic transducers can signal stimulus intensity. The spatial pattern of active transducers in the peripheral tissue encodes stimulus position.

A final consideration is the concept of the receptive field. Every neuron in a sensory system has a receptive field. It is defined in terms of the area of transducer-containing tissue that, when stimulated, affects the neuron in question. The precise receptive field for a neuron often depends on the spatial and temporal properties of the stimulus used to elucidate it. Detailed knowledge of receptive field properties for neurons at many levels within a sensory system leads to insight regarding the functional interactions among these neurons.

■ *Bibliography*

Journal article

Northcutt, R.G.: Evolution of the telencephalon in nonmammals, Ann. Rev. Neuroscience **4**:301, 1981.

Books and monographs

Ariens Kappers, C.U., Huber, G.C., and Crosby, E.C.: The comparative anatomy of the nervous system of vertebrates, including man, New York, 1967, Hafner Press. (Originally published in 1936.)

Nauta, W.J.H., and Karten, H.J.: A general profile of the vertebrate brain, with sidelights on ancestry of cerebral cortex. In Schmitt, F.O., editor: The neurosciences second study program, New York, 1970, Rockefeller University Press.

chapter 8

The visual system

■ *The eye and physiological optics*

Although the visual system begins with the transduction process in the retina, a complete understanding of vision must take into account the transformations in photic stimuli as they pass through the eye's optics to the retina. Thus this section discusses the role of the eye as an optical instrument.*

■ *General structure*

Fig. 8-1 shows a horizontal section through the middle of the human right eye as viewed from above. The eye has three coats. Most of the outside of the eye is covered by the *sclera,* a tough connective tissue that reflects most light striking it. It thus appears white in normal illumination. At the front of the eye, the sclera is replaced by a transparent outer coat known as the *cornea.* The circular junction between sclera and cornea is called the *limbus.* The middle and inner coats occupy only the posterior two thirds of the eye. The middle coat is the *choroid,* which contains a rich blood supply as well as considerable melanin. This pigment absorbs most light striking it. The inner coat is the retina.

The posterior portion of the eye is occupied by the *vitreous body* or *vitreous humor,* a clear, jellylike substance. The vitreous body is bounded at its anterior end by the *ciliary body, zonular fibers,* and *lens;* these structures will be discussed more thoroughly later. The *iris* lies anterior to the lens. It reflects and absorbs most of the light that strikes it, but some light passes through a circular hole, the *pupil,* in the center of the iris. The remaining space between the lens and cornea is occupied by a watery, clear substance called the *aqueous humor.* This fluid is segregated into two compartments: an *anterior chamber* in front of the iris and a *posterior chamber* between the iris and lens.

The *optic disc* and *fovea* appear as depressions in the retina and form two obvious and important landmarks. The fovea is the portion of retina specialized for most acute vision, and the eye is typically directed so that the most important point in the visual world, the *fixation point,* is imaged onto the fovea.

The fovea serves as the center of a frame of reference for locations and directions on the retina. All retina between the fovea and nose is *nasal* retina, and that between the fovea and temple is *temporal* retina. The view from above of the right eye (Fig. 8-1) thus depicts nasal retina to the left of the fovea and temporal retina to the right. If the left eye were also drawn in this figure, the left portion of its retina would be temporal and the right, nasal. The optic disc lies in each nasal retina. The retina above the fovea is *superior,* and below is *inferior* retina. Superior and inferior retinas are roughly equal

*The reader is expected to have a basic knowledge of optics that includes an understanding of (1) how lenses form images and (2) the meanings of terms such as *refraction, focal length, diopter,* and *spherical* or *chromatic aberration.*

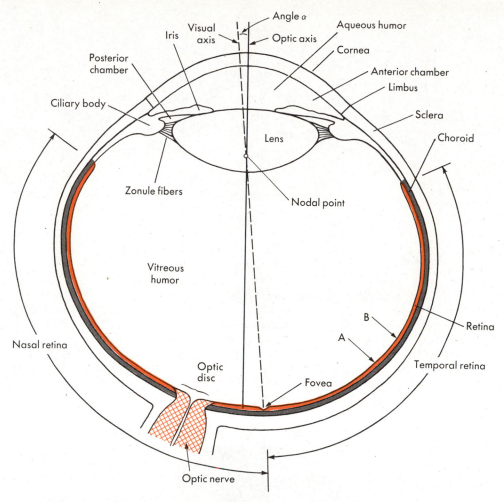

Fig. 8-1 ■ Schematic diagram of a horizontal section through the human right eye as viewed from above. (Redrawn from Walls, G.L.: The vertebrate, eye and its adaptive radiation, Bloomfield Hills, Mich., 1942, Cranbrook Institute of Science.)

in size, but the nasal retina is somewhat greater in extent than the temporal retina. Finally, nasal, temporal, superior, and inferior directions on the retina refer to relative positions in this coordinate system. For instance, the fovea is temporal to the optic disc on the retina; similarly, points *A* and *B* in Fig. 8-1 are both in the temporal retina, although point *A* is nasal to point *B*.

Fig. 8-1 also shows the *visual* and *optic axes*. The visual axis is a line that passes through the fovea, nodal point of the eye, and fixation point. The *nodal point* of an optical system can be defined in terms of the relationship between an object and the reversed, inverted image formed by the optics. Straight lines can be drawn between each point on the object and its corresponding point on the image. All of these lines intersect at the nodal point (Figs. 8-1 and 8-2). The visual axis is the axis along which the eye is directed. It is not to be confused with the eye's optic axis, which is the axis of optical symmetry. More simply, the optic axis can be approximated by the eye's axis of anatomical symmetry, or the line that passes through the center of the cornea, pupil, lens, etc. It, too, passes through the nodal point. Typically, the optic axis intersects the retina nasal to the fovea (Fig. 8-1).

The angle between the optic and visual axes is called the angle α (Fig. 8-1). In humans the angle α averages roughly 5 degrees in each eye so that fixation of a distant object, which leads to parallel visual axes, also leads to about a 10-degree angle be-

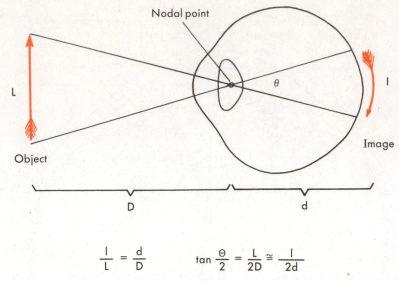

$$\frac{l}{L} = \frac{d}{D} \qquad \tan\frac{\Theta}{2} = \frac{L}{2D} \cong \frac{l}{2d}$$

Fig. 8-2 ■ Image formation by the eye. If straight lines are drawn between equivalent points on the object *(O)* and image *(I),* they intersect at the nodal point. The relationship between *D* (approximately the distance between the object and the eye), *d* (roughly the eye's focal length), *L* (object size), *l* (image size), and θ (the angular subtense of the object and image) is given. Because *d* is a fixed constant, *l* and θ bear a constant trigonometric relationship to each other. Note, however, that the approximation that tan θ/2 = l/2d breaks down for large values of θ or *l,* since *l* represents an arc rather than a chord of a circle.

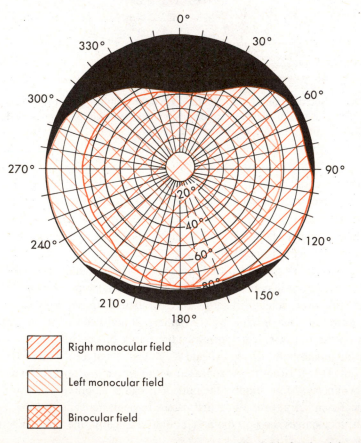

Fig. 8-3 ■ Human visual fields shown in polar coordinates. The fields for the right and left eye are shown separately. The binocular segment of visual field is the overlap region of the two monocular fields. The two monocular segments are the peripheral crescents that can be seen by only one eye.

tween the optic and visual axes. Ignorance of the angle α can lead to misdiagnosis of the clinical condition of *strabismus,* which is defined as the inability to direct both foveas to the same fixation point. When one informally judges the direction in which an eye is aimed, one usually estimates the direction the center of the pupil seems to assume. This refers to the optic and not the visual axis. If the angle between these axes is large enough, one might not diagnose eyes that are pathologically crossed because the optic axes might be parallel. Conversely, eyes might be misdiagnosed as pathologically diverged because the optic axes are divergent when in fact the visual axes are parallel. Although the angle α averages roughly 5 degrees, it is quite variable among humans. In some eyes it is much larger or smaller and can even be reversed in sign; that is, in rare cases, usually involving a high degree of myopia (p. 105), the visual axis intersects the retina temporal to the fovea.

Fig. 8-2 demonstrates a final point regarding the eye's general structure. The size of an image on the retina can be specified in two equivalent fashions: (1) the linear extent of the image or (2) the angle formed by the nodal point and limits of the object or image. For the human retina each millimeter of retina equals slightly more than 3 degrees of visual space. The optic disc occupies 5 degrees of visual space, and the size of the visual field seen by one eye is approximately 150 degrees horizontally by 120 degrees vertically (Fig. 8-3).

■ *Images formed by the eye*

The main role of the nonneural parts of the eye is to form an optically ideal image on the retina. The refractive power of the cornea and lens focuses images on the retina. The transparency of the cornea and lens, as well as that of the aqueous and vitreous humors, preserves the quality of these focused images with little or no intensity loss. The reflective sclera and absorbant choroid serve to prevent improperly focused light from reaching the retina and degrading the image.

Camera optics. In many ways the eye can be compared to a camera (Fig. 8-4). A good camera has a compound lens system (two or more elements to minimize optical distortions) that forms an inverted and reversed image of an object on the photosensitive film plane. The focus can be adjusted by changing the distance between the lens and the film. The image formed can be thought of as a precise, point-to-point map of the visual space within the camera's field of view; each of the points in space can be precisely located in the image once the direction and orientation of the camera are known.

Stray light (i.e., photons that are not imaged by the optics) reduces contrast in the image. Stray light is prevented from reaching the film in two ways. The camera body is opaque, so light can enter only through the lens. Also, internal reflections of stray light are minimized by the camera's interior, which is painted black to absorb such light.

Finally, the camera has an iris/diaphragm system in the lens to control the size of the entrance aperture of the optical system. This serves two purposes. First, the amount of light striking the film is controlled because it is proportional to the cross-sectional area of the aperture. This is relatively unimportant because the shutter speed also controls the amount of light. Second and more importantly, *depth of field* and overall optical quality are controlled by the aperture size. Fig. 8-5 shows how a small aperture increases the depth of field relative to a large one. Each optical system has an aperture size that optimizes the optics. The size is usually intermediate; it is a compromise between smaller sizes, which reduce spherical aberrations and increase depth of field, and larger sizes, which minimize diffraction of the image.

Physiological optics. Each of the optical features considered in the previous section has a counterpart in the eye, which has a compound lens system that forms an inverted and reversed image on the photosensitive retina. The compound lens system has two

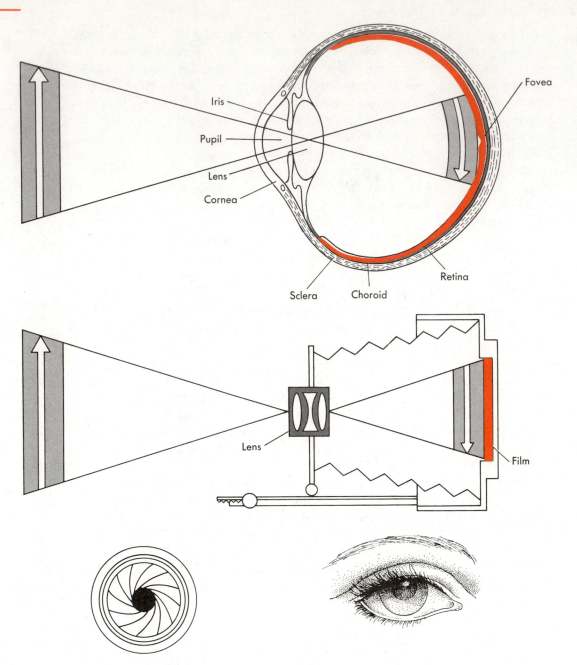

Fig. 8-4 ■ Comparison of eye and camera. Many close parallels can be seen in the optical properties of eyes and cameras, only some of which are illustrated here. See text for details. (Redrawn from Wald, G.: Sci. Am. **183**:32, 1950.)

chief components: the cornea and lens. The cornea is actually responsible for most of the refractive power of the eye, but its power of approximately 43 diopters (D) is fixed. The lens has less power but can be continuously adjusted between approximately 13 and 26 D to change focus of the eye. Most of this change in power, or *accommodation*, results from changing the shape of the lens, although there is also some displacement of the lens relative to the retina.

The retinal image formed can be specified in terms of the fixation point imaged on the fovea; that is, once an observer in a normal, upright position fixates a point, images of other points can be precisely specified in terms of their retinal locations.

Stray light stimulation of the retina is minimized in two ways. First, light is pre-

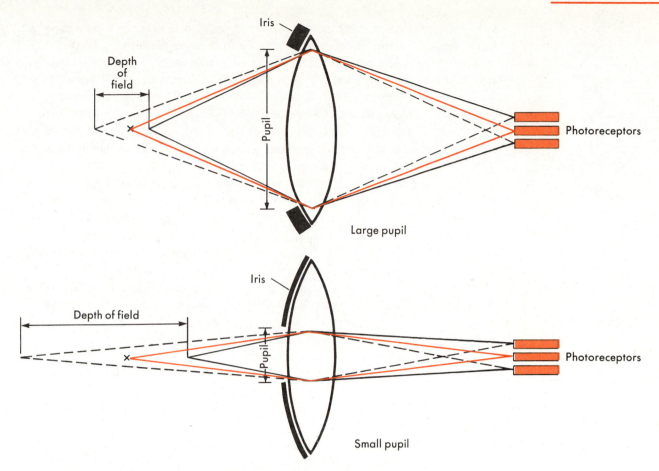

Fig. 8-5 ■ The relationship between pupil size and depth of field. In practical terms the image is as sharp as possible (i.e., unblurred and focused) as long as a single object point is imaged onto a single photoreceptor. When sufficient blur causes some rays from that point to fall onto neighboring photoreceptors, the image will be functionally out of focus. Thus if the eye is accurately focused on a single point *(x)*, there is a range in front of and behind that point within which other points are effectively focused because they are imaged onto a single photoreceptor. This range is called the *the depth of field* or *depth of focus*. As rays from a single point are focused, they describe a cone of light. A larger pupil *(above)* creates cones with larger vertex angles than does a smaller pupil *(below)*. Consequently larger pupils result in smaller depths of field.

vented from reaching the retina along paths that do not pass through the cornea and lens. The sclera reflects most of this light, and the choroid absorbs most of the remainder. Second, light that does enter the eye and is not absorbed by photopigment in the photoreceptors is prevented from reflecting to inappropriate photoreceptors by absorption in the *pigment epithelium*, a layer of retina that lies just beyond the photoreceptors. Pigment in the choroid and pigment epithelium are thus important in the prevention of stray light and reduced contrast. Albinos lack such pigment, and the image of the world they see is thus rather poor. Albinos have other visual deficits as well that seem to involve abnormal visual pathways, but much of their visual deficit can be attributed to poor optics.

Finally, the eye has an iris that creates an adjustable diaphragm or pupil. Part of the adjustment serves to control the light levels striking the retina, but this function is relatively unimportant. The human pupil can be altered in area by approximately a factor of 30 (1.5 to 8 mm in diameter). Since at any one time photoreceptors can respond to roughly a 5 log unit range of intensities but the human visual system can function over a 9 to 10 log unit range, it is clear that photoreceptor adaptation is much more important than is pupil size in the visual system's reaction to different illumination levels.

Much more important is the pupil's role in the optical quality of the eye, particularly with regard to depth of field (Fig. 8-5). In practical terms control of depth of field becomes more important for near than for distant vision. This is because the depth of field is always larger the further away the optical system is focused. Thus pupil size is not of great concern for distance vision. For near vision, such as during threading a needle, the pupil becomes smaller to ensure adequate depth of field. Otherwise too much of the relevant field of view would always be blurred.

There is a *synkinesis* involving the *near reaction*, which involves three related actions for transfer of fixation from a more distant to a nearer object: the lens accomodates, the visual axes converge, and the pupil constricts. This synkinetic pupillary constriction should be distinguished from the reflex constriction to light. The neural pathways involved in the synkinetic constriction are not known. The reflex constriction involves a pathway from the retina to cells in the pretectum, from these cells to preganglionic cells in the Edinger-Westphal portion of the oculomotor nucleus, and from these to postganglionic cells that innervate the pupillary constrictor muscles. A nineteenth century British physician, Argyll Robertson, described patients who lacked the pupillary reflex to light but who still demonstrated pupillary constriction when asked to shift fixation from a distant to a near object. This pathology is still known as the *Argyll Robertson pupil*, although it is rarely seen, and its pathophysiology is not well understood.

■ Dynamic properties of the iris

Pupillary changes are brought about by changes in the configuration of the iris. Two dominant smooth muscle groups are found there. One is circumferentially organized and receives parasympathetic innervation; its contraction causes pupillary constriction. The other is radially organized and receives sympathetic innervation; its contraction causes pupillary dilation.

■ Dynamic properties of the lens and accommodation

Normal accommodation. Accommodation is a process whereby the eye changes its focal length, primarily by changes in lens shape and partially by a shift in lens position. The eye or lens is *relaxed* when the lens is most flat and in its most posterior position. Different *degrees of accommodation* imply a more curved lens that shifts forward slightly. Thus the relaxed eye has a longer focal length and is focused furthest away; the accommodated eye has a shorter focal length that causes the eye to be focused for closer objects.

Fig. 8-6 shows details of the structures responsible for accommodation. The lens itself is a plastic, jellylike substance that is contained within a tough, elastic *lens capsule*. This is analogous to petrolatum placed under pressure inside a balloon. The shape of the lens inside its capsule tends to be spherical. However, tough *zonular fibers* are attached at one end around the equator of the lens capsule, and at their other end they attach to the *ciliary body*. The ciliary body, in turn, is firmly attached at its other end to the sclera. Tension from the zonular fibers pulls the lens and its capsule into a relatively flat shape. The amount of tension in the zonular fibers thus determines the extent of the flattening. The tension is controlled by action of smooth muscles that are located in the ciliary body and receive parasympathetic innervation. These muscles run predominantly in a circumferential fashion and thus act as a sphincter. Contraction of the ciliary muscle reduces tension in the zonular fibers. This allows the lens to assume a more spherical shape as a result of the passive elastic properties of the lens capsule. The entire lens also moves slightly forward during ciliary muscle contraction. Tension on the zonular fibers is maximal and lens curvature minimal during relaxation of the muscle. The lens then is in its most posterior position. Accommodation thus implies ciliary muscle contraction and near vision; this muscle is relaxed to permit distance vision.

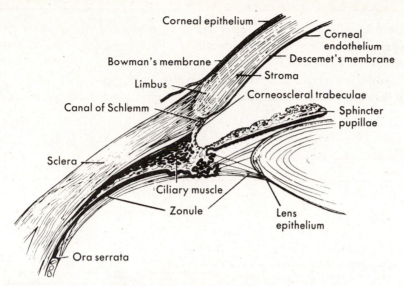

Fig. 8-6 ■ Ciliary muscle, zonule, lens, and associated ocular tissues involved in accommodation. (From Davson, H.: The eye, ed. 2, vol. 1, New York, 1969, Academic Press, Inc.)

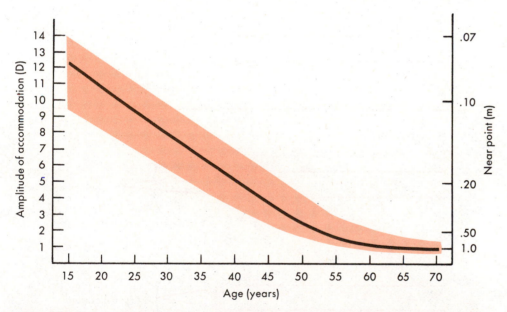

Fig. 8-7 ■ Typical values for emmetropic individuals; normal appearance of presbyopia with age. As a normal consequence of aging, the amplitude of accommodation decreases and the near point (i.e., the closest point that can be focused) increases. *D,* Diopters. (Redrawn from Mountcastle, V.B., editor: Medical physiology, ed. 14, St. Louis, 1980, The C.V. Mosby Co.)

Presbyopia. Accommodation achieves nearer focus largely because of the lens capsule's elasticity. Imagine a lens capsule stretched out of shape and no longer elastic. Ciliary muscle contraction would have a negligible effect on lens shape, and the ability to focus near objects would be severely compromised. This is precisely the condition of *presbyopia,* a normal concomitant of the aging process during which tissue elasticity generally declines throughout the body. This is the reason that focusing ability is lost with age (Fig. 8-7), and older people often wear "reading glasses" or bifocals for close work.

■ *Emmetropia and ametropia*

Probably the most common and easily treated disturbance of vision is improper focus of the eyes. Normal focus is called *emmetropia*. Improper focus is called *ametropia*, and several forms exist. *These conditions are always defined with respect to the relaxed eye.* This is the reason that tests for emmetropia or ametropia usually involve pharmacological paralysis of accommodation (achieved by appropriate eye drops) to ensure that

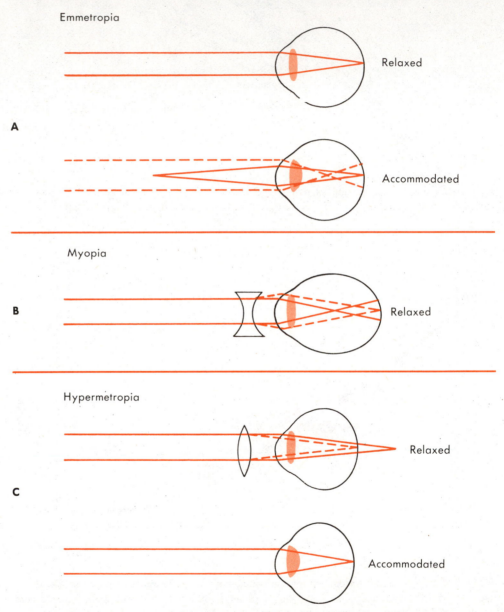

Fig. 8-8 ■ Emmetropia and spherical ametropias. **A,** Emmetropia. When the eye is relaxed *(upper),* rays from an infinitely distant point source, which are effectively parallel, converge at the retina where they are focused. For practical purposes any distance greater than 6 m or 20 feet is considered "infinitely distant" and leads to parallel rays from a point source. Accommodation of the lens *(lower)* permits points from closer objects to be focused onto the retina *(solid lines),* while distant objects are focused in front of the retina *(dashed lines).* **B,** Myopia. Parallel rays are focused in front of the retina *(solid lines),* typically because the eyeball is too long. This is corrected by the addition of a negative lens *(dashed lines).* **C,** Hypermetropia. In the relaxed eye *(upper)* parallel rays are focused behind the retina *(solid lines),* typically because the eyeball is too short. This is corrected by the addition of a positive lens *(dashed lines).* Note that for limited hypermetropia (i.e., within the range of accommodation) lens accommodation can bring distant objects to focus on the retina *(lower).* Thus the eye should always be relaxed to test for hypermetropia.

the relaxed eye is being tested. Because the relaxed eye is tested and is focused to its furthest point, that is, its longest focal length, the test measures its ability to focus distant objects. In practice, anything further than approximately 6 m from the human eye can be treated as infinitely distant for these tests.

The relaxed, emmetropic eye sharply focuses distant objects (Fig. 8-8, *A*). Nearer objects can be focused through accommodation. An ametropic eye cannot focus distant objects when relaxed. Three major forms of ametropia exist.

Myopia (nearsightedness) is illustrated in Fig. 8-8, *B*. Distant objects are focused in front of the retina because the length of the eye is too great for its refractive power. Myopia can be corrected by a simple spectacle or contact lens of appropriate negative power.

Hypermetropia (hyperopia or farsightedness) is illustrated in Fig. 8-8, *C*. Distant objects are brought to focus behind the retina because the eye is too short for its refractive power. The correction in this case is a positive lens. Hypermetropia is often confused with presbyopia for two reasons. First, constant accommodation can compensate for mild hypermetropia so that distant objects are always in focus, and no problem may be evident. Second, less accommodation is thus available for closer focusing with the result that the closest point that can be focused is rather distant. Constant accommodation, which an uncorrected hypermetropic individual might employ, is not desirable. Instead, such an individual should wear an optical corrective device constantly. In contrast, presbyopic persons limit such correction to close work only.

Astigmatism is a more complex form of ametropia. An emmetropic eye or well-designed optical system must be able to image objects clearly regardless of the object orientation in visual space; that is, an emmetropic eye can readily focus lines oriented vertically, obliquely, horizontally, etc. In practice, such an eye is radially symmetrical because the radii of curvature of the cornea, lens, and retina are constant for all meridians. An *astigmatic eye* lacks such symmetry; thus not all orientations at a given distance can be focused (Fig. 8-9). The usual cause of this ametropia is a cornea that has different radii of curvature for different meridians. Thus the corneal surface does not approximate a spherical surface. More rarely, astigmatism can be caused by a lens or retinal shape that is similarly distorted. Astigmatism can be corrected by a spectacle

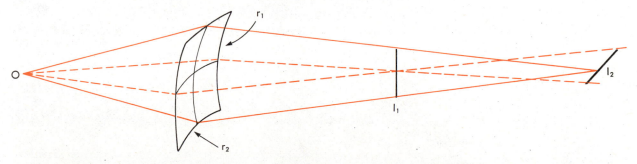

Fig. 8-9 ■ Astigmatism caused by a nonspherical corneal surface. A section of the cornea is shown with different radii of curvature for the vertical and horizontal axes (respectively, r_1 and r_2). In this example, $r_1 > r_2$; thus the horizontal axis of the cornea has more dioptric power and a shorter focal length than does the vertical axis. Because of this, rays (*colored lines*) from a single point object (*O*) cannot be imaged in a single point. Rays that span the vertical axis of the cornea (*solid colored line*) are focused at I_2, but the rays spanning the horizontal axis (*dashed colored line*) are focused closer at I_1. As a consequence, a vertical line is imaged at I_1 and a horizontal line at I_2. In between, the object point is imaged as an oval blur. Because the object point is nowhere imaged as a point, blur is always present. Such astigmatism can be corrected by a nonspherical (or "cylindrical") spectacle lens that creates an opposite astigmatism to cancel out the cornea's astigmatism. Alternatively, a contact lens can optically replace the corneal surface with a spherical refractive surface and eliminate the astigmatism.

lens with appropriately different radii of curvature for different orientations. Contact lenses can compensate for most forms of astigmatism that result from corneal distortion, since the spherical surface of the contact lens optically replaces the distorted corneal surface.

Finally, it should be reemphasized that presbyopia is not a form of ametropia, and it has a different etiology. Indeed, it is common for an individual to have presbyopia and ametropia, and these conditions independently contribute to image formation and need to be corrected separately. Also, astigmatism can occur in addition to myopia or hypermetropia. Finally, although it is most common for an individual to have roughly equivalent emmetropia or ametropia in each eye, exceptions to this occur. When the two eyes differ in this regard, the condition is known as *anisometropia*.

■ *Retina*
■ *Layering and regional variations*

General layering scheme. Fig. 8-10 shows a cross section through the human retina with ten layers indicated. This layering pattern is common to all vertebrates. Layer 1 is the *pigment epithelium*. It is adjacent to the choroid on one side and the photoreceptors on the other. Cells in this layer are nonneural and serve at least two functions. Because they contain pigment that absorbs photons not captured by the photoreceptors, they reduce scattered light. The pigment epithelium also serves a nutritive and metabolic function by acting as an intermediary between the choroidal blood supply and the photoreceptors. Layer 2 is the *receptor layer,* and it is divided into layers comprising the *outer* (2a) and *inner* (2b) segments of rods and cones. Layer 3 is the *external limiting membrane*. This is not a true membrane, but rather it is composed of closely apposed processes from glial elements, known as Müller cells, with somata in layer 6. Layer 4 is the *outer nuclear layer* and contains the somata of the photoreceptors. Those of the cones are in the outer half, and those of the rods are in the inner half. Layer 5 is the *outer plexiform layer* and represents a zone of synaptic interactions among the photoreceptors and certain of the retinal interneurons. Layer 6 is the *inner nuclear layer* and

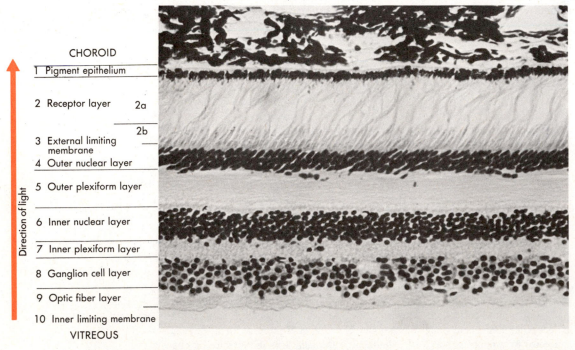

Fig. 8-10 ■ Cross section of a macaque monkey's retina. Note the direction of light as it passes through the retinal layers to reach the photoreceptors. (Nissl stain; courtesy R.E. Weller.)

contains the somata of Müller cells and the interneurons. These interneurons include horizontal cells, bipolar cells, amacrine cells, and interplexiform cells. Layer 7 is the *inner plexiform layer* and represents a zone of synaptic interactions among certain of the interneurons and retinal ganglion cells. Layer 8 is the *ganglion cell layer* and contains the somata of these neurons. Layer 9 is the *optic fiber layer* and contains unmyelinated ganglion cell axons that course toward the optic disc. These axons become myelinated as they penetrate the optic disc, and they represent the only output from the retina. Finally, layer 10 is the *inner limiting membrane,* which is not a true membrane but is formed by apposed processes of the same Müller cells that form the outer limiting membrane. Below layer 10 is the vitreous humor.

Neural input to the retina from the rest of the brain has never been demonstrated unambiguously in any mammal, although such a pathway clearly exists in many species of nonmammalian vertebrates. Details of functional interactions among the retinal neurons are provided later, but first we shall consider variations in the general plan as illustrated in Fig. 8-12.

Changes with eccentricity. In general, the density of neurons in all layers decreases monotonically as the distance from the fovea, or *eccentricity,* increases. Retinal thickness consequently decreases with increasing eccentricity. This is illustrated for rods and cones in Fig. 8-11, but similar functions apply to the other retinal cell types as well (Fig. 8-12). As a result visual performance declines monotonically as the test stimulus is imaged further from the fovea. For example, spatial resolution, or the finest detail that can be resolved, declines with eccentricity from the fovea.

Fovea. As seen in Fig. 8-12, the fovea is a depression, or pit, in the retina. The purpose of this structure probably relates to the curious placement of photoreceptors as seen in Fig. 8-10. Because light enters the retina from the vitreous humor, photons must pass through most of the retina to reach the rods and cones. Although fairly transparent, these retinal layers are not optically ideal, and the resultant image at the photoreceptors is somewhat degraded. The effect would be noticeable only for species with particularly acute vision such as certain birds, reptiles, fish, and mammals, including humans and other primate species. Even in these species the reduced visual performance with reduced retinal cell density away from the fovea suggests that optical degradation by the retinal layers becomes important only where cell density is sufficiently high.

The fovea provides a solution for this problem by sweeping away from the path of light most unnecessary tissue in order to optimize the image at the photoreceptors. As shown in Fig. 8-12 for the human fovea, this excavation leaves only layers 1 to 4, and thus only the photoreceptors are present among the retina's neural components. Because of the excavation, image quality is best at the fovea, which is the only region that can benefit from it.

The fovea of humans contains only cones; rods do not appear until the nearby parafoveal region is reached. Thus cone density peaks in the fovea, whereas rod density peaks in an annulus surrounding the fovea (Fig. 8-11). A simple experiment on a starry evening will illustrate this arrangement. Outside in the darkness the illumination is so dim that only the rods function; the less sensitive cones do not respond. In such illumination the fovea will not function, and the most sensitive retinal region becomes the parafoveal annulus of peak rod density. Thus a very dim, barely detectable star can be seen just beside the fixation point (parafoveally) but not at the fixation point (foveally). In other words, the star disappears when one looks directly at it and reappears when one looks just to one side of it.

A final consequence of the fovea concerns the trajectory of ganglion cell axons toward the optic disc, where they become myelinated and form the optic nerve. As noted earlier, these fibers travel in layer 9, and as a result of the foveal structure, they

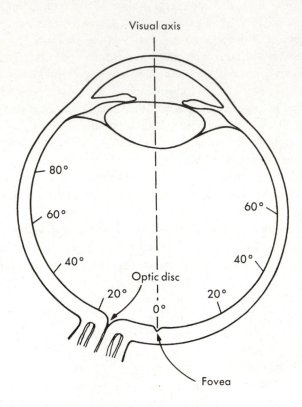

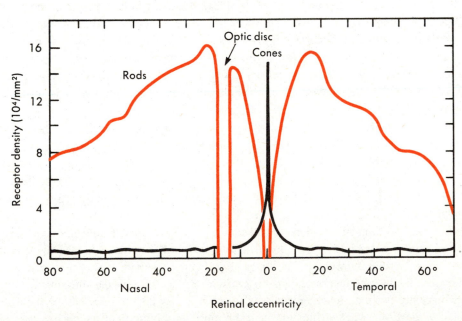

Fig. 8-11 ■ Change with eccentricity in density of rods *(colored curve)* and cones *(black curve)*. Cone density peaks at the fovea, and that of rods peaks parafoveally, but both density functions decrease with eccentricity. (Redrawn from Cornsweet, T.N.: Visual perception, New York, 1970, Academic Press, Inc.)

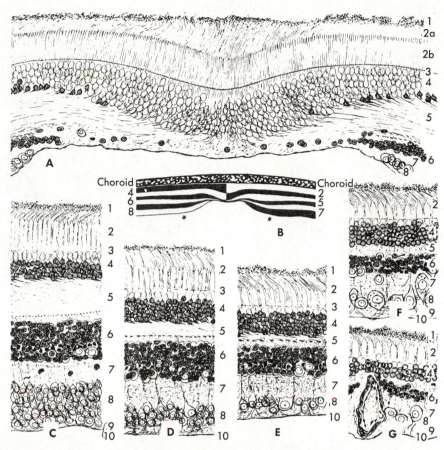

Fig. 8-12 ■ Cross sections of macaque monkey's retina at different locations show changes with eccentricity. Numbers identify the various retinal layers. **A,** Fovea. Note that nearly all elements other than the photoreceptors and their cell bodies are swept aside to create a pit. In this monkey (as in humans), the fovea contains only cones (with open cell bodies in layer *4*) and no rods (with filled cell bodies in layer *4*). Receptor density is increased by virtue of the long, thin cone outer segments. **B,** More schematic view of fovea and layering. The asterisks in the parafoveal region mark the limits of the fovea, and no rods are found between the asterisks. **C** to **G,** Increasingly peripheral retina. Note decrease in retinal thickness and cell numbers with increasing eccentricity. (From Polyak, S.: The vertebrate visual system, Chicago 1957, University of Chicago Press. Copyright © 1957 by University of Chicago Press.)

must detour around the fovea. Fig. 8-13 illustrates the course taken by these fibers. Generally, nasal to the optic disc the course is straight because no fovea must be avoided. Also, between the fovea and optic disc the fibers take a direct route. In the rest of the retina all of these fibers must arch superiorly or inferiorly around the fovea in long, curved courses. A horizontal raphe is formed so that ganglion cells superior to this raphe arch their axons superior to the fovea. The converse is true for inferior ganglion cells. Lesions restricted to a small area of retina but involving all layers produce complicated deficits called *scotomata* (singular, *scotoma*). There are blind regions that can be mapped by *perimetry,* which is a process whereby the visibility is determined for small targets

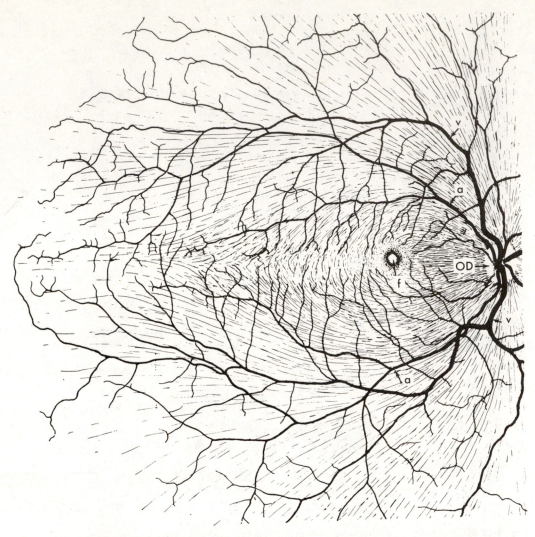

Fig. 8-13 ■ Part of a human ocular fundus or that part of the retina and associated structures visible during ophthalmoscopic examination. The pattern of ganglion cell axons *(fine lines)* and veins *(v)* coursing toward and arteries *(a)* radiating from the optic disc *(OD)* can be seen. The key to this pattern is that they must avoid the fovea *(f)*, since no blood vessels or ganglion cell axons are found there. Thus axons from the temporal retina take looping courses toward the optic disc, looping first away from and then toward the horizontal raphe (a straight line extending from the optic disc through the fovea). In nasal retina, fibers take a more direct, essentially straight route to the optic disc, since no foveal region need be avoided. The blood vessel pattern is also more cruciate in nasal retina. (From Polyak, S.: The vertebrate visual system, Chicago 1957, University of Chicago Press. Copyright © 1957 by University of Chicago Press.)

placed throughout the visual field with reference to a fixation point. Fig. 8-14 shows that such a lesion nasal to the disc produces a wedge-shaped scotoma because the bundle of fibers interrupted by the lesion originates from the corresponding retinal region. A lesion in the temporal retina, however, produces a curious scimitar-shaped scotoma that rests on the horizontal meridian of the visual field. These scotomatal shapes are readily explicable from a knowledge of the course of retinal ganglion cell axons as described earlier.

Optic disc. Because the ganglion cell fibers run along the inside of the retina, they must penetrate all of the retinal layers to exit from the eye (another consequence of the

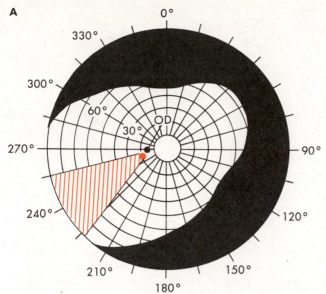

A

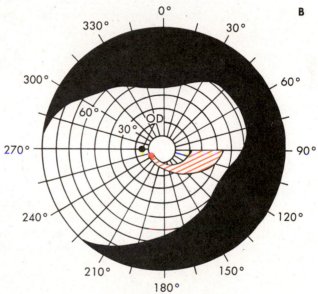

B

Fig. 8-14 ■ Scotomata caused by various retinal lesions shown in polar coordinates as in Fig. 8-3. The lesions are drawn as colored circles and affect all retinal layers. Thus not only are ganglion cells and photoreceptors damaged at the site of the lesion, but also axons from more distant ganglion cells are cut, thereby increasing the scotoma *(colored hatching)*. The shape of the scotoma is determined by the course of ganglion cell fibers through the lesion (see Fig. 8-13). **A,** Lesion nasal to the optic disc *(OD)*. This creates a wedge-shaped lesion that fans out symmetrically toward the periphery. **B,** Lesion temporal to the optic disc. This creates a scimitar-shaped lesion that abuts the horizontal midline of the visual field. (Redrawn from Moses, R.A.: Adler's physiology of the eye, St. Louis, 1970, The C.V. Mosby Co.)

inverted vertebrate retina). They do this at the optic disc (Fig. 8-15). This is also called the *blind spot* because, with no retina at this location, any image formed here is unseen.

Fig. 8-16 offers an experiment to demonstrate and map the blind spot. Close the left eye; fixate on the circle with the right eye, and hold the page so that the cross is horizontal to the circle. Now move the page back and forth at a distance of roughly 15 cm from the eye. The cross disappears and reappears as its image moves in and out of the optic disc. This technique is a rough perimetry test and can be used to map the blind spot fairly accurately. The process can be repeated for the left eye by closing the right eye, fixating on the cross, and noting the disappearance and reappearance of the circle.

Note that with both eyes open no blind spot appears. This is simply because when a target is imaged on one eye's optic disc, it is also imaged in the other eye's temporal retina where no blind spot exists. More interesting is the common observation that viewing the world through one eye does not generally provide the percept of a blind spot; that is, with only a single eye open, one is not aware of a white, gray, or black spot that corresponds to the blind spot in the visual field.

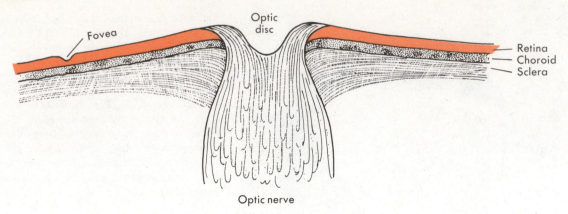

Fovea

Optic
disc

Retina
Choroid
Sclera

Optic nerve

Fig. 8-15 ■ Cross section through the eye at the optic disc. Ganglion cell fibers converge at the disc and penetrate all layers of the eye to enter the cranium and form the optic nerve. Consequently the retina is interrupted here, leading to the physiological blind spot. Note the shape of the optic disc profile. It is sensitive to pressure and becomes more excavated during increased intraocular pressure (e.g., glaucoma) and less excavated during increased intracranial pressure.

Fig. 8-16 ■ Demonstration of the physiological blind spot created by the optic disc. Close one eye and hold the page roughly 15 cm from the open eye. If the right eye is used, fixate on the spot, keep the cross oriented directly to the right of the spot, and move the page slightly closer and further from the open eye. Note that the cross appears and disappears as the image moves out of and into the optic disc. With this technique the shape and location of the optic disc can be accurately plotted. To plot the left optic disc, close the right eye, open the left, and fixate the cross. The spot can be be used to plot the optic disc.

This exemplifies a perceptual phenomenon known as *filling in of the scotoma*. A small enough scotoma is usually undetected unless objective tests, such as visual perimetry, are used. Thus the optic disc creates a blind spot that is rather obvious with perimetry testing and that is ignored or undetected otherwise. The same principle applies to pathological scotomata resulting from lesions of the retina or central visual pathways. Surprisingly large scotomata can be ignored due to the filling-in phenomenon. Perceptual problems usually result from sufficiently large scotomata, but these problems are often not associated with a blind area. Thus objective tests must always be relied on rather than subjective impressions when scotomata are suspected.

■ *Organization*

Circuitry. Fig. 8-17 summarizes a simplified version of the microcircuitry of the vertebrate retina. In the outer plexiform layer, photoreceptors form synapses onto bipolar and horizontal cells. Horizontal cells send processes along the outer plexiform layer to innervate bipolar cells. Horizontal cells also receive input from interplexiform cells. In the inner plexiform layer, bipolar cells synapse onto amacrine and/or ganglion cells, and ganglion cells are also innervated by amacrine cells. Synapses are found as well between amacrine cells and between amacrine and interplexiform cells. Information is carried from the outer to the inner plexiform layer by bipolar cells and in the reverse direction by interplexiform cells.

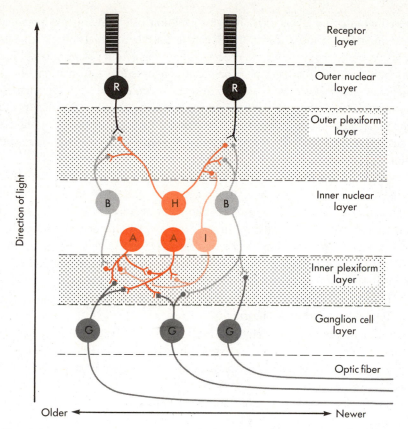

Fig. 8-17 ■ Schematic diagram showing the connections among retinal neurons and the significance of prominent layers. The neurons shown are photoreceptors *(R)*, horizontal cells *(H)*, bipolar cells *(B)*, interplexiform cells *(I)*, amacrine cells *(A)*, and ganglion cells *(G)*. It has been suggested that ganglion cells dominated by amacrine cell inputs represent phylogenetically older circuitry and that ganglion cells dominated by bipolar cell inputs represent newer circuitry. Note the direction of light as it passes through the retina to reach the photoreceptors.

Ganglion cells integrate the synaptic processing that cascades through the retina to innervate the rest of the brain. Fig. 8-17 illustrates variations of neural paths that can be taken to reach ganglion cells from the photoreceptors. One extreme shown by the ganglion cell on the right involves a more direct circuit with ganglion cells dominated by bipolar cell input. The other extreme shown by the ganglion cell on the left is a less direct circuit and involves ganglion cells dominated by amacrine cell input. The central ganglion cell represents probably the most common example of convergent input from both bipolar and amacrine cells. Comparative studies of many vertebrate retinas have shown that the ratio of bipolar-to-amacrine synapses onto ganglion cells increases in a progression from nonmammalian species through mammals to primates. Thus circuits involving ganglion cell innervation by amacrine cells are probably phylogenetically older than those involving bipolar cell innervation.

Functional interactions. To date there is little direct evidence regarding functional properties of the interplexiform cell, but the other retinal neurons have been extensively studied. Consequently, this section focuses on these other cell types.

Two curious, related properties of retinal neurons will be described before their functional properties are discussed. First, only amacrine and ganglion cells generate action potentials. The other cell types communicate via graded, electrotonically conducted potentials. Such electrotonic conduction is sufficient for the short distances of the intraretinal circuitry. This explains why ganglion cells have action potentials, but why amacrine cells should differ in this regard from other retinal interneurons remains a mystery. Second, when a threshold process such as an action potential is not involved, hyperpolarizing and depolarizing potentials can convey equal information. By convention a depolarizing response will be referred to as *excitatory* and a hyperpolarizing one as *inhibitory,* although these designations are somewhat arbitrary. Because photoreceptors hyperpolarize to photic stimulation (Chapter 6), they are inhibited by light according to this convention. Indeed, photic stimulation seems to reduce neurotransmitter release from the photoreceptors.

Most of our understanding of and hypotheses about interactions among retinal ganglion cells are derived from receptive field studies. As noted in Chapter 7, the receptive field approach for visual neurons involves the determination of how they are affected by visual stimuli that vary in contrast, shape, temporal parameters, and retinal location. Fig. 8-18 presents a greatly simplified summary of receptive field properties for some of the retinal neurons as well as the possible synaptic interactions that can be inferred from an analysis of these properties. This exemplifies the value of the receptive field approach in understanding functional circuitry. However, the speculative and incomplete nature of Fig. 8-18 must be emphasized.

Photoreceptors are tonically inhibited by light. They have small, circular receptive fields coextensive with their retinal location and size (Fig. 8-18, *A*). Some weak responses to light applied outside this area result from synaptic contacts between photoreceptors (not shown in Figs. 8-17 and 8-18).

Horizontal cells have larger, more uniform receptive fields and are tonically inhibited by photic stimuli (i.e., they hyperpolarize). They seem to receive convergent excitatory input from several adjacent photoreceptors (Fig. 8-18, *B*).

Bipolar cells also respond tonically to photic stimulation, and there are two complementary types. One is the "depolarizing" or "on-center" type, and the other is the "hyperpolarizing" or "off-center" type. For the former (the left example in Fig. 8-18, *C*), light shone anywhere within a small central region of the receptive field (called the *center*) depolarizes or excites the cell. Light applied to an annular region surrounding the center (called the *surround*) hyperpolarizes or inhibits the cell. This represents an antagonistic organization of the center-surround receptive field, since the response to a stimulus placed in the center of the cell's receptive field is antagonistic to the response elicited by a stimulus placed in the surround. As might be predicted from such an arrangement, a small light spot that just fills the center will maximally depolarize the cell. A larger spot that simultaneously falls on the surround will depolarize the cell less, if at all, because the depolarization from central stimulation sums more or less linearly with the hyperpolarization from surround stimulation. This cell type is called *on center* because light turned on in the center excites the cell. Typically the size of the receptive field center of these bipolar cells is equal to or a little larger than the size of the receptive field of a photoreceptor; the surround is roughly as large as a horizontal cell's receptive field. Thus the circuitry illustrated in Fig. 8-18, *A* to *C,* can be proposed for these on-center, bipolar cells. The central response is subserved by inhibitory synapses onto the bipolar cell from one or a small number of photoreceptors. Such an inhibitory synapse can "invert" the photoreceptor hyperpolarization to postsynaptic depolarization. The surround response is provided by excitatory input from a horizontal cell. The evidence for this explanation for the surround response is weak; thus the question mark is indicated in Fig. 8-18.

The complementary type of bipolar cell, or *off-center* type, is almost the negative image of the on-center type. The off-center cell (shown to the right of Fig. 8-18, *C*) has a similar center-surround receptive field organization, but light in the center hyperpolarizes the cell, whereas light in the surround depolarizes it. Such a cell is called off center

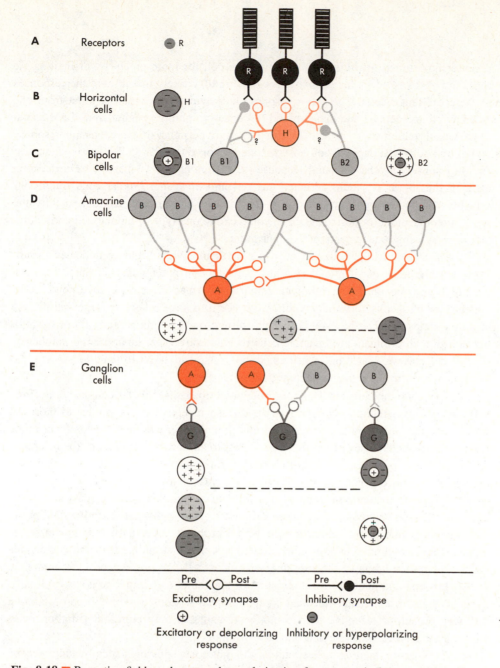

A   Receptors

B   Horizontal cells

C   Bipolar cells

D   Amacrine cells

E   Ganglion cells

Pre ⟶ Post
Excitatory synapse

⊕
Excitatory or depolarizing response

Pre ⟶ Post
Inhibitory synapse

⊖
Inhibitory or hyperpolarizing response

Fig. 8-18 ■ Receptive fields and proposed neural circuitry for many retinal neurons (**A** to **E**), including photoreceptors *(R)*, horizontal cells *(H)*, bipolar cells *(B)*, amacrine cells *(A)*, and ganglion cells *(G)*. Interplexiform cells are not included due to lack of receptive field information specific to these cells. Photoreceptors have small receptive fields equal in size to the retinal area these cells occupy. By convention these fields are considered inhibitory (shown by *minus signs*) because photoreceptors are hyperpolarized by light. Horizontal cells receive convergent excitatory input from several photoreceptors and thus have larger inhibitory fields. Bipolar cells exhibit a center-surround receptive field organization. On-center cells *(B1)* receive inhibitory input from photoreceptors and excitatory input from horizontal cells; the converse holds for off-center cells *(B2)*. Many varieties of amacrine cell receptive fields have been described. These include some that are mostly excitatory due to convergent input from many on-center bipolar cells, some that exhibit excitatory and inhibitory regions diffusely intermixed due to mixed input from on- and off-center bipolar cells, and many that are mostly inhibitory due to convergent input from many off-center bipolar cells. A variety of ganglion cell types has also been described. Some seem to have receptive fields quite similar to those of amacrine cells and presumably reflect nearly exclusive amacrine cell inputs. Some have receptive fields with center-surround organization similar to those of bipolar cells, presumably resulting from nearly exclusive bipolar cell input. Some have more complicated receptive fields that may reflect a combination of amacrine and bipolar cell inputs. Clearly our understanding of how circuitry of the outer plexiform layer results in neuronal receptive fields is more detailed than that of the inner plexiform layer.

because light turned off in the center excites the cell (or more precisely, disinhibits the cell). Fig. 8-18, *A* to *C*, also summarizes the circuitry likely to subserve the responses of these off-center cells. The center response is provided by excitatory input from one or a few photoreceptors and the surround response by inhibitory input from a horizontal cell. Again, inhibition is needed to "invert" a hyperpolarizing presynaptic response (horizontal cell) to a depolarizing postsynaptic response (bipolar cell). The evidence for circuitry subserving the surround response is inferential and is marked by a question mark in Fig. 8-18. It should be noted that on- and off-center bipolar cells could conceivably respond in opposite polarity to the same neurotransmitter. Consequently the same photoreceptors (or horizontal cells) could excite one type of bipolar cell and inhibit the other by using one synaptic neurotransmitter.

Amacrine cells (Fig. 8-18, *D*) form a much more varied and complicated group. These are the most distal retinal cells to exhibit action potentials, and thus excitation and inhibition are no longer arbitrary definitions. Amacrine cells can respond either tonically or phasically to visual stimuli. Most seem to have relatively large and diffuse receptive fields and can be excited by light turned on, by light turned off, or by both light onset and termination. Amacrine cells receive convergent input from a number of bipolar cells as well as from other amacrine cells, and it is difficult to specify the circuitry responsible for their receptive field properties.

Finally, ganglion cells receive input from amacrine cells, bipolar cells, or both (Fig. 8-18, *E*). Many ganglion cells have large, diffuse receptive fields similar to those of some amacrine cells, and these ganglion cells probably receive their predominant input from amacrine cells. They generally respond best to stimuli in the center of the receptive field, and the responsiveness diminishes as the stimulus is placed more eccentrically in the receptive field. Such ganglion cells were only recently discovered in mammals, and now they appear to represent roughly one third of all ganglion cells in cats and monkeys. At the other extreme are ganglion cells with center-surround receptive field organization much like that of bipolar cells, which probably represents their predominant input. These ganglion cells are either excited by photic stimulation in the center and inhibited by stimulation of the surround (on-center cells) or are excited by stimulation of the surround and inhibited by stimulation of the center (off-center cells). Also, many ganglion cells have a center-surround configuration that seems superimposed on a more diffusely organized receptive field component, and this may represent ganglion cells with more balanced amacrine and bipolar cell input.

The bipolar-to-ganglion cell circuit is thought to be phylogenetically more recent than the amacrine-to-ganglion cell circuit. This, in turn, implies that the center-surround receptive field organization evolved relatively recently. Although this seems a likely speculation, a weakness in this inference stems from the fact that center-surround receptive fields for visual neurons were described first for the primitive horseshoe crab *(Limulus)*.

Surround or lateral inhibition

The center-surround receptive field organization seen most distally in bipolar cells and represented by many or most ganglion cells is also called *surround* or *lateral inhibition*. Analogous lateral inhibition is seen in other sensory systems and is generally associated with their more recently evolved components. Fig. 8-19 shows how such receptive field organization can enhance two-point discrimination or spatial resolution compared with that possible without lateral inhibition. This example illustrates on-center cells, but a similar result is derived from off-center cells.

Fig. 8-19, *A*, shows how cells with overlapping center-surround receptive fields respond to a small light spot. The spot is in the center of one cell and excites it; it lies in the surrounds of neighboring cells and inhibits them; it lies completely outside the receptive fields of more distantly located cells and does not affect them. The three-dimensional graph on the right depicts the response profile to a single photic stimulus of a patch of retinal ganglion cells with center-surround stimulation. As indicated, a central zone of active cells is surrounded by a zone of inhibited cells.

Fig. 8-19, *B*, shows that for cells with overlapping but diffusely organized fields, no surround zone of inhibited cells exists to a single stimulus. As noted earlier, each of these cells responds best to a stimulus applied to the center of its receptive field and less so as the stimulus is placed more eccentrically in the receptive field. Thus the active neuronal zone in the presence of a single stimulus peaks at the stimulus location and gradually diminishes without a surrounding zone of inhibited cells.

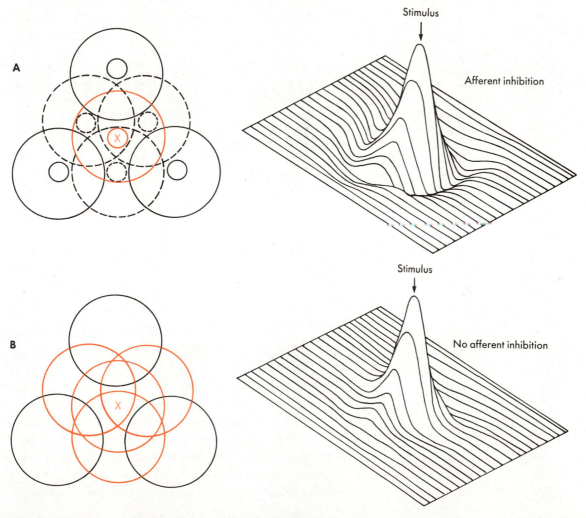

Fig. 8-19 ■ Significance of afferent inhibition to discrimination of two stimuli from a single stimulus. **A,** With afferent inhibition. On the left are shown seven partially overlapping receptive fields, each with a central excitatory zone and an inhibitory surround. A stimulus placed at the point indicated by the colored *X* would fall in the excitatory center of one cell *(solid color),* in the inhibitory surround of three cells *(dashed black),* and outside the field of three cells *(solid black).* The pattern of neuronal activity evoked by such a single stimulus is shown on the right. Neurons representing the stimulus location are maximally excited; a surrounding annulus of neurons is inhibited, and beyond this is no neuronal response. **B,** without afferent inhibition. On the left are shown seven partially overlapping receptive fields. A stimulus placed at the point indicated by the colored *X* would fall within four fields to excite their cells and would fall outside the fields of the other three cells. No cells are inhibited by the stimulus. Thus the pattern of neuronal activity evoked by such a stimulus *(right)* includes a single peak with no surrounding inhibitory trough. *Continued.*

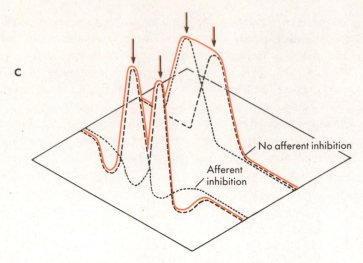

Fig. 8-19 cont'd ■ **C,** Comparison of neuronal activity patterns to paired stimuli with and without afferent inhibition. From the activity patterns shown in **A** and **B,** the pattern of neuronal activity to each stimulus alone is indicated by the dashed black curves. The colored curves indicate the summed activity to the pair of stimuli. With afferent inhibition two peaks of active neurons are evident, and because a single stimulus could not evoke such an activity pattern, a pair of stimuli is unambiguously signaled. Without afferent inhibition, a single broad peak of active neurons results, and such activity could result from a single more intense stimulus. Thus afferent inhibition more clearly encodes fine spatial details of the stimulus than is possible without afferent inhibition.

Fig. 8-19, *C,* shows the consequence of these types of receptive fields for two-point discrimination. With afferent inhibition two nearby stimuli produce two separate zones of activity, a pattern that clearly signals the presence of two stimuli. Without afferent inhibition a single broad zone of activity results, and this can easily be confused with a more intense stimulus placed in the middle.

■ *W-, X-, and Y-cells*

Recent studies of cats have demonstrated a division of retinal ganglion cells into physiologically and morphologically distinct classes called *W-, X-, and Y-cells*. These neuronal classes represent, respectively, roughly 40%, 55%, and 5% of all ganglion cells. More limited data from other mammalian species, including monkeys and probably humans, suggest a similar classification of retinal ganglion cells. We shall refer to these as W-, X-, and Y-cells in primates, although more detailed understanding may emphasize differences between the cat and primates and may require another terminology. Most of the comments concerning W-, X-, and Y-cells derive from experiments on cats, and their application to other species, including monkeys and humans, is as yet somewhat tenuous.

In the cat retina, X- and Y-cells seem to be internally homogeneous classes, but W-cells are not and may well include several further distinct neuronal classes. With this important proviso, we shall refer to these neurons simply as W-cells. X- and Y-cells (and most W-cells) have center-surround receptive field configurations that suggest strong bipolar cell input. Most W-cells have more uniform, diffuse receptive field organization indicative of strong amacrine cell input. Generally, however, we lack precise knowledge concerning the relative strength of amacrine and bipolar cell inputs to these ganglion cell classes.

Many morphological and physiological differences can be briefly noted for W-, X-, and Y-cells. W-cells have small to medium-sized somata and small axons; somata and axons are medium-sized in X-cells; and in Y-cells they are large. Each of these cell classes seems to have distinctive dendritic morphology. Compared with X- and Y-cells, W-cells have large, diffuse receptive fields, respond poorly to visual stimuli, and possess very slowly conducting axons. For these reasons it has been speculated that W-cells are phylogenetically older than X- or Y-cells. The many differences between X- and Y-cells are more subtle, and some of the more obvious differences follow: compared with Y-cells, X-cells generally have smaller receptive fields, more tonic responses to visual stimuli, better responses to small stimuli or fine detail but poorer rseponses to larger or cruder forms, more slowly conducting axons, and more linear summation of responses to visual stimuli. Regarding this last difference, an X-cell generally responds to the combination of two or more stimuli in a fashion that can be predicted from the linear addition of the responses to each of the stimuli alone. Y-cell responses to complex stimuli often exhibit nonlinear summation and thus cannot be readily predicted from knowledge of responses to simpler components of the stimulus. Also, in monkeys at least, X-cells respond differently to different wavelengths or colors of light, whereas W- and Y-cells seem relatively insensitive to color differences. Finally, as noted later, the central projections differ among each of these retinal ganglion cell classes. It seems reasonable to assume that all of these differences signify different roles for W-, X-, and Y-cells in visual perception; these roles are considered later in this chapter.

Axons of retinal ganglion cells exit from the eye at the *optic nerve* and course medially and posteriorly to the *optic chiasm*. A partial decussation of axons occurs at the chiasm, since some cross the midline and project contralaterally, while others do not and project ipsilaterally. The axons from each eye then form the *optic tract,* in which these fibers continue posteriorly toward their targets in the brainstem (Fig. 8-20). No ganglion cell has a bifurcating axon that innervates both hemispheres; each of these axons enters only one optic tract.

■ *Retinofugal projections*
■ *Route of retinofugal fibers*

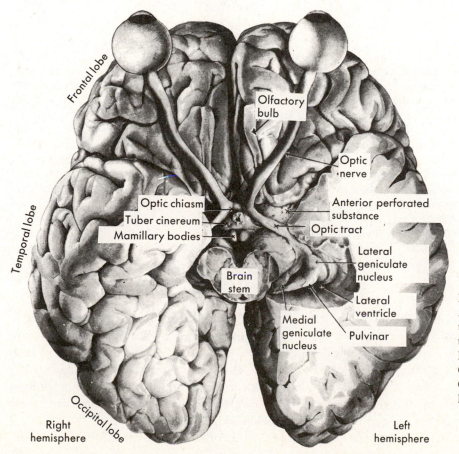

Fig. 8-20 ■ Ventral surface of the human brain. Among various landmarks can be seen the eyes, the optic nerves, the optic chiasm, the optic tracts, and the lateral geniculate nucleus. (From Polyak, S.: The vertebrate visual system, Chicago 1957, University of Chicago Press. Copyright © 1957 by University of Chicago Press.)

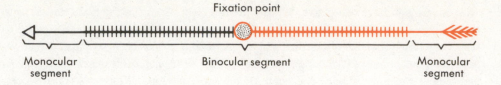

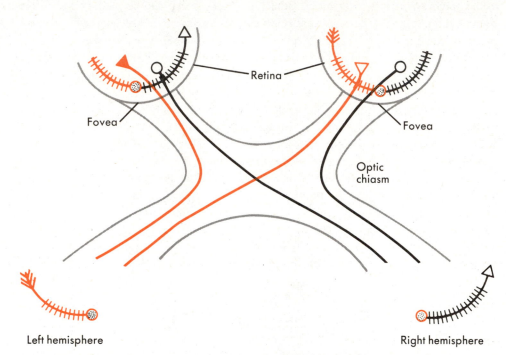

Fig. 8-21 ■ Rules governing optical image formation and the partial decussation of optic nerve fibers at the optic chiasm. In the example the object is an arrow that spans the entire horizontal extent of visual field so that the fixation point is imaged onto each fovea. Optical considerations lead to a reversal of the image onto the retina so that the right half of the arrow *(color)* is mapped onto ganglion cells of the left temporal and right nasal retinas *(triangles),* while the left half *(black)* is mapped onto ganglion cells of the right temporal and left nasal retinas *(circles).* Also, the extreme limits of the arrow (monocular segments) can be viewed by only the nasal retina on that side, and the binocular segments are imaged in each eye. Ganglion cells from the nasal retina *(open symbols)* decussate in the optic chiasm, and ganglion cells from the temporal retina *(filled symbols)* project ipsilaterally. Thus the right half of visual space is mapped onto the left hemisphere and vice versa. Although not shown, ganglion cells within 1 to 2 degrees of a vertical line running through the fovea project to one or the other optic tract regardless of their nasal or temporal location so that portions of the visual field near the vertical midline *(colored stippled circles)* are represented in both hemispheres.

Fig. 8-21 illustrates the simple rules that govern whether or not an axon crosses in the optic chiasm. To a first approximation, axons of ganglion cells in the nasal retina cross to enter the contralateral optic tract, and those of cells in the temporal retina enter the ipsilateral optic tract without crossing. This feature, combined with the optically reversed image of visual objects, leads to a continuous map of visual space in the brain so that each hemifield is mapped in the contralateral hemisphere. Furthermore, these maps contain an input from each eye, and these inputs are in register.

One detail must be added to this description of the decussation at the optic chiasm.

The retinal division between ganglion cells that project to one or the other optic tract is not a precise line. Rather, there is a vertical strip of retina, approximately 1 to 2 degrees across and passing through the fovea, in which ganglion cells that project to each hemisphere are intermingled. This has been termed the *strip of nasotemporal overlap*. The significance of this overlap probably relates to stereopsis, which is described later.

Retinal ganglion cells project to several brainstem sites, but the vast majority of the optic tract fibers terminate in the dorsal lateral geniculate nucleus (LGN) of the thalamus and the superior colliculus of the midbrain. Other sites include the suprachiasmatic nucleus of the hypothalamus, the ventral LGN* of the thalamus, several pretectal nuclei (including the nucleus of the optic tract) in the midbrain, and the accessory optic nucleus in the midbrain tegmentum (Fig. 8-22).

The retinofugal projections differ significantly among W-, X-, and Y-cells (Fig. 8-22). The dorsal LGN receives input from most or all retinal X- and Y-cells. Limited W-cell input to the dorsal LGN has been described for the cat and probably exists as well in primates. The superior colliculus receives a mixture of W- and Y-cell inputs. Each retinofugal Y-cell axon in the cat bifurcates to innervate both the dorsal LGN and the superior colliculus; similar data for primates are not yet available.

Although detailed information is lacking for most of the other sites of retinofugal projections, they probably receive W-cell input and little, if any, X-cell or Y-cell input. In other words, X-cells project to the dorsal LGN, Y-cells project primarily to the dorsal LGN and superior colliculus, and W-cells probably project to all of the retinofugal sites. The dorsal LGN is thus associated with the phylogenetically newest portion of the visual system; the superior colliculus probably represents an intermediate stage of evolution; and the remaining retinofugal sites are much older.

*The ventral LGN is a thalamic structure quite distinct from the dorsal LGN in most mammals. Among other differences, dorsal LGN neurons project exclusively to the visual cortex, whereas ventral LGN neurons project only to subcortical loci. In many monkey species and in humans, however, the ventral LGN homologue is unclear. It probably includes part or all of a band of neurons called the *pregeniculate nucleus,* which lies immediately dorsomedial to the dorsal LGN (Fig. 8-22).

■ *Terminations of retinofugal axons*

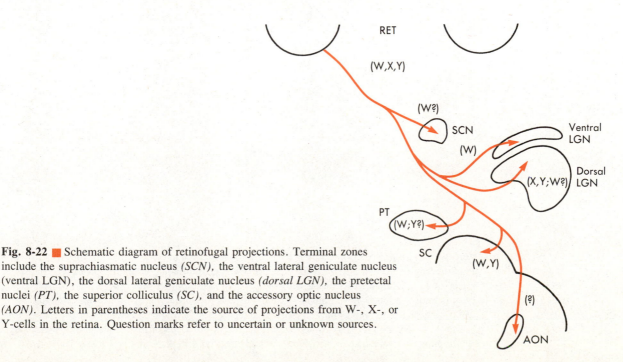

Fig. 8-22 ■ Schematic diagram of retinofugal projections. Terminal zones include the suprachiasmatic nucleus *(SCN)*, the ventral lateral geniculate nucleus (ventral LGN), the dorsal lateral geniculate nucleus *(dorsal LGN)*, the pretectal nuclei *(PT)*, the superior colliculus *(SC)*, and the accessory optic nucleus *(AON)*. Letters in parentheses indicate the source of projections from W-, X-, or Y-cells in the retina. Question marks refer to uncertain or unknown sources.

■ *Geniculostriate system*

■ *Geniculate lamination*

Retinal X-cells and Y-cells from both eyes project to the dorsal LGN, and their precise termination depends largely on the geniculate laminar patterns. Fig. 8-23 illustrates the lamination of the human infant dorsal LGN as seen in a coronal section. The dorsal LGN is kidney shaped with its hilus oriented ventromedially. Typically, six laminae can be distinguished. These are more or less horizontally oriented, but they curve

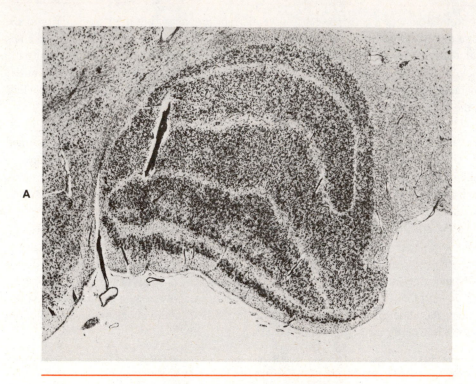

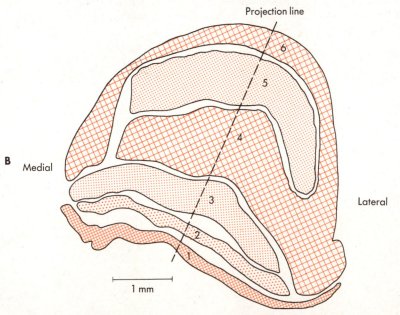

Fig. 8-23 ■ Coronal section through the lateral geniculate nucleus of a human infant. **A,** Photomicrograph of Nissl-stained material. **B,** Explanation of photomicrograph in **A.** The magnocellular laminae are *1* and *2,* and the parvocellular laminae are *3, 4, 5,* and *6.* The contralateral eye innervates laminae *1, 4,* and *6,* and the ipsilateral eye innervates laminae *2, 3,* and *5.* Lines running approximately at right angles to the laminae represent projection lines (one example is shown). Projection lines roughly represent the locus of geniculate neurons that map a single point in visual space. (**A** courtesy T.L. Hickey and R.W. Guillery.)

according to the shape of the dorsal LGN. The laminae are numbered 1 to 6 from the ventral to the dorsal surface (Fig. 8-23, *B*). Laminae 1 and 2 tend to have larger cells than do laminae 3 to 6. Thus laminae 1 and 2 are called the *magnocellular laminae,* and laminae 3 to 6 are called the *parvocellular laminae.*

The significance of these laminae lies in their ocular input, or ''ocular dominance'' (Fig. 8-23, *B*), and probably also in the differential X- and Y-cell inputs. Although fibers from each eye are intermingled in the optic tract, they segregate before terminating in the dorsal LGN so that each lamina receives input exclusively from one or the other eye. Laminae 1, 4, and 6 are innervated by the contralateral eye and laminae 2, 3, and 5 by the ipsilateral eye. The magnocellular laminae thus represent a matched pair innervated by each eye, and the parvocellular laminae represent two pairs innervated by each eye.

Physiological evidence suggests that each dorsal LGN neuron receives a dominant excitatory input from one or very few retinal ganglion cells. Thus, with minor and subtle differences, each dorsal LGN neuron has a receptive field virtually indistinct from that of its retinal input. The parvocellular neurons exhibit many X-cell properties, and the magnocellular neurons display mostly Y-cell properties. For instance, the parvocellular cells receive relatively slow-conducting optic tract input that originates with medium-sized ganglion cells, have small receptive fields, and are sensitive to differences in color. Conversely, the magnocellular neurons receive fast-conducting optic tract input from large ganglion cells, have large receptive fields, and are insensitive to differences in color. Thus the X- and Y-cell pathways remain segregated through the dorsal LGN and project in parallel to the striate cortex. Another significance of the lamination thus seems to be the segregation of X- and Y-cell pathways through the parvocellular and magnocellular laminae.

■ *Nonretinal inputs to the dorsal LGN*

More than half of the synapses found in the dorsal LGN derive from axons that are nonretinal in origin. Unfortunately these various sources of innervation are poorly understood. A major source is the corticogeniculate pathway that originates from neurons in layer VI of visual cortex (Fig. 8-26); Chapter 19 includes a more complete description of cortical layering. Despite the large extent of the corticogeniculate pathway, its function remains unclear, although recent studies suggest a role in stereopsis (pp. 130-132).

Other inputs to the dorsal LGN originate with various brainstem structures; knowledge of these is derived principally from studies of the cat's central visual pathways. Small projections to the dorsal LGN have been traced from various portions of the brainstem reticular formation, particularly the *locus ceruleus,* a small nucleus in the pontine reticular formation at the level of the isthmus. Its neurons contain norepinephrine and have axons that branch extensively to innervate much of the brain. Finally, sparse projections to the dorsal LGN emanate from the superior colliculus and nucleus of the optic tract.

■ *Geniculostriate projection*

Practically every geniculate neuron is a *relay* cell, which means it has an axon that terminates in the cerebral cortex, and practically all of these relay cells innervate the *striate cortex* (area 17). There is no known subcortical site to which these cells project, although axon collaterals within the dorsal LGN are not uncommon. Few, if any, interneurons exist in the dorsal LGN of primates.

Course of geniculostriate axons. The striate cortex is located at the posterior pole of the occipital cortex and extends anteriorly from the pole along the medial surface of the hemisphere. Much of the striate cortex lies buried in a deep sulcus called the *calcarine fissure.* This fissure runs rostrocaudally and divides the striate cortex into roughly equal portions lying above and below the fundus of this sulcus (Fig. 8-24). However, geniculostriate axons do not take a direct route from the dorsal LGN to the striate cortex

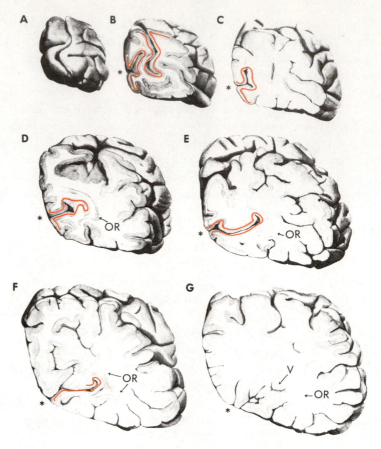

Fig. 8-24 ■ Sections through the right occipital lobe of a human beginning in **A** with the posterior pole and ending in **G** at the level of the lateral horn of the lateral ventricle *(V)*. Asterisks denote the calcarine fissure, and the colored lines indicate the stripe of Gennari that delineates the striate cortex. Note how much of the striate cortex lies buried in the calcarine fissure. Also shown are fibers of the optic radiation *(OR)* as they approach the striate cortex. (From Polyak, S.: The vertebrate visual system, Chicago, 1957, University of Chicago Press. Copyright © 1957 by University of Chicago Press.)

but rather loop around the lateral ventricles before coursing toward their terminal sites. Those which terminate in the cortex inferior to the calcarine fissure take an exaggerated detour well into the temporal lobe to skirt the ventricle. This is known as *Meyer's loop* (Fig. 8-25), and it explains why lesions of the temporal lobe can interfere with the geniculostriate projection.

Recent evidence suggests that, in primates, an extremely small number of scattered dorsal LGN neurons project to cortical areas near the striate cortex. This small extrastriate pathway may involve W-cells. In cats and many other infraprimate mammalian species, however, the geniculocortical projection extends to many cortical areas outside the striate cortex. The significance of this interspecies difference is not clear. Nonetheless, it raises a note of caution because much of the evidence on which our understanding of the central visual pathways rests is derived from studies of cats. As in other areas of the neocortex, the striate cortex has six distinct layers of cells and fibers that run parallel to the pial surface. Layer I is nearest the pia and layer VI is next to the white matter (see Chapter 19 for details).

Termination in striate cortex. The geniculocortical fibers terminate primarily in layer IV of the striate cortex, although a sparse terminal field has also been reported for layer VI. The axon terminal fields in layer IV are particularly dense and can be seen macro-

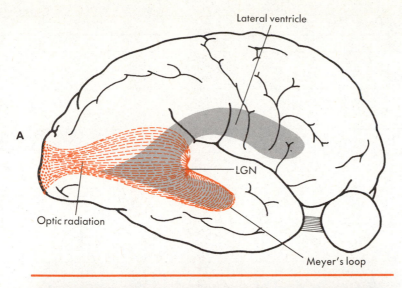

A

Lateral ventricle

LGN

Optic radiation

Meyer's loop

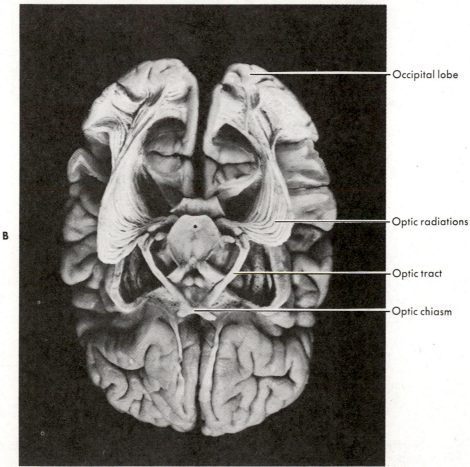

B

Occipital lobe

Optic radiations

Optic tract

Optic chiasm

Fig. 8-25 ■ Course of the geniculocortical fibers in the optic radiation. **A,** Schematic drawing of the optic radiation *(color)*. The geniculocortical fibers start among cell bodies of the lateral geniculate nucleus *(LGN)*, which lies medial to the lateral ventricle *(gray)*. The fibers sweep to the lateral side of the ventricle before turning caudally to reach the striate cortex. This turning is known as *Meyer's loop*. Because of the shape of the lateral ventricle, some of the optic radiation extends well into the temporal lobe. **B,** Photograph of dissection of the ventral surface of the human brain showing the optic chiasm, optic tracts, and optic radiations. (**A** redrawn from Sandford, H.S., and Bair, H.L.: Arch Neurol. Psychiatry **42:**21, 1939; **B** from Gluhbegovic, N., and Williams, T.H.: The human brain: a photographic guide, New York, 1980, Harper & Row, Publishers, Inc.)

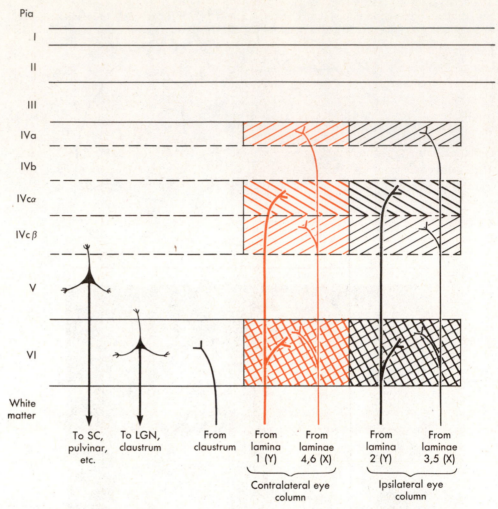

Fig. 8-26 ■ Simplified schema of the subcortical connections involving the striate cortex. Projections to most subcortical loci (e.g., the superior colliculus and pulvinar) derive from layer V pyramidal cells. Although not shown, pulvinar projections to the striate cortex probably exist, and they may terminate largely in layer I. Layer VI pyramidal and fusiform cells innervate the lateral geniculate nucleus and parts of the claustrum, and this region of claustrum projects to cortical layer VI. Geniculate innervation from the magnocellular laminae terminates in laminae IVcα and VI, whereas that from the parvocellular laminae terminates in laminae IVa, IVcβ, and VI. Note also that input representing one or the other eye is segregated into separate "ocular dominance columns" that are roughly 500 μm across. Left and right eye columns (or more precisely, vertically oriented slabs 500 μm wide and several millimeters long) lie next to one another in a fairly regular array.

scopically as a glistening band in gross specimens. This band is often called the *stria of Gennari*, and its stripelike appearance gives rise to the term *striate cortex*.

Layer IV of the striate cortex in primates is not uniformly innervated by dorsal LGN axons. The precise termination patterns depend on the dorsal LGN laminae from which the axons derive (Fig. 8-26). A thin sublayer at the dorsal tier of layer IV, called layer IVa, receives input from the parvocellular laminae. A slightly larger zone just beneath this, layer IVb, receives no geniculate innervation. The ventral region, layer IVc, is approximately half the width of layer IV, and it is divided into two zones of roughly equal width. Layer IVcα is innervated by magnocellular laminae. Immediately ventral to this is a layer, IVcβ, which receives terminals from the parvocellular laminae. Little or no overlap exists between magnocellular and parvocellular terminal zones.

Another pattern of dorsal LGN input relates to ocular dominance. Adjacent and alternating patches of layer IV, approximately 500 μm wide, are innervated by dorsal LGN laminae 1, 4, and 6 or 2, 3, and 5. In other words, these alternate patches represent the ipsilateral or contralateral eye (Fig. 8-26).

Although the dorsal LGN may be regarded as the major subcortical source of innervation of the striate cortex, several other sources will also be considered briefly. The pulvinar, a large nucleus in the thalamus, gives rise to a sparse projection to layer I of the striate cortex, although a larger projection is evident to neighboring cortical areas. The visual cortex projects reciprocally to the pulvinar. The claustrum, a striatal nucleus located just below the cortex, is also reciprocally connected with the striate cortex (Fig. 8-26) as well as other cortical areas. The corticoclaustral projection derives from cells in layer VI, and the claustocortical fibers terminate in all layers, but mostly in layers IV and VI. Finally, the locus ceruleus projects to many brain centers, including all layers of the striate cortex. However, its densest projection is to layer I. The functional significance of these relatively sparse, extrageniculate inputs to the striate cortex is obscure.

■ *Extrageniculate subcortical inputs to striate cortex*

Both the dorsal LGN and the striate cortex exhibit a precise point-to-point, but slightly distorted, map of the retina and thus of visual space; that is, neurons in these structures have well-defined receptive fields, and as one proceeds in an orderly sequence among neurons across the dorsal LGN, striate cortex, etc., the positions of the receptive fields gradually shift in a predictable fashion. This reflects the precise and orderly arrangement of connections along the entire retinogeniculostriate pathway. A general term for this arrangement is *neurotopic organization,* and it is a general feature of newer portions of sensory systems. Older portions often lack such organization. In the visual system the more specific term is *retinotopic organization.*

■ *Retinotopic organization*

There are actually two retinotopic maps in the dorsal LGN and striate cortex. There is one for each eye, and these are in register with one another. For instance, the dorsal LGN laminae are stacked in such a manner that a line passing orthogonally through them corresponds to a small region of visual space (Fig. 8-23). This same region is separately mapped for one or the other eye as the appropriate laminae are considered. Likewise, a line passing perpendicularly through the cortical layers represents a similar region of visual space. As these lines are across the dorsal surface of the dorsal LGN or striate cortex, there is a shift in the small region mapped, as predicted by the retinotopic map.

Fig. 8-27 illustrates the details of these dorsal LGN and cortical maps. A distortion in these maps is evident. Areas of retina nearer the fovea are represented by more neural tissue in the dorsal LGN and striate cortex than are areas more distant from the fovea. This is called the *magnification factor*. To an extent, the magnification factor is a consequence of the ganglion cell density at various retinal locations. For a given retinotopic location, a more or less constant ratio seems to exist among numbers of cortical, dorsal LGN, and retinal ganglion cells, but because only an unspecified subset of retinal ganglion cells projects to the dorsal LGN, this conclusion remains qualified. In order to devote more neurons to the analysis of visual targets closer to the fovea, optical considerations require that retinal ganglion cell density vary with retinal eccentricity. This lack of such optical constraints allows the dorsal LGN and striate cortex to maintain a fairly homogeneous neuron density throughout. Different magnification factors for different retinotopic locations thus result. A concomitant of retinotopic organization and magnification factor is the change with eccentricity of spatial resolution and receptive field size. Fields get smaller and resolution consequently improves as one nears the foveal representation because more neurons are available to map a given region of visual space.

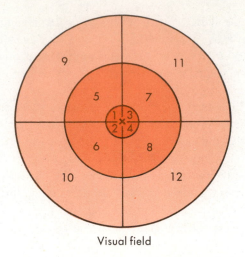

Visual field

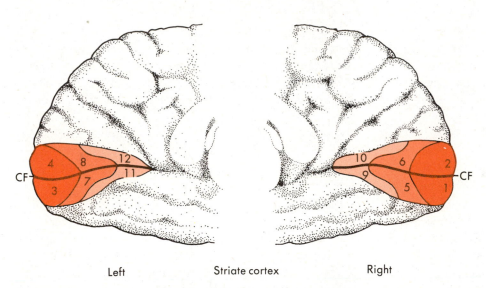

Left   Striate cortex   Right

Fig. 8-27 ■ Location and retinotopic organization of the striate cortex (area 17) in humans. This cortex is located almost exclusively on the medial surface of the hemisphere at the posterior pole, although in some persons it is shifted so that part of it extends variably onto the lateral surface. Much of the cortex lies buried in the calcarine fissure *(CF)*. Numbers indicate the mapping referenced to the fovea (x). Note that each half of the visual field is mapped to the contralateral hemisphere, that upper and lower fields are mapped, respectively, below and above the fundus of the calcarine fissure. Note also that the map for the more central field is greatly magnified with respect to the map for the more peripheral field.

A detailed knowledge of the retinotopic organization of the central visual pathways, particularly of the striate cortex, can be very useful clinically. A small lesion results in a proscribed scotoma. From the location and size of this scotoma one can often localize a lesion precisely.

■ *Cortical receptive field properties*

Most of our contemporary understanding of function in the striate cortex stems from the pioneering studies of Hubel and Wiesel. Many or most neurons in the striate cortex have inhibitory receptive field regions that are analogous to the surround or lateral inhibition in retinal and geniculate cells. However, the receptive field properties of these cortical cells are quite complicated, particularly with respect to those of the geniculate

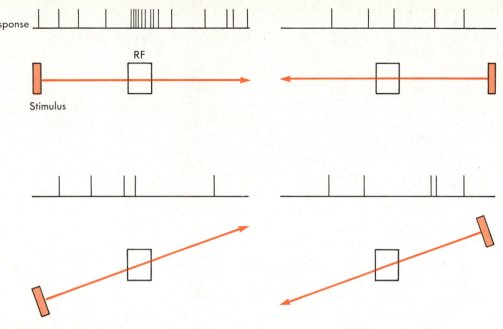

Fig. 8-28 ■ Orientation and direction selectivity for a neuron in the visual cortex. The response of the cell is shown above the drawing of the stimulus *(colored bar)* moving as indicated through the receptive field *(RF; open rectangle)*. The vertically oriented stimulus moving left to right *(upper left)* maximally excites the cell. The same stimulus moving right to left with vertical orientation *(upper right)* or in either direction with an orientation other than vertical *(lower left and right)* fails to discharge the cell.

afferent neurons. As noted earlier, each geniculate neuron has a monocular receptive field (i.e., related to only one eye), and one need specify only the position of a bright or dark spot within the center or surround in order to excite or inhibit the cell in a predictable fashion.

Three additional stimulus parameters must be specified, however, to predict the response properties of most neurons in the striate cortex. One is known as *orientation selectivity* (Fig. 8-28). The stimulus must be elongated in one dimension, that is, a rectangle or straight border between light and dark regions, and the cells are quite sensitive to the orientation of the long axis of this stimulus. For instance, a given neuron in the striate cortex might respond to a vertically oriented stimulus, but a tilt from vertical of as little as 5 to 10 degrees is often sufficient to render the cell unresponsive, or it may even inhibit it. Different cortical cells are "tuned" for different stimulus orientations so that orientations of every angle are represented in the cortex. The manner by which the geniculate inputs that are not orientation sensitive are formed into orientation-sensitive cortical neurons is unclear.

In addition to orientation selectivity, most cortical neurons display *direction selectivity*. If a properly oriented target is moved back and forth through the receptive field, responses for one direction of motion are larger than those for the other, and frequently only one direction of motion evokes a response (Fig. 8-28). As with orientation selectivity, the manner by which direction selectivity is generated remains a mystery.

Finally, most cortical neurons have binocular receptive fields, which means that stimulation of either eye can activate the cell. The receptive field in each eye maps roughly the same region of visual space (but see pp. 130-132). A distinct minority of cortical neurons are monocular, but these tend to be concentrated in layer IV, where the dorsal LGN input remains largely segregated in terms of ocularity. Evidently intracortical circuitry beyond layer IV includes convergence onto single cells of inputs related to each eye. This binocular convergence is crucial to stereopsis and binocular single vision: one is generally aware of a single, integrated view of the world and not separate views for each eye.

■ *Serial versus parallel processing*

As noted earlier, it is far from clear how the geniculostriate pathways and cortical circuitry are organized to produce the functional types of neurons seen in the cortex. Two different hypotheses, referred to as *serial* and *parallel processing,* have been suggested. These hypotheses are not mutually exclusive, and it is quite likely that some synthesis of these ideas is closer to the truth.

Hubel and Wiesel proposed the hypothesis of serial processing in the early 1960s. It was based on seminal research that was recognized with the award of the Nobel Prize in 1981 to these scientists. They argued that retinal ganglion and geniculate cells are fairly homogeneous (the W-, X-, and Y-cell classification was not known at the time their hypothesis was formulated) and that cells in the striate cortex could be classified in groups that reflect receptive field complexity and thus hierarchial order. More complex receptive fields require more specification of the stimulus to discharge the cell. *Simple cells* thus are the first-order cortical cells that receive geniculate input; *complex cells* are innervated by simple cells; *hypercomplex cells* are innervated by complex cells; and so on. Thus visual processing proceeds along a functional neuronal chain. As the chain ascends, more stimulus detail is encoded. The pattern of activity at the pinnacle of each hierarchy can specify the stimulus. These hierarchial neural chains represent a set of homogeneous building blocks for stimulus analysis. Identical neuronal chains are used repeatedly in a point-by-point analysis of the visual scene, and thus each point is analyzed in only this manner. Although there is little direct evidence for or against this hypothesis, it has the virtue of providing a simplifying theoretical framework.

Because it was formulated before knowledge of W-, X-, and Y-cells, however, the hypothesis of serial processing does not account for these cell classes. Proponents of the parallel processing hypothesis have most forcefully emphasized the importance of W-, X-, and Y-cells by suggesting that they represent three distinct, functionally independent pathways to and through the cortex. It is further suggested that these pathways analyze different aspects of the visual scene in parallel.

It is presently not clear how W-, X-, and Y-cell innervation from the dorsal LGN relates to the simple, complex, and hypercomplex cell classes in the striate cortex. This is an active area of research that is replete with new ideas and considerable controversy. Thus present suggestions as to how the geniculostriate pathways are organized should be regarded as nothing more than working hypotheses. The functional implications of the geniculostriate pathway and the W-, X-, and Y-cell pathways are considered later in this chapter.

■ *Stereopsis*

Stereopsis can be defined as binocular depth perception. It is distinguished from the many monocular cues to depth, such as different retinal image sizes for objects at different distances and different rates of object movement at different distances (motion parallax). The cues for stereopsis are based on the slightly different retinal image each eye forms of a given visual scene because each eye has a slightly different perspective. This is the same principle used to calculate distances in range-finder cameras and to generate stereoscopic photographs by cameras with two lenses offset from one another.

The useful range of stereopsis is thus determined by interocular distance; in humans it is best suited for visually guided work with the hands. Human stereopsis is useful only up to a distance of 125 m. *Stereoacuity* (the minimum separation of two objects necessary to be perceived as such) improves dramatically as object distance lessens, and this is a straightforward consequence of trigonometry. Therefore stereopsis is more useful for threading a needle than for landing an airplane.

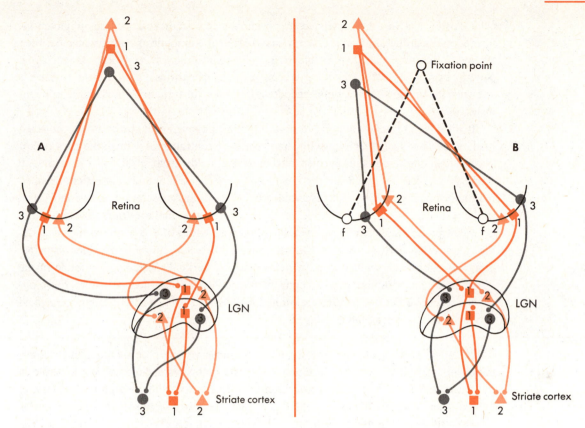

Fig. 8-29 ■ Schematic diagrams illustrating how receptive field disparities might provide the neural basis for stereopsis. A cortical neuron with nondisparate receptive fields in each retina means that the distance and direction (i.e., left, right, up, or down) of each field from the fovea are identical. Receptive field disparities are caused by differences in the distance and/or direction of the fields in each retina. In each drawing the neural circuitry provides cortical cell *1* with nondisparate receptive fields, whereas cells *2* and *3* have disparities in opposite directions. Thus cell *2* encodes farther objects than does cell *1*, and cell *3* encodes nearer objects. The maximum disparities seen in studies of cats and monkeys are roughly 2 degrees. **A,** Objects near the fixation point. The fixation point *(1)* is imaged onto each fovea. Thus the nearer point *(3)* falls onto each temporal retina, and the farther point *(2)* falls onto each nasal retina. **B,** Objects far from the fixation point. The same principles apply as in **A,** except that all of the disparities exist to one side of the fovea.

A possible neural basis for stereopsis has been described in the cat and extended to monkeys. Binocular neurons in the striate cortex do not have receptive fields in precisely homologous locations for each retina. Instead, they usually exhibit *receptive field disparities,* which means disparate receptive field locations in each retina. Fig. 8-29, *A,* illustrates this phenomenon for three cortical neurons with receptive fields near the fovea. Cell *1* has a receptive field in each fovea; cell *2*, in each nasal retina; and cell *3*, in each temporal retina. Cell *1* has no "disparity" because its receptive fields are in precisely corresponding retinal locations, but cells *2* and *3* exhibit disparity. When object *1* is fixated, cell *1* responds best to object *1*; cell *2* responds best to a further object (object *2*); and cell *3* responds best to a nearer object (object *3*). The largest disparities for cortical cells in monkeys are approximately a degree or so; that is, the retinal location of one receptive field may be that far from the precisely corresponding position for the other receptive field.

Receptive field disparities thus provide these cells with sensitivity to the third dimension relative to a fixation point. The range of disparities exhibited by these cortical cells closely matches that predicted by psychophysical studies of stereopsis. Given information about the alignment of the visual axes or the amount of convergence between the eyes, the brain can deduce three-dimensional information from knowledge of the pattern of activity of cells with receptive field disparities. Unfortunately little is known concerning how the brain encodes the critical information of visual axis alignment.

If all ganglion cells temporal to a vertical line passing through the fovea projected ipsilaterally and all cells nasal to this line projected contralaterally, then the presence of cortical cells with binocularly disparate receptive fields from each nasal or temporal retina (e.g., cells *2* and *3,* respectively, in Fig. 8-29, *A*) would require complicated interhemisphere pathways. This is avoided because of the "nasotemporal overlap" seen in the retina. A 1- to 2-degree wide vertical strip of retina includes intermixed ganglion cells that project ipsilaterally or contralaterally. The width of this strip roughly matches the range of disparities seen in cortical cells. This feature thereby provides disparate input to cortical cells with receptive fields near the vertical meridian or midline of the visual field, and complicated interhemisphere pathways are not needed. Neither stereopsis nor receptive field disparities are limited to foveal regions. The entire binocular field of view exhibits these features (Fig. 8-29, *B*).

■ Color

How the visual pathways are organized to subserve color or wavelength discrimination is not well understood. The geniculostriate pathways seem to play a key role in this process. Clearly wavelength discrimination begins with the cone photoreceptors; in humans there are three types. Fig. 8-30 illustrates the *spectral sensitivity* (the sensitivity to different wavelengths) of each type. However, physiological data suggest that most retinal ganglion cells receive both rod and cone input via the retinal interneurons. This implies that these ganglion cells can function at photopic (i.e., cone only) and scotopic (i.e., rod only) levels of illumination. Thus wavelength discrimination is not determined by unique cone inputs to various cells; rather the nature and interactions of these inputs lead to receptive fields of certain neurons (e.g., X-cells) that respond differently to different patterns of cone stimulation.

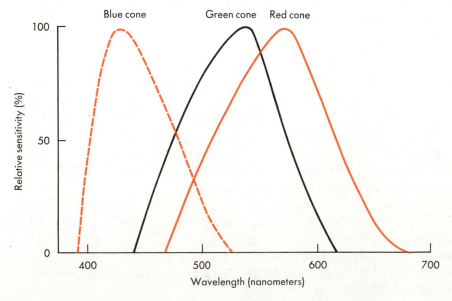

Fig. 8-30 ■ Relative spectral sensitivity of the three cone types in humans.

Retinal ganglion and geniculate cells can be divided into two general categories based on their responses to different wavelengths. One category of response is called *broadband* because the neuronal discharges do not depend significantly on wavelength. The other category is called *color opponent,* and for cells in this category, wavelength is a key stimulus parameter. Most color opponent cells are excited by one color turned on (or off) in the receptive field center and turned off (or on) by another in the surround. For instance, many of these cells are excited by red light turned on in the center or by green light turned off in the surround. Other more complicated patterns of spectral opponency have also been described. Finally, color opponent cells have been described in the striate cortex and in certain extrastriate cortical areas, but details of their spectral sensitivity are currently uncertain.

Within the monkey's geniculostriate pathways, the broadband neurons have been associated with Y-cells and the magnocellular laminae of the dorsal LGN. Color opponent neurons have been associated with X-cells and the parvocellular laminae. Perhaps another functional significance of the parallel processing in X- and Y-cell geniculostriate pathways concerns wavelength discrimination, which might be a unique function of the X-cell pathway.

■ *Superior colliculus pathways*

■ *Lamination and connections*

A large portion of the visual system involves pathways through the superior colliculi, which are bilaterally paired, prominently laminated structures that form the dorsal portion of the rostral midbrain (Fig. 8-31). Seven collicular layers can be distinguished. From the dorsal surface ventrally, these layers are the *stratum zonale, stratum griseum superficiale, stratum opticum, stratum griseum intermedium, the stratum album intermedium, stratum griseum profundum,* and *stratum album profundum.* Below this last layer is the periaqueductal gray matter. The three dorsal layers are exclusively involved in vision, and the four ventral layers are involved in more complex multimodal processing. Therefore only the three dorsal layers are considered here.

Two major visual inputs of roughly equal extent innervate these dorsal layers. One is represented by retinocollicular fibers that pass into the brachium of the superior colliculus, enter the stratum opticum, and terminate in the upper three layers (Fig. 8-31). These axons arise from retinal W- and Y-cells, predominantly from the contralateral nasal retina. The other input is a corticocollicular pathway predominantly from the striate cortex (area 17) and nearby cortex (areas 18 and 19) but also involving other cortical zones as well. Much of this pathway involves the following neuronal chain: retinal Y-cell to geniculate Y-cell to complex cell in layer V of the striate cortex; these complex cells comprise the source of the corticocollicular pathway. There is some evidence that these cells in layer V of striate cortex receive direct geniculocortical input from Y-cells via their dendrites that extend into layer IV and/or layer VI. It follows then that the W- and Y-cell pathways, but not the X-cell pathway, innervate the superior colliculus.

Neurons in the upper layers of the superior colliculus project to the thalamus, and most of this projection terminates in the pulvinar. Data from the cat suggest that some may also terminate in the dorsal LGN. The pulvinar and dorsal LGN innervate virtually all known cortical areas involved in visual processing (pp. 135-136). The functional significance of this relevant to the superior colliculus is also considered later.

Consideration of these innervation patterns suggests a mixed evolutionary age for the superior colliculus. The W-cell pathway is probably old, but the Y-cell pathway and particularly the corticocollicular input can be considered new. In nonmammalian vertebrates the optic tectum (i.e., the homologue of the superior colliculus) is the predominant central visual structure. Presumably as thalamocortical pathways evolved in mammals, the superior colliculus continued to evolve and to play a prominent role in vision.

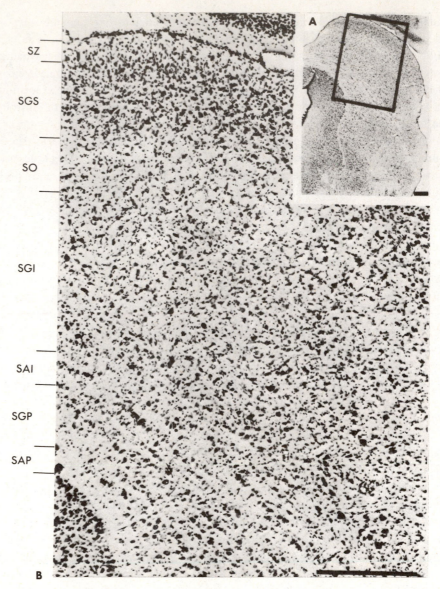

SZ

SGS

SO

SGI

SAI

SGP

SAP

A

B

Fig. 8-31 ■ Coronal section through the superior colliculus in a prosimian primate *(Galago senegalensis)*. **A,** Lower power photomicrograph of midbrain region. **B,** Higher power photomicrograph of area demarcated in **A.** Collicular lamination is poorly differentiated in primates, but the layers are indicated. *SZ,* Stratum zonale; *SGS,* stratum griseum superficiale; *SO,* stratum opticum; *SGI,* stratum griseum intermedium; *SAI,* stratum album intermedium; *SGP,* stratum griseum profundum; *SAP,* stratum album profundum. Scale bars in **A** and **B** are 0.5 mm. (Courtesy D. Raczkowski.)

■ *Retinotopic organization*

As befits a relatively new portion of the visual system, the superior colliculus is organized retinotopically. The fovea is represented slightly lateral to the rostral pole of the superior colliculus. The line that divides the visual fields into superior and inferior halves runs mediocaudally. Rostromedial to this line is represented the superior visual field, and the inferior visual field is represented caudolaterally.

■ *Receptive field properties*

Collicular neurons tend to have fairly large receptive fields that are most sensitive to rapid stimulus motion in a particular direction. These cells seem less concerned with spatial detail than with stimulus motion. In addition to direction selectivity, most possess binocular receptive fields. However, unlike cortical cells, these collicular neurons lack orientation selectivity.

In cats removal of the visual cortex and thus the corticocollicular input dramatically alters the collicular receptive field properties. Postoperatively collicular neurons lose their direction selectively, respond poorly to rapidly moving stimuli, and tend to respond almost exclusively to stimulation of the contralateral eye. These properties presumably reflect the retinocollicular input. Corticocollicular input is thus needed for binocularity, direction selectivity, and responsiveness to rapidly moving targets. Nonetheless, collicular cells still respond well to visual stimuli after cortical removal. Curiously, effects of cortical removal or inactivation on collicular receptive fields in monkeys are somewhat different; cells dorsal to the upper half of layer III are relatively unaffected, whereas more ventral cells lose responsiveness to visual stimuli. The reasons for these differences between cat and monkey are not understood.

■ ***Extrastriate visual cortex***

In the cat at least 13 separate visual areas have been mapped in the cerebral cortex. In primates a large number of separate visual cortical areas also exists; the actual number perhaps varies among species. Each of these areas is retinotopically organized, but receptive field data are generally unavailable for most of these regions. Fig. 8-32 illus-

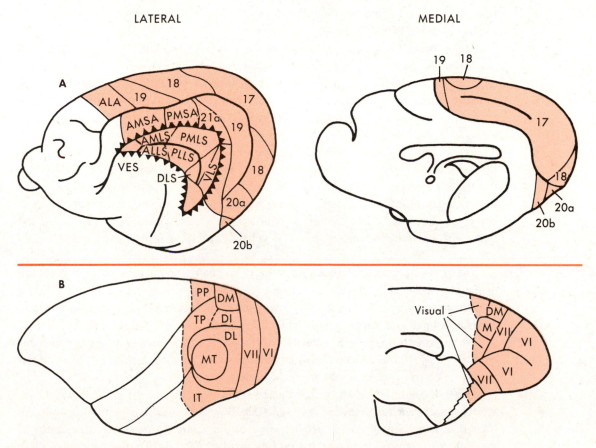

Fig. 8-32 ■ Visual cortical areas in various species. Lateral views are on the left and medial views on the right. Triangles indicate the retraction of a sulcus to visualize the cortex in its depths. **A,** Thirteen or more visual areas of the cat. In addition to the seven numbered areas *(17, 18, 19, 20a, 20b, 21a, and 21b)* are six additional visual areas *(AMLS, PMLS, VLS, ALLS, PLLS, and DLS)*. Areas 17, 18, and 19 are equivalent to VI, VII, and VIII of older literature. Also shown are polysensory areas that contain visually responsive neurons *(ALA, AMSA, PMSA, and VES)*. **B,** Twelve known visual areas of the owl monkey. *VI* and *VII* are equivalent to areas 17 and 18. *M, DM, DI,* and some other visual areas are all found in area 19. *PP* lies in area 7, and the temporal lobe contains areas *TP, MT,* and *IT*. Other visual regions are also shown on the medial surface. *Continued.*

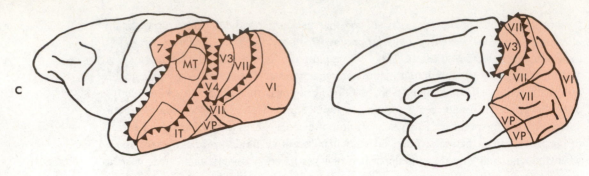

Fig. 8-32, cont'd ■ C, Eight known visual areas of the macaque monkey. *VI* is equivalent to area 17. Area 18 contains *VII, V3, V4,* and *VP. MT* and *IT* are located in the temporal lobe. It should be noted that studies of multiple visual areas of macaque monkeys are recent compared with those of cats and owl monkeys, and more areas may emerge as more detailed maps become available. (Redrawn from Tusa, R.J.: Visual cortex: multiple areas and multiple functions. In Morrison, A.R., and Strick, P.L., editors: Changing concepts of the nervous system, New York, 1982, Academic Press, Inc.)

trates these areas for the cat, the owl monkey (a New World monkey), and the macaque monkey (an Old World monkey).

These visual areas have many complex interconnections with each other and with the thalamus, particularly the pulvinar. Indeed the pulvinar innervates most or all of these areas. Although the functional significance of these multiple extrastriate areas is unknown, their extent relative to the striate cortex raises some questions about the primacy of geniculostriate pathways for vision. This is considered again later in this chapter.

These multiple areas have been clarified only in the last decade. A similar, but presently less complete, picture is emerging for the representation of other sensory systems in the cortex. Previously no specific sensory function was attributed to most of the posterior half of the cortex, and it was designated as *association cortex* by default. The contemporary view is that little or no such association cortex exists in this region and that all of this posterior cortical area is devoted to specific sensory analysis.

■ *Phylogenetically older visual pathways*

The previous discussion describes some of the complex interactions among certain of the phylogenetically newer visual pathways. These are characterized by the presence of retinotopic organization, and often the neurons of these pathways respond tonically to visual stimuli and display lateral inhibition in their receptive fields. As noted earlier, the retina projects sparsely to other brainstem sites that are regarded here as phylogenetically older. These include retinal projections to the suprachiasmatic nucleus, ventral LGN, pretectum, and the accessory optic nucleus.

Generally little is known of these pathways beyond the retinal projection, and for the most part their functions are only superficially known. For those retinofugal projections that have been adequately studied (i.e., to the ventral LGN and to the nucleus of the optic tract), the projection predominantly involves W-cells. The few studies of these pathways have uncovered neither retinotopic organization nor afferent inhibition among neurons.

Where function can be suggested for these structures, it is generally reflexive and subconscious in nature. The suprachiasmatic nucleus is thought to be involved in circadian rhythms involving the light/dark diurnal cycle. The nucleus of the optic tract is involved in pupillary light reflexes, among other possible functions. The accessory optic nucleus seems to participate in reflex eye movements such as optokinetic nystagmus. No obvious function can yet be suggested for the ventral LGN.

The previous sections deal with the functional organization of the various central visual pathways. This section deals with the possible functional role each of these pathways plays in conscious visual perception. Most attention in this context has been focused on the geniculostriate pathways and the extrastriate pathways involving the superior colliculus.

Until recently most investigators emphasized the importance of the geniculostriate pathways to visual perception and relegated the superior colliculus to vestigial or reflex roles such as crude brightness analysis or reflex eye movement. However, the importance of the superior colliculus has been emphasized from recent behavioral studies of animals and humans with brain lesions. Lesions of the visual cortex commonly lead to severe visual dysfunction in many mammalian species, including humans, and this has led to the concept that this cortex and not the superior colliculus is crucial to visual perception. However, there are two flaws in this conclusion. The first is logical. Lesions of the visual cortex interrupt the corticocollicular pathways, and the superior colliculus thus becomes extensively denervated. Visual capacity following such lesions hardly elucidates the functional role of the superior colliculus in normal individuals. Second, more recent and complete behavioral studies have revealed surprising visual capacity in animals and humans with lesions of the visual cortex. Most of this work has been done with cats, but a limited amount has involved monkeys and human patients.

Complete bilateral destruction of the striate cortex in the cat produces remarkably little impairment in the animal's visual performance. There is moderate loss of acuity, but otherwise the cats are indistinguishable before and after such an ablation on tests of visual performance. (As noted later in this chapter, good form vision does not depend on acuity per se.) However, lesions that involve extensive extrastriate areas in the occipitotemporal cortex do dramatically impair form vision.

Because these areas are innervated by both the dorsal LGN and pulvinar (which in turn receives collicular input), it is not clear to what extent geniculocortical or colliculothalmocortical pathways contribute to the cat's spatial vision. Similar data exist for other infraprimate mammalian species. As a general rule, for these species it consequently seems clear that the geniculostriate pathways are not essential for good form vision but may be needed for high acuity vision and perhaps stereopsis.

Behavior-ablation data cannot be unambiguously interpreted, however, because of the complex and poorly understood effects of neural lesions throughout the brain. Fig. 8-33, which illustrates a phenomenon known as the *Sprague effect,* both exemplifies this proviso and also suggests the underestimated role of the superior colliculus in normal form vision. Large unilateral ablations of the occipitotemporal cortex that destroy all known visual areas produce a profound and permanent *hemianopsia* (blindness in a half-field vision) in the hemifield contralateral to the lesion (Fig. 8-33, *B*). In the past such data were used to emphasize the importance of the cortex in vision and to deemphasize subcortical pathways. However, if the superior colliculus contralateral to the earlier cortical lesion is then ablated or if the colliculi are disconnected from one another by splitting their commissure, there is a dramatic restoration of vision for the previously blind hemifield (Fig. 8-33, *C*). We still lack an adequate explanation for this phenomenon of a second lesion that partially restores vision after blindness was produced by an earlier lesion.

Such "collicular vision" is by no means normal. It allows the cat to move about visually and to locate visual objects of interest. However, detailed pattern vision and the ability to discriminate among various spatial patterns are absent or rudimentary. Conversely, removal of the superior colliculus with intact cortex results in rather subtle

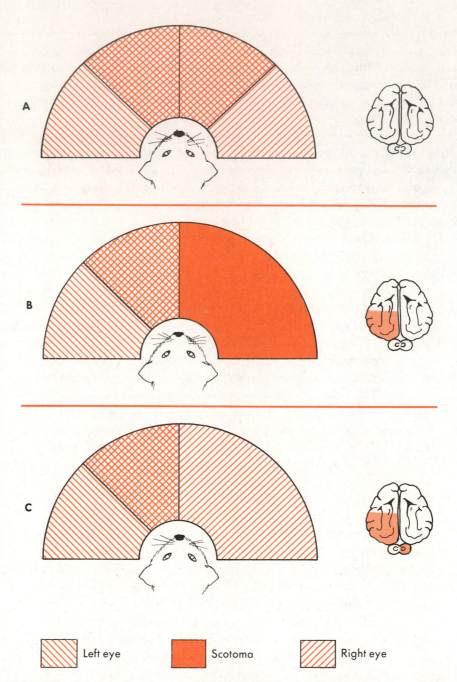

Fig. 8-33 ■ The Sprague effect from studies of cats. **A,** Visual fields in a normal cat. Each eye sees from about 45 degrees across the midline to 90 degrees ipsilaterally. Thus each temporal retina sees from the midline to 45 degrees into the opposite hemifield, and each nasal retina sees the entire ipsilateral hemifield. **B,** Visual fields after left visual cortex removal. The cat now exhibits a right hemianopia, and vision is restricted to the left nasal and right temporal retinas. **C,** Visual fields after subsequent removal of the right superior colliculus. Vision is restored to the right hemifield, although visual capacity here remains subnormal. Note, however, that the left temporal retina remains blind. (Redrawn from Sherman, S.M.: J. Comp. Neurol. **172:**231, 1977.)

deficits that generally do not seem to involve spatial vision. This evidence and similar data from other mammals have led to the concept that the superior colliculus is involved primarily in determining *where* objects are, whereas the cortex is needed to determine *what* objects are. However, it should be reemphasized that the functioning of the superior colliculus after removal of its massive corticocollicular innervation can hardly be expected to be normal.

Analogous ablation-behavior data are much less complete for monkeys and humans. Many older studies have repeatedly emphasized the nearly total loss of spatial vision in monkeys and humans after damage to the striate cortex. Although it certainly seems clear that such damage reduces visual performance more drastically for primates than for cats, two recent series of studies have emphasized certain similarities between primates and cats regarding these ablation-behavior data.

First, it appears that many or all of the earlier experimental studies of monkeys involved lesions that extended well beyond the striate cortex. Thus much of the cortical innervation from the pulvinar, as well as from the dorsal LGN, was interrupted. More recent studies with monkeys that had lesions limited to the striate cortex demonstrated significant remaining spatial vision, although vision in these monkeys was drastically impaired.

Second, in studies of human patients with cortical damage resulting from vascular accidents or penetrating missile wounds, the extents of these lesions are difficult to assess. Nonetheless, surprising visual function has been described in such patients when appropriate measures are used. For example, on a standard perimetry test for which the patient must verbally indicate visualization of the target, a large scotoma might be mapped that can be related to the cortical injury. If nonverbal responses are then employed, such as requiring the patient to point a finger at the verbally denied target, it is clear that visual stimuli can be crudely localized. The cortical lesion seems to interfere with language related to visual stimuli in addition to its effect on visual capacity per se. Even in humans some visual capacity thus seems to survive interruption of the geniculostriate pathways. Although of importance to vision in primates, these pathways are not the sole substrate of conscious vision as was once thought.

■ *Studies of visual function in monkeys and humans*

Because W-, X-, and Y-cells have been appreciated relatively recently, our understanding of the significance of their differences is far from complete. Studies of these cells and their pathways represent an active research area, replete with speculation and controversy. Nonetheless, because an understanding of these pathways will undoubtedly prove crucial to appreciation of visual processing, it is worth considering them in greater detail with emphasis on the tentative and speculative nature of many of the conclusions.

We know a great deal more about these pathways in the cat than in any other species, and even in the cat we know far more about X- and Y-cells than about W-cells. However, because W-cells respond poorly to most visual stimuli and because they seem to play a relatively minor role in geniculocortical innervation (particularly for their homologues in monkeys), most relevant hypotheses have focused on X- and Y-cells.

An appealing hypothesis (among many) is derived from a consideration of both X- and Y-cell response properties and the behavioral consequences of some of the ablation-behavior data noted earlier. Y-cells are most responsive to large targets or cruder forms. X-cells are relatively unresponsive to such stimuli but respond better to smaller targets having finer spatial details. Evidence from behavioral studies in which the cruder and finer forms in a visual scene were separated suggests that basic form analysis is performed on the cruder forms and that the finer ones add details to this basic analysis.* The Y-cell pathway may thus be involved with basic form analysis, whereas the X-cell

■ *W-, X-, and Y-cell pathways*

*In much the same way that complex sounds can be broken down into their pure tones or frequency components (Chapter 10), so can complex visual scenes be analyzed in terms of similar components. These visual components are *sine wave gratings,* which are sinusoidal variations of brightness with position. These are analogous to tones, which are sinusoidal variations of sound level with time, except that the longitudinal dimension is spatial rather than temporal. Thus a visual scene can be described in terms of its component spatial frequencies (i.e., cycles of a sine wave grating per spatial unit). Lower spatial frequencies correspond to the cruder shapes in a visual scene, and higher spatial frequencies correspond to the finer details. Behavioral evidence indicates that basic form information is carried by the lower spatial frequencies and that the higher ones add details and raise acuity. If the higher frequencies are removed from a typical scene (photograph of a landscape, a face, etc.), details are lost, but the basic subject can be recognized.

pathway may be involved in high acuity vision that analyzes the minute details in a visual scene. Also, the X-cell pathway in many primate species may be used to analyze colors.

Behavioral studies of cortically lesioned animals are consistent with this notion. In cats the geniculocortical projection of X-cells is limited nearly exclusively to the striate cortex, whereas the Y-cells directly innervate striate and extrastriate cortical areas, often via branching axons. Thus a bilateral ablation of the striate cortex effectively eliminates the X-cell pathway but leaves much of the Y-cell pathway intact. As noted earlier, such a lesion has remarkably little effect on basic form vision in a cat, except for reduced acuity. Such a lesion in a monkey, in contrast, is much more devastating to form vision. Perhaps this results not from the essential nature of the striate cortex per se but from evidence that in monkeys all geniculocortical homologues of Y-cells, as well as of X-cells, terminate only in the striate cortex. It may be the course of the Y-cell pathway that determines the extent of visual impairment consequent to striate cortex damage.

Anatomical evidence from the cat further supports this notion of X- and Y-cell roles (Fig. 8-34). Compared with X-cells and the X-cell pathway, relatively few Y-cells exist in the retina, but the Y-cell pathway diverges extensively to the dorsal LGN and visual cortex. As a result, a small minority of retinal ganglion cells seem to dominate the geniculocortical innervation patterns. Few Y-cells may be needed in retina because coding of the important crude forms requires no more than a coarse retinal "grain." The importance of this information to basic form vision, however, requires that much of the cortex need be devoted to its analysis. In contrast, many X-cells may exist in retina because a fine grain is needed to code the smallest details in a visual stimulus. Relatively little cortex need be devoted to this analysis performed via the X-cells. To support such a relative divergence in the Y-cell pathway, individual Y-cell axons branch more extensively and issue larger terminal arbors than do those of X-cells, both at the retinogeniculate and geniculocortical levels. Whether or not a similar pattern of connectivity exists for the homologue of X- and Y-cells in primates remains to be determined.

■ ■ ■

In summary, although the visual system begins with the retina, the optics of the eye play a crucial role in preparing visual stimuli for the retina. Important ocular landmarks include the fovea, optic disc, visual axis, and optic axis. The cornea and adjustable lens serve as a compound lens system to focus light, and the adjustable pupil controls depth of field. A reflective outer coat (the sclera) and absorbing inner coats (pigment epithelium and choroid) serve to prevent stray, unfocused or scattered light from striking the retina. Adjustments to focus result from the action of the ciliary muscle passively transmitted to the lens via the tough zonular fibers and elastic lens capsule. The relaxed muscle achieves focus for distant objects; the constricted muscle accommodates the eye and focuses it on nearer objects. *Emmetropia* is defined as normal focus for the relaxed eye. *Ametropia* is abnormal focus for the relaxed eye and includes myopia, hypermetropia, and astigmatism. Presbyopia is not an ametropic condition, but rather it is the failure of the lens to change shape passively during ciliary muscle contraction; it is a normal concomitant of aging and loss of tissue elasticity.

Photic transduction begins with rods and cones in the retina. Rods are more sensitive and are used for black-and-white vision in dim light; the three cone types are less sensitive and are used for color vision in bright light. Light must first travel through all the retinal layers, and this may have led to the evolution of the fovea for detailed vision. Only amacrine and ganglion cells in the retina exhibit action potentials; electrotonically conducted, graded potentials are sufficient for the short conduction distances of the other neurons. The photoreceptors are hyperpolarized by photons. The various synaptic circuitry in the outer and inner plexiform layers involving retinal interneurons (i.e., hori-

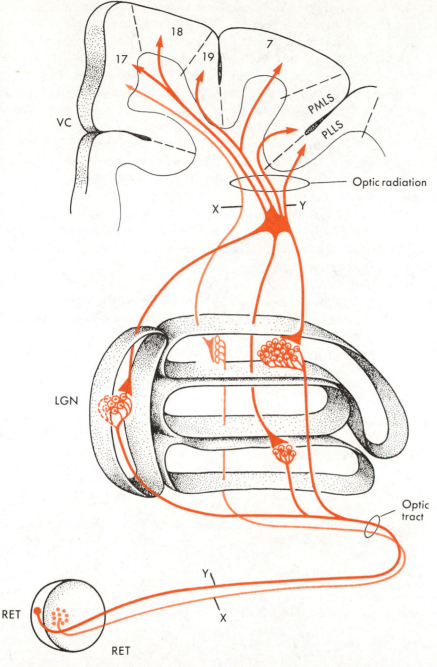

Fig. 8-34 ■ Schematic and hypothetical illustration of the cat's X- and Y-cell pathways, drawn for simplicity only from the contralateral eye. The X-cell pathway is shown in lighter color and the Y-cell pathway in darker color. X-cells outnumer Y-cells in the retina by roughly 10:1, but Y-cell axons of both the optic tract and optic radiation exhibit much more divergence that do X-cell axons. Consequently many more postsynaptic cells are innervated by each Y-cell axon than by each X-cell axon. Single Y-cell retinogeniculate axons innervate all subregions of the lateral geniculate nucleus (LGN). Their geniculocortical axons innervate nearly all known areas of visual cortex (see Fig. 8-32), and although single axons do not innervate all areas, they commonly branch to innervate two or more. Retinogeniculate X-cell axons, by contrast, innervate only one LGN laminar region, and their geniculocortical axons innervate only the striate cortex. The result is that the 10:1 X-to-Y-cell retinal ratio is roughly 1:1 in the LGN and probably becomes dominated by Y-cells in the cortex. (Redrawn from Sherman, S.M., and Spear, P.D.: Neural development of cats raised with deprivation of visual patterns. In Willis, W.D., editor: The clinical neurosciences, section V, Neurobiology, New York, 1983, Churchill Livingstone, Inc.)

zontal, bipolar, interplexiform, and amacrine cells) eventually leads to several classes of ganglion cell output that have been identified as W-, X- and Y-cells.

X- and Y-cells project to the dorsal LGN, which in turn projects to the striate cortex. In monkeys there may be small W-cell input to dorsal LGN neurons, which innervate striate and extrastriate cortical areas. The X- and Y-cell pathways remain separate; X-cells involve parvocellular geniculate laminae and primarily layer IVcβ of the striate cortex; Y-cells involve magnocellular geniculate laminae and cortical layer IVcα. Data from cats suggest that the Y-cell pathway may be used for basic form analysis, whereas the X-cell pathway is used for analysis of fine details. Also, in primates X-cells seem particularly sensitive to wavelength and may represent the neural substrate for color vision. Unlike the straightforward, monocular, center-surround receptive fields of geniculate neurons, those of the striate cortex are binocular and exhibit orientation and often direction selectivity. Also, the binocular receptive fields are slightly disparate in each retina, and this may be the neural substrate for stereopsis. The importance of the geniculostriate pathways for visual capacity is probably not as great as previously thought, particularly for nonprimate mammals.

Visual inputs to the dorsal collicular layers include W- and Y-cell retinocollicular axons and axons from the striate and nearby cortical areas. Collicular neurons are binocular and movement and direction sensitive but are not orientation selective. Cells in these collicular layers project to the pulvinar, which in turn projects extensively to each of the 10 to 15 known cortical visual areas, including the striate cortex.

■ *Bibliography*

Journal articles

Barbur, J.L., Ruddock, K.H., and Waterfield, V.A.: Human visual responses in the absence of a geniculo-calcarine projection, Brain **103:**905, 1980.

Benevento, L.A., and Yoshida, K.: The afferent and efferent organization of the lateral geniculo-prestriate pathways in the macaque monkey, J. Comp. Neurol. **203:**455, 1981.

Berkley, M.A., and Sprague, J.M.: Striate cortex and visual acuity functions in the cat, J. Comp. Neurol. **187:**679, 1979.

Bowling, D.B., and Michael, C.R.: Projection patterns of single physiologically characterized optic tract fibers in cats, Nature **286:**899, 1980.

Carey, R.G., Fitzparick, D., and Diamond, I.T.: Layer I of striate cortex of *Tupaia glis* and *Galago senegalensis:* projections from thalamus and claustrum revealed by retrograde transport of horseradish peroxidase, J. Comp. Neurol. **186:**393, 1979.

Dowling, J.E.: Organization of vertebrate retinas, Invest. Ophthalmol. **9:**655, 1970.

Dreher, B., Fukada, Y., and Rodieck, R.W.: Identification, classification, and anatomical segregation of cells with X-like and Y-like properties in the lateral geniculate nucleus of old-world primates, J. Physiol. (Lond.) **258:**433, 1976.

Ferster, D.: A comparison of depth mechanisms in areas 17 and 18 of the cat visual cortex, J. Physiol. (Lond.) **311:**623, 1981.

Friedlander, M.J., et al.: Morphology of functionally identified X- and Y-cells in the cat's lateral geniculate nucleus, J. Neurophysiol. **46:**80, 1981.

Hubel, D.H., and Wiesel, T.N.: Functional architecture of macaque monkey visual cortex (Ferrier Lecture), Proc. R. Soc. Lond. Biol. **198:**1, 1977.

Lehmkuhle, S., Kratz, K.E., and Sherman, S.M.: Spatial and temporal sensitivity of normal and amblyopic cats, J. Neurophysiol. **48:**372, 1982.

Lennie, P.: Parallel visual pathways, Vision Res. **20:**561, 1980.

LeVay, S., and Sherk, H.: The visual claustrum of the cat. I. Structure and connections, J. Neurosci. **1:**956, 1981.

Miller, M., Pasik, P., and Pasik, T.: Extrageniculostriate vision in the monkey. VII. Contrast sensitivity functions, J. Neurophysiol. **43:**1510, 1980.

Poggio, G.F., and Talbot, W.H.: Mechanisms of static and dynamic stereopsis in foveal cortex of the rhesus monkey, J. Physiol. (Lond.) **315:**469, 1981.

Pöppel, E., Held, R., and Frost, D.: Residual visual function after brain wounds involving the central visual pathways in man, Nature **243:**295, 1973.

Rodieck, R.W.: Visual pathways, Annu. Rev. Neurosci. **2:**193, 1979.

Schiller, P.H., and Malpeli, J.G.: Properties of tectal projections of monkey retinal ganglion cells, J. Neurophysiol. **40:**428, 1977.

Schiller, P.H., and Malpeli, J.G.: Functional specificity of lateral geniculate nucleus laminae of the rhesus monkey, J. Neurophysiol. **41:**788, 1978.

Schiller, P.H., et al.: Response characteristics of single cells in the monkey superior colliculus following ablation or cooling of visual cortex, J. Neurophysiol. **37:**181, 1974.

Shapley, R.M., Kaplan, E., and Soodak, R.: Spatial summation and contrast sensitivity of X- and Y-cells in the lateral geniculate nucleus of the macaque monkey, Nature **292:**543, 1981.

Sherman, S.M.: The effect of superior colliculis lesions upon the visual fields of cats with cortical ablations, J. Comp. Neurol. **172:**211, 1977.

Sherman, S.M., and Spear, P.D.: Organization of visual pathways in normal and visually deprived cats, Physiol. Rev. **62:**738, 1982.

Sherman, S.M., et al.: X- and Y-cells in the dorsal lateral geniculate nucleus of the owl monkey *(Aotus trivirgatus),* Science **192:**475, 1976.

Sprague, J.M.,: Interaction of cortex and superior colliculus in mediation of visually guided behavior in the cat, Science **153:**1544, 1966.

Stone, J., Dreher, B., and Leventhal, A.: Hierarchical and parallel mechanisms in the organization of visual cortex, Brain Res. Rev. **1:**345, 1979.

Sur, M., and Sherman, S.M.: Retinogeniculate terminations in cats: morphological differences between X- and Y-cell axons, Science **218:**389, 1982.

Van Essen, D.C.,: Visual areas of the mammalian cerebral cortex, Ann. Rev. Neurosci. **2:**277, 1979.

Weiskrantz, L.: Behavioural analysis of the monkey's visual nervous system, Proc. R. Soc. Lond. (Biol.) **182:**425, 1972.

Weiskrantz, L., Cowey, A., and Passingham, C.: Spatial responses to brief stimuli by monkeys with striate cortex ablations, Brain **100:**655, 1970.

Yukie, M., and Iwai, E.: Direct projection from the dorsal lateral geniculate nucleus to the prestriate cortex in macaque monkeys, J. Comp. Neurol. **187:**679, 1979.

Books and monographs

Bishop, P.O.: Neurophysiology of binocular single vision and stereopsis. In Jung, R., editor: Handbook of sensory physiology: central processing of visual information, part A, Integrative functions and comparative data, vol. VII/3, New York, 1973, Springer-Verlag New York, Inc.

Boynton, R.M.: Human color vision, New York, 1979, Holt, Rinehart, & Winston.

Cornsweet, T.N.: Visual perception, New York, 1970, Academic Press, Inc.

Diamond, I.T.: Changing views of the organization and evolution of the visual pathways. In Morrison, A.R., and Strick, P.L., editors: Changing concepts of the nervous system, New York, 1982, Academic Press, Inc.

Duke-Elder, W.S.: Textbook of ophthalmology, vol. I, St. Louis, 1942, The C.V. Mosby Co.

Gouras, P.: Color vision. In Kandel, E.R., and Schwartz, J.H., editors: Principles of neural science, New York, 1981, Elsevier North-Holland, Inc.

Graybiel, A.M., and Berson, D.M.: On the relationship between transthalamic and transcortical pathways in the visual system. In Schmitt, F.O., et al., editors: The organization of the cerebral cortex, Cambridge, Mass., 1981, The MIT Press.

Karten, H.J.: Visual lemniscal pathways in birds. In Granda, A.M., and Maxwell, J.H., editors: Neural mechanisms of behavior in the pigeon, New York, 1979, Plenum Publishing Corp.

Lund, J.S.: Intrinsic organization of the primate visual cortex, area 17, as seen in Golgi preparations. In Schmitt, F.O., et al., editors: The organization of the cerebral cortex, Cambridge, Mass., 1981, The MIT Press.

Polyak, S.: The vertebrate visual system, Chicago, 1957, University of Chicago Press.

Ratliff, F.: Mach bands: quantitative studies on neural networks in the retina, San Francisco, 1965, Holden-Day, Inc.

Rodieck, R.W.: The vertebrate retina, San Franscisco, 1973, W.H. Freeman & Co., Publishers.

Rosenquist, A.C., Raczkowski, D., and Symonds, L.: The functional organization of the lateral posterior-pulvinar complex in the cat. In Morrison, A.R., and Strick, P.L., editors: Changing concepts of the nervous system, New York, 1982, Academic Press, Inc.

Sherman, S.M.: Parallel pathways in the cat's geniculocortical system: W-, X-, and Y-cells. In Morrison, A.R., and Strick, P.L., editors: Changing concepts of the nervous system, New York, 1982, Academic Press, Inc.

Sprague, J.M., Berlucchi, G., and Rizzolatti, G.: The role of the superior colliculis and pretectum in vision and visually guided behavior. In Jung, R., editor: Handbook of sensory physiology: central processing of visual information, part B, Visual centers in the brain, vol. VII/3, New York, 1973, Springer-Verlag New York, Inc.

Tusa, R.J.: Visual cortex: multiple areas and multiple functions. In Morrison, A.R., and Strick, P.L., editors: Changing concepts of the nervous system, New York, 1982, Academic Press, Inc.

Walls, G.L.: The vertebrate eye and its adaptive radiation, Bloomfield Hills, Mich., 1942, Cranbrook Institute of Science.

Westheimer, G.: Optical properties of vertebrate eyes. In Fuortes, M.G.F., editor: Handbook of sensory physiology, vol. VII/2, Physiology of photoreceptor organs, New York, 1972, Springer-Verlag New York, Inc.

Westheimer, G.: The eye, including central nervous system control of eye movement. In Mountcastle, V.B., editor: Medical physiology, ed. 14, vol. I, St. Louis, 1980, The C.V. Mosby Co.

Woolsey, C.N., editor: Cortical sensory organization, vol. 2, Multiple visual areas, Clifton, N.J., 1981, Humana Press.

chapter 9

The somatosensory system

Humans perceive at least four distinct body sensations. These are touch (also known as touch-pressure), kinesthesia (or joint sensation), temperature, and nociception (or pain). Most schemata include tactile perception of vibration as a submodality distinct from touch. The following sections discuss the organization of somatosensory pathways represented via the dorsal root ganglia, and a subsequent section deals with the analogous representation of the cranial component, mostly via the trigeminal nerve.

In general, the transducers, or sensory receptors, for somesthesia are poorly understood. Two examples, the pacinian corpuscle and a tonically responsive transducer, are covered in Chapter 6, and they are involved in touch or pressure sensation. In most cases of a tonically responsive transducer we cannot identify particular types of transducers with particular somesthetic properties. A variety of specialized corpuscular and disk-shaped structures have been found near endings of primary afferent fibers in the skin and deeper tissue (joints, fascia, etc.). These presumably act as transducers (e.g., the tonically responsive transducer), or they assist in the transduction process (e.g., the "onion" of the pacinian corpuscle). Many free nerve endings without specialized structures have also been found, and most or all of the transducers for nociception or pain are thought to be free nerve endings. Although the details have yet to be elucidated, there is no reason to doubt that a variety of specialized and uniquely sensitive transducers exist for touch, kinesthesia, temperature, and nociception.

■ *Transducers*

Table 9-1 summarizes the classification of afferent fiber types that enter the spinal cord. From largest to smallest, they are designated group I through group IV, and an alternate designation frequently used is given in brackets. All of these fibers, with the possible exception of some group IV fibers as noted below, have their somata in the dorsal root ganglia. The largest fibers, group I (also known as A-α), originate in muscle spindles and tendon organs; they are not included in Table 9-1. Group II includes large myelinated fibers that convey information about touch and kinesthesia. Group III includes small myelinated fibers that convey information about touch, temperature, and nociception. Group IV represents small unmyelinated fibers that convey information about touch, nociception, and perhaps temperature. The pathways originating from muscle spindles and tendon organs (e.g., all group I and some group II fibers) are discussed in Chapter 14 and will not be considered further in this chapter.

Recent studies have shown that many group IV fibers enter the spinal cord via the *ventral* root. Therefore the classical notion that the dorsal and ventral roots are strictly afferent and efferent, respectively, must be revised. The central pathways of these ventral

■ *Afferent fibers*

TABLE 9-1

Somatosensory afferent fibers

| | Fiber type | | |
Modality	Group II (new)* [A-β and A-γ]	Group III (old)† [A-δ]	Group IV (oldest)† [C]
Touch	X	X	X
Kinesthesia	X		
Temperature		X	?
Pain		X	X

*Dorsal column system.
†Anterolateral system.

root afferent fibers and the location of their somata are presently unclear, although it is thought that their somata lie in the dorsal root ganglia. The possible functional significance of this is considered later in the discussion of nociception or pain.

Table 9-1 also indicates the relationship among modalities, fiber type, and the location within the spinal cord of the fibers that ascend to the brain. Two main tracts are involved, the dorsal columns that participate in the *dorsal column system* and the anterolateral columns that participate in the *anterolateral system* (Fig. 9-1). Group II fibers are related to the dorsal column system, and fibers of groups III and IV relate to the

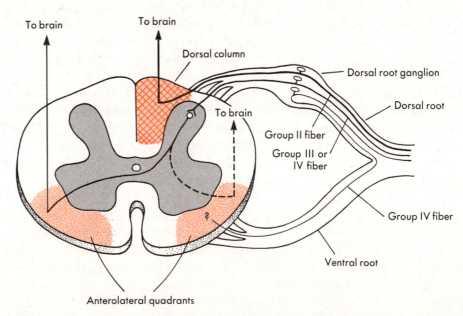

Fig. 9-1 ■ Spinal cord organization of dorsal column and anterolateral systems. Somata of afferent fibers are located in the dorsal root ganglion. Group II afferent fibers enter the cord via the dorsal root. Their branches ascend in the ipsilateral dorsal column to terminate in the dorsal column nuclei of the medulla. Group III and most of group IV afferent fibers also enter the cord via the dorsal root and terminate in the dorsal horn. Secondary neurons project axons across the midline into the contralateral anterolateral quadrant of the spinal cord, although some appear to enter the ipsilateral quadrant. These secondary fibers ascend in the anterolateral quadrants to terminate at many levels of the neuraxis, including brainstem reticular formation and thalamus (see Fig. 9-7). Some group IV afferent fibers enter the cord via the ventral root, and their central connections are presently unknown.

anterolateral system. Consequently touch is the only modality represented by both systems. Kinesthetic information other than that involving muscle spindles and tendon organs is processed almost completely via the dorsal column system. Temperature plus nociception information, however, are processed only via the anterolateral system.

Like the visual system, the somatosensory system can be roughly divided into phylogenetically newer and older components. The dorsal column system is newer than is the anterolateral system. Furthermore, the anterolateral system itself contains older and newer portions. There appears to be a good correlation between fiber size and phylogenetic age such that smaller fibers are older. Touch, then, is represented by older and newer components; kinesthesia is represented only by the newer component, and temperature and nociception are represented by the older one only. However, it is important to recognize the incomplete and sometimes implausible nature of these divisions. For instance, kinesthesia was probably present fairly early in mammalian evolution, yet it is currently seen almost exclusively in the newest dorsal column system.

■ *Dorsal column system*
■ *General anatomical considerations*

Ascending pathways. Fig. 9-2 illustrates the ascending pathways involved in the dorsal column system. A large branch of the primary afferent fiber ascends through the spinal cord in the dorsal columns; other branches are involved in local spinal circuits. Fibers entering at higher spinal segments occupy a progressively lateral position in the dorsal columns so that there is a precise topographical order to these fibers. At higher levels a division of the dorsal columns into two funiculi can be discerned (Fig. 9-2). The more medial *funiculus gracilis* includes fibers from T7 and below; the more lateral *cuneate funiculus* includes fibers from T6 and above. Although not shown in Fig. 9-2, some of the fibers ascending in the dorsal columns originate among neurons in laminae III to V of the dorsal horn; these are sensitive to light touch or hair displacement.

These dorsal column fibers, the vast majority of which are primary afferent fibers, terminate at the level of the lower medulla in either the *nucleus gracilis* or *nucleus cuneatus,* depending on the funiculus in which they ascend (Fig. 9-2). Together these nuclei are called the *dorsal column nuclei*. The secondary fibers issue forth from these nuclei as the *internal arcuate fibers,* which soon cross the midline at the *decussation of the medial lemniscus.* These fibers continue to ascend in the medial lemniscus to their terminus in the *ventralis posterolateralis* (VPL) nucleus of the thalamus. Tertiary fibers then issue from the VPL nucleus to terminate in the somatosensory cortex, mostly in layer IV but also in layer VI. This thalamocortical termination pattern closely resembles that in the visual system.

Historically, two somatosensory areas were defined: SI, or "primary" somatosensory cortex, and SII, or "secondary" somatosensory cortex (Fig. 9-3). SI is located in the postcentral gyrus, and it was thought to be the exclusive terminus of VPL nucleus afferents; SII is located in the parietal cortex just above the sylvian fissure, and it was thought to receive secondary VPL nucleus input from SI. This terminology is outdated by contemporary research for two reasons. First, we know now that SII receives direct VPL nucleus input. Second, recent studies of owl monkeys have established a multiplicity of somatosensory cortical areas, again as in the visual system. SI is actually composed of at least four different zones and SII of at least two. Other separate cortical representations might also be described.

Descending pathways. As in the visual system, the newer somatosensory pathways include a large descending component; that is, corticofugal fibers, originating mostly from the somatosensory cortex, project directly to the thalamic and medullary relay nuclei of the dorsal column system. There is probably also an indirect projection to these nuclei via corticobulbar fibers from the somatosensory cortex to the brainstem reticular formation. The functional significance of this descending component has yet to be elucidated.

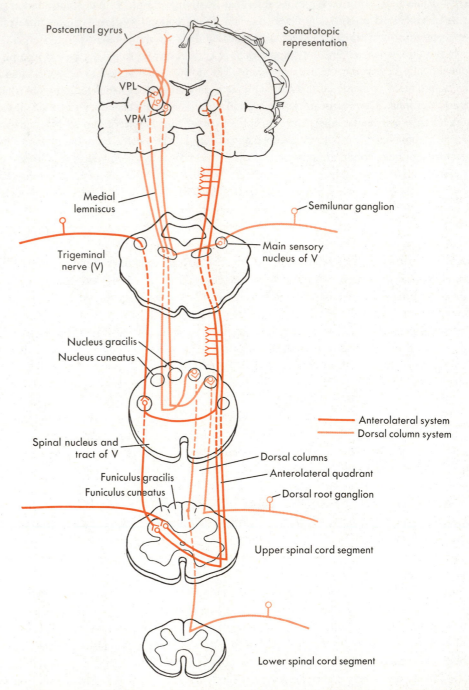

Fig. 9-2 ■ Schematic representation of some central connections in the dorsal column and anterolateral systems. Both trigeminal and spinal components are shown. The dorsal column system is shown in lighter color, and the anterolateral system is shown in darker color. Note the somatotopic arrangement throughout the neuraxis for the dorsal column system. Fibers that enter dorsal roots at higher spinal segments ascend in the dorsal columns more laterally, and this relationship is preserved throughout; furthermore, the trigeminal component merges with the spinal component in the medial lemniscus next to fibers that represent the highest cervical spinal segments. Consequently an accurate map of the body surface exists throughout, including the cortical areas of the postcentral gyrus. *VPL*, Ventralis posterolateralis nucleus; *VPM*, ventralis posteromedialis nucleus.

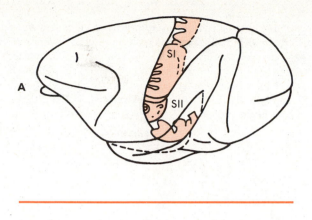

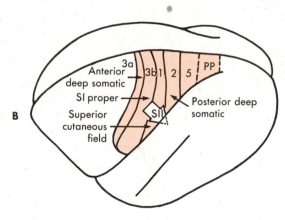

Fig. 9-3 ■ Organization of somatosensory cortical areas in primates. This representation is derived from studies of the owl monkey. **A,** Older, traditional view with a primary *(SI)* and secondary *(SII)* area. **B,** Contemporary view with at least four and probably more separate areas. The four definite areas are the anterior deep somatic area (area *3a*), SI (area *3b*), the posterior cutaneous area (area *1*), and the posterior deep somatic area (area *2*). Areas *SII, 5,* and the posterior parietal *(PP)* cortex probably each represent more than one somatosensory area. (Redrawn from Merzenich, M.M., and Kaas, J.H.: Prog. Psychobiol. Physiol. Psychol. **9:**1, 1980.)

Many neurons throughout the dorsal column system are sensitive to light touch applied to the skin or to pressure at certain deep tissues, for example, fascia, joint capsules, periosteum, and ligaments. We know most about those related to the skin because it is easier to study receptive fields on the body surface than to study those related to deep tissues.

Spatial resolution for touch sensitivity can be defined as the minimum separation needed for two stimuli to be distinguished from a single, more intense stimulus. This refers, for example, to how closely applied to the skin two pinpricks can be to one another and still be distinguished from a single prick. Two-point discrimination varies with position on the body's surface from roughly 2 mm for the fingertips to roughly 40 mm for the back. As expected from knowledge of the visual system, receptive field size and afferent inhibition both play roles in such discrimination. Cutaneous receptive fields are smaller where discrimination is more acute (e.g., the fingertips) and larger where discrimination is poorer (e.g., the surface of the trunk); this is analogous to the smaller visual receptive fields that occur as acuity increases nearer the foveal representation. Furthermore, whereas primary cutaneous afferent fibers do not display lateral inhibition, neurons of the dorsal column nuclei, VPL nucleus, and somatosensory cortex do exhibit

■ *Touch sensation*

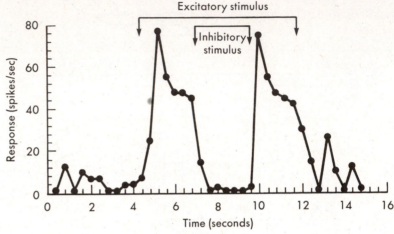

Fig. 9-4 ■ Afferent inhibition in a somatosensory neuron of the postcentral gyrus in a monkey. On the left is the receptive field with an excitatory center *(darker color)* surrounded by an inhibitory region *(lighter color)*. Responses are shown on the right to cutaneous stimuli applied to the receptive field. When a stimulus is applied to the center, the cell discharges. Stimulation of the surround silences the cell even in the presence of an excitatory stimulus. (Redrawn from Mountcastle, V.B., and Powell, T.P.S.: Johns Hopkins Med. J. **105**:201, 1959. Copyright 1959 by The Johns Hopkins University Press.)

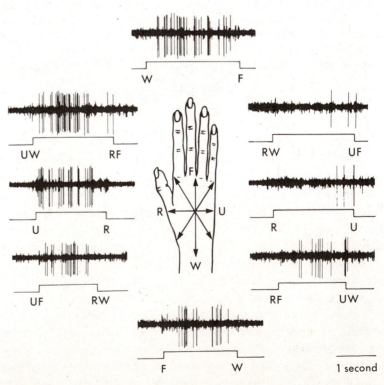

Fig. 9-5 ■ Direction selectivity in a somatosensory neuron of the postcentral gyrus in a monkey. The neuron is most responsive to cutaneous stimuli moving along the axis *UW* to *RF* and least responsive along the axis *RW* to *UF*. (From Costanzo, R.M., and Gardner, E.P.: J. Neurophysiol. **43**:1319, 1980.)

such inhibition (Fig. 9-4); that is, stroking the cutaneous receptive field excites such a neuron for some central cutaneous zone and inhibits it for a surrounding annular zone. This arrangement seems analogous to the center-surround receptive fields of retinal ganglion and lateral geniculate cells.

Another analogy between the geniculocortical and dorsal column systems concerns the specificity of adequate stimuli to excite the neuron. For primary afferent fibers, cells of the dorsal column nuclei, and VPL neurons, only the position of the stimulus is crucial to the cell's response, and this is similar to the limited specificity of retinal and geniculate cells. However, many cells in the somatosensory cortex exhibit a dramatic direction selectivity (Fig. 9-5); that is, they respond when a cutaneous stimulus, such as a brush, is moved in one direction across the skin but not for the reverse or orthogonal directions. This is analogous to the direction and orientation selectivity seen first in the geniculocortical system among cortical neurons.

Many touch-sensitive neurons in the dorsal column system respond tonically to touch or pressure, much like the responses of many neurons in the newer, geniculocortical portion of the visual system. These tonic responses are characterized by a monotonic relationship between firing rate and touch intensity. This applies to neurons throughout the system, including primary afferent neurons (Fig. 6-4) and cortical neurons.

■ *Kinesthesia*

Neurons sensitive to joint position or angle have also been studied throughout the dorsal column system. They generally exhibit tonic responses that encode joint angle in the frequency of firing (Fig. 9-6, *A*). As Fig. 9-6, *B*, shows, some of these cells increase firing with increased flexion, and others increase firing with increased extension.

The plots in Fig. 9-6, *B,* should not be mistaken for receptive fields. They merely represent the relationship between stimulus intensity and response, and they are strictly analogous to the example in Fig. 6-4. A change in joint angle changes the pressure on related tissues (joint capsules, ligaments, etc.), and thus it changes the intensity of the stimulus at the transducer.

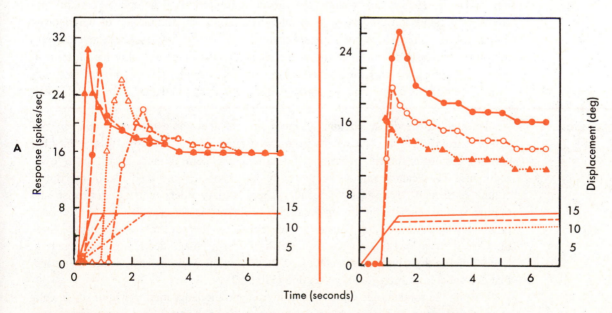

Fig. 9-6 ■ Response properties of kinesthesia-sensitive neurons. **A,** Responses of primary afferent fibers to flexion of the knee of a cat. As shown, the magnitude of the sustained response encodes the extent of flexion, whereas the rate of flexion determines the rate of rise of the response.

Continued.

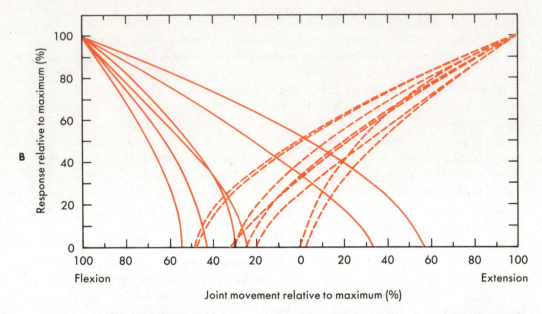

Fig. 9-6, cont'd ■ **B,** Response ranges of 14 separate neurons in the monkey's thalamic VPL nucleus. Each neuron responds monotonically either to increasing flexion or extension of the appropriate joint, but never to both. Note that most neurons respond to stimuli that correspond to half or more of the total range of joint movements. These curves illustrate the frequency code by which neuronal discharges represent joint position. (**A** redrawn from Boyd, I.A., and Roberts, T.D.M.: J. Physiol. [Lond.] **122:**38, 1953; **B** redrawn from Mountcastle, V.B., et al.: J. Neurophysiol. **26:**807, 1963.)

Joint angle thus has little to do with the concept of a receptive field for such a cell. Rather, as defined in Chapter 7, the receptive field of such a cell is determined by the area of appropriate transducer-containing tissue. One might refer to the receptive field of such a cell as "elbow" as opposed to "wrist" or "shoulder." Afferent inhibition, if it existed in such a cell, would be seen not as inhibition of the cell by certain joint angles, but rather it would be seen as inhibition caused by stimulation of neighboring joints. To our knowledge such an experiment has yet to be described, but afferent inhibition for these neurons may be unnecessary. The brain presumably has little difficulty determining which joint has been stimulated, and spatial resolution is of less concern in kinesthesia than is the frequency code that is used to analyze joint angle.

■ *Somatotopic organization*

Throughout the dorsal column system a precise neurotopic or *somatotopic organization* exists (Fig. 9-2). This results from two factors. First, each primary afferent fiber entering the dorsal funiculi of the spinal cord ascends lateral to those arising from lower segments. Thus caudal regions of the body are mapped more medially in the dorsal columns. Second, the primary, secondary, and tertiary fibers maintain a precise and regular spatial relationship among themselves (Fig. 9-2). An orderly map of the body, or *homunculus* (i.e., "little human"), thus exists for the somatosensory cortex. As is the case with vision, this map is distorted by differing magnification factors that relatively enlarge the neural representation of those somatic regions, such as the fingertips, that are capable of high spatial resolution (Fig. 9-2).

Despite the somatotopic organization, there is no modality mixing within single cells at any level of the system from the dorsal funiculi of the spinal cord to the somatosensory cortex; that is, the cells respond either to touch or to joint angle changes but never to both. Because of the somatotopy, kinesthesia and touch-sensitive cells that map similar body parts intermingle. For instance, a neuron that maps the skin over the elbow would be found near one that responds to elbow extension or flexion. This is another example of cells being responsive to a single modality, a feature that seems essential to the ability of the central nervous system to identify the nature of the stimulus.

Mountcastle et al. provided the first physiological evidence for columnar organization of cortex similar to that already described for visual cortex. The evidence was derived from receptive field studies of cells in the postcentral gyrus. These cells are organized into columns from pia to white matter. Cells in each column share many functional features with one another, including modality. Thus columns of cells sensitive to the angle of a particular joint are found intermingled among cell columns sensitive to touch of skin near that joint. However, more recent evidence suggests that submodalities are represented by adjoining slabs of neurons running for several millimeters across the cortex rather than by much smaller discrete columns.

■ *Cortical columnar organization*

The anterolateral system is phylogenetically older than is the dorsal column system. However, the former can be subdivided into newer and older portions that relate to innervation by group III and group IV afferent fibers, respectively. The anterolateral system processes information concerning touch, temperature, and pain. Because of its unique importance, pain is treated separately later in this chapter.

■ *Anterolateral system: touch and temperature*

Fig. 9-1 illustrates the intraspinal circuitry of the anterolateral system. Primary afferent fibers enter lamina V of the dorsal horn of the spinal cord and synapse onto neurons there. In the newer portion of this system, the secondary neuron usually projects an axon across the midline to the anterolateral quadrant of the spinal cord where it ascends to brainstem levels. In older portions there may be interneurons in the dorsal horn so that the ascending fiber is of a higher order than secondary, or in rare cases the fiber may ascend in the anterolateral column ipsilateral to the dorsal horn synapses (Fig. 9-1).

Fig. 9-7 illustrates the diverse brainstem sites innervated by these ascending fibers.

■ *General anatomical considerations*

Fig. 9-7 ■ Schematic representation of terminal zones of fibers ascending in the anterolateral quadrant of the spinal cord. Terminal zones include the medullary, pontine, and mesencephalic reticular formation plus certain areas of the thalamus. Most thalamic terminals are found in midline nuclei, but some reach the VPL and VPM nuclei. Roman numerals, cranial nerve nuclei; *IC*, inferior colliculus; *SC*, superior colliculus; *RF*, reticular formation; *CL*, central lateral nucleus of the thalamus; *VB*, ventrobasal complex (VPL and VPM) of the thalamus. (Redrawn from Mountcastle, V.B., editor: Medical physiology, ed. 14, St. Louis, 1980, The C.V. Mosby Co.; courtesy W.R. Mehler.)

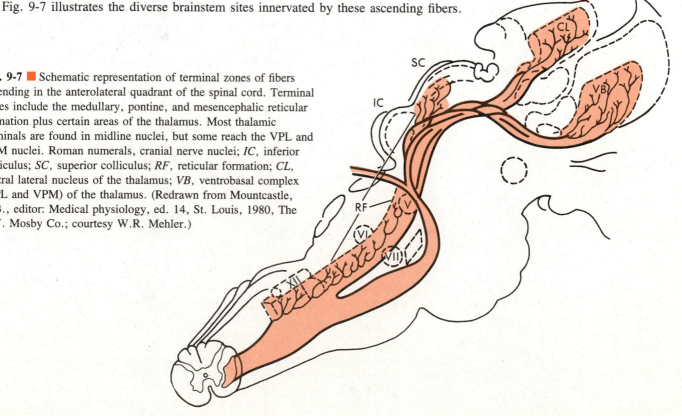

These sites include the medullary and pontine reticular formation, the superior colliculus, the central gray matter of the mesencephalon, some midline thalamic nuclei (such as the nucleus centralis lateralis), and the VPL nucleus of the thalamus. This suggests that the system can be divided into *spinobulbar* (spinoreticular, spinotectal, etc.), *paleospinothalamic* (e.g., involving midline thalamic nuclei), and *neospinothalamic* (involving VPL nucleus) tracts, from oldest to newest. This may be a useful functional division, but there seems to be little evidence for the frequent claim that these tracts are anatomically separated within the anterolateral quadrants of the spinal cord.

■ *Touch sensation*

The neospinothalamic pathway that ascends to the VPL nucleus joins the fibers of the medial lemniscus en route. Because all VPL neurons share basic receptive field features (such as small fields, afferent inhibition, and precise somatotopic organization), touch-sensitive fibers of the neospinothalamic pathway seem to have properties identical to those of fibers emerging from the dorsal column nuclei. Central to their merger in the medial lemniscus, the two populations of fibers from the dorsal column and anterolateral systems are no longer distinguishable.

Little is known about the functional organization of the older portions of the anterolateral system. Limited data, mostly from recordings of cutaneous receptive fields in the midline thalamic nuclei, suggest detailed similarities with older portions of the visual system. These cutaneous receptive fields are large and diffusely organized and lack lateral inhibition. Furthermore, the neurons respond phasically to tactile stimulation and are not arranged in any obvious somatotopic order.

■ *Temperature sensation*

Little is known about the neural substrate of temperature sensation. Recordings of peripheral nerves indicate that many group III afferent fibers have small cutaneous receptive fields that are insensitive to mechanical deformation but respond tonically to small temperture changes in the skin. Fig. 9-8 illustrates the responses of typical thermosensitive fibers. Within a range of roughly 20° to 40° C, these fibers increase their firing rate monotonically with either increasing temperature ("warm" fibers) or decreasing temperature ("cool" fibers). There are roughly 5 to 10 times as many cool fibers as warm fibers. Interestingly, at temperatures below 20° and above 40° C most of these fibers cease responding. At such extreme temperatures, tissue damage will begin, and a different set of neurons that uniquely signal nociception become responsive.

Reports of temperature-sensitive neurons in the brain have been rare. Occasional temperature-sensitive units have been found in the anterolateral quadrants of the spinal cord, VPL nucleus of the thalamus, and somatosensory cortex. Temperature sensation

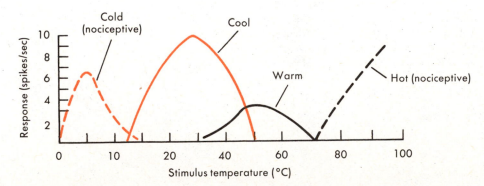

Fig. 9-8 ■ Examples of response of various primary afferent fibers to stimuli of different temperatures. The solid curves represent responses of fibers sensitive to cool *(color)* or warm *(black)* stimuli. The dashed curves represent responses of nociceptive fibers to cold *(color)* or hot *(black)* stimuli. (Modified from Zotterman Y.: Annu. Rev. Physiol. **15:**357, 1953.)

thus seems to be at least partly processed via the neospinothalamic pathways, but a great deal more data are needed. Finally, some indirect evidence, based on differential nerve blockades of myelinated and unmyelinated peripheral fibers, suggests that some group IV afferent fibers might contribute to temperature sensation.

Pain is obviously an important subject to understand clinically. Unfortunately we know surprisingly little about the neural substrates for pain sensation. This is partly a result of the complex and poorly understood central pathways that mediate pain and partly a result of the difficulty in defining pain. Pain is an emotional and changeable perception. What is painful for one person at one time may not be at another time, or it may never be painful for another person. The battlefield has provided many dramatic examples of the difficulty in defining pain. Many soldiers have experienced horrible wounds, but in the excitement of battle, many claim awareness of the injury but experience no great pain. Pain typically begins much later when the excitement associated with the battle has ceased.

For these reasons it is easier to define and describe the concept of *nociception* than to deal with the qualitative nature of pain. Nociception is the sensation associated with tissue damage, and it is the sensation that leads to the percept of pain. Hardy et al. rather elegantly demonstrated a precise relationship between nociception and pain. It is illustrated by Fig. 9-9, which plots the rate of heat transfer to the skin necessary to elicit pain as a function of initial skin temperature. The linear relationship intercepts the abscissa at approximately 45° C. If the skin were maintained at or above this temperature, irreversible tissue damage would soon result. The sensation of pain thus relates rather precisely to actual or threatened tissue damage. In the discussion that follows we shall use the terms *nociception* and *pain* interchangeably.

■ *Anterolateral system: pain or nociception*

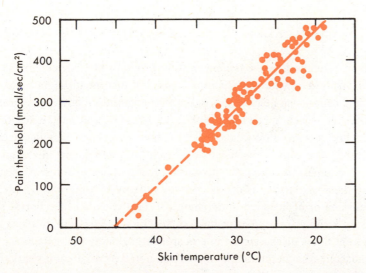

Fig. 9-9 ■ Relationship between skin temperature of forehead and rate of heat application needed to produce a painful sensation in normal humans. The linear relationship intercepts the abscissa at roughly 45° C. (Redrawn from Hardy, J.D., et al.: Pain sensations and reactions, Baltimore; © 1952, Williams & Wilkins Co.)

Before what is known of the central nociceptive pathways is described in detail, it is useful to consider two general theories regarding the neural basis of pain. The first is the *specificity theory* that nociception is processed as any other submodality of somesthesia; that is, specific nociceptive transducers exist that respond only to nociceptive stimuli, and they are functionally connected to specific neural pathways. Nociception

■ *Specificity and pattern theories of pain*

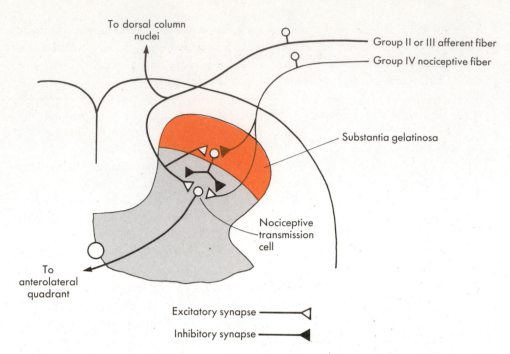

To dorsal column
nuclei

Group II or III afferent fiber

Group IV nociceptive fiber

Substantia gelatinosa

Nociceptive
transmission
cell

To
anterolateral
quadrant

Excitatory synapse ▷

Inhibitory synapse ▶

Fig. 9-10 ■ Schematic representation of one version of the ''gate control'' theory of pain. If the group IV nociceptive fiber activates the transmission cell, pain will be felt. However, a gating mechanism operates through an interneuron in the substantia gelatinosa *(color)* whereby transmission in the nociceptive pathway can be blocked via activation of certain large afferent fibers. These afferent fibers excite the interneuron, which in turn presynaptically inhibits all inputs to the transmission cell.

thus has its own transducers and ''labeled lines,'' much like kinesthesia or touch sensation.

The second hypothesis is the *pattern theory*. It argues that nociception shares transducers and/or pathways with other submodalities, but that the different pattern of activity in the same neuron can be used to signal either nociception or a nonpainful stimulus. For example, light touch applied to the skin might cause transducers and their central pathways to fire at some low frequency, but intense pressure that damages the skin might cause the same cells to discharge at a much higher frequency that signals pain.

Fig. 9-10 summarizes a popular theory that incorporates features of specificity and pattern. The theory is known as the *gate control* theory of pain, and it states that synaptic transmission of nociceptive information can be ''gated'' in the dorsal horn of the spinal cord by activity from other pathways. Specifically, the activity of large somatosensory afferent fibers discharges an interneuron in the substantia gelatinosa, which, in turn, can cause presynaptic inhibition of smaller, nociceptive afferent fibers (Fig. 9-10). Activity in the second-order neuron of the nociceptive pathway can thus be modified by activity in other somatosensory pathways. Theories such as this one have been influential in our thoughts about pain mechanisms. They also serve as plausible explanations for certain phenomena. For instance, minor cutaneous pain often can be abolished by rubbing the affected skin area; that is, rubbing activates the large somatosensory afferent fibers to inhibit transmission of the nociceptive afferent fibers. Also, some have attempted to explain the effects of acupuncture by these theories. There is, however, little direct experimental evidence to support such theories.

Most relevant data point rather convincingly toward the simpler specificity theory. Neurophysiological studies have demonstrated afferent fibers and central neurons that are specifically activated by nociceptive stimuli. Furthermore, many other somatosensory neurons, for example, those sensitive to touch and temperature, actually cease responding when stimulus levels become intense enough to cause tissue damage (Fig. 9-8). Thus our ensuing description of nociceptive pathways will treat these as submodalities, like those for touch or temperature, with unique transducers and labeled lines to the brain.

■ *Peripheral mechanisms*

Afferent fiber types. Nerve block experiments and neurophysiological studies of single neurons both demonstrate that nociception is conveyed centrally by group III and IV afferent fibers. Roughly one fifth of all group III and half of all group IV afferent fibers are nociceptive. All of the group III afferent fibers enter the spinal cord via the dorsal root, but many of the unmyelinated nociceptive afferent fibers probably enter the cord via the ventral root. In the framework of our description of sensory systems, it is likely that the nociceptive pathways include phylogenetically very old (group IV afferent fibers) and moderately old (group III afferent fibers) components.

■ *Fast and slow pain*

Two qualitatively different types of pain can be readily appreciated. These are termed *fast* and *slow pain*. Fast pain is a short, well-localized sensation that is well matched to the stimulus. Examples are a pinprick and strong pinch. Such stimuli are usually distinguishable from one another. The sensation starts and stops abruptly when the stimulus is applied and removed, respectively. Slow pain is a throbbing, burning, or aching sensation that is poorly localized and less specifically related to the stimulus. One cannot usually ascribe a specific nociceptive stimulus (heat, cutting, pinching, etc.) to this sensation. Also, the onset of the sensation has a long latency following application of the stimulus, and this pain continues for hours or days after removal of the stimulus. Fast pain is conveyed by group III afferent fibers and is associated strictly with the skin. Slow pain is conveyed by group IV afferent fibers and involves both cutaneous and deep tissues such as internal organs, joints, and muscles.

■ *Transduction mechanisms*

Our understanding of the transduction process in nociception is far from complete or satisfactory, and the following should be considered as mostly speculative and a basis for further research.

Fast pain. The transducers for fast pain can be considered to be mechanical, thermal, or chemical. These transducers employ the same mechanisms as their nonnociceptive counterparts, but with higher thresholds that require nociceptive stimuli. This would explain the relatively close relationship between the fast pain perceived and the nature of the stimulus.

Slow pain. Fig. 9-11 outlines our current understanding of the less direct transduction process for slow pain. A nociceptive stimulus causes tissue damage that leads to cell death. As some of these cells die, they release certain proteolytic enzymes into the interstitial fluid, and these enzymes cleave certain γ-globulins circulating in the blood. The result is a variety of short-chain polypeptides, similar to and perhaps including *bradykinin,* in the vicinity of the injury. Some of these polypeptides then activate the group IV nociceptive afferent fibers, which can thus be considered chemical transducers, in the same manner that neurotransmitters lead to excitatory postsynaptic potentials (EPSPs) in central neurons. A variety of indirect data is consistent with this hypothesis. For instance, injection of bradykinin or the contents of blister fluid into healthy skin evokes the sensation of slow pain. Also, analysis of blister fluid and certain insect bite venoms has revealed the presence of bradykinin-like polypeptides.

This hypothetical series of processes explains many of the phenomena peculiar to slow pain. The long latency of the sensation can be explained by the time needed for cell death and for the series of chemical reactions that lead to bradykinin-like substances. The prolonged sensation following removal of the stimulus can be explained

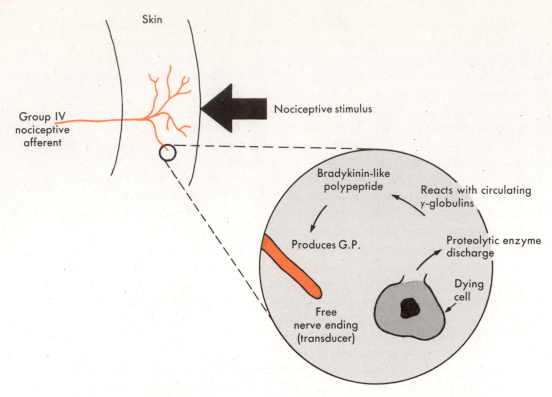

Skin

Group IV
nociceptive
afferent

Nociceptive stimulus

Bradykinin-like
polypeptide

Reacts with circulating
γ-globulins

Produces G.P.

Proteolytic enzyme
discharge

Dying
cell

Free
nerve ending
(transducer)

Fig. 9-11 ■ Hypothetical schema to suggest transduction mechanisms for slow pain. A nociceptive stimulus applied near the free nerve endings of a group IV nociceptive afferent fiber *(color)* causes tissue damage and cell death. As some cells die, they discharge proteolytic enzymes that react with circulating γ-globulins to produce bradykinin-like polypeptides. The free nerve ending is a chemical transducer sensitive to these polypeptides; consequently a generator potential *(GP)* is evoked.

by the continued presence of these substances. Finally, the lack of differential sensations for different nociceptive stimuli (heat, mechanical, etc.) would result from the indirect response of group IV afferent fibers to cell death (and release of proteolytic enzymes, etc.), regardless of which specific nociceptive stimulus caused it.

■ *Central pathways*

We know distressingly little about the central pathways that represent nociception. The diffuse connections, small neurons and fibers, and difficulty in defining painful or nociceptive stimuli render this a particularly difficult problem for experimental analysis. Indeed, much of our understanding stems from clinical observations, particularly from surgical attempts to alleviate intractable pain by interruption of various neuronal pathways.

Some of the nociceptive information is conveyed via the anterolateral system (Fig. 9-7). A small number of axons from the secondary neurons in lamina V of the dorsal horn ascend ipsilaterally rather than contralaterally in the anterolateral columns. There is some reason to believe that a disproportionately large number of these axons are nociceptive rather than touch or temperature sensitive. Nociceptive neurons have proven elusive during attempts to study them neurophysiologically. Limited data from animals suggest that such neurons are found in the anterolateral quadrants of the spinal cord, various brainstem reticular nuclei, and certain midline thalamic nuclei. In humans electrical stimulation of these sites during explorative neurosurgery frequently leads to a painful sensation in the patient. However, because of limited data all of these assignations are speculative. Such nociceptive neurons have not yet been described for either the VPL nucleus or any area of the cerebral cortex.

Unfortunately there may be many pathways other than the anterolateral system that process nociceptive information. For example, many group IV afferent fibers enter the spinal cord via the ventral root, and we presently know virtually nothing regarding the central connections related to this afferent pathway. It is likely that many of these afferent fibers are nociceptive.

The ubiquity of the nociceptive pathways is evident from attempts to interrupt them in the treatment of intractable pain such as often occurs in terminal cancer patients. If the suffering is intense and drugs no longer provide relief, often an attempt will be made to destroy the presumptive nociceptive cells or fibers. The most common procedure is to cut through the anterolateral quadrant of the spinal cord at the appropriate spinal level, thereby interrupting fibers ascending in the anterolateral funiculus as well as causing other damage. Virtually every level of the neuraxis shown in Fig. 9-7 has been interrupted in some patient, and overall the results have been disappointing. Relief, if it occurs at all, is usually incomplete and temporary. This is true even when the appropriate dorsal roots have been sectioned, presumably because of the course taken by many group IV afferent fibers through the ventral roots. Apparently nociceptive pathways are so diffuse and redundant that it is impractical to achieve permanent analgesia surgically.

Often pain is sensed in a region of the body that is inappropriate to the nociceptive stimulus. This is known as *referred pain* because the pain seems to be referred from the damaged region to healthy tissue. Most commonly such pain may be sensed in the skin when internal organs become damaged.

■ *Referred pain*

The explanation for pain referral is not understood, but it is an important clinical phenomenon that generally follows a basic rule called the *dermatome rule;* that is, the area of skin in which pain is felt is usually innervated by the same spinal segment as is the affected organ. Thus heart attack victims often feel cutaneous pain radiating from the left shoulder down the arm because sensory innervation from these skin regions and the affected cardiac tissue pass through the same spinal segments. One should always be suspicious of an internal pathological condition when cutaneous pain is inexplicable. The dermatome rule can be used as a guide to locate the source.

The dermatome rule offers a possible explanation for referred pain, as illustrated in Fig. 9-12. Although most nociceptive afferent fibers have unique pathways to the brain

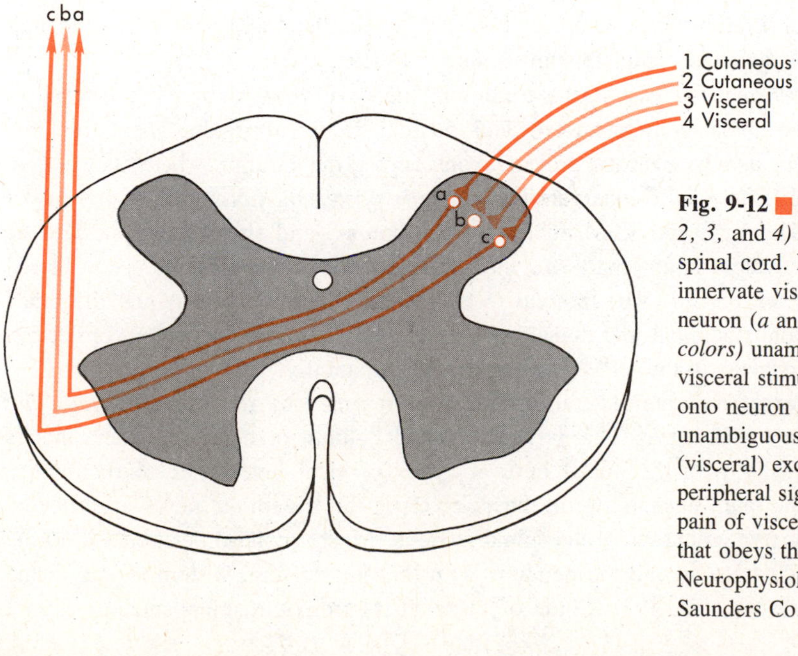

1 Cutaneous
2 Cutaneous
3 Visceral
4 Visceral

Fig. 9-12 ■ Referred pain patterns. Four nociceptive afferent fibers *(1, 2, 3,* and *4)* innervate three secondary neurons *(a, b,* and *c)* in the spinal cord. Fibers *1* and *2* innervate the skin, and fibers *3* and *4* innervate visceral structures. Fibers *1* and *4* each innervate a single neuron *(a* and *c,* respectively). Activity along these pathways *(darker colors)* unambiguously reflects cutaneous stimulation for neuron *a* and visceral stimulation for neuron *c.* However, fibers *2* and *3* converge onto neuron *b (lighter colors)* so that activity of neuron *b* cannot be unambiguously interpreted by the brain. Frequently even if fiber *3* (visceral) excites neuron *b,* the perception is one of pain as if the peripheral signals were transmitted along fiber *2* (cutaneous). Thus pain of visceral origin is often referred to a cutaneous site in a pattern that obeys the dermatome rule. (Modified from Ruch, T.C., et al.: Neurophysiology, ed. 2, Philadelphia, 1965, W.B. Saunders Co.)

(represented by afferent fibers *1* and *4* and neurons *a* and *c* in Fig. 9-12), perhaps some cutaneous and visceral afferent fibers (*2* and *3*) converge on the same second-order neuron (*b*). Activity of neuron *b* cannot distinguish between cutaneous and visceral pain. However, in most persons cutaneous pain is more often experienced than is visceral pain, and the brain might simply have learned to interpret activity in neuron *b* as due to activation of the cutaneous afferent fiber (*2*). When neuron *b* is activated by the visceral afferent fiber (*3*), the brain would therefore misinterpret this as cutaneous in origin. This is a plausible explanation of referred pain and of the dermatome rule, since such convergence of afferent fibers is most probable for fibers innervating the same spinal segment. Nonetheless, there is no direct evidence to support this hypothetical neural basis of referred pain.

■ Action of aspirin and morphine

Probably the two most widely used treatments for pain are aspirin and morphine and their related compounds. Their effects and sites of action are quite different. Aspirin acts peripherally, probably at the level of transduction, to minimize or eradicate the afferent nociceptive signal. Aspirin is thus a true analgesic, since it affects the entire sensation of pain. Morphine acts centrally at many sites in the central nervous system to reduce the suffering caused by the pain. Unlike aspirin, morphine works not as an analgesic to block nociception but at a higher level to change the psychological reaction to the painful stimulus.

In recent years considerable attention has been focused on the discovery that many neurons throughout the mammalian central nervous system have receptors for morphine. These cells are thus probably sensitive to morphine, and they may play a role in the effect morphine has on the reaction to nociceptive stimuli. Not only have such receptors been described, but endogenous morphinelike substances, called *endorphins,* have been discovered throughout the brain. Perhaps these endorphins represent a natural means of controlling or altering the qualitative sensations related to pain. This is an area of particularly active research that could lead to a dramatic breakthrough in our understanding of how pain is processed by the central nervous system.

■ *Trigeminal pathways*

The functional organization of somesthesia for the head is virtually the same as that for the body with the exception that the afferent fibers have cell bodies in the semilunar ganglion, and most enter the neuraxis via the trigeminal nerve; a few enter via the facial and vagus nerves. The trigeminal system is organized into analogues of the dorsal column and anterolateral systems. These pathways from the head and body come together to form a continuous and complete somatotopic map (Fig. 9-2).

The *main sensory nucleus* of the trigeminal nerve is located in the pons and is analogous to the dorsal column nuclei. This nucleus receives afferent fibers from the trigeminal nerve, and its neurons project axons across the midline where they merge with the medial lemniscus to innervate the *ventralis posteromedialis* (VPM) nucleus of the thalamus. The VPL and VPM nuclei are contiguous, and since they appear to be functionally equivalent counterparts for the head and body, they are often considered together as the *ventrobasal* (VB) *complex* (VB = VPL + VPM). One minor difference between the main trigeminal and dorsal column systems is that a very small projection from the main nucleus to the VPM nucleus exists ipsilaterally.

The analogue of the anterolateral system is represented by the *spinal nucleus and tract* of the trigeminal nerve. This is an elongated column of cells and fibers that is continuous caudally with the dorsal horn at upper cervical levels and nearly merges rostrally with the main nucleus. Some fibers entering the trigeminal nerve are directed caudally and innervate neurons of the spinal nucleus. Most second-order fibers from the spinal nucleus cross to ascend the neuraxis with the anterolateral system to end in the same brainstem sites (Fig. 9-7). Some of these fibers from the spinal nucleus ascend ipsilaterally.

The functional divisions in the trigeminal system are thought to mimic those of the ascending spinal pathways. The main nucleus is innervated only by large myelinated fibers that convey information about touch and kinesthesia. The spinal nucleus receives smaller myelinated and unmyelinated fibers that signal touch, temperature, and nociception.

As is the case with the spinal components of the somatosensory system, the somatosensory cortex projects fibers to the trigeminal components. Heavy direct projections have been described to the VPM nucleus and to the main and spinal trigeminal nuclei. Indirect projections via brainstem reticular pathways probably exist as well.

■ *Functional significance of old and new pathways*

In the visual system the newer geniculostriate portion may not be the sine qua non for vision that is often claimed. The same principle applies to the somatosensory pathways. For instance, lesions of the dorsal funiculi of the spinal cord (and thus the newest position of the pathways) lead to only a transient and partial loss of touch sensitivity. (Kinesthesia is permanently and nearly totally lost after such a lesion because this submodality is poorly represented in older portions of the system.) Likewise, lesions of the somatosensory cortex produce a temporary anesthesia with a gradual return of sensation that is only subtly abnormal. One must thus keep in mind the principle that somesthesia results from complex interactions among the various somatosensory pathways, old and new.

The somatosensory submodalities of touch sensation, kinesthesia, temperature sensation, nociception or pain, and perhaps vibration sensation can be distinguished. Each has its own transducers and labeled lines or neuronal pathways. These submodalities are conveyed to the central nervous system by large myelinated group II fibers (touch and kinesthesia), small myelinated group III fibers (touch, temperature, and pain), and small unmyelinated group IV fibers (touch, pain, and perhaps temperature). The central pathways can be divided roughly into phylogenetically newer and older portions. The former are represented by the dorsal column system and main sensory trigeminal nucleus; the latter are represented by the anterolateral system and spinal trigeminal nucleus. The newer pathways begin with group II afferent fibers, whereas the older ones receive smaller afferent fibers. Generally the newer portions display a high degree of somatotopic organization, small receptive fields with lateral inhibition, and many neurons that respond tonically to sensory stimuli. The older pathways lack these features. Older and newer pathways combine in the processing of somesthesia, and the newer pathways appear to be uniquely necessary and important only for the processing of kinesthesia.

In general, the functional organization of the somatosensory pathways shares much in common with that of the visual pathways. Both systems have older and newer components with similar distinguishing characteristics. It is interesting to consider a more detailed comparison, and as an example we shall consider comparisons among the newer portions: the retinogeniculostriate pathways for vision and the dorsal column system and neospinothalamic pathways for somesthesia. Fig. 9-13 outlines this comparison. Because neurons analogous to retinal horizontal and amacrine cells are not evident in somesthesia, these cells are omitted from Fig. 9-13. The somatosensory transducers are both represented by specialized cells that synapse onto the primary afferent fiber (e.g., the tonically responsive transducer described in Chapter 6). In all three pathways, *cell 1* is a transducer (photoreceptor or tonically responsive transducer) that synapses onto a bipolar neuron *(cell 2)*. The bipolar neuron then synapses onto *cell 3* (ganglion cell, dorsal column nucleus cell, or dorsal horn cell), which projects an axon to the thalamus (LGN or VPL nucleus). This projection is such that the sensory environment is mapped onto the contralateral thalamus. Thus the axon from each somatosensory *cell 3* crosses the midline, and that from the retinal ganglion cell does or does not cross in the optic chiasm so that each hemifield is contralaterally mapped in the LGN. The thalamocortical

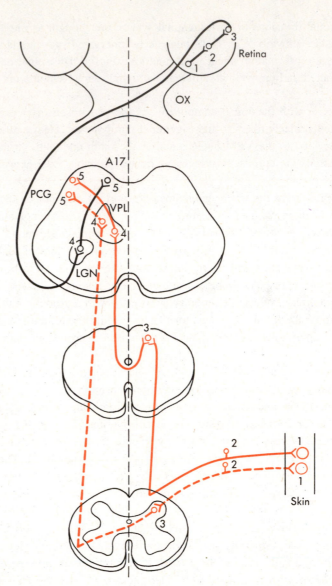

Fig. 9-13 ■ Some similarities among the visual pathways *(solid black)*, dorsal column somatosensory pathways *(solid color)*, and newer portion of the anterolateral somatosensory pathways *(dashed color)*. In all pathways transduction begins with a specialized cell (cell *1*, a rod or cone for vision or a specialized mechanoreceptor in the skin) that synapses onto a bipolar-type neuron (cell *2*, a bipolar cell in retina or primary afferent fiber in the skin). The fiber from cell *2* synapses onto cell *3* (retinal ganglion cell, dorsal column nucleus cell, or dorsal horn cell). In all systems the fiber from cell *3* crosses the midline if needed to map sensory space onto the contralateral half of the neuraxis. Thus both somatosensory fibers decussate, and the ganglion cell axon decussates in the optic chiasm *(OX)* if it derives from nasal but not temporal retina. The fibers from cell *3* terminate onto cell *4* in the thalamus *(VPL* or *LGN)*, and cell *4* projects to the appropriate area of cortex, either area 17 *(A17)* or the postcentral gyrus *(PCG)*. Cell *5* in cortex represents the first stage of direction or orientation specificity in response to stimulus parameters.

projection from *cell 4* terminates mostly in layer IV but also in layer VI of the appropriate cortex. The cortical neurons *(cell 5)* are functionally arranged in columns perpendicular to the layering. Throughout each system neurotopic organization is evident, and the sensory maps are distorted to magnify areas of higher acuity. Neurons display lateral inhibition at the most peripheral site possible (i.e., the first site of synaptic interactions); this occurs in *cell 2* of the visual system (because of the action of horizontal cells that have no somatosensory counterpart) but not until *cell 3* of the somatosensory system. More specific stimulus requirements, such as the shape and direction of movement, are first seen among the cortical neurons *(cell 5)*.

It should be clear that, except for certain anatomical details, the basic organization of the central pathways that underlie vision are remarkably similar to those related to somesthesia. It is these common principles that are most fundamental to sensory processing and, therefore, should be understood most clearly.

■ Bibliography

Journal articles

Bennett, R.E., Ferrington, D.G., and Rowe, M.: Tactile neuron classes within second somatosensory area (SII) of cat cerebral cortex, J. Neurophysiol. **43**:292, 1980.

Blair, R.W., Weber, R.N., and Foreman, R.D.: Characteristics of primate spinothalamic tract neurons receiving viscerosomatic convergent inputs in T_3-T_5 segments, J. Neurophysiol. **46**:797, 1981.

Boivie, J.: An anatomical reinvestigation of the termination of the spinothalamic tract in the monkey, J. Comp. Neurol. **186**:343, 1979.

Coggeshall, R.E., Coulter, J.D., and Willis, W.D.: Unmyelinated axons in the ventral roots of the cat lumbosacral enlargement, J. Comp. Neurol. **153**:39, 1974.

Coggeshall, R.E., and Ito, H.: Sensory fibres in ventral roots L7 and S1 in the cat, J. Physiol. (Lond.) **267**:215, 1977.

Constanzo, R.M., and Gardner, E.P.: A quantitative analysis of responses of direction-sensitive neurons in somatosensory cortex of awake monkeys, J. Neurophysiol. **43**:1319, 1980.

Craig, A.D., Jr., and Burton, E.: Spinal and medullary lamina I projection to nucleus submedius in medial thalamus: a possible pain center, J. Neurophysiol. **45**:443, 1981.

Duclaux, R., and Kenshalo, D.R., Sr.: Response characteristics of cutaneous receptors in the monkey, J. Neurophysiol. **43**:1, 1980.

Duclaux, R., Schafer, K., and Hensel, E.: Response of cold receptors to low skin temperatures in nose of the cat, J. Neurophysiol. **43**:1571, 1980.

Dykes, R.W., et al.: Submodality segregation and receptive-field sequences in cuneate, gracile, and external cuneate nuclei of the cat, J. Neurophysiol. **47**:389, 1982.

Fields, H.L., and Basbaum, A.I.: Brainstem control of spinal pain-transmission neurons, Annu. Rev. Physiol. **40**:217, 1978.

Geisler, G.J., Jr., Spiel, H.R., and Willis, W.D.: Organization of spinothalamic tract axons within the rat spinal cord, J. Comp. Neurol. **195**:243, 1981.

Geisler, G.J., Jr., et al.: Spinothalamic tract neurons that project to medial and/or lateral thalamic nuclei: evidence for a physiologically novel population of spinal cord neurons, J. Neurophysiol. **46**:1285, 1981.

Hensel, H.: Thermoreceptors, Annu. Rev. Physiol. **36**:233, 1974.

Hoffman, D.S., et al.: Neuronal activity in medullary dorsal horn of awake monkeys trained in a thermal discrimination task. I. Responses to innocuous and noxious thermal stimuli, J. Neurophysiol. **46**:409, 1981.

Hyvärinen, J., and Poranen, A.: Movement-sensitive and direction and orientation-selective cutaneous receptive fields in the hand area of the post-central gyrus in monkeys, J. Physiol. (Lond.) **283**:523, 1978.

Juliano, S.L., Hand, P.J., and Whitsel, B.L.: Patterns of increased metabolic activity in somatosensory cortex of monkeys, *Macaca fascicularis,* subjected to controlled cutaneous stimulation: a 2-deoxyglucose study, J. Neurophysiol. **46**:1260, 1981.

Kenshalo, D.R., Jr., et al.: Responses of neurons in primate ventral posterior lateral nucleus to noxious stimuli, J. Neurophysiol. **43**:1594, 1980.

Kerr, F.W.L., and Wilson, P.R.: Pain, Annu. Rev. Neurosci. **1**:83, 1978.

Kosar, E., and Hand, P.J.: First somatosensory cortical columns and associated neuronal clusters of nucleus ventralis posterolateralis of the cat: an anatomical demonstration, J. Comp. Neurol. **198**:515, 1981.

Marfurt, C.F.: The central projections of trigeminal primary afferent neurons in the cat as determined by the transganglionic transport of horseradish peroxidase, J. Comp. Neurol. **203**:785, 1981.

Melzack, R., and Wall, P.D.: Pain mechanisms: a new theory, Science **150**:971, 1965.

Norrsell, U.: Behavioral studies of the somatosensory system, Physiol. Rev. **60**:327, 1980.

Snyder, S.E., and Childers, S.R.: Opiate receptors and opioid peptides, Annu. Rev. Physiol. **34**:315, 1979.

Sur, M., Merzenich, M.M., and Kaas, J.H.: Magnification, receptive-field area, and "hypercolumn" size in areas 3b and 1 of somatosensory cortex in owl monkeys, J. Neurophysiol. **44**:295, 1980.

Tanji, J., and Wise, S.P.: Submodality distribution in sensorimotor cortex of the unanesthetized monkey, J. Neurophysiol. **45**:467, 1981.

Wall, P.D., and Dubner, R.: Somatosensory pathways, Annu. Rev. Physiol. **34**:315, 1972.

Willis, W.D., Kenshalo, D.R., Jr., and Leonard, R.B.: The cells of origin of the primate spinothalamic tract, J. Comp. Neurol. **188**:543, 1979.

Woolsey, C.N., Erickson, T.C., and Gilson, W.E.: Localization in somatic sensory and motor areas of human cerebral cortex as determined by direct recording of evolved potentials and electrical stimulation, J. Neurosurg. **51**:476, 1979.

Books and monographs

Bonica, J.J., editor: International Symposium on Pain: advances in neurology, vol. 4, New York, 1974, Raven Press.

Bonica, J.J., editor: Pain, research publications: Association for Research in Nervous and Mental Disorders, vol. 58, New York, 1980, Raven Press.

Burgess, P.R., and Perl, E.R.: Cutaneous mechanoreceptors and nociceptors. In Iggo, A., editor: Handbook of sensory physiology, somatosensory system, vol. II, New York, 1973, Springer-Verlag New York, Inc.

Darian-Smith, I.: The trigeminal system. In Iggo, A., editor: Handbook of sensory physiology, somatosensory system, vol. II, New York, 1973, Springer-Verlag New York, Inc.

Hardy, J.D., Wolff, H.G., and Goodell, H.: Pain sensations and reactions, Baltimore, 1952, Williams & Wilkins Co.

Iggo, A., and Ramsey, R.A.: Thermosensory mechanisms in the spinal cord of monkeys. In Zotterman, Y., editor: Sensory functions of the skin in primates, with special reference to man, New York, 1976, Pergamon Press, Inc.

Kaas, J.H., et al.: Organization of somatosensory cortex in primates. In Schmitt, F.O., et al., editors: The organization of the cerebral cortex, Cambridge, Mass., 1981, The MIT Press.

Kenshalo, D.R., editor: The skin senses, Springfield, Ill., 1968, Charles C Thomas, Publisher.

Klee, W.A.: Endogenous opiate peptides. In Gainer, H., editor: Peptides in neurobiology, ed. 2, New York, 1979, Plenum Publishing Corp.

Mountcastle, V.B.: Neural mechanisms in somesthesia. In Mountcastle, V.B., editor: Medical physiology, ed. 14, vol. 1, St. Louis, 1980, The C.V. Mosby Co.

Poulos, D.A.: Central processing of peripheral temperature information. In Kornbuber, H.H., and Thieme, G., editors: The somatosensory system, Littleton, Mass., 1975, PSG Publishing Co, Inc.

Ruch, T.C.: Pathophysiology of pain. In Ruch, T.C., and Patton, H.D., editors: Physiology and biophysics: the brain and neural function, Philadelphia, 1979, W.B. Saunders Co.

Skoglund, S.: Joint receptors and kinesthesis. In Iggo, A., editor: Handbook of sensory physiology, somatosensory system, vol. II, New York, 1973, Springer-Verlag New York, Inc.

Snyder, S.H.: Opioid peptides in the brain. In Schmitt, F.O., and Worden, I.G., editors: The Neurosciences Fourth Study Program, Cambridge, Mass., 1979, The MIT Press.

Willis, W.D., and Coggeshall, R.E.: Sensory mechanisms of the spinal cord, New York, 1978, Plenum Publishing Corp.

Wolff, H.G.: Headache and other pain, ed. 2, London, 1963, Oxford University Press.

Woolsey, C.N., editor: Cortical sensory organization, vol. I, Multiple somatic areas, Clifton, N.J., 1981, Humana Press.

The auditory system

The sense of hearing in humans enables us to detect and analyze minute pressure changes that propagate as waves through air (or other media). By way of a complex but elegant series of events, these pressure changes are transduced into neural signals. Before this process is described, however, some salient features of the physical nature of sound will be reviewed briefly.

■ *Sound waves*

Sounds can be characterized as changes in pressure with time. These changes can be described by waveforms that are often quite complex, but each waveform uniquely describes a particular sound. For example, Fig. 10-1 depicts a waveform for a middle C played on a piano. For reasons given later, we consider the basic components of sound, called *pure tones,* to be sinusoidal changes in pressure with time (Fig. 10-2). Each pure tone can be completely characterized by three parameters: *amplitude, temporal frequency,* and *temporal phase* (Fig. 10-2). The amplitude is a measure of the peak-to-trough pressure change. The frequency is the number of cycles of the sine wave that occurs per unit time and is usually expressed in cycles per second or hertz (Hz). The phase refers to the absolute temporal position of the sine wave; two otherwise equal but out-of-phase tones are shown in Fig. 10-2. Phase is expressed in fractions of a cycle, or in degrees, where 360 degrees equals one full cycle. Thus the two tones of equal frequency and amplitude in Fig. 10-2 are 90 degrees out of phase with one another.

In 1822 Fourier produced a general mathematical theorem that can be applied directly to sound and explains why pure tones are considered to be the basic components

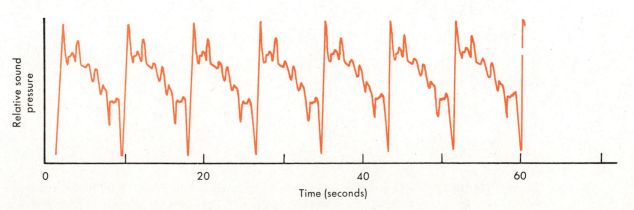

Fig. 10-1 ■ Pressure versus time waveform representing the sound created by striking the C below middle C on a piano. (Redrawn from Cornsweet, T.N.: Visual perception, New York, 1970, Academic Press, Inc.)

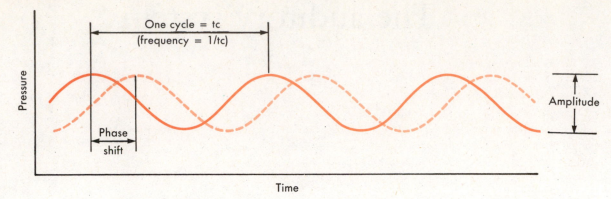

Fig. 10-2 ■ Representation of two pure tones 90 degrees out of phase with one another. A pure tone is a sinusoidal change in pressure with time. The parameters needed to characterize such a tone are the temporal frequency (the inverse of the time needed to encompass one full cycle), amplitude of the pressure change, and the temporal phase.

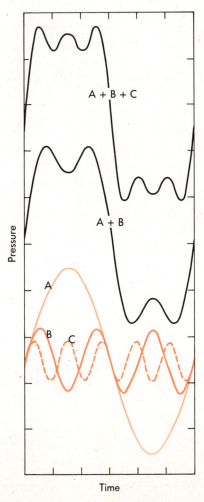

Fig. 10-3 ■ Fourier synthesis and analysis. Three pure tones (*A, B,* and *C*) are shown, and they can be linearly added in various combinations to form more complex sounds (e.g., *A + B* and *A + B + C*). Any complex sound (or other waveform) can likewise be synthesized from the addition of pure tones (or sine waves) appropriately chosen for frequency, amplitude, and phase. (Redrawn from Cornsweet, T.N.: Visual perception, New York, 1970, Academic Press, Inc.)

of sound. Fourier stated that *any* complex waveform can be synthesized by the linear addition of sine waves that have been appropriately selected for amplitude, frequency, and phase. Fig. 10-3 shows that a relatively simple waveform (a square wave) can be created by the addition of appropriately selected sine waves. *Fourier synthesis* is the construction of complex waveforms from sine waves. The converse, *Fourier analysis,* is the description of complex waveforms in terms of their component sine waves. (For a more complete description of Fourier's theorem and its application, suitable mathematical texts should be consulted.) Fourier thus emphasized the elemental nature of sine waves.

From the preceding description it follows that any sound wave, no matter how complicated a function of pressure versus time, can be synthesized and analyzed in terms of pure tones, which are sinusoidal functions of pressure versus time. Fig. 10-4 demonstrates this for the sound illustrated in Fig. 10-1. Fig. 10-4, *B,* shows its "Fourier spectrum," or the relative amplitude of the pure tones that make up this sound. When phase information is added, the Fourier spectrum is a useful and concise means of characterizing a sound.

This analytical approach allows us to determine how a sound will be changed by adding or subtracting certain tones. For instance, an audio amplifier typically does not

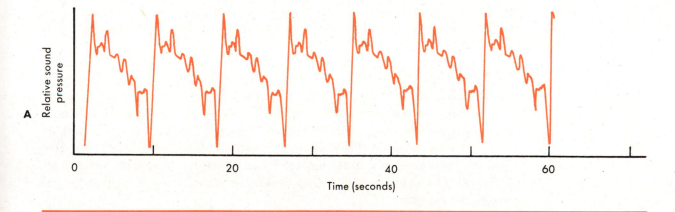

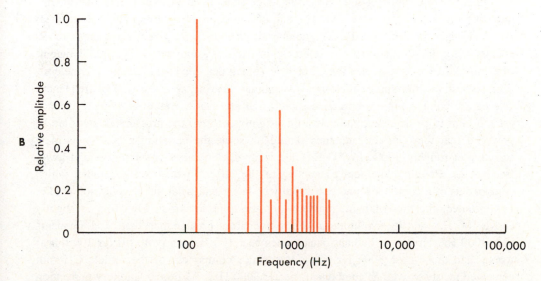

Fig. 10-4 ■ A complex sound and its Fourier spectrum. **A,** Waveform of same sound as depicted in Fig. 10-1. **B,** Fourier spectrum of the sound showing the relative strength of each of the component pure tones. (From Cornsweet, T.N.: Visual perception, New York, 1970, Academic Press, Inc.)

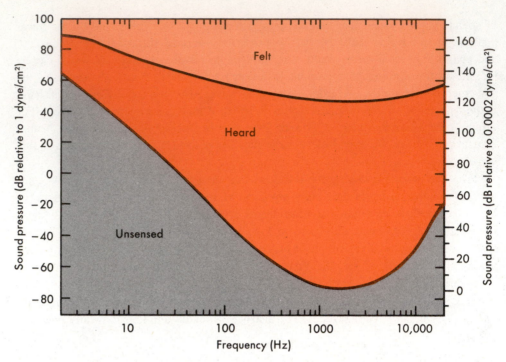

Fig. 10-5 ■ Range of human hearing *(darker color)*. Above this range *(lighter color)* sound pressure waves are sufficiently intense to be felt with somesthesia. Below this range sound pressure waves are too weak to be sensed.

reproduce all frequencies equally, and a particular amplifier might be incapable of reproducing tones of less than 20 Hz or more than 20,000 Hz. If sounds (e.g., music) containing frequencies under 20 or over 20,000 Hz were to be played through such an amplifier, the following could be done. Fourier analysis of the original would show the component tones lower than 20 Hz and higher than 20,000 Hz; these could be removed. Fourier synthesis of the remaining tones would then precisely match the sound reproduced by the amplifier.

Finally, if the range of human hearing is considered, it becomes clear that all frequencies are not equally perceived. The unit of sound pressure commonly used to measure hearing is the decibel (dB). The number of decibels for a given sound pressure (P) equals 20 log P/P_R, where P_R is a reference pressure. P_R most often is 0.002 dyne/cm^2 (the threshold for human hearing), but it is sometimes considered to be 1 dyne/cm^2. Fig. 10-5 shows the approximate range of human hearing as a function of frequency. The darkly colored region represents sounds that are heard. Below this *(gray),* sounds are too faint to be detected; and above this *(lightly colored)* the pressure changes are so powerful that they are felt and thus sensed by somesthesia. Human hearing is most sensitive at roughly 3000 to 4000 Hz, and it rapidly deteriorates for higher and lower frequencies. Fourier's theorem as applied to Fig. 10-5 enables one to predict which components of a complex sound can actually be heard and what these would sound like. For instance, the predominant frequency for the tone shown in Fig. 10-1 is slightly above 100 Hz, and its amplitude is nearly five times greater than that of components near 3000 Hz. However, humans detect tones near 100 Hz nearly 40 dB (or 100 times) worse than those near 3000 Hz. Thus this major component of the middle C is not perceived nearly as well as components nearer 3000 Hz, and human auditory perception distorts actual sounds in a predictable fashion. One can predict this distortion simply by multiplying a sound's Fourier spectrum (Fig. 10-4, *B*) by the relative sensitivity curve of Fig. 10-5.

Fig. 10-6 shows the three major divisions of the ear. These are the *external* or *outer ear*, the *middle ear,* and the *internal* or *inner ear*. Actual neural transduction occurs in the inner ear, but the nature of the stimulus is significantly transformed before reaching the transducers.

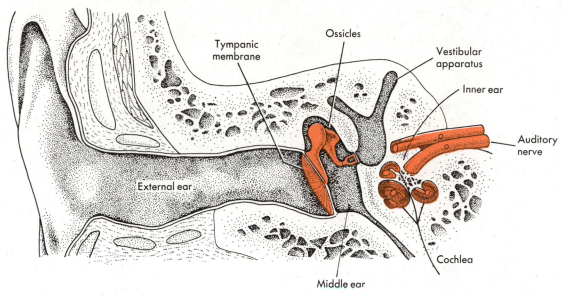

Fig. 10-6 ■ Major divisions of the human ear. Shown in color are the tympanic membrane and ossicles of the middle ear plus the cochlea and auditory nerve of the inner ear. (Redrawn from von Békésy, G.: Sci. Am. **197**:66, 1957.)

The external or outer ear consists of the *pinna* and *external meatus,* or *auditory canal*. The pinna has a complicated shape that probably tends to direct sound waves toward the auditory canal. There is some evidence that the pinna also is important in localizing the source of sounds. The human pinna is relatively immobile, but most animals can readily move the pinna toward the source of sound and thereby improve sensitivity. The auditory canal is an air-filled tube down which sound waves must travel to reach the middle ear. The shape of the canal provides it with a resonant frequency of approximately 3500 Hz, but this is broadly tuned (roughly 800 to 6000 Hz) because the canal surfaces are elastic. Frequencies nearer the resonant frequency arrive at the middle ear with less attenuation than higher and lower ones, and much of the differential sensitivity of human hearing thus is derived from the shape and physical characteristics of the auditory canal.

■ *External or outer ear*

The middle ear consists of the *tympanic membrane* (or *eardrum*) and three bones, or ossicles, known as the *malleus* (or *hammer*), the *incus* (or *anvil*), and the *stapes* (or *stirrup*) (Fig. 10-7, *A*). The rest of the middle ear is air filled. The ossicles are firmly attached to one another and move as a unit. The base of the malleus is attached to the tympanic membrane, and the footplate of the stapes is attached to a membrane covering an opening in the bony shell of the inner ear. This opening is called the *oval window,* and its membranous covering separates the air-filled middle ear from the liquid-filled cochlea (Fig. 10-7, *B*).

Sound waves that travel down the outer ear impinge on the tympanic membrane and

■ *Middle ear*

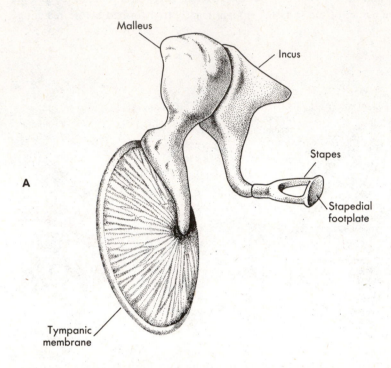

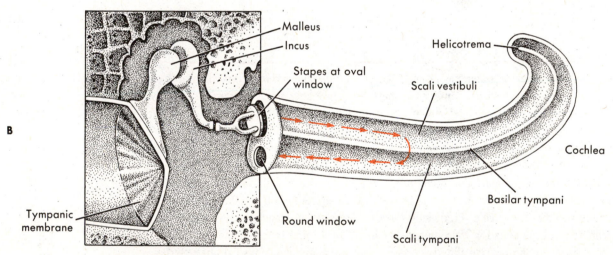

Fig. 10-7 ■ Middle ear and its relationship to the cochlea. **A,** Tympanic membrane (eardrum), malleus (hammer), incus (anvil), and stapes (stirrup). The stapedial footplate rests on the membrane covering the oval window. **B,** Relationship of ossicles and cochlea. The liquid-filled, bony cochlea has elastic membranes covering two apertures, the round and oval windows. Movement of the stapedial footplate causes displacement of the oval window, resulting in displacement of the liquids *(arrows)* and round window. In the process the basilar membrane is displaced. (**A** redrawn from von Békésy, G.: Sci. Am. **197:**66, 1957; **B** redrawn from Kandel, E., and Schwartz, J.H.: Principles of neural science, New York, 1981, Elsevier North-Holland, Inc.)

cause it to vibrate at an amplitude and frequency corresponding to the sound itself. These vibrations are transmitted to the ossicles, which vibrate about an axis running near the juncture of the malleus and incus. This, in turn, causes the footplate of the stapes to vibrate against the oval window, and the sound waves are thereby transmitted to the liquids of the inner ear.

What function does the middle ear serve? Why not bypass the middle ear and have the airborne sound waves directly strike the membrane covering the oval window? The answers to these questions lie in the fact that the *airborne* sound waves in the auditory canal must pass to the watery liquids of the inner ear. The sound-conducting properties of air and water are quite different, and a complicated parameter known as *acoustic impedance* describes much of a medium's sound-conducting qualities. (Acoustic impedance is determined by the amplitude of the sound pressure variations divided by the volume velocity of the sound waves; volume velocity is the product of the velocity of the sound wave and the area through which it passes.) The acoustic impedance of water is considerably higher than that of air. When such an impedance mismatch occurs, most of the sound energy is reflected at the interface between two media, rather than being transmitted from one to the other; that is, most of the sound traveling from air to water would be reflected off the surface back into the air; very little would enter the water. More specifically, the ratio of transmitted energy (E_t) to incident energy (E_i) can be specified as

$$\frac{E_t}{E_i} = \frac{4r}{(r + 1)^2}$$

where *r* is the ratio of impedances. In the case of an air-to-water interface, the impedance mismatch would result in reflection of 99.9% of the sound energy. Because sound energy is proportional to the square of the sound pressure, this represents a 30 dB loss. Indeed such a loss would result without the middle ear if the tympanic membrane were applied directly to the fluids of the inner ear.

The middle ear serves to provide a much better impedance match between the outer and inner ears so that much more of the sound energy is transmitted. In humans the consequence is such that only about 10 to 15 dB are lost. The ossicles of the middle ear perform this impedance-matching function. They act as a transformer that increases the pressures from the tympanic membrane to the oval window. Two factors are most relevant to this function. First, the area of the tympanic membrane is much greater than that of the oval window so that a force applied to the malleus and transmitted to the stapedial footplate exerts more pressure at the oval window than that originated at the tympanic membrane. Second, the ossicular chain moves as a lever that rotates around the malleus-incus junction. The lever action is such that the movements of the stapedial footplate cover less distance and, consequently, exert more force than do movements of the malleus caused by the tympanic membrane. Again this serves to amplify sound pressures between the tympanic membrane and oval window.

The treatment of *otosclerosis* in the recent past serves to illustrate the function of the ossicles. Otosclerosis is a condition in which pathological bone growth around the ossicles tends to cement or freeze them against the skull (Fig. 10-8). In such a condition no vibrations or sound can be transmitted from the outer to the inner ear. This form of deafness was commonly alleviated until recently by a *fenestration operation* (Fig. 10-8, *C*). The ossicles can be removed or disconnected from the tympanic membrane, and an artificial membrane-covered opening in the cochlea can be made to replace the immobile oval window. Now sounds can again be transmitted to the cochlear fluids, but without the benefit of the ossicular impedance matching. The restored hearing suffers from a 15 to 20 dB loss from normal, but given the large dynamic range of human hearing (Fig. 10-5), this loss is compatible with useful hearing. It should be noted that fenestration operations are now rarely needed. Instead otosclerosis is typically treated with artificial ossicular replacement that restores hearing to almost normal levels.

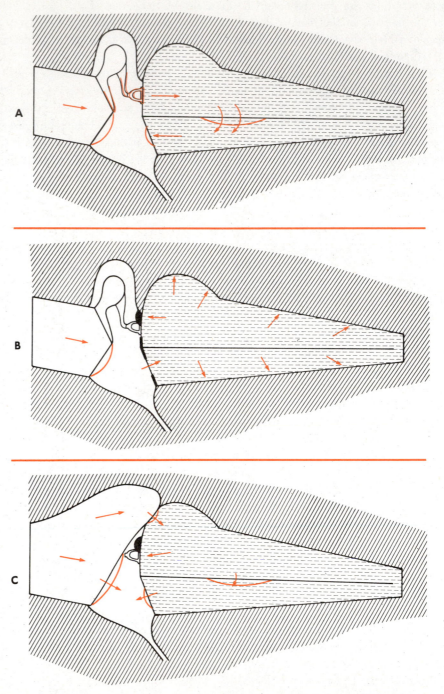

Fig. 10-8 ■ Otosclerosis and its alleviation by a fenestration operation. **A,** Normal ear. Displacements *(color)* caused by sound waves are transmitted via the middle ear to the basilar membrane in the cochlea. **B,** Otosclerosis. Pathological bone formation *(black)* freezes the stapedial footplate and oval window so that they cannot move in response to tympanic membrane displacement. Because the cochlear liquids are incompressible, no round window or basilar membrane movement is possible either. **C,** Fenestration operation. An extra aperture is created in the cochlear wall exposed to the outer ear, and the aperture is covered with a flexible membrane. Sound waves can now cause displacement of the basilar membrane, although the impedance-matching quality of the middle ear is lost. (Redrawn from von Békésy, G.: Sci. Am. **197:**66, 1957.)

The inner ear consists of the cochlea as well as organs for vestibular sensation. The vestibular system is treated in Chapter 11, and only the cochlea is considered here.

Cochlear structure. Fig. 10-9 shows the complicated structure of the cochlea. It is a bony labyrinth filled with liquids and various structural elements. (See Fig. 11-1 for the relationship between the cochlear and vestibular labyrinths in the inner ear.) The *oval* and *round windows* are two membrane-covered openings in the bone that permit

■ *Internal or inner ear*

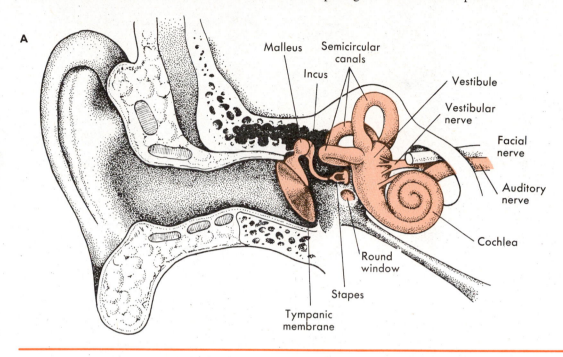

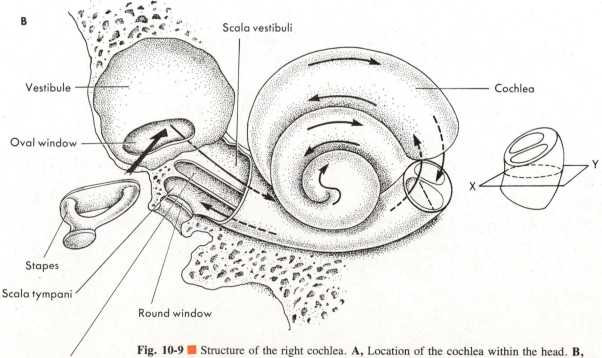

Fig. 10-9 ■ Structure of the right cochlea. **A,** Location of the cochlea within the head. **B,** Coiled structure of cochlea showing the relationships among the oval window, scala vestibuli, scala media, and scala tympani. The arrows indicate the path of continuous fluid flow from the scala vestibula through the helicotrema to the scala tympani. The small drawing at the right illustrates the plane of section for **C.**

Continued.

Fig. 10-9, cont'd ■ **C,** Cross
section through the cochlea. The *X*
and *Y* indicate the orientation vis-a-
vis the drawing on the right in **B.**
The hair cells and sensory fibers
are shown in color. The dashed
rectangle outlines the area shown
in **D. D,** Organ of Corti, including
the hair cells and sensory fibers
(color). The hair cells have fine
hairlike processes embedded in the
tectorial membrane. A single inner
hair cell and several outer hair
cells are separated by the rods of
Corti. (**B** redrawn from Gulick,
W.L.: Hearing: physiology and
psychophysics, New York, 1971,
Oxford University Press; after
Maloney.)

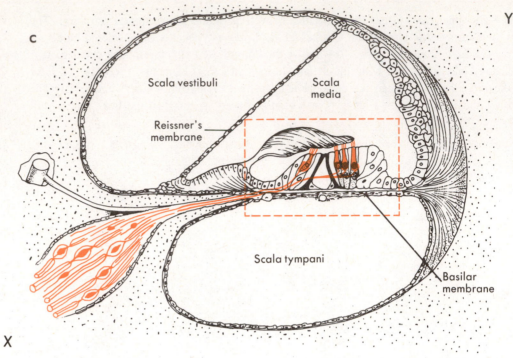

C

Scala vestibuli

Scala media

Reissner's membrane

Scala tympani

Basilar membrane

Y

X

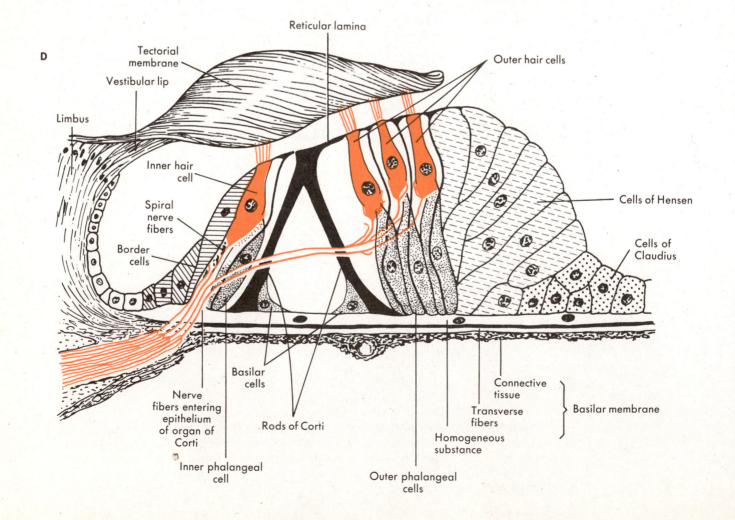

D

Reticular lamina

Tectorial membrane

Vestibular lip

Limbus

Inner hair cell

Spiral nerve fibers

Border cells

Outer hair cells

Cells of Hensen

Cells of Claudius

Nerve fibers entering epithelium of organ of Corti

Inner phalangeal cell

Basilar cells

Rods of Corti

Outer phalangeal cells

Homogeneous substance

Transverse fibers

Connective tissue

Basilar membrane

displacement of the cochlear liquids in response to motion of the stapedial footplate. Like most liquids, those in the cochlea are practically incompressible, and they could not move in response to sound stimuli without these windows. Among the structures inside the cochlea are the *basilar membrane* and *organ of Corti* (Fig. 10-9). The organ of Corti rests on the basilar membrane and contains the transducers, which are known as *hair cells*. The entire structure is complicated by the fact that the cochlea is tightly coiled into a spiral. A cross section through the cochlea shows that the fluids are divided among three compartments (Figure 10-9): the *scala vestibuli* is separated from the scala media by *Reissner's membrane;* the *scala media* lies between the basilar and Reissner's membranes, and it contains the organ of Corti; the final chamber is the *scala tympani.*

Before proceeding with the events leading to transduction, it is useful to consider a simple model of the cochlear plan (Fig. 10-10). The cochlear model begins with a liquid-filled glass container covered at each end by an elastic membrane (Fig. 10-10, *A*). Downward or upward movement of the top membrane (the "oval window") causes

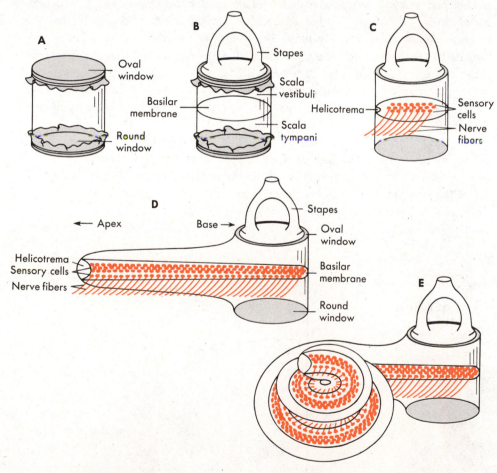

Fig. 10-10 ■ Simplified schema for the construction of a model cochlea. **A,** Glass cylinder filled with liquid and covered at each end with an elastic membrane. The top covering models the oval window, and the bottom covering models the round window. **B,** Addition of the stapes to serve as a piston against the oval window and the basilar membrane stretched across the cylinder. The latter divides the fluids into the scala vestibula and scala tympani (for simplicity the scala media is not represented in this figure). **C,** Addition of sensory hair cells and auditory nerve fibers *(color)*. Also, a gap in the basilar membrane known as the *helicotrema* is added. **D,** Elongation of the cochlea along axis parallel to the basilar membrane. **E,** Coiling of the cochlea. (Redrawn from Kiang, N.Y.S.: Stimulus representation in the discharge patterns of auditory neurons. In Tower, D.B., editor: The human nervous system, vol. 3, Human communication and its disorders, New York, 1975, Raven Press. © 1975.)

displacement of the liquid and a similar movement of the bottom membrane (the "round window"). In Fig. 10-10, *B*, the stapes (to vibrate against the oval window) and the basilar membrane are added. The basilar membrane moves in phase with the oval and round windows, and it divides the fluid into two compartments: the scala vestibuli and scala tympani.

The hair cells and fibers are added to the basilar membrane in Fig. 10-10, *C*. Later we shall see how vibrations of the basilar membrane affect the hair cells. Also shown is the *helicotrema,* which is a small gap in the basilar membrane. It permits the liquids in the scala vestibuli and scala tympani to be continuous. Furthermore, the helicotrema allows the basilar membrane to return to its resting position in the presence of a constant displacement of the oval and round windows. In other words, it dampens out low-frequency movements of the liquids so that only high-frequency displacements (i.e., in the range of sound frequencies) cause appreciable vibration of the basilar membrane. A static change in air pressure is an example of a low-frequency stimulus that the helico-trema serves to filter out as a potential auditory stimulus. In Fig. 10-10, *D*, the cochlea is stretched along a dimension parallel to the basilar membrane, and the cochlea is coiled in Fig. 10-10, *E*. The human cochlea is approximately 35 mm in length and contains slightly more than 2½ turns. For simplicity the scala media is not shown in Fig. 10-10. The scala media is a self-contained, membrane-bound chamber in which the organ of Corti is located (Fig. 10-9). Fig. 10-10, *E,* is a fairly accurate model of the mammalian cochlea, and from it one should readily comprehend the manner whereby sound vibrations are transmitted through the ossicles and cause the basilar membrane to vibrate.

Organ of Corti and transduction. Fig. 10-9, *D,* shows the key structural features of the organ of Corti. The hair cells are supported in a rigid structure, the *reticular lamina,* which in turn is supported on stiff triangular structures called the *rods of Corti.* All of this rests on the basilar membrane. The rods of Corti divide the hair cells into a single row of inner hair cells and three rows of outer hair cells.

The hair cell (Fig. 10-11) is so named because of hairlike cilia that protrude from

Fig. 10-11 ■ Schematic drawing of inner (**A**) and outer (**B**) hair cell. The hair bundles extend into the tectorial membrane (not shown). Each hair cell is postsynaptic to efferent fibers *(light color)* from the olivocochlear bundle and presynaptic to afferent fibers *(dark color)* that comprise the auditory nerve. The efferent fibers may also form synapses onto the terminals of the afferent fibers.

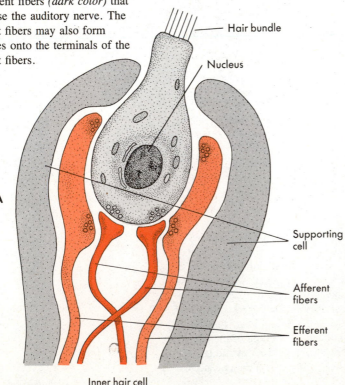

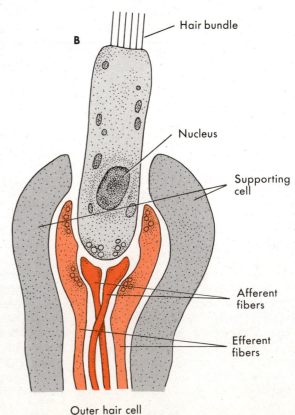

the top of the cell. Many of these cilia are embedded at their distal ends into the *tectorial membrane,* which lies just dorsal to the reticular lamina (Fig. 10-9). As the oval and round windows vibrate to cause displacement of the basilar membrane, the rods of Corti and reticular lamina pivot as a unit around the base of the inner rod. This is shown in Fig. 10-12. This creates a shearing force on the cilia, and this mechanical deformation leads to a generator potential (GP) in the hair cell.

At the base of the hair cell can be found a number of synapses formed between the cell and axons of the eighth nerve (Fig. 10-11). The hair cell is presynaptic to many of these axons, which are thus afferent fibers that innervate the brain, and it is postsynaptic to others. The efferent fibers also seem to synapse onto the afferent ones. The hair cell seems to transmit the GP synaptically to these afferent fibers, and if the GP is sufficiently large, action potentials travel toward the brain. As in other systems, the amplitude of the GP, and thus the frequency of action potentials, is related monotonically to the amplitude of the basilar membrane displacement. This displacement, in turn, relates to the amplitude or intensity of the sound waves. From the foregoing it should be clear how sound stimuli produce coded signals that reach the brain.

A unique feature of the auditory system is the fact that the transducers are under efferent control from the brain. As noted earlier, some of the axons in the organ of Corti form presynaptic terminals onto the hair cells and afferent fibers. These efferent axons run in the *olivocochlear bundle* that originates with cells in the *superior olivary nucleus* of the pons (Fig. 10-15). This pathway is predominantly crossed but has an

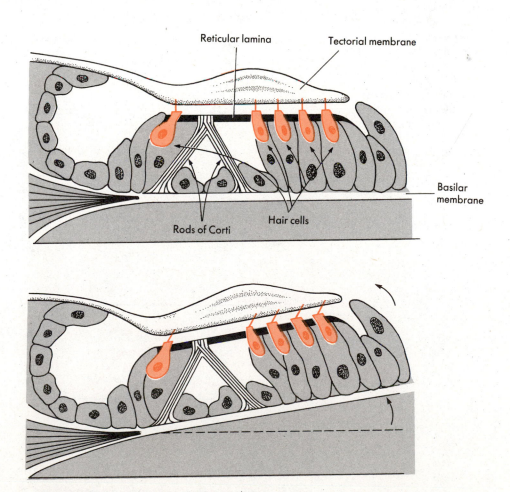

Fig. 10-12 ■ Effect of basilar membrane displacement on hair cells. Note how upward movement of the membrane causes a shearing force to be exerted on the hairs.

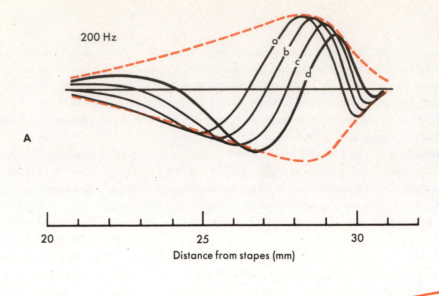

A

Distance from stapes (mm)

Fig. 10-13 ■ Differential frequency responses of the basilar membrane. **A,** Traveling wave evoked along the basilar membrane by a 200 Hz sound. The wave is shown at four successive intervals (*a, b, c,* and *d*), and the dashed line describes the envelope of the peaks. Note how the maximum amplitude varies with distance along the basilar membrane, thereby resulting in the maximum response to a 200 Hz stimulus being evoked from hair cells located approximately 29 mm from the stapes or oval window. **B,** Envelopes of traveling waves for a variety of frequencies. Note that the maximum response varies regularly with frequency so that regions closer to the stapes or oval window are more sensitive to higher frequencies. (Redrawn from von Békésy, G.: Experiments in hearing [research articles from 1928 to 1958], New York, 1960, McGraw-Hill, Inc. Copyright © 1960 by McGraw-Hill Book Co. Used with permission.)

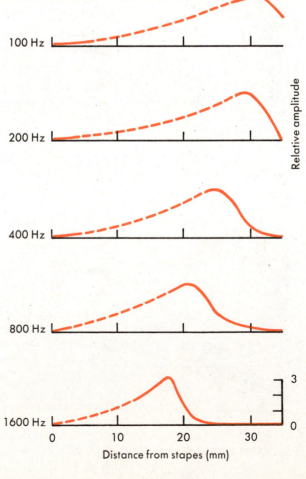

B

Distance from stapes (mm)

ipsilateral component. The function of this efferent pathway is not well understood, but it may affect the hair cell threshold or contribute to afferent inhibition in auditory nerve fibers (pp. 183-185).

Tonotopic organization of the basilar membrane. A final important structural feature of the inner ear is its *tonotopic map.* The basilar membrane is not structurally homogeneous from the stapes to the helicotrema, but rather it gradually changes in width, tension, and viscoelastic properties. The result is that each portion of the basilar membrane vibrates with maximum amplitude only to one specific resonant frequency. In other words, a standing wave is set up for each sound frequency within the range of human hearing (Fig. 10-13, *A*). Those waves for higher frequencies reach peak amplitude closer to the stapes (Fig. 10-13, *B*).

The consequence of this is that each tone causes a limited patch of basilar membrane to vibrate sufficiently to produce large GPs and subsequent action potentials. A tonotopic map of the cochlea is shown in Fig. 10-14. Because of the placement of each hair cell in a limited patch of basilar membrane, each of these cells is "tuned" to a particular sound frequency or tone.

Fig. 10-14 also serves to illustrate a fundamental difference between the auditory

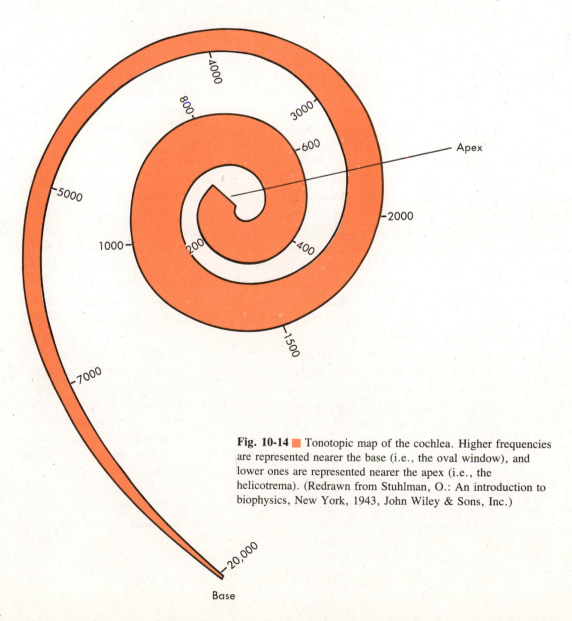

Fig. 10-14 ■ Tonotopic map of the cochlea. Higher frequencies are represented nearer the base (i.e., the oval window), and lower ones are represented nearer the apex (i.e., the helicotrema). (Redrawn from Stuhlman, O.: An introduction to biophysics, New York, 1943, John Wiley & Sons, Inc.)

system and the visual or somatosensory system. Remember that receptive fields are defined in terms of the spatial distribution of relevant transducers in the periphery. The functional significance of this for vision and somesthesia is that receptive fields in these systems relate to space on the retina or skin. Auditory receptive fields, however, are *not* organized along a spatial dimension; rather they are organized in terms of tones or frequencies, since the frequency domain is what is mapped by the auditory periphery. Thus while vision and somesthesia are elegant spatial senses, the auditory system is basically organized to discriminate tones and not their source locations. Note that this concept of the auditory receptive field follows from the same definition and principle as applies to visual or somesthetic receptive fields. The spatial distribution of transducers along the cochlea characterizes the frequency distribution of a sound rather than its spatial location, whereas the spatial distributions of transducers along the retina or cutaneous surface characterize the spatial location of the stimulus. Auditory spatial localization is also performed, but the underlying neural processes are more complicated and less basic than those for tonotopic analysis.

■ *The central auditory pathways*
■ *Anatomical features*

As for other sensory systems, the central auditory pathways include both ascending and descending pathways. These pathways also seem to be organized into phylogenetically older and newer portions, but sufficient data to elucidate these fully are not available.

Ascending pathways. Fig. 10-15 is a simplified version of some of the main features of the ascending pathways, which are complex. Auditory fibers from the auditory or cochlear nerve (part of the eighth nerve) terminate ipsilaterally in the *dorsal* and *ventral cochlear nuclei.* Many fibers from these nuclei cross the midline to innervate the *superior olivary complex,* and others ascend in the contralateral *lateral lemniscus.* The superior olivary complex is composed of a number of distinct nuclei, the most prominent of which are the *lateral* and *medial superior olivary nuclei.* Many of the crossing fibers from the cochlear nuclei decussate in the *trapezoid body,* although some innervate the ipsilateral superior olivary complex. Each superior olivary complex innervates the other across the trapezoid body, and these nuclei also contribute fibers to the lateral lemnisci. Therefore the lateral lemniscus, which represents the major auditory component ascending from the medulla, contains fibers that represent both cochleae; these fibers are a mixture of fibers of different orders (secondary, tertiary, etc.). Unlike the visual and auditory systems, in which various levels of the pathways typically map only the contralateral representation of space and include units of the same order (e.g., secondary, tertiary), the auditory system is much more complicated even at such a peripheral level as the lateral lemniscus.

The lateral lemniscus ascends to the midbrain, where most of its fibers terminate in the *inferior colliculus.* At this level, too, considerable auditory information is relayed across the midline in the *commissure of the inferior colliculus.* Fibers leave the inferior colliculus via the *brachium of the inferior colliculus* to terminate in the *medial geniculate nucleus.* Most of these fibers travel ipsilaterally, but some innervate the contralateral medial geniculate nucleus after crossing the collicular commissure. Neurons of the medial geniculate nucleus project fibers via the auditory radiations to the auditory cortex.

The colliculogeniculocortical projection is even more complex than depicted earlier. It has been divided into *core* and *belt* portions that may relate to phylogenetic age. Fig. 10-15 shows these divisions. The presumably newer *core* portion includes the central nucleus of the inferior colliculus, the laminated portion of the ventral division of the medial geniculate nucleus, and the primary auditory cortex. This cortex has been designated AI, and it corresponds to Brodmann's areas 41 and 42 (Chapter 13). The core

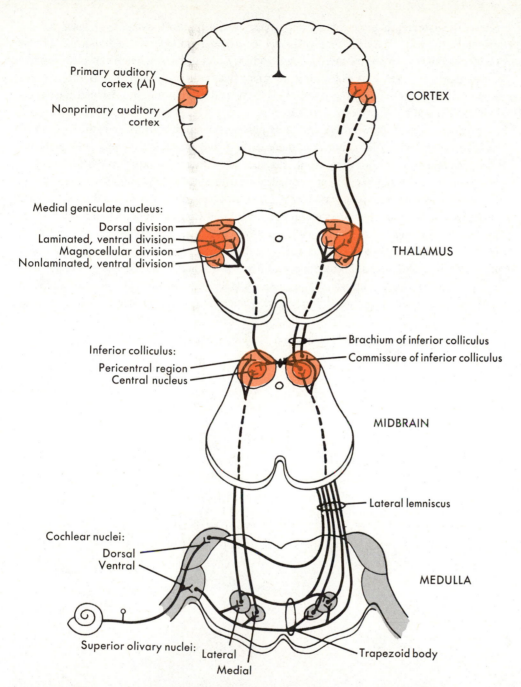

Primary auditory
cortex (AI)

Nonprimary auditory
cortex

CORTEX

Medial geniculate nucleus:

Dorsal division
Laminated, ventral division
Magnocellular division
Nonlaminated, ventral division

THALAMUS

Brachium of inferior colliculus
Commissure of inferior colliculus

Inferior colliculus:

Pericentral region
Central nucleus

MIDBRAIN

Lateral lemniscus

Cochlear nuclei:

Dorsal
Ventral

MEDULLA

Superior olivary nuclei: Lateral
Medial

Trapezoid body

Fig. 10-15 ■ Simplified and highly schematic diagram of some of the central auditory
pathways. The partial division of these pathways into core *(darker color)* and belt *(lighter color)*
regions is shown at the levels of the inferior colliculus, medial geniculate nucleus, and auditory
cortex. Note that as peripherally as the lateral lemniscus the pathways include a mixture of
secondary, tertiary, and other higher level fibers as well as a representation of both cochleae.

geniculocortical projection terminates mainly in layer IV, although a sparse projection
to layer VI has also been described.

The presumably older belt portion includes the pericentral nucleus and tegmentum
of the inferior colliculus, the nonlaminated ventral, dorsal, and magnocellular divisions
of the medial geniculate nucleus, and cortical auditory areas that include AI and neighboring regions. As in the visual and somatosensory systems, there are numerous sepa-

rate cortical representations of sensory space in the auditory system (at least six are presently known). They all receive geniculocortical input via the belt portion of the pathway. Fig. 10-16 shows these known auditory cortical areas. Here the projection is substantially to layer I of the cortex. In many ways the core and belt portions seem analogous, respectively, to the portions of the visual system that are divided into geniculocortical and pulvinar-cortical (Chapter 8).

In addition to the newer core and older belt portions of the auditory pathways shown in Figs. 10-15 and 10-16, there are more diffuse and presumably even older portions that for simplicity are not diagramed. For instance, fibers passing through the medulla at the level of the superior olivary complex emit fine collaterals among neurons of the tegmentum reticular formation and among cells scattered along the trapezoid body. Also, fibers ascending in the lateral lemniscus terminate among neurons of the midbrain reticular formation and among cells scattered along the lateral lemniscus. Although little detailed information is available concerning these diffuse pathways, they are probably functionally analogous to the oldest parts of the visual and somatosensory systems.

The complexity of the ascending auditory pathways not only poses difficulties in analyzing the system functionally in experimental animals, but it also makes it difficult

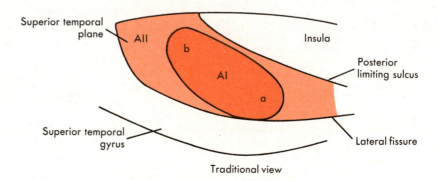

Traditional view

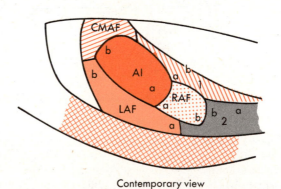

Contemporary view

Fig. 10-16 ■ Organization of auditory cortex in the macaque monkey; view of the surface of the superior temporal plane and dorsolateral surface of the superior temporal gyrus. The representation of the cochlear apex *(a)* and base *(b)*, or low and high frequencies, respectively, are designated for those areas in which tonotopic organization has been verified. Above is the older traditional view with the primary *(AI)* and secondary *(AII)* auditory areas. Below is the contemporary view that illustrates the five or more auditory areas. These include the primary auditory cortex *(AI)*, the nontonotopic caudiomedial auditory field *(CMAF)*, the lateral auditory field *(LAF)*, plus two as yet unnamed areas (designated *1* and *2*) with tonotopic organization. Furthermore, the superior temporal gyrus seems to contain one or more additional auditory areas. (Redrawn from Merzenich, M.M., and Kass, J.H.: Progr. Psychobiol. Physiol. Psychol. **9:**1, 1980.)

to localize lesions in humans on the basis of the resultant hearing deficits. Unilateral lesions produce a unilateral deafness only if they occur in or peripherally to the cochlear nuclei (Fig. 10-15). At all other levels unilateral lesions produce bilateral hearing disorders that are only subtly different from one another. Furthermore, the cerebral cortex does not seem to be essential to hearing, at least in certain experimental animals. In such animals a complete, bilateral removal of the auditory cortical areas produces only a slight reduction of auditory sensitivity. These lesions, however, do create difficulties for the animals in the discrimination of tonal patterns (e.g., tones given in a different sequence such as AABA or ABAA) that are readily discriminated by intact animals. Such bilateral cortical lesions have rarely been seen in humans. However, available evidence from experimental animals indicates that considerable sensory perception occurs subcortically.

Descending pathways. As in other sensory systems, the auditory pathways include a significant descending component that probably interacts with the ascending pathways at many levels to create numerous, complex feedback loops. All of the known auditory cortical areas ipsilaterally innervate all portions of the medial geniculate nucleus. However, no descending fibers from this nucleus to other auditory structures have been described. The auditory cortex also ipsilaterally innervates the inferior colliculus, which in turn ipsilaterally innervates the superior olivary nucleus and the dorsal cochlear nucleus. Finally, as noted earlier, hair cells in the cochlea are innervated bilaterally from the superior olivary nuclei via the olivocochlear bundle.

■ *Functional organization*

As for other sensory systems, the study of individual neuronal receptive fields has done much to elucidate the functional organization of the auditory system. Most of this analysis has been limited to the core or newer portion of the auditory system. Thus afferent inhibition and neurotopic organization are key features that have been described in experimental animals.

Receptive fields. Because receptive fields refer to regions of transducer-containing tissue, the tonotopic organization of the cochlea dictates that auditory receptive fields are described in terms of frequency and not space (pp. 179-180).

Fig. 10-17, *A,* shows a receptive field for a typical auditory neuron. This is the conventional manner in which to describe auditory receptive fields, and it corresponds to a "tuning curve," that is, a curve *(solid line)* that depicts the minimum sound intensity needed to evoke a neural discharge as a function of sound frequency. The dark colored region indicates intensity/frequency combinations that discharge the cell; the light colored region indicates combinations that inhibit the cell; other combinations do not affect the cell. Two important features of these receptive fields are the *characteristic frequency,* which is found at the lowest point of the curve, and the *selectivity for frequency,* which is a function of how steeply the curve rises from the characteristic frequency. To illustrate these two characteristics, Fig. 10-17, *B,* shows auditory receptive fields for cochlear nerve fibers that include a range of characteristic frequencies. Fig. 10-17, *B,* shows relatively highly selective or tightly tuned cells and relatively poorly selective or broadly tuned ones. A tightly tuned cell is analogous to a visual or somatosensory neuron with a small receptive field. Presumably tonal discriminations depend on tightly tuned neurons.

Note that curves for most auditory neurons are asymmetrical (Fig. 10-17, *B*), although this is largely an artifact of the logarithmic scale for the abscissa. The curves tend to rise from the characteristic frequency more steeply toward higher than toward lower frequencies. This is a fairly straightforward consequence of the tuning of the basilar membrane itself. As shown in Fig. 10-13, traveling waves in this membrane set up by pure tones possess a similar asymmetry.

Afferent inhibition. As is discussed in Chapters 8 and 9, *afferent inhibition* means

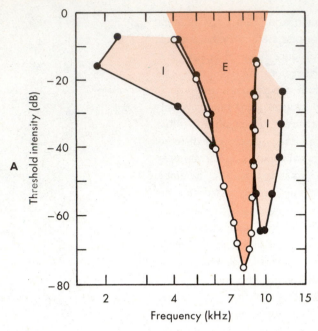

Fig. 10-17 ■ Receptive fields of auditory neurons. These are defined as tuning curves for which the sound intensity needed to evoke a threshold neuronal response is plotted as a function of sound frequency. **A,** Representative auditory receptive field showing afferent inhibition (see Fig. 8-20). A central excitatory region *(E)* is flanked by inhibitory regions *(I)*. An intensity-frequency combination below the curves is without effect on the neuron. Combinations in the inhibitory or excitatory regions, respectively, inhibit or excite the neuron. **B,** Examples of auditory receptive fields, depicted as tuning curves for the excitatory region, for cochlear nerve fibers *(upper left),* neurons in the inferior colliculus *(upper right),* units near the trapezoid body *(lower left),* and neurons in the medial geniculate nucleus *(lower right).* (**A** redrawn from Arthur, R.M., et al.: J. Physiol. [Lond.] **212:**593, 1971; **B** redrawn from Katusi, Y. In Rosenblith, W.A., editor: Sensory communication, Cambridge, Mass., 1961, The MIT Press. © 1961 by the MIT Press.)

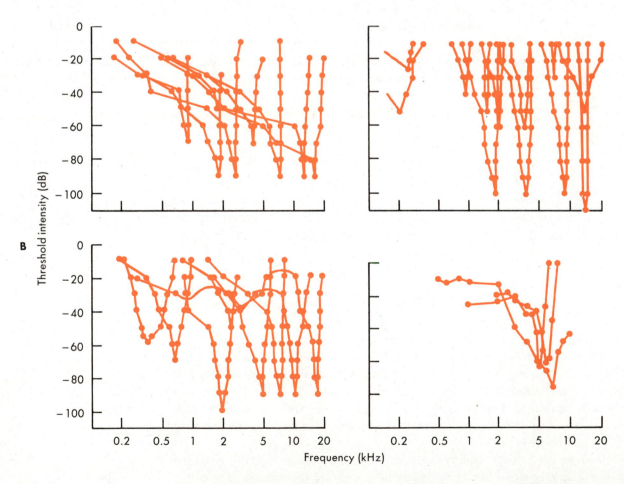

that a stimulus applied to transducers just beyond a neuron's receptive field will inhibit that neuron. Afferent inhibition has the same connotation for auditory neurons; that is, stimuli of intensity/frequency combinations just outside the excitatory receptive field will inhibit an auditory neuron that possesses afferent inhibition. Fig. 10-17, *A,* shows such an inhibitory receptive field *(light colored region).* As is the case in vision and somesthesia, such afferent inhibition enhances the system's ability to discriminate stimuli applied to slightly different regions of the transducer surface. In other words, afferent inhibition in the auditory system serves to sharpen frequency discriminations. Afferent inhibition has been described at all levels of the auditory pathways, including fibers of the cochlear nerve, but it seems to be enhanced at each higher level of the system.

Tonotopic organization. The quality of neurotopic organization for newer portions of a sensory system appears in the newer portions of the auditory system as *tonotopic organization.* Neighboring neurons in a structure exhibit characteristic frequencies that differ only slightly, and these frequencies change systematically as one moves across the structure. This is precisely equivalent to the retinotopic and somatotopic organizations that are found in the visual and somatosensory systems, respectively. Tonotopic organization and neurons that are sharply tuned for sound frequency are general features of the newer portions of the auditory system, including the auditory nerve, cochlear nuclei, superior olivary complex, lateral lemniscus, inferior colliculus, medial geniculate nucleus, and auditory cortex. Fig. 10-16 shows the tonotopic arrangements of the auditory cortex.

Binaural interactions. Input from each cochlea is present at every level of the auditory system central to the cochlear nuclei (Fig. 10-15). Single-neuron studies further show that most cells of the superior olivary complex, inferior colliculus, medial geniculate nucleus, and auditory cortex have binaural receptive fields; that is, a neural response can be generated from stimulation of either cochlea, and the two receptive fields generally have similar tuning characteristics.

Such binaural interactions are thought to play an important role in sound localization. Humans can localize the direction of sounds with an accuracy that depends on frequency, but the accuracy can approach 1 degree. Two cues are thought to be important. A sound source closer to one ear will reach that ear sooner and with greater intensity than it will for the other ear. Thus interaural differences in temporal phase and intensity are key factors in sound localization. Such a difference actually specifies a locus of directions that fall within a vertical plane passing through the head.

Presumably binaural neurons sensitive to differences in phase and/or firing rate (since firing rate encodes intensity) of their monaural inputs could encode certain features of the direction of the sound source. Most attention related to this issue has been focused on the superior olivary complex because this is the first stage of effective binaural integration. (However, the olivocochlear pathway raises the possibility that even hair cells might exhibit binaural interactions, and some evidence of such limited interactions has been found for cells of the dorsal cochlear nucleus.) In particular, many neurons of the medial superior olivary nucleus have two major dendrites, one directed medially and the other directed laterally (Fig. 10-18). Most of the inputs to the medial dendrites are excitatory, and such inputs derive from the contralateral ventral cochlear nucleus. Most inputs to the lateral dendrites are inhibitory, and they derive from the ipsilateral ventral cochlear nucleus. It has been suggested that, in order for these cells to fire, the timing of arrival of excitatory and inhibitory inputs must be precise. One of the factors that governs this timing is the phase difference of the sound arriving at the two ears. Activity transmitted through such a neuron of the medial superior olivary nucleus might thus signal a particular interaural time difference and thus a locus of directions for the sound source.

Cortical architecture and specificity. From the descriptions of the visual and somatosensory systems, one might predict two features that distinguish the auditory cortex from lower levels within the system. First, the cortical neurons appear to be arranged in functional columns of cells perpendicular to the pial surface. Second, there is some evidence that cortical neurons display certain forms of stimulus specificity, such as the direction that a frequency-modulated stimulus is swept (i.e., from high to low frequencies or the reverse) through the auditory receptive field. This is precisely analogous to direction selectivity for visual or somatosensory receptive fields.

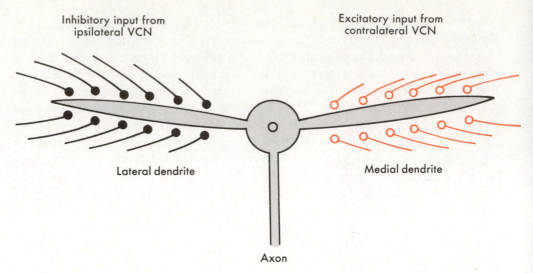

Inhibitory input from
ipsilateral VCN

Excitatory input from
contralateral VCN

Lateral dendrite

Medial dendrite

Axon

Fig. 10-18 ■ Schematic diagram of binaural interactions into neuron of the medial superior olivary nucleus on the left side. The cell's medial dendrite receives excitatory input from the contralateral ventral cochlear nucleus, whereas the lateral dendrite receives inhibitory input from the ipsilateral ventral cochlear nucleus.

■   ■   ■

Consideration of the auditory system begins with a consideration of the nature of sound waves and how they can lead to a generator potential in cochlear hair cells. Sound travels down the auditory canal and causes the tympanic membrane to vibrate. This creates vibrations in the ossicular chain of the middle ear and ultimately leads to vibrations of the basilar membrane in the liquid-filled cochlea. The ossicles permit an effective impedance match between the sounds in the air and the vibrations they cause in the liquids of the cochlea. Vibrations of the basilar membrane mechanically distort the hair cells, and this leads to generator potentials that are synaptically transmitted from the hair cells to auditory nerve fibers.

The viscoelastic properties of the basilar membrane shift gradually along its length. This causes different portions to be uniquely sensitive to particular sound frequencies in such a way that a precise and orderly tonotopic map is formed in the cochlea. Because this receptor-containing tissue maps sound frequency rather than space, auditory receptive fields are defined in terms of these frequencies.

The central auditory pathways are extremely complicated, and the various levels from cochlear nerve to cortex are arranged in series and parallel with one another. Important auditory structures include the cochlear nuclei, trapezoid body, and superior olivary complex of the medulla, the lateral lemniscus, which contains ascending auditory fibers, the inferior colliculus of the midbrain, the medial geniculate nucleus of the thalamus, and several areas of the cerebral cortex. Descending projections are also prominent, and there exists an efferent projection from the superior olivary complex to the hair cells.

The organization of the auditory system is best suited for discrimination of sound frequencies. However, fairly precise sound localization is also possible. This is thought to involve a neural encoding of binaural differences in intensity and temporal phase to a given sound source.

Finally, the functional organization of the auditory system bears many similarities to the visual and somatosensory systems. These similarities include a division into parallel portions that may reflect phylogenetic age. Also, the newer portions exhibit tonotopic organization, afferent inhibition, and several distinct thalamocortical areas. Other more detailed similarities exist as well.

■ *Bibliography*

Journal articles

Allen, J.B.: Cochlear micromechanics—a physical model of transduction, J. Acoust. Soc. Am. **68**:1660, 1981.

Allon, N., Yeshurun, Y., and Wollberg, Z.: Responses of single cells in the medial geniculate body of awake squirrel monkeys, Exp. Brain Res. **41**:222, 1981.

Anderson, R.A., Knight, P.L., and Merzenich, M.M.: The thalamocortical and corticothalamic connections of AI, AII, and the anterior auditory field (AAF) in the cat: evidence for two largely segregated systems of connections, J. Comp. Neurol. **194**:663, 1980.

Anderson, R.A., et al.: The efferent projections of the central nucleus and the pericentral nucleus of the inferior colliculus in the cat, J. Comp. Neurol. **194**:649, 1980.

Bodian, D., and Gucer, G.: Denervation studies of synapses of organ of Corti of old world monkeys, J. Comp. Neurol. **192**:785, 1980.

Brugge, J.F., and Geisler, C.D.: Auditory mechanisms of the lower brainstem, Annu. Rev. Neurosci. **1**:363, 1978.

Brunso-Bechtold, J.K., Thompson, G.C., and Masterton, R.B.: HRP study of the organization of auditory afferents ascending to central nucleus of inferior colliculus in cat, J. Comp. Neurol. **197**:705, 1981.

Calford, M.B., and Webster, W.R.: Auditory representation within principal division of cat medial geniculate body: an electrophysiological study, J. Neurophysiol. **45**:1013, 1981.

Galaburda, A., and Sanides, F.: Cytoarchitectonic organization of the human auditory cortex, J. Comp. Neurol. **190**:597, 1980.

Glendenning, K.K., et al.: Ascending auditory afferents to the nuclei of the lateral lemniscus, J. Comp. Neurol. **197**:673, 1981.

Hebrank, J., and Wright, D.: Spectral cues used in the localization of sound sources on the medial plane, J. Acoust. Soc. Am. **56**:1829, 1974.

Irvine, D.R.F.: Acoustic properties of neurons in posteromedial thalamus of cat, J. Neurophysiol. **43**:395, 1980.

Kitzes, L.M., Wrege, K.S., and Cassady, J.M.: Patterns of responses of cortical cells to binaural stimulation, J. Comp. Neurol. **192**:455, 1980.

Lim, D.J.: Cochlear anatomy related to cochlear micromechanics: a review, J. Acoust. Soc. Am. **67**:1686, 1980.

Middlebrooks, J.C., Dykes, R.W., and Merzenich, M.M.: Binaural response-specific bands in primary auditory cortex (AI) of the cat: topographic organization orthogonal to isofrequency contours, Brain Res. **181**:31, 1980.

Middlebrooks, J.C., and Pettigrew, J.D.: Functional classes of neurons in primary auditory cortex of the cat distinguished by sensitivity to sound location, J. Neurosci. **1**:107, 1981.

Phillips, D.P., and Irvine, D.R.F.: Responses of single neurons in physiologically defined primary auditory cortex (AI) of the cat: frequency tuning and responses to intensity, J. Neurophysiol. **45**:48, 1981.

Reale, R.A., and Imig, T.J.: Tonotopic organization in auditory cortex of the cat, J. Comp. Neurol. **192**:265, 1980.

Zwislocki, J.J.: Five decades of research on cochlear mechanics, J. Acoust. Soc. Am. **67**:1679, 1980.

Books and monographs

Brodal, A.: Neurological anatomy in relation to clinical medicine, ed. 3, New York, 1981, Oxford University Press, Inc.

Flock, A.F.: Physiological properties of sensory hairs. In Evans, E.F., and Wilson, J.P., editors: Psychophysics and physiology of hearing, London, 1977, Academic Press, Inc.

Goldstein, M.H.: The auditory periphery. In Mountcastle, V.B., editor: Medical physiology, ed. 14, vol. I, St. Louis, 1980, The C.V. Mosby Co.

Gulick, W.L.: Hearing: physiology and psychophysics, New York, 1971, Oxford University Press, Inc.

Kiang, N.Y.S.: Stimulus representation in the discharge patterns of auditory neurons. In Tower, D.B., editor: The nervous system, vol. 3, Human communication and its disorders, New York, 1975, Raven Press.

Kiang, N.Y.S., et al.: Discharge patterns of single fibers in the cat's auditory nerve, research monograph No. 35, Cambridge, Mass., 1965, The MIT Press.

Morest, D.K.: Structural organization of the auditory pathways. In Tower, D.B., editor: The nervous system, vol. 3, Human communication and its disorders, New York, 1975, Raven Press.

Neff, W.D., Diamond, I.T., and Casseday, J.H.: Behavioral studies of auditory discrimination: central nervous system. In Keidel, W.D., and Neff, W.D., editors: Handbook of sensory physiology, auditory system: physiology, central nervous system, behavioral studies, and psychoacoustics, vol. V/2, New York, 1975, Springer-Verlag New York, Inc.

Von Békésy, G.: Experiments in hearing, New York, 1960, McGraw-Hill, Inc.

Woolsey, C.N., editor: Cortical sensory organization, vol. 3, Multiple auditory areas, Clifton, N.J., 1981, Humana Press.

chapter II

The vestibular system

The function of the vestibular system is to sense forces of acceleration, both linear and rotational. The most prominent of these forces is the linear force of gravity. Because the vestibular apparatus is part of the inner ear and is thus located inside the head, it is head acceleration that is sensed.

Two general and related functions are served by the vestibular system. In the presence of forces acting on the head, balance of the body is maintained through postural reflexes that are largely determined from vestibular input. Also, because of vestibulo-ocular reflexes, stable eye positions are maintained with respect to visual objects in spite of pronounced head movements. This allows the maintenance of fixation and a stable visual field during activities such as running and jumping.

This discussion of the vestibular system is more superficial than that of vision, somesthesia, and hearing for two reasons. First, the vestibular pathways have nothing analogous to a "lemniscal" thalamocortical system. Second, although we are conscious of forces of acceleration, most of the vestibular information is processed subconsciously to serve postural and oculomotor reflexes (see also Chapter 18). This discussion concerns the peripheral vestibular apparatus and some of the related central pathways.

■ *Vestibular apparatus*

■ *Gross structure of the labyrinth*

The peripheral apparatus of the vestibular system lies in the inner ear. It is adjacent to and continuous with the cochlea (Fig. 11-1). The vestibular apparatus consists of a *bony labyrinth* into which is fitted the *membranous labyrinth*. Between the bony and membranous labyrinths is the *perilymph;* viscous *endolymph* fills the membranous labyrinth. The perilymph and endolymph are continuous with the same fluids contained within the cochlea (Chapter 10).

The labyrinths are functionally divided into five prominent divisions. These include two bulbous enlargements, the *utricle* and *saccule,* and three *semicircular ducts* or *canals,* the *superior, posterior,* and *horizontal.* Fig. 11-2 shows the relationship of these structures to one another and their orientation within the head. Note that the semicircular canals of one side are in planes that are roughly perpendicular to one another. Also, the planes of the two horizontal canals are approximately parallel with one another. This is also true for both the planes of the right superior and left posterior canals as well as those of the left superior and right posterior canals.

Near one end of each canal is an enlargement called the *ampulla* (Fig. 11-1). The epithelium inside each ampulla is thickened in a region called the *crista;* the crista contains the vestibular transducers. These transducers are *hair cells,* which are quite similar to those in the cochlea (Fig. 11-3). The saccule and utricle also have a region of epithelium called the *macula,* which contains a dense aggregate of hair cells. The utricular macula is oriented in the horizontal plane and the saccular macula vertically, approximately in a parasagittal plane.

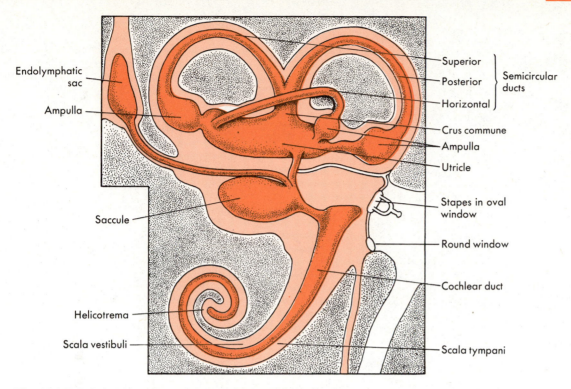

Fig. 11-1 ■ Relationships between the membranous labyrinths (darker color), perilymph (lighter color), and bone (stippling) in the inner ear. (Redrawn from Kandel, E.R., and Schwartz, J.H.: Principles of neural science, New York, 1981, Elsevier.)

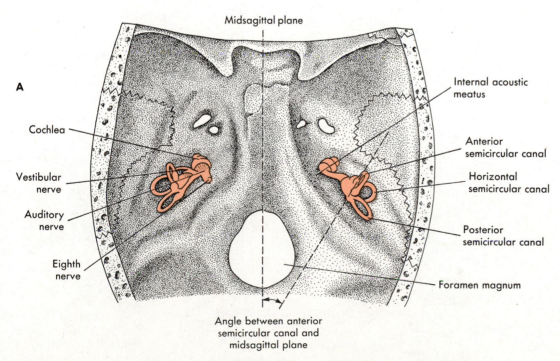

Fig. 11-2 ■ Structure of the inner ear. **A,** View of cranium from above, showing relative location and orientation of the inner ear (color). Note that the horizontal semicircular canals are roughly coplanar and that the anterior canal and contralateral posterior canal lie in roughly parallel planes.

Continued.

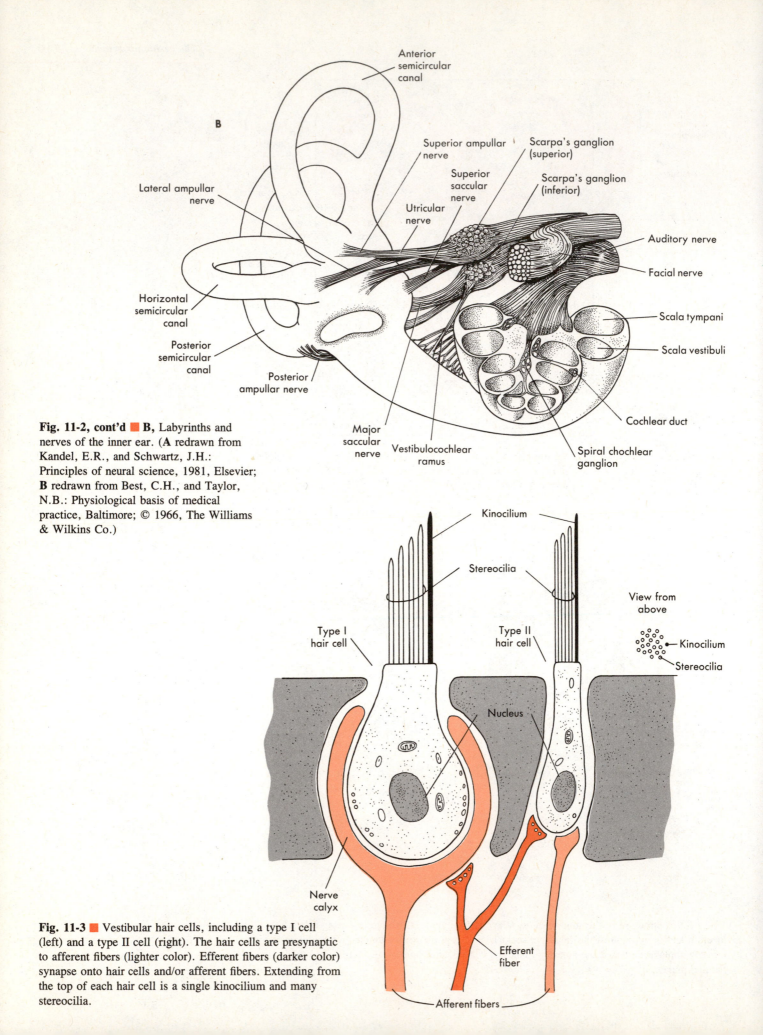

Fig. 11-2, cont'd ■ **B,** Labyrinths and nerves of the inner ear. (**A** redrawn from Kandel, E.R., and Schwartz, J.H.: Principles of neural science, 1981, Elsevier; **B** redrawn from Best, C.H., and Taylor, N.B.: Physiological basis of medical practice, Baltimore; © 1966, The Williams & Wilkins Co.)

Fig. 11-3 ■ Vestibular hair cells, including a type I cell (left) and a type II cell (right). The hair cells are presynaptic to afferent fibers (lighter color). Efferent fibers (darker color) synapse onto hair cells and/or afferent fibers. Extending from the top of each hair cell is a single kinocilium and many stereocilia.

Fig. 11-3 shows the main features of the vestibular hair cells, which form synapses onto sensory fibers of the eighth nerve. Also, as is the case for cochlear hair cells, synapses are formed onto these receptors from efferent fibers. At the top of each hair cell are a number of *stereocilia* of variable length, plus a single *kinocilium* at one edge of the bundle of cilia. Bending of the stereocilia leads to a generator potential that synaptically affects the firing rate of the eighth nerve fiber. The longer stereocilia are closer to the kinocilium; this provides a functional axis of polarization (Fig. 11-4). Bending of stereocilia toward the kinocilium depolarizes the hair cell and increases the firing rate of the afferent vestibular nerve fiber; bending in the reverse direction hyperpolarizes the hair cell and decreases the afferent fiber's firing rate. By definition, the axis and direction of a hair cell are the line from the shortest stereocilium to the kinocilium.

Semicircular canals. The cilia of the hair cells in each ampulla are embedded in the *cupula,* which is a gelatinous capsule that hangs down from the roof of the ampulla (Fig. 11-5). Movements of the viscous endolymph with respect to the walls of the semicircular canal push on and displace the cupula, thereby bending the stereocilia and effecting neural transduction. Because the endolymph has considerable inertia, it lags behind in response to head rotations or angular accelerations. This lagging provides the necessary stimulus for a generator potential. Thus the direction of bending of the stereocilia is opposite to the direction of head rotation (Fig. 11-6).

Within an ampulla the hair cells are all oriented (with regard to their functional axis as just defined) in the same direction. This direction points toward the utricle for the ampulla of the horizontal canal and away from the utricle for the ampullae of the other semicircular canals. This arrangement allows the complementary pairs of semicircular canals on each side (left and right horizontal, left superior and right posterior, left posterior and right superior) (Fig. 11-2) to synchronize their outputs in a heteronymous fashion. Fig. 11-6 shows how this works for the horizontal canals. In this example a clockwise head rotation (as viewed from above) causes a counterclockwise bending of the hair cells. This depolarizes hair cells of the right ampulla and increases the firing

■ *Hair cells*

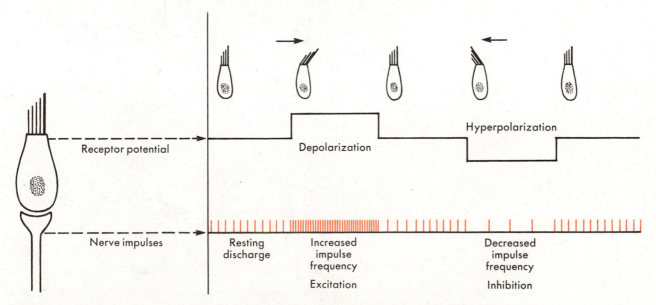

Fig. 11-4 ■ Directional selectivity of hair cells. Bending of the kinocilium away from the stereocilia depolarizes the hair cell, which increases the firing rate in its afferent fiber. Bending of the kinocilium toward the stereocilia hyperpolarizes the hair cell, which decreases the firing rate in its afferent fiber. (Redrawn from Kandel, E.R., and Schwartz, J.H.: Principles of neural science, New York, 1981, Elsevier.)

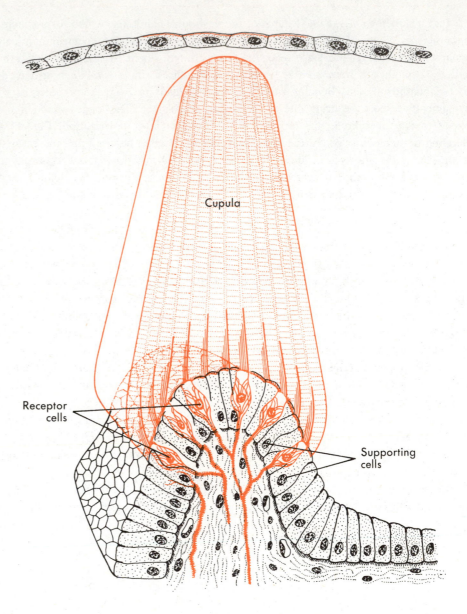

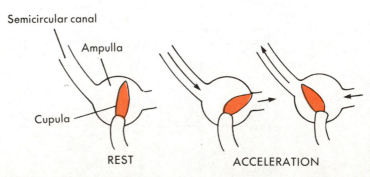

Fig. 11-5 ■ Response of the cupula in the ampulla to angular acceleration. *Above,* Note the hair cells with their cilia embedded in the ampulla. *Below,* Ampullar displacement (and thus bending of the stereocilia and kinocilia) in response to acceleration. (Redrawn from Wersäll, J.: Acta Otolaryngol. [Stock.] Suppl. **126:**1, 1956.)

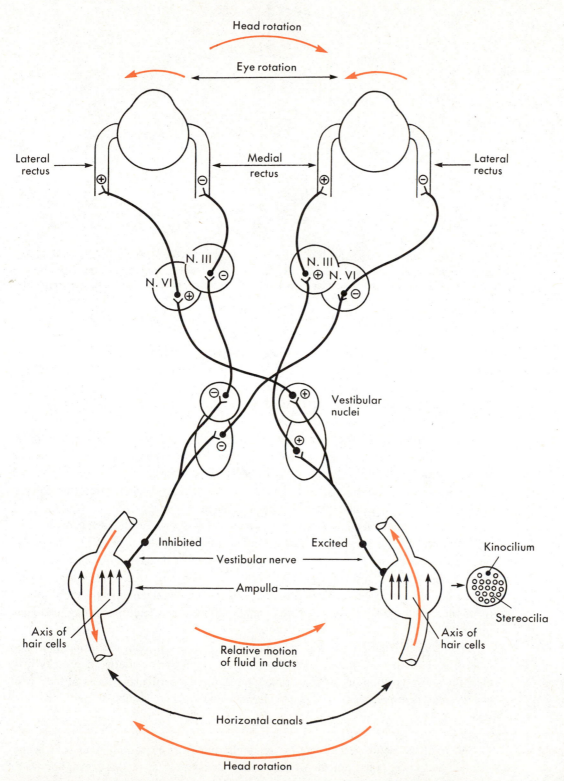

Fig. 11-6 ■ Vestibuloocular reflex in response to head rotation in the horizontal plane. For the head rotation as shown, the right ampulla is excited and the left is inhibited. Because of the organization of the vestibular pathways, this results in an eye movement that compensates for the head rotation, so retinal images are stabilized.

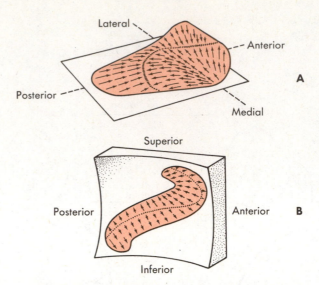

Fig. 11-7 ■ Axes of polarization of macular hair cells. The arrows represent the direction from the stereocilia to the kinocilium. **A,** Utricle. **B,** Saccule. (Redrawn from Spoendlin, H.H. In Wolfson, R.J., editor: The vestibular system and its diseases, Philadelphia, 1966, University of Pennsylvania Press.)

rate of vestibular nerve afferents on the right side; hyperpolarization and decreased firing result on the left. An analogous result can be predicted for head rotations in other planes that involve the superior or posterior canals. Because the semicircular canals respond in this manner to head movements, their responses are most important in the control of eye movements to maintain fixation.

Utricle and saccule. The utricle and saccule respond principally to changes in linear acceleration, particularly to that caused by gravity. Their responses are most important for the control of posture and maintenance of balance. Hair cells in the macula of the utricle and saccule are stimulated by forces similar to those which affect the canals, although some of the details differ. The cilia of macular hair cells extend from the cells into a dense *otolithic membrane,* which is a matrix floating within the utricle or saccule. The utricular macula is oriented horizontally within the head, so head tilt causes the otolithic membrane to deform the stereocilia. The saccular macula is oriented vertically in the sagittal plane. Hence a change in upward or downward acceleration (e.g., jumping or falling) results in an analogous deformation of the stereocilia by the otolithic membrane. Both macular structures are sensitive to linear accelerations in the horizontal plane.

The hair cell polarization axes are orderly but complex in both the utricle and saccule. Fig. 11-7 shows the axes for each structure. From these patterns of polarity it is clear that each direction of linear acceleration results in a particular pattern of increased and decreased responsiveness of hair cells and their related vestibular nerve afferent fibers.

■ *Vestibular pathways*

Fibers of the eighth nerve carry the vestibular information from the hair cells to the medulla and lower pons. Somata of these fibers are located in Scarpa's ganglion (Fig. 11-2). Their axons terminate among the four nuclei of the vestibular nuclear complex: the *superior vestibular nucleus* (Bechterew's nucleus), the *lateral vestibular nucleus* (Deiters' nucleus), the *medial vestibular nucleus* (Schwalbe's nucleus), and the *inferior vestibular nucleus* (Roller's nucleus) (Fig. 11-8). These nuclei receive functionally different afferent inputs and innervate different motor areas (see also Chapter 10).

The medial and superior vestibular nuclei receive most of their vestibular input from

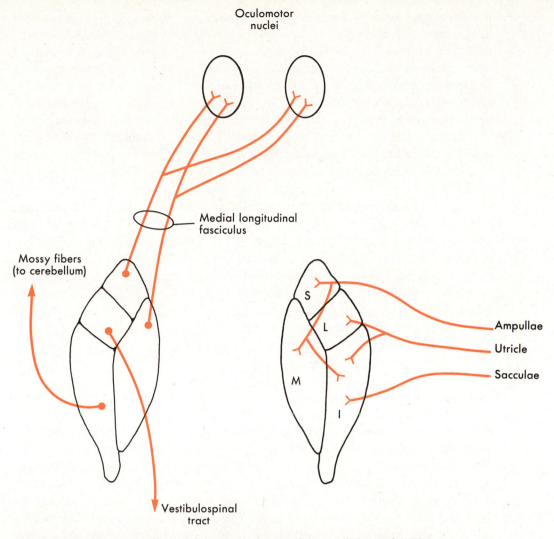

Oculomotor
nuclei

Medial longitudinal
fasciculus

Mossy fibers
(to cerebellum)

S

L

M

I

Ampullae

Utricle

Sacculae

Vestibulospinal
tract

Fig. 11-8 ■ Simplified diagrams of the major pathways of the vestibular nuclei. The subdivisions of the vestibular nuclei are *S*, superior; *L*, lateral; *M*, medial; and *I*, inferior.

the semicircular canals. Axons from cells in these nuclei ascend in the *medial longitudinal fasciculus* and bilaterally innervate motor nuclei of the extraocular muscles (Fig. 11-8). Fig. 11-6 shows how these pathways subserve vestibuloocular reflexes. They maintain visual fixation by stabilizing the position of the eyes in space during head movements.

As described in Fig. 11-6, a clockwise head rotation causes the horizontal ducts to move clockwise relative to the endolymph because of inertia of the endolymph within the ducts. This, in turn, leads to increased firing in the right vestibular nerve and decreased firing in the left nerve. Such head rotation ultimately evokes appropriate changes in firing rates for neurons in the abducens and oculomotor nuclei so that conjugate movement of the eyes is produced to compensate precisely for the head movement.

If the head rotation continues, eventually each eye moves to its extreme position in the orbit and cannot be deviated further. At this point a rapid eye movement in the direction of the head rotation is made, a new fixation point is chosen, and a slow drift to compensate for the further head rotation begins again. These rhythmic eye movements are called *vestibular nystagmus* (nystagmus refers to periodic, conjugate eye movements) and are involuntary. The slow phase of vestibular nystagmus is the eye movement opposite to the direction of head rotation; the fast phase is in the direction of head rotation. Eventually friction between the ducts and endolymph reduces the movement between them. The nystagmus slows down and stops after about 20 seconds of head rotation. However, on abrupt termination of head rotation the endolymph continues to move in the same direction as the prior head rotation. Thus relative movement between the endolymph and duct begins, and this movement is opposite in relative direction to that present during the head rotation. This leads to *postrotary nystagmus,* which is characterized by fast and slow phases opposite in direction to those of the nystagmus observed during the head rotation. Spontaneous nystagmus or a failure to elicit normal nystagmus is often associated with brainstem lesions that interrupt the vestibular pathways.

The lateral vestibular nucleus receives utricular innervation. Neurons of this nucleus also receive inhibitory innervation from cerebellar Purkinje cells (Chapter 16). The output of the lateral vestibular nucleus is largely to spinal motoneurons via the vestibulospinal tract (Fig. 11-8). This portion of the vestibular system is chiefly concerned with postural functions and contributes to medial descending motor pathways (Chapter 18).

Finally, the inferior vestibular nucleus receives input from the utricle, saccule, and semicircular canals, as well as from the cerebellum. This nucleus innervates a number of brainstem neurons via the medial longitudinal fasciculus and also projects the *mossy fibers* to the flocculonodular lobe of the cerebellum (Fig. 11-8).

■   ■   ■

In brief, the vestibular system provides information about linear and rotational head acceleration. Transduction occurs in hair cells of the labyrinths. These transducers are much like cochlear hair cells, and they form synapses onto afferent fibers of the vestibular nerve. Linear acceleration (e.g., gravity) is signaled by the utricle and saccule. This information is necessary for postural reflexes. Rotational movements are signaled by the semicircular canals. This information permits compensatory eye movements to maintain visual fixation during head movements. The vestibular nerve innervates the vestibular nuclei of the medulla and pons. These nuclei innervate extraocular motor nuclei and other brainstem structures via the medial longitudinal fasciculus, the cerebellum, and spinal motoneurons via the vestibulospinal tract. Through these pathways, vestibuloocular and postural reflexes are effected.

■ *Bibliography*

Journal articles

Anastosopoulos, D., and Mergner, T.: Canal-neck interaction in vestibular nuclear neurons of the cat, Exp. Brain Res. **46:**269, 1982.

Anderson, J.H., Blanks, R.H.I., and Precht, W.: Response characteristics of semicircular canal and otolith systems in cats. I. Dynamic responses of vestibular fibers, Exp. Brain Res. **32:**491, 1978.

Boyle, R., and Pompeiano, O.: Responses of vestibulospinal neurons to sinusoidal rotation of neck and macular vestibular inputs on vestibulospinal neurons, J. Neurophysiol. **45:**852, 1981.

Büttner, U., and Waespe, W.: Vestibular nerve activity in the alert monkey during vestibular and optokinetic nystagmus, Exp. Brain Res. **41:**310, 1981.

Corey, D.P., and Hudspeth, A.J.: Ionic basis of the receptor potential in a vertebrate hair cell, Nature **281:**675, 1979.

Donaghy, M.: The cat's vestibulo-ocular reflex J. Physiol. (Lond.) **300:**337, 1980.

Goldberg, J.M., and Fernández, C.: Vestibular mechanisms, Annu. Rev. Physiol. **37:**129, 1975.

Goldberg, J.M., and Fernández, C.: Efferent vestibular system in the squirrel monkey: anatomical location and influence on afferent activity, J. Neurophysiol. **43:**986, 1980.

Hudspeth, A.J.: Extracellular current flow and the site of transduction by vertebrate hair cells, J. Neurosci. **2:**1, 1982.

Hudspeth, A.J., and Jacobs, R.: Stereocilia mediate transduction in vertebrate hair cells, Proc. Natl. Acad. Sci. (U.S.A.) **76:**1506, 1979.

Kubin, L., Manzoni, D., and Pompeiano, O.: Responses of lateral reticular neurons to convergent neck and macular vestibular inputs, J. Neurophysiol. **46:**48, 1981.

Mergner, T., Deecke, L., and Wagner, H.-J.: Vestibulo-thalamic projection to the anterior suprasylvian cortex of the cat, Exp. Brain Res. **44:**455, 1981.

Precht, W.: Vestibular mechanisms, Annu. Rev. Neurosci. **2:**265, 1979.

Raphan, T., and Cohen, B.: Brainstem mechanisms for rapid and slow eye movements, Annu. Rev. Physiol. **40:**527, 1978.

Shotwell, S.L., Jacobs, R., and Hudspeth, A.J.: Direction sensitivity of individual vertebrate hair cells to controlled deflection of their hair bundles, Ann. N.Y. Acad. Sci. **374:**1, 1981.

Tomko, D.L., Peterka, R.J., and Schor, R.H.: Responses to head tilt in cat eighth nerve afferents, Exp. Brain Res. **41:**216, 1981.

Tomko, D.L., et al.: Response dynamics of horizontal canal afferents in barbiturate-anesthetized cats, J. Neurophysiol. **45:**376, 1981.

Books and monographs

Brodal, A.: Neurological anatomy in relation to clinical medicine, ed. 3, New York, 1981, Oxford University Press.

Flock, Å.: Sensory transduction in haircells. In Lowenstein, W.R., editor: Handbook of sensory physiology, vol. 1, Principles of receptor physiology, New York, 1971, Springer-Verlag.

Kornhuber, H.H.: Nystagmus and related phenomena in man: an outline of otoneurology. In Kornhuber, H.H., editor: Handbook of sensory physiology, vol. 6, Vestibular system, part 2, psychophysics, applied aspects, and general interpretations, New York, 1974, Springer-Verlag.

Naunton, R.F., editor: The vestibular system, New York, 1975, Academic Press, Inc.

Wersäll, J., and Bagger-Sjöbäck, D.: Morphology of the vestibular sense organ. In Kornhuber, H.H., editor: Handbook of sensory physiology, vol. 6, Vestibular system, part 1, Basic mechanisms, New York, 1974, Springer-Verlag.

Wilson, V.J., and Melvill Jones, G.: Mammalian vestibular physiology, New York, 1979, Plenum Press.

chapter 12

Chemical senses

Gustation (taste) and olfaction (smell) are the only senses that can identify the chemical nature of stimuli in the external environment. In many animals these important senses not only help to define the environment but also affect social behavior, such as mating, territoriality, and feeding. During the course of primate evolution, however, these chemical senses have become less important to conscious perception as vision, touch, and hearing have dramatically evolved. For this reason, and because of our relative ignorance of the functional organization of the chemical senses, less attention will be devoted to this subject.

■ *Taste*

The elemental qualities of taste are usually divided into four basic groups: salty, sweet, sour, and bitter. This grouping can be demonstrated from behavioral experiments and seems to have a reasonably straightforward neural basis. More complex tastes are thought to be combinations of these four qualities in much the same way that a few primary colors mixed in various combinations can provide the perception of any desired color.

■ *Receptors*

The chemoreceptors involved in taste are the *taste buds*. These are located throughout the mouth but are most concentrated on the dorsal surface of the tongue. They are also found on the palate, faucial pillars, pharynx, and larynx. Taste buds typically occur in groups of one to five on raised *papillae*. Fig. 12-1 includes drawings of a papilla and a taste bud with the transducer cells. The taste bud has 40 to 50 cells in the shape of a bottle; the opening at the top is the *taste pore*.

The transducer cells have microvilli that extend into the taste pore. These microvilli are thought to be chemosensitive, and somehow their contact with the appropriate molecule or molecules leads to a generator potential. Each transducer cell is innervated at its base by fine nerve fibers. Although direct physiological or morphological evidence is lacking, it seems likely that these fibers include afferent axons that conduct taste information toward the brain as well as efferent fibers from the brain. Also, it seems likely that conventional synapses exist between these fibers and the transducer cells, although direct evidence, again, is lacking.

Recordings from individual sensory fibers indicate that individual taste buds and fibers are not typically selective for any of the four elemental taste stimuli. However, these fibers usually are selective for some of these stimuli and not others, and it is thought that the combined activity of many of these partially selective fibers provides the information needed to extract the relative presence of the four elemental stimuli.

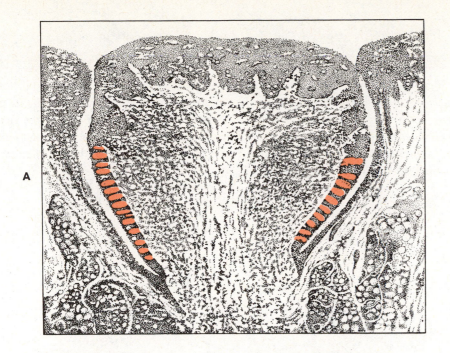

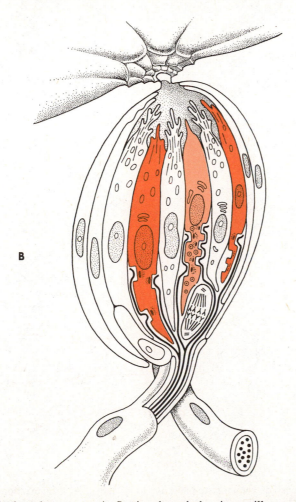

Fig. 12-1 ■ Taste buds and receptors. **A,** Section through the circumvillate papilla of a macaque monkey. The taste buds are shown in color. The top surface of the papilla is roughly 1.5 mm across. **B,** Taste bud, cut away to expose two types of receptor cells. One (lighter color) has presynaptic vesicles with dense cores, and the other (darker color) has standard vesicles. Supporting cells are not colored. The receptor cell innervation is shown in black. (**A** redrawn from Bloom, W., and Fawcett, D.W.: A textbook of histology, ed. 8, Philadelphia, 1962, W.B. Saunders Co.; **B** redrawn from Williams, P.L., and Warwick, R.: Functional neuroanatomy of man, Philadelphia, 1975, W.B. Saunders Co.)

■ *Pathways*

Fig. 12-2 summarizes the taste pathways. Most or all of the innervation of the tongue is supplied by fibers from the *facial* (seventh), *glossopharyngeal* (ninth), and *vagus* (tenth) nerves. The somata of these fibers are located, respectively, in the *geniculate, superior petrosal,* and *nodose ganglia.* These fibers enter the medulla and terminate in the *nucleus of the solitary tract.* Fibers from the left or right half of the tongue innervate the ipsilateral nucleus.

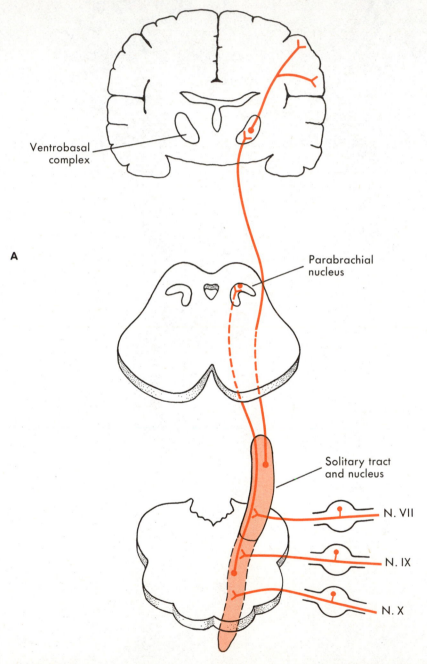

Fig. 12-2 ■ Summary of known taste pathways in monkeys. **A,** Organization of central pathways. The primary afferent fibers enter the medulla via the facial *(N. VII),* glossopharyngeal *(N. IX),* and vagus *(N. X)* nerves. Synapses are formed in the nucleus of the solitary tract (lighter color). Some second-order fibers ascend to terminate in the parabrachial nucleus of the pons, whereas others terminate in the ventrobasal complex of the thalamus. A projection from the parabrachial nucleus to the thalamus has not yet been described for the monkey, although it appears to exist in many other vertebrates (e.g., rats, hamsters, cats, and fish). Finally, neurons in the thalamus project to the cortex.

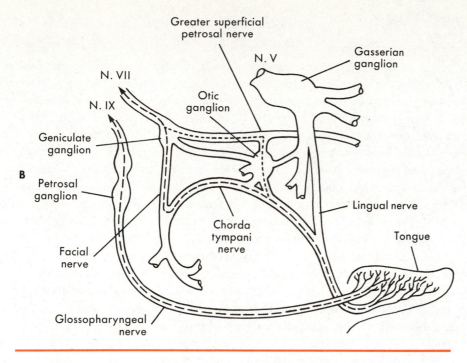

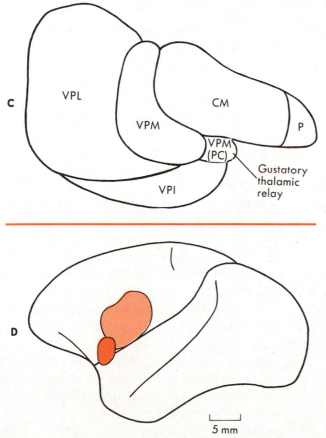

Fig. 12-2, cont'd ■ **B,** Details of the gustatory innervation of the tongue. **C,** Coronal section through the thalamus of a monkey at the level of the gustatory thalamic relay, the parvocellular portion of the ventral posteromedial nucleus [*VPM (PC)*]. Other nuclei shown are ventral posterolateral *(VPL)*, ventral posteromedial *(VPM)*, centromedian *(CM)*, ventral posteroinferior *(VPI)*, and parafascicular *(P)*. **D,** Two separate cortical representations of taste, termed TNI (lighter color) and TNII (darker color), in the monkey. Neurons in TNI are responsive to somatosensory as well as gustatory stimuli, but TNII seems to be a purely gustatory area. (**B** redrawn from Beidler, L.M. In Mountcastle, V.B, editor: Medical physiology, ed. 14, St. Louis, 1980, The C.V. Mosby Co.; **D** redrawn from Benjamin, R.M., and Burton, H.: Brain Res. **7:**221, 1968.)

Our knowledge of the central taste pathways is sketchy at best. In the rat, neurons from the nucleus of the solitary tract project ipsilaterally to a pontine nucleus just lateral to the brachium conjunctivum; this pontine structure is known as the *parabrachial nucleus*. Limited evidence from hamsters, cats, and fish suggests a homologous pathway. Fibers from the parabrachial nucleus terminate in a small nuclear complex medial and adjacent to the ventral posterior medial (VPM) nucleus of the thalamus. In monkeys (and probably humans), however, some neurons from the nucleus of the solitary tract bypass the parabrachial nucleus and project directly to the thalamus. This gustatory thalamic relay may indeed be a part of the ventrobasal (VB) complex described in Chapter 9. The pontothalamic projection is predominantly ipsilateral, although a small crossed component seems also to be present. Finally, thalamocortical gustatory fibers exist, and well-defined cortical taste areas have been described in many mammalian species. At least two have been found in primates—one associated within the somatosensory region on the lateral convexity and another in the opercular insular cortex. Note that, unlike the predominantly crossed somatosensory pathway, each half of the tongue is mapped primarily within the ipsilateral hemisphere.

Finally, gustatory information seems to reach many other brainstem areas. These include the hypothalamus, the amygdaloid complex, the substantia innominata, and the bed nucleus of the stria terminalis.

■ *Smell*

The sense of smell is the least understood of all the major senses, for at least two reasons. First, it is difficult to define adequately either an olfactory stimulus or the response to such a stimulus. There are no "elemental" odors analogous to salty, sweet, sour, or bitter tastes. Thus, even though we have remarkable abilities to detect and identify molecules based on odor (and among mammals, primates are quite poor at this), we lack a detailed, quantitative picture of these abilities. This in turn means that investigations of the olfactory pathways are often carried out without a theoretical framework built on specific functional questions. Second, the olfactory pathways themselves are diffuse and difficult to characterize. These pathways are probably the most primitive of any of the major sensory systems, and this might explain the diffuse nature of the central connections.

■ *Receptors*

The olfactory receptor cells lie in the *olfactory mucosa,* which is located on the upper posterior portion of the nasal septum and the opposite part of the lateral wall (Fig. 12-3). The mucosa occupies a total of only about 2.5 square centimeters in each nostril, and it consists of a number of supporting cells as well as the receptor cells.

The receptor cells act as both transducer and ganglion cells (Fig. 12-4, *A*). At one end of the cell a number of microvilli project into the nasal cavity; at the other end the soma thins to a single, fine unmyelinated axon. These axons pass through fenestrations in the cribriform plate of the ethmoid bone and enter the *olfactory bulbs* of the brain. The axons entering the olfactory bulbs form the *olfactory (first) nerve.*

■ *Pathways*

Historically, two general names have sometimes been used to describe the olfactory system: the *limbic system* and the *rhinencephalon*. However, contemporary evidence makes it clear that the structures conventionally associated with the limbic system are not equivalent to those of the rhinencephalon, and neither is isomorphic with known olfactory pathways. This section deals only with the olfactory pathways.

The olfactory bulbs are located at the anterior ventral surface of the brain. They contain three major neuron types: *tufted* and *mitral cells*, which are second-order projection neurons, and *granule cells*, which are interneurons (Fig. 12-4, *B*). Synapses formed between the olfactory nerve afferent fibers and tufted or mitral cells occur in pronounced morphological specializations known as *glomeruli*. The olfactory bulb also

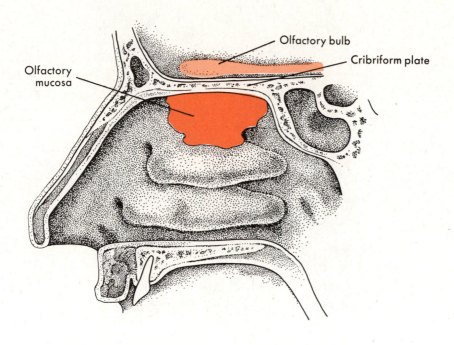

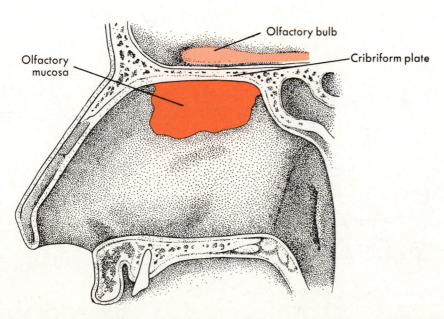

Fig. 12-3 ■ Location of olfactory mucosa (darker color) relative to the olfactory bulb (lighter color). Note the separation of mucosa and bulb by the bony cribriform plate. *Top,* Distribution on lateral nasal wall; *bottom,* distribution on septum.

Fig. 12-4 ■ **A,** Olfactory receptors. Each receptor (lighter color) is continuous with an axon (darker color) that penetrates the cribriform plate and forms the olfactory nerve. Supporting or sustentacular cells are also present in the olfactory mucosa. **B,** Section through the olfactory mucosa and bulb. Axons of olfactory receptors (darker color) penetrate the cribriform plate and enter the bulb, where they terminate in glomeruli, ending on processes of mitral and tufted cells (both black). These latter cells send axons out of the bulb. Also shown are granule cells (lighter color), which are involved in intrinsic circuitry of the bulb. (**A** redrawn from de Lorenzo, A.J.D. In Zotterman, Y., editor: Olfaction and taste, Elmsford, N.Y., 1963, The Pergamon Press; **B** redrawn from House, E.L., and Pansky, B.: A functional approach to neuroanatomy, ed. 2, New York, 1967, McGraw-Hill Book Co. Copyright © 1967 by McGraw-Hill Book Co. Used with permission.)

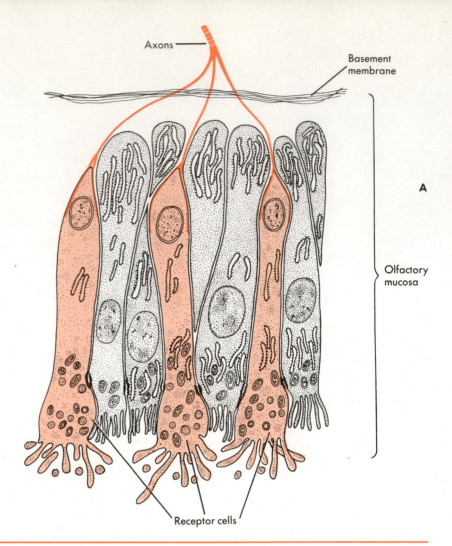

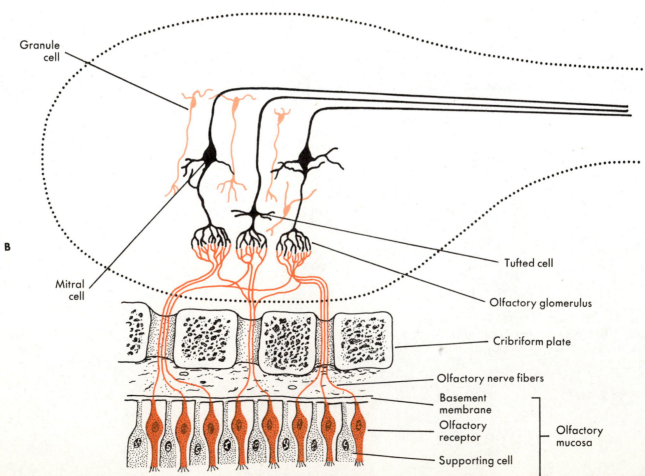

receives inputs from other brain regions, mostly of unknown origin, but many derive from the contralateral olfactory bulb via the *anterior commissure*.

Fig. 12-5 summarizes some of the known connections of the olfactory bulb. Fibers of tufted and mitral cells leave the olfactory bulb posteriorly and form the *olfactory tract*. The tract splits into *lateral* and *medial olfactory striae,* and the enclosed triangular zone is known as the *olfactory trigone*. Central connections of the olfactory striae include the *anterior olfactory nucleus, olfactory tubercle* (or *anterior perforated substance*), parts of the *hypothalamus,* parts of the *amygdaloid nucleus,* the *paleocortex* of the *prepyriform, pyriform,* and *entorhinal* areas, and the contralateral olfactory bulb. No direct connections from the olfactory bulb to the septal areas or hippocampus have been unambiguously demonstrated. Also, no region of the neocortex is devoted to olfaction in a manner analogous to that of the visual or somatosensory cortex.

■ ■ ■

In brief, our description of the chemical senses has been sketchy because we know relatively little of the neural pathways that subserve them. Both taste and smell begin

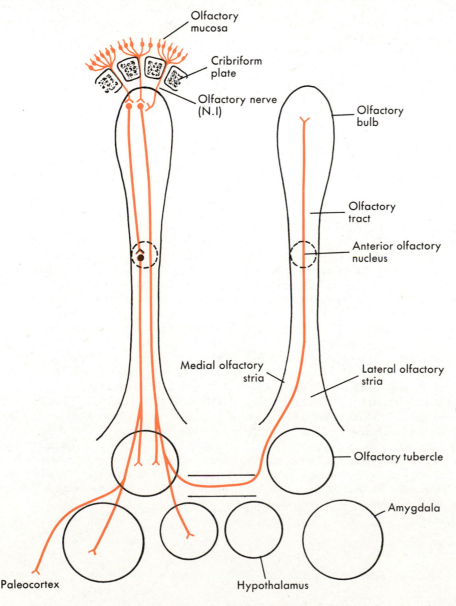

Fig. 12-5 ■ Some of the major olfactory pathways.

with transducers that produce generator potentials in the presence of appropriate chemicals. Limited evidence suggests that taste pathways have a traditional thalamocortical component and may not be organized very differently from visual, somatosensory, or auditory pathways. Olfactory pathways, however, are organized differently from any other sensory system and include no neocortical component. This is thought to reflect the primitive status of these pathways.

■ *Bibliography*

Journal articles

Beckstead, R.M.: Afferent connections of the entorhinal area in the rat as demonstrated by retrograde cell labeling with horseradish peroxidase, Brain Res. **152:**249, 1978.

Beckstead, R.M., Morse, J.R., and Norgren, R.: The nucleus of the solitary tract in the monkey: projections to the thalamus and brainstem nuclei, J. Comp. Neurol. **190:**259, 1980.

Beckstead, R.M., and Norgren, R.: An autoradiographic examination of the central distribution of the trigeminal, facial, glossopharyngeal, and vagal nerves in the monkey, J. Comp. Neurol. **184:**455, 1979.

Davis, B.J., and Macrides, F.: The organization of the centrifugal projections from the anterior olfactory nucleus, ventral hippocampal rudiment, and piriform cortex to the main olfactory bulb in the hamster: an autoradiographic study, J. Comp. Neurol. **203:**475, 1981.

Haberly, L.B., and Price, J.L.: The axonal projection patterns of the mitral and tufted cells of the olfactory bulb in the rat, Brain Res. **129:**152, 1977.

Haberly, L.B., and Price, J.L.: Association and commissural fiber system of the olfactory cortex of the rat. I. Systems originating in the piriform cortex and adjacent areas, J. Comp. Neurol. **178:**711, 1978.

Heimer, L., Van Hoesen, G.W., and Rosene, D.L.: The olfactory pathways and the anterior perforated substance in the primate brain, Int. J. Neurol. **12:**42, 1977.

Newman, R., and Winans, S.S.: An experimental study of the ventral striatum of the golden hamster. I. Neuronal connections of the nucleus accumbens, J. Comp. Neurol. **191:**167, 1980.

Newman, R., and Winans, S.S.: An experimental study of the ventral striatum of the golden hamster. II. Neuronal connections of the olfactory tubercle, J. Comp. Neurol. **191:**193, 1980.

Pfaffmann, C., Frank, M., and Norgren, R.: Neural mechanisms and behavioral aspects of taste, Annu. Rev. Psychol. **30:**283, 1979.

Scott, J.W.: Electrophysiological identification of mitral and tufted cells and distribution of their axons in olfactory system of the rat, J. Neurophysiol. **46:**918, 1981.

Scott, J.W., McBride, R.L., and Schneider, S.P.: The organization of projection from the olfactory bulb to the piriform cortex and olfactory tubercle in the rat, J. Comp. Neurol. **194:**519, 1980.

Shepherd, G.M.: Synaptic organization of the mammalian olfactory bulb, Physiol. Rev. **52:**864, 1972.

Books and monographs

Beidler, L.M.: The chemical senses: gustation and olfaction. In Mountcastle, V.B., editor: Medical physiology, ed. 14, St. Louis, 1980, The C.V. Mosby Co.

Brodal, A.: Neurological anatomy in relation to clinical medicine, ed. 3, New York, 1981, Oxford University Press.

Heimer, L.: The olfactory cortex and the ventral striatum. In Livingston, K.E. and Hornykiewicz, O., editors: Limbic mechanisms: the continuing evolution of the limbic system concept, New York, 1978, Plenum Press.

Pfaffmann, C., et al.: Coding gustatory information in the squirrel monkey chorda tympani. In Sprague, J.M., and Epstein, A.N., editors: Progress in Psychobiology and Physiological Psychology, vol. 6, New York, 1976, Academic Press, Inc.

Shepherd, G.M., Getchell, T.V., and Kauer, J.S.: Analysis of structure and function in the olfactory pathway. In Tower, D.B., editor: The nervous system vol. 1, The basic neurosciences, New York, 1975, Raven Press.

Stoddart, D.M.: The ecology of vertebrate olfaction, London, 1980. Chapman & Hall.

chapter 13

A functional neuroanatomical framework for motor systems

The preceding discussion of sensory systems illustrates the remarkable advances that neurobiology has realized with respect to our understanding of how inputs to the brain are processed. An important substrate for these advances is that in analyzing sensory processing one is able to present precisely controlled stimuli to a peripheral array of receptors and then describe the receptive fields of successively more central neurons along a sensory pathway. When such data are combined with information on the detailed neuronal connectivity of the pathway, it is possible to infer how it is functionally organized to analyze stimulus information.

This capability of analyzing sensory pathways centrally from the periphery has facilitated progress in sensory neurobiology. Unfortunately our understanding of motor systems has not developed as rapidly, in part because this powerful peripheral-to-central approach has been applied only recently. This is not surprising because for the motor systems this requires analysis in a "retrograde" direction, that is, from the muscles centrally. Until recently the techniques available for such an analysis have been limited. However, developments over the past few decades have established a foundation for an organized, functional neuroanatomical treatment of the motor systems beginning, appropriately, at the motor periphery.

■ A functional classification of muscles

There are various ways in which muscles can be classified. For example, anatomically one can define flexors and extensors, abductors and adductors, and pronators and supinators. Physiologically one can classify most muscles as flexors or extensors. *Extensors* are defined as mucles, that resist gravity (if only in part), and *flexors* are defined as muscles that assist gravity or participate in withdrawal reflexes. There is no question as to the usefulness of such classifications.

A somewhat different approach, however, has considerable heuristic value in the analysis of the central control of movement. This approach is based on a medial-lateral or proximal-distal distinction, in which a class of medial or proximal muscles and a class of lateral or distal muscles are defined. The medial musculature would include, for example, the axial and girdle muscles, be they flexors or extensors. The actions of these muscles involve the axis and proximal limbs. Thus they subserve functions that require body and whole limb movement such as posture, progression, and equilibration. Although the medial flexor muscles do not perform an antigravity function, they do act in an organized manner with the medial extensors to control posture. The lateral or distal musculature would include, for example, the intrinsic muscles of the digits and the distal muscles of the extremities, such as those acting around the wrist joint. Such muscles, flexor or extensor, do not serve a primary postural function, and it is more appropriate to consider them as mediating manipulatory activity. This is not a rigorous anatomical formulation because the two classes are not necessarily mutually exclusive. However, it does provide an extremely useful foundation for understanding the functional neuroanatomy of movement.

■ *Topographical organization of motoneuronal pools*

As previously described, the spinal motoneurons located in the ventral horn are arrayed in longitudinal columns having different mediolateral positions (Fig. 6-7). These different positions reflect a highly organized topography with respect to target muscles. The most medially situated motoneurons innervate the most proximal muscle groups, and the most laterally situated motoneurons innervate the most distal muscles of the extremities; for example, the intrinsic muscles of the digits are innervated by the most dorsolaterally located motoneurons (Fig. 13-1). Consequently, this motoneuronal cell column is restricted to the brachial enlargement, and it is found extending only over the lower cervical and first thoracic segments. As one moves ventromedially in the ventral horn, the motoneuronal pools for successively more proximal limb muscles are encountered. For example, near the base of the ventral horn at middle to lower cervical levels is a column of motoneurons innervating the muscles that act at the shoulder and elbow. Located at lumbosacral levels are analogous columns that innervate the muscles of the lower limbs. Situated more medially are the motoneurons that innervate the muscles of the trunk and neck. Thus this cell column extends throughout most of the spinal cord.

This topographical organization, taken in conjunction with the functional muscle classification developed earlier, allows one to consider medial and lateral groups of motoneurons that innervate, respectively, the proximal muscles involved in postural

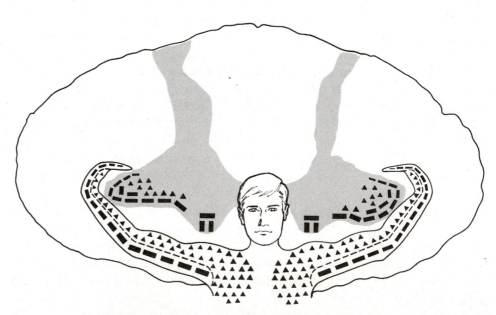

Fig. 13-1 ■ Upper body superimposed on a cross section of the spinal cord with corresponding symbols in the spinal cord and on the upper body. The positions of the motoneurons that innervate the muscles of different portions of the upper extremity are shown. More proximal muscles are indicated by larger symbols and more distal muscles by smaller symbols. Correspondingly, the motoneuronal locations supplying the innervation to more proximal muscles are indicated by larger symbols and the locations supplying the innervation to more distal muscles by smaller symbols. This shows the shift from ventromedial to dorsolateral motoneurons as one moves from more proximal to distal muscles of the extremity. Triangles indicate flexor muscles, and rectangles indicate extensor muscles. (Redrawn from Crosby, E.C., Humphrey, T., and Lauer, E.W.: Correlative anatomy of the nervous system, New York, 1962, The Macmillan Co. Copyright © 1962 by the Macmillan Publishing Co., Inc.)

control and the distal muscles involved in manipulatory activity. From this it can be inferred that inputs that preferentially distribute to the more medial motoneurons will be primarily involved in the control of posture. Conversely, inputs that preferentially distribute to the more lateral motoneuronal cell groups will be principally concerned with manipulatory acitivity.

■ *Topographical organization of spinal interneurons*

It must be appreciated, however, that most influences on the motoneurons are not exerted via direct projections on these cells, but rather through spinal interneurons. It is thus a critical issue as to whether the mediolateral topography of the motoneurons is preserved in the interneuronal cell groups that project on these motoneurons. Anatomical studies have established that this appears to be the case, although the precision of the topography has not been determined. The more dorsolaterally situated interneurons of the spinal gray matter project preferentially on the more dorsolaterally situated motoneurons (Fig. 13-2). Analogously, more ventromedially situated interneurons project perferentially on the more medially located motoneuronal cell groups. Consequently, this inference can be extended, and it can be concluded that any pathway that preferentially projects on the more medial motoneurons or their associated medial interneurons will be primarily involved in controlling posture. Conversely, any pathway that projects preferentially on the more lateral motoneurons or their associated lateral interneurons will be more concerned with manipulatory activity involving the distal limb muscles.

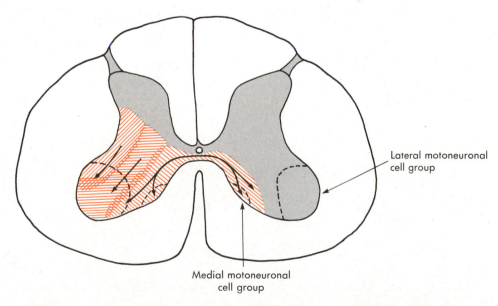

Lateral motoneuronal cell group

Medial motoneuronal cell group

Fig. 13-2 ■ Summary of the relationship of interneurons in the intermediate and ventral spinal gray matter to the motoneuronal cell groups. Note that more medially situated interneurons project preferentially to more medially situated motoneurons (bilaterally), whereas more laterally situated interneurons project preferentially on more laterally situated motoneurons. (Redrawn from Sterling, P., and Kuypers, H.G.J.M.: Brain Res. **7:**419, 1968.)

■   ■   ■

With the preceding information as a peripheral foundation, we are now able to develop an organizational scheme for the motor pathways that is based on their peripheral sites of termination rather than their cells of origin; that is, the descending pathways involved in motor control are divided into (1) those primarily influencing proximal muscles and, therefore, postural function and (2) those influencing distal muscles and, therefore, manipulatory function. This leads to a more coherent functional neuroanatomy of the major motor pathways, a concept that will be developed in later chapters.

■ *Bibliography*

Journal articles

Sterling, P., and Kuypers, H.G.J.M.: Anatomical organization of the brachial spinal cord of the cat. II. The motoneuron plexus, Brain Res., **4:**16, 1967.

Sterling, P., and Kuypers, H.G.J.M.: Anatomical organization of the brachial spinal cord of the cat. IV. The propriospinal connections, Brain Res. **7:**419, 1968.

Books and monographs

Brodal, A.: Neurological anatomy, ed. 3, New York, 1981, Oxford University Press.

Kuypers, H.G.J.M.: The anatomical organization of the descending pathways and their contributions to motor control especially in primates. In Desmedt, J.E., editor: New developments in electromyography and clinical neurophysiology, vol. 3, Basel, 1973, S. Karger.

chapter 14

Spinal organization of motor function

The various descending pathways that control motor behavior must ultimately exert their influences by acting on spinal circuitry (or brainstem circuitry in the case of the cranial nerve motor nuclei). As mentioned previously, only a small proportion of suprasegmental (from above the spinal cord) and segmental (at the level of the spinal cord) inputs that affect the motoneurons terminate monosynaptically on them. Consequently, it is important to gain an understanding of the neuronal circuitry of the spinal cord to appreciate how movements are organized. Indeed, a certain amount of motor organization can be mediated by spinal circuitry, and descending influences can elicit patterns of motor behavior by accessing these circuits via appropriate spinal interneurons.

The experimental approach to investigating motor organization at spinal levels has involved studying animals after spinal cord transection. In such an experimental preparation, one can explore the patterns of motor behavior elicited by activating various segmental inputs and then attempt to delineate the intraspinal circuitry that mediates these patterns. In this chapter we will consider (1) the effects of spinal transection, (2) the classes of segmental inputs that influence the motoneurons, (3) the specific patterns of reflex behavior that are elicited by such inputs, and (4) some general principles of spinal organization.

■ *Effects of spinal transection*

Immediately on transection of the spinal cord there ensues a *paralysis* of all muscles innervated by spinal segments below the lesion. Voluntary contraction of these muscles is permanently abolished. Also, there is a permanent *anesthesia* (loss of sensation) in all parts of the body innervated by segments below the transection. Reflexes mediated at spinal levels are abolished as well, but only temporarily. This *areflexia* is often referred to as *spinal shock*. It includes the loss both of somatic reflexes, such as those mediating muscle tone and the protective withdrawal responses, and of autonomic reflexes, such as those involved in peristalsis, sweat secretion, maintenance of arterial blood pressure, and emptying the bladder and bowels. The level of the transection will, of course, determine the specific pattern of autonomic areflexia.

Although the loss of voluntary muscle contraction and anesthesia are chronic conditions, the areflexia is transient. Over time the spinally mediated reflexes show varying courses of recovery. The evidence is rather compelling that spinal shock results from the loss of suprasegmental influences on the spinal cord. First, the extent and duration of the areflexia are much greater in species that have more extensive suprasegmental control of the spinal cord. For example, in nonmammalian vertebrates, such as amphibia, spinal shock is minimal, and the spinal reflexes reappear in a few minutes. In contrast, recovery in humans has a course of months. Second, if after a transection one waits for recovery from spinal shock and then makes a second transection rostral to the first, the areflexia of the previously affected muscles does not reappear. Because this second transection cannot further compromise suprasegmental influences, this experimental result seems to implicate the loss of descending pathways in the genesis of spinal shock. Third, there is neurophysiological evidence for a depression of motoneuronal excitability after spinal transection.

In humans the time course of recovery from spinal shock is highly variable. Infrequently reflex activity can appear as soon as 24 hours. However, most often no reflex activity appears for 2 to 6 weeks. The first reflexes to reappear are flexion responses to tactile stimulation and reflexes involving the genitals and the anal sphincter. With further recovery the flexor responses generally become hyperactive so that an innocuous tactile stimulus can elicit a mass flexion reflex involving many joints. Finally, there is a slow recovery of tendon reflexes that may ultimately become hyperactive and may be accompanied by clonus (p. 223).

The mechanisms of recovery from spinal shock are not known; however, it is presumed that they involve an increasing effectiveness of segmental influences. It has been shown electrophysiologically that there is an increase in the activity of the motoneurons and the fusimotor neurons, and this parallels recovery in spinal animals. A possible mechanism that might contribute to recovery is axonal sprouting (Chapter 21); that is, sensory fibers that enter the cord or spinal interneurons might sprout additional terminals that occupy the synaptic space vacated by the terminal degeneration of the suprasegmental projections to the spinal gray matter.

■ Segmental inputs, including muscle afferent fibers, that elicit reflex activity

■ General considerations

As a foundation for describing the spinal reflexes, it is first necessary to review the classes of input from the periphery to the spinal cord (Fig. 14-1). Three such classes originate in muscle, and they transmit to the nervous system information on the length or tension of muscles (*proprioception*). One of these classes derives from the *primary (annulospiral) endings* of the muscle spindle. These are stretch receptors that transmit information on the length and rate of change of length of muscles. The fibers from these receptors are large (12 to 20 μm) and myelinated and conduct at 70 to 120 m/second; these fibers are in the Ia group. Another proprioceptive class of fibers derives from the

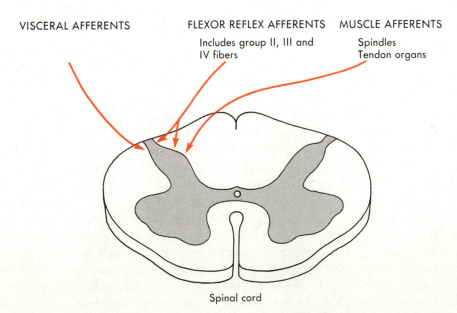

Fig. 14-1 ■ Schematic illustration of the three major classes of segmental input to the spinal cord: muscle afferent fibers, flexor reflex afferent fibers, and visceral afferent fibers.

secondary (flower-spray) endings of the muscle spindle. These, too, are stretch receptors, and they provide information primarily on muscle length. They do so over somewhat smaller myelinated fibers (5 to 12 μm) that conduct at 30 to 70 m/second and are
in the group II category. The third class of proprioceptive inputs is derived from the
Golgi tendon organs. These are specialized stretch receptors located in the region of the
tendon, and they transmit information on muscle tension. Their afferent fibers are in the
group Ib category; such fibers have axonal diameters and conduction velocities that
approximate those of the Ia afferent fibers.

The fourth class of segmental inputs germane to the spinal reflexes consists of somatosensory receptors. It includes a wide variety of receptors that are located throughout
the body and contribute to the group II, III, and IV afferent fiber classes; these have
been described previously. However, from the perspective of the spinal reflexes these
can be aggregated into a single functional class designated *flexor reflex afferent fibers*
because they exert a common reflex action, namely flexor patterns.

The final class of segmental inputs to be considered consists of the *visceral afferent
fibers*. This is a complex and highly differentiated class of inputs that form the afferent
limb of the spinal autonomic reflexes. Included, for example, are stretch receptors that
are located in the urinary bladder; the activation of these receptors elicits reflexes involved in bladder emptying.

■ *Muscle afferent fibers*

Three of the five classes of segmental afferent fibers that constitute the inputs for
spinal reflexes originate in muscle. It is important to elaborate on these before proceeding to the reflexes. These proprioceptors consist of the muscle spindle receptors and the
Golgi tendon organ, and almost all muscles contain this receptor complement. The *muscle spindle* is a fusiform- or spindle-shaped structure (Fig. 14-2) that is several millimeters in length and no more than a few hundred microns in width. It consists of a
connective tissue sheath that encloses 2 to 12 *intrafusal muscle fibers;* each end of the
spindle is attached to extrafusal muscle fibers. The spindles are distributed throughout
the muscle, and their density appears to vary with the degree of control required by a
given muscle. The intrinsic muscles of the digits have a considerably greater density of
spindles per muscle mass than, for example, the large muscles involved in postural
control.

There are at least two types of intrafusal muscle fibers, the *nuclear bag fibers* and
the *nuclear chain fibers* (Fig. 14-3). The nuclear bag fibers, which are longer and wider
than the nuclear chain fibers, have an expanded central or *equatorial* region with densely
aggregated nuclei, hence, the name nuclear bag. Beyond the equatorial region and most
prominently at the distal poles, the nuclear bag fiber contains contractile elements. The
smaller nuclear chain fibers have many fewer nuclei that are serially arrayed within the
central region; they, too, have contractile material distal to the central region.

The primary endings of the muscle spindle tend to coil around the central regions of
both the nuclear bag and nuclear chain fibers, whereas the secondary endings tend to
arise from pericentral regions of the nuclear chain fiber with minimal termination on the
nuclear bag fiber (Fig. 14-3). A typical spindle contains two nuclear bag fibers and
about five nuclear chain fibers. It is innervated by a single Ia afferent fiber and a single
group II afferent fiber. These fibers ramify within the spindle to supply the various
primary and secondary endings.

The intrafusal fibers, in addition to their sensory innervation, receive motor innervation from the central nervous system; they are therefore under *centrifugal* control (Fig.
14-3). This innervation derives from smaller neurons, γ-*motoneurons* or *fusimotor neurons,* that are located in the ventral horn among the α-motoneurons that innervate the
extrafusal muscle fibers. The axons of the γ-motoneurons are of smaller diameter than
those of the α-motoneurons; they travel in muscle nerves, and they consist of two prom-

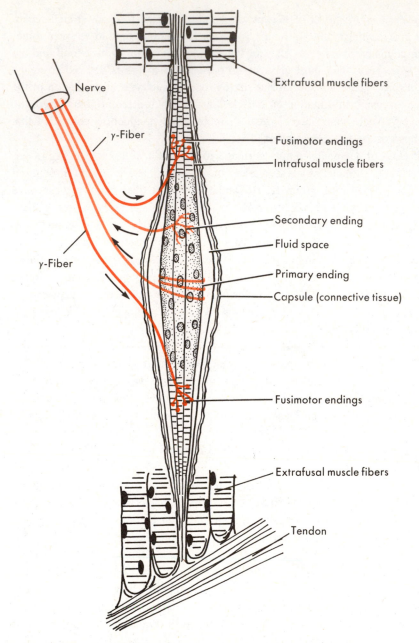

Nerve

γ-Fiber

γ-Fiber

Extrafusal muscle fibers

Fusimotor endings

Intrafusal muscle fibers

Secondary ending

Fluid space

Primary ending

Capsule (connective tissue)

Fusimotor endings

Extrafusal muscle fibers

Tendon

Fig. 14-2 ■ Schematic illustration of the muscle spindle and its principal components. (Redrawn from Brodal, A.: Neurological anatomy, ed. 3. Copyright © 1981 by Oxford University Press, Inc. Reprinted by permission.)

inent types (Fig. 14-3). One terminates in discrete motor endplates *(plate endings)* on the distal (contractile) poles of the nuclear bag fibers. The other terminates as a more extensive network of *trail endings,* primarily on the nuclear chain fibers. There is a class of fibers, Aβ fibers, that innervates both intrafusal and extrafusal muscle fibers via plate endings. The Aβ fibers are rather limited, however, and will not be discussed further.

The third source of muscle afferent fibers, the Golgi tendon organ, is not located in the muscle spindle but at the musculotendinous junction (Fig. 14-4). It consists of a rich arborization of small unmyelinated fibers that extend for perhaps 500 μm and are enclosed in a fine capsule. Tendon organs tend to be fewer in number than spindles in any given muscle, and they are more prominent in slowly contracting muscle.

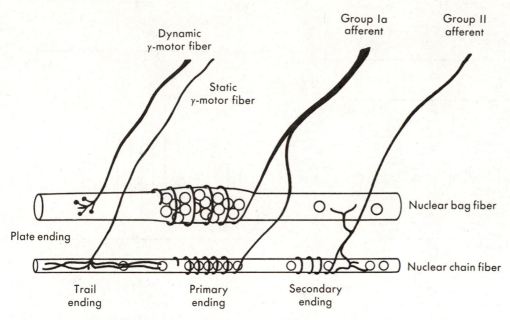

Fig. 14-3 ■ Types of intrafusal fibers in mammalian muscle spindles and their innervation. (Redrawn from Matthews, P.B.C.: Physiol. Rev. **44:**219, 1964).

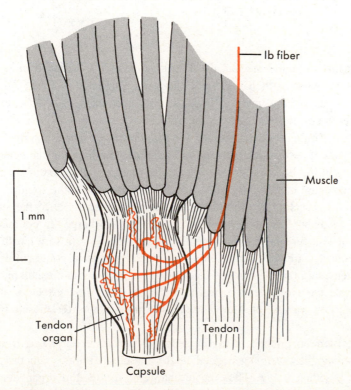

Fig. 14-4 ■ Schematic illustration of a Golgi tendon organ. (Redrawn from Barker, D.: Muscle receptors, Hong Kong, 1962, Hong Kong University Press.)

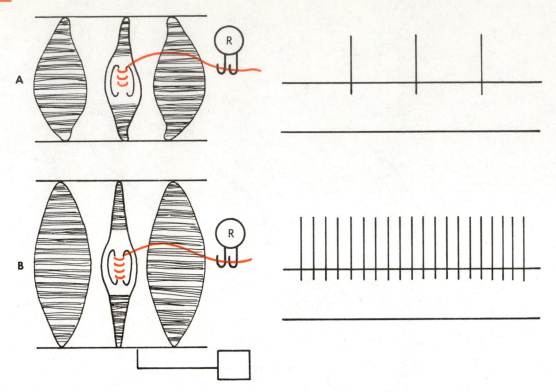

Fig. 14-5 ■ The effects of muscle shortening (contraction) (**A**) and muscle lengthening (stretch) (**B**) on the discharge rate of an afferent fiber from a spindle of the muscle. Contraction of the muscle shortens the spindle and reduces the discharge rate of the afferent fiber. Stretch of the muscle stretches the spindle and increases the discharge rate. *R,* Recording electrode. (Redrawn from Eyzaguirre, C., and Fidone, S.J.: Physiology of the nervous system, ed. 2, Chicago, 1975, Year Book Medical Publishers, Inc.; modified from Ruch, T.C., and Patton, H.D.: Physiology and biophysics, ed. 19, Philadelphia, 1965, W.B. Saunders Co.; Hunt, C.C., and Kuffler, S.W.: J. Physiol. **113:**298, 1951.)

It is essential to appreciate that the muscle spindles are arranged *in parallel* to the extrafusal fibers, whereas the Golgi tendon organs are situated *in series* with the muscle (Fig. 14-5). Because the primary endings, secondary endings, and Golgi tendon organs are stretch receptors, this arrangement is critical for the nature of their adequate stimulation. Being in parallel with the muscle, the spindle will stretch as the muscle lengthens (relaxes). This will deform the primary and secondary endings on the central and pericentral regions of the intrafusal fibers, giving rise to a generator potential and ultimately to an action potential in the group Ia or II fibers that arise from the spindle. Conversely, contraction of a muscle will shorten the muscle spindle and reduce the stretch on the primary and secondary endings, thus decreasing their discharge. In contrast to this behavior of spindle receptors, the Golgi tendon organ, which is in series with the muscle, is most effectively stretched, and hence activated, by muscle contraction (Fig. 14-6). Although stretch of a muscle can generate enough tension at the musculotendinous junction to activate the Golgi tendon organ, this is a much less effective stimulus than muscle contraction because of the relative elastic properties of the muscle fibers and the tendon.

The Golgi tendon organ is a highly sensitive stretch receptor that transmits information regarding the *tension* generated at the musculotendinous junction; indeed, its threshold is less than 0.1 g. Thus the discharge of the tendon organ does not signal muscle length because the tension generated during a muscle contraction will vary with the load against which the muscle is working; for example, considerable tension can be

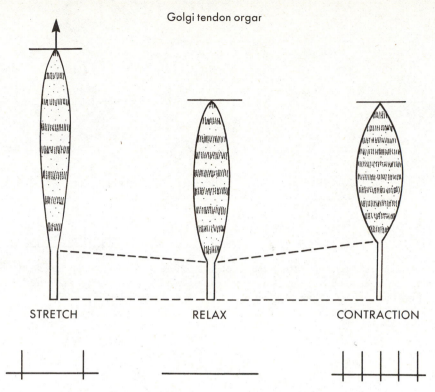

Golgi tendon organ

STRETCH   RELAX   CONTRACTION

Fig. 14-6 ■ Tendon elongation during muscle stretch, relaxation, and active contraction. Contraction is more effective in elongating the tendon than externally applied stretch, where some of the force is dissipated in the elongation of the muscle. Consequently active contraction is more effective in eliciting an increase in discharge from the Golgi tendon organ. (Redrawn with permission from Eyzaguirre, C., and Fidone, S.J.: Physiology of the nervous system, ed. 2. Copyright © 1975 by Year Book Medical Publishers, Inc., Chicago.)

generated during isometric contraction. It should also be appreciated that a tendon organ will reflect the tensions of the specific muscle fibers inserting on it. Therefore these receptors do not signal the average tension of the entire muscle, but they contribute information regarding local tension. They also show both static and dynamic components in their discharge; that is, they are sensitive to both the velocity of tension development as well as to the steady-state tension.

In contrast to the tendon organs, spindle receptors are activated by lengthening of muscle. However, in the discussion thus far, no functional distinction has been made between the primary and secondary endings. For some years it has been known that both receptor types signal muscle length (Fig. 14-7). If the discharge of primary (Ia) and secondary (II) endings at different fixed muscle lengths were recorded, it could be demonstrated that their discharge rates increase as a function of length. Moreover, at any given length these receptors show a *static response* that is maintained as long as the muscle is maintained at that length; that is, little adaptation occurs. It was not until the activity of these receptors was recorded *during the lengthening* of muscle that a distinction between the primary and secondary endings was appreciated (Fig. 14-7). In such an experiment the primary ending discharge is velocity-sensitive during lengthening; that is, it has a *dynamic response component* so that during muscle lengthening its discharge is a function of both length and velocity of lengthening. In contrast, the secondary ending has little or no dynamic response, and it continues to signal principally length during the active phase of stretch. Again, under static or steady-state conditions

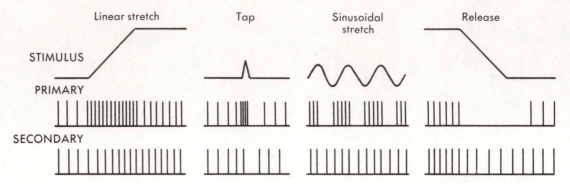

Fig. 14-7 ■ Idealized responses of primary and secondary endings to various manipulations of muscle length. The responses are shown as if the muscle were initially under moderate stretch and with no fusimotor activity. (From Matthews, P.B.C.: Physiol. Rev. **44:**219, 1964.)

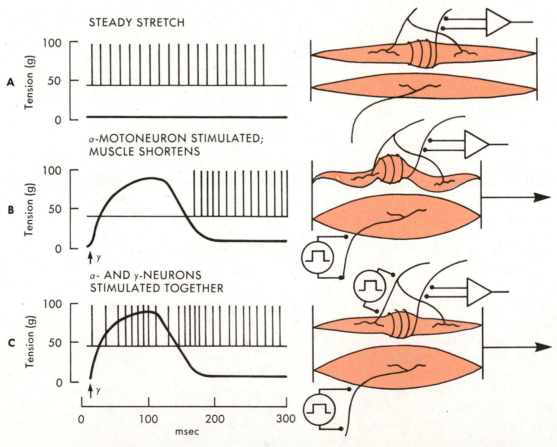

Fig. 14-8 ■ Effect of fusimotor stimulation during muscle contraction. **A,** During steady stretch the discharge of the spindle afferent fiber is maintained. **B,** With muscle contraction the spindle shortens, and the discharge of the afferent fiber ceases. **C,** If the fusimotor innervation is activated together with the innervation of the extrafusal fibers, then the discharge is maintained during shortening of the muscle. (Redrawn from Kuffler, S.W., and Nicholls, J.G.: From neuron to brain, Sunderland, Mass., 1976, Sinauer Associates, Inc.)

(at fixed length) there is little difference between the behavior of the two types of receptors. During release from stretch (as during contraction), the muscle spindle shortens and "unloads" the tension at the receptor regions (Fig. 14-7). During this unloading the secondary ending shows a gradual decrease in discharge proportional to muscle length. However, the primary ending shows a dynamic response during such shortening, providing information on the velocity of shortening. In fact, it may even show a cessation of discharge or *silent period* during the release from stretch (Fig. 14-7).

The mechanisms of the different discharge properties of the primary and secondary endings are not fully understood, but they are thought to reflect differences in the viscoelastic properties of the nuclear bag and nuclear chain fibers. The secondary endings terminate primarily on the nuclear chain fibers (Fig. 14-3). Thus these fibers may have properties that give rise to the static response of the secondary endings. If this is the case, then the static component of the primary ending discharge could derive from its termination on the nuclear chain fiber, whereas the dynamic component would be mediated by its termination on the nuclear bag fiber (Fig. 14-3).

The final aspect of the muscle afferent fibers to be considered is the motor innervation of the muscle spindle (Fig. 14-3). This innervation originates from γ-motoneurons in the ventral horn. The fusimotor innervation is confined to the intrafusal fibers, and it causes the more distal regions of these fibers to contract. If an intrafusal fiber is maintained at a constant length, then contraction or shortening of its distal poles by fusimotor activation would result in a stretch or lengthening of its more central regions (Fig. 14-8). These are the regions where the receptors are located. Thus activation of the fusimotor fibers increases the tension in the receptor regions. This central control of the spindle receptor sensitivity allows these receptors to respond with high resolution over a wide range of muscle lengths. For example, if the receptors are capable of discharge rates of up to 500 per second, then the fusimotor control of tension at the receptor region can be set so that this entire discharge range is dedicated to but a few millimeters of muscle lengthening. This permits resolution of length changes on the order of hundreds of microns. If the fusimotor discharge is then readjusted, this "sensitivity window" can be shifted to another portion of the muscle's possible excursion. Indeed, it is thought that during movements initiated by descending neural pathways, the α- and γ-motoneurons to a given muscle are *coactivated* (Fig. 14-9). With such an arrangement the spin-

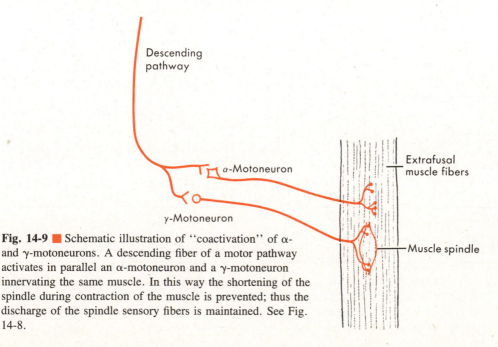

Fig. 14-9 ■ Schematic illustration of "coactivation" of α- and γ-motoneurons. A descending fiber of a motor pathway activates in parallel an α-motoneuron and a γ-motoneuron innervating the same muscle. In this way the shortening of the spindle during contraction of the muscle is prevented; thus the discharge of the spindle sensory fibers is maintained. See Fig. 14-8.

Descending
pathway

α-Motoneuron

γ-Motoneuron

Extrafusal
muscle fibers

Muscle spindle

dle-receptor sensitivity is maintained during muscle contraction because the parallel activation of the fusimotor fibers of that muscle prevents unloading of the spindle.

Given the fusimotor system, it should be clear that the spindle receptors do not actually signal absolute muscle length. A given discharge rate of a spindle receptor may correspond to various muscle lengths, depending on the level of fusimotor discharge. Thus the spindle receptors actually transmit information about the length of the spindle relative to the muscle length. It is necessary to know both the level of fusimotor discharge and the spindle afferent discharge to infer absolute muscle length. How this integrative computation is achieved centrally is not yet understood.

The two morphological types of fusimotor fibers, plate endings and trail endings, may correspond to two functional fusimotor classes (Fig. 14-10). One such class, the *static fusimotor fiber,* when activated increases the static responses of both the primary and secondary endings. It has been suggested that this functional class may correspond to the fusimotor fibers with trail endings. (The trail endings distribute preferentially to the nuclear chain fibers, which may give rise to the static discharge of the primary and secondary endings.) The other functional type, the *dynamic fusimotor fiber,* has little or no effect on the static response of the spindle receptors, but it increases its velocity sensitivity. These fibers may correspond to the fibers with plate endings. (These are distributed preferentially to the nuclear bag fibers, which presumably give rise to the dynamic response of the primary ending.) Some nuclear bag fibers apparently have trail endings, and these fibers have recently been identified as a subtype of nuclear bag fibers having some static properties.

In brief, the muscle spindle is a complex receptor organ that has a dual afferentation and at least a dual efferentation. The receptors respond to muscle stretch or lengthening and provide information to the nervous system regarding muscle length (primary and secondary endings) and rate of change of length (primary endings). The sensitivity of these receptors can be adjusted by the central nervous system via the γ-motoneurons, and the static and dynamic fusimotor innervations allow the static and dynamic receptor discharges to be differentially controlled. The importance of such a complex proprioceptive system can be appreciated if it is recognized that the sequence of muscle contractions required to move the arm to a given position is a function of the starting position of the limb. Information regarding muscle length is sufficient to infer that start-

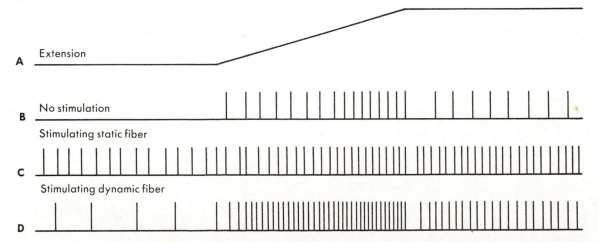

Fig. 14-10 ■ Effects of stimulating single fusimotor fibers on the response of a primary ending to muscle stretch. **A,** Extension of a muscle. **B,** Absence of stimulation. **C,** The effects of stimulating a static fusimotor fiber. **D,** The effects of stimulating a dynamic fusimotor fiber. (From Crowe, A., and Matthews, P.B.C.: J. Physiol. **174:**109, 1964.)

ing position, and given this the appropriate sequence of muscle contractions can be programed. However, in much of our motor behavior the limb will not be static, but moving. Therefore the system must also be capable of predicting the limb's position at any given moment; this requires information on the velocity of change in muscle length. The muscle spindle system with its receptors and their centrifugal control is clearly designed to provide the information needed for the appropriate programing of goal-directed movements.

■ *Spinal reflexes*

A foundation has now been established for discussing the spinal reflexes elicited by each of the afferent classes described earlier. By way of introduction, Fig. 14-11 illustrates a basic spinal reflex circuit. It consists of (1) a receptor and its afferent fiber, (2) one or more spinal interneurons (with the exception of the myotatic reflex), (3) a motoneuron and its axon, and (4) an effector or muscle. The pathways for autonomic reflexes are somewhat more complex and will be described later. The properties of any given reflex are a function of the characteristics of the receptors initiating the reflex, the nature of the involved spinal circuitry, and the anatomical distribution of the involved motoneurons.

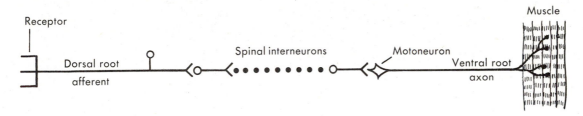

Fig. 14-11 ■ Schematic illustration of a reflex pathway consisting of a receptor and its afferent fiber, spinal interneurons, and a motoneuron and the effector it innervates.

■ *The myotatic reflex*

One of the most important reflexes subserving a postural function is spinally mediated and is elicited by activation of the primary (Ia) endings of the muscle spindle (Fig. 14-12). This is the *myotatic (tendon) reflex* (*myotatic* means muscle stretch).

The Ia afferent fiber, on entering the spinal cord, bifurcates to give rise to a branch that ascends in the dorsal columns and a segmental branch that terminates largely within the segment of entry. The ascending Ia fibers ultimately provide information on muscle length and velocity of lengthening to various central structures such as the cerebellum. The segmental branch not only terminates on various interneurons, but it also terminates *monosynaptically* on motoneurons. The pattern of this monosynaptic termination is quite specific because it is primarily confined to motoneurons innervating the muscle from which the Ia afferent fiber arose, the *homonymous motoneurons*. In fact, a single Ia afferent fiber terminates on most, if not all, cells of the homonymous motoneuronal pool. To a lesser extent the Ia afferent fibers also terminate on motoneurons innervating muscles that act synergistically with their muscles of origin.

Given the properties of the primary endings and this simple monosynaptic circuit, the nature of the myotatic reflex can readily be inferred. The adequate stimulus for the reflex is muscle lengthening or stretch. When this occurs, the primary endings are activated, and the resulting discharge in the rapidly conducting Ia afferent fibers provides a monosynaptic excitatory input to the homonymous motoneurons. Activation of these motoneurons then results in contraction of the stretched muscle. Consequently, the myotatic reflex is a *stretch-evoked reflex,* and its action is to produce muscle contraction or shortening in response to stretch. Because the primary ending discharge has both dynamic and static components, the myotatic reflex can be activated by either of these discharge components. Thus in the case of a rapid muscle stretch, as with a tap of the

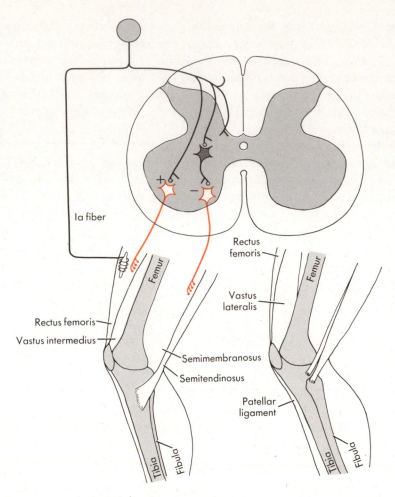

Fig. 14-12 ■ The neural circuitry for the myotatic reflex. The afferent segment of the reflex pathway consists of the group Ia fibers from a spindle of the homonymous muscle. The afferent segment then monosynaptically innervates motoneurons of that muscle (plus sign). Through an inhibitory interneuron, shown in black, it disynaptically inhibits (minus sign) the motoneurons of the antagonistic muscle (reciprocal innervation).

patellar tendon, the resulting myotatic reflex (contraction of the quadriceps muscles and hence extension around the knee) will be driven principally by the dynamic component of the primary ending discharge. With sustained stretch of a muscle, the resulting myotatic reflex will be driven by the static component of the primary ending discharge (Fig. 14-7).

The myotatic reflex circuitry also illustrates a general principle of spinal organization called *reciprocal innervation*. The Ia afferent fibers, in addition to their monosynaptic terminations on homonymous motoneurons, project on a population of spinal interneurons that inhibit the motoneurons of muscles antagonistic to the homonymous muscle (Fig. 14-12). This reciprocal innervation functions to relax the muscles, the contraction of which would oppose the myotatic reflex action. This simple organizational feature is prominent in spinal organization, and it ensures the integrated action of agonists and antagonists at a given joint.

The myotatic reflex, particularly its static component, is most prominent in the medial extensor muscles that serve a critical antigravity function. Thus this reflex is a fundamental and most important postural mechanism. For example, when a person is in an upright posture, gravity produces a stretch influence on the quadriceps muscles. This elicits a sustained myotatic reflex contraction of the quadriceps, and that contraction functions to maintain extension around the knee joint and thereby an upright posture. Because the primary endings show little adaptation, this reflex contraction can be sustained as long as the stretch is imposed. Furthermore, the finely graded nature of the primary ending discharge, its faithful transmission over rapidly conducting afferent fibers, and the involvement of but a single synapse in the reflex circuit all combine to permit rapid and finely graded reflex contractions in response to small increments in muscle length. Recent evidence suggests that the secondary endings of the muscle spindle may also contribute to the myotatic reflex (static component) by means of a monosynaptic connection to the homonymous motoneurons.

It is important to appreciate that the myotatic reflex is the mechanism of muscle tone. *Muscle tone* is defined as the resistance of a muscle to active or passive stretch, and this is precisely the action of the myotatic reflex. This has important clinical implications because alterations in tone frequently occur with lesions at many levels of the neuraxis. In most instances of altered tone, the lesion affects the myotatic reflex by perturbing the fusimotor control of spindle sensitivity. For example, the atonia following spinal transection largely results from a severe reduction or cessation of γ-motoneuronal discharge consequent to the loss of suprasegmental inputs to the γ-motoneurons. The fusimotor system generally maintains a background level of discharge that is modulated up or down. Elimination of this discharge shifts the spindle to one extreme of its dynamic range so that even large muscle stretches may be insufficient to generate enough tension on the intrafusal fibers to activate their receptors. At the other extreme, any lesion that removes major inhibitory influences on the γ-motoneurons will result in an increased background discharge that raises the sensitivity of the spindle receptors and can result in abnormally active myotatic reflexes. This results in *hypertonia* or *spasticity*. *Spastic paralysis* would then refer to the situation in which interruption of descending pathways has eliminated the ability to contract a muscle voluntarily via the α-motoneurons (paralysis), but the spinally mediated myotatic reflexes involving that muscle are hyperactive (spasticity). Interrupting the myotatic reflex circuit by cutting the dorsal roots (dorsal rhizotomy) and therefore the Ia afferent fibers will generally eliminate the hypertonia and transform it to hypotonia.

A common clinical sign of hypertonia is *clonus,* which provides another exercise in understanding the myotatic reflex circuit. In hypertonia resulting from enhanced fusimotor discharge, a tendon tap will elicit an abnormally brisk myotatic reflex, for example, exaggerated extension of the leg in response to a patellar tendon tap. Because the tendon tap is a transient stimulus and the myotatic reflex ends abruptly with stimulus termination, the extension is immediately followed by relaxation or lengthening. Under normal fusimotor control, the leg would return to a resting position; however, with enhanced fusimotor discharge the stretch imposed by the return to a resting position elicits another myotatic reflex contraction, although somewhat weaker.

Although tendon stretch (tap) is an effective clinical means of evaluating the myotatic reflexes of various muscles, it is also possible to monitor the monosynaptic myotatic circuit electrophysiologically in humans. Superficial electrical stimulation of the skin behind the knee (the popliteal fossa) spreads sufficient current to excite the group Ia afferent fibers of the medial popliteal nerve, which carries sensory information from the gastrocnemius muscle. This elicits in that muscle a myotatic response called *Hoffmann's reflex* (H reflex). With recording electrodes on the surface of the gastrocnemius muscle, this myotatic response can be recorded electromyographically.

A final point relates to the dynamic and static components of the primary ending discharge. These can be modulated independently by the dynamic and static fusimotor fibers, respectively. Hence it is possible to affect differentially myotatic responses to rapid muscle stretch and sustained stretch. Because the response to tendon tap is primarily driven by the dynamic component of the Ia fiber discharge, enhanced deep tendon reflexes could result from any lesion that selectively increases the activity of the dynamic fusimotor fibers. In contrast, a lesion that selectively increases the activity of the static fusimotor fibers would not enhance deep tendon reflexes, but it would augment myotatic responses to sustained stretch. This kind of hypertonia would not be velocity sensitive, as in the plastic rigidity of parkinsonism.

In brief, the myotatic reflex is elicited by stretch of a muscle; it is a reflex contraction in response to such stretch. It is the basic mechanism of muscle tone and serves a critical postural function. The afferent segment of the reflex circuit consists of the primary (and possibly secondary) endings of the muscle spindle and their afferent fibers. The spinal circuitry is simple, consisting of a monosynaptic excitatory connection to the homonymous motoneurons. Thus the reflex is local because its effects are largely restricted to the muscle from which the activating afferent fiber arises. Moreover, the reflex action is rapid (short latency), stimulus-locked, finely graded, and capable of being sustained (particularly in medial extensor muscles). The myotatic circuitry also includes a disynaptically mediated inhibition of the antagonistic muscles, producing relaxation of the antagonist muscles, or reciprocal innervation.

■ *The inverse myotatic reflex*

Another important spinal reflex is initiated by the activation of the Golgi tendon organs. In this circuit (Fig. 14-13) the Ib afferent fibers from the tendon organs terminate on spinal interneurons that inhibit the homonymous motoneurons and synergistic motoneurons, including those acting at other joints. Thus the reflex action is a relaxation or lengthening of a muscle in response to its contraction, and the distribution of this reflex effect is most prominent in antigravity muscles. Because the myotatic reflex is a contraction of a muscle in response to lengthening, this tendon organ–mediated reflex has been called the *inverse myotatic reflex*. However, it is important to appreciate that the adequate stimulus for the myotatic reflex is muscle length, whereas that for the inverse myotatic reflex is muscle tension. Therefore, the designation inverse myotatic is not truly appropriate.

Like the myotatic circuit, the inverse myotatic reflex includes a disynaptically mediated reciprocal innervation (Fig. 14-13). However, in contrast to the myotatic reflex, it also includes a crossed-component (Fig. 14-13). If, for example, the inverse myotatic action is to relax the quadriceps muscles of one limb in response to their contraction, the crossed-component will involve an excitation of the motoneurons of the contralateral quadriceps muscles. This crossed-component, called *Phillipson's reflex*, itself includes the circuitry for reciprocal innervation so that the antagonistic hamstring muscles of the contralateral limb are relaxed. This reciprocal innervation in the crossed-component of a reflex is sometimes referred to as *double reciprocal innervation*. A teleological view of the crossed-component is to consider it as postural in function. If the inverse myotatic action is on the antigravity muscles of a limb, leading to their relaxation, then the crossed-component will be extensor in nature. This crossed extension serves to establish postural support.

Unlike the myotatic reflex, the inverse myotatic reflex cannot be demonstrated behaviorally, except in a hypertonic limb. If, for example, one begins to flex a spastic lower limb at the knee, increasing resistance to the stretch of the extensor muscles will be encountered. This resistance will continue to increase as more force is imposed until at some point the extensor contraction abruptly terminates, and the limb passively flexes. This behavior inspired the descriptive name *clasp-knife reaction;* another frequently used designation is the *lengthening reaction*.

It is generally accepted that this reflex is mediated via the Golgi tendon organs. It was originally thought that these tendon organs were high-threshold receptors and that the inverse myotatic reflex served the protective function of preventing the muscle from

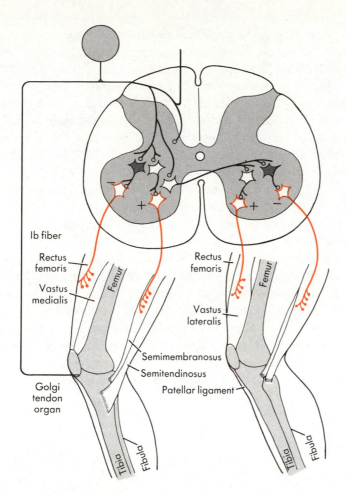

Fig. 14-13 ■ The neural circuitry for the inverse myotatic reflex. The afferent segment of the reflex pathway consists of the group Ib fibers from Golgi tendon organs. These afferent fibers disynaptically inhibit motoneurons innervating the contracting muscle. They disynaptically excite the motoneurons innervating the antagonistic muscle to produce reciprocal innervation. The afferent fibers also excite a commissural neuron that disynaptically mediates a crossed-extensor response that also shows reciprocal innervation. Black interneurons and minus signs are inhibitory, and clear interneurons and plus signs are excitatory.

overloading. However, we now appreciate that tendon organs are, in fact, highly sensitive receptors if the appropriate stimulus, muscle contraction, is applied. It is likely that the inverse myotatic reflex is integrally involved in spinally organized motor patterns. It may, for instance, contribute to the smooth onset and termination of contraction, the rotation among motor units during sustained contraction, the rapid switching between flexion and extension required for such behavior as running, and as a tension feedback system.

■ *Stretch-evoked flexion*

Confusion has prevailed for some years regarding the role in spinal reflex behavior of the group II afferent fibers from the secondary endings of the muscle spindle. Classical descriptions of the spinal effects of the secondary endings were based on studies of spinally transected animals. In this preparation, activating the group II spindle afferent fibers results in an excitation of flexor motoneurons that resembles that seen in the flexor withdrawal reflex (pp. 226-229); extensor motoneurons are inhibited. This finding generated the hypothesis that the secondary endings mediate a stretch-evoked flexion reflex that might contribute to motor patterns such as stepping.

More recently it has been found that in intact or decerebrate preparations, activating the group II spindle afferent fibers produces an effect that is opposite of that seen in the spinal animal. In the intact or decerebrate animal the secondary endings appear to excite extensor and inhibit flexor motoneurons. This effect is more consistent with the recently discovered role of the secondary endings in contributing to the myotatic reflex.

Still more recently it has been suggested that the secondary endings may contribute to the clasp-knife effect (inverse myotatic reflex). This would again implicate the group II spindle afferent fibers in a stretch-evoked flexion. How this relates to their excitatory action on the extensor motoneurons is unclear.

At this point it is necessary to conclude that knowledge regarding the role of the group II spindle afferent fibers in spinally mediated motor behavior is insufficiently developed to allow any firm statements. Indeed, these receptors may have no significant role in this regard, and their primary function may be to transmit information on muscle length to suprasegmental structures.

■ *The flexor withdrawal reflex*

The final spinal reflex to be considered is elicited by activation of a diverse group of receptors that signal somatosensory information, particularly pain. These receptors give rise to afferent fibers in the group II, III, and IV classes. On entering the spinal cord, the fibers course both rostrally and caudally, terminating extensively among the spinal interneurons of many segments (Fig. 14-14). Through complex and incompletely understood spinal circuitry, these afferent fibers ultimately produce excitation of flexor motoneurons innervating muscles around multiple joints (Fig. 14-15).

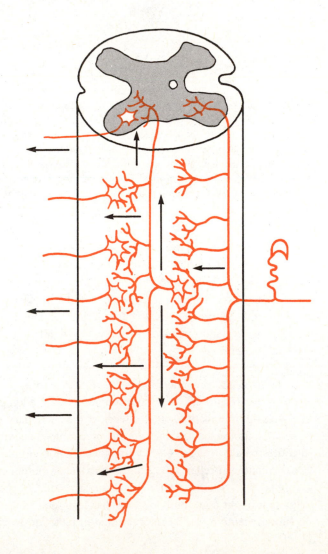

Fig. 14-14 ■ Schematic illustration of the divergence that occurs with the flexor reflex afferent fibers that constitute the afferent segment of the flexor withdrawal reflex. (Redrawn with permission from Eyzaguirre, C., and Fidone, S.J.: Physiology of the nervous system, ed. 2. Copyright © 1975 by Year Book Medical Publishers, Inc., Chicago. Slightly modified from Cajal, S.R.: Histologie du système nerveux, Paris, 1909, Maloine.)

If only the group II somatic afferent fibers (not to be confused with the group II spindle afferent fibers that arise from the secondary endings) are activated, a weak and rather restricted flexor pattern results. With inclusion of the group III afferent fibers, which include fibers from nociceptors that mediate fast pain, a much more active flexion response occurs that involves more joints. The maximal reflex response is evoked when the group IV afferent fibers, which mediate slow pain, are included as well. Because this diverse class of somatic afferent fibers elicits a common reflex pattern of flexion, they have been aggregated for functional purposes as the *flexor reflex afferent fibers*. It is clear from the afferent information that elicits the reflex that it serves the protective function of withdrawing the stimulated body part from potentially damaging stimuli (*nociceptive stimuli*). This contrasts with the other spinal reflexes that have been discussed thus far and that subserve primarily postural functions.

As with the other spinal reflexes, the general principle of reciprocal innervation applies to the organization of the flexor withdrawal reflex (Fig. 14-15). In this case the reciprocal innervation consists of a polysynaptic inhibition of the extensor motoneurons antagonistic to the excited flexor motoneurons. Moreover, the flexor reflex includes a crossed-extensor component (Fig. 14-15) that serves a postural function (as in the inverse myotatic reflex), and the crossed-component includes reciprocal innervation. This results in a complex neural network that mediates a basic motor pattern consisting of flexion of one limb, for example, and extension of the contralateral limb.

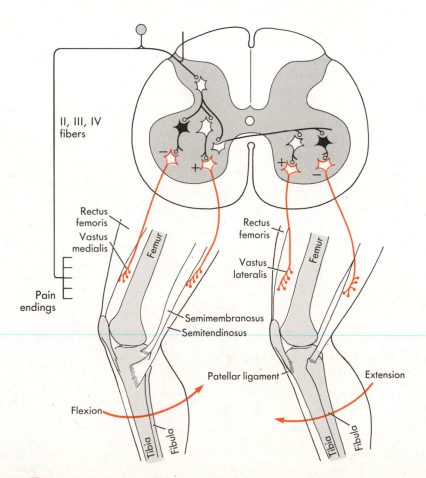

Fig. 14-15 ■ The neural circuitry of the flexor withdrawal reflex. The afferent segment consists of the flexor reflex afferent fibers. Through a chain of spinal interneurons these fibers excite flexor motoneurons and inhibit extensor motoneurons. The opposite pattern occurs contralaterally via commissural interneurons to produce a crossed-extensor response. Black interneurons and minus signs are inhibitory, and clear interneurons and plus signs are excitatory.

The properties of the flexor withdrawal reflex are quite different from those of the myotatic reflex. These differences are derived from the nature of the involved receptors, the degree of divergence of the afferent projections to the spinal cord, the complexity of the intraspinal circuitry, and the distribution of the reflex effects with respect to target muscles. First, the more slowly conducting fibers of the afferent segment and the more extensive intraspinal circuitry combine to make the flexor withdrawal reflex of longer latency. Second, where the myotatic reflex is finely graded as a function of muscle stretch, the flexor withdrawal reflex has a much more nonlinear input-output relationship. At low stimulus intensities there is little response, but as the nociceptive afferent fibers are activated, there rapidly develops a full response that is augmented little by further increases in stimulus intensity. Third, where the myotatic reflex is stimulus locked and terminates abruptly with stimulus termination, the flexor response persists beyond stimulus termination. This is presumed to reflect the persistence of neuronal activity within the complex spinal circuitry that mediates the reflex, a phenomenon designated *afterdischarge*. It has been hypothesized that recurrent circuits are responsible for such afterdischarge. Regardless, it serves the function of maintaining withdrawal from a tissue-damaging stimulus for a sufficient time to allow adjustments that prevent recontacting that stimulus. Fourth, the specificity of the reflex effect is less than that of the myotatic reflex, reflecting the considerable divergence of the sensory inputs, the complexity of the spinal circuitry, and the wide distribution to different motoneuronal pools.

It must be appreciated, however, that despite the extensive divergence in the flexor withdrawal circuitry (Figs. 14-14 and 14-15), there is a definite organization of the output. The extent to which flexors around a given joint are activated is a function of the location of the stimulus, a property called *local sign* (Fig. 14-16). For example, if the nociceptive stimulus is delivered to the plantar surface of the foot, then the resulting reflex action will include considerable flexion at the ankle, somewhat less at the knee, and still less at the hip. Delivering such a stimulus to the skin of the popliteal fossa will evoke a reflex response that includes considerable flexion at the hip and knee, with much less at the ankle. Thus the final limb position will be a function of the site of

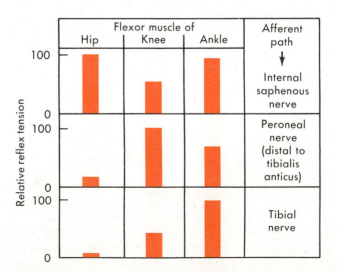

Fig. 14-16 ■ Local sign in the flexor withdrawal reflex. The bars indicate relative reflex tensions developed in three flexor muscles as a result of stimulation of each of three afferent nerves innervating different parts of the hindlimb. Each path activates all three muscles, but the relative participation of each muscle in the reflex movement varies as a function of the afferent path that is stimulated. (Redrawn from Patton, H.D.: Reflex regulation of movement and posture. In Ruch, T., and Patton, H.D., editors: Physiology and biophysics, vol. IV, Philadelphia, 1982, W.B. Saunders Co.)

stimulation, generally involving a flexor pattern that is appropriate to removing the stimulated site from the threatening stimulus.

A final point to recognize is that, although the preceding discussion describes the general flexor withdrawal response, there are specific reflexes of a different nature that can be evoked from various localized regions. For example, in the spinal animal, tactile stimulation of the footpads can elicit an extension of the entire limb, ostensibly a postural response (the *positive supporting reaction*). The most effective stimulus for producing this effect is not, however, activation of the cutaneous afferent fibers but stretch of the interosseus muscles.

■ *Some principles of spinal organization*

Although the application of contemporary methods, such as single-cell marking by dye injection, is now advancing knowledge of the detailed connectivity of the intrinsic spinal circuitry, appreciation of this circuitry is still limited. Nevertheless, there are some established properties of spinal organization that are instructive.

Two such properties are *convergence* and *divergence*. Although these can apply to the connections between any two cell groups, they are readily illustrated by the segmental afferent fibers to the spinal cord. Convergence refers to a many-to-one projection, where axons from a number of neurons converge to terminate on a single neuron. An excellent example of this is the convergence of many Ia afferent fibers from a given muscle on each motoneuron of the motoneuronal pool of that muscle. An example of divergence, a one-to-many projection, is the distribution of group III and IV cutaneous afferent fibers that ramify extensively on entering the spinal cord (Fig. 14-14).

The property of convergence is the anatomical foundation for *spatial summation*. In this phenomenon a neuron receives multiple inputs, none of which alone may be capable of discharging it. However, activity in more than one of these convergent inputs can summate to discharge the target neuron. A related phenomenon is *temporal summation*. In this case a given level of discharge of a single input may be insufficient to discharge the target neuron, but increasing the frequency of that input may permit sufficient summation of postsynaptic potentials to activate the target cell.

Both spatial and temporal summation are examples of *facilitation,* where an input to a neuron increases the probability of its discharging in response to an input from another source (spatial) or a greater number of inputs from the same source (temporal). A basis for this phenomenon is that the distribution of terminals of a given fiber, for example, to a motoneuronal pool, may involve considerable divergence. Some of the motoneurons contacted by that fiber may be insufficiently excited to discharge, whereas others may be activated by that fiber. Thus one can define a *subliminal fringe* and a *discharge zone* for the set of motoneurons contacted by a given fiber (Fig. 14-17). Another fiber projecting onto the motoneuronal pool will also diverge to contact a number of cells; some of these will be in its subliminal fringe and some in its discharge zone. The population of motoneurons contacted by this second fiber may not be identical to that contacted by the first fiber, and thus one can think of these inputs as *fractionating* the motoneuronal pool. However, the neurons contacted by each fiber may overlap, so that convergence occurs on certain of them. If this overlap involves their subliminal fringes, then facilitation by spatial summation can occur, and the effect of stimulating the two fibers simultaneously will be greater than the sum of the effects of stimulating each individually. In contrast, if their discharge zones overlap, then the effects of stimulating the two fibers simultaneously may be less than the sum of the effects of stimulating them individually. This phenomenon is called *occlusion.*

Another principle that can be illustrated with the motoneuronal pool is that of feedback circuits (Fig. 14-18). To this point we have described the ventral horn as containing α- and γ-motoneurons. However, also located in the ventral horn is an intrinsic neuron (interneuron), designated the *Renshaw cell,* that is inhibitory to local motoneu-

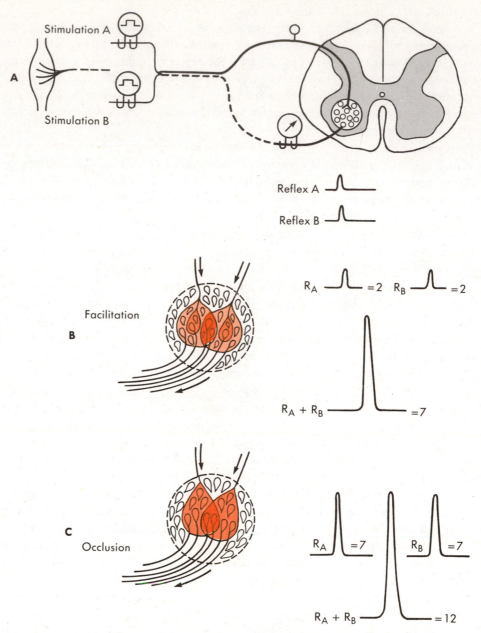

Fig. 14-17 ■ The phenomena of facilitation and occlusion. The monosynaptic reflex is used as an example. **A** shows the preparation, where a muscle nerve is dissected into two branches of equal size (*A* and *B*), which are placed on stimulating electrodes. Stimulating each of these branches will evoke monosynaptic reflexes of equal size (*reflex A* and *reflex B*). The reflexes are recorded from the ventral root, which is cut (*broken line*) distal to the recording electrode. **B** shows facilitation. The discharge zone of each afferent nerve branch (*A* and *B*) is included in the dark circle. Stimulating each branch will activate the motoneurons in its discharge zone. The subliminal fringe of each branch includes those motoneurons within the lighter circle. When both nerves are simultaneously stimulated, the neurons in their overlapping subliminal fringes will be activated as well as the neurons in each discharge zone. Consequently the number of neurons activated with simultaneous stimulation of both branches will be greater than the sum of the neurons activated by stimulating each branch individually. **C** shows occlusion. In this case the discharge zones overlap. Consequently the number of motoneurons activated by stimultaneously stimulating the two branches will be less than the sum of the motoneurons activated by stimulating each branch individually. (Redrawn with permission from Eyzaguirre, C., and Fidone, S.J.: Physiology of the nervous system, ed. 2. Copyright © 1975 by Year Book Medical Publishers, Inc., Chicago.)

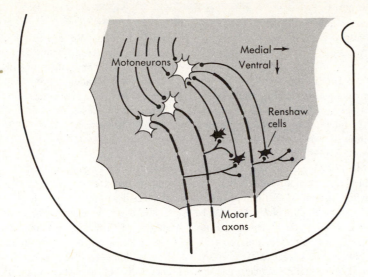

Fig. 14-18 ■ Schematic illustration of recurrent inhibition of motoneurons by inhibitory interneurons *(black),* the Renshaw cells. (Redrawn from Eccles, J.C.: The physiology of synapses, New York, 1964, Academic Press, Inc.)

rons. The Renshaw cell receives inputs from various sources; one important source is the collaterals of motoneuron axons. The Renshaw cell may then project back on a motoneuron that innervates it, and this establishes a negative feedback circuit *(recurrent inhibition)* that limits the discharge of the motoneuron. The spinal cord also contains positive feedback circuits *(recurrent facilitation).* An example is a collateral of a motoneuron axon that terminates on an inhibitory interneuron, which in turn terminates on a neuron that is inhibitory to the motoneuron. Such a circuit "turns off" inhibition of the motoneuron *(disinhibition)* and thereby facilitates it.

Finally, there are some principles, discussed previously, that are worth recalling in this context. These include reciprocal innervation (Fig. 14-12), crossed-components involving a spinal commissural neuron (Fig. 14-13), and double reciprocal innervation (Fig. 14-13). These are characteristics of the spinal circuitry that form the basis for organizing simple motor patterns at spinal levels. In a sense they provide simple "motor programs" that can be activated by descending pathways.

■ *Bibliography*

Journal articles

Burke, D., et al.: Spasticity, decerebrate rigidity and the clasp-knife phenomenon: an experimental study in the cat, Brain **95**:31, 1972.

Crowe, A., and Matthews, P.B.C.: The effects of stimulation of static and dynamic fusimotor fibres on the response to stretching of the primary endings of muscle spindles, J. Physiol. **174**:109, 1964.

Houk, J., and Henneman, E.: Responses of Golgi tendon organs to active contractions of the soleus muscle of the cat, J. Neurophysiol. **30**:466, 1967.

Hunt, C.C., and Perl, E.R.: Spinal reflex mechanisms concerned with skeletal muscle, Physiol. Rev. **40**:538, 1960.

Kuffler, S.W., and Hunt, C.C.: The mammalian small-nerve fibers: a system for efferent nervous regulation of muscle spindle discharge, Res. Publ. Assoc. Nerv. Ment. Dis. **30**:24, 1952.

Kuhn, R.A.: Functional capacity of the isolated human spinal cord, Brain **73**:1, 1950.

Landau, W.M., and Clare, M.H.: Fusimotor function, VI. H reflex, tendon jerk, and reinforcement in hemiplegia, Arch. Neurol. **10**:128, 1964.

Laporte, Y., and Lloyd, D.P.C.: Nature and significance of the reflex connections established by large afferent fibers of muscular origin, Am. J. Physiol. **169**:609, 1952.

Liddell, E.G.T.: Spinal shock and some features in isolation-alteration of the spinal cord in cats, Brain **57**:386, 1934.

Liddell, E.G.T., and Sherrington, C.S.: Reflexes in response to stretch (myotatic reflexes), Proc. R. Soc. Lond. (Biol.) **96**:212, 1924.

Lloyd, D.P.C.: Facilitation and inhibition of spinal motoneurons, J. Neurophysiol. **9**:421, 1946.

Lloyd, D.P.C.: Integrative pattern of excitation and inhibition in two-neuron reflex arcs, J. Neurophysiol. **9**:439, 1946.

Mendell, L.M., and Henneman, E.: Terminals of single Ia fibers: location, density, and distribution within a pool of 400 homonymous motoneurons, J. Neurophysiol. **34**:171, 1971.

Renshaw, B.: Central effects of centripetal impulses in axons of spinal ventral roots, J. Neurophysiol. **9**:191, 1946.

Sahs, A.L. and Fulton, J.F.: Somatic and autonomic reflexes in spinal monkeys, J. Neurophysiol. **3**:258, 1940.

Sherrington, C.S.: Flexion-reflex of the limb, crossed extension-reflex, and reflex stepping and standing, J. Physiol. **40**:28, 1910.

Stein, R.B.: Peripheral control of movement, Physiol. Rev. **54**:215, 1974.

Stuart, D.G. et al.: Stretch responsiveness of Golgi tendon organs, Exp. Brain Res. **10**:463, 1970.

Books and monographs

Creed, R.S., et al.: Reflex activity of the spinal cord, London, 1932, Oxford University Press.

Eccles, J.C.: The physiology of synapses, New York, 1964, Academic Press, Inc.

Harris, D.A., and Henneman, E.: Feedback signals from muscle and their efferent control. In Mountcastle, V.B., editor: Medical physiology, ed. 14, St. Louis, 1980, The C.V. Mosby Co.

Henneman, E.: Organization of the spinal cord and its reflexes. In Mountcastle, V.B., editor: Medical physiology, ed. 14, St. Louis, 1980, The C.V. Mosby Co.

Matthews, P.B.C.: Mammalian muscle receptors and their central actions, Baltimore, 1972, The Williams & Wilkins Co.

Mendell, L.M., and Henneman, E.: Input to motoneuron pools and its effects. In Mountcastle, V.B., editor: Medical physiology, ed. 14, St. Louis, 1980, The C.V. Mosby Co.

Patton, H.D.: Reflex regulation of movement and posture. In Ruch, T., and Patton, H.D., editors: Physiology and biophysics, vol. IV, Philadelphia, 1982, W.B. Saunders Co.

Sherrington, C.S.: The integrative action of the nervous system, ed. 2, New Haven, Conn., 1947, Yale University Press.

chapter 15

Descending pathways involved in motor control

Chapter 14 describes the basic features of spinal organization and establishes a foundation for developing an overview of the principal descending pathways that control movement. The extent to which these pathways contribute to the rich repertoire of motor behavior can be appreciated simply by recalling the severely restricted motor capability of humans after spinal transection. The task now is to develop a basic understanding of how the suprasegmental influences on the spinal cord are organized.

An axon that descends to the spinal cord can influence the local spinal circuitry in a number of ways. Its specific influence will be a function of its discharge properties, the nature of its physiological effect (e.g., excitatory or inhibitory), and the distribution of its terminals. Fig. 15-1 illustrates some of the different possibilities of terminal distribution within a spinal reflex circuit, each resulting in a different influence. The axon can terminate on cells that influence the receptor component of the circuit, as in the centrifugal control of muscle spindle receptor sensitivity. It can terminate on one or more interneurons of the circuit, or it can bypass the intrinsic spinal circuitry and terminate on motoneurons directly to gain greater control over muscle action. Also important is how a fiber system distributes among its population of target neurons. For example, if the fiber projects to motoneurons, its terminal field might be restricted only to certain motoneurons, such as those innervating the intrinsic muscles of the digits. This would restrict the musculature that the fiber is capable of influencing, which in turn shapes its functional role in motor control. Finally, if one considers only a single postsynaptic target of a descending fiber, how the fiber terminates on that neuron contributes to its effect. For example, if the terminals are restricted to the distal dendrites of the postsynaptic cell, the fiber will have less influence on the neuron than if its terminals were close to the spike initiation zone. The various descending pathways use all of these possibilities in controlling motor behavior.

Before the actual pathways involved in motor control are discussed, a definitional issue should be raised. We tend to describe tracts that descend from higher centers as ''motor pathways.'' However, most properly only the final common path should be defined as motor. To appreciate this definitional problem one need only consider that the Ia afferent fibers from the muscle spindles terminate monosynaptically on the motoneurons, yet these are dorsal root fibers that are clearly sensory. Consequently, from a morphological viewpoint the basis for defining any descending pathway as motor is problematic. Nevertheless, we take the license to do so, based on the functional effects of certain suprasegmental systems.

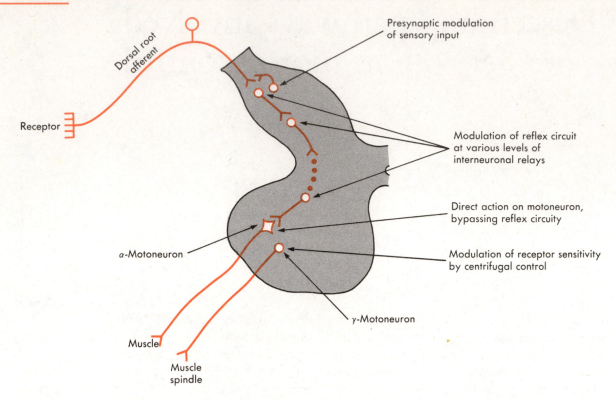

Fig. 15-1 ■ Schematic illustration of a spinal reflex pathway and the sites at which descending pathways can potentially influence such a pathway.

■ *A classification of descending pathways*

The traditional approach to classifying the motor pathways has been to designate them as *pyramidal* or *extrapyramidal*. This classification is derived from the early clinical literature but has little anatomical foundation. Strictly speaking, the pyramidal system consists of all fibers traversing the medullary pyramids, regardless of their cells of origin or sites of termination. Extrapyramidal systems are then defined, largely by exclusion, to include "motor systems" the fibers of which do not travel in the pyramids. Strikingly different motor deficits consequent to lesions of the pyramidal and extrapyramidal systems have reinforced this classification for many years.

In the past two decades, however, there has been a rapid erosion of this concept. With regard to the pyramidal system, Brodal states*:

> There can be little doubt that the classical "pyramidal tract syndrome" does not follow lesions of the pyramidal tract and is accordingly a misnomer. . . . It follows . . . that the term "pyramidal tract syndrome" should be discarded. It is convenient to have a designation of a symptom complex, but the name should not be such that it implies incorrect relations to definite structures.

With regard to the extrapyramidal system, Brodal writes†:

> In spite of the fact that these [extrapyramidal] disorders may have some clinical features in common, it is difficult to see the advantages of considering them under a common heading, the more so since the "system" to which they are referred escapes definition. It is the author's firm conviction that neither theoretically nor practically is a useful purpose served by retaining the term "extrapyramidal."

*From Brodal, A.: Neurological anatomy, Oxford, 1981, Oxford University Press, p. 274.
†From Brodal, A.: Neurological anatomy, Oxford, 1981, Oxford University Press, p. 183.

As suggested by Brodal, the pyramidal-extrapyramidal classification has not contributed to developing an understanding of the descending pathways involved in motor control. The call to abandon this traditional view of the motor pathways is now particularly compelling, since a more useful classification has become available. The basis for this alternative is developed in Chapter 13; it involves considering the *sites of termination* of a given fiber system rather than its cells of origin or trajectory. To review briefly, the spinal cord includes topographically organized columns of motoneurons with the more ventromedially situated motoneuronal cell groups innervating the axial and girdle muscles and the more dorsolaterally located motoneurons innervating the muscles of the distal extremities. This mediolateral topography is maintained in the interneuronal projections on the motoneurons. Thus from a functional perspective one can think in terms of a *lateral system* that is principally concerned with manipulatory activity of the extremities and a *medial system* that is primarily involved in controlling body and whole limb movement.

If the descending pathways are now considered with respect to their spinal sites of termination, then each can be classified with respect to the extent to which it participates in lateral as opposed to medial system types of functions; that is, pathways that terminate primarily on the ventromedial motoneurons or their associated interneurons will be more concerned with controlling posture, equilibration, and progression, whereas those terminating on the lateral motoneurons or their associated interneurons will be more concerned with controlling finer manipulatory activity of the extremities.

■ *Lateral system pathways*

There are two major pathways that terminate on the more laterally situated motoneurons or their associated interneurons: the *lateral corticospinal tract* and the *rubrospinal tract* (Figs. 15-2 and 15-3).

The *corticospinal tract* arises from both the precentral gyrus (including the motor and premotor cortices) and the postcentral gyrus (the somatosensory cortex), deriving from both small and large neurons, primarily in layer V (Fig. 15-9). The tract then descends in the posterior limb of the internal capsule, the medial two thirds of the cerebral peduncle, and the medullary pyramids, where it decussates. The crossed pathway then descends in the lateral funiculus of the spinal cord, where it constitutes the lateral corticospinal tract. A small contingent of fibers does not decussate at the pyramids but remains ipsilateral and descends in the ventral funiculus. This ventral corticospinal tract is a minor pathway that does not exist in all persons. When present, it does not descend beyond thoracic levels. (It should be noted that along its descending brainstem trajectory the descending fibers from the precentral gyrus also project onto the motor cranial nerve nuclei.)

By definition the pyramidal tract consists of all fibers coursing longitudinally in the medullary pyramids, regardless of their sites of origin or termination. The corticospinal tract thus consists of a subset of these fibers that has its cells of origin in the cortex and sites of termination in the spinal cord. The corticospinal tract component that arises from the postcentral gyrus terminates in the dorsal horn and is sensory in function (Fig. 15-3). The motor component arises from the precentral gyrus, and it terminates more ventrally in the spinal gray matter (Figs. 15-2 and 15-4). As they descend, the sensory and motor components of the corticospinal tract issue collaterals at numerous levels of the neuraxis, including the basal ganglia and thalamus.

The motor component of the tract will now be considered in more detail. The precentral gyrus, from which the tract arises, has an orderly topographical representation (homunculus) of the body musculature (Fig. 15-5). This representation was discovered by electrical stimulation experiments in which it was found that localized stimulation of the motor cortex will elicit discrete movements. For example, stimulation of the "wrist area" will evoke a rapid flexion at the wrist. Noteworthy is the disproportionate repre-

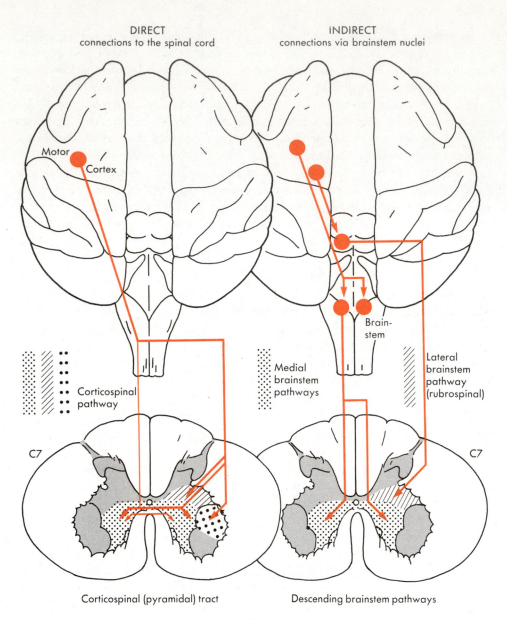

DIRECT
connections to the spinal cord

INDIRECT
connections via brainstem nuclei

Motor Cortex

Brain-stem

Corticospinal pathway

Medial brainstem pathways

Lateral brainstem pathway (rubrospinal)

C7

C7

Corticospinal (pyramidal) tract

Descending brainstem pathways

Fig. 15-2 ■ The descending connections from the cerebral cortex and brainstem to the spinal cord of the monkey. The dorsolateral motoneurons innervating the muscles of the distal extremities are indicated by dots. The dorsolateral interneurons that project onto these motoneurons are shown by hatching, and the ventromedial interneurons that project onto motoneurons innervating more medial muscles are shown by stippling. (Redrawn from Brinkman, C.: Split-brain monkeys: cerebral control of contralateral and ipsilateral arm, hand, and finger movements, doctoral dissertation, Rotterdam, 1974, Erasmus University Rotterdam.)

sentation of certain musculatures, such as those of the face and limbs (Fig. 15-5). The fibers of the corticospinal tract that originate more rostrally in the precentral gyrus, where the more proximal muscles are represented, terminate bilaterally in the more ventral spinal gray matter. Here they influence the interneurons and motoneurons that innervate the more proximal muscles. However, the greatest proportion of corticospinal fibers influence the face and extremities (Fig. 15-5). Consequently, the tract terminates prominently in the more lateral spinal gray matter, where it influences the interneurons and motoneurons associated with the muscles of the extremities. On the basis of this

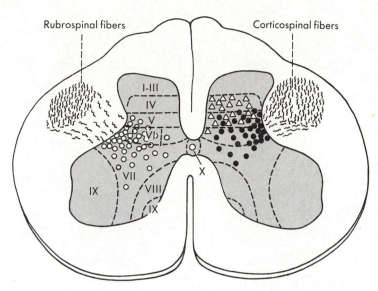

Fig. 15-3 ■ A transverse section of the cervical spinal cord of the cat showing the location in the lateral funiculus and the sites of termination in the spinal gray matter of rubrospinal and corticospinal fibers from "motor" and "sensory" regions of the sensorimotor cortex. Note the similarities in the areas of termination of rubrospinal and corticospinal fibers from the motor cortex. Because the section is from the cat, there are no direct corticospinal projections to mononeurons. *Open circles,* Sites of termination of rubrospinal fibers; *solid circles,* sites of termination of corticospinal fibers from motor cortex; *open triangles,* sites of termination of corticospinal fibers from sensory cortex. (Redrawn from Brodal, A.: Neurological anatomy, ed. 3. Copyright © 1981 by Oxford University Press, Inc. Reprinted by permission.)

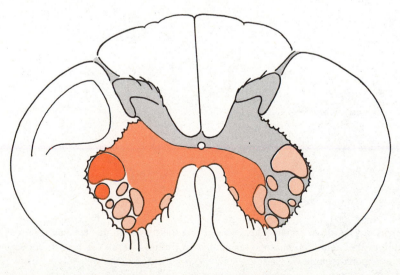

Fig. 15-4 ■ Schematic representation of the termination of the precentral (motor) component of the corticospinal tract in the monkey. These fibers terminate unilaterally in the dorsolateral intermediate zone but bilaterally in the ventromedial intermediate zone. In addition, fibers terminate unilaterally on the dorsolateral motoneurons. (Redrawn from Brinkman, C.: Split-brain monkeys: cerebral control of contralateral and ipsilateral arm, hand and finger movements, doctoral dissertation, Rotterdam, 1974, Erasmus University Rotterdam.)

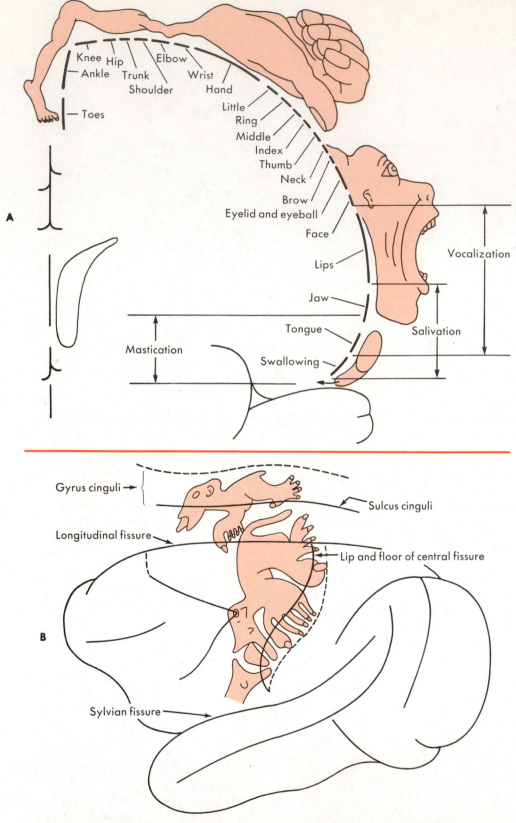

Fig. 15-5 ■ **A,** The motor homunculus of the precentral gyrus. In this schematic coronal section the location of the motor cortical representation of different parts of the body is shown, and size of the various parts is proportional to the amount of cortical surface area serving them. **B,** The somatotopic organization of primary, supplementary, and secondary motor areas in the cerebral cortex of the monkey. The central and longitudinal fissures are shown opened out, with the broken line indicating the floor of the fissure and the solid line the lip of the fissure on the brain's surface. At the bottom is an ipsilateral face area (secondary motor area). (Redrawn with permission from Eyzaguirre, C., and Fidone, S.J.: Physiology of the nervous system, ed. 2. Copyright © 1975 by Year Book Medical Publishers, Inc., Chicago. **A** slightly modified from Penfield, W., and Rasmussen, T.: The cerebral cortex of man, New York, 1950, The Macmillan Co.; **B** slightly modified from Woolsey, C.N., et al.: Res. Publ. Assoc. Res. Nerv. Ment. Dis. **30:**238, 1952.)

prominent termination in the lateral spinal gray matter, the corticospinal tract is classified as a component of the lateral system.

A unique feature of the tract is its direct termination on the motoneurons that innervate the intrinsic muscles of the digits (Figs. 15-2 and 15-4). Indeed, the motor component of the tract is most highly developed in species such as raccoons and most primates, which have evolved independent movement of the digits. In this context it might also be noted that independent finger movement in primates does not occur ontogenetically until development of the corticospinal projection on the motoneurons of the digits. It should be recognized, however, that the size of the entire corticospinal tract does not necessarily correlate with the development of the digits, since the size of the sensory component can vary independently of this. Thus cetaceans, such as whales, have prominent corticospinal tracts, but most of the fibers are sensory. These fibers originate in the postcentral gyrus and terminate in the dorsal horn.

The second major component of the lateral system is the *rubrospinal tract*. It arises from both large and small cells of the caudal part of the red nucleus (Fig. 15-6), a

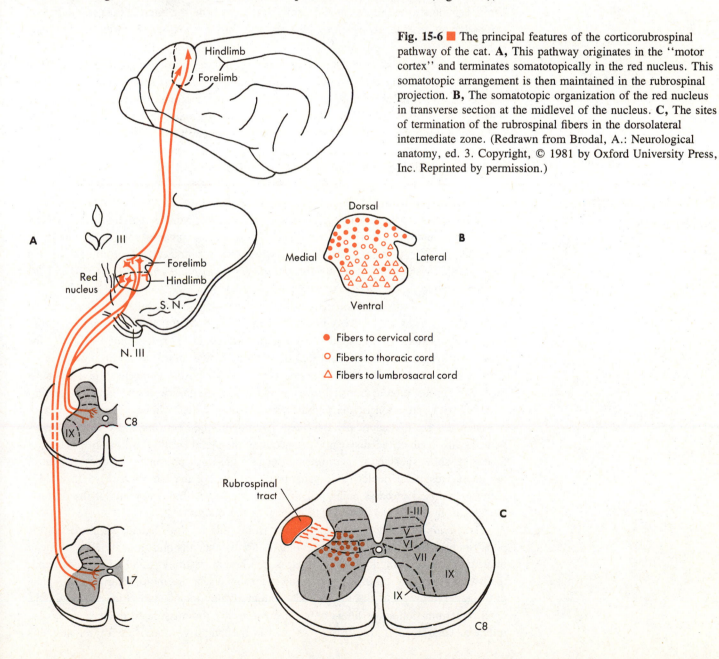

Fig. 15-6 ■ The principal features of the corticorubrospinal pathway of the cat. **A,** This pathway originates in the "motor cortex" and terminates somatotopically in the red nucleus. This somatotopic arrangement is then maintained in the rubrospinal projection. **B,** The somatotopic organization of the red nucleus in transverse section at the midlevel of the nucleus. **C,** The sites of termination of the rubrospinal fibers in the dorsolateral intermediate zone. (Redrawn from Brodal, A.: Neurological anatomy, ed. 3. Copyright, © 1981 by Oxford University Press, Inc. Reprinted by permission.)

● Fibers to cervical cord
○ Fibers to thoracic cord
△ Fibers to lumbrosacral cord

midbrain structure. Although not as discretely organized as the motor cortex, there is a topographical representation of the body musculature in the red nucleus. Immediately on leaving the nucleus, the descending fibers cross the midline and ultimately assume a trajectory in the lateral funiculus of the spinal cord just ventrolateral to that of the corticospinal tract (Fig. 15-3). The terminal field of the rubrospinal tract overlaps to a large extent with that of the corticospinal tract, although it does not appear to have any direct terminations on motoneurons. As in the corticospinal tract, the rubrospinal fibers terminate preferentially on interneurons associated with the more laterally situated motoneurons (Figs. 15-3 and 15-6).

It is probably most appropriate to consider the rubrospinal tract as a corticorubrospinal system, since the motor cortex projects in a topographical fashion onto the cells of origin of the rubrospinal tract (Fig. 15-6). However, the corticorubral projection apparently arises from a population of cortical neurons different from that giving rise to the corticospinal tract. This projection is excitatory, but weakly so because it is restricted primarily to the distal dendrites of the rubrospinal neurons. A more powerful excitatory input to the rubrospinal neurons originates from the intermediate or interposed cerebellar nuclei, the axons of which terminate on the proximal dendrites and somata (Chapter 16).

Thus the lateral system pathways, consisting of the corticospinal and rubrospinal tracts, have spinal trajectories in the lateral funiculus. They terminate most prominently on the interneuronal zone that is associated with the motoneurons that innervate the muscles of the extremities. The corticospinal tract is unique in having monosynaptic connections with the most dorsolaterally located motoneurons that innervate the intrinsic muscles of the digits. It is, in fact, the only descending pathway having access to the final common path for independent movement of the digits.

■ Medial system pathways

There are a number of descending pathways that travel in the ventral funiculus of the spinal cord and terminate preferentially on the more ventromedially situated interneurons and motoneurons (Fig. 15-2). These motoneurons innervate the axial and girdle musculatures, the principal functions of which are postural. Most prominent among these medial system pathways are the *lateral vestibulospinal tract* (Fig. 15-7) and the *pontine reticulospinal tract* (Fig. 15-8).

The *lateral vestibulospinal tract* arises from the lateral vestibular, or Deiters', nucleus. The pathway descends, without crossing, in the ventral funiculus, and it terminates ventromedially in the spinal gray matter throughout the rostrocaudal extent of the spinal cord (Fig. 15-7). This terminal distribution is the substrate for its influence on the extensor motoneurons that innervate the proximal postural muscles. In fact, most medial extensor motoneurons show monosynaptic excitatory postsynaptic potentials (EPSPs) with activation of the lateral vestibulospinal tract, whereas the medial flexor motoneurons are disynaptically inhibited. Moreover, the ventromedially situated interneurons involved in the disynaptic connections of lateral vestibulospinal fibers to motoneurons show substantial recruitment. Because the most prominent source of input to the lateral vestibular nucleus originates in the utricles of the labyrinth (Fig. 15-8), this pathway can be viewed as being principally concerned with adjustments of the postural muscles to linear acceleratory displacements of the body.

The *pontine reticulospinal tract* arises from cells of the medial two thirds of the pons, including the entire nucleus reticularis pontis caudalis and the caudal segment of nucleus reticularis pontis oralis. Similar to the lateral vestibulospinal tract, it descends ipsilaterally in the ventral funiculus, and it terminates in the more ventromedial spinal gray matter throughout the rostrocaudal extent of the spinal cord (Fig. 15-9). At the level of termination some fibers cross in the anterior commissure to innervate the contralateral ventromedial gray matter. Also as in the lateral vestibulospinal tract, it exerts

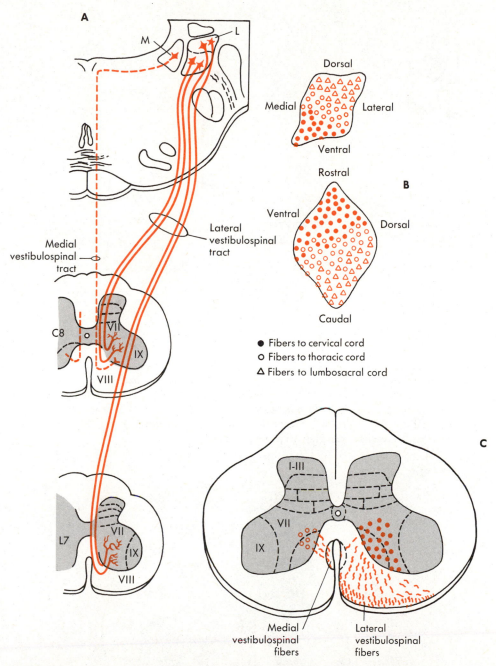

● Fibers to cervical cord
○ Fibers to thoracic cord
△ Fibers to lumbosacral cord

Fig. 15-7 ■ **A,** Principal features of the vestibulospinal pathways of the cat. The lateral vestibulospinal tract arises from the lateral vestibular nucleus of Deiters *(L),* which is somatotopically organized. The medial vestibulospinal tract arises from the medial vestibular nucleus *(M)* and does not project beyond the thoracic spinal cord. **B,** The somatotopic organization of the lateral vestibular nucleus in transverse *(above)* and sagittal *(below)* reconstructions. **C,** The spinal termination patterns of the vestibulospinal pathways. (Redrawn from Brodal, A.: Neurological anatomy, ed. 3. Copyright © 1981 by Oxford University Press, Inc. Reprinted by permission.)

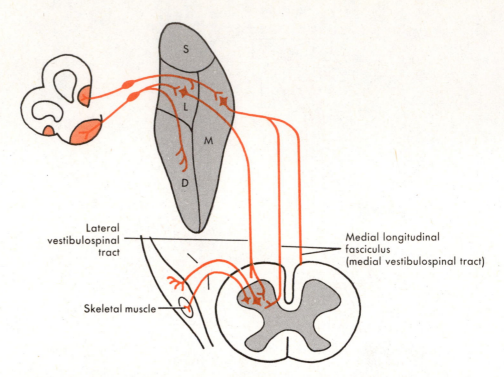

Fig. 15-8 ■ The principal features of the organization of primary vestibular projections on the vestibular nuclei and of the descending projections of these nuclei. *M* and *S* refer to the medial and superior vestibular nuclei, respectively. Note that lateral and descending vestibular nuclei receive projections primarily from the utricle, whereas the medial nucleus receives its input from the semicircular canals. (Redrawn from Brodal, A.: Neurological anatomy, ed. 3. Copyright © 1981 by Oxford University Press, Inc. Reprinted by permission.)

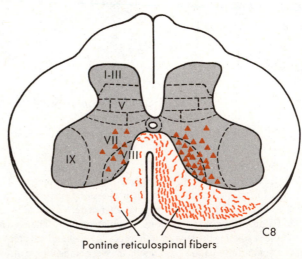

Fig. 15-9 ■ A transverse section of the cervical spinal cord of the cat showing the position and terminal distribution of the pontine reticulospinal tract. (Redrawn from Brodal, A.: Neurological anatomy, ed. 3. Copyright © 1981 by Oxford University Press, Inc. Reprinted by permission.)

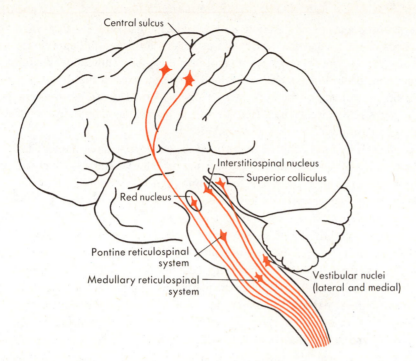

Fig. 15-10 ■ The major fiber systems descending to the spinal cord. Shown are the corticospinal, rubrospinal, tectospinal, interstitiospinal, pontine and medullary reticulospinal, and medial and lateral vestibulospinal tracts. (Redrawn from Brodal, A.: Neurological anatomy, ed. 3. Copyright © by Oxford University Press, Inc. Reprinted by permission.)

an excitatory influence on the more medial motoneurons to the proximal extensor muscles, primarily through interneurons.

Other less prominent pathways that contribute to the medial system are the *medial vestibulospinal tract, interstitiospinal tract,* and *tectospinal tract* (Fig. 15-10). The medial vestibulospinal tract (Figs. 15-7 and 15-8) arises from the medial vestibular nucleus, which receives its primary input from the semicircular canals. The tract descends ipsilaterally in the ventral funiculus to midthoracic levels, and it is concerned with adjustments of the neck and upper limbs to angular acceleration of the body. Ascending fibers from the medial vestibular nucleus mediate adjustments of eye position to angular acceleration. The interstitiospinal tract arises from the interstitial nucleus of Cajal in the rostral midbrain (Fig. 15-10). It descends without crossing to travel in the ventral funiculus, and it terminates in the ventromedial spinal gray matter throughout the rostrocaudal extent of the spinal cord. It is presumed to be principally involved in rotation of the head and body about the longitudinal axis. The tectospinal tract has its cells of origin in the superior colliculus (Fig. 15-10). In contrast to other pathways of the medial system, it crosses the midline before descending in the ventral funiculus. However, it distributes only to upper cervical levels and mediates visually guided head movement.

Thus, in contrast to the lateral system pathways, the tracts constituting the medial system originate primarily from the brainstem. Their spinal trajectories are through the ventral, rather than the lateral, funiculus, and they distribute to the ventromedial spinal gray matter. Their spinal terminal fields thus determine that their influences will be largely restricted to the motoneurons that innervate the more proximal muscle groups. Functionally this implicates these pathways in the control of posture, equilibration and progression, in contrast to the lateral system pathways, the principal function of which is the control of spatially organized movements involving the distal extremities.

■ *Interruption of the descending pathways*

The previous anatomical description of the descending pathways that control movement allows one to predict the distribution of deficits following the interruption of each source of motor control. Thus interruption of lateral system pathways would be expected to compromise control of the extremities and of the muscles of the digits in particular. In contrast, interruption of medial system pathways would be expected to compromise the axial and girdle musculatures. Considerable experimental and clinical evidence supports these predictions.

The *corticospinal tract* can be *interrupted* in monkeys without damage to other ascending or descending systems by a lesion at the medullary pyramids. Such a lesion has suprisingly minimal effect, and the few lasting signs are confined to the distal limb muscles. A mild flexor hypotonia may ensue, consistent with evidence that the corticospinal tract primarily excites flexor motoneurons and inhibits extensor motoneurons of the extremities. The primary persistent deficit is a loss of the ability to move the digits independently, that is, a *loss of independent or fractionated finger movement*. Because the lateral corticospinal tract is the only descending pathway that has direct access to the dorsolateral motoneurons that innervate the intrinsic muscles of the digits, this deficit is not surprising. As expected, there are no deficits involving the more proximal muscles; thus posture, equilibration, and progression are not compromised. Indeed, the monkey can sit, walk, and climb quite normally. Although a deficit restricted to the digits may not appear to be critical, it is important to appreciate that it seriously compromises the ability for fine manipulatory activity, a most important aspect of the human's motor repertoire. With respect to the lower extremities, the toes are affected as well. In a normal adult, stroking the sole of the foot will elicit a plantar flexion of the big toe. Frequently the other toes will flex and adduct as well. In the infant, where the corticospinal innervation of the dorsolateral lumbar motoneurons has not yet developed, such a stimulus elicits instead a dorsoflexion (i.e., an inverted plantar reflex) with fanning of the toes. This is called *Babinski's sign*. In an adult animal in which the corticospinal tract has been interrupted, Babinski's sign also occurs. Because the plantar response is generally considered a flexor response, the shift to an inverted plantar, or extensor, response reflects the loss of flexor motoneuronal innervation consequent to interruption of the lateral corticospinal tract.

Interruption of the other lateral system pathway, the *rubrospinal tract*, has almost no evident effect, and the role of this pathway in humans remains moot. Perhaps the only sequela to interrupting this pathway is a reduction in limb movement, that is, *limb hypokinesia*. However, if one surgically *interrupts both the corticospinal and rubrospinal tracts*, then striking deficits involving movements of the extremities ensue. The monkey has difficulty in flexing its arm and closing its hand. Independent finger movement is, of course, lost, since this follows interruption of the corticospinal tract. After a period of recovery, the capacity for grasping or hand closure is regained, but this tends to occur only as part of a total arm movement. Thus in attempting to guide the affected arm to an object and grasp it, the animal must implement a sweeping arm movement that involves flexion at the elbow and wrist, with a grasp in which the fingers flex in concert. Although predictable, it is still striking that such combined interruption of the two lateral system pathways results in virtually no deficits involving the axial and girdle muscles. Thus posture and equilibration are totally unaffected.

The effects of experimentally *interrupting the medial system pathways* by a brainstem lesion are quite different. Such a procedure produces an immediate reduction in tone in the proximal extensor muscles so that there is a flexor bias of the trunk. It is several weeks before the righting reflexes are sufficiently recovered to allow the animal to maintain an upright posture. After recovery of the righting reflexes, the animal tends to slump forward and frequently falls. This results from impairment of correcting movements that involve the axial and girdle muscles and the proximal muscles of the limbs.

There is extreme difficulty in walking, and when walking does occur, there is severe difficulty in directing the course of progression. However, if the animal is supported properly, it can demonstrate perfectly intact control of the distal extremity muscles, allowing fine manipulatory activity involving fractionated finger movement.

■ *Pyramidal tract signs compared with corticospinal tract interruption*

The traditional pyramidal-extrapyramidal classification of the motor pathways is in part based on clinical phenomenology that includes a group of deficits that are presumed to reflect damage to the corticospinal tract. This set of observations is referred to as the *pyramidal tract syndrome, corticospinal signs,* or sometimes *upper motoneuron disease.* The constellation of abnormalities includes (1) increased deep tendon reflexes, (2) hypertonia (spasticity), (3) paresis, (4) Babinski's sign, and (5) decreased superficial (flexor) reflexes. However, when a lesion restricted to the corticospinal tract is produced experimentally, the consequent abnormalities consist primarily of the loss of independent finger movement. Although it is not generally included among the set of pyramidal tract signs, this deficit does occur. Similarly, Babinski's sign is shared by both the experimental and clinical lesions, yet the increased deep tendon reflexes and spasticity do not follow experimental interruption of the corticospinal tract. If there is any alteration in tone, it is reflected more in the flexor than in the extensor muscles.

The resolution of this discrepancy is reasonably straightforward. The clinician rarely, if ever, sees a case of pure corticospinal tract damage. Corticospinal tract signs constitute a syndrome that generally involves damage to the cortex or internal capsule, where systems beyond the corticospinal tract are compromised as well. For example, a cerebrovascular accident involving the middle cerebral artery characteristically produces pyramidal tract signs. However, such a lesion clearly involves other cortical projections, such as corticoreticular (Fig. 15-2). It is on this basis that Brodal has emphatically stated that the pyramidal tract syndrome is indeed a misnomer, and he has suggested substituting the designation *internal capsule syndrome* or *pure motor hemiplegia.*

It is not even clear that lesions restricted to the motor cortex (area 4) provide a reasonable model of the pyramidal tract syndrome. The literature on experimental lesions in primates suggests that ablating the motor cortex can produce the deficits that follow experimental interruption of the corticospinal tract by pyramidotomy; this would be expected. Additionally, at least a transient paresis ensues, as with stroke in humans. However, producing the spasticity, a sine qua non of the pyramidal tract syndrome, requires a lesion that extends well beyond the motor cortex and includes at least the premotor cortex (area 6) as well. With a lesion restricted to the motor cortex the experimental animal shows a flaccid paralysis that is most severe in the more distal musculature, as one would anticipate from functional neuroanatomical considerations. Some recovery occurs with time, but such improvement is primarily seen in the more proximal muscles. Although some have argued that this discrepancy between the experimental and human clinical literature reflects species differences, it is more likely that a lesion restricted to the motor cortex rarely occurs in humans. Although the genesis of the spasticity that follows cortical lesions in humans is not fully understood, one possible explanation is that it results from destruction of inhibitory corticoreticular projections that release the pontine reticulospinal tract and thus increase medial extensor tone. This subject is discussed in more detail in Chapter 18.

■ *Bibliography*

Journal articles

Asanuma, H.: Cerebral cortical control of movement, Physiologist **16**:143, 1973.

Coulter, J.D., Ewing, L., and Carter, C.: Origin of primary sensorimotor cortical projections to lumbar spinal cord of cat and monkey, Brain Res. **403**:366, 1976.

Kuypers, H.G.J.M., and Brinkman, J.: Precentral projections to different parts of the spinal intermediate zone in the rhesus monkey, Brain Res. **24**:29, 1970.

Kuypers, H.G.J.M., and Lawrence, D.G.: Cortical projections to the red nucleus and the brainstem in the rhesus monkey, Brain Res. **4**:151, 1967.

Landau, W.M., and Clare, M.H.: The plantar reflex in man, with special reference to some conditions where the extensor response is unexpectedly absent, Brain **82**:321, 1959.

Lawrence, D.G., and Hopkins, D.A.: The development of motor control in the rhesus monkey: evidence concerning the role of corticomotoneuronal connections, Brain **99**:235, 1976.

Lawrence, D.G., and Kuypers, H.G.J.M.: The functional organization of the motor system in the monkey. I. The effects of bilateral pyramidal lesions, Brain **91**:1, 1968.

Lawrence, D.G., and Kuypers, H.G.J.M.: The functional organization of the motor system in the monkey. II. The effects of lesions of the descending brainstem pathways, Brain **91**:15, 1968.

Nyberg-Hansen, R.: Functional organization of descending supraspinal fibre systems to the spinal cord: anatomical observations and physiological correlations, Ergeb. Anat. Entwickl. Gesch. **39**(2):1, 1966.

Sterling, P., and Kuypers, H.G.J.M.: Anatomical organization of the brachial spinal cord of the cat. III. The propriospinal connections, Brain Res. **7**:419, 1968.

Books and monographs

Brodal, A.: Neurological anatomy, Oxford, 1981, Oxford University Press.

Ghez, C.: Introduction to the motor systems. In Kandel, E.R., and Schwartz, J.H.: Principles of neural science, New York, 1981, Elsevier North-Holland, Inc.

Kuypers, H.G.J.M.: The anatomical organization of the descending pathways and their contributions to motor control especially in primates. In Desmedt, J.E., editor: New developments in electromyography and clinical neurophysiology, vol. 3, Basel, 1973, S. Karger.

Lundberg, A.: Control of spinal mechanisms from the brain. In Tower, D.B., editor: The nervous system, vol. 1, The basic neurosciences, New York, 1975, Raven Press.

chapter 16

The cerebellum

The cerebellum has been traditionally viewed as a motor structure involved in the regulation rather than in the execution of movements. Situated above the brainstem in the posterior fossa, its afferent and efferent connections are via three pairs of fiber tracts: the inferior cerebellar peduncle or restiform body, the middle cerebellar peduncle or brachium pontis, and the superior cerebellar peduncle or brachium conjunctivum cerebelli (Fig. 16-1). The cerebellum receives a complex array of sensory information from most, if not all, modalities. In fact, it appears to be the principal central target for proprioceptive information. For this reason, it was described by Sherrington as the "head ganglion of the proprioceptive system." The moment-by-moment input of detailed sensory information, particularly concerning body position, muscle length, and muscle tension, is integrated by the cerebellar cortex, and the resultant output is transmitted to the descending motor pathways to modulate ongoing movements. In broad descriptive terms the cerebellum can be thought of as coordinating and smoothing muscular activity on

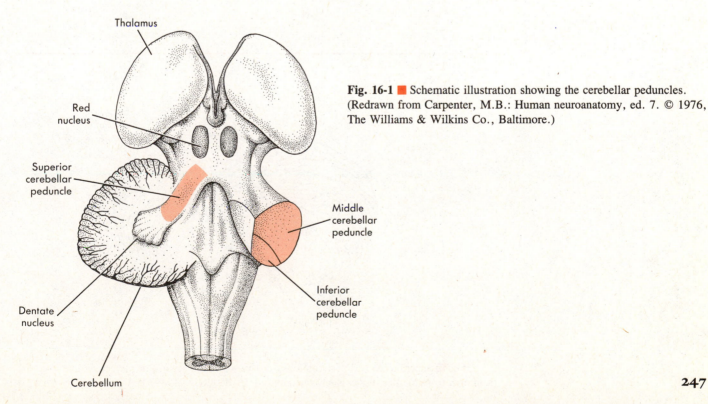

Fig. 16-1 ■ Schematic illustration showing the cerebellar peduncles. (Redrawn from Carpenter, M.B.: Human neuroanatomy, ed. 7. © 1976, The Williams & Wilkins Co., Baltimore.)

Thalamus

Red nucleus

Superior cerebellar peduncle

Middle cerebellar peduncle

Inferior cerebellar peduncle

Dentate nucleus

Cerebellum

the basis of such continuous sensory inputs. More specifically it may be viewed as regulating the rate, range, force, and direction of movements.

The preceding description of the role of the cerebellum in motor control is clearly imprecise. Unfortunately it reflects our presently inadequate understanding of how this structure integrates its inputs and then interacts with descending motor pathways. This inadequacy is somewhat ironic because the microanatomy and microphysiology of the cerebellum are perhaps better characterized than are those for any other structure of the mammalian central nervous system. As we have seen with other central nervous system structures, such elusiveness of functional understanding often prevails when the functional relationships of a given structure with either the sensory or motor peripheries have not been rigorously specified.

■ *Anatomical considerations*
■ *Topography*

A superficial view of the cerebellum (Fig. 16-2) shows its complexity. Immediately striking are the extensive transverse convolutions of the surface and the pattern of fissures that differentiates the cerebellum into lobes and lobules. Indeed, many traditional descriptions of cerebellar structure are long scholarly treatments of the various lobules,

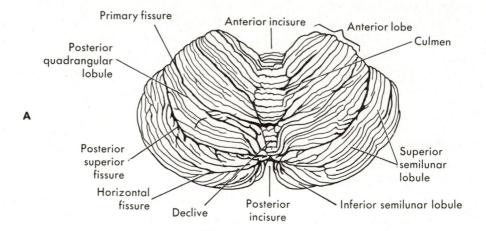

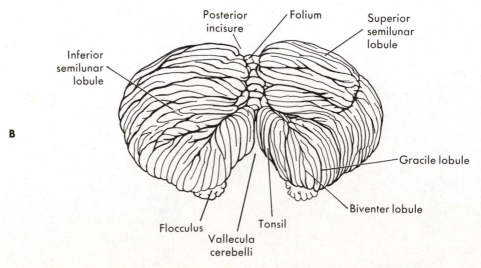

Fig. 16-2 ■ Superior (**A**) and posteroinferior (**B**) views of the surface of the cerebellum. (Redrawn from Mettler, F.A.: Neuroanatomy, ed. 2, St. Louis, 1948, The C.V. Mosby Co.)

the names of which defy recall. Briefly, the most prominent divisions are the anterior, posterior, and flocculonodular lobes, and these are defined by two large fissures (Fig. 16-3). The anterior and posterior lobes are divided by the primary fissure, and the posterior and flocculonodular lobes are separated by the posterolateral fissure. Each of these lobes is then further differentiated into a number of lobules by smaller fissures. Although this lobular organization is meaningful, as will become evident in the discussion of the afferent pathways to the cerebellum, a more effective initial approach is to consider the cerebellum's longitudinal organization.

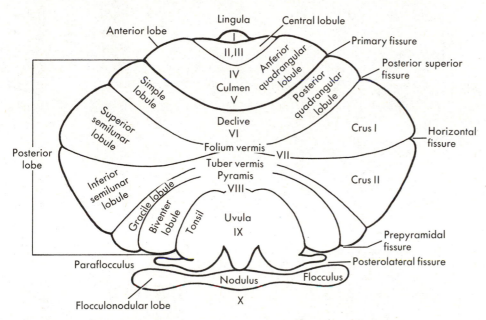

Fig. 16-3 ■ Fissures and lobules of the cerebellum. Portions of the cerebellum caudal to the posterolateral fissure constitute the flocculonodular lobe (archicerebellum). Portions rostral to the primary fissure constitute the anterior lobe (paleocerebellum). The neocerebellum lies between the primary and posterolateral fissures. Roman numerals refer to portions of the cerebellar vermis only. (Redrawn from Carpenter, M.B.: Human neuroanatomy, ed. 7. © 1976, The Williams & Wilkins Co., Baltimore; modified from Larsell, Jansen and Brodal [1958], and Angevine et al.)

The cerebellum consists of a three-layered cortex and deep nuclear groups (Fig. 16-4). The anatomy will be reviewed in more detail later. However, it is helpful to consider first the general relationship between the cortex and the deep nuclei.

The flow of information through the cerebellum begins with inputs arriving at the cerebellar cortex. The information is then processed within the cortical circuitry; its output element is the Purkinje cell (Fig. 16-4). The Purkinje cells then project onto the deep cerebellar nuclei, which relay the cerebellar outflow to various areas of the central nervous system. (There are some exceptions, however, where the Purkinje cell axons exit the cerebellum directly without relaying through the deep nuclei.) There is an orderly topography in this corticonuclear projection (Fig. 16-5). The most medial or *vermal zone* of the cerebellar cortex projects onto the most medial deep nucleus, the *fastigial nucleus*. The most lateral or *hemispheric zone* projects onto the most lateral deep nucleus, the *dentate nucleus*. Between these medial and lateral longitudinal zones is an intermediate or interposed zone, the *paravermal zone*, which projects onto the interposed or intermediate deep nuclei, the *globus* and *emboliform nuclei*. (These nuclei

■ *Longitudinal corticonuclear zones*

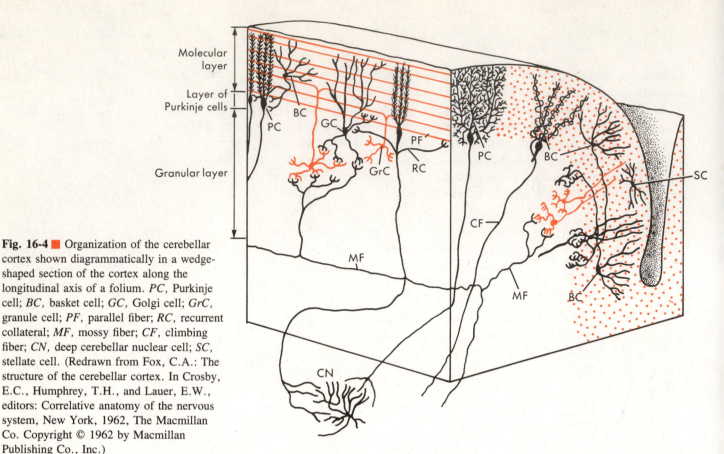

Fig. 16-4 ■ Organization of the cerebellar cortex shown diagrammatically in a wedge-shaped section of the cortex along the longitudinal axis of a folium. *PC,* Purkinje cell; *BC,* basket cell; *GC,* Golgi cell; *GrC,* granule cell; *PF,* parallel fiber; *RC,* recurrent collateral; *MF,* mossy fiber; *CF,* climbing fiber; *CN,* deep cerebellar nuclear cell; *SC,* stellate cell. (Redrawn from Fox, C.A.: The structure of the cerebellar cortex. In Crosby, E.C., Humphrey, T.H., and Lauer, E.W., editors: Correlative anatomy of the nervous system, New York, 1962, The Macmillan Co. Copyright © 1962 by Macmillan Publishing Co., Inc.)

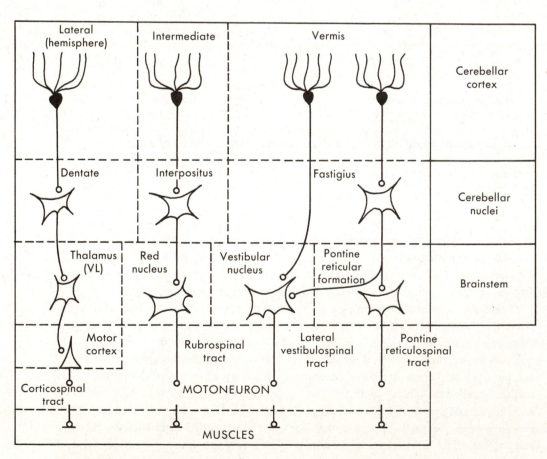

Fig. 16-5 ■ Pattern of outflow from the cerebellar cortex. Inhibitory neurons are shown in black. (Redrawn from Bell, C.C., and Dow, R.S.: Cerebellar circuitry. In Schmitt, F.O., et al., editors: Neurosciences Research Symposium Summaries, vol. 2, Cambridge, Mass., 1967, The MIT Press. Copyright © 1967 by The MIT Press.)

correspond to the *interpositus nucleus* in commonly studied carnivores.) Thus the cerebellum can be viewed as being organized into three *longitudinal corticonuclear zones:* medial or vermal, intermediate or paravermal, and lateral or hemispheric.

■ *Distribution of cerebellar efferent fibers*

A significant advantage in viewing the cerebellum from the perspective of its longitudinal or mediolateral organization is that the outputs of these zones distribute in a predictable fashion with respect to the medial and lateral descending motor pathways (Fig. 16-5). The medial or vermal zone, via the fastigial nucleus, projects prominently onto the pontine reticular formation to access the cells of origin of the pontine reticulospinal tract. This zone also projects onto the lateral vestibular nucleus, which contains the cells of origin of the lateral vestibulospinal tract. This projection is of particular interest because, in addition to fibers from the fastigial nucleus, it includes a complement of Purkinje cell axons that do not relay through the fastigial nucleus (Fig. 16-4). Thus the medial cerebellar zone primarily influences the cells of origin of the principal pathways of the medial descending system: the pontine reticulospinal tract and the lateral vestibulospinal tract.

In contrast to the outflow from the medial zone, the lateral or hemispheric zone projects, via the dentate nucleus, onto the ventral lateral nucleus of the thalamus, which then projects onto the motor cortex (Fig. 16-5). By this route the lateral zone influences the most prominent pathway of the lateral descending system, the lateral corticospinal tract. There is a much less extensive projection from the dentate nucleus to the rostral or parvocellular portion of the red nucleus. Because most of the cells of origin of the rubrospinal tract are located in the caudal two thirds of the nucleus, it is unclear to what extent this projection from the dentate nucleus provides the lateral cerebellar zone with access to the rubrospinal tract.

The intermediate or paravermal zone, via the intermediate deep nuclei, projects primarily onto the red nucleus and to a much lesser extent the ventral lateral thalamus. Thus its principal influence is on the rubrospinal tract, associating it with the lateral descending system.

■ *Distribution of cerebellar afferent fibers*

The pattern of cerebellar afferentation is rather complex, and an appreciation of it requires some consideration of the cerebellum's lobular organization. Vestibular information influences a restricted area of the cerebellum, namely the flocculonodular lobe, with a very small extension into the adjacent area (uvula) of the posterior lobe (Fig. 16-6). Therefore this region of the cerebellum is referred to as the *vestibulocerebellum*. It is also the oldest area of the cerebellum phylogenetically, and thus it is also referred to as the *archicerebellum*.

A later phylogenetic development consists of the anterior lobe and the posterior segment of the vermis. Consequently these areas are referred to as the *paleocerebellum*. It includes segments of both the vermal and paravermal longitudinal zones. Sensory information from the spinal cord distributes over the paleocerebellum and is relayed via a number of direct and indirect pathways. In view of its spinal input, this cerebellar region is also referred to as the *spinocerebellum*.

The body surface is topographically mapped on the paleocerebellum in multiple fashion (Fig. 16-7). One sensory map is found in the anterior lobe. The axial body surface is represented medially, and the limb representation extends laterally. This map is inverted, with the head representation situated posteriorly. A second representation is found in the posterior lobe. Again, the axial surface is represented closest to the midline, and the limb representation extends laterally. This posterior map is longitudinally oriented in a direction opposite of the anterior map so that the head representations of each map are found more centrally on the cerebellar surface. Visual and auditory information is relayed to the cerebellum. Such information influences areas within the head

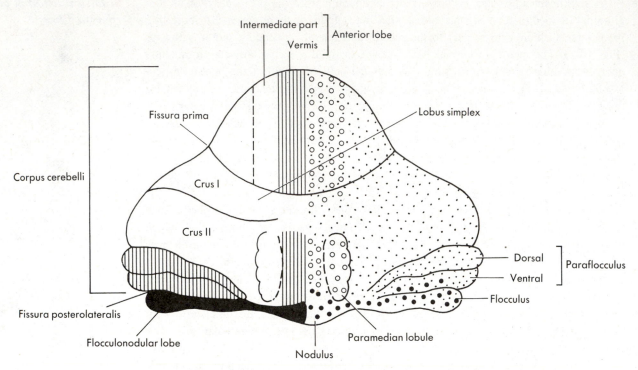

Fig. 16-6 ■ Cerebellar afferentation. On the left are the archicerebellum *(black)*, paleocerebellum *(hatched)*, and neocerebellum *(clear)*. On the right are the distributions of vestibulocerebellar fibers *(closed circles)*, spinocerebellar fibers *(open circles)*, and pontocerebellar fibers *(dots)*. (Redrawn from Brodal, A.: Neurological anatomy, ed. 3. Copyright © 1981 by Oxford University Press, Inc. Reprinted by permission.)

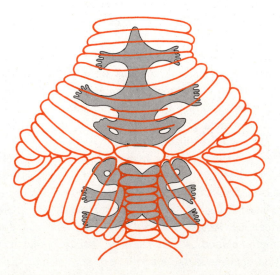

Fig. 16-7 ■ Somatotopic representations of the body on the cerebellar cortex shown in figurine form. (Redrawn from Snider, R.: The cerebellum, Scientific American **199**(6):4, 1958.)

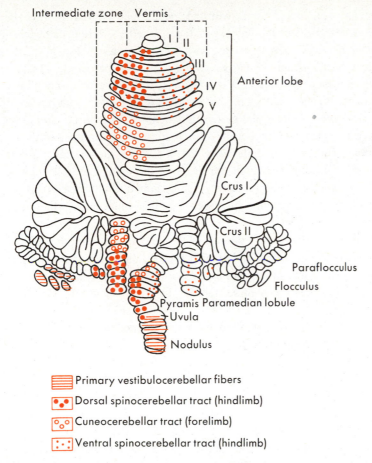

Intermediate zone   Vermis

Anterior lobe

Crus I

Crus II

Paraflocculus

Flocculus

Pyramis Paramedian lobule

Uvula

Nodulus

≡ Primary vestibulocerebellar fibers

Dorsal spinocerebellar tract (hindlimb)

Cuneocerebellar tract (forelimb)

Ventral spinocerebellar tract (hindlimb)

Fig. 16-8 ■ The cerebellar surface showing the distribution of fibers of the dorsal and ventral spinocerebellar tracts and fibers from the external cuneate nucleus. Note the somatotopic pattern. The terminal distribution of primary vestibular fibers is also shown. (Redrawn from Brodal, A.: Neurological anatomy, ed. 3. Copyright © 1981 by Oxford University Press, Inc. Reprinted by permission.)

representations in a zone between the anterior and posterior components of the spinocerebellum (Fig. 16-7). Furthermore, these sensory maps are coincident with motor maps that have been derived by electrical stimulation of the cerebellar surface.

A comprehensive discussion of the spinocerebellar pathways would be inappropriately lengthy. The most prominent direct pathways are the dorsal and ventral spinocerebellar pathways that relay information from thoracic and upper lumbar levels (Fig. 16-8). The corresponding pathways that provide information from cervical levels are the cuneocerebellar and rostral spinocerebellar pathways. The dorsal spinocerebellar tract arises from cells in Clarke's column, and it projects ipsilaterally to the vermal and paravermal zones. The cuneocerebellar pathway arises from the external cuneate nucleus, and it also terminates ipsilaterally in the vermal and paravermal zones. Both of these pathways provide detailed cutaneous and proprioceptive information from very circumscribed regions, and they distribute in a highly localized fashion on the cerebellum. The ventral and rostral spinocerebellar pathways arise from cells that are distributed more diffusely in the spinal gray matter and that project bilaterally onto the spinocerebellum. They relay cutaneous and proprioceptive information from wider regions than the dorsal and cuneocerebellar pathways, and they terminate in a less discrete pattern on the cerebellum. The indirect pathways are complex, but the most prominent pathways include relays in the inferior olive (spino-olivocerebellar pathways) and in the lateral reticular nucleus (spinoreticulocerebellar pathways).

Such spinal input is consistent with the longitudinal pattern of cerebellar organization that was described in the context of the cerebellar efferent pathways (Fig. 16-8). Vestibular inputs, which are associated with the medial descending system, distribute to the flocculonodular lobe and uvula, which constitute that part of the medial longitudinal zone which influences the lateral vestibulospinal tract. The representation of the head, as well as of visual and auditory inputs, is in the medial zone, and head and eye movements are associated with medial system functions. As would be anticipated, the axial body surface is represented within the vermal zone, which influences the pontine reticulospinal tract. Thus the cerebellar zone that influences the medial musculature appropriately receives information about the status of the medial muscles, joints, and body surface. The more distal limb representation extends into the paravermal zone, the outflow of which most prominently influences the rubrospinal tract, which in turn controls the limb muscles.

The phylogenetically newest portion of the cerebellum is the lateral or hemispheric zone; it is referred to as *neocerebellum*. There is no direct sensory representation in this zone. Its principal source of afferent fibers is from the cortex, and this corticocerebellar influence is relayed via the pontine nuclei (Fig. 16-6). Thus the hemispheric zone is also referred to as *pontocerebellum*. This corticopontocerebellar projection arises from widespread areas of the neocortex, including the association cortex. The projection includes collaterals of corticospinal and corticobulbar pathways, which provide information to the cerebellum regarding commands to motoneurons.

■ The position of the cerebellum in the chain of motor command

The cerebellum occupies an interesting position in respect to structures involved in controlling movement. It receives a wealth of sensory information, particularly of a proprioceptive nature. Such information is important for the regulation of motor activity. In addition, the cerebellum receives information regarding the commands being delivered to motoneurons, and this information is provided from many levels of the chain of motor command. For example, collaterals of corticospinal tract fibers influence the cerebellum through the medial and lateral pontine nuclei and inform the cerebellum of the output of the motor cortex. Also, certain of the spinocerebellar inputs originate from spinal interneurons that influence motoneurons, either directly or indirectly, and these interneurons are also recipients of messages from various descending motor pathways. Thus the cerebellum receives inputs about motor commands from cells near the final common path.

The cerebellum is thus one of the best-informed motor structures of the nervous system. It is consequently in a position to process this mass of information and relay its processed output to the various descending systems. Another way of viewing this is to consider the cerebellum as a principal component of a number of major loops involving the descending pathways. For example, the neocerebellum receives extensive cortical input via the corticopontocerebellar pathway. In turn, it relays information back to the cortex via the ventral thalamus, where it also interacts with the outflow of the basal ganglia. The flocculonodular lobe is the primary cerebellar target of vestibular information, and its outflow then plays back on the vestibulospinal pathways.

■ *Functional neuroanatomy*
■ *Functional correlates of the mediolateral longitudinal organization*

On the basis of the efferent distribution of the longitudinal zones it should be possible to predict the distribution of motor deficits consequent to cerebellar lesions. Indeed, it follows the mediolateral organization. Lesions involving the medial zone of the cerebellum produce primarily ipsilateral truncal deficits. If the flocculonodular region is included, then deficits in equilibration will result as well. These deficits appear as disturbances of tone and synergy, where synergy refers to the temporal properties of movements. Thus *asynergia* would involve disturbances in the timing of the onset and termination of contractions, the relationship between the contractions of agonist and antagonistic muscles, and the temporal sequences of muscular contractions.

Lesions confined to the intermediate zone affect principally the ipsilateral proximal limbs, impairing skilled movement of the limb and its participation in postural adjustments as well. Lesions confined to the lateral zone compromise spatially organized movements of the limbs, but they have no effect on posture, progression, or equilibrium. Although the dentate projection to the thalamus is crossed, the corticospinal tract then decussates to return the cerebellar influence to the ipsilateral side. Hence, the effects of lesions are ipsilateral.

Thus the distribution of deficits with restricted cerebellar lesions is entirely understandable, given the longitudinal organization and its efferent pathway distribution with respect to the medial and lateral descending systems. More of the specific nature of the deficits will be considered later.

■ *Cerebellar disorders*

The longitudinal organization of the cerebellum also provides a predictive model for understanding the distribution of cerebellar deficits in disease. Although human cerebellar disease frequently does not respect the mediolateral boundaries of this organization, there are some disorders that provide excellent illustrative confirmation of the basic principles of longitudinal organization.

As a preface it is important to appreciate that cerebellar lesions result in motor disorders that are unique to damage of the cerebellum or to certain of its direct connections. Although the classification is somewhat oversimplified, these disorders can be grouped into three classes: disturbances of *synergy, equilibrium,* or *tone.* Synergy is merely another way of describing coordination, and it refers to the regulation of the rate, range, force, and direction of movements. Cerebellar lesions that disturb synergy and thus produce asynergia are frequently referred to clinically as producing *ataxia.* The asynergia can manifest itself in a number of ways. For example, a common form of asynergia is *dysmetria,* in which there are errors in the direction and force of a movement that compromise an individual's ability to bring a limb smoothly to a desired position. The limb may either overshoot the desired position *(hypermetria)* or undershoot it *(hypometria).* When the lower limbs are involved, it produces an unsteady gait that is frequently broad based. If the desired movement is a more complex one, the asynergia is often expressed as a *decomposition of movement,* where the required sequence of muscle contractions is executed in discrete steps rather than as a smooth movement. The *intention tremor* associated with certain cerebellar lesions is a tremor that is perpendicular to the direction of movement and increases in magnitude toward the end of the movement. It can also be viewed as a disturbance of synergy that involves the temporal relationship between the contraction of agonistic and antagonistic muscles. The ways in which asynergia is expressed are probably as numerous as the clinical tests that can be devised. Depending on the locus of the cerebellar damage, asynergia can involve any muscle group, including the extrinsic eye muscles (producing *nystagmus)* and the muscles of speech (producing *dysarthria).* However, the underlying theme of the disturbance appears to be a loss of the precise temporal regulation of sequences of contractions.

Disorders of equilibration, that is, impairment of the ability to maintain an upright posture, can occur if the lesion involves the vestibulocerebellum. With respect to muscle tone, cerebellar lesions in humans most often produce a hypotonia, which can result in what is known as a *pendular limb.* After elicitation of a patellar tendon reflex, for example, the limb will continue to swing back and forth in a pendular fashion.

It is also instructive to consider specific lesions in the context of longitudinal cerebellar organization. There are two prominent disorders that affect the medial longitudinal zone. One is a childhood tumor, a medulloblastoma, that most commonly involves the flocculonodular lobe. The principal deficit is impaired equilibration with no ataxia or alterations in muscle tone. Another medial disorder is the cerebellar degeneration in alcoholism. This degeneration is largely restricted to the paleocerebellum. The motor deficits are principally postural, and they include a truncal ataxia and broad-based gait. Both of these diseases involve mainly the vermal cerebellum, and the motor abnormalities are thus confined to the more proximal muscles. In contrast to this, various degenerative disorders and tumors of the cerebellum involve principally the neocerebellum or lateral zone. In these disturbances the deficits involve the ipsilateral limbs, and hypotonia, limb ataxia, and tremor are common.

Thus the distribution of motor deficits in cerebellar disease is predictable from the anatomical organization of the cerebellar outflow. The deficits are ipsilateral to the lesion, and the mediolateral position and extent of the lesion will determine the proximodistal distribution of involved muscles.

A final point of interest regarding cerebellar damage is that the patient can frequently compensate well for such disturbed function, particularly if the lesion occurs in childhood. The mechanism of this compensation is not understood.

■ *Cerebellar function*

From the previous discussion it is clear that understanding of cerebellar function is reasonably well advanced with respect to the distribution of deficits. This understanding follows in a straightforward manner from functional neuroanatomical considerations. What is not understood is the nature of the deficits in cerebellar disorders. This, in turn, reflects an inadequate understanding of how the cerebellum participates in normal motor behavior. Considerable detailed data have emerged over the past two decades, and although this new information has not yet provided the answer to the problem of cerebellar participation in movement, it does provide some provocative possibilities.

■ *Organization of the cerebellar cortex*

At this point it is appropriate to review briefly the organization of the cerebellar cortex. As illustrated in Fig. 16-9, it consists histologically of three layers. The deepest, or *granular,* layer contains mainly densely packed *granule cells,* which are the recipient neurons of most of the input to the cerebellar cortex. Also in the granular layer are the *Golgi neurons,* one of three types of interneurons of the cerebellar cortex. Above the granular layer is the *Purkinje layer*. This consists of a single layer of *Purkinje cells,* the output element of the cerebellar cortex. The most superficial layer is the *molecular layer*. The two other interneuronal classes, the *stellate* and *basket cells,* are situated in this layer. In addition, the molecular layer contains the dendrites of the Purkinje and Golgi cells. The final important constituents of this layer are the axons of the granule cells. These axons ascend from the granular layer, and on reaching the molecular layer, they bifurcate to run longitudinally for a few millimeters in the orientation of the folium. These axons constitute the *parallel fiber system*.

Inputs to the cerebellar cortex are transmitted over two fiber systems, both excitatory (Fig. 16-9). These are the *climbing fibers,* which arise from cells of the contralateral inferior olive, and the *mossy fibers;* the cells of origin of the mossy fibers are found in the various precerebellar nuclei of the spinal cord and brainstem. Both systems also innervate the deep cerebellar nuclei *en passage* to the cortex. The climbing fiber system shows very little convergence or divergence. Each Purkinje cell receives input from only a single climbing fiber, and a climbing fiber contacts no more than a dozen Purkinje cells. Furthermore, the climbing fiber makes multiple contacts with the soma and proximal dendrites of its target Purkinje cell. Such multiple contacts lead to a powerful excitatory action where a single discharge along a climbing fiber initiates a burst of impulses from the Purkinje cell (Fig. 16-10). The mossy fiber system, in contrast, shows massive convergence and divergence. This system derives from all sources of input to the cerebellum, except the inferior olive. The mossy fibers terminate on the granule

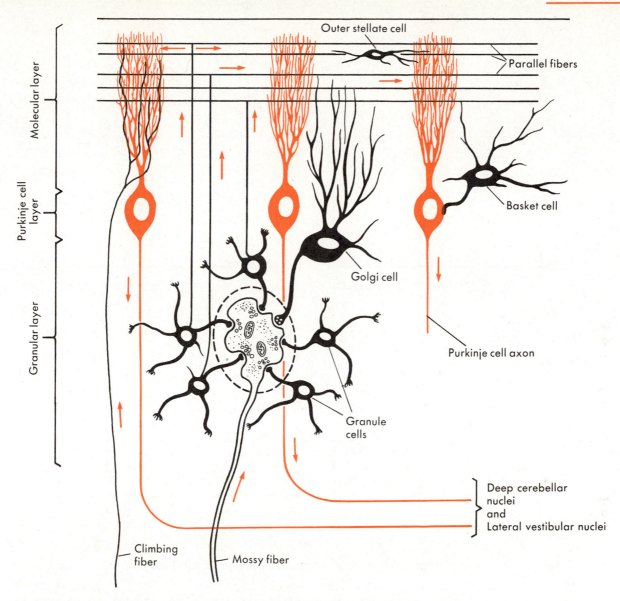

Fig. 16-9 ■ Elements of the cerebellar cortex in the longitudinal axis of a folium. (Redrawn from Carpenter, M.B.: Human neuroanatomy, ed. 7. © 1976, The Williams & Wilkins Co., Baltimore; based on Gray and Eccles et al.)

cells in the deepest layer of the cerebellar cortex. The contacts are axodendritic and form distinctive glomeruli (Fig. 16-9). Thus the mossy fibers activate the parallel fiber system, which consists of the axons of the granule cells (Fig. 16-9).

When the parallel fibers are activated, they excite a band of Purkinje cells. The Purkinje cell dendrites are oriented perpendicularly to the longitudinal axis of the folium, and the parallel fibers course parallel to this axis (Fig. 16-4). Hence each parallel fiber runs perpendicularly through a row of Purkinje cell dendrites, contacting perhaps 50 Purkinje cells. Each Purkinje cell is contacted by perhaps 200,000 parallel fibers.

The parallel fibers also activate the three classes of cerebellar cortical interneurons, all of which have their dendrites in the molecular layer (Fig. 16-9). These interneurons are all inhibitory. The Golgi cell feeds back to inhibit the granule cells, thus shutting off the input via the mossy fiber system. The basket and stellate cells both feed forward to inhibit the Purkinje cells. The stellate cell inhibition occurs on more distal dendrites,

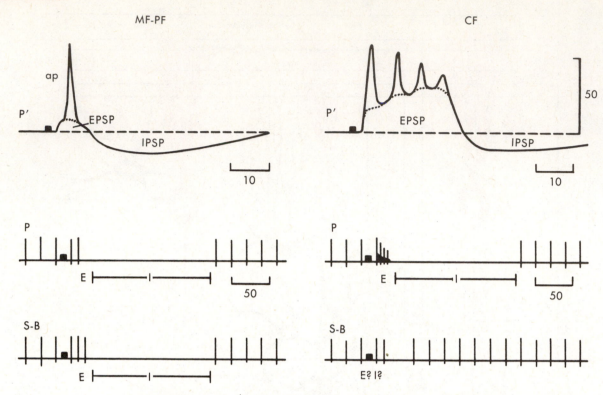

Fig. 16-10 ■ Main types of synaptic actions in the cerebellum. *Right: MF-PF,* The synaptic actions elicited by a volley in the mossy-parallel fiber pathway; *P',* intracellular recording of a simple response in a Purkinje cell; *P,* extracellular recording with a slower time base; excitatory and inhibitory periods, *E* and *I,* respectively; *S-B,* extracellular recording from a stellate or basket cell. *Left: CF,* Synaptic actions elicited by a volley in the climbing fiber pathway; *P'* and *P,* a complex response in a Purkinje cell; *S-B,* ? response in a stellate or basket cell. (Redrawn from Shepherd, G.M.: The synaptic organization of the brain, ed. 2, Oxford, 1979, Oxford University Press.)

and the basket cell inhibition occurs on more proximal dendrites and the soma (Fig. 16-11).

The geometry of these intracortical connections is rather precise. Activation of a narrow band of parallel fibers will activate a narrow beam of Purkinje cells. After a short delay the inhibitory feedback from the Golgi cells will turn off the parallel fiber input over that narrow band. The feed-forward inhibition via the basket and stellate cells is directed toward the bands of Purkinje cells that are arrayed parallel to and on either side of the activated band. The net effect of this is that (1) mossy fiber activation of a small cluster of granule cells excites a well-defined beam of Purkinje cells via the parallel fibers; (2) these same parallel fibers excite basket and stellate cells that inhibit the surrounding beams of Purkinje cells; and (3) after a short delay the Golgi cell feedback inhibition shuts off the activating input. Consequently, mossy fiber activation of a cluster of granule cells produces a well-focused beam of Purkinje cell excitation for a brief period, and concomitantly the off-beam Purkinje cells are inhibited. This is a sophisticated form of lateral inhibition, similar to that seen in sensory systems, and it contributes to a high degree of spatial localization of the cerebellar output. With natural stimuli many such beams become activated in a highly organized spatial pattern. The pattern changes constantly as inputs from the periphery continuously impinge on the cerebellum.

By capitalizing on the Purkinje cells that project directly to the lateral vestibular nucleus, it has been possible to record intracellularly from these large vestibulospinal

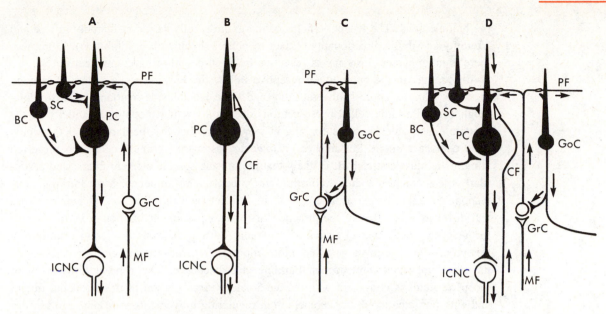

Fig. 16-11 ■ Component neuronal circuits of the cerebellar cortex. Inhibitory neurons are shown in black; excitatory neurons are clear. **A** shows a mossy fiber input that excites the granule cell and parallel fiber system, which in turn excites Purkinje, stellate, and basket cells. Stellate and basket cells inhibit Purkinje cells, and Purkinje cells inhibit deep cerebellar nuclear cells. **B** shows a climbing fiber input exciting a Purkinje cell, which in turn inhibits deep cerebellar nuclear cells. **C** shows a mossy fiber input that excites the granule cell and parallel fiber system, which in turn excites Golgi cells. The Golgi cell axon returns to inhibit the granule cell. **D** shows a combination of the circuits shown in **A** to **C**. *PC,* Purkinje cell; *SC,* stellate cell; *BC,* basket cell; *PF,* parallel fiber; *GrC,* granule cell; *MF,* mossy fiber; *ICNC,* deep cerebellar nuclear cell; *CF,* climbing fiber; *GoC,* Golgi cell. (Redrawn from Eccles, J.C.: Functional organization of the cerebellum in relation to its role in motor control. In Nobel Symposium I: Muscular afferents and motor control, New York, 1966, John Wiley & Sons, Inc. © The Nobel Foundation 1966.)

cells and to demonstrate that activation of the Purkinje cells produces inhibition. This finding has since been generalized, establishing that the output of the cerebellar cortex is inhibitory. Indeed, the only excitatory cell type in the cerebellar cortex is the granule cell. Apparently the deep cerebellar nuclei maintain a high level of background activity, most likely via collateral activation by climbing and mossy fibers. The output of the cerebellar cortex then inhibits this ongoing excitatory activity in a precise geometrical fashion. Thus the maintained excitatory influences of the deep cerebellar nuclei on the various targets of the cerebellum are modulated up or down by the output of the cerebellar cortex. For example, an increase in Purkinje cell activity of the paravermal cortex would then inhibit cells of the nucleus interpositus and thus *disfacilitate* the rubrospinal neurons. This regulatory action occurs with a temporal and spatial precision appropriate for a structure that governs the rate, range, direction, and force of movements.

■ *Cellular studies of cerebellar function*

In recent years the technology has become available to undertake analysis of single cerebellar neurons in response to natural stimulation and during movement. With respect to the receptive fields of cerebellar cells, there is apparently little transformation from the mossy fiber to the granule cell, although considerable transformation of sensory information can occur before mossy fiber input. Most granule cells respond to a single modality, most frequently changes in muscle length or tension. Responses to tactile stimulation are also common, whereas joint movement is a less effective stimulus for the granule cells. The receptive fields of the granule cells vary considerably in size and laterality and in the directionality of their responses (excitation or inhibition). Moreover, there is no apparent clustering of receptive fields in neighboring granule cells.

In contrast to the granule cell receptive fields, the Purkinje cells generally respond to inputs from an entire ipsilateral limb, and there is extensive overlap of fields among neighboring Purkinje cells. These neurons do not show the modality specificity of the granule cells, which is not surprising in view of the fact that perhaps 200,000 granule cells contact a single Purkinje cell. However, as in the granule cells, the preferred stimulus is muscle stretch. It is likely that the muscle spindle afferent fibers, and particularly the secondary endings, constitute the predominant influence. Most Purkinje cells respond to displacement around only a single joint, and the direction of their response can differ as a function of the direction of joint displacement. Although knowledge of the receptive field characteristics of cerebellar cortical neurons is clearly primitive at this point, the data do suggest that information arising from the muscle spindles may be the most important influence on Purkinje cell discharge. The general distribution of receptive fields is consistent with the mediolateral organization of the cerebellar output and with participation of the cerebellum in regulating movements in response to sensory feedback.

Much more data are required on the nature of the sensory information that reaches the cerebellum and the transformation of this information within the cerebellar cortex. This will undoubtedly be a prerequisite to understanding cerebellar function. The solution to this difficult problem may well require the ability to analyze the responses of many neurons simultaneously, However, during the pursuit of this objective another kind of information may be derived by studying the responses of cerebellar neurons during movement. Although such an approach has not yet yielded a definitive answer, it has generated some provocative suggestions. In particular, it has been found that neurons in the interpositus nucleus tend to show altered discharge after a movement, implying a response to proprioceptive feedback. However, neurons in the dentate nucleus frequently discharge before the movement, raising the possibility of cerebellar involvement in the initiation of movement. It remains unclear whether these dentate neurons respond before or after the discharge of the corticospinal neurons that transmit the final command.

Data such as these have led to the hypothesis that the association areas of the neocortex act in concert with the neocerebellum to organize and initiate movements, with the final command being transmitted by the motor cortex. The intermediate zone in this scheme may act primarily in a regulatory fashion through the feedback of sensory information. Another current issue that is emerging is the extent to which the cerebellar outflow influences α- or γ-motoneurons. There is evidence which suggests that cerebellar influences may be most prominent on the γ-motoneurons.

Although final answers are not imminent, contemporary neurophysiological techniques may provide a means of understanding normal cerebellar function. The required studies are indeed challenging, and they will not lead immediately to an understanding of the unique deficits that characterize cerebellar lesions. However, the most productive direction of investigation, as has been seen in the past, may well be to study normal function rather than pathological signs because an understanding of pathology often follows readily when normal function is understood.

■ *Bibliography*

Journal articles

Allen, G.L., and Tsukahara, N.: Cerebrocerebellar communication systems, Physiol. Rev. **54**:957, 1974.

Bloedel, J.R.: Cerebellar afferent systems: a review, Prog. Neurobiol. **2**:1, 1973.

Chambers, W.W., and Sprague, J.M.: Functional localization in the cerebellum. I. Organization in longitudinal cortico-nuclear zones and their contribution to the control of posture, both extrapyramidal and pyramidal, J. Comp. Neurol. **103**:105, 1955.

Evarts, E.V., and Thach, W.T.: Motor mechanisms of the CNS: cerebrocerebellar interrelations, Annu. Rev. Physiol. **31**:451, 1969.

Gray, E.G.: The granule cells, mossy synapses and Purkinje spinal synapses of the cerebellum: light and electron microscopic observations, J. Anat. **95**:345, 1961.

Holmes, G.: The cerebellum of man, Brain **62**:1, 1939.

Jansen, J., and Brodal, A.: Experimental studies on the intrinsic fibers of the cerebellum. II. The corticonuclear projection, J. Comp. Neurol. **73**:267, 1940.

Snider, R.S., and Stowell, A.: Receiving areas of the tactile, auditory and visual systems in the cerebellum, J. Neurophysiol. **7**:331, 1944.

Thach, W.T.: Correlation of neural discharge with pattern and force of muscular activity, joint position, and direction of intended next movement in motor cortex and cerebellum, J. Neurophysiol. **41**:654, 1978.

Books and monographs

Angevine, J.B., et al.: The human cerebellum: an atlas of gross topography in serial sections, Boston, 1961, Little, Brown & Co.

Brodal, A.: Neurological anatomy, Oxford, 1981, Oxford University Press.

Eccles, J.C., Ito, M., and Szentágothai, J.: The cerebellum as a neuronal machine, New York, 1967, Springer-Verlag New York, Inc.

Ghez, C., and Fahn, S.: The cerebellum. In Kandel, E.R., and Schwartz, J.H., editors: Principles of neural science, New York, 1981, Elsevier North-Holland, Inc.

Jansen, J., and Brodal, A.: Das Kleinhirn. In von Möllendorff, W., editor: Handbuch der mikroskopischen Anatomie des Menschen, vol. 3, Berlin, 1958, Julius Springer.

Larsell, O.: Anatomy of the nervous system, ed. 2, New York, 1951, Appleton-Century-Crofts.

Thach, W.T., Jr.: The cerebellum. In Mountcastle, V.B., editor: Medical physiology, ed. 14, St. Louis, 1980, The C.V. Mosby Co.

chapter 17

The basal ganglia

■ *Anatomical overview*
■ *Definition*

Full consensus has not yet been achieved regarding a precise anatomical definition of the basal ganglia. However, there is agreement that the *caudate nucleus, putamen,* and *globus pallidus* constitute its main mass (see Fig. 16-1). The amygdaloid nucleus is generally excluded, and the claustrum is generally included. Other anatomical terms relating to the basal ganglia germane to the literature are the *corpus striatum* (caudate nucleus, putamen, globus pallidus, and claustrum), *neostriatum* (caudate nucleus and putamen), *paleostriatum* (globus pallidus), and *lentiform nucleus* (putamen and globus pallidus). As used here, basal ganglia refers to the caudate nucleus, putamen, and globus pallidus, the major subcortical cell masses of the telencephalon.

Also associated with the basal ganglia are certain subtelencephalic cell groups, the most prominent of which are the *substantia nigra* of the mesencephalon and the *subthalamic nucleus* of the basal caudal diencephalon (Fig. 17-1). The substantia nigra

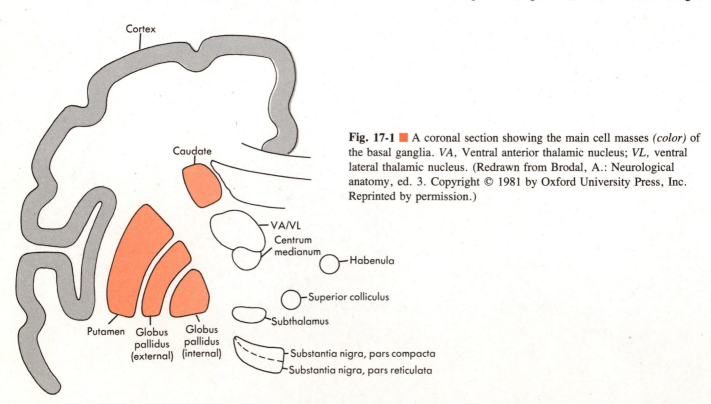

Fig. 17-1 ■ A coronal section showing the main cell masses *(color)* of the basal ganglia. *VA,* Ventral anterior thalamic nucleus; *VL,* ventral lateral thalamic nucleus. (Redrawn from Brodal, A.: Neurological anatomy, ed. 3. Copyright © 1981 by Oxford University Press, Inc. Reprinted by permission.)

consists of a ventral pars reticulata, which appears histologically to be a caudal continuation of the globus pallidus, and a dorsally located pars compacta, which contains highly pigmented (melanin-rich) neurons. The basal ganglia and their associated diencephalic and mesencephalic structures classically have been referred to as the *extrapyramidal system,* a term having little significance in contemporary treatments of the motor system.

■ *Afferentation*

The structures that consitute the basal ganglia, in addition to being extensively interconnected, receive inputs from four major sources: the cortex, thalamus, substantia nigra, and raphe (Fig. 17-2). The neocortex projects widely onto the caudate nucleus and putamen in a topographical fashion. A prominent component of these corticostriatal projections is an input from the precentral gyrus (the motor and premotor cortices). The fibers from the cortical motor areas probably arise from a population of neurons in layer V that is different from the cells of origin of the corticospinal tract.

Perhaps the next most prominent source of striatal afferentation is the substantia nigra (Fig. 17-2). This *nigrostriatal pathway* derives primarily from neurons in the pars compacta, and it terminates in the caudate nucleus and putamen. It is of considerable significance that the neurotransmitter of this nigrostriatal projection is dopamine (pp. 265-266).

A third source of afferent fibers to the basal ganglia is the thalamus; this thalamostriatal projection issues principally from the intralaminar nuclei (Fig. 17-2). Like the corticostriatal projection, the fibers from the intralaminar nuclei distribute topographically on the caudate nucleus and putamen. More recently a thalamic projection has also been

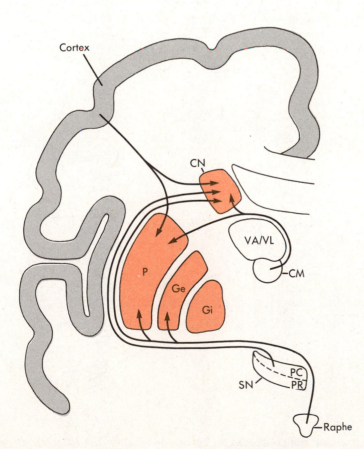

Fig. 17-2 ■ A coronal section showing the major afferent fibers to the basal ganglia. See Fig. 17-1 for identification of structures. (Redrawn from Brodal, A.: Neurological anatomy, ed. 3. Copyright © 1981 by Oxford University Press, Inc. Reprinted by permission.)

identified to the globus pallidus, but it originates in the subthalamic nucleus rather than in the intralaminar thalamus.

Finally, the dorsal raphe nucleus projects extensively onto the basal ganglia and its associated cell groups (Fig. 17-2). Indeed most, if not all, of the serotonin in the basal ganglia is derived from this pathway.

■ *Efferentation*

With respect to the outputs of the basal ganglia, the major outflow is from the globus pallidus, and it terminates in the ventral thalamus (Fig. 17-3). The internal pallidal segment then gives rise to two distinct fiber systems: the ansa lenticularis and the fasciculus lenticularis. These merge in the field of Forel to form the fasciculus thalamicus, which terminates in the ventral thalamus, primarily in the ventral anterior (VA) and ventral lateral (VL) nuclei. These nuclei in turn project onto the motor cortex (Fig. 17-3). Because the motor cortex is a major source of afferent fibers to the caudate nucleus and putamen, this establishes a loop from the motor cortex to the basal ganglia and back to the motor cortex via the ventral thalamus (Figs. 17-2 and 17-3). The neurotransmitters of the relays in this loop are not known, although there is evidence that interneurons in the caudate nucleus are cholinergic.

Another prominent output of the basal ganglia involves a projection from the caudate nucleus and putamen to the substantia nigra. The projection is topographical, and its terminal field includes both the pars compacta and the pars reticulata of the substantia nigra. Moreover, the neurotransmitter of this pathway is thought to be γ-aminobutyric acid (GABA). Together, the striatonigral and nigrostriatal pathways establish a loop involving the caudate nucleus, putamen, and the substantia nigra (Figs. 17-2 and 17-3). Dopamine is apparently involved in the nigrostriatal projection and GABA in the reciprocal striatonigral projection.

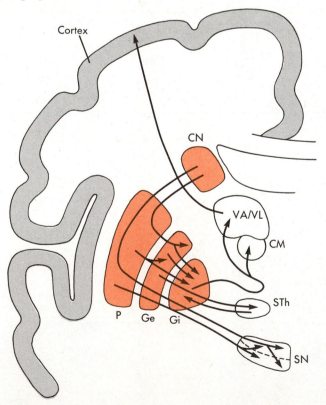

Fig. 17-3 ■ A coronal section showing the intrinsic connections and major efferent projections of the basal ganglia. See Fig. 17-1 for identification of structures. (Redrawn from Brodal, A.: Neurological anatomy, ed. 3. Copyright © 1981 by Oxford University Press, Inc. Reprinted by permission.)

The preceding are the principal outputs of the basal ganglia. However, additional information regarding the connectional anatomy of these structures is emerging with the application of newer neuroanatomical methods. For example, in recent years reciprocal connections between the subthalamic nucleus and the globus pallidus and between the subthalamic nucleus and the substantia nigra have been identified. Thus increased knowledge of the interactions of the basal ganglia with other parts of the brain, as well as their intrinsic connections, can be anticipated.

■ *Lesions of the basal ganglia*

The role of the basal ganglia in motor control is elusive. This, in part, reflects their difficult accessibility and their relative remoteness from both the sensory and motor peripheries. Indeed the basal ganglia constitute the only major motor complex that does not have direct access to the cells of origin of a descending motor pathway. Because of this, it might have been expedient to defer intensive investigation of this system until more was known of its inputs and outputs; however, because diseases of the basal ganglia are common and devastating, this was not advisable. Consequently a sizable literature has developed regarding the effects of both naturally occurring and experimentally induced lesions. The literature, unfortunately, has not satisfactorily clarified the functions of the basal ganglia.

The literature on experimental lesions has been particularly uninformative, which probably reflects the anatomical complexity of the basal ganglia and their interconnections. Such lesions have generally been unsuccessful in reproducing the motor abnormalities of diseases of the basal ganglia. As we begin to understand these disease processes, it is becoming apparent that they frequently involve disturbances of a particular transmitter system. Thus it is possible that traumatic experimental lesions provide an inappropriate approach because they generally are not confined to a single transmitter system.

Clinical disorders involving the basal ganglia generally result in three types of deficits: (1) abnormal movements, or *dyskinesias,* (2) *changes in muscle tone,* and (c) slowness in initiating and executing movements, or *bradykinesia.* The abnormal movements are largely characteristic of disease of the basal ganglia. They include regular involuntary *tremor* at rest, slow writhing movements most prominently involving the distal extremities *(athetosis),* flicklike movements involving the extremity and facial muscles *(chorea),* and violent flailing movements most commonly involving the proximal limb musculature *(ballismus).* Changes in muscle tone can result in rigidity, spasticity, or hypotonus, depending on the particular disease process and the site of the lesion.

■ *Parkinson's disease*

Perhaps the most common disorder involving the basal ganglia is Parkinson's disease, which is characterized by a resting tremor, cogwheel or plastic rigidity, and bradykinesia. The primary lesion site in this disease is the substantia nigra, particularly the dopaminergic neurons of the pars compacta. Most of the dopamine in the central nervous system is found in the basal ganglia, and it has been shown that brain dopamine levels are severely reduced in persons with Parkinson's disease. This would seem to implicate the nigrostriatal pathway. It is presently thought that in Parkinson's disease the dopaminergic neurons of the substantia nigra are destroyed, resulting in degeneration of the nigrostriatal fibers. There is experimental evidence that this dopaminergic system inhibits its target neurons in the caudate nucleus and putamen. Therefore it has been argued that the symptoms of Parkinson's disease reflect a disinhibition or *release* of these striatal neurons. This results in abnormal discharge of the cells in the caudate nucleus and putamen, which in turn produces abnormal discharge of their respective target cells in the globus pallidus. Ultimately this affects the motor cortex via the ventral thalamic relay of the striatocortical system (Fig. 17-4).

This line of reasoning led to the treatment of parkinsonian patients with L-dopa, a

Fig. 17-4 ■ Schematic diagram of the neuronal circuitry of the basal ganglia. Solid neurons are inhibitory. In Parkinson's disease the dopaminergic *(DA)* fibers from the substantia nigra degenerate, releasing the striatothalamocortical pathway from inhibition. In Huntington's disease the GABAergic *(GABA)* fibers from the striatum to the substantia nigra degenerate, disinhibiting the nigral DA neurons, thereby increasing the nigral inhibition of the striatum. The cholinergic interneurons *(ACh)* of the striatum also degenerate, further enhancing the inhibitory effect of the nigra on the striatum.

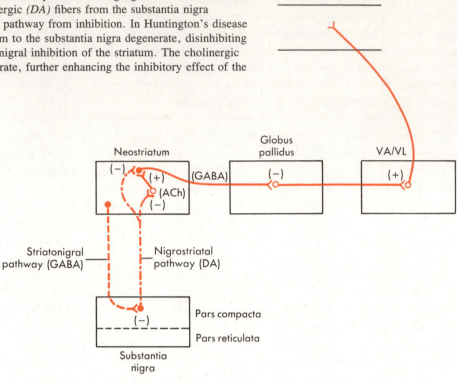

precursor of dopamine that crosses the blood-brain barrier. The rationale of this approach is to restore brain dopamine levels and reverse the disinhibition consequent to the loss of the inhibitory dopaminergic input from the nigrostriatal pathway. This treatment has, in fact, relieved some of the distressing symptoms. Unfortunately, however, it is not presently viewed with the same enthusiasm that prevailed when it was introduced. However, it must be appreciated that any limitations of L-dopa treatment do not necessarily compromise the preceding explanation of the disease process in Parkinson's disease. Also, in this context it is of relevance that anticholineric agents were used in the treatment of Parkinson's disease before the advent of L-dopa therapy. This treatment presumably blocked the excitatory action of the cholinergic interneurons of the striatum, which are disinhibited after degeneration of the dopaminergic input from the substantia nigra (Fig. 17-4). Reversing the release of these interneurons diminishes the enhanced outflow from the striatum. Consistent with this approach is the finding that acetylcholinesterase inhibitors, which prolong the action of acetylcholine, exacerbate the symptoms of Parkinson's disease.

With respect to the functional anatomy of the disease, the tremor and rigidity, *but not the bradykinesia,* have been alleviated by surgical lesions that interrupt the outflow of the basal ganglia either at the globus pallidus or ventral thalamus. Surgical lesions of the motor cortex also alleviate these symptoms. Also, if a person with Parkinson's disease sustains a stroke involving the internal capsule, then the tremor disappears on the hemiplegic side. It is significant that these lesions, which are all on the output side of the basal ganglia, do not alleviate the bradykinesia. This is consistent with the hypothesis that bradykinesia is a primary deficit of certain lesions of the basal ganglia, whereas the tremor and rigidity are release phenomena. It has been suggested that the tremor may result from loss of the serotonergic input from the raphe to the basal ganglia. Whether or not this is correct, it emphasizes the importance of determining the roles of the different afferent systems to the basal ganglia in producing the various symptoms of Parkinson's disease.

A final point regarding this disorder relates to rigidity, which is more evident in flexor than in extensor muscles. In some patients the rigidity is cogwheel in nature, where passive movement of the limb results in an alternation between increased and decreased resistance to the movement. Other persons show a plastic rigidity, in which the resistance to passive movement is independent of the velocity of the movement. In either case neither clonus nor a clasp-knife reaction occurs. Although agreement has not been reached with respect to the mechanism of parkinsonian rigidity, it is probable that enhanced fusimotor discharge participates. Consistent with this hypothesis is that the rigidity can be abolished by dorsal root section. Furthermore, the insensitivity to the velocity of passive movement and the absence of clonus suggest that such enhanced discharge may occur preferentially in the static fusimotor system. Thus the rigidity may be viewed as a release phenomenon in which inhibitory influences that restrain the static fusimotor activity are interrupted.

■ *Huntington's disease*

Huntington's disease is an autosomal dominant disorder characterized by dementia, chorea, and sometimes hypotonia. The chorea results from involvement of the basal ganglia. It has been demonstrated recently that this involvement includes degeneration of the cholinergic interneurons of the caudate nucleus and putamen and of the GABAergic neurons that give rise to the striatonigral projection (Fig. 17-4). The GABAergic striatal neurons inhibit the dopaminergic cells of the substantia nigra. Hence, the effect of this lesion is to disinhibit the dopaminergic nigrostriatal pathway (Fig. 17-4). Loss of the cholinergic interneurons removes a source of excitation to the striatal output neurons. Thus the output of the corpus striatum is reduced through a combination of enhanced inhibition and reduced excitation. This is in marked contrast to Parkinson's disease, where the nigrostriatal pathway degenerates, and the pallidal output is thus enhanced. Supporting this argument are the observations that L-dopa enhances the chorea in Huntington's disease, and excessive L-dopa administration in individuals with Parkinson's disease can produce chorea. Although the identification of the involved transmitter systems is a significant advance in the understanding of this disorder, it still leaves unexplained the specific patterns of abnormal movement that occur in Huntington's disease.

■ *Other disorders*

There are various other diseases that affect the basal ganglia, but only two will be briefly mentioned. Athetosis can follow damage to the basal ganglia at birth (cerebral palsy), and the lesion probably involves both the corpus striatum and the globus pallidus. As in Parkinson's disease, some relief from the motor abnormality can be achieved with lesions of the pallidal outflow or of the motor cortex. Ballismus is a striking abnormality consequent to damage, usually vascular, of the subthalamic nucleus. The ballistic movements can be abolished by lesions of the motor cortex or the corticospinal tract.

■ *Role of the basal ganglia in motor control*

■ *Relationship to medial and lateral systems*

Although the connectional anatomy of the basal ganglia is complex, the principal extrinsic influence of these structures is on the motor cortex. The most prominent source of input to the basal ganglia is the neocortex, which gives rise to projections that terminate on the neostriatum. The caudate nucleus and putamen then project onto the globus pallidus, the output of which relays through the ventral thalamus to the precentral gyrus (motor cortex). Thus the basal ganglia appear to influence motor behavior primarily through pathways that arise in the motor cortex, such as the corticospinal tract. On the basis of these connections, it can be inferred that the basal ganglia act as part of the lateral system and are thus preferentially concerned with organized activity of the limb musculature.

There are a variety of observations consistent with this hypothesis. Most importantly, the motor deficits in disorders of the basal ganglia are more prominent in the extremities. For example, the resting tremor in Parkinson's disease is most evident in the fingers. Athetotic movements principally involve the fingers and hands, sometimes the toes and feet, and less often the more proximal muscle groups. Choreiform movements in Huntington's disease mainly involve the limbs, as in ballismus, and the rigidity in parkinsonism is more prominent in flexor muscles than in the antigravity extensors. Moreover, many of the motor abnormalities of disease of the basal ganglia are attenuated or abolished by destruction of the pallidal outflow, its thalamic relay to the cortex, the motor cortex itself, or the primary descending output of the motor cortex—the corticospinal tract.

The flow of information from the basal ganglia to the cortex converges with cerebellar influences at the VA and VL nuclei of the thalamus. The lateral or hemispheric cerebellar zone accesses the motor cortex and thus the corticospinal tract via this thalamic region. Consequently the VA and VL nuclei represent critical relays for the lateral system because both cerebellar and basal ganglial influences are integrated there before being relayed to the cells of origin of the corticospinal tract. Unfortunately there is presently minimal information about the nature of this interaction. However, it may be significant that tremor can result from disease of either the cerebellum or the basal ganglia. Although cerebellar "intention tremor" is considered to be distinct from the "resting tremor" of basal ganglial disease, it has been suggested that they share a common pathological basis. The disorder may be expressed differently because the background tone in cerebellar disease differs from that in disease of the basal ganglia. Surgical lesions of the VL nucleus can abolish either tremor.

■ *Contribution to motor behavior*

As stated earlier in this chapter, the role of the basal ganglia in motor behavior is still unclear; however, some suggestions can be ventured. Although it is not exclusively associated with the distal musculature, this motor complex appears to act preferentially through the lateral system via its influence on the motor cortex. Strongly supportive of this contention is that disease of the basal ganglia is most prominently expressed in abnormal movements of the extremities, and lesions of the motor cortex can abolish such movements. Furthermore, recording of cellular activity in the globus pallidus during movement indicates that the basal ganglia are concerned with movement of the contralateral extremities and they discharge prior to neurons of the motor cortex. This suggests that the basal ganglia play a role early in the initiation of movement and do not merely have a feedback relationship with the motor cortex.

Although it is not a new concept, it is reasonable to hypothesize that the basal ganglia assist in setting the limb and girdle musculatures as the background for spatially and temporally organized fine manipulatory activity. For example, they might appropriately position the shoulder and elbow as the basis for a goal-directed movement of the wrist and fingers. Furthermore, they might be involved in setting the "postural background" for limb movements. In this sense they would constitute an important bridge between the medial and lateral systems. Viewed in this way, many of the motor abnormalities consequent to disease of the basal ganglia might be interpreted as loss of control in setting the appropriate muscle groups to support an organized, voluntary, manipulatory activity.

■ *Bibliography*

Journal articles

Coyle, J.T., and Schwarcz, R.: Lesion of striatal neurons with kainic acid provides a model for Huntington's chorea, Nature **263**:244, 1976.

Hore, J., Meyer-Lohmann, J., and Brooks, V.B.: Basal ganglia cooling disables learned arm movements of monkeys in the absence of visual guidance, Science **195**:584, 1977.

Kennard, M.: Experimental analysis of functions of the basal ganglia in monkeys and chimpanzees, J. Neurophysiol. **7**:127, 1944.

Books and monographs

Brodal, A.: Neurological anatomy, Oxford, 1981, Oxford University Press.

Côté, L.: Basal ganglia, the extrapyramidal motor system, and disease of transmitter metabolism. In Kandel, E.R., and Schwartz, J.H., editors: Principles of neural science, New York, 1981, Elsevier North-Holland, Inc.

DeLong, M.R.: Motor functions of the basal ganglia: single-unit activity during movement. In Schmitt, F.O., and Worden, F,G., editors: The neurosciences third study program, Cambridge, Mass, 1974, The MIT Press.

DeLong, M.R., and Georgopoulos, A.P.: Functional organization of the substantia nigra, globus pallidus, and subthalamic nucleus in the monkey. In Poirier, L.J., Sourkes, T.L., and Bedard, P.J., editors: The extrapyramidal system and its disorders, New York, 1979, Raven Press.

Hornykiewicz, O.: Neurochemical pathology and pharmacology of brain dopamine and acetylcholine: rational basis for the current drug treatment of parkinsonism. In McDowell, F.H., and Markham, C.H., editors: Recent advances in Parkinson's disease, Philadelphia, 1971, F.A. Davis Co.

Martin, J.P.: The basal ganglia and posture, Philadelphia, 1967, J.B. Lippincott Co.

Yahr, M.D.: The basal ganglia, New York, 1976, Raven Press.

chapter 18

Control of movement and posture

In the preceding chapters on the motor system, motor control was analyzed by beginning at the periphery and proceeding centrally. It has been argued that analysis of the central nervous system is most productive when the structure under study can be linked to the sensory or motor peripheries. Indeed the remarkable progress in our understanding of the sensory systems in large measure reflects the ability to study successive transformations in receptive fields as one proceeds systematically from the receptors to the central nervous system. Advances in understanding the motor system have been realized in recent years, most prominently from efforts to study the control of motor behavior by beginning at the muscles and their innervation. An understanding of the functional neuroanatomical organization of the final common path then establishes a basis for the rational classification of the major descending pathways involved in motor control; Chapters 13 to 15 develop these ideas. Chapters 16 and 17 introduce the cerebellum and basal ganglia within this framework. The objectives of this chapter, which concludes the discussion of the motor system, are (1) to introduce additional material on the control of posture and movement and (2) to present some examples of sensorimotor integration. These examples will illustrate the directions that are required to approach the question of how motor outputs are guided by sensory inputs.

■ *Control of posture*

Posture is defined as the *active muscular resistance to displacement of the body by gravity or acceleration*. The maintenance of an upright position is a critical substrate for the performance of phasic goal-directed movements. This is achieved largely through reflex adjustments of the proximal extensor muscles in response to perturbations that displace the body. It is for this reason that the proximal extensor muscles are often referred to as the *antigravity muscles*. Because the proximal musculature mediates these important functions, the suprasegmental pathways that constitute the output segments of these reflexes are components of the medial system. Therefore an analysis of postural control is equivalent to a further analysis of the pathways that comprise the medial system and the inputs that govern their activities.

■ *The postural reflexes*

There are two broad classes of postural reflexes: the *statokinetic reflexes,* which are elicited by acceleratory displacement of the body, and the *static reflexes,* which are elicited by gravitational displacement. Thus the first step in categorizing the postural reflexes is to segregate them with respect to the origin of their initiating stimuli. Further subdivision within each class can then be made on the basis of the target muscles of the reflex.

By definition the *statokinetic reflexes* are elicited by acceleratory stimuli. Therefore the afferent limb of all such reflex circuits must involve the labyrinths. Based on our knowledge of labyrinthine receptors, it can be inferred that reflexes evoked by linear acceleration, as occur in falling, originate with signals from the utricle. Moreover, based on knowledge of vestibular projections, such reflexes would be expected to be mediated by the lateral vestibulospinal tract because this is the primary target of utricular information.

An example of a statokinetic reflex evoked by linear acceleration is the *vestibular placing reaction*. For example, if a cat is blindfolded and held by the pelvis with its head oriented downward, a sudden lowering of the animal will stimulate the utricular maculae, and this will elicit extension of the forelimbs. This reflex response can be considered as an adaptive reaction that prepares the animal for appropriate support by the limbs on surface contact. Destruction of the utricles abolishes this response. However, if the blindfold is removed, the response can be initiated by visual cues; in this case it is a *visual placing reaction*.

Many postural reflexes that are mediated by the vestibular system can also be elicited by visual stimuli. Thus the visual system frequently compensates for lesions involving the vestibular apparatus or its central pathways. In fact, an animal with a bilateral labyrinthectomy can appear to be normal unless it is deprived of visual cues, in which case the ability to maintain an upright posture is seriously impaired. Indeed, humans are more compromised by abnormal signals from the labyrinths, as in Ménière's disease, than by the loss of labyrinthine information—provided visual cues are available.

Statokinetic reflexes evoked by angular acceleration, as during rotation, originate with signals from the semicircular canals. Information from the canals most prominently influences the medial vestibular nucleus. Hence, the evoked reflex effects are primarily mediated by the ascending (medial longitudinal fasciculus) and descending (medial vestibulospinal tract) pathways from this nucleus (Fig. 18-1). Consider a situation in which an individual is rotated to the right. This will establish a flow to the left of the endolymph of the horizontal canals until the inertia of the endolymph is overcome. This flow to the left deflects the hair cells of the cristae and produces a change in discharge of the vestibular fibers from the horizontal canals (Fig. 18-2) (see Chapter 11). The change in discharge then informs the brain of an angular acceleration to the left in the horizontal plane. This signal will elicit important reflex effects on the eyes, neck, and upper limbs.

The reflex effects on the eyes are called *vestibular nystagmus*. They are mediated by projections from the medial vestibular nucleus, through interneurons of the brainstem, to the cranial nerve nuclei that innervate the extrinsic eye muscles (Fig. 18-1). As the head rotates to the right, the eyes will undergo a slow conjugate deviation to the left, that is, in the direction of the endolymph flow. This functions to maintain the center of gaze. As the rotation continues, the fixation point can no longer be maintained, and the eyes move rapidly to the right to establish a new fixation point. This slow conjugate deviation in the direction of the endolymph flow and the rapid return in the opposite direction occur repeatedly until the inertia of the endolymph is overcome and the endolymph follows the rotation of the canal (Fig. 18-2). The slow deviation to the right is called the *slow component* of the nystagmus, and the rapid return to the left is the *fast component*. By convention, the direction of nystagmus is designated as that of the fast component.

When the endolymph flows with the rotation of the canal, that is, when there is no longer an acceleratory stimulus, and the head is rotating at a constant velocity, vestibular nystagmus ceases. However, if visual cues are available, reflex eye movements will continue. This is *optokinetic nystagmus,* and it is independent of vestibular input. To return to the earlier example of vestibular nystagmus, if the rotation is now terminated, a deceleratory stimulus is imposed. The inertia of the endolymph will cause it to continue to move to the right in the direction of the previous rotation (Fig. 18-2). This

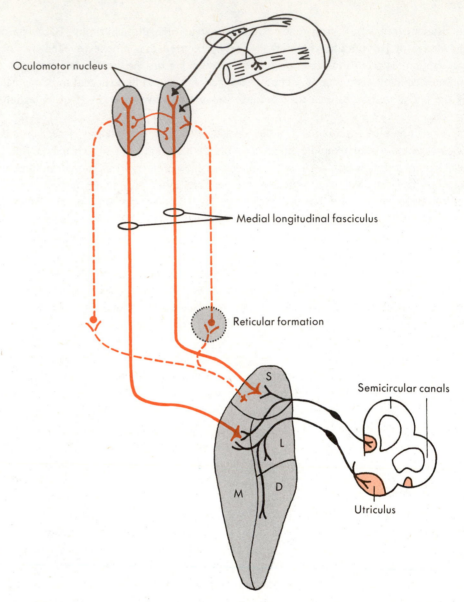

Oculomotor nucleus

Medial longitudinal fasciculus

Reticular formation

Semicircular canals

S

L

M

D

Utriculus

Fig. 18-1 ■ Principal features of the organization of the primary vestibular projections on the vestibular nuclei and the ascending projections of these nuclei. *D, L, M,* and *S,* descending, lateral, medial, and superior vestibular nuclei. (Redrawn from Brodal, A.: Neurological anatomy, ed. 3. Copyright © 1981 by Oxford University Press, Inc. Reprinted by permission.)

will evoke a *postrotatory nystagmus* in which the slow deviation of the eyes is to the right, again in the direction of the endolymph flow, and the rapid return is to the left.

Vestibular nystagmus can also be evoked by *caloric stimulation* of the semicircular canals. This provides a convenient clinical means of assessing the integrity of the brainstem circuitry that mediates vestibulo-ocular reflexes. In this test the head is tilted 60 degrees from the vertical plane to orient the horizontal canal vertically. If the external auditory canal is then irrigated with warm water, a convection current is established, and the endolymph rises toward the ampulla. This evokes a nystagmus with the slow component toward the contralateral side. Irrigating the canal with cold water will establish a convection current in the opposite direction, and it will elicit a nystagmus with the slow component toward the side of irrigation.

Certain diseases, particularly those involving the brainstem, can produce nystagmus. Thus nystagmus in the absence of an appropriate stimulus is a clear pathological sign, and it can be quite useful in the localization of brainstem lesions.

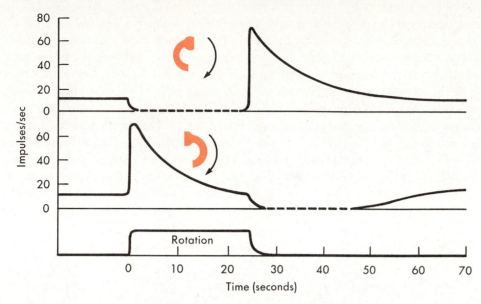

Fig. 18-2 ■ Effects of head rotation on the discharge of the vestibular nerve. There is a steady discharge at rest, which is suppressed by rotation in one direction and increased by rotation in the opposite direction. Cessation of rotation produces an afterdischarge in one case and a silent period in the other. (Redrawn from Adrian, E.D.: J. Physiol. **100**:389, 1943.)

If the earlier example of body rotation is considered in the context of reflex effects on the neck and upper limbs, the rotation to the right will elicit extension of the right arm and relaxation or flexion of the left arm. The effect on the neck muscles is to produce a turning of the head to the left. These reflex responses are mediated via the medial vestibulospinal tract. They may be viewed simply as head and upper limb adjustments that tend to resist the rotation (displacement) of the body. In general, postural reflexes can be understood intuitively as adjustments that resist displacement. This concept can assist in remembering the positions that will be assumed by the head and limbs in the various reflexes. Changes in eye position will generally be in a direction such that the starting fixation point is maintained.

The other major class of postural reactions is the *static reflexes*, which produce adjustments to displacements by gravity. This large class of reactions may be further subdivided with respect to the target of the reflex action. Conventionally the static reflexes are designated as *local, segmental,* or *general static reflexes*. Local static reflexes are those in which the reflex effect is on the same limb from which the stimulus was initiated. The most important of these, the myotatic reflex, is described in detail in Chapter 14. To review briefly, the myotatic reflex is the basis of muscle tone. Its importance as a postural mechanism can be easily appreciated by considering the quadriceps muscles in an individual who is standing in a normal upright posture. In this situation gravity tends to displace the body downward, stretching the quadriceps muscles. The muscle stretch evokes discharge from the muscle spindles of the quadriceps, and this initiates a myotatic reflex contraction that extends the leg at the knee. This contributes to maintaining the leg as a pillar of support and thus counteracts the gravitational displacement of the body.

Segmental static reactions are those reflexes in which the stimulus originates in one limb and the reflex effect is on the contralateral limb. The crossed-extensor components of reflexes, such as the crossed-component of the flexor withdrawal response, fall into this category (Chapter 14).

The largest group of postural reflexes is the general static reactions. These are defined as those postural reflexes in which a stimulus that arises at one site of the body exerts effects on many muscle groups. For example, numerous postural adjustments occur in response to changes in head position. Beyond labyrinthine stimulation, a change in head position relative to body position stimulates the stretch receptors of the neck muscles. Such proprioceptive information from the neck muscles can exert reflex effects on the limbs *(tonic neck reflexes)* and the eyes *(head-on-eyes reflexes),* and it participates in evoking the complex ensemble of *righting reactions* as well. The tonic neck reflexes are most easily illustrated in a quadruped such as the cat (Fig. 18-3). Turning the head to the right, relative to the body, will elicit extension of the right limbs and relaxation of the left limbs; rotation to the left will elicit an opposite limb response. Dorsiflexion of the head will elicit extension of the forelimbs and flexion of the hindlimbs, whereas ventroflexion will elicit flexion of the forelimbs and extension of the hindlimbs. Again, these limb responses compensate for changes in the animal's center of gravity.

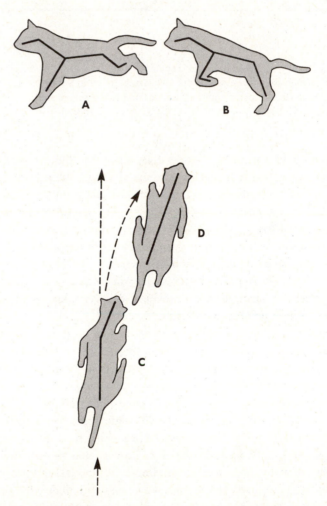

Fig. 18-3 ■ Tonic neck reflexes in locomotion of the cat. During running or leaping movements (**A** and **B**) the animal's head remains in a fixed attitude in relation to the ground. **A,** Dorsiflexion of the head and neck in relation to the back produces forelimb extension and hindlimb flexion. **B,** Ventroflexion of the head and neck in relation to the back produces inhibition of the forelimb extensors and facilitation of the hindlimb extensors. **C** and **D,** The animal is running; turning the head to the right reflexly increases stability during lateral movement to the right. Facilitation of forelimb and hindlimb extensors on the right is accompanied by inhibition of extensors on the left side. The extensor inhibition and flexor facilitation prepare the animal for the first step in the new direction. (Redrawn with permission from Eyzaguirre, C., and Fidone, S.J.: Physiology of the nervous system, ed. 2. Copyright © 1975 by Year Book Medical Publishers, Inc., Chicago.)

The neck proprioceptors also contribute to the reflex adjustments of eye position that contribute to maintaining an invariant visual field. In this reaction a rotation of the head to the right will result in an eye movement to the left; rotation to the left yields an eye movement to the right. Dorsiflexion elicits a downward eye movement, whereas ventroflexion produces an upward movement. This head-on-eyes reflex is demonstrable in a comatose patient whose appropriate brainstem circuitry is still intact, and it is referred to as the *doll's eye phenomenon.*

In review, it is obvious that no effort has been made to treat the topic of the postural reflexes comprehensively. The postural reactions are numerous and often complex, and in most instances their precise circuitry remains unspecified. These reflexes involve virtually every level of the neuraxis, and they occur continuously. They interact to adjust for displacements of the body and to maintain as constant a visual field as possible. This is accomplished through the various pathways of the medial system, and the appropriate occurrence and calibration of these reflexes constitute an essential substrate for phasic, goal-directed motor activity.

■ Control of tone in proximal extensor muscles

Another way of viewing the maintenance of an antigravity posture is from the perspective of the control of tone in the proximal extensor muscles. The maintenance of tone in these muscles is an important element of postural control. Indeed, the continuous inputs to the medial motoneurons originate with stimuli that elicit the postural reflexes, and such inputs contribute significantly to the maintenance of tone. This is produced by activity in the various medial system pathways that descend to excite both the α- and γ-motoneurons that innervate the proximal muscles and their spindles. It is thus readily understandable that spinal transection abolishes the tone of the antigravity muscles that are innervated by segments below the transection because these important suprasegmental contributions to the maintenance of tone are interrupted. The areflexia in spinal shock is a direct manifestation of this loss of medial system input to the spinal cord (see Figs. 15-2 and 15-10).

The two pathways of the medial system that are of the greatest importance in the maintenance of tone are the lateral vestibulospinal and pontine reticulospinal tracts. Thus spinal shock can be produced with any transection caudal to the lateral vestibular nucleus. However, if the transection is just rostral to this nucleus, then in contrast to spinal shock, the tone of the antigravity muscles is actually enhanced and an exaggerated antigravity posture obtains. This phenomenon was discovered by Sherrington, who named it *decerebrate rigidity.* When this occurs in humans (Fig. 18-4), the arms and legs are extended, the back arched, and the head dorsiflexed. The feet are ventroflexed and the arms pronated. However, there is flexion at the wrists and minimal involvement of the digits, indicating that the rigidity is, indeed, a medial system phenomenon. Clinically this can occur after trauma or with compression of the brain by a space-occupying lesion such as a tumor. The explanation for this antigravity posture is that the lateral vestibular nucleus, which exerts a powerful excitatory action on the proximal extensor motoneurons, is released from inhibitory inputs from more rostral structures.

The decerebrate rigidity can be enhanced by making the transection more rostrally so that the cells of origin of the pontine reticulospinal tract remain in continuity with the spinal cord. This tract also provides a powerful tonic excitatory influence on the proximal extensor motoneurons, and the transection releases it from inhibitory control, particularly by corticoreticular projections. Thus the released pontine reticulospinal tract acts synergistically with the lateral vestibulospinal tract to enhance the activity of the α- and γ-motoneurons that innervate the proximal extensor muscles.

Any experimental procedure that reduces these descending excitatory influences will attenuate the rigidity. Thus placing a lesion in the ventral funiculus of the spinal cord or destroying the cells of origin of the pontine reticulospinal or lateral vestibulospinal

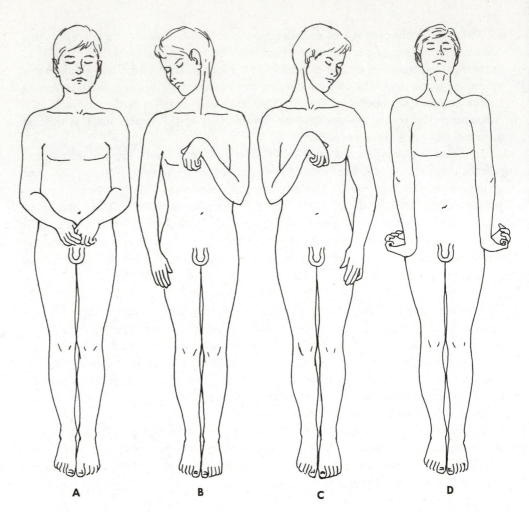

Fig. 18-4 ■ Decorticate rigidity (**A** to **C**) and decerebrate rigidity (**D**). **A,** The patient is lying supine with the head unturned. **B** and **C,** The tonic reflexes are shown by turning the head to the right or left. (Redrawn from Fulton, J.F.: Textbook of physiology, ed. 17, Philadelphia, 1955, W.B. Saunders Co.)

tract will attenuate or abolish the rigidity. Because the cerebellum influences the cells of origin of these two pathways, experimental manipulations involving these cerebellar influences will also affect the rigidity. For instance, lesions of the fastigial nucleus will eliminate an excitatory drive on the pontine reticulospinal and lateral vestibulospinal tracts and will therefore reduce rigidity. Electrical stimulation of the fastigial nucleus will have the opposite effect. In contrast, a lesion of the vermal cortex will eliminate the Purkinje cell inhibition of the fastigial nucleus and the lateral vestibular nucleus. Thus a vermal lesion will enhance the rigidity; vermal stimulation will, of course, attenuate it. Yet another exercise is to remove the labyrinths or transect the vestibular portion of the eighth nerve. The utricles provide the major source of input to the lateral vestibular nucleus. Therefore either of these procedures will remove that excitatory drive and thus reduce the rigidity.

Cutting the appropriate dorsal roots will also abolish decerebrate rigidity. This interrupts the myotatic reflex loop but leaves the α-motoneuron innervation intact. The abolition of the rigidity suggests that it is mediated primarily by enhanced drive on γ-motoneurons to spindles of the antigravity muscles. On this basis decerebrate rigidity is considered a γ-*rigidity* because it largely reflects tonically hyperactive myotatic reflexes that involve the antigravity muscles.

Although many rigidities are γ-mediated, α-*rigidity* also exists. It can be demonstrated experimentally by a rather puzzling phenomenon that involves the anterior cerebellum. If one begins with a preparation in which decerebrate rigidity has been established and then has been eliminated by dorsal rhizotomy, a subsequent lesion of the anterior cerebellum will reinstate the rigidity. Because the dorsal rhizotomy precludes any involvement of the γ-motoneurons, the reestablished rigidity must be mediated by the α-motoneurons. Indeed it has been shown that anterior cerebellectomy increases the activity of the α-motoneurons. This observation has generated the hypothesis that in the intact animal the cerebellum participates in determining the relative bias of descending systems on the α- and γ-motoneurons. Furthermore, this α-rigidity appears to be mediated primarily by the lateral vestibulospinal tract because it is abolished by section of the eighth nerve. This is in contrast to γ-rigidity, in which the pontine reticulospinal pathway plays a particularly prominent role.

The exaggerated antigravity posture of decerebrate rigidity is mediated by the release of medial system pathways from more rostral inhibitory influences. Therefore as more of such inputs are left intact, there will be an increased capability for phasic movements to be superimposed on the rigidity. As transections are made more rostrally, the rigidity can be modified by applying stimuli that elicit various reflexes. For example, in humans if only the cortex is compromised, then a rigidity ensues that is characterized by extension of the lower limbs. However, the position of the upper limbs will be contingent on head position (Fig. 18-4). If the head is rotated to the right, then the left arm will be flexed and the right arm extended; the converse occurs if the head is rotated to the left. This is reminiscent of the tonic neck reflexes, and it suggests that such reflexes are interacting with the *decorticate rigidity* to modify the classic posture. In many mammalian species decorticate rigidity is not prominent. In such animals transections rostral to the midbrain may even permit walking.

The effects of transecting the neuraxis at various levels dominated much of the early literature on the motor system. This approach was based on a "segmental view" of the nervous system that had been successful in helping to understand the spinal cord. However, as it was appreciated that suprasegmental motor control was primarily organized with respect to longitudinal pathways, the transection approach was recognized to be inappropriate in gaining an understanding of normal motor control. (It still, of course, retained value in providing models of certain lesions that occur clinically.) An increasing knowledge of the descending pathways and their effects on spinal mechanisms now provides a reasonable understanding of many of the effects of transections. This serves as yet another example of how an understanding of the fundamental organization of the nervous system is a valuable, if not essential, preface to understanding the pathophysiology of brain lesions.

The control of phasic, goal-directed movement is now considered further to provide an opportunity for a more detailed treatment of the lateral system.

■ Control of movement

■ Motor areas of the cortex

Movements can be elicited by stimulating many areas of the cortex. However, if only those areas from which reasonably localized responses can be evoked at low stimulation thresholds are considered, the "motor regions" of cortex become more limited (Fig. 18-5). These are generally identified as the primary motor, premotor, and supplementary motor areas. A secondary motor area and frontal eye fields are often identified as separate areas as well. Although it is perhaps oversimplified, the cortical motor representation can be viewed in terms of two regions. One region is encompassed by the representation of the musculature on the dorsal surface of the precentral gyrus (see Fig. 15-5). Area 4, the primary motor area, includes the representation of the extremities and is closest to the central sulcus. Area 6, the premotor area, includes the representa-

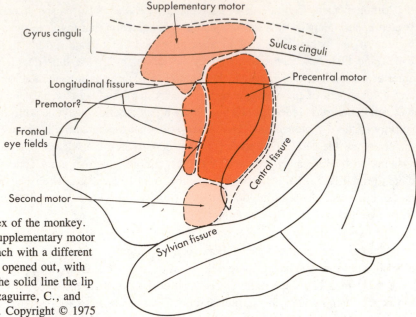

Fig. 18-5 ◼ Principal motor areas of the cerebral cortex of the monkey. Shown are the precentral or primary motor area, the supplementary motor area, the second motor area, and the premotor area, each with a different shade. The central and longitudinal fissures are shown opened out, with the broken line indicating the floor of the fissure and the solid line the lip of the fissure on the brain surface. (Redrawn from Eyzaguirre, C., and Fidone, S.J.: Physiology of the nervous system, ed. 2. Copyright © 1975 by Year Book Medical Publishers, Inc., Chicago.)

tion of the axial musculature and much of the head. Area 8, the frontal eye fields, includes the representation of the eyes within the head. The second motor region is the supplementary motor cortex, which is located on the mesial surface of the precentral cortex; it, too, includes a full representation of the body muscles (see Fig. 15-5).

Stimulation of the primary motor area is described in Chapter 15. Discrete surface stimulation elicits low-threshold, highly localized movements such as wrist flexion. If one stimulates with microelectrodes (microstimulation) deeper in the motor cortex, it is even possible to evoke contraction of single muscles. Moreover, there appear to be localized *efferent cortical zones,* where stimulation consistently contracts the same muscle or muscles. These may be analogous to the columnar organizations of the sensory cortices. Of course, an efferent cortical zone for any given muscle or muscle group will be within the appropriate area of the topographical map of the musculature. Stimulation of the supplementary motor cortex requires higher currents and yields less localized and more complex responses. Stimulating the primary motor area elicits discrete contralateral responses, but activating the supplementary motor cortex evokes postural adjustments that involve many muscles, often bilaterally.

As stated earlier, the corticospinal tract is not the only motor pathway that originates from the cortical motor areas. This can be demonstrated readily for the primary motor cortex by interrupting the corticospinal tract and then stimulating the motor cortex. In this case movements are still elicited, and there remains a topographical map of the musculature. However, the character of the evoked movements changes strikingly. The rapid flicklike movements that primarily involve the contralateral flexor muscles are now replaced by higher threshold, slower movements that are long lasting and frequently involve the entire limb. Fractionated movements of the digits do not occur.

The movements elicited by stimulating the supplementary motor cortex do not require that the corticospinal tract be intact. Moreover, they are not affected by lesions of the primary motor area. Thus the supplementary motor cortex must influence motor activity through cortical projections to the brainstem independently of those derived from the primary motor area.

The multiple cortical representation of the musculature is reminiscent of the multiple sensory representations. It is not yet clear what function is served by such multiple representation in either sensory or motor systems.

Regarding the primary motor area, data have recently become available concerning the aspects of movement that are controlled by the corticospinal tract. Pyramidal tract neurons can be identified in the awake animal by implanting stimulating electrodes in the medullary pyramids. If a cortical neuron from which one is recording responds to pyramidal stimulation with characteristics that indicate it is being activated antidromically, then it can be assumed it is a cell of origin of a pyramidal tract fiber. Such cells can then be studied during various controlled movements, and their discharge characteristics can be correlated with the various features of the movements.

If a monkey is trained in response to a stimulus cue to displace a lever in such a way that the required movement involves flexion of the wrist, one can record from pyramidal tract neurons in a cortical efferent zone associated with the wrist muscles. What one finds is that cells associated with wrist flexors discharge prior to the movement. Some will generate a transient high-frequency burst *(dynamic neurons);* others will continue discharging at increased frequency throughout the wrist flexion *(static neurons);* still others have mixed properties *(mixed neurons).* If the amount of force required to displace the lever is now varied, it can be shown that there exists a class of pyramidal tract neurons the discharge of which is a function of the force of contraction (or rate of change of force) rather than of the intended displacement (Fig. 18-6). There

■ *Activity of pyramidal tract neurons during movement*

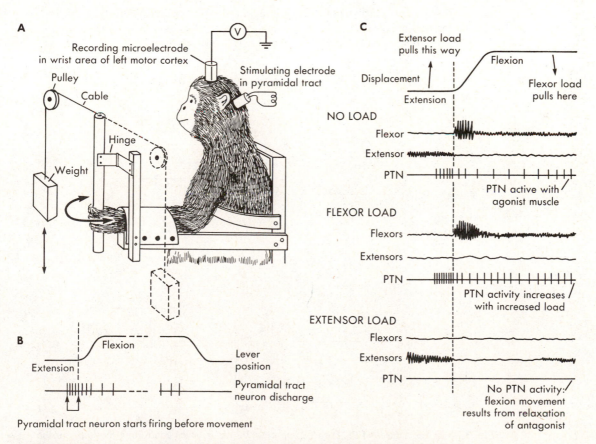

Fig. 18-6 ■ The activity of corticospinal neurons reflects the direction of force exerted. **A** shows the experimental arrangement for recording the discharge of a pyramidal tract neuron while the awake monkey alternately flexes and extends its wrist. **B** demonstrates that the cortical neuron begins firing before the movement *(arrows).* **C** shows a pyramidal tract neuron *(PTN)* that increases its activity with flexion of the wrist. Flexor and extensor electromyograms and PTN discharge records are shown under different load conditions. The absence of neuronal activity with an extensor load indicates that the neuronal output codes for force rather than displacement. (Redrawn from Kandel, E.R., and Schwartz, J.H.: Principles of neural science, New York, 1981, Elsevier North-Holland, Inc.)

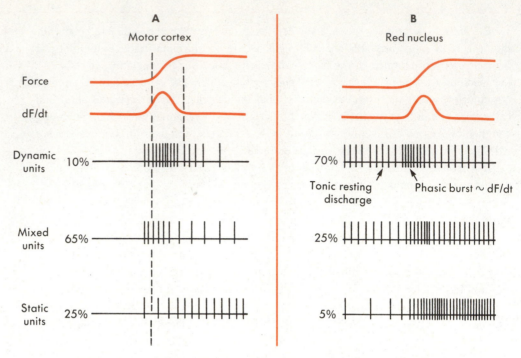

Fig. 18-7 ■ Patterns of activity of dynamic, static, and mixed neurons in the motor cortex **(A)** and red nucleus **(B)** of the cat during voluntary isometric contraction. (Redrawn from Kandel, E.R., and Schwartz, J.H.: Principles of neural science, New York, 1981, Elsevier North-Holland, Inc.)

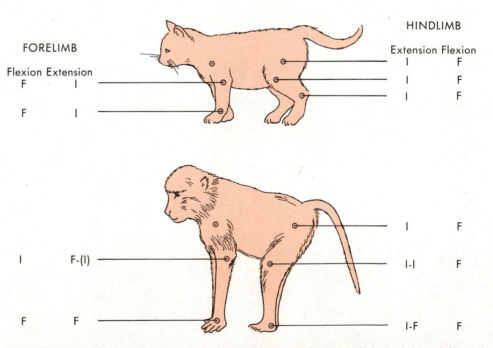

Fig. 18-8 ■ The actions of the corticospinal tract on the muscles controlling the joints of the forelimbs and hindlimbs of the cat and baboon. These animals are shown schematically in their habitual standing postures. Inhibitory *(I)* and facilitative *(F)* effects of the corticospinal tract are shown for the motoneurons innervating the flexor and extensor muscles at each joint. (Redrawn from Preston, J.B., Shende, M.C., and Uemura, K.: The motor cortex-pyramidal system: patterns of facilitation and inhibition on motoneurons innervating limb musculature of cat and baboon and their possible adaptive significance. In Yahr, M.D., and Purpura, D.P., editors: Neurophysiological basis of normal and abnormal motor activities, New York, 1967, Raven Press.)

are other pyramidal tract neurons that influence the wrist flexors, but their discharge properties relative to the movement are more complex and are not yet sufficiently understood. Analogous investigations of the rubrospinal pathway indicate a similar physiological organization but with a higher proportion of dynamic neurons (Fig. 18-7). From such investigations it is now clear that the corticospinal neurons of the primary motor area and the rubrospinal neurons discharge shortly before a movement and that a population of these neurons is involved in controlling the force of contraction.

Other lines of investigation have indicated how this control is distributed with respect to the motoneurons that innervate different muscles. The extent to which the corticospinal tract excites a given motoneuron appears to be determined by the extent to which the muscle it innervates participates in (1) antigravity functions and (2) phasic movements. If the muscle is primarily involved in antigravity or postural function, then its motoneurons are inhibited by the corticospinal tract. If, in contrast, the muscle serves no postural role but is involved in spatially organized movements, then the corticospinal tract will excite its motoneurons (Fig. 18-8). Thus the intrinsic muscles of the digits, both flexor and extensor, are excited by the corticospinal tract. However, for the muscles of the proximal limb, the extensors will be inhibited, and the flexors will be excited. This reinforces the concept that the corticospinal and rubrospinal tracts, the principal components of the lateral system, are mainly concerned with spatially organized, goal-directed movements. To perform this function, their action includes an inhibition of the activity of antigravity muscles, presumably to reduce the excitation of postural muscles that might compete with the desired phasic movement.

■ *Interaction of the corticospinal tract with spinal circuitry*

The corticospinal tract has prominent monosynaptic connections with the motoneurons that innervate the intrinsic muscles of the digits. These connections allow it to bypass the intrinsic spinal circuitry in controlling independent movement of the digits. However, much of the influence of the corticospinal tract on motoneurons is exerted through spinal interneurons. This raises the issue of whether this tract, and descending motor pathways in general, uses independent populations of spinal interneurons or converge on interneurons that participate in spinal reflexes, such as flexor withdrawal responses. There is evidence that the corticospinal projections onto interneurons generally use the existing spinal organization. For example, a corticospinal fiber that excites a flexor motoneuron is most likely to terminate on an interneuron that also receives input from flexor reflex afferent fibers and from the Ib fibers. Thus it would use the existing circuitry for flexor withdrawal and inverse myotatic reflexes, thereby using the intrinsic spinal connections for inhibiting antagonistic extensor motoneurons (reciprocal innervation) and perhaps even for crossed-components (see Fig. 14-15).

Although considerably more investigation is required, the available evidence suggests that use of spinal reflex circuitry by descending motor pathways is a general rule for both the lateral and medial systems. Such an arrangement leads to an economy of organization and emphasizes the need for a comprehensive understanding of the spinal reflex circuitry.

■ *Sensorimotor integration*

When the descending motor pathways are viewed from the perspective of their sites of termination in the spinal cord, it is obvious that they are not far removed from the final common path. Moreover, the physiological data on such pathways as the corticospinal and rubrospinal tracts indicate that they discharge shortly before movement. It is reasonable to ask, then, what distinguishes these pathways from spinal interneurons. Is the corticospinal tract merely a glorified interneuron with a long axon derived from cells of origin in the cortex? These pathways do, indeed, differ from spinal interneurons in an exceedingly important respect. Because their cells of origin are located suprasegmentally, they have access to a rich variety of information that is not available to the spinal circuitry. Thus one of the important challenges in analyzing motor control is to describe how sensory stimuli initiate and guide motor behavior, that is, the problem of sensori-motor integration.

The spinal reflexes described in Chapter 14 represent simpler models of such integration. The myotatic reflex, for example, is a particularly well-described case where much of the circuitry is specified, and the characteristics of the sensory input and motor output are generally understood. The flexor withdrawal reflex provides an illustration of a more complex and less understood circuit for sensorimotor integration. Yet, despite the challenge of understanding the spinal reflexes, these are much simpler than the organization of the complex movements that involve suprasegmental pathways. In the following sections two examples are given of recent investigative efforts to approach this important problem.

■ *Input-output relationships across the motor cortex*

Returning to the pyramidal tract neurons of the primary motor area, one can ask how such neurons respond to various sensory stimuli; that is, one can turn toward the sensory periphery and attempt to determine the receptive field characteristics of the cells of origin of motor pathways. The massive convergence of inputs on these neurons from many sources makes this a difficult task. It is not feasible to hold all these sources of input constant except the one under investigation; this results in rather labile receptive

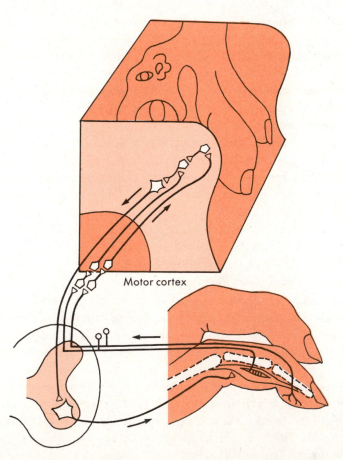

Motor cortex

Fig. 18-9 ■ The input-output organization of the cortical efferent zone controlling the flexor muscle of a digit. See text for further explanation. (Redrawn from Asanuma, H.: Physiologist **16:**143, 1973.)

fields. Nevertheless, recent studies have generated important information, particularly with respect to inputs from the cutaneous, joint, and muscle spindle receptors.

In the motor cortex, pyramidal tract neurons respond to cutaneous input as well as to active and passive movements of a limb. For a specific cortical efferent zone, the cutaneous fields of its neurons lie in the path of the movement elicited by stimulating that zone. For example, if stimulation of the zone produces flexion of a digit, then the cutaneous fields of neurons in that zone will be on the ventral surface of the digit, and the cutaneous input will excite the cortical neurons polysynaptically (Fig. 18-9). Those joint receptors activated by the flexion also will have an excitatory influence on the neurons of the zone. The spindle receptors of the digit flexor will excite the cortical neurons as well (Fig. 18-9). Because these receptors are activated when the flexor muscle lengthens (relaxes), a slowing of the flexion will act to excite the cortical neurons that drive the flexion.

The general rule is that these sensory stimuli influence the motor cortex in a manner that facilitates the movement. This feedback in all likelihood contributes to serial motor acts and assists the motor cortex in tracking objects. For example, the flexor contractions of the muscles involved in grasping an object would be facilitated by the cutaneous and proprioceptive stimuli that are generated by the movement and by its consequent contact with the object.

Another example involves the *tactile placing reaction*, which is mediated by the motor cortex. If a cat is held and moved toward a surface, when the dorsal aspect of its paw contacts that surface, the cat will flex its limb and then extend it to establish the leg as a source of support. In this placing response, the cutaneous stimuli generated by contact of the dorsum of the paw with the surface will facilitate the discharge of pyramidal tract neurons involved in flexing the paw. The cutaneous stimuli subsequently generated by the contact of the ventral aspect of the paw with the surface would then facilitate pyramidal tract neurons activating extension.

■ *Visually guided finger movement*

The preceding section illustrates an approach to sensorimotor integration based on receptive field analysis of pyramidal tract neurons. Although far from fully specifying the sensory control of these cells, it represents an exciting start with respect to cutaneous and proprioceptive inputs. Following is an illustration of another approach that is based on a functional neuroanatomical analysis of the pathways that mediate visual guidance of independent finger movements. A monkey is required to remove a small piece of food from a well in a board, a task that demands fractionated movement of the digits (Fig. 18-10, *A*). This task can be accomplished only via the corticospinal tract that arises from the motor cortex contralateral to the hand that is used (Fig. 18-11). As described more fully in Chapter 19, if the optic chiasm and certain of the transverse commissures of the brain are sectioned, then visual information presented to an eye will reach only the ipsilateral hemisphere (Fig. 18-11). In such a preparation it has been shown that if the left eye is covered, the monkey is able to retrieve food from the board with its right hand but not its left hand (Fig. 18-10); that is, the visual information from the food board is available only to the left hemisphere via the left eye. The left hemisphere can then control the fingers of the right hand because the corticospinal tract decussates at the pyramids. Of course, if the left eye is covered and the visual information is available to the right eye, then retrieval of the food can only be negotiated by the left hand.

This finding suggests that visual information reaching a given hemisphere can access the motor cortex of that hemisphere via intracortical circuitry. If a lesion is now made in the parietal lobe between the visual and motor cortices, then visually guided finger movement is eliminated in this task. This suggests the outline of a pathway whereby visual stimuli access the cells of origin of the corticospinal tract that control fractionated finger movement.

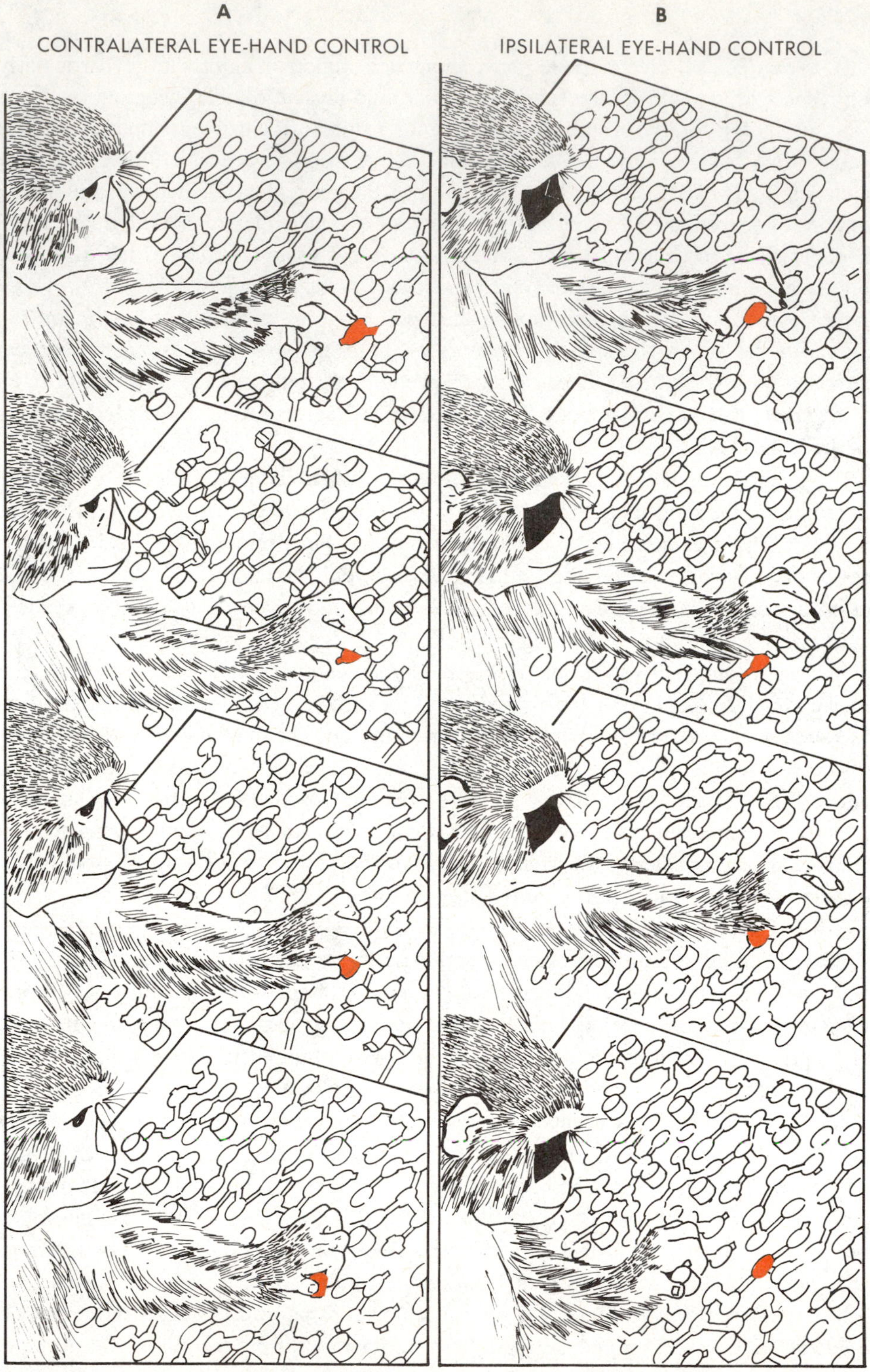

Fig. 18-10 ■ Drawings from a film showing the hand and finger movements of a split-brain monkey with a complete commissurotomy taking a small food pellet (shown in color) from a test board. Under guidance of the contralateral eye **(A)** the hand and fingers in reaching out assume the precision grip posture *(top drawing),* and the index finger and thumb dislodge the pellet from the well. In this situation the visual information can influence the corticospinal neurons controlling the hand and fingers being used. Under the guidance of the ipsilateral eye **(B)** the hand and fingers do not assume the precision grip posture until the hand has touched the board. The hand is brought to the proper place, but the pellet is not taken from the well. Instead the hand and fingers explore the surface of the board as if "blind." In this case visual information cannot gain access to the corticospinal neurons controlling the hand. (Redrawn from Brinkman, C.: Split-brain monkeys: cerebral control of contralateral and ipsilateral arm, hand and finger movements, doctoral dissertation, Rotterdam, 1974, Erasmus University Rotterdam.)

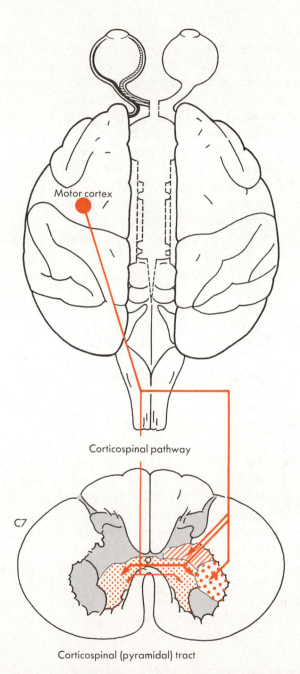

Motor cortex

Corticospinal pathway

C7

Corticospinal (pyramidal) tract

Fig. 18-11 ■ The corticospinal projections in the monkey (see Fig. 19-2). Note that in the split-brain preparation visual information presented to the left eye reaches only the left hemisphere. It can therefore influence the corticospinal fibers controlling the right distal extremity but not those controlling the left. (Redrawn from Brinkman, C.: Split-brain monkeys: cerebral control of contralateral and ipsilateral arm, hand and finger movements, doctoral dissertation, Rotterdam, 1974, Erasmus University Rotterdam.)

■ *Overview of motor control*

In the preceding chapters on the motor system an effort has been made to analyze the control of posture and movement by starting at the periphery and proceeding centrally. Beginning with the topographical organization of the spinal cord with respect to the musculature, an approach to the descending motor pathways is developed that emphasizes their spinal sites of termination. From this foundation it is possible to classify these descending pathways as either concerned primarily with organized movements involving the muscles of the extremities (lateral system) or with postural control involving more proximal muscles (medial system). This distinction is developed for heuristic purposes. However, all movements must occur on a postural background; therefore these two systems cannot operate entirely independently.

Certain muscles, such as the intrinsic muscles of the digits, serve no postural function and thus are principally under the control of the lateral system. Analogously, the axial muscles are almost exclusively postural in function and are therefore primarily under the control of the medial system. However, many muscles, such as the proximal limb muscles, subserve both functions and are therefore under shared control by the two systems. This is seen in the opposing actions of the medial and lateral systems on such muscles; the medial brainstem pathways excite the extensor motoneurons for the proximal limb muscles, and the corticospinal tract inhibits these motoneurons and excites the flexors of the proximal limb.

Beyond this shared control of certain muscles, interaction between the medial and lateral systems is necessary to establish the "postural background" for phasic goal-directed movements. Spatially organized movements of the wrist, for example, require setting the muscles at the elbow and shoulder, and this implies an interaction with the medial system. The basal ganglia may be involved in this context. Thus future study of the motor system must deal not only with the control of specific muscles that implement a movement but also with the problem of how many other muscle groups are controlled to establish the proper "background" for that movement.

We have touched superficially on the question of sensory guidance of movement, a critical area that leads into the broad question of how movements are initiated. This is a fascinating and most challenging question that will probably not be answered satisfactorily for some time. However, recent neurophysiological analyses of the motor system have begun to yield germane information. For example, we know that the corticospinal and rubrospinal tracts discharge just prior to movement. The discharge of these pathways is preceded by discharge changes in the basal ganglia and cerebellum. The outputs of the basal ganglia and the lateral zone of the cerebellum converge at the ventral thalamus and then are relayed to the motor cortex. Thus the basal ganglia and cerebellum participate in the organization of movement and its postural background as a step preceding the actual cortical command for movement. As to the structures that become active before reaching the cerebellum and basal ganglia, these are in all likelihood cortical, but at present there is little information regarding their localization.

The complexity of analyzing such problems may seem overwhelming, but this has been true for various problems in neurobiology that have ultimately yielded to analysis. As a final point it would be useful to recall that, regardless of the complexity of the central integration for organizing a movement, the final command must be expressed through the final common path. There is much information yet to be gained by studying this level of the motor system, for it is at this site that the outputs of all the various segmental and suprasegmental influences converge.

■ Bibliography

Journal articles

Asanuma, H.: Cerebral cortical control of movement, Physiologist **16**:143, 1973.

Asanuma, H., and Rosén, I.: Topographical organization of cortical efferent zones projecting to distal forelimb muscles in the monkey, Exp. Brain Res. **14**:243, 1972.

Asanuma, H., Stoney, S.D., and Abzug, C.: Relationship between afferent input and motor outflow in cat motorsensory cortex, J. Neurophysiol. **31**:670, 1968.

Evarts, E.V.: Pyramidal tract activity associated with a conditioned hand movement in the monkey, J. Neurophysiol. **29**:1011, 1966.

Evarts, E.V.: et al.: Central control of movement, Neurosci. Res. Program Bull. **9**:1, 1971.

Haaxma, R., and Kuypers, H.G.J.M.: Role of the occipito-frontal cortico-cortical connections in visual guidance of relatively independent hand and finger movements in rhesus monkeys, Brain Res. **71**:361, 1974.

Lawrence, D.G., and Kuypers, H.G.J.M.: The functional organization of the motor system in the monkey. II. The effects of lesions of the descending brain-stem pathways, Brain **91**:15, 1968.

Pollack, L.J., and Davis, L.: The reflex activities of a decerebrate animal, J. Comp. Neurol. **50**:377, 1930.

Sherrington, C.S.: Decerebrate rigidity and reflex coordination of movements, J. Physiol. (Lond.) **22**:319, 1898.

Sprague, J.M., and Chambers, W.W.: Regulation of posture in intact and decerebrate cat. I. Cerebellum, reticular formation, vestibular nuclei, J. Neurophysiol. **16**:451, 1953.

Sprague, J.M., and Chambers, W.W.: Control of posture by reticular formation and cerebellum in the intact, anesthetized and unanesthetized and in the decerebrated cat, Am. J. Physiol. **176**:52, 1954.

Tanji, J., and Evarts, E.V.: Anticipatory activity of motor cortex neurons in relation to direction of an intended movement, J. Neurophysiol. **39**:1062, 1976.

Books and monographs

Bard, P., and Macht, M.B.: The behaviour of chronically decerebrate cats. In Neurological basis of behavior, Ciba Foundation Symposium, Boston, 1958, Little, Brown & Co.

Brodal, A.: Neurological anatomy, ed. 3, Oxford, 1981, Oxford University Press.

Evarts, E.V.: Representation of movements and muscles by pyramidal tract neurons of the precentral motor cortex. In Yahr, M.D., and Purpura, D.P., editors: Neurophysiological basis of normal and abnormal motor activities, New York, 1967, Raven Press.

Ghez, C.: Cortical control of voluntary movement. In Kandel, E.R., and Schwartz, J.H., editors: Principles of neural science, New York, 1981, Elsevier North-Holland, Inc.

Granit, R.: The basis of motor control, New York, 1970, Academic Press, Inc.

Lundberg, A.: Control of spinal mechanisms from the brain. In Tower, D.B., editor: The nervous system, vol. 1, The basic neurosciences, New York, 1975, Raven Press.

Magnus, R.: Körperstellung, Berlin, 1924, Springer-Verlag.

Penfield, W., and Rasmussen, T.: The cerebral cortex of man, New York, 1950, The Macmillan Co.

Phillips, C.G., and Porter, R.: Corticospinal neurones: their role in movement, London, 1977, Academic Press, Inc.

Preston, J.B., Shende, M.C., and Uemura, K.: The motor cortex-pyramidal system: patterns of facilitation and inhibition on motoneurons innervating limb musculature of cat and baboon and their possible adaptive significance. In Yahr, M.D., and Purpura, D.P., editors: Neurophysiological basis of normal and abnormal motor activities, New York, 1967, Raven Press.

Ruch, T.C.: Brain stem control of posture and orientation in space. In Ruch, T., and Patton, H.D., editors: Physiology and biophysics, vol. 1, Philadelphia, 1979, W.B. Saunders Co.

Ruch, T.C., and Fetz, E.: The cerebral cortex: its structure and motor functions. In Ruch, T., and Patton, H.D., editors: Physiology and biophysics, vol. 1, Philadelphia, 1979, W.B. Saunders Co.

Sherrington, C.S.: The integrative action of the nervous system, ed. 2, New Haven, Conn., 1947, Yale University Press.

Woolsey, C.N.: Organization of somatic sensory and motor areas of the cerebral cortex. In Harlow, H.F., and Woolsey, C.N., editors: Biological and biochemical bases of behavior, Madison, 1958, University of Wisconsin Press.

chapter 19

The cerebral cortex

Viewed from an evolutionary perspective, the expansion of the mammalian *cerebral cortex* is striking. In all vertebrates, however, it is possible to discern three analogous cortical divisions and more or less infer the evolutionary trend. These divisions are the *archicortex,* the *paleocortex,* and the *neocortex.* Neocortex is often called *isocortex,* and *allocortex* is often used to indicate the contiguous expanse of archicortex and paleocortex. Some authorities consider the *cingulate gyrus* to represent a transitional form between allocortex and neocortex. Among these cortical divisions, it is the neocortex that has most dramatically enlarged during mammalian evolution. Its large surface area (2500 cm^2) is folded among numerous *sulci* and *gyri,* and its overall volume approximates 600 cm^3. Equally impressive is the evolutionary development of the *corpus callosum,* which in humans is a massive cerebral commissure interconnecting neocortical areas between the two hemispheres.

■ *Neocortex*

The human neocortex represents roughly 90% of cortical tissue. Neocortex can be distinguished from archicortex and paleocortex on the basis of lamination. During at least some stage of ontogeny six layers parallel to the surface can be clearly discerned in neocortex. Subsequent development and differentiation obscure some of these layers in certain areas of adult neocortex.

■ *Cell types*

Nearly all neocortical neurons can be placed into one of three broad categories: *pyramidal cells, stellate cells,* and *fusiform cells* (Fig. 19-1). Occasionally other cell types have been described.

Pyramidal cells are the largest, with somata typically 30 to 50 μm across and occasionally over 100 μm in diameter. Each cell has a pyramid-shaped soma, with the apex directed toward the pial surface, a large *apical dendrite* that branches and often extends to the pial surface, and a set of branched *basal dendrites* that fan out from the base of the soma for considerable distances. The dendrites are usually covered with spiny protrusions that are thought to be postsynaptic specializations. A single axon emerges from the base of the soma and projects into white matter to other cortical or subcortical areas. The axon also emits collaterals that contribute to intrinsic cortical circuitry.

Stellate cells are also known as *granule cells,* and they have star-shaped or spherical somata that are roughly 10 μm in diameter. Their short dendritic processes are richly branched, and they emanate in all directions from the soma. Some of these cells have spines on their dendrites, but others do not; these are respectively called spiny and smooth stellate cells. Stellate cell axons branch and ramify locally. Consequently these

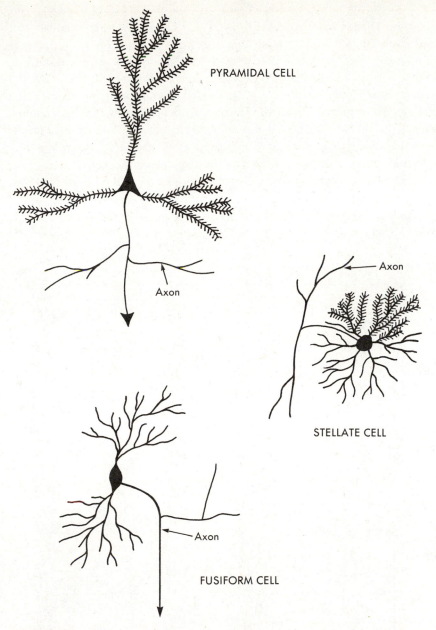

PYRAMIDAL CELL

Axon

Axon

STELLATE CELL

FUSIFORM CELL

Axon

Fig. 19-1 ■ Three major cell types of cerebral cortex. Two major types of stellate cell have been identified: one with no dendritic spines and one with dendrites richly covered by spines. Thus the drawing of one cell shows spiny and nonspiny dendrites.

cells are regarded as cortical interneurons, although some may project axons into the white matter.

Fusiform cells are relatively rare. They have elongated somata, typically 10 to 15 μm wide and up to 50 μm long. Their dendrites emerge as two tufts from either end of the soma, and the cell thus has a bipolar appearance. The cell's orientation is usually more or less perpendicular to the pial surface. Axons from fusiform cells project into white matter, presumably to other cortical or subcortical sites, and branches of the main axon contribute to local intrinsic circuitry.

In brief, pyramidal and fusiform cells are the major efferent neurons of the neocortex, although they also contribute significantly to intrinsic circuitry. Stellate cells form the main population of cortical interneurons.

■ *Functional organization*

Cortical layers. The six main neocortical layers that lie parallel to the pial surface are designated layers I to VI, from the pial surface to white matter (Fig. 19-2). Layer I is the thin *molecular, plexiform,* or *zonal, layer.* It has few somata and is basically a synaptic plexiform layer comprising axonal and dendritic processes. Layer II is the *external granular layer,* comprising mostly stellate cells, although some pyramidal cells are present. Layer III is the *pyramidal* (or *external pyramidal) layer,* and its neurons are predominantly pyramidal cells, with some stellate cells. Layer IV, the *internal granular layer,* consists mostly of stellate cells with some pyramidal cells. Layer V is the *ganglionic (internal pyramidal) layer,* dominated by large pyramidal cells but also including some stellate cells. Finally, layer VI is the *multiform layer.* It contains fusiform cells and a variety of other neuron types, including some stellate and pyramidal cells.

Pyramidal and stellate cell bodies can be found in any layer, but pyramidal cell bodies tend to be concentrated in layers III and IV, and stellate cell bodies in layers II

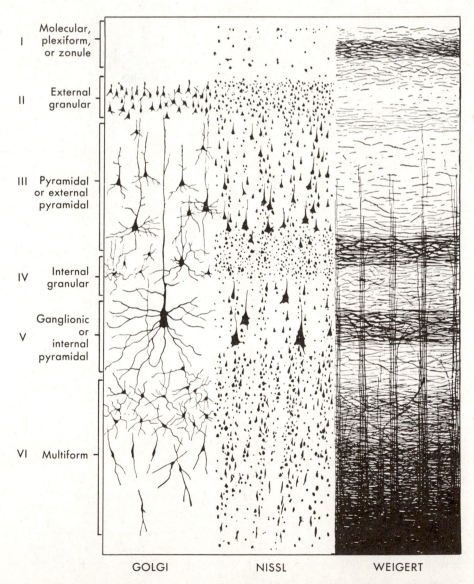

Fig. 19-2 ■ Layers of cerebral cortex as shown by the different staining methods of Golgi (entire structure of a few isolated neurons), Nissl (somata), and Weigert (myelinated axons). (From Brodal, A.: Neurological anatomy, ed. 3. Copyright © 1981 by Oxford Univeristy Press, Inc. Reprinted by permission.)

and IV. Fusiform cell bodies are associated nearly exclusively with layer VI. The dendrites of stellate and fusiform cells are contained mostly or completely within a single layer. In contrast, pyramidal cell dendrites cross layers freely and can receive synaptic inputs through much or all of the cortical depth. For instance, many pyramidal cells with somata in layer V or VI have apical dendrites that reach layer I, and basal dendrites of pyramidal cells commonly descend through one or two layers.

Cortical afferents and efferents. The various afferent fibers to neocortex and sources of corticofugal pathways have distinctly layered arrangements (Fig. 19-3). To a first approximation the supragranular layers (i.e., layers I to III) are involved in corticocortical pathways, and the granular and infragranular layers (i.e., layers IV to VI) participate in subcortical pathways. Exceptions to this generalization exist.

Corticocortical axons enter the gray matter from below and ramify extensively in layer II, although some of these also terminate in layers I and III and may issue collaterals in layer VI. These axons arise from other cortical areas, either in the same hemisphere or contralaterally via commissures, chiefly the *corpus callosum*. Thalamocortical axons terminate chiefly in layer IV, frequently with collaterals in layer VI; however, some thalamic cell groups innervate cortex more diffusely, with terminals in many layers, including layer I. The claustrum innervates all layers, but particularly layers IV and VI of many cortical areas. This pathway is predominantly ipsilateral, but a small crossed

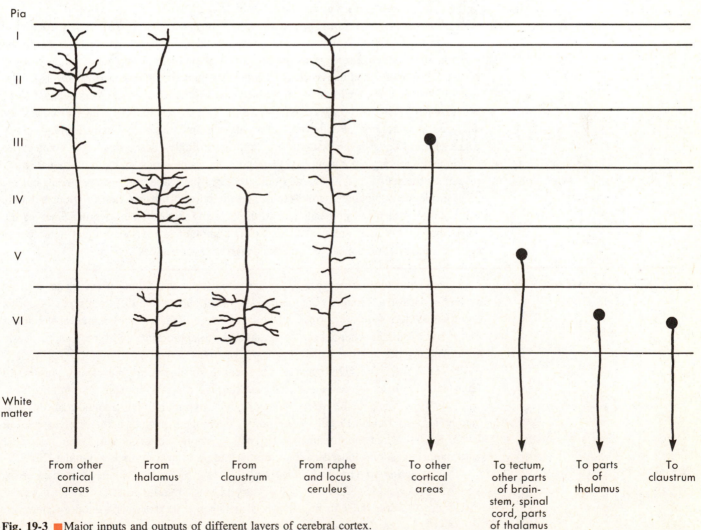

Fig. 19-3 ■ Major inputs and outputs of different layers of cerebral cortex.

component has been demonstrated. Finally, cell groups of the brainstem reticular formation, particularly the *raphe nuclei* and the *locus ceruleus,* project diffusely throughout cortex in all areas and layers. This projection has been identified, by histochemical techniques, as catecholaminergic.

Although many of the afferent fibers to cortex have specific laminar terminations, their target cells are not easy to define. For instance, the lateral geniculate nucleus projects extensively to layers IV and VI of striate cortex, with no appreciable input to layer V. Yet pyramidal cells with somata in layer V typically receive direct geniculate innervation, by virtue of either the apical dendrite that ascends through layer IV or basal dendrites that descend into layer VI.

Cortical efferent fibers also tend to display a layering pattern that depends on the projection target. Corticocortical projections (both ipsilateral and contralateral) originate predominantly with pyramidal cells of layer III, although a few such cells in layers II and V may participate in this pathway. Also, some evidence exists that a sparse projection from supragranular stellate cells also contributes to this pathway. Layer V pyramidal cells are the major source of descending corticofugal pathways, including the corticotectal, corticobulbar, corticospinal, and certain corticothalamic (e.g., to pulvinar) pathways. Layer VI pyramidal cells and fusiform cells innervate certain thalamic targets (e.g., the lateral geniculate nucleus) and the claustrum. Different layer VI cell populations are involved in each pathway. Corticothalamic cells are located throughout the layer; they can be labeled retrogradely by D-aspartate. Corticoclaustral cells are located in the center of layer VI and do not accumulate D-aspartate.

Columnar organization. The functional unit of neocortex appears to be a column of interconnected cells that runs vertically through all the layers. Ample anatomical and physiological evidence for this can be found in visual cortex (Chapter 8), somatosensory cortex (Chapter 9), auditory cortex (Chapter 10), and motor cortex (Chapter 18). This was evident to early anatomists, who noted that cortical cells and fibers seemed to be organized into discrete vertical columns (Fig. 19-2). Presumably, inputs to cortex are analyzed by a column of cells, this information is integrated and analyzed, and appropriate efferent cells in the column discharge accordingly. Also, functional connections within a column are more numerous than those across columns. In many ways neocortex resembles the retina in this regard. Both structures have a laminar arrangement, but the most pronounced functional connections and direction of information flow occur within cell columns organized perpendicular to the layering. Indeed, the layering emerges because the cell columns are stacked next to one another in register, so that homologous cells in each column appear to form the different layers across columns.

■ *Regional variations*

Cytoarchitectonic divisions. A number of distinct variations in the basic layering scheme exist for different cortical areas. Fig. 19-4 shows five different types of cortex and their locations on the surface of the hemisphere. Types *2* to *4* represent variations on the basic laminar scheme because six layers with appropriate cell types can be clearly seen. These types are thus called *homotypic* cortex.

Types *1* and *5* are *heterotypic cortex* because six layers are not clearly present. In fetal life these heterotypical areas also pass through a six-layer stage and are thus neocortical. Type *1* is called *agranular cortex,* since pyramidal cells abound and few stellate cells exist. Layers II and IV are poorly developed, such that the efferent cells in layers III and V are the dominant cell type. Agranular cortex is thus often associated with motor areas, and indeed motor cortex of the precentral gyrus is agranular (Fig. 19-5). Type *5, granular cortex (koniocortex),* has few pyramidal but many stellate cells. The afferent layers II and IV are well developed, and granular cortex is often associated with sensory areas. Indeed, the primary visual, somatosensory, and auditory areas are all granular cortex.

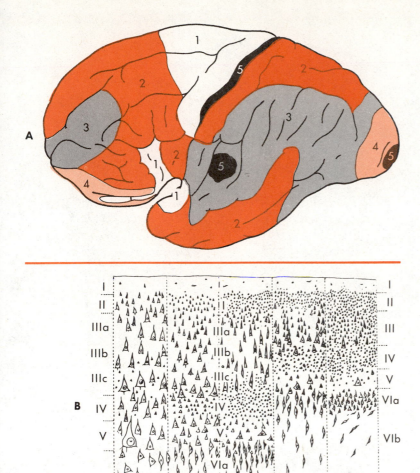

Fig. 19-4 ■ Five major types of cerebral cortex as described by C. von Economo. *1,* Agranular cortex; *2,* frontal type of cortex; *3,* parietal type of cortex; *4,* polar type of cortex; and *5,* granular cortex or koniocortex. **A,** Distribution in the cerebral hemisphere. **B,** Schematic appearance in cross section. (From Kornmüller, A.E., and Janzen, R.: Arch. Psychiatr. Nervenkr. **110:**224, 1939.)

More careful inspection of the different cortical histology indicates that these five cortical types can often be further subdivided. A detailed map that is based on these subtle differences can be produced. Near the turn of the century a number of anatomists produced such *cytoarchitectonic* maps, and the most influential and still widely used is that of Brodmann (Fig. 19-5). Brodmann's map includes roughly 50 discrete areas, the common designations including area 17, which is striate cortex, or VI; areas 1, 2, and 3, which are the primary somatosensory cortex, or SI; areas 41 and 42, which are the primary auditory cortex, or AI; area 4, which is the motor cortex; and area 8, which is the frontal eye field involved in saccadic eye movements. In general, these maps can be correlated well with cortical areas that seem to subserve different functions.

Sensory and motor areas of cortex. Concepts of the functional organization of sensory and motor areas have changed markedly in the past 20 years. Previously it was thought that only areas 17, 18, and 19 were strictly visual, 41 and 42 auditory, and 1, 2, and 3 somatosensory. All cortex in between these areas was designated *association*

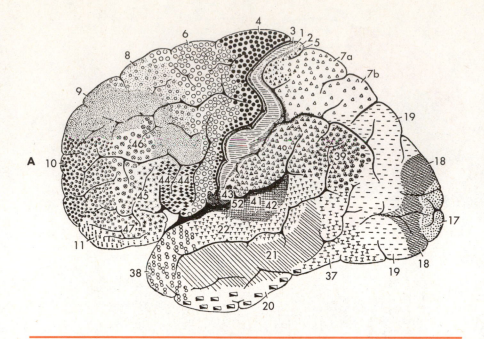

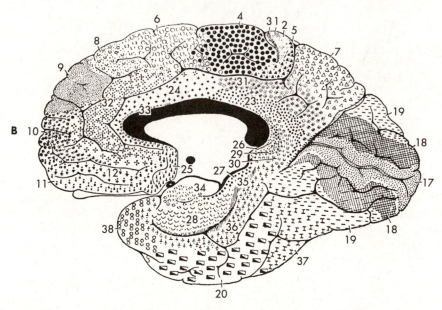

Fig. 19-5 ■ Cytoarchitectonic divisions of human cerebral cortex according to Brodmann. **A,** Lateral surface of left hemisphere. **B,** Medial surface of right hemisphere. (From Crosby, E.C., et al: Correlative anatomy of the nervous system, New York, 1962, Macmillan Publishing Co. Copyright © 1962 by Macmillan Publishing Co., Inc.)

cortex. Association cortex was presumed to be the locus at which all sensory information was analyzed, integrated, and passed on to motor cortex.

Recent studies, however, have made it clear that much or all of what was presumed to be association cortex actually contains multiple, specific sensory representations. For instance, functional mapping of the owl monkey's cortex (Fig. 19-6) shows that there is little room for association cortex. Current thinking suggests that further mapping studies will all but eliminate association cortex and its theoretical consequences from notions of cortical function. Instead, sensory and motor processing seem to involve a bewildering multiplicity of areas that are richly interconnected with each other and with

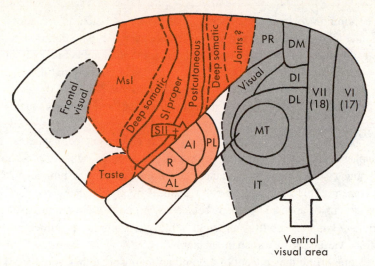

Ventral
visual area

Fig. 19-6 ■ Known sensory areas of cerebral cortex in a New World monkey. After delineation of visual areas *(gray)*, somatosensory areas *(darker color)*, and auditory areas *(lighter color)*, little cortex is unaccounted for. Much of this replaces the outdated notion of "association cortex." Not shown is a medial visual area contiguous with *DM* on the medial surface. *MsI*, Motor-sensory area I; *SI*, primary somatosensory area; *SII +*, other somatosensory areas; *AI*, primary auditory area; *R*, rostral auditory area; *AL*, anterior lateral auditory area; *PL*, posterior lateral auditory area; *VI*, primary visual area; *VII*, secondary visual area; *DM*, dorsal medial visual area; *DI*, dorsal intermediary visual area; *DL*, dorsal lateral visual area; *MT*, middle temporal visual area; and *IT*, inferotemporal visual area. (Redrawn from Merzenich, M.M., and Kaas, J.H. In Sprague, J.M., and Epstein, A.N., editors: Progress in psychobiology and physiological psychology, vol. 9, New York, 1980, Academic Press, Inc.)

the thalamus. Hypotheses as to the significance of these multiple representations are currently limited to speculations. For instance, some of the visual areas are thought to specialize in the analysis of color, movement, stereopsis, or forms. This problem remains an active area of research.

Frontal lobes. The specific sensory and motor areas of cortex are located in the occipital, parietal, and temporal lobes, which are relatively well understood functionally. By contrast, the functional significance of the frontal lobes is poorly grasped. This is an important gap in our knowledge; as one compares cortex among mammals, it is clear that the frontal lobes are larger, proportionally and absolutely, in humans than in any other animal.

Afferent fibers to the frontal lobe arrive directly from other cortical areas, from the thalamus (chiefly the dorsomedial nucleus, or MD), from the locus ceruleus and raphe nuclei, from the amygdala, from the cingulate gyrus, and from the septum. The MD relays further information from some of these areas plus the hypothalamus and olfactory areas. Frontal cortex, in turn, innervates these areas plus the pulvinar and intralaminar thalamic nuclei. Considering some of the areas with which the frontal lobe is extensively connected, such as the amygdala, cingulate gyrus, septum, and hypothalamus, it is thought that the cortical region has important personality and emotional functions.

Behavioral studies of frontal lobe damage in humans and animals generally suggest emotional and intellectual functions of this cortex, but a precise and detailed consensus as to its function and functional subdivisions has not yet emerged. A bilateral frontal lobotomy generally results in the following behavioral changes: although simple intelligence tests reveal no change, the ability to deal with complex problems is reduced; aggressive behavior disappears; the ability to initiate behavior is reduced; and the patient or animal is easily distracted and has trouble concentrating. Unilateral lesions generally produce little or none of these changes. Because of the reduction of aggressive behavior, it was once common to treat certain emotional disorders, such as schizophrenia, with bilateral frontal lobotomies. This form of psychosurgery is highly controversial, since it often fails to produce the desired effect and the resultant intellectual, personality, and emotional changes are sometimes considered worse than the original disorders. Frontal lobotomies are now rarely performed, particularly because powerful drugs are available to achieve more safely, predictably, and reversibly most of the desired personality changes.

■ *Cerebral dominance and language*

Human cortex is divided into two more or less equivalent hemispheres. However, these hemispheres are not strict mirror images of one another. Most of the known hemispheric asymmetry is a result of language functions, which are typically related to a single hemisphere. This language function defines the *dominant hemisphere*. With the possible exceptions of apes and perhaps cetaceans, no nonhuman mammals are believed to have obvious language ability, although subtle hemispheric asymmetries have been described for a variety of mammalian species.

Evidence supporting the concept of a dominant hemisphere has come from processes that affect one hemisphere and interfere with language, such as unilateral trauma (e.g., stroke, tumor, physical injury) or unilateral, temporary anesthesia. Such anesthesia can be induced by injection into one carotid artery of a short-acting anesthetic, such as amobarbital (Amytal). In the vast majority of humans for whom cerebral dominance was determined, the left hemisphere was dominant. In the remainder, either the right hemisphere was dominant or no dominance was seen because each hemisphere subserved language.

Although language function defines cerebral dominance, there exists a strong correlation with handedness. Left hemisphere dominance nearly always results in right-handedness. Right dominance less clearly results in left-handedness because substantial numbers of right-handed or ambidextrous individuals are found with such dominance. Handedness is another example of interhemispheric asymmetry, since each hand is represented predominantly in the contralateral hemisphere. Other examples of functional interhemispheric differences are considered in the discussion of the corpus callosum.

In addition to these examples of functional differences between the hemispheres, there also exist striking anatomical differences. For instance, careful measurements of the *planum temporale,* which is located at the posterosuperior surface of the temporal lobe within the region associated with language function (Fig. 19-7), reveal interhemispheric differences. This region is consistently larger in the dominant hemisphere. This difference can also be found in infants, which suggests a predilection for a specified dominant hemisphere before the development of language. This is interesting, because the effects of a unilateral lesion to the language areas in an adult differ dramatically from these effects in an infant. In an adult such a lesion produces a permanent interference with language ability known as *aphasia.* In infancy such a lesion usually does not result in persistent aphasia, presumably because the immature, nondominant hemisphere retains the ability to develop language capacity. The greater flexibility, plasticity, and adaptability of the young brain is reconsidered in Chapter 21.

From cortical lesions that produce aphasia in humans it is possible to delineate those areas involved in language. Results of electrical stimulation of the cortex in humans undergoing neurosurgical procedures tend to confirm the location and extent of these areas. However, it must be recognized that these methods are imprecise and difficult to control. For instance, aphasic patients suffer from a wide range of cortical lesions that typically produce many deficits in addition to aphasia, and the precise extent and location of the lesions are difficult to assess. For these reasons, the following discussion of cortical language areas and aphasia is tentative.

Fig. 19-7 shows the two major language areas of the dominant hemisphere. *Wernicke's area* is located in the posterior portion of the superior temporal gyrus, near the

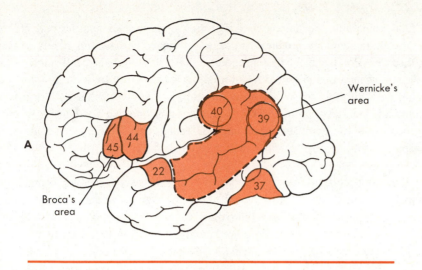

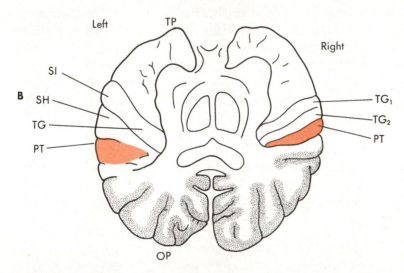

Fig. 19-7 ■ Language areas of cerebral cortex. **A,** Regions of left hemisphere associated with language *(color)*. The numbers represent Brodmann areas (see Fig. 19-5). Wernicke's area is approximately delineated by the dashed line, and Broca's area is roughly coextensive with areas 44 and 45. **B,** Horizontal section through the brain showing interhemispheric asymmetry in the region of the upper temporal lobe. Note the larger planum temporale *(PT)* on the left compared with that on the right. This region is associated with Wernicke's area, which is larger in the dominant, left hemisphere. *TP,* Temporal pole; *OP,* occipital pole; *TG,* transverse gyrus (including primary auditory area); two transverse gyri *(TG₁, TG₂)* are commonly found in the right hemisphere; *SI,* sulcus intermedius; *SH,* sulcus of Heschl. (**A** redrawn from Williams, P.L., and Warwick, R.: Functional neuroanatomy of man, Philadelphia, 1975, W.B. Saunders Co.; **B** redrawn from Geschwind, N. In Mountcastle, V.B., editor: Medical physiology, ed. 14, St. Louis, 1981, The C.V. Mosby Co.)

auditory cortex. The anterior region, *Broca's area*, lies just anterior to the representation of the face area in the motor cortex. Lesions of either area produce language difficulties. However, lesions of Wernicke's area tend to cause more difficulty with understanding written or spoken language than with coherent speech or writing. Conversely, lesions of Broca's area tend to interfere more with speech and writing than with comprehension of written or spoken language. Wernicke's area is thus more or less associated with *sensory aphasia* (difficulty in understanding language) and Broca's area with *motor aphasia* (difficulty in producing coherent language). However, rarely, if ever, does a lesion produce a purely sensory or purely motor aphasia.

A final point must be emphasized regarding aphasia. An aphasic patient may be unable to comprehend or produce language yet typically have no purely sensory or motor disorder. For instance, despite motor aphasia, nonverbal vocalization and humming of complicated melodies is often normal, as is the ability to draw complex pictures or diagrams. Similarly, sensory aphasia may appear with no other auditory or visual symptoms; the identification of complex sounds or tunes and comprehension of complex visual scenes may be completely normal. Thus aphasia is a deficit in language ability quite separate from sensory or motor performance.

■ The electroencephalogram

An important advance in nervous system analysis was the development of electrophysiological methods that permitted the study of ongoing neural activity. The early application of these methods primarily involved recording the summated activity of large populations of neurons. Since neurons are heterogeneous both structurally and functionally, it is difficult to know precisely what is generating such summated potentials. Consequently, their usefulness for the detailed analysis of the organization of the nervous system has been limited, and these methods have been largely replaced in the contemporary experimental literature by recordings of the activity of single nerve cells. However, in studying the human brain, either clinically or experimentally, we are confined to the use of noninvasive techniques in most instances. Thus the use of surface electrodes to record the activity of populations of neurons has considerable value.

By placing a set of disk electrodes at various standardized locations on the scalp, the activity of underlying neural tissue can be recorded as waves that vary in frequency from 1 to over 30 cycles per second (hertz). The record of these waveforms, taken under highly standardized conditions, is called the *electroencephalogram* (EEG), and it is a conventional tool in clinical neurology, particularly in dealing with epileptic patients.

In a normal individual the electroencephalogram will vary as a function of a number

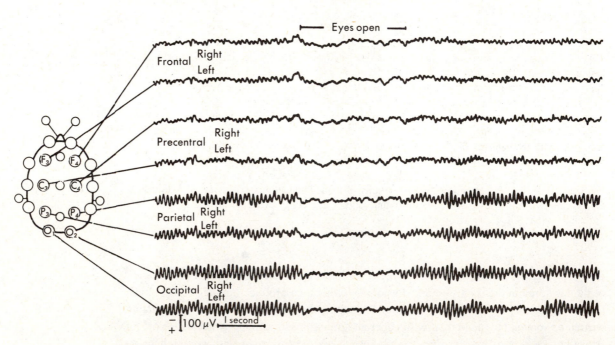

Fig. 19-8 ■ Normal EEG of a resting, awake human. Simultaneous eight-channel unipolar recording from the sites on the skull that are indicated at left. An electrode was attached to each earlobe, and the two together constituted the indifferent electrode. Opening of the eyes *(middle of record)* blocks the alpha rhythm. (From Schmidt, R.F., editor: Fundamentals of neurophysiology, ed. 2, New York, 1978, Springer-Verlag.)

of factors, including the recording site, age, and the state of wakefulness. In an awake relaxed individual with the eyes closed, one observes large slow waves, averaging 10 Hz in frequency, that are most pronounced over the occipital lobe (Fig. 19-8). This *alpha rhythm* represents synchronized neuronal activity in the underlying cortex. During alpha activity any situational change that makes the individual more alert, such as opening the eyes, will block the alpha rhythm and produce less synchronized activity of lower amplitude and higher frequency. This is characterized by *beta waves,* which have an average frequency of 20 Hz. Such waves are characteristic of an alert or aroused state. These potentials recorded in the EEG reflect the activity, both postsynaptic potentials and action potentials, of many hundreds of thousands of neurons in the underlying cortex.

■ *Sleep stages*

The EEG has been particularly useful in the study of sleep, since the different stages of sleep are reflected in different EEG waveforms. Five stages of sleep can be distinguished in the EEG. The first four involve a progressive slowing of the EEG with an increase in wave amplitude (Fig. 19-9). During the early stages of sleep the alpha rhythm diminishes, and low-amplitude *theta waves* appear, with a frequency of 4 to 7 Hz. As sleep deepens, further EEG slowing occurs with the appearance of large *delta waves,* which average 3 Hz in frequency. Early in the appearance of delta waves there are bursts of activity at 12 to 14 Hz—the sleep spindles. Delta waves become more prominent in moderately deep sleep, and K complexes appear. In the deepest stage of this slow-wave sleep the EEG is composed almost entirely of delta waves. The transition through the four stages of slow-wave sleep is smooth, and the ease with which an individual can be wakened varies inversely with the stage.

During the course of sleep a fifth stage occurs. This consists of epochs of EEG desynchronization, accompanied by sympathetic activation and a loss of muscle tone (Fig. 19-9). A prominent feature of this stage are phasic bursts of rapid eye movement

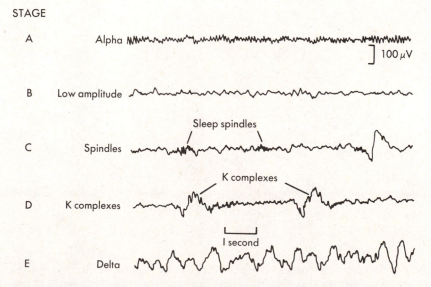

Fig. 19-9 ■ Classification of the stages of human sleep on the basis of the EEG. *Stage A,* Relaxed waking; alpha rhythm prevails. *Stage B,* Falling asleep; alpha rhythm is diminished and shallow theta waves appear. *Stage C,* Light sleep; further decrease in frequency until delta waves appear; occasional "sleep spindles" occur (groups of waves at 12 to 15 Hz). *Stage D,* moderately deep sleep; delta waves and K complexes. *Stage E,* Deep sleep; the trace is composed almost entirely of large, slow delta waves. The REM stage corresponds approximately to stage B of the EEG. There are smooth transitions between the different stages. (It is important to appreciate that other classifications exist.) (From Schmidt, R.F., editor: Fundamentals of neurophysiology, ed. 2, New York, 1978, Springer-Verlag.)

(REM); REM was one of the first observations in the study of this stage of sleep. Of the different stages of sleep, it is most difficult to wake an individual from REM sleep, although spontaneous waking occurs most often in this stage. Moreover, waking an individual during REM sleep leads to the highest frequency of dream recall. Since the EEG is desynchronized in REM sleep and thus resembles the alert state, it is sometimes referred to as paradoxical sleep.

The various theories of the neural mechanisms underlying the sleep-wake cycle concern the involvement of certain neural circuits and different neurotransmitters. However, no single theory has yet gained overwhelming support.

■ Abnormal EEGs

Various diseases of the nervous system, such as tumors, can produce an abnormal EEG. However, most such diseases are more effectively diagnosed by other procedures. Where the EEG has remained most useful is in the diagnosis of epilepsy and sleep disorders. Epilepsy, one of the most common neurological disorders, involves the synchronous discharge of large aggregates of neurons. Since cortical neurons are frequently involved, the EEG is an obvious and effective instrument in diagnosing epileptiform activity. With regard to sleep disorders the EEG is particularly effective, since the stages of sleep are defined on the basis of EEG activity.

■ Corpus callosum

Until about 1950 the corpus callosum represented one of the great mysteries in neuroscience. Its great size and cortical distribution (Fig. 19-10) suggested an important functional role. Yet studies of humans or experimental animals with transection of this commissure consistently failed to reveal any behavioral deficit. As recently as 1951 Lashley jocularly summed up the frustration in these early studies by suggesting that the corpus callosum had only a mechanical function to prevent the hemispheres from sagging. It was left to Sperry and his co-workers in the early 1950s to discover the role of this commissure in interhemispheric integration, and for this work Sperry shared the 1981 Nobel Prize in Medicine.

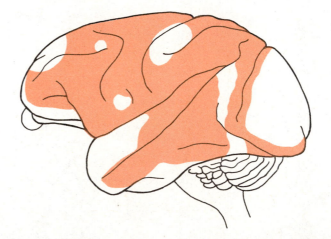

Fig. 19-10 ■ Distribution of fibers of corpus callosum in the cerebral hemisphere of a macaque monkey. The zones of distribution are shown in color. (Redrawn from Ettlinger, E.G., de Reuck, A.V.S., and Porter, R., editors: Functions of the corpus callosum, CIBA Foundation Study Group no. 20, London, 1965, J. & A. Churchill, Ltd.)

■ Interhemispheric transfer

Animal experiments. Fig. 19-11 outlines the basic experiment that uncovered the role of the corpus callosum in interhemispheric transfer. An experimental animal (e.g., a cat or monkey) can be forced to learn visual discrimination through one eye while the

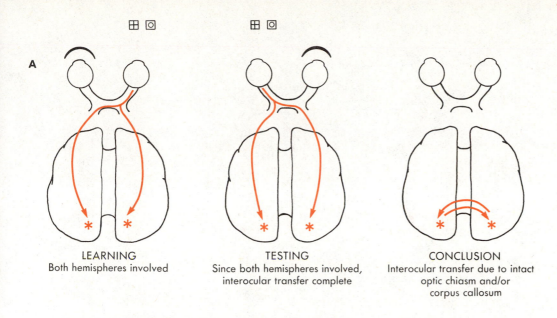

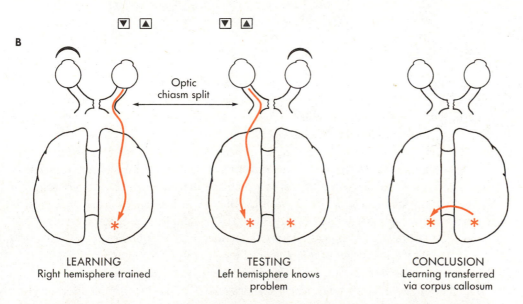

Fig. 19-11 ■ Experiments demonstrating role of corpus callosum in interhemispheric transfer of visual information. In all cases a visual discrimination (e.g., whether a cross or circle indicates the correct door to obtain a food reward) is learned through only one eye. After learning is achieved, the animal is retested through only the naive eye. **A,** Intact animal. Because of the intact optic chiasm, the original learning involves both hemispheres. During testing the naive eye can access both hemispheres that "know" the task, and thus the animal performs as well as with the former eye. Interocular transfer is complete because of the original bilateral learning and/or interhemispheric transfer through the corpus callosum. **B,** Animal with transection of optic chiasm. Now the original learning is limited to the hemisphere ipsilateral to the open eye. Nonetheless, testing with the naive eye shows complete interocular transfer as if that eye knew the task all along. This occurs because information is transferred through the corpus callosum from the originally involved hemisphere to that related to the naive eye. *Continued.*

other eye, which is occluded, never views the test stimuli. For instance, the animal may learn that it can obtain food by pushing a door with a cross on it, but pushing a door with a circle leads to no reward. Once the animal has learned this task through the experienced eye and then is tested monocularly through the naive eye, it continues to perform as if the naive eye had participated in the training (Fig. 19-11, *A*). This is

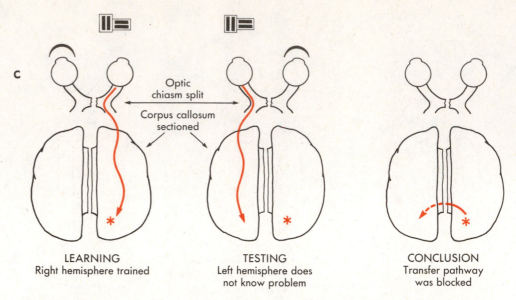

Fig. 19-11, cont'd ■ **C,** Animal with transection of optic chiasm and corpus callosum. Now interhemispheric transfer is blocked, so no interocular transfer occurs. The naive eye must relearn the task as if it were completely new to the animal.

called *interocular transfer* (although the transfer of information occurs between the hemispheres and not between the eyes). Because of the partial decussation of the optic chiasm, which results in both retinas projecting to each hemisphere, such transfer is not surprising. However, if a cat or monkey (after midsagittal transection of the optic chiasm) is put through a training regimen similar to that just described, interocular transfer remains excellent even though each eye now projects only to the ipsilateral hemisphere (Fig. 19-11, *B*). Finally, if an animal has both the optic chiasm and corpus callosum transected, no interocular transfer is seen and the naive eye must relearn the discrimination from the very beginning (Fig. 19-11, *C*). This simple experiment elegantly illustrates the function of the corpus callosum in interhemispheric transfer of information.

Interhemispheric transfer of information is by no means limited to the visual domain. For instance, if a normal cat is taught to discriminate between two somatosensory stimuli applied to one limb, it immediately demonstrates familiarity with the discrimination when tested on a contralateral limb. If the corpus callosum is transected, no such transfer is evident, and the cat must relearn the discrimination for the contralateral limb. Because each limb is represented predominantly in the contralateral hemisphere, information available to contralateral limbs can be compared only in the presence of a corpus callosum.

Two details can be added to this simplified view of corpus callosum function, based on studies of experimental animals. First, information that can be processed subcortically can be transferred between the hemispheres subcortically without a corpus callosum. For instance, a cat with transections of the optic chiasm and corpus callosum exhibits no interocular transfer of pattern discriminations. However, the animal does demonstrate excellent transfer of discriminations based on crude brightness differences. Patterns are processed in cortex, but brightness can be appreciated without cortex. If, however, midbrain commissures (the *habenular, inferior collicular, superior collicular,* and *posterior commissures)* are transected along with the optic chiasm and corpus callosum, interocular transfer of brightness discrimination is blocked. Second, under some conditions the *anterior commissure* seems capable of subserving limited interhemispheric transfer in the absence of a corpus callosum.

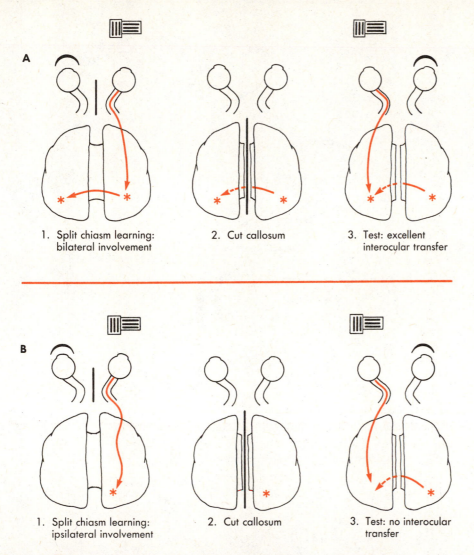

A

1. Split chiasm learning: bilateral involvement

2. Cut callosum

3. Test: excellent interocular transfer

B

1. Split chiasm learning: ipsilateral involvement

2. Cut callosum

3. Test: no interocular transfer

Fig. 19-12 ■ Experiments demonstrating different role of corpus callosum during learning tasks in cats and monkeys with optic chiasm transections. Because of the transections, the visual information is primarily limited to one hemisphere during monocular learning. **A,** Cat. With an intact corpus callosum the visual information is immediately transferred to the other hemisphere. If the corpus callosum is transected between learning and testing of the naive eye, the naive eye thus shows evidence of complete interocular transfer. **B,** Monkey. The original learning is limited to the ipsilateral hemisphere. Thus no interocular transfer is evident when the corpus callosum is transected between learning and testing of the naive eye.

Finally, one can ask whether the corpus callosum immediately involves both hemispheres in newly acquired memories or whether one hemisphere forms the memory and the other can reach it via the commissure. The answer to this seems to depend on the species involved. Cats seem to form memories bilaterally and monkeys do not. For instance, suppose a cat or monkey has only its optic chiasm sectioned and learns a visual task monocularly. Following this, and before the task is presented monocularly to the other eye, the corpus callosum is sectioned. The cat exhibits familiarity with the task when it uses the naive eye, but the monkey does not. Before the callosal section, the cat had already formed the memory bilaterally, but the monkey had not (Fig. 19-12).

Human studies. Sperry and his colleagues expanded their studies to humans and were able to show that the same general principles which apply to animals apply as well

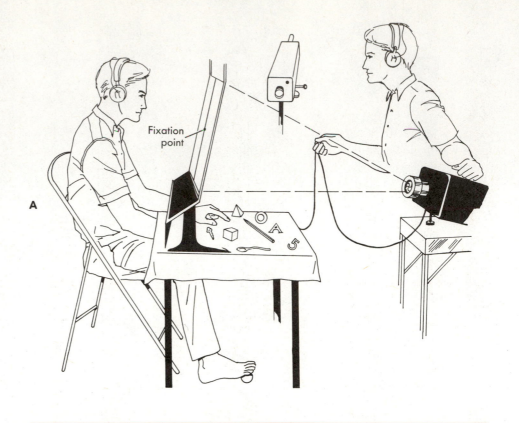

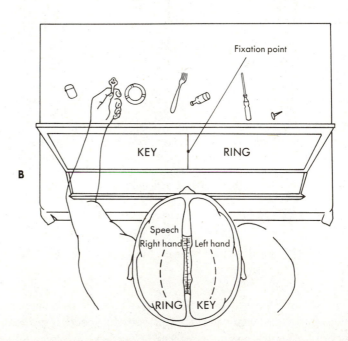

Fig. 19-13 ■ Testing methods used in patients with corpus callosum transection. **A,** Patient looks at fixation point on a rear projection screen while pictures are projected to either side of the point. Out of view are objects that can be palpated for matching to the projected images. **B,** Example of test. A key is selected by the left hand because the picture of a key is represented only in the appropriate (right) hemisphere; the left hand, without an intact corpus callosum, is ignorant of the picture of a ring. Verbally, the patient would deny seeing the key, because the right hemisphere is verbally uncommunicative. Instead, despite reaching with the left hand for a key, the patient would report seeing only a ring. (Redrawn from Sperry, R.W. In Schmitt, F.O., and Worden, F.G., editors: The neurosciences third study program, Cambridge, Mass., 1974, M.I.T. Press. Copyright © 1974 by the M.I.T. Press.)

to the role of the human corpus callosum in interhemispheric transfer. Extensive testing was carried out in human patients with therapeutic transection of this commissure.*

Because these patients had an intact optic chiasm, special procedures were needed to ensure that visual information was provided to only one hemisphere (Fig. 19-13). The subject faced a projection screen and fixated on a small cross in the center. A photograph of a common object was projected to one side of this fixation point. It thus fell within one hemifield and was represented only in the contralateral hemisphere. The subject was then instructed to reach under the screen with one hand to palpate a number of unseen objects and pick out the object projected on the screen. If the hand was ipsilateral to the hemifield exposed, the correct object was consistently chosen, but with the other hand, objects were chosen at random. (A normal subject successfully locates the correct object under these conditions equally well with either hand.) The explanation for this result in the callosum-sectioned patients is straightforward. If the visual and somesthetic information was contained within the same hemisphere (as occurred if the left hand explored for objects after its picture was projected to the left hemifield), the information was successfully integrated. If contained within opposite hemispheres (as occurred if the right hand was used after projection of the picture to the left hemifield), the information could not be integrated without a corpus callosum. Finally, if a separate picture was projected to each hemifield and one hand was used, an object appropriate for that hand was chosen; if two hands were used simultaneously, an appropriate and different object was chosen by each hand.

Because language abilities are typically limited to one dominant hemisphere, interesting and predictable language deficits are seen in patients with corpus callosum transection. In each of Sperry's patients the left hemisphere was dominant. If one of these patients was commanded verbally to raise his right hand, this was readily accomplished because the motor control for that hand is located primarily in the dominant and thus communicative hemisphere. The patient could not, however, raise his left hand on such a simple language command. Likewise, the patient could verbally describe details of somesthetic stimuli applied to the right side of his body, but not to the left side.

Visual testing most dramatically underscored the language asymmetries in the hemispheres of these patients. When the testing condition of Fig. 19-13 was used, the patient could correctly name any object seen in his right hemifield and palpated by his right hand. When the left hemifield was stimulated, although the patient correctly chose the object with his left hand, he verbally denied having seen or palpated anything. Most interestingly, if a different object was projected to each hemifield, such as a comb to the left and a ball to the right, and the patient was allowed to use only his left hand to locate an object, that hand unerringly located the comb. When asked what he saw, the patient consistently claimed to have seen only a ball and denied viewing a comb.

Because language abilities are typically limited to one hemisphere, the question has often arisen as to what special function is performed by areas of the other hemisphere located in the mirror image of the language areas. The callosum-sectioned patients offered a unique opportunity to explore this problem. When asked to solve complicated spatial tasks, such as three-dimensional puzzles, these subjects performed much better with the left hand than with the right. This suggests that the right, nondominant hemisphere specializes for spatial tasks in much the same sense that the other hemisphere specializes for verbal ones. With an intact corpus callosum, of course, the hands exhibit

■ *Integration of different hemispheric functions*

---

*In the 1950s several patients were discovered to have an unusual form of epilepsy. They had an epileptiform focus in one hemisphere that initially produced mild seizures, but when the epileptiform activity was transferred via the corpus callosum to the other hemisphere, serious *grand mal* seizures resulted. To prevent this, each patient had his or her corpus callosum transected, and epileptic symptoms were greatly reduced. Such a procedure is still occasionally performed to treat certain forms of epilepsy.

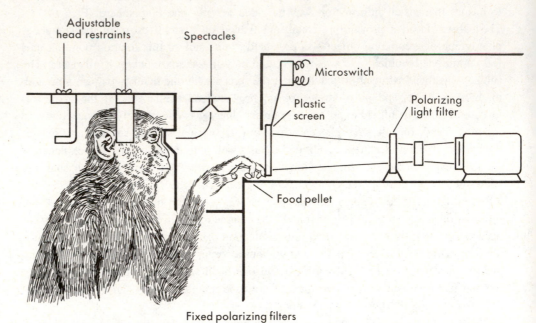

Adjustable head restraints

Spectacles

Microswitch

Plastic screen

Polarizing light filter

Food pellet

Fixed polarizing filters

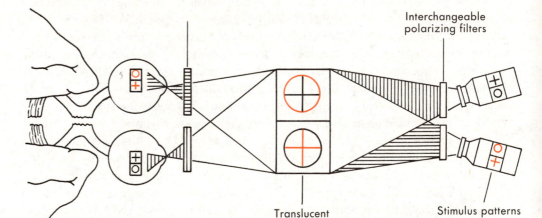

Interchangeable polarizing filters

Stimulus patterns in projectors

Translucent screens

Fig. 19-14 ■ Experimental set-up to demonstrate role of corpus callosum in interhemispheric coordination of behavior. *Above,* Monkey trained to push one of two doors to obtain food reward. The correct door is signaled by visual pattern projected onto it. As the food reward is randomly switched from door to door, the same visual pattern switches with it. *Below,* Optical arrangement to train simultaneously each eye on the opposite task. By the appropriate use of polarizing filters, one eye sees a cross on one door and a circle on the other while the second eye sees the reverse relationship. Thus one eye learns to associate the circle with the correct choice while the other eye learns to attribute the correct choice to the cross. An intact monkey or a monkey with a transected optic chiasm but intact corpus callosum cannot deal with this rivalry, shows signs of frustration, and soon abandons the task. In contrast, a monkey with transections of both structures blissfully performs the task, presumably because no rivalry is evident to it. (Redrawn from Sperry, R.W.: Sci. Am. **210:**42, 1964.)

no such consistent difference in spatial abilities. Perhaps neural substrates of many higher functions in addition to verbal and spatial abilities exhibit interhemispheric asymmetries in humans.

A consequence of interhemispheric asymmetries is that the same function need not require two cortical areas. An intact corpus callosum, then, obviates the need for unnecessary redundancy. For instance, the experiment outlined in Fig. 19-12 shows that cats engage bilateral cortical areas during a visual learning task that in monkeys involves only one hemisphere. Monkeys do not suffer from the consequences of this lateralization as long as the corpus callosum is intact. In a sense the corpus callosum can be viewed as a necessary prerequisite for the release of cortical areas from bilateral redundancy. This release might be the first step in an evolutionary process that led to the development of higher functions such as language.

Not only does the corpus callosum serve as a vital means of communication between the hemispheres, but it also ensures that the hemispheres act in a coordinated fashion in the promulgation of behavior. That is, although we each have two hemispheres capable of considerable neural processing, we normally act as if we have one mind instead of two. Two examples serve to illustrate this point.

Fig. 19-14 illustrates the first example by outlining an experiment carried out in monkeys. One group of monkeys was normal and the other underwent a transection of the optic chiasm and corpus callosum. The monkeys were trained to discriminate between simple targets (a circle and a cross) to select the appropriate panel for an appetitive reward. By a clever optical design involving polarizing filters one eye was exposed to stimuli that indicated the cross was the correct stimulus, and the other eye simultaneously learned that the circle was correct. In intact monkeys, for which this confusing information reached both hemispheres, the situation was most frustrating, and the animals soon ceased working on the problem. The monkeys with optic chiasm and corpus callosum transections evidenced no confusion or frustration. They quickly learned the task and contentedly continued to select the correct panel. However, one hemisphere learned to associate ''cross'' with this panel while the other simultaneously learned to associate ''circle'' with it. The two hemispheres, without a corpus callosum, are capable of learning diametrically opposite tasks. An intact corpus callosum prevents this.

The second example comes from casual observations of patients with therapeutic transection of the corpus callosum. They frequently displayed evidence of poor coordination between the hemispheres in everyday tasks involving the hands. For instance, while one of their hands buttoned a shirt during dressing, the other hand not only failed to cooperate, it even attempted to unbutton the shirt at the same time. Such dramatic examples of lack of cooperation occurred randomly and intermittently, but they underscore the function of the corpus callosum in ensuring interhemispheric coordination.

Sperry eloquently summed up the function of the corpus callosum with respect to the callosum-sectioned patients as follows*:

> Everything we have seen so far indicates that the surgery has left these people with two separate minds, that is, two separate spheres of consciousness. What is experienced in the right hemisphere seems to be entirely outside the realm of awareness of the left hemisphere. This mental division has been demonstrated in regard to perception, cognition, volition, learning, and memory. One of the hemispheres, the left, dominant or major hemisphere, has speech and is normally talkative and conversant. The other, the minor hemisphere, however, is mute and dumb, being able to express itself only through nonverbal reactions.

*From Sperry, R.W.: Brain bisection and mechanisms of consciousness. In Eccles, J.C., editor: Brain and conscious experience, New York, 1966, Springer-Verlag, p. 299.

■ *Allocortex*

Allocortex represents only about 10% of human cortical volume. It can be distinguished from neocortex largely on the basis of its layering: typically no more than three layers can be distinguished in allocortex at any stage of ontogeny as opposed to the six-layered neocortex. Because of the relatively small extent of allocortex and because its significance and functional organization are not well understood, our description of these areas is brief. As noted, allocortex includes both archicortex and paleocortex.

■ *Archicortex: hippocampal formation*

The archicortex is equivalent to a group of structures known as the *hippocampal formation*. This is located in the medial and ventral portion of each hemisphere (Fig. 19-15). The hippocampal formation consists of the *dentate gyrus, hippocampus,* and the *subiculum* (sometimes divided into the *prosubiculum,* the *subiculum,* the *presubiculum,* and the *parasubiculum*) (Figs. 19-16 and 19-17).

Subiculum. The subiculum lies in the *parahippocampal gyrus* adjacent to the *hippocampal sulcus* and merges into the neocortex of the parahippocampal gyrus near the *collateral sulcus.* The *entorhinal area* (Brodmann's area 28), which is located adjacent to the subiculum anteriorly, is thought by some to be a part of paleocortex. Also, because the layering patterns in the subiculum are more variable than those of the re-

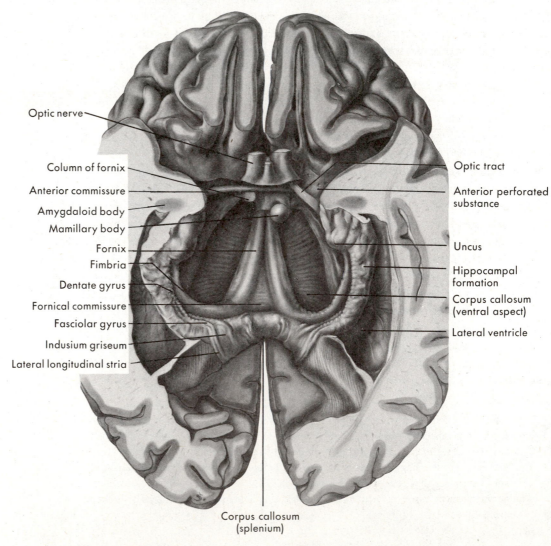

Fig. 19-15 ■ Dissection of base of the brain to show location of hippocampal formation and related structures. (From Mettler, F.A.: Neuroanatomy, St. Louis, 1948, The C.V. Mosby Co.)

maining hippocampal formation, many authorities consider the subiculum to be transitional between allocortex and neocortex.

Hippocampus. The hippocampus derives its name from its resemblance, in coronal sections, to a sea horse. It also resembles a ram's horn and thus is often called *cornu ammonis* or *Ammon's horn* (Ammon was an ancient Egyptian god with a ram's head). Thus separate regions of the hippocampus are designated CA1 through CA4.

The hippocampus bulges into the inferior horn of the lateral ventricle. The *alveus* is a subependymal sheath of afferent and efferent fibers, and the latter collect at the *fimbria* and enter the *fornix*. The three main layers of the hippocampus in a direction from the brain surface to the alveus are the *molecular layer,* which is a neuropil similar to layer I of neocortex; the *pyramidal cell layer,* which is a collection of pyramidal cells oriented with their bases toward the alveus; and the *polymorphic cell layer,* which is similar in appearance to layer VI of neocortex.

Dentate gyrus. The dentate gyrus has a trilayered appearance similar to that of the hippocampus, except that instead of the pyramidal cell layer there is the *granule cell layer* of small cells. The dentate gyrus is C shaped, with the open side directed toward the fimbria, whereas the other side forms the dorsal surface of the hippocampal sulcus. The *molecular layer* is located nearest this sulcus; the *polymorphic cell layer* is located nearest the fimbria.

Hippocampal pathways. Fig. 19-17, *B,* summarizes many of the connections of the hippocampal formation. Inputs derive largely from the parahippocampal gyrus beyond the subiculum, and these derive largely from entorhinal cortex. These inputs enter the hippocampus via either the *perforant path* or the *alvear path.* Other afferent pathways reach all parts of the hippocampal formation via the fornix and arise in the *cingulate gyrus,* the *septal nucleus,* the *indusium griseum,* and the contralateral *hippocampal formation* (via a commissure in the fornix). Because the fornix contains predominantly efferent fibers from the hippocampus, relatively few afferent fibers reach the hippocam-

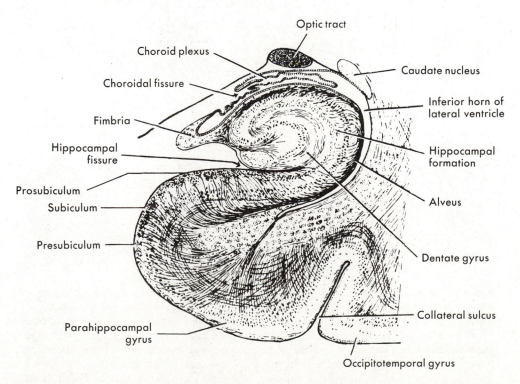

Fig. 19-16 ■ Cross section through the human hippocampus. (From Carpenter, M.B.: Human neuroanatomy, ed. 7, Baltimore; © 1976, The Williams & Wilkins Co.)

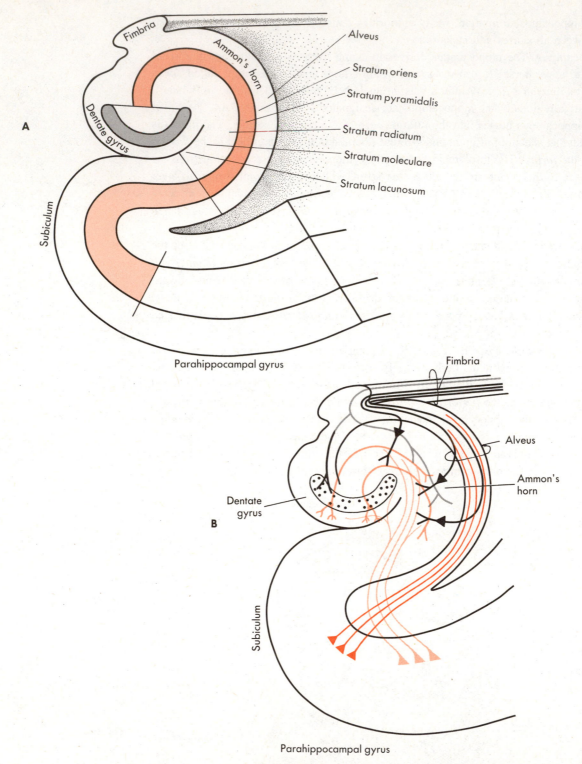

Fig. 19-17 ■ A, Major regions of hippocampus. Cell bodies of the dentate gyrus are in gray. The prominent pyramidal cell bodies of Ammon's horn are in darker color, and this layer merges gradually with neurons of the subiculum *(lighter color)*. Approximate locations of other layers are indicated. **B,** Some connections of the hippocampus. Cells of the dentate gyrus *(medium color)* form the mossy fibers of the hippocampus and innervate the pyramidal cells *(black)*. Pyramidal cell axons course along the alveus and enter the fimbria. Afferent fibers (gray) also enter from the fimbria and innervate the dentate gyrus and/or Ammon's horn. Inputs from the parahippocampal gyrus arrive via the alvear path (dark color) or perforant path (light color). (Redrawn from Williams, P.L., and Warwick, R.: Functional neuroanatomy of man, Philadelphia, 1975, W.B. Saunders Co.)

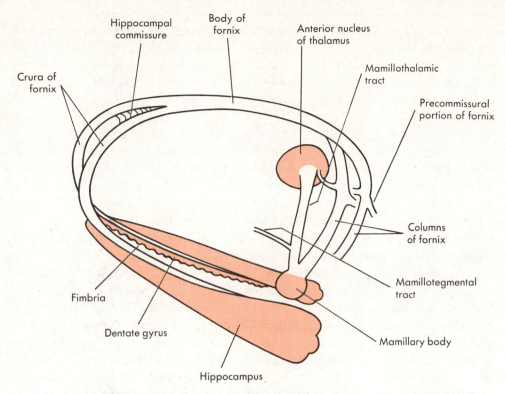

Fig. 19-18 ■ The fornix and related pathways (Redrawn from Barr, M.: The human nervous system: an anatomic viewpoint, ed. 3, New York, 1979, Harper & Row, Publishers.)

pus via this route. Most functional input arrives indirectly via the perforant and alvear paths. In addition, a major intrinsic pathway arises from granule cells of the dentate gyrus, and it terminates on pyramidal cells of the hippocampus.

Nearly all efferent fibers from the hippocampal formation are axons of pyramidal cells in the hippocampus, although some efferent fibers arise from polymorphic cells and from the subiculum. These fibers course along the alveus, collect at the fimbria, and enter the *fornix* (Fig. 19-18) at the posterior tip of the hippocampal formation adjacent to the *splenium of the corpus callosum*. These fibers form the *crus* of the fornix on each side, and they pass over the thalamus, where the two crura merge to form the body of the fornix. Some fibers enter the contralateral crus via a commissure. The body then divides into two ventrally directed *columns* of fibers that pass through the hypothalamus. Some fibers exit the columns in the *precommissural portion* (relative to the anterior commissure), and the remaining columns of fibers constitute the *postcommissural portion*. Axons of this latter portion innervate the *mamillary body*, the *anterior thalamic nucleus*, and the *periaqueductal gray* of the midbrain. Precommissural fibers innervate parts of the *hypothalamus* and the *septal nucleus*, as well as the mamillary body and anterior thalamic nucleus.

Functional considerations. Although the hippocampal formation has received considerable experimental attention, its functional role is not clear. Given its widespread, diffuse connections, it seems likely that no single function can be attributed to it, but rather it participates in many complex behaviors. Among the roles suggested for the hippocampal formation are memory, many emotional responses, and spatial mapping of the environment. However, it should be noted that birds have only a rudimentary hippocampus, and none of the functions suggested for this structure seem prominently absent from the behavioral repertoire of birds. It is at least clear that the hippocampal formation is not essential to olfaction, as was once thought.

■ *Paleocortex: rhinencephalon*

There is no precise knowledge as to which structures form mammalian paleocortex, but one common classification considers the paleocortex and rhinencephalon to be equivalent. In other words, those structures identified with olfaction, including the *olfactory bulb, tract, tubercle,* and *striae,* the *anterior olfactory nucleus, parts of the amygdala,* and parts of the *pyriform cortex,* form the paleocortex (Chapter 12).

■   ■   ■

The most prominent part of the human brain is the cerebral cortex. The vast majority of cortex is the six-layered neocortex, which has shown a pronounced enlargement through mammalian evolution. The layering pattern corresponds to the input/output relationships of cortex, but the functional unit of cortex appears to be a vertical column of interconnected cells. Upward of 50 separate cortical areas have been distinguished based on histological and physiological data. These include multiple representations of various sensory modalities as well as multiple motor areas.

In addition to sensorimotor integration, the cortex plays an obvious role, which is difficult to analyze, in such higher functions as intelligence, personality, spatial relationships, and language. Language, spatial perception, and probably other functions are asymmetrically distributed in the two hemispheres of most people. Typically, the dominant hemisphere (i.e., where language is represented) is the left, and spatial abilities are emphasized in the nondominant hemisphere. The corpus callosum serves to transfer and integrate information between the hemispheres so that processing in unilaterally specialized areas (e.g., language areas) is available to both hemispheres.

In contrast to neocortex, the much older archicortex and paleocortex do not possess six layers. Where layering is clear, usually only three layers can be discerned. Archicortex is represented by the hippocampal formation, and its functional significance is obscure. Paleocortex is represented by the rhinencephalon and has an obvious role in processing of olfactory stimuli.

■ *Bibliography*

Journal articles

Amaral, D.G., and Cowan, W.M.: Subcortical afferents to the hippocampal formation in the monkey, J. Comp. Neurol. **189:**573, 1980.

Asanuma, H.: Recent developments in the study of the columnar arrangement of neurons within the motor cortex, Physiol. Rev. **55:**143, 1975.

Azmitia, E.C.: Bilateral serotonergic projections to the dorsal hippocampus of the rat: simultaneous localization of 3H-5HT and HRP after retrograde transport, J. Comp. Neurol. **203:**737, 1981.

Bentivoglio, M., et al.: Brainstem neurons projecting to neocortex: a HRP study in the cat, Exp. Brain Res. **31:**489, 1978.

Gaarskjaeb, F.B.: The hippocampal mossy fiber system of the rat studied with retrograde tracing techniques: correlation between topographic organization and neurogenetic gradients, J. Comp. Neurol. **203:**717, 1981.

Gatter K.C., and Powell, T.P.S.: The projection of the locus coeruleus upon the neocortex in the macaque monkey, Neuroscience **2:**441, 1977.

Hedreen, J., and Yin, T.C.T.: Homotopic and heterotopic callosal afferents of caudal interior parietal lobule in *Macaca mulatta,* J. Comp. Neurol. **197:**605, 1981.

Hubel, D.H., and Wiesel, T.N.: Functional architecture of macaque monkey visual cortex (Ferrier lecture), Proc. R. Soc. Lond. Biol. **198:**1, 1977.

Imig, T.J., and Brugge, J.F.: Sources and terminations of callosal axons related to binaural and frequency maps in primary auditory cortex of the cat, J. Comp. Neurol. **182:**637, 1978.

Jones, E.G., Coulter, J.D., and Wise, S.P.: Commissural columns in the sensory-motor cortex of monkeys, J. Comp. Neurol. **188:**113, 1979.

Loy, R., et al.: Noradrenergic innervation of the adult rat hippocampal formation, J. Comp. Neurol. **189:**699, 1980.

Markowitsch, H.J., Pritzel, M., and Divoe, I.: The prefrontal cortex of the cat: anatomical subdivisions based on retrograde labeling of cells in the mediodorsal thalamic nucleus, Exp. Brain Res. **32:**335, 1978.

Moruzzi, G.: The sleep-waking cycle, Rev. Physiol. Biochem. Exp. Pharmacol. **64:**1, 1972.

Moruzzi, G., and Magoun, H.W.: Brain stem reticular formation and activation of the EEG, Electroenceph. Clin. Neurophysiol. **1:**455, 1949.

Porrino, L.J., Crane, A.M., and Goldman-Rakic, P.S.: Direct and indirect pathways from the amygdala to the frontal lobe in rhesus monkeys, J. Comp. Neurol. **198:**121, 1981.

Prince, D.A.: Neurophysiology of epilepsy. Annu. Rev. Neurosci. **1:**395, 1978.

Rockel, A.J., Hiorns, R.W., and Powell, T.P.S.: The basic uniformity in structure of the neocortex, Brain **103:**221, 1980.

Rockland, K.S., and Pandya, D.N.: Laminar origins and terminations of cortical connections of the occipital lobe in the rhesus monkey, Brain Res. **179:**3, 1979.

Rosenkilde C.E.: Functional heterogeneity of the prefrontal cortex in the monkey: a review, Behav. Neurol. Biol. **25:**301, 1979.

Sanides, D.: The retinotopic distribution of visual callosal projections in the suprasylvian visual areas compared to the classic visual areas (17, 18, 19) in the cat, Exp. Brain Res. **33:**435, 1978.

Sanides, D., and Albus, K.: The distribution of interhemispheric projections in area 18 of the cat: coincidence with discontinuities of the representation of the visual field in the second visual area (V2), Exp. Brain Res. **38:**237, 1980.

Segraves, M.A., and Rosenquist, A.C.: The afferent and efferent callosal connections of retinotopically defined areas in cat cortex, J. Neurosci. **2:**1090, 1982.

Sperry, R.W.: Mental unity following surgical disconnection of the cerebral hemispheres, Harvey Lecture **62:**293, 1964.

Sperry, R.W.: Some effects of disconnecting the cerebral hemispheres, Science **217:**1223, 1982.

Van Essen, D.C., Newsome, W.T., and Bixby, J.L.: The pattern of interhemispheric connections and its relationship to extrastriate visual areas in the macaque monkey, J. Neurosci. **2:**265, 1982.

Winfield, D.A., Gatter, K.C., and Powell, T.P.S.: An electron microscopic study of the types and proportions of neurons in the cortex of the motor and visual areas of the cat and rat, Brain **103:**245, 1980.

Wyss, J.M.: An autoradiographic study of the efferent connections of the entorhinal cortex in the rat, J. Comp. Neurol. **199:**495, 1981.

Wyss, J.M., Swanson, L.W., and Cowan, W.M.: Evidence for an input to the molecular layer of the *stratum granulosum* of the dentate gyrus from the supramamillary region of the hypothalamus, Anat. Embryol. **156:**165, 1979.

Wyss, J.M., Swanson, L.W.: and Cowan, W.M.: A study of subcortical afferents to the hippocampal formation in the rat, Neuroscience **4:**463, 1979.

Wyss, J.M., Swanson, L.W., and Cowan, W.M.: The organization of the fimbria, dorsal fornix and ventral hippocampal formation in the rat, Anat. Embryol. **158:**303, 1980.

Books and monographs

Benson, D.F.: Aphasia, alexia, and agraphia, New York, 1979, Churchill Livingstone.

Brodal, A.: Neurological anatomy in relation to clinical medicine ed. 3, New York, 1981, Oxford University Press.

Diamond, I.T.: The subdivisions of neocortex: a proposal to revise the traditional view of sensory, motor, and association areas. In Sprague, J.M., and Epstein, A.N., editors: Progress in psychobiology and physiological psychology, vol. 8, New York, 1979, Academic Press, Inc.

Gazzaniga, M.S.: The bisected brain, New York, 1970, Appleton-Century-Crofts.

Geschwind, N.: Some special functions of the human brain: dominance, language, apraxia, memory, and attention. In Mountcastle, V.B., editor: Medical physiology, ed. 14, vol. 1, St. Louis, 1980, The C.V. Mosby Co.

Kleitman, N.: Sleep and wakefulness, ed. 2, Chicago, 1963, University of Chicago Press.

Merzenick, M.M., and Kaas, J.H.: Principles of organization of sensory-perceptual systems of mammals. In Sprague, J.M., and Epstein, A.N., editors: Progress in Psychobiology and Physiological Psychology, vol. 9, New York, 1980, Academic Press, Inc.

Milner, B.: Hemispheric specialization: scope and limits. In Schmitt, F.O., and Worden, F.G., editors: The neurosciences third study program, Cambridge, Mass., 1974, M.I.T. Press.

O'Keefe, J., and Nadel, L.: The hippocampus as a cognitive map, Oxford, 1978, Clarendon Press.

Schmitt, F.O., et al., editors: The organization of the cerebral cortex, Cambridge, 1981, M.I.T. Press.

Sperry, R.W.: Lateral specialization in the surgically separated hemispheres. In Schmitt, F.O., and Worden, F.G., editors: The neurosciences third study program, Cambridge, Mass., 1974, M.I.T. Press.

Woolsey, C.N.: Cortical sensory organization, vols. 1 to 3, Clifton, N.J., 1981, Humana Press.

chapter 20

The autonomic nervous system and its central control

The autonomic (vegetative, visceral) nervous system innervates the internal organs to modulate and control the internal environment; in this context it contributes to maintaining homeostasis. However, it also participates in adjustments to environmental stimuli. For example, a fall of the ambient temperature requires heat production and the prevention of heat loss; exercise necessitates an increase in cardiac output and shunting of blood to the active muscle beds. Such adjustments are mediated at least in part through peripheral components of the autonomic nervous system. However, they also involve complex central circuitry. Thus an important component of central nervous system activity is directed toward the control or modulation of the autonomic outflow.

The significance of the autonomic nervous system and its central control for overall behavior should not be underestimated. By way of illustration, consider the regulation of body temperature. In poikilothermic (cold-blooded) animals much of the regulation of body temperature is behavioral, such as seeking warmer sites when the ambient temperature falls. This places restrictions on both behavior and acceptable environments. Homeothermic (warm-blooded) animals, however, have a greater freedom of movement in environments of varying temperature. This can be strikingly demonstrated by depriving a homeothermic animal of its temperature-regulating capability and then placing it in a cold room with a source of heat in one corner. With thermoregulatory mechanisms intact, the animal will roam the entire room. However, with these mechanisms compromised it will remain close to the heat source, restricting its ambulatory freedom substantially. Thus autonomic regulation of the internal milieu, particularly in response to environmental challenges, provides considerable adaptive advantage.

■ Organization of the autonomic innervation

Most rigorously defined, the autonomic nervous system consists only of the efferent innervation of the viscera, which has two major subdivisions: the *sympathetic* and *parasympathetic systems* (Fig. 20-1). The final common path of the autonomic innervation is conventionally viewed as involving a two-neuron chain. A *preganglionic neuron* is located in the central nervous system, and this projects onto a *postganglionic neuron* located in the periphery (Fig. 20-1). For the sympathetic system the preganglionic neurons are situated in the thoracic and upper lumbar spinal cord. Thus this division of the autonomic nervous system is often referred to as the *thoracolumbar* division. The parasympathetic preganglionic neurons are found in certain cranial nerve nuclei and in the sacral spinal cord; therefore the alternate designation of the parasympathetic system is the *craniosacral* division. The locations of the postganglionic neurons of the two divisions also differ (Fig. 20-1). In the sympathetic system the postganglionic neurons are generally situated in either the *paravertebral (sympathetic) chains* that parallel much of the vertebral column or in *prevertebral ganglia* such as the celiac ganglion (Fig. 20-1).

The autonomic nervous
system and its central
control   315

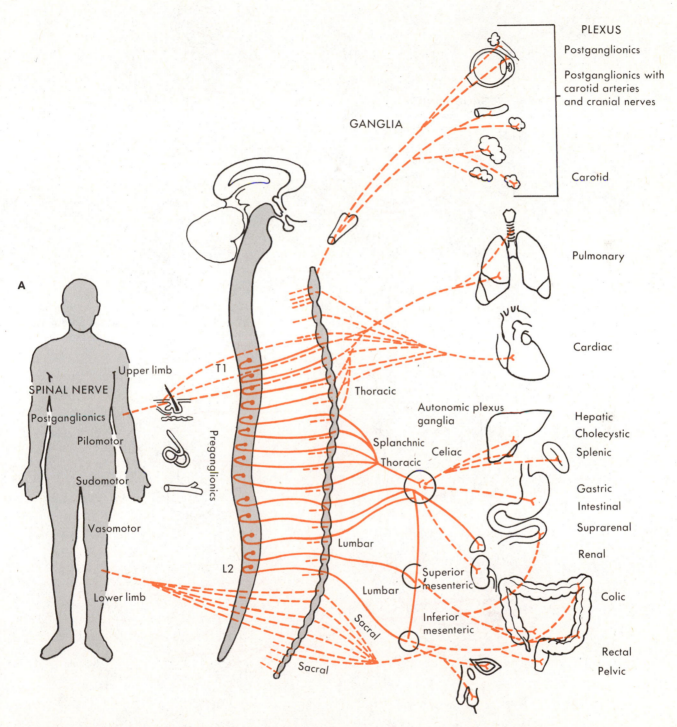

PLEXUS

Postganglionics

Postganglionics with
carotid arteries
and cranial nerves

GANGLIA

Carotid

Pulmonary

Cardiac

Autonomic plexus
ganglia

Hepatic
Cholecystic

Splenic

Gastric

Intestinal

Suprarenal

Renal

Colic

Rectal

Pelvic

SPINAL NERVE

Upper limb

Postganglionics

Pilomotor

Sudomotor

Vasomotor

Lower limb

T1

Preganglionics

L2

Thoracic

Splanchnic
Thoracic

Celiac

Lumbar

Lumbar

Sacral

Sacral

Superior
mesenteric

Inferior
mesenteric

A

Fig. 20-1 ■ **A,** Schematic representation of the sympathetic system.
Continued.

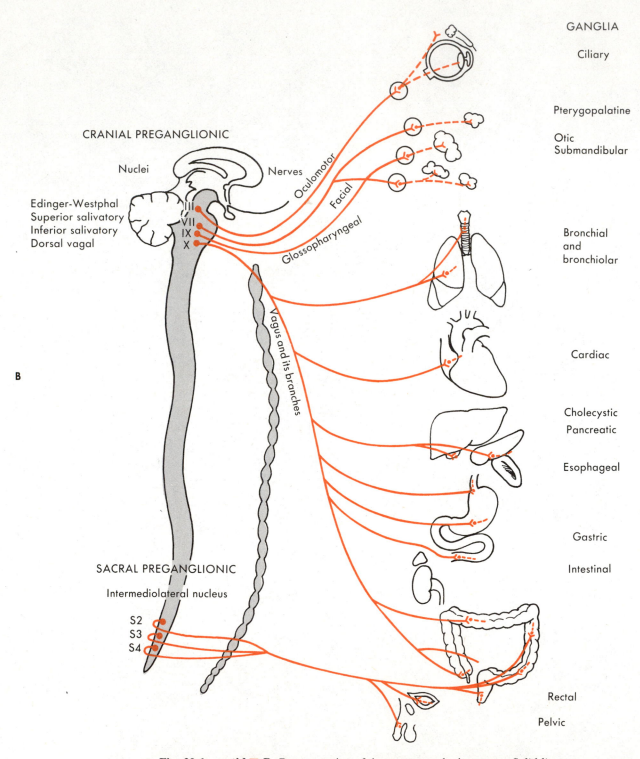

GANGLIA

Ciliary

Pterygopalatine

Otic
Submandibular

CRANIAL PREGANGLIONIC

Nuclei   Nerves

Edinger-Westphal
Superior salivary
Inferior salivary
Dorsal vagal

Bronchial
and
bronchiolar

Cardiac

Cholecystic
Pancreatic

Esophageal

Gastric

Intestinal

SACRAL PREGANGLIONIC

Intermediolateral nucleus

S2
S3
S4

Rectal

Pelvic

Fig. 20-1, cont'd ■ **B,** Representation of the parasympathetic system. *Solid lines,*
Preganglionics; *broken lines,* postganglionics. (Redrawn from Bhagat, B.D., Young, P.A., and
Biggerstaff, D.E.: Fundamentals of visceral innervation, 1977. Courtesy Charles C Thomas,
Publisher, Springfield, Ill.)

In contrast, the postganglionic neurons of the parasympathetic system are located in close proximity to their target organ. (At this point it is important to state the caveat that in the peripheral autonomic innervation exceptions are the rule.)

Many generalities have emerged regarding the structure and function of the autonomic nervous system and its central control. For example, the sympathetic outflow is traditionally viewed as diffuse, whereas the parasympathetic system is described as being more localized. Also, the two divisions are generally conceived as being antagonistic in function. However, as knowledge of autonomic control has advanced, it has become clear that such generalities are oversimplified. Indeed there is considerably greater specificity in sympathetic control than originally envisioned, and the relationship between the two divisions is better characterized as a cooperative integrated action than as adversative. In situations where the sympathetic and parasympathetic innervations of an organ produce apparently opposite responses, it is more productive to view these innervations as acting reciprocally than antagonistically. Yet another generality is that there is no voluntary control of the autonomic innervation, and this, too, has come under challenge in recent years.

■ The sympathetic system

The preganglionic neurons of the sympathetic, or thoracolumbar, division of the autonomic nervous system are located primarily in the *lateral horn* or *intermediolateral cell column* of the thoracic and first two or three lumbar segments of the spinal cord (Figs. 20-1 and 20-2). However, sympathetic preganglionic neurons are also found in lesser numbers more medially, dorsolateral to the central canal. In comparison with the α-motoneurons, the preganglionic neurons are small and distributed in clusters; their axons are of small diameter and are myelinated (group B). These axons exit the spinal cord through the ventral roots along with the axons of α- and γ-motoneurons (Fig. 20-2). However, a preganglionic axon may course for a number of segments within the paravertebral chain before synapsing on postganglionic neurons. Furthermore, the lumbar preganglionic neurons can distribute their axons bilaterally.

The most common trajectory of a sympathetic preganglionic fiber is to leave the ventral root and course through a *white ramus* (Fig. 20-2). The white ramus, so named because the sympathetic preganglionic axons are myelinated, constitutes the pathway to the paravertebral or sympathetic ganglia. This chain of ganglia extends from cervical through sacral levels. However, white rami are found only at segments that contain preganglionic cell bodies, that is, the thoracic and upper lumbar segments. The preganglionic axons ramify extensively in the paravertebral chain and synapse within a number of ganglia. It is via such ramification that the ganglia at cervical, lower lumbar, and sacral levels receive their inputs. It should be noted that at cervical levels there are only three ganglia, each rather large (Fig. 20-1). These are in descending order the *superior, middle,* and *inferior* cervical sympathetic ganglia. Often the inferior cervical ganglion merges with the first thoracic ganglion to form the *stellate ganglion,* an important structure with respect to the innervation of the heart. The postganglionic neurons of the sympathetic ganglia give rise to small unmyelinated axons (group C) that exit the ganglia to join visceral or spinal nerves via the *gray ramus,* so named because these axons are unmyelinated.

Although a preganglionic-postganglionic synapse in the paravertebral chain is the most common organization of the sympathetic outflow, there are two other prominent arrangements. Certain preganglionic axons continue peripherally, with or without synapsing in the paravertebral chain, to terminate on postganglionic neurons in *prevertebral ganglia,* such as the celiac ganglion or mesenteric ganglia (Figs. 20-1 and 20-2). The other exception involves the innervation of the adrenal medulla by the splanchnic nerve. In this case the target cells, the chromaffin cells, are innervated directly by preganglionic axons without relay through a postganglionic neuron.

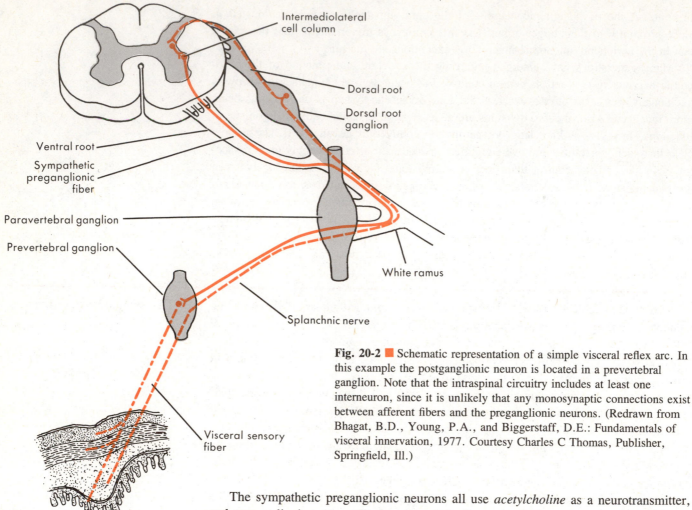

Fig. 20-2 ■ Schematic representation of a simple visceral reflex arc. In this example the postganglionic neuron is located in a prevertebral ganglion. Note that the intraspinal circuitry includes at least one interneuron, since it is unlikely that any monosynaptic connections exist between afferent fibers and the preganglionic neurons. (Redrawn from Bhagat, B.D., Young, P.A., and Biggerstaff, D.E.: Fundamentals of visceral innervation, 1977. Courtesy Charles C Thomas, Publisher, Springfield, Ill.)

The sympathetic preganglionic neurons all use *acetylcholine* as a neurotransmitter, and preganglionic-postganglionic synapses in the sympathetic system have been studied as models of cholinergic transmission. These are *nicotinic cholinergic synapses* because the drug nicotine acts as an agonist at low doses (it blocks transmission at higher doses). Neuromuscular transmission is also nicotinic, although the cholinergic receptors on muscle and on postganglionic neurons do not behave identically with respect to blocking agents such as curare or α-bungarotoxin. The postganglionic sympathetic neurons most commonly use *norepinephrine* as a neurotransmitter. In certain cases, however, such as the innervation of the sweat glands and of certain blood vessels, the junctions are cholinergic. Of interest in this context is the direct innervation of the adrenal medulla by preganglionic axons. The chromaffin cells release catecholamines (in humans approximately 80% is epinephrine and 20% norepinephrine), and hence they may perhaps be viewed as analogous to postganglionic neurons.

Advances in understanding of synaptic transmission in the autonomic nervous system have been explosive in recent years, particularly with respect to the pharmacology of autonomic synapses, and a contemporary pharmacology text should be consulted for further information. Much of this new knowledge involves the postsynaptic receptors at adrenergic postganglionic synapses (Table 20-1). For example, it was noted many decades ago that norepinephrine has an excitatory action at some effectors, such as many blood vessels, and an inhibitory effect at others, such as bronchial muscle. This contributed to the subsequent classification of α-*adrenergic* and β-*adrenergic receptors*. Norepinephrine acting on α-receptors generally produces membrane depolarization, whereas with β-receptors it usually produces hyperpolarization. However, classification of adrenergic receptors is also based on their pharmacological properties. Further subdivisions of adrenergic receptors are being recognized as more is learned of their pharmacology.

TABLE 20-1

Responses of effector organs to autonomic nerve impulses

Effector organs	Receptor type	Adrenergic impulses[1] Responses[2]	Cholinergic impulses[1] Responses[2]
Eye			
Radial muscle, iris	α	Contraction (mydriasis) + +	—
Sphincter muscle, iris		—	Contraction (miosis) + + +
Ciliary muscle	β	Relaxation for far vision +	Contraction for near vision + + +
Heart			
S-A node	β_1	Increase in heart rate + +	Decrease in heart rate; vagal arrest + + +
Atria	β_1	Increase in contractility and conduction velocity + +	Decrease in contractility, and (usually) increase in conduction velocity + +
A-V node	β_1	Increase in automaticity and conduction velocity + +	Decrease in conduction velocity; A-V block + + +
His-Purkinje system	β_1	Increase in automaticity and conduction velocity + + +	Little effect
Ventricles	β_1	Increase in contractility, conduction velocity, automaticity, and rate of idioventricular pacemakers + + +	Slight decrease in contractility
Arterioles			
Coronary	α,β_2	Constriction +; dilatation[3] + +	Dilatation ±
Skin and mucosa	α	Constriction + + +	Dilatation[4]
Skeletal muscle	α,β_2	Constriction + +; dilatation[3,5] + +	Dilatation[6] +
Cerebral	α	Constriction (slight)	Dilatation[4]
Pulmonary	α,β_2	Constriction +; dilatation[3]	Dilatation[4]
Abdominal viscera; renal	α,β_2	Constriction + + +; dilatation[5] +	—
Salivary glands	α	Constriction + + +	Dilatation + +
Veins (systemic)	α,β_2	Constriction + +; dilatation + +	—
Lung			
Bronchial muscle	β_2	Relaxation +	Contraction + +
Bronchial glands	?	Inhibition (?)	Stimulation + + +
Stomach			
Motility and tone	α_2,β_2	Decrease (usually)[7] +	Increase + + +
Sphincters	α	Contraction (usually) +	Relaxation (usually) +
Secretion		Inhibition (?)	Stimulation + + +
Intestine			
Motility and tone	α_2,β_2	Decrease[7] +	Increase + + +
Sphincters	α	Contraction (usually) +	Relaxation (usually) +
Secretion		Inhibition (?)	Stimulation + +
Gallbladder and ducts		Relaxation +	Contraction +
Kidney	β_2	Renin secretion + +	—

From Goodman, L.S., and Gilman, A.: The pharmacological basis of therapeutics, ed. 6, New York, 1980, Macmillan Publishing Co., Inc. Copyright © 1980 by Macmillan Publishing Co., Inc.

[1]A long dash signifies no known functional innervation.

[2]Responses are designated 1+ to 3+ to provide an approximate indication of the importance of adrenergic and cholinergic nerve activity in the control of the various organs and functions listed.

[3]Dilatation predominates *in situ* due to metabolic autoregulatory phenomena.

[4]Cholinergic vasodilatation at these sites is of questionable physiological significance.

[5]Over the usual concentration range of physiologically released, circulating epinephrine, β-receptor response (vasodilatation) predominates in blood vessels of skeletal muscle and liver, α-receptor response (vasoconstriction), in blood vessels of other abdominal viscera. The renal and mesenteric vessels also contain specific dopaminergic receptors, activation of which causes dilatation, but their physiological significance has not been established.

[6]Sympathetic cholinergic system causes vasodilatation in skeletal muscle, but this is not involved in most physiological responses.

[7]It has been proposed that adrenergic fibers terminate at inhibitory β receptors on smooth muscle fibers, and at inhibitory α receptors on parasympathetic cholinergic (excitatory) ganglion cells of Auerbach's plexus.

Continued.

TABLE 20-1—cont'd

Responses of effector organs to autonomic nerve impulses

Effector organs	Receptor type	Adrenergic impulses[1] Responses[2]	Cholinergic impulses[1] Responses[2]
Urinary bladder			
Detrusor	β	Relaxation (usually) +	Contraction + + +
Trigone and sphincter	α	Contraction + +	Relaxation + +
Ureter			
Motility and tone	α	Increase (usually)	Increase (?)
Uterus	α,β$_2$	Pregnant: contraction (α); nonpregnant: relaxation (β)	Variable[8]
Sex organs, male	α	Ejaculation + + +	Erection + + +
Skin			
Pilomotor muscles	α	Contraction + +	—
Sweat glands	α	Localized secretion[9] +	Generalized secretion + + +
Spleen capsule	α,β$_2$	Contraction + + +; relaxation +	—
Adrenal medulla		—	Secretion of epinephrine and norepinephrine
Liver	α,β$_2$	Glycogenolysis, gluconeogenesis[10] + + +	Glycogen synthesis +
Pancreas			
Acini	α	Decreased secretion +	Secretion + +
Islets (β cells)	α	Decreased secretion + + +	—
	β$_2$	Increased secretion +	—
Fat cells	α,β$_1$	Lipolysis[10] + + +	—
Salivary glands	α	Potassium and water secretion +	Potassium and water secretion + + +
	β	Amylase secretion +	
Lacrimal glands		—	Secretion + + +
Nasopharyngeal glands		—	Secretion + +
Pineal gland	β	Melatonin synthesis	—

[8]Depends on stage of menstrual cycle, amount of circulating estrogen and progesterone, and other factors.
[9]Palms of hands and some other sites ("adrenergic sweating").
[10]There is significant variation among species in the type of receptor that mediates certain metabolic responses.

■ The parasympathetic system

For the cranial portion of the parasympathetic system the preganglionic neurons are located in the motor nuclei of cranial nerves III, VII, IX, and X (Fig. 20-1). The sacral portion arises from preganglionic neurons in sacral segments 2 to 4. Their location in the spinal gray matter is analogous to that of the intermediolateral column of the thoracolumbar system. The cranial preganglionic neurons issue myelinated axons that travel in the respective cranial nerves to terminate on postganglionic neurons within the target organs. The sacral preganglionic fibers also issue myelinated axons that emerge via the ventral roots to travel in the pelvic nerves and ultimately terminate on postganglionic cells within the target organs (Fig. 20-1).

As in the sympathetic system, the parasympathetic preganglionic fibers release *acetylcholine* as their neurotransmitter, and the receptors are *nicotinic*. However, unlike the sympathetic innervation, the parasympathetic postganglionic fibers are also cholinergic, and these are *muscarinic* synapses; that is, the drug muscarine acts as an agonist, and transmission is blocked by atropine.

With but a few exceptions, such as the afferent fibers from the carotid sinus and carotid body, our knowledge of visceral afferent fibers is rather limited. There have been recent advances in this regard, but much research is required before we have a comprehensive view of the afferentation of the viscera. Such afferent fibers are extremely important. They participate in basic regulatory reflexes and provide the central nervous system with important information relative to the state of the internal environment. One route by which such information can reach the brain is through sensory components of cranial nerves. For example, information from the carotid sinus regarding arterial blood pressure is transmitted over fibers that are located in the glossopharyngeal nerve (IX) and synapse in the nucleus of the solitary tract. Information from aortic arch receptors is transmitted over vagal afferent fibers that also terminate in the nucleus of the solitary tract. Most visceral afferent fibers, however, have their cell bodies in dorsal root ganglia and enter the spinal cord through dorsal or ventral roots. Recent evidence also suggests that some visceral afferent fibers may synapse directly in autonomic ganglia without entering the spinal cord, providing the possibility for peripheral mediation of autonomic reflexes. A final point is that some visceral receptors, such as the chemoreceptors of the carotid body, receive an efferent innervation from the central nervous system that can modulate their sensitivity; that is, they are under centrifugal control.

■ *Visceral afferent fibers*

The autonomic ganglia are conventionally viewed as simple relays in which preganglionic neurons synapse on postganglionic neurons with varying degrees of convergence and divergence. However, as the peripheral autonomic innervation is studied in greater detail, this generalization has faltered. For example, in the paravertebral chain the possibility of ganglionic interneurons exists. More compelling evidence has been gathered

■ *Autonomic ganglia*

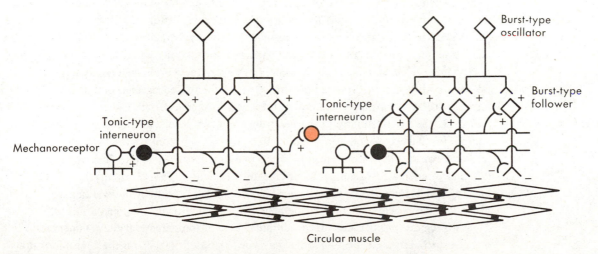

Fig. 20-3 ■ Model of inhibitory neural control of intestinal circular musculature. Circular muscle behaves like a functional syncytium that is activated by myogenic mechanisms. Burst-type oscillators discharge continuously and synaptically drive nonspontaneous follower neurons. Burst-type followers are intrinsic inhibitory neurons that release a nonadrenergic transmitter substance at neuromuscular appositions. Mechanoreceptors activate tonic-type neurons. One kind of tonic-type cell *(black)* forms inhibitory synapses on burst-type followers. These tonic-type cells function to stop the discharge of burst-type followers and release the muscle from inhibition. Sequential activation of mechanoreceptors and release of inhibition account for propagation of a contractile response in the direction of left to right on the diagram. A second kind of tonic-type neuron *(color)* that forms excitatory synapses on burst-type follower cells accounts for descending inhibition of peristaltic propulsion in the direction of left to right on the diagram. Inhibitory synapses, (−); excitatory synapses, (+). This model is presented as an illustration of the complexity of local control in the autonomic innervation. (Redrawn from Wood, J.D.: Physiol. Rev. **55**:307, 1975.)

with respect to certain more peripheral ganglia, such as the mesenteric and myenteric plexuses, where the local circuitry is complex, and substantial local processing occurs (Fig. 20-3). It appears that such peripheral processing is more the rule than the exception in the autonomic innervation. Considerable integrative processing and reflex activity may occur at the periphery. It is also likely that such local peripheral circuits use other neurotransmitters besides acetylcholine and norepinephrine.

■ *Functional considerations regarding the autonomic innervation*

Simplified characterizations of the autonomic innervation describe sympathetic function as concerned with mobilizing the organism for "fight or flight" and parasympathetic functions as "vegetative" and "emptying." Inspection of Fig. 20-1 and Table 20-1 suggests that this is not realistic, however. It now seems more appropriate to abandon such generalizations and to attend to more specific effects at target organs.

■ *Distribution and specificity of the peripheral autonomic innervation*

The distribution of the autonomic efferent nerves is summarized in Fig. 20-1 and Table 20-1. Some structures, such as the adrenal medulla, pilomotor muscles, and most blood vessels, receive only sympathetic innervation. Others, such as the sublingual gland, receive only parasympathetic innervation. However, most structures are *dually innervated*. Although in many instances the sympathetic and parasympathetic innervations have opposite effects on such dually innervated organs, this should not dominate our view of their interaction. For example, in the male sex organs the parasympathetic innervation mediates erection and the sympathetic innervation induces ejaculation; at the salivary glands both innervations produce secretion but with different properties. Such examples cannot truly be considered as illustrative of antagonistic action. In other dually innervated organs the two subdivisions of the autonomic nervous system do appear to have opposite effects. For example, the sympathetic cardiac innervation increases heart rate and contractility, whereas the parasympathetic innervation decreases both. Sympathetic system activation decreases intestinal motility, whereas parasympathetic system activation increases it. However, such antagonistic action would occur only if both innervations were activated concomitantly. In fact, they often behave synergistically. By way of illustration, in situations that demand increased cardiac output, there is an increase in the sympathetic outflow to the heart to increase both rate and contractility. However, the vagal outflow to the heart does not compete with the sympathetic innervation. Instead vagal activity is inhibited, and this also contributes to the increase in rate and contractility. Consequently when one considers (1) singly innervated organs, (2) dually innervated organs without antagonistic action, and (3) dually innervated organs with antagonistic action but where physiologically the two innervations act synergistically, the concept of antagonistic action loses much of its conceptual utility.

With regard to specificity, the traditional view of the parasympathetic innervation is that it exerts restricted local actions. This view is based in part on the low innervation ratio of preganglionic to postganglionic neurons in the parasympathetic system. However, there are exceptions. In contrast, the sympathetic system has been traditionally viewed as discharging diffusely to elicit an ensemble of mobilizing responses. These would include, for instance, increased cardiac output, increased arterial blood pressure, pupillary dilation (mydriasis), decreased gastrointestinal activity, pilomotor responses, and release of catecholamines from the adrenal medulla.

Although these mass responses of the sympathetic system can indeed occur in concert in emergency or stress situations, it is now known that the sympathetic system can mediate highly localized responses as well. For example, blood flow can be increased in but a single muscle. More often patterned responses occur that are generally specific to the stimulus situation. Thus the sympathetic system is more appropriately viewed as having a specificity that is perhaps analogous to local sign in flexor withdrawal reflexes.

*The autonomic nervous
system and its central
control*   **323**

For example, the cardiovascular adjustments to different metabolic or somatic demands are quite specific. In exercise there is an increase in cardiac output and a shunting of blood to the active muscles; arterial blood pressure may not change at all. During temperature changes the alterations in peripheral resistance are largely confined to the cutaneous bed; cutaneous vasodilation occurs with thermal stimulation, and cutaneous vasoconstriction occurs with cold. During orthostasis (moving from a prone to a standing position) there may be no change in cardiac output but an increase in carotid blood flow, which maintains perfusion of the brain.

These patterns are organized centrally and will be discussed further later. However, they do illustrate the capability of the sympathetic system for restricted action. This implies that the postganglionic innervation has a topography that is more organized than traditionally conceived. This can clearly be seen in the cardiac innervation. Not only are there sympathetic postganglionic neurons that innervate only the heart, but cardiac postganglionic fibers arising from different sympathetic ganglia distribute preferentially to different regions of the myocardium. What is not yet established is the extent to which sympathetic preganglionic neurons preferentially distribute to postganglionic neurons of specific function. For example, are there specific ''cardiac preganglionic neurons?'' It would seem that this is likely to be the case.

■ *Autonomic reflexes*

As in the somatomotor system, some important autonomic regulatory actions involve spinally mediated reflexes. For instance, certain important components of control of the urinary bladder, gastrointestinal system, and sexual organs are spinally mediated. The afferent limb of such reflexes frequently consists of visceral afferent fibers, although somatic afferent fibers initiate certain autonomic reflexes. In some cases both visceral and somatic afferent fibers participate. The literature increasingly documents convergence of somatic and visceral afferent nerves onto neurons of the dorsal horn. The afferent fibers that evoke autonomic reflexes generally involve one or more spinal interneurons, and the intraspinal reflex pathway ultimately influences the preganglionic cell column.

An excellent example of segmental reflex control in the autonomic nervous system is provided by the urinary bladder (Fig. 20-4). As the fluid volume of the bladder increases, the intravesical pressure rises. Stretch receptors in the bladder wall transmit information over the pelvic nerves to the sacral spinal cord with respect to the state of bladder distension. This information, via intraspinal circuitry, can activate parasympathetic preganglionic neurons in the second, third, and fourth sacral segments. The axons of these neurons travel in the pelvic nerve to terminate on local postganglionic parasympathetic neurons in the vesical plexuses at the urinary bladder. Activation of these neurons then contracts the detrusor muscle and hence the bladder, contributing to emptying. The afferent information from the bladder also influences sympathetic preganglionic neurons at upper lumbar levels. The axons of these cells relay through postganglionic neurons in prevertebral ganglia (the inferior mesenteric and inferior hypogastric plexuses). The pathway through the inferior mesenteric ganglion inhibits contraction of the internal sphincter, thereby contributing to bladder emptying. As voiding is initiated, the fluid flow through the urethra stimulates an afferent discharge that elicits parasympathetic reflex contraction of the detrusor muscle and contributes to the continuation of micturition.

Spinally mediated reflexes such as the ones just mentioned are what permit a certain degree of bladder control after spinal transection. However, in the intact animal micturition critically involves supraspinal structures, particularly in the medulla and pons. Spinally mediated control of the bladder is normally under massive control by descending pathways, and it is this descending control that allows micturition to be brought under voluntary control. In this regard it is also important to appreciate that control of

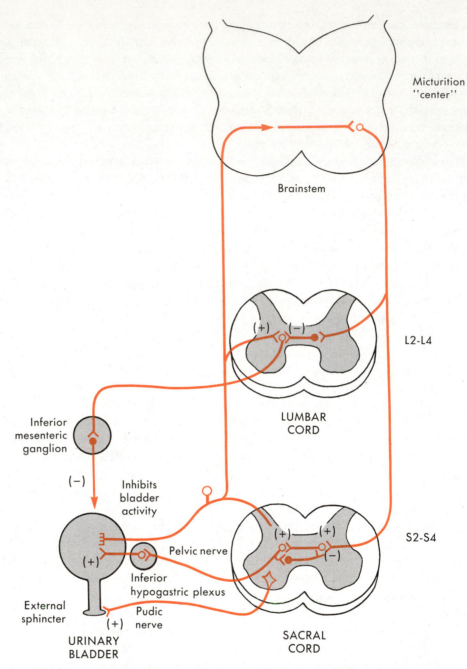

Micturition "center"

Brainstem

L2-L4

LUMBAR
CORD

(+)   (−)

Inferior
mesenteric
ganglion

(−)

Inhibits
bladder
activity

Pelvic nerve

(+)   (+)

(−)

S2-S4

(+)

Inferior
hypogastric plexus

External
sphincter

Pudic
nerve

(+)

URINARY
BLADDER

SACRAL
CORD

Fig. 20-4 ■ Parasympathetic and sympathetic reflex pathways to the urinary bladder of the cat. Inhibitory neurons are black. (Redrawn from De Groat, W.C.: Brain Res. **87:**201, 1975.)

the bladder can be achieved independently of the autonomic innervation because the external sphincter is striated and is innervated by α-motoneurons in the third and fourth sacral segments. Activation of these motoneurons by descending pathways can contract the sphincter and prevent voiding.

Although spinally mediated autonomic reflexes are common, many important reflex adjustments of the internal organs are mediated by suprasegmental reflex pathways. A particularly prominent example is the carotid sinus reflex. Increased pressure in the carotid sinus increases the discharge of afferent fibers that travel in the ninth cranial nerve and synapse in the nucleus of the solitary tract. (This nucleus may be viewed as

The autonomic nervous
system and its central
control

325

a brainstem equivalent of the dorsal horn of the spinal cord.) Through local medullary circuitry, descending fibers are activated that directly or indirectly inhibit the activity of sympathetic preganglionic neurons that participate in the control of peripheral resistance. The action of this circuitry is to decrease peripheral resistance in response to increased arterial blood pressure (Chapter 32). It must be recognized that this is a highly oversimplified treatment of the baroreceptor reflex, but it is sufficient for the present purpose of illustrating the concept of suprasegmental autonomic reflexes.

The autonomic reflexes are relatively simple examples of central control of autonomic activity. However, many critical functions, such as feeding, drinking, reproduction, and temperature regulation, have autonomic components that involve complex central circuitry. Unfortunately such circuitry has not been specified in detail for many of these activities.

As in the somatomotor system, central autonomic control is most productively analyzed by investigating the functional neuroanatomy of the major pathways that influence the autonomic outflow (the autonomic final common path). Analogous to those which control movement, these pathways are primarily organized longitudinally, and it is most effective to analyze them from the periphery centrally. For the autonomic nervous system this entails first specifying the primary descending pathways that terminate on the sympathetic and parasympathetic preganglionic cell groups. As these yield to investigation, one can then approach the problem of "sensorimotor integration" in order to specify how various stimuli (internal and external) guide the motor (autonomic) outflow. Again, as in the somatomotor system, the cells of origin of the major descending pathways are located from the lower brainstem to the neocortex. However, unlike the pathways that control movement, in autonomic control there is a major confluence of such pathways at the hypothalamus. Thus it is appropriate at this point to review briefly the organization of the hypothalamus.

■ *Higher autonomic control*

The hypothalamus is a diencephalic structure bordering the third ventricle (Fig. 20-5). Rostrocaudally it can be divided into *supraoptic, tuberal,* and *mammillary regions,* each consisting of aggregates of nuclear groups. The rostral continuation of the hypothalamus is the *preoptic region* (Fig. 20-5), which continues forward as the *septum.* Both the preoptic and septal regions participate in autonomic control. Major fiber tracts course longitudinally through the hypothalamus, the most prominent being the *medial forebrain bundle* and the *fornix.* The latter tract divides the hypothalamus into *medial* and *lateral regions.* Many of the hypothalamic cell groups are named on the basis of their location with respect to this mediolateral division.

Although there are various well-defined nuclear groups within the hypothalamus, it is difficult to delineate nuclear boundaries in some regions. Thus a cytoarchitectonic description of the hypothalamus does not have the precision it does when applied to other areas of the nervous system, such as the cortex and cerebellum. This and other factors have rendered it more difficult to study afferent and efferent connections. A highly schematic summary of these connections is presented in Fig. 20-6, but it should also be appreciated that there are important intrinsic hypothalamic connections such as local neurons that establish communication between the medial and lateral divisions.

The hypothalamus is of unquestionable importance in autonomic control. Indeed certain hypothalamic nuclei, such as the supraoptic nucleus, constitute final common paths in neuroendocrine control (Chapter 52). However, there is also substantial involvement of extrahypothalamic structures in the control of autonomic function. For example, in respiration, micturition, and emesis most processing involves central structures other than the hypothalamus.

■ *The hypothalamus*

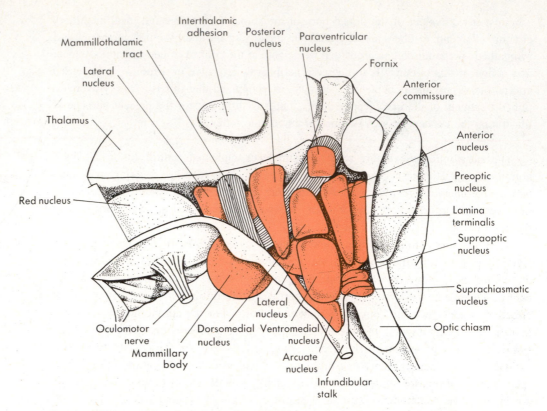

Fig. 20-5 ■ Principal nuclei of the hypothalamus and preoptic area *(color)*. (Redrawn from Nauta, W.J.H., and Haymaker, W.: The hypothalamus, Springfield, Ill., 1969, Charles C Thomas, Publisher.)

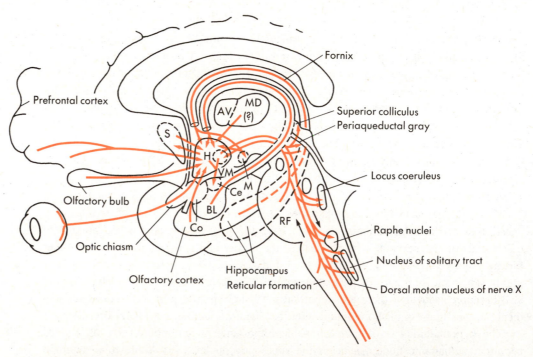

Fig. 20-6 ■ Schematic diagram of the principal afferent and efferent connections of the hypothalamus. *BL, Ce,* and *Co,* Basolateral, central, and cortical amygdaloid nucleus, respectively; *H,* hypothalamus; *M,* mamillary body; *MD,* dorsomedial thalamic nucleus; *RF,* reticular formation; *S,* septum; *SN,* substantia nigra; *VM,* ventromedial hypothalamic nucleus. (Redrawn from Brodal, A.: Neurological anatomy, ed. 3. Copyright © 1981 by Oxford University Press, Inc. Reprinted by permission.)

A concept that has dominated the literature on central autonomic control is that of "centers" that mediate particular autonomic functions. Examples are the vasomotor and vasodilator centers of the medulla, the heat production and heat loss centers of the diencephalon, and the hypothalamic feeding and satiety centers. This notion has some heuristic value, and indeed it reflects a certain localization of autonomic function. However, it is somewhat misleading with respect to the functional neuroanatomy of central autonomic control because it turns attention away from the longitudinal organization. It is perhaps equivalent to describing the motor cortex and the red nucleus as "movement centers" while ignoring the corticospinal and rubrospinal tracts and their connections.

The evolution and dominance of the center concept largely reflect two properties of the organization of central autonomic control. First, the hypothalamus is a major conduit for pathways involved in such control. Thus in early investigations, which primarily involved the lesion approach, major fiber bundles *en passage* through the hypothalamus were destroyed in addition to local cell groups. This compromised both hypothalamic and extrahypothalamic autonomic control and generated an exaggerated view of hypothalamic involvement, thereby reinforcing the center concept. Second, many of the pathways involved in central autonomic control are composed of fine fibers that do not appear as prominent tracts. Thus they were not easily identified by older anatomical methods. This did not encourage the concept of longitudinally organized central autonomic control, which again reinforced the center concept.

These traditional views are now being modified as the central pathways involved in different autonomic functions are being characterized. As will be seen in the sections to follow, we are in a transitional period where a strong residual of the center concept remains for some autonomic activities, but a more realistic description is available for others.

In homeothermic animals, which regulate their body temperatures, a fall in the ambient temperature requires additional heat production and a reduction of heat loss. In contrast, an increase in ambient temperature requires greater heat loss and a reduction of heat production. Such regulation is accomplished by a complex set of responses involving autonomic, somatic, and hormonal systems. For example, when an animal is cooled, shivering (asynchronous contraction of muscle fibers) is elicited, and this is a primary mechanism of heat production. In addition to this somatic response, the output of the thyroid gland increases, which represents a hormonal contribution to heat production through increased metabolism. Heat loss is most prominently restricted by two autonomic responses, piloerection (establishing a dead space) and constriction of the cutaneous vessels (reducing cooling of the blood). If the animal is warmed, heat loss is enhanced by autonomic responses such as sweating (increasing evaporation) and dilation of the cutaneous vessels (promoting cooling of the blood). Hormonal participation is reflected by a decrease in thyroxine release, which lessens heat production by decreasing metabolic activity.

The hypothalamus is critically involved in such thermoregulation. The preoptic region and the anterior hypothalamus participate in mediating responses that result in heat loss. Lesions of this region will prevent sweating and cutaneous vasodilation in response to increased ambient temperature. This led to the designation of this region as a *heat loss center*. Cells within the posterior hypothalamus appear to be involved in heat production and conservation. Posterior hypothalamic lesions dorsolateral to the mamillary body eliminate the responses that serve these functions and result in a chronic hypothermia. Thus this region of the posterior hypothalamus has been viewed as containing a *heat production and conservation center*.

Heating or cooling of the hypothalamus can evoke thermoregulatory responses; pyrogens injected into the hypothalamus can produce fever; and temperature-sensitive neurons have in fact been found in the hypothalamus and preoptic region. Such data support the hypothesis that thermoreceptors in the preoptic region and hypothalamus initiate thermoregulatory responses. However, warming of the heat production and conservation center of the posterior hypothalamus does not abolish the shivering that is elicited by cooling of the skin. Moreover, the center view of thermoregulation contributes little to our understanding of how the posterior hypothalamus mediates shivering, for example. It does not explain how preferential constriction of the cutaneous vasculature is effected. Stated more broadly, it discourages the investigation of extrahypothalamic involvement and thus the delineation of the neuroanatomical pathways that mediate the various thermoregulatory responses. The disturbance of thermoregulatory responses produced by hypothalamic and preoptic lesions is certainly striking. However, such data do not resolve the extent to which the hypothalamus initiates and organizes such responses, as opposed to the extent to which it acts as a relay in a considerably more complex set of pathways.

■ *Regulation of food intake*

The regulation of food intake is even more complex than the regulation of body temperature. The autonomic nervous system does not play a major role in these events and is more prominently involved in digestive than in ingestive function. However, brief mention is made of food intake here because of hypothalamic involvement. This topic is also discussed in Chapter 50.

Some years ago it was observed that lesions of the lateral hypothalamus produce a syndrome that includes *aphagia,* a decrease in food intake that may be of sufficient severity to lead to death by starvation. Electrical stimulation of this region induces eating even after satiety. In contrast, destruction of the ventromedial nucleus of the hypothalamus produces increased food intake, *hyperphagia,* and obesity can result. Electrical stimulation of this hypothalamic area will terminate feeding. Such results generated the hypothesis that the lateral hypothalamus contains a *feeding center* and the ventromedial hypothalamus a *satiety center*. Moreover, these two centers were thought to interact reciprocally. For example, if feeding is induced by lateral hypothalamic stimulation, stimulation of the ventromedial nucleus will terminate the ongoing feeding abruptly.

Analogous to the argument that diencephalic thermoreceptors initiate thermoregulatory responses, a prominent hypothesis of the regulation of food intake involves hypothalamic *glucoreceptors* that sense changes in blood glucose levels. Increased levels are presumed to activate the satiety center and inhibit the feeding center. Decreased blood glucose levels produce the converse effect. Although some evidence is consistent with this hypothesis, it is not definitively established.

As seen with other functions discussed previously, hypothalamic involvement is unquestionable. However, once again it is important to ask to what extent. By way of illustration, the lateral hypothalamic lesions that produce aphagia interrupt many fibers *en passage,* including the nigrostriatal pathway. It has been reported that destruction of this tract outside the hypothalamus can produce aphagia. Similarly, lateral hypothalamic lesions can interrupt the trigeminal lemniscus and abolish ascending sensory information that may participate in ingestion and deglutition; this too can produce aphagia. It now seems that aphagia consequent to lateral hypothalamic lesions can occur if these pathways are spared. However, the simplistic view of a hypothalamic feeding center is challenged by such findings, and a comprehensive examination of extrahypothalamic involvement is dictated.

■ *Central cardiovascular control*

As a final illustration of central autonomic control, cardiovascular function will be considered. It provides a particularly useful case study because it demonstrates (1) the transition from the center concept to longitudinal organization, (2) patterned sympathetic

*The autonomic nervous
system and its central
control*   **329**

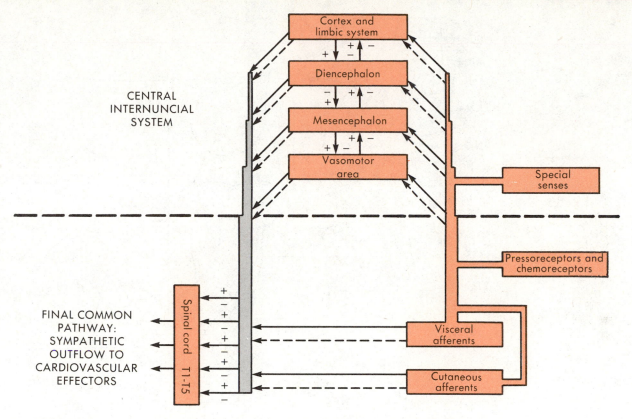

Fig. 20-7 ■ Schematic diagram of proposed central nervous organization for regulation of cardiovascular function. Solid arrows and plus signs represent excitatory connections. Dashed arrows and minus signs represent inhibitory connections. (Redrawn from Peiss, C.N. In Price, H.L., and Cohen, P.J., editors: Symposium on the effects of anesthetics on the circulation, Springfield, Ill., 1964, Charles C Thomas, Publisher.)

activity appropriate to environmental demands, and (3) the integration of reflex activity with such phasic adjustments.

Historically central cardiovascular control has been approached from two directions. One focuses principally on cardiovascular adjustments in response to various peripheral inputs, and this has been primarily a description of reflex activity. Changes in cardiac dynamics and peripheral resistance evoked by activation of the baroreceptors are characteristic of this approach, and indeed a massive literature has evolved in this context. The other approach focuses on the nervous system and is directed primarily toward describing "centers" for the homeostatic regulation of cardiovascular activity. Constructs such as the medullary *vasopressor* and *vasodepressor centers* were derived from this approach, and they generated a hierarchical view of the organization of central cardiovascular control (Fig. 20-7).

This view followed investigations of the regulation of arterial blood pressure in which transections were made at various levels of the neuraxis. The early literature reported that blood pressure was not affected by transections rostral to the medulla. This was supplemented by electrical stimulation experiments which then identified medullary areas where activation increased blood pressure (the vasomotor or vasopressor center) or decreased blood pressure (the vasodilator or vasodepressor center) (Fig. 20-8).

Although these medullary areas are certainly involved in cardiovascular regulation, in the past two decades there has been a shift toward the concept of descending functional pathways that mediate patterns of cardiovascular adjustment specific to behavioral demands. The adjustments required by exercise provide an excellent example. In this case there is a need for increased cardiac output and increased blood flow in the active

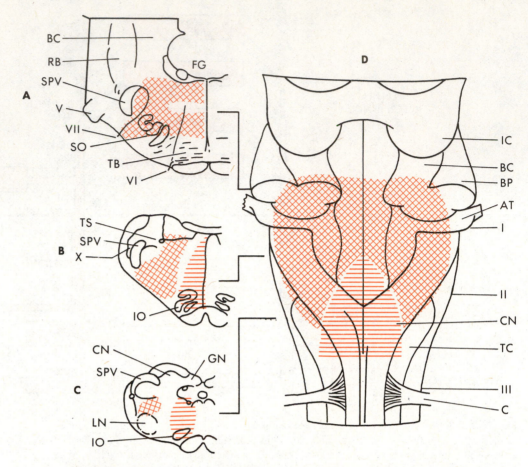

Fig. 20-8 ■ Localization of pressor *(crosshatching)* and depressor *(horizontal lines)* centers of the cat brainstem. **A** to **C,** Cross sections through the medulla at levels indicated by guide lines to **D. D,** Semidiagrammatic projection of pressor and depressor regions onto the dorsal surface of the brainstem viewed with the cerebellar peduncles cut across and the cerebellum removed. *AT,* Auditory tubercle; *BC,* brachium conjunctiva; *BP* brachium pontis; *C,* first cervical nerve; *CN,* cuneate nucleus; *FG,* facial genu; *GN,* gracile nucleus; *IC,* inferior colliculus; *IO,* inferior olivary nucleus; *LN,* lateral reticular nucleus; *RB,* restiform body; *SO,* superior olivary nucleus; *SPV,* spinal trigeminal tract; *TB,* trapezoid body; *TC,* tuberculum cinerum; *TS,* tractus solitarius; *V, VI, VII,* corresponding cranial nerves; *I, II, III,* levels of transection. (Redrawn from Alexander, R.S.: J. Neurophysiol. **9:**205, 1946.)

muscle beds. A pathway originating in the motor cortex will selectively increase the blood flow in muscles that are represented in the activated region. The precise trajectory and connectivity of this pathway have yet to be specified. However, the important point is that viewing the cardiovascular adjustments in exercise in the context of functional neuroanatomical pathways has been considerably more productive than using the center concept.

The so-called exercise pathway illustrates other important points regarding central autonomic control. First, it provides a convincing demonstration that the sympathetic system can behave in a selective and patterned manner. Second, it increases our understanding of the interaction between the sympathetic and parasympathetic innervations in a physiological context. Specifically, the increased cardiac output during exercise involves the synergistic action of the sympathetic and parasympathetic cardiac innervations, with the sympathetic drive on the heart increasing and vagal cardiac inhibition decreasing during exercise. Third, it has allowed investigation of the interaction between patterned responses evoked by environmental demands and reflex activity. During the

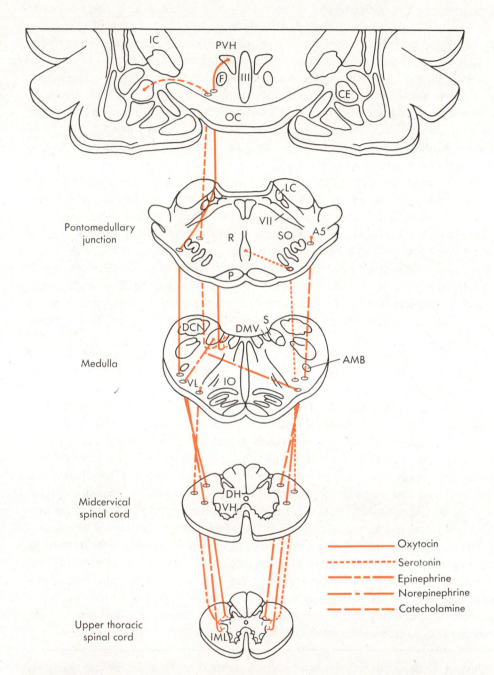

Fig. 20-9 ■ Schematic summary of descending pathways that project directly to the dorsal motor nucleus of the vagus and/or the sympathetic preganglionic cell column. Hypothesized transmitters are indicated. *A5*, A5 catecholamine cell group of Dahlstrom and Fuxe; *AMB*, nucleus ambiguus; *CE*, central nucleus of the amygdala; *DCN*, dorsal column nuclei; *DH*, dorsal horn; *DMV*, dorsal motor nucleus of the vagus; *F*, fornix; *IC*, internal capsule; *III*, third ventricle; *IML*, intermediolateral (sympathetic preganglionic) cell column; *IO*, inferior olive; *LC*, locus coeruleus; *OC*, optic chiasm; *P*, pyramid; *PVH*, paraventricular nucleus of the hypothalamus; *R*, raphe nucleus; *S*, solitary nucleus; *SO*, superior olive; *VH*, ventral horn; *VII*, facial nerve rootlets; *VL*, ventrolateral medullary region including the epinephrine-containing and norepinephrine-containing cell groups. (Redrawn from Cohen, D.H., and Cabot, J.B.: Trends Neurosci. **2:**273, 1979.)

increase in cardiac output the baroreceptor reflex appears to be inhibited, preventing competition via its negative feedback effect. Fourth, a central organization originating in the cortex allows cardiovascular adjustments to occur in anticipation of exercise so that the central control can prepare the organism for impending demands. This is seen, for example, in a runner at the starting line, and it clearly demonstrates that learned control of autonomic outflow plays a functional role, a fifth point. Finally, it stimulates one to view central cardiovascular control in terms of both phasic and tonic functions; that is, there are central pathways that mediate phasic adjustments in response to certain classes of transient demands, such as exercise, and other central pathways that are involved in mediating more tonic or homeostatic functions, such as the moment-by-moment regulation of arterial blood pressure.

We continue to learn more of the specific pathways responsible for specific patterns of cardiovascular adjustment and how they interact. For example, during a shift from a prone to a standing position (orthostasis), it is necessary to effect adjustments that maintain blood flow to the brain, that is, to oppose gravitationally induced pooling of blood in the lower body. It is now established that this is mediated by a reflex pathway, the afferent limb of which involves the vestibular system. This labyrinthine information is then relayed to a localized region of the cerebellum, the rostral third of the medial zone, which then projects back on the brainstem to influence cardiovascular structures that project to the spinal cord. Interruption of this pathway can produce a postural hypotension.

An important contemporary thrust is to describe rigorously the various descending pathways that impinge on autonomic preganglionic neurons (Fig. 20-9). This establishes a foundation for analyzing the different functional ''cardiovascular pathways'' from the periphery to the central nervous system. Moreover, as the neurotransmitters of these pathways are discovered, pharmacological manipulation becomes possible. This is important in dealing with cardiovascular diseases that have a neurogenic etiology, and there is now increasing acceptance of the idea that these may, indeed, be quite common. For example, many now believe that sudden cardiac death may often be initiated neurogenically and that a significant proportion of cases of essential hypertension may involve the central nervous system. It is noteworthy in this regard that many of the most widely used antihypertensive drugs, such as clonidine, have neural sites of action.

Thus more recent investigations of central cardiovascular control are teaching us the merits of reconceptualizing our view of autonomic control. In the case of cardiovascular control this has diminished the historically dominant concept of medullary centers. It has placed them in a more comprehensive context that permits rational investigation of longitudinally organized pathways that mediate specific classes of adjustment, be they homeostatic or in response to environmental demands.

■ Emotional behavior

Just as the hypothalamus was historically designated as the principal suprasegmental structure for autonomic control, the limbic system was assigned the mediation of emotional behavior. Presumed to be in a control position over the hypothalamus, limbic structures then performed higher-order integrative functions involving autonomic control. Again, this reflected the earlier tendency to view the nervous system as being hierarchically organized.

The limbic lobe is phylogenetically the oldest part of the cortex. It was subsequently proposed that a complex circuit involving the limbic lobe (Papez circuit) participated with the hypothalamus in mediating emotional behavior; this became known as the *limbic system* (Fig. 20-10). This circuit involved a flow of information from the cingulate gyrus, via the hippocampus, to the mammillary bodies of the hypothalamus. A loop was closed by information returning from the hypothalamus to the cingulate gyrus via the anterior thalamic nuclei. The concept of this circuit gained popularity for some time,

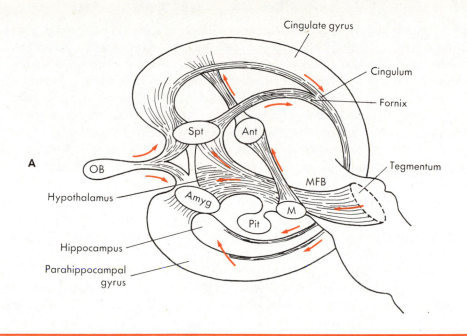

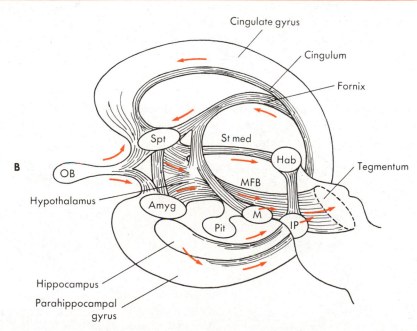

Fig. 20-10 ■ Limbic system circuitry emphasizing the two-way interconnections of most limbic structures. The ascending pathways from the midbrain tegmentum course through the hypothalamus via the medial forebrain bundle **(A)**, analogous to the descending pathways from the rostral limbic system **(B)**. *Amyg,* Amygdala; *Ant,* anterior nucleus of the thalamus; *Hab,* habenula; *Ip,* interpeduncular nucleus; *M,* mammillary body; *MFB,* medial forebrain bundle; *OB,* olfactory bulb; *Pit,* pituitary gland; *Spt,* septum; *St Med,* stria medullaris. (Modified from MacLean, P.D.: Am. J. Med. **25:**611, 1958.)

and as more was learned about autonomic areas of the forebrain, the basic circuit was expanded to include structures such as the amygdala, which is also connected in a loop with the hypothalamus.

The data supporting this approach were derived largely from studying the results of lesions and electrical stimulation because various emotional behaviors could be evoked or perturbed by experimental manipulations of limbic structures. For example, stimulation of certain regions of the amygdala can evoke aggressive behavior, and amygdalar lesions can produce docility. Hypothalamic stimulation can also elicit defensive behavior, and this is modulated by concomitant stimulation of the amygdala. This reinforced the concept of cortical limbic structures that modulate the hypothalamic integration of somatic, autonomic, and endocrine components of emotional behavior. An early finding that stimulated this concept was that surgically separating the hypothalamus from its cortical influences produced an animal that showed *sham rage*. Such a preparation responds to mild stimulation with the extreme aggressive responses that are characteristic of rage.

Another important contribution to the limbic concept is a syndrome caused by temporal lobe lesions. The *Klüver-Bucy syndrome* includes (1) loss of the ability to detect and recognize the meaning of objects on the basis of visual cues (visual agnosia), (2) a tendency to examine objects orally (oral tendencies), (3) an inability to suppress attention to irrelevant stimuli (hypermetamorphosis), (4) hypersexuality with a loss of discriminability of appropriate sexual objects, (5) dramatic changes in dietary habits that include a loss of discriminability of appropriate food objects, and (6) a flatness of emotional responsiveness (taming). These temporal lobe lesions include both the neocortex and limbic cortex, and subsequent studies have allowed separation of the components of the syndrome that are attributable to each of these areas. Thus it is now appreciated that the changes in emotional behavior are primarily the result of damage to the amygdala, whereas such symptoms as visual agnosia reflect damage to visual areas of the temporal neocortex.

Although our knowledge of the organization of emotional behavior remains primitive, it is becoming clear that the concept of the limbic system may not be very useful. As research progresses, it becomes more productive to consider specific pathways that subserve particular functions. Brodal states*:

> It is difficult to see that the lumping together of these different regions under one anatomical heading, "the limbic lobe," serves any purpose. . . .It is even less justifiable to speak of a "limbic system.". . .It is the author's opinion that the use of the terms "limbic lobe" and "limbic system" should be abandoned.

*From Brodal, A.: Neurological anatomy, ed. 3, New York, 1981, Oxford University Press.

■ *Bibliography*

Journal articles

Ánand, B.K., and Brobeck, J.R.: Localization of a "feeding center" in the hypothalamus of the rat, Proc. Soc. Exp. Biol. Med. **77:**323, 1951.

Andersson, B.: Regulation of water intake, Physiol. Rev. **58:**582, 1978.

Cabanac, M.: Temperature regulation, Annu. Rev. Physiol. **37:**415, 1975.

Cohen, D.H., and Cabot, J.B.: Toward a cardiovascular neurobiology, Trends Neurosci. **2:**273, 1979.

De Groat, W.C.: Nervous control of the urinary bladder of the cat, Brain Res. **87:**201, 1975.

Hilton, S.M.: Ways of viewing the central nervous control of the circulation—old and new, Brain Res. **87:**213, 1975.

Kirchheim, H.R.: Systemic arterial baroreceptor reflexes, Physiol. Rev. **56:**100, 1976.

Klüver, H., and Bucy, P.C.: Preliminary analysis of functions of the temporal lobe in monkeys, Arch. Neurol. Psychiatry **42:**979, 1939.

Koizumi, K., and Brooks, C.M.: The integration of autonomic system reactions: a discussion of autonomic reflexes, their control and their association with somatic reactions, Ergeb. Physiol. **67:**1, 1972.

Paintal, A.S.: Vagal afferent fibres, Ergeb. Physiol. **52:**74, 1963.

Papez, J.W.: A proposed mechanism of emotion, Arch. Neurol. Psychiatry **38:**725, 1937.

Petras, J.M., and Faden, A.I.: The origin of sympathetic preganglionic neurons in the dog, Brain Res. **144:**353, 1978.

Smith, O.A.: Reflex and central mechanisms involved in the control of the heart and circulation, Annu. Rev. Physiol. **36**:93, 1974.

Wood, J.D.: Neurophysiology of Auerbach's plexus and control of intestinal motility, Physiol. Rev. **55**:307, 1975.

Books and monographs

Brodal, A.: Neurological anatomy, ed. 3, New York, 1981, Oxford University Press.

Cannon, W.B.: The wisdom of the body, ed. 2, New York, 1939, W.W. Norton & Co., Inc.

Goodman, L.S., and Gilman, A.: The pharmacological basis of therapeutics, ed. 6, New York, 1980, The Macmillan Co.

Jänig, W.: The autonomic nervous system. In Schmidt, R.F., editor: Fundamentals of neurophysiology, ed. 2, New York, 1978, Springer-Verlag New York, Inc.

Pick, J.: The autonomic nervous system: morphological, comparative, clinical and surgical aspects, Philadelphia, 1970, J.B. Lippincott Co.

chapter 21

Neural plasticity

The integrative functioning of the nervous system is based primarily on the connectivity among its basic elements—the neurons. To a large extent these connections are determined genetically and, once established, remain stable. However, it is now becoming increasingly apparent that there are a number of situations in which synaptic connections can be modified, and this modifiability is often referred to as *neural plasticity*.

Considerable contemporary research is focused on this phenomenon. The central questions are: What situations can produce synaptic modification? Is modifiability a property of all neurons or only certain classes of nerve cells? What are the mechanisms of this plasticity, and do the mechanisms mediating neuronal modification in different situations share common properties?

It is beyond the scope of this chapter to treat this captivating and important topic comprehensively. Rather, we present examples from three major classes of plastic change: (1) modifications of neurons and their connections that occur as a result of interactions with the environment during postnatal neural development, (2) the changes in connectivity that occur as a consequence of injury to the brain, and (3) the plasticity during learning or experience. As yet, scientists still lack a detailed understanding of the mechanisms that mediate these various modifications of the nervous system, but ongoing research is providing rapid progress toward such understanding.

■ Postnatal development

Until about 30 or 40 years ago, it was thought that the development of neural connections was essentially under genetic control and that environmental influences had a negligible effect on synapse formation. Thus neural development was thought to be essentially immutable. Scientists now know that the development of many neural connections can be significantly affected by the neuronal environment and are thus *plastic*. Two major environmental manipulations have been used to study these neuroplastic processes: early brain lesions and sensory deprivation. The former directly alters the neuronal environment and the latter does so indirectly by changing the levels of evoked activity among the relevant neurons. As we shall describe later from selected examples limited mostly to the visual system, neonatal lesions and/or deprivation cause many abnormal pathways to develop. Many others develop quite normally, presumably because they are under strict genetic control.

■ Normal development

Before considering the effect of the environment on synaptic development, we shall first very briefly describe a phenomenon that seems common to the development of many or most pathways in a wide range of vertebrate species. Development normally seems to pass through a phase whereby an excess of neurons and pathways develops. The unnecessary or redundant elements are eventually pruned as development proceeds.

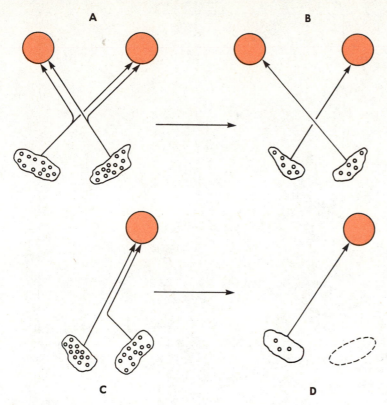

Fig. 21-1 ■ Cell death in the development of the avian isthmo-optic nucleus. Cells of this nucleus project to the retina (color). **A,** Normal early development. A bilateral projection with many cells is present. **B,** Normal adult. The projection is entirely crossed with fewer cells. Presumably, cells not able to sustain a projection (i.e., those projecting ipsilaterally) die and disappear. **C,** Early development with one eye removed. Both nuclei project to the remaining eye. **D,** Later development after early removal of one eye, as in **C.** Only the contralateral projection remains, and cells of the other istho-optic nucleus die and disappear. Note, however, that fewer cells remain in the intact nucleus than is normally the case, as in **B,** presumably because some cells of the ipsilateral nucleus successfully competed for synaptic sites before dying. (Redrawn from Lund, R.D.: Development and plasticity of the brain, New York, 1978, Oxford University Press.)

Numerous examples of this process can be cited. For instance, the adult avian retina (unlike the mammalian retina) receives massive input from the contralateral isthmo-optic nucleus of the midbrain. Neurons of this nucleus are significantly more numerous in the immature brain than in the adult, and the immature pathway is bilateral rather than contralateral, as it is in the adult (Fig. 21-1). Presumably those neurons which fail to make appropriate connections in the retina (i.e., mostly the ipsilaterally projecting neurons) die and disappear. As another example from studies of the monkey, retinogeniculate axons from each eye prenatally innervate all laminae of the lateral geniculate nucleus, and apparently excessive axonal branches are pruned to obtain the laminated ocular input seen in the adult (see Chapter 8). As a final example, the corpus callosum in immature rodents and cats includes fibers from neurons that do not contribute to this commissure in adults, and the immature callosal fibers innervate the entire hemisphere instead of the more limited pattern seen in adults (see Fig. 19-8, *B*).

It should be emphasized that, while the development of cellular and axonal elements appears to be nonmonotonic (i.e., excessive development of these elements is followed by removal of many), this is not the case for synaptic development (Fig. 21-2). That is, the absolute number of synapses increases essentially monotonically during development

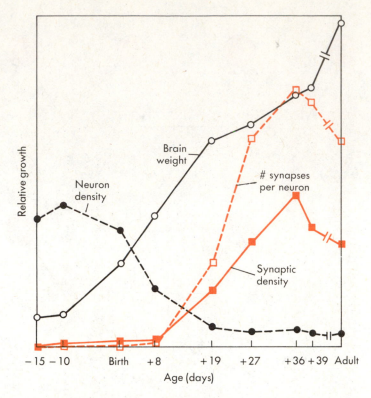

Fig. 21-2 ■ Neuronal development of the kitten's visual cortex. A large postnatal increase in brain weight is mostly due to growth of the synaptic neuropil. Because no new neurons are added postnatally as the brain grows, neuronal density actually decreases. (Redrawn from Cragg, B.G.: Invest. Ophthalmol. **11:**377, 1972.)

until maximum levels are reached approximately asymptotically in adulthood. Apparently the excessive axonal pathways seen during early development do not signify excessive synaptic numbers, since few synapses have been formed anywhere by this stage. It is not yet clear if the pruning of excessive projections and cells is a prelude to or a result of the dramatic increase in synaptic numbers.

■ *Critical period*

The examples of developmental neuroplasticity considered here all exhibit an early *critical period of development*, during which neuronal connections are sensitive or plastic to environmental influences and after which they are not. For instance, a kitten raised with visual deprivation, induced by closing or patching an eye, develops many abnormal connections related to that eye if, and only if, the deprivation occurs during at least part of the first 3 or 4 postnatal months. After this time, deprivation will not produce abnormal connections, and a normal sensory environment will not correct abnormalities that developed during deprivation. Similarly, neonatal lesions can lead to abnormal connections not seen after adult lesions. Adult lesions can lead to plastic changes, but these seem to involve different pathways and processes than are involved in developmental neuroplasticity.

■ *Neonatal lesions*

Animal studies. In a variety of vertebrate species and neuronal pathways, it has been demonstrated that neonatal lesions result in the development of aberrant pathways and cellular development that cannot be produced by later lesions. A critical period is thus evident. The plastic effects of two complementary types of lesions have been studied separately and in combination. In one, a set of afferent fibers to a structure is removed to determine what anomalous afferent fibers might develop in its place. In

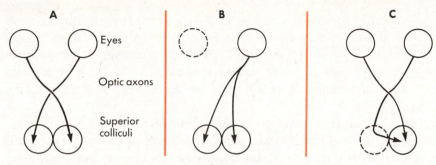

Fig. 21-3 ■ Effect of neonatal lesions on development of the rat's retinocollicular pathway. **A,** Normal development. The pathway is almost entirely crossed. **B,** Removal of one source of afferent input (left eye removed). The remaining eye develops bilateral projections. **C,** Removal of one target (left colliculus removed). The remaining superior colliculus receives input from both eyes.

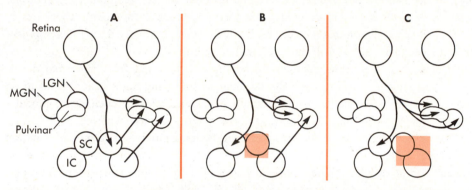

Fig. 21-4 ■ Effect of neonatal lesions on development of the hamster's retinofugal pathways. **A,** Normal development. The retinofugal pathways are practically completely crossed and innervate the lateral geniculate nucleus *(LGN)* and superior colliculus *(SC)*. Also shown is the ipsilateral projection from the SC to the pulvinar and that from the inferior colliculus *(IC)* to the medial geniculate nucleus *(MGN)*. **B,** Removal of the SC. The contralateral retina not only innervates the remaining SC (Fig. 21-3, *C*), but also innervates the denervated pulvinar. **C,** Removal of the SC and IC. The contralateral retina not only innervates the other SC and the denervated pulvinar, but it also innervates the MGN, even though the MGN is normally an auditory structure.

another, a termination site of an afferent pathway is removed to determine if the pathway will develop anomalous connections to another site.

Examples of the former process often involve removal of one eye neonatally to determine the effects on development of the other eye's connections. For instance, the normal adult rat's retinotectal pathway is nearly exclusively crossed (Fig. 21-3, *A*), but if the rat is raised with one eye removed, the remaining eye develops a bilateral retinotectal projection (Fig. 21-3, *B*). Similarly, early monocular enucleation in a kitten or monkey leads to an expanded retinogeniculate projection from the remaining eye into "inappropriate" laminae. There are also complementary examples of plasticity that develop in response to neonatal removal of a normal target zone for a pathway. In rodents, for instance, neonatal removal of one superior colliculus results in the development of binocular input to the remaining colliculus (Fig. 21-3, *C*) instead of the normally crossed retinotectal pattern (Fig. 21-3, *A*).

As might be expected, the most dramatic examples of developmental plasticity result from lesions that remove both a normal input to a structure and a normal termination site of an afferent pathway. For instance, the superior colliculus projects to the pulvinar, which normally receives no retinal afferents (Fig. 21-4, *A*; see also Chapter 8). Neonatal removal of the superior colliculus in rodents leads to growth of retinal afferent fibers into the denervated zone of the pulvinar (Fig. 21-4, *B*). Even more dramatic, if the neonatal tectal lesions involve the inferior colliculus or its brachium, then a major auditory pathway to the medial geniculate nucleus is eliminated (Fig. 21-4, *A*; see also Chapter 11), and aberrant retinal fibers will grow into this auditory structure and form synapses (Fig. 21-4, *C*).

Finally, unusual examples of cell death often result from neonatal lesions. Early removal of one eye in a bird leads to death and disappearance of cells of the contralateral isthmo-optic nucleus (Fig. 21-1, *C* and *D*), presumably because these cells never formed functional connections. More interesting is the result of early visual cortex removal in cats or monkeys. Not only do most geniculate neurons die and disappear, but so do a class of retinal ganglion cells. These are the X-cells; W- and Y-cells remain. As noted in Chapter 8, retinal X-cells project only to the lateral geniculate nucleus, and failure of their normal target cells to develop causes these retinal neurons to die and disappear. Retinal W- and Y-cells survive because they project additionally to neuronal targets, such as the superior colliculus, that survive the cortical removal. These observations support the notion that the cell death which occurs normally during development eliminates neurons that fail to form functional connections.

We emphasize that these examples of neuroplasticity result from abnormal development following neonatal lesions. Plastic changes following adult lesions tend to be much smaller or nonexistent. Also, although many examples of such developmental plasticity can be cited, not all pathways show such plasticity. These may be more strictly under genetic control. Furthermore, even within a single structural pathway that demonstrates developmental plasticity, it is not clear to what extent all cells or some subpopulation of cells form aberrant connections. For instance, there is very little evidence to date regarding the contribution of W-, X-, and Y-cells to plasticity that develops in the retinofugal pathways, although the X-cell pathway is uniquely affected by early cortex removal.

The mechanisms whereby these aberrant connections form are obscure. How much is due to failure of retraction of neonatally exuberant pathways and how much is due to lesion-induced abnormal growth, or "sprouting," is not clear. Both factors may be involved. Whether by failure to retract or by sprouting, two other processes seem to combine to determine the final pattern of connections that develop. First, axons seem to compete with one another for synaptic targets, and a denervated zone thus becomes a particularly attractive target for a developing axon because competition there is reduced. Second, a neuron tends to conserve its axonal terminal arbor, so that if it is prevented from growing at one point it tends to grow more at another. This is the so-called pruning hypothesis because of its analogous horticultural connotation. Removal of one eye means that an aberrant ipsilateral retinotectal pathway develops because less competition for synaptic space is associated with that colliculus. Removal of one colliculus serves to "prune" axons from the contralateral eye, and they then have a tendency to grow aberrant connections (e.g., to the ipsilateral, remaining colliculus). When denervation and pruning coincide, as in aberrant retinal input to pulvinar, the developmental plasticity is particularly evident. Finally, if neonatal lesions prevent neurons from forming functional connections, those neurons may die and disappear. This can even be a retrograde transneuronal process, as for retinal X-cells following neonatal cortex removal.

Clinical implications. Detailed studies of human brains that developed following neonatal lesions are generally unavailable or technically not feasible. However, we can reasonably suppose that the same neuroplastic changes occur. This may be the basis for the common observation that lesions in young children often lead to quite different symptoms than similar lesions in adults. To review an example cited in Chapter 19, a lesion to language areas of cortex in the adult typically leads to permanent aphasia. The same lesion in a child often results in little or no aphasia. Perhaps after the neonatal damage, aberrant pathways can form that subserve language functions. Obviously, such examples of developmental neuroplasticity could be most important clinically, and we need to know a great deal more about the underlying mechanisms.

■ *Monocularly deprived cats*

Sensory deprivation offers a more physiological means of studying developmental neuroplasticity than does the use of neural ablations. The deprivation-induced changes are more subtle and do not involve direct physical trauma. The primary change is limited to different levels of neural activity, and these can have profound effects on the developing brain.

The most thoroughly studied and best understood example of developmental neuroplasticity is probably the monocularly deprived kitten. These animals are raised to adulthood with monocular deprivation by suturing closed the lid over one cornea. The eye can be subsequently reopened to study its ability to influence neural activity or to test the cat's ability to see with that eye. Just a few examples of the effects of such deprivation on visual development will serve to underscore the profound influence of the sensory environment on the developing nervous system.

Geniculostriate abnormalities. Although the deprived retinal ganglion cells seem to develop normally, geniculate and cortical neurons do not (Fig. 21-5). Deprived geniculate cells (i.e., those in laminae innervated by the sutured eye) grow less than in the normal state and exhibit many abnormal response properties. Interestingly, these abnormalities are largely limited to Y-cells; the W- and X-cells are relatively unaffected by the deprivation. To a first approximation, geniculate cells in nondeprived laminae develop normally.

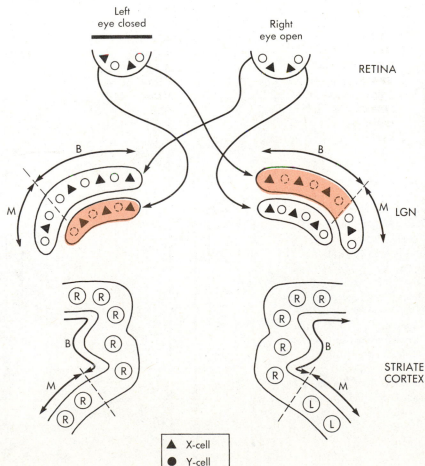

Fig. 21-5 ■ Summary of some of the abnormalities seen in the visual system of cats raised with monocular deprivation due to maintained closure of one eye (the left eye in this example). Ganglion cells in the deprived retina develop normally. In geniculate laminae receiving input from the deprived eye, the X-cells develop fairly normally, but the Y-cells do not. Y-cell abnormalities, however, are limited to the deprived, binocular segment of the nucleus *(color)*. In the binocular segment of striate cortex nearly all cells have purely monocular receptive fields for the right eye (instead of the binocular fields seen in normal cats), but many normal fields can be seen for the deprived eye in the deprived, monocular segment. *LGN,* Lateral geniculate nucleus; *B,* binocular segment; *M,* monocular segment; *R,* cortical cell with receptive field for only the right eye; *L,* cortical cell with receptive field for only the left eye.

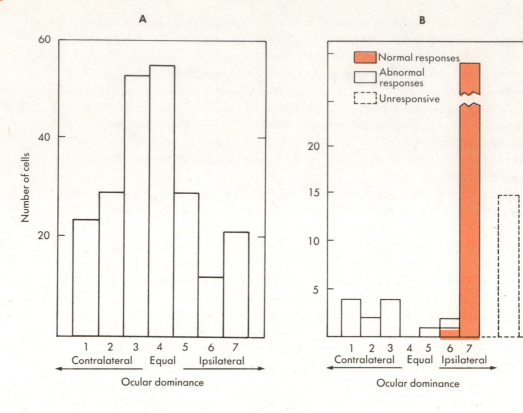

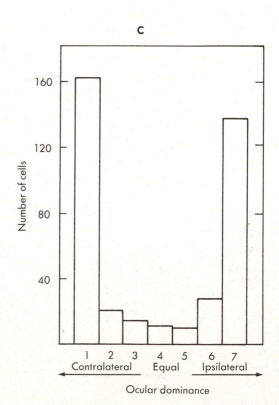

Fig. 21-6 ■ Ocular dominance histograms of cells in the binocular segment of the cat's striate cortex. Seven classes of ocular dominance are shown: *1,* responds only to stimulation of contralateral eye; *2,* responds well to contralateral eye and poorly to ipsilateral eye; *3,* responds somewhat better to contralateral than to ipsilateral eye; *4,* responds equally well to either eye; *5,* responds somewhat better to ipsilateral than to contralateral eye; *6,* responds well to ipsilateral eye and poorly to contralateral eye; and *7,* responds only to ipsilateral eye. **A,** Normal cat. All cells respond to visual stimuli and most have binocular receptive fields. **B,** Cat raised with monocular suture of contralateral eye. Most cells have normal receptive fields only for the (nondeprived) ipsilateral eye. The few with fields for the (deprived) contralateral eye respond abnormally, and many neurons are unresponsive. In the other hemisphere, a mirror-image pattern is evident, since the normal cells are class 1 (i.e., with receptive fields for the nondeprived eye). **C,** Cats raised with exotropia (divergent strabismus). All cells are normally responsive, but they tend to have monocular receptive fields for one or the other eye. (Data from Wiesel, T.N., and Hubel, D.H.: J. Neurophysiol. **26:**1003, 1963; Hubel, D.H., and Wiesel, T.N.: J. Neurophysiol. **28:**1041, 1965.)

Even more dramatic are the effects seen in striate cortex (Fig. 21-6, *B*). Here nearly all cells can be normally driven through the nondeprived eye but fail to respond to stimulation of the deprived eye. Thus, instead of the normal binocular activation of these cells, it seems as if the deprived geniculate cells, even the fairly normal W- and X-cells, fail to develop functional geniculostriate connections.

Critical period. These abnormalities show a distinct critical period. For instance, if one eye is sutured closed at birth and then opened at 4 to 6 months of age, at which time the other eye is sutured closed for an additional several years, the deficits are as just described and are limited to those related to the initially deprived eye. Also, eye closure for just a few days during the critical period can lead to permanent abnormalities.

Binocular competition. There is considerable evidence that much of the neuroplasticity due to monocular deprivation results from competitive interactions among developing axons for synaptic connections. In some form of *binocular competition,* pathways from the open eye develop at the expense of those from the closed eye. This is indicated fairly clearly by a comparison of the development that occurs in the binocular segment with that in the monocular segment (Figs. 21-5 and 21-7).

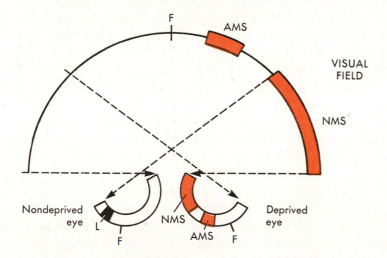

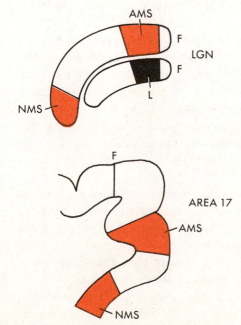

Fig. 21-7 ■ Effects of monocular deprivation on natural and artificial monocular segments (*NMS* and *AMS*) in cats. If a lesion *(L)* is placed in the retina of the nondeprived eye of a neonatal cat, two monocular segments exist for the deprived eye. The NMS corresponds to the extreme nasal retina that views a portion of peripheral visual field beyond the view of the other eye. The AMS corresponds to the retinal region that maps the same portion of visual space as the region lesioned in the other eye. Because of the retinotopic map in the lateral geniculate nucleus *(LGN)* and area 17, the NMS and ALS can be found in these structures. As summarized in Fig. 21-5, the deprived NMS develops relatively normally in the LGN and area 17. Likewise, the deprived AMS develops equally normally, and perimetry tests indicate relatively normal vision for the deprived eye only in the NMS and AMS regions of the visual field. Since the deprived AMS develops fairly normally even though it represents a fairly central portion of visual field, this implies that the deprived monocular segment is spared many of the deleterious consequences of monocular deprivation, not because it represents peripheral visual field but rather because it represents monocular visual field. *F,* Fixation point and its representation in the retina, LGN, and area 17. (Redrawn from Sherman, S.M., and Spear, P.D.: Physiol. Rev. **62:**738, 1982.)

Because the visual field is retinotopically mapped onto the visual parts of the brain, one can define the *binocular segment* of the striate cortex, lateral geniculate nucleus, superior colliculus, etc., as that portion onto which the binocularly viewed, central region of field is mapped. The remaining *monocular segment* corresponds to that extreme lateral portion of the visual field on each side that can be viewed only by the eye on that side. Fig. 21-7 shows that, in addition to this natural monocular segment *(NMS)*, an artificial monocular segment *(AMS)* can be created for one eye by placing a retinal lesion *(L)* in the other.

This division is most interesting because of the different effects of early monocular deprivation on the two segments. Nearly all of the abnormalities just noted for the deprived eye among geniculate and striate cortex neurons are limited to the binocular segment (Figs. 21-5 and 21-7). In the natural or artificial monocular segment innervated by the deprived eye, a segment just as deprived of pattern vision as the binocular segment, geniculate cells grow to normal size; response properties even for Y-cells develop fairly normally; and many striate cortex neurons develop normal receptive field properties for the deprived eye. Subtle abnormalities do develop in the deprived monocular segment, but they are small compared with those seen in the binocular segment. A similar pattern can be found in the superior colliculus and in extrastriate cortical areas that have been studied in these cats. Furthermore, the behavioral capacity with the deprived eye is more normal in response to stimuli placed in the monocular segment than to those in the binocular segment.

Fig. 21-8 diagrammatically illustrates the main conclusion that can be derived from these observations. That is, central pathways related to each eye compete with one another for synaptic connections during development. For illustrative purposes we shall use the geniculostriate connections as an example, although we cannot yet be absolutely certain as to the specific site or sites of this binocularly competitive development. During normal development, pathways from one or the other eye gain no systematic competitive advantage; a balance is struck, and geniculate neurons develop normally with convergent binocular afferentation of cortical cells (Fig. 21-8, *A*). Monocular deprivation upsets the balance (Fig. 21-8, *B*). Deprived geniculate neurons develop at a competitive disadvantage because of their reduced evoked activity. This allows their antagonists from normally innervated laminae to capture the vast majority of synaptic space in the cortex at the expense of the disadvantaged, deprived neurons. As a result, the latter neurons fail to develop normally, and cortical cells become effectively innervated only by the nondeprived geniculate neurons.

However, by definition, a deprived cell can be placed at such a competitive disadvantage only in the binocular segment where such competitive interactions are possible. The deprived cell in the monocular segment may develop more slowly and less completely than normal, but it need not compete against a favored antagonist. Thus many normal connections can be formed in this segment despite the deprivation.

These normal connections in the deprived monocular segment amply illustrate that the deficits which develop in the binocular segment are not due to deprivation per se, but rather are due to the effects of competitive interactions that become unbalanced during monocular deprivation.

Neural basis of amblyopia. The rather serious disruption in the development of the Y-cell pathway during deprivation may explain some of the profound amblyopia (i.e., reduced visual capacity) exhibited by the deprived eye. In Chapter 8 we suggested that the Y-cell pathway, because of its sensitivity to lower spatial frequencies, seems sufficient and perhaps even necessary for normal pattern vision. Its disruption during deprivation results in poor sensitivity to low spatial frequencies and thus amblyopia.

Fig. 21-9 compares the spatial contrast sensitivity among several groups of cats. These include normally reared cats before and after bilateral removal of striate cortex,

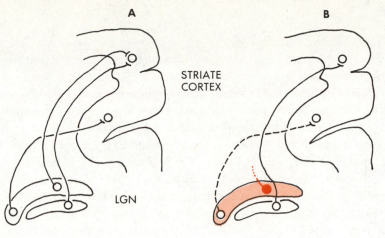

Fig. 21-8 ■ Hypothesis of binocular competition used to explain observations summarized in Figs. 21-5 and 21-7. The hypothesis suggests that development largely proceeds via a process of competitive interactions between the pathways related to each eye. For the sake of clarity, the hypothesis is described as if these interactions occur among geniculostriate synapses, although the actual site of the interactions is currently unknown. **A,** Normal development. Normally, there is an enormous postsynaptic proliferation of geniculocortical synapses, and they compete with one another for space onto and control of the cortical cell. With a normal binocular environment, a competitive advantage in this process is conferred to neither geniculate lamina's neurons, a balance is struck, and binocular cortical cells develop. **B,** Monocular deprivation. The cell in the binocular segment of the deprived lamina *(colored circle)* is somehow placed at a competitive disadvantage with respect to its nondeprived rival. Perhaps its reduced level of activity leads to this disadvantage. The result is that the nondeprived geniculate neuron develops exclusive synaptic control of the cortical cell. However, by definition, the deprived cell in the monocular segment *(open circle)* cannot be placed at a competitive disadvantage, since no nondeprived cells are available to innervate its target neuron. Thus, although deprived as much as its neighbor in the binocular segment, the monocular segment cell is free to develop synaptic control of the cortical cell. Thus, to the extent that development in the deprived monocular segment is less affected than that in the binocular segment, a mechanism of binocular competition is implicated. If, on the other hand, equal abnormalities were seen in both segments, there would be no reason to infer such a competitive process of synaptic development. (Redrawn from Sherman, S.M. In Freeman, R.D., editor: Developmental neurobiology of vision, New York, 1979, Plenum Press.)

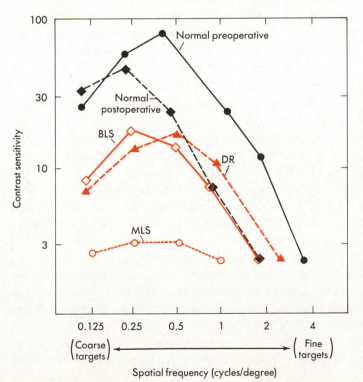

Fig. 21-9 ■ Behavioral contrast sensitivity functions of cats. These functions plot the inverse of the minimum contrast necessary for the animal to detect a sine wave grating of various spatial frequencies. Low spatial frequencies correspond to coarse or large targets, and high frequencies represent fine or small targets. The function for normally reared cats is shown in black before *(solid line)* and after *(dashed line)* bilateral removal of striate cortex, which also effectively eliminates the X-cell pathway. Functions for cats raised with visual deprivation are shown in color and include binocular lid suture *(BLS)*, total dark rearing *(DR)*, and monocular lid suture *(MLS)*. The MLS data are for the deprived eye only; the nondeprived eye's function is normal. Where sensitivity to low spatial frequencies is good (normal and destriate cats), spatial vision is good; where sensitivity to these frequencies is poor (BLS, DR, and MLS cats), spatial vision is poor. (Redrawn from Lehmkuhle, S., et al.: J. Neurophysiol. **48:**372, 1982.)

monocularly deprived cats, and binocularly deprived cats. (Binocular deprivation also leads to somewhat abnormal development of the Y-cell pathway.) Removal of striate cortex completely interrupts the X-cell pathway but not the Y-cell pathway. The destriate cats exhibit far better spatial vision and better low-frequency sensitivity than do the deprived cats, even though there may be no obvious differences in spatial acuity. This suggests that abnormal development of the striate cortex is not sufficient to explain the amblyopia of visually deprived cats, since its total removal in normally raised cats leads to less of a visual deficit than does deprivation rearing. A more likely explanation for the amblyopia seems to be the Y-cell abnormalities, since geniculate Y-cells in the cat innervate many cortical areas. Consequently, a vast expanse of cortex develops abnormal geniculate input. In any case, the neuroplastic changes in the visual system caused by visual deprivation have rather dramatic effects on final visual capacity.

■ *Strabismic and contour-deprived cats*

We know far more about development during monocular suture than any other form of deprivation, but the limited results of studying related forms of visual deprivation offer further insights into the developmental mechanisms at work. Two examples that employ more limited and specific forms of deprivation in kittens are described.

Strabismus. First are a series of experiments whereby kittens are raised with an artificial strabismus, usually an *exotropia* created by transection of one or both medial recti. An exotropia is an outward deviation of the visual axes such that they cross behind the eyes. These exotropic kittens tend not to develop a favored or dominant eye; rather, they alternate fixation between the eyes and acquire good vision with either eye. These kittens are not deprived of spatial or temporal patterns, but they are deprived of normal synchronous binocular vision. The exotropia prevents bifixation, and thus a target of interest cannot simultaneously be brought onto homonymous regions of each retina. The consequence for development is that cortical cells fail to develop normal binocular inputs. Instead, they can be driven only by one or the other eye (Fig. 21-6, *C*). These monocular receptive fields are otherwise normal. Patches of "right eye" or "left eye" cells seem to become randomly distributed through the cortex.

This pattern of monocular cells is thought to be another consequence of binocular competition. During the binocularly competitive development for geniculocortical synapses, the asynchronous stimulation somehow prevents a binocular balance from being struck, and one eye's pathway comes to win the competition. These victories and losses appear to occur in patches that might bear some relation to ocular dominance patches described in Chapter 8. If the cell did not lose one of the eye's inputs, information about the target location would be lost, since each homonymous retinal area now points to different targets. The competition ensures that the inputs to the cortical cell are synchronous and not confusing.

Contour deprivation. The second example is the consequence of raising a kitten in an environment that is limited to contours of a single orientation, such as vertical. Normal cortex would include cells that are nearly all orientation selective to a single small range of orientations, and all preferred orientations are represented by the whole neuronal population (Fig. 21-10, *A*). In cats raised with only vertical contours, nearly all of the orientation selective cortical cells that develop have a preferred orientation for vertically oriented stimuli (Fig. 21-10, *B*); many other cells fail to develop orientation selectivity and respond to a stimulus of any orientation. Presumably, cells in columns destined to develop preferences for vertical orientations receive normal stimulation and develop normally. Cells in other columns are deprived of normal visual stimulation and thereby fail to develop normal orientation selectivity. A similar result is seen for rearing with horizontal contours (Fig. 21-10, *C*).

Limits of developmental plasticity. These examples serve to emphasize the limited consequence for development of the sensory environment. There seem to be definite

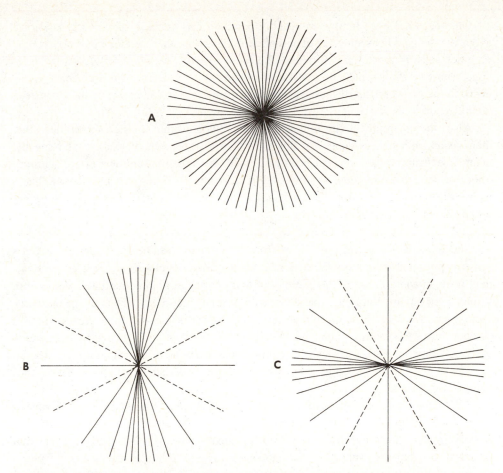

Fig. 21-10 ■ Schematic distribution of preferred orientations for population of neurons in cat striate cortex. The orientation of each line represents for each cell the stimulus orientation that evokes the best response. Solid lines indicate normal orientation selectivity; dashed lines indicate cells with poor orientation selectivity, although a preferred orientation can be determined. **A,** Normal cats. All orientations are equally represented among the neuronal population. **B** and **C,** Cats raised with environment restricted to viewing vertical orientations **(B)** or horizontal ones **(C).** This can be achieved by fitting the animals with special goggles with parallel lines focused onto the retina. With experience of only vertical contours **(B),** nearly all normal cells prefer vertically oriented stimuli, and likewise for horizontal experience **(C).** If both eyes view the same contours, the cells tend to be binocular. If one eye (e.g., the left) views vertical and the other (e.g., the right) views horizontal, the cells tend to be monocular with an orientation preference equivalent to its rearing experience. Not shown are numerous cells without orientation selectivity.

genetic constraints within which the environment can affect development. For instance, exotropia would not prevent binocular activation of cortical cells and the development of binocularity if the retinotopic map were sufficiently plastic. That is, if the map from each eye could be shifted enough to compensate for the strabismus, cortical cells could receive synchronous binocular activation and normal binocularity might develop. However, the visual system is not so plastic.

Likewise, not all cortical cells develop vertical preferred orientations during rearing limited to vertical contours. This observation indicates that cortical circuitry is not a plastic template waiting to be formed by the visual environment. Instead, it seems that genetically predisposed circuits are validated or discarded based on the environment. Only those orientation-selective circuits designed for vertical targets can develop in these cats, and representations of other orientations cannot.

■ *Other studies of sensory deprivation*

Our focus has been limited to visually deprived cats because rather little is known about the detailed consequences of visual deprivation in other species or of deprivation of other sensory modalities. Limited studies of visually deprived monkeys, galagos, dogs, and tree shrews indicate that many of the same processes that are seen for cats occur in the development of these species. For instance, binocular competition has been identified in all these species.

Also, the environment can influence the developing brain through modalities other than vision, although evidence for this is limited to a handful of studies, and specific neural mechanisms have not yet been addressed. Nonetheless, animals raised with auditory or somatosensory deprivation develop clear abnormalities in those senses. There is every reason to believe that the total sensory environment contributes significantly but within limits to neural development.

■ *Clinical significance*

The message for human pediatrics from these animal studies is clear: the infant's sensory environment is a crucial factor in normal brain development. This is most evident from an appreciation of the etiology of many forms of amblyopia that develop from visual deprivation. Numerous parallels with visually deprived cats are apparent, although by no means is it clear that cats and humans share the same developmental mechanisms. Nonetheless, if an infant is raised with any ocular abnormality that distorts the visual environment, a permanent amblyopia is likely to result. Such distortions can be caused by corneal scarring, cataracts, ametropia, strabismus, ptosis of the eyelid, and even therapeutic patching.

All of the available evidence suggests that the deficits caused by these forms of deprivation exhibit a critical period that extends through the first 5 to 10 years of life. One example of this comes from studies of astigmatic patients. Astigmatism means that all target orientations cannot be simultaneously focused. The patients studied focused one meridian and failed to focus the orthogonal one. This astigmatism was corrected optically with an appropriate cylindrical lens. If the correction was made in infancy or early childhood, no substantive amblyopia developed. If the correction was not made until adulthood, an amblyopia specific to the previously defocused meridian developed that was evident despite optical correction. Astigmatism of late onset (i.e., after the critical period) did not lead to such a meridional amblyopia, and vision was normal following appropriate optical correction. These results indicate both that a critical period of development exists and also that the resultant amblyopia is fairly specifically related to the precise nature of the visual deprivation.

Because of this and similar observations, it is now abundantly clear that the causes of visual distortion should be removed as soon as possible in infants to avoid permanent amblyopia. That is, a cataract should be removed, strabismus corrected, and myopia eliminated as soon as diagnosed. This current view is in stark contrast to the prevailing opinion of 20 or 30 years ago, when children were given a chance to outgrow these symptoms, and attempts to correct them were postponed until late in the critical period. In 1970 von Noorden addressed this issue*:

> Strabismus, when neglected in childhood, may cause severe disturbances or loss of binocular functions, and a decrease of visual acuity in the deviated eye, even below the level of legal blindness. Most of these serious consequences can be prevented by early diagnosis and appropriate medical or surgical therapy. It is a tragic fallacy for parents of strabismic children to be told by their family physician that strabismus is usually outgrown as the child gets older, or that, if surgery needs to be done, there will be time enough to do it before the child enters school.

The importance of the critical period also dictates caution in the early treatment of visual disorders. For instance, when many young children begin to show the first signs

*From von Noorden, G.K.: Surg. Clin. North Am. **50**:885, 1970.

of a weak or amblyopic eye, this can often be reversed through forced usage of that eye by patching the good or dominant eye. This is still regarded as an excellent therapy, but it must be used carefully with regard to the effects of even a few days' deprivation of a healthy eye. (Remember that a few days' monocular occlusion in a kitten produces permanent defects in the central pathways related to that eye.) In 1973, von Noorden wrote about patch therapy for a deviating, amblyopic eye in infant strabismus*:

> . . . we no longer occlude the dominant eye of children under two years with strabismus for more than one week without re-examining the patient. If longer periods of occlusion are required to improve the . . . deviated eye, we advocate switching the patch to the amblyopic eye for two days each week to enforce use of the occluded eye. Likewise, the advisability of treating external eye disease with an eye patch in infants should be re-evaluated. . . .

Finally, although our best understanding of the clinical consequences of the critical period involve the visual system, we have every reason to believe that these observations can be extrapolated in a general sense to neural development. Anything that affects the developing neuronal environment, including sensory stimuli, motor activity, learning and motivational experience, and nutrition, may have an important influence on the final product of neural development.

■ *Response to injury of the adult brain*

If axons in the central nervous system of poikilothermic vertebrates are damaged, they have the capacity to regenerate. For instance, if the optic nerve of a fish or frog is transected, the distal fragments will degenerate, as in homeothermic vertebrates. However, the proximal axonal stumps will then regenerate and reestablish connections with the appropriate retinofugal targets, such as the optic tectum. These regenerated connections eventually become functional, such that a nearly normal pathway is reestablished.

In contrast, homeothermic vertebrates show such regeneration only in the peripheral nervous system. With injury of central pathways, regeneration does not proceed to functional completion. Why successful regeneration does not occur centrally in homeothermic vertebrates is a question of obvious importance that is currently receiving considerable investigative attention.

Although regeneration does not occur in the adult mammal, there are plastic changes that follow injury to the central nervous system. The most thoroughly studied example of such injury-induced modification is the phenomenon of *axonal sprouting*. Liu and Chambers first demonstrated such sprouting in adult mammals in the 1950s (Fig. 21-11). They unilaterally transected the dorsal roots for several segments above and below a dorsal root that was left intact. After several months they then used anatomical techniques to compare the projection pattern of the fibers from the remaining dorsal root on that side with the projection pattern of the corresponding dorsal root on the control side of the spinal cord. The projections of the dorsal root on the experimental side were found to have extended dramatically, presumably reflecting the growth or sprouting of the axons into nearby denervated zones that were left vacant by the previous dorsal root transections.

Such axonal sprouting has now been shown to occur in many structures of the mammalian central nervous system, including retinofugal targets, the hippocampus, the septum, and the red nucleus. In some cases ultrastructural and electrophysiological studies have demonstrated that the new connections resulting from such sprouting have indeed formed functional connections. For example, the red nucleus receives prominent projections from both the cerebellum and motor cortex. The cerebellar projection is preferentially distributed to the soma and proximal dendrites of the rubral neurons, whereas the cortical projection is primarily on more distal dendrites. If the cerebellar projection is

*From von Noorden, G.K.: Invest. Ophthalmol. **12:**721, 1973.

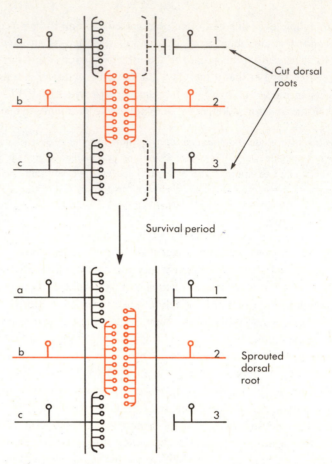

Fig. 21-11 ■ Lesion-induced sprouting in the mammalian spinal cord. Normally fibers from dorsal roots *b* and *2* have similar central projections *(upper)*. If fibers from dorsal roots *1* and *3* (neighboring segments to *2*) are cut, their central connections will degenerate. After a survival period, fibers from dorsal root *2* "sprout" to occupy some of the denervated space so that their terminal extent is greater than that of fibers from dorsal root *b (lower)*.

destroyed, then the cortical projection sprouts to establish more proximal connections at the synaptic sites that have been vacated by degeneration of the cerebellorubral pathway. These new connections are functional, and they increase the influence of the motor cortex on the red nucleus. In this case the evidence suggests that such injury-induced plasticity may contribute to the recovery of motor behavior.

The phenomenon of axonal sprouting implies that denervated portions of dendrites and somata are capable of attracting collateral inputs from nearby intact axons. Moreover, such plasticity occurs in many areas of the adult mammalian brain. However, it is important to recognize that not all pathways seem equally capable of sprouting, and there are several examples of failure to demonstrate sprouting.

The adult mammalian nervous system is thus capable of at least a limited plasticity in response to injury. To what extent axonal sprouting operates as a mechanism for establishing new functional connections after injury is not yet clear. Moreover, the mechanisms involved in axonal sprouting are not fully understood, and it is unclear why the capacity of the developing nervous system to sprout in response to injury diminishes with maturity.

■ *Learning and memory*

The capacity of at least certain classes of neurons to change their behavior as a function of environmental inputs is an exceedingly important property of the nervous system, since it subserves learning and therefore adaptive behavior. In the broadest sense *learning* refers to a process that mediates a change in behavior as a result of experience. *Memory* is the end point of the process and refers to the actual storage of information representing the experience-induced modification. It is generally agreed that such storage of information involves actual physical changes in neurons. The mechanisms of such change have commanded intense interest for many decades, and in recent years exciting progress has been realized in our understanding of the cellular changes that mediate information storage.

Controversies have raged over the years among psychologists as to whether there are different kinds of learning and, if so, what the different classes of learned behavior might be. This is not the appropriate forum to review such arguments, but it is worthwhile to distinguish between two very broad classes of learned behavior—*nonassociative* and *associative*. Nonassociative learning does not require that an organism learn a predictive relationship between stimuli. The simplest example of such learning is *habituation* during which a particular stimulus elicits some unlearned response that declines in magnitude (habituates) with repeated presentation of that stimulus. In teleological terms the organism is learning that the stimulus is not biologically meaningful in a given context. Therefore the organism becomes increasingly unresponsive to it (i.e., it learns not to respond or to inhibit responding). A commonplace example would be a dog or a cat that orients its head and ears toward a novel auditory stimulus. With repetition of that stimulus, such orienting behavior will attenuate or habituate if no meaningful event is associated with the occurrence of the stimulus.

Another kind of nonassociative learning is *sensitization*. If, for example, some noxious event occurs, this may increase an organism's responsiveness to other stimuli that would normally be neutral. If a dog or cat experiences a painful stimulus of some sort, then the subsequent occurrence of the auditory stimulus might elicit more vigorous orienting behavior than normally occurs. Telelogically one could view this kind of learning as an adaptive behavior that prepares the animal for responding to a potentially threatening environment.

In associative learning an organism acquires information about the relationship among stimuli. In *classical conditioning*, described extensively by Pavlov, the temporal pairing of a neutral stimulus, the *conditioned stimulus*, with a stimulus that elicits an unlearned response, the *unconditoned stimulus*, can endow the conditioned stimulus with the novel capacity to elicit a response. Often this conditioned response is similar to that evoked by the unconditioned stimulus, although that need not be the case. A traditional example of such learning is provided by Pavlov's observation that his experimental dogs salivated on the appearance of the caretaker who fed them. Salivation is an unconditioned response to ingestion of food, the unconditioned stimulus. The visual and auditory stimuli associated with the caretaker normally would not elicit salivation, but these became conditioned stimuli because they were associated with the delivery of food. Consequently, these stimuli gained the capacity to elicit a conditioned response, salivation, which in this example resembled the unconditioned response to food intake.

The other major form of associative learning is *instrumental* or *operant conditioning*. In this case responses that are immediately followed by an event that has reinforcing value will show an altered probability of occurrence. If the reinforcing event is positive, such as presentation of food, then the probability of response will increase. If the reinforcing event is negative, such as pain, then the probability of response will decrease. As in classical conditioning, a predictive relationship is being learned.

A vast amount of literature has been produced over the years describing these forms of learning and the rules that govern them. Although it remains unclear as to whether these kinds of learning exhaustively describe all learned behavior, habituation, sensitization, classical conditioning, and instrumental conditioning have provided extremely useful models for the experimental analysis of learning and memory and the mechanisms that mediate such training-induced modification of the nervous system.

■ Stages of memory

An important concept that has emerged from experimental analyses of learning is that the storage of information, or memory, may involve different stages. For instance, early in learning the information may be represented in short-term memory and over time is then transferred in some way to long-term memory. The number of stages, their relationships to each other, and the mechanisms underlying each remain an area of debate. Indeed, the usefulness of even distinguishing among different phases of memory has been questioned by some. However, at this time a distinction at least between short-term and long-term memory retains some heuristic value in understanding a broad range of experimental and clinical findings. For example, head trauma often results in a *retrograde amnesia* where the memory of recently learned events is lost, whereas memory for older events remains unimpaired. A similar phenomenon occurs after electroconvulsive shock therapy, which is used in the treatment of severe depression disorders.

Such observations suggest a more fragile short-term memory that ultimately is transformed into a more robust long-term form of storage, and it is generally agreed that long-term storage must be mediated by physical changes in the involved neurons. However, whether there is a qualitative difference between these forms of storage has not been resolved, and further understanding will most likely derive from the investigation of the cellular mechanisms of learning.

■ Localization of information storage

Neuroscientists have been concerned for many years with what parts of the brain are involved in storing information. Lashley devoted much of his career to investigating this question by using the approach of ablating areas of the brain and assaying the effects on memory. He coined the term *engram* to describe the storage of information and ultimately concluded that there were no specific sites of engram formation; rather, such sites are widely distributed throughout the nervous system.

Since Lashley's time significant progress has been made, and it is now clear that for any given learning task there are specific neural pathways that participate in the performance of that task. Moreover, these pathways may differ, depending on the nature of the specific learning task. Thus there is some localization of storage, since in a given learning situation not all areas of the brain are necessarily involved. However, it is likely that in any specific situation changes occur along much of the involved pathways, though not necessarily in all segments of those pathways. Thus it is unlikely that there exist only a few structures where memory is stored. A related question that remains to be resolved is whether all neurons have the capability of storing information on a long-term basis, or whether there are specific classes of neurons that are specialized for this function.

The neuropathology of learning and memory disorders in humans has not contributed substantially to our understanding of these phenomena. A persistent finding in human investigations is that lesions involving the temporal lobe and hippocampus and certain midline diencephalic structures, such as the dorsomedial nucleus of the thalamus, result in a generalized *anterograde amnesia*. That is, previously stored information is preserved, but the ability to store new information is impaired. In Korsakoff's syndrome these lesions result from a thiamine deficiency following inappropriate nutrition in chronic alcoholism. Although experimental lesions of these structures in animals can produce similar effects, what remains puzzling is why these are the only structures that have been persistently associated with memory deficits in humans.

One difficulty in basing conclusions on lesion data is the important distinction between learning and performance. When we measure learning by any organism behaviorally, what is actually being measured is performance. Performance depends on learning, but it is also a function of factors such as appropriate processing of sensory information, the capability to express appropriately the required motor responses, and motivation. For example, in a learning task that requires a movement, interrupting the motor nerves to the involved muscles would abolish the learned response. However, such a lesion clearly does not compromise memory but merely prevents the expression of it. Thus the deficit is in performance and not learning. Separating learning and performance when lesions of the central nervous system are involved is often not a simple task, and this confounds much of the literature. Indeed, many of the so-called learning disabilities in children may not necessarily involve impairment of learning or memory, since they could well represent rather subtle performance deficits. Analogously, the occurrence of anterograde amnesia after temporal lobe lesions does not necessarily imply that information is stored in that region of the brain.

The experimental literature on animals is moving into an exciting and productive period with respect to mechanisms of memory. The study of human memory is, in many respects, much more difficult, and information will accumulate more slowly. However, even in this context there have been exciting developments. For example, recent studies of dementia are progressing toward an identification of relevant neural pathways. In Alzheimer's disease, characterized by a severe memory loss, evidence is pointing toward a compromise of the cholinergic input to the cortex, which originates in large measure in the nucleus basalis.

For many years investigators were seriously frustrated in their efforts to describe the cellular mechanisms of learning. However, in the past decade there have been significant advances. These have derived mainly from the development of experimental animal models in which the relevant circuitry for some well-specified learned response has been delineated. Describing such circuitry in the complex brains of vertebrates has been quite difficult, and there are but a few experimental models involving vertebrate species. However, the use of simpler invertebrate systems reduces the complexity of mapping the relevant pathways, and there has been significant progress when such systems have been used.

The available invertebrate and vertebrate models have generated exciting information. For example, results from vertebrate models suggest that training-induced modification in a given learning task is probably distributed extensively over the involved neural pathways, including the sensory pathways. The invertebrate models are progressing to a stage of development where a true cell biology of learning is at hand. Some of these models involve reduced systems consisting of a monosynaptic connection between a single sensory neuron and a single motoneuron (Fig. 21-12). In such systems short-term habituation and sensitization have now been studied intensively. Such investigations have shown that the neuronal changes are presynaptic, involving modification of

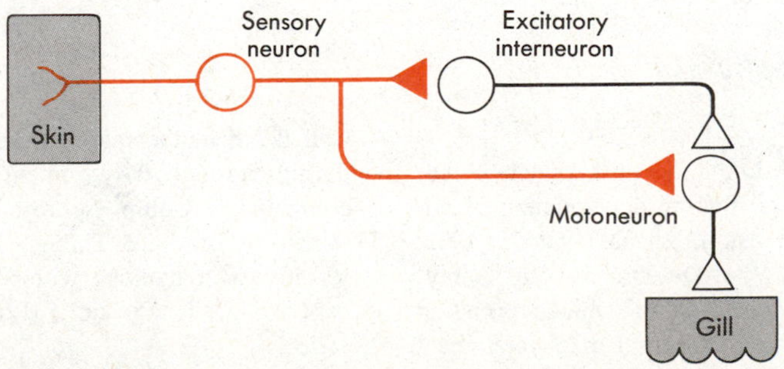

■ *Cellular mechanisms of learning*

Fig. 21-12 ■ Simplified neuronal circuit for the gill-withdrawal reflex in the marine snail *Aplysia*. The site of plasticity in this reflex that underlies its habituation is at the terminals of the sensory neurons on the central target cells *(color)*—the interneurons and motoneurons. (Redrawn from Kandel, E.R., and Schwartz, J.H.: Principles of neural science, New York, 1981, Elsevier/North Holland.)

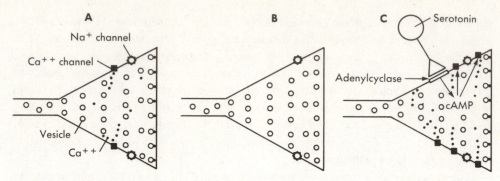

Fig. 21-13 ■ Model of short-term habituation and sensitization. **A,** In the control state an action potential in the terminal membranes of the sensory neurons opens a number of Ca^{++} channels *(black squares)* in parallel with the Na^+ channels *(hexagons)*. As a result some Ca^{++} flows into the terminals and allows a certain number of synaptic vesicles to bind to release sites and be released. **B,** Repeated action potentials in the terminals decrease the number of open Ca^{++} channels in the sensory terminal and, in the limit, may shut them down altogether. The resulting depression in Ca^{++} influx functionally inactivates the synapse by preventing synaptic vesicles from binding. **C,** Sensitization is produced by cells thought to be serotonergic. In the terminals serotonin acts on adenylate cyclase, which stimulates the synthesis of cAMP. cAMP in turn acts on a cAMP-dependent protein kinase to phosphorylate a membrane protein thought to be the K^+ channel, and this leads to a decrease in the repolarizing K^+ current and a broadening of the action potential. The increase in the duration of the action potential increases the time during which Ca^{++} channels can open, leading to a greater influx of Ca^{++} and greater binding of vesicles to release sites, and therefore to an increased release. (Redrawn from Kandel, E.R., and Schwartz, J.H.: Principles of neural science, New York, 1981, Elsevier/North Holland.)

transmitter release at the terminals of the sensory neuron. In habituation there is a decrease in transmitter release, whereas in sensitization there is an increase. Most recently, studies of nonassociative learning in *Aplysia,* a gastropod mollusc, have permitted a biophysical analysis of the presynaptic depression and facilitation that mediate habituation and sensitization, respectively. These studies have implicated changes in the Ca^{++} current at the presynaptic terminals (Fig. 21-13).

We have yet to determine whether the kinds of changes that occur during short-term habituation and sensitization in invertebrates also constitute the mechanisms of long-term nonassociative learning, whether the mechanisms of associative learning are similar, and whether the mechanisms that prevail in invertebrates are used by vertebrates. The important point, however, is that the problem is beginning to yield to cellular analysis. We are clearly entering an era where the cellular analysis of learning and memory is a reality. As this proceeds, it will contribute substantially to our understanding of disorders of learning and memory.

■ *Bibliography*

Journal articles

Banks, M.S., Aslin, R.N., and Letson, R.D.: Sensitive period for the development of human binocular vision, Science **190:**675, 1975.

Blakemore, C., Vital-Durand, F., and Garey, C.J.: Recovery from monocular deprivation in the monkey. I. Reversal of physiological effects in the visual cortex, Proc. R. Soc. Biol. [Lond.] **213:**399, 1981.

Bunt, S.M., and Lund, R.D.: Development of a transient retino-retinal pathway in hooded and albino rats, Brain Res. **211:**399, 1981.

Finlay, B.A., Berg, A.T., and Sengelaub, D.R.: Cell death in the mammalian visual system during normal development. II. Superior colliculus, J. Comp. Neurol. **204:**318, 1982.

Freeman, R.D., and Thibos, L.N.: Contrast sensitivity in humans with abnormal visual experience, J. Physiol. [Lond.] **247:**687, 1975.

Fujito, Y., et al.: Formation of functional synapses in the adult cat red nucleus from the cerebrum following cross-innervation of forelimb flexor and extensor nerves. II. Analysis of newly appeared synaptic potentials, Exp. Brain Res. **45:**13, 1982.

Garey, L.J., and Vital-Durand, F.: Recovery from monocular deprivation in the monkey. II. Reversal of morphological effects in the lateral geniculate nucleus, Proc. R. Soc. Biol. [Lond.] **213**:425, 1981.

Hamburger, V.: Trophic interactions in neurogenesis: a personal historical account, Annu. Rev. Neurosci. **3**:269, 1980.

Harwerth, R.S., et al.: Behavioral studies of stimulus deprivation amblyopia in monkeys, Vision Res. **21**:779, 1981.

Hess, R.F., France, T.D., and Tulunay-Keesey, U.: Residual vision in humans who have been monocularly deprived of pattern stimulation in early life, Exp. Brain Res. **44**:295, 1981.

Hess, R.F., and Howell, E.R.: The threshold contrast sensitivity function in human amblyopia: evidence for a two type classification, Vision Res. **17**:1049, 1977.

Innocenti, G.M.: Growth and reshaping of axons in the establishment of visual callosal connections, Science **212**:824, 1981.

Ivy, G.O., Akers, R.M., and Killackey, H.P.: Differential distribution of callosal projection neurons in the neonatal and adult rat, Brain Res. **173**:532, 1979.

Jen, L.S., and Lund, R.D.: Experimentally induced enlargement of the uncrossed retinotectal pathway in rats, Brain Res. **211**:37, 1981.

Kasamatsu, T., Pettigrew, J.D., and Ary, M.: Cortical recovery from effects of monocular deprivation: acceleration with norepinephrine and suppression with 6-hydroxydopamine, J. Neurophysiol. **45**:253, 1981.

Killackey, H.P., and Belford, G.R.: Central correlates of peripheral pattern alterations in the trigeminal system of the rat, Brain Res. **183**:205, 1980.

Klein, M., Shapiro, E., and Kandel, E.R.: Synaptic plasticity and the modulation of the Ca^{2+} current, J. Exp. Biol. **89**:117, 1980.

Landmesser, L.T.: The generation of neuromuscular specificity, Annu. Rev. Neurosci. **3**:279, 1980.

Lashley, K.S.: In search of the engram, Symp. Soc. Exp. Biol. **4**:454, 1950.

Lehmkuhle, S., Kratz, K.E., and Sherman, S.M.: Spatial and temporal sensitivity of normal and amblyopic cats, J. Neurophysiol. **48**:372, 1982.

LeVay, S., Wiesel, T.N., and Hubel, D.H.: The development of ocular dominance columns in normal and visually deprived monkeys, J. Comp. Neurol. **191**:, 1980.

Manny, R.E., and Levi D.M.: Pyschophysical investigations of the temporal modulation sensitivity function in amblyopia: spatiotemporal interactions, Invest. Ophthalmol. Vis. Sci. **22**:425, 1982.

Manny, R.E., and Levi, D.M.: Psychophysical investigations of the temporal modulation sensitivity function in amblyopia: uniform field flicker, Invest. Ophthalmol. Vis. Sci. **22**:515, 1982.

Movshon, J.A., and Van Sluyters, R.C.: Visual neural development, Annu. Rev. Psychol. **32**:477, 1981.

Pilar, G., Landmesser, L., and Burstein, L.: Competition for survival among developing ciliary ganglion cells, J. Neurophysiol. **43**:233, 1980.

Raisman, G.: Formation of synapses in the adult rat after injury: similarities and differences between a peripheral and a central nervous site, Phil. Trans. R. Soc. [Lond] **278**:349, 1977.

Schneider, G.E.: Early lesions of superior colliculus: factors affecting the formation of abnormal retinal projections, Brain Behav. Evol. **8**:73, 1973.

Sengelaub, D.R., and Finlay, B.L.: Early removal of one eye reduces normally occurring cell death in the remaining eye, Science **212**:573, 1981.

Sengelaub, D.R., and Finlay, B.L.: Cell death in the mammalian visual system during normal development. I. Retinal ganglion cells, J. Comp. Neurol. **204**:311, 1982.

Sherman, S.M.: The effect of cortical and tectal lesions upon the visual fields of binocularly deprived cats, J. Comp. Neurol. **172**:231, 1977.

Sherman S.M., and Spear, P.D.: Organization of visual pathways in normal and visually deprived cats, Physiol. Rev. **62**:738, 1982.

Sherman, S.M., and Sprague J.M.: Effects of visual cortex lesions upon the visual fields of monocularly deprived cats, J. Comp. Neurol. **188**:291, 1979.

Squire, L.R., Slater, P.C., and Chace, P.M.: Retrograde amnesia: temporal gradient in very long-term memory following electroconvulsive therapy, Science **187**:77, 1975.

Sur, M., Humphrey, A.C., and Sherman, S.M.: Monocular deprivation affects X- and Y-cell retinogeniculate terminations in cats, Nature **300**:183, 1982.

Swindale, N.V., Vital-Durand, F., and Blakemore, C.: Recovery from monocular deprivation in the monkey. III. Reversal of anatomical affects in the visual cortex, Proc. R. Soc. Biol. [Lond.] **213**:435, 1981.

Tees, R.C.: Effects of early auditory restriction in the rat on adult pattern discrimination, J. Comp. Physiol. Psychol. **63**:389, 1967.

Tsukahara, N.: Synaptic plasticity in the mammalian central nervous system, Annu. Rev. Neurosci. **4**:351, 1981.

Tsukahara, N., et al.: Formation of functional synapses in the adult cat red nucleus from the cerebrum following cross-innervation of forelimb flexor and extensor nerves, Exp. Brain Res. **45**:1, 1982.

von Noorden, G.K., and Maumenee, A.E.: Clinical observations on stimulus deprivation amblyopia *(amblyopia ex anopsia)*, Am. J. Ophthalmol. **65:**220, 1968.

Woolsey, T.A. et al.: Effects of early vibrissal damage on neurons in the ventrobasal (VB) thalamus of the mouse, J. Comp. Neurol. **184:**363, 1979.

Books and monographs

Black, I.B., and Patterson, P.H.: Developmental regulation of neurotransmitter phenotype. In Moscona, A.A., and Monroy, A., Current topics in developmental biology, vol. 15, part 1, New York, 1980, Academic Press, Inc.

Blakemore, C.: Maturation and modification in the developing visual system. In Held, R., Leibowitz, H.W., and Teuber, H.-L., editors: Handbook of sensory physiology, vol. 8, New York, 1978, Springer-Verlag.

Cohen, D.H.: The functional neuroanatomy of a conditioned response. In Thompson, R.F., Hicks, L.H., and Shvyrkov, V.B., editors: Neural mechanisms of goal-directed behavior and learning, New York, 1980, Academic Press, Inc.

Cotman, C.W., editor: Neuronal plasticity, New York, 1978, Raven Press.

Cowan, W.M.: Neuronal cell death as a regulative mechanism in the control of cell number in the nervous system. In Rockstein, M., editor: Development and aging in the nervous system, New York, 1973, Academic Press, Inc.

Cowan, W.M., Stanfield, B.B., and Kishi, K.: The development of the dentate gyrus. In Moscona, A.A., and Monroy, A., editors: Current topics in developmental biology, vol. 15, part 1, volume 15: Neural development. Part 1 Histogenesis, (Volume Ed. R.K. Hunt) New York, 1980, Academic Press, Inc.

Freeman, R.D., editor: Developmental neurobiology of vision, New York, 1979, Plenum Press.

Goldman-Rakic, P.: Development and plasticity of primate frontal association cortex. In Schmitt, F.O., et al., editors: The organization of the cerebral cortex, Cambridge, Mass., 1981, M.I.T. Press.

Hirsch, H.V.B., and Leventhal, A.G.: Functional modification of the developing visual system. In Jacobson, M., editor: Handbook of Sensory Physiology, vol. 9, New York, 1978, Springer-Verlag.

Hollyday, M.: Motoneuron histogenesis and the development of limb innervation. In Moscona, A.A., and Monroy, A., editors: Current topics in developmental biology, vol. 15, part 1, New York, 1980, Academic Press, Inc.

Innocenti, G.M.: Two types of brain plasticity? In Cuénod, M., Kreutzberg, G.W., and Bloom, F.E., editors: Progress in brain research, vol. 51, Amsterdam, 1979, Elsevier/North-Holland.

Jacobson, M.: Developmental neurobiology, ed. 2, New York, 1978, Plenum Press.

Kandel, E.R.: Cellular basis of behavior, San Francisco, 1976, W.H. Freeman & Co., Publishers.

Kuffler, S.W., and Nicholls, J.G.: From neuron to brain, Sunderland, Mass., 1976, Sinauer Associates, Inc.

Kupferman, I.: Learning. In Kandel, E.R., and Schwartz, J.H., editors: Principles of neural science, New York, 1981, Elsevier-North-Holland.

LeVay, S., Wiesel, T.N., and Hubel, D.H.: The postnatal development and plasticity of ocular-dominance columns in the monkey. In Schmitt, F.O., et al., editors: The organization of the cerebral cortex, Cambridge, Mass., 1981, M.I.T. Press.

Lund, R.D.: Development and plasticity of the brain, New York, 1978, Oxford University Press.

Mistretta, C.M., and Bradley, R.M.: Effect of early sensory experience on brain and behavior development. In Gottlieb, G., editor: Studies on the development of behavior and the nervous system, vol. 4, New York, 1978, Academic Press, Inc.

Rubel, E.W.: Ontogeny of structure and function in the vertebrate auditory system. In Jacobson, M., editor: Handbook of sensory physiology, vol. 9, New York, 1978, Springer-Verlag.

Silver, J.: Cell death during development of the nervous system. In Jacobson, M., editor: Handbook of sensory physiology, vol. 9, New York, 1978, Springer-Verlag.

Steward, O.: Events within the sprouting neuron and the denervated neuropil during lesion-induced synaptogenesis. In Morrison, A.R., and Strick, P.L., editors: Changing concepts of the nervous system, New York, 1982, Academic Press, Inc.

MUSCLE

Richard A. Murphy

chapter 22

Contraction of muscle cells

Skeletal muscle expresses the conscious mind. This description emphasizes the role of skeletal muscle as an effector through which voluntary actions (and some reflex movements), ranging from speaking to running, are accomplished. Cardiac and smooth muscles are effectors subserving the cardiovascular, respiratory, digestive, and genito-urinary systems. Their activity is usually controlled or modified by the autonomic nervous system. In many ways cardiac and smooth muscles express the unconscious mind illustrated by the classic "fright, fight, flight" reflexes. To accomplish these functions, muscle comprises about 45% to 50% of the body mass. Even the muscle in the walls of blood vessels amounts to about 2 kg in an average adult male. Whereas muscle subserves other organs, it is also responsible for the major demands on many of these systems by consuming most of the body's metabolic output. The enormous increase in energy expenditure during sustained exercise dictates many of the characteristics of the cardiovascular and respiratory systems.

This chapter focuses on the cellular processes of contraction and the mechanisms whereby they are controlled. Emphasis is placed on those characteristics that enable a basic mechanism that economically converts chemical energy into mechanical energy (chemomechanical transduction) to fulfill diverse functions in various tissues. Chapter 23 is devoted to an examination of muscle function in various types of tissues and organs. The capacity of different muscles to adapt to changing requirements is also described, along with various proliferative, synthetic, and secretory roles.

■ Structure of the contractile apparatus

Muscle is usually classified as either smooth or striated. This distinction is anatomical and is based on the organization of the contractile apparatus. The structural classification of muscle into smooth and striated types has few functional implications. Striated skeletal and cardiac muscles are excellent examples of the extremes of the functional spectrum (voluntary neural control of cells versus an electrically and mechanically coupled tissue with an inherent genesis of contractile activity). Different smooth muscles exemplify most of the functional spectrum between these extremes.

A skeletal muscle is composed of bundles of enormous, multinucleated cells up to 80 μm in diameter and sometimes many centimeters long (Fig. 22-1). These cells, like those in cardiac muscle, have a striking banding pattern responsible for their classification as *striated* muscles. The striations arise from a highly organized arrangement of subcellular structures. There are few instances in biology where ultrastructure provides a clearer basis for understanding cell function than in striated muscle. Therefore careful study of Fig. 22-1 is warranted. Electron micrographs reveal bundles of filaments running along the long axis of the cell. These bundles are termed *myofibrils*. The gross

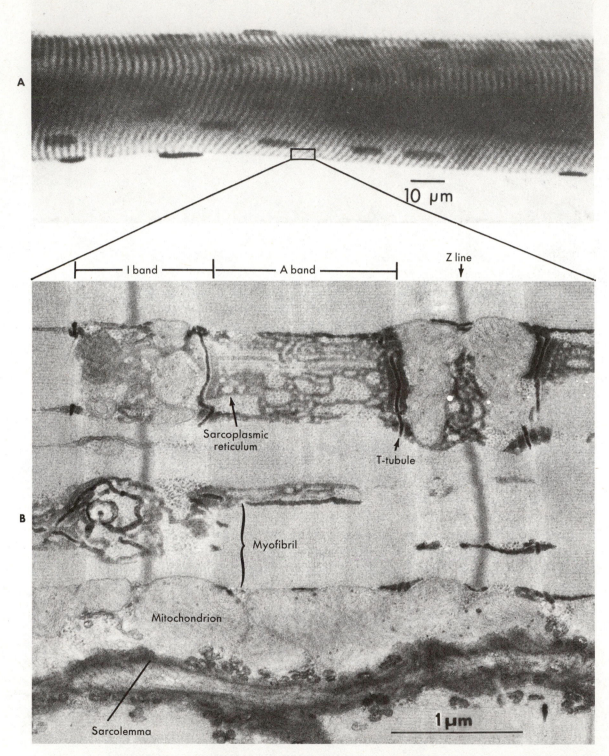

I band   A band   Z line

Sarcoplasmic reticulum

T-tubule

Myofibril

Mitochondrion

Sarcolemma

10 μm

1 μm

Fig. 22-1 ◼ **A,** Segment of a single cell from human gastrocnemius muscle. Cross striations of alternating dark (A) and light (I) bands are visible along with nuclei located peripherally. **B,** Electron micrograph of a longitudinal section through a mouse skeletal muscle cell. The tissue is stained to show the internal membrane system of the sarcoplasmic reticulum as a very dark network of interconnecting tubules. Invaginations of the cell membranes (T-tubules) are visible as dark elements extending into the cell at the level of the junction between the A and I bands. **C,** Drawing illustrating the three-dimensional relationships between membrane elements and the filament lattice. (**A** and **C** redrawn from Leeson, C.R., and Leeson, T.S.: Histology, ed. 3, Philadelphia, 1976, W.B. Saunders Co.; **B** courtesy Dr. Michael S. Forbes.)

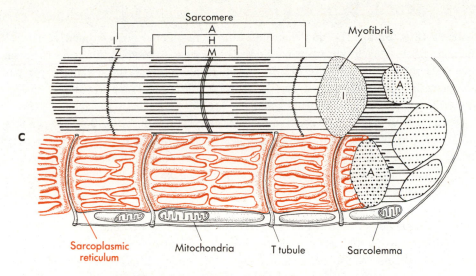

Fig. 22-1, cont'd ■ For legend see opposite page.

striation pattern of the cell arises from a repeating pattern in the myofibrils that is in transverse register across the whole cell (Fig. 22-1, *C*).

The banding pattern arises from two sets of filaments in the myofibrils. Their organization is most clearly seen in a drawing that exaggerates the cross-sectional dimensions (Fig. 22-2). The dark striations are a region containing a lattice of *thick filaments*. This region of the myofibril is termed the *A band* because it appears dark (anisotropic) when viewed in a microscope using polarized light. A second lattice consists of *thin filaments* that are attached to a transverse, darkly staining structure termed the *Z line* or *Z disk*. The thin filaments extend from two adjacent Z lines to interdigitate with the thick filaments. Areas of the myofibril or cell containing only thin filaments and Z disks are termed *I bands* as they are light (isotropic) when viewed under polarized light. The thick and thin filament arrays form the contractile system, and the repeating unit in each myofibril defined by the Z disks is the basic contractile unit called a *sarcomere*. Each sarcomere contains half of two I bands with a central A band. The latter has a less dense central region (H zone) where there is no overlap of thin filaments. The H zone is bisected by a darkly staining M line containing proteins that link the thick filaments together. Cross sections of the myofibril reveal the relationship between thin and thick

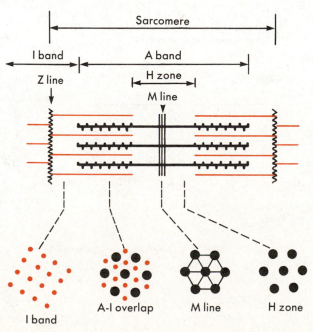

Fig. 22-2 ■ Longitudinal (top) and cross-sectional (bottom) diagrams of a sarcomere showing the relationships between thick (black) and thin (color) filaments. (Redrawn from Squire, J.M.: The structural basis of muscular contraction, New York, 1981, Plenum Press.)

filaments (Fig. 22-2). The thin filaments form a hexagonal array around each thick filament, whereas each thin filament is equidistant from three thick filaments. This arrangement reflects the presence of two thin filaments for each thick filament per half sarcomere in vertebrate striated muscle. The complex membrane systems associated with each sarcomere are considered later.

The comparatively small *smooth muscle* cells are approximately 100 to 300 μm long and 2 to 5 μm in diameter in the center where the nucleus is located. The cells taper toward each end and may be very irregular in cross section. Thick and thin filaments of the contractile apparatus are aligned in the long axis of the cell. The filaments resemble those in striated muscle when observed in electron micrographs. However, there are fewer thick filaments than in striated muscle, and the thin filaments occupy most of the cytoplasm. Bundles of thin filaments are anchored to rather amorphous ellipsoid *dense bodies* distributed through the cytoplasm or in special *dense areas* forming bands along the cell membrane. No structure equivalent to the sarcomere can be seen in smooth muscle. It seems likely that the lack of transverse registration of filaments obscures identification of a functional equivalent to the sarcomere. This organization does not produce a gross striping pattern, and the cells appear rather featureless or ''smooth'' when viewed in the light microscope.

■ The thin filament

Composition and structure. Thin filaments are ubiquitous constituents of all nucleated cells and are a dominant feature in muscle. All thin filaments contain two major proteins. A globular protein called *actin,* with a molecular weight of 45,000 daltons, polymerizes under conditions existing in the cytoplasm to form twisted, two-stranded filaments (Fig. 22-3). Rod-shaped molecules of *tropomyosin* stretch along each strand of the thin filament. Each molecule of tropomyosin is associated with six or seven actins in one strand. Tropomyosin is composed of two separate polypeptide chains. The indi-

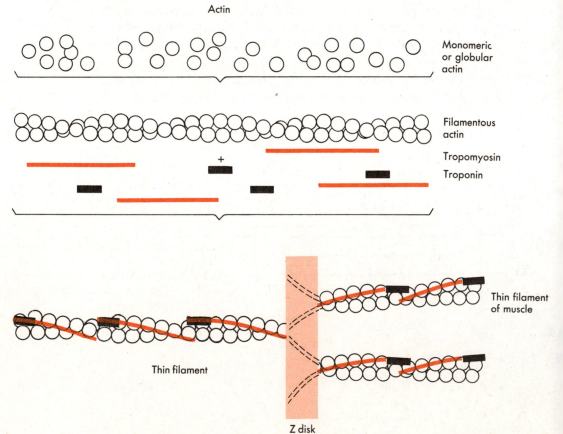

Fig. 22-3 ■ Composition and structure of thin filaments in muscle. Globular actin monomers (top) polymerize into a two-stranded helical filament. The thin filament structure is completed with the addition of stiff, rod-shaped tropomyosin molecules (colored area in center). Troponin (black rectangles) is a regulatory protein bound to the tropomyosin component of the thin filament in vertebrate striated muscles (bottom). Thin filaments are anchored to Z lines in striated muscles. Filaments on each side of a Z line have opposite polarities. (Redrawn from Murray, J.M., and Weber, A.: The cooperative action of muscle proteins, Sci. Am. **230:**58-71, 1974.)

Actin

Monomeric or globular actin

Filamentous actin

Tropomyosin

Troponin

Thin filament of muscle

Thin filament

Z disk

vidual polypeptides have a basic α-helical structure, and the two helical peptides are wound around each other to form a supercoil. Molecules of this type are long, rigid, and insoluble. Other proteins, present in smaller amounts, are associated with the thin filaments. These minor proteins are involved in the attachment of thin filaments to Z disks or dense bodies and probably contribute to the remarkably uniform thin filament length of 1 μm in vertebrate striated muscles. Finally, additional thin filament proteins are involved in regulation of the interaction of actin with the thick filament. The most important regulatory protein in striated muscles is troponin, which is linked to tropomyosin (Fig. 22-3). The regulatory proteins are considered later.

Lattice organization and polarity. Thin filaments are anchored at one end. The filaments have a polarity, although this is not apparent in electron micrographs. Thus thin filaments on each side of the Z disk ''point'' in opposite directions. Fig. 22-3 indicates one view of how this may arise from a splitting of the two strands of the thin filament at the Z disk so that each strand forms half a thin filament of opposite polarity. In cross section the thin filaments can be seen to form a hexagonal lattice around each thick filament (Fig. 22-2). Each thin filament lies equidistant from three thick filaments.

Composition. A very large protein (about 470,000 daltons) called *myosin* forms the thick filament, although small amounts of other proteins are present. The myosin molecule is formed by the association of six different polypeptides. These peptides, which are not covalently linked, can be dissociated by detergents or denaturing agents (Fig. 22-4) and separated into three pairs: one set of large *heavy chains* and two sets of *light chains*. In the intact molecule most of the heavy chain has an α-helical structure, and

■ **The thick filament**

Fig. 22-4 ■ The myosin molecule (top) contains a long tail and two heads. The tail regions of many myosin molecules associate in the cytoplasm to form a thick filament with the heads projecting from the surface. The individual myosin molecule can be dissociated into three pairs of identical polypeptides by denaturing agents (bottom). The molecule can also be cleaved by proteolytic enzymes (at sites shown by the dashed lines) into two rod-shaped fragments plus one pair of globular fragments. Studies of the denatured peptides or of the proteolytic fragments, which retain biological activity, provide information about the function and interactions of myosin.

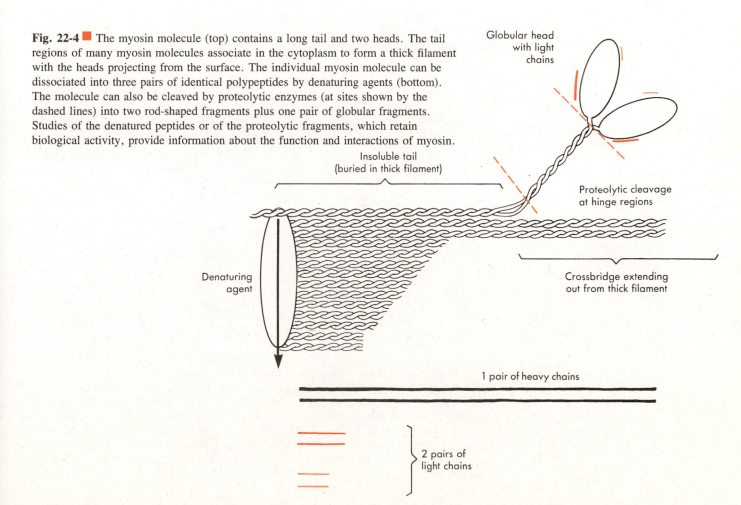

Globular head with light chains

Insoluble tail (buried in thick filament)

Proteolytic cleavage at hinge regions

Crossbridge extending out from thick filament

Denaturing agent

1 pair of heavy chains

2 pairs of light chains

the two strands are twisted around each other in a supercoil that forms a long, rigid, insoluble "tail." However, one end of the heavy chains has a globular tertiary structure. Thus each myosin molecule has two "heads" attached to one end of the long tail (Fig. 22-4). One polypeptide of each set of light chains is associated with each head of the molecule.

If myosin is briefly exposed to proteolytic enzymes such as trypsin or papain, it is preferentially cleaved into four segments (Fig. 22-4). The points of cleavage occur at regions where the basic supercoiled structure is perturbed, exposing the peptide backbone to enzymatic attack. Cleavage points indicate flexible regions that serve as hinges for the molecule. Studies of these proteolytic fragments show the functions of each part of the myosin molecule as described later.

Filament formation and structure. Myosin aggregates in the cytoplasm to form thick filaments. The basic filament represents aggregates of the tail segment of the molecules, and smooth filaments can form from the tail segments obtained by proteolytic cleavage. The remainder of the molecule, including the globular heads and the rod portion between the hinges, projects laterally from the filaments. These projections are visible in electron micrographs and are termed *crossbridges* because they can link adjacent thick and thin filaments. The process of filamentogenesis is extraordinarily spe-

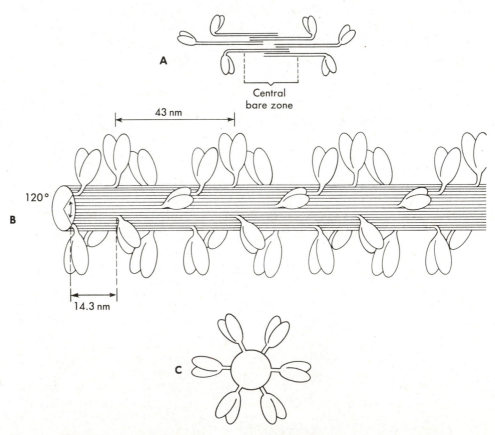

Fig. 22-5 ■ Structure of the thick filament **A,** Initiation of filament. Filament formation begins with an end-to-end association of the tails of myosin molecules giving a central bare zone lacking crossbridge projections. **B,** Segment of filament. "Crowns" of three crossbridges project at angles of 120 degrees relative to each other at intervals of 14.3 nm along the thick filament. Each crown is rotated 60 degrees passing down the filament. Within a row, 43 nm separate each crossbridge. The result is a thick filament with six rows of crossbridges along its length, as can be visualized in a diagram of the filament viewed from one end (**C**). (Redrawn from Murray, J.M., and Weber, A.: The cooperative action of muscle proteins, Sci. Am. **230:**58-71, 1974.)

cific. Each filament begins with an end-to-end association of the tails of the myosin molecules (Fig. 22-5, *A*). This produces a central thick filament segment lacking crossbridge projections. The crossbridges in each half of the filament are consequently oriented in the opposite direction. Therefore the molecules of the thick and thin filaments in each half of the sarcomere have the same relative orientation as a result of the thin filament polarity. The crossbridges project in groups of three from the filament (Fig. 22-5, *B*). Successive crowns of crossbridges are rotated 60 degrees, producing a helical arrangement of crossbridges along the thick filament. In a striated muscle the crossbridges project toward the six thin filaments surrounding each thick filament. It is not known whether crossbridges in thick filaments of smooth muscle have a comparable symmetry. Thick filaments are 1.6 μm long and contain an estimated 300 to 400 crossbridges. Minor protein constituents probably contribute to the highly ordered and regular structure of thick filaments and their basic triangular lattices (Fig. 22-2) in cells. Adjacent thick filaments in the sarcomere are linked by protein connections between the central bare zones, contributing to the stability of the filament lattice.

■ *Intermediate filaments*

A third type of filament with a diameter between that of thin (7 nm) and thick (15 nm) filaments serves as part of the cytoskeleton in muscle cells. These *intermediate filaments* (12 nm diameter) serve a variety of cytostructural roles in all cells. Several types of intermediate filaments composed of different proteins, but having a similar ultrastructure, are known. They are prominent in smooth muscles, where they appear to link the dense bodies serving as thin filament attachment points. Intermediate filaments also link Z disks of adjacent myofibrils in striated muscles.

■ *Crossbridge interactions with the thin filament*
■ *Crossbridge properties*

Each crossbridge consists of two identical heads that exhibit a remarkable set of properties. Most studies suggest that the two heads act largely independently in the reactions discussed in this section. However, the behavior of one head may be somewhat influenced by reactions involving the other (cooperativity). The properties of the myosin head depend on the intact globular part of the heavy chain and the two associated light chains. One light chain on each myosin head appears to have special significance and is termed the *regulatory light chain*.

The adenosine triphosphatase *(ATPase)* activity of myosin can catalyze the hydrolysis of adenosine triphosphate (ATP), producing adenosine diphosphate (ADP) and inorganic phosphate (P_i). However, the ATPase activity of myosin is inhibited by the high magnesium concentrations existing in cells. Myosin can also bind to actin, forming an *actomyosin* complex. Actomyosin is a very active ATPase in the muscle cells. The interaction between actin and myosin, associated with ATP hydrolysis, represents the fundamental *chemomechanical transduction* process considered later in which chemical energy is converted into mechanical energy by the muscle.

The splitting of ATP by actomyosin is a complex cycle involving many steps. The more important steps are illustrated in Fig. 22-6. In a relaxed muscle, regulatory systems prevent actin-myosin interactions (Fig. 22-6, top). This inhibition is overcome on stimulation by Ca^{++} ions and will be considered later. In the presence of ATP, myosin has ADP and P_i bound to each head and, in this state, exhibits a high affinity for actin. The ADP and P_i are released when the myosin heads bind to actin. Product release allows an ATP molecule to bind to myosin, and the affinity of myosin for actin (as A-M) is greatly reduced. The bound ATP is only hydrolyzed after dissociation of myosin and actin although the products (ADP and P_i) remain bound to myosin in the succeeding step. The energy released by splitting of ATP is stored in the myosin molecule, which is now in a high energy state and has a renewed high affinity for actin. The rate-limiting step in this cycle is that involving binding of ATP and dissociation of actomyosin.

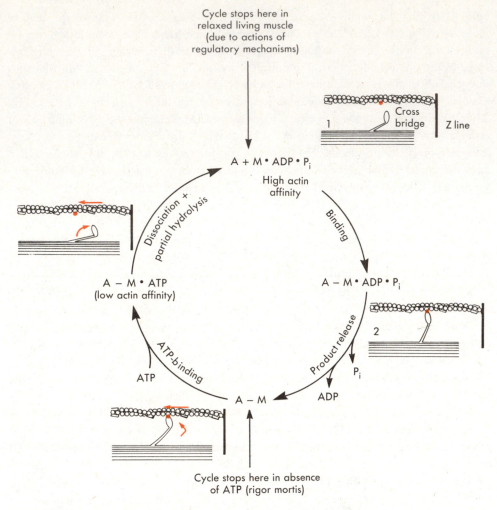

Fig. 22-6 ■ The mechanism of ATP hydrolysis by actomyosin and the major steps in the crossbridge cycle. Actin in the thin filament *(A)* and the myosin crossbridge projecting from the thick filament *(M)* interact cyclically. This interaction involves a number of steps during which ATP is hydrolyzed, and the energy released is harnessed to induce conformational changes in the crossbridge. Each cycle causes the thick and thin filaments to interdigitate by about 10 nm (note position of colored actin monomer). See text for further details.

■ *The crossbridge cycle*

The cycle associated with ATP hydrolysis by isolated actomyosin releases the energy in ATP without any conversion into mechanical work. Chemomechanical transduction involves conformational changes in the crossbridge, which lead to filament movements with shortening and force development (Fig. 22-6). The details of this process are uncertain, but considerable evidence favors the general model illustrated for the crossbridge interacting with the thin filament. In a resting muscle the crossbridge is not attached to the thin filament and is oriented perpendicularly to the myosin filament. When a muscle is stimulated, a rise in myoplasmic Ca^{++} concentrations produces changes in the myofilament structure, allowing crossbridge binding to the thin filament (considered in the section on control mechanisms). The hinge regions in the crossbridge permit the head to swing toward the thin filament. After attachment myosin heads change their conformation (tilt), using the energy stored in the high energy myosin-ADP-P_i complex. This conformational change in the crossbridge generates a force moving the thin filament relative to the thick filament. It is likely that the conformational change also leads to release of ADP and P_i, setting the stage for crossbridge detachment when another ATP is bound. Each cycle can move the filaments about 10 nm relative to each other. The way in which enormous numbers of such cycles generate muscular contraction is considered in the following section. The illustrated cycle will continue until interrupted in the detached state by control systems (which remove Ca^{++} from the myoplasm and produce relaxation) or until the ATP is exhausted. ATP depletion is an abnormal situation in muscle cells and arrests the cycle with the formation of permanent actomyosin complexes as shown in Fig. 22-6. Permanent crossbridge attachment resulting from ATP depletion in a muscle following death causes muscular rigidity *(rigor mortis)*.

■ *Biophysics of the contractile system*

Quantitative estimates of the mechanical output of muscles have provided important inferences about the mechanism of chemomechanical transduction and its control mechanisms. The measurement of muscle contraction also provides a way of assessing the effects of neurotransmitters, drugs, and hormones and of quantifying pathological changes. The important mechanical variables are force, length, and shortening velocity. The basic approach in muscle mechanics is to control all of these factors except the measured dependent variable. Such measurements should employ muscle preparations in which the cells are aligned in the axis in which force and length are determined.

■ *Force-length relationships and the sliding filament mechanism*

The dependence of force generation on the length of the muscle is shown in Fig. 22-7. Points on these curves are determined under conditions in which the length of the muscle is fixed, and force is measured during a contraction. Such a contraction at constant length is termed *isometric*. Force will vary with the size of the muscle and is best expressed as a stress (force/cross-sectional area of the muscle) to allow comparisons with other tissues. A relaxed muscle is elastic, and it requires a force to stretch it to an increased length. The resulting passive force-length or stress-length relationship primarily reflects the properties of connective tissue in intact muscles. Connective tissue is a much greater component in the walls of hollow organs, such as the heart and gastrointestinal tract, than in striated muscle, the in vivo length of which is constrained by the skeletal attachments. When a muscle is stimulated, the total isometric stress is greater than the passive stress at any length. The difference between the stress-length curves for contracting and relaxed muscles is the *active stress-length relationship* that characterizes the contractile system (Fig. 22-7, *B*). There is a family of such curves for any muscle when it is not maximally activated (Fig. 22-7, *C*).

The correlation of structural and mechanical information reveals that active stress is proportional to the overlap between thick and thin filaments in a sarcomere (Fig. 22-7, *D*). As the muscle is stretched beyond the length (L_o) at which maximum active force (F_o) is developed, active stress decreases linearly with the overlap between thick and thin filaments. This result shows that stress is proportional to the number of active crossbridges that can interact with the thin filament. This number includes all the interacting crossbridges in each half sarcomere. There is a small segment at the peak of the stress-length curve at which stress does not change with length. At these lengths the thin filaments move past the central thick filament bare zone that lacks crossbridges. Force declines at sarcomere lengths less than L_o. This is partially a result of disturbances in the filament lattice geometry as indicated in Fig. 22-7, *D*. However, the control mechanisms for the contractile machinery become less sensitive to the stimulus at short muscle lengths, and partial inactivation contributes to the low stress at short lengths. Slight variations in sarcomere lengths in whole muscles obscure the inflection points detected in the stress-length curves for sarcomeres.

Force-generation and muscle type. The maximum stress generated depends on muscle length, the number of crossbridges that are turned on (the level of *activation*), and the type of muscle. Because the lengths of the thick and thin filaments and their packing

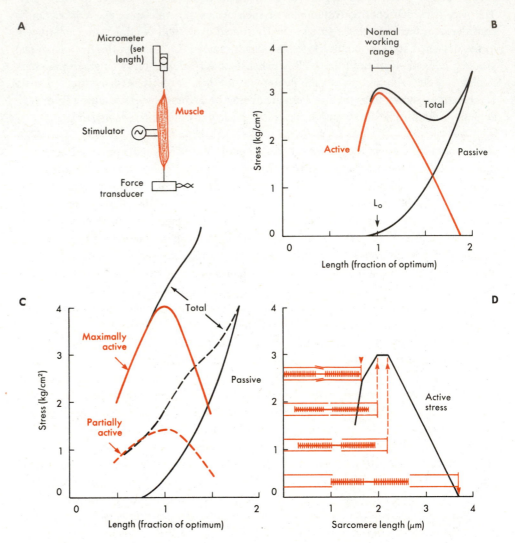

Fig. 22-7 ■ Stress-length relationships in muscle. **A,** Schematic diagram of experimental apparatus in which the tissue is attached to a micrometer to set the length and a transducer to measure force, which is normalized as a stress to allow comparisons of muscles of differing size, that is, force/cross-sectional area of the muscle cells. Values of stress at different lengths are obtained and plotted as shown in **B** to **D. B,** Three stress-length curves are shown for a skeletal muscle: (1) the passive stress exerted as a function of the length of the relaxed muscle, (2) the total stress exerted by the maximally stimulated muscle (passive + active), and (3) the active stress-length curve for the contractile machinery obtained as the difference between the total and passive stresses at any length (colored curve). All muscles have an inherent maximum force-generating capacity, which is obtained at the optimum length (L_o). **C,** In cardiac and smooth muscles, which are not attached to the skeleton, connective tissue limits how far the tissue can be stretched. The characteristics of the connective tissue alter the passive, but not the active, stress-length relationship. Active stress will vary with the extent to which the contractile system is activated in these tissues. **D,** Precise studies of the stress-length behavior of the sarcomeres in single skeletal muscle cells reveal the dependence of stress on the overlap of thick and thin filaments. Diagrams of four sarcomeres at lengths where the slope of the stress-length curve changes show how filament interactions and active stress depend on sarcomere length. (Data from Gordon, A.M., Huxley, A.F., and Julian, F.J.: J. Physiol. [London] **184:**170-192, 1966.)

density are similar in vertebrate striated muscles, they all generate similar maximum stresses at L_o. The maximum stress is about 3 kg/cm^2 of cell cross-sectional area.

The shape of the cellular stress-length curve for smooth muscles is similar to that of skeletal and cardiac muscles. This observation suggests that the basic *sliding filament–crossbridge mechanism* functions in smooth muscle, although anatomical verification of this hypothesis is lacking. Most smooth muscles develop maximum active stresses that are comparable to those developed by striated muscles, and some smooth muscles can generate considerably higher stresses. The mechanical output of all smooth muscles is striking when the comparatively low myosin (and therefore thick filament and crossbridge) content of smooth muscle is considered. Differences in the structure of the smooth muscle filament lattice and in the crossbridge interactions with the thin filament may underlie the comparatively high stress-generating capacities of smooth muscles.

Velocity-stress relationships and the crossbridge cycle. The second basic relationship characterizing the output of a muscle is obtained when the load or stress on a muscle is held constant and the shortening velocity is measured (Fig. 22-8, *A*). A contraction at constant load is termed *isotonic*. Shortening velocity is determined by the load on the muscle as illustrated by the hyperbolic velocity-stress curve. Muscles can lift a heavy load slowly or shorten rapidly when lightly loaded. Fig. 22-8, *A*, also shows that a contracting muscle can withstand (briefly) a heavier stress when forcibly stretched than it can develop isometrically. The strength of the crossbridge attachment to the thin filament is greater than the force generated by its movement. Consequently, muscles can bear a load about 1.6 times F_o before the crossbridge attachment is mechanically broken and rapid lengthening occurs. This situation is important physiologically when a muscle is contracted to decelerate the body as when a person is walking downhill.

The contractile system operates most efficiently (i.e., the greatest mechanical work output for the chemical energy used) with optimal loading. This is illustrated by the power curve (Fig. 22-8, *A*). Power, or work/time, is simply the product of force and velocity, and power can be calculated by multiplying these values at any point on the velocity-force curve. Maximum power is obtained with a load of about 0.3 of the maximum force that can be developed. A muscle contracting isometrically does no work (force × distance = 0), nor does an isotonically shortening muscle with no load. The mechanical efficiency of such contractions is necessarily zero. Inefficient contractions are often physiologically important when either high speed or maximum force is appropriate. The optimal efficiency of the contractile system in converting chemical energy into mechanical work is about 40% to 45%.

If a sarcomere is to shorten more than the 10 nm associated with a single crossbridge cycle, each crossbridge must detach and then reattach at a new site on the thin filament closer to the Z line. The crossbridges must cycle asynchronously to maintain a constant force and permit continuous shortening. This is probably facilitated by the fact that the distances between successive crossbridges along a thick filament are different from the repeat distances on the thin filament.

The number of active crossbridges is the primary variable subject to control. Crossbridge cycling rates depend primarily on the load. A partially activated skeletal muscle developing a low maximum force (F_o) will have the same maximum rate (V_o) as that of a fully activated muscle (Fig. 22-8, *B*).

Shortening velocities in a muscle depend on several factors. First, the velocity varies with the number of sarcomeres in a cell. Total shortening and shortening velocity are the sum of the movements of thin filaments past thick filaments times the number of half sarcomeres in the cell. Velocities can be calculated in terms of μm per second per half sarcomere to allow comparisons between different muscles. However, it is more common to normalize shortening velocities by reporting values in terms of optimal muscle lengths (L_o) per second, which accounts for the number of sarcomeres in a cell. Shortening velocities for smooth muscle can only be normalized this way. Second, the

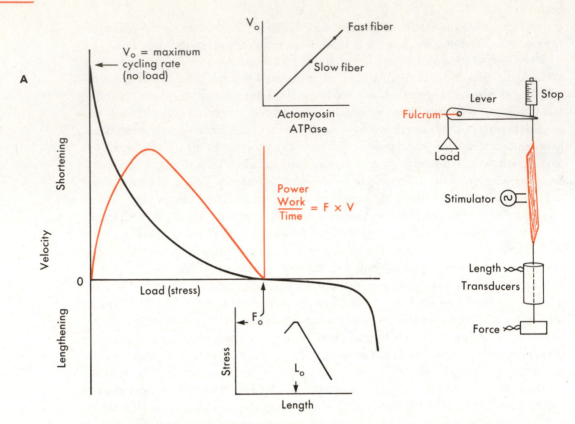

Fig. 22-8 ■ The dependence of shortening velocity on the load (stress) on a muscle. **A,** Shortening velocity may be measured using a lever system, which permits a muscle to shorten against a constant load. Velocity is measured using a length transducer (velocity = Δ length/Δ time). The velocity-load curve is constructed from points obtained in a series of contractions against different loads. (The initial length in these experiments is at L_o, and only a slight shortening is permitted.) If the load placed on the muscle is greater than that which the active crossbridges can bear, the muscle will lengthen. Consequently, the velocity-load curve can be extended to describe this situation. Insets emphasize that the maximum stress (F_o) that a muscle can develop depends on the number of interacting crossbridges, whereas the maximum shortening velocity (V_o) is limited by the rate at which a particular isoenzymatic form of myosin synthesized in a cell can interact with actin and release the energy stored in ATP. The power output of a muscle (colored curve) is the mechanical work (force times distance shortened) per unit of time and can be calculated as the product of load times shortening velocity. Determination of velocity-stress relationships characterizes the output of the contractile machinery and the role of the intracellular regulatory mechanisms for the contractile proteins. These involve mechanisms determining the number of crossbridges that are interacting with the thin filament (**B**) and other mechanisms that appear capable of modulating the rate at which active crossbridges can interact with the thin filament (**C**).

velocity depends on the load on the muscle; the velocity-stress relationship shows that crossbridge cycling rates fall as the stress on the crossbridges increases. This effect can be understood by referring to Fig. 22-6. The conformational change in the crossbridge, which causes shortening between attachment and detachment, is opposed by the load. Heavier loads progressively increase the average time for this to occur and allow a new cycle to take place. Also, an unloaded crossbridge can cycle at a maximum rate, indicated by V_o. This maximum rate depends on the molecular properties of the myosin synthesized within a cell. The direct proportionality between the ATPase activity of myosin isolated from a cell and V_o for that cell (inset, Fig. 22-8) illustrates this molecular diversity, which is responsible for physiological differences in the speed of contraction of muscle cells from different sources. Enzymatically produced modifications of the myosin by certain regulatory mechanisms have recently been found to alter V_o (Fig. 22-8, *C*). These will be considered in the following section.

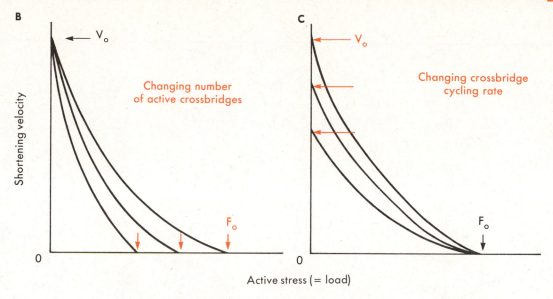

Fig. 22-8, cont'd ■ For legend see opposite page.

Two elements are involved in regulation of the contraction-relaxation cycle. The first involves cell membrane events that alter the cellular concentration of a *second messenger,* that is an ion or molecule that diffuses to the fibrillar contractile apparatus and regulates the crossbridge cycle. The second messenger in the myoplasm of all types of muscle is Ca^{++}. The second aspect of regulation concerns the mechanisms whereby Ca^{++} produces its effects on the contractile machinery. Three different Ca^{++}-binding proteins with regulatory roles are now recognized in different types of muscle.

■ *Intracellular control mechanisms for the contractile system*

Vertebrate skeletal and cardiac muscles contain a regulatory protein in the thin filament termed *troponin.* One molecule of troponin is bound to the end of each tropomyosin molecule (Figs. 22-3 and 22-9). Troponin consists of three polypeptides that are not covalently linked. One of these peptides, troponin C, binds up to four Ca^{++} ions in a cooperative manner. In the presence of micromolar concentrations of Ca^{++} the binding sites on troponin C are occupied, and the conformation of the thin filament changes (Fig. 22-9). This shift allows crossbridges adjacent to that portion of the thin filament to attach and cycle. In effect, Ca^{++} binding to troponin acts as a switch that regulates the number of crossbridges that are interacting with the thin filament.

In vertebrate skeletal and cardiac muscles the contraction-relaxation cycle involves five steps: (1) a stimulus at the cell membrane causes the myoplasmic Ca^{++} to rise above 0.1 μm; (2) binding of Ca^{++} to troponin exposes the sites at which crossbridges attach to actin; (3) crossbridges now bind and cycle at all exposed sites; (4) cessation of stimulation is followed by Ca^{++} removal from the myoplasm and dissociation of Ca^{++} from troponin; (5) the thin filament returns to a configuration in which further crossbridge interactions are blocked.

The binding of Ca^{++} ions to troponin exhibits a steep response curve so that relatively small changes in Ca^{++} produce large changes in the number of active crossbridges and force development (Fig. 22-9). Each crossbridge cycle is associated with ATP hydrolysis; therefore the myofibrillar ATPase activity is also proportional to the $[Ca^{++}]$.

Smooth muscles appear to lack troponin. Whereas there is some evidence for a different Ca^{++}-dependent regulatory system associated with the thin filament, other mechanisms involving myosin appear to play critical roles.

■ *Troponin and thin filament regulation*

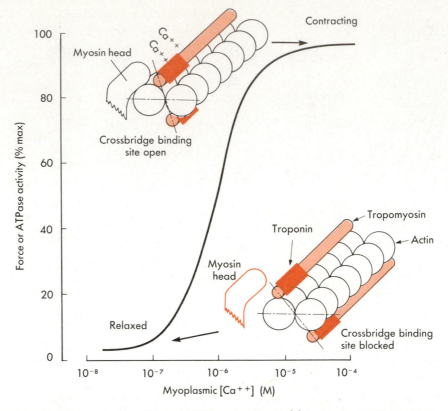

Fig. 22-9 ■ Regulation of the crossbridge cycle by Ca^{++} and troponin. Force development by muscle cells or ATP splitting by actomyosin containing troponin and tropomyosin requires very small concentrations of Ca^{++} in the myoplasm. Ca^{++} activates contraction by binding to the regulatory protein troponin in the thin filament of striated muscle. Troponin is associated with tropomyosin as a 1:1 complex in the thin filament. A conformational change occurs in the thin filament when Ca^{++} binds to troponin. This change may involve movement of the tropomyosin rods into the groove of the two-stranded actin helix (here shown as though it were straight for clarity). This shift exposes the myosin-binding site and allows the crossbridges to interact with all seven actins associated with each Ca^{++}-troponin-tropomyosin complex. (Redrawn from Hartshorne, D.J.: Calcium dependence of contractile apparatus. In Lapedes, D.N., editor: Yearbook of science and technology, New York, 1976, McGraw-Hill, Inc. Copyright 1976 by McGraw-Hill Book Co. Used with the permission of McGraw-Hill Book Co.)

■ *Myosin-linked regulation*

The regulatory light chain of all myosins contains one Ca^{++} binding site. Myosin from many invertebrate muscles cannot interact with the thin filament unless this Ca^{++} binding site on the regulatory light chain is occupied. The Ca^{++}-dependence curve for such myosin-linked regulatory systems is similar to that shown in Fig. 22-9 for thin filament regulation. In muscles exhibiting myosin-linked regulation, Ca^{++} again acts as a simple switch that regulates the number of active crossbridges. This system coexists with troponin and thin filament regulation in some invertebrate muscles. Although vertebrate striated muscle myosins have a Ca^{++} binding site on their regulatory light chains, the crossbridges interact with the thin filament, with no requirement for Ca^{++} binding to the light-chain component of the crossbridge. This shows that Ca^{++} binding by light chains is not the primary Ca^{++} regulatory mechanism. Nevertheless, Ca^{++} binding to myosin may have important effects on the actin-myosin interaction in living cells. Ca^{++} binding to myosin may play a significant regulatory role in smooth muscles, but the evidence is conflicting.

The regulatory light chain of all myosins can be phosphorylated reversibly by endogenous enzymes. Phosphorylation is catalyzed by a specific enzyme, *myosin light-chain kinase* (MLCK), and is reversed by *myosin light chain phosphatase* (MLCP) (Fig. 22-10). Myosin phosphorylation depends on Ca^{++} binding to a separate protein cofactor called *calmodulin*. Calmodulin mediates many Ca^{++}-dependent enzymatic reactions in cells when it activates an enzyme such as MLCK by forming a Ca^{++}-calmodulin-enzyme complex. Troponin C is effectively a form of calmodulin that is permanently associated with two other polypeptides to form the troponin molecule. In contrast, calmodulin is a soluble cytoplasmic protein and does not bind to MLCK unless its Ca^{++} binding sites are occupied.

Smooth muscles exhibit a dependence of force on the myoplasmic Ca^{++} concentration resembling that illustrated for striated muscle in Fig. 22-9. Ca^{++} may indirectly regulate the crossbridge cycle in smooth muscles through activation of MLCK because the ATPase activity of smooth muscle actomyosin increases with crossbridge phosphorylation. Studies of living smooth muscles have shown that V_o is the physiological parameter that is proportional to crossbridge phosphorylation (Fig. 22-8, *C*). Crossbridge phosphorylation appears to play an important role in controlling the crossbridge cycle during brief, phasic contractions in smooth muscle. Other unidentified Ca^{++}-dependent systems determine the force that is maintained during prolonged tonic contractions, which are characterized by greatly reduced cycling rates and low levels of myosin phosphorylation. This poorly understood system is obviously adaptive because it allows smooth muscles to maintain force with low ATP consumption rates against externally imposed loads such as blood pressure.

In summary, crossbridge phosphorylation is an important regulatory mechanism in smooth muscle, acting in concert with another Ca^{++}-dependent system. Enzymatic modification of the crossbridge via Ca^{++} and the MLCK/MLCP system during a contraction can modulate crossbridge cycling rates. This has physiologically significant effects on shortening velocities and the energetics of contraction. Phosphorylation may modify the crossbridge cycle in striated muscles. However, this comparatively slow enzymatic reaction causes little myosin phosphorylation during brief contractions of striated muscles, and the effects of phosphorylation on contraction are much less obvious than in smooth muscles.

■ *Myosin phosphorylation and modulation of crossbridge cycling*

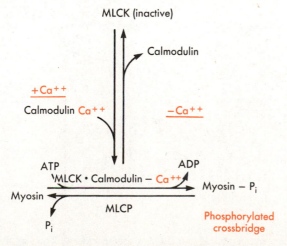

Fig. 22-10 ■ Enzymatic modifications of the crossbridge by Ca^{++}-stimulated phosphorylation of the regulatory light chain of myosin. Ca^{++} acts indirectly in this system by binding to four sites on calmodulin. Phosphorylation produced by myosin light-chain kinase *(MLCK)* is reversed in the absence of Ca^{++} by myosin light-chain phosphatase *(MLCP),* also present in muscle cells. Myosin phosphorylation greatly enhances crossbridge cycling rates in smooth muscles.

■ *Other second messengers and crossbridge regulation*

Agonists that combine with β-receptors in cardiac and smooth muscle can modify contraction. Part of this effect is a result of stimulation of adenylate cyclase and an increase in cellular cyclic AMP. Cyclic AMP acts as a second messenger to activate protein kinase. The active form of protein kinase can phosphorylate troponin in cardiac muscle. Phosphorylated troponin has a lowered affinity for CA^{++}. Protein kinase can also phosphorylate MLCK in smooth muscle, reducing its affinity for the CA^{++}-calmodulin complex. These are potential modulatory mechanisms that can affect the crossbridge cycle, although the physiological significance of these reactions is now uncertain. It may be a general principle that Ca^{++}-dependent regulatory mechanisms mediating the crossbridge cycle may themselves be modulated.

■ *Regulation of cellular Ca^{++}*

The myoplasmic Ca ion concentration in relaxed muscle is very low (less than 0.1 μm). This is maintained by active transport mechanisms against very large concentration gradients. Contraction is initiated by release of Ca^{++} into the myoplasm, and relaxation follows Ca^{++} removal. The Ca^{++} involved in initiating contraction is localized in two pools (Fig. 22-11). One pool is the extracellular space and Ca^{++} bound to the cell membrane in equilibrium with the extracellular Ca^{++}. The other pool is an internal compartment called the *sarcoplasmic reticulum*.

The *sarcolemma*, or plasma membrane, separates the extracellular space from the *myoplasm*, or the intracellular space, which contains the contractile apparatus. The sarcolemma is an excitable membrane and propagates an action potential, which is the first event in the excitation of striated muscle cells (Chapter 3). Small *transverse tubules* (T-tubules) open to the extracellular space at the sarcolemma and form a reticulum in the interior of the cell. This network of T-tubules around the myofibrils forms a grid across the cell at the level of the junction between the A and I bands in each sarcomere of

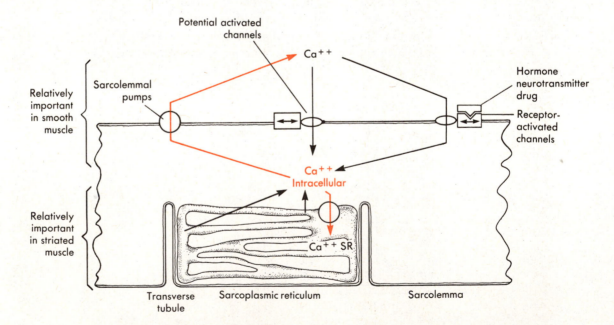

Fig. 22-11 ■ Cellular compartments and Ca^{++} pools in muscle. The sarcolemma and its invaginations (termed T-tubules) separate the myoplasm containing the myofibrils from the extracellular space. The sarcoplasmic reticulum is an internal membrane system delineating an intracellular compartment in which Ca^{++} is present in high concentrations. Ca^{++} can enter the myoplasm to activate the contractile apparatus from the sarcoplasmic reticulum or through the sarcolemma. Relaxation follows Ca^{++} movement back into the sarcoplasmic reticulum or extracellular space by active transport processes. See Fig. 22-1 for anatomical details.

mammalian striated muscles (Figs. 22-1 and 22-11). The lumen of the T-tubules is continuous with the extracellular space. Depolarization of the sarcolemma spreads down the T-tubule into the interior of the cell. The T-tubules are in close association with the fenestrated sheath of sarcoplasmic reticular membranes surrounding the myofibrils. This sheath consists of repeating units with (a) expanded elements (or lateral cisternae) that surround the T-tubules and with (b) narrower elements that run along the myofibrils.

The membranes of the sarcoplasmic reticulum form an intracellular compartment containing large amounts of calcium. The sarcoplasmic reticular membrane is highly specialized and consists almost entirely of pumps that have a higher affinity for Ca^{++} than does troponin. Through the action of this active transport system, 2 moles of Ca^{++} are sequestered in the sarcoplasmic reticulum for each mole of ATP hydrolyzed. These pumps maintain the low resting myoplasmic Ca^{++} concentration. The transit of an action potential along the sarcolemma causes Ca^{++} release from the sarcoplasmic reticulum into the myoplasm. The nature of the coupling between the sarcolemma T-tubular system and the sarcoplasmic reticulum is poorly understood, but apparently it does not involve depolarization of the sarcoplasmic reticular membrane. The net effect is the release of a pulse of Ca^{++} into the myoplasm by the transit of an action potential. The Ca^{++} released into the myoplasm binds to troponin or to other Ca^{++} regulatory sites to initiate contraction.

The interior of the sarcoplasmic reticulum contains a protein called *calsequestrin*. Each molecule of calsequestrin can bind about 43 Ca ions with a moderate affinity. Calsequestrin plus another protein with similar properties acts to reduce the sarcoplasmic reticular concentration of Ca^{++} from perhaps 20 mM, if all were free, to an estimated 0.5 mM. This action greatly reduces the concentration gradient against which the pump must act.

Smooth muscles lack a T-tubular system, and the sarcoplasmic reticulum is limited to a network of narrow tubules near the sarcolemma. The extent of this network and the importance of the sarcoplasmic reticulum in regulating myoplasmic Ca^{++} is quite variable and, in some cases, is negligible.

■ *Ca^{++} regulation by the sarcoplasmic reticulum*

The overall process by which depolarization of the sarcolemma causes Ca^{++} release into the myoplasm and Ca^{++} binding to regulatory sites to initiate crossbridge cycling is termed *excitation-contraction coupling* (E-C coupling). The response of the contractile system depends on the amount of Ca^{++} released into the myoplasm. Ca^{++} release from the sarcoplasmic reticulum depends on the value of the membrane potential, and it occurs when the potential becomes less negative than about -50 mV. Maximal Ca^{++} release with full activation of the cell occurs when the cell is depolarized to a level of about -20 mV (Fig. 22-12). The more negative value is termed the *mechanical threshold* and is similar whether it is achieved during a normal action potential or when the membrane is experimentally depolarized (Fig. 22-12). Agents that lower the mechanical threshold or prolong the action potential increase the magnitude of the Ca^{++} pulse. Release of a greater amount of Ca^{++} potentiates the contractile response to a single action potential. Caffeine is a drug that exhibits both of these effects.

A single action potential releases sufficient Ca^{++} to fully activate the contractile machinery in skeletal muscles. However, the Ca^{++} is very rapidly pumped back into the sarcoplasmic reticulum before the muscle has time to develop its maximum force. The resulting submaximal response to a single action potential is termed a *twitch*. The time relationships among the action potential, the Ca^{++} pulse (expressed in terms of the number of active crossbridges), and isometric force development are illustrated in Fig. 22-13 for a skeletal muscle fiber. Repetitive action potentials can cause summation of twitches, producing a partial or complete *tetanus* as the Ca^{++} pulses sum to maintain saturating Ca^{++} concentrations in the myoplasm (Figs. 22-13 and 22-14).

■ *Excitation-contraction coupling*

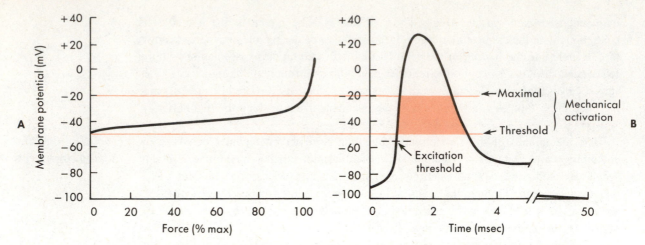

Fig. 22-12 ■ **A,** The relationship between steady state force development and membrane potential (E_m) obtained by increasing the K^+ concentration in the bathing solution. The mechanical threshold for force development occurs at an E_m of about -50 mV when Ca^{++} release from the sarcoplasmic reticulum leads to myoplasmic concentrations exceeding threshold values of about 0.1 μM. At an E_m of -20 mV, myoplasmic Ca^{++} concentrations are maximal for activation. **B,** Under physiological conditions the transit of an action potential along a muscle fiber leads to release of a Ca^{++} pulse, the magnitude of which is proportional to the product of membrane potential and time, as indicated by the colored area in the illustrated action potential. (Redrawn from Sandow, A.: Arch. Phys. Med. Rehabil. **45:**62-81, 1964.)

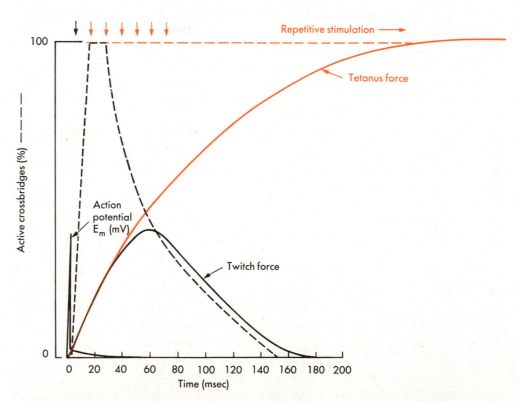

Fig. 22-13 ■ Time relationships between the action potential, activation of crossbridges (dashed lines), and force development (solid lines) in skeletal muscle. The twitch response to a single stimulus and action potential is shown in black. Repetitive stimulation (color) leads to a tetanus. The dashed curve depicting the fraction of active crossbridges is estimated by applying quick stretches to a muscle to determine the stiffness resulting from attached crossbridges. This relationship is often termed an *active state curve*. (Redrawn from Kutchai, H.C.: Muscle cells. In Flickinger, C.J., editor: Medical cell biology, Philadelphia, 1979, W.B. Saunders Co.)

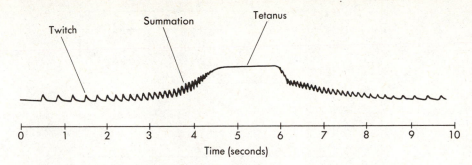

Fig. 22-14 ■ The force developed by a skeletal muscle fiber to repetitive stimulation at gradually increasing frequencies, followed by gradually decreasing frequencies. Individual twitch responses start to fuse and sum as the frequency increases, leading to a complete or fused tetanus. (From Buchthal, F.: Dan. Biol. Med. **17**:1, 1942.)

Force development is much slower than Ca^{++} release and binding to troponin because a number of crossbridge cycles must occur with some sarcomere shortening before maximal force is registered at the tendons. This behavior is a result of the fact that the structural elements of the muscle, including the crossbridges, are somewhat elastic and lengthen when a force is generated. This elasticity (termed *series elasticity*) is seen only in a contracting muscle and is distinct from the *passive (parallel) elasticity* observed in a relaxed muscle (Fig. 22-7).

In the heart the rate of Ca^{++} release into the myoplasm from the less extensive sarcoplasmic reticulum is slow and is the limiting factor for the rate of shortening. The number of active crossbridges is reflected by the twitch force when the rate of Ca^{++} release is slow relative to the rate of shortening of the sarcomeres. Obviously any factor potentiating Ca^{++} influx into the cardiac cell myoplasm will significantly enhance the contractile response.

The action potential in skeletal muscles is caused by the opening of *fast channels* in the membrane. This process permits an initial rapid inward Na^+ current, followed by an outward K^+ movement and repolarization. There is no significant movement of Ca^{++} ions through fast channels from the extracellular space to the myoplasm. The role of the fast channels in the generation and propagation of the action potential is described in Chapter 3. In involuntary muscles two other types of channels in the cell membrane are important in regulating myoplasmic Ca^{++} and are described next.

Potential-sensitive Ca^{++} channels. In addition to the fast channels, cardiac muscle has membrane potential–dependent *slow channels,* which open during the transit of an action potential. Ca^{++} ions are the major inward current-carrying ions in the slow channels. The inward Ca^{++} current is responsible for the maintained depolarization characterized by the plateau phase of the action potential in cardiac muscle (Chapters 3 and 27). Although only a small amount of Ca^{++} enters a cell during a cardiac action potential, this influx contributes significantly to the maintenance of the normal levels of Ca^{++} that are handled by the sarcoplasmic reticulum. The force of the rhythmical cardiac contractions rapidly declines if extracellular Ca^{++} is reduced, and it is enhanced when extracellular Ca^{++} is increased. Similarly, higher frequencies of cardiac contraction increase the magnitude of the Ca^{++} pool available for E-C coupling over the course of several beats. This produces the *staircase phenomenon* or *Treppe* characterized by progressive increases in the force of contraction over several beats (Chapter 33). Thus both the sarcoplasmic reticulum and the extracellular Ca^{++} pools are involved in E-C coupling in the heart (Fig. 22-11).

Smooth muscles usually exhibit some dependence on extracellular Ca^{++} for contraction. The sarcolemma and extracellular space can function as the equivalent of a

■ *Ca^{++} regulation by the sarcolemma*

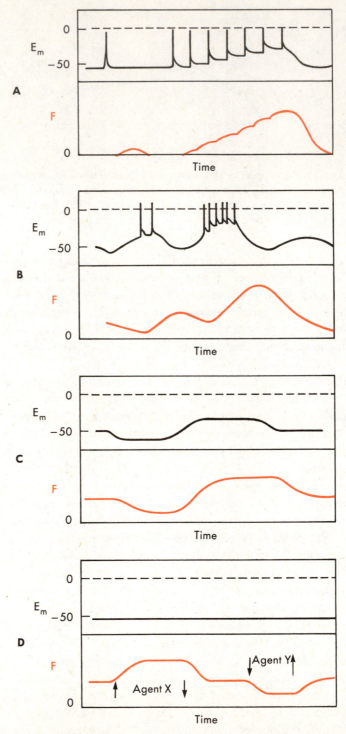

Fig. 22-15 ■ Relationships between membrane potential (E_m) and force generation (F) characteristic of different types of smooth muscle. **A,** Action potentials may be generated and lead to a twitch or larger summed mechanical responses. Action potentials are characteristic of single-unit smooth muscles (many visceral). Gap junctions permit the spread of action potentials throughout the tissue. **B,** Rhythmical activity produced by slow waves that trigger action potentials. The contractions are usually associated with a burst of action potentials. Slow oscillations in membrane potential usually reflect the activity of electrogenic pumps in the cell membrane. **C,** Tonic contractile activity may be related to the value of the membrane potential in the absence of action potentials. Graded changes in E_m are common in multiunit smooth muscles (many vascular), where electrical activity is not propagated from cell to cell. **D,** Pharmacomechanical coupling: changes in force produced by the addition or removal *(arrows)* of drugs or hormones that have no significant effect on the membrane potential.

sarcoplasmic reticulum in these small cells, which have short diffusion distances and comparatively slow rates of contraction and relaxation. Pumps in the smooth muscle cell membrane actively extrude Ca^{++} from the myoplasm. Ca^{++} influx depends on two types of channels (Fig. 22-11). The first is the potential-sensitive slow channel, which responds to the membrane potential. Smooth muscle cell membranes do not contain fast channels. Enough Ca^{++} crosses the membrane to initiate a small contraction, which can be summed to produce larger contractions with trains of action potentials (Fig. 22-15, *A* and *B*).

Not all smooth muscles generate slow action potentials when depolarized. This is probably the result of a simultaneous rise in K^+ permeability, which prevents a regenerative response. Variations in Ca^{++} influx through potential-sensitive slow channels in the absence of action potentials produce graded contractions (Fig. 22-15, *C*).

The "resting" membrane potential (E_m) reflects ionic concentration gradients and conductances as in striated muscle. However, there is a significant and variable contribution to E_m by electrogenic pumps that extrude Na^+ in exchange for K^+ (Chapter 2). Changes in pump activity produce alterations in the membrane potential (Fig. 22-15, *C*). Sufficient depolarization may produce trains of action potentials, most typically in smooth muscles involved in rhythmical contractile activity, as in the gastrointestinal tract (Fig. 22-15, *B*).

Receptor-activated Ca^{++} channels. Many neurotransmitters, hormones, and drugs can combine with highly specific receptors on smooth muscle cell membranes (Fig. 22-11). Such agents can elicit contractions in some cases without evoking significant changes in the membrane potential (Fig. 22-15, *D*). The Ca^{++} sources can involve both extracellular or membrane-bound Ca^{++} and Ca^{++} released internally from the sarcoplasmic reticulum. This process is termed *pharmacomechanical coupling*. Depolarization may also occur in response to the binding of neurotransmitters, hormones, or drugs to the cell membrane. Additional Ca^{++} influx through potential-activated Ca^{++} channels may occur in such cells during depolarization.

The complexity of the processes involved in E-C coupling or pharmacomechanical coupling in smooth muscle reflects the diverse and complex physiology of these tissues. In addition, tissue function can be modified by a wide range of neural and hormonal influences. A broad comparison of how contractile output is graded in the different muscle types is given in the next section.

■ *Grading contractile force*

The mechanisms described earlier for regulating the myoplasmic Ca^{++} concentration and the crossbridge interaction with the thin filament determine contraction and relaxation in muscle cells. The physiologically important ways in which various muscle tissues vary their output are summarized in Table 22-1.

In skeletal muscle contraction only occurs in response to activity in the motor nerves, and semiquantal Ca^{++} release from the sarcoplasmic reticulum is produced by the action potential. Contractile force can be graded by two mechanisms. One mechanism is to recruit more cells in a muscle. Recruitment involves increments of groups of cells, each group consisting of all the muscle fibers innervated by a single motor nerve (*motor unit*). The second mechanism is to sum the twitch responses by increasing the frequency of stimulation to obtain a tetanus. The gradation of contraction is considered in greater detail in the following chapter and in Chapter 14. Force will vary somewhat with shortening according to the stress-length relationship. However, skeletal muscles are attached to the skeleton in arrangements that usually limit the normal working range to lengths near the peak of the stress-length curve (Figs. 22-7, *B,* and 23-3).

Cardiac muscle cells are all electrically coupled, and the cells contract almost synchronously as the action potential propagates rapidly from cell to cell. Furthermore, the prolonged action potentials and subsequent refractory period of cardiac cells prevent

TABLE 22-1

Mechanisms for grading contractile force

General mechanism	Skeletal	Occurrence Cardiac	Smooth*
Recruit more cells (motor units)	+	−	+
Sum twitches by increasing stimulation frequency (tetanus)	+	−	+
Alter filament overlap by stretch	(+)	+	+
Vary twitch via potential-activated Ca^{++} channels	−	+	+
Tonic depolarization and activation of potential-dependent channels without action potentials	−	−	+
Receptor-activated channels (pharmaco-mechanical coupling)	−	−	+

*The relative importance of these mechanisms varies greatly with the type of smooth muscle.

tetanization. Thus different mechanisms are used to grade the force of contraction. One mechanism involves the action of neurotransmitters or hormones to alter Ca^{++} influx from the extracellular space during the action potential or to change the sensitivity of the regulatory proteins to the myoplasmic Ca^{++}. The other mechanism grading contraction involves stretching the cells to lengths more satisfactory for force generation when the cardiac filling pressure increases (the Starling mechanism, Chapter 29).

Some smooth muscles resemble skeletal muscle, with extensive innervation and little electrical coupling between smooth muscle cells. Such tissues grade contractile force in ways that are analogous to those described for skeletal muscle: recruitment of more cells and summation of twitch responses. Other smooth muscles are more like the heart, where action potentials are propagated between electrically coupled muscle cells. Graduation of output is a function of (1) action potential frequency and mechanical summation, (2) the level of the membrane potential and opening of potential-dependent Ca^{++} channels, and (3) alterations in length of smooth muscle cells by variations in the volume of the hollow organs. However, all smooth muscle cells are sensitive to a variety of neurotransmitters and hormones, and the responses can be further modified by receptor-activated Ca^{++} channels. Force development by most smooth muscles reflects a balance of neural, hormonal, mechanical, and local metabolic factors, some of which may be inhibitory and others excitatory. Much is known concerning specific mechanisms for altering force, but there are few tissues about which there is a reasonable understanding of all the physiological factors that determine, for example, the resistance in a vascular bed or motility in the gastrointestinal tract.

■ Energy use and supply

Muscular contraction demands a constant supply of ATP at a rate proportional to consumption. In this section the ATP requirements for various types of contractions (muscle energetics) will be considered. The ways in which cells can adequately provide ATP are then reviewed. Finally, the matching of metabolism and energetics is described for particular fiber types. Such matching illustrates the high degree of specialization of muscle cells for specific contractile activities and forms the basis for an understanding of many physiological aspects of muscle function in the body.

■ *Energetics*

The mechanical output of muscle can be readily determined in terms of work performed (force × distance shortened) or the tension-time integral (force × time) in isometric contractions. Measurement of energy use by the contractile system is more difficult. Heat production can be determined very accurately. Heat release reflects ATP use on the basis of assumptions concerning the total energy released during ATP hydrolysis and the efficiency of its conversion to mechanical work without being lost as heat. There are potential errors in these assumptions, and the method is not well suited for estimating rapid events. The same is true for metabolic measurements in which oxygen consumption (or lactate production when oxidative metabolism is blocked) is used to estimate ATP use. However, more difficult direct chemical measurements of ATP and creatine phosphate use (in muscles where metabolic resynthesis of these compounds is inhibited) have confirmed the basic relationships in muscular energetics.

Muscle has an ATP consumption or metabolism associated with basal cellular activities such as maintaining ion gradients and synthesizing and degrading cellular constituents. Muscles are unusual in that this *resting metabolism* represents only a small fraction of the maximum ATP use that is associated with contraction. The energy requirements for activation associated with action potentials and Ca^{++} release into the myoplasm are also comparatively small.

Most energy consumption is a consequence of the fact that each complete cycle of the enormous numbers of crossbridges in a muscle cell requires one ATP molecule (Fig. 22-6). Cycling rates in skeletal muscle depend on two factors. One is the load on the muscle (Fig. 22-8). A muscle contracting isometrically has a low crossbridge cycling rate and a comparatively moderate rate of ATP use. This rate is reduced even more in a muscle subjected to an imposed stretch. This situation is termed *negative work,* as it is characterized by the product of force times distance stretched. Such an imposed stretch occurs when muscles are used to decelerate the body. When a muscle can shorten, crossbridge cycling rates increase, and ATP consumption rises with the mechanical work done. Low forces and high velocities are associated with extremely high rates of ATP consumption. Speed is costly as in other aspects of life. Shortening velocity is ultimately limited by the myosin isoenzyme present. Fast fibers, which contain a form of myosin that is capable of high rates of ATP hydrolysis, have much greater requirements for rapid ATP synthesis than do slow fibers. Thus the type of myosin is the second factor that dictates metabolic requirements.

Relaxation is also associated with above resting levels of ATP use. In part this increased ATP use provides energy for the Ca^{++} pump in the sarcoplasmic reticulum or the sarcolemma. The creatine pool is also rephosphorylated if it was depleted during the contraction, and glycogenesis, which acts to restore cellular glycogen stores, contributes to ATP use.

The energetics of crossbridge cycling and work performance during shortening in smooth muscles are similar to those described for striated muscle. Nevertheless, many mammalian smooth muscles can maintain force tonically (steadily) with a rate of ATP use that is several hundredfold less than that in striated muscles. Part of this performance may be attributed to a form of myosin present in smooth muscles that has intrinsically lower rates of crossbridge cycling. However, much of the economy in maintaining a steady force is a consequence of the regulatory system that can greatly slow the crossbridge cycle. More and more of the crossbridges in tonic smooth muscles go into a "slowly cycling" state during the course of a maintained contraction in which the muscle can withstand distending pressure and yet expend relatively little energy.

■ *Metabolism*

Muscle shares the same ATP-generating mechanisms that are found in all nucleated cells (Fig. 22-16), although the relative importance of the different mechanisms varies in different muscle cell types.

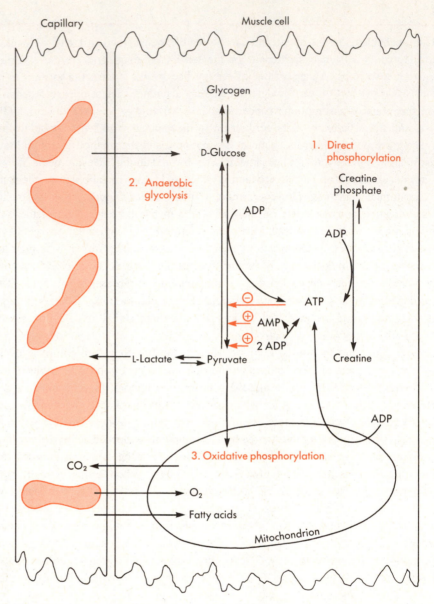

Fig. 22-16 ■ Metabolic pathways in muscle. ATP is supplied via direct phosphorylation of ADP*(1)*, glycolysis *(2)*, and oxidative phosphorylation *(3)*.

1. *Direct phosphorylation* of ADP to regenerate ATP from creatine phosphate is an extremely rapid reaction. This pathway usually functions as a buffer to maintain the normal ATP levels of 3 to 5mM at the beginning of contraction while other systems for regenerating ATP are being turned on. Myoplasmic creatine phosphate concentrations are 5 (smooth) to 20 (skeletal) mM, which are only sufficient to provide the energy for a few twitches. In the second direct phosphorylation reaction, *adenylate kinase* (often termed *myokinase* in muscle) transfers a phosphate group from one ADP to another to form ATP and AMP. This reaction has little metabolic significance, but it plays an important regulatory role in glycolysis (colored arrows, Fig. 22-16). Phosphofructokinase, the rate-limiting enzyme in glycolysis, is inhibited by ATP and stimulated by ADP and AMP. Thus ADP and AMP formed on activation of the contractile apparatus will stimulate glycolysis.

2. *Anaerobic glycolysis* is very rapid and readily meets the ATP demands, even of very fast muscle cells. Consequently it is important in cells of this type and in all muscle cells when the oxygen supply is inadequate. However, this pathway has a net yield of

only 2 moles of ATP per mole of glucose (or 3 if the glucose is derived from cell glycogen), and it is comparatively inefficient. In the presence of oxygen, pyruvate is converted to CO_2 instead of to lactate (aerobic glycolysis), and the net yield of ATP is improved threefold. ATP production by glycolysis may also be limited by the cellular stores of glycogen, which can be rapidly depleted.

3. *Oxidative phosphorylation* of fatty acids is the primary source of energy in muscles that are frequently active. Oxidative phosphorylation is not only efficient (36 moles ATP/mole glucose), but it can operate continuously when the circulation is adequate. However, oxidative phosphorylation is a slow process. It cannot meet the maximum ATP consumption rates of rapidly contracting skeletal muscle fibers unless a major fraction of the fibers consists of mitochondria in close proximity to a capillary.

■ *Matching of ATP production and consumption*

The vertebrate genome contains at least two genes that code for the synthesis of myosin in skeletal muscle in an animal species. One of the two isoenzymatic forms of myosin has a higher rate of ATP hydrolysis than the other, and cells in which the "fast" myosin is synthesized have faster shortening velocities. With appropriate histochemical methods, thin sections of frozen muscles can be cut and incubated with ATP under conditions where only the slow myosin is enzymatically very active. The phosphate released is trapped on the section, "staining" the slow fibers. Fig. 22-17, *A,* illustrates how fast fibers (types IIA and IIB) and slow fibers (type I) are typically intermixed in most mammalian skeletal muscles. Table 22-2 provides a summary of the fiber types and the nomenclature in general use.

Similar histochemical reactions can also be applied to serial sections cut from the same muscle in order to estimate the activities of enzymes in the oxidative (Fig. 22-17, *B*) and glycolytic (Fig. 22-17, *C*) metabolic pathways. The metabolic capacities in muscle fibers can vary continuously over a wide range. However, most fast (type II) fibers show high activities of glycolytic enzymes and low activities of oxidative enzymes, a fact confirmed by the comparatively few mitochondria that can be observed in electron micrographs. Fast fibers with high contraction velocities have a much more extensive sarcoplasmic reticulum with high pumping rates to quickly activate and inactivate the contractile machinery than do slow fibers, cardiac muscle, and smooth muscle. Because the ATP consumption rates in the myofibrils and sarcoplasmic reticulum of such fast fibers can most readily be matched by glycolysis, a metabolism based primarily on glycolysis is appropriate, provided the fibers are recruited only occasionally for brief efforts. This is true, as will be shown in the next chapter. Theoretically, a fiber of small diameter with very high mitochondrial and surrounding capillary densities could oxidatively synthesize ATP at rates used by the fast myosin isoenzyme. Some fast fibers with high glycolytic and oxidative capacities are found in mammals and have led to the subclassification of fast, type II fibers shown in Table 22-2.

Slow, type I fibers can meet relatively modest metabolic demands with oxidative phosphorylation. Molecules associated with oxygen binding (e.g., hemoglobin, myoglobin, cytochromes) contain iron and are red. Therefore the red color of an oxidative muscle cell or a tissue containing mostly type I (or IIA) fibers has led to the designation of such types as *red fibers*.

Cardiac muscle is slow and contains myosin isoenzymes with low ATPase activities. As might be expected, this continuously active muscle is almost entirely oxidative and is highly sensitive to interruptions in its blood supply. Smooth muscles as a group have few mitochondria and derive a significant portion of their ATP from aerobic glycolysis. This is somewhat surprising because ATP consumption rates are comparatively low.

The oxidation of fatty acids provides most of the ATP used by the muscles in the body. However, all muscles have a significant capacity to use other substrates, including carbohydrates, certain amino acids, and ketone bodies.

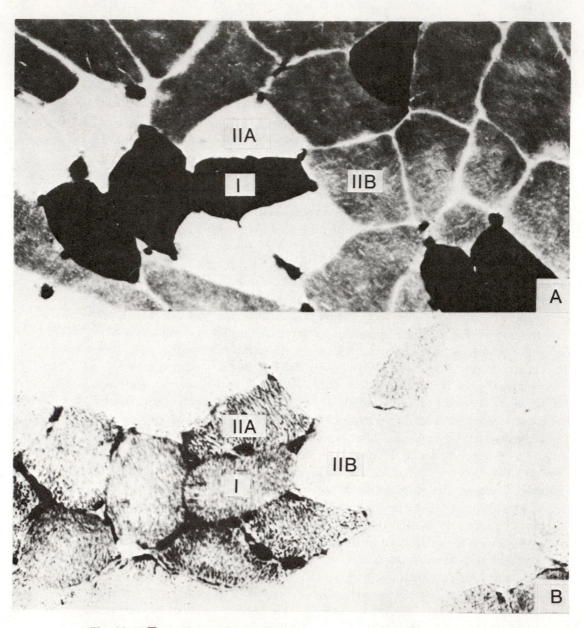

Fig. 22-17 ■ Histochemical staining of cross sections of the feline semitendinosus skeletal muscle. **A,** Staining for the ATPase activity of the slow myosin isoenzyme of type I fibers. Type IIB fibers stain more than type IIA fibers in this species, and the myosins in these two types of fast fibers may differ. **B,** Staining for succinic dehydrogenase activity, an enzyme associated with oxidative phosphorylation.

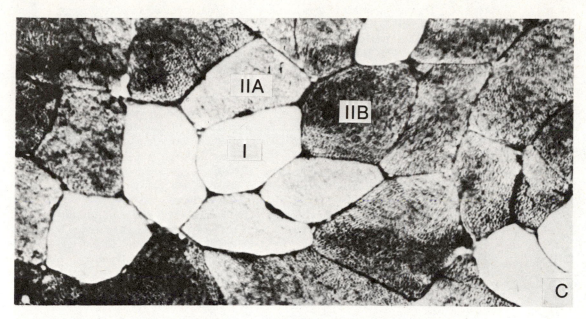

Fig. 22-17, cont'd ■ **C,** PAS stain indicating glycolytic capacity. Three distinct fiber types can be differentiated in these serial sections: slow oxidative or type I, fast glycolytic or type IIB, and fast oxidative (and glycolytic) or type IIA fibers. (×450.) (From Hoppeler, H., et al.: Respir. Physiol. **44:**94, 1981.)

TABLE 22-2

Basic classification of skeletal muscle fiber types

	Type I: slow oxidative (red)	*Type IIB: fast glycolytic (white)*	*Type IIA*: fast oxidative (red)*
Myosin isoenzyme (ATPase rate)	Slow	Fast	Fast
Sarcoplasmic reticular Ca^{++} pumping capacity	Moderate	High	High
Diameter (diffusion distance)	Moderate	Large	Small
Oxidative capacity: mitochondrial content, capillary density, myoglobin	High	Low	Very high
Glycolytic capacity	Moderate	High	High

*Comparatively infrequent in humans and other primates. In the text a simple designation of type II fiber refers to a fast-glycolytic (type IIB) fiber.

■ *Bibliography*

Journal articles

Cohen, C.: The protein switch of muscle contraction, Sci. Am. **233:**36, 1975.

Huxley, H.E.: The mechanism of muscular contraction, Sci. Am. **213:**18, 1965.

Lester, H.A.: The response to acetylcholine, Sci. Am. **236:**106, 1977.

Margaria, R.: The sources of muscular energy, Sci. Am. **226:**84, 1972.

Murray, J.M., and Weber, A.: The cooperative action of muscle proteins, Sci. Am. **230:**58, 1974.

Porter, K.R., and Franzini-Armstrong, C.: The sarcoplasmic reticulum, Sci. Am. **212:**72, 1965.

Staehelin, L.A., and Hull, B.E.: Junctions between cells, Sci. Am. **238:**140, 1978.

Books and monographs

Berne, R.M., and Sperelakis, N., editors: Handbook of physiology, section 2, The cardiovascular system, vol. 1, The heart, Bethesda, Md., 1979, American Physiological Society.

Bohr, D.F., Somlyo, A.P., and Sparks, H.V., Jr.; editors: Handbook of physiology, section 2, The cardiovascular system, vol. 2, Vascular smooth muscle, Bethesda, Md., 1980, American Physiological Society.

Goldman, R., Pollard, T., and Rosenbaum, J.: Cell motility, book A, Motility, muscle and non-muscle cells, Cold Spring Harbor, N.Y., 1972, Cold Spring Harbor Laboratory.

Huddart, H.: The comparative structure and function of muscle, Oxford, 1975, Pergamon Press, Ltd.

Huddart, H., and Hunt, S.: Visceral muscle: its structure and function, New York, 1975, Halsted Press.

Needham, D.M.: *Machina carnis:* the biochemistry of muscular contraction in its historical development, Cambridge, 1971, Cambridge University Press.

Squire, J.: The structural basis of muscular contraction, New York, 1981, Plenum Press.

Stein, R.B.: Nerve and muscle: membranes, cells, and systems, New York, 1980, Plenum Press.

Muscle as a tissue

Individual muscle cells are encased in a connective tissue layer called the *endomysium*. Groups of skeletal muscle cells form *fascicles* that are bounded by the *perimysium*. These fascicles are grouped into the definitive muscle, which is covered by the *epimysium*. The three connective tissue layers are composed primarily of elastin and collagen fibrils. An estimated 250 million cells are found in the more than 400 skeletal muscles in humans. Each of these muscles exerts specific movements via tendons that are attached to the skeleton in most cases. This chapter considers muscular function at the tissue level.

■ *Voluntary muscle: physiological control and function*

The output of a muscle depends on the size of the muscle's cells and their anatomical arrangement. Increasing the diameter of a fiber by synthesis of new myofibrils *(hypertrophy)* will increase the force-generating capacity of a cell (Fig. 23-1). The formation of more cells *(hyperplasia)* also increases tissue force output. However, differentiated skeletal muscle has only a limited capacity to form new cells. The force-generating capacity is unchanged if the length of the cells is increased by adding more sarcomeres in series without increasing the cross-sectional area of the cells. However, the absolute velocity of contraction and the total shortening capacity of the cell increase with the addition of more sarcomeres (Fig. 23-1).

■ *Musculoskeletal relationships*

The output of muscles not only depends on the number of sarcomeres and myofibrils in the cells but also on the orientation of the cells within the tissue. Measurements on intact muscles directly reflect cellular properties only when the cells are arranged parallel with each other and also with the axes of the tendons that transmit the force (Fig. 23-2). This parallel arrangement maximizes shortening capacity and velocity for a muscle. However, force-generating capacity can be enhanced at the expense of shortening capacity and velocity by arrangements placing the muscle fibers at an angle to the tendons. Examples of such *pennate* (featherlike) arrangements are illustrated in Fig. 23-2.

Some skeletal muscles, including those surrounding the mouth and anus, serve as sphincters. Striated muscles are also found in the upper portions of the esophagus, and they are active during swallowing. Others may be attached to the skin. However, most skeletal muscles are attached to the skeleton. The attachments involve the highly inextensible protein *collagen,* which forms *tendons* or flattened sheets termed *aponeuroses*. Muscles are described in terms of an origin at the proximal or relatively fixed point and an insertion on the bone that is moved. (The distinction between origin and insertion is sometimes arbitrary.) Muscles bridge one or, more frequently, two joints. The contraction of an individual muscle can consequently lead to movement of more than one bone.

Specific coordinated movements require the actions of two or more muscles. These

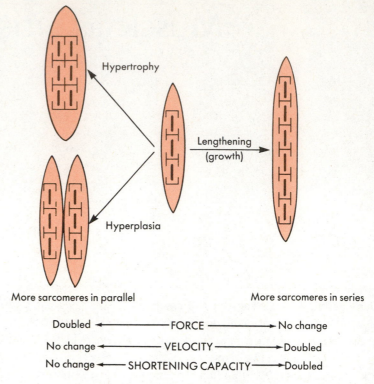

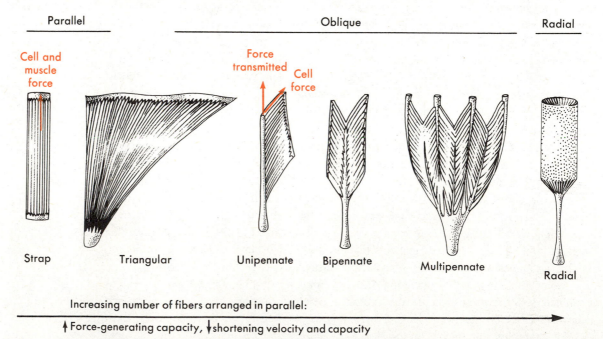

Fig. 23-1 ■ Effects of growth on the mechanical output of a muscle cell. Growth may consist of adding new myofibrils (depicted as a series of model sarcomeres) within a cell (hypertrophy), formation of new cells (hyperplasia), or adding more sarcomeres in series as the muscle cells lengthen along with skeletal growth. The effects of the illustrated cell growth on the absolute force (kg), shortening velocity (m/sec), and shortening capacity (m) of the muscle are summarized in the table.

Fig. 23-2 ■ Some arrangements of skeletal muscle fibers. The force generated by all of the individual cells is fully transmitted to the skeleton in only a few straplike muscles. The cells in most muscles are arranged at an angle to the axis of the muscle. This allows more fibers to be attached to a tendon and increases the total force-generating capacity. However, not all of the force generated by each cell is usefully transmitted to the tendon with an oblique geometry, and the overall shortening velocity and shortening capacity of the muscle are diminished compared to that of the individual cells. (Redrawn from Gray's anatomy, ed. 35 [British], Philadelphia, 1973, W.B. Saunders Co.)

may be *synergists* when the muscles act together or *antagonists* when the actions of the muscles are opposed. *Kinesiology* is the study of the interactions of groups of muscles. The complex nature of coordinated movements has been revealed by electromyographic techniques in which the summed electrical activity of a muscle is detected from electrodes placed on the overlying skin. Another technique can detect the activity of a single cell with needle electrodes inserted into the muscle. These techniques reveal that a specific movement may involve contractions of antagonistic muscles and may not involve contractions of some synergistic muscles. The activity patterns are often unpredictable and can only be determined by direct recordings. In general, the interactions of the muscles serve not only to move a specific bone but also to fix or stabilize another bone or joint.

The skeleton also acts as a lever system with significant mechanical consequences. As illustrated in Fig. 23-3, the biceps muscle of a person holding a 20 kg weight at approximately right angles must develop an isometric force of 140 kg. Enormous forces can be placed on tendons by large muscles such as the gastrocnemius. This can lead to rupture of the tendon. The greatest stress on a tendon occurs upon gravitational loading of a contracting muscle because the crossbridges can bear more force than they can develop. Injuries usually occur as the result of a fall, which subjects a contracting muscle to a sudden large increase in stress. Another consequence of the skeletal lever system is that large movements of the limbs can occur with far less shortening of the muscles (Fig. 23-3). With this arrangement sarcomere lengths remain close to their optimum for force development in most movements.

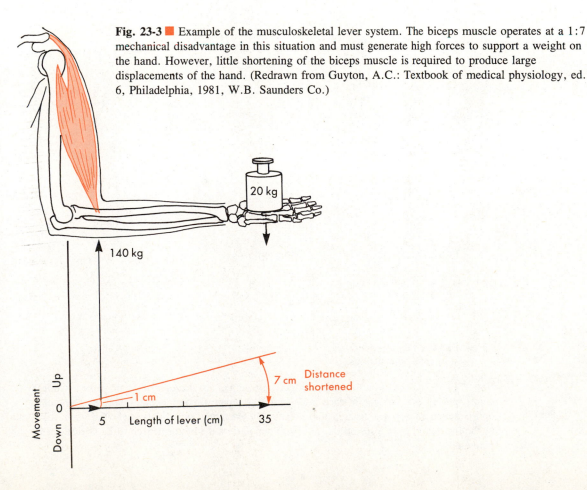

Fig. 23-3 ■ Example of the musculoskeletal lever system. The biceps muscle operates at a 1:7 mechanical disadvantage in this situation and must generate high forces to support a weight on the hand. However, little shortening of the biceps muscle is required to produce large displacements of the hand. (Redrawn from Guyton, A.C.: Textbook of medical physiology, ed. 6, Philadelphia, 1981, W.B. Saunders Co.)

■ *Coordination of muscular activity*

Motor nerves and motor units. The cell bodies of the *motor nerves* (α-motor axons) are in the ventral horn of the spinal cord (Fig. 23-4). The axon exits via the ventral root and reaches the muscle through a mixed peripheral nerve. The motor nerves branch in the muscle, with each terminal innervating a single muscle cell in mammals. The specialized cholinergic synapse that forms the neuromuscular junction and the neuromuscular transmission process that generates an action potential in the muscle fiber are described in Chapter 4. A *motor unit* consists of the motor nerve and all of the muscle fibers innervated by that nerve. The motor unit (and not the individual cell) is the functional contractile unit because all of the cells within a motor unit contract synchronously when the motor nerve fires. The muscle cells of a motor unit are not segregated anatomically into distinct groups, and considerable intermixture of cells occurs among neighboring motor units.

Motor units exhibit considerable specialization. They may consist of only two or three muscle fibers, or there may be over a thousand cells in some motor units in large muscles. The cell bodies and axons of the motor nerve increase in size with the number of muscle fibers in the unit. This relationship is understandable in terms of the metabolic requirements for synthesis and release of acetylcholine. A distinction between slow ox-

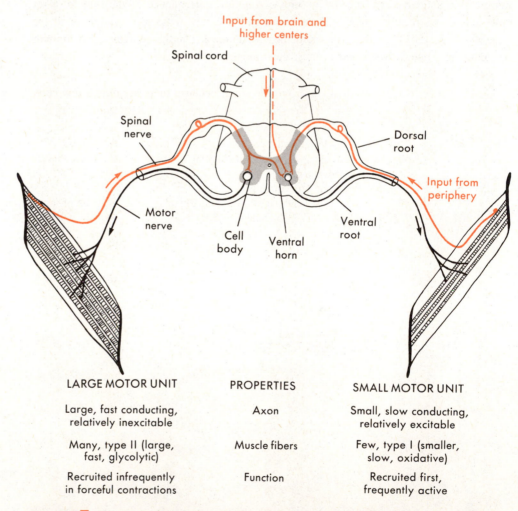

LARGE MOTOR UNIT	PROPERTIES	SMALL MOTOR UNIT
Large, fast conducting, relatively inexcitable	Axon	Small, slow conducting, relatively excitable
Many, type II (large, fast, glycolytic)	Muscle fibers	Few, type I (smaller, slow, oxidative)
Recruited infrequently in forceful contractions	Function	Recruited first, frequently active

Fig. 23-4 ■ The motor unit and some inputs. Large and small motor units are mixed within a single muscle. An example of each is illustrated in a pair of contralateral muscles. The motor unit contracts in response to an action potential in the motor axon. This is elicited when the sum of the inputs from synapses on the cell body depolarizes the motor neuron to its critical firing potential.

idative (type I) and fast glycolytic (type II) muscle fibers is found in Table 22-2. This classification also applies to motor units because all the fibers in one motor unit are of the same type (Table 23-1). The smaller motor units normally consist of type I cells (Table 23-1) (Fig. 23-4). An important point is that only the contraction velocity or myosin ATPase activity clearly distinguishes the fiber types. These two parameters reflect the presence of different isoenzymatic forms of myosin in the cells. Metabolic differences arise from different cellular contents of mitochondria, for example, and may vary considerably between cells of the same type.

The inputs to the motor nerve are both excitatory and inhibitory. The inputs involve (1) neurons from the brain, (2) neurons from elsewhere in the spinal cord, and (3) neurons originating from a variety of receptors within a muscle, from its antagonists and synergists, and from the same group of contralateral muscles (Fig. 23-4). These inputs to the motor neuron and their integrated activity are described in Chapter 14. A motor neuron fires when the sum of the excitatory and inhibitory inputs depolarizes the cell to its critical membrane potential.

Recruitment of motor units. The functional importance of the variations among motor units can be illustrated by considering how the motor units in a muscle are progressively activated in graded contractions. Increasing excitatory or decreasing inhibitory input to the motor neuron pool in the ventral horn will depolarize the cell bodies. However, a given level of excitatory input will produce more depolarization of the smallest neurons because of their smaller membrane areas. Thus the first axons to fire are those of the smallest motor units. The conduction velocity of the action potential will be relatively low, reflecting the cable properties of the axons of small diameter. The total force developed by the muscle will be small because there are only a few cells of moderate diameter in these units. Fig. 23-5 shows the basic relationships between motor unit recruitment and total force generated by a muscle. In the left column *31* indicates the number of motor units among those sampled in the muscle that developed up to 10 g force (average, 5 g) when tetanized. If all 31 motor units fired together, the total force generated would be about 0.15 kg, assuming an average force of 5 g per unit. The important point is that these units are recruited first, and they remain active as long as any part of the muscle is contracting. Many such small motor units permit fine gradations of delicate movements. With increasing excitatory input somewhat larger axons fire. In this sample there were fewer motor units that could generate between 10 and 20 g force, but their summed contributions to the total force in the muscle were comparable, that is, 10 units × 15 g (average force) = 150 g, shown by the height of

TABLE 23-1

Properties of motor units

Characteristics	Motor unit classification	
	Type I	Type II
Properties of nerve		
Cell diameter	Small	Large
Conduction velocity	Fast	Very fast
Excitability	High	Low
Properties of muscle cells		
Number of fibers	Few	Many
Fiber diameter	Moderate	Large
Force of unit	Low	High
Metabolic profile	Oxidative	Glycolytic
Contraction velocity	Moderate	Fast
Fatigability	Low	High

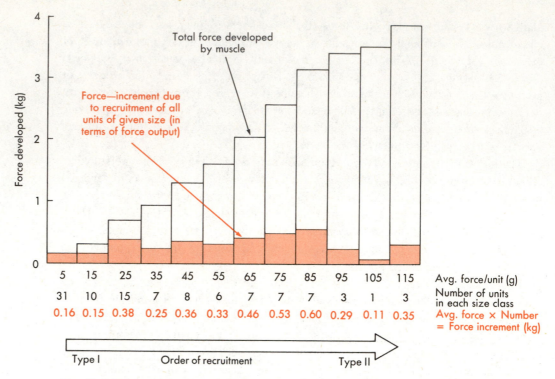

Fig. 23-5 ■ Relationships between the size of a motor unit, that is, those producing up to 10 g for an average of 5 g, etc., the number of motor units of different size classes in a muscle, the contributions of each class of motor units to force development, and the order of recruitment of motor units of different sizes. See text for further explanations. (Data from Henneman, E., and Olson, C.B.: J. Neurophysiol. **28:**581-598, 1965.)

the colored column. The total force generated by the muscle with the increased level of excitatory input is now the sum of 31 units averaging 5 g force plus 10 units averaging 15 g force or 0.31 kg. A smooth gradation in contractile force is still maintained because the percent increase in force produced by a larger motor unit remains small when added to the force already being generated. Fig. 23-5 shows that smaller numbers of larger motor units are successively recruited until all motor units are contracting. There is some evidence that the largest motor units are so inexcitable that most persons cannot recruit them voluntarily. They may account for the exceptional displays of strength exhibited by persons under stress when increased excitatory activity occurs in the central nervous system.

The pattern of increasing force illustrated in Fig. 23-5 is a consequence of recruitment by the *size principle*. Recruitment by size and the simultaneous increase in firing rates, which allows each unit to increase its force by tetanization, are responsible for gradations in contractile force. Because the larger motor units are also the faster units, whole muscle contraction velocities can be increased by recruiting more motor units. Recruiting more units also contributes to the increase in speed by reducing the effective load on each cell, allowing faster crossbridge cycling rates.

The size principle also explains the metabolic profile pattern of the motor units. Highly oxidative units are those which are used most. Maximal efforts, in which fast motor units are recruited, cannot be sustained because of the rapid depletion of glycogen. This intracellular source of energy is needed to supply the glycolytic pathway, which is the principal way in which the high rates of ATP consumption can be met.

Slow or tonic muscle fibers. In amphibians, reptiles, and birds the function of the slow, type I fiber is served by an entirely different type of skeletal muscle cell. This type is confusingly termed a *slow* or sometimes a *tonic* fiber. It differs from the slow, type I fibers described earlier in that it has many motor end plates that receive branches from one motor neuron. These slow muscle fibers generally do not generate action potentials, and they are depolarized only by the decremental spread of the closely spaced end-plate potentials. The slow or tonic fibers of lower vertebrates exhibit only a small response to a single neural stimulus, but the force output can be continuously graded by increasing the frequency of motor nerve firing. Tonic fibers are very rare in mammals, but they are found in the extraocular muscles that control eye movement.

Tone in skeletal muscle. The skeletal system supports the body mass efficiently when the posture is normal. The amount of energy expended for the muscular contraction that is required to maintain a standing posture is remarkably small. However, muscles normally exhibit some level of contractile activity. Isolated, unstimulated muscles are relaxed and quite flaccid. "Relaxed" muscles in the body are comparatively firm. This *tone* is apparently a result of low levels of contractile activity in some motor units driven by reflex arcs from receptors in the muscles because tone is abolished by dorsal root section. Tone in skeletal muscles should be distinguished from "tone" in vascular smooth muscles. The latter refers to the normal tonic or continuous contraction that maintains vascular resistance.

Trophic factors are responsible for the long-term development and maintenance of specific characteristics of a tissue. Skeletal muscle exhibits considerable plasticity (variability in the phenotype or properties of the cells) as shown by a variety of experimental studies. The trophic factors associated with the development and maintenance of fiber phenotypes are the topic of this section.

Growth and development. Skeletal muscle fibers differentiate before innervation (neuromuscular junctions may be formed well after birth). Before innervation the fibers physiologically resemble slow, type I cells. These uninnervated fibers have acetylcholine receptors distributed throughout the sarcolemma and are supersensitive to that neurotransmitter. An end plate is formed when the first growing nerve terminal establishes contact. The fiber forms no further association with nerves, and the receptors to acetylcholine become concentrated in the end-plate membranes. Fibers that are innervated by a small motor neuron form slow oxidative motor units. Fibers innervated by large motor nerves develop all the characteristics of fast, type II motor units. Thus innervation produces major changes in the cells, including the synthesis of the fast and slow myosin isoenzymes, which replace the embryonic variant.

An increase in muscle strength and size occurs during maturation. The cells must lengthen with skeletal growth. This is accomplished by formation of additional sarcomeres at the ends of the cells. This process is reversible. The cells shorten by destruction of terminal sarcomeres if a limb is immobilized such that the muscle is fixed in a shortened position or if an improperly set fracture leads to a shortened limb segment. The gradual increase in strength and diameter of a muscle during growth is achieved mainly by hypertrophy (Fig. 23-1). Skeletal muscles have a limited capacity to form new fibers (hyperplasia) by differentiation of satellite cells that are present in the tissues. Injured cell segments that contain nuclei can grow and fuse with other segments to regenerate a cell. However, major cellular destruction leads to replacement with scar tissue. Overall, skeletal muscle cells exhibit remarkable dynamic adjustments of their morphology to the demands of the organism.

Denervation, reinnervation, and cross-innervation. A variety of studies have shown the importance of innervation for the skeletal muscle phenotype (Fig. 23-6). If the motor nerve is cut, muscle *fasciculation* occurs. This term is used to describe small, irregular contractions caused by the release of acetylcholine from the terminals of the degenerating distal portion of the axon. Several days after denervation, muscle *fibrillation* begins.

■ *Trophic responses of skeletal muscle*

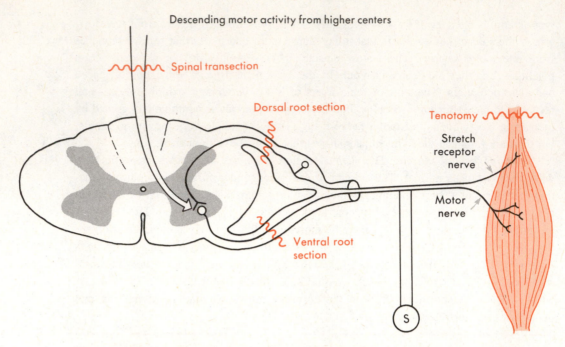

Fig. 23-6 ■ Experimental or pathological situations that modify the phenotype of skeletal muscles. Normal contractile activity in the muscle can be abolished by ventral root section that cuts the motor nerves (peripheral denervation). Contractile activity can be increased in a muscle with normal innervation by pacemakers *(S)*, which are implanted on the motor nerves. Activity can be decreased by blocking excitatory pathways from higher centers (spinal transection) or by interrupting the reflex arcs that modify muscle activity through dorsal root section. If the muscle tendon is severed (tenotomy), stretch receptors within the tendon and muscle become inactive. Tenotomy also lowers excitatory input to the motor nerves.

Fibrillation is characterized by spontaneous, repetitive contractions. These originate after the spread of cholinergic receptors occurs over the entire cell membrane (a return to the preinnervation embryonic characteristic) and are a result of supersensitivity to acetylcholine. Atrophy also occurs, with a decrease in the size of the muscle and its cells. Atrophy is progressive in humans, with degeneration of some cells after 3 or 4 months. Most of the muscle fibers are replaced by fat and connective tissue after 1 to 2 years. These changes are all reversed if reinnervation occurs within a few months. Reinnervation is normally achieved by growth of the peripheral stump of the motor nerve axons along the old nerve sheath.

Reinnervation of a formerly fast, type II fiber by a small motor axon causes that cell to redifferentiate into a slow, type I fiber and vice versa. Such observations suggest that there are qualitative differences in large and small motor nerves and that the nerves have specific "trophic" effects on the muscle fibers. Although it is increasingly recognized that nerve terminals release substances in addition to the primary neurotransmitter at synapses, an alternative explanation of the trophic effects of innervation explains most observations. Simply, it is the frequency of contraction that determines fiber development and phenotype. Electrical stimulation via electrodes implanted in the muscle or its motor nerves can ameliorate denervation atrophy. More strikingly, chronic low-frequency stimulation of fast motor units (by a pacemaker with electrodes placed on their motor nerves) causes fast motor units to be converted to slow units. Some shifts toward a typical fast-fiber phenotype can be observed when the frequency of contraction in slow units is greatly decreased by reducing the excitatory input in the ventral horn. This may be achieved through appropriate spinal or dorsal root section or by severing the tendon, which functionally inactivates peripheral mechanoreceptors (Fig. 23-6) (Chapter 14). In

TABLE 23-2

Effects of exercise

Type of training	Example	Major adaptive response
Learning/coordination	Typing	Increases the rate and accuracy of motor skills (central nervous system)
Endurance (submaximal, sustained efforts)	Marathon running	Increased oxidative capacity in all involved motor units with limited cellular hypertrophy
Strength (brief, maximal efforts)	Weight lifting	Hypertrophy and enhanced glycolytic capacity of motor units employed

brief, fibers that undergo frequent contractile activity form many mitochondria and synthesize the slow myosin isoenzyme. Fibers innervated by large, inexcitable axons contract infrequently. Such relatively inactive fibers form few mitochondria and have large concentrations of glycolytic enzymes. The gene for the fast myosin isoenzyme is expressed in such cells.

Response to exercise. Exercise physiologists identify three categories of training regimens and responses (Table 23-2). In practice, most athletic endeavors involve elements of all three. The learning aspect can involve motivational factors as well as neuromuscular coordination. This aspect of training does not involve adaptive changes in the muscle fibers per se. However, motor skills can persist for years without regular training, unlike the responses of muscle cells to exercise.

All healthy persons can maintain some level of continuous muscular activity that is supported by oxidative metabolism. This level can be greatly increased by a regular exercise regimen that is sufficient to induce adaptive responses. The adaptive response of skeletal muscle fibers to endurance exercise is mainly an increase in the metabolic capacity of the motor units involved. This demand places an increased load on the cardiovascular and respiratory systems, and this demand leads to increases in the oxidative capacity of the heart and respiratory muscles. The latter effects are responsible for the principal physiological benefits associated with endurance exercise.

Muscle strength can be increased by regular massive efforts that involve all motor units. Such efforts recruit fast glycolytic motor units and are brief. The blood supply may be interrupted as the tissue pressures rise above the intravascular pressures during maximal contractions, further limiting the duration of the contraction. Regular maximal strength exercise such as weight lifting induces synthesis of more myofibrils and hypertrophy of the active muscle cells. The increased stress also induces growth of tendons and bones.

The precise cellular adaptive responses to exercise are difficult to determine because experimental animals are unsuited for certain exercise regimens. However, most studies indicate that the effects of exercise on muscle are quantitative. Endurance exercise does not cause fast motor units to become slow with the synthesis of the slow myosin isoenzyme, nor do maximal muscular efforts produce a shift from slow to fast motor units. Experiments involving cross-innervation and other techniques show that such shifts are possible. It seems likely that any practical exercise regimen, when superimposed on normal daily activities, does not provide sufficient changes in the pattern of activation of motor units to induce shifts in the myosin isoenzyme distribution.

Two additional factors that affect the motor unit phenotype and athletic performance are hormones and the genetic endowment of the person. Androgens stimulate muscle hypertrophy and may account for the greater average muscle mass in males.

■ *Fatigue*

Remarkably little is known about the factors responsible for fatigue. Fatigue may potentially involve any of the steps involved in muscular contraction, from the brain to the muscle cells, as well as the systems involved in maintaining energy supplies, including cardiovascular and respiratory functions.

Cellular fatigue. Fatigue of motor units can be assessed experimentally by recording the maximum stress maintained during prolonged contraction or during a series of brief tetani elicited by direct stimulation of the motor nerve to the muscle. The latter regimen allows adequate perfusion when the circulation is intact. Tetanic stress decays rapidly to a level that can be maintained for long periods (Fig. 23-7). This decay represents the rapid and almost total failure of fast motor units. The decline in tetanic stress is paralleled by glycogen and creatine phosphate depletion and by lactic acid production. This implies that ATP depletion leads to the failure of contraction. However, the decline in stress occurs when the ATP pool is not greatly reduced, and the muscle fibers do not go into rigor. Slow motor units, which can meet the energy demands of the stimulus regimen, do not exhibit significant fatigue for many hours. Evidently some factor associated with energy metabolism can inhibit contraction, but this factor has not been clearly identified. There is some evidence that *neuromuscular fatigue* can occur in the largest fast motor units, in which the ability of the motor nerve to synthesize and release acetylcholine may be limiting.

General fatigue. Most persons tire and cease exercise long before there is any motor unit fatigue of the type illustrated in Fig. 23-7. *General physical fatigue* may be defined as a state of disturbed homeostasis produced by work. The basis for the perceived discomfort or even pain probably involves many factors. These factors may include a lowering of plasma glucose levels and accumulation of metabolites. Motor system function in the central nervous system is not impaired. Highly motivated and trained athletes are able to bear the discomfort and will exercise to the point where some motor unit fatigue occurs. Part of the enhanced performance observed after training involves motivational factors.

Recovery. Muscle blood flow and oxygen uptake remain elevated for some time after

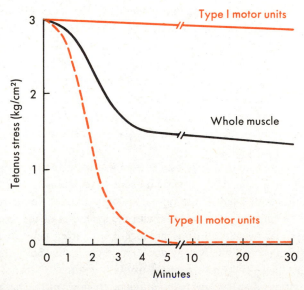

Fig. 23-7 ■ Fatigue of a skeletal muscle in which half of the cross-sectional area is composed of slow, oxidative type I motor units and half of fast, glycolytic type II motor units. The muscle was briefly tetanized once every second by stimulation of its motor nerve in situ. With this regimen, type II motor units exhibit rapid cellular fatigue and failure to contract, while type I motor units maintain almost normal contractile responses.

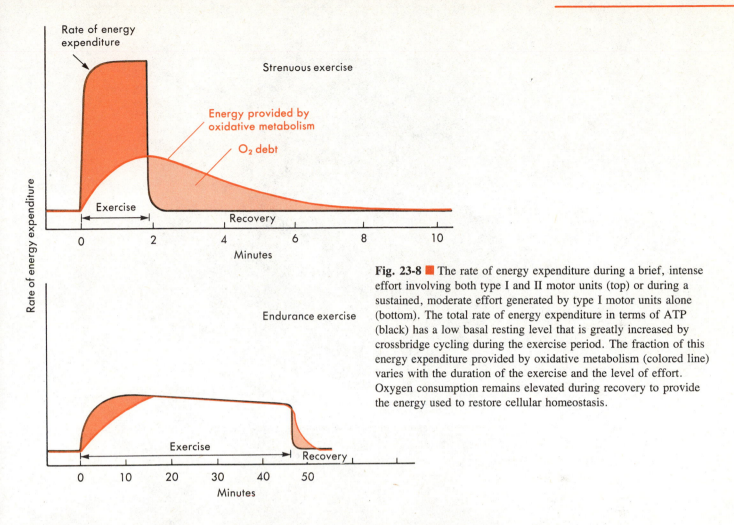

Rate of energy expenditure

Rate of energy
expenditure

Strenuous exercise

Energy provided by
oxidative metabolism

O₂ debt

Exercise

Recovery

0   2   4   6   8   10

Minutes

Endurance exercise

Exercise

Recovery

0   10   20   30   40   50

Minutes

Fig. 23-8 ■ The rate of energy expenditure during a brief, intense effort involving both type I and II motor units (top) or during a sustained, moderate effort generated by type I motor units alone (bottom). The total rate of energy expenditure in terms of ATP (black) has a low basal resting level that is greatly increased by crossbridge cycling during the exercise period. The fraction of this energy expenditure provided by oxidative metabolism (colored line) varies with the duration of the exercise and the level of effort. Oxygen consumption remains elevated during recovery to provide the energy used to restore cellular homeostasis.

exercise. *Oxygen debt* (Fig. 23-8) is the excess amount of oxygen consumed over that required for resting metabolism when energy use by the contractile system has ceased. Some oxygen debt occurs even at low levels of exercise because slow oxidative motor units consume considerable ATP (derived from creatine phosphate or glycolysis) before oxidative metabolism can increase ATP production to the steady state requirements. The oxygen debt is much greater in strenuous exercise during which fast glycolytic motor units are recruited. The oxygen debt is approximately equal to the energy consumed during exercise minus that supplied by oxidative metabolism (i.e., the dark and light colored areas in Fig. 23-8 are roughly equal). The additional oxygen used during recovery represents the energy requirements for restoring normal cellular metabolite levels.

■ *Involuntary
muscles: physiological
control and function*

Involuntary muscles include a highly diverse group of tissue ranging from striated muscle in the heart to many types of smooth muscles. Some general functional characteristics of muscles surrounding hollow organs are outlined in this section. Specific characteristics of various smooth muscle types are given in the chapters devoted to the different organ systems.

■ *Cell-to-cell contacts*

A variety of specialized contacts between involuntary muscle cells serves two functions: mechanical linkages and communication. Smooth and cardiac muscle cells are connected to each other rather than spanning the distance between two tendons. Thus cells anatomically arranged in series should not only be linked mechanically but should also be activated simultaneously and to the same degree. If this were not true, contraction in one region would simply stretch another region without a substantial decrease in

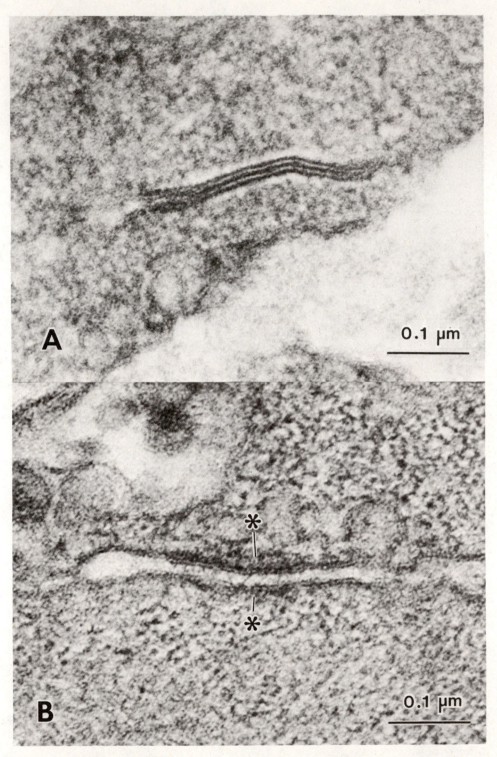

Fig. 23-9 ■ Junctions between smooth muscle cells. **A,** Gap junction (or nexus) joins two vascular smooth muscle cells in the rat aorta. **B,** Intermediate junction between smooth muscle cells of the squirrel monkey coronary artery. Note the accumulation of opaque material or "plaque" in the cytoplasm adjacent to the junction (*) and the presence of filamentous material connecting the opposed cell membranes. (**A** courtesy Susan Purdy-Ramos; **B** courtesy Dr. Michael S. Forbes.)

radius or increase in pressure. The mechanical connections are provided by sheaths of connective tissue and by specific junctions between muscle cells. One example is the intercalated disk in cardiac muscle (Chapter 29).

In smooth muscle there are two basic types of junctions. One is the *gap junction* (or nexus) in which adjacent plasma membranes are separated by only 2 to 3 nm. Gap junctions often exhibit a typical five-layered appearance in electron micrographs (Fig. 23-9, *A*). The other specialized contact is the *intermediate junction* (often termed *attachment plaque* or *desmosome-like attachment*), the membranes of which are separated by a space of 20 to 60 nm, often marked by a central dense line (Fig. 23-9, *B*). A variety of simple appositions between areas of cell membranes that are separated more widely are also common between smooth muscle cells. Some of these are quite elaborate with protrusions of one cell into another.

There is evidence that all these junctions may subserve the roles of both mechanical linkage and cell-to-cell communication. However, gap junctions are of particular importance in forming low-resistance pathways through which currents generated by action potentials in one cell can cause adjacent cells to fire. In certain tissues such as in the heart or the outer longitudinal layer of smooth muscle in the intestine there are large numbers of such junctions. A wave of depolarization is readily propagated from cell to cell through such tissues. Intermediate junctions are presumed to mainly subserve mechanical functions.

Some smooth muscles show little or no electrical coupling via junctions, and these cells may not normally fire action potentials. Synchrony of cellular activation occurs by diffusion of hormones or neurotransmitters through the tissue. Neural control of contraction in involuntary muscle is more complex than in skeletal muscle. Three factors must be considered: (1) the types of innervation and neurotransmitters, (2) the proximity of the nerves to the muscle cells, and (3) the type and distribution of the receptors for neurotransmitters in the muscle cell membranes (Fig. 23-10).

The types of innervation can be divided into three categories. *Extrinsic innervation* is derived from axons of the autonomic nervous system. In arteries this is usually limited to sympathetic nerves, but both sympathetic and parasympathetic innervation commonly are present in other tissues. *Intrinsic nerves* contained in plexuses may occur within the smooth muscle tissue, particularly in the gastrointestinal tract. Finally, *afferent sensory neurons* that mediate various reflexes may be found in the plexuses. A few smooth muscle tissues have no innervation.

Neuromuscular junctions in involuntary muscle are functionally comparable to those in skeletal muscle: presynaptic transmitter release, diffusion across the "junction," and combination with a postsynaptic receptor. However, elaborate neuromuscular contacts at axon terminals are not found. Autonomic nerves have a series of swollen areas or varicosities that are spaced at intervals along the axon. These varicosities contain the vesicles in which the neurotransmitters are found (Fig. 23-10). Each varicosity functions as a neuromuscular junction, although the adjacent muscle membranes exhibit little specialization. The varicosities are closely apposed to the muscle cell membranes with a gap of 6 to 20 nm in tissues with a large degree of neural regulation. The average gap may be 80 to 120 nm and, occasionally, considerably more in tissues in which neural control is less extensive. In arteries the nerves tend to be concentrated in the outer layer of smooth muscle cells just under the adventitial sheath of connective tissue. In such cases neurally mediated activation will be greatest at the periphery, and it may be nonexistent for smooth muscle cells in the luminal layers.

The neurotransmitters that are released and the presynaptic and postsynaptic effects of the neurohormones are highly variable. Marked individuality of the responses of different tissues that contain involuntary muscle is achieved by differences in their in-

■ *Neuromuscular relationships*

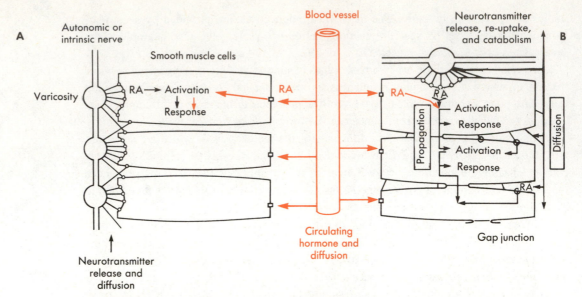

Fig. 23-10 ■ Control of smooth muscle. Contraction (or inhibition of contraction) of smooth muscles can be initiated by (1) intrinsic activity of pacemaker cells, (2) neurally released transmitters, or (3) circulating hormones. The combination of a neurotransmitter, hormone, or drug with specific receptors activates contraction *(RA)* by increasing cell Ca^{++}. The response of the cells depends on the concentration of the transmitters or hormones at the cell membrane and the nature of the receptors present. Hormone concentrations depend on diffusion distance, release, re-uptake, and catabolism. Consequently, cells lacking close neuromuscular contacts will have a limited response to neural activity unless they are electrically coupled so that depolarization is transmitted from cell-to-cell. **A,** Multiunit smooth muscles resemble striated muscles in that there is no electrical coupling, and neural regulation is important. **B,** Single-unit smooth muscles are like cardiac muscle, and electrical activity is propagated throughout the tissue. Most smooth muscles probably lie between the two ends of the single-unit–multiunit spectrum. (Redrawn from Ljung, B.: Physiological patterns of neuroeffector control mechanisms. In Bevan, J.A., et al., editors: Vascular neuroeffector mechanisms, Basel, 1976, S. Karger.)

nervation, the types of transmitter released, and the nature of the receptors for each transmitter.

Relationships with other cell types. Contractile function in smooth muscle can be modulated by cells other than nerves. For example, specialized junctional regions can be observed between the membranes of the smooth muscle cells and the endothelial cells that line blood vessels. The actions of some circulating hormones or drugs can be mediated by the endothelial cells. For example, acetylcholine, acting on endothelial cell receptors, causes release of an unidentified hormone that subsequently produces arterial smooth muscle relaxation. Acetylcholine causes contraction when it combines with cholinergic receptors in the arterial smooth muscle cell membrane.

■ *Patterns of function*

In a classic grouping of involuntary muscles, a distinction is made between *multiunit* and *single-unit* tissues (Fig. 23-10). Multiunit tissues are those in which each cell does not communicate with other muscle cells through junctions. Contraction of multiunit smooth muscle is controlled by extrinsic innervation or hormonal diffusion. Skeletal muscles are an example of this pattern, which is also typical of a few smooth muscles (i.e., those in the vas deferens and iris). However, all smooth muscle cells that are anatomically arranged in series must be part of the same functional motor unit, although many nerves may be involved. Single-unit tissues (Fig. 23-10, *B*) are exemplified by the heart and by a number of smooth muscles that undergo rhythmical contractile activity. All the cells in such tissues are electrically coupled through cell-to-cell junctions.

Single-unit tissues can maintain fairly normal contractile activity without extrinsic innervation. The activity pattern in denervated single-unit muscles is caused by pacemaker cells and intrinsic reflex pathways.

The distinction between single-unit and multiunit tissues is overly simplified. Most smooth muscles are controlled and coordinated by a combination of neural elements and some degree of coupling. One generalization is that tonic muscles such as arteries and sphincters, which maintain more or less continuous levels of tone, approach the multiunit end of the spectrum. Characteristically, such tissues do not exhibit action potentials when stimulated. Tissues that undergo phasic (rhythmical) activity, such as peristalsis, usually generate action potentials that are propagated from cell to cell. Such tissues more fully meet the criteria characterizing single-unit muscles.

■ *Biophysics of hollow organs*

The mechanical output of a skeletal muscle is readily calculated from the stress-length and velocity-stress relationships of the fibers (Figs. 22-7 and 22-8) and the angle formed between the fibers and the axis of the tendon. However, additional factors have considerable significance in the mechanics of hollow organs.

Structural factors. Many arteries have a comparatively simple geometry; the smooth muscle cells run circumferentially around the vessel. However, the heart and organs that contain smooth muscle have multiple layers of cells with different orientations. Such structural arrangements can have a profound effect on the pressure-volume relationships of the tissue. Even with a comparatively simple geometry a moderate shortening of smooth muscle can produce a disproportionately large volume change. This is illustrated for an arteriole with a lumen diameter of 18 μm and a wall composed of a 1 μm thick endothelium surrounded by a 2 μm thick smooth muscle cell (Fig. 23-11). A 50% shortening will halve the mean radius of the smooth muscle cylinder. However, the cells are not compressible so wall thickening reduces the inner radius of the smooth muscle layer to 7 μm. Contraction of the smooth muscle cells produces bulging of the

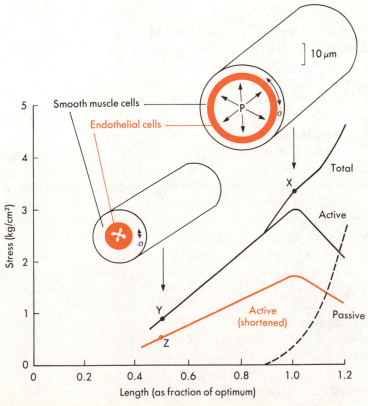

Fig. 23-11 ■ The mechanical properties of smooth muscle in hollow organs. In this simple example the smooth muscle of a small arteriole shortens by 50% from its optimum length for force generation. The stress-length relationships for the arteriolar wall before constriction are illustrated in black. The actual wall stress (σ) in the arteriole is governed by the Laplace relationship (text). If the pressure remained constant, σ would fall from a value proportional to *X* to a value proportional to *Z* during the constriction. Thus the smooth muscle needs to develop considerably less force to maintain the constriction. This geometrical factor more than compensates for the fall in active stress (*Y*) due to the length-stress relationship of the smooth muscle cells.

endothelial cell layer. The overall volume reduction is 36-fold for an arteriolar segment that undergoes a cell shortening of 50%. This reduction is sufficient to stop red cell flow.

Physical factors. The stress borne by the organ wall resulting from transmural pressure (e.g., the internal pressure minus the pressure outside a hollow organ) can be calculated from a derivation of the *law of Laplace* (considered further in Chapter 31). It states that the stress (σ) in the wall of the vessel (double-headed arrows, Fig. 23-11) produced by the transmural pressure (P) is equal to the product of the pressure and the mean radius (r) of the wall divided by the wall thickness (w), or $\sigma = P (r/w)$ with units of force per unit area. Obviously a given pressure in the lumen produces a smaller stress on shortened smooth muscle cells because of the reduced vessel radius and the increased wall thickness. At larger radii the stress on the arterial wall produced by a given transmural pressure is withstood in part by the passive elasticity of the connective tissue and in part by active contraction of the smooth muscle. This is illustrated by point *X* in Fig. 23-11, where the muscle cells are at their optimum length for force development. After the muscle shortens by 50%, all the load must be borne by the smooth muscle cells. Their ability to withstand the distending pressure is reduced (point *Y*) because of their force-length properties. However, the actual wall stress on the smooth muscle cells (neglecting endothelial cell compression) is now significantly lower because of the Laplace relationship (point *Z*). The smooth muscle cells would have to partially "relax" to maintain the reduced diameter if the compressive forces were insignificant and would be operating on a lower stress-length curve (colored curve, Fig. 23-11).

The relatively simple example illustrated in Fig. 23-11 shows the general relationship between muscle function and hollow organ mechanics. Pressure-volume relationships rather than stress-length relationships are normally used to describe the mechanical properties of hollow organs. The pressure-volume curve of an organ depends on the stress-length properties of the smooth muscle. However, structural factors and the Laplace relationship also influence pressure-volume behavior.

■ Functional adaptations in involuntary muscle

Neurotrophic relationships. Unlike skeletal muscles, involuntary muscles show little dependence on their extrinsic innervation for maintaining a normal phenotype. This may be related to the fact that the heart and smooth muscles can maintain contractile activity by other mechanisms when denervated. However, supersensitivity to neurotransmitters occurs in the heart and smooth muscle after denervation.

Development and hypertrophy. During development and growth there is an appropriate increase in the number of smooth and cardiac muscle cells (hyperplasia). In adults an increase in tissue mass occurs if an organ is subjected to a sustained increase in mechanical work. This is termed *compensatory hypertrophy.* A striking example of this occurs in the ventricles and arterial media in hypertension. The increased mechanical load on the muscle cells appears to be the common factor inducing hypertrophy in involuntary muscles. Recent studies of myocardial and arterial smooth muscle cells indicate that tissue hypertrophy, at least in part, reflects cellular hypertrophy. Chromosomal replication (which may or may not be followed by nuclear replication) results in significant numbers of polyploid muscle cells. The polyploid cells, with multiples of the normal sets of chromosomes, synthesize more contractile proteins and increase in size with each chromosomal replication. It is not known whether polyploidy is associated with exercise-induced cellular hypertrophy in striated muscles.

The capacity of skeletal muscles to synthesize different myosin isoenzymes is demonstrated by cross-innervation or reinnervation experiments. However, such changes have only been "physiologically" demonstrated in cardiac muscle after hypertrophy resulting from disease or from experimental procedures to induce cardiac overload. The myosin isoenzyme synthesized in the hypertrophied cells has a lower specific ATPase activity, which results in a reduced contraction velocity of the cells and changes in the myocardial energetics. The overloaded heart is subjected to a continuous stress, unlike the activity pattern in skeletal muscle associated with an exercise regimen. This may be responsible for the appearance of new myosin isoenzymes in cardiac muscle.

The myometrium, which is the smooth muscle component of the uterus, exhibits another striking example of hypertrophy as parturition approaches. Hormones play an important role in this response. During pregnancy (when the hormone progesterone predominates) the smooth muscle is quiescent, and there are few gap junctions electrically coupling the smooth muscle cells. At term, under the dominant influence of estrogen, there is a marked hypertrophy of the myometrium. A great increase in the number of gap junctions occurs just before birth, converting the myometrium into a single-unit tissue to coordinate contraction during parturition.

Growth and development of tissues containing smooth muscles are associated with increases in the connective tissue matrix. Smooth muscle cells can synthesize and secrete the materials that make up this matrix. These constituents include collagen, elastin, and proteoglycans. The synthetic and secretory capacities are evident when smooth muscle cells with extensive contractile filament arrays are isolated and placed in tissue culture. The cells rapidly "dedifferentiate" and lose thick myosin filaments and much of the thin filament lattice. Their places are filled by a greatly expanded rough endoplasmic reticulum and Golgi apparatus (cellular structures that are associated with protein synthesis and secretion). The dedifferentiated cells multiply and lay down connective tissues in the culture plate. This process is reversible, and some degree of redifferentiation with formation of thick filaments is possible after cessation of cell replication. Such studies suggest that the synthetic, secretory, and proliferative smooth muscle cell is phenotypically and functionally distinguishable from the differentiated contractile cell. However, these functions may all be present in varying degrees. The determinants of the smooth muscle cell phenotype are largely unknown, but hormones and growth factors in the blood, as well as the mechanical loads on the cells, are implicated in the control of differentiation and dedifferentiation.

■ *Synthetic and secretory functions*

■ *Bibliography*

Journal articles

Chapman, C.B., and Mitchell, J.H.: The physiology of exercise, Sci. Am. **212:**88, 1965.

Evarts, E.V.: Brain mechanisms in movement, Sci. Am. **229:**96, 1973.

Merton, P.A.: How we control the contraction of our muscles, Sci. Am. **227:**30, 1972.

Books and monographs

Alexander, R.M.: Animal mechanics, Seattle, 1968, University of Washington Press.

Åstrand, P.O., and Rodahl, K.: Textbook of work physiology, ed. 2, New York, 1977, McGraw-Hill Book Co.

Bohr, D.F., Somlyo, A.P., and Sparks, H.V., Jr., editors: Handbook of physiology, section 2, The cardiovascular system, vol. II, Vascular smooth muscle, Bethesda, Md., 1980, American Physiological Society.

Brooks, V.R., editor: Handbook of physiology, The nervous system, vol. II, Motor control, Bethesda, Md., 1981, American Physiological Society.

Bülbring, E., et al., editors: Smooth muscle: an assessment of current knowledge, Austin, 1981, University of Texas Press.

Code, C.F., editor: Handbook of physiology, Alimentary canal, vol. IV, Motility, Bethesda, Md., 1968, American Physiological Society.

Daniel, E.E., and Paton, D.M.: Methods in pharmacology, vol. 3, Smooth muscle, New York, 1975, Plenum Press.

Pette, D., editor: Plasticity of muscle, Berlin, 1980, Walter de Gruyter & Co.

BLOOD

Oscar D. Ratnoff

chapter 24

Blood components

Blood is the vehicle of transportation that makes possible the specialization of structure and function that is characteristic of all but the lowest organisms. It is a suspension of various types of cells in a complex aqueous medium, the plasma. The elements of blood serve multiple functions essential for metabolism and the defense of the body against injury, of which only a portion will be described.

In the normal human adult, plasma makes up approximately 55% to 60% of blood. Almost innumerable substances are dissolved in plasma including, among others, oxygen, carbon dioxide, nitrogen, electrolytes, proteins, lipids, carbohydrates (particularly glucose), amino acids, vitamins, hormones, and nitrogenous breakdown products of metabolism such as urea and uric acid. The concentrations of these constituents may vary as the result of such diverse influences as diet, metabolic demand, and the levels of hormones and vitamins. Normally the composition of the blood is maintained at biologically safe and useful levels by a variety of homeostatic mechanisms.

Atmospheric oxygen diffuses from the pulmonary alveoli into the circulating plasma in the pulmonary capillaries and thence into red blood cells to combine with hemoglobin, the major carrier of oxygen in blood. Similarly, carbon dioxide, elaborated by the tissues through the oxidation of carbon-containing compounds, diffuses into peripheral capillaries. It is then carried by the blood to the lung, where it is excreted. Carbon dioxide is transported in several ways including the formation of bicarbonate ions in plasma and of carbaminohemoglobin in red blood cells.

The ionic constituents of plasma maintain the osmolarity and pH of blood within physiological limits. The chief inorganic cation of plasma is sodium, present normally at an average concentration of 142 mEq/L. Plasma also contains small amounts of ionic potassium, calcium, and magnesium. The principal anion of plasma is chloride; its average concentration is 104 mEq/L. Ionic equilibrium is maintained by the presence of other anions including bicarbonate, plasma protein, and, to a lesser degree, phosphate, sulfate, and organic acids. Elaborate devices preserve the normal ionic composition of plasma. For example, the level of sodium is carefully regulated by the kidney, the function of which is in turn modulated by circulating hormones such as adrenocorticosteroids and vasopressin. Vasopressin is a polypeptide hormone that is secreted by the posterior pituitary gland, and it regulates the reabsorption of water by the kidney tubules (Chapters 48 and 52).

Literally hundreds of different proteins are dissolved in plasma. The bulk of protein is of two types: albumin and the various immunoglobulins. Albumin, which is synthesized in the liver, is present at a concentration of about 4 g/dl. Because albumin does not diffuse freely through intact vascular endothelium, it provides the critical *colloid*

■ *Plasma*

osmotic or *oncotic pressure* that regulates passage of water and diffusible solutes through the capillaries. Water and its solutes diffuse out of the capillaries in their proximal portions, where intravascular pressure exceeds oncotic pressure, and into the capillaries in their distal portions, where the situation is reversed. Albumin also serves as a carrier for substances that are adsorbed to it, both normal components of blood or exogenous agents such as drugs, and it furnishes some of the anions needed to balance the cations of plasma. Second to albumin in concentration are antibodies (immunoglobulins) that are synthesized by plasma cells in the lymphoid organs and are critical in defense against infection. Five classes of immunoglobulins are recognized: IgG, IgM, IgA, IgD, and IgE; they differ in amino acid composition and structure and serve different functions. The specificity of antibodies is determined by differences in the amino acid composition of the amino-terminal "variable" region of the immunoglobulin molecules.

Other plasma proteins include the "clotting factors" needed for the coagulation of blood: fibrinogen, the most plentiful and the precursor of the fibrin clot (Chapter 25); complement, a group of proteins involved in immune responses; many enzymes or their precursors; enzyme inhibitors; specific carriers of such constituents as iron and copper; and scavengers of agents inadvertently released into plasma, for example, haptoglobin, which binds free hemoglobin. Plasma lipids, the chief of which are triglycerides, cholesterol, phospholipids, and fatty acids, are transported as complexes with plasma proteins.

■ *Blood cells*

The cellular constituents of blood include red blood cells (erythrocytes), a variety of white blood cells (leukocytes), and platelets. In normal adults the red cells occupy, on the average, about 48% of the volume of blood in men and about 42% in women; the percent of erythrocytes in blood is defined as the *hematocrit*. The red cells are biconcave disks, each with a diameter of about 8 μm, a thickness of 2 μm, and a volume of about 87 μm^3. The "red blood cell count" normally averages 5.2 million/μl in men and 4.8 million/μl in women. Through a succession of divisions, red cells mature in the bone marrow from primitive hematopoietic "stem" cells, normally losing their nuclei before they enter the bloodstream, so that peripheral blood erythrocytes are anuclear. Maturation of the red cell is dependent on the presence of adequate supplies of vitamin B$_{12}$, folic acid, and iron; the last is needed for the synthesis of hemoglobin, the principal constituent of the cytoplasm of the erythrocyte. Hemoglobin is a protein synthesized by the nucleated precursors of the red cells in the marrow. It is a complex molecule composed of four polypeptide "globin" subunits. Each is attached to a prosthetic group consisting of a tetrapyrrole *heme*. Each heme unit, in turn, surrounds an atom of iron. Atmospheric oxygen, diffusing into the red cell in the lungs, attaches to this iron atom. The oxygen is then released to the tissues in the peripheral capillaries. Some carbon dioxide is also transported from the tissues to the lungs by hemoglobin, attached to its globin portion.

Anemia, a decrease in the circulating mass of erythrocytes, may result from decreased generation of these cells or their premature destruction or loss through hemorrhage. Failure of synthesis can arise in many ways: as the result of hypocellularity of the marrow, from deficiencies of agents essential for the formation of hemoglobin such as vitamin B$_{12}$ or iron, from replacement of the marrow by tumor cells, or by suppression of hematopoiesis as in renal failure. Sickle cell anemia is particularly instructive because it is the result of the substitution of valine for the glutamic acid residue that is the sixth amino acid from the amino-terminal end of the hemoglobin β-chain. This single amino acid substitution, the first of many molecular hemoglobin defects discovered, brings about the aggregation of hemoglobin molecules when the concentration of oxygen in the red cell is low.

The life span of erythrocytes, which is normally about 120 days, may be shortened by damage that is brought about by antibodies directed against erythrocytes or by other endogenous or exogenous agents that may rupture the red cell membrane, a process described as *hemolysis*. Alternatively, excessively rapid destruction of erythrocytes may reflect an inherent defect in their structure, as in hereditary disorders of hemoglobin synthesis such as sickle cell anemia.

Therapy for anemia varies widely, depending on its pathogenesis. In some cases replacement of red cells by transfusion may be needed. Blood transfusion has been made feasible by the recognition that there are hereditary differences in the chemical structure of the red cell membranes. For example, persons can be separated into those with blood groups O, A, B, or AB, depending on saccharide groups of the membrane glycoprotein. Furthermore, those with blood group O have antibodies in their plasma directed against group A and B cells; those of group A have antibodies directed against B cells and vice versa, while those of group AB do not have antibodies directed at any of the blood groups. For transfusion to be successful, blood must be transfused into a recipient whose plasma does not contain antibodies directed against the infused cells. Thus persons of blood group O are universal donors, and those of blood group AB are universal recipients. In common practice, however, blood of the donor is carefully matched to that of the recipient. The A-B-O system is only one of many inherited blood group systems. Of special note are the complex Rh blood groups. Ordinarily a person who lacks the Rh factor does not have antibodies against this substance in his plasma. A woman who is Rh negative, however, may develop such antibodies if she carries a fetus who has inherited the Rh factor from its father. The maternal antibodies can cross the placenta to the fetus and result in a devastating destruction of its erythrocytes, which is described as *hemolytic disease of the newborn*.

Normal peripheral blood contains between 4000 and 10,000 leukocytes/μl. Five classes of leukocytes are recognized: neutrophils, eosinophils, basophils, monocytes, and lymphocytes. The cells are differentiated in blood smears by their morphological and tinctorial characteristics upon staining with a mixture of dyes (Wright-Romanovsky stain).

Neutrophils, eosinophils, and basophils, described collectively as *granulocytes,* are distinguished by the nature of the granules in the cytoplasm. These cells, averaging about 12 to 15 μm in diameter, have small multilobed nuclei in their mature forms and abundant cytoplasm. The neutrophils, which account for about 40% to 75% of blood leukocytes, have nuclei with two to five lobes and in stained blood smears exhibit fine purple granules in a pink cytoplasm. They provide a major defense against infection by bacteria, which they can ingest and destroy. They also accumulate at sites of noninfectious injury, where their enzymes may participate in inflammatory reactions. Eosinophils, 1% to 6% of peripheral blood leukocytes, most often have bilobed nuclei; in stained preparations the cytoplasm contains large red granules. The eosinophils appear to protect against parasitic infestation. They are increased in number in many situations including ''allergic'' states, such as asthma, in which the host responds to the presence of abnormal exogenous or endogenous agents with an immunological reaction. Neutrophils and eosinophils are formed in the marrow through maturation of cells derived from the primitive hematopoietic stem cell.

Basophils, normally less than 1% of the white blood cell count, have multilobulated nuclei and in stained blood smears show large, deep blue cytoplasmic granules. The origin of peripheral blood basophils is uncertain. One view holds that they are formed in the marrow from the primitive stem cells. Alternatively, they are thought to be derived from morphologically similar mast cells found scattered throughout extravascular tissues. The mast cells may be responsible for some of the phenomena associated with localized immunological reactions. On appropriate stimulation, mast cells release histamine from their granules. One result is a local increase in vascular permeability that brings about the formation of wheals.

A fourth class of leukocytes, the monocytes, makes up 2% to 10% of peripheral white blood cells. Larger than other leukocytes, they have an average diameter of about 15 to 20 μm. They have indented, often kidney-shaped nuclei and fine pink cytoplasmic granules in stained blood smears that are readily differentiated from those of the granulocytes. The monocytes appear to be derived from the primitive marrow stem cells. They are closely related to similar monocytic macrophages that are scattered throughout the tissues and comprise the reticuloendothelial system. One view is that the reticuloendothelial cells are peripheral blood monocytes that have migrated into the tissues. The monocytes and macrophages are actively phagocytic and ingest particulate matter including microorganisms and injured or dead cells. Monocytes also participate in the formation of specific antibodies by lymphocytes. They appear to do this by processing antigens, that is, agents that bring about the synthesis of specific antibodies, so that they can be used by the lymphocytes.

The fifth class of leukocytes, the lymphocytes, comprises 20% to 45% of white blood cells. Lymphocytes are a heterogenous group of cells with large nuclei and more or less cytoplasm, depending on their size, which varies from 6 to 20 μm in diameter. The cytoplasm is devoid of granules in stained preparations. Three groups of lymphocytes can be differentiated on functional grounds although they cannot be distinguished in stained preparations. Some lymphocytes, described as B-cells, are transformed on stimulation with antigen into plasma cells that synthesize the specific immunoglobulin antibodies. Others, called T-cells, participate in cell-mediated delayed hypersensitivity reactions, such as sensitivity to tuberculin, that are not dependent on the presence of circulating plasma antibodies. Certain T-cells also either abet or inhibit the transformation of lymphocytes to antibody-producing cells. Other lymphocytes have characteristics of neither T- nor B-lymphocytes and have been described as null cells. Some null cells seem capable of destroying tissue cells that have been coated with antibody and have therefore been described as "killer cells."

In adults lymphocytes are derived from the primitive hematopoietic stem cells in marrow. Most lymphocytes are situated not in peripheral blood but in the lymphoid organs, principally the thymus, spleen, lymph nodes, and the submucosal Peyer's patches of the intestines. On the basis of animal experiments, T-cells are thought to be derived from lymphocytes that have migrated from the marrow to the thymus gland. The origin of human B-cells is less certain; their designation derives from the observation that in birds antibodies are synthesized by lymphocytes derived from the bursa of Fabricius. In mammals B-cells probably mature in bone marrow. Synthesis of immunoglobins by lymphocytes takes place in those cells which have migrated from peripheral blood to the lymphoid organs, where they are transformed on antigenic stimulation to plasma cells.

Platelets are anuclear cytoplasmic fragments of megakaryocytes, large polyploid cells found in the adult in bone marrow. They have an important role in the control of bleeding and the genesis of thrombosis, that is, the formation of clots within blood vessels. The platelets are discussed in detail in Chapter 25.

Hemostasis and blood coagulation

The life of every organism depends on the preservation of its internal environment. Even one-cell creatures can seal disruptions of their external membranes. Invertebrates protect themselves from loss of vital hemolymph by processes that are analogous to vertebrate clotting and that use circulating blood cells, hemolymph, or both.

Vertebrates have evolved complex mechanisms to stem hemorrhage after injury. Failure of the hemostatic mechanisms may lead to fatal exsanguination. In mammals an early event after trauma is transitory contraction of damaged blood vessels, but this furnishes little toward *hemostasis,* that is, the arrest of bleeding. Within seconds after vascular injury, however, *platelets,* small anuclear circulating blood cells, adhere to the site of damage and pile up, one on another, to provide a mechanical plug that effectively stops bleeding from minor injuries. Hemorrhage from more formidable wounds is staunched by the coagulation of blood. When the skin is unbroken, bleeding may be checked by external compression of the injured vessels by the extravasated blood. This back pressure is highly effective where the skin is tightly bound to the underlying tissues, as in the finger tip, but it is essentially useless where the skin is distensible. For example, hemorrhage around the eye is unchecked by the counterpressure of extravasated blood, the pathogenesis of a "black eye." After childbirth hemostasis is aided by contraction of the uterine musculature, which compresses the blood vessels that feed the site from which the placenta was wrenched.

Incredibly, that clots may provide a mechanical barrier to loss of blood after vascular injury was appreciated only at the beginning of the eighteenth century. A clot is a network of protein fibers, the meshes of which trap blood cells and serum, that is, plasma that has undergone coagulation. Speculation about the source of the fibers of the blood (now called *fibrin*) goes back at least as far as Plato and his pupil Aristotle in the fourth century BC. About 200 years ago William Hewson, an English physician, proposed that fibrin was derived during clotting from plasma, the liquid portion of blood. Over the succeeding years it became clear that this came about through the action of a proteolytic enzyme *thrombin* that catalyzed the transformation of the plasma protein *fibrinogen* (factor I) to fibrin.

Thrombin is not normally present in circulating blood, but when blood is shed, it is cleaved from its precursor in plasma, *prothrombin* (factor II). The release of thrombin is the final step in one or both of two convergent and intertwined chains of chemical reactions, the *extrinsic* and *intrinsic pathways*. When trauma disrupts the vascular endothelial lining, blood comes into contact with subendothelial structures and other exposed injured tissues. This sets in motion a succession of catalytic events through either or both pathways. At each step a proenzyme *clotting factor* is transformed to its enzymatic form in which it can activate the next proenzyme in the chain. The series of enzymatic steps magnifies the original disturbance of the blood until, in the end, thrombin is released explosively.

The enzymes that are activated via the extrinsic or intrinsic pathway are endopeptidases; that is, they cleave specific peptide bonds that are not located at the extreme ends of the substrate molecules. Cleavage exposes the site in the substrate clotting factor that is responsible for its biological function. In most cases the bond that is broken is located within a folded portion of the polypeptide chain that is held together by internal disulfide cross-links. Thus with disruption of the peptide bond, the polypeptide chain is split into two parts that are held together by a disulfide bridge. The clotting enzymes of the intrinsic and extrinsic pathways are serine proteases; that is, the catalytic site contains a serine residue. Three clotting factors, *high molecular weight (HMW) kininogen, antihemophilic factor* (AHF, factor VIII), and *proaccelerin* (factor V), have not yet been shown to have catalytic activity. These clotting factors probably act as nonenzymatic cofactors.

Reactions of the *extrinsic pathway of thrombin formation* are initiated by contact of blood with injured tissues. The damaged cells furnish a clot-promoting agent, *tissue thromboplastin* (tissue factor, factor III), that was first studied extensively by a German-Estonian physiologist Alexander Schmidt. Schmidt, whose studies of coagulation dominated the last century, thought that clotting took place in two steps. First, tissue thromboplastin transformed prothrombin to thrombin; then this enzyme converted fibrinogen into a fibrin clot. We now know that generation of thrombin is much more complex.

Studies of the *intrinsic pathway of thromboplastin,* whereby blood clots in the absence of tissue thromboplastin, were pioneered by John Lister, the great British surgeon and a contemporary of Schmidt. Lister observed that blood clotted rapidly when it was drawn into a cup but much more slowly when it was placed in an india-rubber tube. He concluded that clotting came about by contact of blood with a "foreign surface" that differed from the normal vascular endothelial lining. The cup, in this regard, was more foreign than the india-rubber tube.

The successive steps of the intrinsic pathway are now known to begin with activation of a plasma proenzyme, Hageman factor (factor XII), that is brought about by contact with a suitable foreign surface. Activated Hageman factor (factor XIIa) then initiates a succession of enzymatic events that involve at least eight other plasma proteins and that lead to the elaboration of thrombin. The final steps of the intrinsic and extrinsic pathways are identical.

Schmidt appreciated that clotting might be modulated by inhibitory agents within the plasma. The first such substance to be delineated was an inhibitor of thrombin now called *antithrombin III,* the action of which is greatly potentiated by *heparin,* a mucopolysaccharide that is now widely used therapeutically (p. 428).

Blood also has the capacity to redissolve clots that might inadvertently form within blood vessels. This property is attributed for the most part to the elaboration of the plasma proteolytic enzyme *plasmin.* Plasmin can be generated from its precursor, *plasminogen,* in many ways; one of the most interesting is the process of coagulation itself.

Mention has already been made of the important role of *platelets* in hemostasis. The integrity of the *blood vessel walls* is also critical. When the supporting structures surrounding the blood vessels are defective as may occur, for example, in scurvy (vitamin C deficiency), a hemorrhagic tendency may ensue. Our understanding of the role of vessels in the control of bleeding is still primitive. Superimposed on the contributions to hemostasis by the coagulation system, platelets, and blood vessel walls are the influences of blood flow; the formation of an intravascular clot is fostered by stasis of flow and impeded by rapid blood flow.

As each new clotting factor was discovered, it acquired a trivial name that reflected its supposed function or was derived from the name of a patient whose plasma appeared to be deficient in that particular agent. Often several names were used to designate the same substance. To bring order out of this chaos, the International Committee for the Nomenclature of Clotting Factors has assigned a Roman numeral to most of the clotting factors and has recommended that each author refer to this number along with the trivial name he preferred. Currently, most of the scientific literature uses only the numerical nomenclature. Inevitably this has led to numerous errors, which are of little consequence in a published article but are of great moment if the patient or physician is confused. Table 25-1 provides a key to commonly used synonyms. This chapter will avoid the pitfalls of pied type by retaining the admittedly archaic trivial nomenclature, followed where clarity is needed, by the appropriate Roman numeral.

■ *Blood coagulation*
■ *Nomenclature*

TABLE 25-1

Blood clotting factors

Roman numeral	Trivial names	Activated or altered state
Factor I	Fibrinogen	Fibrin
Factor II	Prothrombin	Thrombin
Factor III	Tissue thromboplastin Tissue factor	—
Factor IV	Calcium ions	—
Factor V	Proaccelerin Ac globulin	Altered proaccelerin (factor Va)
Factor VII	—	Factor α-VIIa
Factor VIII	Antihemophilic factor (AHF)	Altered antihemophilic factor
Factor IX	Christmas factor	Factor IXa
Factor X	Stuart factor	Factor Xa
Factor XI	Plasma thromboplastin antecedent (PTA)	Factor XIa
Factor XII	Hageman factor (HF)	Factor XIIa
Factor XIII	Fibrin-stabilizing factor (FSF)	Factor XIIIa (fibrinoligase)
—	Plasma prekallikrein (Fletcher factor)	Plasma kallikrein
—	High–molecular weight (HMW) kininogen (Fitzgerald, Williams, or Flaujeac factor)	—

With the exception of antihemophilic factor (factor VIII), the clotting factors in plasma are synthesized, primarily but not exclusively, in the liver. *Antihemophilic factor* is a complex molecule that can be dissociated into two subcomponents of unequal size (p. 422). The larger subcomponent is most probably synthesized in vascular endothelial cells and megakaryocytes, whereas the site of synthesis of the smaller subcomponent is unknown. Megakaryocytes also synthesize the a- (or α) chains of *fibrin-stabilizing factor* (factor XIII) and a form of fibrinogen whose identity with plasma fibrinogen is disputed.

■ *The synthesis of clotting factors*

Hepatocytes synthesize four clotting factors, *Christmas factor* (factor IX), *factor VII, Stuart factor* (factor X), and *prothrombin* (factor II), only if vitamin K is present. This vitamin was discovered by Henrik Dam of Copenhagen during studies of cholesterol metabolism in chicks. Vitamin K is the generic name for a group of fat-soluble quinone derivatives that are plentiful in leafy vegetables. In the newborn infant vitamin K is provided by milk; cow's milk is a much richer source than human milk. Within a few days after birth, bacteria that have begun to grow within the lumen of the gut become an important source of the vitamin. Because it is fat soluble, absorption of vitamin K from the gut depends on the presence of bile salts that are excreted by the liver into the duodenum and on normal fat digestive and absorptive mechanisms.

These physiological considerations help to explain the commonplace occurrence of combined deficiencies of the vitamin K–dependent clotting factors. In *hemorrhagic disease of the newborn,* for example, the infant ingests inadequate amounts of vitamin K before bacterial flora are established. Vitamin K deficiency also results from *suppression of intestinal flora* by orally administered antibiotics, particularly if little or none of the vitamin is ingested. Deficiency of the four clotting factors may result from its malabsorption, as may occur if bile salts are excluded from the gut by *obstruction of the common bile duct* or if *pancreatic* or *bowel disease* impairs absorption of lipids. The hemorrhagic symptoms and laboratory abnormalities in all these states can be corrected by parenteral administration of vitamin K. Deficiencies of the vitamin K–dependent factors may also complicate *hepatic disease,* but in this situation the administration of vitamin K is usually without benefit.

Recent studies have elucidated the long puzzling function of vitamin K. Synthesis of the vitamin K–dependent clotting factors proceeds in two stages. First, the hepatocytes synthesize polypeptide progenitors of each factor, a process that does not require the vitamin. In the second step, vitamin K, reduced to its hydroquinone form in the liver, acts as a cofactor for a specific microsomal carboxylase (Fig. 25-1). This carboxylase inserts a second carboxyl group into the γ-carbon of certain glutamic acid residues in the polypeptide chains. These unique tricarboxylic glutamic acid residues serve as points of attachment for the calcium ions that are needed during transformation of the vitamin K–dependent factors to their enzymatically active states. During carboxylation

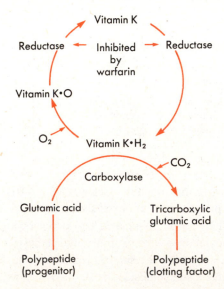

Fig. 25-1 ■ The role of vitamin K is the insertion of a second carboxyl group into the γ-carbon of glutamic acid residues of the vitamin K–dependent clotting factors. *Vitamin K·O,* vitamin K epoxide.

the reduced vitamin K is oxidized to an epoxide form, from which vitamin K is then regenerated by enzymatic reduction.

Synthesis of the vitamin K–dependent factors is inhibited by dicumarol and its congeners, notably warfarin. These agents have had long usage as anticoagulants in the prevention and treatment of thrombosis. Dicumarol was first extracted by Karl Paul Link almost 50 years ago from spoiled sweet clover that induced a mysterious hemorrhagic disorder in cattle. Dicumarol and similar agents act by competitive inhibition of the enzymes that reduce vitamin K and its epoxide (Fig. 25-1). In this way they suppress the carboxylation of the progenitors of the vitamin K–dependent factors so that the synthesis of these factors is not completed.

That four distinct clotting factors require vitamin K to complete their synthesis suggests that they have a common evolutionary origin. In agreement with this, extensive similarities in amino acid sequence have been found. A reasonable explanation is that the different factors arose as the result of the duplication and subsequent mutation of genes responsible for the synthesis of an ancestral protein.

For many years the common wisdom held that vitamin K was needed only for the synthesis of clotting factors. Now we have learned that it is also required for synthesis of other proteins found, for example, in plasma, bone, kidney, lung, spleen, and placenta. Presumably in each case the functioning of these proteins requires the attachment of calcium ions to the tricarboxylic glutamic acid residues formed through the action of vitamin K. Of pertinent interest is the vitamin K–dependent plasma agent *protein C* (autoprothrombin II-A) that, when activated by thrombin, inhibits the coagulant properties of antihemophilic factor (factor VIII:C) and proaccelerin (factor V) (p. 423).

■ Plasma fibrinogen and the formation of fibrin

Blood clotting is the visible result of the conversion of the soluble plasma protein *fibrinogen* (factor I) into an insoluble meshwork of *fibrin*. Fibrinogen is a dimeric glycoprotein with a molecular weight of 340,000. Each half of the dimer is composed of three polypeptide chains, designed Aα, Bβ, and γ, respectively. The six chains are held together by disulfide bonds.

The conversion of fibrinogen to fibrin takes place in three stages (Fig. 25-2). First, fibrinogen undergoes limited proteolysis by thrombin that has evolved during the earlier steps of the coagulation process. The partially digested fibrinogen molecules, now called *fibrin monomers*, polymerize into insoluble strands of fibrin. Finally, the constituent fibrin monomers of fibrin are cross-linked covalently by a plasma enzyme, *activated fibrin-stabilizing factor* (fibrinoligase, factor XIIIa).

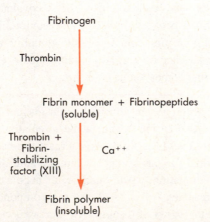

Fig. 25-2 ■ The formation of fibrin. (Redrawn from Ratnoff, O.D.: Hemorrhagic disorders: coagulation defects. In Beeson, P.B., McDermott, W., and Wyngaarden, J.B., editors: Cecil textbook of medicine, ed. 15, Philadelphia, 1979, W.B. Saunders Co.)

In the first step of fibrin formation, *thrombin* cleaves four small polypeptide fragments, each with a molecular weight of about 1500, from each fibrinogen molecule, reducing its weight by about 2%. One fragment, *fibrinopeptide A*, is released from the amino-terminal end of each Aα-chain, and another, *fibrinopeptide B*, from the amino-terminal end of each Bβ-chain. The residue (fibrin monomer) retains its dimeric structure, each half of which has three chains, designated α, β, and γ. Fibrinopeptide A is released by thrombin more rapidly than fibrinopeptide B, and with its separation polymerization begins. Polymerization depends primarily on the separation of fibrinopeptide A.

As the first fibrin monomers are generated, they are surrounded by unaltered fibrinogen molecules with which they form loose complexes. As more fibrin monomers accumulate, the equilibrium shifts so that fibrin monomers now polymerize with each other. At the same time, the complexes of fibrinogen and fibrin monomer dissociate, making more fibrin monomers available for polymerization. The monomers aggregate both side-to-side and end-to-end and gradually build insoluble polymers that thicken and lengthen to form visible fibrous strands. Some of the polymerization appears to be branched so that the fibrin strands build into a complicated meshwork that traps blood cells and serum. Additionally, there is evidence that platelets bind to the polymerizing strands of fibrin. Calcium ions are not required for the transformation of fibrinogen to fibrin monomer, but at the concentration present in plasma these ions greatly accelerate the polymerization process.

Clots formed in purified mixtures of thrombin and fibrinogen are held together by noncovalent forces and have low tensile strength. Such clots dissolve readily in solutions of urea or monochloroacetic acid, which dissociate noncovalent bonds. In contrast, clots formed in normal human plasma are insoluble in these dispersing agents and have high tensile strength. Plasma contains a proenzyme, *fibrin-stabilizing factor* (factor XIII), which, when activated, acts as a transamidase to forge covalent links between the γ-carboxyl groups of glutamic acid residues of one fibrin monomer and the ε-amino groups of lysyl residues of another. Fibrin-stabilizing factor is a tetramer composed of two a- (or α) chains and two b- (or β) chains; half of the fibrin-stabilizing factor in blood is found in platelets, where it is composed only of α-chains.

Activation of fibrin-stabilizing factor is brought about by thrombin, which cleaves a small polypeptide from each a-chain. In the presence of calcium ions, fibrin-stabilizing factor then induces covalent links among fibrin monomers, dimers between the γ-chains of two adjacent monomers, and at a slower rate, polymers among several α-chains. Resistance to dissolution by urea or monochloroacetic acid occurs only when these α-chain fibrin polymers are formed.

Fibrin-stabilizing factor has other actions besides covalent cross-linking of fibrin. For example, it forges links between molecules of contractile proteins of muscle or platelets, and it bonds fibronectin (cold insoluble globulin) to itself and to fibrin or collagen (p. 433). In the absence of fibrin-stabilizing factor, experimental wound healing is retarded, but how this comes about is uncertain.

■ Disorders of fibrin formation

The concentration of fibrinogen in normal human plasma averages 270 to 300 mg/dl. Thus there are about 10 g of fibrinogen in the circulating blood; perhaps an additional 5 g is present in extravascular fluid. An increased concentration of fibrinogen is commonplace during normal pregnancy and in innumerable disease states, particularly those associated with inflammation, tissue damage, or neoplasm, in which it is thought to be an "acute phase reactant," that is, a plasma component, the concentration of which is increased under the stress of disease. An increased concentration of fibrinogen accelerates the settling of red blood cells when blood is allowed to stand. This increased *erythrocyte sedimentation rate* was described by Hippocrates, who interpreted the rapid settling of red cells as a basic cause of disease.

TABLE 25-2

Molecular weight, concentration, and biological half-life of plasma hemostatic factors

Factor	Molecular weight (× 1000)	Approximate plasma concentration (mg/dl)	Biological half-life*
Fibrinogen (I)	340	200-400	3-4 days
Prothrombin (II)	72	10	3 days
Proaccelerin (V)	330	0.5-0.7	15-30 hours
Factor VII	50	0.05-0.06	4-5 hours
Antihemophilic factor (VIII)	1,000-12,000	0.5-1	10-12 hours
Christmas factor (IX)	57	0.5-0.6	18-24 hours
Stuart factor (X)	59	1	24-36 hours
Plasma thromboplastin antecedent (XI)	120-200	0.4-0.6	2-3 days
Hageman factor (XII)	80	1.5-4.5	2 days
Plasma prekallikrein	80-100	3.5-4.5	35 hours
HMW kininogen	110-225	8-9	6 days
Fibrin-stabilizing factor (XIII)	300-350	1-2	11-12 days
Plasminogen	87	20	2-2½ days
Antithrombin III	65	12-15	45-60 hours
α_1-Antitrypsin	55	300	5-6 days
α_2-Macroglobulin	820	285-350	5 days
α_2-Plasmin inhibitor	65-70	5-7	2½ days
C1-esterase inhibitor (CĪ-INH)	78-105	18-25	3-4 days
Protein C	62	0.4	—

*The biological half-life of a clotting factor is measured by infusing it into a patient with a congenital deficiency of the factor or by infusing radio-labeled factor into a normal individual. Ordinarily the curve describing the rate of disappearance of the factor from plasma has two components. First, the concentration of the infused substance decreases relatively rapidly as it diffuses into extravascular spaces. Thereafter, the concentration decreases more slowly, reflecting the catabolism of the factor. The biological half-life, that is, the length of time until the titer of the infused factor decreases by 50%, is calculated from the second component of the disappearance curve.

The biological half-life of fibrinogen under normal conditions is about 3 to 4 days; that is, half of the fibrinogen present in the plasma at any given time will have disappeared in this interval (Table 25-2). Thus about 15% of plasma fibrinogen must be replaced daily by its continual synthesis, which takes place in the parenchymal cells of the liver. Among the agents proposed as stimuli to synthesis are growth hormone, thyroxin, corticotropin, hormones released from inflammatory lesions, and the degradation products released by the digestion of fibrinogen and fibrin by *plasmin,* a plasma proteolytic enzyme (p. 425).

The normal site and mechanism of the catabolism of fibrinogen are unknown. Fibrinogen is used through its conversion to fibrin both in hemostasis and in the formation of inflammatory lesions. Perhaps some fibrinogen is digested by plasmin, particularly when this enzyme has been activated by severe stress. But these mechanisms participate only marginally in the normal catabolism of fibrinogen, and the hypothesis that this process is the result of continual slow intravascular clotting has little to support it.

Hypofibrinogenemia, an abnormally low concentration of fibrinogen in plasma, or afibrinogenemia, the absence of detectable fibrinogen, may be hereditary or acquired. *Congenital afibrinogenemia* is a rare autosomal recessive disorder in which affected persons do not synthesize fibrinogen; the blood does not clot even when large amounts of thrombin are added. Although patients with this disorder bleed uncontrollably from sites of injury, they have surprisingly little difficulty if unchallenged. The lesson learned from these patients is that hemostasis from minor wounds is apparently provided by platelets when the generation of thrombin is normal. Impaired synthesis of fibrinogen, resulting in hypofibrinogenemia, is said to occur in catastrophic liver disease, but this is an uncommon phenomenon.

Hypofibrinogenemia and afibrinogenemia must be distinguished from *dysfibrinogenemia,* a group of rare autosomal dominant disorders in which plasma contains abnormal variants of fibrinogen that clot unusually slowly on addition of thrombin.

The most frequent disorder of fibrinogen is *disseminated intravascular coagulation,* a serious and sometimes fatal complication of many disease processes. Widespread thrombosis, that is, intravascular clotting, within small blood vessels occurs because agents that can bring about the elaboration of thrombin have gained access to the bloodstream. Much or all of the plasma fibrinogen may be consumed in the formation of thrombi, resulting in hypofibrinogenemia or afibrinogenemia. Other clotting factors may be depleted and the platelet count reduced, while the proteolytic enzyme plasmin, (p. 425) may become active and may dissolve the intravascular clots. Among the incitants of disseminated intravascular clotting are such complications of pregnancy and childbirth as *retention of a dead fetus* within the uterus for several weeks, *premature separation of the placenta,* and *amniotic fluid embolism,* in which amniotic fluid and its contents enter the maternal bloodstream during parturition. Other causes of disseminated intravascular coagulation include such diverse events as *envenoming* by the bite of snakes with clot-promoting venoms, *sepsis* with a variety of organisms (an all too frequent cause of death when abortion was illegal), *transfusion of incompatible blood* in which damaged red blood cells appear to foster the evolution of thrombin, and the presence of certain *tumors,* in which clot-promoting products of the neoplasm gain entrance into the bloodstream.

Functional *deficiency of fibrin-stabilizing factor* (factor XIII) is a rare hereditary hemorrhagic disorder accompanied by severe bleeding. Presumably hemorrhage occurs because covalent cross-links between fibrin monomers cannot be established. Usually the disorder is caused by synthesis of a functionally incompetent variant of fibrin-stabilizing factor rather than by deficient synthesis of this protein. The diagnosis is suspected when clots of the patient's plasma dissolve in 6M urea or in 1% monochloroacetic acid.

■ The formation of thrombin

Thrombin, the enzyme ultimately responsible for the formation of fibrin monomers, is generated by the catalytic scission of prothrombin by *activated Stuart factor* (factor Xa). Under physiological conditions, activation of Stuart factor (factor X) takes place through two series of enzymatic steps: the intrinsic and extrinsic pathways (Fig. 25-3). These reactions are so intertwined that the reader may find it helpful to read the next few sections twice to appreciate the nature of the feedback mechanisms that are involved.

■ Surface-mediated reactions initiating the intrinsic pathway

Activation of Stuart factor (factor X) by the intrinsic pathway is the end result of a series of reactions that begin when blood comes into contact with negatively charged surfaces. When venous blood is drawn into a polystyrene or silicone-coated tube and centrifuged to sediment essentially all its cells, the plasma thus separated clots readily in glass tubes but much more slowly in polystyrene or silicone-coated tubes. Such experiments suggest that glass induces a change in plasma that leads to the elaboration of thrombin. The clot-promoting properties of glass have been related to its negative surface charge, and many similarly charged insoluble substances such as kaolin (clay), diatomaceous earth (Celite), and talc have a similar effect. These substances are foreign

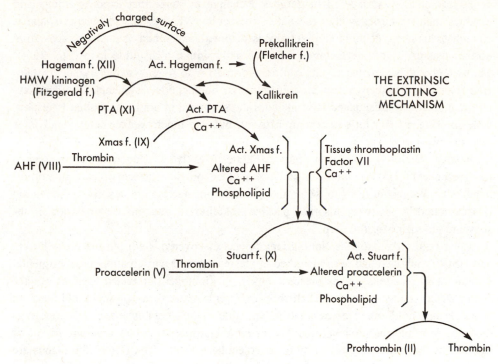

THE INTRINSIC
CLOTTING MECHANISM

Fig. 25-3 ■ The intrinsic and extrinsic pathways for the formation of thrombin. Omitted from the diagram are inhibitors of the various steps. The phospholipid portion of tissue thromboplastin may function in the activation and action of Stuart factor (factor X). The phospholipid for the intrinsic pathway is furnished by platelets and by the plasma itself. Augmentation of the action of factor VII by thrombin and by the activated forms of Hageman factor (factor XIIa), Christmas factor (factor IXa), and Stuart factor (factor Xa) and the activation of Christmas factor by the factor VII–tissue thromboplastin complex are not depicted. *Act,* activated; *HMW,* high molecular weight; *PTA,* plasma thromboplastin antecedent; *Xmas,* Christmas; *AHF,* antihemophilic factor. (Redrawn from Ratnoff, O.D.: Hemorrhagic disorders: coagulation defects. In Beeson, P.B., McDermott, W., and Wyndgaarden, J.B., editors: Cecil textbook of medicine, ed. 15, Philadelphia, 1979, W.B. Saunders Co.)

to the body, but drug addicts may inject themselves with drugs adulterated with kaolin or talc and in this way may self-induce thrombosis.

Uncertainty exists about the natural analogues of the clot-promoting solids. Sebum, the oily secretion of skin, some forms of collagen or basement membrane proteins, disrupted vascular endothelial cells, and endotoxins derived from gram-negative bacilli have all been implicated.

The events triggered by negatively charged surfaces include not only the generation of thrombin but also the mediation of other defense mechanisms of the body against injury including inflammation, the immune response, and fibrinolysis, that is, the dissolution of clots. Four plasma proteins are involved in the initiation of one or another of these surface-mediated defense reactions: *Hageman factor* (factor XII), *plasma thromboplastin antecedent* (PTA, factor XI), *plasma prekallikrein* (Fletcher factor), and *HMW kininogen* (Fitzgerald, Williams, or Flaujeac factor). All four factors are readily adsorbed from normal plasma to negatively charged surfaces such as glass, and it is apparently in this adsorbed state that they exert their clot-promoting functions.

The first step in the various surface-mediated defense reactions, including the initiation of events of the intrinsic pathway, is conversion of Hageman factor to an enzymatic form that can activate PTA. When Hageman factor is adsorbed to clot-promoting, negatively charged surfaces, it undergoes a change in shape that exposes amino acid residues that are buried within the native molecule. Whether this change in shape is enough to induce clot-promoting activity is disputed. Adsorbed to such surfaces from normal plasma, Hageman factor is split internally within a disulfide loop (Fig. 25-4). The two-chain species that results can transform PTA to its enzymatically active state. Hageman factor is then further cleaved into two polypeptide fragments. The carboxy-terminal fragment (designated HF$_f$) includes the sequence of amino acids that is responsible for Hageman factor's enzymatic activity, but is a much weaker activator of PTA than the two-chain species.

The conversion of PTA to its activated state by activated Hageman factor requires the presence of HMW kininogen and takes place on the surface of clot-promoting, negatively charged surfaces to which the three clotting factors are adsorbed. Activation is accelerated by similarly adsorbed plasma prekallikrein, but this factor is not an absolute requirement for this step.

Long before their role as clotting factors was discovered, both plasma prekallikrein and HMW kininogen had been identified in plasma by their participation in experimental inflammatory reactions. In the presence of HMW kininogen, activated Hageman factor (factor XIIa) changes plasma prekallikrein to its enzymatic form (plasma kallikrein). In turn, plasma kallikrein releases small polypeptide *kinins,* notably the nonapeptide bradykinin, from plasma proteins described as *kininogens.* Plasma contains two groups of kininogens that are distinguished by their molecular weights. The HMW kininogens are much more avid substrates for plasma kallikrein than those of lower molecular weight; only HMW kininogens have clot-promoting properties. Bradykinin, released through the action of plasma kallikrein, can dilate small blood vessels, increase their permeability, and induce pain. In this way, bradykinin can bring about the warmth, redness, swelling, and pain of inflammatory lesions. Bradykinin also contracts certain smooth muscles, a property useful in its biological assay. Human (but not bovine) HMW kininogen still retains its clot-promoting properties after the bradykinin sequence has been removed.

Activated Hageman factor participates in at least three other surface-mediated defense reactions. Directly or indirectly, it augments the clot-promoting properties of factor VII (p. 423); it converts plasminogen to plasmin (p. 425), and it transforms Cl, the first component of complement, a group of proteins participating in immune reactions, to its enzymatically active state (Cl̄).

Persons with hereditary deficiency of one of the four plasma proteins participating in surface-mediated defense reactions have been described. In each case the affected persons have impaired blood coagulation in vitro. Paradoxically, those lacking Hageman factor (said to have Hageman trait), plasma prekallikrein (Fletcher trait), or HMW kininogen (Fitzgerald, Williams, or Flaujeac trait) are curiously free of symptoms of a bleeding disorder, and those with PTA deficiency have only a mild bleeding tendency. Each of these disorders is ordinarily inherited as an autosomal recessive trait.

Activation of Christmas factor (factor IX) via the intrinsic pathway. The function of activated PTA (factor XIa) in the intrinsic pathway is the activation of Christmas factor, the first step that requires calcium ions. Christmas factor is a vitamin K–dependent protein that is synthesized by the liver under the direction of a gene on the X chromosome (p. 414). Activated PTA cleaves Christmas factor at two points (Fig. 25-5). First, the single chain of Christmas factor is split within an internal disulfide loop. The two-chain molecule that results acquires enzymatic properties only after activated PTA then severs a small "activation" polypeptide from the longer of the two chains.

In the test tube, activation of Christmas factor can also be brought about by plasma kallikrein, activated Stuart factor (factor Xa), and factor VII. Two of these agents,

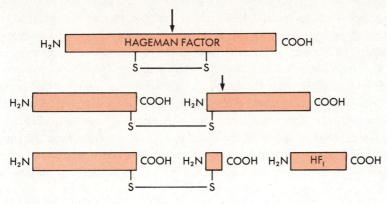

Fig. 25-4 ■ The activation of Hageman factor. What brings about scission *(arrows)* is uncertain. HF_f, carboxy-terminal fragment.

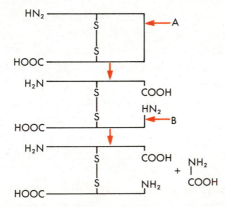

Fig. 25-5 ■ The activation of Christmas factor (factor IX). Activated plasma thromboplastin antecedent (factor XIa) splits Christmas factor successively at points *A* and *B*. Christmas factor acquires enzymatic properties only with the release of the "activation polypeptide."

activated Stuart factor and factor VII, participate in the extrinsic pathway of thrombin formation (p. 423). Perhaps this explains the benign nature of deficiencies of Hageman factor, PTA, plasma prekallikrein, or HMW kininogen. Activation of the extrinsic pathway may circumvent the steps of the intrinsic pathway before the participation of Christmas factor.

Christmas factor is functionally deficient in the plasma of patients with Christmas disease (hemophilia B), a bleeding syndrome that mimics classic hemophilia both in its mode of inheritance and its symptomatology and is distinguished from this disorder only by specific assays. The affected males in some families synthesize reduced amounts of Christmas factor; other males synthesize functionally incompetent variants of this protein.

Activation of Stuart factor (factor X) via the intrinsic pathway. The role of activated Christmas factor (factor IXa) is the conversion of Stuart factor to its activated state. Under physiological conditions, the activation of Stuart factor depends on the interaction of activated Christmas factor with *antihemophilic factor* (AHF, factor VIII), *phospholipids* (derived principally from platelets but also from plasma), and *calcium ions*. Antihemophilic factor is a complex molecule with a variety of defined properties. It is functionally deficient in several hereditary bleeding disorders, the most common of which are *classic hemophilia* and *von Willebrand's disease*.

Antihemophilic factor exists in plasma in multiple forms that range in molecular weight from about 1 to 12 million. It is readily dissociated into two subcomponents of unequal size by disrupting loose, noncovalent bonds. The larger subcomponent, most probably synthesized in vascular endothelial cells and megakaryocytes, contains the bulk of the protein of antihemophilic factor and varies in molecular weight from about 860,000 to 12 million. This subcomponent forms precipitates when it is mixed with antiserum raised in goats or rabbits that have been immunized with antihemophilic factor. For this reason, it has been described as factor VIII:Ag; Ag is the abbreviation for antigen, that is, an agent that induces the formation of antibodies. The larger subcomponent enhances the adhesion of platelets to subendothelial structures and in this way may be important for hemostasis when these structures are exposed by vascular injury. In vitro the higher molecular weight species of the larger subcomponent clump (agglutinate) platelets in the presence of the antibiotic ristocetin; because this property is deficient in von Willebrand's disease but not in classic hemophilia, it has been designated as factor VIII:VWF. Additionally, when normal blood is filtered through a column of glass beads, the majority of its platelets are retained within the column, a phenomenon that is dependent on the presence of the larger subcomponent of antihemophilic factor.

The second, smaller subcomponent of antihemophilic factor contains the amino acid sequences needed for blood coagulation and has therefore been designated factor VIII:C. The subcomponent has a molecular weight estimated to be 200,000 to 260,000 and is synthesized at some unknown site.

Classic hemophilia (hemophilia A) is the most common hereditary disorder of the intrinsic pathway. In this disease the coagulant titer of antihemophilic factor (factor VIII:C) is reduced, but the concentration of antihemophilic factor–like antigens (factor VIII:Ag) and the capacity to agglutinate platelets in the presence of ristocetin (factor VIII:VWF or factor VIIIR:RC) are normal.

Classic hemophilia varies in severity in parallel to the degree of the defect measured in clotting tests; within a family affected members are closely similar. In severe hemophilia, in which the titer of coagulant antihemophilic factor is less than 1% of that of the normal person, the patient may have repeated bleeding into joints (hemarthrosis) that ultimately leads to crippling. Apparently spontaneous bleeding into the skin or soft tissues is frequent, and hematuria (blood in the urine) is an important clinical feature. Injuries and surgical procedures, including dental extraction, may result in devastating hemorrhage. Death from exsanguination is unusual and is more likely to come from bleeding into a vital area or infection at the site of bleeding. In milder cases significant bleeding may occur only after injury or surgical procedures. To control bleeding, the patient is transfused with fractions of normal plasma rich in antihemophilic factor.

Classic hemophilia is the prototype of X chromosome–linked disorders, which affect only males, who pass the abnormal gene to their daughters, all of whom are carriers. In turn, half of a carrier's sons have hemophilia, and half of her daughters are carriers. Women who carry the abnormal gene are usually asymptomatic but can be recognized in about 90% of cases because their plasma contains relatively less coagulant antihemophilic factor than antihemophilia-like antigen.

Von Willebrand's disease, a hereditary disorder of both sexes, is inherited as an autosomal dominant trait. Affected persons have symptoms suggestive of mild classic hemophilia; in women menorrhagia and bleeding after childbirth may be troublesome. The plasma is deficient in antihemophilic factor, as measured in specific clotting assays (factor VIII:C), in immunological tests for antihemophilic factor–like antigen (factor VIII:Ag), and in tests for ristocetin-induced platelet agglutination (factor VIII:VWF). Several variants of von Willebrand's disease have been described. Episodes of bleeding may be treated by infusion of fractions of plasma rich in antihemophilic factor.

Stuart factor (factor X) is a vitamin K–dependent plasma proenzyme (p. 414). Its activation is brought about by proteolytic separation of a polypeptide fragment from one of its two chains. In the intrinsic pathway this cleavage is brought about by a complex of activated Christmas factor (factor IXa), antihemophilic factor (factor VIII), and calcium ions, all adsorbed to phospholipid micelles. The protease responsible for cleavage of Stuart factor is activated Christmas factor, whereas antihemophilic factor appears to act as a nonenzymatic cofactor. The effectiveness of antihemophilic factor is greatly enhanced by partial digestion of its low molecular weight, clot-promoting subcomponent (factor VIII:C) by thrombin or activated Stuart factor. Further digestion by thrombin destroys its coagulating activity.

Activated Stuart factor is the enzyme immediately responsible for the release of thrombin from prothrombin.

Activation of Stuart factor (factor X) via the extrinsic pathway. The early events of the intrinsic pathway are short-circuited when plasma is exposed to injured tissues. Such tissues contain one or more powerful agents known generically as *tissue thromboplastin* (tissue factor). Tissue thromboplastin is found principally in cell membranes of almost every cell including peripheral blood monocytes; platelets are an important exception. Tissue thromboplastin is a lipoprotein complex composed of heat-stable phospholipids and a heat-labile glycoprotein with a molecular weight that varies from 50,000 to 330,000, depending on its source. Removal of the phospholipid inactivates the clot-promoting properties of tissue thromboplastin, which can be restored by re-addition of the phospholipid portion.

The clot-promoting properties of tissue thromboplastin are mediated through its action on a trace plasma protein *(factor VII),* a single-chain protein that is synthesized when vitamin K is available (p. 414). By itself factor VII has no activity. Complexed stoichiometrically with tissue thromboplastin, it cleaves Stuart factor enzymatically, converting it to its activated state (factor Xa). The subsequent steps of the extrinsic and intrinsic pathways are identical.

The enzymatic properties of factor VII are greatly enhanced if it is first altered by one or another of many plasma proteases including thrombin, plasmin, and the activated forms of Hageman factor (factor XIIa), Christmas factor (factor IXa), and Stuart factor (factor Xa). Thus altered, however, factor VII will not activate Stuart factor unless tissue thromboplastin is also present.

As was noted earlier, the factor VII–tissue thromboplastin complex also activates Christmas factor. *Thus reactions of the extrinsic pathway affect the intrinsic pathway and vice versa.*

The generation of thrombin. Thrombin, the protease that is ultimately responsible for the formation of a fibrin clot, is separated from its parent molecule *prothrombin* by activated Stuart factor (factor Xa) through reactions that are augmented by *proaccelerin* (factor V), calcium ions, and phospholipid.

Prothrombin (factor II) is a vitamin K–dependent glycoprotein (p. 414). The molecule has been dissected enzymatically into several distinct parts (Fig. 25-6). Prothrombin is split by thrombin into an amino-terminal segment (fragment-1) and a carboxy-terminal portion (prethrombin-1); fragment-1 contains the tricarboxylic glutamic acid residues that serve as points of attachment for calcium ions. The prethrombin-1 segment can be further cleaved by activated Stuart factor into an amino-terminal portion (fragment-2) and a carboxy-terminal fragment (prethrombin-2) that contains an internal disulfide loop.

Thrombin formation proceeds through several stages. The first activated Stuart factor that forms via reaction of the intrinsic or extrinsic pathways binds via calcium links to tricarboxylic glutamic acid residues in fragment-1 of prothrombin (Fig. 25-6). There it slowly cleaves prothrombin in two places: between fragment-2 and prethrombin-2 and within the internal disulfide loop in prethrombin-2. This second split converts prethrombin-2 to thrombin.

As thrombin is liberated, it alters *proaccelerin* (factor V), a glycoprotein that in its

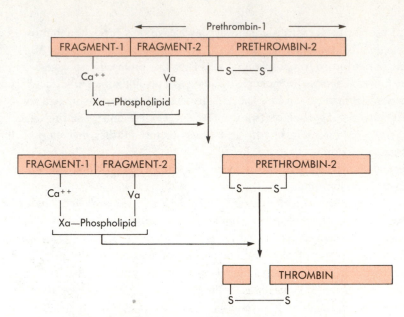

Fig. 25-6 ■ The release of thrombin from prothrombin.

native state has no clot-promoting properties, by splitting one of its polypeptide chains. In this altered form proaccelerin attaches to the fragment-2 portion of as yet unaltered prothrombin molecules. There it is a nonenzymatic accelerator of the action of activated Stuart factor, bringing about a rapid release of thrombin. Both activated Stuart factor and proaccelerin are adsorbed to phospholipid, which appears to augment their activity. Phospholipid for the intrinsic pathway is furnished by platelets and, to a lesser extent, by plasma and for the extrinsic pathway by tissue thromboplastin. The parallel between the activation of Stuart factor and the formation of thrombin is evident.

Thrombin converts fibrinogen to fibrin monomer, activates fibrin-stabilizing factor (factor XIII) and factor VII, augments the action of antihemophilic factor (factor VIII) and proaccelerin (factor V), and aggregates platelets.

■ *Laboratory studies of the clotting mechanism*

Delineation of the nature of defects of the clotting mechanism is largely a laboratory exercise. The tests described in everyday use in clinical laboratories depend on the physiological principles that have been reviewed.

The whole blood clotting time measures the period elapsing until blood withdrawn from a vein clots in glass test tubes under standardized conditions. The clotting time reflects the integrity of the intrinsic pathway of thrombin formation and is abnormally long with deficiencies or qualitative defects of any clotting factor except factor VII and fibrin-stabilizing factor (factor XIII). It is also long if abnormal inhibitors of clotting are present. The clotting time is insensitive to partial deficiencies of clotting factors and is now used almost exclusively to measure the anticoagulant effect of therapeutically administered heparin (p. 428). Sometimes the test is performed in polystyrene tubes, which are deficient in negative surface charges; under these conditions the clotting time may be long with more minor abnormalities of the intrinsic pathway and in severe thrombocytopenia (p. 434).

The *partial thromboplastin time* (PTT), a test devised by Langdell, Wagner, and Brinkhous, has largely supplanted the whole blood clotting time as a measure of the intrinsic pathway. Blood is drawn into citrate solution to reduce the concentration of the calcium ions needed for coagulation. Plasma separated from citrated blood is mixed in glass tubes with phospholipid. The mixture is then recalcified, and the clotting time is measured. The PTT is prolonged under the same conditions as the whole blood clotting time but is much more sensitive to minor abnormalities. Usually kaolin, diatomaceous earth (Celite), or solutions of ellagic acid are added to the mixture to ensure brisk activation of Hageman factor (factor XII). Like the whole blood clotting time, the PTT is normal in deficiencies of factor VII and fibrin-stabilizing factor.

The *prothrombin time,* introduced by Armand J. Quick, assesses the integrity of the extrinsic pathway. Tissue thromboplastin, usually in the form of rabbit brain tissue, is added to citrated plasma. The mixture is recalcified, and the clotting time is measured. The prothrombin time is abnormally long if any of the plasma clotting factors of the extrinsic pathway are deficient or qualitatively defective or if inhibitors of this pathway are present. The prothrombin time is normal in deficiencies of fibrin-stabilizing factor or of the components of the intrinsic pathway that act before the participation of Stuart factor (factor X).

The *thrombin time* is the clotting time of a mixture of thrombin and plasma. It is abnormally long when the concentration of fibrinogen is abnormally low or this protein is qualitatively defective. The thrombin time is also lengthened by the presence of endogenous or exogenous inhibitors of the formation of thrombin, for example, therapeutically administered heparin.

Specific assays for each of the clotting factors are available. Uniquely, the concentration of fibrinogen can be assayed chemically by converting it to fibrin and measuring the protein content of the fibrin. The clotting factors of the extrinsic and intrinsic pathways are assayed in a sample by measuring its ability to shorten either the abnormal prothrombin time or the PTT of plasma known to be deficient solely in the specific factor under study. *Fibrin-stabilizing factor* (factor XIII) can be estimated qualitatively by observing the stability of clots in 6M urea or 1% monochloroacetic acid, which dissolve fibrin that has not been covalently linked. Quantitative assays of fibrin-stabilizing factor rely on its transamidation function. Immunological assays have been devised to quantify each clotting factor.

■ *Fibrinolysis and related phenomena*

That clotted blood may reliquefy was recognized in the early nineteenth century by the French physiologist Denis. The agent responsible for fibrinolysis, which is the dissolution of fibrin, is *plasmin,* a typical serine protease that is the activated form of the plasma glycoprotein *plasminogen.* Plasminogen is a single-chain polypeptide that is synthesized in the liver and perhaps elsewhere as well. It can be changed to plasmin by many activators, not all of which are physiological.

An activator of plasminogen of particular interest because it has been used clinically to dissolve thrombi and emboli is *streptokinase,* a protein elaborated by certain β-hemolytic streptococci. Streptokinase combines stoichiometrically with plasminogen to form a complex that transforms plasminogen to plasmin by cleavage within an internal disulfide loop. Staphylococci synthesize staphylokinase, a similar activator of plasminogen.

Streptokinase and staphylokinase are foreign to the body, but many physiological activators exist. Normal urine, for example, contains a serine protease *(urokinase)* that is elaborated and excreted by the kidney. This enzyme converts plasminogen to plasmin through the same cleavages as the streptokinase-plasminogen complex. Like streptokinase, it has been introduced for the clinical dissolution of thrombi and emboli. Activators have also been identified in human milk, tears, saliva, seminal fluid, many tissues including vascular endothelium, and blood monocytes.

In normal blood, plasminogen may be converted to plasmin "spontaneously" through one or another of several pathways. Three enzymes activated through surface-mediated reactions (p. 418), plasma kallikrein, activated PTA (factor XIa), and, much more weakly, activated Hageman factor (factor XIIa), can activate plasminogen. Thus activation of the intrinsic pathway of thrombin formation sets into motion reactions that lead to fibrinolysis.

Another intrinsic activator of plasminogen can be demonstrated in blood drawn from a vein distal to a tourniquet that has been applied to the upper arm for several minutes. Clots formed from the plasma of such blood dissolve at an increased rate. Activation of plasminogen under these circumstances appears to be initiated by an agent released from venous endothelium. Fibrinolysis in vitro is greatly enhanced by emotional stress, strenuous physical activity, or the injection of epinephrine, vasopressin, or pyrogens; it is likely that this phenomenon is the consequence of the release of the venous endothelial activator. Although the temptation is great to assume that these test tube observations have biological significance, the evidence for this is inferential.

Plasmin is an enzyme of broad specificity. It digests not only fibrin but numerous other substances as well including fibrinogen, antihemophilic factor (factor VIII), proaccelerin (factor V), and other clotting factors. Plasmin is one of the plasma enzymes that can alter factor VII, augmenting its clot-enhancing properties (p. 423). Plasmin may participate in the immune defenses of the body, for it converts the first component of complement (C1) to its enzymatically active state (C$\bar{1}$), and it separates a peptide fragment from the fifth component of complement (C5) that has chemotactic properties, that is, the capacity to attract leukocytes. It fragments Hageman factor, liberating its enzymatically active carboxy-terminal fragment (HF$_f$), and it releases kinins from kininogens, both directly and possibly by converting prekallikrein to kallikrein (p. 420). Additionally, plasmin digests casein and certain synthetic amino acid esters and small polypeptide amides, all of which have been used to examine and assay its function.

The property of plasmin that has received the most attention is the digestion of fibrinogen and fibrin. In purified systems, plasmin appears to attack these two proteins with equal avidity. In plasma, however, generation of plasmin occurs more readily on the surface of a clot. One explanation for this phenomenon is that plasminogen and its activators adhere to fibrin. In this situation, plasmin can digest fibrin relatively unhampered by the inhibitors of this enzyme in plasma (p. 428). The practical result is that the therapeutic injection of streptokinase or urokinase into human subjects may bring about dissolution of intravascular clots without a major effect on circulating fibrinogen. Any plasmin that leaks into the circulation after its activation on the surface of a clot is readily inactivated by its inhibitors in plasma, notably α_2-antiplasmin (p. 428).

The digestion of fibrinogen takes place in steps (Fig. 25-7). First, plasmin separates fragments sequentially from the carboxy-terminal ends of the Aα-chains and the amino-terminal ends of the Bβ-chains of fibrinogen. The principal residue (fragment X) is still coagulable by thrombin. Fragment X is then further digested by plasmin into fragments designated Y, D, and E. The degradation products of fibrinogen are inhibitors of clotting and interfere with the formation and action of thrombin and the polymerization of fibrin monomers. In part this may be a result of the formation of soluble complexes of fragments X and Y with fibrinogen or fibrin monomer. The fragmentation of fibrin proceeds along similar lines, but fragment X is not coagulable.

The digestion of fibrinogen and fibrin can be brought about not only by plasmin but also by proteases released by leukocytes. The proteases responsible have been identified as elastases, enzymes that can digest elastin fibers, collagenases that digest collagen, and chymotrypsin-like enzymes.

Rarely persons undergoing severe stress, such as surgical procedures or trauma, develop *primary fibrinolytic purpura*. The blood of such persons may coagulate, but the clots dissolve within a few minutes. This hemorrhagic syndrome is probably caused by the entrance of activators of plasminogen into the bloodstream; a similar syndrome may follow the therapeutic infusion of activators of plasminogen. The titers of antihemophilic factor (factor VIII) and proaccelerin (factor V) are usually low, but in contrast to the secondary fibrinolytic states that accompany disseminated intravascular coagulation, the

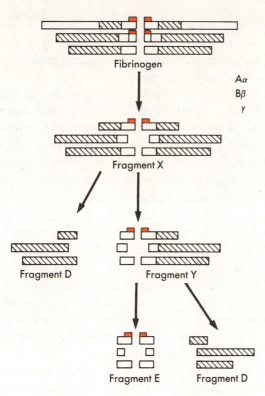

Fig. 25-7 ■ The digestion of fibrinogen by plasmin. The three chains of fibrinogen are Aα, Bβ, and γ; the solid black areas represent fibrinopeptides A and B. The first cleavage of fibrinogen removes a carboxy-terminal segment of the Aα chain; the residue (fragment X) is still coagulable. Fragment X is then cleaved successively to fragment Y, two fragments D, and fragment E. (Redrawn from Sherry, S.: Mechanisms of fibrinolysis. In Williams, W.J., et al., editors: Hematology, ed. 2, New York, 1977, McGraw-Hill, Inc. Copyright 1977 by McGraw-Hill Book Co. Used with the permission of McGraw-Hill Book Co.)

platelet count is normal. The patient has a severe bleeding tendency, probably exaggerated by the anticoagulant properties of the products of fibrinogenolysis and fibrinolysis.

Spontaneous activity of the *fibrinolytic system* of plasma can be assessed with increasing sensitivity by measuring the time required for dissolution of clots formed from whole blood, plasma, or the euglobulin fraction of plasma, that is, the fraction that is insoluble when plasma is diluted in acidified water. The rate of fibrinolysis reflects the combined effects of activators, plasminogen, and inhibitors of the activation and action of plasmin (p. 426). Tests of fibrinolysis measure what takes place after fibrinogen has been converted to fibrin in vitro; their significance in vivo is not clear. The plasma content of plasminogen can be estimated by converting this proenzyme to plasmin by addition of urokinase or streptokinase. The enzymatic activity of plasmin can then be assayed on protein or synthetic ester or amide substrates. Inhibitors of plasmin, particularly α_2-antiplasmin (p. 428), are readily estimated by testing the inhibitory effect of plasma on the fibrinolytic or amidolytic activity of plasmin.

That fibrinolysis has occurred in vivo is often suggested by the presence in *serum* of proteins that are either immunologically related to fibrinogen or that gel upon addition of ethanol or protamine sulfate. Although these proteins are usually described as fibrin(ogen) split or degradation products, they are more accurately called fibrin(ogen)-related antigens because the tests do not distinguish between the presence in serum of digestion products of fibrinogen or fibrin and soluble intermediates in the formation of fibrin such as complexes of fibrinogen and fibrin monomer.

■ *Laboratory examination of the fibrinolytic system*

■ *Inhibitors of clotting and fibrinolysis*

Plasma is rich in substances that can inhibit *activated* clotting factors and plasmin. Presumably they serve to restrict the growth of intravascular clots and minimize the deleterious effects of unbridled fibrinolysis.

Antithrombin III is the principal inhibitor of thrombin and activated Stuart factor (factor Xa). To a lesser extent it blocks the action of the enzymatically active forms of each of the other serine proteases involved in the formation and dissolution of clots; it is without effect on fibrin-stabilizing factor (factor XIII). Inhibition results from the formation of a stoichiometric complex between antithrombin III and the activated enzymes. The importance of antithrombin III is evident from the frequency of thrombosis in persons with a partial deficiency of this inhibitor.

The inhibitory properties of antithrombin III are greatly potentiated by the addition of *heparin,* with which it forms a complex. Heparin is a negatively charged sulfated polysaccharide that is synthesized principally in mast cells. It is found in many tissues including lung, liver, and intestinal mucosa, but it is not a normal constituent of blood. Without antithrombin III, heparin is not a significant anticoagulant. Heparin is widely used in the prevention and treatment of thrombotic states, in which its enhancement of the action of antithrombin III limits the growth of intravascular clots. Overdosage by heparin is readily controlled by administration of *protamine sulfate,* a highly positively charged polypeptide derived from fish sperm.

α_2-*Macroglobulin* is a plasma protein that slowly inactivates the proteolytic properties of plasmin, thrombin, and plasma kallikrein; it is a major inhibitor of plasma kallikrein, and the chief backup system for the inhibition of plasmin when supplies of α_2-plasmin inhibitor are exhausted. Its mode of action is unusual. The enzymes bind to and partially digest the inhibitor. As a consequence, the shape of the enzymes is so changed that their ability to attack protein substrates is greatly reduced. The catalytic sites of the enzymes remain active so that the enzyme-inhibitor complexes can still hydrolyze small synthetic substrates. α_2-Macroglobulin attaches to the surface of vascular endothelial cells, fibroblasts, and macrophages, where it may protect these cells from proteolytic digestion.

α_2-*Plasmin inhibitor* is the major inhibitor of plasmin, with which it combines stoichiometrically and probably covalently in an irreversible link. A patient with a hereditary deficiency of this inhibitor has a severe hemorrhagic disorder in which hemostasis appears to be impeded by rapid dissolution of clots by the unchecked action of plasmin. α_2-Plasmin inhibitor inactivates the activated forms of Hageman factor, plasma kallikrein, activated Stuart factor, and thrombin, but these reactions are probably of little significance. Plasma also contains inhibitors of the *activation* of plasminogen that are less well defined.

α_1-*Antitrypsin* (α_1-protease inhibitor) is a glycoprotein with broad specificity that is synthesized in the liver. Among enzymes that it inactivates in the test tube are activated PTA (factor XIa), thrombin, plasma kallikrein, and plasmin. The biological signficance of these properties is dubious because the hereditary deficiency of α_1-antitrypsin is not associated with any disturbances of clotting or fibrinolysis. Rather, persons with this defect may have a peculiar form of chronic obstructive pulmonary disease or cirrhosis of the liver.

C1 esterase inhibitor ($\overline{C1}$-INH) is a plasma protein that inhibits the activated form of the first component of complement ($\overline{C1}$). In hereditary angioneurotic edema, a disorder in which the affected person has recurrent bouts of soft tissue swelling that are unaccompanied by pain or itching, the concentration of $\overline{C1}$-INH is very low, as determined in functional assays. $\overline{C1}$-INH also inhibits activated HF, plasma kallikrein, activated PTA, and plasmin, but these properties are probably of no clinical importance because patients with hereditary angioneurotic edema do not appear to have defective hemostasis or fibrinolysis.

Protein C (autoprothrombin II-A) is a two-chain vitamin K–dependent plasma proenzyme (p. 415) that can be changed to an enzymatically active state by thrombin. In this form it inhibits antihemophilic factor (factor VIII) and proaccelerin (factor V), perhaps explaining in part their low titers in some cases of widespread intravascular coagulation.

Other devices are available to limit intravascular clotting and fibrinolysis besides the presence of circulating plasma inhibitors. Passage of blood through the liver removes activated Stuart factor (factor Xa), thrombin, soluble fibrin complexes, and probably activators of plasminogen from plasma. This "clearance" of activated factors is thought to be a property of the Kupffer cells, part of the reticuloendothelial system. In agreement with this interpretation, peritoneal or peripheral blood macrophages are known to bind fibrin and thrombin.

■ **Platelets**

The *platelets* are small, disk-shaped anuclear cells with an average diameter of about 2 to 4 μm that serve multiple functions in the hemostatic and defense mechanisms of the body. They arise by budding from the cytoplasm of their progenitors the *megakaryocytes,* large polyploid cells with 4, 8, or 16 nuclei that are derived from the primitive hematopoietic stem cells. In the normal adult, megakaryocytes are found almost exclusively in bone marrow and in the newborn and under certain pathological conditions in the liver, lungs, and other organs as well. Maturation of megakaryocytes to the point at which they shed their platelets takes 4 or 5 days.

Normal blood contains between 150,000 and 350,000 platelets/μl. Not all platelets are in the circulating blood; as many as a third are sequestered elsewhere, principally in the spleen. The life of an individual platelet is about 8 to 12 days; their destruction appears to be a consequence of senescence. Aged or damaged platelets are removed from the circulation by the reticuloendothelial system; the spleen is an important site of removal. In the person whose spleen has been removed, the platelet count may be modestly elevated.

Platelet production is probably regulated at least partially by the total number of platelets destroyed rather than by their concentration in circulating blood. The signal for platelet production is uncertain. Maturation of megakaryocytes and shedding of their platelets are stimulated by a poorly defined hormone *thrombopoietin,* but this substance is probably not required for platelet production.

■ **Platelet structure**

The anatomy of the platelet is complex (Fig. 25-8). The exterior coat or glycocalyx is rich in glycoproteins that may be responsible for adhesion of platelets to subendothelial structures and aggregation of platelets one to another when these cells are appropriately stimulated. Separated from plasma, the platelets have a loose covering of plasma proteins that adhere more or less tightly to the surface and can be removed by careful washing. Among these proteins are fibrinogen, antihemophilic factor (factor VIII), and proaccelerin (factor V). The surface of the platelets has receptors for many agents including fibrinogen, thrombin, adenosine diphosphate (ADP), catecholamines such as epinephrine, serotonin (5-hydroxytryptamine), collagen, and immune complexes, that is, complexes of antigen and antibody.

Just within the exterior coat is a phospholipid-rich "unit membrane" that is the principal source of clot-promoting phospholipids. This layer is the site of many enzymes, among them phospholipases, glycogen synthetase, and adenylate cyclase, the enzyme that converts adenosine triphosphate (ATP) to cyclic adenosine monophosphate (cAMP). The innermost layer of the platelet membrane contains fibrillar microtublar structures and microfilaments of actin and myosin that maintain the shape of the platelet.

The cytoplasm of the platelet is equally complex. At its periphery is an open canalicular system, the channels of which appear to be invaginations of the external platelet

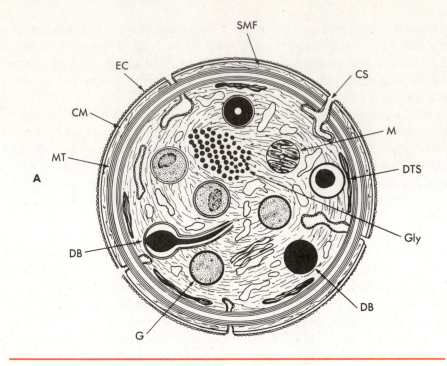

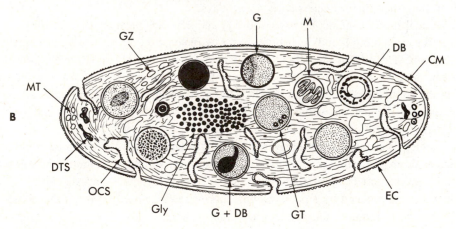

Fig. 25-8 ■ Diagrammatic representation of a platelet in an equatorial plane **(A)** and in cross section **(B)**. Components of the peripheral zone include the exterior coat *(EC)*, trilaminar unit membrane *(CM)*, and submembrane area containing specialized filaments *(SMF)*, which form the platelet wall and line channels of the surface-connected canalicular system *(CS)*. The matrix of platelet cytoplasm contains active microfilaments, structured filaments, the circumferential band of microtubules *(MT)*, and glycogen *(Gly)*. Organelles embedded in the sol-gel zone include mitochondria *(M)*, granules *(G)*, and electron-dense bodies *(DB)*; some α-granules appear to contain tubular structures *(GT)*. The membrane systems include the surface-connected open canalicular system *(CS or OCS)* and the dense tubular system *(DIS)*. An occasional Golgi apparatus *(GZ)* is found in some platelets. (Redrawn from White, J.G.: Am. J. Clin. Pathol. **71:**363-378, 1979.)

membranes. The contents of platelet granules can be discharged through these tubular structures to the exterior of the cell. Scattered through the cytoplasm is a second series of channels, a dense tubular system that does not connect with the open canalicular system. These structures, which represent smooth endoplasmic reticulum, are probably the site of synthesis of prostaglandins and are a site of sequestration of calcium ions. The cytoplasm also contains a potent contractile system that includes proteins resembling muscle actin and myosin and an actin-binding protein. Within the cytoplasm is a metabolic pool of ATP and ADP that furnishes energy for the cell's metabolism and the a- or α-chains of fibrin-stabilizing factor (factor XIII).

Numerous organelles are dispersed throughout the cytoplasm including small electron-dense bodies, lysosomes, α-granules, mitochondria, glycogen granules, and, inconstantly, a Golgi apparatus (see following).

Platelet organelles*

> Electron-dense bodies
> > Serotonin
> > Catecholamines
> > Nonmetabolic storage pool of ATP and ADP
> > Calcium ions
> α-Granules
> > Platelet factor 4
> > β-Thromboglobulin
> > Fibronectin
> > Platelet fibrinogen
> > Platelet growth factor
> > Proaccelerin (factor V)
> > Antihemopilic factor–related antigen (factor VIII:Ag)
> Lysosomes
> > Acid hydrolases
> Mitochondria
> Golgi apparatus (inconstant)
> Glycogen granules

*The localization of certain of the substances in the specific granules is only tentative.

■ **Platelet adhesion, aggregation, and the release reaction**

Within seconds after the endothelial lining of blood vessels is disrupted, some circulating platelets adhere to the exposed subendothelial structures such as collagen; factor VIII:VWF (p. 422), which is probably synthesized in endothelial cells, appears to enhance platelet adhesion. The adherent platelets assume a more spherical shape and, by active contraction that involves polymerization of microfibrillar actin, send out pseudopods or spicules that spread out like the legs of a spider along the subendothelial fibrils. Simultaneously, contraction of the platelet's microfibrils forces the cytoplasmic granules to migrate toward the center of the cell and to discharge their contents into the surrounding blood via the open canalicular system, the so-called *release reaction*. Agents extruded from the granules, including the nonmetabolic pool of ADP in the dense bodies, attract other platelets to those adherent to the injured vessel walls. At the same time, the clotting mechanisms are activated as blood comes into contact with the damaged vascular surfaces. Thrombin, generated locally in this way, promotes further platelet aggregation and release. It also brings about the formation of strands of fibrin that bind the platelet aggregates into a firm *hemostatic plug* that may control bleeding from small vascular injuries but may also serve as a nidus for a more formidable intravascular clot or thrombus.

How platelets stick to subendothelial structures and to each other is now under intensive study. Platelet aggregation and the release reaction can be induced in the test tube by addition of such diverse agents as collagen, thrombin, ADP, catecholamines, serotonin, arachidonic acid, aggregated immunoglobulins, antigen-antibody complexes, and antibodies directed against platelets.

Aggregation by ADP takes place only if the surrounding milieu contains fibrinogen, which adheres to platelets stimulated by the aggregating agent. The platelets of subjects who have ingested aspirin within the preceding few days do not undergo the release reaction upon addition of collagen, ADP, or epinephrine, and platelet aggregation is incomplete and reversible. In normal persons the effect of aspirin is reflected only by a slight prolongation of the bleeding time (p. 434), but in patients with disorders of hemostasis the tendency to hemorrhage may be enhanced.

Recent studies provide a partial explanation for the mechanisms underlying platelet aggregation and release (Fig. 25-9). Collagen, thrombin, ADP, or the catecholamines bind to receptors in the platelet membrane, where they activate membrane phospholipases, perhaps by releasing membrane-bound calcium ions. The activated phospholipases hydrolyze membrane phosphatidylcholine and phosphatidylinositol. Arachidonic acid, released from these phosphatides, is converted to cyclic endoperoxides (prostaglandin G_2 and H_2) by a cyclo-oxygenase that uses molecular oxygen. The endoperoxides are transformed by a microsomal enzyme (thromboxane synthetase) to a short-lived compound (thromboxane A_2) that can aggregate platelets, bring about the release reaction, and constrict small blood vessels. The release of arachidonic acid from membrane phospholipids is blocked by elevated concentrations of cAMP. Aspirin inhibits platelet aggregation and release by inactivating the cyclo-oxygenase, which is irreversibly acetylated, preventing the formation of thromboxane A_2. Not yet clear is the relationship between thromboxane A_2 and the release of ADP from platelet granules.

Like platelets, endothelial cells convert arachidonic acid to cyclic endoperoxides, but these are changed by the microsomal enzyme prostacyclin synthetase to prostacyclin (prostaglandin I_2). Prostacyclin inhibits platelet aggregation, perhaps by increasing the level of cAMP, and dilates blood vessels. Endothelial cells can use endoperoxides released from stimulated platelets to synthesize prostacyclin. Perhaps this phenomenon limits platelet aggregation at sites of vascular damage. Aspirin, at somewhat higher doses than are needed to block the synthesis of thromboxane A_2 in platelets, inhibits prostacyclin formation, suggesting that this drug may have a double-edged effect.

Thrombin brings about aggregation of platelets and discharge of their granular contents through the thromboxane pathway. But thrombin also aggregates platelets and induces the release reaction when this pathway is blocked, suggesting that some additional mechanism may be involved, perhaps the elaboration of another mediator.

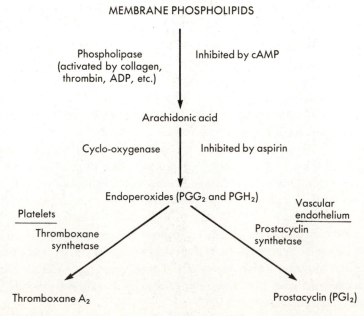

Fig. 25-9 ■ Pathways to the formation of thromboxane A_2 in platelets and prostacyclin in vascular endothelial cells.

A minor constituent of the α-granules that is discharged during the release reaction is *fibronectin,* which is also synthesized by fibroblasts, endothelial cells, and other tissues. Fibronectin is localized to many cellular surfaces, where it may serve as a glue that fosters the adhesion of cells to one another, but it seems to play no role in platelet aggregation. It is an important component of the matrix of connective tissue, where it is bound to collagen by an enzyme similar to activated fibrin-stabilizing factor (factor XIIIa). Fibronectin is also present in plasma, where it has been called *cold-insoluble globulin.* Plasma fibronectin may participate during disseminated intravascular coagulation in reactions that inhibit the polymerization of fibrin monomers. During in vitro clotting it may be covalently linked to fibrin by activated fibrin-stabilizing factor, but whether this has biological importance is not clear.

■ *Platelets and blood coagulation*

In the test tube stimulation of platelets by ADP or other agents or contact of platelets with such negatively charged substances as kaolin makes available membrane phospholipids (platelet factor 3, PF3) for the intrinsic pathway of thrombin formation. Presumably similar stimulation of platelets occurs in vivo. Additionally, activated Stuart factor (factor Xa) binds to receptor sites on the surface of thrombin-stimulated platelets. In this situation, activated Stuart factor converts prothrombin to thrombin much more effectively. The receptor for activated Stuart factor has been identified as proaccelerin (factor V) that has been altered to its clot-promoting state by thrombin. The function of platelet fibrinogen is not known, although this clotting factor makes up as much as 15% of platelet protein. Both proaccelerin and fibrinogen are found in the platelet's α-granules and are secreted during the release reaction.

■ *Clot retraction*

After normal blood coagulates, the clot gradually shrinks and extrudes clear serum. This phenomenon of clot retraction depends on the presence of viable, metabolically active platelets and is initiated by an action of thrombin on these cells, but the subsequent steps are only vaguely appreciated. Perhaps 15% or more of platelet protein consists of actin and myosin similar, but not identical to, that in muscle. One concept of the mechanism of clot retraction is that thrombin releases intracellular calcium ions, which then inactivate regulatory proteins similar to muscle troponin. When this takes place, platelet myosin, an adenosine triphosphatase (ATPase), is phosphorylated and combines with actin through crossbridges. The complex then contracts in a manner similar to the contraction of muscle (Chapter 22); the energy required comes from the splitting of ATP by myosin ATPase. The contractile process is responsible for extrusion of pseudopods and migration of granules to the center of the cell. The platelet spicules stick to polymerizing fibrin in the surrounding blood, whereupon the contractile process pulls the fibrin strands toward the platelet. This decreases the volume of the clot so that the entrapped serum is extruded; the fibrin strands themselves are not shortened. Clot retraction draws the platelets together, and these cells gradually assume an amorphous appearance described as *platelet metamorphosis,* a term now seldom used. Clot retraction is not inhibited by aspirin and is thus not dependent on the release reaction.

The utility of clot retraction has been puzzling. Perhaps the retraction of fibrin strands can bring together the edges of a wound filled with clot, or the shrinkage of the clot allows room for vascular walls to approximate, but these are teleological thoughts rather than scientifically based explanations.

■ *Miscellaneous platelet functions*

Besides their role in the formation of a hemostatic plug and in clot formation and retraction, platelets seem to serve myriad other functions. They have weak phagocytic and bactericidal properties that have not been demonstrated to be clinically important. They possess inhibitory activity against plasmin. During the release reaction a small polypeptide ''growth factor'' is discharged from the α-granules. This growth factor stimulates the proliferation of vascular endothelial cells and may therefore be important in maintaining the integrity of the vascular lining. It also stimulates the growth of fibroblasts and vascular smooth muscle cells and has been implicated in the pathogenesis of atherosclerotic plaques.

Platelets adhere to glass, a phenomenon related to the retention of these cells when blood is filtered through columns of glass beads (p. 422).

■ Laboratory studies of platelet function

Platelets can be counted with reasonable accuracy in a hemocytometer (counting chamber), but currently they are more likely to be enumerated with an electron particle counter. The latter technique is suitable for most purposes but gives erratic results when the platelet count is very low, for example, 10,000 to 20,000/μl, the range at which accuracy is most critical in patients with severe thrombocytopenia.

The *bleeding time* measures the duration of bleeding from a deliberately incised wound. Currently popular is the use of a template to provide a standardized incision of the forearm, but this technique sometimes leaves permanent scars. This complication can be avoided by measuring the bleeding time in the ball of the finger. The bleeding time is long when the platelet count is abnormally low ($< 80,000/\mu$l) or high ($> 800,000/\mu$l) or the platelets are qualitatively abnormal; the paradox of a long bleeding time in association with high platelet counts is poorly understood. It is also long in von Willebrand's disease (p. 422) and in the presence of appreciable amounts of abnormal plasma proteins as in multiple myeloma.

Clot retraction can be observed qualitatively in blood allowed to clot in glass tubes and incubated for 24 hours. Normally the clot shrinks, expressing clear serum, but impaired retraction is observed when the platelet count is less than about 80,000/μl or when the platelets have the qualitative abnormality described as thrombasthenia or Glanzmann's disease. Semiquantitative variations of this test take into account the amount of serum that can theoretically be expressed from a clot; the higher the hematocrit (the percent of red blood cells in a volume of blood) the less serum is available to be expressed.

Platelet aggregation is usually measured by adding the substances to be tested to platelet-rich plasma that is stirred continuously. Aggregation is recognized turbidimetrically by the increase in light transmission that occurs as platelets clump. The aggregating agents used usually include collagen, ADP, epinephrine, and thrombin. Impaired aggregation is observed in many herditary or acquired qualitative platelet disorders. The same technique is used to assay *platelet agglutination* by ristocetin, a measure of the titer of the von Willebrand factor subcomponent of antihemophilic factor (factor VIII:VWF).

Other tests of platelet function measure the release of the components of platelet membranes or granules such as PF3, PF4, serotonin, or the adenine nucleotides.

■ Platelet disorders

Thrombocytopenia, a reduced number of circulating platelets, may reflect their impaired production, increased destruction or use, or sequestration into a space such as the spleen outside the effective circulation. The causes of thrombocytopenia are legion. The hemorrhagic tendency in these patients is proportional to the degree of thrombocytopenia and is characterized by the presence of both petechiae and ecchymoses, menorrhagia, and bleeding from mucous membranes and into the central nervous system. Besides the low platelet count, the patient may have a long bleeding time and impaired clot retraction. Differentiation among the causes of thrombocytopenia nearly always requires careful examination of the bone marrow to see if megakaryocytes are present; their absence implies a failure of platelet production.

Thrombocytosis, an abnormally high number of circulating platelets, is seen in patients with a wide variety of disorders including iron deficiency anemia, chronic infection, and malignancy. Thrombocytosis is also a feature of some disorders of blood production such as polycythemia vera, a disease in which the numbers of red and white blood cells are also increased. Paradoxically, when the platelet count is above about 800,000/µl, the patient may experience a bleeding tendency. Under these conditions, described as *thrombocythemia,* the platelets are often qualitatively abnormal.

Many *qualitative abnormalities of platelets* have been described. In different cases, clot retraction is reduced; the release of clot-promoting activity (PF3) from platelets is inadequate; aggregation of platelets by collagen or ADP is impaired; or the electron-dense bodies are absent or discharge ADP poorly. Often combinations of these or other defects are detected. In some instances the platelet abnormality is inherited; in others it is a complication of some other disorder, notably renal failure. The bleeding time is nearly always prolonged.

■ *Thrombosis*

A *thrombus* is a clot within a blood vessel. The conditions that foster the formation of thrombi were outlined more than a century ago by the German pathologist Rudolf Virchow, who implicated damage to the vascular wall, impaired blood flow (stasis), and alteration in the blood that rendered it more readily coagulable. The initial event is believed to be a disruption of the normal vascular surface that induces localized platelet adhesion and aggregation and localized blood coagulation; such lesions are more evident in arterial than in venous thrombosis. Thereafter, a blood clot extends from this nidus into the lumen of the vessel. The anatomy of the thrombus is conditioned by its location. In arteries, in which blood flow is swift, the adherent portion of the thrombus is rich in platelets, whereas in veins, in which flow is sluggish, the base of the thrombus may resemble a typical blood clot. In veins thrombi are particularly likely to begin at the site of valves, where the flow of blood is slowed.

The degree to which changes in the blood itself foster thrombosis is much debated. Experimentally the injection of activated clotting factors results in the formation of intravascular clots, but only in areas in which blood flow has been halted by the subsequent application of ligatures. Thus stasis and the introduction of the activated form of clotting factors foster this type of experimental thrombosis. The role of increased concentrations of clotting factors is less certain.

Embolism is the process through which thrombi break off from their vascular attachments and are carried by the flowing blood until they lodge in a blood vessel too small to allow their passage. The most frequent example is the commonplace and often fatal disorder pulmonary embolism, in which a venous thrombus originating in a deep vein of the legs or in a pelvic vein becomes wedged in the pulmonary artery.

■ *Bibliography*

Journal articles

Davie, E.W., et al.: The role of serine proteases in the blood coagulation cascade, Adv. Enzymol. **48:**277, 1979.

Jackson, C.M., and Nemerson, Y.: Blood coagulation, Ann. Rev. Biochem. **49:**765, 1980.

Shattil, S.J., and Bennett, J.S.: Platelets and their membranes in hemostasis: physiology and pathophysiology, Ann. Int. Med. **94:**108, 1981.

Suttie, J.W.: The metabolic role of vitamin K, Fed. Proc. **39:**2730, 1980.

Suttie, J.W., and Jackson, C.M.: Prothrombin structure, activation and biosynthesis, Physiol. Rev. **57:**1, 1977.

White, J.G.: Current concepts of platelet structure, Am. J. Clin. Pathol. **71:**373, 1979.

Books and monographs

Gordon, J.L., editor: Platelets in biology and pathology, ed. 2, New York, 1981, Elsevier North-Holland, Inc.

Kline, D.A., and Reddy, K.N.N.: Fibrinolysis, Boca Raton, Fla., 1980, CRC Press, Inc.

Marcus, A.J.: The role of prostaglandins in platelet function. In Brown, E.B., editor: Progress in hematology, vol. 11, New York, 1979, Grune & Stratton, Inc.

Mosher, D.F.: Fibronectin. In Spaet, T.H., editor: Progress in hemostasis and thrombosis, vol. 5, New York, 1980, Grune & Stratton, Inc.

Poller, L., editor: Recent advances in blood coagulation, New York, No. 2, 1977, No. 3, 1981, Churchill Livingstone, Inc.

Ratnoff, O.D.: Hereditary disorders of hemostasis. In Stanbury, J.B., Wyngaarden, J.B., and Fredrickson, D.S., editors: The metabolic basis of inherited disease, ed. 3, New York, 1978, McGraw-Hill, Inc.

Ratnoff, O.D., and Saito, H.: Surface-mediated reactions. In Piomelli, S., and Yachnin, S., editors: Current topics in hematology, vol. 2, New York, 1979, Alan R. Liss, Inc.

THE CARDIOVASCULAR SYSTEM

Robert M. Berne
Matthew N. Levy

chapter 26

The circuitry

The circulatory, endocrine, and nervous systems constitute the principal coordinating and integrating systems of the body. Whereas the nervous system is primarily concerned with communications and the endocrine glands with regulation of certain body functions, the circulatory system serves to transport and distribute essential substances to the tissues and to remove by-products of metabolism. The circulatory system also shares in such homeostatic mechanisms as regulation of body temperature, humoral communication throughout the body, and adjustments of oxygen and nutrient supply in different physiological states.

The cardiovascular system that accomplishes these chores is made up of a pump, a series of distributing and collecting tubes, and an extensive system of thin vessels that permit rapid exchange between the tissues and the vascular channels. The primary purpose of this section is to discuss the function of the components of the vascular system and the control mechanisms (with their checks and balances) that are responsible for alteration of blood distribution necessary to meet the changing requirements of different tissues in response to a wide spectrum of physiological and pathological conditions.

Before considering the function of the parts of the circulatory system in detail, it is useful to consider it as a whole in a purely descriptive sense. The heart consists of two pumps in series: one to propel blood through the lungs for exchange of oxygen and carbon dioxide (the *pulmonary circulation*) and the other to propel blood to all other tissues of the body (the *systemic circulation*). Unidirectional flow through the heart is achieved by the appropriate arrangement of effective flap valves. Although the cardiac output is intermittent in character, continuous flow to the periphery occurs by virtue of distension of the aorta and its branches during ventricular contraction (*systole*) and elastic recoil of the walls of the large arteries with forward propulsion of the blood during ventricular relaxation (*diastole*). Blood moves rapidly through the aorta and its arterial branches, which become narrower and whose walls become thinner and change histologically toward the periphery. From a predominantly elastic structure—the aorta—the peripheral arteries become more muscular in character until at the arterioles the muscular layer predominates (Fig. 26-1). As far out as the beginning of the arterioles, frictional resistance to blood flow is relatively small, and, despite a rapid flow in the arteries, the pressure drop from the root of the aorta to the point of origin of the arterioles is relatively small (Fig. 26-2). The arterioles, the stopcocks of the vascular tree, are the principal points of resistance to blood flow in the circulatory system. The large resistance offered by the arterioles is reflected by the considerable fall in pressure from arterioles to capillaries. Adjustment in the degree of contraction of the circular muscle of these small vessels permits regulation of tissue blood flow and aids in the control of arterial blood pressure.

In addition to a sharp reduction in pressure across the arterioles, there is a change

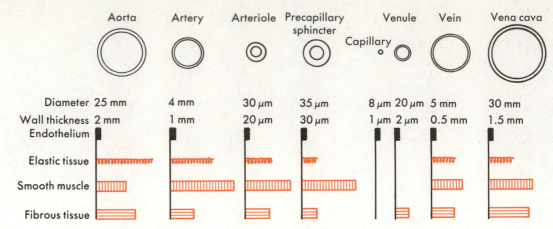

Fig. 26-1 ■ Internal diameter, wall thickness, and relative amounts of the principal components of the vessel walls of the various blood vessels that compose the circulatory system. Cross sections of the vessels are not drawn to scale because of the huge range from aorta and venae cavae to capillary. (Redrawn from Burton, A.C.: Physiol. Rev. **34:**619, 1954.)

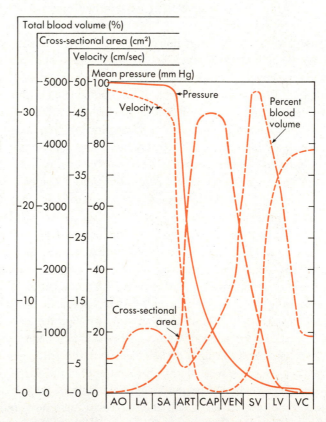

Fig. 26-2 ■ Pressure, velocity of flow, cross-sectional area, and capacity of the blood vessels of the systemic circulation. The important features are the inverse relationship between velocity and cross-sectional area, the major pressure drop across the arterioles, the maximal cross-sectional area and minimal flow rate in the capillaries, and the large capacity of the venous system. The small but abrupt drop in pressure in the venae cavae indicates the point of entrance of these vessels into the thoracic cavity and reflects the effect of the negative intrathoracic pressure. To permit schematic representation of velocity and cross-sectional area on a single linear scale, only approximations are possible at the lower values. *AO,* Aorta; *LA,* large arteries; *SA,* small arteries; *ART,* arterioles; *CAP,* capillaries; *VEN,* venules; *SV,* small veins; *LV,* large veins; *VC,* venae cavae.

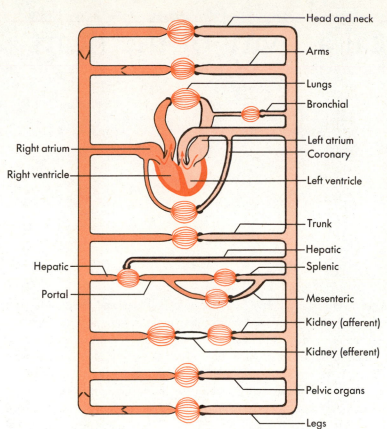

Head and neck
Arms
Lungs
Bronchial
Left atrium
Coronary
Right atrium
Right ventricle
Left ventricle
Trunk
Hepatic
Hepatic
Splenic
Portal
Mesenteric
Kidney (afferent)
Kidney (efferent)
Pelvic organs
Legs

Fig. 26-3 ■ Schematic diagram of the parallel and series arrangement of the vessels composing the circulatory system. The capillary beds are represented by thin lines connecting the arteries (on the right) with the veins (on the left). The crescent-shaped thickenings proximal to the capillary beds represent the arterioles (resistance vessels). (Redrawn from Green, H.D.: Circulation: physical principles. In Glasser, O., editor: Medical physics, vol. 1, Chicago, 1944, Year Book Medical Publishers, Inc. Reproduced with permission. Copyright © 1944 by Year Book Medical Publishers, Inc.)

from pulsatile to steady flow. The pulsatile character of arterial blood flow, caused by the intermittency of cardiac ejection, is damped at the capillary level by the combination of distensibility of the large arteries and frictional resistance in the arterioles. Many capillaries arise from each arteriole so that the total cross-sectional area of the capillary bed is very large, despite the fact that the cross-sectional area of each individual capillary is less than that of each arteriole. As a result, blood flow becomes quite slow in the capillaries, analogous to the decrease in flow rate seen at the wide regions of a river. Since the capillaries consist of short tubes whose walls are only one cell thick and since flow rate is slow, conditions in the capillaries are ideally suited for the exchange of diffusible substances between blood and tissue.

On its return to the heart from the capillaries, blood passes through venules and then through veins of increasing size. As the heart is approached, the number of veins decreases, the thickness and composition of the vein walls change (Fig. 26-1), the total cross-sectional area of the venous channels is progressively reduced, and velocity of blood flow increases (Fig. 26-2). Also note that the greatest proportion of the circulating blood is located in the venous vessels (Fig. 26-2). The cross-sectional area of the venae cavae is larger than that of the aorta (although not evident from Fig. 26-2 because cross-sectional areas of venae cavae and aorta are so close to zero with a scale that includes the capillaries), and hence the flow is slower than that in the aorta. Blood entering the right ventricle via the right atrium is then pumped through the pulmonary arterial system at mean pressures about one-seventh those developed in the systemic arteries. The blood then passes through the lung capillaries, where carbon dioxide is released and oxygen taken up. The oxygen-rich blood returns via the pulmonary veins to the left atrium and ventricle to complete the cycle. Thus in the normal intact circulation the total volume of blood is constant, and an increase in the volume of blood in one area must be accompanied by a decrease in another. However, the velocity at which the blood circulates through the different regions of the body is determined by the output of the left ventricle and by the contractile state of the arterioles (resistance vessels) of these regions. The circulatory system is composed of conduits arranged in series and in parallel, as schematized in Fig. 26-3.

chapter 27

Electrical activity of the heart

The experiments on "animal electricity" conducted by Galvani and Volta in the last half of the eighteenth and early nineteenth centuries prepared the stage for the discovery that electrical phenomena were involved in the spontaneous contractions of the heart. In 1855 Kölliker and Müller observed that when they placed the nerve of a nerve-muscle preparation in contact with the surface of a frog's heart, the muscle twitched with each cardiac contraction. This phenomenon may be observed in the laboratory by allowing the phrenic nerve of an anesthetized dog to lie across the exposed surface of the heart: the diaphragm will contract with each heart beat. Precise measurement of this electrical activity was not feasible until the end of the past century, when the construction of sensitive, high-fidelity galvanometers permitted registration of the changes in electrical potential during the various phases of the cardiac cycle, which led to the science of electrocardiography.

■ Transmembrane potentials

The electrical behavior of single cardiac muscle cells has been investigated by inserting microelectrodes into the interior of cells from various regions of the heart. The potential changes recorded from a typical ventricular muscle fiber are illustrated schematically in Fig. 27-1. When two electrodes are situated in an electrolyte solution near a strip of quiescent cardiac muscle, there will be no potential difference measurable between the two electrodes (from point A to point B, Fig. 27-1). At point B one of the electrodes, a microelectrode with a tip diameter less than 1 μm, was inserted into the interior of a cardiac muscle fiber. Immediately the galvanometer recorded a potential difference across the cell membrane, indicating that the potential of the interior of the cell was about 90 mV lower than that of the surrounding medium. Such electronegativity of the interior of the resting cell with respect to the exterior is also characteristic of skeletal and smooth muscle, of nerve, and indeed of most cells within the body. At point C a propagated action potential was transmitted to the cell impaled with the microelectrode. Very rapidly the cell membrane became depolarized; actually, the potential difference was reversed (positive overshoot), so that the potential of the interior of the cell exceeded that of the exterior by about 20 mV. The rapid upstroke of the action potential is designated phase 0. Immediately after the upstroke, there was a brief period of partial repolarization (phase 1), followed by a *plateau* (phase 2) that persisted for about 0.1 to 0.2 second. The potential then became progressively more negative (phase 3), until the resting state of polarization was again attained (at point E). The process of rapid repolarization (phase 3) proceeds at a much slower rate of change than does the process of depolarization (phase 0). The interval from the completion of repolarization until the beginning of the next action potential is designated phase 4.

The time relationships between the electrical events and the actual mechanical con-

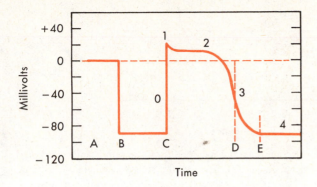

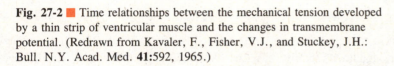

Fig. 27-1 ■ Changes in potential recorded by an intracellular microelectrode. From time *A* to *B* the microelectrode was outside the fiber; at *B* the fiber was impaled by the electrode. At time *C* an action potential began in the impaled fiber. Time *C* to *D* represents the effective refractory period, and *D* to *E* represents the relative refractory period.

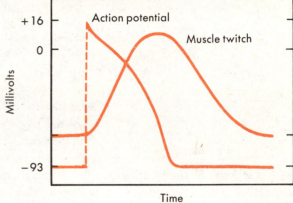

Fig. 27-2 ■ Time relationships between the mechanical tension developed by a thin strip of ventricular muscle and the changes in transmembrane potential. (Redrawn from Kavaler, F., Fisher, V.J., and Stuckey, J.H.: Bull. N.Y. Acad. Med. **41**:592, 1965.)

traction are shown in Fig. 27-2. It can be seen that rapid depolarization (phase 0) precedes tension development and that completion of repolarization coincides approximately with peak tension development. The duration of contraction tends to parallel the duration of the action potential. Also, as the frequency of cardiac contraction is increased, there is a progressive reduction in the duration of both the action potential and the mechanical contraction.

Two main types of action potentials are observed in the heart, as shown in Fig. 27-3. One type, the so-called *fast response*, occurs in the normal myocardial fibers in the atria and ventricles and in the specialized conducting fibers *(Purkinje fibers)* in these chambers. The action potentials shown in Figs. 27-1 and 27-2 are also typical fast responses. The other type of action potential, the so-called *slow response*, is found in the *sinoatrial (SA) node*, the natural pacemaker region of the heart, and in the *atrioventricular (AV) node*, the specialized tissue involved in conducting the cardiac impulse from atria to ventricles. Furthermore, fast responses may be converted to slow responses either spontaneously or under certain experimental conditions. For example, in a myocardial fiber a gradual shift of the resting membrane potential from its normal level of about −80 to −90 mV to a value of about −60 mV will cause a conversion of subsequent action potentials to the slow response. Such conversions may occur spontaneously in patients with severe coronary artery disease, in those regions of the heart in which the blood supply has been severely curtailed.

As shown in Fig. 27-3, not only is the resting membrane potential of the fast response considerably more negative than that of the slow response, but also the slope of the upstroke (phase 0), the amplitude of the action potential, and the extent of the overshoot of the fast response are greater than in the slow response. It will be explained later that the magnitude of the resting potential is largely responsible for these other distinctions between the fast and slow responses. Furthermore, the amplitude of the action potential and the rate of rise of the upstroke are important determinants of propagation velocity. Hence, in cardiac tissue characterized by the slow response, conduction velocity is very much slower and there is a much greater tendency for impulses to

■ *Principal types of cardiac action potentials*

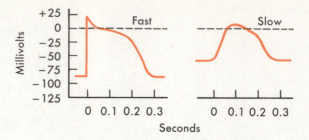

Fig. 27-3 ■ A fast and a slow response action potential recorded from the same canine Purkinje fiber. In the left panel the Purkinje fiber bundle was perfused with a solution containing K^+ at a concentration of 4mM. In the right panel epinephrine was added and the K^+ concentration was raised to 16mM. (Redrawn from Wit, A.L., Rosen, M.R., and Hoffman, B.F.: Am. Heart J. **88:**515, 1974.)

be blocked than in tissues displaying the fast response. Slow conduction and tendency toward block are conditions that increase the likelihood of certain rhythm disturbances in the heart.

■ *Ionic basis of the resting potential*

The various phases of the cardiac action potential are associated with changes in the permeability of the cell membrane, mainly to Na, K, and Ca ions. These changes in permeability produce alterations in the rate of passage of these ions across the membrane. Just as with all other cells in the body, the concentration of potassium ions inside a cardiac muscle cell, $[K^+]_i$, greatly exceeds the concentration outside the cell, $[K^+]_o$, as shown in Fig. 27-4. The reverse concentration gradient exists for Na ions and for unbound Ca ions. Furthermore, the resting cell membrane is relatively permeable to K^+, but much less so to Na^+ and Ca^{++}. Because of the high permeability to K^+, there tends to be a net diffusion of K^+ from the inside to the outside of the cell, in the direction of the concentration gradient, as shown on the right side of the cell in Fig. 27-4. Many of the anions (labeled A^-) inside the cell, such as the proteins, are not free to diffuse out with the K^+. Therefore, as the K^+ diffuses out of the cell and leaves the A^- behind, the deficiency of cations causes the interior of the cell to become electronegative, as shown on the left side of the cell in Fig. 27-4.

Therefore two opposing forces are involved in the movement of K^+ across the cell membrane. A chemical force, based on the concentration gradient, results in the net outward diffusion of K^+. The counter force is an electrostatic one; the positively charged K ions are attracted to the interior of the cell by the negative potential that exists there. If the system came into equilibrium, the chemical and the electrostatic forces would be equal. As has already been explained in Chapter 2, this equilibrium is expressed by the Nernst equation for potassium:

$$E_k = -61.5 \log ([K^+]_i/[K^+]_o)$$

The right-hand term represents the chemical potential difference at the body temperature of 37° C. The left-hand term, E_K, represents the electrostatic potential difference that would exist across the cell membrane if K^+ were the only diffusible ion. E_K is called the *potassium equilibrium potential*. When the measured concentrations of $[K^+]_i$ and $[K^+]_o$ for mammalian myocardial cells are substituted into the Nernst equation, the calculated value of E_K equals about -90 to -100 mV. This value is close to, but slightly more negative than, the resting potential actually measured in myocardial cells. Therefore there is a small potential of about 10 to 15 mV tending to drive K^+ out of the resting cell.

The balance of forces acting on the Na ions is entirely different in resting cardiac cells. The intracellular Na^+ concentration, $[Na^+]_i$, is much lower than the extracellular

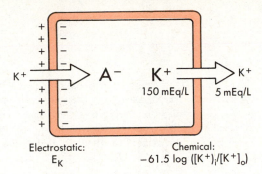

Electrostatic:
E_K

Chemical:
$-61.5 \log ([K^+]_i/[K^+]_o)$

Fig. 27-4 ■ The balance of chemical and electrostatic forces acting on a resting cardiac cell membrane, based on a 30:1 ratio of the intracellular to extracellular K^+ concentrations, and the existence of a nondiffusible anion (A^-) inside but not outside the cell.

concentration $[Na^+]_o$. At 37° C the *sodium equilibrium potential*, E_{Na}, expressed by the Nernst equation is $-61.5 \log ([Na^+]_i [Na^+]_o)$. For cardiac cells E_{Na} is about $+40$ to $+60$ mV. At equilibrium, therefore, an electrostatic force of 40 to 60 mV, oriented with the inside of the cell more positive than the outside, would be necessary to counterbalance the chemical potential for Na^+. However, the actual polarization of the resting cell membrane is just the opposite. The resting membrane potential of myocardial fibers is about -80 to -90 mV. Hence both chemical and electrostatic forces act to pull extracellular Na^+ into the cell. The influx of Na^+ through the cell membrane is small, however, because the permeability of the resting membrane to Na^+ is very low. Nevertheless, it is mainly this small inward current of positively charged Na ions that causes the potential on the inside of the resting cell membrane to be slightly less negative than the value predicted by the Nernst equation for K^+.

The steady inward leak of Na^+ would cause a progressive depolarization of the resting cell membrane were it not for the metabolic pump that continuously extrudes Na^+ from the cell interior and pumps in K^+. The metabolic pump involves the enzyme, Na^+, K^+ -activated ATPase, which is located in the cell membrane itself. Because the pump must move Na^+ against both a chemical and an electrostatic gradient, operation of the pump requires the expenditure of metabolic energy. Increases in $[Na^+]_i$ or in $[K^+]_o$ accelerate the activity of the pump. The quantity of Na^+ extruded by the pump exceeds the quantity of K^+ transferred into the cell (Chapter 2). Therefore the pump itself tends to create a potential difference across the cell membrane, and thus it is termed an *electrogenic pump*. If the pump is partially inhibited, as by large doses of digitalis, the concentration gradients for Na^+ and K^+ are partially dissipated, and the resting membrane potential becomes less negative than normal.

The dependence of the transmembrane potential, V_m, on the intracellular and extracellular concentrations of K^+ and Na^+ and on the conductances (g_K and g_{Na}) of these ions is described by the chord conductance equation. Equation 12 in Chapter 2 is the relevant form of this equation:

$$V_m = \frac{g_K}{g_K + g_{Na}} E_K + \frac{g_{Na}}{g_K + g_{Na}} E_{Na}$$

It is apparent from the chord conductance equation that it is the relative conductances to Na^+ and K^+, and not the absolute magnitude of each conductance, that determine the resting potential. In the resting cardiac cell, because g_{Na} is so much less than g_K (that is, $g_{Na}/g_K \cong 0.01$), the chord conductance equation reduces essentially to the Nernst equation for K^+. When the ratio $[K^+]_i/[K^+]_o$ is decreased experimentally by raising $[K^+]_o$, the measured value of V_m (curved line, Fig. 27-5) approximates that predicted by the Nernst equation for K^+ (straight line). For extracellular K^+ concentra-

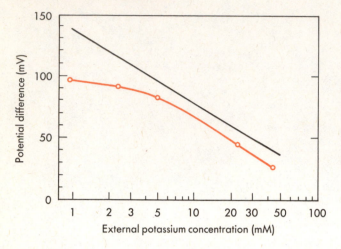

Fig. 27-5 ■ Transmembrane potential of a cardiac muscle fiber varies inversely with the potassium concentration of the external medium *(curve)*. The straight line represents the change in transmembrane potential predicted by the Nernst equation for E_K. (Redrawn from Page, E.: Circulation **26**:582, 1962. By permission of the American Heart Association, Inc.)

Fig. 27-6 ■ Concentration of sodium in the external medium is a critical determinant of the amplitude of the action potential in cardiac muscle *(upper curve)* but has relatively little influence on the resting potential *(lower curve)*. (Redrawn from Weidmann, S.: Elektrophysiologie der Herzmuskelfaser, Bern, 1956, Verlag Hans Huber.)

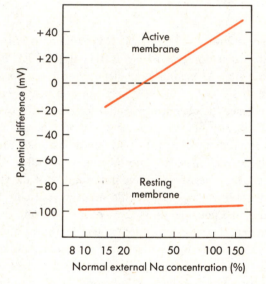

tions of about 5mM and above, the measured values correspond closely with the predicted values. The measured levels are slightly less than those predicted by the Nernst equation because of the small but finite value of g_{Na}. For values of $[K^+]_o$ below about 5mM, it has been found that the membrane properties become altered, such that there is a progressive reduction in g_K as $[K^+]_o$ is diminished. As g_K is decreased, the effect of the Na^+ gradient on the transmembrane potential becomes relatively more important, as predicted by the constant-field equation. This change in g_K accounts for the greater deviation of the measured V_m from that predicted by the Nernst equation for K^+ at low levels of $[K^+]_o$ (Fig. 27-5). Also, in accordance with the chord conductance equation, changes in $[Na^+]_o$ have relatively little effect on resting V_m (Fig. 27-6) because of the low value of g_{Na}.

■ Ionic basis of the fast response

Any process that abruptly changes the resting membrane potential to a critical value (called the *threshold*) will result in a propagated action potential. The characteristics of fast response action potentials resemble those shown on the left side of Fig. 27-3. The rapid depolarization that takes place during phase 0 is related almost exclusively to the inrush of Na^+ by virtue of a sudden increase in the permeability of the cell membrane to Na^+. The amplitude of the action potential (the magnitude of the potential change during phase 0) varies linearly with the logarithm of $[Na^+]_o$, as shown in Fig. 27-6. When $[Na^+]_o$ is reduced from its normal value of about 140mM to about 10 to 30mM, the cell is no longer excitable.

The physical and chemical forces responsible for these transmembrane movements

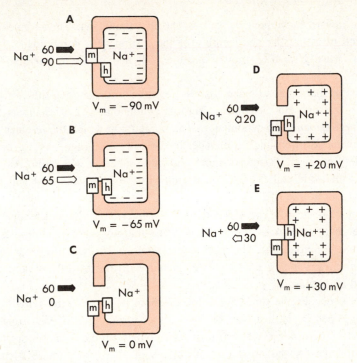

Fig. 27-7 ■ The Na$^+$ permeability of a cardiac cell membrane during phase 4 (panel *A*) and during various stages of phase 0, when the transmembrane potential (V$_m$) had attained values of −65 mV (panel *B*), 0 mV (panel *C*), +20 mV (panel *D*), and +30 mV (panel *E*). The positions of the *m* and *h* electrostatic gates in the fast Na$^+$ channels are shown at the various levels of V$_m$. The electrostatic forces are represented by the white arrows, and the chemical (diffusional) forces by the black arrows.

of Na$^+$ are explained in Fig. 27-7. When the resting membrane potential, V$_m$, is suddenly changed to the threshold level of about −60 to −70 mV, there is a dramatic change in the properties of the cell membrane. It is believed that *fast channels* for Na$^+$ exist in the membrane and that the flux of Na$^+$ through these channels is controlled by two polar components, referred to as "gates." The opening and closing of these gates are governed principally by the electrostatic charge, V$_m$, across the membrane. One of these gates, the *m* gate, tends to open the channel as V$_m$ becomes less negative and is therefore called an *activation gate*. The other, the *h* gate, tends to close the channel as V$_m$ becomes less negative and hence is called an *inactivation gate*. The *m* and *h* designations were originally employed by Hodgkin and Huxley in their mathematical model of conduction in nerve fibers (Chapter 3).

With the cell at rest, V$_m$ is about −80 to −90 mV. At this level the *m* gates are closed and the *h* gates are wide open, as shown in Fig. 27-7, panel *A*. The concentration of Na$^+$ is much greater outside than inside the cell, and the interior of the cell is electrically negative with respect to the exterior. Hence both chemical and electrostatic forces are oriented to draw Na$^+$ into the cell. The electrostatic force in Fig. 27-7, *A*, is a potential difference of 90 mV, and it is represented by the white arrow. The chemical force, based on the difference in Na$^+$ concentration between the outside and inside of the cell, is represented by the black arrow. For a Na$^+$ concentration difference of about 130 mM, a potential difference of 60 mV (inside more positive than the outside) would be necessary to counterbalance the chemical, or diffusional, force, according to the Nernst equation for Na$^+$. Therefore we may represent the net chemical force favoring the inward movement of Na$^+$ in Fig. 27-7 (black arrows) as being equivalent to a potential of 60 mV. With the cell at rest, therefore, the total electrochemical force favoring the inward movement of Na$^+$ is 150 mV (panel *A*). The *m* gates are closed, however, and therefore the permeability of the resting cell membrane to Na$^+$ is very low. Hence virtually no Na$^+$ moves into the cell; that is, the *inward Na$^+$ current* is negligible.

Any process that tends to make V$_m$ less negative tends to open the *m* gates, thereby "activating" the fast Na$^+$ channels. The activation of the fast channels is therefore called a *voltage-dependent* phenomenon. The potential at which the *m* gates swing open varies somewhat from channel to channel in the cell membrane. As V$_m$ becomes progressively less negative, therefore, more and more *m* gates will open. As the *m* gates open, Na$^+$ enters the cell (Fig. 27-7, *B*), by virtue of the chemical and electrostatic forces referred to before.

The entry of positively charged Na$^+$ into the interior of the cell tends to neutralize some of the negative charges inside the cell and thereby tends to diminish further the transmembrane potential, V$_m$. The resultant reduction in V$_m$, in turn, tends to open more *m* gates, thereby producing a still greater increase in the inward Na$^+$ current. Hence this is called a *regenerative process*.

As V$_m$ approaches the threshold value of about -65 mV, the remaining *m* gates rapidly swing open in the fast Na$^+$ channels, until virtually all of the *m* gates are open (Fig. 27-7, *B*). There is a rapid inrush of Na$^+$, which produces an abrupt reduction of V$_m$. This accounts for the rapid rate of change of V$_m$ during phase 0 of the action potential (Fig. 27-1). The maximum rate of change of V$_m$ (that is, the maximum dV$_m$/dt) has been found to be from 100 to 200 V/second in myocardial cells and from 500 to 1000 V/second in Purkinje fibers. Although the quantity of Na$^+$ that enters the cell during one action potential alters V$_m$ by over 100 mV, it is too small to change the intracellular Na$^+$ concentration measurably. Therefore the chemical force remains virtually constant, and only the electrostatic force changes throughout the action potential. Hence the lengths of the black arrows in Fig. 27-7 remain constant at 60 mV, whereas the white arrows change in magnitude and direction.

As the Na$^+$ rushes into the cardiac cell during phase 0, the negative charges inside the cell are neutralized, and V$_m$ becomes progressively less negative. When V$_m$ becomes zero (Fig. 27-7, *C*), there is no longer an electrostatic force pulling Na$^+$ into the cell. As long as the fast Na$^+$ channels are open, however, Na$^+$ continues to enter the cell because of the large concentration gradient that still exists. This continuation of the inward Na$^+$ current causes the inside of the cell to become postively charged with respect to the exterior of the cell (Fig. 27-7, *D*). This reversal of the membrane polarity is the so-called overshoot of the cardiac action potential, which is evident in Fig. 27-1. Such a reversal of the electrostatic gradient would, of course, tend to repel the entry of Na$^+$ (Fig. 27-7, *D*). However, as long as the inwardly directed chemical forces exceed these outwardly directed electrostatic forces, the net flux of Na$^+$ will still be inward, although the rate of influx will be diminished. The inward Na$^+$ current finally ceases when the *h* (inactivation) gates close (Fig. 27-7, *E*).

The activity of the *h* gates is governed by the value of V$_m$ just as is that of the *m* gates. However, whereas the *m* gates tend to open as V$_m$ becomes less negative, the *h* gates tend to close under this same influence. Furthermore, the opening of the *m* gates occurs very rapidly (in about 0.1 to 0.2 msec), whereas the closure of the *h* gates is a relatively slow process, requiring 1 msec or more. Phase 0 is finally terminated when the *h* gates have closed and have thereby "inactivated" the fast Na$^+$ channels.

The *h* gates then remain closed until the cell has partially repolarized during phase 3 (at about point *D* in Fig. 27-1). Until these gates do reopen partially, the cell is refractory to further excitation. This mechanism therefore prevents a sustained, tetanic contraction of cardiac muscle, which would of course be inimical to the intermittent pumping action of the heart.

In cardiac cells that have a prominent plateau and especially in Purkinje fibers, phase 1 constitutes an early, brief period of limited repolarization between the end of the upstroke and the beginning of the plateau. This early phase of repolarization has been ascribed to a transient, inward Cl^- current, although recent work has tended to discredit this explanation.

During the plateau (phase 2) of the action potential, there is a weak flow of Ca^{++} and Na^+ into the cell through *slow channels,* which appear to be entirely different from the fast Na^+ channels that are operating during phase 0. The activation, inactivation, and recovery processes are much slower for the slow than for the fast channels. The fast channels may be blocked by tetrodotoxin, whereas the slow channels may be blocked by Mn^{++} or verapamil, agents that are known to impede the movement of Ca^{++} into the cell. The slow channels are activated when V_m reaches their threshold voltage of about -30 to -40 mV. Activation probably represents the opening of electrostatic gates in the slow channels, similar to the process that occurs in the fast channels. Other gates in the slow channels slowly begin to close, thereby initiating the process of inactivation, which helps terminate the plateau.

The slow inward current is increased by catecholamines, such as epinephrine and norepinephrine. This is probably one of the principal mechanisms whereby the catecholamines enhance the contractile process in cardiac muscle. The Ca^{++} that enters the myocardial cell during the plateau is involved in excitation-contraction coupling (Chapter 22.) Slow channel blocking drugs are now widely used in the treatment of various cardiac disorders.

During the plateau of the action potential the concentration gradient for K^+ between the inside and outside of the cell is virtually the same as it is during phase 4, but V_m is close to 0 mV. Therefore the chemical forces acting on K^+ greatly exceed the electrostatic forces during the plateau, and K^+ tends to diffuse out of the cell. The efflux of this positively charged ion would tend to make the interior of the cell membrane more negative; that is, it would tend to repolarize the cell membrane, thereby terminating the plateau. In nerve fibers, g_K increases when the neuron is depolarized, and the resultant outward current of K^+ promotes rapid repolarization. In cardiac cells, conversely, the permeability of the cell membrane to K^+ in the outward direction diminishes during phase 2, although the conductance in the inward direction is considerably greater. This unidirectional decrease in g_K during the plateau has been called *anomalous rectification.* As a consequence of the reduction in g_K in the outward direction, there is only a small outward current of K^+ during the plateau. It tends to balance the slow inward currents of Ca^{++} and Na^+ and thereby contributes to the maintenance of a prolonged plateau at a level of V_m close to 0 mV. The roles of both the slow inward Ca^{++} and Na^+ currents and of the reduction in g_K in the production of the plateau have been demonstrated by the administration of verapamil. If the slow inward currents are completely blocked with this agent, a plateau still exists, but it occurs at more negative voltages than when the slow channels are not blocked.

The process of final repolarization (phase 3) appears to depend on two principal processes, namely (1) an increase in g_K and (2) inactivation of the slow inward Ca^{++} and Na^+ currents. The increase in g_K may be induced by the elevation in intracellular Ca^{++}, consequent to the inward Ca^{++} current during the plateau. The enhancement of g_K leads to an efflux of K^+ from the cell. The outward current of K^+ is no longer balanced by the slow inward currents of Ca^{++} and Na^+. This efflux of positive K ions therefore causes the charge on the inside of the cell membrane to become progressively more negative. The increase in g_K is voltage dependent; as the inside of the cell becomes more negative, g_K increases and the outward flux of K^+ is accelerated. Hence this rapid phase of repolarization (phase 3) can be considered to be a *regenerative process,* just as is the inward current of Na^+ during phase 0. The efflux of K^+ during phase 3 rapidly restores the resting level of membrane potential. During the subsequent rest period (phase 4), and probably throughout the action potential as well, the active membrane pump eliminates the excess Na^+ that had entered the cell principally during phases 0 and 2, in exchange for the K^+ that had exited chiefly during phases 2 and 3.

■ Ionic basis of the slow response

Fast response action potentials may be considered to consist of two principal components, a spike (phases 0 and 1) and a plateau (phase 2). In the slow response the first component is absent or inoperative, and the second component accounts for the entire action potential. In the fast response the spike is produced by the inward Na^+ current through the fast channels. These channels can be blocked by certain interventions, such as the administration of tetrodotoxin. When the fast Na^+ channels are blocked, slow responses may be generated in the same fibers under appropriate conditions.

The Purkinje fiber action potentials shown in Fig. 27-8 clearly exhibit the two components. In the control tracing (*A*) a prominent notch separates the spike from the plateau. In *B* to *E,* progressively larger quantities of tetrodotoxin were added to the bathing solution to produce a graded blockade of a larger and larger fraction of the fast Na^+ channels. It is evident that the spike becomes progressively less prominent in *B* to *D,* and it disappears entirely in *E.* Thus the tetrodotoxin had a pronounced effect on the spike, and only a negligible influence on the plateau. With elimination of the spike (*E*), the action potential resembles a typical slow response.

Certain cells in the heart, notably those in the SA and AV nodes, are normally slow response fibers. In such fibers, depolarization is achieved by the inward currents of Ca^{++} and Na^+ through the slow channels. These ionic events closely resemble those that occur during the second component of fast response action potentials. The slow channels in nodal cells can be blocked by Mn^{++} or verapamil, just as in fast response fibers.

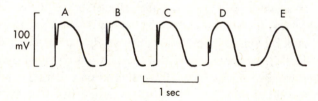

Fig. 27-8 ■ Effect of tetrodotoxin on the action potential recorded in a calf Purkinje fiber perfused with a solution containing epinephrine and 10.8mM K^+. The concentration of tetrodotoxin was 0M in *A*, 3×10^{-8}M in *B*, 3×10^{-7}M in *C*, and 3×10^{-6} M in *D* and *E; E* was recorded later than *D*. (Redrawn from Carmeliet, E., and Vereecke, J.: Pflügers Arch. **313:**300, 1969.)

■ Conduction in cardiac fibers

An action potential traveling down a cardiac muscle fiber is propagated by local circuit currents, similar to the process that occurs in nerve and skeletal muscle fibers (Chapter 3). In Fig. 27-9, consider that the left half of the cardiac fiber has already been depolarized, whereas the right half is still in the resting state. The fluids normally in contact with the external and internal surfaces of the membrane are essentially solutions of electrolytes and thus are good conductors of electricity. Hence current (in the abstract sense) will flow from regions of higher to those of lower potential, denoted by the plus and minus signs, respectively. In the external fluid, current will flow from right to left between the active and resting zones, and it will flow in the reverse direction intracellularly. In electrolyte solutions, the true current is carried by a movement of cations in one direction and anions in the opposite direction. At the cell exterior, for example, cations will flow from right to left, and anions from left to right (Fig. 27-9). In the cell interior the opposite migrations will occur. These local currents at the border

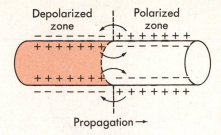

Fig. 27-9 ■ The role of local currents in the propagation of a wave of excitation down a cardiac fiber.

between the depolarized and polarized sections of the fiber will tend to depolarize the region of the resting fiber adjacent to the border.

■ *Conduction of the fast response*

In the fast response the fast Na^+ channels will be activated when the transmembrane potential is suddenly brought to the threshold value of about -70 mV. The inward Na^+ current will then depolarize the cell very rapidly at that site. This portion of the fiber will become part of the depolarized zone, and the border will be displaced accordingly (to the right in Fig. 27-9). The same process will then begin at the new border. This process will be repeated over and over and the border will move continuously down the fiber as a wave of depolarization.

At any given point on the fiber, the greater the amplitude of the action potential and the greater the rate of change of potential (dV_m/dt) during phase 0, the more rapid the conduction down the fiber. The amplitude of the action potential equals the difference in potential between the fully depolarized and the fully polarized regions of the cell interior (Fig. 27-9). The magnitude of the local currents is proportional to this potential difference. Since these local currents shift the potential of the resting zone toward the threshold value, they are the local stimuli that depolarize the adjacent resting portion of the fiber to its threshold potential. The greater the potential difference between the depolarized and polarized regions (that is, the greater the amplitude of the action potential), the more efficacious the local stimuli, and the more rapidly the wave of depolarization is propagated down the fiber.

The rate of change of potential (dV_m/dt) during phase 0 is also an important determinant of the conduction velocity. The reason can be appreciated by referring again to Fig. 27-9. If the active portion of the fiber depolarizes very gradually, the local currents across the border between the depolarized and polarized regions would be very small. Thus the resting region adjacent to the active zone would be depolarized very slowly, and consequently each new section of the fiber would require more time to reach threshold.

The level of the resting membrane potential is also an important determinant of conduction velocity. This factor operates through its influence on the amplitude and maximum slope of the action potential. The level of the resting potential may vary for a variety of reasons: (1) it can be altered experimentally by varying $[K^+]_o$ (Fig. 27-5); (2) in cardiac fibers that are intrinsically automatic, V_m becomes progressively less negative during phase 4 (Fig. 27-11, *B*); (3) during a premature contraction, repolarization may not have been completed before the beginning of the next excitation (Fig. 27-10). In general, the less negative the level of V_m, the less the velocity of impulse propagation, regardless of the reason for the alteration of the level of V_m.

The reason for the effect of the level of V_m on conduction velocity resides in the fact that the inactivation, or *h*, gates (Fig. 27-7) in the fast Na^+ channels are voltage dependent. The less negative the value of V_m, the greater the number of *h* gates that tend to close. During the normal process of excitation, depolarization proceeds so rap-

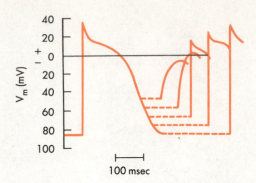

Fig. 27-10 ■ The changes in action potential amplitude and slope of the upstroke as action potentials are initiated at different stages of the relative refractory period of the preceding excitation. (Redrawn from Rosen, M.R., Wit, A.L., and Hoffman, B.F.: Am. Heart J. **88**:380, 1974.)

idly during phase 0 that the comparatively slow h gates do not close until the end of that phase. If partial depolarization is produced by a more gradual process, however, such as by elevating the level of external K^+, then the h gates do have ample time to close. When the cell is in a partial state of depolarization, therefore, many of the fast Na^+ channels will already be inactivated; when this situation exists, only a fraction of these channels will be available to conduct the inward Na^+ current during phase 0.

■ Conduction of the slow response

Local circuits are also responsible for propagation of the slow response (Fig. 27-9). However, the characteristics of the conduction process are entirely different from those of the fast response. The threshold potential is between -45 and -35 mV for the slow response, and the conduction velocity is of course very much less than for the fast response. Slow responses are more likely to be blocked than are fast responses. Also, the former are not able to be conducted at as rapid repetition rates. Fast responses are easily conducted in either an antegrade or a retrograde direction. The velocity of conduction is virtually the same in both directions. The velocity of conduction of the slow response in one direction may be much greater than in the opposite direction. Not uncommonly, the slow response will be conducted in one direction only, but it will be blocked in the opposite direction. This condition of *undirectional* block is a sine qua non for reentry. This is the basis for many arrhythmias.

■ *Cardiac excitability*

The excitability characteristics of cardiac cells depends on whether the action potentials are fast or slow responses.

■ Fast response

Once the fast response has been initiated, the depolarized cell will no longer be excitable until about the middle of the period of final repolarization (phase 3). The interval from the beginning of the action potential until the fiber is able to conduct another action potential is called the *effective refractory period*. In the fast response this period extends from the beginning of phase 0 to a point in phase 3 where repolarization has proceeded to a value of about -50 mV (period C to D in Fig. 27-1). It is at about this value of V_m that the electrochemical gates (m and h) for some of the fast Na^+ channels have been reset.

Full excitability is not regained until the cardiac fiber has been fully repolarized (point E in Fig. 27-1). During period D to E in the figure an action potential may be evoked, but only with a stimulus that is stronger than that which would be capable of eliciting a response during phase 4. Period D to E is called the *relative refractory period*.

The characteristics of a fast response evoked during the relative refractory period of an antecedent excitation vary with the membrane potential that exists at the time of stimulation. The nature of this voltage dependency is illustrated in Fig. 27-10. It is evident that as the fiber is stimulated later and later in the relative refractory period, the amplitude of the response and the rate of rise of the upstroke increase progressively. Presumably, the number of fast Na^+ channels that have recovered from inactivation increase as repolarization proceeds during phase 3. As a consequence of the increase in amplitude and slope of the evoked response, the propagation velocity becomes greater the later the cell is stimulated in the relative refractory period. Once the fiber is fully repolarized, the response is constant at no matter what time in phase 4 the stimulus is applied. By the end of phase 3, the m and h gates of all channels are in their final positions, and therefore there is no further change in excitability with time.

■ *Slow response*

The effective refractory period during the slow response frequently extends well beyond phase 3. Even after the cell has completely repolarized, it may not be possible to evoke a propagated response for some time. The relative refractory period then extends into phase 4, during which V_m is virtually constant in nonpacemaker cells. During the relative refractory period there is a progressive recovery of excitability despite the presence of this constant level of V_m. The recovery of full excitability usually requires considerably more time than for the fast response. The process might involve a few seconds, as compared with a few tenths of a second for recovery in the fast response. Until full recovery of excitability is achieved, conduction velocity varies with excitability. Impulses arriving at a slow response fiber early in its relative refractory period are conducted much more slowly than those arriving late in that period. The lengthy refractory periods also account for the observed tendency toward conduction blocks. Even when slow responses recur at a relatively low repetition rate, the fiber may be able to conduct only a fraction of those impulses.

■ *Natural excitation of the heart*

The properties of *automaticity* (the ability to initiate its own beat) and of *rhythmicity* (the regularity of such pacemaking activity) are intrinsic to cardiac tissue. The heart will continue to beat even when it is completely removed from the body. If the coronary vasculature is artificially perfused, rhythmic cardiac contraction will persist for considerable periods of time. Apparently, at least some cells in the walls of all four cardiac chambers are capable of initiating beats; such cells probably reside in the nodal tissues or specialized conducting fibers of the heart. The region of the mammalian heart that ordinarily displays the highest order of rhythmicity is the *sinoatrial*, or SA, node; it is called the *natural pacemaker* of the heart. Other regions of the heart that initiate beats under special circumstances are called *ectopic foci* or *ectopic pacemakers*. Ectopic foci may become pacemakers when (1) their own rhythmicity becomes enhanced, (2) the rhythmicity of the higher order pacemakers becomes depressed, or (3) all conduction pathways between the ectopic focus and those regions with a higher degree of rhythmicity become blocked.

When the SA node is suddenly excised or destroyed, pacemaker cells in the AV junction usually have the next highest order of rhythmicity, and they become the pacemakers for the entire heart. After some time, which may vary from minutes to days, automatic cells in the atria usually become dominant. Purkinje fibers in the specialized conduction system of the ventricles also possess the property of automaticity. Characteristically, they fire at a frequency of only 30 to 40 beats per minute.

■ *Sinoatrial node*

The SA node, which is the phylogenetic remnant of the sinus venosus of lower vertebrate hearts, was first described for mammalian hearts by Keith and Flack in 1906. In humans it is a crescent-shaped structure, approximately 15 mm long, 5 mm wide, and 2 mm thick. It lies in the sulcus terminalis on the posterior aspect of the heart, where the superior vena cava joins the right atrium.

A typical transmembrane action potential recorded from a cell in the SA node is depicted in Fig. 27-11, *B*. In comparison with the action potential recorded from a ventricular myocardial cell (Fig. 27-11, *A*), the resting potential of the SA nodal cell is usually less, the upstroke of the action potential (phase 0) has a much slower velocity, a plateau is absent, and repolarization (phase 3) is more gradual. These are all characteristic of the slow response. Under ordinary conditions tetrodotoxin has no influence on the SA nodal action potential. This indicates that the upstroke of the action potential is not produced by an inward current of Na^+ through the fast channels. However, the principal distinguishing feature of a pacemaker fiber resides in phase 4. In nonautomatic cells the potential remains constant during this phase, whereas in a pacemaker fiber there is a slow, steady depolarization, called the *pacemaker potential*. Depolarization proceeds at a steady rate during phase 4 until a threshold is attained, and then an action potential is triggered.

The frequency of discharge of pacemaker cells may be varied by a change in (1) the rate of depolarization during phase 4, (2) the threshold potential, or (3) the resting potential (Fig. 27-12). With an increase in the rate of depolarization (*b* to *a* in Fig. 27-12, *A*) the threshold potential will be attained earlier, and heart rate will increase. A rise in the threshold potential (from *TP-1* to *TP-2* in Fig. 27-12, *B*) will delay the onset of phase 0 (from time *b* to time *c*), and heart rate will be reduced accordingly. Similarly, when the magnitude of the resting potential is increased (from *a* to *d*), more time will be required to reach threshold *TP-2* when the slope of phase 4 remains unchanged, and the heart rate will diminish.

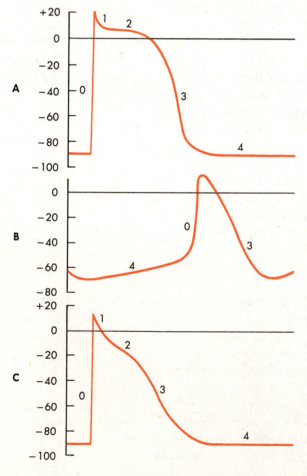

Fig. 27-11 ■ Typical action potentials (in millivolts) recorded from cells in the ventricle (**A**), sinoatrial node (**B**), and atrium, (**C**). Sweep velocity in **B** is half that in **A** or **C**. (Redrawn from Hoffman, B.F., and Cranefield, P.F.: Electrophysiology of the heart, New York, 1960, McGraw-Hill Book Co., Inc. Copyright © 1960 by McGraw-Hill Book Co. Used with permission of McGraw-Hill Book Co.)

The frequency of pacemaker firing is controlled by the activity of both divisions of the autonomic nervous system. Increased sympathetic nervous activity, through the release of norepinephrine, raises the heart rate principally by increasing the slope of the pacemaker potential. Increased vagal activity, through the release of acetylcholine, diminishes the heart rate by hyperpolarizing the pacemaker cell membrane and by reducing the slope of the pacemaker potential. This hyperpolarization is ascribable to a significant increase in g_K evoked by the action of acetylcholine on the pacemaker cell membrane. Fig. 27-13 shows the changes in transmembrane potential in an SA nodal cell in response to a brief vagal stimulus (at the arrow). The immediately ensuing hyperpolarization caused the cardiac cycle length to increase to 1250 msec from a basic cycle length of 700 msec. The next several cycles were also lengthened. However, these longer cycles were associated with a decreased slope of the pacemaker potential, but no significant hyperpolarization. Vagal stimulation frequently also evokes a *pacemaker shift,* wherein the true pacemaker cells are inhibited more than some of the latent pacemakers within the node, and these then become the true pacemakers.

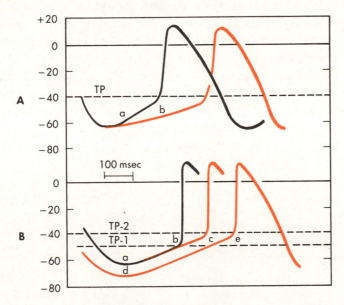

Fig. 27-12 ■ Mechanisms involved in changes of frequency of pacemaker firing. In **A** a reduction in the slope of the pacemaker potential from *a* to *b* will diminish the frequency. In **B** an increase in the threshold (from *TP-1* to *TP-2*) or an increase in the magnitude of the resting potential (from *a* to *d*) will also diminish the frequency. (Redrawn from Hoffman, B.F., and Cranefield, P.F.: Electrophysiology of the heart, New York, 1960, McGraw-Hill Book Co., Inc. Copyright © 1960 by McGraw-Hill Book Co. Used with permission of McGraw-Hill Book Co.)

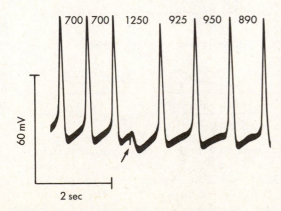

Fig. 27-13 ■ Effect of a brief vagal stimulus *(arrow)* on the transmembrane potential recorded from an SA nodal pacemaker cell in an isolated cat atrium preparation. The cardiac cycle lengths, in milliseconds, are denoted by the numbers at the top of the figure. (Modified from Jalife, J., and Moe, G.K.: Circ. Res. **45**:595, 1979. By permission of the American Heart Association, Inc.)

■ *Ionic basis of automaticity*

On the basis of the chord conductance equation (p. 445), the gradual phase 4 depolarization characteristic of pacemaker cells could be accounted for either by a progressive increase in g_{Na} or by a progressive reduction in g_K. Measurement of the membrane resistance of automatic Purkinje fibers and certain other types of pacemaker cells has disclosed a gradual increase in resistance during phase 4. This suggests that a progressive reduction in g_K is more likely than an increase in g_{Na} as the mechanism of the pacemaker potential in such cells. As a consequence of the changing g_K, there is a progressive reduction in outward K^+ current during phase 4. Also, there is a small, but steady, inward Na^+ current, reflecting the concentration and electrostatic gradients for this ion (Fig. 27-7, *A*). The imbalance between the steady inward Na^+ current and the gradually diminishing outward K^+ current produces the slow diastolic depolarization that is characteristic of such automatic cells. Because of these two opposing currents during phase 4, raising $[K^+]_o$ or lowering $[Na^+]_o$ diminishes the firing frequency of automatic Purkinje fibers.

Considerably less is known about the ionic mechanisms of automaticity in nodal cells, including the pacemaker cells in the SA node itself. There is a progressive reduction in K^+ conductance that accompanies the slow diastolic depolarization during phase 4, just as in automatic Purkinje fibers. However, lowering $[Na^+]_o$ or raising $[K^+]_o$ has relatively little influence on such cells, in contrast to the pronounced effects of such changes on automatic Purkinje fibers. Instead, changes in the external Ca^{++} concentration (Fig. 27-14) and the addition of slow channel blocking agents, such as verapamil (Fig. 27-15), decrease the firing rate and alter the characteristics of the action potentials of SA nodal pacemaker cells. Hence a steady inward Ca^{++} current is probably responsible for the pacemaker potential in these cells.

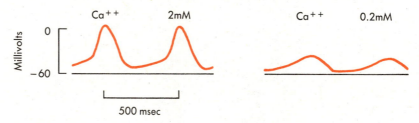

Fig. 27-14 ■ Transmembrane action potentials recorded from an SA nodal pacemaker cell in an isolated rabbit atrium preparation. The concentration of Ca^{++} in the bath was changed from 2mM to 0.2mM. (Modified from Kohlhardt, M., Figulla, H.-R., and Tripathi, O.: Basic Res. Cardiol. **71**:17, 1976.)

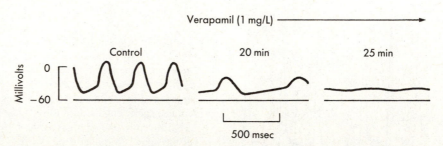

Fig. 27-15 ■ Transmembrane action potentials recorded from an SA nodal pacemaker cell in an isolated rabbit atrium preparation. After a control record was obtained, verapamil (1 mg/L) was added to the tissue bath, and records were taken 20 and 25 minutes later. (Modified from Kohlhardt, M., Figulla, H.-R., and Tripathi, O.: Basic Res. Cardiol. **71**:17, 1976.)

The automaticity of pacemaker cells becomes depressed after a period of excitation at a frequency greater than their intrinsic firing rate. This phenomenon is known as *overdrive suppression*. Because of the greater intrinsic rhythmicity of the SA node than of the other latent pacemaking sites in the heart, the firing of the SA node tends to suppress the automaticity in the other loci. If an ectopic focus in one of the atria suddenly began to fire at a rate of 150 impulses per minute in an individual with a normal heart rate of 70 beats per minute, the ectopic center would become the pacemaker for the entire heart. When that rapid ectopic focus suddenly stopped firing, the SA node might remain quiescent briefly, by virtue of overdrive suppression.

■ *Overdrive suppression*

From the SA node the cardiac impulse spreads radially throughout the right atrium (Fig. 27-16) along ordinary atrial myocardial fibers, at a conduction velocity of approximately 1 m/second. The configuration of the atrial transmembrane potential is depicted in Fig. 27-11. There is a special pathway, the *anterior interatrial myocardial band* (or *Bachmann's bundle*), which conducts the impulse from the SA node directly to the left atrium. There are also three tracts, the *anterior, middle,* and *posterior internodal pathways,* which conduct the cardiac impulse directly from the SA to the AV node; of these, the anterior pathway is probably the most important. These pathways consist of a mixture of ordinary myocardial cells and of specialized conducting fibers similar to those that exist in the ventricles.

■ *Atrial conduction*

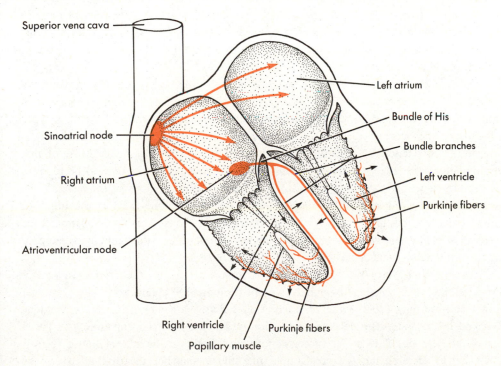

Superior vena cava

Sinoatrial node

Right atrium

Atrioventricular node

Left atrium

Bundle of His

Bundle branches

Left ventricle

Purkinje fibers

Right ventricle   Purkinje fibers

Papillary muscle

Fig. 27-16 ■ Schematic representation of the conduction system of the heart.

The cardiac action potential proceeds along the internodal pathways in the atrium and ultimately reaches the AV node. This node (described by Tawara in 1906) is approximately 22 mm long, 10 mm wide, and 3 mm thick. It is situated posteriorly on the right side of the interatrial septum near the ostium of the coronary sinus.

The AV node has been divided into three functional regions: (1) the AN region, the transitional zone between the atrium and the remainder of the node; (2) the N region, the midportion of the AV node; and (3) the NH region, the zone in which the nodal

■ *Atrioventricular
conduction*

fibers gradually merge with the *bundle of His,* which is the upper portion of the specialized conducting system for the ventricles. Normally, the AV node and bundle of His constitute the only pathways for conduction from atria to ventricles.

Several features of atrioventricular conduction are of physiological and clinical significance. It is in the AN region of the AV node that the principal delay occurs in the passage of the impulse from the atria to the ventricular myocardial cells. The conduction velocity is actually less in the N region than in the AN region. However, the path length is substantially greater in the latter than in the former zone, which accounts for the difference in the total conduction time through the two regions. The conduction times through the AN and N zones account for a considerable fraction of the time interval between the onsets of the atrial and ventricular systoles and between the onsets of the *P wave* (the electrical manifestation of the spread of atrial excitation) and the *QRS complex* (spread of ventricular excitation) in the electrocardiogram. The period of time between the initiation of the P wave and the QRS complex is called the *PR interval* (see Fig. 27-21). Functionally, this delay between atrial and ventricular excitation permits optimal ventricular filling during atrial contraction.

In the N region the action potentials display many of the characteristics of the slow response. The resting potential is about -50 to -60 mV, the upstroke velocity is very low (about 5 V/second), and the conduction velocity is about 0.05 m/second. Tetrodotoxin, which blocks the fast Na^+ channels, has virtually no effect on the action potentials in this region. Conversely, Mn^{++} and verapamil, which are slow channel blocking agents, have a depressant effect on AV conduction. The action potentials in the AN region are intermediate in configuration between those in the N region and atria. Similarly, the action potentials in the NH region are transitional between those in the N region and His bundle.

The relative refractory period of the cells in the N region extends well beyond the period of complete repolarization. Such refractoriness is said to be *time dependent,* in contrast to the *voltage-dependent* refractoriness that characterizes most of the other types of cells in the heart (Fig. 27-10). As the repetition rate of atrial depolarizations is increased, conduction through the AV junction tends to be prolonged. Most of that prolongation takes place in the N region of the AV node. Impulses tend to be blocked at stimulus repetition rates that are easily conducted in other regions of the heart. If the atria are depolarized at a high frequency, only one half or one third of the impulses might be conducted through the AV junction to the ventricles. This tends to protect the ventricles from excessive contraction frequencies, wherein the filling time between contractions might be inadequate.

The autonomic nervous system plays an important role in the regulation of AV conduction. Weak vagal activity may simply prolong AV conduction time. Stronger vagal activity may cause some or all of the impulses arriving from the atria to be blocked in the node. The delayed conduction or block tends to occur largely in the N region of the node. The cardiac sympathetic nerves, however, have a facilitatory effect. They act to decrease the AV conduction time and to enhance the rhythmicity of the latent pacemakers in the AV junction.

■ *Ventricular conduction*

The bundle of His (described by His in 1893) passes subendocardially down the right side of the interventricular septum for approximately 12 mm and then divides into the right and left *bundle branches* (Fig. 27-16). The right bundle branch is a direct continuation of the bundle of His and proceeds down the right side of the interventricular septum. The left bundle branch, which is considerably thicker than the right, arises almost perpendicularly from the bundle of His and perforates the interventricular septum. On the subendocardial surface of the left side of the interventricular septum the main left bundle branch splits into a thin *anterior division* and a thick *posterior division.*

The right bundle branch and the two divisions of the left bundle branch ultimately subdivide into a complex network of conducting fibers called *Purkinje fibers* (described by Purkinje in 1839), which ramify over the subendocardial surfaces of both ventricles. In certain mammalian species, such as cattle, the Purkinje fiber network is arranged in discrete, encapsulated bundles.

The conduction velocity for propagation of the action potential over the Purkinje fiber system is the fastest of any tissue within the heart; estimates vary from 1 to 4 m/second. This permits a rapid activation of all regions of the endocardial surface of the ventricles.

The configuration of the action potentials recorded from Purkinje fibers is quite similar to that from ordinary ventricular myocardial fibers (Fig. 27-11, *A*). In general, phase 1 is more prominent in Purkinje fiber action potentials than in those recorded from ventricular fibers, and the duration of the plateau (phase 2) is longer.

The intimate details of the spread of the action potential over the ventricles are of major concern in clinical cardiology. Numerous studies have been conducted to determine the precise course of the wave of excitation under normal and various abnormal conditions. Such knowledge serves as a basis for the interpretation of the electrocardiogram. However, only the elementary, salient features of ventricular activation will be considered here. The first portions of the ventricles to be excited are the interventricular septum (except the basal portion) and the papillary muscles. The wave of activation spreads into the substance of the septum from both its left and its right endocardial surfaces. Early contraction of the septum tends to make it more rigid and allows it to serve as an anchor point for the contraction of the remaining ventricular myocardium. Also, early contraction of the papillary muscles serves to prevent eversion of the AV valves during ventricular systole.

The endocardial surfaces of both ventricles are activated rapidly, but the wave of excitation spreads from endocardium to epicardium at a slower velocity (about 0.3 to 0.4 m/second). Because the right ventricular wall is appreciably thinner than the left, the epicardial surface of the right ventricle is activated earlier than that of the left ventricle. Also, apical and central epicardial regions of both ventricles are activated somewhat earlier than their respective basal regions. The last portions of the ventricles to be excited are the posterior basal epicardial regions and a small zone in the basal portion of the interventricular septum.

■ Reentry

Under appropriate conditions a cardiac impulse may reexcite some region through which it has passed previously. This phenomenon, known as *reentry,* is responsible for many clinical disturbances of cardiac rhythm. The conditions necessary for reentry are illustrated in Fig. 27-17. In each of the four panels a single bundle *(S)* of cardiac fibers is seen to divide into a left *(L)* and a right *(R)* branch. A connecting bundle *(C)* runs between the two branches.

Normally, the impulse coming down bundle *S* is conducted along the *L* and *R* branches (panel *A*). As the impulse reaches connecting link *C,* it enters from both sides and becomes extinguished at the point of collision. The impulse from the left side cannot proceed further because the tissue beyond is absolutely refractory, since it had just undergone depolarization from the other direction. The impulse cannot pass through bundle *C* from the right either, for the same reason.

It is obvious from panel *B* that the impulse cannot make a complete circuit if antegrade block exists in the two branches (*L* and *R*) of the fiber bundle. Furthermore, if bidirectional blocks exists at any point in the loop (for example, branch *R* in panel *C*), the impulse will not be able to reenter.

A necessary condition for reentry is that at some point in the loop the impulse is able to pass in one direction but not in the other. This phenomenon is called *unidirec-*

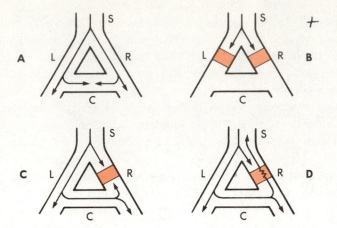

Fig. 27-17 ■ The role of unidirectional block in reentry. In panel *A* an excitation wave traveling down a single bundle *(S)* of fibers continues down the left *(L)* and right *(R)* branches. The depolarization wave enters the connecting branch *(C)* from both ends and is extinguished at the zone of collision. In panel *B* the wave is blocked in the *L* and *R* branches. In panel *C* bidirectional block exists in branch *R*. In panel *D* unidirectional block exists in branch *R*. The antegrade impulse is blocked, but the retrograde impulse is conducted through, and reenters bundle *S*.

tional block. As shown in panel *D*, the impulse may travel down branch *L* normally and may be blocked in the antegrade direction in branch *R*. The impulse that had been conducted down branch *L* and through the connecting branch *C* may be able to penetrate the depressed region in branch *R* from the retrograde direction, even though the antegrade impulse had been blocked previously at this same site. Such unidirectional block is commonly observed with slow response action potentials, and it is frequently a temporal phenomenon. It is evident from panel *D* that the antegrade impulse will arrive at the depressed region in branch *R* earlier than the impulse coming from the opposite direction. The antegrade impulse may be blocked simply because it happens to arrive at the depressed region during its effective refractory period. If the retrograde impulse is delayed sufficiently, the refractory period may have ended, and the impulse will be conducted back into bundle *S*.

Unidirectional block is a necessary condition for reentry, but not a sufficient one. It is also essential that the effective refractory period of the reentered region be less than the propagation time around the loop. In panel *D*, if the retrograde impulse is conducted through the depressed zone in branch *R* and if the tissue just beyond is still refractory from the antegrade depolarization, branch *S* will not be reexcited. Therefore the conditions that promote reentry are those that prolong conduction time or shorten the effective refractory period.

The functional components of reentry loops responsible for specific arrhythmias are protean. Some loops are very large and involve entire specialized conduction bundles. Others are microscopic in size. The loop may include myocardial fibers, specialized conducting fibers, nodal cells, and junctional tissues, in almost any conceivable arrangement. Also, the cardiac cells in the loop may be normal or deranged. Some of the important arrhythmias that occur on the basis of reentry are described at the end of this chapter.

■ *Basis of electrocardiography*

The electrocardiograph is a valuable instrument, for it enables the physician to infer the course of the cardiac impulse simply by recording the variations in electrical potential at various loci on the surface of the body. By analyzing the details of these potential fluctuations, the physician gains valuable insight concerning (1) the anatomical orientation of the heart, (2) the relative sizes of its chambers, (3) a variety of disturbances of rhythm and of conduction, (4) the extent, location, and progress of ischemic damage to the myocardium, (5) the effects of altered electrolyte concentrations, and (6) the influence of certain drugs (notably digitalis and its derivatives).

Electrocardiograms are usually recorded from *indirect leads* (recording electrodes located on the skin), which are at some distance from the heart, the source of potential. However, *direct leads* have been applied to the surface of the heart in experimental animals and in humans during thoracic surgical procedures. Electrocardiograms recorded by direct leads are discussed first because they are simpler to comprehend.

The principles are illustrated by describing the potential changes recorded directly from the surface of a single, long myocardial fiber. In Fig. 27-18 the changes in potential in one localized region (in contact with electrode A) will first be considered. Electrode A is connected to the lower vertical deflecting plate of a cathode ray oscilloscope; electrode C, far to the right of A, is connected to the upper deflecting plate. If the strip of muscle is stimulated at its left end, the action potential travels from left to right along the strip. Before excitation reaches region A, however, the external surface of the fiber at A is at the same potential as the surface at C, and no difference in potential is recorded between A and C (section *1*). When the action potential reaches region A, there

■ *Direct leads*

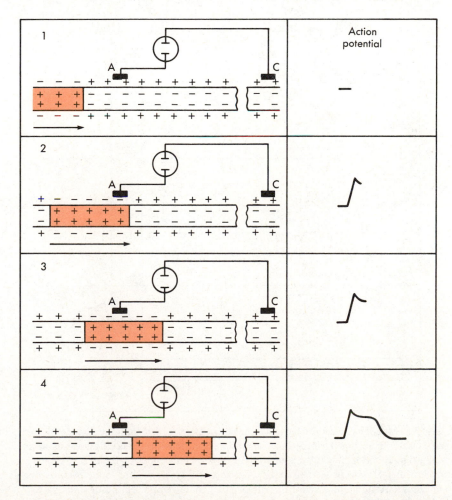

Fig. 27-18 ■ Sequential changes of potential recorded from an external electrode, *A*, as an action potential travels along a strip of cardiac muscle. Electrode *C* is located far to the right of *A;* the entire cycle of depolarization and repolarization is completed under *A* before the action potential reaches *C*.

is a rapid reversal of the transmembrane potential in that region, and the external surface at *A* becomes electronegative with respect to the surface at *C* (section *2*). This difference in potential persists as long as region *A* remains depolarized (section *3*). As region *A* repolarizes, the potential at *A* again becomes equal to the potential at *C* (section *4*). Repolarization proceeds much more slowly than depolarization; hence the slope of phase 3 is more gradual than that of phase 0.

Fig. 27-19 depicts a similar situation but illustrates the potential differences that would be recorded when the wave of activation is recorded simultaneously from two regions of the same fiber. In section *1* of Fig. 27-19 the lead locations and oscilloscope connections are identical to those represented in Fig. 27-18. Hence the recorded action potential is the same. Section *2* of Fig. 27-19 shows the record that is obtained under identical conditions, except that the potential difference between electrodes *B* and *C* is registered. In this case *B* is connected to the upper and *C* to the lower vertical deflecting plate. Relative to the action potential recorded in section *1* of Fig. 27-19, the record is inverted and is displaced in time, depending on the distance between electrodes *A* and *B* and on the propagation velocity of the action potential.

Finally, if electrodes *A* and *B* are connected to the lower and upper deflecting plates, respectively, then the potential difference recorded between them has the configuration shown at the right of section *3*. This action potential represents the algebraic sum of the potentials recorded separately under *A* and *B* (with respect to resting region *C*). The

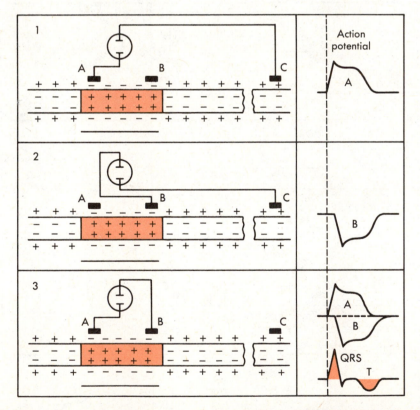

Fig. 27-19 ■ Monophasic and biphasic action potentials recorded from the surface of a strip of cardiac muscle. In section **1** electrodes *A* and *C* are connected to the oscilloscope as in the preceding figure, and the recorded action potential is therefore identical with that shown in Section **4** of Fig. 27-18. In section **2** electrodes *B* and *C* are connected to the oscilloscope as shown, and the resulting deflection, *B*, will be inverted and displaced in time relative to deflection *A* in section **1**. In section **3** electrodes *A* and *B* are connected to the oscilloscope, and the resulting QRS and T waves represent the algebraic sum of the individual deflections, *A* and *B*.

externally recorded action potential consists of an initial upright spike, the *R wave,* followed by a small downward deflection, the *S wave.* The second major wave, termed the *T wave,* is inverted and occurs during repolarization. In electrocardiograms the analogous deflection occurring during ventricular depolarization is often triphasic (see Fig. 27-21) and therefore is designated the *QRS complex.* The T wave is of lesser amplitude but of greater duration than the QRS complex because the rate of repolarization is slower than the rate of depolarization (as is evident in the monophasic tracings shown in sections *1* and *2*). The portion of the tracing between the end of the QRS complex and the start of the T wave is called the *ST segment.* During this interval the regions under both electrodes are depolarized. Since they are of equal negativity, the ST segment lies along the *isoelectric line,* or line of zero potential difference.

In most normal electrocardiograms the T wave is deflected in the same direction as the QRS complex from the isoelectric line, contrary to the oppositely directed waves seen in section *3* of Fig. 27-19. The explanation may be illustrated by subjecting a strip of cardiac muscle to a temperature gradient (temperature increasing from left to right), as shown in Fig. 27-20. If a stimulus is applied to the left end, then the wave of depolarization is, of course, propagated from left to right. However, because the right end of the strip is warmer than the left end, the repolarization process may actually proceed from right to left, that is, the reverse of the direction of propagation of the wave of depolarization. Stated in another way, the action potential duration under electrode *A* (near the cooler end) will exceed that under *B* (near the warmer end). Therefore the T wave will be deflected in the same direction as the QRS complex under such conditions, as shown in Fig. 27-20. Thus, when the spread of depolarization and of repolarization occur in the same direction, the T wave is inverted with respect to the QRS complex; when these processes proceed in opposite directions, the deflections occur in the same direction.

As stated on p. 459, the wave of depolarization in the ventricles proceeds from endocardium to epicardium. However, the wave of repolarization normally travels in the opposite direction across the ventricular walls, producing the concordant QRS and T deflections. Stated in another way, the duration of the action potentials in the subendocardial cells is greater than that in the subepicardial cells (analogous to action potentials *A* and *B* in section *2* of Fig. 27-20). It has been postulated that the pressure developed in the ventricular chambers during systole retards the repolarization process more in the subendocardial than in the subepicardial region.

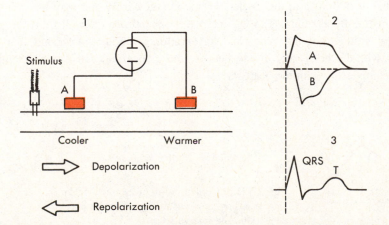

Fig. 27-20 ■ When the wave of depolarization progresses in one direction along a cardiac muscle strip and the wave of repolarization travels in the opposite direction, then the QRS and T waves will be deflected in the same direction relative to the isoelectric line.

■ *Scalar electrocardiography*

At any given moment in the cardiac cycle there is a complex pattern of electrical charges across the membranes of the myriad cells that comprise the heart. The charge at any given locus in the heart has a magnitude and a direction, that is, it is a vector quantity. The *resultant cardiac vector* is the sum of all the individual vectors that exist at any given time within the heart. The changes in the resultant cardiac vector throughout the cardiac cycle may be recorded as a *vectorcardiogram.*

More commonly, however, lead systems are used that record certain projections of the resultant cardiac vector. A system of leads oriented in a given plane detects only the projection of the three-dimensional vector on that plane. Furthermore, the potential difference between two recording electrodes represents the projection of the vector on the line between the two leads. Components of vectors projected on such lines are not vectors but are *scalar quantities* (having magnitude, but not direction). Hence a recording of the changes with time of the differences of potential between two points on the surface of the skin is called a *scalar electrocardiogram.*

Configuration of the scalar electrocardiogram. The scalar electrocardiograph detects the changes with time of the electrical potential between some point on the surface of the skin and an indifferent electrode or between pairs of points on the skin surface. The cardiac impulse progresses through the heart in an extremely complex three-dimensional pattern. Hence the precise configuration of the electrocardiogram varies from individual to individual, and in any given individual the pattern varies with the anatomical location of the leads.

In general, the pattern consists of P, QRS, and T waves (Fig. 27-21). The PR interval is a measure of the time from the onset of atrial activation to the onset of ventricular activation; it normally ranges from 0.12 to 0.20 second. A considerable fraction of this time involves passage of the impulse through the AV conduction system. Pathological prolongations of this interval are associated with disturbances of AV conduction produced by inflammatory, circulatory, pharmacological, or nervous mechanisms.

The configuration and amplitude of the QRS complex vary considerably among individuals. The duration is usually between 0.06 and 0.10 second. Abnormal prolongation may indicate a block in the normal conduction pathways through the ventricles (such as a block of the left or right bundle branch). The duration of the ST segment is of little clinical significance and is usually not measured in routine analyses of electrocardiograms. During this interval the entire ventricular myocardium is depolarized. Therefore the ST segment lies on the isoelectric line under normal conditions. Any appreciable deviation from the isoelectric line is noteworthy and may indicate ischemic damage of the myocardium. The QT interval is sometimes referred to as the period of "electrical systole" of the ventricles. Its duration is about 0.4 second, but it varies inversely with the heart rate, in part because the myocardial cell action potential duration varies inversely with the heart rate.

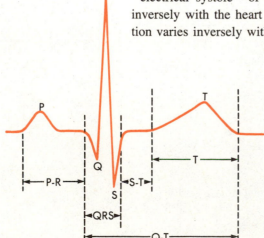

Fig. 27-21 ■ Configuration of a typical scalar electrocardiogram, illustrating the important deflections and intervals.

In most leads the T wave is deflected in the same direction from the isoelectric line as the major component of the QRS complex, although biphasic or oppositely directed T waves are perfectly normal in certain leads. When the T wave and QRS complex deviate in the same direction from the isoelectric line, it indicates that the repolarization process does not follow the same route as the depolarization process, as explained previously (Fig. 27-20). T waves that are abnormal either in direction or amplitude may indicate myocardial damage, electrolyte disturbances, or other abnormal conditions.

Standard limb leads. The original electrocardiographic lead system was devised by Willem Einthoven (1860-1927), who was professor of physiology at the University of Leiden. In his lead system the resultant cardiac vector was considered to lie in the center of a triangle (assumed to be equilateral) formed by the left and right shoulders and the pubic region (Fig. 27-22). This triangle, called the *Einthoven triangle,* is oriented in the frontal plane of the body. Hence only the projection of the resultant cardiac vector on the frontal plane will be detected by this system of leads. For convenience the electrodes are connected to the right and left forearms rather than to the corresponding shoulders, since the arms are considered to represent simple extensions of the leads from the shoulders; this assumption has been validated experimentally. Similarly, the leg (the left leg, by convention) is taken as an extension of the lead system from the pubis, and the third electrode is therefore connected to the left leg.

Certain conventions prevail in the manner in which these so-called *standard limb leads* are connected to the galvanometer. Lead I records the potential difference between the left arm (LA) and the right arm (RA). The galvanometer connections are such that when the potential at LA (V_{LA}) exceeds the potential at RA (V_{RA}), the galvanometer will be deflected upward from the isoelectric line. In Figs. 27-22 and 27-23 this arrangement of the galvanometer connections for lead I is designated by a (+) at LA and a (−) at RA. Lead II records the potential difference between RA and LL (left leg) and

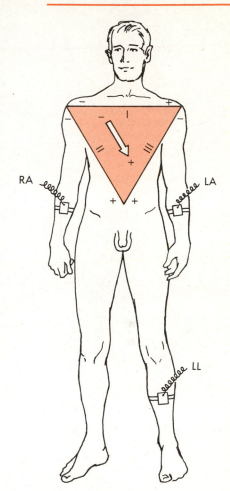

Fig. 27-22 ■ Einthoven's triangle, illustrating the galvanometer connections for standard limb leads I, II, and III.

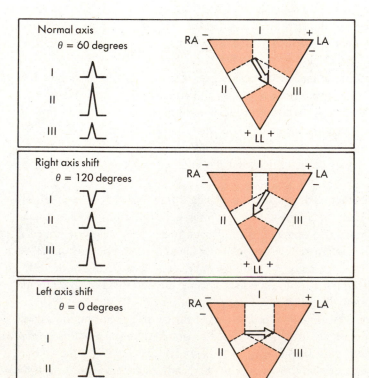

Fig. 27-23 ■ Magnitude and direction of the QRS complexes in limb leads I, II, and III, when the mean electrical axis (θ) is 60° *(top section)*, 120° *(middle section)*, and 0° *(bottom section)*.

yields an upward deflection when L_{LL} exceeds V_{RA}. Finally, lead III registers the potential difference between LA and LL and yields an upward deflection when V_{LL} exceeds V_{LA}. It will become evident from the following discussion that these galvanometer connections were arbitrarily chosen so that the QRS complexes will be upright in all three standard limb leads in the majority of normal individuals.

Let the frontal projection of the resultant cardiac vector at some moment in time be represented by an arrow (tail negative, head positive), as in Fig. 27-22. Then the potential difference, $V_{LA} - V_{RA}$, recorded in lead I will be represented by the component of the vector projected along the horizontal line between LA and RA, as shown in Fig. 27-23. If the vector makes an angle (θ) of 60° with the horizontal (as in the top section of Fig. 27-23), the magnitude of the potential recorded by lead I will equal the vector magnitude times cosine 60°. The deflection recorded in lead I will be upward, since the positive arrowhead lies closer to LA than to RA. The deflection in lead II will also be upright, since the arrowhead lies closer to LL than to RA. The magnitude of the lead II deflection will be greater than that in lead I, since in this example the direction of the vector parallels that of lead II; therefore the magnitude of the projection on lead II exceeds that on lead I. Similarly, in lead III the deflection will be upright, and in this example, where $\theta = 60°$, its magnitude will equal that in lead I.

If the vector in the top section of Fig. 27-23 happens to represent the resultant of the electrical events occurring during the peak of the QRS complex, then the orientation of this vector is said to represent the *mean electrical axis* of the heart in the frontal plane. The positive direction of this axis is taken in the clockwise direction from the horizontal (contrary to the usual mathematical convention). For normal individuals the average mean electrical axis is approximately + 60° (as in the top section of Fig. 27-23). Therefore the QRS complexes are usually upright in all three leads and largest in lead II.

Changes in the mean electrical axis may occur with alterations in the anatomical position of the heart or with changes in the relative preponderance of the right and left ventricles. For example, the axis tends to shift toward the left (more horizontal) in short, stocky individuals and toward the right (more vertical) in tall, thin persons. Also, with left or right ventricular hypertrophy (increased myocardial mass), the axis will shift toward the hypertrophied side.

With appreciable shift of the mean electrical axis to the right (middle section of Fig. 27-23, where $\theta = 120°$), the displacements of the QRS complexes in the standard leads will change considerably. In this case the largest upright deflection will be in lead III and the deflection in lead I will be inverted, since the arrowhead will be closer to RA than to LA. With left axis shift (bottom section of Fig. 27-23, where $\theta = 0°$), the largest upright deflection will be in lead I, and the QRS complex in lead III will be inverted.

As is evident from the above discussion, the standard limb leads, I, II, and III, are oriented in the frontal plane at 0°, 60°, and 120°, respectively, from the horizontal. Other limb leads, which are also oriented in the frontal plane, are usually recorded in addition to the standard leads. These leads (principally the unipolar limb leads of Wilson or the augmented unipolar limb leads of Goldberger) lie along axes at angles of +90°, −30°, and −150° from the horizontal. Such lead systems are described in all textbooks on electrocardiography and will not be considered further here.

To obtain information concerning the projections of the cardiac vector on the sagittal and transverse planes of the body in scalar electrocardiography, the so-called *precordial leads* are usually recorded. Most commonly, each of six selected points on the anterior and lateral surfaces of the chest in the vicinity of the heart is connected in turn to the galvanometer. The other galvanometer terminal is usually connected to a *central terminal,* which is composed of a junction of three leads from LA, RA, and LL, each in series with a 5000-ohm resistor. It can be shown that the voltage of this central terminal remains at a theoretical zero potential throughout the cardiac cycle.

The frequency of pacemaker discharge varies by the mechanisms described earlier in this chapter (Fig. 27-12). Changes in SA nodal discharge frequency are usually produced by the cardiac autonomic nerves. Examples of electrocardiograms of sinus tachycardia and sinus bradycardia are shown in Fig. 27-24. The P, QRS, and T deflections are all normal, but the duration of the cardiac cycle (the so-called *PP interval*) is altered. Characteristically, when sinus bradycardia or tachycardia occurs under natural conditions, the cardiac frequency changes gradually and requires several beats to attain its new steady-state value. Electrocardiographic evidence of *respiratory cardiac arrhythmia* is common and is manifested as a rhythmic variation in the PP interval at the respiratory frequency (Chapter 33).

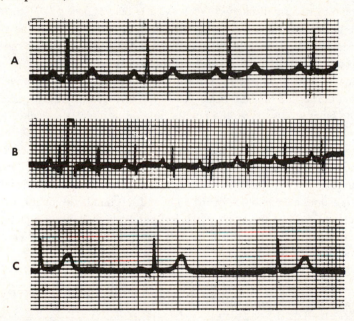

Fig. 27-24 ■ Sinoatrial rhythms. **A,** Normal sinus rhythm. **B,** Sinus tachycardia. **C,** Sinus bradycardia.

Various physiological, pharmacological, and pathological processes can impede impulse transmission through the atrioventricular conduction tissue. The site of block can now be localized more precisely by recording the *His bundle electrogram* (Fig. 27-25). To obtain such tracings, an electrode catheter is introduced into a peripheral vein and is threaded centrally until the tip containing the electrodes lies in the junctional region between the right atrium and ventricle. When the electrodes are properly positioned, a

Fig. 27-25 ■ His bundle electrogram (*lower tracing,* retouched) and lead II of the scalar electrocardiogram (*upper tracing*). The deflection, *H,* which represents the impulse conduction over the bundle of His, is clearly visible between the atrial, *A,* and ventricular, *V,* deflections. The conduction time from the atria to the bundle of His is denoted by the *AH* interval; that from the His bundle to the ventricles, by the *HV* interval. (Courtesy Dr. J. Edelstein.)

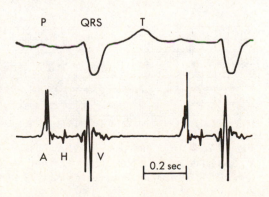

distinct deflection (Fig. 27-25, *H*) is registered, which represents the passage of the cardiac action potential down the His bundle. The time intervals required for propagation from the atrium to the His bundle (*AH interval*) and from the His bundle to the ventricles (*HV interval*) may be measured accurately. Abnormal prolongation of the former or latter interval indicates block above or below the His bundle, respectively.

Three degrees of AV block can be distinguished, as shown in Fig. 27-26. *First-degree AV block* is characterized by a prolonged PR interval. In Fig. 27-26, *A*, the PR interval is 0.28 second; an interval greater than 0.2 second is usually considered to be abnormal. In most cases of first-degree block the AH interval of the His bundle electrogram is prolonged, whereas the HV interval is normal. Hence the delay is located above the bundle of His, that is, in the AV node.

In *second-degree AV block* all QRS complexes are preceded by P waves, but not all P waves are followed by QRS complexes. The ratio of P waves to QRS complexes is usually the ratio of two small integers (such as, 2:1, 3:1, 3:2). Fig. 27-26, *B*, illustrates a typical 2:1 block. His bundle electrograms have demonstrated that the site of block may be above or below the bundle of His. When the block arises above the bundle, the H deflections are absent after the A waves of the blocked beats. When the block occurs below the bundle, the H deflection is readily apparent after each A wave, but the V wave is absent during the blocked beats. This type of block implies a graver prognosis than when the block exists above the bundle, and an artificial pacemaker is frequently required.

Third-degree AV block is often referred to as *complete heart block* because the impulse is unable to traverse the AV conduction pathway from atria to ventricles. His bundle electrograms reveal that the most common sites of block are distal to the His bundle; that is, simultaneous block of the right and left bundle branches, or of the right bundle branch and the two divisions of the left bundle branch. In complete heart block the atrial and ventricular rhythms are entirely independent. A classic example is dis-

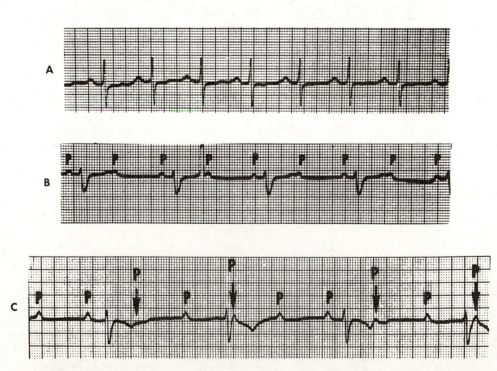

Fig. 27-26 ■ Atrioventricular blocks. **A,** First-degree heart block; PR interval is 0.28 second. **B,** Second-degree heart block (2:1). **C,** Third-degree heart block; note that there is a dissociation between the P waves and the QRS complexes.

played in Fig. 27-26, *C*, where the QRS complexes bear no fixed relationship to the P waves. Because of the slow ventricular rhythm (32 beats per minute in this example), circulation is often inadequate, especially during muscular activity. Third-degree block is often associated with syncope (so-called Stokes-Adams attacks) caused principally by insufficient cerebral blood flow. It is for such a condition that artificial pacemakers are most often installed to ensure a more nearly normal ventricular frequency.

■ *Premature systoles*

Premature contractions occur at times in most normal individuals but are more common under certain abnormal conditions. They may originate in the atria, AV junction, or ventricles (Fig. 27-27). Two principal mechanisms are believed to be responsible for premature beats. One type of premature beat is coupled to a normally conducted beat. If the normal beat is suppressed in some way (for example, by vagal stimulation), the premature beat will also be abolished. Such premature beats, called *extrasystoles,* probably reflect the *reentry* of the cardiac impulse around some slow conduction pathway, such that the impulse reenters previously activated tissue after it has regained its excitability.

The other type of premature beat occurs as the result of enhanced automaticity in some ectopic focus. This ectopic center may fire regularly and be protected in some way from depolarization by the normal cardiac impulse. If this premature beat occurs at a regular interval or at a simple multiple of that interval, the disturbance is called *parasystole*.

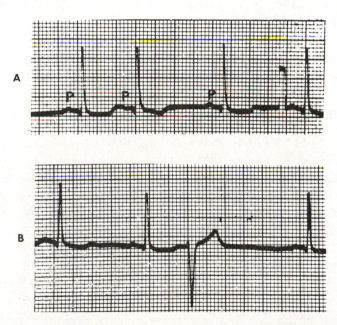

Fig. 27-27 ■ A premature atrial systole, **A,** and a premature ventricular systole, **B,** recorded from the same patient. The premature atrial systole (the second beat in the top tracing) is often characterized by an inverted P wave and usually normal QRS and T waves. The interval following the premature beat is not much longer than the usual interval between beats. The brief rectangular deflection just before the last beat is a standardization signal. The premature ventricular systole, **B,** is characterized by bizarre QRS and T waves and is followed by a pause.

■ *Ectopic tachycardias*

When a tachycardia originates from some ectopic site in the heart, the onset and termination are typically abrupt, as distinguished from the more gradual changes in heart rate in sinus tachycardia. Because of the sudden appearance and abrupt reversion to normal, such ectopic tachycardias are usually referred to as *paroxysmal tachycardias*. Episodes of ectopic tachycardia may persist for only a few beats or for many hours or

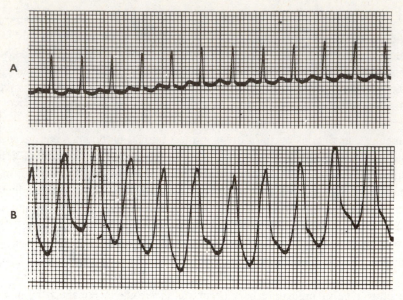

Fig. 27-28 ■ Paroxysmal supraventricular (**A**) and ventricular (**B**) tachycardia.

days, and the episodes are often recurrent. Paroxysmal tachycardias may occur either as the result of the rapid firing of an ectopic pacemaker or an impulse circling a reentry loop repetitively.

Paroxysmal tachycardias originating in the atria or in the AV junctional tissues are usually indistinguishable, and therefore both are included in the term *paroxysmal supraventricular tachycardia*. An electrocardiogram illustrating this arrhythmia is shown in Fig. 27-28, *A*. The QRS complexes are normal, since ventricular activation proceeds over the normal pathways. When the supraventricular frequency is excessively rapid, the AV conduction tissue may be incapable of conducting all impulses, and second-degree AV blocks (for example, 2:1 block) may be a concomitant of the paroxysmal tachycardia.

Paroxysmal ventricular tachycardia originates from an ectopic focus in the ventricles. The electrocardiogram is characterized by the rapid repeated, bizarre QRS complexes that reflect the aberrant intraventricular impulse conduction (Fig. 27-28, *B*). Paroxysmal ventricular tachycardia is much more ominous than supraventricular tachycardia because it is frequently a precursor of ventricular fibrillation, a lethal arrhythmia that will be described in the following section. The paroxysmal ventricular tachycardia illustrated in Fig. 27-28, *B*, was recorded from a patient immediately after resuscitation from ventricular fibrillation.

■ Fibrillation

Under certain conditions cardiac muscle undergoes an extremely irregular type of contraction that is entirely ineffectual in propelling blood. Such an arrhythmia is termed *fibrillation* and may involve either the atria or ventricles. Fibrillation probably represents a reentry phenomenon, in which the reentry loop fragments into multiple, irregular circuits.

The tracing in Fig. 27-29, *A*, illustrates the electrocardiographic changes in *atrial fibrillation*. In this condition, which occurs quite commonly in various types of chronic heart disease, the atria do not contract and relax sequentially during each cardiac cycle and hence do not contribute to ventricular filling. Instead, the atria undergo a continuous, uncoordinated, rippling type of activity. On the electrocardiogram there are no P waves; they are replaced by continuous irregular fluctuations of potential, called *f* waves. The AV node is activated at intervals that may vary considerably from cycle to cycle.

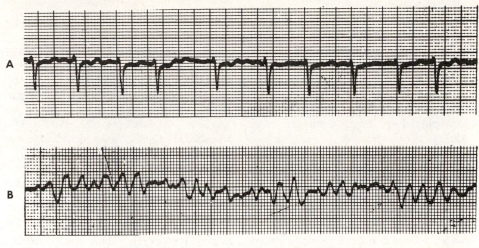

Fig. 27-29 ■ Atrial (**A**) and ventricular (**B**) fibrillation.

Hence there is no constant interval between QRS complexes and therefore between ventricular contractions.

Although atrial fibrillation is compatible with life and even with full activity, the onset of *ventricular fibrillation* leads to loss of consciousness within a few seconds. The irregular, continuous, uncoordinated twitchings of the ventricular muscle fibers result in no output of blood. Death ensues except when immediate, effective resuscitation is achieved or when ventricular fibrillation reverts to a more normal rhythm spontaneously, which rarely occurs. Ventricular fibrillation may supervene when the entire ventricle, or some portion of it, is deprived of its normal coronary blood supply. It may also occur as a result of electrocution or in response to certain drugs and anesthetics. In the electrocardiogram (Fig. 27-29, *B*) large, irregular fluctuations of potential are manifest.

Fibrillation is often initiated when a premature impulse arrives during the so-called *vulnerable period*. In the ventricles this period coincides with the downslope of the T wave. During this period there is some variability in the excitability of the cardiac cells. Some fibers are still in their effective refractory periods; others have almost fully recovered their excitability; and still others are able to conduct impulses, but only at very slow conduction velocities. As a consequence the action potentials are propagated over the chambers in multiple wavelets that travel along circuitous paths and at widely varying conduction velocities. As a region of cardiac cells becomes excitable again, it will ultimately be reentered by one of the wave fronts traveling about the chamber. The process tends to be self-sustaining.

Atrial fibrillation may be reverted to a normal sinus rhythm by means of certain depressant drugs, such as quinidine, which act in part by prolonging the refractory period. Therefore the cardiac impulse, on retracing its path, may find the myocardial fibers no longer excitable. However, much more dramatic therapy is required in ventricular fibrillation. Conversion to a normal sinus rhythm is accomplished by means of a strong electric current that places the entire myocardium in a refractory state. Originally it was necessary to open the patient's chest and apply the electrodes directly to the walls of the heart. However, techniques have now been developed so that the current can be administered safely through the intact chest wall. In successful cases the SA node again takes over as the normal pacemaker for the entire heart. Direct current shock has been found to be more effective than alternating current shock, and it is now widely used clinically to treat not only ventricular fibrillation but atrial fibrillation and certain other arrhythmias as well. Electrical abolition of such disturbances of rhythm is often called *cardioversion*.

■ *Bibliography*

Journal articles

Armstrong, C.M.: Sodium channels and gating currents, Physiol. Rev. **61:**644, 1981.

Brown, H.F.: Electrophysiology of the sino-atrial node, Physiol. Rev. **62:**505, 1982.

Childers, R.: The AV node: normal and abnormal physiology, Prog. Cardiovasc. Dis. **19:**361, 1977.

Coraboeuf, E.: Ionic basis of electrical activity in cardiac tissues, Am. J. Physiol. **234:**H101, 1978.

Cranefield, P.F.: Action potentials, afterpotentials, and arrhythmias, Circ. Res. **41:**415, 1977.

Elharrar, V., and Zipes, D.P.: Cardiac electrophysiologic alterations during myocardial ischemia, Am. J. Physiol. **233:**H329, 1977.

Ferrier, G.R.: Digitalis arrhythmias: role of oscillatory afterpotentials, Prog. Cardiovasc. Dis. **19:**459, 1977.

Glitsch, H.G.: Characteristics of active Na transport in intact cardiac cells, Am. J. Physiol. **236:**H189, 1979.

Hauswirth, O., and Singh, B.N.: Ionic mechanisms in heart muscle in relation to the genesis and the pathological control of cardiac arrhythmias, Pharmacol. Rev. **30:**5, 1979.

Irisawa, H.: Comparative physiology of the cardiac pacemaker mechanism, Physiol. Rev. **58:**461, 1978.

Jalife, J., and Moe, G.K.: Phasic effects of vagal stimulation on pacemaker activity of the isolated sinus node of the young cat, Circ. Res. **45:**595, 1979.

Lee C.O.: Ionic activities in cardiac muscle cells and application of ion-selective microelectrodes, Am. J. Physiol. **241:**H459, 1981.

Martin, P.: The influence of the parasympathetic nervous system on atrioventricular conduction, Circ. Res. **41:**593, 1977.

Singer, D.H., Baumgarten, C.M., and Ten Eick, R.E.: Cellular electrophysiology of ventricular and other dysrhythmias: studies on diseased and ischemic heart, Prog. Cardiovasc. Dis. **24:**97, 1981.

Strauss, H.C., Prystowsky, R.N., and Scheinman, M.M.: Sino-atrial and atrial electrogenesis, Prog. Cardiovasc. Dis. **19:**385, 1977.

Ten Eick, R.E., Baumgarten, C.M., and Singer, D.H.: Ventricular dysrhythmia: membrane basis, or of currents, channels, gates, and cables, Prog. Cardiovasc. Dis. **24:**157, 1981.

Vassalle, M.: Cardiac automaticity and its control, Am. J. Physiol. **233:**H625, 1977.

Wit, A.L., and Cranefield, P.F.: Reentrant excitation as a cause of cardiac arrhythmias, Am. J. Physiol. **235:**H1, 1978.

Books and monographs

Bonke, F.I.M.: The sinus node: structure, function, and clinical relevance, The Hague, 1978, Martinus Nijhoff.

Carmeliet, E., and Vereecke, J.: Electrogenesis of the action potential and automaticity. In Handbook of physiology, Section 2: Cardiovascular system, vol. 1, Bethesda, Md., 1979, American Physiological Society, pp. 269-334.

Cranefield, P.F.: The conduction of the cardiac impulse, Mount Kisco, N.Y., 1975, Futura Publishing Co., Inc.

Fozzard, H.A.: Conduction of the action potential. In Handbook of physiology, Section 2: Cardiovascular system, vol. 1, Bethesda, Md., 1979, American Physiological Society, pp. 335-356.

Levy, M.N., and Vassalle, M.: Excitation and neural control of the heart, Bethesda, Md., 1982, American Physiological Society.

Noble, D.: The initiation of the heartbeat, ed. 2, Oxford, 1979, Oxford University Press.

Scher, A.M., and Spach, M.S.: Cardiac depolarization and repolarization and the electrocardiogram. In Handbook of physiology, Section 2: Cardiovascular system, vol. 1, Bethesda, Md., 1979, American Physiological Society, pp. 357-392.

Sperelakis, N.: Origin of the cardiac resting potential. In Handbook of physiology, Section 2: Cardiovascular system, vol. 1, Bethesda, Md., 1979, American Physiological Society, pp. 187-267.

chapter 28

Hemodynamics

The problem of treating the pulsatile flow of blood through the cardiovascular system in precise mathematical terms is virtually insuperable. The heart is an extremely complicated pump, the behavior of which is affected by a large variety of physical and chemical factors. The blood vessels are multibranched, elastic conduits of continuously varying dimensions. The blood itself is not a simple fluid but is composed of red and white corpuscles, platelets, and lipid globules suspended in a colloidal solution of proteins. Despite these complicating factors, considerable insight may be gained from an understanding of the more elementary principles of fluid mechanics as they pertain to simpler physical systems. Such principles are expounded in the following sections to explain the interrelationships among velocity of blood flow, blood pressure, and the dimensions of the various components of the systemic circulation.

■ *Velocity of the bloodstream*

In describing the variations in blood flow in different vessels it is first essential to distinguish between the terms *velocity* and *flow*. The former term, sometimes designated as linear velocity, refers to the rate of displacement with respect to time and has the dimensions of distance per unit time, for example, centimeters per second. The latter term frequently is designated as volume flow and has the dimensions of volume per unit time, for example, cubic centimeters per second. In a conduit of varying cross-sectional dimensions velocity (v), flow (Q), and cross-sectional area (A) are related by the equation:

$$v = Q/A$$

The interrelationships among velocity, flow, and area are portrayed in Fig. 28-1. In the case of an incompressible fluid flowing through rigid tubes the flow past successive cross sections must be the same. For a given constant flow the velocity varies inversely as the cross-sectional area. Thus for the same volume of fluid per second passing from section *a* into section *b*, where the cross-sectional area is five times greater, the velocity of flow diminishes to one fifth of its previous value. Conversely, when the fluid proceeds from section *b* to section *c*, where the cross-sectional area is one tenth as great, the velocity of each particle of fluid must increase tenfold. The velocity at any point in

Fig. 28-1 ■ As fluid flows through a tube of variable cross-sectional area *(A)*, the linear velocity *(v)*, varies inversely as the cross-sectional area.

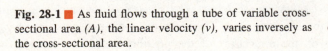

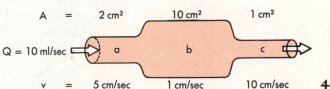

473

the system is dependent not only on area but also on the magnitude of the flow, Q. This in turn depends on the pressure gradient, the properties of the fluid, and the dimensions of the entire hydraulic system, as discussed in the following section. For any given flow, however, the ratio of the velocity past one cross section relative to that past a second cross section depends only on the inverse ratio of the respective areas, that is,

$$v_1/v_2 = A_2/A_1$$

This rule pertains regardless of whether a given cross-sectional area applies to a single large tube or to several smaller tubes in parallel.

As shown in Fig. 26-2, there is a progressive reduction in velocity as the blood traverses the aorta, its larger primary branches, the smaller secondary branches, and the arterioles. Finally, a minimum value is reached in the capillaries. As the blood then passes through the venules and continues centrally toward the venae cavae, the velocity progressively increases again. The relative velocities in the various components of the circulatory system are not related directly to the pressure gradients or to any other physical factors but only to the cross-sectional area. For this reason each point on the curve representing the total cross-sectional area is inversely proportional to the corresponding point on the curve representing velocity for any given subdivision of the vascular bed (Fig. 26-2).

■ *Relationship between velocity and pressure*

In that portion of a hydraulic system in which the total energy remains virtually constant, changes in velocity may be accompanied by appreciable alterations in the measured pressure. Consider three sections (*A, B,* and *C*) of such a hydraulic system (Fig. 28-2). Six pressure probes, or *pitot tubes,* have been inserted. The openings of three of these (*2, 4,* and *6*) are tangential to the direction of flow and thus measure the *lateral,* or *static,* pressure within the tube. The openings of the remaining three pitot tubes (*1, 3,* and *5*) face upstream. Therefore they detect the *total pressure,* which is the lateral pressure plus a pressure component ascribable to the kinetic energy of the flowing fluid. This dynamic component, P_d, of the total pressure may be calculated from the following equation:

$$P_d = \rho v^2/2$$

where ρ is the density of the fluid, and *V* the velocity. If the midpoints of segments *A, B,* and *C* are at the same hydrostatic level, then the corresponding total pressure, P_1, P_3, and P_5, will be virtually equal, if the energy loss from viscosity in these segments is negligible. However, with the changes in cross-sectional area the consequent alteration in velocity induces variations in the dynamic component.

In sections *A* and *C,* let $\rho = 1$ g/cm^3, and v = 100 cm/second. Therefore

$$P_d = 5000 \text{ dyne/cm}^2$$
$$= 3.8 \text{ mm Hg}$$

since 1330 dyne/cm^2 = 1 mm Hg. In the narrow section, *B,* let the velocity be twice as great as in sections *A* and *C*. Therefore

$$P_d = 20,000 \text{ dyne/cm}^2$$
$$= 15.0 \text{ mm Hg}$$

Thus in the wide sections of the conduit the lateral pressures (P_2 and P_6) will be only 3.8 mm Hg less than the respective total pressures (P_1 and P_5). However, in the narrow section the lateral pressure (P_4) is 15 mm Hg less than the total pressure (P_3).

The peak velocity of flow in the ascending aorta of normal dogs has been found to be in the range of 100 to 200 cm/second. Therefore the measured pressure may vary significantly, depending on the orientation of the pressure probe. In the descending thoracic aorta the peak velocity is only half as great as in the ascending aorta, and lesser

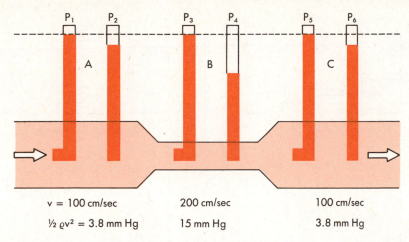

v = 100 cm/sec　　　200 cm/sec　　　100 cm/sec

$\frac{1}{2}\varrho v^2 = 3.8$ mm Hg　　15 mm Hg　　　3.8 mm Hg

Fig. 28-2 ■ In a narrow section *(B)* of a tube the linear velocity *(v)* and hence the dynamic component of pressure *($\frac{1}{2}\rho v^2$)* are greater than in the wide sections *(A and C)* of the same tube. If the total energy is virtually constant throughout the tube (that is, if the energy loss due to viscosity is negligible), the total pressures *(P_1, P_3, and P_5)* will not be detectably different, but the lateral pressure *(P_4)*, in the narrow section will be less than the lateral pressures *(P_2 and P_6)* in the wide sections of the tube.

magnitudes have been recorded in other major arterial sites. The dynamic component thus will be a negligible fraction of the total pressure, and the orientation of the pressure probe will not materially influence the magnitude of the pressure recorded. However, at the site of a constriction, such as a stenotic heart valve or coarctation of the aorta, the dynamic component may attain substantial values.

The most fundamental law governing the flow of fluids through cylindrical tubes was derived empirically in 1842 by Jean Léonard Marie Poiseuille. This French physician was primarily interested in the physical determinants of blood flow but substituted simpler liquids for blood in his measurements of flow through glass capillary tubes. His work was precise and of such importance that his observations have been designated *Poiseuille's law*. Subsequently this same law has been derived on certain theoretical bases.

Poiseuille's law is applicable to the flow of fluids through cylindrical tubes only under special conditions. It applies to the case of steady, laminar flow of newtonian fluids. The term *steady flow* signifies the absence of variations of flow in time, that is, a nonpulsatile flow. *Laminar flow* is the type of motion in which the fluid moves as a series of individual layers, with each stratum moving at a different velocity from its neighboring layers. In the case of flow through a tube the fluid consists of a series of infinitesimally thin concentric tubes sliding past one another. Laminar flow is described in greater detail later, where it is distinguished from turbulent flow. Also a *newtonian fluid* is defined more precisely. For the present discussion it will suffice to consider it as a homogeneous fluid, such as air or water, in contradistinction to a suspension, such as blood.

Pressure is one of the principal determinants of the rate of flow. The pressure, P, in dynes per square centimeter, at a distance h centimeters below the surface of a liquid is

$$P = h\rho g$$

where ρ is the density of the liquid in grams per cubic centimeter, and g is the acceleration of gravity in centimeters per square second. For convenience, however, pressure is frequently expressed simply in terms of height, h, of the column of liquid above some arbitrary reference point.

■ *Relationship between pressure and flow*

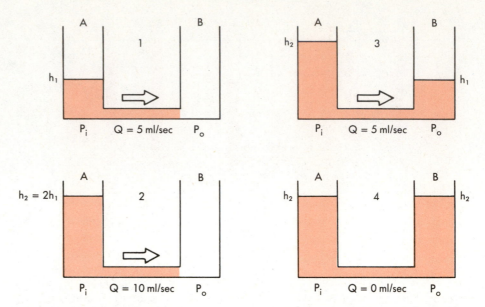

Fig. 28-3 ■ The flow *(Q)* of fluid through a tube connecting two reservoirs *(A* and *B)* is proportional to the difference between the pressure *(P_i)* at the inflow end and the pressure *(P_o)* at the outflow end of the tube.

Consider the tube connecting reservoirs *A* and *B* in Fig. 28-3. Let reservoir *A* be filled with liquid to height h_1, and let reservoir *B* be empty, as in section *1* of Fig. 28-3. The outflow pressure, P_o, is therefore equal to the atmospheric pressure, which shall be designated as the zero, or reference, level. The inflow pressure, P_i, is then equal to the same reference level plus the height, h_i, of the column of liquid in reservoir *A*. Under these conditions let the flow, *Q*, through the tube be 5 ml/second. If reservoir *A* is filled to height h_2, which is twice h_1, and reservoir *B* is again empty (as in section *2*), the flow will be twice as great, that is, 10 ml/second. Thus with reservoir *B* empty the flow will be directly proportional to the inflow pressure, P_i. If reservoir *B* is now allowed to fill to height h_1 and the fluid level in *A* is maintained at h_2 (as in section *3*), the flow will again become 5 ml/second. Thus flow is directly proportional to the difference between inflow and outflow pressures:

$$Q \propto P_i - P_o$$

If the fluid level in *B* attains the same height as in *A*, flow will cease (section *4*).

For any given pressure difference between the two ends of a tube the flow will be dependent on the dimensions of the tube. Consider the tube connected to reservoir *1* in Fig. 28-4. With length l_1 and radius r_1, the flow Q_1 is observed to be 10 ml/second. The tube connected to reservoir *2* has the same radius but is twice as long. Under these conditions the flow Q_2 is found to be 5 ml/second, or only half as great as Q_1. Conversely, for a tube half as long as l_1 the flow would be twice as great as Q_1. In other words, flow is inversely proportional to the length of the tube:

$$Q \propto 1/l$$

The tube connected to reservoir *3* in Fig. 28-4 is the same length as l_1, but the radius is twice as great. Under these conditions the flow Q_3 is found to increase to a value of 160 ml/second, which is 16 times greater than Q_1. The precise measurements of Poiseuille revealed that flow varies directly as the fourth power of the radius:

$$Q \propto r^4$$

Thus in the example above, since $r_3 = 2r_1$, Q_3 will be proportional to $(2r_1)^4$, or $16r_1^4$; therefore Q_3 will equal $16Q_1$.

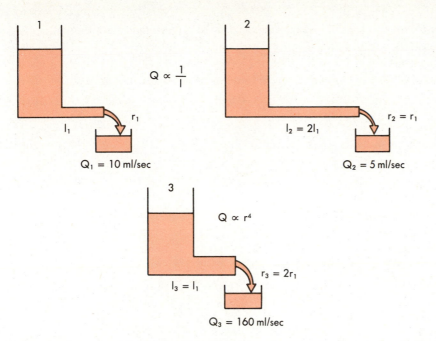

Fig. 28-4 The flow *(Q)* of fluid through a tube is inversely proportional to the length *(l)* and directly proportional to the fourth power of the radius *(r)* of the tube.

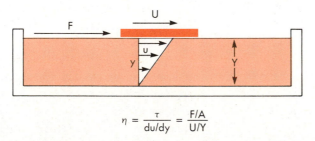

Fig. 28-5 ■ For a newtonian fluid the viscosity (η) is defined as the ratio of shear stress (τ) to rate of strain *(du/dy)*. For a plate of contact area *(A)* moving across the surface of a liquid, τ equals the ratio of the force *(F)* applied in the direction of motion to the contact area, and *du/dy* equals the ratio of the velocity of the plate *(U)* to the depth of the liquid *(Y)*.

Finally, for a given pressure difference and for a cylindrical tube of given dimensions the flow will vary, depending on the nature of the fluid itself. This flow-determining property of fluids is termed *viscosity* (η), which has been defined by Newton as the ratio of *shear stress* to the *rate of strain* of the fluid. These terms may be comprehended most clearly by considering the flow of a homogeneous fluid between parallel plates. In Fig. 28-5 let the bottom plate (the bottom of a large basin) be stationary, and let the upper plate move at a constant velocity along the upper surface of the fluid. The *shear stress* (τ) is defined as the ratio of F/A, where F is the force applied to the upper plate in the direction of its motion along the upper surface of the fluid, and A is the area of the upper plate in contact with the fluid. The *rate of strain* is du/dy, where u is the velocity in the direction parallel to the motion of the upper plate for any minute fluid element contained between the plates, and y is the distance of that element above the bottom, stationary plate.

With regard to the flow of fluids through cylindrical tubes, the flow will vary inversely as the viscosity. Thus in the example of flow from reservoir *1* in Fig. 28-4, if the viscosity of the fluid in the reservoir were doubled, then the flow would be halved (5 ml/second instead of 10 ml/second).

In summary, for the steady, laminar flow of a newtonian fluid through a cylindrical tube, the flow, Q, varies directly as the pressure difference, $P_i - P_o$, and the fourth power of the radius (r) of the tube and varies inversely as the length (l) of the tube and the viscosity (η) of the fluid. The full statement of Poiseuille's law is

$$Q = \frac{\pi(P_i - P_o)r^4}{8\,\eta\,l}$$

where $\pi/8$ is the constant of proportionality.

■ *Resistance to flow*

In direct-current electrical circuit theory it has been useful to employ the concept of resistance (R), which is defined as the ratio of voltage drop (E) to current flow (I). Similarly in fluid mechanics the analogous hydraulic resistance (R) may be defined as the ratio of pressure drop $(P_i - P_o)$ to flow (Q). For the steady, laminar flow of a newtonian fluid through a cylindrical tube the physical components of hydraulic resistance may be perceived readily by rearranging Poiseuille's law to give the hydraulic resistance equation:

$$R = \frac{P_i - P_o}{Q} = \frac{8\,\eta\,l}{\pi\,r^4}$$

Thus, when Poiseuille's law applies, the resistance to flow is dependent only on the dimensions of the tube and on the characteristics of the fluid.

In the circulatory system the length of any given vessel is virtually constant, and ordinarily the viscosity of the blood does not vary appreciably. Thus the major alterations in resistance are produced, physiologically and pathologically, by virtue of variations in radius. From Fig. 26-2 it may be noted that the greatest pressure drop occurs across the arterioles. Since the total flow is the same through the various series components of the circulatory system, it follows that the greatest resistance to flow resides in the arterioles. The arterioles are vested with a thick coat of circularly arranged smooth muscle fibers, by means of which variations in lumen radius may be produced. From the hydraulic resistance equation, wherein R varies inversely as r^4, it is clear that small changes in radius will result in relatively great alterations in resistance.

In the cardiovascular system the various types of vessels listed in Fig. 26-2 lie in series with one another. Furthermore the individual members of each category of vessels are ordinarily arranged in parallel with one another (Fig. 26-3). For example, the capillaries throughout the body are in most instances parallel elements, with the notable exceptions of the renal vasculature (wherein the peritubular capillaries are in series with the glomerular capillaries) and the splanchnic vasculature (wherein the intestinal and hepatic capillaries are aligned in series). Formulas for the total hydraulic resistance of components arranged in series and in parallel have been derived in a manner analogous to those employed for similar combinations of electrical resistances.

Three hydraulic resistances (R_1, R_2, and R_3) are arranged in series in the schema depicted in Fig. 28-6. The pressure drop across the entire system—that is, the difference

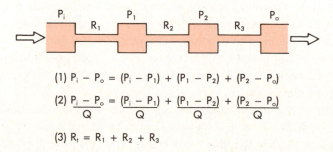

(1) $P_i - P_o = (P_i - P_1) + (P_1 - P_2) + (P_2 - P_o)$

(2) $\dfrac{P_i - P_o}{Q} = \dfrac{(P_i - P_1)}{Q} + \dfrac{(P_1 - P_2)}{Q} + \dfrac{(P_2 - P_o)}{Q}$

(3) $R_t = R_1 + R_2 + R_3$

Fig. 28-6 ■ For resistances (R_1, R_2, and R_3) arranged in series the total resistance (R_t) equals the sum of the individual resistances.

between inflow pressure (P_i) and outflow pressure (P_o)—consists of the sum of the pressure drops across each of the individual resistances (equation *1*). Under steady-state conditions the flow (Q) through any given cross section must equal the flow through any other cross section. By dividing each component in equation *1* by Q (equation *2*), it becomes evident from the definition of resistance that the total resistance (R_t) of the entire system equals the sum of the individual resistances, that is,

$$R_t = R_1 + R_2 + R_3$$

For resistances in parallel, as illustrated in Fig. 28-7, the inflow and outflow pressures are the same for all tubes. Under steady-state conditions the total flow, Q_t, through the system equals the sum of the flows through the individual parallel elements (equation *1*). Since the pressure gradient ($P_i - P_o$) is identical for all parallel elements, each term in equation *1* may be divided by the pressure gradient to yield equation *2*. From the definition of resistance, equation *3* may be derived. This states that the reciprocal of the total resistance, R_t, equals the sum of the reciprocals of the individual resistances, that is,

$$\frac{1}{R_t} = \frac{1}{R_1} + \frac{1}{R_2} + \frac{1}{R_3}$$

Stated in another way, if we define hydraulic *conductance* as the reciprocal of resistance, in analogy to the practice in electrical theory, it becomes evident that, for tubes in parallel, the total conductance is the sum of the individual conductances.

By considering a few simple illustrations, some of the fundamental properties of parallel hydraulic systems become apparent. For example, if the resistances of the three parallel elements in Fig. 28-7 were all equal, then

$$R_1 = R_2 = R_3$$

Therefore

$$1/R_t = 3/R_1$$

and

$$R_t = R_1/3$$

Thus the total resistance is less than any of the individual resistances. After further consideration it becomes evident that for any parallel arrangement the total resistance must be less than that of any individual component. For example, consider a system in which a very high-resistance tube is added in parallel to a low-resistance tube. The total resistance must be less than that of the low-resistance component by itself, since the high-resistance component affords an additional pathway, or conductance, for fluid flow.

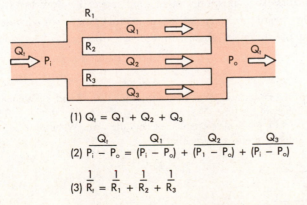

(1) $Q_t = Q_1 + Q_2 + Q_3$

(2) $\dfrac{Q_t}{P_i - P_o} = \dfrac{Q_1}{(P_i - P_o)} + \dfrac{Q_2}{(P_1 - P_o)} + \dfrac{Q_3}{(P_i - P_o)}$

(3) $\dfrac{1}{R_t} = \dfrac{1}{R_1} + \dfrac{1}{R_2} + \dfrac{1}{R_3}$

Fig. 28-7 ■ For resistances (R_1, R_2, and R_3) arranged in parallel the reciprocal of the total resistance (R_t) equals the sum of the reciprocals of the individual resistances.

■ *Laminar and turbulent flow*

Under certain conditions the flow of a fluid in a cylindrical tube will be *laminar* (sometimes called *streamlined*), as illustrated in Fig. 28-8. At the entrance of the tube all the fluid elements will have the same linear velocities, regardless of their radial positions. In progressing along the tube, however, the thin layer of fluid in contact with the wall of the tube remains adherent to the wall and thus is motionless. The layer of fluid just central to this external lamina must shear against this motionless layer and therefore moves slowly but with a finite velocity. Similarly the adjacent, more central layer travels still more rapidly. Close to the tube inlet the fluid layers near the axis of the tube still move with the same velocity and do not shear against one another. However, at a distance-from the tube inlet equal to several tube diameters, laminar flow becomes *fully developed;* that is, the velocity profiles do not change with longitudinal distance along the tube. In fully developed laminar flow the longitudinal velocity profile is that of a paraboloid. The velocity of the fluid adjacent to the wall is zero, whereas the velocity at the center of the stream is maximal and equal to twice the mean velocity of flow across the entire cross section of the tube. In laminar flow, fluid elements in one lamina, or streamline, remain in that streamline as the fluid progesses longitudinally along the tube.

Irregular motions of the fluid elements may develop in the flow of fluid through a tube; such a flow is called *turbulent*. Under such conditions fluid elements do not remain confined to definite streamlines, but rapid, radial mixing occurs. A considerably greater pressure difference is required to force a given flow of fluid through the same tube under conditions of turbulence as compared with the pressure difference required for laminar flow. In turbulent flow the pressure drop is approximately proportional to the square of the flow rate, whereas in laminar flow the pressure drop is proportional to the first power of the flow rate. Therefore, a pump such as the heart would have to do considerably more work for a given flow if turbulence developed.

Whether turbulent or laminar flow will exist in a tube under given conditions may be predicted on the basis of a dimensionless number called *Reynold's number* (N_R). This number represents the ratio of inertial to viscous forces and equals $\rho D \bar{v} / \eta$, where D is the tube diameter, $\bar{v}$ is the mean velocity, ρ is the density, and η is the viscosity. For $N_R < 2000$ the flow usually will be laminar; for $N_R > 3000$, turbulence usually will exist. Various possible conditions may develop in the transition range of N_R between 2000 and 3000. Since flow tends to be laminar at low N_R and turbulent at high N_R, it is evident from the definition of N_R that large diameters, high velocities, and low viscosities predispose the flow to the development of turbulence. In addition to these factors, abrupt variations in tube dimensions or irregularities in the tube walls will produce turbulence.

Turbulence usually is accompanied by vibrations in the auditory frequency ranges. When turbulent flow exists within the cardiovascular system, it usually is detected as a *murmur*. The factors listed previously that predispose the flow to turbulence may account for murmurs heard clinically. In severe anemia *functional cardiac murmurs* (murmurs not caused by structural abnormalities) are frequently detectable. The physical

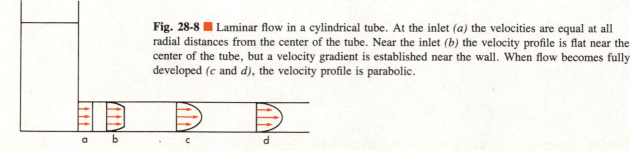

Fig. 28-8 ■ Laminar flow in a cylindrical tube. At the inlet *(a)* the velocities are equal at all radial distances from the center of the tube. Near the inlet *(b)* the velocity profile is flat near the center of the tube, but a velocity gradient is established near the wall. When flow becomes fully developed *(c and d)*, the velocity profile is parabolic.

basis for such murmurs resides in (1) the reduced viscosity of blood caused by the low red cell content, and (2) the high flow velocities associated with marked augmentation of cardiac output that usually occurs in anemic patients.

In Fig. 28-5 an external force was applied to a plate floating on the surface of a volume of liquid contained in a large basin. This force, exerted parallel to the surface, caused a shearing stress on the liquid below, thereby producing a differential motion of each layer of liquid relative to the adjacent layers. At the bottom of the basin the flowing liquid produced a shearing stress on the surface of the basin in contact with the liquid. By rearranging the formula for viscosity stated in Fig. 28-5, it is apparent that the shear stress (τ) equals η (du/dy); that is, the shear stress equals the product of the viscosity and the rate of strain. Thus the greater the rate of flow, the greater the shear stress that the liquid exerts on the walls of the container in which it flows.

For precisely the same reasons the rapidly flowing blood in a large artery tends to pull the endothelial lining of the artery along with it. This force, or *viscous drag,* is proportional to the rate of strain (du/dy) of the layers of blood near the wall. For Poiseuille flow

$$\tau = 4\eta Q/\pi r^3$$

The greater the rate of blood flow (Q) in the artery, the greater will be du/dy near the arterial wall, and the greater will be the viscous drag (τ).

In certain types of arterial disease, particularly with hypertension, the subendothelial layers tend to degenerate locally, and small regions of the endothelium may lose their normal support. The viscous drag on the arterial wall may cause a tear between a normally supported and unsupported region of the endothelial lining. Blood then may enter the rift in the lining from the vessel lumen and dissect between the various layers of the artery. Such a lesion is called a *dissecting aneurysm.* It occurs most commonly in the proximal portions of the aorta and is extremely serious. One reason for its predilection for this site is the very high velocity of blood flow, with the associated high values of du/dy at the endothelial wall. A major aim of treatment is to depress the heart to reduce the velocity of blood flow and hence the viscous drag on the aortic endothelium.

The viscosity of a newtonian fluid, such as water, may be determined by measuring the rate of flow of the fluid at a given pressure gradient through a cylindrical tube of known length and radius. As long as the fluid flow is laminar, the viscosity may be computed by substituting these values into the Poiseuille equation. With careful measurement the viscosity of a given newtonian fluid at a specified temperature will be constant over a wide range of tube dimensions and flow rates. However, for a nonnewtonian fluid the viscosity calculated by substituting into Poiseuille's equation may vary considerably as a function of tube dimensions and flow rates. Therefore in considering the rheological properties of a suspension such as blood the term *viscosity* does not have a unique meaning. The terms *anomalous viscosity* and *apparent viscosity* are frequently applied to the value of viscosity obtained for blood under the particular conditions of measurement.

Rheologically blood is a suspension, principally of erythrocytes in a relatively homogeneous liquid, the blood plasma. For this reason the apparent viscosity of blood varies as a function of the *hematocrit ratio* (ratio of volume of red blood cells to volume of whole blood). In Fig. 28-9 the upper curve represents the ratio of the apparent viscosity of whole blood to that of plasma over a range of hematocrit ratios from 0% to 80%, measured in a tube 1 mm in diameter and 250 mm in length. The viscosity of plasma is 1.2 to 1.3 times that of water. From the graph it may be seen that blood, with a normal hematocrit ratio of 45%, has an apparent viscosity 2.4 times that of plasma. With severe anemia therefore there is a considerable reduction in blood viscosity. With

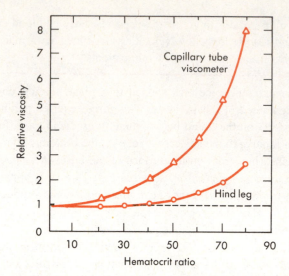

Fig. 28-9 ■ Viscosity of whole blood, relative to that of plasma, increases at a progressively greater rate as hematocrit ratio increases. For any given hematocrit ratio the apparent viscosity of blood is less when measured in a biological viscometer (such as the hind leg of a dog) than in a conventional capillary tube viscometer. (Redrawn from Levy, M.N., and Share, L.: Circ. Res. **1:**247, 1953.)

increasing hematocrit ratios the slope of the curve increases progressively; it is especially steep at the upper range of erythrocyte concentrations. A rise in hematocrit ratio from 45% to 70%, which occurs in *polycythemia,* results in more than a twofold increase in apparent viscosity, with a proportionate effect on the resistance to blood flow. The magnitude of the effect on peripheral resistance may be appreciated when it is recognized that even in the most severe cases of essential hypertension, in which peripheral resistance is augmented by arteriolar constriction rather than by greater blood viscosity, the total peripheral resistance rarely increases by more than a factor of two.

For any given hematocrit ratio the apparent viscosity of blood depends on the dimensions of the tube employed in estimating the viscosity. In 1931 Fåhraeus and Lindqvist demonstrated that the apparent viscosity of blood diminishes appreciably in glass capillary tubes approaching the radial dimensions of the microscopic blood vessels. Fig. 28-10 illustrates the changes in viscosity of blood relative to that of water as a function of tube diameter. The graph demonstrates that the apparent viscosity of blood diminishes progressively as tube diameter decreases below a value of about 0.3 mm. Since the highest resistance blood vessels, the arterioles, possess diameters considerably less than this critical value, it was pointed out by these investigators that the consequent reduction of apparent viscosity "allows the heart to drive a given volume of blood through the arterioles at a much lower pressure than would be the case if the blood behaved as a [newtonian] fluid." It was demonstrated subsequently that the apparent viscosity of blood, when measured in living tissues, is considerably less than when measured in a conventional capillary tube viscometer with a diameter greater than 0.3 mm. In the lower curve of Fig. 28-9 apparent relative viscosity of blood was assessed by using the hind leg of a dog as a biological viscometer. It is evident that over the entire range of hematocrit ratios the apparent viscosity is less as measured in the living tissue than in the capillary tube viscometer (*upper curve*) and that the disparity becomes greater with higher hematocrit ratios. Undoubtedly this phenomenon may be ascribed to the so-called Fåhraeus-Lindqvist effect, since the diameter of the high-resistance blood vessels is considerably less than the diameter of the glass-tube viscometer employed in the same study.

The influence of tube diameter on apparent viscosity is ascribable in part to the

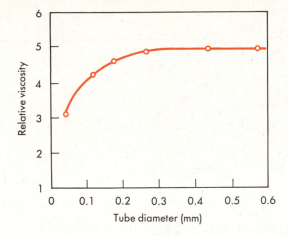

Fig. 28-10 ■ Viscosity of blood, relative to that of water, increases as a function of tube diameter up to a diameter of about 0.3 mm. (Redrawn from Fåhraeus, R., and Lindqvist, T.: Am. J. Physiol. **96:**562, 1931.)

Fig. 28-11 ■ The "relative hematocrit" of blood flowing from a feed reservoir through capillary tubes of various calibers, as a function of the tube diameter. The relative hematocrit is the ratio of the hematocrit of the blood in the tubes to that of the blood in the feed reservoir. (Redrawn from Barbee, J.H., and Cokelet, G.R.: Microvasc. Res. **3:**6, 1971.)

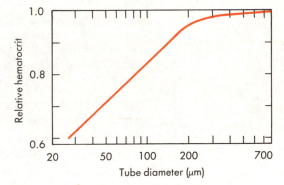

change in actual composition of the blood as it flows through small tubes. The alteration in composition results from the tendency for the red blood cells to accumulate in the faster axial stream, whereas largely plasma flows in the slower marginal layers. To illustrate this phenomenon, a reservoir such as *A* in Fig. 28-3 could be filled with blood possessing a given hematocrit ratio. If the blood in *A* was constantly agitated to prevent settling and was permitted to flow through a narrow capillary tube into reservoir *B*, the hematocrit ratio of the blood in *B* would not be detectably different from that in *A*. Surprisingly, however, if the capillary tube joining the two reservoirs were suddenly disconnected and the hematocrit ratio of the blood contained within the tube determined, it would be found to be considerably lower than the hematocrit ratio of the blood in either reservoir. In actual experiments it has been found that the smaller the tube, the more pronounced the change in hematocrit ratio in the tube. In Fig. 28-11, the relative hematocrit is the ratio of the hematocrit in the tube to that in the reservoir at either end of the tube. For tubes of 500 μm diameter or greater the relative hematocrit was observed to be close to 1.0. However, with reductions in the tube diameter there was a progressive decrease in the relative hematocrit, such that for a tube diameter of 30 μm the relative hematocrit was only 0.6.

The anomalous rheological behavior of blood also becomes manifest as a consequence of changes in the rate of shear (Fig. 28-12). The observed reduction in apparent viscosity with increasing shear rate has been called *shear thinning*. This non-newtonian behavior has been explained in part on the basis of a greater tendency of the erythrocytes to accumulate in the axial laminae at higher flow rates. However, a more important factor is that at very slow rates of shear there is a distinct tendency for the suspended cells to form aggregates, which would increase viscosity. This tendency toward aggregation becomes progressively less as the rate of flow is augmented, producing the diminution in apparent viscosity that is exhibited in Fig. 28-12. The tendency toward aggregation at low flows is dependent on the concentration in the plasma of the larger protein molecules, such as the globulins and particularly fibrinogen. For this reason the changes in blood viscosity with shear rate are much more pronounced when the plasma

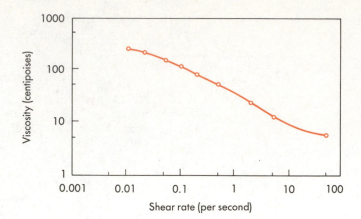

Fig. 28-12 ■ Decrease in apparent viscosity of blood (hematocrit ratio, 51.7%) at increasing rates of shear, both plotted on logarithmic scales. The shear rate refers to the relative velocity of one layer of fluid with respect to that of the adjacent layers and is directionally related to the rate of flow. (Redrawn from Chien, S.: J. Appl. Physiol. **21**:81, 1966.)

contains a high concentration of fibrinogen than when the content is low. In addition, there is a tendency at low flow rates for leukocytes to adhere to the endothelial walls of the microvasculature, thereby increasing the apparent viscosity.

The deformability of the erythrocytes is also a factor in shear thinning, especially at high hematocrit ratios. The mean diameter of human red blood cells is about 8.5 μm, yet they are able to pass through openings with a diameter of only 3.0 μm. As blood, densely packed with erythrocytes, is caused to flow at progressively greater rates, the erythrocytes become more and more deformed. The deformation is such that the apparent viscosity of the blood diminishes. If the red blood cells become hardened, as they are in certain spherocytic anemias, shear thinning may become much less prominent, or the apparent viscosity may actually increase slightly at greater rates of shear.

■ *Bibliography*

Journal articles

Badeer, H.S., and Rietz, R.R.: Vascular hemodynamics: deep-rooted misconceptions and misnomers, Cardiology **64**:197, 1979.

Fåhraeus, R., and Lindqvist, T.: The viscosity of blood in narrow capillary tubes, Am. J. Physiol. **96**:562, 1931.

Levy, M.N., and Share, L.: The influence of erythrocyte concentration upon the pressure-flow relationships in the dog's hind limb, Circ. Res. **1**:247, 1953.

Prokop, E.K., Palmer, R.F., and Wheat, M.W., Jr.: Hydrodynamic forces in dissecting aneurysms, Circ. Res. **27**:121, 1970.

Schmid-Schönbein, H., and Wells, R.E., Jr.: Rheological properties of human erythrocytes and their influence upon the "anomalous" viscosity of blood, Ergeb. Physiol. **63**:146, 1971.

Schmid-Schönbein, G.W., et al.: Cell distribution in capillary networks, Microvasc. Res. **19**:18, 1980.

Yen, R.T., and Fung, Y.C.: Inversion of Fåhraeus effect and effect of mainstream flow on capillary hematocrit, J. Appl. Physiol. **42**:578, 1977.

Zamir, M.: The role of shear forces in arterial branching, J. Gen. Physiol. **67**:213, 1976.

Books and monographs

Cokelet, G.R., Meiselman, H.J., and Brooks, D.E., editors: Erythrocyte mechanics and blood flow, New York, 1980, Alan R. Liss, Inc.

Dintenfass, L.: Rheology of blood in diagnostic and preventive medicine, London, 1976, Butterworths.

Hwang, N.H.C., and Normann, N.A.: Cardiovascular flow dynamics and measurements, Baltimore, 1977, University Park Press.

Milnor, W.R.: Hemodynamics, Baltimore, 1981, The Williams & Wilkins Co.

Noordergraaf, A.: Circulatory system dynamics, New York, 1979, Academic Press, Inc.

chapter 29

The cardiac pump

It is nearly impossible to contemplate the pumping action of the heart without being struck by its simplicity of design, its wide range of activity and functional capacity, and the staggering amount of work it performs relentlessly over the lifetime of an individual. To understand how the heart accomplishes its important task, it is first necessary to consider the relationships between the structure and function of its various components.

A number of important morphological and functional differences exist between myocardial and skeletal muscle cells (see Chapter 22). However, the contractile elements within the two types of cells are quite similar; each skeletal and cardiac muscle cell is made up of sarcomeres (from Z line to Z line) containing thick filaments composed of myosin (in the A band) and thin filaments composed of actin. The thin filaments extend from the Z line (through the I band) to interdigitate with the thick filaments. The sliding filament hypothesis appears to explain contraction satisfactorily in both. Furthermore, skeletal and cardiac muscle show similar length/tension relationships. The sarcomere length has been determined with electron microscopy in papillary muscles and intact ventricles rapidly fixed during systole or diastole. Maximal developed tension is observed at resting sarcomere lengths of 2 to 2.3 μm for cardiac muscle. At such lengths there is overlap of thick and thin filaments and a maximal number of crossbridge attachments. Developed tension of cardiac muscle is less than the maximum value when the sarcomeres are stretched beyond the optimum length, probably because of less overlap of the filaments and thus less interaction between crossbridges on the thick and thin filaments. At resting sarcomere lengths shorter than the optimum values, the thin filaments are either compressed, or they overlap each other; this may interfere in some manner with the development of contractile force. At short sarcomere lengths there may be failure of excitation to release Ca^{++} from the sarcoplasmic reticulum, as is the case in skeletal muscle.

In general, the fiber length/tension relationship for the papillary muscle also holds true for fibers in the intact heart. This relationship may be expressed graphically (Fig. 29-1) by substituting ventricular systolic pressure for tension and end-diastolic ventricular volume for myocardial resting fiber (and hence sarcomere) length. The lower curve in Fig. 29-1 represents the pressure that exists at each volume when the heart is in diastole. The upper curve represents the peak pressure developed by the ventricle during systole at each degree of filling and illustrates the *Frank-Starling relationship* of initial myocardial fiber length (or initial volume) to tension (or pressure) development by the ventricle. Note that the pressure/volume curve in diastole is initially quite flat, indicating that large increases in volume can be accomodated with only small increases in pressure, yet systolic pressure development is considerable at the lower filling pressures. How-

■ *Structure of the heart in relation to function*
■ *Myocardial cell*

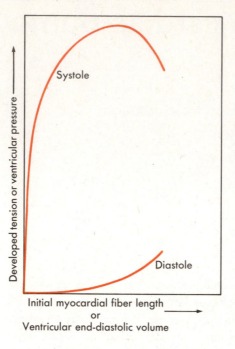

Fig. 29-1 ■ Relationship of myocardial resting fiber length (sarcomere length) or end-diastolic volume to developed tension or peak systolic ventricular pressure during ventricular contraction in the intact dog heart. (Redrawn from Patterson, S.W., Piper, H., and Starling, E.H.: J. Physiol. **48:**465, 1914.)

ever, the ventricle becomes much less distensible with greater filling, as evidenced by the sharp rise of the diastolic curve at large intraventricular volumes. In the normal intact heart, peak tension may be attained at a filling pressure of 12 mm Hg. At this intraventricular diastolic pressure, which is about the upper limit observed in the normal heart, the sarcomere length is 2.2 μm. However, developed tension has been shown to peak at filling pressures as high as 30 mm Hg in the isolated heart. Even at higher diastolic pressures (>50 mm Hg) the sarcomere length is not greater than 2.6 μm in cardiac muscle, whereas in many skeletal muscles sarcomere lengths as great as 3.65 μm can be obtained with stretch. This resistance to stretch of the myocardium at high filling pressures probably resides in the noncontractile constituents of the tissue and may serve as a safety factor against overloading of the heart in diastole. Usually ventricular diastolic pressure is about 0 to 7 mm Hg and the average diastolic sarcomere length about 2 μm. Thus the normal heart operates on the ascending portion of the Frank-Starling curve depicted in Fig. 29-1.

A striking difference in the appearance of cardiac and skeletal muscle is the semblance in cardiac muscle of a syncytium with branching interconnecting fibers. However, the myocardium is not a true anatomical syncytium, since, laterally, the myocardial fibers are separated from adjacent fibers by their respective sarcolemmas, and the end of each fiber is separated from its neighbor by dense structures, *intercalated disks,* that are continuous with the sarcolemma. Nevertheless cardiac muscle functions as a syncytium, since a wave of depolarization followed by contraction of the entire myocardium (an all-or-none response) occurs when a suprathreshold stimulus is applied to any one focus. Graded contraction, as seen in skeletal muscle by activation of different numbers of fibers, does not occur in heart muscle. Whether the intercalated disks represent a high-resistance or a low-resistance conduction pathway is controversial. The *tight junctions,* where there appears to be fusion of the intercalated disks (Fig. 29-2), are thought by most investigators to represent low-resistance pathways between myocardial cells.

Another difference between cardiac and fast skeletal muscle fibers is in the abundance of mitochondria *(sarcosomes)* in the two tissues. Fast skeletal muscle, which is called on for relatively short periods of repetitive or sustained contraction and which can metabolize anaerobically and build up a substantial oxygen debt, has relatively few

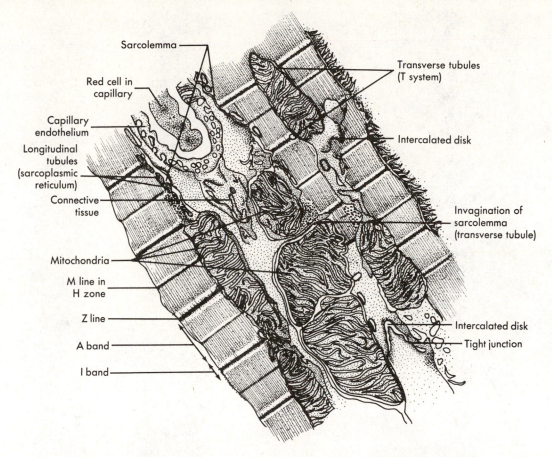

Sarcolemma

Red cell in
capillary

Capillary
endothelium

Longitudinal
tubules
(sarcoplasmic
reticulum)

Connective
tissue

Mitochondria

M line in
H zone

Z line

A band

I band

Transverse tubules
(T system)

Intercalated disk

Invagination of
sarcolemma
(transverse tubule)

Intercalated disk

Tight junction

Fig. 29-2 ■ Diagram of an electron micrograph of cardiac muscle showing large numbers of mitochondria, the intercalated disks with tight junctions (nexi), the transverse tubules, and the longitudinal tubules. (Approximately ×30,000.)

mitochondria in the muscle fibers. In contrast, cardiac muscle, which is required to contract repetitively for a lifetime and which is incapable of developing a significant oxygen debt, is very rich in sarcosomes (Figs. 29-2 and 29-3). Rapid oxidation of substrates with the synthesis of ATP can keep pace with the myocardial energy requirements by virtue of the large numbers of sarcosomes containing the respiratory enzymes necessary for oxidative phosphorylation.

To provide adequate oxygen and substrate for its metabolic machinery, the myocardium also is endowed with a rich capillary supply, about one capillary per fiber. Thus diffusion distances are short, and oxygen, carbon dioxide, substrates, and waste material can move rapidly between myocardial cell and capillary. With respect to exchange of substances between the capillary blood and the myocardial cells, electron micrographs of myocardium show deep invaginations of the sarcolemma into the fiber at the Z lines (Figs. 29-2 and 29-3). These sarcolemmal invaginations constitute the transverse or T-tubular system. The lumina of these T-tubules are continuous with the bulk interstitial fluid, and they play a key role in excitation-contraction coupling (Chapter 22). They are thought to provide a pathway for the rapid transmission of the electrical signal from the surface sarcolemma to the inside of the fiber, thus enabling nearly simultaneous activation of all myofibrils, including those deep within the interior of the fiber. In mammalian ventricular cells adjacent transverse tubules are interconnected by longitudinally running or axial tubules, thus forming an extensively interconnected lattice of "intracellular" tubules. This T system is open to the interstitial fluid, is lined with a basement membrane continuous with that of the surface sarcolemma, and contains micropinocytotic-

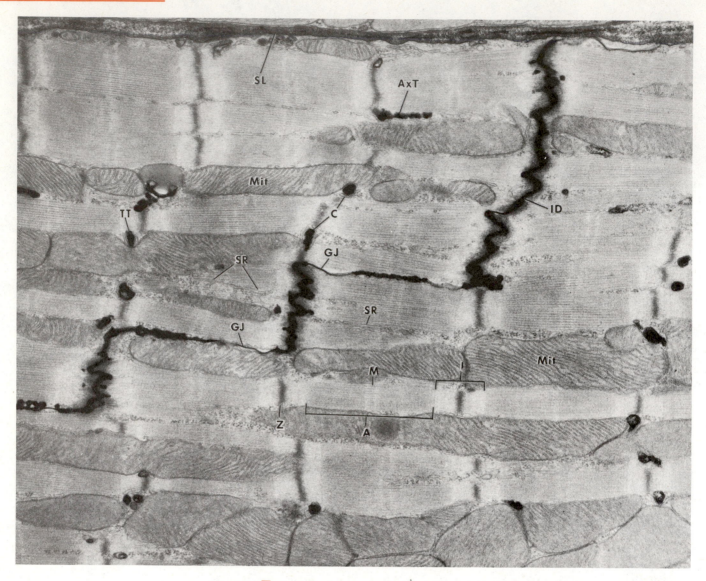

Fig. 29-3 ■ Electron micrograph of cardiac muscle in longitudinal section. Surface structures shown include the sarcolemma *(SL)*, intercalated disk *(ID)* with its associated gap junctions *(GJ)*, and sarcolemmal invaginations, which may be oriented either transverse to the cell axis (transverse tubules, *TT*) or parallel to it (axial tubules, *AxT*). The sarcoplasmic reticulum is seen in the form of tubules *(SR)* or saccules, which form couplings *(C)* with the sarcolemma or its invaginations. Mitochondria *(Mit)* are large and elongated, lying between the myofibrils. Z, Z line of sarcomere; *A,* A band; *I,* I band; *M,* M line in H zone of sarcomere. (×30,000.) (Electron micrograph courtesy Dr. Michael S. Forbes.)

like vesicles (Fig. 29-4). A network of sarcoplasmic reticulum consisting of small-diameter sarcotubules is also present surrounding the myofibrils; these sarcotubules are believed to be "closed," since colloidal tracer particles (20 to 100 Å in diameter) do not enter them. They do not contain basement membrane. Flattened elements of the sarcoplasmic reticulum are often found in close proximity to the T system as well as to the surface sarcolemma, forming "diads" (Fig. 29-4). The sarcoplasmic reticulum se-

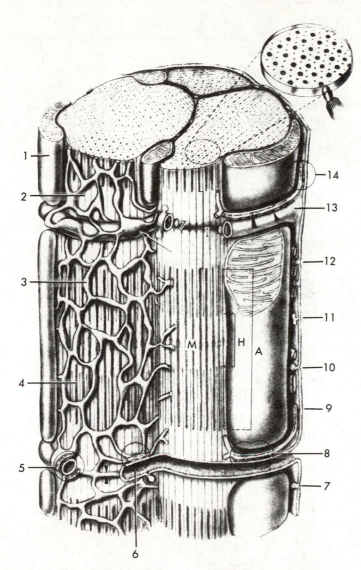

Fig. 29-4 ■ Model of mammalian cardiac muscle showing a portion of a muscle fiber: *1*, mitochondrion; *2*, actin (thin) filament; *3*, sarcoplasmic reticulum; *4*, myosin (thick) filament; *5*, triad, made up of two couplings at a transverse tubule; *6*, diad, made up of one coupling at a transverse tubule. The couplings in *5, 6,* and *8* are interior couplings. *7*, Pit (or coated vesicle) also commonly seen at transverse tubules; *8*, interior coupling; *9*, sarcolemma; *10*, pinocytotic vesicle or caveola; *11*, branched caveola; *12*, caveola containing dense granules; *13*, transverse tubule; *14*, peripheral coupling. *Large arrowhead:* junctional granules (or central membrane), only within junctional sarcoplasmic reticulum. *Small arrowhead:* junctional processes. *Arrow:* junctional sarcoplasmic reticulum with connections of the sarcoplasmic reticulum across the Z line. Also shown are the M line, A band, H band, and I band. Note the narrow junction of the sarcoplasmic reticulum of the couplings at *5, 6, 8, 14.* (Reproduced by permission from Z. Zellforsch. **98:**437, 1969.)

questers calcium (during diastole) and releases some of the calcium involved in activation of the contractile proteins in myocardial contraction. The other source of calcium in cardiac muscle contraction is that present in the extracellular fluid. Thus in ventricular cells the myofibrils and mitochondria have close access to a space that is continuous with the interstitial fluid. The T system is absent or poorly developed in atrial cells of many mammalian hearts, but these atrial cells do contain a small sarcotubular system.

■ *Myocardial contractile machinery and contractility*

Studies on the mechanisms of muscle contraction for the most part have been carried out on skeletal muscle. However, in recent years a great deal of attention has been focused on cardiac muscle in an attempt to determine to what extent the information gained about skeletal muscle applies to the myocardium. Despite the fact that none of the existing models for cardiac muscle contraction are compatible with all the data on muscle fiber length, force, and velocity, the model originally proposed for skeletal muscle by A.V. Hill has been found to be of some use in the consideration of cardiac muscle. It consists of a *contractile element,* a *series elastic element* (an elastic component in series with the contractile element), and a *parallel elastic element* (an elastic component in parallel with the contractile and series elastic elements). The elastic elements are defined for the present only in functional terms, since their anatomical counterparts have not been established.

The true meaning of the *active state* is not fully understood. In general, it refers to the ability of the contractile system to bear a load after activation of the muscle. Its magnitude and duration depend on the concentration and persistence of free Ca^{++} in the myoplasm as well as that bound to active sites. However, it cannot be measured experimentally, and therefore its value as a means of providing a better understanding of the contractile process is questionable. Velocity and force of contraction are a function of the intracellular concentration of free calcium ions as well as the mechanical load on the contractile system. Force and velocity are inversely related, so that with no load, force is negligible, and velocity is maximal, whereas with an isometric contraction, where no external shortening occurs, force is maximal, and velocity is zero.

The sequence of events in an afterloaded isotonic contraction of a papillary muscle is illustrated in Fig. 29-5. Point *A* represents the resting state in which the preload is responsible for the existing initial stretch. With stimulation the contractile element begins to shorten, and at point *B* the series elastic element has been stretched, but the load has not yet been lifted (the isometric phase of the contraction). This stretch of the series elastic element (an expression of the muscle extensibility) consumes a certain amount

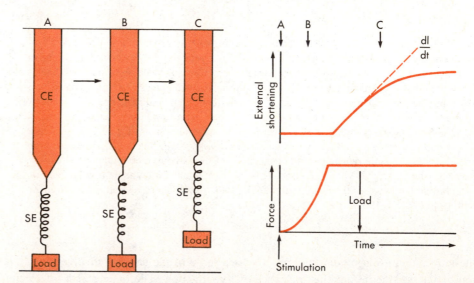

Fig. 29-5 ■ Model for an afterloaded isotonic contraction of a papillary muscle. *A,* At rest; *B,* partial contraction of the contractile element *(CE)* with stretch of the series elastic element *(SE)* but without external shortening (the isometric phase of the contraction); *C,* further contraction of the *CE* with external shortening and lifting of the afterload. The tangent *(dl/dt)* to the initial slope of the shortening curve on the right is the velocity of initial shortening. (Redrawn from Sonnenblick, E.H.: The myocardial cell, Philadelphia, 1966, University of Pennsylvania Press.)

of energy. Therefore the energy used for shortening of the muscle is actually less than the total energy expenditure in a single contraction. Stretch of the series elastic element is represented in the diagram at the lower right as a progressive rise in force with no external shortening. At point *C* the force developed by the contractile element has equaled the load, and the load has been raised without further stretch of the series elastic element. This is represented in the diagram on the right as external shortening of the muscle without a further increase in force.

The initial slope (dashed tangent) of the shortening curve (Fig. 29-5, upper right section) depicts the initial rate of shortening (change in length with change in time—dl/dt). Since the initial velocity is dependent on the magnitude of the afterload, a series of initial velocities can be obtained from a papillary muscle by varying the afterload. Furthermore the onset of shortening is delayed (longer isometric phase), but the time from stimulation to maximal shortening is unchanged when afterload is increased.

If the initial velocity of shortening is plotted against the afterload, the force/velocity curves shown in Fig. 29-6, *A,* are obtained. The maximal velocity *(V_max)* may be estimated by extrapolation of the force/velocity curve back to zero load (as indicated by the *dashed lines* in Fig. 29-6, *A)* and represents the maximum rate of cycling of the cross-bridges.

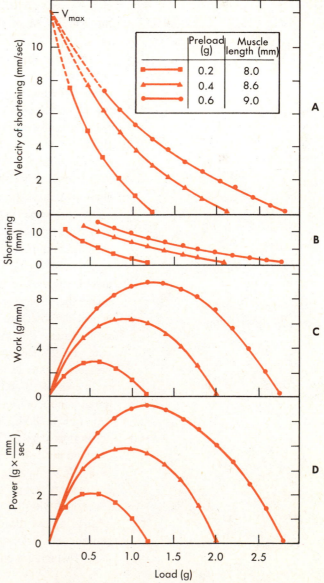

Fig. 29-6 ■ The effect of increasing initial length of a cat papillary muscle on the force-velocity relationship, degree of shortening, muscle work, and muscle power. (Redrawn and reproduced by permission from Sonnenblick, E.H.: Am. J. Physiol. **202:**931, 1962.)

Contractility can be defined as a change in developed tension at a given resting fiber length. However, it also may be defined in terms of a change in V_{max}. Augmentation of contractility is observed with certain drugs, such as norepinephrine or digitalis, and with an increase in contraction frequency (*tachycardia,* when applied to the whole heart). The increase in contractility *(positive inotropic effect)* produced by any of the preceding interventions is reflected by increments in developed tension and V_{max}.

An increase in initial fiber length produces a more forceful contraction, as shown in Fig. 29-1. However, this greater tension development is not associated with any change in contractility, as estimated by V_{max} (Fig. 29-6, *A*). Fig. 29-6 also illustrates that at any given load there is an increase in the degree of shortening, the work (shortening $\times$ load, or distance $\times$ force), and the power (velocity $\times$ load, or work/time) when the initial length of the papillary muscle is increased. It is apparent that with an increase in initial fiber length greater force may be developed, but the estimated V_{max} is the same for all three initial lengths. Thus changes in resting length may alter tension development but not contractility. This conclusion is based of course on the assumption that the displayed extrapolation of the force/velocity curves back to the vertical axis provides the true value for V_{max}. This assumption has recently been challenged; some investigators have failed to obtain a hyperbolic force/velocity relationship and have observed changes in V_{max} with changes in initial length (that is, V_{max} is *length dependent*). For this and several other reasons estimates of V_{max} may not be a reliable index of contractility.

Although the experiments from which the length-tension and force-velocity relationships have been derived were carried out on papillary muscles (essentially one dimensional), the findings are applicable to some extent to the intact heart (three dimensional); for the left ventricle the preload is reflected by the ventricular pressure at the end of diastole (end-diastolic pressure), and the afterload is reflected by the aortic pressure. However, the complex changes in ventricular shape that occur in systole and the fact that fibers branch at different angles and therefore do not all contract in their optimal longitudinal axis (Fig. 29-9) make accurate determinations of contractility in the whole heart very difficult.

A reasonable index of myocardial contractility can be obtained from the contour of ventricular pressure curves (Fig. 29-7). A hypodynamic heart is characterized by an elevated end-diastolic pressure, a slowly rising ventricular pressure, and often a lower peak pressure (curve *C,* Fig. 29-7), whereas a normal ventricle under adrenergic stimulation shows a reduced end-diastolic pressure, a fast rising ventricular pressure, and a higher peak pressure (curve *B,* Fig. 29-7). The slope of the ascending limb of the ventricular pressure curve indicates the maximum rate of force development by the ventricle (maximum rate of changes in pressure with time; maximum *dP/dt,* as illustrated by the tangents to the steepest portion of the ascending limbs of the ventricular pressure curves in Fig. 29-7). The slope is maximal during the isovolumetric phase of systole (p. 499) and at any given degree of ventricular filling provides an index of the initial contraction velocity and hence of contractility. Similarly one can obtain an indication of the initial velocity of blood flow in the ascending aorta (the initial slope of the aortic flow curve, Fig. 29-12). The *ejection fraction,* which is the ratio of the volume of blood ejected from the left ventricle per beat (stroke volume) to the volume of blood in the left ventricle at the end of diastole, is widely used clinically as an index of contractility. Other measurements or combinations of measurements that in general are concerned with the magnitude or velocity of the ventricular contraction also have been used to assess the contractile state of the cardiac muscle. There is no index that is entirely satisfactory at present, and this undoubtedly accounts for the large number of indices which are currently in use.

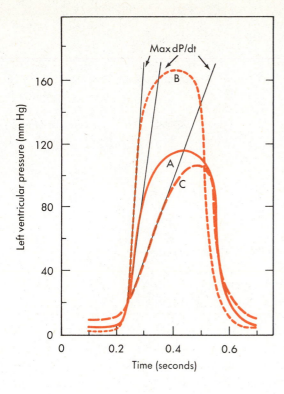

Fig. 29-7 ■ Left ventricular pressure curves with tangents drawn to the steepest portion of the ascending limbs to indicate maximum dP/dt values. *A,* Control; *B,* hyperdynamic heart, as with norepinephrine administration; *C,* hypodynamic heart, as in cardiac failure.

Excitation-contraction coupling. In addition to influences by neural, neurohumoral (for example, norepinephrine), and exogenous agents (for example, cardiac glycosides), myocardial contractility also is affected by the concentration of cations in the plasma. The earliest studies on isolated hearts perfused with isotonic salt solutions indicated the need for optimal concentrations of sodium, potassium, and calcium. In the absence of sodium the heart will not beat because the action potential is dependent on extracellular sodium ions. In contrast, the resting membrane potential is independent of the sodium ion gradient across the membrane (Fig. 27-6). Under normal conditions the extracellular potassium concentration is low (4mM), and further reduction in extracellular potassium has little effect on myocardial excitation and contraction. However, increases in extracellular potassium, if great enough, produce depolarization and loss of excitability of the myocardial cells and cardiac arrest in diastole. In addition to effects on membrane excitability, calcium has a direct effect on myocardial contractility. Increases in extracellular calcium concentrations enhance myocardial contractility and, at unphysiologically high levels, produce cardiac arrest in systole (rigor), whereas decreases in extracellular calcium concentrations reduce contractility. That free intracellular calcium is the agent which initiates and sustains the contractile state of the myocardium is well documented (Chapter 22). As indicated in Chapter 27, the electrical excitation alters the membrane permeability to Ca^{++}, thereby allowing Ca^{++} to enter the cell from the interstitial fluid (especially the T-tubules) and also releasing Ca^{++} from its storage sites in the sarcoplasmic reticulum within the cell. This calcium binds to troponin and brings about some conformational change in the thin filament, which permits interaction between the actin and myosin filaments through the crossbridges (myosin heads) (Chapter 22). ATP is dephosphorylated to ADP by the ATPase present in the crossbridges and provides the energy for movement of the crossbridges in a ratchetlike fashion. This action increases the overlap of the actin and myosin filaments, resulting in shortening of the sarcomeres and contraction of the myocardium. Relaxation occurs when free calcium is taken up by the sarcoplasmic reticulum or is bound to the cell membrane and extruded from the cell. This oversimplification of excitation-contraction coupling is diagrammed in Fig. 29-8.

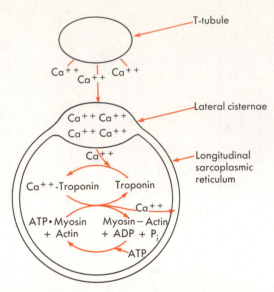

Fig. 29-8 ■ Ca^{++} from the interstitial fluid (primarily from T-tubules that penetrate the cell as invaginations of the sarcolemma but do not directly communicate with the intracellular compartment) enters the myoplasm of the cell. Ca^{++} is also released from the sarcoplasmic reticulum. The Ca^{++} reaches the affinity sites of troponin, which enables the actin and myosin filaments to interact and bring about contraction. Relaxation is initiated by the removal of Ca^{++} from the myoplasm by the sarcoplasmic reticulum. (Reproduced by permission from Morkin, E., and LaRaia, P.S.: N. Engl. J. Med. **290**:445, 1974. Reprinted, by permission of the New England Journal of Medicine.)

■ *Cardiac chambers*

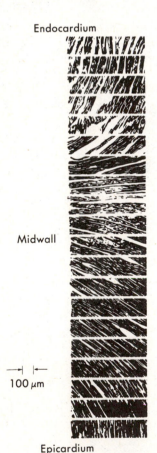

Endocardium

Midwall

⊢ ⊣
100 μm

Epicardium

The atria are thin-walled, low-pressure chambers that function more as large reservoir conduits of blood for their respective ventricles than as important pumps for the forward propulsion of blood. The ventricles once were thought to be made up of bands of muscle. However, it now appears that they are formed by a continuum of muscle fibers that take origin from the fibrous skeleton at the base of the heart (chiefly around the aortic orifice). These fibers sweep toward the apex at the epicardial surface and also pass toward the endocardium as they gradually undergo a 180-degree change in direction to lie parallel to the epicardial fibers and form the endocardium and papillary muscles (Fig. 29-9). At the apex of the heart the fibers twist and turn inward to form papillary muscles, whereas at the base and around the valve orifices they form a thick powerful muscle that not only decreases ventricular circumference for ejection of blood but also narrows the AV valve orifices as an aid to valve closure. In addition to a reduction in circumference, ventricular ejection is accomplished by a decrease in the longitudinal axis with descent of the base of the heart. The earlier contraction of the apical part of the ventricles coupled with approximation of the ventricular walls propels the blood toward the outflow tracts. The right ventricle, which develops a mean pressure about one seventh that developed by the left ventricle, is considerably thinner than the left.

Fig. 29-9 ■ Sequence of photomicrographs showing fiber angles in successive sections taken from the middle of the free wall of the left ventricle from a heart in systole. The sections are parallel to the epicardial plane. Fiber angle is +90 degrees at the endocardium, running through 0 degrees at the midwall to −90 degrees at the epicardium. (From Streeter, D.D., Jr., et al.: Circ. Res. **24**:339, 1969. By permission of The American Heart Association, Inc.)

■ *Cardiac valves*

The cardiac valves consist of thin flaps of flexible, tough endothelium-covered fibrous tissue firmly attached at the base to the fibrous valve rings. Movements of the valve leaflets are essentially passive, and the orientation of the cardiac valves is responsible for unidirectional flow of blood through the heart. There are two types of valves in the heart: the *atrioventricular valves,* or AV valves, and the *semilunar valves* (Figs. 29-10 and 29-11).

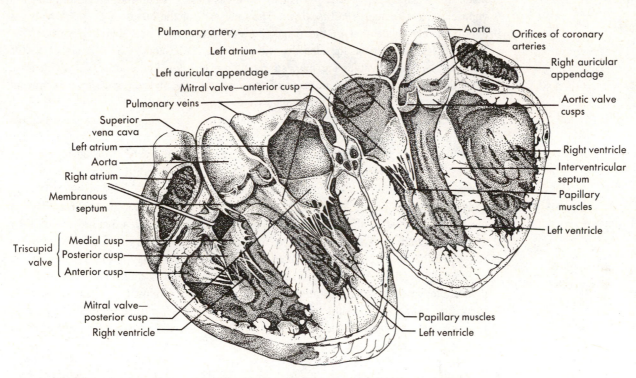

Fig. 29-10 ■ Drawing of a heart split perpendicular to the interventricular septum to illustrate the anatomical relationships of the leaflets of the AV and aortic valves.

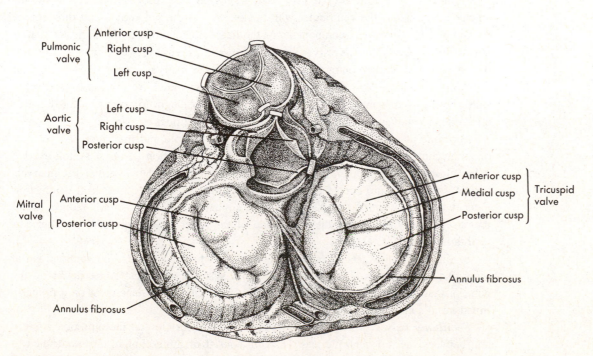

Fig. 29-11 ■ Drawing of the four cardiac valves as viewed from the base of the heart. Note the manner in which the leaflets overlap in the closed valves.

Atrioventricular valves. The valve between the right atrium and right ventricle is made up of three cusps *(tricuspid valve),* whereas that between the left atrium and left ventricle has two cusps *(mitral valve).* The total area of the cusps of each AV valve is approximately twice that of the respective AV orifice, so that there is considerable overlap of the leaflets in the closed position (Figs. 29-10 and 29-11). Attached to the free edges of these valves are fine, strong filaments *(chordae tendineae),* which arise from the powerful papillary muscles of the respective ventricles and serve to prevent eversion of the valves during ventricular systole.

The mechanism of closure of the AV valves has been the subject of considerable investigation, and a number of factors are thought to play a role in approximating the valve leaflets. In the normal heart the valve leaflets are relatively close during ventricular filling and provide a funnel for the transfer of blood from atrium to ventricle. This partial approximation of the valve surfaces during diastole is believed to be caused by eddy currents behind the leaflets and possibly also by some tension on the free edges of the valves, exerted by the chordae tendineae and papillary muscles that are stretched by the filling ventricle. The finding of muscle fibers of atrial origin in the mitral valve has suggested that contraction of these fibers during atrial systole may aid in this positioning of the valve leaflets. With atrial systole the abrupt ejection of additional blood into the distensible ventricles that are already filled with blood, coupled with the subsequent relaxation of the atria, results in a small reversal of the AV pressure gradient sufficient to close the AV valves. Thus the AV valves may actually close before the onset of ventricular systole.

The rapid rise in ventricular pressure in early ventricular systole bulges the closed valves toward the atria and presses the leaflets tightly together. Simultaneous contraction of the papillary muscles restrains the edges of the valves from being everted into the atria. The concept that the AV valves close prior to the onset of ventricular contraction has been challenged by studies employing *echocardiography* in humans. This technique consists of sending short pulses of high-frequency sound waves (ultrasound) through the chest tissues and the heart and recording the echoes reflected from the various structures. The timing and the pattern of the reflected waves provide information such as the diameter of the heart, the ventricular wall thickness, and the magnitude and direction of the movements of various components of the heart. Echocardiography indicates that at normal PR intervals the AV valves are closed by the increase in ventricular pressures at the onset of ventricular contraction. Only when the PR interval exceeds 0.18 second do the AV valves close as a result of atrial relaxation (and pressure reversal) prior to the start of ventricular contraction. If the ventricles do not begin to contract shortly thereafter, the AV valves are reopened as atrial pressure continues to rise.

The time interval between atrial and ventricular systole and the vigor and velocity of atrial contraction and relaxation determine the contribution of atrial systole to AV valve closure. If ventricular systole follows atrial systole by the normal PR interval, then AV valve closure occurs by ventricular systole. A strong, rapid atrial contraction induces a larger ventricular pressure elevation, and a quick relaxation reduces atrial pressure, thereby increasing the magnitude of the reversed AV pressure gradient and facilitating AV valve closure.

Studies with intravascular radiopaque dyes, in which no perceptible ventriculoatrial regurgitation occurred in the absence of atrial systole, have cast some doubt on the concept that closure of the AV valves without regurgitation is dependent on a properly timed atrial contraction.

Semilunar valves. The valves between the right ventricle and the pulmonary artery and between the left ventricle and the aorta consist of three cuplike cusps attached to the valve rings (Figs. 29-10 and 29-11). At the end of the reduced ejection phase of

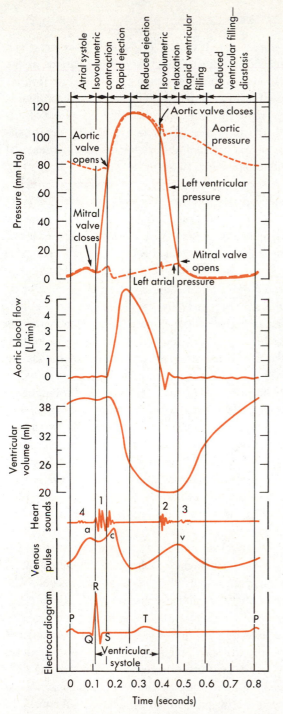

Fig. 29-12 ■ Left atrial, aortic, and left ventricular pressure pulses correlated in time with aortic flow, ventricular volume, heart sounds, venous pulse, and electrocardiogram for a complete cardiac cycle in the dog.

ventricular systole there is a brief reversal of blood flow toward the ventricles (shown as a negative flow in the phasic aortic flow curve in Fig. 29-12) that snaps the cusps together and prevents regurgitation of blood into the ventricles. During ventricular systole the cusps do not lie back against the walls of the pulmonary artery and aorta but float in the bloodstream approximately midway between the vessel walls and their closed position. Behind the semilumar valves are small outpocketings of the pulmonary artery and aorta *(sinuses of Valsalva),* where eddy currents develop that tend to keep the valve cusps away from the vessel walls. The orifices of the right and left coronary arteries are located behind the right and the left cusps, respectively, of the aortic valve. Were it not for the presence of the sinuses of Valsalva and the eddy currents developed therein, the coronary ostia could be blocked by the valve cusps.

■ *The pericardium*

The pericardium is an epithelized fibrous sac. It closely invests the entire heart and the cardiac portion of the great vessels and is reflected onto the cardiac surface as the epicardium. It normally contains a small amount of fluid, which provides lubrication for the continuous movement of the enclosed heart. The distensibility of the pericardium is small, so that it strongly resists a large, rapid increase in cardiac size. By virtue of this characteristic, the pericardium plays a role in preventing sudden overdistension of the heart. However, in congenital absence of the pericardium or after its surgical removal cardiac function appears to be well within physiological limits.

■ *Heart sounds*

There are usually four sounds produced by the heart, but only two are ordinarily audible through a stethoscope. With the aid of electronic amplification the less intense sounds can be detected and recorded graphically as a *phonocardiogram*. This means of registering heart sounds that may be inaudible to the human ear aids in delineating the precise timing of the heart sounds relative to other events in the cardiac cycle.

The first heart sound is initiated at the onset of ventricular systole (Fig. 29-12) and consists of a series of vibrations of mixed, unrelated, low frequencies (a noise). It is the loudest and longest of the heart sounds, has a crescendo-decrescendo quality, and is heard best over the apical region of the heart. The tricuspid valve sounds are heard best in the fifth intercostal space just to the left of the sternum, and the mitral sounds are heard best in the fifth intercostal space at the cardiac apex. The first heart sound is caused chiefly by the oscillation of blood in the ventricular chambers and vibration of the chamber walls. The vibrations are engendered in part by the abrupt rise of ventricular pressure with acceleration of blood back toward the atria but primarily by sudden tension and recoil of the AV valves and adjacent structures with deceleration of the blood by closure of the AV valves. The vibrations of the ventricles and the contained blood are transmitted through surrounding tissues and reach the chest wall, where they may be heard or recorded. The intensity of the first sound is primarily a function of the force of ventricular contraction but also of the interval between atrial and ventricular systoles (p. 496).

The second heart sound, which occurs on closure of the semilunar valves (Fig. 29-12), is composed of higher frequency vibrations (higher pitch), is of shorter duration and lower intensity, and has a more snapping quality than the first heart sound. The second sound is caused by abrupt closure of the semilunar valves, which initiates oscillations of the columns of blood and the tensed vessel walls by the stretch and recoil of the closed valve. The second sound, caused by closure of the pulmonic valve, is heard best in the second thoracic interspace just to the left of the sternum, whereas that caused by closure of the aortic valve is heard best in the same intercostal space but to the right of the sternum. Conditions that bring about a more rapid closure of the semilunar valves, such as increases in pulmonary artery or aortic pressure (for example, pulmonary or systemic hypertension), will increase the intensity of the second heart sound. In the adult the aortic valve sound is usually louder than the pulmonic, but in cases of pulmonary hypertension the reverse is often true.

A normal phonocardiogram taken simultaneously with an electrocardiogram is illustrated in Fig. 29-13. Note that the first sound, which starts just beyond the peak of the R wave, is composed of irregular waves and is of greater intensity and duration than the second sound, which appears at the end of the T wave. A third and fourth heart sound do not appear on this record.

The third heart sound, which is more frequently heard in children with thin chest walls or in patients with left ventricular failure, consists of a few low-intensity, low-frequency vibrations heard best in the region of the apex. It occurs in early diastole and is believed to be the result of vibrations of the ventricular walls caused by abrupt accel-

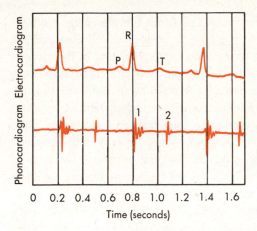

Fig. 29-13 ■ Phonocardiogram illustrating the first and second heart sounds and their relationship to the P, R, and T waves of the electrocardiogram. (Time lines = 0.04 second.)

eration and deceleration of blood entering the ventricles on opening of the atrioventricular valves (Fig. 29-12).

A fourth or atrial sound, consisting of a few low-frequency oscillations, is heard occasionally in normal individuals. It is caused by oscillation of blood and cardiac chambers created by atrial contraction (Fig. 29-12).

Since the onset and termination of right and left ventricular systoles are not precisely synchronous, differences in time of vibration of the two atrioventricular valves or of the two semilunar valves sometimes can be detected with the stethoscope. Such asynchrony of valve vibrations, which may sometimes indicate abnormal cardiac function, is manifested as a *split sound* over the apex of the heart for the atrioventricular valves and over the base for the semilunar valves. The heart sounds also may be altered by deformities of the valves; *murmurs* may be produced, and the character of the murmur serves as an important guide in the diagnosis of valvular disease. When the third and fourth (atrial) sounds are accentuated, as occurs in certain abnormal conditions, triplets of sounds may occur, resembling the sound of a galloping horse. These *gallop rhythms* are essentially of two types: *presystolic gallop* caused by accentuation of the atrial sound and *protodiastolic gallop* caused by accentuation of the third heart sound.

Isovolumetric contraction. The onset of ventricular contraction coincides with the peak of the R wave of the electrocardiogram and the initial vibration of the first heart sound. It is indicated on the ventricular pressure curve as the earliest rise in ventricular pressure after atrial contraction. The interval of time between the start of ventricular systole and the opening of the semilunar valves (when ventricular pressure rises abruptly) is termed *isovolumetric contraction,* since ventricular volume is constant during this brief period (Fig. 29-12).

The increment in ventricular pressure during isovolumetric contraction is transmitted across the closed valves and is evident in Fig. 29-12 as a small oscillation on the aortic pressure curve. Isovolumetric contraction also has been referred to as isometric contraction. However, some fibers shorten, and others lengthen, as evidenced by changes in ventricular shape; therefore it is not a true isometric contraction.

Ejection. Opening of the semilunar valves marks the onset of the ejection phase, which may be subdivided into an earlier, shorter phase *(rapid ejection)* and a later, longer phase *(reduced ejection).* The rapid ejection phase is distinguished from the reduced ejection phase by the sharp rise in ventricular and aortic pressures that terminates at the peak ventricular and aortic pressures, a more abrupt decrease in ventricular volume, and a greater aortic blood flow (Fig. 29-12). The sharp decrease in the left atrial pressure curve at the onset of ejection results from the descent of the base of the heart and stretch of the atria. During the reduced ejection period, runoff of blood from the

■ ***Cardiac cycle***
■ *Ventricular systole*

aorta to the periphery exceeds ventricular output, and therefore aortic pressure declines. Throughout ventricular systole the blood returning to the atria produces a progressive increase in atrial pressure. Note that during approximately the first third of the ejection period left ventricular pressure slightly exceeds aortic pressure, and flow accelerates (continues to increase), whereas during the last two thirds of ventricular ejection the reverse holds true. This reversal of the ventricular/aortic pressure gradient in the presence of continued flow of blood from left ventricle to aorta (caused by the momentum of the forward blood flow) is the result of the storage of potential energy in the stretched arterial walls, which produces a deceleration of blood flow into the aorta. The peak of the flow curve coincides in time with the point at which the left ventricular pressure curve intersects the aortic pressure curve during ejection. Thereafter flow decelerates (continues to decrease) because the pressure gradient has been reversed. With right ventricular ejection there is shortening of the free wall of the right ventricle (descent of the tricuspid valve ring) in addition to lateral compression of the chamber. However, with left ventricular ejection there is very little shortening of the base-to-apex axis, and ejection is accomplished chiefly by compression of the left ventricular chamber.

The venous pulse curve shown in Fig. 29-12 has been taken from a jugular vein, and the *c* wave is caused by impact of the adjacent common carotid artery. Note that except for the *c* wave the venous pulse closely follows the atrial pressure curve.

At the end of ejection a volume of blood approximately equal to that ejected during systole remains in the ventricular cavities. This *residual volume* is fairly constant in normal hearts but is smaller with increased heart rate or reduced outflow resistance and larger when the opposite conditions prevail. An increase in myocardial contractility may decrease residual volume (or increase stroke volume and ejection fraction), especially in the depressed heart. With severely hypodynamic and dilated hearts, as in *heart failure,* the residual volume can become many times greater than the stroke volume. In addition to serving as a small adjustable blood reservoir, the residual volume to a limited degree can permit transient disparities between the outputs of the two ventricles.

■ *Ventricular diastole*

Isovolumetric relaxation. Closure of the aortic valve produces the incisura on the descending limb of the aortic pressure curve and the second heart sound (with some vibrations evident on the atrial pressure curve) and marks the end of ventricular systole (Fig. 29-12). The period between closure of the semilunar valves and opening of the AV valves is termed *isovolumetric* (or *isometric*) relaxation and is characterized by a precipitous fall in ventricular pressure without a change in ventricular volume.

Rapid filling phase. The major part of ventricular filling occurs immediately on opening of the atrioventricular valves when blood that had returned to the atria during the previous ventricular systole is abruptly released into the relaxing ventricles. This period of ventricular filling is called the *rapid filling phase.* In Fig. 29-12 the onset of the rapid filling phase is indicated by the decrease in left ventricular pressure below left atrial pressure, resulting in the opening of the mitral valve. The rapid flow of blood from atria to relaxing ventricles produces a decrease in atrial and ventricular pressures and a sharp increase in ventricular volume.

The decrease in pressure from the peak of the *v* wave of the venous pulse is caused by transmission of the pressure decrease incident to the abrupt transfer of blood from the right atrium to the right ventricle with opening of the tricuspid valve. Rapid ventricular filling is responsible for the few small vibrations that constitute the third heart sound. Elastic recoil of the previous ventricular contraction may aid in drawing blood into the relaxing ventricle when residual volume is small, but it probably does not play a significant role in ventricular filling under most normal conditions.

Diastasis. The rapid filling phase is followed by a phase of slow filling, called *diastasis.* During diastasis, blood returning from the periphery flows into the right ven-

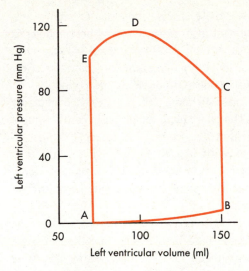

Fig. 29-14 ■ Pressure/volume loop of the left ventricle for a single cardiac cycle *(ABCDE)*.

tricle and blood from the lungs into the left ventricle. This small, slow addition to ventricular filling is indicated by a gradual rise in atrial, ventricular, and venous pressures and in ventricular volume (Fig. 29-12).

Pressure/volume relationship. The changes in left ventricular pressure and volume throughout the cardiac cycle are summarized diagrammatically in Fig. 29-14. The element of time is not considered in this pressure/volume loop. Diastolic filling starts at *A* and terminates at *B,* when the mitral valve closes. Note that there is only a small increase in pressure with the increase in ventricular volume during diastole. With isovolumetric contraction (*B* to *C*) there is a steep rise in pressure with no change in ventricular volume. At *C* the aortic valve opens, and during the first phase of ejection (rapid ejection—*C* to *D*) there is a large reduction in volume associated with a continued but lesser increase in ventricular pressure than that which occurred during isovolumetric contraction. This volume reduction is followed by reduced ejection (*D* to *E*) and a small decrease in ventricular pressure. The aortic valve closes at *E,* and this event is followed by isovolumetric relaxation (*E* to *A*), which is characterized by a sharp drop in pressure with no change in volume. The mitral valve opens at *A* to complete one cardiac cycle.

Atrial systole. The onset of atrial systole occurs soon after the beginning of the P wave of the electrocardiogram (curve of atrial depolarization), and the transfer of blood from atrium to ventricle made by the peristalsis-like wave of atrial contraction completes the period of ventricular filling. Atrial systole is responsible for the small increases in atrial, ventricular, and venous (*a* wave) pressures as well as in ventricular volume shown in Fig. 29-12. Throughout ventricular diastole, atrial pressure barely exceeds ventricular pressure, indicating a low-resistance pathway across the open AV valves during ventricular filling. A few small vibrations produced by atrial systole constitute the fourth, or atrial, heart sound.

Since there are no valves at the junctions of the venae cavae and right atrium or of the pulmonary veins and left atrium, atrial contraction can force blood in both directions. Actually little blood is pumped back into the venous tributaries during the brief atrial contraction, mainly because of the inertia of the inflowing blood.

Atrial contraction certainly is not essential for ventricular filling, as can be observed in atrial fibrillation or complete heart block. However, its contribution is governed to a great extent by the heart rate and the structure of the atrioventricular valves. At slow heart rates filling practically ceases toward the end of diastasis, and atrial contraction contributes little additional filling. During tachycardia diastasis is abbreviated and the atrial contribution can become substantial, especially if atrial contraction follows immediately after the rapid filling phase when the atrioventricular pressure gradient is maximal. Should tachycardia become so great that the rapid filling phase is encroached on, atrial contraction assumes great importance in rapidly propelling blood into the ventricle during this brief period of the cardiac cycle. Of course, if the period of ventricular relaxation is so brief that filling is seriously impaired, even atrial contraction cannot prevent inadequate ventricular filling. The consequent reduction in cardiac output may result in syncope. Obviously, if atrial contraction occurs simultaneously with ventricular contraction, no atrial contribution to ventricular filling can occur. In certain disease states the atrioventricular valves may be markedly narrowed *(stenotic)*. Under such conditions atrial contraction may play a much more important role in ventricular filling than it does in the normal heart.

Ventricular contraction has been shown to aid indirectly in right ventricular filling by its effect on the right atrium. Descent of the base of the heart stretches the right atrium downward, and pressure measurements indicate a sharp reduction in right atrial pressure, which results in acceleration of blood flow in the venae cavae toward the heart. Enhancement of venous return by ventricular systole provides an additional supply of atrial blood for ventricular filling during the subsequent rapid filling phase of diastole. However, this mechanism is probably of little physiological importance, except possibly at rapid heart rates.

■ *Bibliography*

Journal articles

Abbott, B.C., and Mommaerts, W.F.H.M.: A study of inotropic mechanisms in the papillary muscle preparation, J. Gen. Physiol. **42:**533, 1959.

Alpert, N.R., Hamrell, B.B., and Mulieri, L.A.: Heart muscle mechanics, Annu. Rev. Physiol. **41:**521, 1979.

Armour, J.A., and Randall, W.C.: Structural basis for cardiac function, Am. J. Physiol. **218:**1517, 1970.

Brutsaert, D.L., Claes, V.A., and Sonnenblick, E.H.: The velocity of shortening of unloaded heart muscle relative to the length-tension relation, Circ. Res. **29:**63, 1971.

Burggraf, G.W., and Craige, E.: The first heart sound in complete heart block; phono-echocardiographic correlations, Circulation **50:**17, 1974.

Fabiato, A., and Fabiato, F.: Calcium and cardiac excitation—contraction coupling, Annu. Rev. Physiol. **41:**473, 1979.

Jewell, B.R.: A reexamination of the influence of muscle length on myocardial performance, Circ. Res. **40:**221, 1977.

Lau, V.K., and Sagawa, K.: Model analysis of the contribution of atrial contraction to ventricular filling, Ann. Biomed. Eng. **7:**167, 1979.

Little, R.C.: The mechanism of closure of the mitral valve: a continuing controversy, Circulation **59:**615, 1979.

Nayler, W.G., and Seabra-Gomes, R.: Excitation-contraction coupling in cardiac muscle, Prog. Cardiovasc. Dis. **18:**75, 1976.

Noble, M.I.M.: The contribution of blood momentum to left ventricular ejection in the dog, Circ. Res. **23:**663, 1968.

Ross, J., Jr., and Sobel, B.E.: Regulation of cardiac contraction, Annu. Rev. Physiol. **34:**47, 1972.

Sagawa, K.: The ventricular pressure-volume diagram revisited, Circ. Res. **43:**677, 1978.

Sonnenblick, E.: Force-velocity relations in mammalian heart muscle, Am. J. Physiol. **202:**931, 1962.

Streeter, D.D., Jr., et al.: Fiber orientation in the canine left ventricle during diastole and systole, Circ. Res. **24:**339, 1969.

Weber, K.T., and Janicki, J.S., editors: Cardiac mechanics (symposium), Fed. Proc. **39:**131, 1980.

Zaky, A., Stenimentz, E., and Feigenbaum, H.: Role of atrium in closure of mitral valve in man, Am. J. Physiol. **217:**1652, 1969.

Books and monographs

Brady, A.J.: Mechanical properties of cardiac fibers. In Handbook of physiology; Section 2: The cardiovascular system—the heart, vol. I, Washington, D.C., 1979, American Physiological Society.

Braunwald, E., Ross, J., Jr., and Sonnenblick, E.H.: Mechanisms of contraction of the normal and failing heart, ed. 2, Boston, 1976, Little, Brown & Co.

Katz, A.M.: Physiology of the heart, New York, 1977, Raven Press.

Langer, G.A., and Brady, A.J.: The mammalian myocardium, New York, 1974, John Wiley & Sons, Inc.

McKusick, V.A.: Cardiovascular sound in health and disease, Baltimore, 1958, Williams & Wilkins Co.

Mirsky, I., Ghista, D.N., and Sandler, H.: Cardiac mechanics: physiological, clinical, and mathematical considerations, New York, 1974, John Wiley & Sons, Inc.

Parmley, W.W., and Talbot, L.: Heart as a pump. In Handbook of physiology; Section 2: The cardiovascular system—the heart, vol. I, Washington, D.C., 1979, American Physiological Society.

Sommer, J.R., and Johnson, E.A.: Ultrastructure of cardiac muscle. In Handbook of physiology; Section 2: The cardiovascular system—the heart, vol. I, Washington, D.C., 1979, American Physiological Society.

chapter 30

The arterial system

■ *Hydraulic filter*

The principal function of the systemic and pulmonary arterial systems is to distribute blood to the capillary beds throughout the body. The arterioles, the terminal components of this system, serve to regulate the fractional distribution among the various capillary beds. Between the arterioles and the heart, the aorta and pulmonary artery and their major branches constitute a system of conduits of considerable volume and distensibility. An arterial system composed of elastic conduits and high-resistance terminals constitutes a *hydraulic filter* analogous to the resistance-capacitance (R-C) filters of electrical circuits.

Hydraulic filtering enables the intermittent output of the heart to be converted to a steady flow through the capillaries. This important function of the large elastic arteries first was recognized in 1834 by Ernst Weber of Leipzig. He explained the similarity between the distensible arteries and the *Windkessels* of the fire engines of that day. In those devices there was a large volume of air trapped in a container between the inflow and outflow ends of the manually operated engine. The *Windkessel* converted the intermittent inflow to a steady outflow of water at the nozzle of the fire hose.

The analogous function of the large elastic arteries is illustrated in Fig. 30-1. The entire stroke volume is discharged into the arterial system during systole, which occupies approximately one third the duration of the cardiac cycle at normal heart rates. In fact, as described in Chapter 29, most of the stroke volume is pumped during the rapid ejection phase, which constitutes only a fraction of total systole. Part of the energy of cardiac contraction is dissipated as forward capillary flow during systole; the remainder is stored as potential energy, in that much of the stroke volume is retained by the distensible arteries. During diastole the elastic recoil of the arterial walls converts this potential energy into capillary blood flow. If the arterial walls were rigid, then capillary flow would cease during diastole.

Hydraulic filtering also plays an important role in minimizing the work load of the heart. More work is required to pump a given flow intermittently than steadily, and the more effective the filtering, the less the magnitude of this excess work. This principle was well recognized by Weber over a century ago. A simple example will illustrate this point.

Consider first the steady flow of a fluid at a rate of 100 ml/second through a hydraulic system with a resistance of 1 mm Hg/ml/second. This would result in a constant pressure of 100 mm Hg (Fig. 30-2, *A*). Neglecting any inertial effect, hydraulic work, W, may be defined as

$$W = \int_{t_1}^{t_2} P dV$$

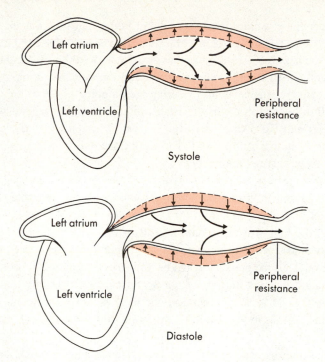

Fig. 30-1 ■ During ventricular systole the stroke volume ejected by the ventricle results in some forward capillary flow, but most of the ejected volume is stored in the elastic arteries. During ventricular diastole the elastic recoil of the arterial walls maintains capillary flow throughout the remainder of the cardiac cycle.

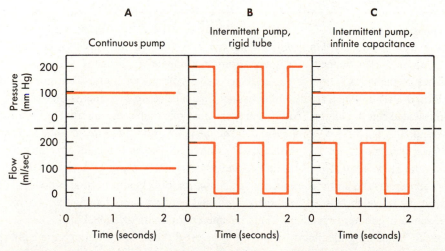

Fig. 30-2 ■ The relationships between pressure and flow for three systems, in each of which the flow is 100 ml/second and the resistance is 1 mm Hg/ml/second. In system *A* the flow is steady, and the distensibility of the conduit is immaterial. In systems *B* and *C* the flow is intermittent; it is steady for half the cycle and ceases for the remainder of the cycle. In system *B* the conduit is rigid, whereas in system *C* the conduit is infinitely distensible, resulting in perfect filtering of the pressure. In systems *A* and *C* the work per second is 10,0000 mm Hg-ml (or 1.33×10^7 dyne-cm); in system *B* the work per second is twice as great.

That is, each small increment of volume pumped is multiplied by the pressure existing at the time, and the products, dW, are integrated over the time interval of interest, $t_2 - t_1$, to give the total work, W. For steady flow, W = PV. In the example in Fig. 30-2, *A*, the work done in pumping the fluid for 1 second would be 10,000 mm Hg-ml (or 1.33×10^7 dyne-cm).

Next consider the example of an intermittent pump that puts out the same volume per second but pumps the entire volume at a steady rate over 0.5 second and then pumps nothing during the next 0.5 second. It pumps at the rate of 200 ml/second for 0.5 second (Fig. 30-2, *B* and *C*). In *B* the conduit is rigid, and the fluid is incompressible, but the system has the same resistance as in *A*. During the pumping phase of the cycle (systole) the flow of 200 ml/second through a resistance of 1 mm Hg/ml/second would produce a pressure of 200 mm Hg. During the filling phase of the pump (diastole) the pressure would be 0 mm Hg in this rigid system. The work done during systole would be 20,000 mm Hg-ml, or twice that required in *A*.

If the system were very distensible, hydraulic filtering would be very effective, and the pressure would remain virtually constant throughout the entire cycle (Fig. 30-2, *C*). Of the 100 ml of fluid pumped during the 0.5 second of systole, only 50 ml would be emitted through the high-resistance outflow end of the system during systole. The remaining 50 ml would be stored by the distensible conduit during systole and would flow out during diastole. Thus the pressure would be virtually constant at 100 mm Hg throughout the cycle. The fluid pumped during systole would be ejected at only half the pressure that prevailed in Fig. 30-2, *B*, and therefore the work would be only half as great. With nearly perfect filtering, as in Fig. 30-2, *C*, the work would be identical to that for steady flow (Fig. 30-2, *A*).

Naturally the filtering accomplished by the systemic and pulmonic arterial systems is at some level intermediate between the examples in Fig. 30-2, *B* and *C*. It has been estimated that under average normal conditions the additional work imposed by intermittency of pumping, in excess of that for the steady-flow case, is about 35% for the right ventricle and about 10% for the left ventricle. These fractions change, however, with variations in heart rate, peripheral resistance, and arterial distensibility.

◼ *Arterial elasticity*

The elastic properties of the arterial wall may be appreciated by considering first the *static pressure/volume relationship* for the aorta. To obtain the curves shown in Fig. 30-3, aortas were obtained at autopsy from individuals in different age groups. All branches of the aorta were ligated, and successive volumes of liquid were injected into this closed elastic system in the same manner that successive increments of water might be introduced into a balloon. After each increment of volume the internal pressure was measured. In Fig. 30-3 it is apparent that the curve relating pressure to volume for the youngest age group (curve *a*) is sigmoidal. Although it is quite linear over most of its extent, the slope decreases at the upper and lower ends. At any given point the slope (dV/dP) represents the aortic *capacitance* (or *compliance*). Thus in normal individuals the aortic capacitance is least at extremely high and low pressures and greatest over the usual range of pressure variations. This resembles the familiar capacitance changes encountered in inflating a balloon, where the greatest difficulty in introducing air is experienced at the beginning of inflation and again at near-maximum volume, just prior to rupture of the balloon.

It is also apparent from Fig. 30-3 that the curves become displaced downward and the slopes diminish as a function of advancing age. Thus for any given pressure above about 80 mm Hg the capacitance decreases with age, a manifestation of increased rigidity caused by progressive atherosclerosis. The heart is unable to eject its stroke volume into a rigid arterial system as rapidly as into a more compliant system. As capacitance diminishes, peak arterial pressure occurs progressively later in systole. Thus there is a significant prolongation of the rapid ejection phase of systole.

A complete representation of arterial elasticity cannot be derived from such static pressure/volume curves. When *dynamic curves* are obtained, that is, when pressure is recorded during continuous injection or withdrawal of fluid, it is found that the pressure is a function not only of volume but also of the rate of change of volume. Therefore

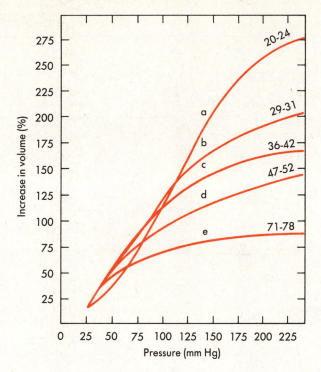

Fig 30-3 ■ Pressure/volume relationships for aortas obtained at autopsy from humans in different age groups (denoted by the numbers at the right end of each of the curves). (Redrawn from Hallock, P., and Benson, I.C.: J. Clin. Invest. **16**:595, 1937.)

the arterial wall has *viscoelastic* rather than purely elastic properties (where pressure would be a function of volume alone).

■ *Determinants of the arterial blood pressure*

The determinants of the pressure that may exist at any moment within the arterial system cannot be evaluated with great precision at present. Yet the arterial blood pressure is a quantitative measurement routinely obtained for diagnosis in most patients, and it provides a useful clue to their cardiovascular status. A simplified approach therefore is undertaken in an attempt to gain a general understanding of the principal determinants of the arterial blood pressure. To accomplish this, the determinants of the *mean arterial pressure* (defined in the following section) are analyzed. The *systolic* and *diastolic arterial pressures* are considered then as the upper and lower limits of periodic oscillations about this mean pressure. Finally, the more complex aspect of the arterial impedance is considered to explain the changes in arterial pressure as the pulse wave progresses from the origin of the aorta toward the capillaries.

The determinants of the arterial blood pressure are subdivided arbitrarily into "physical" and "physiological" factors. For the sake of simplicity the arterial system is considered as a static, elastic system, and the only two "physical" factors to be considered are the blood volume within the arterial system and the elastic characteristics (capacitance) of the system. Several "physiological" factors are considered, such as heart rate, stroke volume, cardiac output, and peripheral resistance. However, such physiological factors are shown to operate through one or both of the physical factors.

■ *Mean arterial pressure*

The *mean arterial pressure* is the average pressure during a given cardiac cycle that exists in the aorta and its major branches. It may be obtained from an arterial pressure tracing by measuring the area under the curve and dividing this area by the time interval involved (Fig. 30-4). During the registration of an arterial pressure curve the tracing

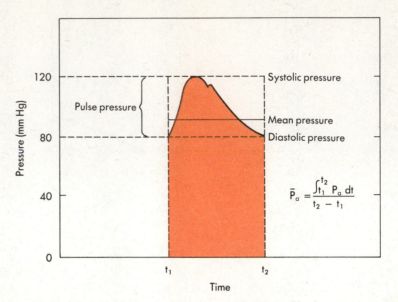

Fig. 30-4 ■ Arterial systolic, diastolic, pulse, and mean pressures. The mean arterial pressure ($\overline{P}_a$) represents the area under the arterial pressure curve *(shaded area)* divided by the cardiac cycle duration $(t_2 - t_1)$.

may be damped electronically to record the mean pressure. With a mercury manometer connected to a peripheral artery of an experimental animal, the mean pressure may be obtained by hydraulically damping the oscillations by means of a screw clamp on the tubing between the artery and manometer. The mean arterial pressure ($\overline{P}_a$) usually can be approximated satisfactorily from the measured values of the systolic (P_s) and diastolic (P_d) pressures by means of the following formula:

$$\overline{P}_a \cong P_d + \frac{1}{3}(P_s - P_d) \tag{1}$$

For the purposes of the present discussion the mean pressure is considered to be dependent only on the mean volume of blood in the arterial system and the elastic properties of the arterial walls. The arterial volume, V_a, in turn, is dependent on the rate of inflow, Q_i, from the heart into the arteries *(cardiac output)* and the rate of outflow, Q_o, from the arteries through the capillaries *(peripheral runoff)*, expressed mathematically as:

$$dV_a/dt = Q_i - Q_o \tag{2}$$

If arterial inflow exceeds outflow, then arterial volume increases, the arterial walls are stretched more, and pressure rises. The converse happens when arterial outflow exceeds inflow. When inflow equals outflow, then arterial pressure remains constant.

The change in pressure in response to an alteration of cardiac output can be better appreciated by considering some simple examples. Under control conditions, let cardiac output be 5 L/minute and mean arterial pressure ($\overline{P}_a$) be 100 mm Hg. From the definiton of total peripheral resistance

$$R = (\overline{P}_a - \overline{P}_{ra})/Q_o \tag{3}$$

If $\overline{P}_{ra}$ (right atrial pressure) is close to zero,

$$R \cong \overline{P}_a/Q_o \tag{4}$$

Therefore, in the example, R is 100/5, or 20 mm Hg/L/minute.

Now let cardiac output (Q_i) suddenly increase to 10 L/minute. Instantaneously, $\overline{P}_a$ will be unchanged. Since the outflow (Q_o) from the arteries depends on $\overline{P}_a$ and R,

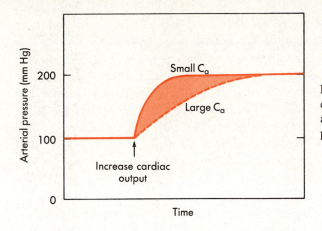

Fig. 30-5 ■ When cardiac output is suddenly increased, the arterial capacitance *(Cₐ)* determines the *rate* at which the mean arterial pressure will attain its new, elevated value but does not determine the *magnitude* of the pressure.

instantaneously Q_o also will remain unchanged. Therefore Q_i, now 10 L/minute, will exceed Q_o, still only 5 L/minute. This will result in an increase in the mean arterial blood volume $(\overline{V}_a)$. From equation 2, when $Q_i > Q_o$, then $d\overline{V}_a/dt > 0$; that is, volume is increasing. When $Q_i < Q_o$, the converse is true. When $Q_i = Q_o$, then $d\overline{V}_a/dt = 0$, and $\overline{V}_a$ remains constant.

Since $\overline{P}_a$ is essentially dependent only on the mean arterial blood volume, $\overline{V}_a$, and the arterial capacitance, C_a, an increase in $\overline{V}_a$ will be accompanied by a rise in $\overline{P}_a$. By definition

$$C_a = d\overline{V}_a/d\overline{P}_a \tag{5}$$

Therefore

$$d\overline{V}_a = C_a d\overline{P}_a \tag{6}$$

and

$$\frac{d\overline{V}_a}{dt} = C_a \frac{d\overline{P}_a}{dt} \tag{7}$$

From equation 2

$$\frac{d\overline{P}_a}{dt} = \frac{Q_i - Q_o}{C_a} \tag{8}$$

Hence P_a will rise when $Q_i > Q_o$, will fall when $Q_i < Q_o$, and will remain constant when $Q_i = Q_o$.

In the previous example, where Q_i is suddenly doubled, $\overline{P}_a$ will continue to rise as long as Q_i exceeds Q_o. It is evident from equation 4 that Q_o will not attain a value of 10 L/minute until $\overline{P}_a$ reaches a level of 200 mm Hg, as long as R remains constant at 20 mm Hg/L/minute. It is apparent then that, as $\overline{P}_a$ approaches 200, Q_o will almost equal Q_i, and $\overline{P}_a$ will rise very slowly. When Q_i is first raised, however, Q_i is greatly in excess of Q_o, and therefore $\overline{P}_a$ will rise sharply. The pressure/time tracing in Fig. 30-5 shows that, regardless of the value of C_a, the initial slope is relatively steep; the slope diminishes as pressure rises, to approach a final value asymptotically.

Furthermore the height to which $\overline{P}_a$ will rise is entirely independent of the elastic characteristics of the arterial walls. $\overline{P}_a$ must rise to a level such that $Q_o = Q_i$. It is apparent from equation 4 that Q_o depends only on pressure gradient and resistance to flow. So C_a determines only the rate at which the new equilibrium value of $\overline{P}_a$ will be approached (Fig. 30-5). When C_a is small (rigid vessels), a relatively slight increase in $\overline{V}_a$ (associated with the transient excess of Q_i over Q_o) produces a relatively large increase in $\overline{P}_a$. Thus $\overline{P}_a$ attains its new equilibrium level quickly. Conversely, when C_a is large, then considerable volumes can be accommodated with relatively small pressure changes, and the new equilibrium value of $\overline{P}_a$ is reached at a slower rate.

Similar reasoning may be applied now to explain the changes in $\overline{P}_a$ that accompany alterations in peripheral resistance. Let the control conditions be identical to those of the preceding example, that is, $Q_i = 5$, $\overline{P}_a = 100$, and $R = 20$. Then let R suddenly be increased to 40. Instantaneously, $\overline{P}_a$ will be unchanged. With $\overline{P}_a = 100$ and $R = 40$, $Q_o = \overline{P}_a/R = 2.5$ L/minute. If Q_i remains constant at 5 L/minute, $Q_i > Q_o$, and V_a will increase; thus $\overline{P}_a$ will rise. $\overline{P}_a$ will continue to rise until it reaches 200 mm Hg. At this level $Q_o = 200/40 = 5$ L/minute, which equals Q_i. $\overline{P}_a$ will then remain at this new elevated equilibrium level as long as Q_i and R do not change.

It is clear therefore that *the level of the mean arterial pressure is dependent only on cardiac output and peripheral resistance*. It is immaterial whether any change in cardiac output is accomplished by an alteration of heart rate, of stroke volume, or of both. Any change in heart rate that is balanced by a concomitant, oppositely directed change in stroke volume will not alter Q_i; thus $\overline{P}_a$ will not be affected.

■ Pulse pressure

If we assume that the arterial pressure, P_a, at any moment in time is dependent primarily on arterial blood volume, V_a, and arterial capacitance, C_a, then it can be shown that the arterial *pulse pressure* (difference between systolic and diastolic pressures) is principally a function of stroke volume and arterial capacitance.

Stroke volume. The effect of a change in stroke volume on pulse pressure may be analyzed under conditions in which C_a remains virtually constant over the range of pressures under consideration. In this situation the curve relating P_a to V_a is linear (Fig. 30-6). This curve would correspond fairly closely with the curve for the 20- to 24-year age group in Fig. 30-3, especially over the pressure range between 75 and 150 mm Hg.

In an individual with such a P_a:V_a curve the arterial pressure would oscillate about some mean value ($\overline{P}_A$ in Fig. 30-6) that depends entirely on cardiac output and peripheral resistance, as explained previously. This mean pressure corresponds to some mean arterial blood volume, $\overline{V}_A$, and the coordinates $\overline{P}_A$, $\overline{V}_A$ define point $\overline{A}$ on the graph. During diastole peripheral runoff occurs in the absence of ventricular ejection of blood, and P_a and V_a diminish to minimum values, P_1 and V_1, just prior to the next ventricular ejection. P_1 is then, by definition, the *diastolic pressure*.

During ventricular ejection there is rapidly introduced into the arterial system a volume of blood that greatly exceeds the peripheral runoff during this same portion of the cardiac cycle. Arterial pressure and volume therefore rise from point A_1 toward point A_2 in Fig. 30-6. As described in Chapter 29, the normal heart discharges most of its stroke volume during the early part of systole, the so-called rapid ejection phase. The

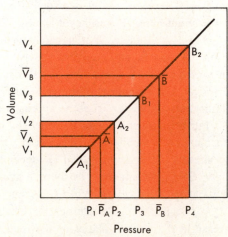

Fig. 30-6 ■ Effect of a change in stroke volume on pulse pressure in a system in which arterial capacitance is constant over the range of pressures and volumes involved. A larger volume increment $(V_4 - V_3$ as compared with $V_2 - V_1)$ results in a greater mean pressure $(\overline{P}_B$ as compared with $\overline{P}_A)$ and a greater pulse pressure $(P_4 - P_3$ as compared with $P_2 - P_1)$.

maximum arterial volume, V_2, is reached at the end of the rapid ejection phase, and this volume corresponds to a peak pressure, P_2, which is the *systolic pressure*. The mean arterial pressure is ordinarily somewhat less than the arithmetic average of the systolic and diastolic pressures, as illustrated in Fig. 30-4.

The *pulse pressure* is the difference between systolic and diastolic pressures ($P_2 - P_1$ in Fig. 30-6), and it corresponds to some *volume increment*, $V_2 - V_1$. This increment equals the volume of blood discharged by the left ventricle during the rapid ejection phase minus the volume that has run off to the periphery during this same phase of the cardiac cycle. When a normal heart beats at a normal frequency, this volume increment is a large fraction of the stroke volume (about 80%). It is this increment which will raise arterial volume rapidly from V_1 to V_2 and thus will cause the arterial pressure to rise from the diastolic to the systolic level (P_1 to P_2 in Fig. 30-6). During the remainder of the cardiac cycle peripheral runoff will greatly exceed cardiac ejection. The resultant arterial blood volume decrement will cause volumes and pressures to fall from point A_2 back to point A_1.

If stroke volume is now doubled, while heart rate and peripheral resistance remain constant, the mean arterial pressure will be doubled, to $\overline{P}_B$ in Fig. 30-6. Thus the arterial pressure now will oscillate about this new value of the mean arterial pressure. A normal, vigorous heart will eject this greater stroke volume during a fraction of the cardiac cycle approximately equal to the fraction observed at the lower stroke volume. Therefore the volume increment, $V_4 - V_3$, will be a large fraction of the new stroke volume and therefore will be approximately twice as great as the previous volume increment ($V_2 - V_1$). With a linear $P_a : V_a$ curve the greater volume increment will be reflected by a pulse pressure ($P_4 - P_3$) that will be approximately twice as great as the original pulse pressure ($P_2 - P_1$). With a rise in both mean and pulse pressures it is evident from inspection of Fig. 30-6 that the rise in systolic pressure (from P_2 to P_4) exceeds the rise in diastolic pressure (from P_1 to P_3).

Arterial capacitance. To assess arterial capacitance as a determinant of pulse pressure, the relative effects of the same volume increment ($V_2 - V_1$ in Fig. 30-7) in a young person (curve A) and in an elderly person (curve B) can be compared. Let cardiac output and total peripheral resistance be the same. It is apparent from Fig. 30-7 that the same volume increment will result in a greater pulse pressure in the less distensible arteries of the elderly individual ($P_4 - P_1$) than in the more compliant arteries of the young one ($P_3 - P_2$). For the reasons given on p. 504, this will impose a greater work load on the left ventricle of the elderly than of the young person, even if the stroke volumes, total peripheral resistances, and mean arterial pressures are equivalent.

Fig. 30-8 displays the effects of a change in arterial capacitance on the arterial

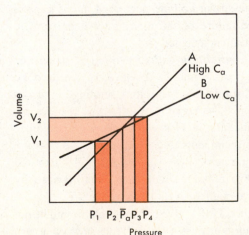

Fig. 30-7 ■ For a given volume increment ($V_2 - V_1$) a reduced arterial capacitance (curve B as compared with curve A) results in an increased pulse pressure ($P_4 - P_1$ as compared with $P_3 - P_2$).

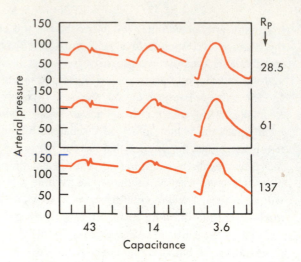

Fig. 30-8 ■ The changes in aortic pressure with changes in arterial capacitance and peripheral resistance *(R$_p$)* in an isolated cat heart preparation. (Modified from Elizinga, G., and Westerhof, N.: Pressure and flow generated by the left ventricle against different impedances, Circ. Res. **32**:178, 1973. By permission of the American Heart Association, Inc.)

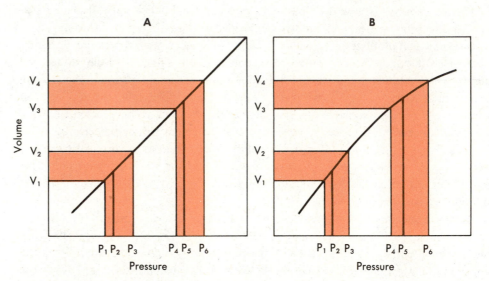

Fig. 30-9 ■ Effect of a change in total peripheral resistance (volume increment remaining constant) on pulse pressure when the pressure/volume curve for the arterial system is rectilinear (**A**) or curvilinear (**B**).

pressure in an isolated cat heart preparation. As the capacitance was reduced from 43 to 14 to 3.6 units (left to middle to right column, respectively), the pulse pressure increased significantly. In this preparation the stroke volume decreased as the capacitance was diminished. This accounts for the failure of the mean arterial pressure to remain constant at the different levels of arterial capacitance.

Total peripheral resistance and arterial diastolic pressure. It often is stated that increased total peripheral resistance (TPR) affects primarily the level of the diastolic arterial pressure. The validity of such an assertion deserves close scrutiny. First, let TPR be increased in an individual with a linear $P_a:V_a$ curve (Fig. 30-9, *A*). If heart rate and stroke volume remain constant, then an increase in TPR will evoke a proportionate increase in $\overline{P}_a$ (from P_2 to P_5). If the volume increments ($V_2 - V_1$ and $V_4 - V_3$) are equal at both levels of TPR, then the pulse pressures ($P_3 - P_1$ and $P_6 - P_4$) also will be equal. Thus systolic (P_6) and diastolic (P_4) pressures will have been elevated by exactly the same amounts from their respective control levels (P_3 and P_1).

It is certainly conceivable that with a higher TPR a given stroke volume might not be ejected as rapidly as the same stroke volume against a lower TPR. In such a case the volume increment might not be as great for the same stroke volume, since a greater fraction of the peripheral runoff might occur during ejection. A somewhat smaller volume increment would result in a proportionaltely smaller pulse pressure; in Fig. 30-9, A, $V_4 - V_3$ would be less, and so also would $P_6 - P_4$. Under such circumstances an augmentation of TPR would be associated with a somewhat greater rise in diastolic than in systolic pressure.

In the experiment illustrated in Fig. 30-8, as the peripheral resistance (R_p) was raised from 28.5 to 61 to 137 units (top to middle to bottom row, respectively), the mean arterial pressure increased. The heart responded to the augmentation of resistance by pumping a smaller stroke volume. Thus the volume increment undoubtedly decreased as the resistance was augmented. This was reflected by a progressive reduction in the pulse pressure, for any given level of arterial capacitance. Note that in the left and middle columns of Fig. 30-8 the pulse pressure decreased as R_p was raised.

Chronic hypertension, a condition characterized by a persistent elevation of TPR, occurs more commonly in middle-aged and elderly individuals than in younger persons. The $P_a:V_a$ curve for a hypertensive patient therefore would possess the configuration shown in Fig. 30-9, B (which resembles curves b to e in Fig. 30-3), rather than that displayed in Fig. 30-9, A. The type of curve in Fig. 30-9, B, reveals that C_a is less at higher than at lower pressures. As before, if cardiac output were to remain constant, an increase in TPR would cause a proportionate rise in $\overline{P}_a$ (from P_2 to P_5). For equivalent increases in TPR the elevation of pressure from P_2 to P_5 will be the same in Fig. 30-9, A and B, for reasons discussed on p. 509. Assuming the volume increment ($V_4 - V_3$ in Fig. 30-9, B) at elevated TPR to be equal to the control increment ($V_2 - V_1$), it is evident that the pulse pressure ($P_6 - P_4$) in the hypertensive range will greatly exceed that ($P_3 - P_1$) at normal pressure levels. In other words, a given volume increment will produce a greater pressure increment when the tube is more rigid than when it is more compliant. Thus the rise in systolic pressure ($P_6 - P_3$) will greatly exceed the increase in diastolic pressure ($P_4 - P_1$). These changes in arterial pressure closely resemble those seen in patients with hypertension. Diastolic pressure is indeed elevated in such individuals, but ordinarily not more than 10 to 40 mm Hg above the average normal level of 80 mm Hg, whereas it is not uncommon for systolic pressures to be elevated by 50 to 150 mm Hg above the average normal level of 120 mm Hg.

It is apparent therefore that the *pulse pressure is principally dependent on volume increment* (a function primarily of stroke volume) and *arterial capacitance*. Interventions that alter pulse pressure usually do so by changing one or both of these factors. If a change in heart rate, for example, is not accompanied by an alteration of stroke volume, then pulse pressure may or may not be modified, depending on the value of C_a at the new level of P_a. If C_a remains constant, then an increased heart rate will cause a rise in $\overline{P}_a$ but no change in pulse pressure. The situation is represented by Fig. 30-9, A, where $V_2 - V_1$ represents the volume increment at the normal heart rate and $V_4 - V_3$ the volume increment during tachycardia. If stroke volumes are equal, the volume increments will be approximately equal. Thus the pulse pressure ($P_6 - P_4$) during tachycardia will be similar to the pulse pressure ($P_3 - P_1$) at the control heart rate. If the increased P_a during tachycardia is associated with a reduction in arterial compliance, as in Fig. 30-9, B, the pulse pressure will be augmented ($P_6 - P_4$, as compared with the control pulse pressure, $P_3 - P_1$) as long as stroke volume remains constant. However, changes in heart rate usually are accompanied by changes in stroke volume, and this in turn will affect pulse pressure in accordance with the principles illustrated in Fig. 30-6.

■ *Peripheral arterial pressure curves*

The radial stretch of the ascending aorta brought about by left ventricular ejection initiates a pressure wave that is propagated down the aorta and its branches with a finite velocity that is considerably faster than the actual forward movement of the blood itself. It is this propagated pressure wave which one perceives in counting the pulse rate by palpating the radial artery.

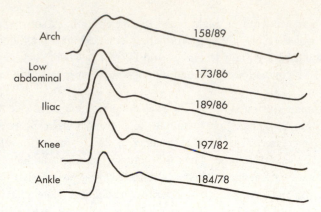

Arch   158/89
Low abdominal   173/86
Iliac   189/86
Knee   197/82
Ankle   184/78

Fig. 30-10 ■ Arterial pressure curves recorded from various sites in an anesthetized dog. (From Remington, J.W., and O'Brien, L.J.: Am. J. Physiol. **218**:437, 1970.)

The velocity of transmission of the pressure wave varies inversely with the vascular capacitance. With accurate measurement of the transmission velocity, valuable information has been derived concerning the elastic characteristics of the arterial tree. In general, transmission velocity increases with age, confirming the observation that the arteries become less compliant with advancing age. Also velocity increases progressively as the pulse wave travels from the ascending aorta toward the periphery. This indicates that vascular capacitance diminishes in the more distal portions of the arterial system, a fact that also has been confirmed by direct measurement.

The arterial pressure contour becomes progressively more distorted as the wave is transmitted down the arterial system; the changes in configuration of the pulse with distance are shown in Fig. 30-10. Aside from the increasing delay in the time of onset of the initial pressure rise associated with the transmission delay, three major changes occur in the arterial pulse contour as the pressure wave travels away from the aortic arch. First, the high-frequency components of the pulse, such as the incisura, are damped out and soon disappear. Second, the systolic portions of the pressure wave become narrowed and attain greater peak values. In the curves shown in Fig. 30-10 the systolic pressure at the level of the knee was 39 mm Hg greater than that recorded in the aortic arch. Third, a hump may become prominent on the diastolic portion of the pressure wave. These changes in contour of the pulse wave are pronounced in young individuals, but the magnitude of the alterations diminishes with age. In elderly patients with severe atherosclerosis the pulse wave may be transmitted virtually unchanged from the ascending aorta to the periphery.

■ Blood pressure measurement in humans

In certain instances needles or catheters are introduced into peripheral arteries of patients, and arterial blood pressure is measured *directly* by means of strain gauges. In the vast majority of cases, however, the blood pressure is estimated *indirectly* by means of a *sphygmomanometer*. This instrument consists of an inextensible cuff containing an inflatable bag. The cuff is wrapped around the extremity (usually the arm, occasionally the thigh) so that the inflatable bag lies between the cuff and the skin, directly over the artery to be compressed. The artery is occluded by inflating the bag, by means of a rubber squeeze bulb, to a pressure in excess of arterial systolic pressure. The pressure in the bag is measured by means of a mercury manometer or an aneroid manometer. Pressure is released from the bag at a rate of 2 or 3 mm Hg per heartbeat by means of a needle valve in the inflating bulb.

When blood pressure readings are taken from the arm, the systolic pressure may be estimated by palpating the radial artery at the wrist (*palpatory method*). When pressure

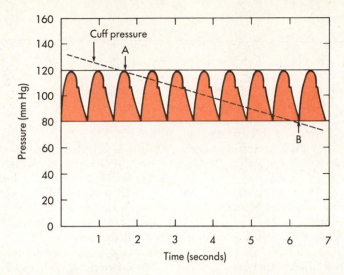

Fig. 30-11 ■ Principle of measurement of arterial blood pressure with a sphygomanometer. The oblique line represents pressure in the inflatable bag in the cuff. At cuff pressures greater than the systolic pressure (to the left of *A*) no blood progresses beyond the cuff, and no sounds can be detected below the cuff. At cuff pressures between the systolic and diastolic levels (between *A* and *B*) spurts of blood traverse the arteries under the cuff and produce the Korotkoff sounds. At cuff pressures below the diastolic pressure (to the right of *B*) arterial flow past the region of the cuff is continuous, and no sounds are audible.

in the bag exceeds the systolic level, no pulse will be perceived. As the pressure falls just below the systolic level (Fig. 30-11, *A*), a spurt of blood will pass through the brachial artery under the cuff during the peak of systole, and a slight pulse will be felt at the wrist.

The *auscultatory method* is a more sensitive and therefore a more precise method for measuring systolic pressure, and it permits the estimation of the diastolic level as well. The physician listens with a stethoscope applied to the skin of the antecubital space over the brachial artery. When the pressure in the bag exceeds the systolic pressure, the brachial artery is occluded, and no sounds are heard. When the inflation pressure falls just below the systolic level (point *A* in Fig. 30-11), the small spurt of blood escapes through the cuff, and a slight tapping sound is heard. This represents the systolic pressure. It usually corresponds closely with the systolic pressure when it is measured directly and exceeds by a few millimeters of mercury the pressure estimated by the palpatory method (because the auscultatory method is more sensitive than the palpatory method). As inflation pressure continues to fall, more blood escapes under the cuff per beat, and the sounds (called *Korotkoff sounds*) are heard as louder thuds. As the inflation pressure approaches the diastolic level, the Korotkoff sounds become muffled. As they fall just below the diastolic level (point *B* in Fig. 30-11), the sounds disappear; this indicates the diastolic pressure. The origin of the Korotkoff sounds is related to the spurt of blood passing under the cuff and meeting a static column of blood; the impact and turbulence generate vibrations, some of which are in the audible range of frequencies. Once the inflation pressure is less than the diastolic pressure, flow is continuous in the brachial artery, and sounds are no longer audible.

■ *Bibliography*

Journal articles

Attinger, E.O., and Attinger, F.M.: Frequency dynamics of peripheral vascular blood flow, Annu. Rev. Biophys. Bioeng. **4:**7, 1973.

Dobrin, P.B.: Mechanical properties of arteries, Physiol. Rev. **58:**397, 1978.

Elzinga, G., and Westerhof, N.: Pressure and flow generated by the left ventricle against different impedances, Circ. Res. **32:**178, 1973.

Hallock, P., and Benson, I.C.: Studies on the elastic properties of isolated human aorta, J. Clin. Invest. **16:**595, 1937.

Milnor, W.R.: Arterial impedance as ventricular afterload, Circ. Res. **36:**565, 1975.

Murgo, J.P., et al.: Aortic input impedance in normal man: relationship to pressure wave forms, Circulation **62:**105, 1980.

O'Rourke, M.F.: Vascular impedance in studies of arterial and cardiac function, Physiol. Rev. **62:**570, 1982.

Pepine, C.J., Nichols, W.W., and Conti, C.R.: Aortic input impedance in heart failure, Circulation **58:**460, 1978.

Van den Bos, G.C., et al.: Reflection in the systemic arterial system: effects of aortic and carotid occlusion, Cardiovasc. Res. **10:**565, 1976.

Books and monographs

Attinger, E.O.: Pulsatile blood flow, New York, 1964, McGraw-Hill Book Co.

Bader, H.: The anatomy and physiology of the vascular wall. In Handbook of physiology; Section 2; Circulation, vol. II, Washington, D.C., 1963, American Physiological Society.

Bauer, R.D., and Busse, R., editors: Arterial system: dynamics, control theory and regulation, Heidelberg, 1978, Springer-Verlag.

McDonald, D.A.: Blood flow in arteries, London, 1960, Edward Arnold, Ltd.

Spencer, M.P., and Denison, A.B., Jr.: Pulsatile blood flow in the vascular system. In Handbook of physiology; Section 2: Circulation, vol. II, Washington, D.C., 1963, American Physiological Society.

chapter 31

The microcirculation and lymphatics

The entire circulatory system is geared to supply the body tissues with blood in amounts commensurate with their requirements for oxygen and nutrients. The capillaries, consisting of a single layer of endothelial cells, permit rapid exchange of water and solutes with interstitial fluid. The muscular arterioles, which are the major *resistance vessels,* regulate regional blood flow to the capillary beds, and the venules and small veins serve primarily as collecting channels and storage, or *capacitance, vessels.*

The arterioles, which range in diameter from about 5 to 100 μm, have a thick smooth muscle layer, a thin adventitial layer, and an endothelial lining (Fig. 26-1). The arterioles give rise directly to the capillaries (5 to 10 μm diameter) or in some tissues to metarterioles (10 to 20 μm diameter), which then give rise to capillaries (Fig. 31-1). The metarterioles can serve either as thoroughfare channels to the venules, bypassing the capillary bed, or as conduits to supply the capillary bed. There are often cross connections between arterioles and between venules as well as in the capillary network. Arterioles that give rise directly to capillaries regulate flow through the capillaries by constriction or dilation. At the points of origin of the capillaries in some tissues there is a small cuff of smooth muscle called the *precapillary sphincter* that controls the blood flow through the capillaries (Fig. 31-1). The capillaries form an interconnecting network of tubes of different lengths, with an average length of 0.5 to 1 mm.

Capillary distribution varies from tissue to tissue. In metabolically active tissues, such as cardiac and skeletal muscle and glandular structures, capillaries are numerous, whereas in less active tissues, such as subcutaneous tissue or cartilage, *capillary density* is low. Also all capillaries are not of the same diameter, and since some capillaries have diameters less than those of the erythrocytes, it is necessary for the cells to become temporarily deformed in their passage through these capillaries. Fortunately the normal red cells are quite flexible and readily change their shape to conform with that of the small capillaries.

Blood flow in the capillaries is not uniform and is chiefly dependent on the contractile state of the arterioles and, where present, precapillary sphincters. The average velocity of blood flow in the capillaries is approximately 1 mm/second; however, it can vary from zero to several millimeters per second in the same vessel within a brief period. These changes in capillary blood flow may be of random type or may show rhythmical oscillatory behavior that is caused by contraction and relaxation *(vasomotion)* of the precapillary vessels. This vasomotion is to some extent an intrinsic contractile behavior of the vascular smooth muscle and is independent of external input. Furthermore changes in *transmural pressure* (intravascular minus extravascular pressure) at the precapillary vessels influence their contractile state; an increase in transmural pressure, whether produced by an increase in venous pressure or by dilation of arterioles, results

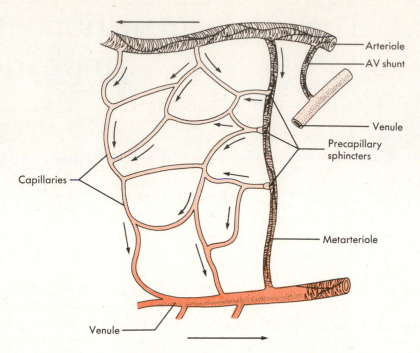

Fig. 31-1 ■ Composite schematic drawing of the microcirculation (modified from Zweifach). The circular structures on the arteriole and venule represent smooth muscle fibers, and the branching solid lines represent sympathetic nerve fibers. The arrows indicate the direction of blood flow.

in contraction of the terminal arterioles at the points of origin of the capillaries. In addition, humoral and possibly neural factors also affect vasomotion. For example, when the precapillary sphincters contract in response to increased transmural pressure, the contractile response can be overridden and vasomotion abolished. This effect is accomplished by metabolic (humoral) factors (p. 534) when the oxygen supply is reduced below the requirements of the parenchymal tissue, as occurs in muscle during exercise.

Reduction of transmural pressure will induce relaxation of the terminal arterioles, but blood flow through the capillaries obviously cannot increase if the reduction in intravascular pressure is caused by severe constriction of the parent arterioles or metarterioles. Large arterioles and metarterioles also exhibit vasomotion, but in the contraction phase they usually do not completely occlude the lumen of the vessel and arrest blood flow, contrary to the effect of contraction of the terminal arterioles and precapillary sphincters. Thus flow rate may be altered by arteriolar and metarteriolar vasomotion.

Since blood flow through the capillaries provides for exchange of gases and solutes between blood and tissue, it has been termed *nutritional flow,* whereas blood flow that bypasses the capillaries in traveling from the arterial to the venous side of the circulation has been termed *nonnutritional* or *shunt flow* (Fig. 31-1). In some areas of the body (such as skin) true arteriovenous (AV) shunts exist (p. 605). However, in many tissues, such as muscle, evidence of anatomical shunts is lacking. Nevertheless nonnutritional flow does occur and can be demonstrated in muscle by stimulation of the adrenergic sympathetic fibers to the muscle vessels during constant flow perfusion or by stimulation of the sympathetic cholinergic fibers to muscle (p. 540). With adrenergic sympathetic fiber stimulation at constant flow the capillary surface area available for solute exchange is reduced, as measured by decrease in intraarterially administered ^{86}Rb uptake by the tissue. With sympathetic cholinergic fiber stimulation, blood flow increases, but tissue clearance (washout) of radioactive isotope deposited in the tissue is unchanged. With

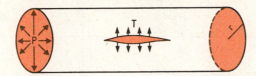

Fig. 31-2 ■ Diagram of a small blood vessel to illustrate the law of Laplace: *T = Pr*, where *P* = intraluminal pressure, *r* = radius of the vessel, and *T* = wall tension as the force per unit length tangential to the vessel wall, tending to pull apart a theoretical longitudinal slit in the vessel.

this technique for the estimation of nutritional blood flow a radioactive substance such as $^{22}Na^+$ is injected into the tissue in a small volume of fluid to minimize tissue damage. The rapidity with which it is removed by the perfusing blood is indicative of the rate of nutritional blood flow. Increased flow through open capillaries in the absence of morphological AV shunts has been termed a *physiological shunt*. It is the result of a greater flow of blood through open capillaries (shunts), with either no change or an increase in the number of closed capillaries. Adrenergic sympathetic fiber stimulation elicits arteriolar constriction, but when flow is set constant by a pump, all the perfusing blood passes through the decreased number of open channels during the sympathetic nerve stimulation. Cholinergic sympathetic fiber stimulation dilates arterioles, and the increased volume of blood flow caused by the decrease in vascular resistance occurs through the same capillaries that were open before nerve stimulation. In tissues that have metarterioles, shunt flow may be continuous from arteriole to venule during low metabolic activity when many precapillary vessels are closed. When metabolic activity increases in such tissues and precapillary vessels open, blood passing through the metarterioles is readily available for capillary perfusion.

The true capillaries are devoid of smooth muscle and therefore are incapable of active changes in caliber. Changes in capillary diameter are passive and are caused by alterations in precapillary and postcapillary resistance. The thin-walled capillaries can withstand high internal pressures without bursting because of their narrow lumen. This can be explained in terms of the law of Laplace and is illustrated in the following comparison of wall tension of a capillary with that of the aorta. The Laplace equation is

$$T = Pr$$

where

      T = tension in the vessel wall
      P = transmural pressure
      r = radius of the vessel

Wall tension is the force per unit length tangential to the vessel wall that opposes the distending force (Pr), which tends to pull apart a theoretical longitudinal slit in the vessel (Fig. 31-2). Transmural pressure is essentially equal to intraluminal pressure, since extravascular pressure is negligible. The Laplace equation applies to very thin-walled vessels, such as capillaries. Wall thickness must be taken into consideration when the equation is applied to thick-walled vessels, such as the aorta (Chapter 23). This is done by dividing Pr (pressure × radius) by wall thickness (w). The equation now becomes

$$\sigma \text{ (wall stress)} = Pr/w$$

To convert pressure in mm Hg (height of mercury column) to dynes per square centimeter, $P = h\rho g$, where h = height of mercury column in centimeters, ρ = density of mercury in grams per cubic centimeter, g = gravitational acceleration in centimeters per square second (Chapter 28), σ = force per unit area, and w = wall thickness.

	Aorta	*Capillary*
Radius (r)	1.5 cm	5×10^{-4} cm
Height of Hg column (h)	10 cm Hg	2.5 cm Hg
ρ	13.6 g/cm^3	13.6 g/cm^3
g	980 cm/second^2	980 cm/second^2
$\overline{P}$	$10 \times 13.6 \times 980 =$ 1.33×10^5 dyne/cm^2	$2.5 \times 13.6 \times 980 =$ 3.33×10^4 dyne/cm^2
w	0.2 cm	1×10^{-4} cm
$T = Pr$	$(1.33 \times 10^5)(1.5) =$ 2×10^5 dyne/cm	$(3.33 \times 10^4)(5 \times 10^{-4}) =$ 16.7 dyne/cm
$\sigma = \dfrac{Pr}{w}$	$\dfrac{2 \times 10^5}{0.2} =$ 1×10^6 dyne/cm^2	$\dfrac{16.7}{1 \times 10^{-4}} =$ 1.67×10^5 dyne/cm^2

Thus at normal aortic and capillary pressures the wall tension of the aorta is about 12,000 times greater than that of the capillary. In a person standing quietly capillary pressure in the feet may reach 100 mm Hg. Under such conditions capillary wall tension increases to 66.5 dyne/cm, a value that is still only one three-thousandth that of the wall tension in the aorta at the same internal pressure. Because T (tension) applies only to very thin-walled vessels such as capillaries, it is not applicable to the aorta. However, σ (stress), which takes wall thickness into consideration, is only about tenfold greater in the aorta than in the capillary.

In addition to providing an explanation for the ability of capillaries to withstand large internal pressures, the preceding calculations also point out that, in dilated vessels, wall tension increases even when internal pressure remains constant and under certain circumstances (for example, syphilitic aneurysm of the aorta) may be an important factor in rupture of the vessel. The preceding equation also indicates that, as the wall of the vessel becomes thicker, the wall stress decreases. In *hypertension* (high blood pressure) the arterial vessel walls thicken (hypertrophy of the vascular smooth muscle), thereby minimizing the arterial wall stress and hence the possibility of vessel rupture.

The diameter of the resistance vessels is determined by the balance between the contractile force of the vascular smooth muscle and the distending force produced by the intraluminal pressure. The greater the contractile activity of the vascular smooth muscle of an arteriole, the smaller its diameter, until a point is reached, in the case of small arterioles, when complete occlusion of the vessel occurs, in part because of infolding of the endothelium and the cells trapped in the vessel. With progressive reducton in intravascular pressure, vessel diameter decreases, as does tension in the vessel wall (law of Laplace). When perfusion pressure is reduced, a point is reached where blood flow ceases even though there is still a positive pressure gradient. This phenomenon has been referred to as the *critical closing pressure,* and its mechanism is still controversial. This critical closing pressure is low when vasomotor activity is reduced by inhibition of sympathetic nerve activity to the vessel, and it is increased when vasomotor tone is enhanced by activation of the vascular sympathetic nerve fibers. It has been suggested that flow stops because of vessel collapse when vascular smooth muscle contractile stress exceeds the stress associated with vessel radius and wall thickness and intraluminal pressure (law of Laplace).

■ *Transcapillary exchange*

Solvent and solute move across the capillary endothelial wall by three processes: diffusion, filtration, and via endothelial cell vesicles (pinocytosis). By far the greatest number of molecules traverse the capillary endothelium by diffusion.

Under normal conditions only about 0.06 ml of water per minute moves back and forth across the capillary wall per 100 g of tissue as a result of filtration and absorption, whereas 300 ml of water per minute transfer across the endothelium per 100 g of tissue by diffusion, a 5000-fold difference. Relating filtration and diffusion to blood flow, we find that about 2% of the plasma passing through the capillaries is filtered, whereas the diffusion of water is 40 times greater than the rate at which it is brought to the capillaries by blood flow. The transcapillary exchange of solutes also is governed primarily by diffusion. Thus diffusion is the key factor in providing exchange of gases, substrates, and waste products between the capillaries and the tissue cells.

The process of diffusion is described by Fick's law (p. 8):

$$J = -DA\frac{dc}{dx}$$

where

J = quantity of a substance moved per unit of time (t)
D = free diffusion coefficient for a particular molecule
    (the value is inversely related to the square root of the molecular weight)
A = cross-sectional area of the diffusion pathway
$\frac{dc}{dx}$ = concentration gradient

Fick's law is also expressed as:

$$J = -PS\,(C_o - C_i) \text{ or } J = -kpA\,(C_o - C_i) \text{ (p. 12)}$$

where

P = capillary permeability of the substance
S = capillary surface area
C_i = concentration of the substance inside the capillary
C_o = concentration of the substance outside the capillary

Hence the PS product provides a convenient expression of available capillary surface, since permeability rarely is altered under physiological conditions.

In the capillaries, diffusion of lipid-insoluble molecules is not free but is restricted to the pores, whose mean size can be calculated by measurement of the diffusion rate of an uncharged molecule whose free diffusion coefficient is known. Movement of solutes across the endothelium is quite complex and involves corrections for attractions between solute and solvent molecules, interactions between solute molecules, pore configuration, and charge on the molecules relative to charge on the endothelial cells (as observed in the kidney). It is not simply a question of random thermal movements of molecules down a concentration gradient.

For small molecules, such as water, NaCl, urea, and glucose, the capillary pores offer little restriction to diffusion (low reflection coefficient), and diffusion is so rapid that the mean concentration gradient across the capillary endothelium is extremely small. However, with lipid-insoluble molecules of increasing size, diffusion through muscle capillaries becomes progressively more restricted, until diffusion becomes minimal with molecules of a molecular weight above about 60,000. With small molecules the only limitation to net movement across the capillary wall is the rate at which blood flow transports the molecules to the capillary surfaces *(flow limited)*, whereas with larger molecules diffusion across the capillaries becomes the limiting factor *(diffusion limited)*. The rate of diffusion of small lipid-insoluble molecules is so great that it is uninfluenced by filtration in the direction opposite to the concentration gradient of the diffusible substance. In fact filtration *or* absorption accelerates the movement of tracer ions from interstitial fluid to blood (tissue clearance). The reason for this enhanced tissue clearance is not known, but it may be the result of a stirring effect on the interstitial fluid or of changes in its structure (for example, gel to sol transformation or "canals" in a gel matrix).

Movement of lipid-soluble molecules across the capillary wall is not limited to capillary pores (only about 0.02% of the capillary surface), since such molecules can pass directly through the lipid membranes of the entire capillary endothelium. Consequently lipid-soluble molecules move with great rapidity between blood and tissue. The degree of lipid solubility (oil-to-water partition coefficient) provides a good index of the ease of transfer of lipid molecules through the capillary endothelium.

Oxygen and carbon dioxide are both lipid soluble and readily pass through the endothelial cells. Calculations based on the diffusion coefficient for oxygen, capillary density and diffusion distances, blood flow, and tissue oxygen consumption indicate that the oxygen supply of normal tissue at rest and during activity is not limited by diffusion or the number of open capillaries. Recent measurements of oxygen tension and saturation of blood in the microvessels indicate that in many tissues oxygen saturation at the entrance of the capillaries already has decreased to about 80% as a result of diffusion of oxygen from arterioles. Such studies also have shown that CO_2 loading and the resultant intravascular shifts in the oxyhemoglobin dissociation curve occur in the precapillary vessels. These findings reflect not only the movement of gas to respiring tissue at the precapillary level but also the direct flux of O_2 and CO_2 between adjacent arterioles, venules, and possibly arteries and veins (contercurrent exchange). This exchange of gas represents a diffusional shunt of gas around the capillaries, and at low blood flow rates it may limit the supply of oxygen to the tissue.

■ *Capillary filtration*

The direction and the magnitude of the movement of water across the capillary wall are determined by the algebraic sum of the hydrostatic and osmotic pressures existing across the membrane. An increase in intracapillary hydrostatic pressure favors movement of fluid from the vessel to the interstitial space, whereas an increase in the concentration of osmotically active particles within the vessels favors movement of fluid into the vessels from the interstitial space.

■ *Hydrostatic forces*

The hydrostatic pressure (blood pressure) within the capillaries is not constant and is dependent on the arterial pressure, the venous pressure, and the precapillary (arteriolar and precapillary sphincter) and postcapillary (venules and small veins) resistances. A gain in arterial or venous pressure increases capillary hydrostatic pressure, whereas a reduction in each has the opposite effect. Increase in arteriolar resistance or closure of precapillary sphincters reduces capillary pressure, whereas greater venous resistance (venules and veins) increases capillary pressure (Fig. 31-3).

The effect of precapillary and postcapillary pressure and resistance on capillary pressure can be expressed by the equation:

$$P_c = \frac{(R_v/R_a)\, P_a + P_v}{1 + (R_v/R_a)}$$

where

P_c = capillary hydrostatic pressure
P_a = arterial pressure
P_v = venous pressure
R_a = precapillary resistance (on the arterial side)
R_v = postcapillary resistance (on the venous side)

A given increment in venous pressure produces a fivefold to tenfold greater effect on capillary hydrostatic pressure than the same increment in arterial pressure, and about 80% of the increase in venous pressure is transmitted back to the capillaries. The im-

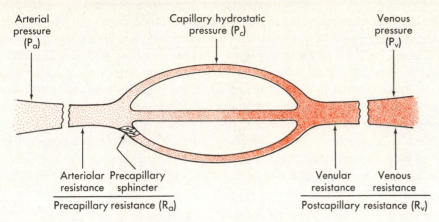

Fig. 31-3 ■ Diagram of the terminal vascular bed, illustrating precapillary and postcapillary pressures and resistances used in the calculation of capillary hydrostatic pressure.

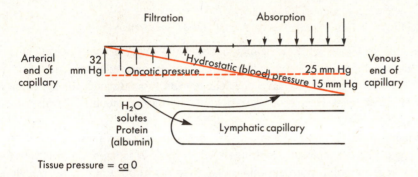

Fig. 31-4 ■ Schematic representation of the factors responsible for filtration and absorption across the capillary wall and the formation of lymph.

portant variable in the preceding equation is the ratio of postcapillary resistance to precapillary resistance (R_v/R_a), where R_a is much greater than R_v (about 4 to 1). Thus the effect of an increase in one resistance (R_a) may be offset by a proportional increase in the other (R_v).

Despite the fact that capillary hydrostatic pressure is variable from tissue to tissue, even within the same tissue, average values, obtained from many direct measurements in human skin, are about 32 mm Hg at the arterial end of the capillaries and 15 mm Hg at the venous end of the capillaries at the level of the heart (Fig. 31-4). The hydrostatic pressure in capillaries of the lower extremities will be higher and that of capillaries in the head will be lower in an individual in the standing position than when he is recumbent. Measurements of pressure at the arteriolar and venous ends of the capillaries in different tissues in several species of mammals give values that are reasonably close to the ones found in human skin capillaries. This hydrostatic pressure is the principal force in filtration across the capillary wall.

Tissue pressure, or more specifically interstitial fluid pressure (P_i) outside the capillaries, opposes capillary filtration, and it is $P_c - P_i$ that constitutes the driving force for filtration. The true value of P_i is still controversial. For years it was assumed to be close to zero in the normal (nonedematous) state. However, recent studies, using perforated, plastic capsules implanted in the subcutaneous tissue or wicks inserted through the skin, indicate a negative P_i of about -7 mm Hg. If the pressures recorded by these techniques are representative of interstitial fluid pressure in undisturbed tissue, then the hydrostatic driving force for capillary filtration is 7 mm Hg greater than the value of P_c.

■ *Osmotic forces*

The key factor that restrains fluid loss from the capillaries is the osmotic pressure of the plasma proteins—usually termed the *colloid osmotic pressure* or *oncotic pressure* (π_p) (Chapter 1). The total osmotic pressure of plasma is about 6000 mm Hg, whereas the oncotic pressure is only about 25 mm Hg. However, this small oncotic pressure plays an important role in fluid exchange across the capillary wall, since the plasma proteins essentially are confined to the intravascular space, whereas the electrolytes that are responsible for the major fraction of plasma osmotic pressure are practically equal in concentration on both sides of the capillary endothelium. The relative permeability of solute to water influences the actual magnitude of the osmotic pressure. The *reflection coefficient* (σ) is the index of the deviation from Van't Hoff's law caused by permeability of the solute. Therefore true osmotic pressure is defined by

$$\pi = \sigma RT\,(C_i - C_o)$$

where

σ = reflection coefficient
R = gas constant
T = absolute temperature
C_i and C_o = solute concentration inside and outside the capillary

For albumin, to which the endothelium is essentially impermeable, the reflection coefficient has a value of 1, whereas with small molecules the reflection coefficient is less than 1. Also different tissues have different reflection coefficients for the same molecule, and therefore movement of a given solute across the endothelial wall varies with the tissue.

Of the plasma proteins, albumin is preponderant in determining oncotic pressure. The average albumin molecule (molecular weight 69,000) is approximately half the size of the average globulin molecule (molecular weight 150,000) and is present in almost twice the concentration as the globulins (4.5 g/dl versus 2.5 g/dl of plasma). Albumin also exerts a greater osmotic force than can be accounted for solely on the basis of the number of molecules dissolved in the plasma, and for these reasons it cannot be completely replaced by inert substances of appropriate molecular size such as dextran. This additional osmotic force becomes disproportionately greater at high concentrations of albumin (as in plasma) and is weak to absent in dilute solutions of albumin (as in interstitial fluid). One reason for this behavior of albumin is its negative charge at the normal blood pH and the attraction and retention of cations (principally Na^+) in the vascular compartment (the *Gibbs-Donnan effect*). Furthermore albumin binds a small number of chloride ions, which increases its negative charge and hence its ability to retain more sodium ions inside the capillaries. The small increase in electrolyte concentration of the plasma over that of the interstitial fluid produced by the negatively charged albumin enhances its osmotic force to that of an ideal solution containing a solute of molecular weight of 37,000. If albumin did indeed have a molecular weight of 37,000, it would not be retained by the capillary endothelium because of its small size and obviously could not function as a counterforce to capillary hydrostatic pressure. If, however, albumin did not have an enhanced osmotic force, it would require a concentration of about 12 g of albumin per 100 ml of plasma to achieve a plasma oncotic pressure of 25 mm Hg. Such a high albumin concentration would greatly increase blood viscosity and the resistance to blood flow through the vascular system. The other factors that contribute to the nonlinearity of the relationship of albumin concentration to osmotic force are not known. About 65% of plasma oncotic pressure is attributable to albumin, about 15% to the globulins, and the remainder to other ill-defined components of the plasma.

Small amounts of albumin escape from the capillaries and enter the interstitial fluid, where they exert a very small osmotic force (0.1 to 5 mm Hg). This force (π_i) is small because of the low concentration of albumin in the interstitial fluid and because at low concentrations the osmotic force of albumin becomes simply a function of the number of albumin molecules per unit volume of interstitial fluid.

The relationship between hydrostatic pressure and oncotic pressure and the role of these forces in regulating fluid passage across the capillary endothelium were expounded by Starling in 1896 and constitute the *Starling hypothesis*. It can be expressed by the equation:

$$\text{Fluid movement} = k[(P_c + \pi_i) - (P_i + \pi_p)]$$

where

P_c = capillary hydrostatic pressure
P_i = interstitial fluid hydrostatic pressure
π_p = plasma protein oncotic pressure
π_i = interstitial fluid oncotic pressure
k = filtration constant for the capillary membrane

Filtration occurs when the algebraic sum is positive, and absorption occurs when it is negative.

Classically it has been thought that filtration occurs at the arterial end of the capillary and absorption at its venous end because of the gradient of hydrostatic pressure along the capillary. This is true for the idealized capillary, as depicted in Fig. 31-4, but direct observations have revealed that many capillaries show filtration for their entire length, whereas others show only absorption. In some vascular beds (for example, the renal glomerulus) hydrostatic pressure in the capillary is high enough to result in filtration along the entire length of the capillary. In other vascular beds, such as in the intestinal mucosa, the hydrostatic and oncotic forces are such that absorption occurs along the whole capillary. As discussed earlier in this chapter, capillary pressure is quite variable and depends on several factors, the principal one being the contractile state of the pre-capillary vessel. In the normal steady state, arterial pressure, venous pressure, postcapillary resistance, interstitial fluid hydrostatic and oncotic pressures, and plasma oncotic pressure are relatively constant, and change in precapillary resistance is the determining factor with respect to fluid movement across the wall for any given capillary. Since water moves so quickly across the capillary endothelium, the hydrostatic and osmotic forces are nearly in equilibrium along the entire capillary. Thus filtration and absorption in the normal state occur at very small degrees of imbalance of pressures across the capillary wall. Only a small percentage (2%) of the plasma flowing through the vascular system is filtered, and of this about 85% is absorbed in the capillaries and venules. The remainder returns to the vascular system as lymph fluid, which also contains the albumin that escapes from the capillaries.

In the lungs the mean capillary hydrostatic pressure is only about 8 mm Hg. Since the plasma oncotic pressure is 25 mm Hg, the forces across the capillary membrane favor absorption. However, pulmonary lymph is formed and consists of fluid that is osmotically drawn out of the capillaries by the small amount of plasma protein that escapes through the capillary endothelium. Only in pathological conditions, such as left ventricular failure or stenosis of the mitral valve, does pulmonary capillary hydrostatic pressure exceed plasma oncotic pressure. When this occurs, it may lead to pulmonary edema, a condition that can seriously interfere with gas exchange in the lungs; hence it is a condition that is often fatal.

The permeability of the capillary endothelial membrane is not the same in all body tissues. For example, the liver capillaries are quite permeable, and albumin escapes at a rate severalfold greater than from the less permeable muscle capillaries. Also there is not uniform permeability along the whole capillary; the venous ends are more permeable than the arterial ends, and permeability is greatest in the venules.

■ *Balance of hydrostatic and osmotic forces: Starling hypothesis*

■ *Capillary pores*

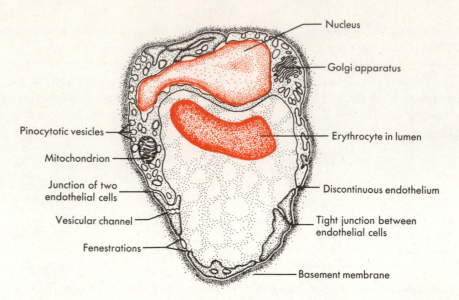

Fig. 31-5 ■ Diagrammatic sketch of an electron micrograph of a composite capillary in cross section.

The sites where filtration occurs have been a controversial subject for a number of years. Some investigators believe that a fraction of the water flows through the capillary endothelial cell membranes. However, most investigators believe that water flows through apertures in the endothelial wall of the capillaries. Calculations based on the transcapillary movement of solutes of small molecular size led to the prediction of pore diameters of about 40 Å. However, electron microscopy failed to reveal pores, and the clefts at the junctions of endothelial cells appeared to be fused at the tight junctions (Fig. 31-5). Nevertheless studies with horseradish peroxidase, a protein with a molecular weight of 40,000, have demonstrated that in cardiac and skeletal muscle many of the clefts between adjacent endothelial cells are open. Electron microscopy revealed filling of the clefts with peroxidase from the lumen side of the capillaries with a gap at the narrowest point of about 40 Å *(gap junctions),* providing morphological support of the physiological evidence for the existence of capillary pores. The clefts (pores) are sparse and represent only about 0.02% of the capillary surface area. In cerebral capillaries, where a blood-brain barrier to many small molecules exists, peroxidase studies do not reveal any pores. Some studies have failed to reveal the presence of interendothelial pores or clefts, even when microperoxidase (molecular weight of 1900) was used. Transcapillary movement of solute (large and small molecules) is thought to occur through channels formed by the fusion of vesicles (vesicular channels) across the endothelial cells (Fig. 31-5). The basement membrane appears to restrict passage of large molecules (greater than 100 Å radius).

In addition to clefts, some of the more porous capillaries (for example, kidney, intestine) contain fenestrations 200 to 1000 Å wide, whereas others (such as in the liver) have a discontinuous endothelium (Fig. 31-5). The fenestrations that appear to be sealed by a thin diaphragm are quite permeable to horseradish peroxidase as well as to a number of other tracers. Therefore larger molecules can penetrate capillaries with fenestrations or discontinuous endothelium than can pass through the intercellular clefts of the endothelium.

■ *Capillary filtration coefficient*

The rate of movement of fluid across the capillary membrane (Q_f) is dependent not only on the algebraic sum of the hydrostatic and osmotic forces across the endothelium (ΔP) but also on the area of the capillary wall available for filtration (A_m), the distance across the capillary wall (Δx), the viscosity of the filtrate (η), and the filtration constant of the membrane (k). These factors may be expressed by the equation:

$$Q_f = \frac{kA_m\Delta P}{\eta\Delta x}$$

The dimensions are units of flow per unit of pressure gradient across the capillary wall per unit of capillary surface area. It should be apparent that this expression, which describes the flow of fluid through a membrane (pores), is essentially Poiseuille's law for flow through tubes (Chapter 28).

Since the thickness of the capillary wall and the viscosity of the filtrate are relatively constant, they can be included in the filtration constant, k. Also if the area of the capillary membrane is not known, the rate of filtration can be expressed per unit weight of tissue. Thus the equation can be simplified to

$$Q_f = k_t\Delta P$$

where k_t is the capillary filtration coefficient (CFC) for a given tissue, and the units for Q_f are milliliters per minute per 100 g of tissue per millimeter of mercury.

The rate of filtration and absorption is determined by measuring the rate of change in tissue weight or volume at different mean capillary hydrostatic pressures, which are altered by adjustment of arterial and venous pressures. At the isogravimetric or isovolumetric point (constant weight or constant volume, respectively, as continuously measured with an appropriate scale or volume recorder) the hydrostatic and osmotic forces are balanced across the capillary wall, and hence there is neither net filtration nor absorption. An abrupt increase in arterial pressure will increase capillary hydrostatic pressure, and fluid will move from the capillaries to the interstitial fluid compartment. Since the pressure increment, the weight increase of the tissue per unit of time, and the total weight of the tissue are known, the capillary filtration coefficient (CFC or k_t) in milliliters per minute per 100 g of tissue per millimeter of mercury can be calculated. With the isogravimetric and isovolumetric techniques it is assumed that 80% of increments in venous pressure are transmitted back to the capillaries, that precapillary and postcapillary resistances are constant when venous pressure is changed, and that the weight or volume change that occurs immediately after raising venous pressure is the result of vascular distension and not filtration. These assumptions may not always be correct; nevertheless the filtration coefficient constitutes a useful index of capillary permeability and surface area.

In any given tissue the filtration coefficient per unit area of capillary surface, and hence capillary permeability, is not changed by different physiological conditions, such as arteriolar dilation and capillary distension, or by adverse conditions such as hypoxia, hypercapnia, or reduced pH. With capillary injury (toxins, severe burns) capillary permeability increases greatly, as indicated by the filtration coefficient, and significant amounts of fluid and protein leak out of the capillaries into the interstitial space.

Since capillary permeability is constant under normal conditions, the filtration coefficient can be used to determine the relative number of open capillaries (total capillary surface area available for filtration in tissue). For example, increased metabolic activity of contracting skeletal muscle induces relaxation of precapillary resistance vessels, with opening of more capillaries (*capillary recruitment*, resulting in an increased filtering surface).

Some quantity of plasma protein apparently is required to maintain the integrity of the endothelial membrane. If the plasma proteins are replaced by nonprotein colloids so as to give the same oncotic pressure, the filtration coefficient is doubled, and edema occurs. However, if as little as 0.2% protein is added, normal permeability is restored. Possibly the protein molecules are adsorbed on the endothelial membrane and alter pore dimensions.

■ *Disturbances in hydrostatic-osmotic balance*

Changes in arterial pressure per se may have little effect on filtration, since the change in pressure may be countered by adjustments of the precapillary resistance vessels (autoregulation, p. 532), so that hydrostatic pressure in the open capillaries remains the same. However, with severe reduction in arterial pressure, as may occur in hemorrhage, there may be arteriolar constriction mediated by the sympathetic nervous system and a fall in venous pressure resulting from the blood loss. These changes will lead to a decrease in the capillary hydrostatic pressure. Furthermore the low blood pressure causes a decrease in blood flow (and hence oxygen supply) to the tissue with the result that vasodilator metabolites accumulate and induce relaxation of arterioles. Precapillary vessel relaxation also is engendered by the reduced transmural pressure. As a consequence of these several factors, absorption predominates over filtration and occurs at a larger capillary surface area. This is one of the compensatory mechanisms employed by the body to restore blood volume (Chapter 36).

An increase in venous pressure alone, as occurs in the feet when one changes from the lying to the standing position, would elevate capillary pressure and enhance filtration. However, the increase in transmural pressure causes precapillary vessel closure, so that the capillary filtration coefficient actually decreases. This reduction in capillary surface available for filtration serves to protect against the extravasation of large amounts of fluid into the interstitial space (edema). With prolonged standing, particularly when associated with some elevation of venous pressure in the legs (such as that caused by tight garters or pregnancy), or with sustained increases in venous pressure, as seen in congestive heart failure, filtration is greatly enhanced and exceeds the capacity of the lymphatic system to remove the capillary filtrate from the interstitial space.

A large amount of fluid can move across the capillary wall in a relatively short time. In a normal individual the filtration coefficient (k_t) for the whole body is about 0.0061 ml/minute/100 g of tissue/mm Hg. For a 70 kg man elevation of venous pressure of 10 mm Hg for 10 minutes would increase filtration from capillaries by 342 ml. This would not lead to edema formation, since the fluid is returned to the vascular compartment by the lymphatic vessels. When edema does develop, it usually appears in the dependent parts of the body, where the hydrostatic pressure is greatest, but its location and magnitude also are determined by the type of tissue. Loose tissues, such as the subcutaneous tissue around the eyes or in the scrotum, are more prone to collect larger quantities of interstitial fluid than are firm tissues, such as muscle, or encapsulated structures, such as the kidney.

The concentration of the plasma proteins also may change in different pathological states and thus alter the osmotic force and movement of fluid across the capillary membrane. The plasma protein concentration is increased in dehydration (for example, water deprivation, prolonged sweating, severe vomiting, and diarrhea), and water moves by osmotic forces from the tissues to the vascular compartment. In contrast, the plasma protein concentration is reduced in nephrosis (a renal disease in which there is loss of protein in the urine), and edema may occur. Leaks occur when there is extensive capillary injury, as in burns, and plasma protein escapes into the interstitial space along with fluid and increases the oncotic pressure of the interstitial fluid. This greater osmotic force outside the capillaries leads to additional fluid loss and possibly to severe dehydration of the patient.

■ Pinocytosis

Some transfer of substances across the capillary wall can occur in tiny pinocytotic vesicles (pinocytosis). These vesicles (Fig. 31-5), formed by a pinching off of the surface membrane, can take up substances on one side of the capillary wall, move across the cell, and deposit their contents at the other side. The amount of material that can be transported in this way is very small relative to that moved by diffusion. However, pinocytosis may be responsible for the movement of large lipid-insoluble molecules between blood and interstitial fluid.

■ Lymphatics

The terminal vessels of the lymphatic system consist of a widely distributed closed-end network of highly permeable lymph capillaries that are similar in appearance to blood capillaries. However, they generally are lacking in tight junctions between endothelial cells and possess fine filaments that anchor them to the surrounding connective tissue. With muscular contraction these fine strands may distort the lymphatic vessel and thereby open spaces between the endothelial cells. This would permit the entrance of protein, large particles, and cells present in the interstitial fluid. The lymph capillaries drain into larger vessels that finally enter the right and left subclavian veins at their junctions with the respective internal jugular veins. Only cartilage, bone, epithelium, and the tissues of the central nervous system are devoid of lymphatic vessels. The plasma capillary filtrate is returned to the circulation by virtue of tissue pressure. It is also facilitated by intermittent skeletal muscle activity, contractions of the lymphatic vessels, and an extensive system of one-way valves. In this respect the lymphatics resemble the veins, although even the larger lymphatic vessels have thinner walls than the corresponding veins and contain only a small amount of elastic tissue and smooth muscle.

The volume of fluid transported through the lymphatics in 24 hours is about equal to the animal's total plasma volume, and the protein returned by the lymphatics to the blood in a day is about one fourth to one half the circulating plasma proteins. This is the only means whereby protein (albumin) that leaves the vascular compartment can be returned to the blood, since back diffusion into the capillaries cannot occur against the large albumin concentration gradient. Were the protein not removed by the lymph vessels, it would accumulate in the interstitial fluid and act as an oncotic force to draw fluid from the blood capillaries to produce increasingly severe edema. In addition to returning fluid and protein to the vascular bed, the lymphatic system filters the lymph at the lymph nodes and removes foreign particles, such as bacteria. The largest lymphatic vessel, the *thoracic duct,* in addition to draining the lower extremities, serves to return protein lost through the permeable liver capillaries and to carry substances absorbed from the gastrointestinal tract, principally fat in the form of chylomicrons, to the circulating blood.

Lymph flow varies considerably, being almost nil from resting skeletal muscle and increasing during exercise in proportion to the degree of muscular activity. It is increased by any mechanism that enhances the rate of blood capillary filtration, for example, increased capillary pressure or permeability or decreased plasma oncotic pressure. When either the volume of interstitial fluid exceeds the drainage capacity of the lymphatics or the lymphatic vessels become blocked, as may occur in certain disease states, interstitial fluid accumulates, chiefly in the more compliant tissues (for example, subcutaneous tissue) and gives rise to clinical edema.

■ *Bibliography*

Journal articles

Bundgaard, M.: Transport pathways in capillaries—in search of pores, Annu. Rev. Physiol. **42:**325, 1980.

Duling, B.R., and Berne, R.M.: Longitudinal gradients in periarteriolar oxygen tension, Circ. Res. **27:**669, 1970.

Duling, B.R., and Klitzman, B.: Local control of microvascular function: role in tissue oxygen supply, Annu. Rev. Physiol. **42:**373, 1980.

Garlick, D.G., and Renkin, E.M.: Transport of large molecules from plasma to interstitial fluid and lymph in dogs, Am. J. Physiol. **219:**1595, 1970.

Gauer, O.H., et al.: Proceedings of a symposium on capillary exchange and the interstitial space, Pfluegers. Arch. **336:**S1, 1972.

Gore, R.W., and McDonagh, P.F.: Fluid exchange across single capillaries, Annu. Rev. Physiol. **42:**337, 1980.

Karnovsky, M.J.: The ultrastructural basis of transcapillary exchanges, J. Gen. Physiol. **52:**645, 1968.

Krogh, A.: The number and distribution of capillaries in muscles with calculations of the oxygen pressure head necessary for supplying the tissue, J. Physiol. (London) **52:**409, 1919.

Leak, L.V.: Electron microscopic observations on lymphatic capillaries and the structural components of the connective tissue–lymph interface, Microvasc. Res. **2:**361, 1970.

Lewis, D.H., editor: Symposium on lymph circulation, Acta Physiol Scand. (Suppl.) **463:**9, 1979.

Palade, G.E.: Blood capillaries of the heart and other organs, Circulation **24:**368, 1961.

Rosell, S.: Neuronal control of microvessels, Annu. Rev. Physiol. **42:**359, 1980.

Starling, E.H.: On the absorption of fluids from the connective tissue spaces, J. Physiol. (London) **19:**312, 1896.

Books and monographs

Guyton, A.C., Taylor, A.E., and Granger, H.J.: Circulatory physiology II: dynamics and control of the body fluids, Philadelphia, 1975, W.B. Saunders Co.

Johnson, P.C.: The microcirculation, and local and humoral control of the circulation. In Guyton, A.C., and Jones, C.E., editors: Cardiovascular physiology, series 1, London, 1974, Butterworths.

Johnson, P.C., editor: Peripheral circulation, New York, 1978, John Wiley & Sons, Inc.

Kaley, G., and Altura, A., editors: Microcirculation, vols. 1 and 2, Baltimore, 1977, 1978, University Park Press.

Krogh, A.: The anatomy and physiology of capillaries, New York, 1959, Hafner Co.

Yoffey, J.M., and Courtice, F.C.: Lymphatics, lymph and the lymphomyeloid complex, London, 1970, Academic Press, Inc. (London), Ltd.

chapter 32

The peripheral circulation and its control

The peripheral circulation is essentially under dual control: centrally through the nervous system and locally in the tissues by the environmental conditions in the immediate vicinity of the blood vessels. The relative importance of these two control mechanisms is not the same in all tissues. In some areas of the body, such as the skin and the splanchnic regions, neural regulation of blood flow predominates, whereas in others, such as the heart and brain, this mechanism plays a minor role.

The vessels chiefly involved in regulating the rate of blood flow throughout the body are referred to as the *resistance vessels,* since these blood vessels offer the greatest resistance to the flow of blood pumped to the tissues by the heart and thereby are important in the maintenance of arterial blood pressure. Smooth muscle fibers constitute a large percentage of the composition of the walls of the resistance vessels (Fig. 26-1). Therefore the vessel lumen can be varied from one that is completely obliterated by strong contraction of the smooth muscle, with infolding of the endothelial lining, to one that is maximally dilated as a result of full relaxation of the vascular smooth muscle. Some resistance vessels are closed at any given moment in time, and partial contraction (or *tone*) of the vascular smooth muscle exists in essentially all of the other arterioles. Were all the resistance vessels in the body to dilate simultaneously, blood pressure would fall precipitously to very low levels.

■ *Vascular smooth muscle*

Vascular smooth muscle is the tissue responsible for the control of total peripheral resistance, arterial and venous tone, and the distribution of blood flow throughout the body. The smooth muscle cells are small, mononucleate, spindle shaped, and generally arranged in one or more helical or circular layers around the blood vessels (Chapter 22). In general, the close association between action potentials and contraction observed in skeletal and cardiac muscle cells cannot be demonstrated in vascular smooth muscle. However, graded changes in membrane potential often are associated with increases or decreases in force. Contractile activity generally is elicited by neural or humoral stimuli. The behavior of smooth muscle in different vessels is quite variable. For example, some vessels, particularly in the portal or mesenteric circulation, contain longitudinally oriented smooth muscle, which is spontaneously active and which shows action potentials that are correlated with the contractions and the electrical coupling between cells.

The cells contain large numbers of thin, actin filaments and comparatively small numbers of thick, myosin filaments. These filaments are not arranged in transverse register to form visible sarcomeres, although the sliding filament mechanism is believed to operate in this tissue. Compared with skeletal muscle, the smooth muscle contracts very slowly, but it can develop high forces and operates over a considerable range of lengths under physiological conditions.

The interaction between myosin and actin, leading to contraction, is controlled by the myoplasmic Ca^{++} concentration, as in other muscles, but the molecular mechanism whereby Ca^{++} regulates contraction appears to differ. The increased myoplasmic Ca^{++} that initiates contraction can come from intracellular stores in the sarcoplasmic reticulum, be displaced from the plasma membrane, or pass into the cell following an increase in membrane Ca^{++} permeability. The relative importance of intracellular and extracellular Ca^{++} for activation varies with different vascular smooth muscles and different agonists.

Most of the arteries and veins of the body are supplied to different degrees solely by fibers of the sympathetic nervous system. These nerve fibers exert a tonic effect on the blood vessels, as evidenced by the fact that cutting or freezing the sympathetic nerves to a vascular bed (such as muscle) results in an increase in blood flow. Activation of the sympathetic nerves either directly or reflexly (see pp. 536 and 542) enhances vascular resistance. In contrast to the sympathetic nerves, the parasympathetic nerves tend to decrease vascular resistance, but they innervate only a small fraction of the blood vessels in the body, mainly in certain viscera and pelvic organs. Vascular smooth muscle also responds to humoral stimulation (hormones and drugs) without evidence of electrical excitation. This has been referred to as *pharmacomechanical coupling* and is mediated by Ca^{++} influx or release. In the category of pharmacological stimuli are substances such as catecholamines, histamine, acetylcholine, serotonin, angiotensin, adenosine, and the prostaglandins. Local environmental changes alter the contractile state of vascular smooth muscle, and alterations such as increased temperature or increased CO_2 levels induce relaxation of this tissue.

In studies on this interesting and important type of muscle, great care should be taken in extrapolating results from one tissue to another or from the same tissue under different physiological conditions. For example, some agents elicit vasodilation in some vascular beds and vasoconstriction in others.

■ Intrinsic or local control of peripheral blood flow

In a number of different tissues the blood flow appears to be adjusted to the existing metabolic activity of the tissue. Furthermore, imposed changes in the perfusion pressure (arterial blood pressure) at constant levels of tissue metabolism, as measured by oxygen consumption, are met with vascular resistance changes that tend to maintain a constant blood flow. This mechanism is commonly referred to as *autoregulation* of blood flow (Fig. 32-1). In the skeletal muscle preparation from which these data were gathered, the muscle was completely isolated from the rest of the animal and was in a resting state. From a control pressure of 100 mm Hg the pressure was abruptly increased or decreased, and the blood flows observed immediately after changing the perfusion pressure are represented by the closed circles. Maintenance of the altered pressure at each new level was followed within 30 to 60 seconds by a return of flow to or toward the control levels; the open circles represent these steady-state flows. Over the pressure range from 20 to 120 mm Hg the steady-state flow is relatively constant. Calculation of resistance across the vascular bed (pressure/flow) during steady-state conditions indicates that with elevation of perfusion pressure the resistance vessels constricted, whereas with reduction of perfusion pressure, dilation occurred.

The mechanism responsible for this constancy of blood flow in the presence of an altered perfusion pressure is not known. However, three explanations have been suggested: the *tissue pressure hypothesis,* the *myogenic hypothesis,* and the *metabolic hypothesis*.

According to the tissue pressure concept an increase in perfusion pressure produces an increase in blood volume of the tissue and a net transfer of fluid from the intravascular to the extravascular compartments. The resultant increase in tissue pressure (turgor) is believed to compress the very thin-walled vessels such as the capillaries, venules, and veins and thereby to reduce the flow of blood into the tissue. A reduction in per-

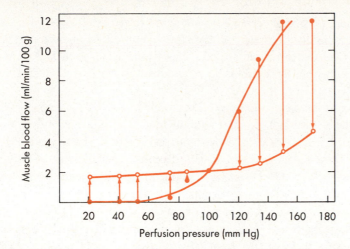

Fig. 32-1 ■ Pressure/flow relationship in the skeletal muscle vascular bed of the dog. The closed circles represent the flows obtained immediately after abrupt changes in perfusion pressure from the control level (point where lines cross). The open circles represent the steady-state flows obtained at the new perfusion pressure. (Redrawn from Jones, R.D., and Berne, R.M.: Circ. Res. **14**:126, 1964. By permission of the American Heart Association, Inc.)

fusion pressure would elicit the opposite response. Such a mechanism can operate only in an encapsulated structure where expansion of the tissue is restricted. However, even in the kidney, which possesses a fairly rigid connective tissue capsule and shows a high degree of autoregulation of blood flow, conclusive evidence for a tissue pressure mechanism is lacking.

The myogenic hypothesis states that the vascular smooth muscle contracts in response to stretch and relaxes with a reduction in tension. Therefore the initial flow increment produced by an abrupt increase in perfusion pressure that passively distends the blood vessels would be followed by a return of flow to the previous control level by contraction of the smooth muscles of the resistance vessels. One difficulty with the myogenic hypothesis is that, for flow to remain constant following an increase in perfusion pressure, it is necessary for the caliber of the resistance vessels to be less than it was prior to the elevation of pressure. Should this occur, the stretch stimulus for the maintenance of increased resistance would be removed. However, if an increase in vessel wall tension, and not stretch, triggers a contractile (constrictor) response, then it is possible, by consideration of the Laplace equation ($T = Pr$, see p. 519) to explain autoregulation in terms of the myogenic mechanism. Raising perfusion pressure increases the pressure across the vessel wall and by passive stretch increases the radius, resulting in a large increment in wall tension. Constriction of the vessel in response to the increase in tension results in a reduction of its diameter to less than the original value so that the product of pressure (increased) and radius (decreased) is restored to the control level. Since the resistance vessels also show intermittent contraction and relaxation, it is possible to avoid the paradox encountered when a simple maintained contractile response to stretch is postulated, by proposing an increase in the frequency of contractile responses (via stretch-induced increase in frequency of action potentials) evoked by elevation of perfusion pressure. If such a mechanism were operative, the resistance vessels as a whole would spend more time in the contracted state when pressure was raised than they did prior to elevation of perfusion pressure.

The myogenic mechanism has been demonstrated in certain tissues and in isolated arterioles. Since blood pressure is reflexly maintained at a fairly constant level under normal conditions, operation of a myogenic mechanism would be expected to be minimized. However, when one changes position (from a lying to a standing position) a large change in transmural pressure occurs in the lower extremities. Were it not for the fact that the precapillary vessels constrict in response to this imposed stretch, the hydrostatic pressure in the lower parts of the legs would reach such high levels that large volumes of fluid would pass from the capillaries into the interstitial fluid compartment and produce edema.

According to the metabolic hypothesis the blood flow is governed by the metabolic activity of the tissue, and any intervention that results in an oxygen supply that is inadequate for the requirements of the tissue gives rise to the formation of vasodilator metabolites. These metabolites are released from the tissue and act locally to dilate the resistance vessels. When the metabolic rate of the tissue increases or the oxygen delivery to the tissue decreases, more vasodilator substance is released, and the metabolite concentration in the tissue increases. When metabolic activity at constant perfusion pressure decreases or perfusion pressure at constant metabolic activity increases, the tissue concentration of the vasodilator agent falls. A decrease in metabolite production or an increase in washout and/or inactivation of the metabolite elicits an increase in precapillary resistance. An attractive feature of the metabolic hypothesis is that in most tissues blood flow closely parallels metabolic activity. Thus, although blood pressure is kept fairly constant, metabolic activity and blood flow in the different body tissues vary together under physiological conditions (for example, exercise).

Many substances have been proposed as mediators of metabolic vasodilation. Some of the earliest ones suggested are lactic acid, CO_2, and hydrogen ion. However, the decrease in vascular resistance induced by supernormal concentrations of these dilator agents falls considerably short of the dilation observed under physiological conditions of increased metabolic activity.

Changes in oxygen tension can evoke changes in the contractile state of vascular smooth muscle; an increase in Po_2 elicits contraction, and a decrease in Po_2, relaxation. If significant reductions in the intravascular Po_2 occur before the arterial blood reaches the resistance vessels (diffusion through the arterial and arteriolar walls, see p. 522), small changes in oxygen supply and/or consumption could elicit contraction or relaxation of the resistance vessels. However, direct measurements of Po_2 at the resistance vessels indicate that over a wide range of Po_2 (11 to 343 mm Hg) there is no correlation between oxygen tension and arteriolar diameter. Furthermore, if Po_2 were directly responsible for vascular smooth muscle tension, one would not expect to find a parallelism between the duration of arterial occlusion and the duration of the reactive hyperemia observed on release of the occlusion (Fig. 32-2). With either short occlusions (5 to 10 seconds) or long occlusions (1 to 3 minutes) the venous blood becomes bright red (well oxygenated) within 1 to 2 seconds after release of the arterial occlusion, and therefore the smooth muscle of the resistance vessels must be exposed to a high Po_2 in each instance. Nevertheless the longer occlusions result in longer periods of reactive hyperemia. These observations are more compatible with the release of a vasodilator metabolite from the tissue than with a direct effect of Po_2 on the vascular smooth muscle.

The potassium ion, inorganic phosphate, and interstitial fluid osmolarity also can induce vasodilation; since potassium and phosphate are released and osmolarity is increased during skeletal muscle contraction, it has been proposed that these factors contribute to *active hyperemia* (increased blood flow caused by enhanced tissue activity). However, significant increases of phosphate concentration and osmolarity are not consistently observed during muscle contraction, and they may produce only transient increases in blood flow. Therefore they are not likely candidates for mediators of the vasodilation observed with muscular activity. Potassium release occurs with the onset of skeletal muscle contraction or an increase in cardiac activity and could be responsible for the initial decrease in vascular resistance observed with exercise or increased cardiac work. However, potassium release is not sustained, despite continued arteriolar dilation

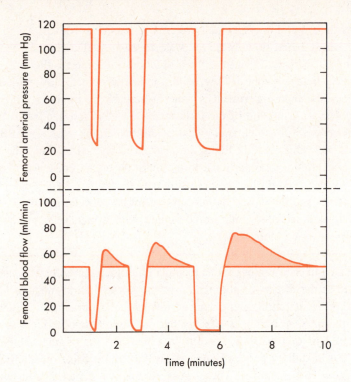

Fig. 32-2 ■ Reactive hyperemia in the hind limb of the dog after 15-, 30-, and 60-second occlusions of the femoral artery.

throughout the period of enhanced muscle activity. Therefore some other agent must serve as mediator of the vasodilation associated with the greater metabolic activity of the tissue. Reoxygenated venous blood obtained from active cardiac and skeletal muscles under steady-state conditions of exercise does not elicit vasodilation when infused into a test vascular bed. It is difficult to see how oxygenation of the venous blood could alter its potassium or phosphate content or its osmolarity and thereby destroy its vasodilator effect.

Recent evidence indicates that adenosine, which plays a role in the regulation of coronary blood flow, also may participate in the control of the resistance vessels in skeletal muscle; also some of the prostaglandins have been proposed as important vasodilator mediators in certain vascular beds.

Thus there are a number of candidates for the mediator of metabolic vasodilation, and the relative contribution of each of the various factors remains the subject for future investigation. Several factors may be involved in any given vascular bed, and different factors preponderate in different tissues.

Metabolic control of vascular resistance via the release of a vasodilator substance is predicated on the existence of basal vessel tone. This tonic activity, or *basal tone,* of the vascular smooth muscle is readily demonstrable, but in contrast to tone in skeletal muscle it is independent of the nervous system. The factor responsible for basal tone in blood vessels is not known, but one or more of the following factors may be involved: (1) an expression of myogenic activity in response to the stretch imposed by the blood pressure, (2) the high oxygen tension of arterial blood, (3) the presence of calcium ions, or (4) some unknown factor in plasma, since addition of plasma to the bathing solution of isolated vessel segments evokes partial contraction of the smooth muscle.

A phenomenon that is mechanistically linked to autoregulation of blood flow is *reactive hyperemia.* If arterial inflow to a vascular bed is stopped for a few seconds to several minutes, the blood flow, on release of the occlusion, immediately exceeds the flow before occlusion and only gradually returns to the control level. This is illustrated in Fig. 32-2, where blood flow to the leg was stopped by clamping the femoral artery for 15, 30, and 60 seconds. Release of the 60-second occlusion resulted in a peak blood flow 70% greater than the control flow, with a return to control flow within about 110 seconds. When this same experiment is done in humans by inflating a blood pressure cuff on the upper arm, dilation of the resistance vessels of the hand and forearm, immediately after release of the cuff, is evident from the bright red color of the skin and the fullness of the veins. Within limits the peak flow and particularly the duration of the reactive hyperemia are proportional to the duration of the occlusion (Fig. 32-2). If the extremity is exercised during the occlusion period, reactive hyperemia is increased. These observations and the close relationship that exists between metabolic activity and blood flow in the unoccluded limb are consonant with a metabolic mechanism in the local regulation of tissue blood flow.

■ *Extrinsic control of peripheral blood flow*
■ *Neural sympathetic vasoconstriction*

There are a number of regions in the medulla oblongata that influence cardiovascular activity. Some of the effects of stimulation of the dorsal lateral medulla are vasoconstriction, cardiac acceleration, and enhanced myocardial contractility. Caudal and ventromedial to the *pressor* region is a zone that on stimulation produces a decrease in blood pressure. This *depressor area* exerts its effect by direct spinal inhibition and by inhibition of the medullary pressor region. However, the precise mechanism of its depressor actions is still unknown. These areas comprise a center not in an anatomical sense in that a discrete group of cells is discernible, but in a physiological sense in that stimulation of the pressor region produces the responses mentioned previously. From the vasoconstrictor regions, fibers descend in the spinal cord and synapse at different levels of the thoracolumbar region (T1 to L2 or L3). Fibers from the intermediolateral gray matter of the spinal cord emerge with the ventral roots but leave the motor fibers to join the paravertebral sympathetic chains through the white rami communicantes (Chapter 20). These preganglionic white (myelinated) fibers may pass up or down the sympathetic chains to synapse in the various ganglia within the chains or in certain outlying ganglia. Postganglionic gray rami (unmyelinated) then join the corresponding segmental spinal nerves and accompany them to the periphery to innervate the arteries and veins. Postganglionic sympathetic fibers from the various ganglia join the large arteries and accompany them as an investing network of fibers to the resistance and capacitance vessels.

The vasoconstrictor regions are tonically active, and reflexes or humoral stimuli which enhance this activity result in an increase in frequency of impulses reaching the terminal branches to the vessels, where a constrictor neurohumor (norepinephrine) is released and elicits constriction (α-adrenergic effect) of the resistance vessels. Inhibition of the vasoconstrictor areas reduces their tonic activity and thus diminishes the frequency of impulses in the efferent nerve fibers, resulting in vasodilation. In this manner neural regulation of the peripheral circulation is accomplished primarily by alteration of the number of impulses passing down the vasoconstrictor fibers of the sympathetic nerves to the blood vessels. The vasomotor regions may show rhythmic changes in tonic activity manifested as oscillations of arterial pressure that occur at the frequency of respiration *(Traube-Hering waves)* or that are independent of and at a lower frequency than respiration *(Mayer waves)*.

■ *Sympathetic constrictor influence on resistance and capacitance vessels*

The vasoconstrictor fibers of the sympathetic nervous system supply the arterioles and the smaller veins. The arteries and larger veins also receive sympathetic innervation, but neural influence on the larger vessels is of far less functional importance than it is on the microcirculation. Capacitance vessels are apparently more responsive to sympathetic nerve stimulation than are resistance vessels, since they reach maximal constric-

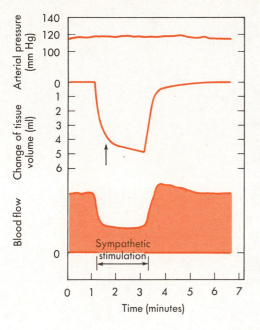

Fig. 32-3 ■ Effect of sympathetic nerve stimulation (2 impulses per second) on blood flow and tissue volume in the hindquarters of the cat. The arrow denotes the change in slope of the tissue volume curve where the volume decrease caused by emptying of capacitance vessels ceases, and loss of extravascular fluid becomes evident. (Redrawn from Mellander, S.: Acta Physiol. Scand. **50**[Suppl. 176]:1-86, 1960.)

tion at a lower frequency of stimulation than do the resistance vessels. Norepinephrine is the neurotransmitter released at the sympathetic nerve terminals at the blood vessels, and many factors, such as circulating hormones and particularly locally released substances, modify the liberation of norepinephrine from the vesicles of the nerve terminals. The response of the resistance and capacitance vessels to stimulation of the sympathetic fibers is illustrated in Fig. 32-3. At constant arterial pressure, sympathetic fiber stimulation evoked a reduction of blood flow (constriction of the resistance vessels) and a decrease in blood volume of the tissue (constriction of the capacitance vessels). The initial abrupt decrease in tissue volume was caused by movement of blood out of the capacitance vessels and out of the hindquarters of the cat, whereas the late, slow progressive decline in volume (to the right of the *arrow*) was caused by movement of extravascular fluid into the capillaries and then away from the tissue. The loss of tissue fluid was a consequence of the lowered capillary hydrostatic pressure brought about by constriction of the resistance vessels, with establishment of a new equilibrium of the forces responsible for filtration and absorption across the capillary wall.

In addition to active changes (contraction and relaxation of the vascular smooth muscle) in vessel caliber, there are also passive changes caused solely by alterations in intraluminal pressure; an increase in intraluminal pressure produces distension of the vessels, and a decrease produces a reduction in caliber by recoil of the elastic components of the vessel walls. The relative effects of the passive and active forces on the volume changes of the tissues are depicted in Fig. 32-4. With occlusion of the arterial blood supply to the tissue, perfusion pressure dropped almost to zero, blood flow became nil, and there was an abrupt 2 ml decrease in tissue volume. This rapid reduction in tissue volume was a passive response to the reduction in arterial pressure, with expulsion of blood from the tissue. The slow decrease in tissue volume (between the first two *arrows*) was quite small, relative to that shown in Fig. 32-3, because of the brevity of the period between arterial occlusion and the start of sympathetic stimulation. This slight reduction in volume was caused by the decrease in capillary hydrostatic pressure incident to the reduction of the perfusion pressure. It represents loss of extravascular fluid via absorption from the capillaries and movement out through the venous channels. When the sympathetic nerve fibers were stimulated during the period of arterial occlusion, a large decrease in tissue volume was observed. Since blood flow had already been stopped by arterial occlusion, this change in tissue volume cannot be a passive

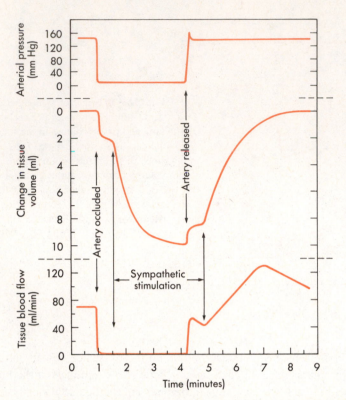

Fig. 32-4 ■ Effect of arterial occlusion and sympathetic nerve stimulation (6 impulses per second) on the tissue volume in the hindquarters of the cat. Note the small changes in tissue volume with arterial occlusion and release compared with the large volume change obtained with sympathetic nerve stimulation. (Redrawn from Mellander, S.: Acta Physiol. Scand. **50**[Suppl. 176]:1-86, 1960.)

response secondary to arteriolar constriction but must represent active constriction of the capacitance vessels. Note that once again a brief slow decrease in tissue volume followed the rapid change and indicates a continuation of the loss of extravascular fluid as a result of the reduction in capillary hydrostatic pressure. If the capillary pressure is held at low levels for 2 to 3 minutes, either by arterial occlusion or by constriction of the resistance vessels, a new pressure equilibrium is established across the capillary wall, and further loss of tissue fluid becomes negligible. The protein concentration of the blood leaving the tissue during the period of slow decrease in tissue volume was found to be reduced, indicating dilution of the plasma proteins by the absorption of low-protein tissue fluid by the capillaries. When the arterial occlusion was released during continuous sympathetic nerve stimulation, a small abrupt increase in intravascular pressure occurred because of a passive stretch of the capacitance vessels by the increased intraluminal pressure. This rapid increment in tissue volume was followed by a small gradual increase (between pair of *arrows* to the right) that represented movement of fluid from the capillaries into the interstitial spaces in response to the elevation of capillary hydrostatic pressure. Cessation of sympathetic stimulation resulted in relaxation of the smooth muscle of the capacitance vessels and a restoration of their preexisting blood volume. Note that blood flow increased above, and gradually returned toward, the control level, illustrating a typical reactive hyperemia response to the period of ischemia. From these observations it can be seen that the passive changes of the capacitance vessels are small, relative to those induced by active contraction and relaxation of vascular smooth muscle of the capacitance vessels. Under physiological conditions, in which such drastic reduction in perfusion pressure would not occur, the contribution

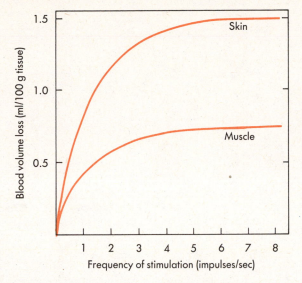

Fig. 32-5 ■ Comparison of the blood volume loss from skin and muscle with sympathetic nerve stimulation. (Redrawn from Mellander, S.: Acta Physiol. Scand. **50**[Suppl. 176]:1-86, 1960.)

of a passive factor to changes in volume of the capacitance vessels would be even less than shown in Fig. 32-4.

At basal tone approximately one third the blood volume of a tissue can be mobilized on stimulation of the sympathetic nerves at physiological frequencies. The basal tone is very low in capacitance vessels; with veins denervated only small increases in volume are obtained with maximal doses of the vasodilator acetylcholine. Therefore the blood volume at basal tone is close to the maximal blood volume of the tissue. The amount of blood that can be mobilized from skin and muscle at different frequencies of stimulation of the sympathetic nerves is depicted in Fig. 32-5. The greater mobilization of blood from the skin than from the muscle capacitance vessels may in part be caused by a greater sensitivity of these vessels to sympathetic stimulation, but it also is caused by the fact that basal tone is lower in skin than in muscle. Therefore, in the absence of neural influence, the skin capacitance vessels contain more blood than do the muscle capacitance vessels.

Blood is mobilized from capacitance vessels in response to physiological stimuli. In exercise, activation of the sympathetic nerve fibers produces constriction of veins and thus augments the cardiac filling pressure. Also, in arterial hypotension (as in hemorrhage) the capacitance vessels constrict, which aids in overcoming the decreased central venous pressure associated with this condition. In addition, the resistance vessels constrict in shock, thereby assisting in the maintenance or restoration of arterial pressure. With arterial hypotension the enhanced arteriolar constriction also leads to a small mobilization of blood from the tissue by virtue of recoil of the postarteriolar vessels when intraluminal pressure is reduced. Further, there is mobilization of extravascular fluid because of greater absorption of fluid into the capillaries in response to the lowered capillary hydrostatic pressure.

Clear dissociation between responses of the resistance and capacitance vessels can be demonstrated with nerve stimulation or with the use of epinephrine and acetylcholine. In Fig. 32-6, *A*, a small intravenous dose of epinephrine elicited an increase in blood flow (active dilation of resistance vessels) and a small abrupt increase in tissue volume (passive expansion of the capacitance vessels by the increased intraluminal pressure). Following the passive increase in tissue volume (to the left of the first *arrow*) a steady gradual tissue volume increase occurred (between *arrows*), attributable to movement of fluid from the capillaries to the interstitial space as a result of the elevated capillary hydrostatic pressure. Fig. 32-6, *B*, represents the effect of a dose of acetylcholine that elicited the same degree of resistance vessel dilation as that obtained with epinephrine

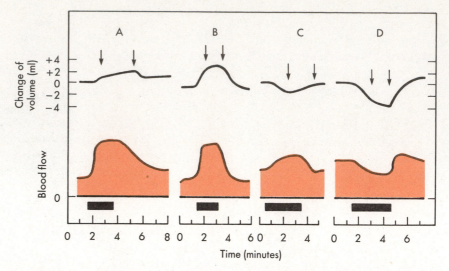

Fig. 32-6 ■ Effect of acetylcholine and different doses of epinephrine on resistance and capacitance vessels in skeletal muscle. *A*, Epinephrine 0.3 μg/kg/minute. *B*, Acetylcholine 1.7 μg/kg/minute (to give same degree of dilation of resistance vessels as in *A*). *C*, Epinephrine 0.7 μg/kg/minute. *D*, Epinephrine 3.0 μg/kg/minute. Pairs of arrows indicate periods of filtration and absorption. (Redrawn from Mellander, S.: Acta Physiol. Scand. **50**[Suppl. 176]:1-86, 1960.)

in Fig. 32-6, *A*. Note that the acetylcholine produced a large increment in tissue volume (blood volume) and only a small increase in tissue volume attributable to increased capillary filtration (between *arrows*). In this case the concomitant dilation of the capacitance and resistance vessels resulted in a smaller increment in capillary pressure and thus a slower rate of capillary filtration. In Fig. 32-6, *C*, the dose of epinephrine was about twice that given in *A;* it elicited dilation of the resistance vessels but constriction of the capacitance vessels, resulting in a small decrease in tissue blood volume. These opposite effects on the resistance and capacitance vessels served to increase capillary hydrostatic pressure and enhance the rate of capillary filtration, as evidenced from the slowly rising portion of the tissue volume curve (between *arrows*). A still larger dose of epinephrine (Fig. 32-6, *D*) produced constriction of both resistance and capacitance vessels, with a decrease in tissue blood volume (to left of first *arrow*) and movement of fluid from the extravascular to the intravascular compartment (between *arrows*) because of a decrease in net capillary hydrostatic pressure.

From these observations it becomes apparent that neural and humoral stimuli can exert similar or dissimilar effects on different segments of the vascular tree and in so doing can alter blood flow, tissue blood volume, and extravascular volume to meet the physiological requirements of the organism.

■ *Active sympathetic vasodilation*

Although the major neural control of the peripheral vessels is provided by the adrenergic vasoconstrictor fibers of the sympathetic nervous system, there are also *sympathetic cholinergic fibers* innervating the resistance vessels in skeletal muscle and skin. Electrical stimulation of the sympathetic nerves to blood vessels produces vasoconstriction. However, if the adrenergic constrictor effect is blocked by a suitable adrenergic receptor blocking agent, or if the neural stores of norepinephrine are depleted by prior treatment with reserpine, the stimulation results in *active dilation,* which can be blocked by the administration of atropine. There is no evidence that these cholinergic sympathetic dilator fibers innervate the capacitance vessels. However, a small fraction of the vasodilation observed with activation of the baroreceptors may occur through the sympathetic cholinergic neurons.

Fibers of the cholinergic sympathetic dilator system arise in the motor cortex of the cerebrum and pass through the hypothalamus and the ventral medulla oblongata before joining the other sympathetic outflow in the spinal cord. Activation of the cholinergic sympathetic dilator system produces a relatively large transient initial increase in blood flow, followed by a smaller sustained increment in flow during the period of nerve stimulation. There is no evidence for any tonic activity of these fibers. Since excitement or apprehension seems to activate this system and induce vasodilation in skeletal muscle, it has been suggested that the role of the sympathetic vasodilators is to provide the muscles with an increased blood flow in anticipation of the use of the muscles (for example, flight or fight). These cholinergic sympathetic fibers have been demonstrated in dog and cat and may also be present in humans.

In addition to the cholinergic sympathetic vasodilator system, there are β-adrenergic receptors on the resistance vessels that can be demonstrated by sympathetic nerve stimulation after the administration of atropine and an α-adrenergic receptor blocker. The vasodilation observed in muscle with intraarterial injection of small doses of epinephrine or with isoproterenol is caused by stimulation of these β-adrenergic receptors. However, conclusive evidence of neural reflex activation of the vascular β-receptor is lacking. Finally, two other active vasodilator mechanisms mediated by the sympathetic nerve fibers have been suggested. One produces a transitory active vasodilation in response to baroreceptor or direct sympathetic nerve stimulation, and the other evokes a slow sustained vasodilation on sympathetic nerve stimulation that can be demonstrated only when the overriding constrictor response is blocked by adrenergic receptor blocking agents. The neurohumor mediating the transient response is reputed to be histamine, whereas that mediating the sustained vasodilation is unknown.

The efferent fibers of the cranial division of the parasympathetic nervous system supply blood vessels of the head and viscera, whereas fibers of the sacral division supply blood vessels of the genitalia, bladder, and large bowel. Skeletal muscle and skin do not receive parasympathetic innervation. Thus only a small proportion of the resistance vessels of the body receives parasympathetic fibers, and therefore the effect of these cholinergic fibers on total vascular resistance is small. Stimulation of the parasympathetic fibers to the salivary glands induces marked vasodilation. However, the increase in submaxillary gland blood flow seen with chorda tympani stimulation is believed by some investigators to be secondary to the increase in metabolic activity of the gland and by others to be a primary action on the arterioles of the gland. A vasodilator polypeptide, *bradykinin*, formed locally from the action of an enzyme on a plasma protein substrate present in the glandular lymphatics has been considered to be the metabolic mediator of the vasodilation produced by chorda tympani stimulation. Whether vasodilation in salivary glands results from the release of a cholinergic neurohumor from nerve endings, from the formation and release of bradykinin, or from both is unsettled. Bradykinin also has been reported to be formed in other exocrine glands, such as the lacrimal glands and the sweat glands. Its presence in sweat is thought to be partly responsible for the dilation of cutaneous blood vessels with sweating.

■ *Parasympathetic neural influence*

Vasomotor impulses at one time were thought to travel antidromically in the spinal sensory nerves. Other than the antidromic impulses seen with the axon reflex (p. 608), there is no good evidence that the spinal nerves transmit impulses from the spinal cord to the peripheral blood vessels antidromically via the dorsal roots.

■ *Dorsal root dilators*

Epinephrine and norepinephrine exert a profound effect on the peripheral blood vessels. In skeletal muscle, epinephrine in low concentrations dilates resistance vessels (β-adrenergic effect). In skin only vasoconstriction is elicited by epinephrine, whereas in

■ *Humoral factors*

all vascular beds the primary effect of norepinephrine is vasoconstriction. The adrenal gland, when stimulated, can release epinephrine and norepinephrine into the systemic circulation. However, the concentrations reached are relatively low, and the effect of catecholamines released from the adrenal medulla is negligible compared with the effect produced by direct sympathetic innervation.

■ *Vascular reflexes*

In addition to the vasomotor areas in the medulla oblongata that initiate sympathetic nerve activity, the dorsal motor nucleus of the vagus and the nucleus ambiguus are regions that initiate cardioinhibition (Chapter 33). These areas are tonically active and serve essentially as brakes on the heart; interruption of fibers from these areas results in tachycardia, whereas stimulation of them produces bradycardia. Areas of the medulla that mediate sympathetic and vagal influences are under the influence of afferent neural impulses arising in the baroreceptors, peripheral chemoreceptors, hypothalamus, cerebral cortex, and skin and also can be altered by changes in the blood concentrations of CO_2 and O_2.

Baroreceptors. The *baroreceptors* (or *pressoreceptors*) are stretch receptors located in the carotid sinuses (slightly widened areas of the internal carotid arteries at their points of origin from the common carotid arteries) and in the aortic arch (Figs. 32-7 and 32-8). Impulses arising in the carotid sinus travel up the carotid sinus nerve (nerve of Hering) to the glossopharyngeal nerve and, via the latter, to the medulla, whereas impulses arising in the pressoreceptors of the aortic arch reach the medulla via afferent fibers in the vagus nerves. The pressoreceptor nerve terminals in the walls of the carotid sinus and aortic arch respond to the stretch and deformation of the vessel induced by the arterial pressure. With increase in blood pressure the frequency of firing is enhanced,

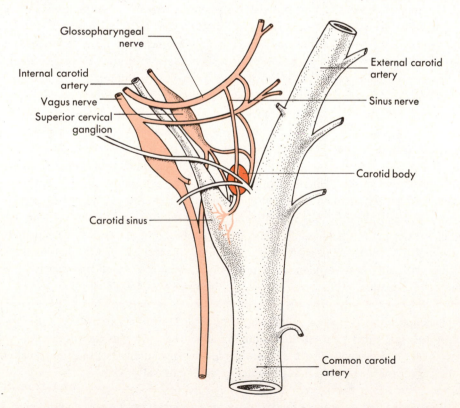

Fig. 32-7 ■ The carotid sinus and carotid body and their innervation in the dog. (Redrawn from Adams, W.E.: The comparative morphology of the carotid body and carotid sinus, Springfield, Ill., 1958. Courtesy of Charles C Thomas, Publisher.)

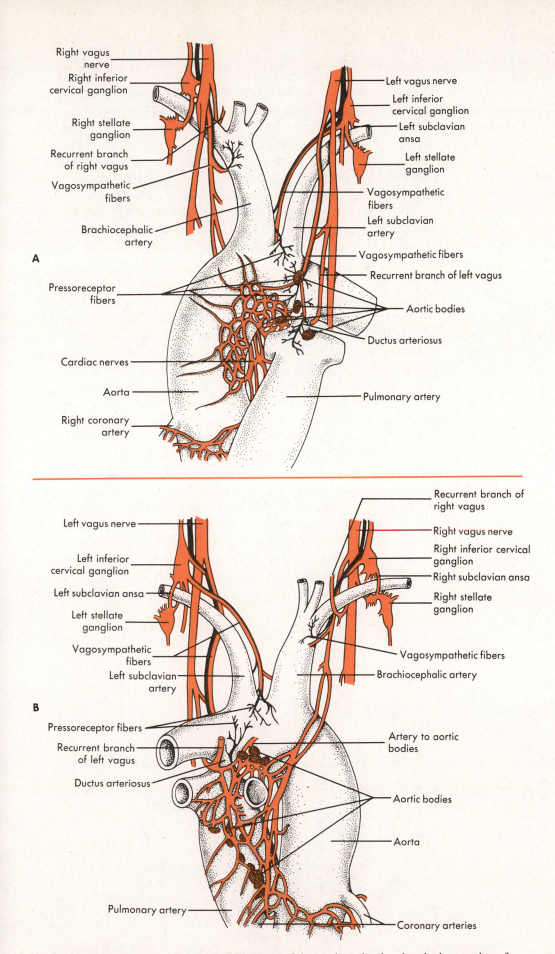

Right vagus nerve

Right inferior cervical ganglion

Right stellate ganglion

Recurrent branch of right vagus

Vagosympathetic fibers

Brachiocephalic artery

A

Pressoreceptor fibers

Cardiac nerves

Aorta

Right coronary artery

Left vagus nerve

Left inferior cervical ganglion

Left subclavian ansa

Left stellate ganglion

Vagosympathetic fibers

Left subclavian artery

Vagosympathetic fibers

Recurrent branch of left vagus

Aortic bodies

Ductus arteriosus

Pulmonary artery

Left vagus nerve

Left inferior cervical ganglion

Left subclavian ansa

Left stellate ganglion

Vagosympathetic fibers

Left subclavian artery

B

Pressoreceptor fibers

Recurrent branch of left vagus

Ductus arteriosus

Pulmonary artery

Recurrent branch of right vagus

Right vagus nerve

Right inferior cervical ganglion

Right subclavian ansa

Right stellate ganglion

Vagosympathetic fibers

Brachiocephalic artery

Artery to aortic bodies

Aortic bodies

Aorta

Coronary arteries

Fig. 32-8 ■ Anterior (**A**) and posterior (**B**) views of the aortic archs showing the innervation of the aortic bodies and pressoreceptors in the dog. (Modified from Nonidez, J.F.: Anat. Rec. **69:**299, 1937.)

whereas the converse is true with a reduction of blood pressure. An increase in frequency of impulses, as occurs with a rise in arterial pressure, acts to inhibit the vasoconstrictor regions, resulting in peripheral vasodilation and a lowering of blood pressure. Contributing to a lowering of the blood pressure is a bradycardia brought about by stimulation of the vagal regions. The carotid sinus and aortic baroreceptors are not equipotent in their effects on peripheral resistance in response to nonpulsatile alterations in blood pressure. The carotid sinus baroreceptors are more sensitive than are those in the aortic arch; changes in mean pressure in the carotid sinus evoke greater alterations in perfusion pressure, and thus in resistance, than do equivalent mean changes in aortic arch pressure. However, with pulsatile changes in blood pressure the two sets of baroreceptors respond similarly.

The carotid sinus with the sinus nerve intact can be isolated from the rest of the circulation and perfused by either a donor dog or an artificial perfusion system. Under these conditions changes in the pressure within the carotid sinus are associated with reciprocal changes in the blood pressure of the experimental animal. The receptors in the walls of the carotid sinus show some adaptation and therefore are more responsive to constantly changing pressures than to sustained constant pressures. This is illustrated in Fig. 32-9, where at normal levels of blood pressure a barrage of impulses from a single fiber of the sinus nerve is initiated in early systole by the pressure rise, and only a few spikes are observed during late systole and early diastole. At lower pressures these phasic changes are even more evident, but the overall frequency of discharge is reduced. The blood pressure threshold for eliciting sinus nerve impulses is about 50 mm Hg, and a maximal sustained firing is reached at around 200 mm Hg.

Since the pressoreceptors show some degree of adaptation, it is to be expected that their response at any level of mean arterial pressure would be greater with a large than with a small pulse pressure. This is indeed the case and is diagramatically illustrated in Fig. 32-10. At any level of mean arterial pressure the lower the pulse pressure, the greater the systemic vascular resistance. This modulation of the baroreceptor reflex by pulse pressure can play a significant role in compensatory adjustments of the circulatory system. For example, if mean blood pressure decreases (as in hemorrhage) from P_1 to

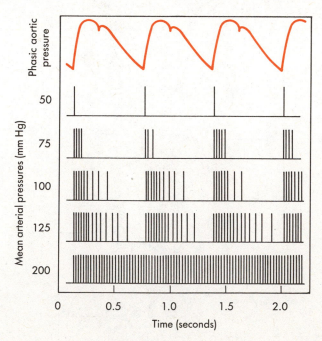

Fig. 32-9 ■ Relationship of phasic aortic blood pressure to the firing of a single afferent nerve fiber from the carotid sinus at different levels of mean arterial pressure.

P_2 (Fig. 32-10) and pulse pressure remains at 60 mm Hg, systemic vascular resistance increases reflexly from point *A* to point *B*. However, pulse pressure decreases in hypotension caused by hemorrhage, and if it decreased to 40 mm Hg (point *C*), there would be a greater reflex constriction of the systemic resistance vessels. These changes in resistance would tend to restore blood pressure more closely toward normal levels and constitute an important protective mechanism in maintaining blood flow to vital tissues such as the brain and heart (Chapter 36). The resistance increases that occur in the peripheral vascular beds in response to a reduced pressure in the carotid sinus vary from one vascular bed to another and thereby produce a redistribution of blood flow. For example, in some vessels studied, the resistance changes elicited by altering carotid sinus pressure around the normal operating sinus pressure were greatest in the femoral vessels, less in the renal vessels, and least in the mesenteric and celiac vessels. The sensitivity of the carotid sinus reflex can be altered by changing the distensibility of the sinus. Local application of epinephrine causes contraction of the smooth muscle in the sinus wall and increases the response to a given rise in intravascular pressure.

In some individuals the carotid sinus is quite sensitive to pressure. Therefore tight collars or other forms of external pressure over the region of the carotid sinus may elict marked hypotension and fainting. In some patients with severe coronary artery disease and chest pain *(angina pectoris)* symptoms have been relieved temporarily by stimulation of the sinus nerve by means of a chronically implanted stimulator than can be activated externally. The reduction in blood pressure achieved by sinus nerve stimulation decreases the pressure work of the heart and hence the myocardial ischemia responsible for the pain. As would be expected, denervation of the carotid sinus can produce temporary, and in some instances prolonged, hypertension.

The baroreceptors play a key role in short-term adjustments of blood pressure when relatively abrupt changes in blood volume, cardiac output, or peripheral resistance (as in exercise) occur. However, long-term control of blood pressure—that is, over days, weeks, and longer—is determined by the fluid balance of the individual, namely the balance between fluid intake and fluid output. By far the single most important organ in the control of body fluid volume, and thus blood pressure, is the kidney. With overhy-

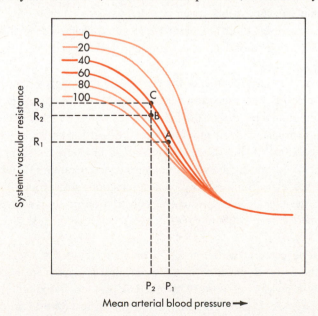

Fig. 32-10 ■ The family of curves relating the mean arterial (carotid sinus and aortic arch) pressure and systemic vascular resistance at pulse pressures of 0, 20, 40, 60, 80, and 100 mm Hg. (Redrawn from Angell James, J.E., and Daly, M. de B.: J. Physiol. [London] **209:**257, 1970.)

dration excessive fluid intake is excreted, whereas with dehydration there is a marked reduction in urine output.

Cardiopulmonary baroreceptors. In addition to the carotid sinus and aortic baroreceptors, there are also cardiopulmonary receptors that show low levels of continuous activity. These receptors are located in the atria, ventricles, and pulmonary vessels and, when stimulated by increases in intracardiac or pulmonary vascular pressures, inhibit vasoconstrictor tone of the resistance vessels. The atrial receptors are of two types: those which are stimulated by atrial contraction (A receptors) and those stimulated by atrial filling and distension (B receptors). The atrial receptors also play a role in the regulation of vasomotor tone and of urine output by means of stimulation or inhibition of angiotensin and of vasopressin secretion. The cardiopulmonary receptors as well as the carotid and aortic baroreceptors are necessary for the full expression of blood pressure regulation. The major responses elicited by stimulation of these vary in different vascular beds. For example, activation of the carotid pressoreceptors has a large effect on the splanchnic vascular bed but only a little influence on forearm vascular resistance. The reverse is true for the cardiopulmonary receptors.

Peripheral chemoreceptors. The peripheral chemoreceptors consist of small, highly vascular bodies in the region of the aortic arch and just medial to the carotid sinuses (Fig. 32-8). They are sensitive to changes in the P_{O_2}, P_{CO_2}, and pH of the blood. Although they are concerned primarily with the regulation of respiration, they reflexly influence the vasomotor regions to a minor degree. A reduction in arterial blood oxygen tension (Pa_{O_2}) stimulates the chemoreceptors, and the increase in the number of impulses in the afferent nerve fibers from the carotid and aortic bodies stimulates the vasoconstrictor regions, resulting in increased tone of the resistance and capacitance vessels. The chemoreceptors also are stimulated by increased arterial blood carbon dioxide tension (Pa_{CO_2}) and reduced pH, but the reflex effect induced is quite small compared with the direct effect of hypercapnia and hydrogen ions on the vasomotor regions in the medulla. When hypoxia and hypercapnia occur at the same time, the stimulation of the chemoreceptors is greater than the sum of the two stimuli when they act alone. When the chemoreceptors are stimulated simultaneously with a reduction in pressure in the baroreceptors, the chemoreceptors potentiate the vasoconstriction observed in the peripheral vessels. However, when the baroreceptors and chemoreceptors are both stimulated (for example, high carotid sinus pressure and low arterial P_{O_2}), the effects of the baroreceptors predominate.

Hypothalamus. Optimal function of the cardiovascular reflexes requires the integrity of pontine and hypothalamic structures. Furthermore these structures are responsible for behavioral and emotional control of the cardiovascular system. Stimulation of the anterior hypothalamus produces a fall in blood pressure and bradycardia, whereas stimulation of the posterolateral region of the hypothalamus produces a rise in blood pressure and tachycardia. The hypothalamus also contains a temperature-regulating center that affects the skin vessels. Stimulation by cold applications to skin or by cooling of the blood perfusing the hypothalamus results in constriction of the skin vessels and heat conservation, whereas warm stimuli result in cutaneous vasodilation and enhanced heat loss.

Cerebrum. The cerebral cortex also can exert a significant effect on blood flow distribution in the body. Stimulation of the motor and premotor areas can affect blood pressure; usually a pressor response is obtained. However, vasodilation and depressor responses may be evoked, as in blushing or fainting in response to an emotional stimulus.

Skin and viscera. Painful stimuli can elicit either pressor or depressor responses, depending on the magnitude and location of the stimulus. Distension of the viscera often evokes a depressor response, whereas painful stimuli on the body surface usually evoke a pressor response. In the anesthetized animal strong electrical stimulation of a sensory nerve will produce a strong pressor response. However, it is sometimes possible to obtain a depressor response with low-intensity and low-frequency stimulation. Furthermore all vascular beds do not exhibit the same response; in some, resistance increases, whereas in others it decreases. In addition, muscle contractions can elicit reflex changes in the magnitude of vasoactivity in the muscle. For the most part these reflexes are mediated through the vasomotor areas in the medulla, but there are also spinal areas that can aid in the regulation of peripheral resistance.

■ *Pulmonary reflexes*

Inflation of the lungs reflexly induces systemic vasodilation and a decrease in arterial blood pressure. Conversely, collapse of the lungs evokes systemic vasoconstriction. Afferent fibers mediating this reflex run in the vagus nerves and possibly to a limited extent in the sympathetic nerves. Their stimulation by stretch of the lungs inhibits the vasomotor areas. The magnitude of the depressor response to lung inflation is related directly to the degree of inflation and to the existing level of vasoconstrictor tone.

■ *Chemosensitive regions of the medulla*

Increases of Pa_{CO_2} stimulate the medullary vasoconstrictor regions (central chemoreceptors, Chapter 40), thereby increasing peripheral resistance, whereas reduction in Pa_{CO_2} below normal levels (as with hyperventilation) decreases the degree of tonic activity of these areas, thereby decreasing peripheral resistance. The central chemoreceptors also are affected by changes of pH. A lowering of blood pH stimulates and a rise in blood pH inhibits these receptors. These effects of changes in Pa_{CO_2} and blood pH possibly operate through changes in cerebrospinal fluid pH, as appears to be the case for the respiratory center. Whether there are special hydrogen ion chemoreceptors mediating pH-induced vasomotor effects has not been established.

Oxygen tension has relatively little direct effect on the vasomotor region. The primary effect of hypoxia is reflexly mediated via the carotid and aortic chemoreceptors. Moderate reduction of Pa_{O_2} will stimulate the vasomotor region, but severe reduction will depress vasomotor activity in the same manner that other areas of the brain are depressed by very low oxygen tensions.

Cerebral ischemia, which may occur because of excessive pressure exerted by an expanding intracranial tumor, results in a marked increase in vasoconstriction. The stimulation probably is caused by a local accumulation of CO_2 and reduction of O_2 and possibly by excitation of intracranial baroreceptors. With prolonged, severe ischemia, central depression eventually supervenes, and the blood pressure falls.

■ *Balance between extrinsic and intrinsic factors in regulation of peripheral blood flow*

Dual control of the peripheral vessels by virtue of intrinsic and extrinsic mechanisms makes possible a number of vascular adjustments that enable the body to direct blood flow to areas where it is needed in greater supply and away from areas whose immediate requirements are less. In some tissues a more or less fixed relative potency of extrinsic and intrinsic mechanisms exists, and in other tissues the ratio is changeable, depending on the state of activity of that tissue.

In the brain and the heart, both vital structures with very limited tolerance for a reduced blood supply, intrinsic flow-regulating mechanisms are dominant. For instance, massive discharge of the vasoconstrictor region over the sympathetic nerves, which might occur in severe, acute hemorrhage, has negligible effects on the cerebral and cardiac resistance vessels, whereas skin, renal, and splanchnic blood vessels become greatly constricted.

In the skin the extrinsic vascular control is dominant. Not only do the cutaneous vessels participate strongly in a general vasoconstrictor discharge, but they also respond selectively through hypothalamic pathways to subserve the heat loss and heat conservation function required in body temperature regulation. However, intrinsic control can

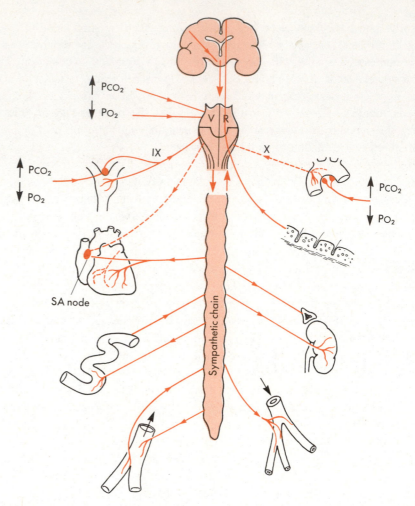

Fig. 32-11 ■ The neural input and output of the vasomotor region *(VR)*. *IX,* Glossopharyngeal nerve; *X,* vagus nerve.

be demonstrated by local changes of temperature that can modify or override the central influence on resistance and capacitance vessels.

In skeletal muscle the interplay and changing balance between extrinsic and intrinsic mechanisms can be clearly seen. In resting skeletal muscle, neural control (vasoconstrictor tone) is dominant, as can be demonstrated by the large increment in blood flow that occurs immediately after section of the sympathetic nerves to the tissue. In anticipation of and at the start of exercise, such as running, there is an increase in blood flow in the leg muscles, possibly mediated by activation of the cholinergic sympathetic dilator system. After the onset of exercise the intrinsic flow-regulating mechanism assumes control, and because of the local increase in metabolites, vasodilation occurs in the active muscles. Vasoconstriction occurs in the inactive tissues as a manifestation of the general sympathetic discharge, but constrictor impulses reaching the resistance vessels of the active muscles are overridden by the local metabolic effect. Operation of this dual control mechanism thus provides increased blood where it is required and shunts it away from relatively inactive areas. Similar effects may be achieved with an increase in Pa_{CO_2}. Normally the hyperventilation associated with exercise keeps Pa_{CO_2} at normal levels. However, were Pa_{CO_2} to increase, a generalized vasoconstriction would occur by virtue of stimulation of the vasoconstrictor region by CO_2, but in the active muscles, where the CO_2 concentration is highest, the smooth muscle of the arterioles would relax in response to the local P_{CO_2}. Factors affecting and affected by the vasomotor region are summarized in Fig. 32-11.

■ *Bibliography*

Journal articles

Abboud, F.M., et al.: Carotid and cardiopulmonary baroreceptor control of splanchnic and forearm vascular resistance during venous pooling in man, J. Physiol. (London) **286**:173, 1979.

Angell James, J.E., and Daly, M. de B.: Comparison of the reflex vasomotor responses to separate and combined stimulation of the carotid sinus and aortic arch baroreceptors by pulsatile and nonpulsatile pressures in the dog, J. Physiol. (London) **209**:257, 1970.

Belloni, F.L., and Sparks, H.V.: The peripheral circulation, Annu. Rev. Physiol. **40**:67, 1978.

Bohr, D.F.: Vascular smooth muscle updated, Circ. Res. **32**:665, 1973.

Brown, A.M.: Receptors under pressure—an update on baroreceptors, Circ. Res. **46**:1, 1980.

Browse, N.L., Lorenz, R.R., and Shepherd, J.T.: Response of capacity and resistance vessels of dog's limb to sympathetic nerve stimulation, Am. J. Physiol. **210**:95, 1966.

Coleridge, H.M., and Coleridge, J.C.G.: Cardiovascular afferents involved in regulation of peripheral vessels, Annu. Rev. Physiol. **42**:413, 1980.

Donald, D.E., and Shepherd, J.T.: Reflexes from the heart and lungs: physiological curiosities or important regulatory mechanisms, Cardiovasc. Res. **12**:449, 1978.

Donald, D.E., and Shepherd, J.T.: Autonomic regulation of the peripheral circulation, Annu. Rev. Physiol. **42**:429, 1980.

Duling, B.R.: Oxygen sensitivity of vascular smooth muscle. II. In vivo studies, Am. J. Physiol. **227**:42, 1974.

Guyton, A.C., et al.: A systems analysis approach to understanding long-range arterial blood pressure control and hypertension, Circ. Res. **35**:159, 1974.

Hilton, S.M., and Spyer, K.M.: Central nervous regulation of vascular resistance, Annu. Rev. Physiol. **42**:399, 1980.

Kirchheim, H.R.: Systemic arterial baroreceptor reflexes, Physiol. Rev. **56**:100, 1976.

Mellander, S., and Johansson, B.: Control of resistance, exchange, and capacitance functions in the peripheral circulation, Pharmacol. Rev. **20**:117, 1968.

Somlyo, A.P., and Somlyo, A.V.: Vascular smooth muscle, Pharmacol. Rev. **20**:197, 1968.

Books and monographs

Bevan, J.A., Bevan, R.D., and Duckles, S.P.: Adrenergic regulation of vascular smooth muscle. In Handbook of physiology; Section 2: The cardiovascular system—smooth muscle, vol. II, Bethesda, Md., 1980, American Physiological Society.

Brown, A.M.: Cardiac reflexes. In Handbook of physiology; Section 2: The cardiovascular system—the heart, vol. I, Bethesda, Md., 1979, American Physiological Society.

Johnson, P.C.: The myogenic response. In Handbook of physiology; Section 2: The cardiovascular system—vascular smooth muscle, vol. II, Bethesda, Md., 1980, American Physiological Society.

Korner, P.I.: Central nervous control of autonomic cardiovascular function. In Handbook of physiology; Section 2: The cardiovascular system—the heart, vol. I, Bethesda, Md., 1979, American Physiological Society.

Rhodin, J.A.G.: Architecture of the vessel wall, In Handbook of physiology; Section 2: The cardiovascular system—vascular smooth muscle, vol. II, Bethesda, Md., 1980, American Physiological Society.

Sparks, H.V., Jr.: Effect of local metabolic factors on vascular smooth muscle. In Handbook of physiology; Section 2: The cardiovascular system—vascular smooth muscle, vol. II, Bethesda, Md., 1980, American Physiological Society.

chapter 33

Control of the heart

The quantity of blood pumped by the heart may be varied by changing the frequency of its beats or the volume ejected per stroke. A discussion of the control of cardiac activity therefore may be subdivided into a consideration of the regulation of pacemaker activity and the regulation of myocardial performance. However, in the intact organism a change in the behavior of one of these features of cardiac activity almost invariably produces an alteration in the other.

Experimentally it has been shown that certain local factors, such as temperature changes and tissue stretch, can affect the discharge frequency of the sinoatrial node. Under natural conditions, however, the principal control of heart rate is relegated to the autonomic nervous system, and this discussion is restricted to this aspect of heart rate control. Relative to the performance of the myocardium, both intrinsic and extrinsic factors warrant extensive considerations.

■ *Nervous control of heart rate*

In normal adults the average heart rate at rest is approximately 70 beats per minute, but it is significantly greater in children. During sleep the heart rate diminishes by 10 to 20 beats per minute, but during emotional excitement or muscular activity it may accelerate to rates considerably above 100. In well-trained athletes at rest the rate is usually only 50 to 60 beats per minute. During diving the heart decelerates, and bradycardia persists even during the vigorous exercise of underwater swimming. This bradycardia associated with diving is considerably more pronounced in certain aquatic mammals, such as the seal.

Under most conditions the sinoatrial (SA) node is under the tonic influence of both divisions of the autonomic nervous system. The sympathetic system exerts a facilitatory influence on the rhythmicity of the pacemaker, whereas the parasympathetic system imposes an inhibitory effect. It was first demonstrated by the Webers in 1845 that changes in heart rate usually involve a reciprocal action of the two divisions of the autonomic nervous system. Thus an acceleration of the heart rate is produced by a diminution of parasympathetic activity and a concomitant increase in sympathetic activity; deceleration usually is evoked by a reversal of these mechanisms. Under certain conditions changes in heart rate may be achieved by selective action of just one division of the autonomic nervous system rather than by reciprocal changes in both divisions.

Ordinarily in healthy, resting individuals the parasympathetic tone is predominant. Abolition of parasympathetic influences by transection of the vagus nerves or by the administration of atropine usually elicits a pronounced tachycardia, whereas abrogation of sympathetic activity usually results in only slight slowing of the heart. When both divisions of the autonomic nervous system are blocked, the heart rate of young adults has been found to average about 105 beats per minute. The rate that prevails after complete autonomic blockade is called the *intrinsic heart rate*.

The cardiac parasympathetic fibers originate in the medulla, in a column of cells that lies in the dorsal motor nucleus or in the region of the nucleus ambiguus. The precise location varies from species to species. The medullary center on a given side projects to both the ipsilateral and contralateral vagus nerves. Centrifugal vagal fibers pass inferiorly through the neck in close proximity to the common carotid arteries and through the mediastinum to synapse with postganglionic cells located within the heart itself. Most of the cardiac ganglion cells are located near the SA node and atrioventricular (AV) conduction tissue. Stimulation of the cardiac ends of the severed vagus nerves of experimental animals has revealed that the right and left vagi usually are distributed differentially to the various cardiac structures. The right vagus nerve affects the SA node predominantly, to produce sinus bradycardia or even complete cessation of SA nodal activity for several seconds. The left vagus nerve exerts its greatest influence on the AV conduction tissue, to produce various degrees of AV block. However, there is a considerable overlap of distribution, so that left vagal stimulation also depresses the SA node, and right vagal stimulation evokes some impairment of AV conduction.

The SA and AV nodes are rich in cholinesterase. Thus the effects of any given vagal impulse are ephemeral because the acetylcholine released at the nerve terminals is rapidly hydrolyzed. Also parasympathetic influences preponderate over sympathetic effects at the SA node (Fig. 33-1). As the frequency of vagal stimulation in an anesthetized dog was increased from 0 to 8 pulses per second *(left panel)*, the heart rate decreased by 75 beats per minute in the absence of concomitant cardiac sympathetic stimulation (S = 0). Sympathetic stimulation at 4 pulses per second (S = 4) caused an increase in heart rate of 80 beats per minute in the absence of vagal stimulation. However, when vagal stimulation at 8 pulses per second was combined with sympathetic stimulation at 4 pulses per second (S = 4), the heart rate declined by 155 beats per minute. From the right panel it is evident that increasing the sympathetic stimulation frequency from 0 to 4 pulses per second produced a substantial cardioacceleration in the absence of vagal stimulation (V = 0). However, when the vagi were stimulated at 8 pulses per second (V = 8), increasing the sympathetic stimulation frequency from 0 to 4 pulses per second had a negligible influence on heart rate.

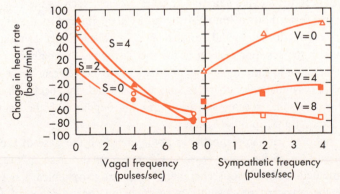

Fig. 33-1 ■ The changes in heart rate in an anesthetized dog when the vagus and cardiac sympathetic nerves were stimulated simultaneously. The sympathetic nerves were stimulated at 0, 2, and 4 pulses per second; the vagus nerves at 0, 4, and 8 pulses per second. The symbols represent the observed changes in heart rate; the curves represent a derived regression equation. (Modified from Levy, M.N., and Zieske, H.: J. Appl. Physiol. **27:**465, 1969.)

The cardiac sympathetic fibers originate in the intermediolateral columns of the upper five or six thoracic and lower one or two cervical segments of the spinal cord. They emerge from the spinal column through the white rami communicantes and enter the paravertebral chain of ganglia. The anatomical details of the sympathetic innervation of

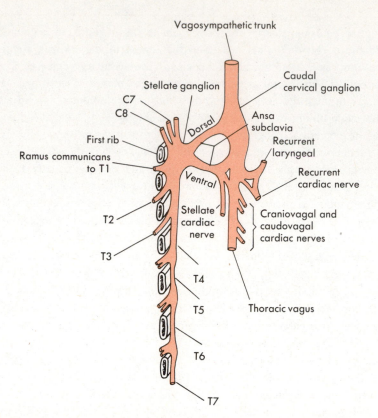

Fig. 33-2 ■ Upper thoracic sympathetic chain and the cardiac autonomic nerves on the right side in the dog. (Modified from Mizeres, N.J.: Anat. Rec. **132:**261, 1958. A publication of the Wistar Press.)

the heart vary among mammalian species; the innervation has been elaborated in greatest detail in the dog (Fig. 33-2). In that species virtually all the preganglionic neurons ascend in the paravertebral chain and funnel through the stellate ganglia. In many species, including the cat, the synapse between preganglionic and postganglionic neurons takes place mainly in the stellate ganglia. In other species, such as the dog, the preganglionic neurons traverse the two limbs of the ansa subclavia and then synapse with the postganglionic neurons in the caudal cervical ganglia. These latter ganglia lie close to the vagus nerves in the superior portion of the mediastinum. Sympathetic and parasympathetic fibers then join to form a complex network of mixed efferent nerves to the heart (Fig. 33-2). Other sympathetic fibers ascend to the head and neck—as the cervical sympathetic chain in some species and in a common bundle with the vagus nerve (as the vagosympathetic trunk) in other species.

The postganglionic cardiac sympathetic fibers approach the base of the heart along the adventitial surface of the great vessels. On reaching the base of the heart, these fibers are distributed to the various chambers as an extensive epicardial plexus. They then penetrate the myocardium, usually accompanying the branches of the coronary vessels. The adrenergic receptors in the nodal regions and in the myocardium are predominantly of the β type; that is, they are responsive to β-adrenergic agonists, such as isoproterenol, and are inhibited by specific β-adrenergic blocking agents, such as propranolol.

As with the vagus nerves, there tends to be a differential distribution of the left and right sympathetic fibers. In the dog, for example, the fibers on the left side have more pronounced effects on myocardial contractility (*augmentor fibers*) than on heart rate (*accelerator fibers*). In some dogs left cardiac sympathetic nerve stimulation may produce no detectable effects on heart rate, even though it may exert pronounced facilita-

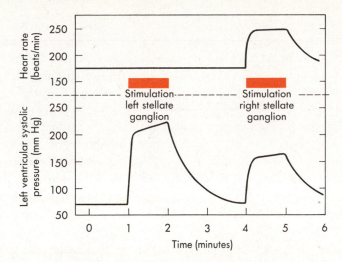

Fig. 33-3 ■ In the dog, stimulation of the left stellate ganglion produces a greater effect on ventricular contractility than does right-sided stimulation, but it has a lesser effect on heart rate. In this example, traced from an original record, left stellate ganglion stimulation had no detectable effect at all on heart rate but had a considerable effect on ventricular performance in an isovolumetric left ventricle preparation.

tory effects on ventricular performance (Fig. 33-3). Conversely, right cardiac sympathetic nerve stimulation elicits considerably greater cardiac acceleration and less augmentation of contractile force than does stimulation of equivalent strength of sympathetic fibers on the left side. This bilateral asymmetry probably also exists in humans. It has been found recently that right stellate ganglion blockade caused a mean reduction in heart rate of 14 beats per minute, whereas left-sided blockade elicited an average reduction of only 2 beats per minute.

It is evident from Fig. 33-3 that the effects of sympathetic stimulation decay very gradually after the cessation of stimulation, in contrast to the abrupt termination of the response after vagal activity (not shown). Most of the norepinephrine released during sympathetic stimulation is taken up again by the nerve terminals, and much of the remainder is carried away by the bloodstream. Relatively little of the released norepinephrine is degraded in the tissues, in contrast to the fate of the acetylcholine released at the vagus nerve endings.

■ *Control by higher centers*

Dramatic alterations in cardiac rate, rhythm, and contractility have been induced experimentally by stimulation of various regions of the brain. In the cerebral cortex the centers regulating cardiac function are located mostly in the anterior half of the brain, principally in the frontal lobe, the orbital cortex, the motor and premotor cortex, the anterior part of the temporal lobe, the insula, and the cingulate gyrus. In the thalamus tachycardia may be induced by stimulation of the midline, ventral, and medial groups of nuclei. Variations in heart rate also may be evoked by stimulating the posterior and posterolateral regions of the hypothalamus. Stimuli applied to the H_2 fields of Forel in the diencephalon elicit a variety of cardiovascular responses, including tachycardia; such changes simulate closely those observed during muscular exercise. Undoubtedly the cortical and diencephalic centers are responsible for initiating the cardiac reactions that occur during excitement, anxiety, and other emotional states. The hypothalamic centers also are involved in the cardiac response to alterations in environmental temperature. Recent studies have shown that localized temperature changes in the preoptic anterior hypothalamus evoke pronounced changes in heart rate and peripheral resistance.

Stimulation of the parahypoglossal area of the medulla produces a reciprocal activation of cardiac sympathetic and inhibition of cardiac parasympathetic pathways. In certain dorsal regions of the medulla distinct cardiac accelerator and augmentor points have been detected in animals with transected vagi. The accelerator regions were found to be more abundant on the right and the augmentor sites more prevalent on the left. A similar distribution also has been reported to exist in the hypothalamus. It appears therefore that for the most part the sympathetic fibers descend the brainstem ipsilaterally.

■ The Bainbridge reflex and atrial receptors

In 1915 Bainbridge reported that infusions of blood or saline resulted in cardiac acceleration. This increase in heart rate occurred whether arterial pressure was unaffected or became somewhat elevated. Acceleration was observed whenever central venous pressure rose sufficiently to distend the right side of the heart, and the effect was abolished by bilateral transection of the vagi. Bainbridge postulated that increased cardiac filling elicited tachycardia reflexly and that the afferent impulses were conducted by the vagi.

Numerous investigators have confirmed the acceleration of the heart in response to the intravenous administration of fluid. However, the magnitude and direction of the response are dependent on the prevailing heart rate. At relatively slow rates intravenous infusions usually result in cardiac acceleration. At more rapid initial rates, however, infusions ordinarily will slow the heart. Increases in blood volume not only evoke the Bainbridge reflex, but they also activate other reflexes, notably the baroreceptor reflex, which tend to elicit oppositely directed heart rate changes. The actual change in heart rate that occurs in response to an alteration of blood volume under a given set of conditions is therefore the result of these antagonistic reflex effects.

In a recent study on unanesthetized dogs, volume loading with blood resulted in increases in heart rate that were proportional to the augmentations of cardiac output (Fig. 33-4). Consequently stroke volume remained virtually constant. Conversely, reductions in blood volume diminished the cardiac output but also evoked a tachycardia.

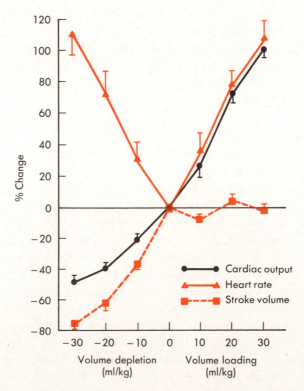

Fig. 33-4 ■ Effects of blood transfusion and of bleeding on cardiac output, heart rate, and stroke volume in unanesthetized dogs. (From Vatner, S.F., and Boettcher, D.H.: Circ. Res. **42**:557, 1978. By permission of the American Heart Association, Inc.)

Undoubtedly the Bainbridge reflex was prepotent over the baroreceptor reflex when the blood volume was raised, but the baroreceptor reflex prevailed over the Bainbridge reflex under conditions of hypovolemia.

Receptors that influence heart rate have been detected in both atria. They are located principally in the venoatrial junctions—in the right atrium at its junctions with the venae cavae and in the left atrium at its junctions with the pulmonary veins. Distension of these receptors sends impulses centripetally in the vagi. The efferent impulses are carried by fibers from both autonomic divisions to the SA node, and they produce an increase in heart rate. The cardiac response appears to be highly selective. Even when the reflex increase in heart rate is quite large, changes in ventricular contractility have been negligible. Furthermore the increase in heart rate is unattended by any augmentation of sympathetic activity to the peripheral arterioles. In fact the only observed change in sympathetic activity to the peripheral vasculature is a significant reduction in sympathetic tone in the renal vessels.

Stimulation of the atrial receptors causes an increase in urine volume, in addition to the cardiac acceleration. The reduction in activity in the renal sympathetic nerve fibers might be partially responsible for this diuresis. However, the principal mechanism appears to be a neurally mediated reduction in the secretion of vasopressin (antidiuretic hormone) by the posterior pituitary gland.

The inverse relationship between arterial blood pressure and heart rate was first described in 1859 by the French physician Étienne Marey and often is referred to as *Marey's law of the heart*. Largely through the work of Hering, Koch, and Heymans it was demonstrated that the alterations in heart rate evoked by changes in blood pressure were dependent on the baroreceptors located in the aortic arch and carotid sinuses. These two sets of baroreceptors appear to be about equally potent in the regulation of heart rate. An example of the effects on heart rate elicited by a stepwise elevation of pressure in an isolated aortic arch preparation is illustrated in Fig. 33-5. An abrupt rise

■ *Baroreceptor reflex*

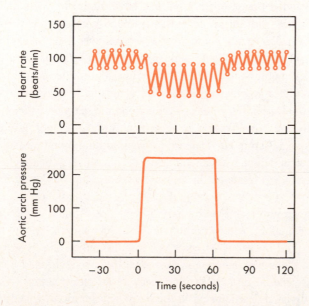

Fig. 33-5 ■ Effect on heart rate of a stepwise pressure change in the isolated aortic arch. When pressure is raised, the mean heart rate decreases, and there is an increase in the magnitude of the rhythmic fluctuations of heart rate at the frequency of respiratory movements. (Redrawn from Levy, M.N., Ng, M.L., and Zieske, H.: Circ. Res. **19:**930, 1966. By permission of the American Heart Association, Inc.)

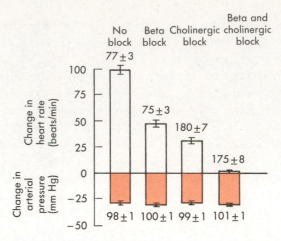

Fig. 33-6 ■ The effects of β-adrenergic receptor blockade, cholinergic blockade, and combined blockade on the change in heart rate in a group of 13 resting dogs when the mean arterial blood pressure was decreased by about 30 mm Hg by means of nitroglycerin. The control values of heart rate and blood pressure prior to hypotension are shown in each column. (From Vatner, S.F., Higgins, C.B., and Braunwald, E.: Cardiovasc. Res. **8**:153, 1974.)

in pressure in the aortic arch from 0 to 250 mm Hg resulted in bradycardia and an exaggeration of the respiratory cardiac arrhythmia (defined in the following section). The subsequent decline of pressure was followed by the return of heart rate to control levels.

It generally has been assumed that the changes in heart rate which occur in response to alterations in baroreceptor stimulation are mediated by means of reciprocal changes in activity in the two divisions of the autonomic nervous system. Recent investigations have challenged this concept. Currently the predominant idea is that reciprocal reflex changes in sympathetic and vagal activity do occur for small deviations in blood pressure within the normal range of pressures. However, when blood pressure is gradually increased to high levels, cardiac sympathetic tone is completely suppressed when only a fraction of the blood pressure elevation has been attained. Thereafter the additional reduction in heart rate with further increases in blood pressure is evoked entirely by an augmentation of vagal activity. The converse applies during the development of severe hypotension. Vagal tone virtually disappears after the initial, relatively small drop in blood pressure. As the pressure continues to decline, the further acceleration of the heart is ascribable solely to a progressive increase in sympathetic neural activity.

The participation of both divisions of the autonomic nervous system in the heart rate response to a moderate hypotension is shown in Fig. 33-6. In a group of 13 normal unanesthetized dogs a 30 mm Hg decrease in mean arterial blood pressure was produced by nitroglycerin, which dilates peripheral arterioles. With efferent vagal and sympathetic pathways both intact this reduction in pressure was accompanied by a reflex increase in heart rate of about 100 beats per minute. After β-adrenergic receptor blockade with propranolol the heart rate increased by only 50 beats per minute in response to the same blood pressure reduction. After cholinergic blockade with atropine, hypotension increased the heart rate by about 30 beats per minute. When the two drugs were given simultaneously to block both autonomic divisions, there was virtually no change in heart rate in response to the 30 mm Hg decline in mean arterial pressure. It is apparent that both autonomic divisions participated in the cardioacceleration evoked by a 30 mm Hg blood pressure reduction in these experiments. Had more severe degrees of hypotension been induced, any additional cardiac acceleration probably would have been mediated solely by efferent sympathetic pathways. Once the mean arterial pressure has declined to about 30 mm Hg below the normal level, vagal tone is usually negligible. Thus no additional increase in heart rate can be achieved through a further reduction in vagal tone.

■ *Respiratory cardiac arrhythmia*

Rhythmic variations in heart rate, occurring at the frequency of respiration, are detectable in most individuals and tend to be especially pronounced in children. Typically the cardiac rate accelerates during inspiration and decelerates during expiration. The vagus nerves are principally responsible for mediating this respiratory cardiac arrhythmia, and the extent of the arrhythmia varies with the degree of vagal tone. Respiratory sinus arrhythmia becomes more pronounced when vagal tone is enhanced. An example is shown in Fig. 33-5, where vagal tone was increased by elevating pressure in the aortic arch. Under these conditions the amplitude of the cyclical variations in heart rate was twice as great as when the pressure in the aortic arch was low.

Both reflex and central factors contribute to the genesis of the respiratory cardiac arrhythmia. During inspiration the intrathoracic pressure decreases, and venous return to the right side of the heart is accelerated. It has been postulated that this elicits the Bainbridge reflex. After the time delay required for the increased venous return to reach the left side of the heart, left ventricular output is increased and produces a rise in arterial blood pressure. This in turn will reduce heart rate through baroreceptor stimulation. Vasomotor tone also varies periodically at the frequency of respiration and results in rhythmic fluctuations in arterial blood pressure, with consequent baroreceptor reflex effects on heart rate. Stretch receptors located in the lungs are capable of affecting heart rate. With moderate degrees of pulmonary inflation cardioacceleration may be evoked reflexly. The afferent and efferent limbs of this reflex are located in the vagus nerves.

It was postulated by Traube in 1865 that respiratory cardiac arrhythmia was ascribable to an influence of the brainstem respiratory center on the cardiac autonomic centers. Abundant experimental evidence subsequently has been provided to demonstrate that central as well as reflex mechanisms play an important role. In the experiment displayed in Fig. 33-5 a total heart-lung bypass was employed to eliminate variations in respiration, venous return, or arterial blood pressure. Yet in that experiment a respiratory cardiac arrhythmia was still prominent.

■ *Chemoreceptor reflex*

In intact animals stimulation of the carotid chemoreceptors elicits a consistent augmentation of ventilatory rate and depth but ordinarily evokes only slight increases or decreases in heart rate. The directional change in heart rate is related to the magnitude of the enhancement of pulmonary ventilation (Fig. 33-7). When respiratory stimulation is relatively mild, heart rate usually diminishes; when the increase in pulmonary ventilation is more pronounced, heart rate usually accelerates.

The cardiac response to peripheral chemoreceptor stimulation represents the result of primary and secondary reflex mechanisms. The primary reflex effect of carotid chemoreceptor excitation on the SA node is inhibitory. Secondary effects are largely related to the concomitant stimulation of respiration.

An example of the primary inhibitory influence is displayed in Fig. 33-8. In this experiment on an anesthetized dog the lungs were completely collapsed, and blood oxygenation was accomplished by an artificial oxygenator. When the carotid chemoreceptors were stimulated, an intense bradycardia ensued that usually was accompanied by some degree of AV block. Such effects are mediated primarily by efferent vagal fibers.

The identical primary inhibitory effect recently has been shown to operate in humans. The electrocardiogram in Fig. 33-9 was recorded from a quadriplegic patient who could not breathe spontaneously but required tracheal intubation and artificial respiration. When the tracheal catheter was disconnected briefly to permit nursing care, the patient quickly developed a profound bradycardia. His heart rate was 65 beats per min-

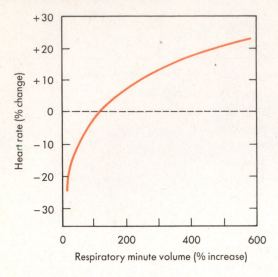

Fig. 33-7 ■ Relationship between the change in heart rate and the change in respiratory minute volume during carotid chemoreceptor stimulation in spontaneously breathing cats and dogs. When respiratory stimulation was relatively slight, heart rate usually diminished; when respiratory stimulation was more pronounced, heart rate usually increased. (Modified from Daly, M. de B., and Scott, M.J.: J. Physiol. [London] **144:**148, 1958.)

Fig. 33-8 ■ Changes in heart rate during carotid chemoreceptor stimulation when the lungs are deflated and respiratory gas exchange is accomplished by an artificial oxygenator. The lower tracing represents the oxygen saturation of the blood perfusing the carotid chemoreceptors. The blood perfusing the remainder of the animal, including the myocardium, was fully saturated with oxygen throughout the experiment. (Modified from Levy, M.N., DeGeest, H., and Zieske, H.: Circ. Res. **18:**67, 1966. By permission of the American Heart Association, Inc.)

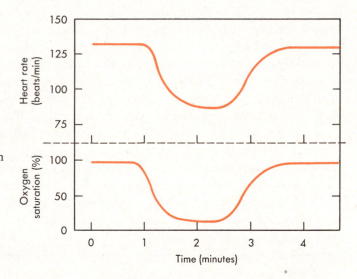

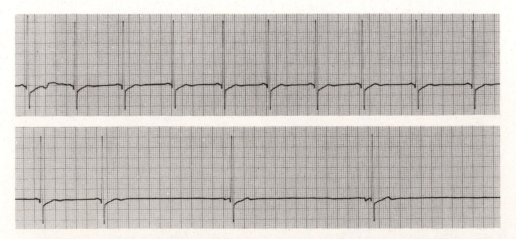

Fig. 33-9 ■ Electrocardiogram of a 30-year-old quadriplegic man who could not breathe spontaneously and required tracheal intubation and artificial respiration. The two strips are continuous. The tracheal catheter was temporarily disconnected from the respirator at the beginning of the top strip, at which time his heart rate was 65 beats per minute. In less than 10 seconds his heart rate decreased to about 20 beats per minute. (Modified from Berk, J.L., and Levy, M.N.: Eur. Surg. Res. **9:**75, 1977.)

ute just before the tracheal catheter was disconnected. In less than 10 seconds after cessation of artifical respiration his heart rate dropped to about 20 beats per minute. This bradycardia could be prevented with atropine, and its onset could be delayed considerably by hyperventilating the patient prior to disconnecting the tracheal catheter.

The potent enhancement of pulmonary ventilation that ordinarily is evoked by carotid chemoreceptor stimulation influences heart rate secondarily, both by initiating more pronounced pulmonary inflation reflexes and by producing hypocapnia. Each of these influences has been found to accelerate the heart per se and to depress the primary cardiac response to chemoreceptor stimulation. Therefore, when pulmonary ventilation is not controlled experimentally, carotid chemoreceptor stimulation produces only a slight change (in either direction) in heart rate instead of the profound bradycardia that occurs when ventilation is held constant.

■ Intrinsic regulation of cardiac performance

Just as the heart has the inherent ability to initiate its own beat in the absence of any nervous or hormonal control, so also does it possess the capacity to adapt to changing hemodynamic conditions by virtue of mechanisms that are intrinsic to cardiac muscle itself. Experiments on completely denervated hearts reveal that this organ is able to adjust remarkably well to a variety of stressful circumstances. In racing greyhounds, for example, it was found that animals with denervated hearts perform essentially as well as those with intact innervation. Their maximal running speed was reduced on the average by only 5% after complete cardiac denervation. In these dogs exercise induced a threefold to fourfold increase in cardiac output, achieved principally by means of an increase in stroke volume. In normal dogs the increase of cardiac output with moderate exercise is accompanied by a proportionate increase of heart rate; stroke volume remains remarkably constant. It is unlikely that the cardiac adaptation in the denervated animals is achieved entirely by intrinsic mechanisms; circulating catecholamines undoubtedly play a role. If the β-adrenergic receptors are blocked in greyhounds with denervated hearts, their racing performance is severely impaired.

The heart is partially or completely denervated in a variety of clinical situations: (1) the surgically transplanted heart is totally denervated; (2) atropine is capable of blocking vagal effects on the heart, and propranolol and other β-adrenergic receptor blocking agents can abrogate the sympathetic influences; (3) certain drugs, such as reserpine, deplete cardiac norepinephrine stores and thereby restrict or abolish sympathetic control; and (4) in severe, chronic congestive heart failure cardiac norepinephrine stores are often severely diminished, thereby attenuating any sympathetic influences.

Historically the intrinsic cardiac adaptive mechanism that has received the greatest attention involves changes in the resting length of the myocardial fibers. This type of adaptation frequently is designated *Starling's law of the heart*, or the *Frank-Starling mechanism*. The mechanical and structural bases for this mechanism are explained in Chapter 29. The term *heterometric autoregulation* has been coined by Sarnoff and his collaborators to refer to those adaptive mechanisms which involve changes in myocardial fiber length. They also suggested the term *homeometric autoregulation* to be applied to those other intrinsic adjustments of cardiac performance which are independent of changes in myocardial fiber length.

■ Heterometric autoregulation

Studies on isolated hearts. In 1895 the German physiologist Otto Frank described the response of the isolated heart of the frog to alterations in the tension of the myocardial fibers just prior to contraction—the *initial tension*. He recognized, however, that such changes in the initial tension were probably accompanied by changes in the resting fiber length. Representative isovolumetric pressure curves from his experiments are reproduced in Fig. 33-10. These curves illustrate the response of the heart to increased filling. It is evident that the initial tension increases with greater degrees of filling, and

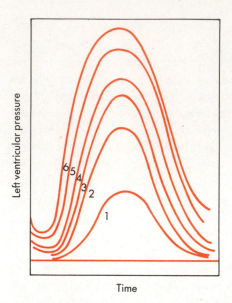

Fig. 33-10 ■ Response of the frog ventricle to increased filling. The initial tension (intraventricular pressure at the onset of each contraction) increases as the filling volume is increased (denoted by the successively higher numbers in the figure). As initial tension is raised, the peak pressure developed during systole is also augmented. (Redrawn from Frank, O.: Z. Biol. **32:**370, 1895.)

at each succeeding level, contraction produces a progressively greater peak pressure. Frank recognized that such behavior of cardiac muscle was similar to that of skeletal muscle when it is stretched to progressively greater initial lengths prior to contraction.

In 1914 the noted English physiologist Ernest Starling and his collaborators described the intrinsic response of the heart to changes in venous return and arterial pressure in the canine heart-lung preparation (Fig. 33-11). In this preparation the right atrium is filled with blood from the reservoir that is connected to a cannula tied into the atrial end of the ligated superior vena cava. Venous return is varied either by altering the height of the reservoir or by adjusting a screw clamp on the connecting tube. Right atrial pressure is recorded by means of a water manometer in the atrial end of the ligated inferior vena cava. From the right atrium, blood enters the right ventricle, which then pumps it through the pulmonary vessels. The trachea is cannulated, and the lungs are artificially ventilated by means of an intermittent respiration pump. The pulmonary venous return enters the left atrium, in which the pressure is recorded by a water manometer. The aorta is ligated distal to the arch, and a cannula is inserted into the brachiocephalic artery. Blood is pumped by the left ventricle through this cannula and through artificial tubing that ultimately conducts the blood back through a heating coil to the right atrial reservoir. Arterial pressure is recorded by means of a mercury manometer. The volume of the right and left ventricles is recorded by means of a cardiometer. Peripheral resistance is adjusted by means of a pressure-limiting device, which has become known as a Starling resistance. This device consists of a piece of collapsible tubing in a rigid chamber. Any desired air pressure is applied to the collapsible tubing through an inlet and is measured with a manometer. The cardiac output is measured by temporarily diverting the flow, which is returning to the venous reservoir, into a graduated cylinder for a measured interval of time.

The response of the isolated heart to a sudden augmentation of filling pressure is shown in Fig. 33-12. Right atrial pressure increased rapidly and appreciably, whereas aortic pressure was permitted to increase only slightly. The top tracing in Fig. 33-12 reveals the changes in ventricular volume that resulted from the abrupt increase of filling

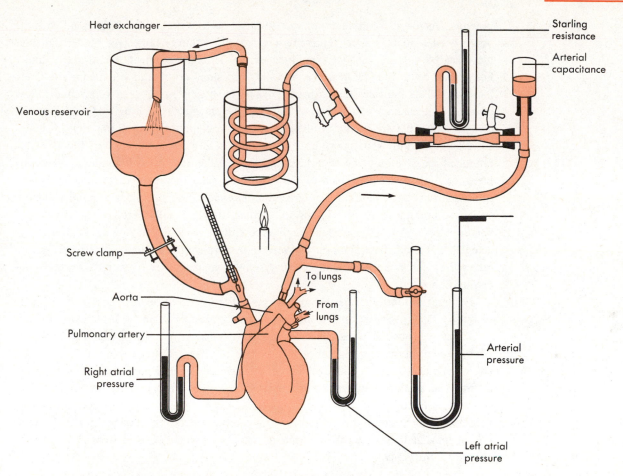

Heat exchanger

Starling
resistance

Arterial
capacitance

Venous reservoir

Screw clamp

Aorta

To lungs

From
lungs

Pulmonary artery

Right atrial
pressure

Arterial
pressure

Left atrial
pressure

Fig. 33-11 ■ Heart-lung preparation. (Redrawn from Patterson, S.W., and Starling, E.H.: J. Physiol. [London] **48:**357, 1914.)

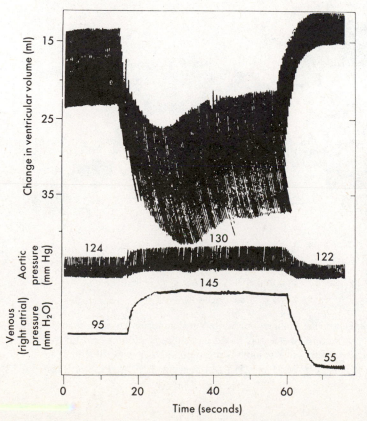

Change in ventricular volume (ml)

15

25

35

Aortic
pressure
(mm Hg)

124

130

122

Venous
(right atrial)
pressure
(mm H₂O)

145

95

55

0   20   40   60

Time (seconds)

Fig. 33-12 ■ Changes in ventricular volume in a heart-lung preparation when the venous reservoir was suddenly raised (right atrial pressure increased from 95 to 145 mm H_2O) and subsequently lowered (right atrial pressure decreased from 145 to 55 mm H_2O). Note that an increase in ventricular volume is registered as a downward shift in the tracing. (Redrawn from Patterson, S.W., Piper, H., and Starling, E.H.: J. Physiol. [London] **48:**465, 1914.)

pressure. Increased ventricular volume is registered as a downward deflection. Thus the upper border of the tracing represents the systolic volume, the lower border indicates the diastolic volume, and the amplitude of the deflections reflects the stroke volume. For several beats after augmentation of filling pressure, ventricular volume progressively increased. This indicates that a disparity must have existed between ventricular inflow during diastole and ventricular output during systole; that is, during systole the ventricles did not expel an amount of blood equal to that which entered during diastole, until a new equilibrium was attained. It is this progressive accumulation of blood which produced the dilation of the ventricles and the lengthening of the individual myocardial fibers that constitute the walls of the ventricles. The increased diastolic fiber length somehow facilitates ventricular contraction and enables the ventricles to pump a greater stroke volume, so that, at equilibirum, cardiac output exactly matches the augmented venous return. An optimal fiber length apparently exists, beyond which contraction is actually impaired. Therefore excessively high levels of filling pressure may depress rather than enhance the pumping capacity of the ventricles by overstretching the myocardial fibers.

Changes in diastolic fiber length also permit the isolated heart to compensate for an increase of peripheral resistance. In the experiment depicted in Fig. 33-13 the arterial resistance was abruptly raised in three steps, whereas venous inflow was held constant. Each step of increase in resistance was associated with an increase in arterial pressure and ventricular volume. With each abrupt elevation of arterial pressure the left ventricle at first was unable to pump a normal stroke volume. Since venous return was held constant, diminution of stroke volume was attended by a rise in ventricular diastolic volume and therefore in the length of the myocardial fibers. This change in end-diastolic fiber length finally enabled the ventricle to pump a given stroke volume against a greater peripheral resistance.

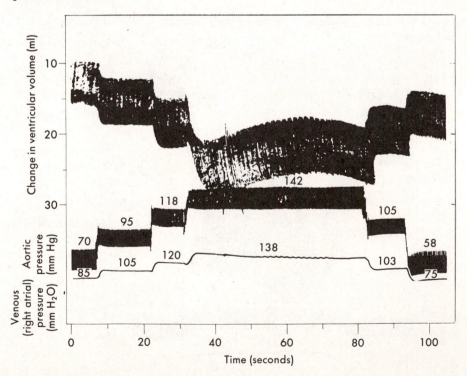

Fig. 33-13 ■ Changes in ventricular volume, aortic pressure, and right atrial pressure in a heart-lung preparation when peripheral resistance was raised and subsequently lowered in several steps. Note that an increase in ventricular volume is registered as a downward shift in the tracing. (Redrawn from Patterson, S.W., Piper, H., and Starling, E.H.: J. Physiol. [London] **48:**465, 1914.)

Changes in ventricular volume also have been shown to be involved in the cardiac adaptation to alterations in heart rate. During bradycardia, for example, the increased duration of diastole permits greater ventricular filling. The consequent augmentation of myocardial fiber length results in an increased stroke volume. With constant venous return therefore the reduction in heart rate is fully compensated by virtue of an increase in stroke volume, such that cardiac output remains constant.

When cardiac compensation involves ventricular dilation, the force required by each myocardial fiber to generate a given intraventricular systolic pressure must be appreciably greater than that developed by the fibers in a ventricle of normal size. The relationship between wall stress and cavity pressure resembles that for cylindrical tubes (Chapter 31) in that, for a constant internal pressure, wall stress varies directly with the radius. As a consequence, the dilated heart has been found to have a considerably greater oxygen requirement to perform a given amount of external work than does the normal heart.

In the intact animal, of course, the heart is enclosed in the pericardial sac. Thus the relatively rigid pericardium determines the pressure/volume relationship at the higher levels of pressure and volume. It is likely that this limitation by the pericardium is exerted even under normal conditions, when an individual is at rest and the heart rate is slow. In the cardiac dilation and hypertrophy that usually accompany chronic heart failure, the pericardium is stretched considerably, and its limitation is exerted at pressures and volumes that are entirely different from those in normal individuals.

Studies on more intact preparations. The major problem involved in assessing the role of the Frank-Starling mechanism in intact animals and humans resides in the difficulty of obtaining a representative measure of end-diastolic myocardial fiber length. The Frank-Starling mechanism has been represented graphically, with some index of ventricular performance usually plotted along the ordinate and some index of fiber length along the abscissa. The most common indices of ventricular performance that have been assessed are cardiac output, stroke volume, and stroke work. The indices of fiber length include ventricular end-diastolic volume, ventricular end-diastolic pressure, ventricular circumference, and mean atrial pressure.

The Frank-Starling mechanism usually is represented by a family of so-called ventricular function curves, rather than by a single curve. To construct a given ventricular function curve, blood volume is altered over a wide range of values, and stroke work and end-diastolic pressure are measured at each step. A similar series of observations then is made during the desired experimental intervention. For example, the ventricular function curve obtained during a norepinephrine infusion lies above and to the left of a control ventricular function curve (Fig. 33-14). It is evident that, for a given level of left ventricular end-diastolic pressure, the left ventricle performs more work during a norepinephrine infusion than during control conditions. Thus a shift of the ventricular function curve to the left usually signifies an improvement of ventricular contractility (Chapter 29); a shift to the right usually indicates an impairment of contractility and a consequent tendency toward *cardiac failure*. A shift in a ventricular function curve cannot be interpreted uniformly as an indication of a change in contractility, however. Contractility is a measure of cardiac performance at a given level of preload and afterload. The end-diastolic pressure is ordinarily a good index of the preload. In assessing changes in myocardial contractility the cardiac afterload must be held constant as the end-diastolic pressure is varied over a range of values.

It is very difficult at present to determine the precise position on a Frank-Starling curve at which the heart of an intact, conscious person or animal may be operating. Recent studies on instrumented, conscious dogs indicate that, when the dog is reclining and relaxed, the left ventricle seems to be at nearly its maximal size at the end of diastole; it probably is limited by the pericardium. When blood volume is reduced, the

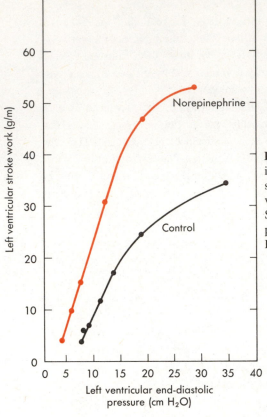

Fig. 33-14 ■ A constant infusion of norepinephrine in the dog causes the ventricular function curve to shift to the left, signifying an enhancement of ventricular contractility. (Redrawn from Sarnoff, S.J., et al.: Circ. Res. **8**:1108, 1960. By permission of the American Heart Association, Inc.)

end-diastolic volume diminishes, and the Frank-Starling mechanism is undoubtedly involved in cardiac output regulation. Conversely, when blood volume is acutely expanded, end-diastolic volume apparently cannot increase appreciably. Thus extrinsic mechanisms probably play a more crucial role in cardiac output regulation. In the experiment illustrated in Fig. 33-4, for example, the transfusion of blood caused a parallel increase in heart rate and cardiac output; stroke volume did not change appreciably. The Bainbridge reflex seemed to play the dominant role in this adaptation to an increased blood volume.

The Frank-Starling mechanism is the one that is certainly most ideally suited for matching the cardiac output to the venous return. Any sudden, excessive output by one ventricle soon results in a greater venous return to the other ventricle. The consequent increase in diastolic fiber length serves as the stimulus to increase the output of the second ventricle to correspond with that of its mate. For this reason it is the Frank-Starling mechanism that maintains a precise balance between the outputs of the right and left ventricles. Since the two ventricles are arranged in series in a closed circuit, it is apparent that even a small, but maintained, imbalance in the outputs of the two ventricles would have catastrophic consequences.

■ *Homeometric autoregulation*

Pressure-induced regulation. An increase in afterload may evoke a type of homeometric autoregulation, which sometimes is referred to as the *Anrep effect.* In the experiment shown in Fig. 33-15 arterial and left ventricular diastolic pressure rose when peripheral resistance was increased abruptly. However the rise (phase *1*) in left ventricular diastolic pressure was only temporary. After the peak diastolic pressure was reached, the diastolic pressure diminished progressively toward the control level (phase *2*), while aortic pressure remained elevated. Finally a new equilibrium level of ventricular diastolic pressure was reached (phase *3*); in many cases this pressure was actually at or

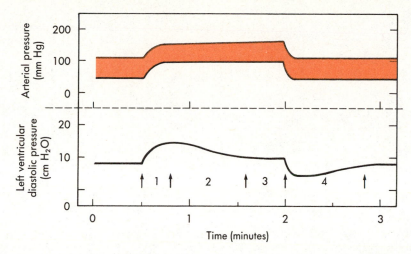

Fig. 33-15 ■ Homeometric autoregulation in the dog heart in response to a sustained increase in peripheral resistance. Initially there was an appreciable rise in left ventricular end-diastolic pressure (phase *1*), but this subsequently returned toward the control level (phases *2* and *3*), even though peripheral resistance was still augmented. (Redrawn from Sarnoff, S.J., and Mitchell, J.H.: Am. J. Med. **34**:440, 1963.)

slightly below the control value. Measurement of myocardial segment length revealed changes during diastole that paralleled the changes in ventricular diastolic pressure. With abrupt return of peripheral resistance to the control level, left ventricular diastolic pressure declined to a value appreciably below control and then gradually returned to the control level (phase *4*). Coronary blood flow was controlled in these experiments, so that the adaptation of the heart cannot be attributed to an improvement in the myocardial circulation. More recent evidence indicates that these autoregulatory changes may be caused in part by the release of myocardial catecholamines.

This evidence demonstrates that the mammalian ventricle possesses the intrinsic capabilities of adapting to changes in filling pressure and arterial resistance without a continued increase in resting fiber length. The conditions under which homeometric autoregulation, rather than maintained heterometric autoregulation, would prevail have not been delineated. Recent studies suggest that homeometric autoregulation may be relatively unimportant in the normal heart but may be more prominent when contractility is depressed. It is likely that transient heterometric adaptation usually precedes homeometric regulation, as in the previous examples, but ventricular distension is not a necessary prelude to homeometric autoregulation. This mechanism therefore averts the mechical disadvantage that dilation of the ventricles places on the myocardial fibers during heterometric autoregulation.

Rate-induced regulation. Approximately one century ago Bowditch directed attention to the modifications of myocardial performance that depend on the time interval between beats. He demonstrated the occurrence of *Treppe* (the *staircase phenomenon*) in the frog ventricle; that is, after a period of rest the ventricle repsonded to repetitive stimuli with progressively greater contractions until a plateau was attained.

The effects of contraction frequency on the tension developed in an isometrically contracting cat papillary muscle are shown in Fig. 33-16, *B*. Initially the strip of cardiac muscle was stimulated to contract only once every 20 seconds. When the muscle was made to contract once every 0.63 second, the developed tension increased progressively over the next several beats. At the new steady state the developed tension was more than five times as great as it was at the lower contraction frequency. A return to the slower rate had the opposite influence on developed tension, with an ultimate return to the initial value.

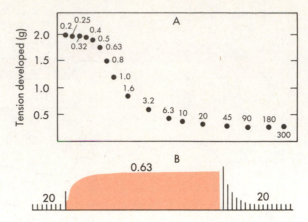

Fig. 33-16 ■ Changes in tension development in an isolated papillary muscle from a cat as the interval between contractions is varied. The numbers in both sections of the record denote the interval (in seconds) between beats. In section *A* the points represent the steady-state tensions developed at the intervals indicated. (Redrawn from Koch-Weser, J., and Blinks, J.R.: Pharmacol. Rev. **15**:601, 1963. © 1963, American Society for Pharmacology and Experimental Therapeutics.)

The effect of the interval between contractions on the steady-state level of developed tension is shown in panel *A* (Fig. 33-16) for a wide range of intervals. It is evident that, as the interval is diminished from 300 seconds to about 10 to 20 seconds, there is little change in developed tension. As the interval is reduced further to a value of about 0.5 second, tension increases steeply. Further reduction of the interval to 0.2 second has little additional effect on developed tension.

The progressive rise in developed tension observed in *B* as the contraction frequency is increased is ascribable to a gradual rise in intracellular calcium. During the plateau of the action potential, Ca^{++} enters the cell. As the contraction frequency is increased, there is greater influx of Ca^{++} into the cell. The duration of the cardiac action potential diminishes as contraction frequency is raised, and so less Ca^{++} actually enters per contraction. The influx of Ca^{++} per minute equals the influx per beat times the number of beats per minute. As the frequency of contraction is increased, the increment in the number of beats per minute exceeds the decrement in Ca^{++} influx per beat, and so there is a net rise in the Ca^{++} influx per minute. The intracellular content of Ca^{++} increases, and as a consequence, developed tension rises as the interval between contractions is diminished, as shown in *A* (Fig. 33-16).

■ *Extrinsic regulation of cardiac performance*

Although the completely isolated heart is capable of a remarkable degree of adaptation to changes in filling pressure and peripheral resistance, in the intact animal there are various extrinsic factors that also exercise a potent regulatory influence on the myocardium. Under many natural conditions these extrinsic mechanisms may dwarf in importance the intrinsic regulatory mechanisms. These extrinsic regulatory factors may be subdivided into nervous and chemical components.

■ *Nervous control*

Sympathetic influences. The sympathetic division of the autonomic nervous system has a profound facilitatory effect on the atrial and ventricular myocardium. The concentration of norepinephrine in the various regions of the heart reflects the relative density of the sympathetic innervation to those areas. In the normal heart the norepinephrine concentration in the atria is about three times that in the ventricles. The norepinephrine concentration in the SA and AV nodes is no greater than that in the surrounding atrial

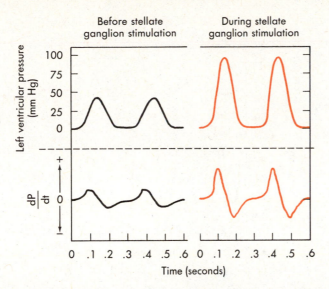

Fig. 33-17 ■ In an isovolumetric left ventricle preparation, stimulation of cardiac sympathetic nerves evokes a substantial rise in peak left ventricular pressure and in the maximum rates of intraventricular pressure rise and fall (dP/dt).

regions. When the heart is denervated, the tissue concentrations of this neurotransmitter approach zero.

The alterations in ventricular contraction evoked by electrical stimulation of the left stellate ganglion in a canine isovolumetric left ventricle preparation are shown in Fig. 33-17. The peak pressure developed by the ventricle during systole is considerably augmented, and the maximum rate at which pressure is generated (dP/dt) is markedly increased. Also the duration of systole is reduced, and the rate of ventricular relaxation is increased during the early phases of diastole. In the dog sympathetic nerve fibers from the left side exert a much more potent effect on left ventricular performance than do those from the right side. On the average a given stimulus applied to the left stellate ganglion produces a rise in peak left ventricular pressure which is more than twice as great as that evoked by right-sided stimulation (Fig. 33-3).

Sympathetic nervous activity facilitates myocardial performance principally by enhancing the contractility of the individual cardiac muscle cells. The manner in which this is accomplished is not fully understood, but the stimulation of adenylate cyclase appears to be a critical step in the process. The consequent elevation in the level of cyclic AMP leads to the activation of a protein kinase that is involved in the phosphorylation of proteins in the sarcolemma and sarcoplasmic reticulum. These reactions are believed to alter Ca^{++} fluxes across the sarcolemma and sarcoplasmic reticulum. Catecholamines are known to increase the permeability of the cell membrane to Ca^{++}, resulting in a greater inward Ca^{++} current during the plateau of the action potential. Such a change in Ca^{++} flux across the sarcolemma, and probably also across the sarcoplasmic reticulum, makes more Ca^{++} available for interaction with the contractile proteins, which thereby affects the enhancement of myocardial contracility.

The facilitatory sympathetic influences on the heart of intact animals can best be appreciated in terms of families of ventricular function curves. When stepwise increases in the frequency of electrical stimulation are applied to the left stellate ganglion, the ventricular function curves are shifted progressively to the left. The changes parallel those produced by catecholamine infusions (Fig. 33-14). Thus for any given left ventricular end-diastolic pressure the ventricle is capable of performing more work as the level of sympathetic nervous activity is progressively raised. During cardiac sympathetic

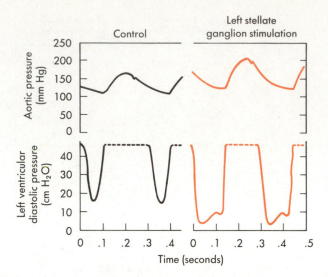

Fig. 33-18 ■ Stimulation of the left stellate ganglion of a dog increases arterial pressure, stroke volume, and stroke work despite a concomitant reduction in ventricular end-diastolic pressure. Note also the abridgment of systole, thereby allowing more time for ventricular filling. In the ventricular pressure tracings the pen excursion is limited at 45 mm Hg; actual ventricular pressures during systole can be estimated from the aortic pressure tracings. (Redrawn from Mitchell, J.H., Linden, R.J., and Sarnoff, S.J.: Circ. Res. **8:**1100, 1960. By permission of the American Heart Association, Inc.)

stimulation the shift from curve to curve usually occurs in such a manner that an increase in work is achieved despite a reduction in mean left ventricular end-diastolic pressure. An example of such a reduction in left ventricular end-diastolic pressure during stellate ganglion stimulation in a paced heart is shown in Fig. 33-18. In this experiment stroke work increased by about 50%, despite a 7 cm H_2O reduction in the left ventricular end-diastolic pressure. The pronounced abridgment of the duration of ventricular systole, with the consequent lengthening of the filling period, is also evident in the same tracing. The reason for the reduction in ventricular end-diastolic pressure is explained in Chapter 34.

Parasympathetic influences. It is universally recognized that the vagus nerves exert profound depressant effects on the cardiac pacemaker, the atrial myocardium, and the atrioventricular conduction tissue. The ventricles generally have been considered to be essentially devoid of parasympathetic innervation. However, considerable evidence has been accumulated over the past two decades which demonstrates that the vagus nerves do indeed depress ventricular contractility. Fig. 33-19 reveals that the effects of vagal stimulation on the isovolumetric left ventricle preparation are just the reverse of those induced by sympathetic stimulation (Fig. 33-17). Vagal stimulation produces a reduction in peak left ventricular pressure, in the maximum rate of pressure development (dP/dt), and in the maximum rate of pressure decline during diastole. In pumping heart preparations the ventricular function curve is shifted to the right. No differences are detectable between the effects of the right and left vagus nerves on ventricular performance, in contrast to the distinct differences between the two sides observed in the case of the cardiac sympathetic nerves.

The depressant effect of vagal activity on the ventricular myocardium appears to be achieved by at least four mechanisms (Fig. 33-20). One process is direct, and the others involve an interaction with the sympathetic nervous system. With respect to the direct mechanism it has been found that vagal stimulation or acetylcholine infusions raise the intracellular levels of cyclic GMP (arrow *a*). This nucleotide in turn may depress myocardial contractility through some process that has not yet been elucidated.

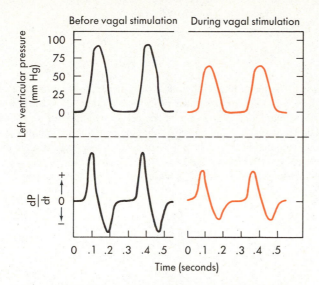

Fig. 33-19 ■ In an isovolumetric left ventricle preparation, when the ventricle is paced at a constant frequency, vagal stimulation decreases the peak left ventricular pressure and diminishes the maximum rates of pressure rise and fall ($^{dP}/_{dt}$).

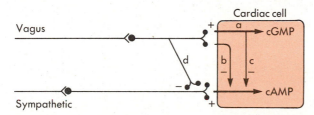

Fig. 33-20 ■ The interneuronal and intracellular mechanisms responsible for the vagal-sympathetic interactions. (Modified from Levy, M.N. In Baan, J., Noordergraaf, A., and Raines, J., editors: Proceedings of the International Conference on Cardiovascular System Dynamics, Cambridge, Mass., 1976, The M.I.T. Press. © 1976 by The M.I.T. Press.)

With respect to the mechanisms involving an interaction with the sympathetic nervous system, it has been observed that, when the existing level of cardiac sympathetic nervous activity is low, the depressant effect of increased parasympathetic activity on the ventricular myocardium is relatively feeble. However, against a background of tonic sympathetic activity the negative inotropic effect produced by vagal stimulation is considerably more prominent. A similar phenomenon occurs with respect to the chronotropic responses to autonomic neural activity. In the left half of Fig. 33-1, for example, vagal stimulation produced a much greater reduction in the heart rate during simultaneous cardiac sympathetic stimulation at 4 pulses per second (S = 4) than in the absence of sympathetic stimulation (S = 0).

This accentuated antagonism between the parasympathetic and sympathetic systems may be accomplished in three different ways (Fig. 33-20). First, the acetylcholine released at the vagal endings causes a direct reduction in the intracellular levels of cyclic AMP (arrow *b*). As stated in the preceding section, cyclic AMP probably mediates the enhancement of contractility produced by sympathetic neural activity. Second, increased vagal activity raises intracellular levels of cyclic GMP, as explained previously. This nucleotide accelerates the hydrolysis of cyclic AMP (arrow *c*), thereby lowering its concentration in the myocardial cell.

The third mechanism responsible for the accentuated antagonism between the two divisions of the autonomic nervous system involves extracellular processes. As shown

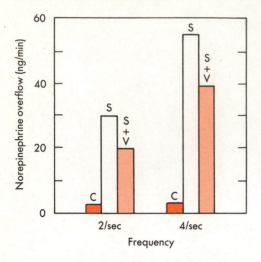

Fig. 33-21 ■ The mean rates of overflow of norepinephrine into the coronary sinus blood in a group of seven dogs under control conditions *(C)*, during cardiac sympathetic stimulation *(S)* at 2 or 4 cycles per second and during combined sympathetic and vagal stimulation *(S + V)*. The combined stimulus consisted of sympathetic stimulation at 2 or 4 cycles per second and vagal stimulation at 15 cycles per second. (Redrawn from Levy, M.N., and Blattberg, B.: Circ. Res. **38:**81, 1976. By permission of the American Heart Association, Inc.)

Fig. 33-22 ■ As the pressure in the carotid sinus is progressively raised, there is a shift to the right in the ventricular function curves. The numbers at the tops of each curve represent the systolic/diastolic perfusion pressures (in millimeters of mercury) in the carotid sinus regions of the dog. (Redrawn from Sarnoff, S.J., et al.: Circ. Res. **8:**1123, 1960. By permission of the American Heart Association, Inc.)

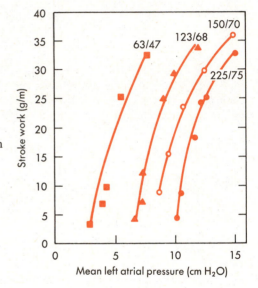

in Fig. 33-20, some postganglionic vagal terminals *(d)* end near the postganglionic sympathetic terminals in the heart. The acetylcholine released at these vagal endings inhibits the release of norepinephrine from the sympathetic fibers. Fig. 33-21 demonstrates that stimulation of the cardiac sympathetic nerves *(S)* results in the overflow of substantial amounts of norepinephrine into the coronary sinus blood, in comparison with the rates of overflow during the control state *(C)*. Concomitant vagal stimulation *(S + V)* causes a reduction of about 30% in the rate of overflow of norepinephrine produced by sympathetic stimulation alone at a given frequency. The amount of norepinephrine overflowing into the coronary sinus blood probably parallels the amount released at the sympathetic terminals.

Baroreceptor reflex. Just as stimulation of the carotid sinus and aortic arch baroreceptors evokes marked changes in heart rate (p. 555), so also does it elicit reflex alterations of myocardial performance. Evidence of reflex alterations of ventricular contractilty is presented in Fig. 33-22. Ventricular function curves were obtained at four different levels of carotid sinus perfusion pressure. With each successive rise in perfusion pressure the ventricular function curves were displaced further and further to the right, denoting a progressively greater reflex depression of ventricular performance.

In normal, resting individuals and animals the tonic level of sympathetic activity is usually very low. Recent studies have shown that under such conditions mild to moderate changes in baroreceptor activity have little reflex influence on myocardial contractility. In states of augmented sympathetic neural activity, however, the effects of the baroreceptor reflex on contractility may play an important role. In the adaptation to blood loss, for example, a reflex change in myocardial contractility may constitute an important means of compensation. Blood loss diminishes cardiac output. The associated

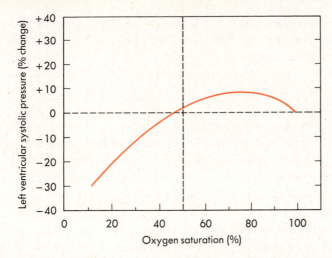

Fig. 33-23 ■ In the isovolumetric left ventricle preparation a reduction in the oxygen saturation of the coronary arterial blood has a mild stimulatory effect on ventricular contractility when the oxygen saturation is between 45% and 100% but a depressant effect when the oxygen saturation falls below 45%. (Redrawn from Ng, M.L., et al.: Am. J. Physiol. **211:**43, 1966.)

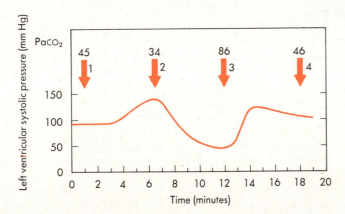

Fig. 33-24 ■ Decrease in arterial blood carbon dioxide tension (Pa_{CO_2}) increases left ventricular systolic pressure (arrow *2*) in an isovolumetric left ventricle preparation, whereas a rise in Pa_{CO_2} (arrow *3*) has the reverse effect.

reduction in arterial blood pressure alters the intensity of baroreceptor stimulation, thereby evoking not only an acceleration of heart rate but also an improvement of myocardial contractility. Thus this type of feedback mechanism tends to minimize the extent of the reduction in cardiac output that would be induced by the loss of a given volume of blood.

Blood gases

Oxygen. The Pa_{O_2} of the blood perfusing the myocardium influences myocardial performance directly. The effect of hypoxia is actually biphasic, with moderate degrees being stimulatory and more severe degrees being depressant. As shown in Fig. 33-23, when the O_2 saturation is reduced to levels below 50% in isolated hearts, the effect is predominantly depressant in that peak left ventricular pressures are less than the control levels. However, with less severe degrees of hypoxia (O_2 saturation >50%) the effects are predominantly facilitatory. Furthermore recent studies have shown that moderate degrees of hypoxia may enhance the contractile response of the heart to circulating catecholamines.

Carbon dioxide. The alterations in myocardial performance elicited by changes of Pa_{CO_2} in the coronary arterial blood are illustrated in Fig. 33-24. In this experiment on an isolated, isovolumetric left ventricle preparation the control Pa_{CO_2} was 45 mm Hg. (arrow *1*). Decreasing the Pa_{CO_2} to 34 mm Hg (arrow *2*) had a pronounced stimulatory effect, whereas increasing Pa_{CO_2} to 86 mm Hg (arrow *3*) was severely depressant. In intact animals, however, systemic hypercapnia also activates the sympathoadrenal system, which tends to compensate for the direct depressant effect on the heart of the increased Pa_{CO_2}.

The changes in Pa_{CO_2} just described were achieved by varying the CO_2 content of the gas mixture in the oxygenator of the perfusion system. Therefore changes in Pa_{CO_2} of the blood were accompanied by inverse changes in blood pH. By analysis of the effects of experimental alterations of Pa_{CO_2} and pH in various combinations, it has become apparent that neither the Pa_{CO_2} nor the blood pH is actually a primary determinant of myocardial behavior. It is more likely that the resultant change in intracellular pH is the critical factor. The precise mechanisms whereby an intracellular acidosis depresses myocardial contractility are not known. Recent studies do suggest that a reduced intracellular pH diminishes the amount of Ca^{++} released from the sarcoplasmic reticulum in response to excitation of a myocardial cell. Furthermore the diminished pH depresses the myofilaments directly. When they are exposed to a given concentration of Ca^{++}, the lower the prevailing pH, the less the developed tension.

■ *Bibliography*

Journal articles

Berk, J.L., and Levy, M.N.: Profound reflex bradycardia produced by transient hypoxia or hypercapnia in man, Eur. Surg. Res. **9**:75, 1977.

Boettcher, D.H., et al.: Extent of utilization of the Frank-Starling mechanism in conscious dogs, Am. J. Physiol. **234**:H338, 1978.

Coleman, T.G.: Arterial baroreflex control of heart rate in the conscious rat, Am. J. Physiol. **238**:H515, 1980.

DeGeest, H., et al.: Depression of ventricular contractility by stimulation of the vagus nerves, Circ. Res. **17**:222, 1965.

Donald, D.E., and Shepherd, J.T.: Cardiac receptors: normal and disturbed function, Am. J. Cardiol. **44**:873, 1979.

Downing, S.E., and Lee, J.C.: Myocardial and coronary vascular responses to insulin in the diabetic lamb, Am. J. Physiol. **237**:H514, 1979.

Frank, O.: On the dynamics of cardiac muscle (translated by Chapman, C.B., and Wasserman, E.), Am. Heart J. **58**:282, 467, 1959.

Geis, G.S., and Wurster, R.D.: Cardiac responses during stimulation of the dorsal motor nucleus and nucleus ambiguus in the cat, Circ. Res. **46**:606, 1980.

Higgins, C.B., Vatner, S.F., and Braunwald, E.: Parasympathetic control of the heart, Pharmacol. Rev. **25**:119, 1973.

Kirchheim, H.R.: Systemic arterial baroreceptor reflexes, Physiol. Rev. **56**:100, 1976.

Koch-Weser, J., and Blinks, J.R.: The influence of the interval between beats on myocardial contractility, Pharmacol. Rev. **15**:601, 1963.

Kollai, M., and Koizumi, K.: Reciprocal and non-reciprocal action of the vagal and sympathetic nerves innervating the heart, J. Auton. Nerv. Syst. **1**:33, 1979.

Korner, P.I.: Integrative neural cardiovascular control, Physiol. Rev. **51**:312, 1971.

Kunze, D.L.: Regulation of activity of cardiac vagal motoneurons, Fed. Proc. **39**:2513, 1980.

Lakatta, E.G., and Spurgeon, H.A.: Force staircase kinetics in mammalian cardiac muscle: modulation by muscle length, J. Physiol. (London) **299**:337, 1980.

Levy, M.N.: Sympathetic-parasympathetic interactions in the heart, Circ. Res. **29**:437, 1971.

Levy, M.N., and Blattberg, B.: Effect of vagal stimulation on the overflow of norepinephrine into the coronary sinus during cardiac sympathetic nerve stimulation in the dog, Circ. Res. **38**:81, 1976.

Loewy, A.D., and McKellar, S.: Neuroanatomical basis of central cardiovascular control, Fed. Proc. **39**:2495, 1980.

Morad, M., and Goldman, Y.: Excitation-contraction coupling in heart muscle: membrane control of development of tension, Prog. Biophys. Mol. Biol. **27**:257, 1973.

Noble, M.I.M.: The Frank-Starling curve, Clin. Sci. Mol. Med. **54**:1, 1978.

Patterson, S.W., Piper, H., and Starling, E.H.: The regulation of the heart beat, J. Physiol. (London) **48**:465, 1914.

Sagawa, K.: The ventricular pressure-volume diagram revisited, Circ. Res. **43**:677, 1978.

Sarnoff, S.J., et al.: Homeometric autoregulation in the heart, Circ. Res. **8:**1077, 1960.

Thorén, P.N., Donald D.E., and Shepherd, J.T.: Role of heart and lung receptors with non-medullated vagal afferents in circulatory control, Circ. Res. **38**(Suppl. II):II-2, 1976.

Vanhoutte, P.M., and Levy, M.N.: Prejunctional cholinergic modulation of adrenergic neurotransmission in the cardiovascular system, Am. J. Physiol. **238:**H275, 1980.

Vatner, S.F., and Boettcher, D.H.: Regulation of cardiac output by stroke volume and heart rate in conscious dogs, Circ. Res. **42:**557, 1978.

Books and monographs

Braunwald, E., and Ross, J., Jr.: Control of cardiac performance. In Handbook of physiology; Section 2: Cardiovascular system—the heart, vol. I, Bethesda, Md., 1979, American Physiological Society.

Brown, A.M.: Cardiac reflexes. In Handbook of physiology; Section 2: Cardiovascular system—the heart, vol. I, Bethesda, Md., 1979, American Physiological Society.

Coleridge, J.C.G., and Coleridge, H.M.: Chemoreflex regulation of the heart. In Handbook of physiology; Section 2: Cardiovascular system—the heart, vol. I, Bethesda, Md., 1979, American Physiological Society.

Downing, S.E.: Baroreceptor regulation of the heart. In Handbook of physiology; Section 2: Cardiovascular system—the heart, vol. I, Bethesda, Md., 1979, American Physiological Society.

Hainsworth, R., Kidd, C., and Linden, R.J., editors: Cardiac receptors, Cambridge, 1979, Cambridge University Press.

Korner, P.I.: Central nervous control of autonomic cardiovascular function. In Handbook of physiology; Section 2: Cardiovascular system—the heart, vol. I, Bethesda, Md., 1979, American Physiological Society.

Levy, M.N., and Martin, P.J.: Neural control of the heart. In Handbook of physiology; Section 2: Cardiovascular system—the heart, vol. I, Bethesda, Md., 1979, American Physiological Society.

Levy, M.N., and Vassalle, M.: Excitation and neural control of the heart, Bethesda, Md., 1982, American Physiological Society.

Randall, W.C., editor: Neural regulation of the heart, New York, 1976, Oxford University Press, Inc.

Stull, J.T., and Mayer, S.E.: Biochemical mechanisms of adrenergic and cholinergic regulation of myocardial contactility. In Handbook of physiology; Section 2: Cardiovascular system—the heart, vol. I, Bethesda, Md., 1979, American Physiological Society.

chapter 34

Cardiac output and the venous system

In 1870 a German physiologist, Adolph Fick, contrived the first method for measuring cardiac output that would be applicable in intact animals and humans. The basis for this method has been called the *Fick principle,* and it is simply an application of the law of conservation of mass. It is derived from an algebraic statement of the fact that the quantity of oxygen per minute delivered to the pulmonary capillaries via the pulmonary artery plus the quantity of oxygen per minute that enters the pulmonary capillaries from the alveoli must equal the quantity of oxygen per minute that is carried away by the pulmonary veins.

This is depicted schematically in Fig. 34-1. The rate (q_1) at which O_2 is delivered to the lungs by the pulmonary artery equals the oxygen concentration in that blood, $[O_2]_{pa}$, times the pulmonary arterial blood flow, Q, which is, in fact, the cardiac output, that is

$$q_1 = Q[O_2]_{pa} \qquad (1)$$

Let q_2 be the net rate of oxygen uptake by the pulmonary capillaries from the alveoli and distributed to the tissues, the so-called *oxygen consumption* of the body. The rate (q_3) at which oxygen is carried away by the pulmonary veins equals the O_2 concentration in that blood, $[O_2]_{pv}$, times the total pulmonary venous blood flow, Q, which is virtually equal to the pulmonary arterial blood flow at equilibrium, that is

$$q_3 = Q[O_2]_{pv} \qquad (2)$$

From conservation of mass

$$q_1 + q_2 = q_3 \qquad (3)$$

Therefore

$$Q[O_2]_{pa} + q_2 = Q[O_2]_{pv} \qquad (4)$$

Solving for cardiac output

$$Q = q_2 /([O_2]_{pv} - [O_2]_{pa}) \qquad (5)$$

Equation (5) is the statement of the Fick principle.

In the clinical determination of cardiac output, the rate of oxygen consumption is computed from measurements of the volume and oxygen content of expired air over a given interval of time. Because the oxygen concentration of peripheral arterial blood is

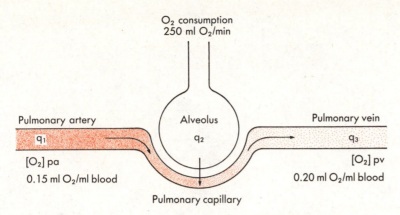

Fig. 34-1 ■ Schema illustrating the Fick principle for measuring cardiac output. The change in color intensity from pulmonary artery to pulmonary vein represents the change in color of the blood as venous blood becomes fully oxygenated.

essentially identical to that in the pulmonary veins, $[O_2]_{pv}$ is determined on a sample of peripheral arterial blood withdrawn by needle puncture. Pulmonary arterial blood actually represents mixed systemic venous blood. Samples for oxygen analysis are obtained from the pulmonary artery or right ventricle through a catheter. For many years the catheter was a relatively stiff, radiopaque tube that had to be introduced into the pulmonary artery under fluoroscopic guidance. Now a very flexible catheter with a small balloon near the tip may be inserted into a peripheral vein. As the tube is advanced, it is carried by the flowing blood toward and then into the heart. By simply following the pressure changes, the physician is able to know when the catheter tip has been passed through the tricuspid valve into the right ventricle and then through the pulmonic valve into the pulmonary artery.

An example of the results ordinarily obtained in a normal, resting adult is illustrated by the values of Fig. 34-1. With an oxygen consumption of 250 ml/minute, an arterial (pulmonary venous) oxygen content of 0.20 ml O_2/ml blood (or 20 vol%), and a mixed venous (pulmonary arterial) oxygen content of 0.15 ml O_2/ml blood (or 15 vol%), the cardiac output would equal 250 ÷ (0.20 − 0.15) = 5000 ml/minute.

The Fick principle is also used for estimating the oxygen consumption of organs in situ when blood flow and the oxygen contents of the arterial and venous blood can be determined. Algebraic rearrangement reveals that oxygen consumption equals the blood flow times the arteriovenous oxygen concentration difference. For example, if the blood flow through one kidney is 700 ml/minute, arterial oxygen content is 0.20 ml O_2/ml blood and renal venous oxygen content is 0.18 ml O_2/ml blood, then the rate of oxygen consumption by that kidney must be 700 (0.20 − 0.18) = 14 ml O_2/minute.

The principle of the indicator dilution technique for measuring cardiac output is also based on the law of conservation of mass and is illustrated by the model in Fig. 34-2. Let a liquid flow through a tube at a rate of Q ml/second, and let q mg of dye be injected as a slug into the stream at point *A*. Let mixing occur at some point downstream. If a small sample of liquid is continually withdrawn from point *B* farther downstream and passed through a densitometer, then a curve of the dye concentration (c) may be recorded as a function of time (t) as shown in the lower half of the figure.

If there is no loss of dye between points *A* and *B*, the amount of dye (q) passing point *B* between times t_1 and t_2 will be

$$q = \bar{c}Q (t_2 - t_1) \tag{1}$$

■ *Indicator dilution techniques*

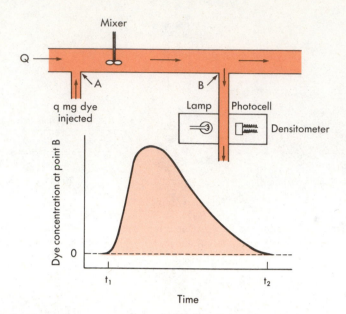

Fig. 34-2 ■ Schema illustrating the indicator dilution technique for measuring cardiac output. In this model, in which there is no recirculation, q mg of dye is injected instantaneously at point *A* into a stream flowing at Q ml/minute. A mixed sample of the fluid flowing past point *B* is withdrawn at a constant rate through a densitometer. The resulting dye concentration curve at point *B* has the configuration shown in the lower section of the figure.

where $\bar{c}$ is the mean concentration of dye. The value of $\bar{c}$ may be computed by dividing the area of the dye concentration curve by the time duration ($t_2 - t_1$) of that curve, that is

$$\bar{c} = \int_{t_1}^{t_2} c \ dt/(t_2 - t_1) \tag{2}$$

Substituting this value of $\bar{c}$ into equation (1) and solving for Q yields

$$Q = \frac{q}{\int_{t_1}^{t_2} c \ dt} \tag{3}$$

Thus flow may be measured by dividing the amount of indicator injected upstream by the area under the downstream concentration curve.

This technique has been widely applied for the estimation of cardiac output in humans. An accurately measured quantity of some indicator (a dye or isotope that remains within the circulation) is injected rapidly as a slug into a large central vein or into the right side of the heart through a cardiac catheter. Arterial blood is continuously drawn through a detector (densitometer or isotope rate counter), and a curve of indicator concentration is recorded as a function of time.

Because some of the indicator recirculates and reappears at the site of arterial withdrawal before the entire curve is inscribed, the concentration curve is not as simple as that shown in the lower half of Fig. 34-2. Instead, on the downstroke of the curve there is a secondary increase in concentration (Fig. 34-3) as the recirculated dye becomes mixed with the last portions of dye still undergoing its primary passage past the site of withdrawal. To compute the area under the concentration curve, the downslope of the curve is extrapolated to zero concentration beyond the beginning of recirculation (dashed line, Fig. 34-3). The extrapolation, of course, introduces some error into the estimation of cardiac output. In model systems without recirculation it has been found that, for the descending portion of the curve, the logarithm of the concentration varies linearly with

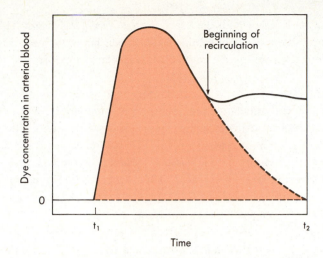

Fig. 34-3 ■ Typical dye concentration curve recorded from a human subject. Because of recirculation of the dye, the concentration does not return to zero as in the model in Fig. 34-2. The dashed line on the descending limb represents the semilogarithmic extrapolation of the upper portion of the descending limb, before the beginning of recirculation.

time. Usually, in the intact circulatory system, this same relationship holds for the downstroke of the concentration curve up till the time of recirculation. Therefore in practice this initial portion of the descending limb of the concentration curve is plotted on semilogarithmic paper; the curve is extrapolated to concentrations approaching zero, and the extrapolated curve is then replotted on the original linear tracing. Cardiac output equals the amount of indicator injected divided by the area of the curve to the point of recirculation plus the area under the extrapolated portion of the curve after the beginning of recirculation. The total area is represented by the colored region in Fig. 34-3.

Over the past several years, the most popular indicator dilution technique has been that of *thermodilution*. The indicator used is cold saline. The temperature and volume of the saline are measured accurately before injection. A very flexible double-lumen catheter is introduced into a peripheral vein and advanced so that the tip lies in the pulmonary artery. A small thermistor attached to the catheter tip permits rapid changes in temperature to be recorded. One of the openings of the catheter lies a few inches proximal to the tip. When the tip is in the pulmonary artery, the proximal opening lies in or near the right atrium. The cold saline is injected rapidly through the catheter, and it enters the circulation through this upstream opening. The resultant change in temperature downstream is recorded by the thermistor in the pulmonary artery.

The thermodilution technique has the following distinct advantages: (1) an arterial puncture is not necessary; (2) the small volumes of saline used in each determination are innocuous so repeated determinations may be made; and (3) there is virtually no problem with recirculation. Temperature equilibration takes place as the cooled blood flows through the pulmonary and systemic capillary beds before it flows by the thermistor in the pulmonary artery the second time. Therefore the curve of temperature change resembles that shown in Fig. 34-2, and the errors associated with extrapolation are largely averted.

The various factors that determine cardiac performance have been discussed in detail in Chapters 29 and 33. There are usually considered to be four factors that affect the heart directly, namely, preload, afterload, heart rate, and myocardial contractility. The last two of these factors (heart rate and contractility) are characteristics of the cardiac tissues per se, although they are subject to modulation by various neural and humoral mechanisms. The first two factors (preload and afterload), however, are depen-

■ *Control of cardiac output*

dent on the characteristics of both the heart and the vascular system. The preload and afterload are critical determinants of cardiac performance, but at the same time the preload and afterload are determined by the cardiac and vascular components of the circulatory system.

Over the past two decades we have become increasingly aware of the fact that changes in the peripheral circulation are often just as important in determining the level of the cardiac output as are changes in the cardiac tissues themselves. Therefore in order to understand the regulation of cardiac output, it is important to appreciate the nature of the coupling between the heart and the vascular system. Guyton and his colleagues have made important contributions to this field. They have developed useful graphic techniques that we shall use in modified form for analyzing the interactions between the cardiac and vascular components of the circulatory system.

The graphic analysis involves two simultaneous functional relationships between the cardiac output and the *central venous pressure,* that is, the pressure in the right atrium and thoracic venae cavae. The curve defining one of these relationships will be called the *cardiac function curve.* It is an expression of the well-known Frank-Starling relationship and reflects the fact that the cardiac output depends, in part, on the preload, that is, the central venous or right atrial pressure. The cardiac function curve is a characteristic of the heart itself and has been studied in hearts completely isolated from the rest of the circulatory system (Chapter 33).

The second functional relationship between the central venous pressure and the cardiac output is defined by a second curve, which we shall call the *vascular function curve.* This relationship depends only on certain critical characteristics of the vascular system: the peripheral resistance, the arterial and venous capacitances, and the blood volume. It is entirely independent of the characteristics of the heart, and it can be studied even if the heart were replaced by a mechanical pump.

■ *Vascular function curve*

The vascular function curve defines the change in central venous pressure that occurs as a consequence of a change in cardiac output; that is, central venous pressure is the dependent variable (or response), and cardiac output is the independent variable (or stimulus). This contrasts with the cardiac function curve (or Frank-Starling mechanism), for which the central venous pressure (or preload) is the independent variable, and the cardiac output is the dependent variable.

The simplified model of the circulation illustrated in Fig. 34-4 will be used to explain how the cardiac output determines the level of the central venous pressure. For the sake of simplicity the essential components of the cardiovascular system may be lumped into four elements as illustrated in Fig. 34-4. The right and left sides of the heart as well as the pulmonary vascular bed will be considered simply as a pump-oxygenator, much like that employed during open heart surgery. The high-resistance microcirculation is designated the peripheral resistance. Finally, the entire *capacitance* of the system is subdivided into two components: the total arterial capacitance (C_a) and

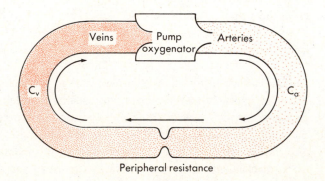

Fig. 34-4 ■ Simplified model of the cardiovascular system consisting of a pump-oxygenator, an arterial capacitance (C_a), a peripheral resistance, and a venous capacitance (C_v).

the total venous capacitance (C_v). As defined in Chapter 30, *capacitance* (C) is the increment of volume (dV) accomodated per unit change of pressure (dP); that is, C = dV/dP. The venous capacitance is approximately twenty times as great as the arterial capacitance. In the example to follow, the ratio of $C_v:C_a$ will be set at 19:1; this will simplify the mathematics as will be evident below. Thus for a 1 mm Hg increment in venous pressure, nineteen times as much blood would be stored on the venous side of the circuit as would be stored on the arterial side for an equivalent rise in arterial pressure. It must be emphasized that these statements apply strictly to the usual pressure ranges that exist on both sides of the vascular circuit because capacitance varies as a function of transmural pressure (Chapter 30).

With the model system at rest, pressures are the same throughout the entire circuit, and there is no flow. The pressure that exists at rest is a function of only the total volume of blood contained within the system and the elastic characteristics of the walls, that is, the overall capacitance of the system. This equilibrium pressure has been termed the *mean circulatory pressure* by Guyton. At normal blood volumes and with normal vessels, the magnitude of the mean circulatory pressure has been estimated to be about 7 mm Hg. To the left of arrow *1* in Fig. 34-5, the arterial pressure (P_a) and the venous pressure (P_v) are both equal to 7 mm Hg when the cardiac output (C.O.) is zero.

Let the pump-oxygenator (or simply the pump) in Fig. 34-4 start suddenly to deliver a constant flow of 1 L/minute (at arrow *1*, Fig. 34-5), and let peripheral resistance remain constant at 20 mm Hg/L/minute. Because of the arrangement of the valves, the direction of transfer will be from the venous to the arterial side of the circuit. Hence, pressure will begin to fall on the venous side and rise on the arterial side. The arterial pressure (P_a) will continue to rise until a pressure of 20 mm Hg above the venous pressure (P_v) is attained. From the definition of peripheral resistance, as explained in Chapter 30, the following is derived:

$$R = (P_a - P_v)/Q \qquad (1)$$

$$P_a = P_v + QR = P_v + 20 \qquad (2)$$

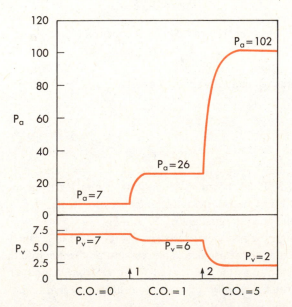

Fig. 34-5 ■ The changes in arterial (P_a) and venous (P_v) pressures in the circulatory model shown in the preceding figure. The total peripheral resistance is 20 mm Hg/L/minute, and the ratio of C_v to C_a is 19:1. The cardiac output (C.O.) is zero to the left of arrow *1*. It is increased to 1 L/minute at arrow *1* and to 5 L/minute at arrow *2*.

Hence, P_a will continue to be 20 mm Hg above P_v as long as the pump output is maintained at 1 L/minute and the peripheral resistance remains at 20 mm Hg/L/minute. The arterial volume increment required to achieve this new, elevated level of P_a is entirely dependent on the arterial capacitance, C_a. For a rigid arterial system (low capacitance) this volume will be small; for a distensible system the volume will be large. Whatever the magnitude of this volume, however, it represents the translocation of some quantity of blood from the venous to the arterial side of the circuit. For a given total blood volume, any increment in arterial volume (ΔV_a) must be equal to the decrement in venous volume (ΔV_v), that is

$$\Delta V_a = -\Delta V_v \tag{3}$$

From the definition of capacitance

$$C_a = \Delta V_a / \Delta P_a \tag{4}$$

and

$$C_v = \Delta V_v / \Delta P_v \tag{5}$$

By substitution into equation (3)

$$\frac{\Delta P_v}{\Delta P_a} = -\frac{C_a}{C_v} \tag{6}$$

Given that C_v is nineteen times as great as C_a, then the increment in P_a will be nineteen times as great as the decrement in P_v; that is

$$\Delta P_a = -19\Delta P_v \tag{7}$$

Let ΔP_a represent the difference between the prevailing P_a and the mean circulatory pressure (P_{mc}); that is, let

$$\Delta P_a = P_a - P_{mc} \tag{8}$$

and let ΔP_v represent the difference between the prevailing P_v and the mean circulatory pressure:

$$\Delta P_v = P_v - P_{mc} \tag{9}$$

Substituting these values for ΔP_a and ΔP_v into equation (7):

$$P_a - P_{mc} = -19 (P_v - P_{mc}) \tag{10}$$

By solving equations (2) and (10) simultaneously:

$$P_a = P_{mc} + 19 \tag{11}$$

and

$$P_v = P_{mc} - 1 \tag{12}$$

Hence, in Fig. 34-5, P_v is shown to decrease to 6 mm Hg from the mean circulatory pressure of 7 mm Hg. Concomitantly, P_a increases to 26 mm Hg, resulting in the required arteriovenous pressure gradient of 20 mm Hg.

The flow through the peripheral resistance (Fig. 34-4) is a function of the pressure gradient ($P_a - P_v$) and the resistance. The pressure gradient across the peripheral resistance is often referred to as the *vis a tergo*, or force from behind. *It is the single most important factor responsible for the venous return and is directly ascribable to the pumping action of the heart itself.*

If the pump output is abruptly increased to a constant level of 5 L/minute (Fig. 34-5, arrow 2) and peripheral resistance remains constant at 20 mm Hg/L/minute, an additional volume of blood will again be translocated from the venous to the arterial side

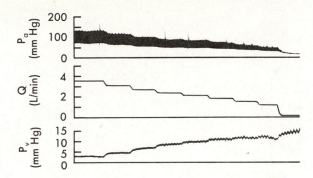

Fig. 34-6 ■ The changes in arterial (P_a) and central venous (P_v) pressures produced by changes in systemic blood flow (Q) in a canine right-heart bypass preparation. Stepwise changes in Q were produced by altering the rate at which blood was mechanically pumped from the right atrium to the pulmonary artery. (Redrawn from Levy, M.N.: Circ. Res. **44**:739, 1979. By permission of the American Heart Association, Inc.)

of the circuit. It will progressively accumulate in the arteries until P_a reaches a level of 100 mm Hg above P_v. By substitution into equation (1)

$$P_a = P_v + QR = P_v + 100 \qquad (13)$$

By solving equations (10) and (13) simultaneously, we find that P_a rises to a value of 95 mm Hg above P_{mc}, and P_v falls to a value 5 mm Hg below P_{mc}. In Fig. 34-5, therefore, P_v declines to 2 mm Hg and P_a rise to 102 mm Hg. The resultant pressure gradient of 100 mm Hg is that value which will force a cardiac output of 5 L/minute through a constant peripheral resistance of 20 mm Hg/L/minute.

Experiments in animals and observations in human patients have demonstrated that alterations in cardiac output do indeed evoke the directional changes in P_a and P_v that have been predicted above for our simplified model. In the experiment on an anesthetized dog shown in Fig. 34-6, a mechanical pump was substituted for the right ventricle. As the cardiac output (Q) was diminished in a series of small steps, there were concomitant reductions in P_a and elevations of P_v. Similarly, a major coronary artery may suddenly become occluded in a human subject. The resultant *acute myocardial infarction* often leads to a substantial reduction in cardiac output, which is attended by a fall in the arterial pressure and a rise in the central venous pressure.

In animal experiments, if the output of the heart or of a substitute mechanical pump were varied over a range of values and the resultant values of P_v were recorded, a *vascular function curve* could be constructed. Such a curve is shown in Fig. 34-7, and it is derived from the theoretical data shown in Fig. 34-5. Contrary to the usual convention, the independent variable, cardiac output in this instance, is plotted along the ordinate, and the dependent variable, venous pressure, is plotted along the abscissa. The customary arrangement of axes is reversed for reasons of convenience that will become apparent later in this chapter. In the vascular function curve depicted in Fig. 34-7, point A represents the mean circulatory pressure, point B the value of P_v at a cardiac output of 1 L/minute, and point C the value of P_v at a cardiac output of 5 L/minute. From equations (1), (6), (8), and (9) earlier, the following equation for the linear portion of the vascular function curve can be derived:

$$P_v = -\frac{RC_a}{C_a + C_v} Q + P_{mc} \qquad (14)$$

From this equation, it is evident that when Q = 0, $P_v = P_{mc}$. Also, the slope of the vascular function curve depends only on R, C_v, and C_a.

Fig. 34-7 also shows that there is a limit to the reduction of P_v that can be produced

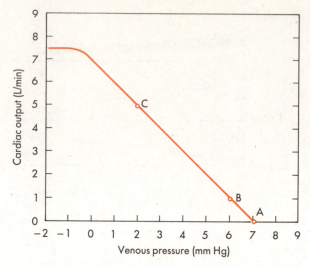

Fig. 34-7 ■ Changes in venous pressure as cardiac output is varied over a range from 0 to over 7 L/minute. Point *A* is the mean circulatory pressure, which is the equilibrium pressure throughout the cardiovascular system when cardiac output is 0. Points *B* and *C* represent the values of venous pressure at cardiac outputs of 1 and 5 L/minute, respectively. Contrary to the usual convention, the independent variable (in this instance, cardiac output) is plotted along the ordinate, whereas the dependent variable (venous pressure) is plotted along the abscissa.

by an increase in cardiac output. At some critical maximum value of cardiac output, sufficient fluid will be translocated from the venous to the arterial side of the circuit such that P_v will drop below the ambient pressure. In a system of distensible tubes, the venous system will be collapsed by this negative transmural pressure (P_v minus ambient pressure). This will, of course, limit the maximum value of cardiac output, regardless of the capabilities of the pump.

In the simplified schema in Fig. 34-4, the venous system was considered to be without resistance. In the body, however, there is a continuous pressure gradient from the venules to the right side of the heart. In an open-chest animal, therefore, if an artificial pump is substituted for the heart and if the pump output is progressively increased, the site of collapse will be at the junction of the venae cavae with the inflow side of the pump. In the normal, closed-chest animal, venous collapse occurs at the points of entry of the veins into the chest for reasons to be described later in this chapter.

■ Blood volume

The vascular function curve is affected by variations in the total blood volume. During circulatory standstill (cardiac output zero), the mean circulatory pressure is simply a function of vascular capacitance and blood volume, as stated previously. Thus for a given vascular capacitance, the mean circulatory pressure will be increased during *hypervolemia,* when the blood volume is expanded (as by transfusion), and decreased during *hypovolemia,* when the blood volume is diminished (as by hemorrhage). This is illustrated by the X-axis intercepts in Fig. 34-8, where the mean circulatory pressure is 5 mm Hg with hemorrhage and 9 mm Hg with transfusion, as compared with the value of 7 mm Hg at the normal blood volume *(normovolemia)*.

Furthermore, the differences in P_v during hypervolemia, normovolemia, and hypovolemia in the static system are preserved at each level of cardiac output so that the venous pressure curves parallel each other (Fig. 34-8). To illustrate, consider the example of hypervolemia, in which the mean circulatory pressure is 9 mm Hg. In Fig. 34-5 both P_a and P_v would be 9 mm Hg instead of 7 mm Hg, when the cardiac output was zero. With a sudden increase in cardiac output to 1 L/minute (at arrow *1*), if the peripheral resistance were still 20 mm Hg/L/minute, an arteriovenous pressure gradient

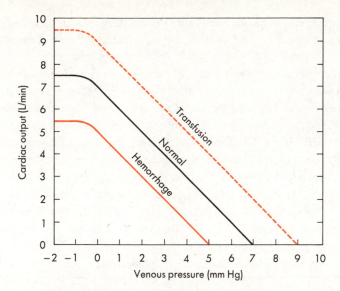

Fig. 34-8 ■ Effects of increased blood volume (*transfusion* curve) and of decreased blood volume (*hemorrhage* curve) on the vascular function curve. Similar shifts in the vascular function curve are produced by increases and decreases, respectively, in venomotor tone.

of 20 mm Hg would still be necessary for 1 L/minute to flow through the resistance vessels. This does not differ from the example for normovolemia. Assuming the same ratio of C_v to C_a of 19:1, the pressure gradient would be achieved by a 1 mm Hg decline in P_v and a 19 mm Hg rise in P_a. Hence, a change in cardiac output from 0 to 1 L/minute would evoke the same 1 mm Hg reduction in P_v irrespective of the blood volume, as long as the C_v/C_a ratio and the peripheral resistance were independent of blood volume. Equation (14) also discloses that the slope of the vascular function curve remains constant as long as there is no change in R, C_v, or C_a.

From Fig. 34-8 it is also apparent that the value for cardiac output at which P_v equals zero varies directly with the blood volume. Therefore the maximum value of cardiac output becomes progressively more limited as the total blood volume is reduced, but the pressure at which the veins collapse (sharp change in slope on the vascular function curve) is not altered appreciably by changes in blood volume.

■ *Venomotor tone*

The vascular function curves representing changes in venomotor tone closely resemble those for changes in blood volume. In Fig. 34-8, for example, the transfusion curve could just as well represent increased venomotor tone, whereas the hemorrhage curve could represent decreased tone. During circulatory standstill, for a given blood volume, the pressure within the vascular system will rise as the tension exerted by the smooth muscle within the vascular walls increases. Present evidence indicates that it is principally the arteriolar and venous smooth muscle that is under any appreciable nervous or humoral control. Because the fraction of the blood volume located within the arterioles is small (Fig. 26-2), only changes in venomotor tone are of any practical importance in altering the magnitude of the mean circulatory pressure. Hence, mean circulatory pressure rises with increased venomotor tone and falls with diminished tone. Changes in venomotor tone may alter the elastic characteristics of the veins. The consequent change in C_v may therefore alter the slope of the vascular function curve as predicted by equation (14).

Experimentally, the pressure attained shortly after abrupt circulatory standstill is usually above 7 mm Hg, even when blood volume is normal. This is attributable to the generalized venoconstriction elicited by cerebral ischemia, activation of the chemoreceptors, and reduced stimulation of the baroreceptors. If resuscitation is not achieved, then this reflex response subsides as central nervous activity ceases, and at normal blood volumes the mean circulatory pressure usually approaches a value close to 7 mm Hg.

■ *Blood reservoirs*

The extent of venoconstriction is considerably greater in certain regions of the body than in others. In effect, vascular beds that undergo appreciable venoconstriction constitute blood reservoirs. The vascular bed of the skin is one of the major blood reservoirs in humans. During blood loss, profound subcutaneous venoconstriction occurs, giving rise to the characteristic pale appearance of the skin. The resultant redistribution of blood thereby liberates several hundred milliliters of blood to be perfused through more vital regions. The vascular beds of the liver, lungs, and spleen also serve as important blood reservoirs. In the dog the spleen is packed with red blood cells and is capable of constricting to a small fraction of its normal size. During hemorrhage this mechanism serves to autotransfuse blood of high erythrocyte content into the general circulation. However, in humans the volume changes of the spleen are considerably less extensive and play only a minor role in compensating for blood loss.

■ *Peripheral resistance*

The modification of the vascular function curve induced by changes in arteriolar tone is shown in Fig. 34-9. It has been estimated that the arterioles contain only about 3% of the total blood volume (Fig. 26-2). Hence, changes in the contractile state of these vessels do not appreciably alter the pressure-volume relationship in the static system. Therefore the mean circulatory pressure is virtually independent of arteriolar tone, and the family of venous pressure curves representing a range of peripheral resistances will converge at a common point on the abscissa.

At any given level of cardiac output, P_v will vary inversely with the arteriolar tone if all other factors remain constant. Vasoconstriction sufficient to double the peripheral resistance will result in a twofold rise in P_a. In the example shown in Fig. 34-5, a change in the cardiac output from 0 to 1 L/minute caused P_a to rise from 7 to 26 mm Hg, an increment of 19 mm Hg. If peripheral resistance had been twice as great, the same change in cardiac output would have evoked twice as great an increment in P_a. To achieve this greater rise in P_a, twice as great an increment in blood volume would be required on the arterial side of the circulation, assuming a constant arterial capacitance. Given a constant total blood volume, this larger arterial volume signifies a cor-

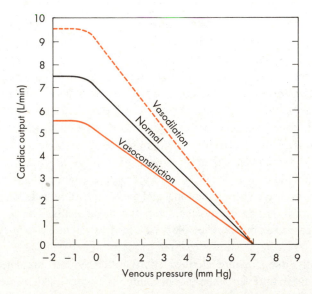

Fig. 34-9 ■ Effects of arteriolar vasodilation and vasoconstriction on the vascular function curve.

responding reduction in venous volume. Hence, the decrement in venous volume would be twice as great when the peripheral resistance is doubled. With a constant venous capacitance, a twofold reduction in venous volume would be reflected by a twofold decline in P_v. Therefore, in Fig. 34-5, an increase in cardiac output to 1 L/minute (arrow *1*) would have caused a 2 mm Hg decrement in P_v, to a level of 5 mm Hg instead of the 1 mm Hg decrement that occurred with the normal peripheral resistance. Similarly, greater increases in cardiac output would have evoked proportionately greater decrements in P_v under conditions of increased peripheral resistance than with normal levels of resistance. This inverse relationship between the peripheral resistance and the decrement in P_v, together with the failure of peripheral resistance to affect the mean circulatory pressure, accounts for the counterclockwise rotation of the venous pressure curves with increased peripheral resistance that is observed in Fig. 34-9. Conversely, arteriolar vasodilation produces a clockwise rotation from the same horizontal axis intercept. The direct proportionality between the peripheral resistance and the slope of the vascular function curve is also evident from the algebraic equation for that curve (equation 14).

■ *Interrelationships between cardiac output and venous return*

Cardiac output and venous return are inextricably interdependent. Clearly, except for small, transient disparities, the heart will be unable to pump any more blood than is delivered to it through the venous system. Similarly, because the circulatory system is a closed circuit, the rate of venous return must equal the cardiac output over any appreciable time interval. The flow around the entire closed circuit depends on the capability of the pump, the characteristics of the circuit, and the total volume of fluid in the system. Cardiac output and venous return are simply two terms for expressing the flow around the closed circuit. *Cardiac output* is the volume of blood being pumped by the heart per unit of time. *Venous return* is the volume of blood returning to the heart per unit of time. At equilibrium these two flows are equal.

The techniques of circuit analysis will be applied in an effort to gain some insight into the control of flow around the circuit. Acute changes in cardiac contractility, peripheral resistance, or blood volume may transiently exert differential effects on cardiac output and venous return. Except for such brief disparities, however, such factors simply alter flow around the entire circuit, and it is irrelevant whether one thinks of that flow as "cardiac output" or "venous return." Not uncommonly, authors have ascribed the reduction in cardiac output during hemorrhage, for example, to a decrease in venous return. It will become clear that such an explanation is a blatant example of circular reasoning in its most literal sense. Hemorrhage reduces flow around the entire circuit. To attribute the reduction in cardiac output to a curtailment of venous return is equivalent to ascribing the decrease in total flow to a decrease in total flow.

■ *Coupling between the heart and the vasculature*

In accordance with Starling's law of the heart, or so-called heterometric autoregulation, cardiac output is intimately dependent on P_v, since right atrial pressure is virtually identical to central venous pressure. Furthermore, the right atrial pressure is approximately equal to the right ventricular end-diastolic pressure because the normal tricuspid valve constitutes a low-resistance junction between the right atrium and ventricle. In the discussion to follow, graphs of cardiac output as a function of P_v will be called *cardiac function curves*. Supervention of other types of intrinsic regulatory mechanisms, for example, homeometric autoregulation (Chapter 33), will modify such curves appreciably. For completeness such changes must be taken into consideration, but for purposes of simplicity the cardiac function curves will be drawn to represent only the Frank-Starling relationship. Extrinsic regulatory influences may be expressed as shifts in such curves, as has previously been indicated (Chapter 33).

A typical cardiac function curve is plotted on the same coordinates as a normal

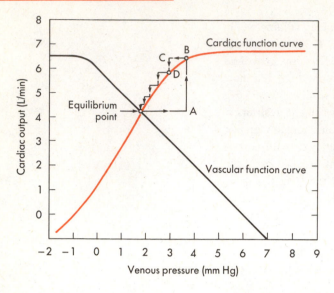

Fig. 34-10 ■ Typical vascular and cardiac function curves plotted on the same coordinate axes. The coordinates of the equilibrium point, at the intersection of these curves, represent the stable values of cardiac output and central venous pressure at which the system tends to operate. Any perturbation (such as when venous pressure is suddenly increased to point *A*) institutes a sequence of changes in cardiac output and venous pressure such that these variables are returned to their equilibrium values.

vascular function curve in Fig. 34-10. Contrary to the vascular function curve, the cardiac function curve will be plotted according to the usual convention; that is, the independent variable (P_v) will be plotted along the abscissa, and the dependent variable (cardiac output) will be plotted along the ordinate. In accordance with the Frank-Starling mechanism, the cardiac function curve reveals that a rise in P_v is associated with an increase in cardiac output. Conversely, the vascular function curve describes an inverse relationship between cardiac output and P_v. The *equilibrium point* of such a system is defined by the point of intersection of these two curves. The coordinates of this equilibrium point represent the values of cardiac output and P_v at which such a system tends to operate. Only transient deviations from such values for cardiac output and P_v are possible as long as the given cardiac and vascular function curves accurately describe the system.

The tendency for the cardiovascular system to operate about such an equilibrium point may best be illustrated by examining its response to a sudden perturbation. Consider the changes elicited by a sudden rise in P_v from the equilibrium point to point *A* in Fig. 34-10. Such a change might be induced experimentally by the rapid injection, during ventricular systole, of a given volume of blood on the venous side of the circuit, accompanied by the withdrawal of an equal volume from the arterial side so that total blood volume would remain constant. Because of the Frank-Starling mechanism, this elevated P_v would result in an enhanced cardiac output (point *B*) during the very next ventricular systole. The increased cardiac output, in turn, would result in the net transfer of blood from the venous to the arterial side of the circuit, with a consequent reduction in P_v (to point *C*). During the next systole, cardiac output would therefore be less (point *D*), although still above the equilibrium point. This process would continue in ever-diminishing steps until the equilibrium values for P_v and cardiac output were reestablished.

■ Enhanced myocardial contractility

The effect of alterations in ventricular contractility may be comprehended on the basis of a similar graphic representation. In Fig. 34-11 the lower cardiac function curve represents the control state, whereas the upper curve represents a condition of improved contractility, analogous to the family of ventricular function curves described in Chapter 33. Such enhancement of ventricular contractility might be achieved by electrical stimulation of the cardiac sympathetic nerves. Because the effects of such stimulation would be restricted to the heart, a single vascular function curve would be appropriate to both control and experimental conditions as shown in Fig. 34-11.

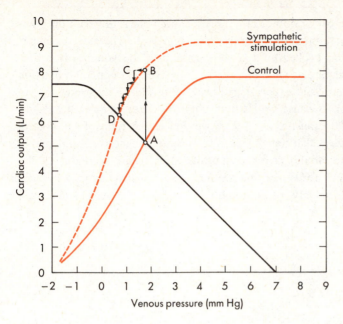

Fig. 34-11 ■ Enhancement of myocardial contractility, as by cardiac sympathetic nerve stimulation, causes the equilibrium values of cardiac output and P_v to shift from the intersection (point *A*) of the control vascular and cardiac function curves (continuous lines) to the intersection (point *D*) of the same vascular function curve with that cardiac function curve (dashed line) representing enhanced myocardial contractility.

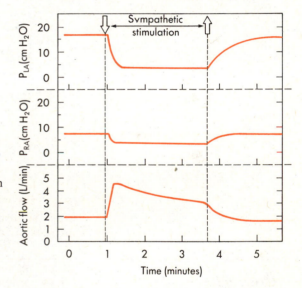

Fig. 34-12 ■ During electrical stimulation of the left stellate ganglion (containing cardiac sympathetic nerve fibers), aortic blood flow increased while pressures in the left atrium (P_{LA}) and right atrium (P_{RA}) diminished. These data conform with the conclusions derived from Fig. 34-11, in which the equilibrium values of cardiac output and venous pressure are observed to shift from point *A* to point *D* during cardiac sympathetic nerve stimulation. (Redrawn from Sarnoff, S.J., et al.: Circ. Res. **8**:1108, 1960. By permission of the American Heart Association, Inc.)

During the control state the equilibrium values for cardiac output and P_v are designated by point *A*. With the onset of cardiac sympathetic nerve stimulation (assuming the effects to be instantaneous and constant), the prevailing level of P_v would elicit an abrupt rise in cardiac output to point *B* because of the enhanced contractility. However, this level of cardiac output would result in the net transfer of blood from the venous to the arterial side of the circuit, and consequently P_v will fall (to point *C*). Cardiac output will therefore continue to diminish until a new equilibrium point *(D)* is reached, which is located at the point of intersection of the vascular function curve with the new cardiac function curve (representing the effects of increased sympathetic activity on the heart). The new equilibrium point *(D)* lies above and to the left of the control equilibrium point *(A)*, revealing that sympathetic stimulation evokes a greater cardiac output at a lower level of P_v. That such a change accurately describes the actual experimental situation is illustrated by the tracings reproduced in Fig. 34-12. In this experiment the left stellate ganglion was stimulated between the two arrows at the top of Fig. 34-12. During stimulation there was a substantial rise in aortic flow accompanied by reductions in right and left atrial pressures (P_{RA} and P_{LA}).

■ *Blood volume*

Changes in blood volume per se do not affect myocardial contractility but do influence the vascular function curve in the manner shown in Fig. 34-8. Therefore to understand the circulatory alterations evoked by a given change in blood volume, it is necessary to plot the appropriate cardiac function curve along with the vascular function curves that represent the control and experimental states. Fig. 34-13 illustrates the response to a blood transfusion. Because equilibrium point *B*, which denotes the values for cardiac output and P_v after transfusion, lies above and to the right of the control equilibrium point *A*, it is evident that transfusion results in an enhanced cardiac output and elevated P_v. Hemorrhage would have the opposite effect. Pure increases or decreases in venomotor tone elicit responses that are analogous to those evoked by augmentations or reductions, respectively, of the total blood volume, for reasons that were discussed on p. 583.

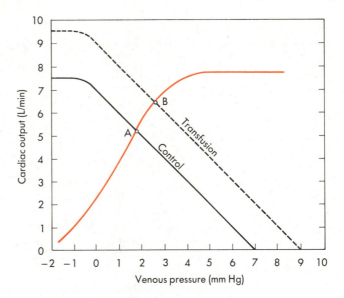

Fig. 34-13 ■ After a blood transfusion the vascular function curve is shifted to the right (dashed curve). Therefore cardiac output and venous pressure are both increased, as denoted by the translocation of the equilibrium point from *A* to *B*.

■ *Heart failure*

Heart failure may be acute or chronic. Acute heart failure may be caused by toxic quantities of certain drugs and anesthetics or by certain pathological conditions such as coronary artery occlusion. Chronic heart failure may occur in such conditions as essential hypertension or rheumatic heart disease. In these various forms of heart failure, myocardial contractility is impaired. Consequently, the cardiac function curve is shifted to the right as depicted in Fig. 34-14.

In acute heart failure there has not been sufficient time for blood volume to change. Therefore the equilibrium point will shift from the intersection of the normal curves (Fig. 34-14, point *A*) to the intersection of the normal vascular function curve with a depressed cardiac function curve (point *B* or *C*).

In chronic heart failure there is not only a shift in the cardiac function curve but also an increase in blood volume caused in part by fluid retention by the kidneys. The fluid retention is related to the concomitant reduction in glomerular filtration rate and to the increased secretion of aldosterone by the adrenal cortex. The resultant hypervolemia causes the vascular function curve to be shifted to the right as shown in Fig. 34-14. Hence, with moderate degrees of heart failure, P_v will be elevated, but cardiac output will be approximately normal (point *D*). With more severe degrees of heart failure, P_v will be still higher, and cardiac output will be subnormal (point *E*).

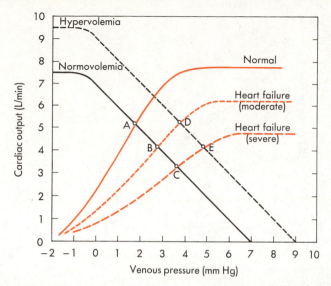

Fig. 34-14 ■ With moderate or severe heart failure, the cardiac function curves are shifted to the right. With no change in blood volume, cardiac output decreases and venous pressure rises (from control equilibrium point *A* to point *B* or point *C*). With the increase in blood volume that usually occurs in heart failure, the vascular function curve is shifted to the right. Hence, venous pressure may be elevated with no reduction in cardiac output (point *D*) or (in severe heart failure) with some diminution in cardiac output (point *E*).

Classically heart failure has been explained on the basis of the descending limb of the Frank-Starling curve. As explained in Chapter 33, when myocardial fibers are stretched beyond an optimal point, then further degrees of stretch impair rather than improve performance. Beyond the optimal point, cardiac output diminishes as venous pressure is raised; this portion of a cardiac function curve beyond the optimal point is designated the *descending limb*. In the normal heart the descending limb does not become manifest until the ventricular end-diastolic pressure reaches very high levels, but with myocardial failure it may appear at much lower end-diastolic pressures. Not uncommonly, patients in severe heart failure improve when blood is withdrawn and become worse when blood volume is expanded. However, whether such clinical observations indicate that the heart is operating on the descending limb of a cardiac function curve or whether other mechanisms are responsible remains controversial.

Predictions concerning the precise alterations evoked by changes in peripheral resistance are complex because shifts in both the cardiac and vascular function curves are involved. With increased peripheral resistance (Fig. 34-15) the vascular function curve is displaced downward but converges to the same X-axis intercept as the control curve as described on p. 584. The cardiac function curve is also shifted downward because at any given P_v the heart is able to pump less blood against a greater resistive load. Because both the vascular and cardiac function curves are displaced downward by an increase of peripheral resistance, it is clear that the new equilibrium point *(B)* will fall below the control point *(A)*

Whether point *B* will fall directly below point *A* or will lie somewhat to the right or left depends on the magnitude of the shift in each of the two curves. For example, if a given increase in peripheral resistance resulted in a relatively greater downward shift of the vascular function curve than of the cardiac function curve, then equilibrium point *B* would fall not only below *A* but also to the left of *A;* that is, both cardiac output and P_v would diminish. Conversely, if the cardiac function curve is displaced more than the vascular function curve, then point *B* will fall below and to the right of point *A;* that is, cardiac output would decrease, but P_v would rise.

■ *Peripheral resistance*

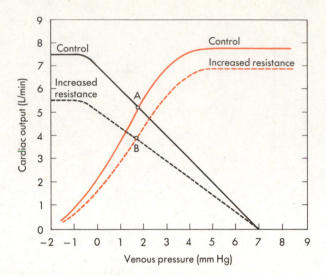

Fig. 34-15 ■ With increased peripheral resistance, both the cardiac and vascular function curves are displaced downward.

■ *Ancillary factors affecting the venous system and cardiac output*

The interrelationships between central venous pressure and cardiac output have been discussed in terms of an oversimplified schema. The effects evoked by changes in single variables have been described. However, it must be recognized that, partly because of the multitude of feedback control loops that regulate the cardiovascular system, an isolated change in a single variable rarely occurs. A change in blood volume, for example, would rapidly elicit reflex alterations in cardiac function, peripheral resistance, and venomotor tone. Aside from the complications involved in analyzing the effects of multiple variables acting simultaneously, several ancillary factors must also be considered for a complete understanding of the control of cardiac output. Such ancillary factors may be considered to modulate the operation of the simplified schema described previously.

■ *Gravity*

Experience has shown that gravitational forces may exert dramatic effects on the cardiac output. Among soldiers standing at attention for protracted periods, particularly in warm weather, it is not unusual for some persons to faint because of reduced cardiac output. Gravitational effects are exaggerated in cases of airplane pilots executing pull-outs from dives, where a centrifugal force several times greater than the force of gravity may be exerted briefly in the footward direction; such persons characteristically black out during the maneuver as blood is drained from the cephalic regions and pooled in the lower parts of the body.

Specious reasoning is sometimes applied to explain the curtailment of cardiac output under such conditions. It is often argued that when a person is oriented in a vertical, footdown position, the forces of gravity act counter to those forces that ordinarily promote venous return from the dependent regions of the body. This statement is incomplete, however, because it ignores the fact that any impedimentary gravitational force on the venous side of the circuit is exactly balanced by a facilitatory counterforce on the arterial side of the same circuit.

In this sense, therefore, the vascular system may be considered to be a U tube. To comprehend the action of gravity on flow through such a system, the models depicted in Figs. 34-16 and 34-17 will be analyzed. In Fig. 34-16 all the U tubes represent rigid cylinders of constant diameter. With both limbs of the U tube oriented horizontally (*A*) the flow is dependent only on the pressures at the inflow and outflow ends of the tube (P_i and P_o, respectively), the viscosity of the fluid, and the length and radius of the tube

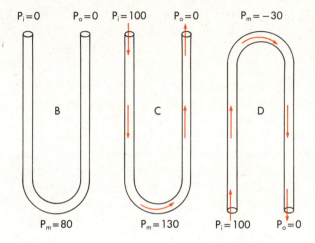

Fig. 34-16 ■ Pressure distributions in rigid U tubes with constant internal diameters, all with the same dimensions. For a given inflow pressure (P_i = 100) and outflow pressure (P_o = 0), the pressure at the midpoint (P_m) depends on the orientation of the U tube, but the flow through the tube is independent of the orientation.

Fig. 34-17 ■ In U tubes with a distensible section at the bend, even when inflow (P_i) and outflow (P_o) pressures are the same, the resistance to flow and the fluid volume contained within each tube vary with the orientation of the tube.

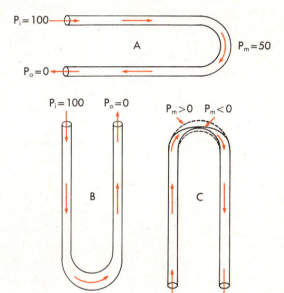

in accordance with Poiseuille's equation (Chapter 28). With a constant cross section the pressure gradient will be uniform; hence, the pressure midway down the tube (P_m) will be the average of the inflow and outflow pressures.

When the U tube is oriented vertically *(B to D)*, hydrostatic forces must be taken into consideration. In tube *B* both limbs are open to atmospheric pressure at the same hydrostatic level; hence, there is no flow. The pressure at the midpoint of the tube will simply be ρhg; that is, it will depend on the density of the fluid (ρ), the height of the U tube (h), and the acceleration of gravity (g). In the example the length of the U tube is such that the pressure at the midpoint is 80 mm Hg.

Now consider case *C*, where the U tube is oriented the same as tube *B*, but where a 100 mm Hg pressure difference is applied across the two ends. The flow will precisely equal that in *A*, since the pressure gradient, the tube dimensions, and the fluid viscosity are all the same. Gravitational forces are precisely equal in magnitude but opposite in direction in the two limbs of the U tube. Therefore, since the flow will be the same as that in *A*, there will be a pressure drop of 50 mm Hg at the midpoint because of the viscous losses resulting from flow. Furthermore, there will be an increased pressure of 80 mm Hg at the midpoint because of gravitational effects, just as in *B*. The actual pressure at the midpoint of tube C then will be the resultant of the viscous loss and hydrostatic gain, or 130 mm Hg in this example.

In D a pressure gradient of 100 mm Hg is applied to the same U tube but oriented in the opposite direction. Gravitational forces will be so directed that the pressure at the midpoint will tend to be 80 mm Hg less than that at the ends of the U tube. Viscous losses will still produce a 50 mm Hg pressure drop at the midpoint relative to P_i. Hence, with this orientation, pressure at the midpoint of the U tube will be 30 mm Hg below ambient pressure. Flow will, of course, be the same as in tubes A and C for the reasons stated in relation to C.

Therefore in a system of rigid tubes, gravitational effects will not alter the rate of blood flow. Because experience shows that gravity does affect the cardiovascular system, it may be suspected that the reason resides in the fact that the cardiovascular system is composed of distensible rather than rigid vessels. Therefore the pressures in a set of U tubes with distensible components (at the bends in the tubes of Fig. 34-17) will be examined. In tubes A and B the pressure distributions will resemble those in tubes A and C, respectively, of Fig. 34-16. Because the pressure is higher at the bend of tube B than at the bend of tube A and the segment is distensible in this region, the tubing at the bend of B will be distended more than at the bend of A. The extent of the distension will depend on the elastic characteristics of these tube segments. Because flow is directly related to the tube diameter, the flow through B will exceed the flow through A for a given pressure difference applied at the ends.

Because orienting a U tube with its bend downward actually increases rather than diminishes flow for any given pressure difference, how then is the observed impairment of cardiovascular function explained when the body is similarly oriented? Of course the explanation resides in the fact that the cardiovascular system is a closed circuit of constant volume, whereas in the example the U tube is conceived of as an open conduit supplied by a source of unlimited volume. In the cardiovascular system, most of the distension will occur on the venous rather than the arterial side of the circuit in the dependent regions of the body because the venous capacitance is so much greater than the arterial capacitance. Such venous distension is readily observed on the back of the hands when the arms are allowed to hang down below heart level. Whatever the volume of blood involved in such venous distension *(venous pooling)*, the hemodynamic effects resemble those caused by the loss of an equivalent volume of blood from the body during hemorrhage. It has been estimated that when a person shifts from a supine to a relaxed standing position from 300 to 800 ml of blood is pooled in the legs. This results in a decrease in cardiac output of about 2 L/minute and a 40% reduction in stroke volume.

The compensatory adjustments that enable humans to adapt to the erect position are largely identical to those which permit them to adapt to blood loss. For example, the diminution in baroreceptor excitation reflexly initiates acceleration of the heart, a more vigorous cardiac contraction, and increased arteriolar and venular constriction. The baroreceptor reflex per se apparently has a much more pronounced effect on the resistance than on the capacitance vessels. Warm ambient temperatures tend to interfere with the compensatory vasomotor reactions, and the absence of muscular activity exaggerates the effects. Therefore fainting is not uncommon during prolonged standing in hot weather.

The reverse situation prevails when the U tube is rotated so that the bend is directed upward. In Fig. 34-17 the pressure at the bend of tube C would tend to be -30 mm Hg, just as in tube D of the preceding figure. Because the ambient pressure exceeds the internal pressure, however, the tube will collapse. Flow will then cease, and there will no longer be any viscous decline of pressure associated with flow. In U tube C, when flow stops, the pressure at the top of each limb will be 80 mm Hg less than at the bottom (the hydrostatic pressure difference). Hence, in the left or inflow limb the pressure will approach +20 mm Hg. As soon as this pressure exceeds ambient pressure, the collapsed tubing will be forced open, and flow will begin. With the onset of flow, however, pressure will again drop below the ambient pressure. Thus the tubing at the bend will flutter; that is, it will fluctuate between the open and closed states.

When an arm is raised, the cutaneous veins in the hand and forearm are seen to collapse, for the reasons described previously. Fluttering does not occur here because the deeper veins are protected from collapse by surrounding structures; these deeper veins therefore accommodate the flow. The situation would be analogous to adding a rigid tube (representing the deeper veins) in parallel with the collapsible tube (representing the superficial veins) at the bend of tube *C* in Fig. 34-17. The collapsible tube would no longer flutter but would remain closed. All flow would occur through the rigid tube, just as in tube *D* in Fig. 34-16.

It may not be appreciated why the superficial veins in the neck are ordinarily partially collapsed in the normal person in the upright position. Venous return from the head is conducted largely through the deeper cervical veins. However, when *central venous pressure* (pressure in the thoracic venae cavae and right atrium) is abnormally elevated, the superficial neck veins are distended, and they do not collapse even when the subject assumes the upright position. Such cervical venous distension is an important clinical sign, often indicative of congestive heart failure.

■ *Muscular activity and venous valves*

When a person assumes the upright position but remains at rest, the pressure rises in the veins in the dependent regions of the body. The venous pressure in the legs increases gradually and does not reach an equilibrium value until almost 1 minute after standing. The gradual nature of the rise in P_v is attributable to the valves located within the veins, which permit flow only in the direction toward the heart. When a person stands up, the valves prevent blood in veins at higher levels from actually falling toward the feet. Hence, the column of venous blood is supported at numerous levels by these valves; the venous column temporarily consists of many separate segments. However, blood continues to enter the column at many points from the venules and smaller tributary veins, and pressure continues to rise. As soon as the pressure in one segment exceeds that in the segment just above it, the valve is forced open. Ultimately all the valves are opened, and the column is then continuous, similar to the outflow limbs of the U tubes shown in Figs. 34-16 and 34-17.

Precise measurement reveals that the final level of P_v in the feet during standing is only slightly greater than that which would have existed in a static column of blood extending from the right atrium to the feet. This indicates that the pressure drop caused by flow from foot veins to the right atrium is very small. This is one justification for lumping all the veins as a common venous capacitance in the model illustrated in Fig. 34-4.

When a person who has been standing quietly begins to walk, the venous pressure in his legs decreases appreciably during this muscular activity. Because of the intermittent venous compression produced by the contracting muscles and because of the orientation of the venous valves, blood is forced from the veins toward the heart. Hence, muscular contraction lowers the mean venous pressure and serves as an auxiliary pump to assist venous return. Furthermore, it prevents venous pooling and lowers capillary hydrostatic pressure (which must always exceed P_v), thereby reducing the tendency toward the formation of tissue edema in the dependent regions.

■ *Respiratory activity*

During quiet, normal breathing, the periodic activity of the respiratory muscles results in rhythmic variations in vena caval flow and serves as an auxiliary pump to promote venous return. Other activities involving the muscles of respiration, such as coughing and straining at stool, usually exert more profound influences on cardiac output and venous return; substantial changes in the cardiac output may also occur during artificial respiration.

The changes in blood flow in the superior vena cava associated with normal spontaneous respiration are shown in Fig. 34-18. During inspiration the reduction in intrathoracic pressure is transmitted to the lumina of the blood vessels located within the thoracic cavity. This is reflected by the diminution of right atrial pressure accompanying the intrathoracic pressure change. This reduction in central venous pressure during inspiration increases the pressure gradient between extrathoracic and intrathoracic veins. The consequent acceleration of venous return to the right atrium is displayed in the figure as an increase in superior vena caval blood flow from 5.2 ml/second during expiration to 11 ml/second during inspiration.

An exaggerated reduction in intrathoracic pressure achieved by a strong inspiratory effort against a closed glottis (called the *Müller maneuver*) does not evoke a proportionately greater acceleration of venous return. This is largely caused by the tendency for many of the extrathoracic veins to collapse near their points of entry into the chest when their pressures fall below the atmospheric level. As the veins collapse, flow into the chest momentarily stops. The cessation of flow results in a rise in pressure upstream, forcing the collapsed segment to open again. The process is repetitive so that the segments of veins adjacent to the chest will actually flutter rapidly between the open and closed states.

During expiration there is, of course, a deceleration of the flow into the central veins. However, it has been shown that the mean rate of venous return during normal respiration exceeds the rate in the temporary absence of respiratory activity. Hence, normal inspiration apparently exerts a greater effect toward promoting venous return than does normal expiration toward impeding it. In part this must be attributable to the presence of valves in the veins of the extremities and neck, which would prevent any reversal of flow during expiration. Thus the respiratory muscles and venous valves constitute an auxiliary pump for venous return.

Sudden sustained increases in intrathoracic pressure tend to impede venous return. Straining against a closed glottis (termed the *Valsalva maneuver*) regularly occurs during coughing, defecation, and lifting. Intrathoracic pressures in excess of 100 mm Hg have

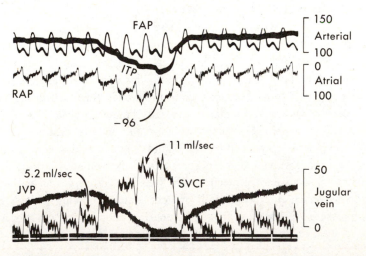

Fig. 34-18 ■ During a normal inspiration, intrathoracic *(ITP)*, right atrial *(RAP)*, and jugular venous *(JVP)* pressures decrease, and flow in the superior vena cava *(SVCF)* increases, in this case from 5.2 to 11 ml/second. All pressures are in mm H₂O, except for femoral arterial pressure *(FAP)*, which is in mm Hg. (Redrawn from Brecher, G.A.: Venous return, New York, 1956, Grune & Stratton, Inc. By permission.)

been recorded in trumpet players, and pressures over 400 mm Hg have been observed during paroxysms of coughing. Arterial pressure may rise considerably above the control level during the cough by virtue of a direct transmission of the greatly increased intrathoracic pressure to the lumina of the aorta and its branches within the chest. However, after cessation of coughing the arterial blood pressure might drop transiently to very low levels because of the severe impediment to venous return during the preceding paroxysm of coughing.

■ Bibliography

Journal articles

Carneiro, J.J., and Donald, D.E.: Blood reservoir function of dog spleen, liver, and intestine, Am. J. Physiol. **232:**H67, 1977.

Gauer, O.H., and Thron, H.L.: Properties of veins in vivo: integrated effects of their smooth muscle, Physiol. Rev. **42**(suppl. 5):283, 1962.

Grodins, F.S.: Integrative cardiovascular physiology: a mathematical synthesis of cardiac and blood vessel hemodynamics, Q. Rev. Biol. **34:**93, 1959.

Grodins, F.S., Stuart, W.H., and Veenstra, R.L.: Performance characteristics of the right heart bypass preparation, Am. J. Physiol. **198:**552, 1960.

Lautt, W.W., and Greenway, C.V.: Hepatic venous compliance and role of liver as a blood reservoir, Am. J. Physiol. **231:**292, 1976.

Levy, M.N.: The cardiac and vascular factors that determine systemic blood flow, Circ. Res. **44:**739, 1979.

Levy, M.N.: The cardiovascular physiology of the critically ill patient, Surg. Clin. North Am. **55:**483, 1975.

Ludbrook, J.: The musculovenous pumps of the human lower limb, Am. Heart J. **71:**635, 1966.

Rothe, C.F.: Reflex control of the veins in cardiovascular function, Physiologist **22:**28, 1979.

Sagawa, K.: Critique of a large-scale organ system model: guytonian cardiovascular model, Ann. Biomed. Engin. **3:**386, 1975.

Samar, R.E., and Coleman, T.G.: Measurement of mean circulatory filling pressure and vascular capacitance in the rat, Am. J. Physiol. **234:**H94, 1978.

Shoukas, A.A., and Sagawa, K.: Control of total systemic vascular capacity by the carotid sinus baroreceptor reflex, Circ. Res. **33:**22, 1973.

Shoukas, A.A., et al.: Importance of the spleen in blood volume shifts of the systemic vascular bed caused by the carotid sinus baroreceptor reflex in the dog, Circ. Res. **49:**759, 1981.

Weisel, R.D., Berger, R.L., and Hechtman, H.B.: Measurement of cardiac output by thermodilution, N. Engl. J. Med. **292:**682, 1975.

Books and monographs

Alexander, R.S.: The peripheral venous system. In Handbook of physiology; Section 2: Circulation, vol. II, Washington, D.C., 1963, American Physiological Society.

Bloomfield, D.A.: Dye curves: the theory and practice of indicator dilution, Baltimore, 1974, University Park Press.

Brecher, G.A.: Venous return, New York, 1956, Grune & Stratton, Inc.

Green, J.F.: Determinants of systemic blood flow; International Review of Physiology III: Cardiovascular physiology, vol. 18, Baltimore, 1979, University Park Press.

Guyton, A.C., Jones, C.E., and Coleman, T.G.: Circulatory physiology: cardiac output and its regulation, ed. 2, Philadelphia, 1973, W.B. Saunders Co.

Shepherd, J.T., and Vanhoutte, P.M.: Veins and their control, Philadelphia, 1975, W.B. Saunders Co.

Zierler, K.L.: Circulation times and the theory of indicator-dilution methods for determining blood flow and volume. In Handbook of physiology; Section 2: Circulation, vol. I, Washington, D.C., 1962, American Physiological Society.

chapter 35

Special circulations

■ *Coronary circulation*

■ *Anatomy of coronary vessels*

The right and left coronary arteries, which arise at the root of the aorta behind the right and left cusps of the aortic valve, respectively, provide the entire blood supply to the myocardium. The right coronary artery supplies principally the right ventricle and atrium; the left coronary artery, which divides near its origin into the anterior descendens and the circumflex branches, supplies principally the left ventricle and atrium, but there is some overlap. In the dog the left coronary artery supplies about 85% of the myocardium, whereas in humans the right coronary artery is dominant in 50% of persons; the left coronary artery is dominant in another 20%, and the flow delivered by each main artery is about equal in the remaining 30%. Coronary blood flow is measured in humans by (1) the nitrous oxide technique as described for cerebral blood flow measurement on p. 612, (2) measurement of rubidium 84 (a positron emitter) uptake by the myocardium, by using a procedure known as coincident counting over the chest, or (3) the clearance rate of an inert radioactive gas (for example, xenon 133 or krypton 85) injected directly into a coronary artery by means of a catheter passed from a peripheral artery (e.g., brachial) into the coronary artery. With the last method the rate of washout of the injected radioactivity, as monitored by a radiation detector placed over the precordium, is proportional to the blood flow through the myocardium supplied by the injected vessel.

After passage through the capillary beds, most of the venous blood returns to the right atrium through the coronary sinus, but some reaches the right atrium by way of the anterior coronary veins. There are also vascular communications directly between the vessels of the myocardium and the cardiac chambers; these comprise the *arteriosinusoidal,* the *arterioluminal,* and the *thebesian vessels*. The arteriosinusoidal channels consist of small arteries or arterioles that lose their arterial structure as they penetrate the chamber walls and divide into irregular, endothelium-lined sinuses (50 to 250 μm). These sinuses anastomose with other sinuses and with capillaries and communicate with the cardiac chambers. The arterioluminal vessels are small arteries or arterioles that open directly into the atria and ventricles. The thebesian vessels are small veins that connect capillary beds directly with the cardiac chambers and also communicate with cardiac veins and other thebesian veins. On the basis of anatomical studies, intercommunication appears to exist among all the minute vessels of the myocardium in the form of an extensive plexus of subendocardial vessels. It has been suggested that some myocardial nutrition can be derived from the cardiac cavities through those channels. Isotope-labeled blood in the cardiac chambers does penetrate a short distance into the endocardium but does not constitute a significant source of oxygen and nutrients to the myocardium. In the dog a major fraction of the left coronary inflow returns to the right atrium via the

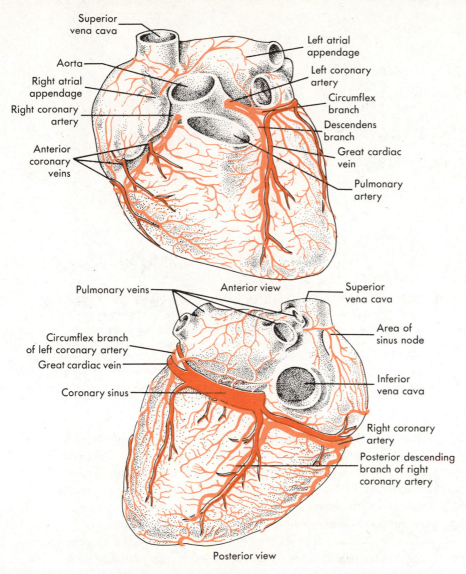

Superior
vena cava

Aorta

Right atrial
appendage

Right coronary
artery

Anterior
coronary
veins

Left atrial
appendage

Left coronary
artery

Circumflex
branch

Descendens
branch

Great cardiac
vein

Pulmonary
artery

Anterior view

Pulmonary veins

Circumflex branch
of left coronary artery

Great cardiac vein

Coronary sinus

Superior
vena cava

Area of
sinus node

Inferior
vena cava

Right coronary
artery

Posterior descending
branch of right
coronary artery

Posterior view

Fig. 35-1 ▪ Anterior and posterior surfaces of the heart, illustrating the location and distribution of the principal coronary vessels.

coronary sinus, and a small fraction supplying the interventricular septum returns directly to the right ventricular cavity. Right coronary artery drainage is primarily via the anterior cardiac veins to the right atrium. The epicardial distribution of the coronary arteries and veins is illustrated in Fig. 35-1.

Physical factors. The primary factor responsible for perfusion of the myocardium is the aortic pressure, which is, of course, generated by the heart itself. Changes in aortic pressure generally evoke parallel directional changes in coronary blood flow. However, alterations of cardiac work, produced by an increase or decrease in aortic pressure, have a considerable effect on coronary resistance. By means of mechanisms that have not yet been fully elucidated, increased metabolic activity of the heart results in a decrease in coronary resistance, and a reduction in cardiac metabolism produces an increase in coronary resistance. If a cannulated coronary artery is perfused by blood from a pressure-controlled reservoir, perfusion pressure can be altered without changing aortic pressure and cardiac work. Under these conditions abrupt variations in perfusion pressure produce equally abrupt changes in coronary blood flow in the same direction. However, maintenance of the perfusion pressure at the new level is associated with a return of

▪ *Factors that influence coronary blood flow*

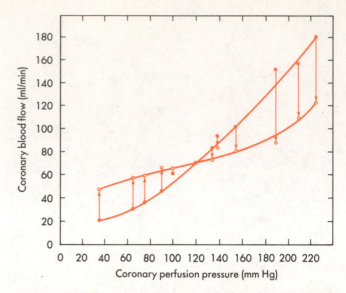

Fig. 35-2 ■ Pressure-flow relationships in the coronary vascular bed. At constant aortic pressure, cardiac output, and heart rate, coronary artery perfusion pressure was abruptly increased or decreased from the control level indicated by the point where the two lines cross. The closed circles represent the flows that were obtained immediately after the change in perfusion pressure, and the open circles represent the steady state flows at the new pressures. There is a tendency for flow to return toward the control level (autoregulation of blood flow), and this is most prominent over the intermediate pressure range (about 60 to 180 mm Hg).

blood flow toward the level observed before the induced change in perfusion pressure (Fig. 35-2). This phenomenon is an example of autoregulation of blood flow and is discussed in Chapter 32. Under normal conditions blood pressure is kept within relatively narrow limits by the baroreceptor reflex mechanisms so that changes in coronary blood flow are primarily caused by caliber changes of the coronary resistance vessels in response to metabolic demands of the heart.

In addition to providing the head of pressure to drive blood through the coronary vessels, the heart also influences its blood supply by the squeezing effect of the contracting myocardium on the blood vessels that course through it *(extravascular compression)*. This force is so great during early ventricular systole that blood flow, as measured in a large coronary artery supplying the left ventricle, is briefly reversed. Maximal left coronary inflow occurs in early diastole, when the ventricles have relaxed and extravascular compression of the coronary vessels is virtually absent. This flow pattern is seen in the phasic coronary flow curve for the left coronary artery (Fig. 35-3). After an initial reversal in early systole, left coronary blood flow follows the aortic pressure until early diastole, when it rises abruptly and then declines slowly as aortic pressure falls during the remainder of diastole. Left ventricular myocardial pressure (pressure within the wall of the left ventricle) is greatest near the endocardium and lowest near the epicardium.

However, under normal conditions this pressure gradient does not result in impairment of endocardial blood flow because the greater diastolic flow in the endocardium compensates for the greater systolic flow in the epicardium. In fact, when 10 μm diameter radioactive microspheres are injected into the coronary arteries, their distribution indicates that the blood flow to the epicardial and endocardial halves of the left ventricle are approximately equal under normal conditions. Because extravascular compression is greatest at the endocardial surface of the ventricle, equality of epicardial and endocardial blood flow must mean that the tone of the endocardial resistance vessels is less than that of the epicardial vessels.

Under abnormal conditions, when diastolic pressure in the coronary arteries is low,

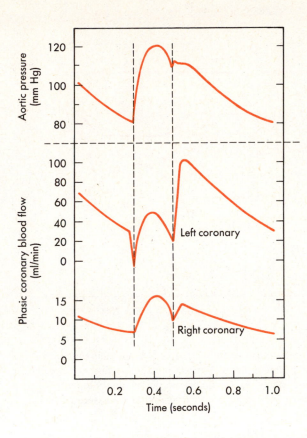

Fig. 35-3 ■ Comparison of phasic coronary blood flow in the left and right coronary arteries.

such as in severe hypotension, partial coronary artery occlusion, or severe aortic stenosis, the ratio of endocardial to epicardial blood flow falls below a value of 1.0. This indicates that the blood flow to the endocardial region is more severely impaired than that to the epicardial regions of the left ventricle. This redistribution of coronary flow is also reflected in an increase in the gradient of myocardial lactic acid and adenosine concentrations from epicardium to endocardium. For this reason, the myocardial damage observed in arteriosclerotic heart disease, for example, following coronary occlusion, is greatest in the inner wall of the left ventricle.

Flow in the right coronary artery shows a similar pattern (Fig. 35-3), but because of the lower pressure developed during systole by the thin right ventricle, reversal of blood flow does not occur in early systole, and systolic blood flow constitutes a much greater proportion of total coronary inflow than it does in the left coronary artery. The extent to which extravascular compression restricts coronary inflow can be readily seen when the heart is suddenly arrested in diastole or on the induction of ventricular fibrillation. Fig. 35-4 depicts mean left coronary flow when the vessel was perfused with blood at a constant pressure from a reservoir. At the arrow in record *A*, ventricular fibrillation was electrically induced, and an immediate and substantial increase in blood flow occurred. Subsequent increase in coronary resistance over a period of many minutes reduced myocardial blood flow to below the level existing before the induction of ventricular fibrillation (record *B*, before stellate ganglion stimulation).

Tachycardia and bradycardia have dual effects on coronary blood flow. A change in heart rate is accomplished chiefly by shortening or lengthening of diastole. With tachycardia the proportion of the cardiac cycle spent in systole and consequently the period of restricted inflow increases. However, this mechanical reduction in mean coronary flow is overridden by the coronary dilation associated with the increased metabolic activity of the more rapidly beating heart. With bradycardia the opposite is true; restriction of coronary inflow is less (more time in diastole) but so are the metabolic (oxygen) requirements of the myocardium.

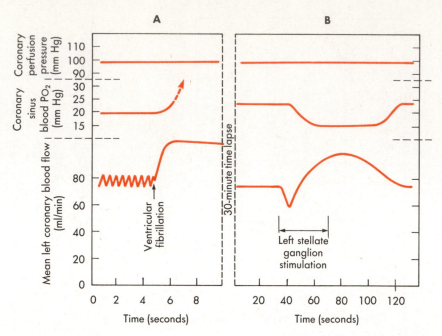

Fig. 35-4 ■ **A,** Unmasking of the restricting effect of ventricular systole on mean coronary blood flow by induction of ventricular fibrillation during constant pressure perfusion of the left coronary artery. **B,** Effect of cardiac sympathetic nerve stimulation on coronary blood flow and coronary sinus blood oxygen tension in the fibrillating heart during constant pressure perfusion of the left coronary artery.

Neural and neurohumoral factors. Stimulation of the sympathetic nerves to the heart elicits a marked increase in coronary blood flow. However, the increase in flow is associated with cardiac acceleration and a more forceful systole. The stronger myocardial contractions and the tachycardia (with the consequence that a greater proportion of time is spent in systole) tend to restrict coronary flow, whereas the increase in myocardial metabolic activity, as evidenced by the rate and contractility changes, tends to evoke dilation of the coronary resistance vessels. The increase in coronary blood flow observed with cardiac sympathetic nerve stimulation is the algebraic sum of these factors. In perfused hearts in which the mechanical effect of extravascular compression is eliminated by cardiac arrest or ventricular fibrillation, an initial coronary vasoconstriction is often observed with cardiac sympathetic nerve stimulation before the vasodilation attributable to the metabolic effect comes into play (Fig. 35-4, *B*). Such observations suggest that the primary action of the sympathetic nerve fibers on the coronary resistance vessels is vasoconstriction.

In contrast to skeletal muscle, sympathetic cholinergic innervation of the coronary vessels does not exist, and whether there are sympathetic fibers to β-adrenergic receptors on the coronary arterioles is questionable. However, experiments with the use of α- and β-adrenergic drugs and their respective blocking agents reveal the presence of α(constrictor)- and β(dilator)-receptors on coronary vessels. Recent studies have indicated that the coronary resistance vessels participate in the baroreceptor and chemoreceptor reflexes and that there appears to be sympathetic constrictor tone of the coronary arterioles that can be reflexly modulated. Nevertheless coronary resistance is predominantly under local nonneural control.

Vagus nerve stimulation has little direct effect on the caliber of the coronary arterioles. In the fibrillating heart with the coronary arteries perfused at a constant pressure or in the beating, paced heart with flow measured in late diastole when there is essentially no extravascular compression, small increments in coronary blood flow can be observed with stimulation of the peripheral ends of the vagi. In addition, activation of the carotid and aortic chemoreceptors can elicit a decrease in coronary resistance via the vagus nerves to the heart. However, it is likely that the vagi exert less effect on coronary resistance in the normal animal because of the overriding influence of metabolic mechanisms. The failure of strong vagal stimulation to evoke a large increase in coronary blood flow is not because of insensitivity of the coronary resistance vessels to acetylcholine because intracoronary administration of this agent elicits marked vasodilation.

Reflexes originating in the myocardium and altering vascular resistance in peripheral systemic vessels, including the coronary vessels, have been conclusively demonstrated. However, the existence of extracardiac reflexes, with the coronary resistance vessels as the effector sites, has not been established.

Metabolic factors. One of the most striking characteristics of the coronary circulation is the close parallelism between the level of myocardial metabolic activity and the magnitude of the coronary blood flow. This relationship is also found in the denervated heart or the completely isolated heart, whether in the beating or the fibrillating state. The ventricles will continue to fibrillate for many hours when the coronary arteries are perfused with arterial blood from some external source. With the onset of ventricular fibrillation, an abrupt increase in coronary blood flow occurs because of the removal of extravascular compression (Fig. 35-4). Flow then gradually returns toward and often decreases below the prefibrillation level. The increase in coronary resistance, which occurs despite the elimination of extravascular compression, is a manifestation of the heart's ability to adjust its blood flow to meet its energy requirements. The fibrillating heart uses less oxygen than the pumping heart, and blood flow to the myocardium is reduced accordingly.

The link between cardiac metabolic rate and the coronary blood flow remains unsettled. Numerous agents, generally referred to as *metabolites,* have been suggested as the mediator of the vasodilation observed with increased cardiac work. Accumulation of vasoactive metabolites may also be responsible for reactive hyperemia because the magnitude and, to a greater degree, the duration of the coronary flow following release of the briefly occluded vessel is, within certain limits, proportional to the duration of the period of occlusion. Among the substances implicated are carbon dioxide, oxygen (reduced oxygen tension), lactic acid, hydrogen ions, histamine, potassium ions, increased osmolarity, polypeptides, and adenine nucleotides. None of these has satisfied all the criteria for the physiological mediator. Although potassium release from the myocardium can account for about half of the initial decrease in coronary resistance, it cannot be responsible for the increased coronary flow observed with prolonged enhancement of cardiac metabolic activity, because its release from the cardiac muscle is transitory. Recent evidence suggests that adenosine plays the role of metabolic vasodilator.

According to the adenosine hypothesis, a reduction in myocardial oxygen tension produced by low coronary blood flow, hypoxemia, or increased metabolic activity of the heart leads to the formation of adenosine, which reaches the interstitial fluid space and induces dilation of the coronary resistance vessels. The adenosine concentration of the myocardium increases in response to enhanced cardiac work in the absence of a reduced oxygen supply, and it is rapidly formed. (In a single cardiac cycle the myocardial adenosine level increases signficantly during ventricular systole over that observed during diastole.) Coronary dilation results in an increase in coronary blood flow that enhances the washout of adenosine and reduces its formation. Under normal resting conditions only small amounts of adenosine are released by the heart and exert a minimal dilating effect. The schema for adenosine regulation of coronary blood flow is illustrated in Fig. 35-5. The enzyme 5'-nucleotidase is located at the cell membranes, transverse tubules, and to some extent inside the myocytes and dephosphorylates AMP (formed by degradation of ATP) to adenosine. The adenosine enters the interstitial space and produces arteriolar dilation. From the interstitial space the adenosine may reenter

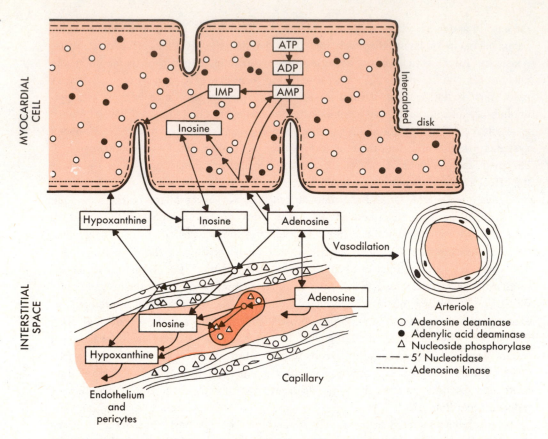

MYOCARDIAL CELL

INTERSTITIAL SPACE

ATP
ADP
AMP
IMP
Inosine
Intercalated disk

Hypoxanthine   Inosine   Adenosine

Vasodilation

Adenosine

Inosine

Hypoxanthine

Endothelium and pericytes

Capillary

Arteriole

○ Adenosine deaminase
● Adenylic acid deaminase
△ Nucleoside phosphorylase
– – – 5' Nucleotidase
- - - - - Adenosine kinase

Fig. 35-5 ■ Schematic diagram of myocardial tissue illustrating the formation, fate, and site of action of adenosine arising from intracellular ATP and the distribution of the enzymes involved in the metabolism of adenosine. (Reproduced by permission from Rubio, V.R., Wiedmeier, T., and Berne, R.M.: Am. J. Physiol. **222**:550, 1972.)

the myocardial cell and (1) be rephosphorylated to AMP by adenosine kinase, (2) be deaminated to inosine by adenosine deaminase in the myocardial cell, or (3) enter the capillaries where it can be deaminated to inosine in the vessel wall or red cells or further degraded to hypoxanthine by the enzyme nucleoside phosphorylase, which is located in pericytes (perivascular fibroblasts), endothelial cells, or red cells. Relative to adenosine deaminase, the concentration of adenylic acid deaminase in the myocardium is low so that only small amounts of inosinic acid (IMP) are formed.

■ *Coronary collateral circulation and vasodilators*

In the normal human heart there are virtually no functional intercoronary channels, whereas in the dog there are a few small vessels that link branches of the major coronary arteries. Abrupt occlusion of a coronary artery or one of its branches in a human or dog leads to ischemic necrosis and eventual fibrosis of areas of myocardium supplied by the occluded vessel. However, if narrowing of a coronary artery occurs slowly and progressively over a period of days, weeks, or longer, collateral vessels develop and may furnish sufficient blood to the ischemic myocardium to prevent or reduce the extent of the necrosis. The development of collateral coronary vessels has been extensively studied in dogs, and the clinical picture of coronary atherosclerosis, as it occurs in humans, can be simulated by gradual narrowing of the normal dog's coronary arteries. Collateral vessels develop between branches of occluded and nonoccluded arteries. Recent work indicates that they originate from preexisting small vessels that undergo proliferative changes of the endothelium and smooth muscle, possibly in response to wall stress and chemical agents released by the ischemic tissue.

Numerous surgical attempts have been made to enhance the development of coronary collateral vessels. However most of the techniques used do not increase the collateral circulation over and above that produced by coronary artery narrowing alone. When discrete occlusions occur in coronary arteries (even vessels as small as 1 mm in diameter), the lesions can be bypassed with a vein graft connecting the aorta to a point on the coronary artery distal to the site of the occlusion.

A number of drugs are available that induce coronary vasodilation, and they are used in patients with coronary artery disease to relieve *angina pectoris,* the chest pain associated with myocardial ischemia. Many of these compounds are nitrites. They are not selective dilators of the coronary vessels, and the mechanism whereby they accomplish their beneficial effects has not been established. The arterioles that would dilate in response to the drugs are undoubtedly already maximally dilated by the ischemia responsible for the symptoms. It has been suggested that the relief of angina pectoris by nitrites is brought about by a reduction in cardiac work and myocardial oxygen requirement caused by the moderate hypotension these drugs produce. In short, the reduction in pressure work and oxygen requirement must be greater than the reduction in coronary blood flow and oxygen supply consequent to the lowered coronary perfusion pressure. It has also been demonstrated that nitrites dilate large coronary arteries and coronary collateral vessels, thus increasing blood flow to ischemic myocardium and alleviating precordial pain.

■ *Cardiac oxygen consumption and work*

The volume of oxygen consumed by the heart is determined by the amount and the type of activity the heart performs. Under basal conditions, myocardial oxygen consumption is about 8 to 10 ml/minute/100 g of heart. It can increase severalfold with exercise and decrease moderately under conditions such as hypotension and hypothermia. The cardiac venous blood is normally quite low in oxygen (about 5 vol%), and the myocardium can receive little additional oxygen by further oxygen extraction from the coronary blood, a situation that also exists in contracting skeletal muscle. Therefore increased oxygen demands of the heart must be met primarily by an increase in coronary blood flow. When the heart beat is arrested, as with administration of potassium, but coronary perfusion is maintained experimentally, the oxygen consumption falls to 2 ml/minute/100 g or less, which is still six or seven times greater than that for resting skeletal muscle.

Left ventricular work per beat (*stroke work,* Chapter 30) is generally considered to be equal to the product of the stroke volume and the mean aortic pressure against which the blood is ejected by the left ventricle. At resting levels of cardiac output the kinetic energy component is negligible (Chapter 28). However, at high cardiac outputs as in severe exercise, the kinetic component can account for up to 50% of total cardiac work. One can simultaneously halve the aortic pressure and double the cardiac output, or vice versa, and still arrive at the same value for cardiac work. However, the oxygen requirements are greater for any given amount of cardiac work when a major fraction is so-called pressure work as opposed to volume work. An increase in cardiac output at a constant aortic pressure (volume work) is accomplished with a small increase in left ventricular oxygen consumption, whereas increased arterial pressure at constant cardiac output (pressure work) is accompanied by a large increment of myocardial oxygen consumption. Thus myocardial oxygen consumption may not correlate well with overall cardiac work. The magnitude and duration of left ventricular pressure do correlate with left ventricular oxygen consumption. The area under the systolic portion of the left ventricular pressure curve has been termed the *tension-time index,* and this index correlates reasonably well with myocardial oxygen consumption in many different physiological states. Because this index does not take into consideration the velocity of contraction and because the velocity factor influences myocardial oxygen consumption, there are conditions (for example, exercise, sympathetic activation, epinephrine administration) in which the tension-time index fails to reflect accurately the oxygen consumption. The tension development (ventricular wall tension), the velocity of shortening, and, to a lesser extent, the degree of shortening of the myocardial fibers constitute the chief determinants of myocardial oxygen consumption. The greater energy demand of pressure work over volume work can be readily demonstrated in the heart-lung preparation and may be associated with a decrease in myocardial creatine phosphate levels. It is also of great clinical importance, especially in aortic stenosis, in which left ventricular oxygen consumption is increased because of the high intraventricular pressures developed during systole, but coronary perfusion pressure is normal or reduced because of the pressure drop across the diseased aortic valve. Because mean pulmonary artery pressure is about one seventh that of aortic pressure and the outputs of the two ventricles are equal, work of the right ventricle is one seventh that of the left ventricle.

■ Cardiac efficiency

As with an engine, the efficiency of the heart is the ratio of the work accomplished to the total energy used. Assuming an average oxygen consumption of 9 ml/minute/100 g for the two ventricles, a 300 g heart consumes 27 ml O_2/minute, which is equivalent to 130 small calories at a respiratory quotient of 0.82. Together the two ventricles do about 8 kg-m of work per minute, which is equivalent to 18.7 small calories. Therefore the gross efficiency is 14% ($18.7/130 \times 100 = 14\%$). The net efficiency is slightly higher (18%) and is obtained by subtracting the oxygen consumption of the nonbeating (asystolic) heart (about 2 ml/minute/100 g) from the total cardiac oxygen consumption in the calculation of efficiency. It is thus evident that the efficiency of the heart as a pump is relatively low and is comparable to the efficiency of many mechanical devices used in everyday life. With exercise, efficiency improves because mean blood pressure shows little change, whereas cardiac output and work increase considerably without a proportional increase in myocardial oxygen consumption. The energy expended in cardiac metabolism that does not contribute to the propulsion of blood through the body appears in the form of heat. The energy of the flowing blood is also dissipated as heat, chiefly in passage through the arterioles.

■ Substrate utilization

The heart is quite versatile in its use of substrates, and within certain limits the uptake of a particular substrate is directly proportional to its arterial concentration. The utilization of one substrate is also influenced by the presence or absence of other substrates. For example, the addition of lactate to the blood perfusing a heart metabolizing glucose leads to a reduction in glucose uptake and vice versa. At normal blood concentrations glucose and lactate are consumed at about equal rates, whereas pyruvate uptake is very low, but so is its arterial concentration. For glucose the threshold concentration is about 4mM, and below this blood level no myocardial glucose uptake occurs. Insulin reduces this threshold and increases the rate of glucose uptake by the heart. A very low threshold exists for cardiac utilization of lactate; insulin does not affect its uptake by the myocardium. Wth hypoxia, glucose use is facilitated by an increase in the rate of transport across the myocardial cell wall, whereas lactate cannot be metabolized by the hypoxic heart and is in fact produced by the heart under anaerobic conditions. Associated with lactate production by the hypoxic heart is the breakdown of cardiac glycogen.

Of the total cardiac oxygen consumption, only 35% to 40% can be accounted for by the oxidation of carbohydrate. Thus the heart derives the major part of its energy from oxidation of noncarbohydrate sources. The chief noncarbohydrate fuel used by the heart is esterified and nonesterified fatty acid, which accounts for about 60% of myocardial oxygen consumption in the postabsorptive state. The various fatty acids show different thresholds for myocardial uptake but are generally used in direct proportion to their arterial concentration. Ketone bodies, especially acetoacetate, are readily oxidized by the heart and contribute a major source of energy in diabetic acidosis. As is true of carbohydrate substrates, utilization of a specific noncarbohydrate substrate is influenced by the presence of other substrates, both noncarbohydrate and carbohydrate. Therefore, within certain limits, the heart uses preferentially that substrate which is available in the largest concentration. Most evidence indicates that the contribution to myocardial energy expenditure provided by the oxidation of amino acids is small.

Normally the heart derives its energy by oxidative phosphorylation in which each mole of glucose yields 36 moles of ATP. However, with hypoxia glycolysis supervenes and 2 moles of ATP are provided by each mole of glucose; this is in part compensated by a greater glycolytic flux as a result of several factors. β-Oxidation of fatty acids is also curtailed. If hypoxia is prolonged, there is initial depletion of cellular creatine phosphate and eventually ATP. It is also conceivable that the enhanced adenosine release that occurs with an inadequate myocardial oxygen supply contributes to the reduction in myocardial contractility. In contrast to hypoxia, ischemia results in accumulation of lactic acid (lack of washout) with a decrease in intracellular pH. This condition leads to inhibition of glycolysis and to greater inhibition of fatty acid use and protein synthesis resulting in membrane impairment, cellular damage, and eventually necrosis of myocardial cells. Detailed analysis of myocardial metabolism and its alteration in hypoxia and ischemia is an important subject but lies beyond the scope of this book. A number of procedures, some specific and some nonspecific, have been extensively studied in an effort to reduce myocardial infarct size following coronary occlusion and to restore and preserve marginally affected cardiac muscle during and after impairment of its oxygen supply.

■ *Cutaneous circulation*

The oxygen and nutrient requirements of the skin are relatively small and, in contrast to most other body tissues, the supply of these essential materials is not the chief governing factor in the regulation of cutaneous blood flow. The primary function of the cutaneous circulation is the maintenance of a constant body temperature. Consequently the skin shows wide fluctuations in blood flow, depending on the need for loss or conservation of body heat; mechanisms responsible for alterations in skin blood flow are primarily activated by changes in ambient and internal body temperatures.

■ *Regulation of skin blood flow*

There are essentially two types of resistance vessels in skin—*arterioles,* similar to those found elsewhere in the body, and *arteriovenous anastomoses,* which shunt blood from arterioles to venules and venous plexuses and therefore bypass the capillary bed. The former are distributed over most of the body surface, whereas the latter are found primarily in the fingertips, palms of the hands, toes, soles of the feet, ears, nose, and lips. The arteriovenous anastomoses differ morphologically from the arterioles in that they are generally short, coiled vessels about 20 to 40 μm in lumen diameter, with thick muscular walls richly supplied with nerve fibers. These vessels are almost exclusively under sympathetic neural control and become maximally dilated when their nerve supply is interrupted. Conversely, reflex stimulation of the sympathetic fibers to these vessels may produce constriction to the point of complete obliteration of the vascular lumen. Although the arteriovenous anastomoses do not exhibit *basal tone* (tonic activity of the vascular smooth muscle independent of innervation), they are nevertheless very sensitive to vasoconstrictor agents like epinephrine and norepinephrine. In fact, the absence of basal tone in these vessels and their greater sensitivity to catecholamines have been put forth as evidence against the concept that the basal tone displayed by all other resistance vessels is caused by circulating catecholamines. Furthermore, the arteriovenous anastomoses do not appear to be under metabolic control, and they fail to show reactive hyperemia or autoregulation of blood flow. Thus the regulation of blood flow through these anastomotic channels is governed principally by the nervous system in response to reflex activation by temperature receptors or from higher centers of the central nervous system.

The bulk of the skin resistance vessels exhibits some basal tone and is under dual control of the sympathetic nervous system and local regulatory factors, in much the same manner as are other vascular beds. However, in the case of skin, neural control plays a more important role than local factors. Stimulation of sympathetic nerve fibers to skin blood vessels (arteries and veins as well as arterioles) induces vasoconstriction, and severance of the sympathetic nerves, vasodilation. With chronic denervation of the cutaneous blood vessels, the degree of tone that existed before denervation is gradually regained over a period of several weeks. This is accomplished by an enhancement of basal tone that compensates for the degree of tone previously contributed by sympathetic nerve fiber activity. Epinephrine and norepinephrine elicit only vasoconstriction in cutaneous vessels. Whether the increased basal tone following denervation of the skin vessels is the result of their enhanced sensitivity to circulating catecholamines *(denervation hypersensitivity)* has not been established.

Parasympathetic vasodilator nerve fibers do not supply the cutaneous blood vessels. However, stimulation of the sweat glands, which are innervated by cholinergic fibers of the sympathetic nervous system, results in dilation of the skin resistance vessels. Sweat contains an enzyme that acts on a protein moiety in the tissue fluid to produce *bradykinin,* a polypeptide with potent vasodilator properties. It is thought that the bradykinin formed in the tissue acts locally to dilate the arterioles and increase blood flow to the skin. This polypeptide has also been found in the secretion of several other glandular structures and may be the mediator of the vasodilation associated with increased metabolic activity of the glands. Finally, the skin vessels of certain regions, particularly the head, neck, shoulders, and upper chest, are under the influence of the higher centers of the central nervous system. Blushing, as with embarassment or anger, and blanching, as with fear or anxiety, are examples of cerebral inhibition and stimulation, respectively, of the sympathetic nerve fibers to the affected regions. Whether blushing is caused in part by activation of the sweat glands and the local formation of bradykinin remains to be proved.

In contrast to the arteriovenous anastomoses in the skin, the cutaneous resistance vessels show autoregulation of blood flow and reactive hyperemia. If the arterial inflow to a limb is stopped with an inflated blood pressure cuff for a brief period, the skin shows a marked reddening below the point of vascular occlusion when the cuff is deflated. This increased cutaneous blood flow (reactive hyperemia) is also manifested by the distension of the superficial veins in the erythematous extremity. With respect to autoregulation of blood flow in the skin, the relatively low metabolic activity of the tissue favors a myogenic rather than a metabolic mechanism (Chapter 32).

Ambient and body temperature in regulation of skin blood flow. Because the primary function of the skin is to preserve the internal milieu and protect it from adverse changes in the environment and because ambient temperature is one of the most important external variables the body must contend with, it is not surprising that the vasculature of the skin is chiefly influenced by environmental temperature. Exposure to cold elicits a generalized cutaneous vasoconstriction that is most pronounced in the hands and feet. That this reponse is chiefly mediated by the nervous system is evident from the fact that when the circulation to a hand is arrested by a pressure cuff, immersion of the hand in cold water results in vasoconstriction in the skin of the other extremities that are exposed to room temperature. With the circulation to the chilled hand unoccluded, the reflex vasoconstriction is caused in part by the cooled blood returning to the general circulation and stimulating the temperature-regulating center in the anterior hypothalamus. Direct application of cold to this region of the brain produces cutaneous vasoconstriction.

The skin vessels of the cooled hand also show a direct response to cold. Moderate cooling or exposure for brief periods to severe cold (0° to 15° C) results in constriction of the resistance and capacitance vessels including arteriovenous anastomoses. However, prolonged exposure of the hand to severe cold has a secondary vasodilator effect. Prompt vasoconstriction and severe pain are elicited by immersion of the hand in water near 0° C, but these are soon followed by dilation of the skin vessels with reddening of the immersed part and alleviation of the pain. With continued immersion of the hand, alternating periods of constriction and dilation occur, but the skin temperature rarely drops to as low a degree as it did with the initial vasoconstriction. Prolonged severe cold, of course, results in tissue damage. The rosy faces of people working or playing in a cold environment are examples of cold vasodilation. However, the blood flow through the skin of the face may be greatly reduced despite the flushed appearance. The red color of the slowly flowing blood is in large measure the result of the reduced oxygen uptake by the cold skin and the cold-induced shift to the left of the oxyhemoglobin dissociation curve.

Direct application of heat produces not only local vasodilation of resistance and capacitance vessels as well as arteriovenous anastomoses but also reflex dilation in other parts of the body. The local effect is independent of the vascular nerve supply, whereas the reflex vasodilation is a combination of anterior hypothalamic stimulation by the returning warmed blood and of stimulation of receptors in the heated part. However, evidence for a reflex from peripheral temperature receptors is not as definitive for warm stimulation as it is for cold stimulation.

With exposure of the extremities to cold or heat, it must be remembered that the close proximity of the major arteries and veins to each other permits considerable heat exchange (countercurrent) between artery and vein. If cold blood flows from a cooled hand toward the heart, heat is taken up by the blood in the arm veins from that in the adjacent arteries, resulting in warming of the venous blood and cooling of the arterial blood. Heat exchange is of course in the opposite direction with exposure of the extremity to heat. Thus heat conservation is enhanced, and heat gain is minimized during exposure of extremities to cold and warm environments, respectively, and the temperature of the blood returning to the core of the body is brought closer to the existing core temperature.

The color of the skin is of course caused in large part by pigment, but in all but very dark skin, the degree of pallor or ruddiness is primarily a function of the amount of blood in the skin. With little blood in the venous plexuses the skin appears pale, whereas with moderate to large quantities of blood in the venous plexuses, the skin shows color. Whether this color is bright red, blue, or some shade between is determined by the degree of oxygenation of the blood in the subcutaneous vessels. For example, a combination of vasoconstriction and reduced hemoglobin can produce an ashen gray color of the skin, whereas a combination of venous engorgement and reduced hemoglobin can result in a dark purple hue. The skin color provides little information about the rate of cutaneous blood flow. There may coexist rapid blood flow and pale skin when the arteriovenous anastomoses are open and slow blood flow and red skin when the extremity is exposed to cold.

■ *Skin color and special reactions of the skin vessels*

White reaction and triple response. If the skin of the forearm of many individuals is lightly stroked with a blunt instrument, a *white line* appears at the site of the stroke within 20 seconds. The blanching becomes maximal in about 30 to 40 seconds and then gradually disappears within 3 to 5 minutes. This response is known as a *white reaction* and has been attributed by Sir Thomas Lewis to capillary contraction because it occurs in the denervated limb and is unaffected by arrest of the limb circulation. Because all direct evidence indicates that capillaries do not contract and skin color is primarily a

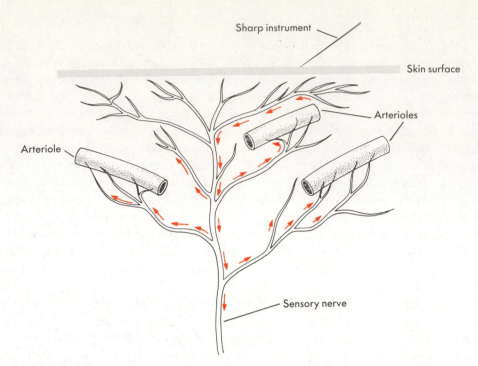

Fig. 35-6 ▪ Schematic representation of the axon reflex in response to a scratch on the skin surface with a sharp instrument. Arrows indicate the pathways of impulses in a sensory nerve from the site of stimulation to adjacent arterioles to produce local vasodilation (flare).

result of blood content of venous plexuses, venules, and small veins, it seems logical to attribute the white reaction to venous constriction induced by mechanical stimulation.

If the skin is stroked more strongly with a sharp pointed instrument, a *triple response* is elicited. Within 3 to 15 seconds a thin *red line* appears at the site of the stroke, followed in about 15 to 30 seconds by a red blush or *flare* extending out 1 to 2 cm from either side of the red line. This in turn is followed in 3 to 5 minutes by an elevation of the skin along the red line, with gradual fading of the red line as the elevation, a *wheal,* becomes more prominent. The red line is probably caused by dilation of the vessels because of mechanical stimulation. The flare, however, is the result of dilation of neighboring arterioles caused to relax by an *axon reflex* originating at the site of mechanical stimulation. In an axon reflex, the nerve impulse travels centripetally in the cutaneous sensory nerve fiber and then antidromically down the small branches of the afferent nerve to adjacent arterioles to elicit vasodilation (Fig. 35-6). The flare is not affected by acute section or anesthetic block of the sensory nerve central to the point of branching, whereas it is abolished when the nerve degenerates after section. The wheal is caused by increased capillary permeability induced by the trauma. Fluid containing protein leaks out of the capillaries locally and produces edema at the site of injury. Because the triple response can be elicited by an intradermal injection of histamine, Lewis attributed the response to histamine or a histamine-like substance *(H-substance)*. Whether it is histamine, ATP, a vasoactive polypeptide like bradykinin, or some yet unidentified substance remains to be determined. In summary, slight trauma is believed to release a substance that (1) produces the red line, (2) stimulates the local sensory nerve endings, and (3) increases capillary permeability.

▪ Skeletal muscle circulation

The rate of blood flow in skeletal muscle varies directly with the contractile activity of the tissue and the type of muscle; blood flow in red (slow-twitch, high-oxidative) muscle is greater than it is in white (fast-twitch, low-oxidative) muscle. In resting muscle the precapillary arterioles exhibit asynchronous intermittent contractions and relaxations so that at any given moment in time a very large percent of the capillary bed is not perfused. Consequently, total blood flow through quiescent skeletal muscle is low (1.4 to 4.5 ml/minute/100 g). With exercise the resistance vessels relax, and the muscle blood flow may increase manyfold (up to fifteen to twenty times the resting level), the magnitude of the increase depending largely on the severity of the exercise.

■ Regulation of skeletal muscle blood flow

Control of muscle circulation is achieved by neural and local factors; the relative contribution of these factors is dictated by the state of activity of the muscle. At rest neural and myogenic regulations are predominant, whereas in exercise metabolic control supervenes. As with all tissues, physical factors such as arterial pressure, tissue pressure, and blood viscosity influence muscle blood flow. However, another physical factor comes into play during exercise—the squeezing effect of the active muscle on the vessels. With intermittent contractions, inflow is restricted and venous outflow is enhanced during each brief contraction. The presence of the venous valves prevents backflow of blood in the veins between contractions, thereby aiding in the forward propulsion of the blood. With strong sustained contractions the vascular bed can be compressed to the point where blood flow actually ceases temporarily.

Neural factors. Although the resistance vessels of muscle possess a high degree of basal tone, they also display tone attributable to continuous low frequency activity in the sympathetic vasoconstrictor nerve fibers. Evidence for this sympathetic tone is the increase in blood flow observed with local anesthetic block of the sympathetic fibers.

The basal frequency of firing in the sympathetic vasoconstrictor fibers is quite low (about 1 to 2 per second), and maximal vasoconstriction is observed at frequencies as low as 8 to 10 per second. Stimulation of the sympathetic nerve fibers to skeletal muscle elicits vasoconstriction that is caused by the release of norepinephrine at the fiber endings. Intraarterial injection of norepinephrine elicits only vasoconstriction, whereas low doses of epinephrine produce vasodilation in muscle and large doses cause vasoconstriction. The vasodilator effect of epinephrine is believed to be the result of stimulation of β-adrenergic receptors on the resistance vessels or possibly to stimulation of muscle metabolism by epinephrine. However, attempts to identify the mediator of such a metabolic response have not met with success. Evidence that lactic acid serves as the mediator is controversial.

The tonic activity of the sympathetic nerves is greatly influenced by reflexes from the baroreceptors. An increase in carotid sinus pressure results in dilation of the vascular bed of the muscle, whereas a decrease in carotid sinus pressure elicits vasoconstriction (Fig. 35-7). When the existing sympathetic constrictor tone is high, as in the experiment

Fig. 35-7 ■ Evidence for participation of the muscle vascular bed in vasoconstriction and vasodilation mediated by the carotid sinus baroreceptors after common carotid artery occlusion and release. In this preparation the sciatic and femoral nerves constituted the only direct connection between the hind leg muscle mass and the rest of the dog. The muscle was perfused by blood at a constant pressure that was completely independent of the animal's arterial pressure. (Modified from Jones, R.D., and Berne, R.M.: Am. J. Physiol. **204:**461, 1963.)

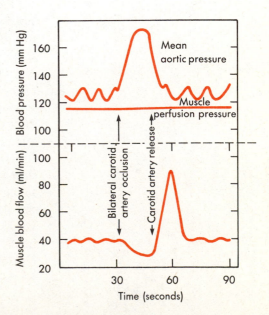

illustrated in Fig. 35-7, the decrease in blood flow associated with common carotid artery occlusion is small, but the increase following the release of occlusion is large.

The vasodilation produced by baroreceptor stimulation is caused only in part by inhibition of sympathetic vasoconstrictor activity. Usually a biphasic dilator response (a decrease in flow followed by an increase) is obtained. A large initial transient dilation is attributed to active vasodilation, whereas a lesser and more sustained effect is the result of release from vasoconstrictor tone. There is some evidence to indicate that the active vasodilation is caused by histamine release from sympathetic fibers innervating the resistance vessels. This dilator effect is abolished by tissue depletion of histamine or by histamine antagonists. Furthermore, histamine release into venous blood during this dilator response has been demonstrated with the aid of isotope-labeled tissue-bound histamine. Because muscle is the major body component on the basis of mass and thereby represents the largest vascular bed, the participation of its resistance vessels in vascular reflexes plays an important role in the maintenance of a constant arterial blood pressure.

Reference has previously been made to the sympathetic cholinergic vasodilator pathway from the cortex and hypothalamus to the muscle resistance vessels (Chapter 32). These nerve fibers are believed to induce vasodilation of the muscle vessels in anticipation of exercise. Whether these nerve fibers in truth serve this function and whether they exist in humans has not been firmly established. There is some evidence that the increased muscle blood flow observed in fainting or with acute emotional upsets is in part mediated by these cholinergic fibers.

Participation of the sympathetic cholinergic vasodilator fibers in the active vasodilation induced by carotid sinus stimulation is minimal because atropine administration reduces only slightly the degree of vasodilation obtained. From studies on the clearance of ^{131}I injected into muscle, it appears that the vasodilation produced by stimulation of sympathetic cholinergic fibers is limited to the *thoroughfare or nonnutrient* channels because increases in blood flow through the muscle are not accompanied by increases in the rate of ^{131}I removal. Hence, activation of the sympathetic dilator nerves permits more blood to flow through open channels, thereby creating a functional shunt of blood through the muscle and bypassing the inactive capillary beds. Other procedures that increase muscle blood flow (e.g., injection of dilator agents, inhibition of sympathetic vasoconstrictor tone, or increased metabolic activity of the muscle) affect nutrient blood flow through capillary beds because they produce proportionate increases in ^{131}I clearance.

Finally, there has recently been described a slowly evoked, sustained, active vasodilation obtained with sympathetic nerve stimulation when the masking effect of the sympathetic vasoconstrictors is pharmacologically blocked. The neurohumor released by these nerve fibers is neither acetylcholine nor histamine. Its identity and the physiological role of these nerve fibers must await further investigation.

A comparison of the vasoconstrictor and vasodilator effects of the sympathetic nerves to blood vessels of muscle and skin is summarized in diagrammatic form in Fig 35-8. Note the lower basal tone of the skin vessels, their greater constrictor response, and the absence of active cutaneous vasodilation.

Local factors. It has already been stressed that neural regulation of muscle blood flow is superseded by metabolic regulation when the muscle changes from the resting to the contracting state. However, local control is also demonstrable in innervated resting skeletal muscle of the dog's hind leg when reflex stimulation of the vasomotor nerves is minimal and local thermal or mechanical stimulation of the muscle is eliminated. Such preparations show autoregulation of blood flow (a prime example of local control) to the same degree as do actively contracting muscle and denervated muscle. If, however, vascular reflexes are activated when the vasomotor nerves to the muscle

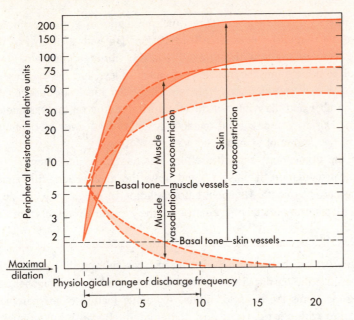

Fig. 35-8 ■ Diagrammatic representation of basal tone and the range of response of the resistance vessels in muscle and skin to sympathetic nerve stimulation. Peripheral resistance plotted on a logarithmic scale. (Redrawn from Celander, O., and Folkow, B.: Acta Physiol. Scand. **29:**241, 1953.)

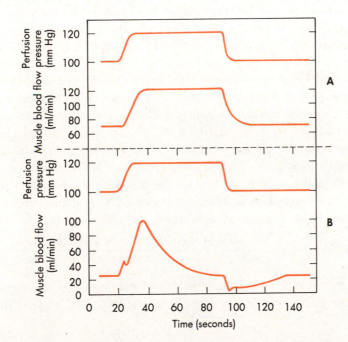

Fig. 35-9 ■ **A,** Absence of autoregulation of blood flow. Skeletal muscle blood flow follows changes in perfusion pressure. **B,** Autoregulation of skeletal muscle blood flow. Blood flow returns to control level despite maintenance of elevated perfusion pressure.

are intact or if the muscle receives mechanical or thermal stimulation (with or without the nerves intact), blood flow increases and autoregulation is either lost or diminished (Fig. 35-9). In resting muscle an autoregulating preparation is characterized by low blood flow and low venous blood oxygen saturation, whereas in contracting muscle venous blood oxygen saturation is low but the flow is high. Thus the common denominator for autoregulation of blood flow is a low venous blood oxygen level and presumably a low muscle oxygen tension. Direct measurements of P_{O_2} in resting skeletal muscle cells yield a range of values from 0 to 10 mm Hg with a mean P_{O_2} of 4 mm Hg. Studies on humans also reveal that at rest blood flow through muscle is low and oxygen saturation of blood draining the muscles is frequently less than 50%. These observations indicate that, at rest, muscle blood flow is under either local or neural control, depending on the number and the nature of the stimuli impinging on the vasomotor regions of the medulla. During exercise local metabolic factors take over blood flow regulation, regardless of the degree of sympathetic nerve activity.

Despite the tremendous scientific effort that has gone into attempts to clarify the nature of local regulation of muscle blood flow, the mechanism still remains obscure. Some of the current hypotheses about autoregulation of blood flow and about the identity of the mediator of metabolic control of the resistance vessels are considered in Chapter 32.

■ *Cerebral circulation*

Blood reaches the brain through the internal carotid and the vertebral arteries. The latter join to form the basilar artery, which, in conjunction with branches of the internal carotid arteries, forms the *circle of Willis*. A unique feature of the cerebral circulation is that it all lies within a rigid structure, the cranium. Because intracranial contents are incompressible, any increase in arterial inflow, as with arteriolar dilation, must be associated with a comparable increase in venous outflow. The volume of both blood and extravascular fluid can show considerable variations in most tissues, whereas in the brain the volume of blood and extravascular fluid is relatively constant; changes in either of these fluid volumes must be accompanied by a reciprocal change in the other. In contrast to most other organs, the rate of total cerebral blood flow is held within a relatively narrow range; in humans it averages 55 ml/minute/100 g of brain.

■ *Estimation of cerebral blood flow*

Total cerebral blood flow is usually measured in humans by the nitrous oxide (N_2O) method, which is based on the Fick principle (Chapter 34). The subject breathes a gas mixture of 15% N_2O, 21% O_2, and 64% N_2 for a period of 10 minutes, which is sufficient time to permit equilibration of the N_2O between the brain tissue and the blood leaving the brain. Simultaneous samples of arterial (any artery) blood and mixed cerebral venous (internal jugular vein) blood are taken at the start of N_2O inhalation and at 1-minute intervals throughout the 10-minute period of N_2O administration. From these data the cerebral blood flow can be calculated by the Fick equation.

$$\text{Cerebral blood flow} = \frac{\text{Amount of } N_2O \text{ taken up by brain during time } (t_2 - t_1)}{\text{A-V difference of } N_2O \text{ across brain during time } (t_2 - t_1)}$$

Because the arterial and venous concentrations are continuously changing with time, it is necessary to get the true A-V difference during the period of N_2O inhalation by

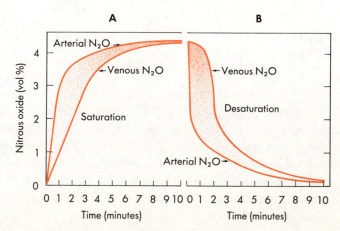

Fig. 35-10 ■ Concentrations of nitrous oxide in arterial and cerebral venous blood during saturation (**A**) and desaturation (**B**). The colored areas represent the arteriovenous differences of nitrous oxide during the 10-minute period of N_2O inhalation and the 10 minutes after discontinuance of the N_2O administration.

integration of the A-V difference over the 10-minute period. This is represented in Fig. 35-10, *A,* by the colored area between the arterial and venous N_2O concentration curves constructed from the blood concentrations observed at successive 1-minute intervals during N_2O administration. The amount of N_2O removed by the brain, as well as the concentration of N_2O in the brain tissue, is unknown. Because the partition coefficient between brain and blood is about 1 and equilibrium of N_2O between the brain and the blood leaving the brain is reached by the end of 10 minutes, the concentration of N_2O in the brain tissue closely approximates that of the cerebral venous blood in the 10-minute sample. The total weight of the brain is not known, and so for convenience the concentration in brain tissue is multiplied by 100 to express the cerebral blood flow (CBF) in ml/minute/100 g of brain tissue. The equation is

$$CBF = \frac{V_{10} \times S \times 100}{\int_0^{10} (A\text{-}V)dt}$$

where

V_{10} = venous concentration of N_2O at equilibrium (at 10 minutes)

S = partition coefficient of N_2O between brain and blood = 1

$A\text{-}V$ = arteriovenous difference of N_2O

Cerebral blood flow can also be calculated from the desaturation A-V curves, which are constructed from N_2O concentrations of simultaneously drawn arterial and venous blood samples taken each minute for 10 minutes, starting when equilibrium is reached between brain tissue and cerebral venous blood (Fig. 35-10, *B*). In this procedure the subject breathes the N_2O mixture for 10 minutes, and sampling starts at the moment N_2O inhalation is stopped. The only difference from the preceding equation is that the denominator becomes

$$\int_{10}^{20} (V\text{-}A)dt$$

The nitrous oxide method is also used for determination of coronary blood flow in humans. The procedure is the same except that the venous blood samples are taken from the coronary sinus through a cardiac catheter. Because the coronary sinus drains the left ventricle, blood flow is expressed as ml/minute/100 g of left ventricle.

Cerebral blood flow and its distribution to different areas of the brain can be measured in animals by injection into the internal carotid artery of microspheres (about 15 μm) labeled with radioactive substances. The microspheres become lodged in the arterioles and capillaries; the brain tissue is sampled, and the radioactivity of the tissue is determined. Blood flow to each tissue sample is proportional to the radioactivity in that sample. By the use of microspheres labeled with different radioactive isotopes, several measurements of cerebral blood flow can be made with a gamma counter that can measure each isotope independently of the other isotopes in the sample. This method is also used for measurements of blood flow in other tissues such as the myocardium. One can also measure cerebral blood flow in animals with the use of ^{14}C-antipyrine, which is taken up by the brain in proportion to the blood flow. The brain is then sliced, and the radioactivity of the slice is determined by radioautography. The advantage of these methods over the N_2O method or the direct measurement of venous outflow from the brain is that blood flow to different regions of the brain *(regional blood flow)* can be determined. The obvious disadvantage is that the animals must be killed to obtain the samples of brain tissue.

Recently the development of multiple collimated scintillation detectors built into a helmet that fits over the cranium has made possible the measurement of regional blood flow (cortical blood flow) in animals and humans. An inert radioactive gas (such as xenon 133) is injected into an internal carotid artery, and from its rate of washout from the brain, regional cerebral blood flow can be determined. The radioactive gas may also be given by inhalation, but more sophisticated techniques are required to eliminate errors introduced by noncerebral blood flow and to distinguish between blood flow to cortical (gray matter) and deep cerebral (white matter) tissue.

■ *Regulation of cerebral blood flow*

Of the various body tissues, the brain is the least tolerant of ischemia. Interruption of cerebral blood flow for as short a time as 5 seconds results in loss of consciousness, and ischemia lasting just a few minutes results in irreversible tissue damage. Fortunately, regulation of the cerebral circulation is primarily under direction of the brain itself. Local regulatory mechanisms and reflexes originating in the brain tend to maintain cerebral circulation relatively constant in the presence of fluctuations in arterial blood pressure. Under certain conditions the brain also regulates its blood flow by initiating changes in systemic blood pressure. For example, elevation of intracranial pressure results in an increase in systemic blood pressure. This response, first described by Cushing, is apparently caused by ischemic stimulation of vasomotor regions of the medulla. It aids in maintaining cerebral blood flow in such conditions as expanding intracranial tumors.

Neural factors. The cerebral vessels receive innervation from the cervical sympathetic nerve fibers that accompany the internal carotid and vertebral arteries into the cranial cavity. Until recently it was generally accepted that stimulation of the sympathetic nerves to the cerebral vessels elicited, at most, minimal vasoconstriction. However, the role of the sympathetic nerves in the regulation of the cerebral circulation has become quite controversial because of reports of severe constriction with either sympathetic nerve stimulation or the application of catecholamines to the pial arterioles and arteries. Despite these new findings of marked cerebral vasoconstriction with sympathetic nerve stimulation, the prevalent belief is that relative to other vascular beds sympathetic control of the cerebral vessels is weak and that the contractile state of the cerebral vascular smooth muscle is primarily dependent on local metabolic factors. There are no known sympathetic vasodilator nerves to the cerebral vessels, but the vessels do receive parasympathetic fibers from the facial nerve, which produce a slight vasodilation on stimulation.

Local factors. Until quite recently it was thought that cerebral blood flow was relatively constant despite wide variations in cortical activity. However, with the advent of methods for measurement of regional cortical blood flow, it became apparent that there is a tight coupling between regional metabolic activity and regional blood flow. For example, movement of one hand results in increased blood flow only in the projected hand area of the contralateral sensory-motor and premotor cortex, whereas stimulation of the retina with flashes of light increases blood flow only in the visual cortex. The reason this parallelism between cerebral metabolism and blood flow went unrecognized is that changes in the regional distribution of cerebral blood flow were masked when only total cerebral blood flow was measured. It has also been recently demonstrated that glucose uptake is closely coupled to regional cortical neuronal activity. For example, when the retina is stimulated by light, uptake of ^{14}C-2-deoxyglucose is enhanced in the visual cortex. This analogue of glucose is taken up and phosphorylated by cerebral neurons but cannot be metabolized further. The magnitude of its uptake is determined from radioautographs of slices of the brain. The mediator of the link between cerebral metabolism and blood flow has not been established, but there are currently three principal candidates: pH, potassium, and adenosine.

It is well known that the cerebral vessels are very sensitive to carbon dioxide tension. Increases in arterial blood carbon dioxide tension (Pa_{CO_2}) elicit marked cerebral vasodilation; inhalation of 7% CO_2 results in a two fold increment in cerebral blood flow. By the same token decreases in Pa_{CO_2}, such as produced by hyperventilation, produce a decrease in cerebral blood flow. Carbon dioxide produces changes in arteriolar resistance by altering perivascular (and probably intracellular vascular smooth muscle) pH. By independently changing P_{CO_2} and bicarbonate concentration, it has been demonstrated that pial vessel diameter (and presumably blood flow) and pH are inversely related, regardless of the level of the P_{CO_2}. Carbon dioxide can diffuse to the vascular smooth muscle from the brain tissue or from the lumen of the vessels, whereas hydrogen ions in the blood are prevented from reaching the arteriolar smooth muscle by the "blood-brain barrier." Hence, the cerebral vessels dilate when the hydrogen ion concentration of the cerebrospinal fluid is increased but show only minimal dilation in response to an increase of the hydrogen ion concentration of the arterial blood. Despite the responsiveness of the cerebral vessels to pH changes, the precise role of hydrogen ions in the regulation of cerebral blood flow remains obscure. The initiation of increases in cerebral blood flow produced by seizures has been reported to be associated with transient increases rather than decreases in perivascular pH. Also the intracellular and extracellular decreases in pH that occur with electrical stimulation of the brain or hypoxia often occur after cerebral blood flow has increased in response to the stimulus. Furthermore, with prolonged hypocapnia, cerebrospinal fluid pH may return to control levels in the face of a persistent reduction in cerebral blood flow.

With respect to potassium ions, such stimuli as hypoxia, electrical stimulation of the brain, and seizures elicit rapid increases in cerebral blood flow and are associated with increases in perivascular K^+. The increments in K^+ are similar to those which produce pial arteriolar dilation when applied topically to these vessels. However, the increase in K^+ may not be sustained throughout the period of stimulation. Hence, only the initial increment in cerebral blood flow may be attributed to the release of K^+.

Adenosine levels of the brain increase with ischemia, hypoxemia, hypotension, hypocapnia, electrical stimulation of the brain, or induced seizures. When it is applied topically, adenosine is a potent dilator of the pial arterioles. In short, any intervention that either reduces the oxygen supply to the brain or increases the oxygen need of the brain results in rapid (within 5 seconds) formation of adenosine in the cerebral tissue. Unlike pH or K^+, the adenosine concentration of the brain increases with the initiation of the stimulus and remains elevated throughout the period of oxygen imbalance. The adenosine released into the cerebrospinal fluid during conditions associated with inadequate brain oxygen supply is available to the brain tissue for reincorporation into cerebral tissue adenine nucleotides.

In all likelihood, all three factors—pH, K^+, and adenosine—act in concert to adjust the cerebral blood flow to the metabolic activity of the brain, but how these factors interact in accomplishing this regulation of cerebral blood flow remains to be elucidated.

The cerebral circulation shows reactive hyperemia and excellent autoregulation between pressures of about 60 and 160 mm Hg. Mean arterial pressures below 60 mm Hg result in reduced cerebral blood flow and syncope, whereas mean pressures above 160 may lead to increased permeability of the blood-brain barrier and cerebral edema. Autoregulation of cerebral blood flow is abolished by hypercapnia or any other potent vasodilator, and none of the candidates for metabolic regulation of cerebral blood flow has been shown to be responsible for this phenomenon. To what extent autoregulation of the cerebral vessels is attributable to a myogenic mechanism has not been established. However, the bulk of evidence favors a metabolic rather than a myogenic mechanism for autoregulation of cerebral blood flow.

■ *Splanchnic circulation*

The splanchnic circulation consists of the blood supply to the gastrointestinal (GI) tract, liver, spleen, and pancreas. There are several features that distinguish the splanchnic circulation, the most noteworthy of which is that there are two large capillary beds partially in series with one another. The splanchnic arterial branches supply the capillary beds in the GI tract, spleen, and pancreas. From these capillary beds the venous blood ultimately flows into the portal vein, which normally provides most of the blood supply to the liver. However, there is also a hepatic arterial blood supply. In the next two subsections, we shall deal almost exclusively with the intestinal and hepatic circulations; the blood supply to the spleen and pancreas will not be discussed.

■ *Intestinal circulation*

Anatomy. The GI tract is supplied by three large vessels: the celiac, superior mesenteric, and inferior mesenteric arteries. The superior mesenteric artery is the largest of all the branches of the aorta and carries over 10% of the cardiac output. Small mesenteric arteries form an extensive vascular network in the submucosa (Fig. 35-11). Branches of these small arteries penetrate the longitudinal and circular muscle layers and give rise to third- and fourth-order arterioles. Some third-order arterioles in the submucosa penetrate to the tip of the villus as its main arteriole.

The direction of the blood flow in the capillaries and venules in a villus is opposite to that in the main arteriole. This arrangement presents the anatomical potential for a countercurrent exchange system. Experiments in certain species suggest that there is an effective countercurrent multiplier in the villus that is involved in the absorption of sodium and water. There is probably also a countercurrent exchange of O_2 from arterioles to venules. At low flow rates especially, a substantial fraction of the O_2 may be shunted from arterioles to venules near the base of the villus, thereby curtailing the supply of O_2 to the mucosal cells at the tip of the villus. When intestinal blood flow is reduced, the shunting of oxygen is exaggerated; this could lead to extensive necrosis of the intestinal villi.

Neural regulation. The neural control of the mesenteric circulation is almost exclusively by the sympathetic nervous system. Increased sympathetic activity elicits a pronounced constriction of the mesenteric arterioles, precapillary sphincters, and capacitance vessels. These responses are mediated by α-receptors, which are prepotent in the mesenteric circulation; however, β-receptors are also present. Infusion of a β-receptor agonist, such as isoproterenol, causes vasodilation.

In spontaneously occurring fighting behavior or in response to artificial stimulation

Fig. 35-11 ■ The distribution of small blood vessels in the rat intestinal wall. *SA*, small artery; *SV*, small vein; *1A, 2A, . . ., 5A*, first-, second-, . . ., fifth-order arterioles; *IV, . . ., 4V*, first-, . . ., fourth-order venules; *CC* and *LC*, capillaries in circular and longitudinal muscle layers; *MA* and *CV*, main arteriole and collecting venule of a villus; *DA*, distributing arteriole; *2VM*, second-order mucosal venule; *PC*, precapillary sphincter; *MC*, mucosal capillary. (From Gore, R.W., and Dohlen, H.G.: Am. J. Physiol. **233**:H685, 1977.)

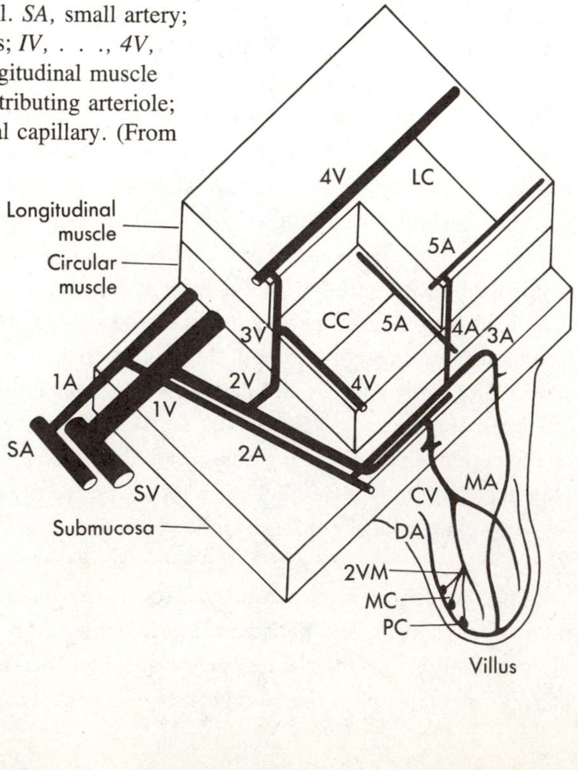

of the hypothalamic "defense" area, pronounced vasoconstriction occurs in the mesenteric vascular bed. This serves to promote a shift of blood flow from the temporarily less important intestinal circulation to the more crucial skeletal muscles, heart, and brain.

Autoregulation. Autoregulation exists in the intestinal circulation, but it is not as well developed as it is in certain other vascular beds such as those in the brain and kidney. However, the oxygen consumption of the small intestine is more rigorously controlled. The oxygen uptake of the small intestine has been found to remain constant when the arterial perfusion pressure was varied over the range of 30 to 125 mm Hg. The principal mechanism responsible for autoregulation appears to be metabolic, although a myogenic mechanism probably also participates. A fourfold rise in the adenosine concentration in the mesenteric venous blood was observed after a brief arterial occlusion. Adenosine is a potent vasodilator in the mesenteric vascular bed and may be the principal metabolic mediator of autoregulation. However, potassium and altered osmolality might also contribute to the overall response.

Functional hyperemia. The ingestion of food leads to an appreciable increase in intestinal blood flow. Several mechanisms contribute to this hyperemia, but the secretion of certain gastrointestinal hormones is probably involved. Gastrin and cholecystokinin have been shown to augment intestinal blood flow, and they are, of course, secreted in response to food ingestion (Chapter 43).

The absorption of food is also closely correlated with the rate of intestinal blood flow. Undigested food has no vasoactive influence, whereas several products of digestion are potent vasodilators. Among the various constituents of chyme, the principal mediators of the mesenteric hyperemia seem to be glucose and long-chain fatty acids.

■ *Hepatic circulation*

Anatomy. The blood flow to the liver normally is about 25% of the cardiac output. It is derived from two sources, the portal vein and the hepatic artery. Ordinarily, about three fourths of the blood flow is delivered by the portal vein. The portal venous blood has already passed through the gastrointestinal capillary bed; therefore much of the O_2 has already been extracted. The remaining one fourth of the blood supply is delivered by the hepatic artery. Ordinarily this blood is fully saturated with O_2. Hence, about three fourths of the O_2 used by the liver is derived from the hepatic arterial blood, although the blood flow rate in the hepatic artery is only about one fourth that in the portal vein.

The small branches of the portal vein and hepatic artery give rise to terminal portal venules and hepatic arterioles (Fig. 35-12). These terminal vessels enter the hepatic acinus (the functional unit of the liver) at its center. Blood flows from these terminal vessels into the sinusoids, which constitute the capillary network of the liver. The sinusoids radiate toward the periphery of the acinus, where they connect with the terminal hepatic venules. Blood from these terminal venules drains into progressively larger branches of the hepatic veins, which are tributaries of the inferior vena cava.

Hemodynamics. The mean blood pressure in the portal vein is about 10 mm Hg, and that in the hepatic artery is about 90 mm Hg. The resistance of the vessels upstream to the hepatic sinusoids is considerably greater than that of the downstream vessels. Consequently, the pressure in the sinusoids is only 2 or 3 mm Hg greater than that in the hepatic veins and inferior vena cava. The ratio of presinusoidal to postsinusoidal resistance in the liver is much greater than is the ratio of precapillary to postcapillary resistance for almost any other vascular bed. Hence, drugs and other interventions that tend to alter the presinusoidal resistance have only a negligible effect on the pressure in the sinusoids; consequently such changes in presinusoidal resistance have very little effect on the fluid exchanges across the sinusoidal wall. Conversely, changes in hepatic venous (and in central venous) pressure are transmitted almost quantitatively to the

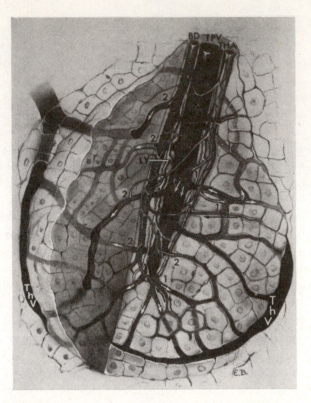

Fig. 35-12 ■ Microcirculation to an hepatic acinus. *THA,* terminal hepatic arteriole; *TPV,* terminal portal venule; *BD,* bile ductule; *THV,* terminal hepatic venule; *LY,* lymphatic. The hepatic arterioles empty either directly *(1)* or via the peribiliary plexus *(2)* into the sinusoids that run from the terminal portal venule to the terminal hepatic venules. (From Rappaport, A.M.: Microvasc. Res. **6:**212, 1973.)

hepatic sinusoids. Such venous pressure changes, therefore, have a profound effect on the transsinusoidal exchange of fluids. When central venous pressure is elevated, as in congestive heart failure, there may be a large transudation of plasma water from the liver into the peritoneal cavity, leading to the development of *ascites*.

Regulation of flow. There is a reciprocal relation between the rates of blood flow in the portal venous and hepatic arterial systems. When blood flow is curtailed in one system, the flow increases in the other system. However, the resultant increase in flow in one system usually does not fully compensate for the initiating reduction in flow in the other system.

The portal venous system does not display any significant autoregulation. As the portal venous pressure and flow are raised, resistance either remains constant or it decreases. The hepatic arterial system does display some capacity to autoregulate. However, the tendency is prominent only in denervated preparations for some unknown reason.

Despite the weak autoregulation of hepatic blood flow, the liver does show a remarkable capacity to maintain a constant O_2 consumption. This is achieved by a very efficient mechanism for extraction of O_2 from the hepatic blood supply. As the rate of O_2 delivery to the liver is varied, the liver compensates by an appropriate change in the fraction of O_2 extracted from each unit volume of blood. This extraction process is facilitated by the distinct separation of the presinusoidal vessels at the acinar center from the postsinusoidal vessels at the periphery of the acinus (Fig. 35-12). There is little opportunity for a countercurrent exchange of O_2, contrary to the condition that exists in an intestinal villus, for example (Fig. 35-11).

The sympathetic nerves are capable of constricting the presinusoidal resistance vessels in the portal venous and hepatic arterial systems. However, neural effects on the capacitance vessels are probably of greater importance. The effects are mediated mainly via α-receptors.

Capacitance vessels. The liver contains about 15% of the total blood volume of the body. Under appropriate conditions, such as in response to hemorrhage, about half of this volume of blood can be rapidly expelled. Hence, the liver constitutes an important blood reservoir. It is probably the most important blood reservoir in humans. In certain other species, such as the dog, the spleen is an important blood reservoir. Smooth muscle in the capsule and trabeculae of the spleen contract in response to increased sympathetic neural activity, such as occurs during exercise or hemorrhage (Chapter 36). However, this mechanism does not exist in humans.

■ *Fetal circulation*

The circulation of the fetus shows a number of differences from that of the postnatal infant. The fetal lungs are functionally inactive, and the fetus is completely dependent on the placenta for oxygen and nutrient supply. Oxygenated fetal blood from the placenta passes through the umbilical vein to the liver. A major fraction passes through the liver, and a small fraction bypasses the liver to the inferior vena cava through the *ductus venosus* (Fig. 35-13). In the inferior vena cava, blood from the ductus venosus joins blood returning from the lower trunk and extremities and this combined stream is in

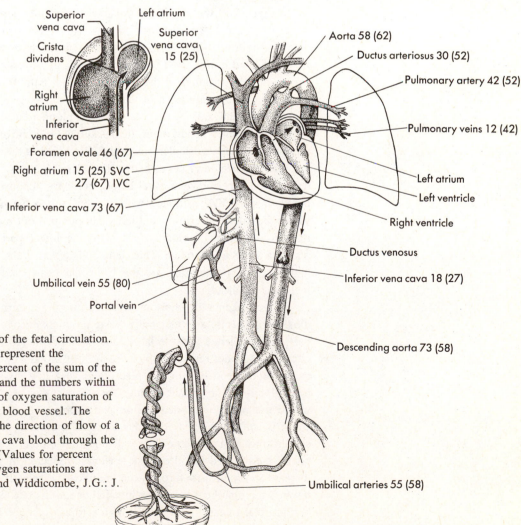

Fig. 35-13 ■ Schematic diagram of the fetal circulation. The numbers without parentheses represent the distribution of cardiac output in percent of the sum of the right and left ventricular outputs, and the numbers within parentheses represent the percent of oxygen saturation of the blood flowing in the indicated blood vessel. The insert at the upper left illustrates the direction of flow of a major portion of the inferior vena cava blood through the foramen ovale to the left atrium. (Values for percent distribution of blood flow and oxygen saturations are from Dawes, G.S., Mott, J.C., and Widdicombe, J.G.: J. Physiol. **126:**563, 1954.)

turn joined by blood from the liver through the hepatic veins. The streams of blood tend to maintain their identity in the inferior vena cava and are divided into two streams of unequal size by the edge of the interatrial septum *(crista dividens)*. The larger stream, which is primarily blood from the umbilical vein, is shunted through the *foramen ovale,* which lies between the inferior vena cava and the left atrium (inset, Fig. 35-13), to the left atrium. The other stream passes into the right atrium, where it is joined by superior vena caval blood returning from the upper parts of the body and blood from the myocardium. In contrast to the adult, in whom the right and left ventricles pump in series, in the fetus the ventricles operate essentially in parallel. Because of the large pulmonary resistance, less than one third of the right ventricular ouptut goes through the lungs. The remainder passes through the *ductus arteriosus* from the pulmonary artery to the aorta at a point distal to the origins of the arteries to the head and upper extremities. Flow from pulmonary artery to aorta occurs because pulmonary artery pressure is about 5 mm Hg higher than aortic pressure in the fetus. The large volume of blood coming through the foramen ovale into the left atrium is joined by blood returning from the lungs and is pumped out by the left ventricle into the aorta. About one third of the aortic blood goes to the head, upper thorax, and arms and the remaining two thirds to the rest of the body and the placenta. The amount of blood pumped by the left ventricle is about 20% greater than that pumped by the right ventricle, and the major fraction of the blood that passes down the descending aorta flows by way of the two umbilical arteries to the placenta.

In Fig. 35-13 the distribution of fetal blood flow is given in percentage of the combined right and left ventricular outputs. Note that over half of the combined cardiac output is returned directly to the placenta without passing through any capillary bed. Also indicated in Fig. 35-13 are the oxygen saturations of the blood (numbers in parentheses) at various points of the fetal circulation. Fetal blood leaving the placenta is 80% saturated, but the saturation of the blood passing through the foramen ovale is reduced to 67% by some mixing with desaturated blood returning from the lower part of the body and the liver. Addition of the desaturated blood from the lungs reduces the oxygen saturation of left ventricular blood to 62%, which is the level of saturation of the blood reaching the head and upper extremities. The blood in the right ventricle, a mixture of desaturated superior vena caval blood, coronary venous blood, and inferior vena caval blood, is only 52% saturated with oxygen. When the major portion of this blood traverses the ductus arteriosus and joins that pumped out by the left ventricle, the resultant oxygen saturation of blood traveling to the lower part of the body and back to the placenta is 58% saturated. Thus it is apparent that the tissues receiving blood of the highest oxygen saturation are the liver, the heart, and the upper parts of the body, including the head.

At the placenta the chorionic villi dip into the maternal sinuses, and oxygen, carbon dioxide, nutrients, and metabolic waste products exchange across the membranes. The barrier to exchange is quite large, and equilibrium of oxygen tension between the two circulations is not reached at normal rates of blood flow. Therefore the oxygen tension of the fetal blood leaving the placenta is very low. Were it not for the fact that fetal hemoglobin has a greater affinity for oxygen than does adult hemoglobin, the fetus would not receive an adequate oxygen supply. The fetal oxyhemoglobin dissociation curve is shifted to the left so that at equal pressures of oxygen fetal blood will carry significantly more oxygen than will maternal blood. If the mother is subjected to hypoxia, the reduced blood oxygen tension is reflected in the fetus by tachycardia and an increase in blood flow through the umbilical vessels. If the hypoxia persists or if flow through the umbilical vessels is impaired, fetal distress occurs and is first manifested as bradycardia. In early fetal life the high cardiac glycogen levels that prevail (which gradually decrease to adult levels by term) may protect the heart from acute periods of hypoxia.

The umbilical vessels have thick muscular walls that are very reactive to trauma, tension, sympathomimetic amines, bradykinin, angiotensin, and changes in oxygen tension. In animals in which the umbilical cord is not tied, hemorrhage of the newborn is prevented by constriction of these large vessels in response to one or more of these stimuli. Closure of the umbilical vessels produces an increase in total peripheral resistance and of blood pressure. When blood flow through the umbilical vein ceases, the ductus venosus, a thick-walled vessel with a muscular sphincter, closes. What initiates closure of the ductus venosus is still unknown. The asphyxia, which starts with constriction or clamping of the umbilical vessels, and the cooling of the body activate the respiratory center. With the filling of the lungs with air, pulmonary vascular resistance decreases to about one tenth of the value existing before lung expansion. This resistance change is not caused by the presence of oxygen in the lungs because the change is just as great if the lungs are filled with nitrogen. However, filling the lungs with liquid does not reduce pulmonary vascular resistance.

The left atrial pressure is raised above that in the inferior vena cava and right atrium by (1) the decrease in pulmonary resistance, with the resulting large flow of blood through the lungs to the left atrium, (2) the reduction of flow to the right atrium caused by occlusion of the umbilical vein, and (3) the increased resistance to left ventricular output produced by occlusion of the umbilical arteries. This reversal of the pressure gradient across the atria abruptly closes the valve over the foramen ovale, and fusion of the septal leaflets occurs over a period of several days.

With the decrease in pulmonary vascular resistance, the pressure in the pulmonary artery falls to about half its previous level (to about 35 mm Hg), and this change in pressure, coupled with a slight increase in aortic pressure, reverses the flow of blood through the ductus arteriosus. However, within several minutes the large ductus arteriosus beings to constrict, and this constriction produces turbulent flow, which is manifest as a murmur in the newborn. Constriction of the ductus arteriosus is progressive and usually is complete within 1 to 2 days after birth. Closure of the ductus arteriosus appears to be initiated by the high oxygen tension of the arterial blood passing through it, since pulmonary ventilation with oxygen or with air low in oxygen induces, respectively, closure and opening of this shunt vessel. Whether oxygen acts directly on the ductus or through the release of a vasoconstrictor substance is not known. Similarly, in a heart-lung preparation made from a newborn lamb, the ductus arteriosus may be made to close with high Pao_2 and to open with low Pao_2. However, recent studies on fetal lambs suggest that bradykinin formed from kininogen in the lungs when they fill with air contributes to the closure of the ductus arteriosus. The bradykinin may act directly or by inducing a release of catecholamines from the adrenal medulla. In addition, the prostaglandins may play a role in the closure of the ductus arteriosus. Permanent closure usually takes several weeks.

At birth the walls of the two ventricles are approximately of the same thickness, with possibly slight preponderance of the right ventricle. There is also present in the newborn thickening of the muscular layers of the pulmonary arterioles, which is apparently responsible in part for the high pulmonary vascular resistance of the fetus. After birth the thickness of the walls of the right ventricle diminishes, as does the muscle layer of the pulmonary arterioles, whereas the left ventricular walls increase in thickness. These changes are progressive over a period of weeks after birth.

Failure of the foramen ovale or ductus arteriosus to close after birth is occasionally observed and constitutes some of the more common congenital cardiac abnormalities that are now amenable to surgical correction.

■ *Circulatory changes that occur at birth*

■ *Bibliography*

Journal articles

Belloni, F.L.: The local control of coronary blood flow, Cardiovasc. Res. **13**:63, 1979.

Bergofsky, E.H.: Mechanisms underlying vasomotor regulation of regional pulmonary blood flow in normal and disease states, Am. J. Med. **57**:378, 1974.

Berne, R.M.: Regulation of coronary blood flow, Physiol. Rev. **44**:1, 1964.

Berne, R.M.: The role of adenosine in the regulation of coronary blood flow, Circ. Res. **47**:807, 1980.

Berne, R.M., Winn, H.R., and Rubio, R.: The local regulation of cerebral blood flow, Prog. Cardiovasc. Dis. **24**:243, 1981.

Carneiro, J.J., and Donald, D.E.: Blood reservoir function of dog spleen, liver, and intestine, Am. J. Physiol. **1**:H67, 1977.

Feigl, E.O.: Sympathetic control of coronary circulation, Circ. Res. **20**:262, 1967.

Feigl, E.O.: Parasympathetic control of coronary blood flow in dogs, Circ. Res. **25**:509, 1969.

Fox, R.H., and Hilton, S.M.: Bradykinin formation in human skin as a factor in heat vasodilatation, J.. Physiol. (London) **142**:219, 1958.

Granger, D.N., and Kvietys, P.R.: The splanchnic circulation; intrinsic regulation, Ann Rev. Physiol. **43**:409, 1981.

Granger, D.N., et al.: Intestinal blood flow, Gastroenterology **78**:837, 1980.

Greenway, C.V., and Stark, R.D.: Hepatic vascular bed, Physiol. Rev. **51**:23, 1971.

Gregg, D.E.: The natural history of coronary collateral development, Circ. Res. **35**:335, 1974.

Heymann, M.A., Iwamoto, H.S., and Rudolf, A.M.: Factors affecting changes in the neonatal systemic circulation, Ann. Rev. Physiol. **43**:371, 1981.

Heymann, M.A., and Rudolph, A.M.: Control of the ductus arteriosus, Physiol. Rev. **55**:62, 1975.

Jones, R.D., and Berne, R.M.: Intrinsic regulation of skeletal muscle blood flow, Circ. Res. **14**:126, 1964.

Kontos, H.A.: Regulation of the cerebral circulation, Ann. Rev. Physiol. **43**:397, 1981.

Lanciault, G., and Jacobsen, E.D.: Gastrointestinal circulation, Gastroenterology **71**:851, 1976.

Lautt, W.W.: Hepatic vasculature: a conceptual review, Gastroenterology **73**:1163, 1977.

Lautt, W.W.: Hepatic nerves: a review of their functions and effects, Canad. J. Physiol. Pharmacol. **58**:105, 1980.

Murray, P.A., Belloni, F.L., and Sparks, H.V.: The role of potassium in the metabolic control of coronary vascular resistance in the dog. Circ. Res. **44**:767, 1979.

Olsson, R.A.: Local factors regulating cardiac and skeletal muscle blood flow, Ann. Rev. Physiol. **43**:385, 1981.

Rappaport, A.M.: The microcirculatory hepatic unit, Microvasc. Res. **6**:212, 1973.

Rappaport, A.M., and Schneiderman, J.H.: The function of the hepatic artery, Rev. Physiol. Biochem. Pharmacol. **76**:129, 1976.

Rubio, R., and Berne, R.M.: Regulation of coronary blood flow, Progr. Cardiovasc. Dis. **18**:105, 1975.

Wearn, J.T., et al.: The nature of the vascular communications between the coronary arteries and the chambers of the heart, Am. Heart J. **9**:143, 1933.

Books and monographs

Berne, R.M.: The coronary circulation. In Langer, G.A., and Brady, A.J., editors: The mammalian myocardium, New York, 1974, John Wiley & Sons, Inc.

Berne, R.M., and Rubio R.: Coronary circulation. In Handbook of physiology; Section 2: The cardiovascular system—the heart, vol. I, Bethesda, Md., 1979, American Physiological Society.

Dawes, G.S.: Foetal and neonatal physiology, Chicago, 1968, Year Book Medical Publishers, Inc.

Lewis, T.: Blood vessels of the human skin and their responses, London, 1927, Shaw & Son, Ltd.

Longo, L.D., and Reneau, D.D., editors: Fetal and newborn cardiovascular physiology; vol. 1, Developmental aspects, New York, 1978, Garland STPM Press.

Morgan, H.E., Rannels, D.E., and McKee, E.E.: Protein metabolism of the heart. In Handbook of physiology; Section 2: The cardiovascular system—the heart, vol. I, Bethesda, Md., 1979, American Physiological Society.

Randle, P.J., and Tubbs, P.K.: Carbohydrate and fatty acid metabolism. In Handbook of physiology; Section 2: The cardiovascular system—the heart, vol. I, Bethesda, Md., 1979, American Physiological Society.

Schaper, W.: The collateral circulation of the heart, New York, 1971, North-Holland Publishing Co.

chapter 36

Interplay of central and peripheral factors in the control of the circulation

The primary function of the circulatory system is to deliver the supplies needed for tissue metabolism and growth and to remove the products of metabolism. To explain how the heart and blood vessels serve this function, it has been necessary to analyze the system morphologically and functionally and to discuss the mechanisms of action of the component parts in their contribution to the maintenance of adequate tissue perfusion under different physiological conditions. Once the functions of the various components are understood, it is essential that their interrelationships in the overall role of the circulatory system be considered. Tissue perfusion is dependent on arterial pressure and local vascular resistance, and arterial pressure in turn is dependent on cardiac output and total peripheral resistance. Arterial pressure is maintained within a relatively narrow range in the normal person, a feat that is accomplished by reciprocal changes in cardiac output and total peripheral resistance. However, cardiac output and peripheral resistance are each influenced by a number of factors, and it is the interplay among these factors that determines the level of these two variables. The autonomic nervous system and the baroreceptors play the key role in the regulation of blood pressure. However, from the long-range point of view the control of fluid balance by the kidney, by the adrenal cortex, and by the central nervous system, with maintenance of a constant blood volume, is of the greatest importance.

In a well-regulated system one way to study the extent and the sensitivity of the regulatory mechanism is to disturb the system and observe its response to restore the preexisting equilibrium state. With respect to the circulatory system, disturbances in the form of physical exercise and hemorrhage will be used to illustrate the effects of the various factors that go into its regulation.

■ *Exercise*

Over the years ideas about the cardiovascular changes that take place in exercise have undergone considerable revision, and the adjustments that were once explained in a simple, seemingly logical sequence are now known to be incorrect. It was formerly thought that the chain of events was initiated in the active muscles, where vasodilation led to a decrease in blood pressure, which activated the baroreceptors to induce an increase in heart rate and myocardial contractility and an enhancement of peripheral vasoconstriction in the inactive tissues. Evidence to substantiate the baroreceptor mechanism is lacking; in fact, the baroreceptor reflex is diminished in exercise. Furthermore, attempts to demonstrate that either peripheral, neural, or humoral input alone triggers the increase in cardiac output have not been supported by experimental data. With regard to humoral stimuli, receptors that are known to respond to reduced P_{O_2} and increased P_{CO_2} are all located on the arterial side of the circulation and the Pa_{O_2} and Pa_{CO_2} are within normal limits during exercise. Furthermore, in exercise experiments in which venous blood from the active muscles was prevented from returning to the central veins, subjects still displayed an increase in heart rate and cardiac output, indicating the lack of need of a humoral stimulus to initiate the cardiovascular adjustments in exercise. With respect to the role of the nervous system, it has been observed that reflexes originating in the contracting muscles result in activation of the sympathetic nerves to the heart and peripheral blood vessels. Hence, neural input is in large part responsible for the cardiovascular changes observed with exercise. The earliest circulatory adjustments, particularly those occurring before the onset of exercise, apparently take origin from the cerebral cortex.

■ Preparation for exercise

In humans or in trained animals, anticipation of and preparation for physical activity evoke primarily inhibition of vagal nerve activity to the heart and generalized sympathetic discharge. The concerted effects of inhibition of parasympathetic and activation of sympathetic areas of the medulla on the heart are an increase in heart rate and in myocardial contractility (a greater ventricular emptying at the same or at reduced ventricular filling pressure), which leads to an increase in cardiac output. This would be only transient were it not for the fact that the enhanced output of the pump also increases blood flow back to the heart *(vis a tergo)*.

At the same time that cardiac stimulation occurs, the sympathetic nervous system also elicits vascular resistance changes in the periphery. In cats and dogs and possibly in humans, the cholinergic sympathetic vasodilator system is activated and produces dilation of the larger resistance vessels (arterioles and metarterioles) in the muscles. In the skin, kidneys, and splanchnic regions, the sympathetic vasoconstrictor fibers increase vascular resistance, which in effect diverts blood away from these areas. The increased renal and splanchnic vascular resistance persists throughout the period of exercise. The renal vasoconstriction is probably an autoregulatory response to the rise in blood pressure because it occurs even in the denervated kidney. Blood flow to the renal and splanchnic regions remains at rest levels during exercise. Only with impaired cardiac function, as in heart failure, does the blood flow to the renal and splanchnic regions decrease with exercise. Under conditions of poor cardiac performance, this reduction in blood flow is caused by intense sympathetically mediated vasoconstriction.

■ Moderate exercise

Local factors in muscle. With the onset of exercise, several other circulatory adjustments come into play; the major one involves the vasculature of the active muscles. The local formation of vasoactive metabolites induces marked dilation of the resistance vessels, which progresses rapidly to maximal degrees at moderate levels of exercise. Potassium is one of the vasodilator substances released by contracting muscle, and it is at least in part responsible for the initial decrease in vascular resistance in the active muscles. Other contributing factors may be an increase in interstitial fluid osmolarity during the initial phase of exercise and the release of adenosine during sustained exercise. Afferent impulses from receptors in the muscle travel to the medulla oblongata, and an increase in sympathetic nerve activity ensues. The result is an increase in heart rate and myocardial contractility. Peripheral resistance increases in nonactive tissues. The local accumulation of metabolites induces relaxation of the terminal arterioles. Hence, blood flow through the muscle may increase to fifteen to twenty times the resting level, although the increment in blood volume in the muscle increases only about 50%. This metabolic vasodilation of the precapillary vessels in the active muscles occurs very soon after the onset of exercise, and the decrease in total peripheral resistance enables the heart to pump more blood at a lesser load and more efficiently (less pressure work, Chapter 35) than if peripheral resistance were unchanged. Because only a small percent of the capillaries are perfused at rest, whereas in exercise all or nearly all of the capillaries contain flowing blood *(capillary recruitment)*, the surface available for exchange

*Interplay of factors in
the control of the
circulation*

625

of gases, water, and solutes is increased many times in exercise. Furthermore, the hydrostatic pressure in the capillaries is increased by virtue of the relaxation of the resistance vessels, and there is a net movement of water and solutes into the muscle tissue. Tissue pressure rises and remains elevated during exercise as fluid continues to move out of the capillaries and is carried away by the lymphatics. Lymph flow is increased as a result of the increase in capillary hydrostatic pressure and the massaging effect of the contracting muscles on the valve-containing lymphatic vessels.

The contracting muscle avidly extracts needed oxygen from the perfusing blood (increased arteriovenous O_2 difference, Fig. 36-1), and the venous blood leaving the active muscles has a low oxygen content (about 5 vol%). The removal of oxygen is facilitated by the nature of oxyhemoglobin dissociation (Chapter 39). The reduction in pH caused by the high concentration of CO_2 and the formation of lactic acid and the increase in temperature in the contracting muscle contribute to shifting the oxyhemoglobin dissociation curve to the right. Hence at any given partial pressure of oxygen, less oxygen is held by the hemoglobin in the red cells; consequently there is a more effective oxygen removal from the blood. Oxygen consumption may increase as much as sixtyfold with only a fifteenfold increase in muscle blood flow. Some investigators believe that the muscle myoglobin may serve as an important oxygen store in exercise. Myoglobin will release attached oxygen only at very low partial pressures, which are probably reached in actively contracting muscle. However, relative to the rate of oxygen consumption of the active muscle, the amount of oxygen bound to myoglobin is negligible.

Arterial pressure. If the exercise involves a large proportion of the body musculature, such as in running or swimming, the reduction in total vascular resistance can be considerable (Fig. 36-1). Nevertheless arterial pressure starts to rise with the onset of exercise, and the increase in blood pressure roughly parallels the severity of the exercise performed (Fig. 36-1). Therefore the increase in cardiac output is proportionally greater than the decrease in total peripheral resistance. The vasoconstriction produced in the inactive tissues by the sympathetic nervous system (and to some extent by the release of catecholamines from the adrenal medulla) is important for the maintenance of normal or increased blood pressure because sympathectomy or drug-induced block of the adrenergic sympathetic nerve fibers results in a decrease in arterial pressure (*hypotension*) during exercise. In general, systolic pressure increases more than diastolic pressure, resulting in an increase in pulse pressure (Fig. 36-1). The latter may be in part attributable to a greater stroke volume and in part to a more rapid ejection of blood by the left ventricle, with less peripheral runoff during the brief ventricular ejection period.

As body temperature rises, the skin vessels dilate in response to thermal stimulation of the heat-regulating center in the hypothalamus, and total peripheral resistance decreases further. This would result in a decline in blood pressure were it not for the increasing cardiac output and constriction of arterioles in the renal, splanchnic, and other inactive tissues. Coronary blood flow increases as heart work increases, but flow to the brain remains constant throughout the period of exercise.

■ *Heart rate and stroke volume—cardiac output*

The enhanced sympathetic drive and the reduced parasympathetic inhibition of the sinoatrial node continue during exercise, and consequently tachycardia persists. If the work load is moderate and constant, the heart rate will reach a certain level and remain there throughout the period of exercise. However, if the work load increases, a concomitant increase in heart rate occurs until a plateau is reached in severe exercise at about 180 beats per minute (Fig. 36-1). In contrast to the large increment in heart rate, the increase in stroke volume is only about 10% to 35% (Fig. 36-1), the larger values occurring in trained persons. (In very well-trained distance runners, whose cardiac outputs can reach six to seven times the resting level, stroke volume reaches about twice

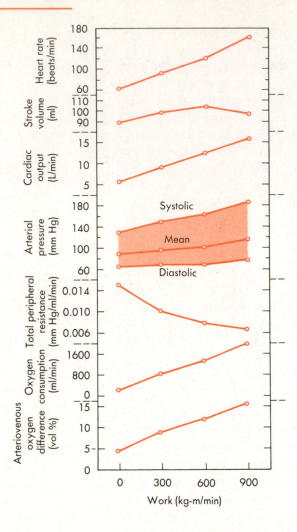

Fig. 36-1 ■ Effect of different levels of exercise on several cardiovascular variables. (Data from Carlsten, A., and Grimby, G.: The circulatory response to muscular exercise in man, Springfield, Ill., 1966, Charles C Thomas, Publisher.)

the resting value.) Thus it is evident that the increase in cardiac output observed with exercise is accomplished principally by an increase in heart rate. If the baroreceptors are denervated, the cardiac output and heart rate responses to exercise are sluggish when compared to the changes in animals with normally innervated baroreceptors. However, in the absence of autonomic innervation of the heart, as produced experimentally in dogs by total cardiac denervation, exercise still elicits an increment in cardiac output comparable to that observed in normal animals, but chiefly by means of an elevated stroke volume. However, if a β-adrenergic receptor blocking agent is given to dogs with denervated hearts, exercise performance is impaired. The β-adrenergic receptor blocker apparently prevents the cardiac acceleration and enhanced contractility caused by increased amounts of circulating catecholamines and hence limits the increase in cardiac output necessary for maximal exercise performance.

Auxiliary pumps. In addition to the contribution made by sympathetically mediated constriction of the capacitance vessels in both exercising and nonexercising parts of the body, an auxiliary pumping action is provided by the working skeletal muscles and the muscles of respiration. As pointed out in Chapter 34, the intermittently contracting muscles compress the vessels that course through them and, in the case of veins with their valves oriented toward the heart, pump blood back toward the right atrium. The flow of venous blood to the heart is also aided by the increase in the pressure gradient developed by the more negative intrathoracic pressure produced by deeper and more frequent inspirations. In humans, with the exception of the skin, lungs, and liver, there is little evidence that blood reservoirs contribute much to the circulating blood volume.

*Interplay of factors in
the control of the
circulation*

627

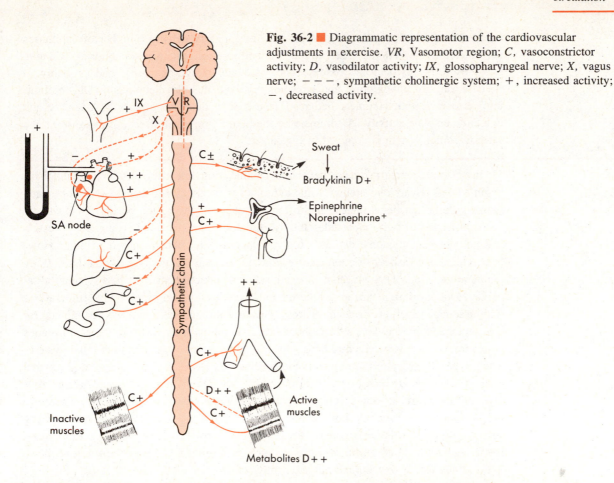

Fig. 36-2 ■ Diagrammatic representation of the cardiovascular adjustments in exercise. *VR*, Vasomotor region; *C*, vasoconstrictor activity; *D*, vasodilator activity; *IX*, glossopharyngeal nerve; *X*, vagus nerve; − − −, sympathetic cholinergic system; +, increased activity; −, decreased activity.

In fact, blood volume is usually slightly reduced in exercise, as evidenced by a rise in the hematocrit ratio, because of loss of water externally by perspiration and enhanced ventilation, and internally into the contracting muscle. The fluid loss into contracting muscle reaches a plateau as interstitial fluid pressure rises and opposes the increased hydrostatic pressure in the capillaries of the active muscle. The fluid loss is partially offset by movement of fluid from the splanchnic regions and inactive muscle into the bloodstream. This influx of fluid occurs as a result of a decrease of hydrostatic pressure in the capillaries of these tissues and an increase in plasma osmolarity caused by movement of osmotically active particles into the blood from the contracting muscle. In addition, reduced urine formation by the kidneys helps to conserve body water.

The large volume of blood returning to the heart is so rapidly pumped through the lungs and out into the aorta that central venous pressure remains essentially constant. Thus the Frank-Starling mechanism of a greater initial fiber length does not account for the greater stroke volume in moderate exercise. Chest x-ray films of persons at rest and during exercise reveal a decrease in heart size in exercise, a finding contrary to old beliefs but in harmony with the observations of a constant ventricular filling pressure. However, in maximal or near maximal exercise, right atrial pressure and end-diastolic ventricular volume do increase. Thus the Frank-Starling mechanism contributes to the enhanced stroke volume in very vigorous exercise. In trained persons the resting heart rate is slow, and stroke volume is large. With exercise the cardiac output of the trained person can reach much greater values than can that of the untrained person as a result of greater increments in stroke volume. Heart rates will reach similar peak values with maximal exercise stress in trained and untrained subjects. Hence, there is greater oxygen delivery to the active muscles and a greater oxygen consumption in the trained person. In addition to an enhanced maximal oxygen consumption, physical conditioning is associated in skeletal muscle with an increase in capillary density (number of capillaries per unit of cross-sectional area), in the concentration of certain mitochondrial oxidative enzymes (e.g., cytochrome oxidase and succinate dehydrogenase), and probably in ATPase activity, myoglobin, and enzymes involved in lipid metabolism. Deconditioning, as occurs with complete bed rest, results in a decrease in capillary density and oxidative enzyme activity. A summary of the neural and local effects of exercise on the cardiovascular system is depicted in Fig. 36-2.

■ *Severe exercise*

In severe exercise taken to the point of exhaustion, the compensatory mechanisms begin to fail. Heart rate attains a maximum level of about 180 beats per minute, and stroke volume reaches a plateau and often decreases. Dehydration occurs. Sympathetic vasoconstrictor activity supersedes the vasodilator influence on the cutaneous vessels and has the hemodynamic effect of a slight increase in effective blood volume. However, cutaneous vasoconstriction also results in a decrease in the rate of heat loss. Body temperature is normally elevated in exercise, and reduction of heat loss through cutaneous vasoconstriction can, under these conditions, lead to very high body temperatures with associated feelings of acute distress. Because the arterial blood oxygen concentration remains at normal levels throughout exercise, the limiting factor in the performance of severe exercise is believed to be the functional capacity of the cardiovascular system. The oxygen delivery to the tissues is the cardiac output times the arterial blood oxygen content. The venous blood oxygen reaches very low levels in active muscle (large A-V O_2), and the heart rate tends to reach a plateau at about 180 beats per minute. Therefore the limiting factor in whole body exercise is the extent to which the heart can increase stroke volume and thereby increase oxygen delivery to the active muscles. In exercise involving only a small group of muscles, such as in finger exercise, the limiting factor is the tissue use of oxygen and not the supply of oxygen.

The tissue and blood pH decrease as a result of increased lactic acid and CO_2 production, and the reduced pH is probably the key factor in determining the maximal amount of exercise a given individual can tolerate because of muscle pain, subjective feeling of exhaustion, and inability or loss of will to continue.

■ *Postexercise recovery*

When exercise stops, there is an abrupt decrease in heart rate and cardiac output; the sympathetic drive to the heart is essentially removed. In contrast, total peripheral resistance remains low for some time after the exercise is ended, presumably because of the accumulation of vasodilator metabolites in the muscles during the exercise period. As a result of the reduced cardiac output and persistence of vasodilation in the muscles, arterial pressure falls, often below pre-exercise levels, for brief periods. Blood pressure then stabilizes at normal levels as a result of activation of the baroreceptor reflexes.

■ *Hemorrhage*

In a person who has lost a large quantity of blood, the principal findings, as might be anticipated, are related to the cardiovascular system. The arterial systolic, diastolic, and pulse pressures are diminished, and the pulse is rapid and feeble. The cutaneous veins are collapsed and fill slowly when compressed centrally. The skin is pale, moist, and slightly cyanotic. Respiration is rapid, but the depth of respiration may be shallow or deep.

■ *Course of arterial blood
pressure changes*

Cardiac output decreases as a result of blood loss for the reasons described in Chapter 34. This reduction in cardiac output produces the characteristic fall in arterial blood pressure. The changes in mean arterial pressure that ensue after an acute hemorrhage in experimental animals are illustrated in Fig. 36-3. If sufficient blood is withdrawn rapidly to bring mean arterial pressure to 50 mm Hg, it is found that in almost all animals there is a tendency for the pressure to rise spontaneously toward control levels over the sub-

*Interplay of factors in
the control of the
circulation*

629

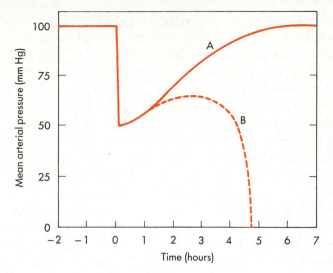

Fig. 36-3 ■ The changes in mean arterial pressure after a rapid hemorrhage. At time zero, the animal is bled rapidly to a mean arterial pressure of 50 mm Hg. After a period in which there is a return of pressure toward the control level, some animals will continue to improve until the control pressure is attained (curve *A*). However, in other animals the pressure will begin to decline until death ensues (curve *B*).

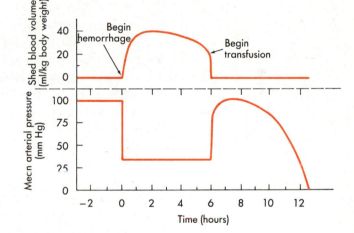

Fig. 36-4 ■ Changes in shed blood volume during a 6-hour period of hemorrhage sufficient to hold mean arterial pressure at 35 mm Hg and the changes in mean arterial pressure after transfusion of the shed blood.

sequent 20 or 30 minutes. In some animals (curve *A*, Fig. 36-3) this trend continues with the result that normal pressures are regained within a few hours. Conversely, in other animals (curve *B*) after the initial rise of pressure above 50 mm Hg to some peak value (usually significantly below the control level), the arterial pressure begins to decline and continues to fall at an accelerating rate until death ensues.

In other experimental studies of the response to hemorrhage, animals are bled to a given hypotensive level, for example, to 35 mm Hg, and then held at that level for a fixed time. This is usually accomplished by connecting a peripheral artery to a reservoir elevated to an appropriate height above the animal; the volume of blood in the reservoir is continuously monitored. The results of such a procedure are shown in Fig. 36-4. The arterial blood runs rapidly into the reservoir until the pressures become equilibrated and then continues to flow into the reservoir at a progressively slower rate for about 2 hours. This gradual increase in shed blood volume is a manifestation of the same compensatory mechanisms that produced the tendency for the arterial blood pressure to rise back toward normal after hemorrhage in the experiment depicted in Fig. 36-3. However, under the experimental conditions portrayed in Fig. 36-4, as the arterial pressure tends to rise to a level higher than the hydrostatic pressure in the blood reservoir, blood flows from the cannulated vessel into the reservoir.

After the peak shed volume is attained about 2 hours after the beginning of hemorrhage, blood begins to flow in the opposite direction—from the reservoir to the animal. This is a manifestation of the tendency for arterial pressure to diminish below the established level, thereby resulting in an uptake of blood from the reservoir. It has been found that once 40% to 50% of the maximum shed volume has spontaneously returned to the animal, rapid reinfusion of the remainder of the shed blood results only in a transient improvement in arterial pressure. There follows an accelerating decline of arterial pressure, eventuating in death (Fig. 36-4). This progressive deterioration of cardiovascular function is termed *shock*. At some point the deterioration appears to become irreversible; that is, a lethal outcome can be retarded only temporarily by any known therapy, including massive transfusions of donor blood.

■ *Compensatory mechanisms*

The changes in arterial pressure immediately after an acute blood loss (Fig. 36-3) and in shed blood volume during the initial stages of sustained hemorrhage (Fig. 36-4) indicate that certain compensatory mechanisms that combat the effects of blood loss must be operating. Any mechanism that tends to return the arterial pressure toward normal in response to the reduction in pressure produced by hemorrhage may be designated a *negative feedback mechanism*. It is termed "negative" because the secondary change in pressure is opposite in direction to the initiating change. In response to hemorrhage the following negative feedback mechanisms are evoked: (1) the baroreceptor reflexes, (2) the chemoreceptor reflexes, (3) cerebral ischemia responses, (4) reabsorption of tissue fluids, (5) release of endogenous vasoconstrictor substances, and (6) renal conservation of salt and water.

Baroreceptor reflexes. The reduction in mean arterial pressure and in pulse pressure during hemorrhage results in diminished stimulation of the baroreceptors located in the carotid sinuses and aortic arch. As a consequence multiple cardiovascular responses are evoked, all of which tend to return the arterial pressure toward its normal level. Reduction of vagal tone and enhancement of sympathetic tone result in tachycardia and a positive inotropic effect on the atrial and ventricular myocardium. The increased sympathetic discharge also produces generalized venoconstriction, which has the same hemodynamic consequences as a transfusion of blood (Chapter 34). Venoconstriction is probably not elicited exclusively by the baroreceptor reflexes because experimental evidence indicates that these reflexes exert a rather feeble influence on venomotor tone. Contraction of certain blood reservoirs in response to sympathetic activation provides an autotransfusion of blood into the circulating bloodstream. In the dog considerable quantities of blood are mobilized by contraction of the spleen. In humans the spleen does not have the same relative importance as a blood reservoir. Instead, the cutaneous, pulmonary, and hepatic vasculatures probably constitute the principal blood reservoir sites.

Generalized arteriolar vasoconstriction is a prominent component of the response to the diminished baroreceptor stimulation during hemorrhage. The reflex increase in total peripheral resistance minimizes the extent of the fall in arterial pressure resulting from the reduction of cardiac output. Fig. 36-5 shows the changes in mean aortic pressure in a group of dogs in response to an 8% blood loss. With both vagi cut and only the carotid sinus baroreceptors operative (left panel), this degree of hemorrhage produced a 14% reduction in mean aortic pressure. This did not differ significantly from the pressure reduction (12%) that occurred with all baroreceptor reflexes intact. When the carotid sinuses were denervated and the aortic baroreceptor reflexes were intact, the same percentage reduction in blood volume caused mean aortic pressure to decline by 38% (middle panel). Hence, it is apparent that the carotid sinus baroreceptors are more efficacious than the aortic baroreceptors in attenuating the fall in pressure. The aortic baroreceptors did play some role, however, because when both sets of afferent baroreceptor pathways were interrupted, an 8% blood loss produced a 48% decline in arterial pressure.

Although the arteriolar vasoconstriction is widespread during hemorrhage, it is by no means uniform in intensity. Vasoconstriction is most severe in the cutaneous, skeletal muscle, and splanchnic vascular beds and is slight or absent in the cerebral and coronary circulations. In many instances the vascular resistance of the coronary circulation is

*Interplay of factors in
the control of the
circulation*

631

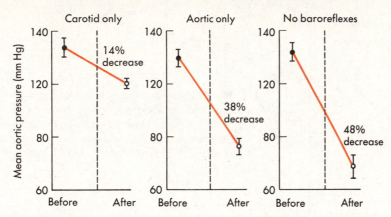

Fig. 36-5 ■ The changes in mean aortic pressure in response to an 8% blood loss in a group of eight dogs. In the left panel the carotid sinus baroreceptor reflexes were intact and the aortic reflexes were interrupted; in the middle panel the aortic reflexes were intact, and the carotid sinus reflexes were interrupted; in the right panel all sinoaortic reflexes were abrogated. (Redrawn from Shepherd, J.T.: Circulation **50**:418, 1974. By permission of the American Heart Association, Inc.; derived from the data of Edis, A.J.: Am. J. Physiol. **221**:1352, 1971.)

found to be diminished. Thus a drastic redistribution of the reduced cardiac output in favor of flow through the brain and the heart occurs.

The severe cutaneous vasoconstriction accounts for the characteristic pale, cold skin of patients suffering from blood loss. Warming the skin of such patients improves their appearance considerably, much to the satisfaction of well-meaning persons rendering first aid; however, it also inactivates an effective, natural compensatory mechanism—to the detriment of the patient.

In the early stages of mild to moderate hemorrhage, the changes in renal resistance are usually slight. The tendency for increased sympathetic activity to produce renal vasoconstriction is counteracted by intrinsic autoregulatory mechanisms. With more prolonged and more severe hemorrhages, however, intense renal vasoconstriction does occur. The reductions in renal circulation are most intense in the outer layers of the renal cortex. The inner zones of the cortex and outer zones of the medulla tend to be spared.

The severe renal and splanchnic vasoconstriction during hemorrhage undoubtedly produces more favorable conditions for the heart and the brain. However, if such constriction persists too long, it may have serious consequences. Not infrequently patients have survived the acute hypotensive period only to die several days later of kidney failure resulting from renal ischemia during hypotension. Intestinal ischemia also may have dire effects. In the dog, for example, intestinal bleeding and extensive sloughing of the mucosa occur after only a few hours of hemorrhagic hypotension. Furthermore, the low splanchnic flow produces swelling of the centrilobular cells in the liver. The resultant obstruction of the hepatic sinusoids produces an elevation of the portal venous pressure, which intensifies the intestinal blood loss. Fortunately the pathological changes in the liver and intestine are usually much less severe in humans.

Chemoreceptor reflexes. Once mean arterial pressure has dropped to about 60 mm Hg, further reductions of pressure do not evoke any additional responses through the baroreceptor reflexes. However, extremely low levels of arterial pressure result in peripheral chemoreceptor stimulation because of anoxia of the chemoreceptor tissue consequent to inadequate local blood flow. Chemoreceptor excitation enhances the already existent peripheral vasoconstriction associated with baroreceptor reflexes. Also respiratory stimulation provides an auxiliary pumping mechanism as described in Chapter 34.

Cerebral ischemia. At very low levels of arterial pressure (below 40 mm Hg) extensive sympathetic nervous discharge occurs in response to inadequate cerebral blood flow.

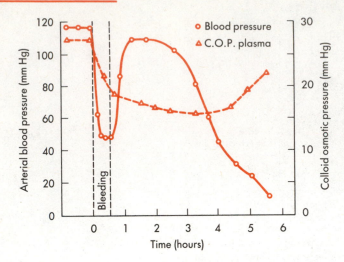

Fig. 36-6 ■ The changes in arterial blood pressure and plasma colloid osmotic pressure in response to withdrawal of 45% of the estimated blood volume over a 30-minute period, beginning at time zero. The data are the average values for 23 cats. (Redrawn from Zweifach, B.W.: Anesthesiology **41:**157, 1974.)

The intensity of the nervous discharge is severalfold greater than the maximum that occurs when the baroreceptors cease to be stimulated. Therefore severe vasoconstriction and facilitation of myocardial contractility occur. With more severe degrees of cerebral ischemia, however, the vagal centers become active, and pronounced bradycardia may ensue, which would tend to aggravate the hypotension.

Reabsorption of tissue fluids. As a consequence of arterial hypotension, arteriolar constriction, and reduced venous pressure, the hydrostatic pressure in the capillaries is low during and after hemorrhage. The balance of forces, therefore, is disturbed in the direction of net reabsorption of interstitial fluid into the vascular compartment. An example of the rapidity of this response is shown in Fig. 36-6. In a group of cats a single hemorrhage of 45% of the estimated blood volume was carried out over a 20- to 30-minute period, beginning at time zero on the graph. The mean arterial blood pressure declined rapidly to about 45 mm Hg during the bleeding. The pressure then returned rapidly, but only temporarily, to near the control level. The colloid osmotic pressure declined markedly during the bleeding and continued to decrease at a more gradual rate for several hours. The reduction in plasma colloid osmotic pressure is a reflection of the extent of dilution of the blood by the ingress of tissue fluids into the vascular compartment.

Considerable quantities of fluid may thus be drawn into the circulation during hemorrhage. Values in the region of 0.25 ml fluid reabsorbed per minute per kilogram body weight have been reported. Thus approximately 1 L of fluid per hour might be autoinfused into the circulatory system of an average person from his own interstitial spaces after an acute blood loss. This mechanism therefore tends to restore the depleted blood volume, and it is manifested as a reduction in hematocrit ratio soon after the onset of hemorrhage.

A slower mechanism probably also comes into play, involving the translocation of considerable quantities of fluid from intracellular to extracellular spaces. This fluid exchange is probably mediated in part by the increased secretion of cortisol by the adrenal cortex in response to hemorrhage. This hormone does cause a shift of fluid from the cells to the extracellular compartment, and it appears to be essential for a full restoration of the plasma volume after hemorrhage.

Endogenous vasoconstrictors. The *catecholamines,* epinephrine and norepinephrine, are released from the adrenal medulla in response to the same stimuli that evoke widespread sympathetic nervous discharge. For all the reasons cited previously, therefore, blood levels of catecholamines are high during and after hemorrhage. In experiments in which animals were bled to an arterial pressure level of 40 mm Hg, it was found that there was a fiftyfold increase in blood levels of epinephrine and a tenfold increase in levels of norepinephrine. The epinephrine comes almost exclusively from the adrenal medulla, whereas the norepinephrine is derived both from that source and from the peripheral sympathetic nerve endings. These humoral substances reinforce the effects of sympathetic nervous activity listed previously.

Vasopressin, which is a potent vasoconstrictor, is actively secreted by the posterior pituitary gland in response to hemorrhage. Removal of about 20% of the blood volume in experimental animals causes an increase in the rate of vasopressin secretion to about 40 times the normal rate. The stimuli responsible for the accelerated release are the diminished pressures on both the arterial and the venous sides of the vascular system. It has been shown that the sinoaortic baroreceptors and receptors in the left atrium are involved in the regulation of vasopressin secretion.

The diminished renal perfusion during hemorrhagic hypotension leads to the secretion of *renin* from the juxtaglomerular apparatus. This enzyme acts on a plasma protein, *angiotensinogen,* to form *angiotensin,* a very powerful vasoconstrictor substance.

Renal conservation of water. Fluid and electrolytes are conserved by the kidneys during hemorrhage for several reasons, including the effect of the increased secretion of vasopressin (antidiuretic hormone) noted previously. The lower arterial pressure per se leads to a diminished glomerular filtration rate, with a consequent reduction in the rate of excretion of water and electrolytes. Also the diminished renal blood flow results in elevated blood levels of angiotensin as described above. This polypeptide accelerates the release of *aldosterone* from the adrenal cortex. Aldosterone in turn stimulates sodium reabsorption by the renal tubules, and the sodium that is actively reabsorbed is accompanied by water. Therefore this constitutes a mechanism for the conservation of extracellular fluid.

In contrast to the negative feedback mechanisms just described, there are also latent *positive feedback mechanisms* that are evoked as a result of hemorrhage. Such mechanisms tend to exaggerate any change that occurs. Specifically, positive feedback mechanisms aggravate the hypotension induced by blood loss and tend to initiate *vicious cycles,* which may lead to death. The operation of positive feedback mechanisms is manifest in curve *B* of Fig. 36-3 and in the uptake of blood from the reservoir in Fig. 36-4.

■ ***Decompensatory mechanisms***

Whether a positive feedback mechanism of itself will lead to a vicious cycle depends on the *gain* of that mechanism. For a positive feedback system, gain may be defined as the ratio of the magnitude of the secondary change evoked by the mechanism in question to the magnitude of the initiating change itself. A gain greater than 1 would lead to a vicious cycle; a gain less than 1 would not. For example, consider a positive feedback mechanism with a gain of 2. If, for any reason, mean arterial pressure decreased by 10 mm Hg, the positive feedback mechanism would then evoke a secondary reduction of pressure of 20 mm Hg, which in turn would cause a further decrement of 40 mm Hg; that is, each change would induce a subsequent change that was twice as great. Hence, mean arterial pressure would decline at an ever-increasing rate until death supervened, much as is depicted by curve *B* in Fig. 36-3.

Conversely, a positive feedback mechanism with a gain of 0.5 would indeed exaggerate any change in mean arterial pressure but would not necessarily lead to death. For example, if arterial pressure suddenly decreased by 10 mm Hg, the positive feedback mechanism would initiate a secondary, additional fall of 5 mm Hg. This in turn would provoke a further decrease of 2.5 mm Hg, and the process would continue in ever-diminishing steps, with the arterial pressure approaching an equilibrium value asymptotically.

Some of the more important positive feedback mechanisms include (1) cardiac failure, (2) acidosis, (3) inadequate cerebral blood flow, (4) aberrations of blood clotting, and (5) depression of the reticuloendothelial system.

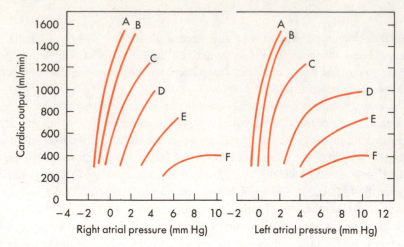

Fig. 36-7 ■ Ventricular function curves for the right and left ventricles during the course of hemorrhagic shock. Curves *A* represent the control function curve; curves *B*, 117 minutes; curves *C*, 247 minutes; curves *D*, 280 minutes; curves *E*, 295 minutes; and curves *F*, 310 minutes after the initial hemorrhage. (Redrawn from Crowell, J.W., and Guyton, A.C.: Am. J. Physiol. **203**:248, 1962.)

Cardiac failure. Considerable controversy exists at present concerning the role of cardiac failure in the progression of shock during hemorrhage. All investigators agree that the heart fails terminally, but opinions differ concerning the relative importance of cardiac failure during earlier stages of hemorrhagic hypotension. Shifts to the right in ventricular function curves (Fig. 36-7) constitute experimental evidence of a progressive depression of myocardial contractility during hemorrhage.

The hypotension induced by hemorrhage reduces the rate of coronary blood flow and therefore tends to depress ventricular function. The consequent reduction in cardiac output leads to a further decline in arterial pressure, a classical example of a positive feedback mechanism. Furthermore, the reduced tissue blood flow leads to an accumulation of vasodilator metabolites, which serves to decrease peripheral resistance and therefore to aggravate the fall in arterial pressure.

In addition to the curtailment of coronary blood flow, other mechanisms contribute to the development of cardiac failure during hemorrhagic hypotension. As described in the following section, acidosis develops in the course of hemorrhagic shock. This may depress the myocardium directly and also indirectly by impairing the responsiveness of the heart to sympathetic stimulation and to circulating catecholamines.

Certain substances that impair cardiac function have been isolated from the blood of animals in various types of experimental shock. Such substances include the amino acid, leucine, and a small, unidentified peptide. They are apparently released from the pancreas and other splanchnic viscera as a consequence of the curtailed blood flow. The role of such *myocardial depressant factors* in the pathogenesis of cardiac failure during shock is controversial at present.

Subendocardial necrosis and hemorrhage are frequently found in the course of hemorrhagic shock. Such pathological changes probably result from a combination of inadequate coronary perfusion and excessive sympathoadrenal activity. These anatomical alterations undoubtedly accentuate the functional impairment of the heart.

Acidosis. Because of inadequate blood flow during hemorrhage, the metabolism of all cells in the body is affected. The resultant stagnant anoxia leads to increased production of lactic acid and other acid metabolites by the tissues. Furthermore, impaired kidney function prevents adequate excretion of H^+, with the result that generalized metabolic acidosis ensues. The resulting depressant effect of acidosis on the heart causes a further reduction in tissue perfusion, with an intensification of the metabolic acidosis. Acidosis also diminishes the reactivity of the resistance vessels to neurally released and circulating catecholamines, thereby intensifying the hypotension.

Inadequate cerebral blood flow. The cerebral ischemic response was shown to result in pronounced sympathetic nervous stimulation of the heart, arterioles, and veins. With very severe degrees of hypotension, however, the cardiac and vasomotor centers eventually become depressed because of inadequate cerebral perfusion. The resultant loss of sympathetic tone then causes a reduction in cardiac output and peripheral resistance. The resulting reduction in mean arterial pressure intensifies the severity of the inadequate cerebral perfusion.

Aberrations of blood clotting. The alterations of blood clotting after hemorrhage are typically biphasic—an initial phase of hypercoagulability followed by a secondary phase of hypocoagulability and fibrinolysis. In the initial phase intravascular clots or *thrombi* have been detected within a few minutes of the onset of severe hemorrhage, and intravascular coagulation may be extensive throughout the minute blood vessels. The death rate from certain standard shock-provoking procedures has been reduced considerably by the administration of anticoagulants such as heparin.

In the later stages of hemorrhagic hypotension, the clotting time is prolonged, and fibrinolysis is prominent. It was mentioned previously that in the dog hemorrhage into the intestinal lumen is common after several hours of hemorrhagic hypotension. Blood loss into the intestinal lumen would, of course, aggravate the effects of the original hemorrhage.

Reticuloendothelial depression. During the course of hemorrhagic hypotension, reticuloendothelial system (RES) function becomes depressed. The phagocytic activity of the RES is modulated by an opsonic protein. It has been found that the opsonic activity in plasma diminishes during the course of shock. This probably accounts in part for the depression of RES function. As a consequence, the antibacterial and antitoxic defense mechanisms of the body are impaired. Endotoxins from the normal bacterial flora of the intestine constantly enter the intestinal circulation. Ordinarily they are inactivated by the RES, principally in the liver. When the RES is severely depressed, these endotoxins invade the general circulation. Endotoxins produce a form of shock that resembles in many respects that produced by hemorrhage. Therefore depression of the RES leads to an intensification of the hemodynamic changes caused by blood loss. Sterilization of the intestine by means of antibiotics significantly reduces the mortality from certain standard shock-provoking procedures including hemorrhage.

It is obvious from the preceding discussion that a multitude of circulatory and metabolic derangements are produced by hemorrhage. Some of these changes are compensatory; others are decompensatory. Some of these feedback mechanisms possess a relatively high gain; others, a relatively low gain. Furthermore, with regard to any specific mechanism, the magnitude of the gain varies with the severity of the hemorrhage. For example, with only a slight loss of blood, mean arterial pressure will still be within the range of normal, and the gain of the baroreceptor reflexes is appreciable. With greater losses of blood, when mean arterial pressure is below about 60 mm Hg, that is, below the threshold for the baroreceptors, then further reductions of pressure will have no additional influence through the baroreceptor reflexes. Hence, below this critical pressure level the gain of the baroreceptor reflexes will be zero or near zero.

As a general rule, with minor degrees of blood loss, the gains of the negative feedback mechanisms are relatively high, whereas those of the positive feedback mechanisms are low. The converse is true with more severe hemorrhages. The gains of the various mechanisms are additive algebraically. Therefore whether a vicious cycle de-

■ ***Interactions of positive and negative feedback mechanisms***

velops depends on whether the sum of the various gains exceeds +1. Total gains in excess of +1 are, of course, more likely to occur with severe losses of blood. Therefore to avert a vicious cycle, serious hemorrhages must be treated quickly and intensively, preferably by whole blood transfusions, before the process has become irreversible.

■ Bibliography

Journal articles

Bevegård, B.S., and Shepherd, J.T.: Regulation of the circulation during exercise in man, Physiol. Rev. **47**:178, 1967.

Chapman, C.B., editor: Symposium on physiology of muscular exercise, Circ. Res. **20** (suppl. 1):1, 1967.

Chien, S.: Role of the sympathetic nervous system in hemorrhage, Physiol. Rev. **47**:214, 1967.

Clausen, J.P.: Circulatory adjustments to dynamic exercise and effect of physical training in normal subjects and in patients with coronary artery disease, Prog. Cardiovasc. Dis. **18**:459, 1976.

Fredholm, B.B., Farnebo, L.O., and Hamberger, B.: Plasma catecholamines, cyclic AMP and metabolic substrates in hemorrhagic shock of the rat. The effect of adrenal demedullation and 6-OH-dopamine treatment, Acta Physiol. Scand. **105**:481, 1979.

Goldfarb, R.D.: Cardiodynamics following shock: role of circulating cardiodepressant substances, Circ. Shock **9**:317, 1982.

Kaijser, L.: Limiting factors for aerobic muscle performance, Acta Physiol. Scand., Suppl. 346, p. 1, 1970.

Lefer, A.M.: Properties of cardioinhibitory factors produced in shock, Fed. Proc. **37**:2734, 1978.

Liang, C., and Hood, W.B., Jr.: Afferent neural pathway in the regulation of cardiopulmonary responses to tissue hypermetabolism, Circ. Res. **38**:209, 1976.

McCloskey, D.I., and Mitchell, J.H.: Reflex cardiovascular and respiratory responses originating in exercising muscles, J. Physiol. **224**:173, 1972.

Pinardi, G., et al.: Contribution of adrenal medulla, spleen and lymph, to the plasma levels of dopamine β-hydroxylase and catecholamines induced by hemorrhagic hypotension in dogs, J. Pharmacol. Exp. Ther. **209**:176, 1979.

Pirkle, J.C., Jr., and Gann, D.S.: Restitution of blood volume after hemorrhage: mathematical description, Am. J. Physiol. **228**:821, 1975.

Rowell, L.B.: Human cardiovascular adjustments to exercise and thermal stress, Physiol. Rev. **54**:75, 1974.

Saltin, B., and Rowell, L.B.: Functional adaptations to physical activity and inactivity, Fed. Proc. **39**:1506, 1980.

Smith, E.E., et al.: Integrated mechanisms of cardiovascular response and control during exercise in the normal human, Prog. Cardiovasc. Dis. **18**:421, 1976.

Zweifach, B.W., and Fronek, A.: The interplay of central and peripheral factors in irreversible hemorrhagic shock, Prog. Cardiovasc. Dis. **18**:147, 1975.

Books and monographs

Carlsten A., and Grimby, G.: The circulatory response to muscular exercise in man, Springfield, Ill., 1966, Charles C Thomas, Publisher.

Lefer, A.M., Saba, T., and Mela, L.M., editors: Advances in shock research, vol. 1, New York, 1979, Alan R. Liss, Inc.

Lefer, A.M., and Schumer, W., editors: Metabolic and cardiac alterations in shock and trauma, New York, 1979, Alan R. Liss, Inc.

Schumer, W., Spitzer, J.J., and Marshall B.E., editors: Advances in shock research, vol. 2, New York, 1979, Alan R. Liss, Inc.

Wiggers, C.J.: Physiology of shock, New York, 1950, Commonwealth Fund.

THE RESPIRATORY SYSTEM

Neil S. Cherniack

Murray D. Altose

Steven G. Kelsen

chapter 37

Respiratory system mechanics

Normal cellular metabolism requires a continuous supply of oxygen (O_2) and continuous disposal of carbon dioxide (CO_2). The major responsibility of the respiratory system is to ventilate gas-exchanging surfaces and thus add O_2 and remove CO_2 from the blood passing through the lungs. The system is so designed that the rate of ventilation can be adjusted as environmental conditions change, metabolic demands are altered, or the physical characteristics of the ventilatory apparatus are modified by growth, senescence, or disease. Feedback control is involved in such ventilatory adjustments.

From a functional standpoint the respiratory system can be considered to be a control system made up of a controller and a plant (Fig. 37-1). The controller consists of (1) chemoreceptors that measure the level of O_2 and CO_2 in the blood and in the brain, (2) mechanoreceptors in the lung and chest wall that monitor movements and pressure, and (3) networks of neurons in the brain, especially in the medulla and pons, that set the rhythm of respiration and allow appropriate times for inspiration and expiration. Information from chemoreceptors and mechanoreceptors relayed back to the brainstem determines the rate and level of respiratory neuronal output to the muscles of the ventilatory apparatus. The plant includes the lung and chest wall, the respiratory muscles that drive the chest bellows, and blood components, such as hemoglobin, plasma proteins, and buffers, that transport O_2 from the lungs to the body tissues and CO_2 from the tissues to the lungs.

Breathing consists of cyclic, alternating inspiratory and expiratory movements of the chest bellows. During inspiration contractions of the diaphragm and the external intercostal muscles, which are the major inspiratory muscles, expand the chest cage and cause the pressure within the lungs to become subatmospheric. Because of the pressure difference between the external surfaces of the body and the interior of the lung, air passes from the atmosphere along the tracheobronchial tree into terminal airspaces or alveoli, where exposure to pulmonary capillary blood takes place. Inspired air is distributed within the lungs by convection (bulk movement of gas), diffusion (a physical process that depends on the activity of the molecules of gas), and the churning action of the cardiac contractions on the lungs. Whereas inspiration is an active event involving muscle contractions, expiration during quiet breathing occurs passively by means of the elastic recoil of the lung and chest wall. However, at increased levels of breathing, muscles of expiration, including the internal intercostal muscles and the muscles of the anterior abdominal wall, are engaged to assist in expelling air from the lungs and returning the chest bellows to its resting end-expiratory position.

Mixed venous blood from the systemic circulation is pumped by the right ventricle through the pulmonary arteries to the pulmonary capillary bed, which consists of a fine

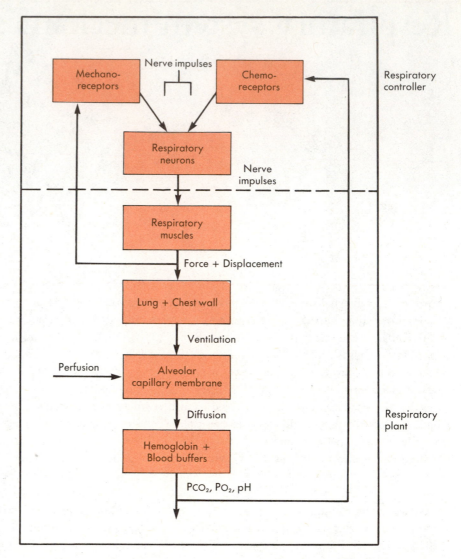

Fig. 37-1 ■ Schematic of the respiratory control system. The control system is made up of a controller that includes the chemoreceptors, mechanoreceptors, and respiratory neurons. The plant comprises the respiratory muscles, chest wall and lungs, and blood components.

network of thin-walled vessels. Exchange of O_2 and CO_2 between alveolar air and pulmonary capillary blood occurs passively by diffusion. O_2-enriched blood is returned to the left atrium and is, in turn, delivered to peripheral tissues by the contraction of the left ventricle. O_2 is transported in the blood primarily in chemical combination with hemoglobin and to a small extent in physical solution in plasma. CO_2 is carried in the blood in physical solution in plasma, in chemical combination with hemoglobin (carbaminohemoglobin), but mainly in the form of bicarbonate.

Rates of O_2 delivery to and CO_2 removal from peripheral body tissues depend on the activity of both the respiratory and circulatory systems (Fig. 37-2). They are determined by the intrinsic sensitivity of the responses of each system to metabolic demands and by the interactions between the systems. Respiratory-circulatory interactions are of two types: (1) mechanical interactions between the chest bellows and the circulatory pump and (2) neural interactions among the receptors and neurons that regulate respiration and circulation and serve to coordinate the responses of the two systems.

An example of mechanical interaction is the pumping action of changes in intrathoracic pressure during breathing on venous return to the heart. Fluctuations in blood pres-

Respiratory system

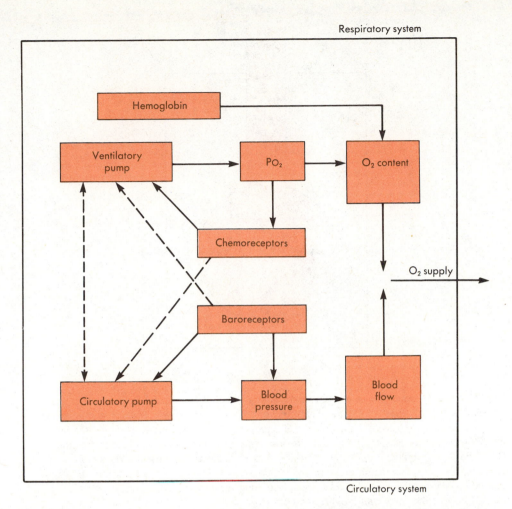

Circulatory system

Fig. 37-2 ■ O_2 delivery to body tissues depends on the function and activity of the ventilatory apparatus and the circulatory pump and important neural and mechanical interactions between the two systems.

sure and heart rate that occur synchronously with breathing are also, in part, a result of changes in the activity of intrapulmonary and airway mechanoreceptors. Pulmonary receptors that are excited by lung inflation (stretch receptors) may dilate systemic blood vessels.

Activation of peripheral arterial chemoreceptors in the carotid and aortic bodies by hypoxia not only stimulates respiration but also alters the sympathetic and parasympathetic output to the heart and peripheral blood vessels, thereby raising systemic blood pressure and altering the heart rate (Chapter 32). Conversely, stimulation of arterial baroreceptors by increasing systemic arterial blood pressure may suppress respiratory activity.

Thus the respiratory system is but one element in a larger integrated network that maintains O_2 and CO_2 homeostasis in the body. Although one component of the network can at least partially compensate for defects in another, optimal operation depends on the coordinated activity of all parts of the network.

■ *Mechanics*

The ventilatory apparatus consists of the lungs and the surrounding chest wall, which includes the rib cage, intercostal muscles, and diaphragm. The lungs fill the chest cavity so that the visceral pleura of the lungs is in contact with the parietal pleura of the chest cage. During breathing the lungs and chest wall move in unison, and from a mechanical standpoint they can be regarded as elements of a pump operating in series.

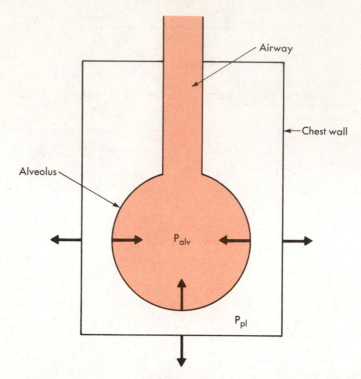

Fig. 37-3 ■ An airway and an alveolus representing the tracheobronchial tree and lungs are enclosed within the chest wall. The pleural pressure (P_{pl}) is the pressure in the potential space between the lungs and chest wall. The alveolar pressure (P_{alv}) is the pressure within the lung. The transpulmonary pressure (P_L) is the pressure difference between the interior of the lungs (P_{alv}) and that surrounding the lungs (P_{pl}). The pressure difference across the chest wall is the difference between the pleural pressure and the pressure at the body surface. At the end of a normal exhalation, the elastic recoil forces of the lung are directed inward, while the elastic recoil forces of the chest wall are directed outward.

At the end of a normal expiration, the respiratory muscles are at rest. The volume of air in the lungs at the resting end-expiratory position is determined by the balance between the elastic recoil forces of the lung, which are directed inward and favor collapse, and the recoil forces of the chest wall, which at end-expiration are directed outward and favor expansion. These opposing forces produce a subatmospheric pressure of about 5 cm H_2O in the potential space between the visceral and parietal pleura.

The elastic recoil of the lungs is measured from the transpulmonary pressure, which is the pressure difference between the interior of the lungs, that is, alveolar pressure (P_{alv}), and that surrounding the lungs, that is, pleural pressure (P_{pl}), under static conditions when airflow is arrested (Fig. 37-3). Similarly, when the respiratory muscles are relaxed the elastic recoil of the chest wall is measured from the pressure gradient across the chest wall, that is, the difference between pleural pressure and the pressure at the external surface of the chest.

Chest expansion produced by inspiratory muscle contraction causes the pleural pressure to become more subatmospheric. The alveolar pressure also temporarily becomes subatmospheric, producing a pressure difference that induces air to flow into the lung from the atmosphere. The difference between alveolar pressure (P_{alv}) and the pressure (P_m) at the airway opening or mouth, that is, the pressure gradient from the distal to the proximal ends of the airways, divided by the rate of airflow ($\dot{V}$) is a measure of the resistance (R) to flow in the tracheobronchial tree (Fig. 37-4).

$$R = \frac{P_m - P_{alv}}{\dot{V}} \tag{1}$$

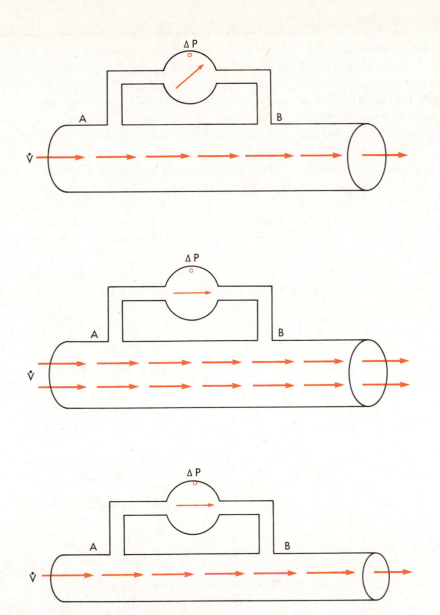

Fig. 37-4 ■ The ratio of the pressure drop (ΔP) across a segment of a tube (from point *A* to point *B*) and the rate of airflow (V̇) represent the resistance of that segment. Increasing the rate of airflow (middle panel) results in a greater pressure drop across the segment. Resistance, however, is unchanged. When the caliber of the tube is narrowed (bottom panel), resistance increases, and the pressure drop for a given airflow rate is greater.

Air moves into the lungs during inspiration until the alveolar pressure again reaches atmospheric levels and the pressure gradient between the alveoli and the airway opening is eliminated.

At the end of inspiration, the volume of air in the lungs and consequently the transpulmonary or recoil pressure of the lung are greater than those at the end of the preceding expiration. The change in transpulmonary pressure for a given change in volume is a measure of lung *elastance* or stiffness. *Compliance,* the reciprocal of elastance, is more commonly used in describing lung elastic properties.

The forces required to overcome lung elastic recoil are stored during inspiration. When these forces are released by relaxation of the inspiratory muscles, expiration occurs passively. The recoil of the lungs causes the alveolar pressure to exceed the pressure at the nose and mouth; consequently air flows out of the lungs.

■ Lung volumes

Lung volumes can be subdivided into a number of compartments (Fig. 37-5). The volume of air in the lungs at the normal, resting end-expiratory position is termed the *functional residual capacity* (FRC).

The volume of air that is inhaled during inspiration with quiet breathing and leaves the lungs passively during expiration is the *tidal volume* (VT). The maximum volume of air that can be drawn into the lungs from functional residual capacity is termed the *inspiratory capacity* (IC). The inspiratory capacity is made up of the tidal volume plus the inspiratory reserve volume (IRV). The volume of air contained in the lungs at the end of a maximum inspiration is the *total lung capacity* (TLC).

The maximum volume of air that can be forcibly exhaled from the functional residual capacity, that is, after the completion of a quiet expiration, is the *expiratory reserve volume* (ERV). The volume of air remaining in the lungs after such a maximum expiratory effort is the *residual volume* (RV).

The *vital capacity* (VC), the maximum volume of air that can be exhaled following a maximum inspiration, is the difference between total lung capacity and residual volume.

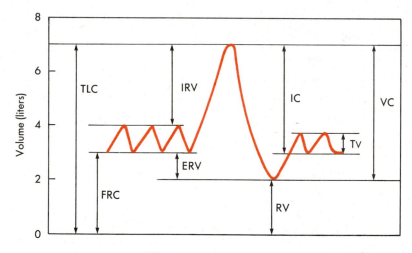

Fig. 37-5 ■ Lung volumes and subdivisions.

■ Elastic properties of the lung

Volume-pressure relationships. If the opposing forces of the chest wall on the lung are eliminated by opening the thorax, for example, the lungs will collapse to a virtually airless state. The lungs may be reexpanded by progressively increasing the transpulmonary pressure.

In practice, measurements of lung volume-pressure relationships in human subjects are made under static conditions, when airflow is temporarily arrested at successive lung volumes during the course of an expiration from total lung capacity (Fig. 37-6). At each lung volume transpulmonary pressure is determined from the difference between alveolar and pleural pressure. Pleural pressure is determined indirectly with a balloon-tipped catheter positioned in the midportion of the esophagus. Alveolar pressure is the same as the pressure at the mouth when airflow along the tracheobronchial tree has ceased, and the glottis is held open.

At lung volumes around functional residual capacity (approximately 40% of total lung capacity), lung distensibility or compliance, as measured from the slope of the volume-pressure curve, is relatively high. However, the volume-pressure characteristics of the lung are nonlinear; compliance progressively falls as the total lung capacity is approached, and increasingly greater applied pressures are required to produce a given volume change (Fig. 37-7).

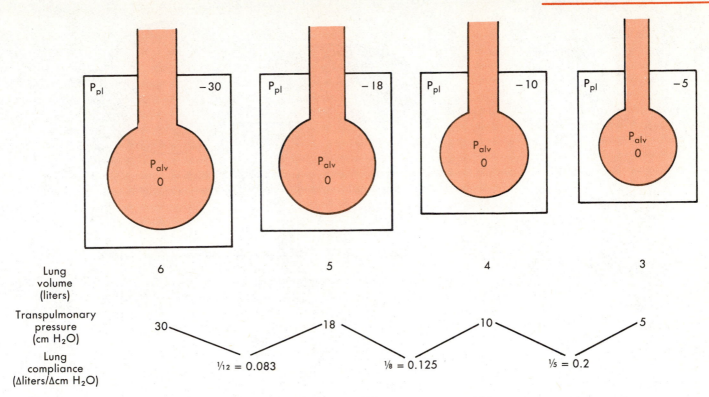

Fig. 37-6 ■ Volume-pressure relationships of the lung. At FRC the transpulmonary pressure is 5 cm H_2O. As lung volume increases (from right to left), there is a progressively greater and greater increase in transpulmonary pressure. Lung compliance, that is, the ratio changes in lung volume to changes in transpulmonary pressure, progressively falls.

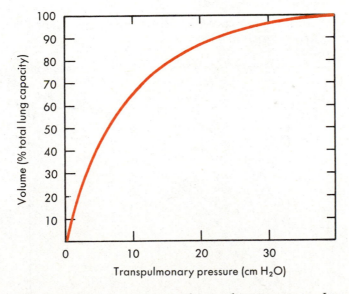

Fig. 37-7 ■ Relationship between lung volume and transpulmonary pressure. Lung compliance, that is, the ratio of change in lung volume to change in transpulmonary pressure, is high at low lung volumes but progressively falls as total lung capacity is approached.

Tissue forces. The elastic properties of the lungs depend on the physical properties of the lung tissues and the surface tension of the film lining the alveolar walls. Lung tissue elasticity arises from the elastin and collagen fibers in the alveolar walls and surrounding bronchioles and pulmonary capillaries. Whereas the elastin fibers can be stretched to approximately double their resting length, the collagen fibers are poorly

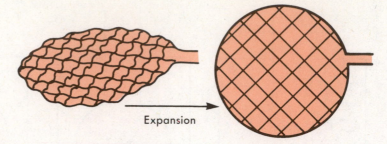

Fig. 37-8 ■ Lung expansion occurs as a result of an unfolding of elastin and collagen fibers in the alveolar walls. The actual lengths of the individual fibers change little.

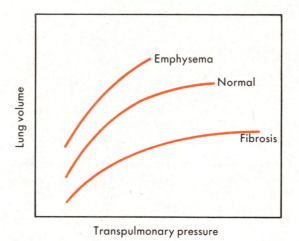

Fig. 37-9 ■ A comparison of lung volume-pressure relationships in a normal person and in patients with emphysema and pulmonary fibrosis.

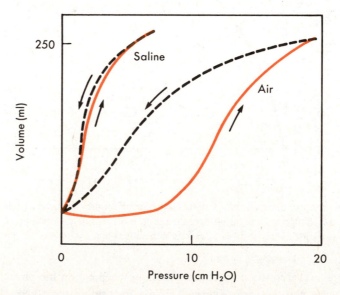

Fig. 37-10 ■ Pressure-volume relationships of an air-filled and a saline-filled lung. In the saline-filled lung, there is no air-liquid interface, and surface forces are abolished. The compliance of the saline-filled lung is considerably greater than that of the air-filled lung. Hysteresis, which is marked in the air-filled lung, is greatly reduced in the saline-filled lung.

extensible and act primarily to limit further expansion at large lung volumes. Lung expansion during breathing seems to occur through an unfolding and geometrical rearrangement of fibers in the alveolar walls. The process is analogous to the manner in which a nylon stocking is stretched, without much change in the length of individual fibers (Fig. 37-8). Changes in the arrangement and physicochemical properties of the elastin and collagen fibers in the lungs account for the increasing distensibility of the lungs with advancing age. Disease processes can also alter lung tissue elasticity. *Emphysema,* which is characterized by a destruction of alveolar walls, decreases the elastic recoil and increases the compliance of the lungs. Conversely, *pulmonary fibrosis,* which involves the interstitial tissues, decreases the distensibility of the lungs (Fig. 37-9).

Surface forces. At an air-liquid interface of a spherical structure, the strong intermolecular forces in the liquid lining cause the area of the lining to shrink. These surface forces acting at the air-liquid interface in the alveoli increase the pressure required to distend the lung and contribute considerably to the recoil pressure. The effects of surface forces on lung elastic recoil are clearly seen by comparing pressure-volume relationships of air-filled and saline-filled lungs of experimental animals (Fig. 37-10). Filling the lung with saline eliminates the air-liquid interface and abolishes surface forces without affecting the elasticity of the pulmonary tissues. At any given volume, the transpulmonary pressure of the liquid-distended lung is about half of that of the lung inflated with air. The contribution of surface forces to the elastic recoil of the lungs is somewhat greater at low than at high lung volumes.

Surfactant. The film lining the interior of alveolar walls is termed *surfactant.* This material is produced by the type II granular pneumocyte and consists of dipalmitoyl lecithin conjugated to protein. Surfactant has a number of interesting and important characteristics. First, the surface tension of surfactant is very low. This serves to minimize the surface forces. Second, as the surface area of the film is reduced, the surface tension decreases yet further (Fig. 37-11). This is critical in maintaining the stability of

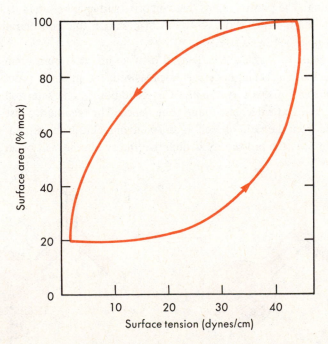

Fig. 37-11 ■ Surface tension–surface area relationships of surfactant. The surface tension of surfactant is extremely low. As the surface area of a surfactant film is reduced, surface tension is further reduced. The surface tension–surface area relationship of surfactant is characterized by marked hysteresis.

alveoli and preventing their collapse. The alveoli are essentially spherical structures and thus behave according to the Laplace equation (Chapter 31) for a thin-walled sphere:

$$P = \frac{2T}{r} \qquad (2)$$

which states that the pressure (P) inside the alveolus is directly proportional to the tension (T) in the wall and inversely proportional to the radius (r). As lung volume falls and hence as the radii of the alveoli become smaller, transpulmonary pressure would progressively increase if the surface tension remained constant. As shown in Fig. 37-12, because of the differences in transpulmonary pressure between parallel large and small alveoli, small alveoli would empty into larger alveoli and collapse. Because of the decreasing surface tension of surfactant, however, the transpulmonary pressure of parallel large and small alveoli remains virtually the same. Small alveoli, which would otherwise empty into larger alveoli with which they communicate, are stabilized and maintain their volume.

When the surface area of surfactant is kept small, the surface tension progressively increases with time as a result of a rearrangement of the molecules of the surface lining. This seems to account for the tendency of peripheral airspaces to collapse and for lung compliance to fall during prolonged periods of breathing at small tidal volumes. A single large breath or sigh, which reopens alveoli and expands the surface, results in a lowering of surface tension and a restoration of the elastic properties of the lung.

The surface tension–surface area relationship of surfactant (Fig. 37-11) is characterized by marked hysteresis, that is, the relationships between tension and area differ during expansion and contraction. Although all viscoelastic structures manifest some degree of hysteresis, the hysteresis of surfactant largely accounts for the differences in the pressure-volume relationships that are noted during the alternate expansions and contractions of the air-filled lung (Fig. 37-10).

Type II granular pneumocytes first appear in the alveolar epithelium of the fetal lung at about 24 weeks of gestation. They begin to produce surfactant between 28 and 32 weeks of gestation. The phospholipids that comprise surfactant contain lecithin and pass into the amniotic fluid. After about 35 weeks of gestation, the concentration of lecithin in the amniotic fluid begins to rise.

If the production of surfactant is delayed or if an infant is born prematurely before adequate amounts of surfactant have been synthesized, the infant is prone to a disorder known as *respiratory distress syndrome*. This condition is characterized by a marked instability of alveoli and a decreased lung compliance. The respiratory distress syndrome can be predicted from the ratio of lecithin to sphingomyelin in the amniotic fluid. If the ratio is less than 2:1, the probability is high that the syndrome will occur.

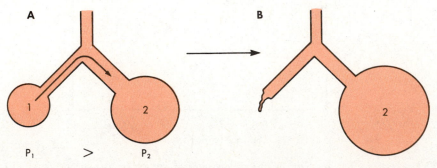

Fig. 37-12 ■ Communicating alveoli of different sizes. If the surface tension in both alveoli were the same, the transpulmonary pressure of the smaller alveolus would exceed that of the larger alveolus **(A)**. The smaller alveolus would collapse and empty into the larger alveolus **(B)**.

Surfactant synthesis in the fetus is related most closely to gestational age, but it is also under humoral influences. The administration of a glucocorticoid to the mother before delivery may accelerate surfactant synthesis in the fetus.

■ *Elastic properties of the chest wall*

The elastic recoil of the chest wall is such that the chest, were it unopposed by the lungs, would enlarge to about 70% of the total lung capacity (Fig. 37-13). This volume represents the equilibrium or resting position of the chest wall, where the pressure gradient across the chest wall, that is, the difference between the pleural pressure and the pressure at the body surface, is zero. If the volume of gas in the lungs is increased to further expand the thorax, the chest wall, like the lung, would tend to recoil inward. This would tend to resist expansion and would favor a return to the equilibrium position. At volumes less than 70% of the total lung capacity, the recoil of the chest is directed outward and is opposite to that of the lung. As residual volume is approached, not only is the outward recoil of the chest wall large, but the compliance of the chest wall falls to very low levels. The stiffness of the chest wall at low lung volumes is one of the major determinants of residual volume.

The compliance of the chest wall can be markedly deranged by abnormalities of bones and joints of the thorax such as kyphoscoliosis and ankylosing spondylitis.

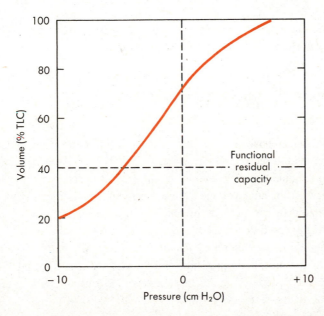

Fig. 37-13 ■ Pressure-volume relationships of the chest wall. The equilibrium position of the chest wall, that is, the volume at which the pressure difference across the chest wall is zero, when the respiratory muscles are at rest, is approximately 70% of total lung capacity. At volumes above the equilibrium position, the chest wall tends to recoil inward, whereas at volumes below the equilibrium position, the tendency of the chest wall is to recoil outward.

■ *Elastic properties of the lung–chest wall system*

The lung and chest wall are considered to be in series with one another so that the algebraic sum of the pressures exerted by the recoil of the lung (P_L) and recoil of the chest wall (P_{CW}) make up the recoil pressure of the total respiratory system (P_{RS}):

$$P_{RS} = P_L + P_{CW} \qquad (3)$$

The recoil of the lung is determined from the difference between alveolar pressure (P_{alv}) and pleural pressure (P_{pl}). Also, the recoil of the chest wall (when the respiratory muscles are completely at rest) is determined from the difference between pleural pres-

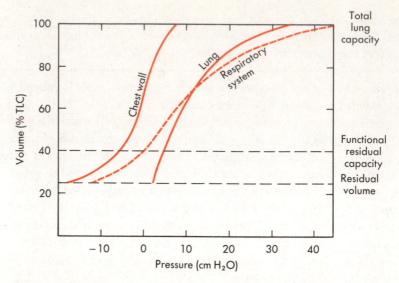

Fig. 37-14 ■ Pressure-volume relationships of the lung, chest wall, and combined system. At any lung volume the elastic recoil of the total respiratory lung–chest wall system is determined from the algebraic sum of the recoil pressures of the lung and chest wall.

sure (P_{pl}) and the pressure at the surface of the body (P_{BS}). Therefore the recoil of the entire respiratory system can be expressed as follows:

$$P_{RS} = (P_{alv} - P_{pl}) + (P_{pl} - P_{BS}) = P_{alv} - P_{BS} \tag{4}$$

When the pressure at the surface of the chest is at atmospheric levels (i.e., $P_{BS} = 0$),

$$P_{RS} = P_{alv} \tag{5}$$

Functional residual capacity represents the equilibrium position of the respiratory system where the overall recoil pressure is zero (Fig. 37-14). At any given lung volume above functional residual capacity, the net recoil pressure exceeds atmospheric pressure and favors a decrease in lung volume. Total lung capacity is reached when the maximum force generated by the muscles of inspiration matches the inward passive elastic recoil pressure of the respiratory system.

At lung volumes below functional residual capacity, the respiratory system recoil pressure is less than atmospheric pressure, and the net effect is directed toward increasing lung volume. At residual volume, the chest wall is virtually solely responsible for the outward recoil of the respiratory system.

■ *Resistance to airflow*

Whereas elastic forces of the lung oppose lung expansion, resistance impedes airflow into and out of the lung. The total resistance of the ventilatory apparatus consists of the flow resistance of the airways and the frictional resistance to the displacement of lung and chest wall tissues during breathing. Tissue resistance normally comprises only a small fraction (about 10%) of the total resistance, but it may increase considerably with diseases of the lung parenchyma.

Airway resistance. Airway resistance is defined as the ratio of the pressure gradient along the tracheobronchial tree to the rate of airflow. Pressure-flow relationships in the lung are particularly complicated because the airways comprise a system of dichotomous branching tubes. To illustrate the range and variety of pressure-flow relationships that may occur along the airways of the lung, it is convenient to consider patterns of flow through rigid circular tubes (Fig. 37-15). In many respects the fluid mechanical features resemble those observed in the cardiovascular system (Chapter 28).

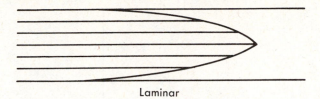

Laminar

Fig. 37-15 ■ Patterns of flow through a tube.

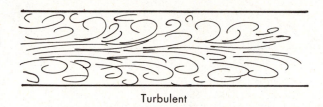

Turbulent

Laminar flow is characterized by organized streamlines that run parallel to the sides of the tube. The streamlines are capable of sliding over one another, and those at the center of the tube tend to move faster than those closest to walls. Consequently, the flow profile is parabolic. The driving pressure (ΔP) producing laminar flow is linearly related to flow rate ($\dot{V}$) and is dependent on the length (l) and radius (r) of the tube and the viscosity of the gas (η), according to Poiseuille's equation:

$$\Delta P = \frac{8\eta l \dot{V}}{\pi r^4} \tag{6}$$

It is apparent from the equation that the driving pressure for a given laminar flow is critically dependent on tube radius. If the radius of the tube is reduced to half, the pressure required to maintain a given flow rate is increased sixteenfold.

Turbulent flow is characterized by a complete disorganization of streamlines. Molecules of gas move laterally, change their velocities, and collide with one another to produce eddies and swirls. The driving pressure is not linearly related to the flow rate, but rather it is proportional to the square of flow, and it increases with gas density.

At branches in the tracheobronchial tree, where air from two separate channels comes together, the parabolic profile of laminar flow may become blunted; the streamlines may separate from the walls of the tube, and minor eddy formation may develop to produce a mixed or transitional flow pattern. Under these conditions the driving pressure to produce a given flow is dependent on the ratio of gas viscosity to density (the *kinematic viscosity*), and the pressure needed to produce a given flow rate is proportional to the flow and to its square:

$$P = K\dot{V} + K\dot{V}^2 \tag{7}$$

Reynold's number (N_R), an empirically derived formulation, can be used to predict the pattern of airflow. Reynold's number depends on flow rate ($\dot{V}$), gas density (ρ) and viscosity (η), and tube diameter (D):

$$N_R = \dot{V}D\frac{\rho}{\eta} \tag{8}$$

When the Reynold's number is less than 2000, flow is fully laminar. However, when gas density is high and viscosity is low and when rates of airflow are high, the Reynold's number is large, and flow may become turbulent.

In the normal lung laminar flow occurs only in very small peripheral airways, where the overall cross-sectional area is great. Consequently flow through any given airway is

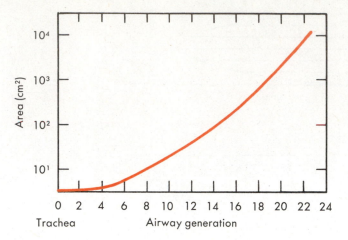

Fig. 37-16 ■ Total cross-sectional area of all airways in a given airway generation increases progressively from the trachea to the level of the alveoli.

extremely slow. In general, airflow through the trachea is turbulent, but through most of the remainder of the branching tracheobronchial tree, flow is transitional or mixed.

The driving pressure that produces flow overcomes friction but also accelerates the air. Acceleration is important during expiration because as air moves from the alveoli toward the airway opening, the total cross-sectional area of the airways decreases. Hence, the molecules of air must accelerate through the converging channels, although the overall flow rate does not change. The pressure required to produce this convective acceleration is proportional to the gas density and the square of the flow rate.

Distribution of airway resistance. All portions of the tracheobronchial tree do not contribute equally to the total airway resistance. A large proportion of the resistance to airflow is provided by the airways of the upper respiratory tract. The resistance of the nasal passages is extremely high; it may comprise 50% of the total airway resistance during nose breathing. The mouth, pharynx, larynx, and trachea account for 20% to 30% of airway resistance during quiet mouth breathing, but this may increase to about 50% at increased levels of ventilation such as those encountered during exercise.

The tracheobronchial tree is a system of irregular dichotomous branching tubes. The total cross-sectional area of daughter branches tends to exceed the cross-sectional area of parent segments. Thus, although the surface area over which air must pass increases, the resistance decreases progressively with successive generations of airways (Fig. 37-16). The major sites of resistance in intrapulmonary airways are in medium-sized lobar, segmental, and subsegmental airways up to about the seventh generation. The small peripheral airways, less than 2 mm in diameter, contribute less than 30% to 40% of the total airway resistance of the normal lung. Certain clinical disorders preferentially affect small bronchi and bronchioles, significantly reducing their caliber. Because these airways contribute relatively little to the overall airway resistance, significant disease at these sites may not be detected by routine tests of pulmonary function.

Effects of lung volume on airway caliber. Like the lung parenchyma, the airways are elastic and are capable of being compressed or distended. Airway caliber varies with the transmural airway pressure, that is, the difference between the pressure within the airway and that surrounding the airway. The airways are exposed to the expansive forces involved in overcoming the elastic recoil of the lung. Accordingly, the pressures surrounding intrathoracic airways are related to and may approximate pleural pressure. The airways can be viewed as being tethered to parenchymal lung tissue. There is an inverse hyperbolic relationship between lung volume and airway resistance (Fig. 37-17). At large volumes, lung elastic recoil is high, and the traction applied to the walls of intra-

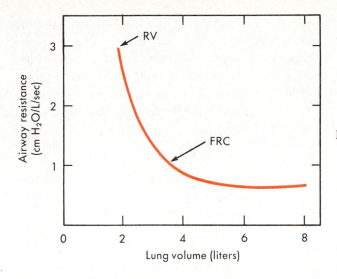

Fig. 37-17 ■ Relationship between lung volume and airway resistance.

thoracic airways is great. The airways widen because of the large transmural airway pressure, and the resistance to airflow falls. At small lung volumes, transmural airway pressure is low so that the airways narrow, and resistance increases.

If the elastic recoil of the lung is reduced by destruction of alveolar walls, as in pulmonary emphysema, the transmural airway pressure at any given lung volume will be correspondingly less, and the airways will be narrowed. Although there may be no disease of the airways per se, airway resistance will be greater.

The effects of changes in transmural pressure on airway caliber depends on the compliance of the airways, which is determined by their structural composition. The trachea is almost completely surrounded by cartilaginous rings, which tend to prevent complete collapse even when the pressure surrounding the trachea is greater than the intraluminal pressure. The bronchi are less well supported by incomplete cartilaginous rings and cartilaginous plates, and the bronchioles lack any cartilaginous support. These structures are much more compliant and may narrow considerably when the extraluminal pressure exceeds that within the airways. Additionally, all airways can be stiffened, although to different degrees, by contraction of the smooth muscle in their walls.

Control of airway smooth muscle. The tone of the smooth muscle cells that encircle the airways affects their caliber and hence the airflow resistance. Airway smooth muscle is reactive, responding to autonomic nervous activity, circulating hormones, and chemicals released by cells located near the tracheobronchial tree (Table 37-1). The reactivity of the airways can be assessed in humans by determining the changes in resistance as graded doses of a constricting agent, for example, methacholine or histamine, are inspired. These bronchial challenges or bronchoprovocation tests suggest that the airways

TABLE 37-1

Control of airway smooth muscle

Stimulus	*Contraction*	*Relaxation*
Nervous	Cholinergic	Adrenergic
Neurohumoral	Acetylcholine	Norepinephrine
Chemical	Histamine SRS-A Prostaglandin F-2α	Prostaglandin E
Physical	Smoke, dust, SO_2	

of asthmatic patients are hyperreactive; that is, smaller doses of the provocative drug are required to produce increases in airway resistance in such patients than in normal subjects. Airway constriction not only increases airway resistance, but it also diminishes expiratory and inspiratory flow rates and can increase functional residual capacity. Moreover, the airways are not all equally reactive; hence, airway constriction tends to occur nonuniformly. Airway smooth muscle tone depends on the balance between the mechanisms that tend to contract and relax the smooth muscles. In patients with hyperreactive airways, this balance appears to be altered so that the constricting influences dominate.

Nerve fibers from three different systems (parasympathetic, sympathetic, and nonadrenergic inhibitory) converge on airway smooth muscle. Stimulation of parasympathetic nerve fibers, carried in the vagal trunks, causes release of acetylcholine, which contracts airway smooth muscle. These effects can be blocked by atropine and other cholinergic antagonists. In humans the activity of the parasympathetic nervous system is probably the most important of the neural influences on airway smooth muscle tone. Airway pollutants such as SO_2 incite coughing and bronchial narrowing. These reactions are initiated by stimulating irritant receptors that are located in the airway submucosa. The afferent fibers are all carried in the vagi. Irritant receptor stimulation appears to trigger reflex bronchoconstriction via the release of acetylcholine from efferent vagal nerve endings (Fig. 37-18). Large doses of atropine block some asthmatic attacks, suggesting that they were induced by such acetylcholine release.

In dogs stimulation of the sympathetic system releases norepinephrine and causes bronchodilation. Norepinephrine can react with β-adrenergic receptors in the smooth muscle cell membrane to cause bronchodilation and with α-receptors in the cell mem-

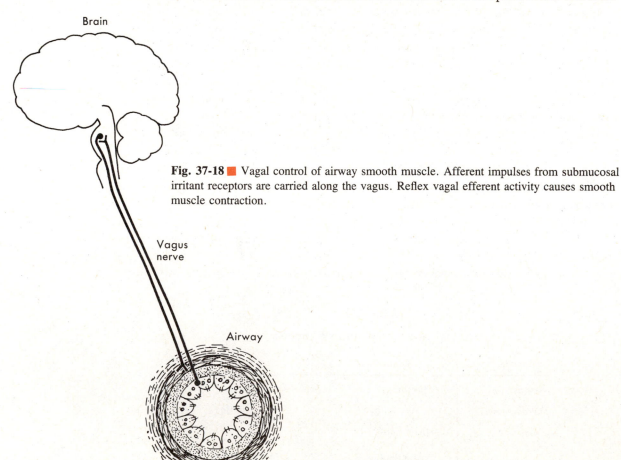

Fig. 37-18 ■ Vagal control of airway smooth muscle. Afferent impulses from submucosal irritant receptors are carried along the vagus. Reflex vagal efferent activity causes smooth muscle contraction.

brane to constrict the airways. Its major effect is bronchodilation because there are many more β-receptors than α-receptors in the airways. Histochemical studies have demonstrated that there are few, if any, sympathetic fibers in the airways of humans. However, the β-receptors present in airway smooth muscle membranes can be stimulated by catecholamines released from the adrenal medulla or exogenously administered.

The nonadrenergic inhibitory system decreases the tone of gastrointestinal smooth muscle, which has an embryological origin similar to the smooth muscle in the airways. In the gastrointestinal tract, the neural mediator of this system may be a purine derivative, and a similar mediator may be released in the airways. In vitro studies on human airway smooth muscle have demonstrated that the nonadrenergic inhibitory system relaxes constricted muscle. Failure of the nonadrenergic bronchodilator system may be one of the factors contributing to increased airway reactivity in asthmatics.

Chemicals released from mast cells by a number of physical and chemical stimuli and immunological reactions can substantially alter airway caliber. Mast cells are located in the connective tissue underlying the smooth muscle and, in lesser numbers, in the walls of the airways. Histamine and slow reacting substance of anaphylaxis (SRS-A) are the most important bronchoconstrictor mediators released from mast cells. Histamine causes rapid contraction of airway smooth muscle by binding to H_1 receptors. However, histamine also causes relaxation of airway muscle by binding to H_2 receptors, which are located more peripherally in the lungs. In humans the H_1-mediated bronchoconstriction appears to be the predominant effect of histamine. Histamine release from mast cells can be blocked by cromolyn sodium.

Histamine and SRS-A are two agents that can increase the production of prostaglandins. Not all prostaglandins can affect smooth muscle tone in the same way. Certain prostaglandins, such as prostaglandin E, have bronchodilator properties. Other prostaglandins, such as F-2α, produce bronchoconstriction.

It is currently believed that the tone of smooth muscle depends largely on the ratio of cyclic AMP (cAMP) to cyclic GMP (cGMP) within the cell. Increased ratios of cAMP/cCMP are associated wtih greater smooth muscle relaxation. Cyclic AMP is formed by the action of adenylate cyclase on ATP. β-Receptor stimulation enhances adenylate cyclase activity and results in an increased cAMP/cGMP ratio within the cell. Cyclic AMP is degraded by the enzyme phosphodiesterase. This enzyme is inhibited by theophylline and its derivatives, which are commonly used in the treatment of asthma.

Smooth muscle tone seems also to depend on membrane potential and membrane permeability to calcium. According to current concepts of electromechanical coupling, contraction of airway smooth muscle is initiated by calcium influx. Uptake of calcium by the sarcoplasmic reticulum is associated with relaxation.

The airway smooth muscle cells, the irritant receptors, and the majority of intrapulmonary mast cells lie beneath the airway epithelium, which acts as a barrier between noxious agents within the airstream and these cellular components. This barrier may be important in the stabilization of airway smooth muscle tone. Freeze-fracture techniques have demonstrated that cell-to-cell connections within the epithelial layer are "tight junctions." Agents such as cigarette smoke or infectious viruses can disrupt the epithelial barrier by loosening these junctions, thereby permitting reactions that induce airway constriction.

Flow-volume relationships. A maximum expiratory flow-volume curve is constructed by plotting airflow against lung volume when an individual, after inspiring to total lung capacity, exhales as forcefully, rapidly, and completely as possible to residual volume. The rate of airflow reaches a peak during a forced expiration at a lung volume close to total lung capacity. As lung volume decreases, there is a reduction in lung elastic recoil; the intrathoracic airways narrow, and the airway resistance increases. Consequently, the rate of airflow progressively falls (Fig. 37-19).

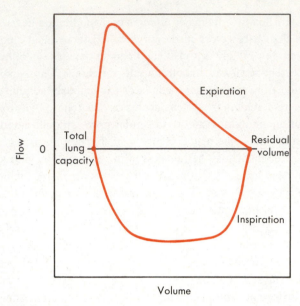

Fig. 37-19 ■ A maximum flow-volume loop. Lung volume is plotted on the horizontal axis; total lung capacity is shown on the extreme left and residual volume on the extreme right. Airflow during inspiration is directed down, whereas airflow during expiration is directed upward on the vertical axis.

In a similar fashion, a maximum inspiratory flow-volume curve can be generated during a rapid, forceful inspiratory maneuver from residual volume to total lung capacity. The pattern of changes in airflow with changes in lung volume differs during maximum inspiratory and expiratory maneuvers. During a maximum inspiratory maneuver, inspiratory flow reaches a high level while lung volume is still low, and the flow remains high over a wide range of the vital capacity. During a maximum inspiratory effort, the pleural pressure is markedly subatmospheric; the transmural airway pressure is large, and the intrathoracic airways are held open. As lung volume increases in the course of the inspiratory maneuver, the muscles of inspiration shorten, and according to the length-tension properties, the forces generated by those muscles fall. The decrease in inspiratory muscle force with increasing lung volume would favor a progressive reduction in inspiratory flow. However, as the lung expands, the airway widens so that airway resistance becomes less and less. Consequently, inspiratory flow is maintained at relatively constant levels until total lung capacity is approached (Fig. 37-19).

A family of flow-volume loops may be obtained by repeating the expiratory and inspiratory maneuvers for a full vital capacity at different degrees of effort (Fig. 37-20). The greater the inspiratory effort, the greater the rate of airflow over the entire vital capacity. Hence, at any lung volume from residual volume to total lung capacity, inspiratory flow rates depend on the forces generated by the muscles of inspiration.

With increasing expiratory efforts, there is also a progressive increase in expiratory airflow rates. At lung volumes less than 75% of the vital capacity, maximum rates of airflow are achieved with less than maximum effort. Thereafter airflow rates increase no further despite further increases in expiratory effort. Accordingly, at volumes less than 75% of the vital capacity, maximum expiratory flow rates are considered to be relatively effort independent. At lung volumes greater than 75% of the vital capacity, expiratory airflow rates continue to increase with increasing effort until a maximum effort is expended.

Isovolume pressure-flow curves. The influence of lung volume on the pressure-flow relationships during expiration is further illustrated by isovolume pressure-flow curves (Fig. 37-21). During repeated expiratory vital capacity maneuvers, each performed with

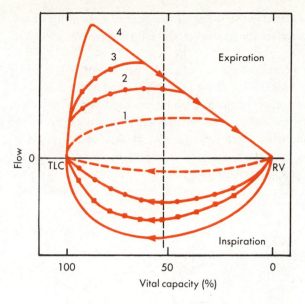

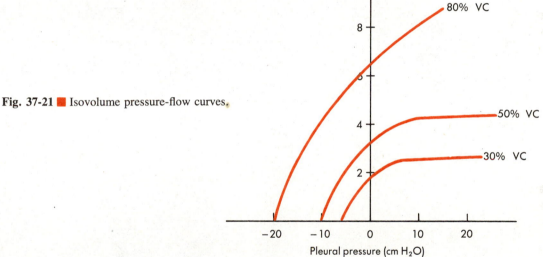

Fig. 37-20 ■ A family of flow-volume loops. Each of the four inspiratory and expiratory vital capacity maneuvers is performed at a different level of effort. The maneuver with maximal effort is designated by the number *4*. Maneuvers *3, 2,* and *1* are performed with progressively less and less effort.

Fig. 37-21 ■ Isovolume pressure-flow curves.

a different degree of effort, as shown in Fig. 37-20, simultaneous measurements are made of airflow rates, lung volume changes, and pleural pressure. At given lung volumes (80%, 50%, and 30% of the vital capacity are shown in Fig. 37-21) during the course of repeated expiratory vital capacity maneuvers, the instantaneous airflow rate is plotted against the instantaneous pleural pressure. When airflow is zero, the pleural pressure is subatmospheric, reflecting the elastic recoil pressure of the lung; the greater the lung volume, the more negative is the pleural pressure. At any given lung volume, as pleural pressure is allowed to become less subatmospheric and the alveolar pressure becomes positive, the rate of expiratory airflow increases. With increasing expiratory efforts, the pleural pressure exceeds atmospheric pressure. At lung volumes greater than 75% of the vital capacity, airflow rates increase progressively as pleural pressure becomes more and more positive; flow is considered to be effort dependent.

In contrast, at lung volumes below 75% of the vital capacity, the rate of airflow levels off and reaches a plateau as pleural pressure becomes positive. Thereafter, further increases in effort and in pleural pressure fail to increase the rate of airflow. Over this range of lower lung volumes, expiratory flow is virtually independent of effort. Because airflow rates remain constant despite increases in expulsive force, it follows that the resistance to airflow increases in direct proportion to the increase in the pressure gradient along the entire airway. This is thought to occur because of compression and narrowing of large intrathoracic airways by the positive pleural and extraluminal airway pressure.

Equal pressure point theory. An analysis of some of the determinants of maximum expiratory flow rates during forced expiratory maneuvers is facilitated by considerations of the equal pressure point theory.

The concept of equal pressure points during forced expiratory efforts is illustrated by using a model of the lung (Fig. 37-22). The alveoli are represented by an elastic sac and the intrathoracic airways by a compressible tube, both enclosed within the chest cage. When a lung volume, for example 50% of the vital capacity, is maintained by the action of the inspiratory muscles and when there is no airflow, pleural pressure is subatmospheric, and it counterbalances the elastic recoil pressure of the lung. When the airway is open to the atmosphere and when there is no airflow, the pressure along the entire airway, down to and including the alveoli, is equal to atmospheric pressure (Fig. 37-22, *A*).

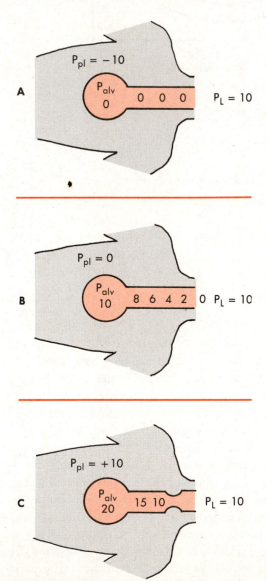

Fig. 37-22 ■ A model of the lung where the alveoli are represented by an elastic sac and the intrathoracic airways by a compressible tube, both enclosed within the chest cage. In each of the three conditions, the lung volume is the same; consequently the elastic recoil pressure of the lung (P_L) is the same. Pleural pressure is designated P_{pl}, and alveolar pressure is designated P_{alv}. In each condition, the pressure at points along the airway is shown. In **C** the intrathoracic airways downstream from the equal pressure point are compressed because the extraluminal pressure exceeds that within the airway lumen.

During quiet expiration at the same lung volume (50% of vital capacity), pleural pressure is less subatmospheric, and alveolar pressure is positive. Pressure is gradually dissipated along the airway in overcoming resistance, and the pressure at the airway opening is zero. However, all along the intrathoracic airway, airway pressure exceeds pleural pressure. The transmural airway pressure is positive, and the airways remain open (Fig. 37-22, *B*).

The pleural pressure exceeds atmospheric pressure during a forceful expiratory effort, and alveolar pressure becomes yet more positive (Fig. 37-22, *C*). Airway pressure falls more steeply from the alveolus to the airway opening because of the greater rate of airflow. At some point along the airways, referred to as the *equal pressure point,* the drop in airway pressure equals the elastic recoil pressure of the lung. At the equal pressure point, the intraluminal pressure equals the pressure surrounding the airway, that is, the pleural pressure. Downstream from the equal pressure point toward the airway opening, the transmural airway pressure is negative because the intraluminal airway pressure falls below the pleural pressure. Consequently the airways are compressed. Once maximum expiratory flow is achieved, any further increase in pleural pressure, achieved by increasing expiratory force, simply produces more compression of the downstream segment, but it does not affect the airflow rate.

At maximum expiratory airflow rates, the equal pressure point is located at the level of segmental bronchi. The equal pressure point divides the airways into the following two components arranged in series: an upstream segment from the alveoli to the equal pressure point and a downstream segment from the equal pressure point to the airway opening. The driving pressure of the upstream segment, that is, the pressure gradient along the airways of that segment, equals the elastic recoil of the lung. Accordingly, maximum expiratory flow ($\dot{V}_{max}$) can be expressed in terms of the elastic recoil pressure of the lung (P_L) and the resistance (R_{us}) of the upstream segment. This expression, $\dot{V}_{max} = P_L/R_{us}$, indicates that lung elasticity and the resistance of the upstream airways are important determinants of maximum expiratory flow.

Changes in the velocity of airflow that occur in the tracheobronchial tree contribute to dynamic compression of the airways. During forced expiration, as air moves downstream from the alveoli toward the airway opening, the molecules of air accelerate, and their velocity increases as the total cross-sectional airway diameter decreases. According to the Bernoulli principle (Chapter 28), which is based on the first law of thermodynamics, the intraluminal pressure diminishes inversely as the square of the velocity. When the intraluminal airway pressure falls below the pleural pressure, the airways just downstream from the equal pressure points are compressed, and their caliber is reduced. This results in a further increase in linear velocity and hence a further fall in intraluminal pressure. These changes in pressure lead to the development of a flow-limiting segment that prevents a rise in flow rate despite any increase in expiratory effort.

Wave-speed limitation theory. Using wave-speed limitation theory, mathematical equations can be written to show the interrelationships between the various factors that contribute to flow limitation during forced expiration. The velocity (v) of propagation of pressure waves along a tube varies directly with the cross-sectional area of the tube (A) and the elastance of the tube walls, $\frac{(dP)}{(dA)}$, and is inversely related to gas density (ρ) according to the equation

$$v = \frac{A^3}{q\rho} \times \frac{dP}{dA} \qquad (9)$$

where q is a conversion factor for departure from a blunt flow profile.

In a tube with elastic walls, flow is limited at a choke point, which occurs when the flow rate equals the velocity of propagation of pressure waves at that point. In a system of branching tubes, there can be several choke points. In the lung the location of choke points depends on the pulmonary volume. At large lung volumes, a choke point is situated in the vicinity of the lower trachea, but at lower lung volumes, choke points develop closer to the alveolar level. Neck extension tends to increase maximum expiratory flow rates at large lung volumes. Longitudinal tension on the trachea, produced by neck extension, stiffens the trachea, decreases tracheal collapsibility, and increases wave velocity.

Regional distribution of lung volume and ventilation. In the upright posture, there is a vertical pleural pressure gradient of approximately 0.25 cm H_2O per centimeter of vertical distance along the lung. Because of the weight of the lung and the effects of gravity, the pleural pressure is more negative at the apex of the lung, and it becomes progressively less negative toward the lung base. Consequently, the transpulmonary pressure, that is, alveolar minus pleural pressure, is greater at the apex than at the base of the lung. The elastic properties of the lung appear to be the same everywhere in the lung; hence, the pressure-volume relationships in all areas of the lung are considered to be identical. At any given overall lung volume, alveoli at the apex and at the base will fall on different locations of the same pressure-volume curve by virtue of their differing transpulmonary pressures (Fig. 37-23). Near total lung capacity, the pressure-volume curve is flat so that despite the differences in transpulmonary pressure between the top and bottom of the lung, alveoli at the lung apex and base are uniformly expanded and are approximately the same size. At intermediate lung volumes (e.g., functional residual capacity) the pressure-volume curve is steep, and because of the regional differences in transpulmonary pressure, alveoli at the lung apices are more expanded and larger than those at the lung bases. At low lung volumes near residual volume, the pleural pressure at the bottom of the lung may actually exceed the pressure inside the airways. The negative transmural airway pressure leads to closure of airways in the gravity-dependent zones of the lung base.

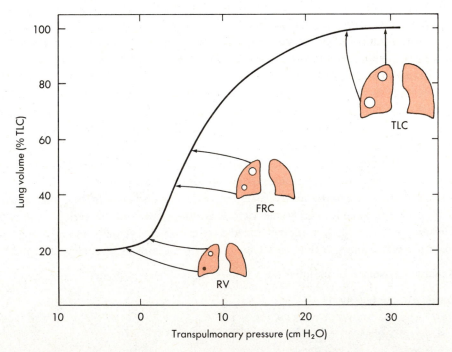

Fig. 37-23 ■ Regional distribution of lung volume. Because of the differences in pleural pressure from the apex to the base of the lung, the transpulmonary pressure of alveoli at the lung apex will differ from that of alveoli at the lung base. At a given lung volume, alveoli at the apex and at the base will fall on different locations of the same pressure-volume curve.

During a slow inspiration from residual volume, the alveoli of the upper lung zones start to fill first, but trapped units at the lung bases receive no ventilation until the transpulmonary pressure exceeds a critical level and the airways reopen. In contrast, during an inspiration from functional residual capacity, ventilation to alveoli in the lung bases is greater than to alveoli at the apex of the lung because for a given change in transpulmonary pressure, alveoli at the lung bases will enlarge more than those at the lung apex.

Distribution of ventilation. The distribution of ventilation in the lungs and the volume at which the airways in the lung bases begin to close can be assessed by the single-breath nitrogen washout test (Fig. 37-24). The subject first takes a single full inspiration of 100% O_2 from residual volume to total lung capacity. The initial portion of the breath goes to the upper lung zones, but the remainder of the breath, containing only O_2, is preferentially distributed to the lung bases. The pattern of distribution of the inspired gas is such that at the end of the maximum inspiration of O_2 the concentration of nitrogen will be lower in the alveoli of the lung bases than in the alveoli of the upper lung zones.

During the subsequent slow but complete exhalation from total lung capacity to residual volume, the concentration of nitrogen in the exhaled air at the mouth is plotted against the expired volume. The initial portion of the exhaled air consists only of O_2 that filled the airways; this dead space air contains no nitrogen (phase I). As the alveolar air that does contain nitrogen begins to be washed out, the concentration of nitrogen in the expired air rises steeply (phase II) to reach a plateau (phase III). Provided that ventilation is distributed evenly and leaves all regions of the lung synchronously during expiration, phase III will be flat. In lung units that fill poorly during inspiration, the nitrogen concentration will be relatively high. These units also tend to empty more slowly and later than well-ventilated units. Consequently, a nonuniform distribution of ventilation will result in a progressive rise in the nitrogen concentration the greater the expired volume during phase III. At low lung volumes, when the airways at the lung bases close, only alveoli at the top of the lung will continue to empty. Because the concentration of nitrogen in the alveoli of the upper lung zones is higher, there will be an abrupt increase in the slope of the nitrogen–lung volume curve (phase IV). The volume at which this deflection occurs is known as the *closing volume*.

Dynamic pressure-volume relationships. The changes in lung volume and pleural

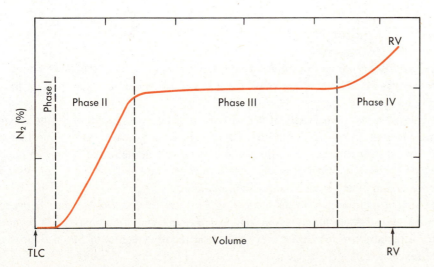

Fig. 37-24 ■ Single-breath nitrogen washout curve. Lung volume is plotted on the horizontal axis. The concentration of nitrogen in the exhaled air is plotted on the vertical axis.

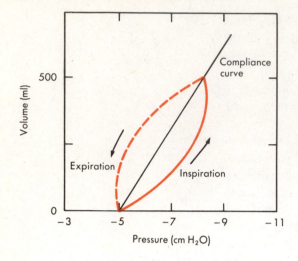

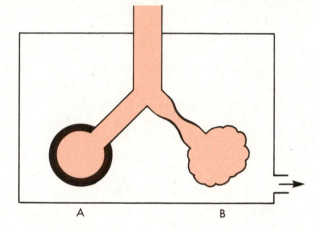

Fig. 37-25 ■ Dynamic pressure-volume relationships during a single breath. Pleural pressure is plotted on the horizontal axis, and lung volume is plotted on the vertical axis. Inspiration is designated by the solid line and expiration by the interrupted line.

Fig. 37-26 ■ Two lung units are arranged in parallel. Unit *A* with a low resistance and low compliance has a short time constant and fills quickly. Unit *B* with a large resistance and a high compliance has a long time constant. It fills more slowly and less completely on application of a distending force.

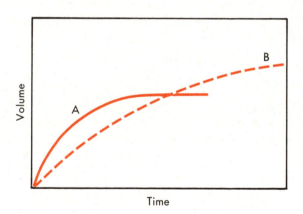

pressure during a normal breathing cycle may be displayed as a pressure-volume loop. Such changes in pressure and volume are affected by the elastic properties of the lung and the resistance of the airways (Fig. 37-25).

Airflow ceases momentarily at the end of expiration and inspiration. The change in pleural pressure from end-expiration to end-inspiration is a measure of the increasing elastic recoil of the lungs. The ratio of the change in volume to the change in pressure, the slope of the line connecting the end-inspiratory and end-expiratory points on the pressure-volume loop is the dynamic compliance.

During inspiration, when air is flowing into the lung, the difference between the actual pleural pressure at any given lung volume and that at end-expiration reflects the forces required to overcome not only the elastic recoil of the lungs but also airway and tissue resistance.

Normally the dynamic compliance closely approximates the inspiratory static compliance. It remains essentially unchanged even when breathing frequency is increased

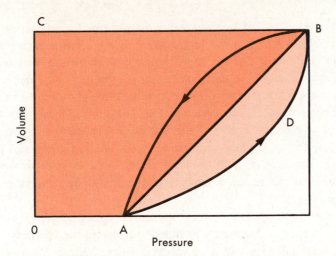

Fig. 37-27 ■ Dynamic pressure-volume curve of the lung illustrating the mechanical work required to overcome elastic forces (dark color: *0ABC0*) and flow and tissue resistance (light color: *ABDA*).

up to 60 breaths per minute. This indicates that airspaces normally fill evenly and empty synchronously even when rates of airflow are high and changes in lung volume are rapid.

As shown in Fig. 37-26, lung units consisting of airways and alveoli are arranged in parallel. The distribution of ventilation to parallel lung units depends on the resistance and compliance of the respective airways and airspaces. In order for ventilation to be distributed evenly at high breathing frequencies and rates of airflow, the resistance and compliance of parallel units must be distributed such that the products of resistance and compliance, that is, the time constants, must be approximately the same.

If the airway resistance of one unit is relatively great, it will not fill as completely, nor will it empty as well as other parallel units. Similarly, lung units that have a greater compliance than other parallel units may not fill as completely if the time for inspiration is limited. Also, high-compliance alveoli tend to empty more slowly during expiration (Fig. 37-26).

Fenestrations in the lung tissue between alveoli (pores of Kohn) and between alveoli and respiratory bronchioles (canals of Lambert) allow air to reach alveoli even if the bronchioles that supply them directly are blocked. In healthy persons they help to minimize further the effects of unequal time constants within the lung.

Also, the parenchymal network of the lung promotes an even distribution of ventilation by resisting distortion. This effect is known as *interdependence*. Adjacent alveoli may have common walls. Also, the traction exerted by the tissue surrounding an alveolus with an obstructed airway will increase as the lung around it expands. In addition, there is interdependence between the movement of the lung and the chest wall. Failure of a region of the lung to enlarge when the chest wall expands will make the local pleural pressure more negative. This will increase the difference in pressure across the region and thereby promote expansion.

Work of breathing. The respiratory muscles perform mechanical work in overcoming the elastic recoil of the lung and the nonelastic resistance of the airways and tissues during breathing. Work (W) is defined as the product of pressure (P) and volume (V) according to the equation:

$$W = \int PdV \qquad (10)$$

The work of breathing is determined from the dynamic pressure-volume loop of the lung. Fig. 37-27 shows the changes in pleural pressure and lung volume over the course of a breath. The dark colored area represents the work performed during inspiration in overcoming the elastic forces of the lung, and the light colored area represents the work required of the respiratory muscles to overcome flow and tissue resistance. The work needed to expand the lung from functional residual capacity is provided not only by the muscles of inspiration but also in part by the outward recoil of the passive chest wall.

The mechanical work used to overcome the elastic recoil and to expand the lung is partially stored as potential energy. That energy is released during expiration and is used to overcome airway and tissue resistance. Accordingly, expiration is passive during quiet breathing. However, at high levels of ventilation and when airway resistance is increased, additional mechanical work may be required during expiration to overcome these nonelastic forces.

It is important to distinguish between mechanical work and energy expenditure. For a given change in pressure there may be little expansion of the thorax when the various respiratory muscles fail to act in a coordinated manner. Similarly, when agonist and antagonist muscle groups contract simultaneously, considerable energy may be expended, but little mechanical work is performed.

The energy expenditure during breathing can be determined from the O_2 cost of breathing. In order to determine the O_2 cost of breathing, the O_2 consumption is measured at rest and at an increased level of ventilation produced by voluntary hyperventilation or CO_2 breathing. Provided there are no other factors acting to increase O_2 consumption, the added O_2 uptake at the higher level of ventilation represents that used by the muscles of respiration. The O_2 cost of breathing is normally about 1 ml/L of ventilation. Ordinarily it constitutes less than 5% of the total O_2 consumption of the body. At high levels of ventilation, however, the O_2 cost of breathing becomes progressively greater.

The efficiency of the respiratory muscles is defined as the ratio of mechanical work to the energy expenditure, that is

$$\text{Efficiency} = \frac{\text{Mechanical work}}{\text{Energy expenditure}} \tag{11}$$

The efficiency of the respiratory muscles is normally only about 5% to 10%, considerably less than the efficiency of limb skeletal muscle. In the presence of lung or chest wall disease, the respiratory muscles are even less efficient.

At any given level of ventilation, the work of breathing depends on the breathing pattern. Large tidal volumes increase the elastic work of breathing, whereas rapid breathing frequencies increase the work against resistive forces. Tidal volume and breathing frequency during spontaneous breathing tend to assume values that minimize the work of breathing (Fig. 37-28).

Fig. 37-28 ■ Relationship of work of breathing to breathing frequency at a constant level of alveolar ventilation. During ventilation at low breathing frequencies with large tidal volumes, the resistive work of breathing is low, but the elastic work of breathing is high. In contrast, rapid, shallow breathing is characterized by a higher resistive but a lower elastic work of breathing. During spontaneous breathing, tidal volume and breathing frequency are regulated to minimize the total work of breathing.

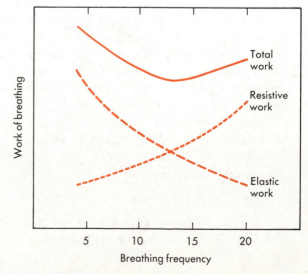

Lung volumes. The subdivisions of thoracic gas volume can be determined using a spirometer, a gas volume recorder. There are many different types of spirometers, but the simplest and one of the most commonly used is the water spirometer, which consists of a double-walled drum into which is fitted a bell. The bell is attached by means of a pulley to a pen that writes on a rotating cylinder (Fig. 37-29).

The residual volume, the air remaining in the lung after a maximum expiration, cannot be determined directly by a spirometer, and alternative methods must be employed. In practice, functional residual capacity (FRC) is measured by either the closed-circuit helium dilution, the open-circuit nitrogen washout, or the body plethysmograph technique. The residual volume is then ascertained by subtracting the expiratory reserve volume from the functional residual capacity.

The closed-circuit helium dilution method involves a period of rebreathing from a spirometer containing a mixture of helium and air. During rebreathing, the expired carbon dioxide is absorbed by soda lime, and the O_2 that is consumed is replaced by adding O_2 to the spirometer. As the helium in the spirometer mixes with the air in the lung, the concentration of helium in the circuit falls. Helium is inert and essentially insoluble so there is no significant transfer of helium across the alveolar-capillary membrane and no significant change in the quantity of helium in the lung-spirometer system. Accordingly, the product of the initial helium concentration (He_I) and the volume of gas in the spirometer (Vs) at the start of the test will equal the product of the final concentration of helium (He_F) at the end of the test and the total volume of gas in the spirometer (Vs) and in the lungs at the end of a normal expiration (FRC). After rearrangement

$$FRC = Vs \, (He_I - He_F)/He_F. \qquad (12)$$

The open-circuit nitrogen washout method involves a period of 100% O_2 breathing for a number of minutes in order to completely displace all nitrogen from the lungs. The quantity of nitrogen displaced from the lungs during the test is determined from the

■ *Assessment of mechanical function of the lung*

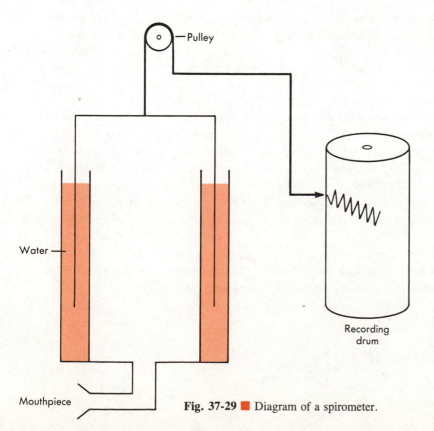

Pulley

Water

Recording
drum

Mouthpiece

Fig. 37-29 ■ Diagram of a spirometer.

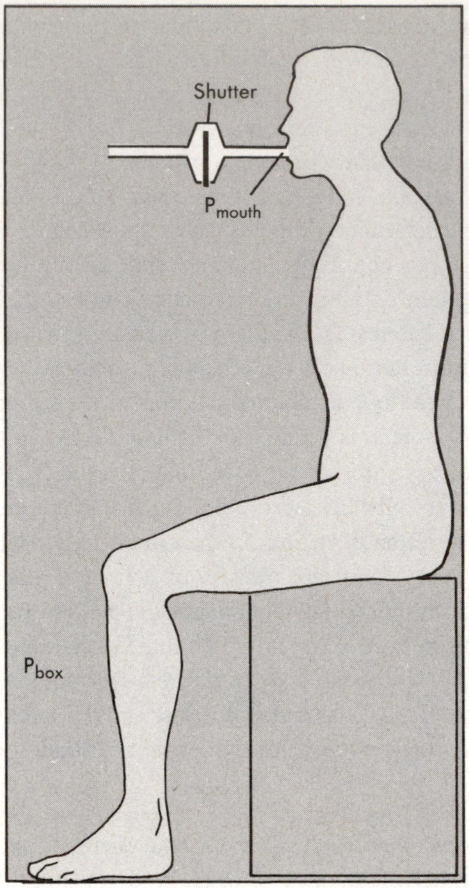

Fig. 37-30 ■ Diagram of a body plethysmograph.

product of the total volume of expired air (VE) and the nitrogen concentration in the mixed expired air (FEN$_2$). This is equal to the quantity of nitrogen in the lungs at the end of a normal expiration before the start of O$_2$ breathing, the product of the FRC and the concentration of nitrogen in the lung (FIN$_2$). After rearrangement

$$FRC = V_E \times \frac{FE_{N_2}}{FI_{N_2}} \tag{13}$$

The volume of gas in the lungs can also be determined using a body plethysmograph, a sealed enclosure within which an individual may be seated. The subject makes gentle inspiratory and expiratory efforts against a closed shutter at the mouth (Fig. 37-30). During inspiratory efforts against the closed airway, air in the lungs undergoes decompression; lung volumes increase slightly, and the alveolar pressure falls to subatmospheric levels. Because the plethysmograph is sealed, the increase in lung volume produces a compression of the air and a consequent increase in pressure within the plethysmograph. Expiratory efforts against an obstructed airway produce the opposite effects.

The volume of air in the lungs is determined by applying Boyle's law, which states that the product of pressure and volume is constant for a given quantity of gas at a constant temperature. Boyle's law can also be expressed as follows:

$$P_i \times V_i = (P_i + \Delta P) \times (V_i + \Delta V) \tag{14}$$

where

P_i = initial pressure
V_i = initial volume
ΔP = change in pressure
ΔV = corresponding change in volume

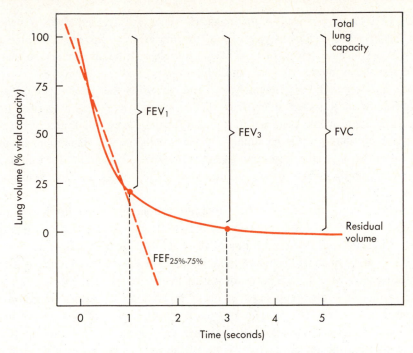

Fig. 37-31 ■ Spirometric tracing of a maximum expiratory vital capacity maneuver. Lung volume expressed as a percentage of the vital capacity is plotted on the vertical axis. Time from the onset of the expiratory maneuver is plotted on the horizontal axis. The forced vital capacity (FVC) and the forced expiratory volumes in 1 second (FEV$_1$) and 3 seconds (FEV$_3$) are shown. The forced midexpiratory flow rate (FEF$_{25\%-75\%}$) is the average airflow rate over the middle half of the vital capacity. It is determined from the slope of the line connecting the points on the curve that correspond to 75% and 25% of the vital capacity.

With respect to the respiratory system, V_i is the lung volume at the end of a normal expiration, and P_i is the pressure in the airways and alveoli (atmospheric pressure at the end of expiration). Changes in intrapulmonary pressure (ΔP) during breathing efforts against the closed shutter are determined from changes in pressure at the mouth; in the absence of flow, pressures are equal throughout the system. Changes in lung volume (ΔV) consequent to gas expansion or compression are reflected by changes in pressure within the plethysmograph.

Airflow rates. The flow resistive properties of the airways are commonly assessed from measurements of airflow rates out of the lungs during a rapid, forceful expiratory maneuver from total lung capacity to residual volume (Fig. 37-31). When a spirometer is used, the rates of airflow are determined from the volume of air exhaled during particular time intervals. Measurements are made of the volume exhaled during the first second, the forced expiratory volume in 1 second (FEV$_1$), and over the first 3 seconds, the forced expiratory volume in 3 seconds (FEV$_3$). FEV$_1$ and FEV$_3$ are also expressed as a percent of the forced vital capacity (FVC). An additional measurement is that of the mean airflow rate over the middle half of the forced vital capacity, that is, between 25% and 75% of the vital capacity. This is called the *maximum midexpiratory flow rate* (MMF) or the forced midexpiratory flow rate (FEF$_{25\%-75\%}$).

■ *Respiratory muscles*

The respiratory muscles act together in purposeful coordinated patterns to move the chest bellows and produce ventilation. Because of intricate anatomical and mechanical interrelationships, the actions of the individual muscles are quite complicated. Muscles that are active during inspiration include the diaphragm, parasternal muscles between the cartilaginous portions of the ribs, external intercostal muscles, as well as the scalene and sternocleidomastoid muscles. Expiration during quiet breathing is generally passive, resulting from the elastic recoil of the lungs. However, at increased levels of ventilation and in the presence of airway obstruction, the muscles of expiration become active. These muscles include the internal intercostal muscle and the abdominal muscles, such as the external and internal obliques and the transverse and rectus muscles.

The diaphragm. The diaphragm, which is supplied by the phrenic nerves, consists of a thin dome-shaped sheet of muscles that is inserted into the lower ribs. It is the principal muscle of inspiration, and it accounts for more than two thirds of the air that enters the lungs during quiet breathing. During anesthesia, the diaphragm usually accounts for an even greater proportion of the inspiratory air volume. Two thirds of the fibers making up the human diaphragm are slow twitch fibers that are resistant to fatigue.

During inspiration the dome of the diaphragm flattens as the myofibrils shorten. The diaphragm increases the vertical dimensions of the thoracic cavity and the anteroposterior and lateral dimensions of the thorax (Fig. 37-32). With diaphragm contraction, pressure within the abdomen becomes more positive. This positive pressure is important in displacing the lower rib cage outward. As the ribs move upward, rotations about their spinal attachments cause expansions to occur in both the lateral and anteroposterior directions.

External intercostal muscles. The external intercostal muscles extend downward and forward from the lower surface of the rib above to the upper surface of the rib below. These muscles are innervated by the intercostal nerves from the first through the eleventh thoracic spinal cord segments. Contraction of the external intercostal muscles raises the ribs during inspiration. The upward displacement of the upper ribs results in an increase in the anteroposterior dimensions of the chest, while elevation of the lower ribs produces an increase in the transverse dimension of the thorax. Intercostal muscle contraction tenses the intercostal spaces, thereby stabilizing the chest and preventing retraction of those spaces as the pleural pressure becomes more negative during inspiration.

During quiet breathing the intercostal muscles probably act mainly to optimize the action of the diaphragm. When the diaphragm is paralyzed or inactivated, the external intercostal muscles become essential to breathing.

Accessory muscles of inspiration. At high levels of ventilation or when movement of air into the lungs is obstructed, the scalene and sternocleidomastoid muscles contribute to inspiration. The scalene muscles arise from the transverse processes of the lower five cervical vertebrae and insert into the upper aspects of the first and second ribs. Contraction of these muscles elevates and enlarges the upper parts of the chest. Simi-

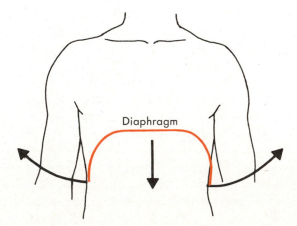

Fig. 37-32 ■ Contraction and descent of the diaphragm during inspiration increases the vertical, anteroposterior, and lateral dimensions of the thorax.

larly, the sternocleidomastoid muscles elevate the sternum and enlarge slightly the anteroposterior and longitudinal dimensions of the chest.

Muscles of expiration. The most important muscles of expiration are the abdominal muscles, which are innervated by nerves from the lower six thoracic and first lumbar segments of the spinal cord. Abdominal muscle contraction depresses the lower ribs, pulls down the anterior part of the lower chest, and compresses the abdominal contents during expiration. The increased abdominal pressure forces the relaxed diaphragm upward, causing it to lengthen.

The abdominal muscles are assisted during expiration by the internal intercostal muscles, which depress the ribs and move them downward and inward. Contraction of the internal intercostals also stabilizes the rib cage and prevents bulging of the intercostal spaces during forceful expiratory maneuvers.

The muscles of expiration are critical in reducing thoracic volume below functional residual capacity and are also important in forceful expiratory maneuvers such as coughing. The abdominal muscles also seem to play a role in maximum inspiratory maneuvers. At lung volumes approaching total lung capacity, abdominal muscles are activated, and this serves to augment the action of the diaphragm in expanding the rib cage.

Assessment of respiratory muscle function. The strength of the respiratory muscles can be determined from the maximum pressures generated during forced inspiratory or expiratory efforts against a closed airway (Fig. 37-33). As with other skeletal muscles, the force of respiratory muscle contraction depends on the velocity of shortening. The less the velocity of shortening, the greater the force exerted. During maneuvers against a closed airway, the respiratory muscles are prevented from shortening and hence contract virtually isometrically. In fact, there are some small changes in muscle length resulting from changes in lung volume consequent to gas expansion or compression, but the changes are negligible. The forces that can be developed during an isometric contraction of muscle depend on the precontraction length of the muscle. The muscles of expiration, for example, are lengthened at total lung capacity and are at their most advantageous position mechanically. It is at that thoracic volume that the expiratory muscles can exert their greatest force. At lower and lower lung volumes there is a progressive decrease in maximum expiratory airway pressure. The muscles of inspiration are longer, and the maximal inspiratory force is greater at lung volumes near resid-

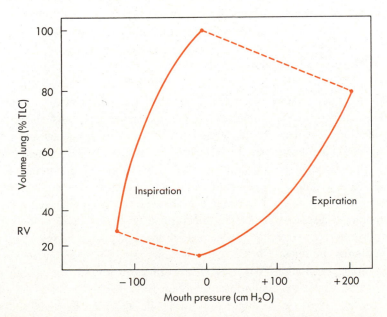

Fig. 37-33 ■ Effects of lung volume on maximum inspiratory and expiratory pressures.

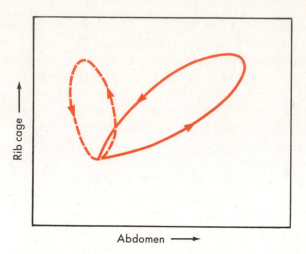

Fig. 37-34 ■ A plot on X-Y coordinates of rib cage and abdomen displacement during a breath. Normally there is synchronous outward displacement of both the rib cage and abdomen during inspiration and inward movement of both the rib cage and abdomen during expiration *(solid line)*. When the diaphragm is paralyzed, the abdomen moves inward during inspiration as the rib cage expands *(dashed line)*.

ual volume. The precontraction length of the muscles of inspiration decreases as thoracic volume increases; consequently the maximum inspiratory airway pressure falls.

The earliest abnormalities associated with respiratory muscle weakness are reduction in maximum inspiratory and expiratory pressures. These changes occur even before there are any significant alterations in lung volumes. More marked weakness of the inspiratory muscles will reduce the inspiratory capacity and lower the vital capacity and total lung capacity. Expiratory muscle weakness, on the other hand, will elevate the residual volume.

Neuromuscular disorders may not affect all muscle groups uniformly. Because the diaphragm is the major muscle of inspiration, disorders that preferentially involve the diaphragm are likely to have the most serious consequences. Diaphragm function can be evaluated directly by measuring transdiaphragmatic pressure either during normal breathing or during maximum inspiratory maneuvers against a closed airway. Transdiaphragmatic pressure is the difference between intrathoracic and intraabdominal pressure. Balloon- or transducer-tipped catheters positioned in the esophagus are used to measure intrathoracic pressure, and catheters in the stomach are used to measure intraabdominal pressure.

A less direct, noninvasive method of evaluating the actions of the diaphragm and intercostal muscles is to record the movements of the rib cage and abdomen during breathing. Devices that measure displacement, such as magnetometers and impedance plethysmographs, have been used for this purpose. As the diaphragm descends during inspiration, the abdomen is displaced outward. The extent of abdominal movement can be used as a measure of diaphragm contribution to tidal volume. Although outward movement of the lower portion of the rib cage can be produced by the actions of the diaphragm, displacement of the upper rib cage at the level of the second or third anterior ribs results from contractions of the intercostal muscles and perhaps also the accessory muscles. Normally the rib cage and abdomen move nearly synchronously during inspiration. When the diaphragm is paralyzed, the abdomen moves inward as the rib cage expands during inspiration (Fig. 37-34). In paralysis of the intercostal muscles, the rib cage moves inward as the abdomen expands with increasing lung volume.

■ *Bibliography*

Journal articles

Agostoni, E.: Mechanics of the pleural space, Physiol. Rev. **42:**57, 1972.

Bolton, T.B.: Mechanisms of action of transmitters and other substances on smooth muscle, Physiol. Rev. **59:**606, 1979.

Briscoe, W.A., and DuBois, A.B.: The relationship between airway resistance, airway conductance, and lung volume in subjects of different age and body size, J. Clin. Invest. **37:**1279, 1958.

Dawson, S.V., and Elliott, E.A.: Wave-speed limitation on expiratory flow—a unifying concept, J. Appl. Physiol. **43:**498, 1977.

DuBois, A.B., Botelho, S.Y., and Comroe, J.H.: A new method for measuring airway resistance in man using a body plethysmograph: values in normal subjects and in patients with respiratory disease, J. Clin. Invest. **35:**327, 1956.

DuBois, A.B., et al.: A rapid plethysmographic method for measuring thoracic gas volume: a comparison with a nitrogen washout method for measuring functional residual capacity in normal subjects, J. Clin. Invest. **35:**322, 1956.

Ferris, B.G., Jr., Mead, J., and Opie, L.H.: Partitioning of respiratory flow resistance in man, J. Appl. Physiol. **19:**653, 1964.

Gibson, G.J., and Pride, N.B.: Lung distensibility: the static pressure-volume curve of the lungs and its use in clinical assessment, Br. J. Dis. Chest **70:**143, 1976.

Glazer, J.B., et al.: Vertical gradients of alveolar size in lungs of dogs frozen intact, J. Appl. Physiol. **23:**694, 1967.

Henderson, R., Hosfield, K. and Cumming, G.: Intersegmental collateral ventilation in the human lung, Respir. Physiol. **6:**128, 1968.

Hyatt, R.E., and Black, L.F.: The flow-volume curve: a current perspective, Am. Rev. Respir. Dis. **107:**191, 1973.

Konno, K., and Mead, J.: Measurement of the separate volume changes of rib cage and abdomen during breathing, J. Appl. Physiol. **22:**407, 1967.

Macklem, P.T.: Airway obstruction and collateral ventilation, Physiol. Rev. **51:**365, 1971.

Macklem, P.T., and Mead J.: Resistance of central and peripheral airways measured by a retrograde catheter, J. Appl. Physiol. **22:**395, 1967.

McCarthy, D.S., et al.: Measurement of ''closing volume'' as a simple and sensitive test for early detection of small airway disease, Am. J. Med. **52:**747, 1972.

Mead, J., et al.: Significance of the relationship between lung recoil and maximum expiratory flow, J. Appl. Physiol. **22:**95, 1967.

Meneely, G.R., et al.: A simplified closed circuit between dilution method for the determination of residual volume of the lungs, Am. J. Med. **28:**824, 1960.

Menkes, H.A., and Traystman, R.J.: Collateral ventilation, Am. Rev. Respir. Dis. **116:**287, 1977.

Milic-Emili, J., et al.: Improved technique for estimating pleural pressure from esophageal balloons, J. Appl. Physiol. **19:**207, 1964.

Milic-Emili, J., et al.: Regional distribution of inspired gas in the lung, J. Appl. Physiol. **21:**749, 1966.

Morgan, T.E.: Pulmonary surfactant, N. Engl. J. Med. **284:**1185, 1971.

Otis, A.B., Fenn, W.O., and Rahn, H.: Mechanics of breathing in man, J. Appl. Physiol. **2:**592, 1950.

Pride, N.B.: The assessment of airflow obstruction: role of measurement of airways resistance and of tests of forced expiration, Br. J. Dis. Chest **65:**135, 1971.

Rahn, H., et al.: The pressure-volume diagram of the thorax and lung, Am. J. Physiol. **146:**161, 1946.

Sharp, J.T., et al.: Relative contributions of rib cage and abdomen to breathing in normal subjects, J. Appl. Physiol. **39:**608, 1975.

Stubbs, S.E., and Hyatt, R.E.: Effect of increased lung recoil on maximal expiratory flow in normal subjects, J. Appl. Physiol. **32:**325, 1972.

Vincent, N.J., et al.: Factors influencing pulmonary resistance, J. Appl. Physiol. **29:**236, 1970.

Woolcock, A.J., Vincent, N.J., and Macklem, P.T.: Frequency dependence of compliance as a test for obstruction in the small airways, J. Am. Invest. **48:**1097, 1969.

Books and monographs

Campbell, E.J.M., Agostoni, E., and Newsom Davis, J.: The Respiratory muscles: mechanics and neural control, ed. 2, Philadelphia, 1970, W.B. Saunders, Co.

Clements, J.A., and King, R.J.: Composition of the surface active material. In Crystal, R.G., editor: The biochemical basis of pulmonary function, New York, 1976, Marcel Dekker, Inc.

Clements, J.A., and Tierney, D.F.: Alveolar instability associated with altered surface tension, In Fenn, W.O. and Rahn, H., editors: Handbook of physiology, Section 3; Respiration, vol. III, Bethesda, Md., 1964, American Physiological Society.

Comroe, J.H.: Physiology of respiration, Chicago, 1974, Year Book Medical Publishers, Inc.

Murray, J.F.: The normal lung: the basis for diagnosis and treatment of pulmonary disease, Philadelphia, 1976, W.B. Saunders Co.

chapter 38

The pulmonary circulation

The pulmonary vascular system is a low-resistance network of highly distensible vessels. The main pulmonary artery is much shorter than the aorta. The walls of the pulmonary artery and its branches are much thinner than the walls of the aorta, and they contain less smooth muscle and elastin. Contrary to systemic arterioles, which have very thick walls composed mainly of circularly arranged smooth muscle, the pulmonary arterioles are very thin and contain little smooth muscle. The pulmonary arterioles certainly do not have the same capacity for vasoconstriction as do their counterparts in the systemic circulation. The pulmonary veins are also thin and possess little smooth muscle.

The pulmonary capillaries also differ markedly from the systemic capillaries. Whereas the systemic capillaries are usually arranged as a network of tubular vessels with some interconnections, the pulmonary capillaries are sandwiched between adjacent alveoli in such a manner that the blood flows as if in thin sheets. This provides for maximum exposure of the capillary blood to the alveolar gases. The total surface area for exchange between alveoli and blood has been estimated to be about 50 to 70 square meters. Only thin layers of vascular endothelium and alveolar epithelium separate the blood and alveolar gas. The thickness of the sheets of blood between adjacent alveoli depends on the intervascular pressure and the intraalveolar pressure. Ordinarily the width of an interalveolar sheet of blood is about equal to the diameter of a red blood cell. During pulmonary vascular congestion, as when the left atrial pressure becomes elevated, the width of the sheet may increase severalfold. Conversely, when the local alveolar pressure exceeds the adjacent capillary pressure, the capillaries are compressed and may collapse. Hydrostatic factors play a crucial role in this phenomenon, particularly with respect to the distribution of blood flow to the various regions of the lungs as described later.

The total cross-sectional area of the pulmonary vascular bed markedly increases from the level of the pulmonary artery to the level of the capillary bed. Despite this increase in cross-sectional area, the resistance of all of the smaller blood vessels, arranged in parallel, is greater than it is in the major arteries, for the same reason that prevails in the systemic circulation (Chapter 28). However, the relative increase in resistance with diminishing vessel size is far less than in the systemic circulation.

In addition to functioning in gas exchange, the pulmonary capillaries also act as filters of systemic venous blood. These filters cleanse the blood of small clots and other particulate materials before they can reach the capillary beds of other vital organs such as the brain. Finally, the capillary endothelium has metabolic pharmacokinetic effects that influence the circulating level of certain vasoactive substances.

The lungs also receive blood through bronchial vessels from the systemic circulation. These vessels nourish the walls of the tracheobronchial tree down to the terminal bronchioles, the supporting vessels, and the outer layers of the pulmonary arteries and veins.

In normal persons the average systolic and diastolic pressures in the pulmonary artery are about 25 and 10 mm Hg, respectively, and the mean pressure is about 15 mm Hg. These pressures are much lower than those in the aorta and reflect the much lower resistance of the pulmonary vascular bed as compared to the systemic vascular bed (Fig. 38-1). The mean pressure in the left atrium is normally about 5 to 8 mm Hg so the total pulmonary arteriovenous pressure gradient is only about 10 mm Hg. The mean hydrostatic pressure in the pulmonary capillaries lies between the pulmonary arterial and pulmonary venous values, but it is somewhat closer to the latter. Mean pulmonary capillary pressure is therefore about 10 mm Hg.

Vascular pressure, the difference between pressure in the lumen of a blood vessel and atmospheric pressure, varies in similar-sized vessels, depending on their position in the lung. In the erect posture, vascular pressure is greater at the base of the lung than at the apex. The pressure differences in the lung vessels are caused by gravitational effects. The pressure at a point at the top of the lung (the apex of the lung in the erect posture) will be equal to the inlet pressure of the pulmonary artery less the height of a

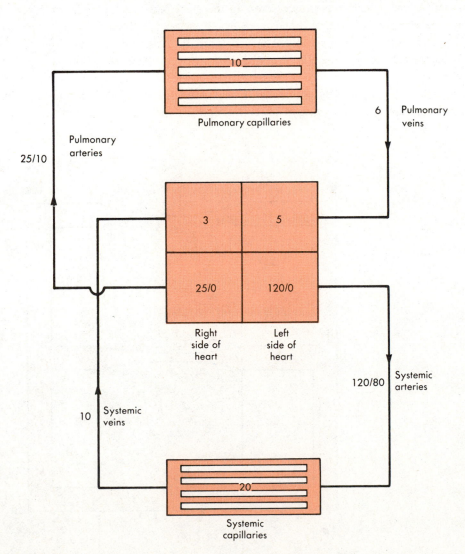

Fig. 38-1 ■ Systolic and diastolic pressures (in mm Hg) in the systemic and pulmonary arteries and mean pressures in the systemic and pulmonary capillaries and veins.

column of blood extending from the inlet to that point. The pressure in capillaries at the bottom of the lung will be greater than inlet pressure by the height of such a column extending from the inlet to that point. When an individual is recumbent, the effects of gravity on vascular pressure are less than when that person is in the erect position because the lungs are not as wide as they are long. Because arterial pressures are low in the pulmonary circulation, these hydrostatic effects substantially influence the driving pressure.

The flow between two points in a system of vessels depends on the driving pressure (the difference between the vascular pressure at these locations). If the pulmonary vessels had rigid walls, there would be no effect of gravity on the total driving pressure, that is, the difference between arterial and venous pressures, because gravity would affect both arterial and venous pressures equally (Chapter 34). The smaller blood flow that is actually observed at the top of the lung than at the bottom occurs because (1) the walls of the pulmonary capillaries are collapsible and not rigid and (2) alveolar pressure can be greater than capillary vascular pressure at the top of the lung but can be much less than capillary pressure at the bottom of the lung.

As shown in Fig. 38-2, there is a potential region at the top of the lung *(zone 1)* where pulmonary arterial pressure may fall so low that it is less than alveolar pressure (normally close to atmospheric pressure). In the erect person, capillaries in this zone would be completely collapsed, and there would be no blood flow. Zone 1 does not exist under normal conditions, but it may be present if pulmonary arterial pressure is reduced or if alveolar pressure is increased (by positive-pressure ventilation).

Pulmonary arterial pressure increases further down the lung because of the greater hydrostatic effects of gravity. Pulmonary artery pressure exceeds alveolar pressure, but venous pressure, which is always lower than arterial pressure, is less than alveolar

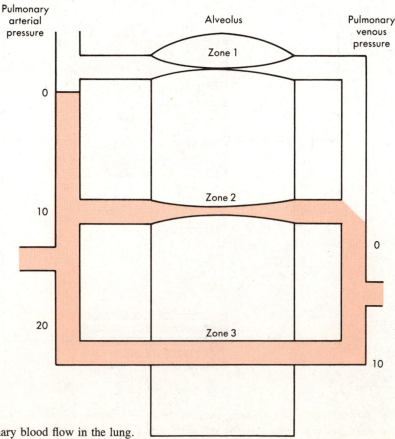

Fig. 38-2 ■ Distribution of pulmonary blood flow in the lung.

pressure *(zone 1)*. Blood flow in *zone 2* depends on the difference between arterial and alveolar pressures and not the arterial-venous pressure difference. This condition corresponds to that of a waterfall and can be simulated by a Starling resistor, a flexible tube inside a rigid chamber (Chapter 33). When chamber pressure is greater than outflow pressure, the tube collapses at its downstream end, and the pressure in the tube at this point limits flow. The pulmonary capillary bed behaves similarly to the Starling resistor. The lower the level within *zone 2,* the greater will be the arterial pressure. However, alveolar pressure is unaffected by gravity, and it therefore remains constant throughout the entire lung. Because of the increasing gradient between arterial and alveolar pressure, blood flow progressively increases down this zone from top to bottom.

In *zone 3* venous pressure exceeds alveolar pressure, and flow is regulated by the arterial-venous pressure difference. The capillary pressure (which is between arterial and venous pressure) increases down the zone, while the alveolar pressure outside the vessels remains constant. The increased intravascular pressure enlarges capillaries and reduces resistance, increasing blood flow in the apex to base direction within this region.

The resistance of larger pulmonary arteries and veins can also influence the distribution of blood flow. The pressure surrounding these vessels approximates pleural pressure. At the very bottom of the lung, because pleural pressure is less subatmospheric than it is near the top of the lung, the transmural vascular pressure is smaller, and the blood vessels narrow. This results in a decrease in regional blood flow at the very bottom of the lung. When the lung is at residual volume, rather than at total lung capacity, the distribution of perfusion is much more uniform. The distribution is determined by the large vessel resistance, which increases from the top to the bottom of the lung, thereby counteracting the hydrostatic effect on flow through the pulmonary capillaries.

The driving pressure across the entire pulmonary vascular bed is the difference between the pressures in the pulmonary artery and the left atrium. It is possible to place catheters to measure pressure at these specific points in the pulmonary circulation. In normal humans the driving pressure across the pulmonary vascular bed is 10 to 15 mm Hg. Obstruction to flow into the left ventricle, as occurs in mitral stenosis, can increase all the pulmonary vascular pressures without affecting the total driving pressure. However, this can result in sufficient elevation in capillary pressure so as to force fluid from the capillaries into the lung. Capillary pressures cannot be measured directly, nor can they be reliably deduced from measurements of pulmonary arterial pressure. For example, obstruction to pulmonary artery flow raises mean pulmonary artery pressure proximal to the obstruction, but pulmonary capillary pressure and left atrial pressure may be normal.

Passive effects on pulmonary vascular resistance

It is useful for purposes of comparison to describe the resistance of a system of blood vessels as the ratio of driving pressure to blood flow. The high resistance of the systemic circulation is largely caused by the muscular arterioles (Chapter 32), which control the distribution of blood flow to various organs of the body. The pulmonary circulation does not have such thick-walled narrow vessels, and the total resistance of the pulmonary circulation in healthy subjects is only one tenth that in the systemic circulation. Pulmonary resistance, however, is not constant, but it varies passively with changes in vascular pressure and lung volume.

The relationship between pulmonary artery pressure and pulmonary vascular resistance is shown in Fig. 38-3. Pulmonary vascular resistance decreases as the pressure within the vessels rises. As the left atrial or pulmonary artery pressure rises, some capillaries that are usually closed begin to conduct blood, thus lowering the overall resistance. In addition, as vascular pressures rise, more and more capillary segments become distended. Because the distension of the pulmonary vascular bed has a limit, the effects of pulmonary arterial pressure changes on resistance are greater when left atrial pressure is low.

Changes in lung volume affect pulmonary vascular resistance by altering transmural pressure in the blood vessels. Transmural pressure is the difference in pressure inside and outside the vessel wall. Increased transmural pressure tends to enlarge blood vessel diameter. Because vascular pressures are so low in the pulmonary circulation, small changes in the pressure surrounding the vessel can produce relatively large percentage changes in the transmural pressure.

From a functional standpoint, it is useful to think of two categories of pulmonary vessels: the extraalveolar vessels (e.g., the major arteries and veins) and the alveolar vessels (e.g., the capillaries and some arterioles and venules), which behave as if they are surrounded by alveolar pressure (Fig. 38-4).

At large lung volumes the pressure surrounding extraalveolar blood vessels is highly subatmospheric. There is some evidence that this pressure is somewhat less than the pressure around the whole lung (pleural pressure). Because of the large transmural vas-

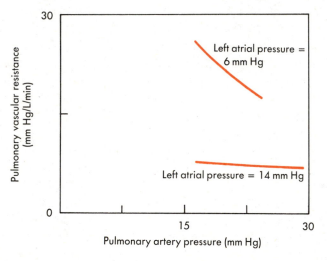

Fig. 38-3 ■ The effects of changes in pulmonary artery pressure on pulmonary vascular resistance. As pulmonary artery pressure increases, vascular resistance falls. This effect is attenuated when left atrial pressure is high.

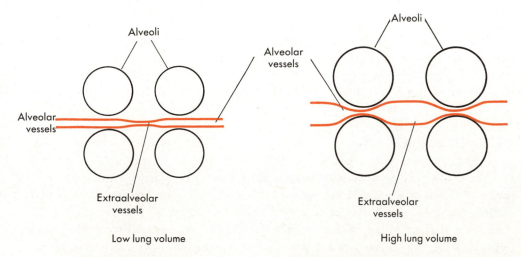

Fig. 38-4 ■ Effects of lung volume on the caliber of alveolar and extraalveolar vessels.

cular pressure when the lung is distended, the extraalveolar vessels are pulled open. As the lung expands, the transmural pressure of both the pulmonary arteries and veins increases, as does their caliber. This tends to lower pulmonary vascular resistance.

On the other hand, alveolar vessels tend to be compressed as the lung expands and alveoli enlarge. This tends to increase pulmonary vascular resistance. Because of these opposing effects of lung volume on the caliber of alveolar and extraalveolar vessels, total pulmonary vascular resistance usually is at its minimum value at functional residual capacity. As lung volume decreases, the resistance of the extraalveolar vessels increases. If the lung is completely collapsed, the tone of the smooth muscle and the elastic tissue in the walls of extraalveolar vessels tend to obstruct blood flow. Pulmonary artery pressure must be raised before blood flow begins; this pressure is called the *critical opening pressure*. When vascular pressures are high, the effects of lung volume changes on transmural pressure are less.

■ *Active regulation of pulmonary vascular resistance*

Although changes in pulmonary resistance are achieved mainly by passive factors, resistance can be actively modified by neural, chemical, and humoral influences. The pulmonary blood vessels are innervated by sympathetic fibers but more sparsely than systemic vessels. Unlike most systemic vessels, the pulmonary blood vessels receive some parasympathetic innervation. The density of nerve fibers is greatest in the largest vessels and decreases with vessel size. Vessels less than 3 mm in diameter have few or no demonstrable nerve fibers. In the adult, stimulation of the stellate ganglion increases the stiffness of the large blood vessels but produces a less consistent increase in vascular resistance. The fetal pulmonary circulation responds to both parasympathetic and sympathetic nerve stimulation. The influence of baroreceptor and chemoreceptor stimulation on resistance seems to be weak, although systemic hypoxia does increase the resistance of the large vessels in the lung.

Local alveolar hypoxia produces vasoconstriction, which increases vascular resistance even in the denervated lung. This reaction seems to be important in shifting blood away from poorly ventilated alveoli. The reaction is accentuated by hypercapnia, acidosis, and hypertrophy of the blood vessel (Fig. 38-5). The mechanism by which local

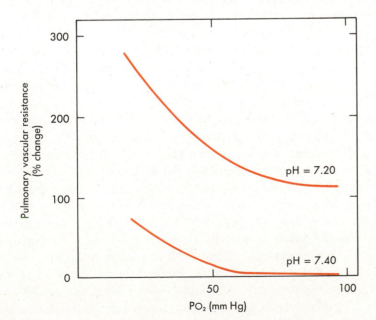

Fig. 38-5 ■ Effects of hypoxia on pulmonary vascular resistance. Hypoxia causes pulmonary vasoconstriction and an increase in pulmonary vascular resistance. This effect is accentuated by acidosis.

hypoxia produces its effects is still under investigation. Some studies suggest that hypoxia acts by altering ion transport across the cell membrane of the smooth muscle in the vessel wall or by affecting the contractile mechanism of the cells. Hypoxia does not appreciably constrict isolated vessels but has a much greater effect when the vessel exposed to hypoxia is surrounded by a cuff of lung tissue. This suggests that hypoxia releases some mediator from the tissue cells, and the mediator then produces the vasoconstriction. Mast cells, which are found in the interstitial space, degranulate when exposed to hypoxia. They contain vasoactive substances that, when released, could produce vasoconstriction. Prostaglandins, angiotensin, epinephrine, norepinephrine, and serotonin are pulmonary vasoconstrictors that have been implicated in the hypoxic vascular response. Histamine is the pulmonary vasoconstrictor that has been most extensively studied. Although there is considerable evidence supporting the idea that histamine released by mast cells mediates the hypoxic response or at least modulates that response, all experimental data do not support this idea.

■ *Fluid movement through the lung*

Efficient gas exchange requires that the alveolocapillary membrane be kept free of excess water. The alveolocapillary barrier has thick and thin portions. In the thin part, the barrier consists of the alveolar epithelium, the capillary endothelium, and their fused basement membranes. In the thick parts, a layer of interstitial tissue separates the two basement membranes. The alveolar epithelium is made of membranous pneumocytes (type I), which have few organelles, and granular pneumocytes (type II), which contain vesicles and multilaminar bodies that are believed to contain the precursors of surfactant. Type I and type II cells are held tightly together so that molecules can pass into the alveoli only by going through cells and not between them. The pulmonary capillary endothelium is continuous and nonfenestrated. The junctions between endothelial cells are only partially obliterated, and between some endothelial cells there is a gap through which molecules can pass.

It is generally believed that fluid transfer across pulmonary capillaries, as well as systemic capillaries, depends on hydrostatic and osmotic forces and obeys the Starling hypothesis (Chapter 31). The capillary hydrostatic pressure favors movement of fluid out of the capillary, and it is opposed by the hydrostatic pressure in the interstitial fluid. The colloid osmotic pressure of the plasma proteins tends to draw fluid into the capillary, but it is opposed by the osmotic pressure of the proteins in the interstitial fluid.

The colloid osmotic pressure within the capillary is about 25 mm Hg. The capillary hydrostatic pressure is probably about halfway between the arterial and venous pressures, and it is higher at the bottom of the lung than at the top. The colloid osmotic pressure of the interstitial fluid is not known precisely but seems to be similar to that of the lung lymph, the colloid osmotic pressure of which is about 20 mm Hg. The interstitial hydrostatic pressure is unknown, but it is thought to approximate pleural pressure.

Based on these estimated values, there seems to be a net force that favors a continuous leak of fluid out of the capillaries into the thick portion of the interstitial space. This fluid travels through the interstitital space to the perivascular and peribronchial spaces within the lung; then it passes into lymphatic channels. Although lymphatic channels do not extend into the alveolar septum but instead end near the terminal bronchioles, they are strategically placed to drain fluid that has passed into the interstitial space. They are held open by tissue tethers that connect their external surfaces to the surrounding connective tissue. Pulmonary lymph is propelled toward the hilum of the lung by the active contraction of smooth muscle in the walls of lymphatics as well as by pulsations arising from ventilatory movement and circulatory contractions. Unicuspid, funnel-shaped valves in the small lymphatics prevent backward flow. Lymph flow from the lung, which is normally about 20 ml/minute, greatly increases if the capillary pressure is raised over a long period and helps to prevent edema formation.

Pulmonary edema can occur with an increase either in hydrostatic pressure or in capillary permeability such that both water and proteins can easily transverse the capillary wall. For example, release of endotoxin by bacteria infecting the lung can cause pulmonary edema by increasing capillary permeability. Greater hydrostatic pressure may also make the capillary more permeable by widening the endothelial cell pores through which water and protein exit. Increased left atrial pressure can produce pulmonary edema by increasing capillary hydrostatic pressure.

Interstitial edema occurs before alveolar edema. It has been suggested that the pressure-volume relationship in the interstitital space is such that even a small amount of excess fluid makes interstitial pressure much more positive, which would tend to retard fluid flow from the capillary. When large amounts of fluid are present in the interstitium, the compliance of the interstitial space becomes much larger so that more fluid can be accepted with only small increases in pressure. This allows the interstitial space to serve as a fluid reservoir that protects the alveolus. The alveolus is further protected by the tight junctions between the epithelial lining cells. Surface-active forces in the alveoli, which would tend to draw fluid into the airspace, are diminished by the layer of surfactant that coats the alveolus. It is only when edema is severe that fluid ultimately enters the alveolus and interferes significantly with gas exchange.

Water accidently entering the alveoli through the airway can be reabsorbed because capillary oncotic pressure is much greater than capillary hydrostatic pressure. For example, large quantities of water may be absorbed from the alveoli in drowning.

■ *Metabolic function of the pulmonary endothelium*

The pulmonary circulation has a unique position in the vascular system because all of the venous blood from the body tissues must pass through the lung before it is recycled. This allows the lung to play an important role in regulating the level of vasoactive substances in the circulation. Attention has been focused on the capillary endothelial cell as a possible site of this metabolic activity. The surface area of the endothelium is large because of the dense network of capillaries in the lung and because endothelial cells have many projections and indentations (caveolae) that seem to contain many enzyme-rich sites. Some prostaglandins of the E and F types, serotonin, norepinephrine, and histamine, are removed from the blood while passing through the lung. Uptake of some substances such as serotonin seems to depend on a specific enzymatic carrier that can be saturated.

Angiotensin I is activated to the more powerful vasoconstrictor, angiotensin II, on passage through the lung (Chapter 47). Angiotensin I, a decapeptide, is formed by the action of renin (secreted by the kidney) on angiotensinogen (produced by the liver). In the lung two peptides are enzymatically cleaved from angiotensin I to form the octapeptide, angiotensin II. The same converting enzyme that is responsible for the conversion of angiotensin I to angiotensin II also inactivates the nonapeptide, bradykinin, which lowers blood pressure. Conversion of angiotensin I to angiotensin II seems to be delayed by hypoxia.

■ *The bronchial vessels*

The bronchial arteries usually originate from the midthoracic aorta, or they may arise from the branches of the intercostal, internal mammary, or subclavian arteries. Two or three arteries accompany the bronchi and form a peribronchial plexus. Venous blood from the large bronchi and pleural space drains into the azygous and hemiazygous systems. Blood from small bronchi enters the pulmonary veins, adding some poorly oxygenated blood to the blood entering from the lungs. A network of arterioles, venules, and capillaries, which form on either side of the muscular layer of the bronchi, is interposed between the arteries and veins. The submucosal network intermingles with pulmonary capillaries so that changes in the permeability of the small bronchial vessels may affect fluid exchange across the pulmonary capillaries.

Under normal conditions, the blood furnished by the bronchial artery does not reach gas-exchanging surfaces. Because the pressure in the bronchial arteries is the same as systemic pressure, it is sufficiently great to perfuse bronchi, bronchioles, and supporting tissue everywhere in the lungs, regardless of body position. Although in humans bronchial blood flow is only 1% to 2% of pulmonary blood flow, this is appropriate to the mass of the lung that is nourished (about 475 g). If the pulmonary circulation becomes obstructed, the bronchial arteries enlarge and open connections with precapillary vessels in the pulmonary circuit. In these circumstances, little or no O_2 may be extracted from the alveoli that receive bronchial blood because the bronchial arterial blood is already well oxygenated. However, this bronchial blood supplies nutritive substrates and maintains the viability of the tissues comprising the alveolar ducts and alveoli.

■ Bibliography

Journal articles

Anthonisen, N.R., and Milic-Emili, J.: Distribution of pulmonary perfusion in erect man, J. Appl. Physiol. **21**:761, 1966.

Borst, H.G., et al.: Influence of pulmonary arterial and left atrial pressures on pulmonary vascular resistance, Circ. Res. **4**:393, 1956.

Dollery, C.T., and Glazer, J.B.: Pharmacological effects of drugs on the pulmonary circulation in man, Clin. Pharmacol. Ther. **7**:807, 1966.

Glazer, J.B., and Murray, J.F.: Sites of pulmonary vasomotor reactivity in the dog during alveolar hypoxia and serotonin and histamine infusion, J. Clin. Invest. **50**:2550, 1971.

Hughes, J.M.B., et al.: Effect of lung volume on the distribution of pulmonary blood flow in man, Respir. Physiol. **4**:58, 1968.

Ingram, R.H., et al.: Effects of sympathetic nerve stimulation on the pulmonary arterial tree of the isolated lobe in-situ, Circ. Res. **22**:801, 1968.

Kato, M., and Staub, N.C.: Response of small pulmonary arteries to unilobar hypoxia and hypercapnia, Circ. Res. **19**:426, 1966.

Lee, G. deJ., and DuBois, A.B.: Pulmonary capillary blood flow in man, J. Clin. Invest. **34**:1380, 1955.

Permutt, S., and Riley, R.L.: Hemodynamics of collapsible vessels with tone: the vascular waterfall, J. Appl. Physiol. **18**:924, 1963.

Szidon, J.P., and Flint, J.F.: Significance of sympathetic innervation of pulmonary vessels in response to acute hypoxia, J. Appl. Physiol. **43**:65, 1977.

West, J.B., Dollery, C.T., and Naimark, A.: Distribution of blood flow in isolated lung: relation to vascular and alveolar pressures, J. Appl. Physiol. **19**:713, 1964.

Books and monographs

Fishman, A.P., and Hecht, H.H.: The pulmonary circulation and interstitial space, Chicago, 1969, University of Chicago Press.

Murray, J.F.: The normal lung: the basis for diagnosis and treatment of pulmonary disease, Philadelphia, 1976, W. B. Saunders Co.

chapter 39

Gas exchange and gas transport

An appreciation of the gas-exchanging function of the lung requires an understanding of some of the fundamental properties of gases. A gas consists of molecules that are in a state of random motion. The molecules fill any container in which they are enclosed and exert a pressure that is caused by the collision of molecules with one another and with the container walls.

The respiratory gases, which include oxygen, carbon dioxide, and nitrogen, follow the law of perfect gases, which can be expressed as follows:

$$PV = nRT \tag{1}$$

where

P = pressure
V = volume
n = number of gas molecules
R = gas constant
T = absolute temperature

This expression indicates that if temperature is kept constant, the volume occupied by an aliquot of gas varies inversely with the pressure to which it is subjected (Boyle's law). Also, at a constant pressure the volume of a gas is proportional to the absolute temperature (Charles' law). Similarly, at a constant volume the pressure exerted by a gas varies directly with the absolute temperature (Gay-Lussac's law).

At sea level the total pressure of atmospheric air is about 1000 cm H_2O or 760 mm Hg (torr). More and more commonly pressure is expressed in units called kilopascals. One kilopascal is equal to 10 cm H_2O.

Air is a mixture of constituents including nitrogen, oxygen, carbon dioxide, and certain inert gases such as argon and neon. Each gas in the mixture exerts a partial pressure that is proportional to its concentration (Dalton's law). The partial pressure of each gas in the mixture is the same as it would be if that gas alone occupied the entire volume. The sum of the partial pressures of each of the gases in the mixture equals the total pressure.

Inspired atmospheric air is warmed and humidified as it passes over the nasal turbinates and mucous membranes of the airways. The water vapor pressure of a saturated gas varies with temperature. At body temperature, water vapor pressure is 47 mm Hg, regardless of the barometric pressure. The total pressure of dry gases in the airways is consequently equal to the barometric pressure (760 mm Hg at sea level) minus water vapor pressure (47 mm Hg), or 713 mm Hg. The partial pressures of the gases in inspired air are proportional to their concentrations. The concentration of O_2 is about 21%, and the partial pressure of O_2 (P_{O_2}) is 149 mm Hg. The partial pressure of CO_2

■ *Properties of gases*

(P_{CO_2}) is less than 1 mm Hg, and the partial pressure of nitrogen (P_{N_2}), the concentration of which is about 79%, is 563 mm Hg.

A gas such as O_2 or CO_2, when exposed to a liquid, will dissolve in that liquid. According to Henry's law, the amount of gas dissolved is proportional to the partial pressure. The greater the concentration of O_2 or CO_2 in the gas phase, the greater the volume that will go into solution. At equilibrium the partial pressures in the gas and liquid phases will be the same. If, however, the partial pressure in the gas phase is then reduced, some of the dissolved molecules of gas would leave solution and reenter the gas phase.

The volume of gas dissolved in a liquid also depends on the solubility of the gas in the solvent and on the temperature. CO_2 is more soluble in water than is O_2 so that at a given partial pressure, greater quantities of CO_2 than of O_2 will be dissolved.

■ Gas exchange

■ Ventilation, O_2 uptake, and CO_2 output

Ventilation is measured as the volume of gas that is inspired or expired in a unit of time. Minute ventilation is the product of tidal volume and the breathing frequency.

In the lung, gas exchange occurs in the alveoli. Blood passing through the pulmonary capillary bed removes O_2 from the alveolar air and adds CO_2. Ventilation of the alveoli, on the other hand, adds O_2 to the alveolar air and removes CO_2. The composition of alveolar gas thus depends on the balance between the levels of alveolar ventilation and pulmonary capillary blood flow.

The volume of O_2 that is taken up by the lungs may be calculated from the difference in the amount of O_2 in the inspired and expired air, according to the equation:

$$\dot{V}_{O_2} = (\dot{V}_I \times F_{IO_2} - (\dot{V}_E \times F_{EO_2}) \qquad (1)$$

where

$\dot{V}_{O_2}$ = O_2 uptake per minute
$\dot{V}_I$ = inspired volume of ventilation per minute
F_{IO_2} = concentration of O_2 in inspired air
$\dot{V}_E$ = expired volume of ventilation per minute
F_{EO_2} = concentration of O_2 in expired gas

The volume of CO_2 that is eliminated can be determined in a similar fashion. Because the concentration of CO_2 in the inspired air is negligible, CO_2 output per minute ($\dot{V}_{CO_2}$) is calculated simply as the product of the expired volume of ventilation ($\dot{V}_E$) and the concentration of CO_2 in the expired gas (F_{ECO_2}):

$$\dot{V}_{CO_2} = \dot{V}_E \times F_{ECO_2} \qquad (2)$$

In the steady state, when metabolic activity is constant, there is a fixed relationship between tissue CO_2 production and O_2 consumption. The *respiratory quotient* is the ratio of CO_2 production to O_2 consumption in the tissues. The quotient depends on whether carbohydrate, protein, or fat is being metabolized; it may range from 0.7 to 1.0, indicating that O_2 consumption equals or exceeds CO_2 production.

The ratio of the amount of CO_2 eliminated to the amount of O_2 taken up by the lungs is called the *respiratory exchange ratio*. Under steady-state conditions, the respiratory exchange ratio equals the respiratory quotient. Accordingly, the amount of O_2 taken up by the capillary blood is greater than the amount of CO_2 added by the blood to the alveolar gas. As a result, the expired volume is somewhat less than the inspired volume. Nitrogen is not exchanged in the lung. Hence, the volume of nitrogen in the inspired air, that is, the product of the inspired volume of ventilation ($\dot{V}_I$) and the concentration of nitrogen in the inspired air (F_{IN_2}), will equal the volume of nitrogen in the expired gas, that is, the product of the expired volume of ventilation ($\dot{V}_E$) and the concentration of nitrogen in the expired gas (F_{EN_2}):

$$\dot{V}_I \times F_{IN_2} = \dot{V}_E \times F_{EN_2} \qquad (3)$$

Differences between the inspired and the expired volumes of ventilation will be reflected in differences in the concentrations of nitrogen in the inspired and expired gas.

■ *Dead space*

The entire volume of each tidal breath does not participate in gas exchange. A portion of each breath just fills the pharynx, larynx, and conducting airways and will not enter the alveoli to come in contact with capillary blood. This volume is termed the *anatomical dead space*. During normal tidal breathing it is approximately equal in milliliters to the person's body weight in pounds. The anatomical dead space is about 150 ml in an average adult male.

Although the dead space of the conducting airways is wasted as far as gas exchange is concerned, it does play an important role in heat and water exchange. It is in the dead space that the inspired air is humidified and brought to body temperature. Also, harmful constituents such as bacteria that are inhaled from the atmosphere can be removed in the dead space before they reach the more delicate gas-exchanging surfaces.

The volume of the anatomical dead space can be determined from the single-breath nitrogen washout method (Fig. 39-1). The concentration of nitrogen in the expired gas, after O_2 has been inspired to total lung capacity, is plotted against expired volume. The initial portion of the exhaled air consists of dead space only and contains no nitrogen (phase I). The next part of the exhalation, which is characterized by a steep rise in nitrogen concentration (phase II), is made up of a mixture of dead space and alveolar gas. Because some mixing occurs, the boundary between the dead space and the alveolar gas is not square. As shown in Fig. 39-1, the volume of phase II can be divided so that the areas of the colored triangles are equal. The total anatomical dead space is the volume of phase I plus the volume of phase II up to the point of division.

Each tidal volume (V_T) can be considered to consist of a dead space volume (V_D) and an alveolar volume (V_A), that is

$$V_T = V_D + V_A \tag{4}$$

The concentration of CO_2 in the dead space is the same as that in the inspired air; it is virtually zero. During expiration the alveolar gas is diluted by the dead space air so that the concentration of CO_2 in the mixed expired gas is lower than that in alveolar gas.

The volume of the dead space can be determined by means of the Bohr equation, which states that the volume of CO_2 in mixed expired gas, that is, the product of the expired tidal volume (V_T) and the concentration of CO_2 in the mixed expired gas (F_{ECO_2}), equals the sum of the volume of CO_2 in the dead space (the dead space volume, V_D, multiplied by the concentration of CO_2 in the inspired air, F_{ICO_2}) and in the alveolar component of the tidal volume (the alveolar volume, V_A, multiplied by the concentration of CO_2 in alveolar gas, F_{ACO_2}). Hence

$$V_T \times F_{ECO_2} = (V_D \times F_{ICO_2}) + (V_A \times F_{ACO_2}) \tag{5}$$

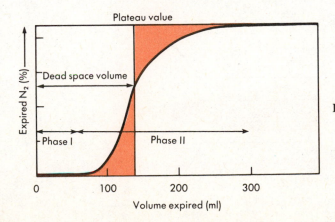

Fig. 39-1 ■ Single-breath nitrogen washout curve.

The amount of air entering the alveolar space is equal to the tidal volume minus the dead space volume. Hence

$$V_A = V_T - V_D \tag{6}$$

Combining equations (5) and (6) and assuming that $FI_{CO_2} \cong 0$,

$$V_T \times FE_{CO_2} = (V_T - V_D) \times FA_{CO_2} \tag{7}$$

Solving for V_D, the expression becomes

$$V_D = V_T \frac{(FA_{CO_2} - FE_{CO_2})}{FA_{CO_2}} \tag{8}$$

When FI_{CO_2} is not zero, the equation becomes

$$V_D = V_T \frac{(FA_{CO_2} - FE_{CO_2})}{(FA_{CO_2} - FI_{CO_2})} \tag{9}$$

The Bohr equation enables the determination of the physiological dead space, that is, the anatomical dead space plus the volume of air that reaches the alveoli but does not come into contact with pulmonary capillary blood and does not participate in gas exchange. The additional alveolar dead space is small in normal persons. In some patients with lung disease, however, the alveolar dead space can become large and can significantly affect gas exchange. Because the alveolar and arterial P_{CO_2} are usually virtually the same, dead space may be calculated from the following equation, in which Pa_{CO_2} is substituted for FA_{CO_2} in equation 8:

$$V_D = V_T \frac{(Pa_{CO_2} - PE_{CO_2})}{Pa_{CO_2}} \tag{10}$$

The dead space volume in an individual is not constant, but it enlarges with inspiration as the conducting airways widen because of the traction exerted on the air passages by the surrounding lung parenchyma. This increase in dead space during inspiration is not totally detrimental because it is associated with a lowering of airway resistance. It has been suggested that certain regulatory systems operate to control the volume of the conducting passages so as to obtain the best compromise between the need to minimize the airway resistance and to minimize the dead space itself.

Even when the tidal volume is less than the anatomical dead space (during very shallow respiration), some of the inspired gas may reach the gas-exchanging surfaces of the alveoli. One reason is that air tends to travel through smaller airways with a parabolic profile (Fig. 39-2). Hence, the more central streamlines will travel further than those nearer the walls. It has been shown that when the breathing frequencies are very high (a cycle or more per second), the lungs can be ventilated even with very low tidal volumes. Nonetheless, it is clear that the amount of alveolar ventilation does depend on the pattern of breathing as well as on the total ventilation. Shallow, rapid breathing

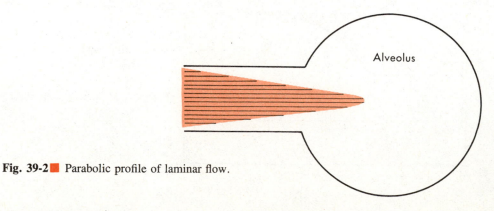

Fig. 39-2 ■ Parabolic profile of laminar flow.

(small tidal volumes and high breathing frequencies) will provide less alveolar ventilation than the same minute ventilation produced by slow and deep breathing. The pattern of breathing also affects the work performed by the respiratory muscles. Therefore the breathing pattern actually used, particularly in patients with lung disease, may reflect an attempt by the respiratory controller to maintain adequate ventilation at a minimal energy cost.

■ *Alveolar ventilation*

Alveolar ventilation ($\dot{V}_A$) is total ventilation minus dead space ventilation. Because the dead space air does not participate in gas exchange, the entire output of CO_2 in the expired gas comes from the alveolar gas. Accordingly, CO_2 output can be expressed in terms of alveolar ventilation ($\dot{V}_A$), that is, the product of the alveolar component of the tidal volume and the breathing frequency, and the partial pressure of CO_2 in the alveolar air (P_{ACO_2}), that is

$$\dot{V}_{CO_2} = K_1 \times \dot{V}_A \times P_{ACO_2} \tag{11}$$

where K_1 is a constant of proportionality. Rearranging the equation, it becomes

$$P_{ACO_2} = K_2 \times \frac{\dot{V}_{CO_2}}{\dot{V}_A} \tag{12}$$

where K_2 is the reciprocal of K_1. This equation indicates that, at any given level of metabolic activity, the P_{ACO_2} varies inversely with the level of alveolar ventilation. This equation is graphically represented in Fig. 39-3.

Similarly, the O_2 uptake, which occurs only in the alveoli, can also be considered in terms of alveolar ventilation ($\dot{V}_A$) and the difference in the partial pressures of O_2 between the inspired air and the alveolar gas:

$$\dot{V}_{O_2} = K_1 \times \dot{V}_A (P_{IO_2} - P_{AO_2}) \tag{13}$$

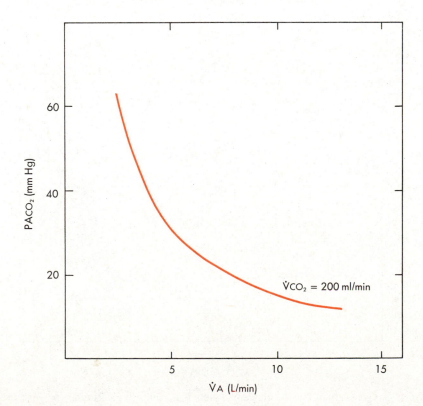

Fig. 39-3 ■ At a constant level of CO_2 production, the alveolar P_{CO_2} varies inversely with alveolar ventilation.

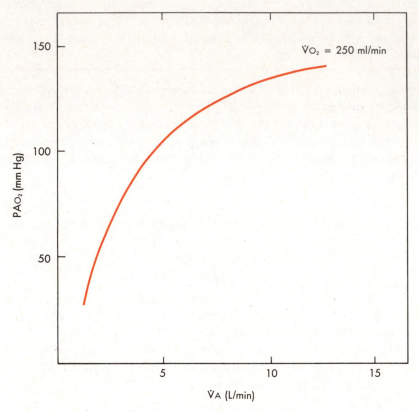

Fig. 39-4 ■ At a constant level of O_2 consumption and a fixed inspired O_2 concentration, there is a direct but nonlinear relationship between alveolar PO_2 and the level of alveolar ventilation.

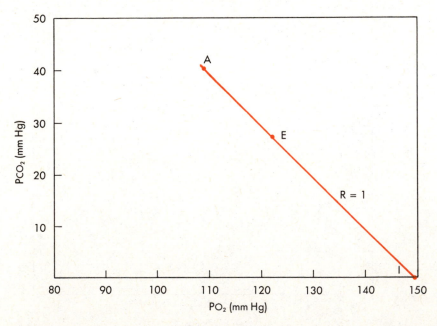

Fig. 39-5 ■ Oxygen–carbon dioxide diagram. Points *I, E,* and *A* represent the composition of inspired, expired, and alveolar gas, respectively.

Rearranging the equation, it becomes

$$PA_{O_2} = PI_{O_2} - K_2 \times \frac{\dot{V}_{O_2}}{\dot{V}_A} \tag{14}$$

This equation is graphically illustrated in Fig. 39-4. When equations (11) and (13) are combined, PA_{O_2} can be expressed

$$PA_{O_2} = PI_{O_2} - \frac{PA_{CO_2}}{R} \tag{15}$$

since $\dfrac{\dot{V}_{O_2}}{\dot{V}_{CO_2}} = \dfrac{1}{R}$, where R is the respiratory exchange ratio.

This equation is a simplified form of the alveolar air equation. It indicates that the alveolar P_{O_2} is dependent on the partial pressure of O_2 in the inspired air (PI_{O_2}) and on the alveolar P_{CO_2}, with a correction (R) for the respiratory exchange ratio. Also, at a given inspired O_2 concentration and a constant value of R, PA_{O_2} and PA_{CO_2} are inversely related to one another.

In its complete form the alveolar air equation is:

$$PA_{O_2} = PI_{O_2} - \frac{PA_{CO_2}}{R} + PA_{CO_2} \times FI_{CO_2} \times \frac{(1-R)}{(R)} \tag{16}$$

The relationship between PA_{O_2} and PA_{CO_2} can be illustrated by the $O_2 - CO_2$ diagram (Fig. 39-5). During room air breathing ($PI_{O_2} = 150$ mm Hg; $PI_{CO_2} = 0$) and when *R* is 1, all possible values for alveolar P_{O_2} and P_{CO_2} will fall on a line that has a slope of -1 and intersects the horizontal axis at a P_{O_2} of 150 mm Hg (point *I*). Point *I* represents the P_{O_2} and P_{CO_2} of inspired air. The location on the line of point *A*, representing alveolar P_{O_2} and P_{CO_2}, will depend on the level of alveolar ventilation. The higher the alveolar ventilation, the higher will be the PA_{O_2} and the lower will be the PA_{CO_2}. Conversely, the lower the alveolar ventilation, the lower will be the PA_{O_2} and the higher will be the PA_{CO_2}. The location of point *E*, representing the composition of mixed expired gas, will depend on the size of the dead space. The larger the dead space, the greater the difference between points *A* and *E*.

Less than 5% of the total amount of O_2 and CO_2 carried by the blood is present in physical solution. Primarily, the ability of the blood to carry O_2 and CO_2 depends on reversible chemical reactions. These reactions require the presence of specialized compounds and catalysts so that these gases can be transported between the lungs and tissues within the time constraint imposed by the circulation.

■ Gas transport by the blood

The amount of O_2 carried in the blood in physical solution and as O_2 chemically bound to hemoglobin in the erythrocytes is shown in Table 39-1.

■ Oxygen in physical solution

TABLE 39-1

Forms of O_2 transport in the blood

	Arterial	*Venous*	*A-V difference*
Physically dissolved (vol%*)	0.3	0.1	0.2
Bound to hemoglobin (vol%)	19.5	14.4	5.1
Total (vol%)	19.8	14.9	5.3

*Vol% = milliliters of gas per deciliter blood.

The volume of O_2 in solution in a given volume of plasma follows Henry's law and is proportional to its partial pressure. At a Po_2 of 100 mm Hg, 0.3 ml of O_2 is dissolved in each liter of blood. In resting humans, with an O_2 requirement of about 250 ml/minute and a normal cardiac ouput of about 6 L/minute, extraction of dissolved O_2 by the body tissues supplies only about 0.7% of the body's O_2 requirement. The O_2 requirements of the body at rest could be satisfied by complete extraction by the tissues of dissolved O_2 from the blood only if the cardiac output were to exceed 80 L/minute.

■ Oxygen combined with hemoglobin

O_2 combines rapidly and reversibly with hemoglobin in the red blood cell to form *oxyhemoglobin* (HbO_2). Hemoglobin that is not combined with O_2 is called *reduced hemoglobin* (Hb).

The hemoglobin molecule consists of *heme* (a pigment) and *globin* (a protein) (Chapter 24). The globin portion of the molecule is composed of four chains; two are identical α-chains, each with 141 amino acids, while the other two are identical β-chains, each with 146 amino acids. In normal human hemoglobin (hemoglobin A), the component amino acids, the sequence in which they are arranged, and their spatial relationships are precisely known. In addition, the different amino acid compositions that characterize more than 100 variants of normal hemoglobin have also been determined. Changes in the amino acid sequence may affect the ability of hemoglobin to combine with O_2. Each of the four globin chains in hemoglobin contains one heme unit. Iron in the ferrous state is present at the center of each heme group; it can combine with one molecule of O_2. Only when heme, iron, and globin are together in their proper spatial relationship can the combination with O_2 occur.

The amount of O_2 in chemical combination with hemoglobin depends on the partial pressure of O_2. Each hemoglobin molecule can bind up to four molecules of O_2. O_2, similarly, can combine with myoglobin, a hemoglobin-like compound in muscle that has only one globin chain. Oxygenation of the iron atoms in one of the chains in hemoglobin accelerates oxygenation in the remaining chains. Similarly, release of O_2 by any of the iron atoms makes the remaining iron atoms release their O_2 molecules more readily.

One gram of hemoglobin is capable of combining chemically with 1.34 ml of O_2. In a normal person with a hemoglobin concentration of 15 g/dl of blood, the blood is capable of carrying about 20 vol%, that is, 20 ml O_2/dl blood, of O_2 as HbO_2. This maximum amount is referred to as the *oxygen capacity*. Normally, arterial Po_2 is not sufficiently high that all the hemoglobin is completely saturated with O_2. The actual amount of O_2 in combination with hemoglobin is less than capacity; the ratio of the O_2 content to the O_2 capacity, expressed as percent, is called the *oxygen saturation*.

The O_2 content and saturation of blood increase progressively with increasing Po_2. However, the relationship between Po_2 and O_2 saturation (the HbO_2 dissociation curve) is not linear, but rather it is sigmoid in shape (Fig. 39-6). The sigmoid shape of the HbO_2 dissociation curve indicates that the amount of O_2 bound to hemoglobin is relatively constant over a fairly wide range of high O_2 tensions. However, the changes are much more pronounced for a given change in Po_2 at the lower ranges of O_2 tension. In a normal person breathing room air, the O_2 saturation of arterial blood is 97%. Small changes in Po_2 produced by alterations in alveolar ventilation will ordinarily have little effect on the amount of O_2 in the arterial blood.

In order for the tissues to extract O_2 from peripheral capillary blood, the Po_2 in the tissues must be lower than the Po_2 of the blood perfusing them. Because of the steep slope of the HbO_2 curve when Po_2 is below 55 mm Hg, large quantities of O_2 can be unloaded from blood to the tissues with relatively small decreases in blood Po_2. Hence, increased metabolic demands for O_2 can be satisfied by small reductions in tissue Po_2.

When healthy humans with an O_2 saturation of 97% breathe gas mixtures containing

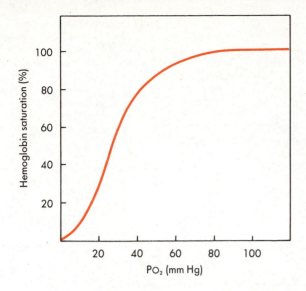

Fig. 39-6 ■ Oxyhemoglobin (HbO$_2$) dissociation curve.

more than 21% O$_2$, the alveolar Po$_2$ rises above 100 mm Hg. However, there will be a very small change in the arterial O$_2$ content. However, when lung disease results in a reduction in arterial Po$_2$ and a decrease in O$_2$ saturation, increasing the O$_2$ concentration in the inspired air even slightly can cause a significant increase in arterial O$_2$ content.

■ *Factors altering the combination of oxygen with hemoglobin*

The amount of O$_2$ bound to hemoglobin depends not only on the O$_2$ tension but also on the partial pressure of CO$_2$, the pH, and the temperature of the blood (Fig. 39-7). The HbO$_2$ dissociation curve is shifted to the right by an increase in temperature, a rise in the CO$_2$ tension, or a decrease in pH (termed the *Bohr shift*). Therefore, at any given partial pressure of O$_2$, there is less HbO$_2$ under such conditions.

An increase in metabolism decreases the Po$_2$ and increases the Pco$_2$ and the hydrogen ion (H$^+$) concentration in the tissues. The shift in the HbO$_2$ dissociation curve resulting from the changes in Pco$_2$ and pH facilitates the unloading of O$_2$ from the

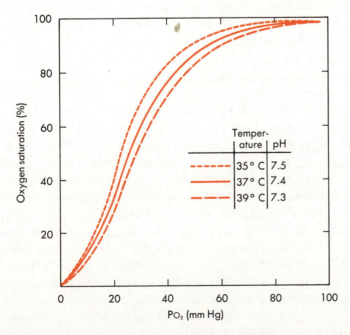

Temperature	pH
35° C	7.5
37° C	7.4
39° C	7.3

Fig. 39-7 ■ Effect of changes in temperature and pH on O$_2$ binding to hemoglobin. An increase in temperature or a decrease in pH produces a shift to the right of the HbO$_2$ dissociation curve. The HbO$_2$ dissociation curve is shifted to the left by a fall in temperature or an increase in pH.

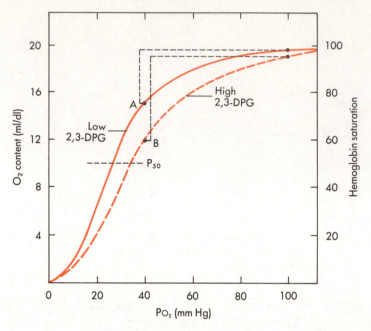

Fig. 39-8 ■ Effect of 2,3-DPG on O_2 binding to hemoglobin. Increasing concentrations of 2,3-DPG *(dashed curve)* shifts the HbO_2 dissociation curve to the right. The P_{50}, that is, the Po_2 required for 50% hemoglobin saturation, is increased. Also, the amount of O_2 released at the tissues from the arterial blood as the Po_2 falls from 100 mm Hg to 40 mm Hg is greater (*B* versus *A*).

blood to the tissues. On the other hand, the fall in blood CO_2 tension as the blood passes through the lungs causes the dissociation curve to shift to the left so that the hemoglobin is capable of combining with more O_2 at a given Po_2.

Certain drugs and chemicals, such as nitrates and sulfonamides, oxidize the iron in hemoglobin from the ferrous to the ferric form. Hemoglobin with iron in ferric form *(methemoglobin)* cannot take up oxygen. Some Hb is oxidized slowly to methemoglobin in a healthy human, even if such drugs are not being used. The methemoglobin is reduced back to Hb by reduced diphosphopyridine nucleotide (DPNH) with the aid of the enzyme methemoglobin reductase. Methemoglobinemia also occurs when this enzyme is deficient in the red blood cell. Conversion of Hb to sulfhemoglobin by certain drugs irreversibly renders it inactive for O_2 transport.

A specific chemical (2,3-diphosphoglycerate, or 2,3-DPG) that is present in high concentrations in red blood cells regulates the release of O_2 from HbO_2. This substance is formed in the erythrocyte by anaerobic glycolysis from 1,3-DPG. When added to red cells, 2,3-DPG shifts the HbO_2 curve to the right. The effect on the dissociation curve can be measured by determining the Po_2 that is required for a hemoglobin-O_2 saturation of 50% at a temperature of 37° C and a pH of 7.40; that Po_2 is called the P_{50} (Fig. 39-8). Increased concentrations of 2,3-DPG raise the P_{50}. At any given Po_2, the O_2 content of blood is less, but the availability of O_2 to the tissues is increased. Conversely, lowered concentrations of 2,3-DPG reduce the P_{50} so that O_2 saturation is higher at a given Po_2, and less O_2 is available to the tissues. Increased amounts of 2,3-DPG have been shown to acidify red blood cells, and this acidification is one reason this substance increases the P_{50}. Also, 2,3-DPG binds more tightly to reduced hemoglobin and interferes with the changes in the shape of the molecule that are required for its combination with O_2. An excess of 2,3-DPG is not always advantageous. When a person breathes O_2-poor gas mixtures or resides at high altitude, excess 2,3-DPG can reduce the O_2 binding to hemoglobin in the lung.

Carbon monoxide (CO) has more than 200 times the affinity for hemoglobin than

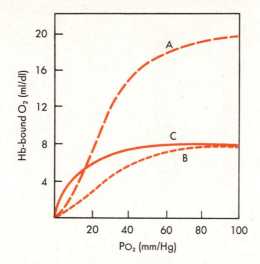

Fig. 39-9 ■ Oxyhemoglobin dissociation curves. *A*, A normal HbO_2 dissociation curve with a hemoglobin concentration of about 15 g/dl. *B*, An HbO_2 dissociation curve in a person with anemia and a hemoglobin concentration of about 6 g/dl. *C*, An HbO_2 dissociation curve after exposure to CO. The hemoglobin concentration is 15 g/dl, but the O_2-carrying capacity is markedly reduced, and the dissociation curve is shifted to the left.

does O_2. Because of this greater affinity for hemoglobin, even very small concentrations of CO may produce appreciable levels of carboxyhemoglobin (HbCO). Because CO competes with O_2 for binding sites on the hemoglobin molecule, HbO_2 will consequently be reduced. Combination of CO with hemoglobin also affects O_2 binding and causes the O_2 dissociation curve to shift to the left and become more hyperbolic. This causes a greater deficit in O_2 uptake than might be expected just from the decrease in carrying capacity of hemoglobin for O_2 (Fig. 39-9). Even when HbCO levels are high and consequently HbO_2 levels are reduced, the P_{O_2} in arterial blood remains normal. Hence, the chemoreceptors, which respond to low P_{O_2}, are not stimulated and ventilation fails to increase. CO poisoning is treated by administering high concentrations of inspired O_2 to favor the combination of hemoglobin with O_2 rather than with CO.

Variations in the globin chains may affect the P_{50}. Certain variants of hemoglobin are designated by letters of the alphabet, such as Hb A for normal adult hemoglobin and Hb F for fetal hemoglobin. Other variants bear the name of the location where they were first described, such as Hb Seattle. Usually, in these variants there is a substitution of one amino acid. In only one, Hb M, does the structural change involve the heme group.

The globin chains in Hb F and Hb A differ. Because of decreased binding with 2,3-DPG in Hb F erythrocytes, Hb F has a greater affinity for O_2 than does Hb A. Fetal red blood cells contain Hb F; therefore they are able to combine with more O_2 as they flow through the capillaries in the placenta, where the P_{O_2} is low. Shortly after birth, Hb F begins to disappear from the red blood cell and is replaced by Hb A.

■ *Tissue oxygenation*

Respiration must be coordinated with circulatory activity to provide the body tissues with enough O_2 to prevent tissue hypoxia. The amount of O_2 carried to the tissues by the capillaries depends on the O_2 content of the arterial blood and the cardiac output. The O_2 content of the blood in turn depends on the level of P_{O_2} and the amount of hemoglobin in the blood. Even in normal persons, the total amount of O_2 carried in the blood is less than a liter so that there is very little O_2 reserve if breathing should stop. Tissue hypoxia may result from anemia or from excessive reductions in arterial P_{O_2} or blood flow.

Transport of O_2 from the capillaries to the tissues takes place passively by diffusion.

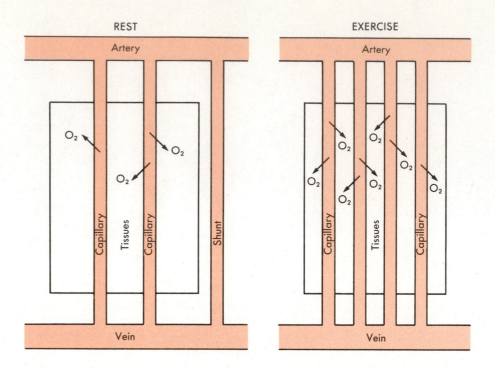

Fig. 39-10 ■ Tissue oxygenation. At rest not all capillaries are open; some blood is shunted through arteriovenous anastomoses. During exercise more capillaries open, which facilitates diffusion and allows more O_2 to be extracted by the tissues.

Hence, tissue P_{O_2} must be less than capillary P_{O_2}. The difference between the tissue P_{O_2} and capillary P_{O_2} will depend on the surface area available for diffusion and the thickness of the tissue layer through which O_2 must diffuse. Under resting conditions, as shown in Fig. 39-10, not all the tissue capillaries are open at any one instant. Also, some blood passing through the tissues may be shunted through arteriovenous anastomoses and thereby bypasses the gas-exchanging capillaries. More capillaries open when the metabolic rate increases (e.g., during exercise). This facilitates diffusion and allows more O_2 to be extracted by the tissues. Arterial hypoxemia also tends to open increased numbers of tissue capillaries and to close shunts, thereby acting to improve tissue perfusion.

The use of O_2 in tissue metabolism involves a series of coupled chemical processes that take place in the mitochondria and require enzymatic facilitation. Various cytoplasmic poisons can interfere with enzymatic function and reduce the O_2 uptake and use by cells. Under conditions in which the cell cannot use O_2, the O_2 consumption is negligible, and the capillary and venous P_{O_2} will be abnormally high in the face of tissue hypoxia.

■ *Types of hypoxia*

Hypoxia has been classified into four main types: hypoxic, anemic, circulatory, and histotoxic.

In *hypoxic hypoxia* the tissues are supplied by blood with an abnormally low O_2 tension, and thus the hemoglobin is poorly saturated with O_2. This condition occurs when the partial pressure of O_2 in the inspired air is lower than normal (e.g., at altitude) or in patients with respiratory diseases that interfere with gas exchange in the lung.

In *anemic hypoxia* the tissues are supplied by arterial blood with a normal O_2 tension but a low hemoglobin concentration. Hence, the O_2 content and the O_2 capacity of the blood are both lower than normal. This variety of hypoxia can occur in anemia as well as when toxic substances combine with hemoglobin and interfere with its capacity to combine with O_2.

Circulatory hypoxia may occur when the cardiac output is low (e.g., in shock or congestive heart failure) or when there is a local obstruction to arterial blood flow.

Histotoxic hypoxia occurs when a toxic substance such as cyanide interferes with the ability of the tissues to use O_2.

The CO_2 produced in body tissues is carried in the blood to the lungs in three forms: in physical solution, as bicarbonate (HCO_3^-), and as a carbamino compound combined with blood proteins (Table 39-2).

■ Carbon dioxide transport and its effects on acid-base balance

Physically dissolved carbon dioxide. The amount of CO_2 dissolved in blood depends on the partial pressure of CO_2. Because CO_2 is about 20 times more soluble than O_2, CO_2 in the dissolved form is more important in CO_2 transport than is dissolved O_2 in O_2 transport. As with most other gases, an increase in body temperature decreases the volume of CO_2 in physical solution. At normal body temperature, an additional 0.6 ml of CO_2 is dissolved in each liter of plasma for every 1 mm Hg increase in P_{CO_2}. Dissolved CO_2 accounts for about 5% of the CO_2 in the blood.

Bicarbonate. Some of the dissolved CO_2 reacts with water to form carbonic acid (H_2CO_3). This hydration of CO_2 takes place according to the equation:

$$CO_2 + H_2O \rightleftarrows H_2CO_3$$

In the plasma this reaction takes place extremely slowly, and the concentration of dissolved CO_2 is about 1000 times greater than the concentration of H_2CO_3. A small portion of the H_2CO_3 so formed subsequently ionizes, according to the equation:

$$H_2CO_3 \rightleftarrows H^+ + HCO_3^-$$

Although a minute amount of bicarbonate (HCO_3^-) is produced in the plasma, much larger amounts are formed in the red blood cells. The increase in red cell HCO_3^- ultimately leads to increased plasma HCO_3^-, and this is the major form of CO_2 transport in the blood.

Red blood cells accelerate the formation of HCO_3^- from CO_2. Red blood cells contain the enzyme *carbonic anhydrase*, which catalyzes the combination of H_2O with CO_2.

$$CO_2 + H_2O \underset{\text{carbonic anhydrase}}{\rightleftarrows} H_2CO_3 \rightleftarrows HCO_3^- + H^+$$

Also, the hemoglobin in the red blood cell provides basic groups, which combine with the hydrogen ions formed by the dissociation of H_2CO_3. This increases the formation of HCO_3^- by shifting the following reaction to the right:

$$H_2CO_3 + Hb^- \rightleftarrows Hb + H^+ + HCO_3^- \rightleftarrows HHb + HCO_3^-$$

Reduced hemoglobin is a better H^+ acceptor than HbO_2.

TABLE 39-2

Forms of CO_2 transport in blood

	Venous	*Arterial*	*Difference*
Dissolved CO_2 (vol%*)	3.1	2.7	0.4
HCO_3^- (vol%)	47.0	43.9	3.1
Carbamino CO_2 (vol%)	3.9	2.4	1.5
Total (vol%)	54.0	49.0	5.0

*Vol% = milliliters of gas per deciliter of blood.

Carbamino compounds. CO_2 can also combine with amine groups in proteins to form carbamino compounds, as shown in the following reaction:

$$Hb-N\left\langle {}_H^H \right. + CO_2 \rightleftarrows Hb-N\left\langle {}_{COO^-}^H \right. + H^+$$

Because more amine groups are available in hemoglobin than in plasma proteins, more CO_2 combines with hemoglobin to form carbamino compounds. Reduced hemoglobin binds CO_2 more readily than does HbO_2. About 30% of the arteriovenous difference in CO_2 is accounted for by the difference in the carbamino CO_2 contents in the arterial and venous blood.

■ *Carbon dioxide dissociation curve*

The CO_2 dissociation curve relates the CO_2 content of the blood to the blood P_{CO_2} (Fig. 39-11). It should be noted that the relationship between CO_2 content and P_{CO_2} is much steeper than the relationship between O_2 content and P_{O_2}. The total quantity of CO_2 in the blood is normally more than twice that of O_2. In the physiological range of CO_2 content and tension, the CO_2 dissociation curve is almost linear.

The degree of oxygenation affects the CO_2 dissociation curve. The greater the saturation of hemoglobin with O_2, the less will be the CO_2 content for a given P_{CO_2}. This effect, called the *Haldane effect,* is caused by the greater ability of reduced hemoglobin to buffer H^+ and to form carbamino hemoglobin. The Haldane effect facilitates CO_2 exchange in both the lungs and the tissues.

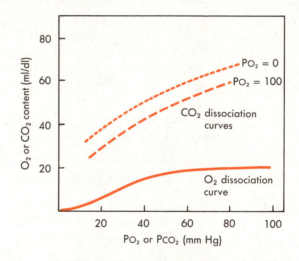

Fig. 39-11 ■ Comparison of HbO_2 and CO_2 dissociation curves. The CO_2 content of blood is more than twice the O_2 content of blood. Also, the CO_2 dissociation curves are much steeper than the O_2 dissociation curve. The CO_2 content at any given P_{CO_2} depends on the level of oxygenation (P_{O_2} of 0 versus P_{O_2} of 100). As the O_2 saturation of hemoglobin is reduced, the CO_2 content at a given P_{CO_2} will increase.

■ *Carbon dioxide exchange in the lung*

CO_2 exchange in the lungs is summarized in Fig. 39-12. Because systemic venous P_{CO_2} is higher than alveolar P_{CO_2}, the physically dissolved CO_2 diffuses from the pulmonary capillary plasma into the alveoli. By mass action, CO_2 is also released from carbamino binding and from HCO_3^- in the plasma. Because of the absence of carbonic anhydrase in plasma, this last reaction is slow. As CO_2 leaves the plasma, it is replaced by CO_2 from the red cell. This CO_2 comes mainly from the HCO_3^- in the red cell, but it also comes from CO_2 in the carbamino form. Carbonic anhydrase is present within the red cell, and as a result, CO_2 formation from both reactions proceeds rapidly. As the blood passes through the pulmonary capillaries, the HCO_3^- in the red blood cell decreases more rapidly than it does in the plasma. Hence, HCO_3^- moves into the red

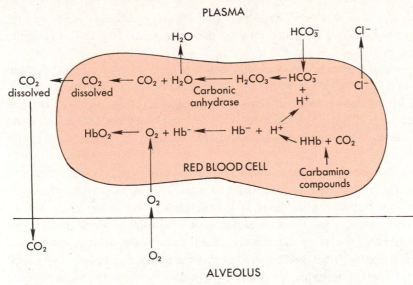

Fig. 39-12 ■ CO_2 exchange in the lungs. CO_2 is carried in solution in the plasma, in combination with hemoglobin in the red cell ($HHbCO_2$), in the form of carbonic acid (H_2CO_3) in the red cell, and as bicarbonate (HCO_3^-) in the red cell and plasma.

blood cell from the plasma as the HCO_3^- in the red blood cell interior falls. This increases the negative charge inside the red cell, and consequently some chloride exits from the red cell into the plasma as a result of electrostatic forces. This transfer of chloride ions between the red blood cells and the plasma is referred to the *chloride* (or *Hamburger*) *shift*.

In the pulmonary capillary bed, hemoglobin is oxygenated at the same time that CO_2 is eliminated and the P_{CO_2} falls. HbO_2 is less able to combine with CO_2 and is more acidic. The consequent increase in the concentration of H^+ as HbO_2 is formed (Haldane effect) facilitates the conversion of HCO_3^- to CO_2. It has been estimated that the Haldane effect accounts for almost half of the transfer of CO_2 in the lung. The change in HCO_3^- and H^+ in the red blood cell decreases the osmotic pressure. Therefore water leaves the red cells, and the red cells shrink.

None of the chemical and physical processes involved in CO_2 transfer occur instantaneously. For example, 0.16 second is required for the chloride shift. The rate of pulmonary blood flow limits the time the erythrocyte is present in the lungs, and hence the time available for CO_2 liberation from the venous blood is also limited. Normally the time required for chemical reactions to reach completion does not effect CO_2 transfer. However, if there were no carbonic anhydrase available in the red cells, the uncatalyzed hydration of CO_2 would be so slow that only 10% of the available CO_2 would be removed from the blood during its passage through the pulmonary capillaries.

Changes in H^+ and HCO_3^- concentrations in the plasma are slow because of the absence of carbonic anhydrase. Moreover, because the red cell does not allow the passage of cations across its membrane (although it allows free passage of anions), the H^+ in the plasma cannot equilibrate with the H^+ in the red cell directly by diffusion. Hence, H^+ concentrations and P_{CO_2} may continue to change even after the red cell has left the pulmonary capillary. The magnitude of these additional changes is not settled, but it could affect the responses of the chemoreceptors, which sense changes in H^+ concentrations and P_{CO_2} in the arterial blood.

■ *Exchange of carbon dioxide in tissue*

Processes that take place in the tissue are the reverse of those which occur in the lung. In the tissue, uptake of CO_2 by the red blood cell occurs because the tissue P_{CO_2} is greater than the P_{CO_2} in the capillary blood. CO_2 uptake is facilitated by the reduction of hemoglobin.

CO_2 is transported across the tissue-capillary membrane by diffusion, as is O_2. It is generally held that tissue and capillary P_{CO_2} rapidly reach equilibrium. CO_2 transfer from the tissues to the blood is limited only by perfusion and not by diffusion or the rate of chemical reactions.

The ratio of tissue perfusion to tissue volume affects the speed at which the P_{CO_2} in different tissues reach equilibrium with P_{CO_2} in the blood. The greater the perfusion, the faster the rate of equilibration. In some tissues that are poorly perfused, such as resting muscle, tissue P_{CO_2} will continue to change for very long periods (minutes and even hours) after a change in Pa_{CO_2} produced, for instance, by breathing air with added CO_2.

The rapidity with which changes in whole body metabolism can produce changes in P_{CO_2} also depends on the ratio of cardiac output to tissue volume. Because chemoreceptor responses are influenced by the level of Pa_{CO_2}, cardiac output can affect the responses to environmental or metabolic disturbances in CO_2 balance. With the onset of exercise, large increases in cardiac output occur, which theoretically could produce rather abrupt changes in Pa_{CO_2}. It has been suggested that the abrupt hyperpnea observed at the beginning of exercise is in part produced by a sudden increase in cardiac output, which rapidly washes out the CO_2 produced metabolically in the tissues, elevates Pa_{CO_2}, and therefore stimulates the chemoreceptors. The term *cardiodynamic hyperpnea* has been applied to this possible mechanism for the increase of ventilation with exercise.

■ Effects of ventilation on acid-base balance

The pH of blood and H^+ homeostasis are regulated by the HCO_3^-/H_2CO_3 buffer system. According to the Henderson-Hasselbalch equation:

$$pH = pK + \log \frac{[HCO_3^-]}{[H_2CO_3]} \qquad (17)$$

It is important to note that H_2CO_3 is in equilibrium with dissolved CO_2:

$$CO_2 + H_2O \rightleftarrows H_2CO_3$$

The equilibrium position of this reaction is far to the left; that is, for every molecule of H_2CO_3, there are about 500 molecules of CO_2 in solutions. The denominator in the Henderson-Hasselbalch equation is in fact the sum of the dissolved CO_2 and H_2CO_3, but this H_2CO_3 pool is overwhelmingly made up of dissolved CO_2. According to Henrys' law, the dissolved CO_2 is proportional to the P_{CO_2} so that the Henderson-Hasselbalch equation can be rewritten:

$$pH = pK + \log \frac{HCO_3^-}{0.03 \, P_{CO_2}} \qquad (18)$$

where 0.03 is the solubility constant.

The HCO_3^- concentration in the blood is regulated chiefly by the kidney and the blood P_{CO_2} by the lungs.

Blood buffers other than the HCO_3^-/H_2CO_3 system prevent wide swings in blood pH when P_{CO_2} changes. These blood buffers include the $HPO_4^-/H_2PO_4^-$ system and numerous proteins, particularly hemoglobin, as explained in Chapter 48. The ability of Hb to buffer changes in H^+ is greater than that of plasma proteins, mainly because there is much more Hb than plasma protein per liter of blood.

■ Acid-base disturbances

The pH of the body can be disturbed in four ways: respiratory acidosis, respiratory alkalosis, metabolic acidosis, and metabolic alkalosis (Fig. 39-13). (See also Chapter 48.)

Respiratory acidosis is caused by hypoventilation, which increases the P_{CO_2}, reduces the HCO_3^-/P_{CO_2} ratio and thus depresses the pH. As the P_{CO_2} increases, H_2CO_3 and H^+ concentrations increase. In acute respiratory acidosis buffering is primarily by cellular buffers. Excess hydrogen ions tend to enter cells in exchange for Na^+ and K^+

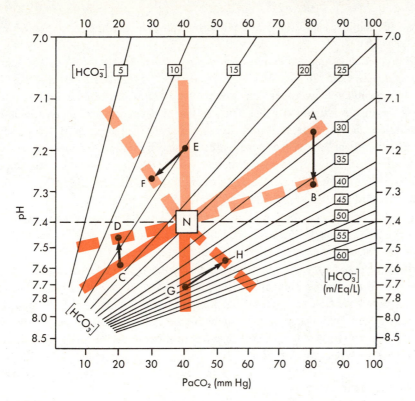

Fig. 39-13 ■ Acute and chronic acid-base relationships. The Henderson-Hasselbalch equation (equation 18) was used to describe interrelations among pH *(ordinate)*, Paco$_2$ *(abscissa)*, and plasma bicarbonate *(solid black radiating lines)*. Conditions of acidosis are designated by light colored lines *(top)*, and conditions of alkalosis are designated by dark colored lines *(bottom)*. The solid colored lines describe average relationships between pH and Paco$_2$ in humans when acids or bases are added to the blood. The slopes of the lines are determined by the body's buffering capacity. The normal plasma values in the human: (pH, 7.4; Paco$_2$, 40 mm Hg; and [HCO$_3^-$], 24 mEq/L) are indicated by the point *N*. The points describing deviations from the normal, which are induced by various acid-base disturbances, are designated by the letters *A* through *H*. The change in pH induced by any particular perturbation is determined by the size of the acid or base load and the amount of extracellular and intracellular buffering. The dashed colored lines represent conditions that might be found after renal or respiratory compensatory processes have returned the pH toward normal. Note that compensation is incomplete. The arrows show the transition from the acute to the compensated situation. *A*, Respiratory acidosis; *B*, compensated respiratory acidosis; *C*, respiratory alkalosis; *D*, compensated respiratory alkalosis; *E*, metabolic acidosis; *F*, compensated metabolic acidosis; *G*, metabolic alkalosis; *H*, compensated metabolic alkalosis.

and are buffered by cellular proteins. HCO$_3^-$ ions, however, are left in the extracellular fluid. Buffering also occurs in red blood cells. CO$_2$ enters the red blood cells and is hydrated to H$_2$CO$_3$, which in turn dissociates. The hydrogen ions that are released are buffered by hemoglobin, but the HCO$_3^-$ returns to the extracellular fluid in exchange for Cl$^-$. The magnitude of the acute compensation by cellular buffers during hypoventilation is small. As the Pco$_2$ increases acutely from 40 to 80 mm Hg (Fig. 39-13, *A*), the HCO$_3^-$ concentration increases from a normal value of 24 mEq/L at *N* to 28 mEq/L at *A* so that the pH falls from a normal value of about 7.40 to approximately 7.15. If respiratory acidosis persists, the body increases the renal excretion of H$^+$ and increases HCO$_3^-$ reabsorption by the kidney. The resulting increase in plasma HCO$_3^-$ returns the HCO$_3^-$/Pco$_2$ ratio toward normal. The renal compensation for respiratory acidosis is delayed for some hours, and a steady state is not reached for several days. Renal compensation is not complete, so the pH is not fully restored to its normal level of 7.40, but instead stablilizes at 7.25 if arterial carbon dioxide tension is held at 80 mm Hg (Fig. 39-13, *B*).

Hyperventilation leads to *respiratory alkalosis* by reducing the P_{CO_2} and increasing the HCO_3^-/P_{CO_2} ratio. In Fig. 39-13 the Pa_{CO_2} at *C* is reduced to 20 mm Hg. During acute hyperventilation hydrogen ions are released from intracellular buffers and from hemoglobin in the red blood cells. This results in a decrease in extracellular HCO_3^- concentration and to a slight degree limits the overall change in H^+ concentration, stabilizing the pH at 7.6. When hyperventilation and hypocapnia persist, renal compensation occurs as a result of decreased tubular H^+ secretion and reduced reabsorption of HCO_3^- by the kidney. Plasma HCO_3^- concentration falls, and the HCO_3^-/P_{CO_2} ratio increases toward normal, stabilizing the pH at 7.5 (Fig. 39-13, *D*).

Metabolic acidosis can be produced by a variety of disease processes either through the loss of HCO_3^- ions or the overproduction and retention of nonvolatile acids by the body. If no respiratory response occurs, extracellular fluid pH is stabilized by buffering the excess hydrogen ions with extracellular fluid HCO_3^-, and with intracellular buffers following transmembrane exchange of H^+ with Na^+ or K^+ (Fig. 39-13, *E*). The total buffering capacity of HCO_3^- and nonbicarbonate buffers, however, is limited, but the respiratory system provides important compensation for metabolic acidosis. The increase in the H^+ concentration stimulates the respiratory chemoreceptors and raises ventilation. The increase in ventilation lowers the P_{CO_2} and raises the depressed HCO_3^-/P_{CO_2} ratio toward normal (Fig. 39-13, *F*). In severe metabolic acidosis the P_{CO_2} can be reduced to values as low as 10 to 15 mm Hg.

Metabolic alkalosis is characterized by an increase in serum HCO_3^- concentration and an increase in the HCO_3^-/P_{CO_2} ratio (Fig. 39-13, *G*). This can be due to a variety of causes including loss of gastric acid by vomiting and excessive ingestion of alkali. The normal compensation by the respiratory system involves hypoventilation, which raises the P_{CO_2} and returns the pH toward 7.4 (Fig. 39-13, *H*). The degree of hypoventilation for any given level of alkalosis varies considerably among individuals and is clearly limited by the development of hypoxia as ventilation is reduced. It is unusual for the P_{CO_2} to increase above 55 to 60 mm Hg in compensation for metabolic alkalosis.

■ *Ventilation-perfusion relationships*

Maximal efficiency of gas exchange occurs when both ventilation and blood flow are distributed uniformly to all regions of the lung. However, this ideal situation does not exist even in the normal lung. The amount of CO_2 delivered and the amount of O_2 extracted in each alveolus is in proportion to its blood flow. If blood flow increases to an alveolus but ventilation remains constant, more CO_2 is delivered and more O_2 is removed; consequently alveolar P_{CO_2} rises, and alveolar P_{O_2} falls. If perfusion remains the same but the alveolar ventilation is increased, more CO_2 will be expelled from the alveolus, and alveolar P_{CO_2} will fall. P_{O_2} in the alveolus will be increased because more O_2 is carried into the alveolus as a result of the increased ventilation. Thus the P_{O_2} and P_{CO_2} in each alveolus will depend on the ratio of *ventilation to perfusion*.

When a subject is in the upright posture, more ventilation is distributed to the bottom than to the top of the lung because of the effects of gravity. Similarly, the distribution of pulmonary blood flow also decreases from the bottom to the top of the lung. The degree of change in blood flow with vertical distance along the lung is greater than the corresponding change in ventilation so that the ventilation/perfusion ratio progressively increases from the bottom to the top of the lung (Fig. 39-14). Overall, the ratio of alveolar ventilation in liters per minute to pulmonary blood flow in liters per minute ($\dot{V}A/\dot{Q}$) is about 0.8.

In alveoli at the top of the lung, alveolar ventilation is about three times greater than blood flow. O_2 uptake by the capillary blood is limited by the relatively low pulmonary blood flow, but the blood passing through the pulmonary capillary bed is well oxygen-

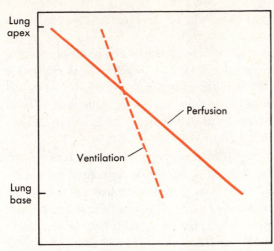

Ventilation or perfusion

Fig. 39-14 ■ Distribution of ventilation and perfusion in the lungs. In the upright position, there is greater ventilation at the bottom than at the top of the lung. Similarly, pulmonary blood flow is greater at the bottom than at the top of the lung, but the difference in blood flow exceeds the difference in ventilation between the lung apex and lung base.

ated. CO_2 is readily extracted from each unit of blood flowing through the capillary bed, and ventilation is high so that regional CO_2 output is relatively high. Because ventilation exceeds blood flow, the alveolar P_{CO_2} is low.

At the bottom of the lung, alveolar ventilation is less than pulmonary capillary blood flow, and $\dot{V}_A/\dot{Q}$ is approximately 0.6. Because of the relative hypoventilation of alveoli of the bottom of the lung, the alveolar P_{O_2} will be lower and the P_{CO_2} will be higher at the lung bases than at the top of the lungs.

The range of ventilation/perfusion ratios throughout the normal lung is relatively narrow so that regional differences in arterial blood gas tensions are small. However, in pulmonary diseases there are often extreme variations in ventilation/perfusion ratios in the lung because of patchy abnormalities in airway resistance, lung compliance, and blood vessel caliber. This may result in ventilation/perfusion ratios in the lung ranging from zero to infinity, independent of the effects of gravity.

A given lung unit may receive no ventilation at all because of airway obstruction; an example is shown in Fig. 39-15, *B*. This produces a unit with a $\dot{V}_A/\dot{Q}$ of zero. Under

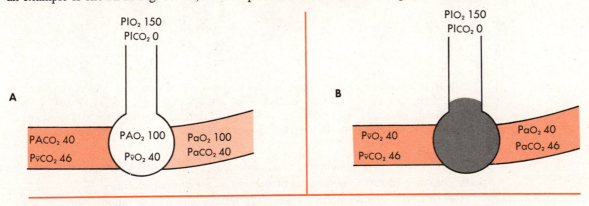

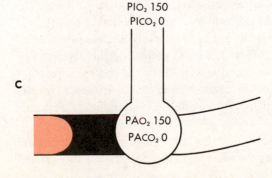

Fig. 39-15 ■ Schematic diagram of gas exchange in lung units. **A,** A normal lung unit. **B,** A lung unit with normal perfusion but receiving no ventilation. **C,** A lung unit with normal ventilation but no perfusion. *P,* Partial pressure; *I,* inspired air; *A,* alveolar gas; *a,* arterial blood; *v̄,* mixed venous blood.

these conditions the end-capillary P_{O_2} and P_{CO_2} will be the same as those in mixed venous blood (P_{O_2} = 40 mm Hg; P_{CO_2} = 46 mm Hg).

The ventilation/perfusion ratio in a lung unit can be increased as a result of constriction, obstruction, or obliteration of pulmonary blood vessels, which reduces pulmonary blood flow. The alveolar air in poorly perfused lung units has a relatively high P_{O_2} and a low P_{CO_2}. With complete obstruction to blood flow, the $\dot{V}_A/\dot{Q}$ of the unit rises to infinity, and the P_{O_2} and P_{CO_2} in the alveoli of that unit approach those of inspired air; that is, P_{O_2} = 150 mm Hg, and P_{CO_2} = 0 mm Hg (Fig. 39-15, *C*).

The consequences of uneven ventilation/perfusion ratios on arterial gas tensions can be illustrated by a model consisting of two lung units, equally well perfused but receiving different ventilation rates (Fig. 39-16). In this example the inspired gas is room air with a P_{O_2} of 150 mm Hg and a P_{CO_2} of zero. Alveolar P_{CO_2} will be lower in the unit with the greater ventilation, and alveolar P_{O_2} will be higher. Assume that the ventilation rates are such that the $P_{A_{CO_2}}$ in the poorly ventilated unit (*A*) is 44 mm Hg and the $P_{A_{O_2}}$ is only 60 mm Hg, whereas in the well-ventilated unit (*B*) $P_{A_{CO_2}}$ is 20 mm Hg, and the $P_{A_{O_2}}$ is 125 mm Hg. The mean arterial P_{O_2} must be ascertained by averaging the O_2 contents or saturations of the mixed end-capillary blood rather than by averaging the end-capillary O_2 tensions of the two units. The blood from the poorly ventilated unit with a P_{O_2} of 60 mm Hg is about 90% saturated, whereas the blood from the normal unit with a P_{O_2} of 125 mm Hg is about 99% saturated. The mixture of blood from the

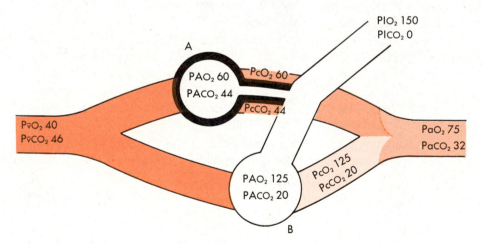

Fig. 39-16 ■ Schematic diagram of two parallel lung units. Both lung units are equally well perfused. Lung unit *A* has a reduced ventilation and a low $\dot{V}_A/\dot{Q}$ ratio. Lung unit *B* has excessive ventilation and a high $\dot{V}_A/\dot{Q}$ ratio. *P*, Partial pressure; *I*, inspired air; *A*, alveolar gas; *c*, end-capillary; *a*, arterial blood; $\bar{v}$, mixed venous blood.

two units will be 94.5% saturated, and according to the HbO_2 dissociation curve, the partial pressure of O_2 in the arterial blood will be about 75 mm Hg.

The CO_2 dissociation curve between the arterial and mixed venous points is essentially linear. As a result, the P_{CO_2} of mixed arterial blood will be equal to the average of the CO_2 tensions of the end-capillary blood from the units, or about 32 mm Hg.

Because the HbO_2 dissociation curve is flat at high P_{O_2} levels, increasing the ventilation of a lung unit with an already high $\dot{V}_A/\dot{Q}$, despite increasing the alveolar P_{O_2}, will not appreciably alter the O_2 content of blood leaving that unit, and consequently it will have little effect on the overall P_{O_2} of the mixed arterial blood. However, increasing the ventilation to high $\dot{V}_A/\dot{Q}$ units will significantly decrease arterial P_{CO_2}, since the CO_2 dissociation curve is linear.

Alterations in the distribution of blood flow so that more blood now goes to the unit with the low $\dot{V}_A/\dot{Q}$ ratio produce a further decrease in the arterial P_{O_2} for two reasons:

(1) the $\dot{V}A/\dot{Q}$ ratio of the unit is even lower than before; and (2) the low $\dot{V}A/\dot{Q}$ unit now provides a larger contribution to the blood leaving the lungs.

Similarly, the PO_2 of the mixed alveolar air collected from the different alveoli will depend on the alveolar PO_2 in each and the relative amount that each alveolus contributes to the total alveolar ventilation. Because alveoli with a high $\dot{V}A/\dot{Q}$ ratio have both higher levels of PO_2 and more ventilation, whereas alveoli with a low $\dot{V}A/\dot{Q}$ ratio have both lower O_2 tensions and relatively more perfusion, abnormally wide distribution of $\dot{V}A/\dot{Q}$ ratios will always cause the PO_2 of mixed alveolar gas to be higher than the average PO_2 of the blood in the pulmonary veins (a value usually close to arterial PO_2). Likewise, a wide range of $\dot{V}A/\dot{Q}$ ratios of alveoli will cause the arterial PCO_2 to be higher than the alveolar PCO_2. The normal difference between alveolar and arterial O_2 tensions is about 5 mm Hg. The difference between alveolar and arterial PCO_2 tensions is normally almost indiscernible. However, with lung diseases that increase the spread of $\dot{V}A/\dot{Q}$ ratios in the lung, the difference between the alveolar and arterial blood gas tensions will become greater.

Ventilation-perfusion inequalities interfere with the efficiency of gas exchange. For the same level of ventilation, arterial PCO_2 is higher and arterial PO_2 is lower the greater the spread in $\dot{V}A/\dot{Q}$ ratios in the alveoli.

The lowering of arterial PO_2 produced by low $\dot{V}A/\dot{Q}$ alveoli cannot be overcome by increasing ventilation to normal regions of the lung. Low $\dot{V}A/\dot{Q}$ regions are the most common cause of hypoxemia in lung disease. Elevations in arterial PCO_2 caused by $\dot{V}A/\dot{Q}$ abnormalities give rise, by stimulating chemoreceptors, to a compensatory increase in ventilation and hence to a fall in alveolar PCO_2. Because of the linearity of the CO_2 dissociation curve, this can restore arterial PCO_2 to normal levels. Ultimately, when $\dot{V}A/\dot{Q}$ abnormalities become very severe, ventilation may not increase sufficiently to prevent chronic elevation of arterial PCO_2.

The ventilation/perfusion ratio of an alveolus can be calculated if the alveolar PCO_2, alveolar PO_2, mixed venous PO_2, and R are known and if it is assumed that there is no gradient for O_2 across the alveolar-capillary membrane so that PO_2 in the alveolus and in its capillary blood are the same.

$$\dot{V}A/\dot{Q} \cong R \times \frac{(PAO_2 - P\bar{v}O_2)}{PACO_2} \tag{19}$$

All the possible values for alveolar PCO_2 and PO_2 that occur in lung alveoli with different $\dot{V}A/\dot{Q}$ ratios can be incorporated in an O_2/CO_2 diagram (Fig. 39-17).

In the steady state the respiratory exchange ratio for the whole lung is the same as the respiratory quotient. We can plot on this diagram the R lines for blood and air (where R is the respiratory quotient of the body). The lines for air and blood intersect at a point on the $\dot{V}A/\dot{Q}$ curve. The PCO_2 and PO_2 at the point of intersection represent the alveolar gas tensions that would exist under ideal circumstances if the $\dot{V}A/\dot{Q}$ ratios in every alveolus were all the same. It is evident from the shape of the $\dot{V}A/\dot{Q}$ curve that low $\dot{V}A/\dot{Q}$ ratios depress alveolar PO_2 more than they raise arterial PCO_2, whereas high $\dot{V}A/\dot{Q}$ ratios lower alveolar PCO_2 proportionally more than they increase alveolar PO_2.

A wide range of ventilation/perfusion ratios in the lung will produce a difference between mixed alveolar PO_2 and the PO_2 of the mixed pulmonary venous and systemic arterial blood. Arterial PO_2 can be further decreased by systemic venous blood, which, for anatomical reasons, fails to come into contact with alveoli but then mixes with blood in the pulmonary veins.

Normally about 5% of the blood ejected by the left ventricle has not been oxygenated. Blood from bronchial veins, which empty into pulmonary veins, and blood from thebesian veins of the ventricular myocardium, which empty into the left ventricle, constitute a normal right-to-left shunt of underoxygenated blood. In some congenital

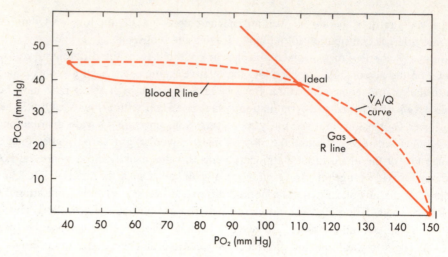

Fig. 39-17 ■ Oxygen–carbon dioxide diagram. P_{O_2} is plotted on the horizontal axis, and P_{CO_2} is plotted on the vertical axis. The interrupted line extending from the inspired air point (I) to the mixed venous point (v̄) is the ventilation/perfusion ratio ($\dot{V}_A/\dot{Q}$) curve. The $\dot{V}_A/\dot{Q}$ line represents all possible P_{O_2} and P_{CO_2} combinations that can result from the equilibration of mixed venous blood and inspired air at $\dot{V}_A/\dot{Q}$ ranging from zero to infinity. At a given respiratory exchange ratio (R), there is a fixed relationship between alveolar P_{O_2} ($P_{A_{O_2}}$) and alveolar P_{CO_2} ($P_{A_{CO_2}}$). The potential combinations of $P_{A_{O_2}}$ and $P_{A_{CO_2}}$ are described by the solid line (*gas R line*) originating at the inspired air point (I). At the same constant R value, the potential combinations of P_{O_2} and P_{CO_2} in the blood are shown by the blood R line, which extends from the mixed venous point (v̄). The intersection of the gas and blood R lines on the $\dot{V}_A/\dot{Q}$ line represents the hypothetical P_{O_2} and P_{CO_2} at this R value when no ventilation/perfusion abnormalities exist.

heart diseases, blood may pass directly from the right to the left side of the heart through intracardiac defects or through pulmonary arteriovenous communications.

The fraction of the cardiac output that fails to reach gas-exchanging surfaces in the lung can be calculated from the *shunt equation*. The total cardiac output ($\dot{Q}_T$) is made up of the shunt flow ($\dot{Q}_S$) plus the blood flow past the ventilated alveoli ($\dot{Q}_T - \dot{Q}_S$). The amount of O_2 in the blood pumped by the left ventricle is the product of the cardiac output and the O_2 content of the arterial blood (Ca_{O_2}). This total quantity of O_2 is made up of the contribution from the shunted blood, the O_2 content of which is the same as that of mixed systemic venous blood ($C\bar{v}_{O_2}$), and the contribution from the blood that has perfused ventilated alveoli. The amount of O_2 in that blood is the product of the blood flow past the ventilated alveoli ($\dot{Q}_T - \dot{Q}_S$) and the O_2 content of the end-capillary blood (Cc_{O_2}). This can be expressed as follows:

$$\dot{Q}_T \times Ca_{O_2} = (\dot{Q}_S \times Q\bar{v}_{O_2}) + (\dot{Q}_T - \dot{Q}_S) \times Cc_{O_2} \qquad (20)$$

By rearranging this equation,

$$\dot{Q}_S/\dot{Q}_T = \frac{Cc_{O_2} - Ca_{O_2}}{Cc_{O_2} - C\bar{v}_{O_2}} \qquad (21)$$

The O_2 content of the arterial and mixed venous blood can be measured directly. The O_2 content of the end-capillary blood must be estimated by considering the mixed alveolar P_{O_2} and the end-capillary P_{O_2} to be identical. If there are many low $\dot{V}_A/\dot{Q}$ alveoli in the lung, this assumption will be incorrect. The usual procedure then is to have the subject breathe 100% O_2. After a sufficient period of O_2 breathing, nitrogen is washed out of the lung, and even the most poorly ventilated alveoli will contain only O_2 and CO_2. The end-capillary P_{O_2} of even poorly ventilated lung units will be the same as the P_{O_2} in the alveolar gas of those units.

By the time inspired air has reached the gas-exchanging units of the lung, all the energy imparted by inspiratory muscle contraction has been dissipated, and the bulk movement of gas is virtually nil. Hence, diffusion is the principal process that accounts for the transfer of O_2 and CO_2 across the alveoli and the intervening interstitial tissue and into the capillary blood.

The rates of diffusion of molecules in a gaseous phase are inversely proportional to the square root of their densities (Graham's Law). Large differences in molecular weight are therefore needed to produce substantial differences in diffusion within the alveolus. For O_2 (molecular weight, 32) and CO_2 (molecular weight, 44) the relative rates of diffusion in the alveolus are not too different.

$$\frac{O_2 \text{ rate}}{CO_2 \text{ rate}} = \frac{\sqrt{44}}{\sqrt{32}} = \frac{6.6}{5.6}$$

In normal alveoli (diameter, approximately 20 μm) CO_2 and O_2 appear to mix evenly in less than 10 msec. Consequently, diffusion in the alveolus itself does not limit gas exchange. Rarely, in severe emphysema, a disease in which alveolar walls are destroyed and large air sacs are formed, equilibration times may be sufficiently prolonged so that gas tensions within the terminal lung units are not uniform, and gas exchange is impaired.

Gas transfer through tissue sheets, such as the alveolar-capillary membrane, into a liquid medium, such as the blood, depends on the gas tension gradient, the solubility of the gas in the liquid, the molecular weight of the gas, and the properties of the membrane. This relationship is expressed by Fick's law (Chapter 1):

$$J = -DA\frac{\Delta c}{\Delta x} \qquad (23)$$

where

- J = flux of the gas across the membrane
- D = diffusion coefficient
- A = area of the membrane
- Δx = thickness of the membrane
- Δc = concentration gradient across the membrane

According to Henry's law:

$$\Delta c = \alpha\Delta P \qquad (24)$$

where
- α = solubility of the gas in the tissue
- ΔP = gas tension gradient across the membrane

Therefore

$$J = -DA\alpha\frac{\Delta P}{\Delta x} \qquad (25)$$

The alveolar capillary membrane averages about 0.5 μm in thickness (Fig. 39-18) and consists of the following layers from the alveolus to the capillary side: surfactant, alveolar epithelium, its basement membrane, a connective tissue layer, and the basement membrane of the capillary and its endothelium.

Capillaries are situated in the lung interstitium between alveoli. One side of each capillary is thin and the other thick. On the thin side, the alveolar and capillary basement membranes are fused, and there is no intervening connective tissue layer. Gas exchange probably takes place mainly through this thin side. The thick side seems to be important in fluid exchange (Chapter 38). The area of the blood-gas barrier in the lung is enormous, 50 to 100 m^2, and over 80% of this area is covered by a thin layer of

■ *Diffusion of gases in the lung*
■ *Diffusion in the alveolus*

■ *Diffusion through the alveolar-capillary membrane*

Fig. 39-18 ■ Schematic diagram of the alveolocapillary membrane.

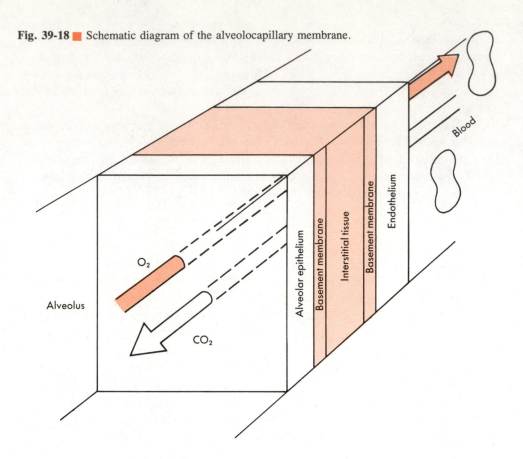

blood. The pulmonary capillaries and associated alveolar-capillary membranes may be conceptualized as a thin layer of blood with interposed connective tissue struts that separate two alveolar sheets.

Although the potential area for gas exchange is large, the effective surface may be much more limited, particularly in the presence of lung disease. Areas of the lung that are neither ventilated nor perfused will not participate in gas exchange. As a consequence, diseases (e.g., emphysema) that affect the distribution of inspired air or of pulmonary blood flow may influence gas diffusion.

In a resting individual, the red blood cells remain within the pulmonary capillary for only about 0.75 second, which imposes a limit on the time available for gas transfer. Depending on the specific gas in the alveoli, transfer of the gas from the alveoli to the blood may be limited by the rates of perfusion or diffusion (Fig. 39-19). Transfer of a gas is said to be *diffusion-limited* when the gas tensions in the alveoli and the capillary fail to reach equilibrium during the time of red cell transit through the capillary. For example, the transfer of CO is diffusion-limited. CO moves rapidly across the extremely thin blood-gas barrier from the alveolar gas into the red blood cell. However, because of the high affinity of CO for hemoglobin in the red blood cell, a large amount of CO can be taken up by the cell, and there is almost no increase in partial pressure in the blood. Thus no appreciable back pressure of CO develops. Gas continues to cross the alveolar wall rapidly, and equilibration is not achieved even if flow through the capillaries is extremely low. Therefore the amount of CO that enters the blood is limited by the diffusion properties of the blood-gas barrier and not by the rate of perfusion.

On the other hand, the transfer of an inert gas, such as nitrous oxide, is *perfusion-limited*. When nitrous oxide moves across the alveolar wall, it dissolves in the blood, but it does not combine with any substance in the blood. As a result, the partial pressure of nitrous oxide in the blood increases rapidly. Hence, by the time the red cell is only one tenth of the way along the capillary, the partial pressures of nitrous oxide on both sides of the membrane are the same, preventing further gas transfer.

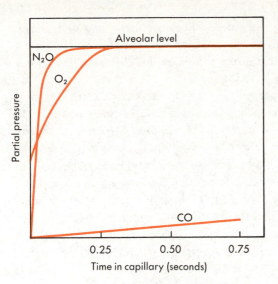

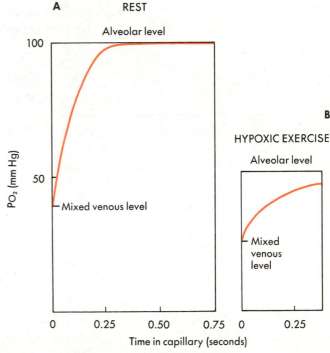

Fig. 39-19 ■ Changes in the partial pressure of carbon monoxide (CO), nitrous oxide (N_2O), and oxygen (O_2) in blood during transit through the pulmonary capillary bed. During transit through the pulmonary capillaries, blood P_{CO} increases only slightly. P_{N_2O} of blood reaches equilibrium with that in the alveolar gas very rapidly. The P_{O_2} of blood and alveolar gas become virtually the same after the blood has travelled only about one third of the way along the capillary bed.

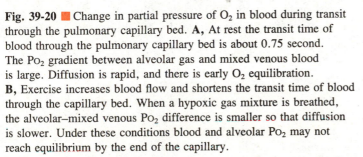

Fig. 39-20 ■ Change in partial pressure of O_2 in blood during transit through the pulmonary capillary bed. **A,** At rest the transit time of blood through the pulmonary capillary bed is about 0.75 second. The P_{O_2} gradient between alveolar gas and mixed venous blood is large. Diffusion is rapid, and there is early O_2 equilibration. **B,** Exercise increases blood flow and shortens the transit time of blood through the capillary bed. When a hypoxic gas mixture is breathed, the alveolar–mixed venous P_{O_2} difference is smaller so that diffusion is slower. Under these conditions blood and alveolar P_{O_2} may not reach equilibrium by the end of the capillary.

Continued transfer of nitrous oxide can occur only if new blood is supplied. Thus the amount of nitrous oxide taken up by the blood depends almost entirely on the rate of perfusion.

In healthy persons breathing quietly, the transfer of O_2 is mainly perfusion-limited. O_2, like CO, combines with the hemoglobin, but it does so less avidly. Therefore, when a given volume of O_2 enters a red blood cell, the rise in partial pressure is much greater than when the same volume of CO is transferred. Under resting conditions, the capillary P_{O_2} is nearly the same as the alveolar P_{O_2} by the time the red cell is about one third of the way along the capillary.

When cardiac output is increased during exercise, the time spent by the red blood cell in the capillary is reduced to one third of normal (approximately 0.25 second). Even under these conditions, the P_{O_2} in the blood will become equal to that in the alveoli before the end of the capillary is reached.

Breathing a hypoxic gas mixture at sea level or ambient air at altitude will decrease alveolar P_{O_2} and reduce the partial pressure difference that drives O_2 across the alveolar-capillary membrane. Hence, P_{O_2} will increase more slowly in the capillary blood. Also, because of the shape of the HbO_2 dissociation curve, there is a smaller increase in blood P_{O_2} for a given increase in O_2 content at lower levels of P_{O_2} than at higher levels. For both these reasons, hypoxia decreases the likelihood that blood P_{O_2} and alveolar P_{O_2} will reach equilibrium by the end of the capillary. But in healthy subjects, a limitation of O_2 transfer by diffusion probably only occurs when exhausting exercise is performed under hypoxic conditions (Fig. 39-20).

In some forms of lung disease (e.g., pulmonary fibrosis), there may be sufficient thickening of the pulmonary interstitium so that diffusion limits O_2 transfer under less severe conditions. Even when marked pulmonary fibrosis occurs, diffusion limitation is not likely to be the cause of hypoxemia in patients at rest.

Although CO_2 has a greater molecular weight than O_2, it is much more soluble in water. For a given partial pressure difference, CO_2 will diffuse 20 times more rapidly than O_2. However, the capacity of blood to retain CO_2 (Δ total gas content/Δ partial pressure of the gas) is also considerably greater than that for O_2. As a consequence, it will take nearly as long for P_{CO_2} as for P_{O_2} to reach equilibrium between the capillary blood and the alveolus. This means that any abnormality in diffusion that affects O_2 equilibration may also affect the equilibration of CO_2. In addition, the chemical reactions between O_2 and hemoglobin and those involved in CO_2 transfer (the carbamino reaction and CO_2 buffering in the red cell) take time, which could affect the equilibration time in the capillary. Nonetheless, diffusion abnormalities rarely cause CO_2 retention. When arterial P_{CO_2} increases, ventilation is stimulated and restores the normal level of arterial P_{CO_2}.

■ *Measurement of diffusing capacity of the lung*

The amount of gas transferred across the alveolocapillary membrane is directly proportional to the area of the membrane and is inversely proportional to the thickness of the membrane. It is not possible to measure during life either the total area (A) or thickness of the alveolocapillary membrane (Δx). Instead a simplified version of equation 25 shown earlier is used:

$$D_L = \frac{DA\alpha}{\Delta x} = -\frac{J}{\Delta P} = \frac{J}{P_A - P_{cap}} \tag{26}$$

where

D_L = diffusing capacity of the lung
J = volume of gas transferred per unit of time
P_A = alveolar tension of the gas
P_{cap} = capillary tension of the gas

D_L includes area, thickness, and diffusion properties of the lung.

The equation for the diffusing capacity of the lung for oxygen is

$$D_L O_2 = \frac{\dot{V}O_2}{PA_{O_2} - Pcap_{O_2}} \tag{27}$$

where

$D_L O_2$ = diffusing capacity in ml O_2/minute/mm Hg P_{O_2}
$\dot{V}O_2$ = volume of O_2 (in ml) transferred per minute from alveolar gas to the blood
PA_{O_2} = mean alveolar P_{O_2} in mm Hg
$Pcap_{O_2}$ = mean pulmonary capillary P_{O_2} in mm Hg

A decrease in $D_L O_2$ may be the result of a decreased area or an increased thickness of the alveolocapillary membrane, or both. $D_L O_2$ may be unaffected when the thickness is increased if the surface area available for diffusion is simultaneously increased.

O_2 diffusing capacity is difficult to determine because the uptake of O_2 is largely blood-flow dependent, and the instantaneous partial pressure gradient of O_2 from the alveolus to the capillary blood cannot be measured. It is much simpler to use CO to measure the diffusing capacity.

The diffusing capacity for CO ($D_L CO$) is given by

$$D_L CO = \frac{\dot{V}_{CO}}{PA_{CO} - Pcap_{CO}} \tag{28}$$

where

PA_{CO} = partial pressure of CO in alveolar gas
$Pcap_{CO}$ = partial pressure of CO in capillary blood
$\dot{V}_{CO}$ = volume of CO transferred (ml/min)

Except in habitual cigarette smokers, the partial pressure of CO in capillary blood is so small that it can be neglected. The equation then becomes

$$D_L{CO} = \frac{\dot{V}_{CO}}{PA_{CO}} \qquad (29)$$

Single-breath and steady-state techniques have been devised for measuring $D_L{CO}$. In the *single-breath method,* the patient inspires a vital capacity–size breath of a dilute mixture of CO (approximately 0.3%). The breath is held for 10 seconds at the end of that inspiration, and the rate of disappearance of CO from the alveolar gas during that period is calculated. The uptake of CO is calculated from the difference between the concentration of CO in alveolar air at the beginning and end of the period of breath-holding multiplied by the alveolar volume during the period of breath-holding. An inert insoluble gas, such as helium or neon, is added to the inspired gas so that the alveolar volume and initial alveolar CO concentration can be estimated by dilution.

$$V_A = \frac{He_I \times \dot{V}_I}{He_E} \qquad (30)$$

and

$$FA_{CO} = \frac{He_E \times FI_{CO}}{He_I} \qquad (31)$$

where

V_A = alveolar volume
He_I = concentration of helium in inspired air
$\dot{V}_I$ = inspired volume
He_E = concentration of helium in expired air
FA_{CO} = concentration of CO in alveolar air
FI_{CO} = concentration of CO in inspired air

This approach to measuring alveolar volume and initial alveolar CO concentration is based on the assumption that the helium and CO are distributed within the lung similarly and mix uniformly during the period of breath-holding. However, in diseased lungs with increased numbers of poorly ventilated alveolar units, the single-breath dilution test may seriously underestimate the alveolar volume. To circumvent this problem, alveolar volume may be measured plethysmographically rather than from the dilution of an inert gas.

The calculation of the volume of CO taken up during a single breath assumes that diffusion occurs uniformly throughout the lung and that alveolar P_{CO} decreases exponentially during the breath-holding period. In normal lungs alveolar P_{CO} does fall exponentially during breath-holding. In the diseased lung, however, there are areas where alveolar ventilation or capillary blood flow is substantially reduced so that CO will not be transferred uniformly.

In the steady-state methods, the subject breathes a very low concentration of CO (about 0.1%) for approximately 1 minute. The rate of uptake of CO from alveolar gas is determined from differences in the concentration of CO in inspired and expired air. Alveolar P_{CO} is either directly measured or indirectly determined from arterial P_{CO} values and the dead space equation (8). Steady-state measurements of PA_{CO} are affected by unevenness in the distribution of the inspired CO even more than are single-breath estimations. Newer techniques that involve rebreathing minimize the problems caused by uneven distribution of ventilation but have other theoretical difficulties. For these reasons, some investigators prefer not to apply the term diffusing capacity to these measurements and instead use the term *transfer capacity*. In practice these tests are useful in detecting patients with subtle abnormalities in lung function that are not apparent with other measurements, although none of them actually measures only the ability of gas to diffuse across the alveolocapillary membrane.

The binding of O_2 or CO to hemoglobin is not instantaneous. The time required for the chemical reactions with hemoglobin appreciably delays the uptake of O_2 and CO by the red cell. Hence, the transfer of O_2 (or CO) across the alveolocapillary membrane can be considered as occurring in two stages: (1) diffusion through the blood-gas barrier (including the plasma and red cell interior) and (2) reaction of the gas with hemoglobin. Each stage can be considered to offer a resistance to gas uptake by the blood, and these resistances can be summed to produce an overall "diffusion" resistance. The reciprocal of D_L is the total resistance to uptake of gas by diffusion. The resistance of the blood-gas barrier to diffusion can be considered to be $1/D_M$, where M signifies membrane. The reciprocal of the product of the rate of reaction (θ) of O_2 (or CO) with hemoglobin in milliliters of gas/minute/mm Hg/milliliters of blood and the volume of capillary blood (V_C), that is, $1/(\theta \times V_C)$, describes the "resistance" of this chemical reaction to the uptake of gas. The two resistances to uptake offered by diffusion across the membrane and the uptake of O_2 or CO by hemoglobin are added to obtain the total diffusion resistance. Thus the complete expression is

$$\frac{1}{D_L} = \frac{1}{D_M} + \frac{1}{\theta \times V_C} \tag{32}$$

Most experimental data suggest that the resistances offered by the membrane and blood components are approximately equal. It is apparent from the equation that a reduction of capillary blood volume by disease or a reduction in the hemoglobin content of the blood by anemia, for example, lowers the diffusing capacity.

Normal values for $D_L CO$ measured by the single-breath test are 25 ml/minute/mm Hg in resting subjects. Values obtained by the single-breath method are greater than those obtained by the steady-state test. Because the single-breath test is performed at a lung volume close to total lung capacity, the higher values obtained most likely reflect the greater surface area of the lung available for diffusion. $D_L CO$ rises during exercise when pulmonary blood flow and pressure increase, probably because the area available for diffusion increases as more pulmonary capillaries open.

CO diffusing capacity also varies with body size, probably because of differences in the area available for gas exchange. The diffusing capacity is about 25% greater in the supine than in the standing position because more capillaries are open in the upper lobe when the subject is supine. The CO diffusing capacity also depends on the P_{O_2}. When the arterial P_{O_2} is reduced, the $D_L CO$ is increased because of changes in the rate of reaction of CO with hemoglobin.

In smokers a considerable portion of the hemoglobin may be combined with CO (as much as 12%); hence, there is an appreciable back pressure of CO in the blood. The values obtained for diffusing capacity for CO in smokers should be corrected by measurements of actual levels of HbCO.

■ *Bibliography*

Journal articles

Bunn, H.F., and Jandl, J.H.: Control of hemoglobin function within the red cell, N. Engl. J. Med. **282:**1414, 1970.

Dantzker, D.R., Wagner, P.D., and West, J.B.: Instability of lung units with low $\dot{V}_A/\dot{Q}$ ratios during O_2 breathing, J. Appl. Physiol. **38:**886, 1975.

Finch, C.A., and Lenfant, C.: Oxygen transport in man, N. Engl. J. Med. **286:**407, 1972.

Forster, R.E., and Crandall, E.D.: Time course of exchanges between red cells and extracellular fluid during CO_2 uptake, J. Appl. Physiol. **38:**710, 1975.

Forster, R.E., and Crandall, E.D.: Pulmonary gas exchange, Ann. Rev. Physiol. **38:**69, 1976.

Goldberg, M., Green, S.B., and Moss, M.L.: Computer-based instruction and diagnosis of acid-base disorders, J.A.M.A. **223:**240, 1973.

Heinemann, H.O., and Goldring, R.M.: Bicarbonate and the regulation of ventilation, Am. J. Med. **57:**361, 1974.

Klocke, R.A.: Mechanisms and kinetics of the Haldane effect in human erythrocytes, J. Appl. Physiol. **35:**673, 1973.

Mellemgaard, K.: The alveolar-arterial oxygen difference: its size and components in normal man, Acta Physiol. Scand. **67:**10, 1966.

Riley, R.L., and Permutt, S.: Venous admixture component of the A-aPo_2 gradient, J. Appl. Physiol. **35:**430, 1973.

Roughton, F.J.W., and Forster, R.E.: Relative importance of diffusion and chemical reaction rates in determining rate of exchange of gases in the human lung with special reference to true diffusing capacity of pulmonary membrane and volume of blood in the lung capillaries, J. Appl. Physiol. **11:**290, 1957.

Siggaard-Andersen, O.: The acid-base status of the blood, Scand. J. Clin. Lab. Invest. **15** (suppl. 70):1, 1963.

Wagner, P.D., Saltzman, H.A., and West, J.B.: Measurement of continuous distributions of ventilation-perfusion ratios: theory, J. Appl. Physiol. **36:**588, 1974.

West, J.B.: Ventilation-perfusion inequalities and overall gas exchange in computer models of the lung, Respir. Physiol. **7:**88, 1969.

West, J.B.: Ventilation-perfusion relationships, Am. Rev. Respir. Dis. **116:**919, 1977.

Books and monographs

Filley, G.F.: Acid-base and blood gas regulation, Philadelphia, 1971, Lea & Febiger.

Forster, R.E.: Diffusion of gases. In Fenn, W.O., and Rahn, H., editors: Handbook of physiology; Section 3: Respiration, vol. I, Bethesda, Md., 1964, American Physiological Society.

Maxwell, M.H., and Kleeman, C.R.: Clinical disorders of fluid and electrolyte metabolism, New York, 1980, McGraw-Hill, Inc.

Rahn, H., and Farhi, L.E.: Ventilation, perfusion, and gas exchange—the $\dot{V}_A/\dot{Q}$ concept. In Fenn, W.O., and Rahn, H., editors: Handbook of physiology; Section 3: Respiration, vol. I, Bethesda, Md., 1964, American Physiological Society.

Rahn, H., and Fenn, W.O.: A graphical analysis of the respiratory gas exchange: the O_2-CO_2 diagram, Bethesda, Md., 1955, American Physiological Society.

West, J.B.: Ventilation/blood flow and gas exchange, ed. 3, Oxford, 1977, Blackwell Scientific Publications.

West, J.B., and Wagner, P.D.: Pulmonary gas exchange. In West, J.B., editor: Bioengineering aspects of the lung, New York, 1977, Marcel Dekker, Inc.

chapter 40

Control of respiration

Breathing is continuously monitored by *chemoreceptors*, which detect changes in P_{CO_2} and P_{O_2}, and by mechanoreceptors, which survey thoracic mechanics. The activity of these receptors allows ventilation to be adjusted automatically so that arterial blood gases are kept within acceptable limits, despite changing internal and external conditions.

Ventilation is affected by signals from the thermal, vascular pressure and mechanoreceptors situated throughout the body and by the activity of higher brain centers. The multiplicity of inputs to the respiratory neurons permits satisfactory levels of ventilation to be maintained even when disease affects one or more afferent pathways or when the response to some sensory cues is blunted by a depressed state of consciousness.

■ Central organization of respiratory neurons
■ *Supraspinal organization*

The usual techniques of ablation, stimulation, or recording electrical activity of neurons with microelectrodes have been used to explore the brain and to locate respiratory areas. Two types of areas have been found: (1) *intrinsic respiratory areas,* that is, centers whose major, if not only, function is the regulation of respiratory activity, and (2) *subsidiary respiratory areas* concerned primarily with other functions but whose outputs project to the intrinsic respiratory areas and modify respiratory activity. At least two regions in the brainstem function as intrinsic respiratory centers: (1) the *medullary respiratory centers,* and (2) the *pneumotaxic center* (or nucleus parabrachialis medialis [NPBM]), which is in the upper third of the pons. Other subsidiary respiratory areas of the brain, with projections to one or more of these intrinsic centers, are the cortex (volitional inputs to respiratory centers), the cerebellum, and the medullary vasomotor areas.

Many investigators have demonstrated that surgical lesions placed in certain regions of the brainstem have profound effects on breathing, whereas decortication or decerebration do not. The neurons responsible for rhythmic breathing reside in a discrete location in the medulla, but the behavior of these neurons is modified by influences arising in the pons.

The precise anatomical and functional organization of the medullary neurons that are responsible for the respiratory rhythm remains a matter for investigation. However, substantial evidence indicates that networks of neurons in two separate nuclei in the medulla generate the respiratory rhythm. These areas include the *nucleus tractus solitarius* (NTS), located dorsally near the exit of the ninth cranial nerve, and the ventrally located *nucleus retroambiguus* (NRA), which extends caudally from the first cervical spinal segment (Fig. 40-1).

Some of the neurons in the NTS project to the phrenic motor neurons, whereas others project to neurons in the NRA. Afferent fibers from the peripheral chemorecep-

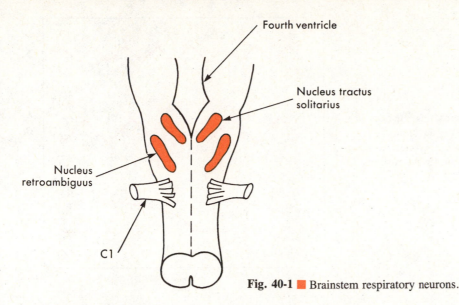

Fourth ventricle

Nucleus tractus
solitarius

Nucleus
retroambiguus

C1

Fig. 40-1 ■ Brainstem respiratory neurons.

tors, baroreceptors, and pulmonary mechanoreceptors synapse in the NTS. Thus the
NTS may be the site of the initial processing of information from peripheral respiratory
sensors. The abundant interconnections within the nucleus appear to provide the basis
for a system of self-excitation, which probably explains the progressive increase in
phrenic nerve activity seen during inspiration. The NTS also may regulate the behavior
of neurons in the NRA.

Some neurons in the NRA fire during expiration, and others discharge during inspiration. Many of the NRA neurons innervate intercostal and abdominal muscle motor
neurons in the spinal cord. Cranial motor nerves that innervate pharyngeal or laryngeal
muscles receive signals from neurons in the nucleus ambiguus, located adjacent to the
NRA. The activity in the nucleus ambiguus waxes and wanes with respiratory activity,
thereby helping to synchronize the activity of the upper airway and chest wall muscles.

The mechanism by which these medullary networks of neurons cause switching between inspiration and expiration is unclear. Studies to date have failed to demonstrate
pacemaker cells, that is, neurons, analogous to cardiac muscles cells, that demonstrate
oscillation in membrane potential in the absence of excitatory and inhibitory inputs. The
bulk of available evidence supports the concepts that rhythmic respiratory activity requires the presence of continuous (tonic) or intermittent (phasic) excitatory inputs and
that respiratory oscillations are the result of reciprocal inhibition of interconnected networks of neurons.

Examination of the respiratory controller in its rudimentary state in deeply anesthetized animals indicates that inputs from thoracic mechanoreceptors, chiefly stretch receptors in the walls of the airways, are of critical importance in setting the breathing
frequency. Interruption of pulmonary stretch receptor input by section of the vagus
nerves leads to a marked prolongation in the duration of inspiration and expiration (Fig.
40-2). Networks of neurons in the pons also influence the switching between inspiration
and expiration. When lesions are made in the NPBM in the rostral pons, inspiration
becomes even more prolonged, and *apneusis* (deep and prolonged inspirations lasting
tens of seconds) ensues. Some studies have suggested that there is an "apneustic center" in the lower pons, because apneustic breathing may disappear when the pons and
medulla are completley separated.

Input from *lung stretch receptors*, traveling in the vagi, affects the frequency of
breathing. However, this input appears to have little effect on the rate of rise of the
motor activity of the respiratory muscles during inspiration. In anesthetized animals with
intact vagi, inflation of the lung during the inspiratory phase can terminate inspiration

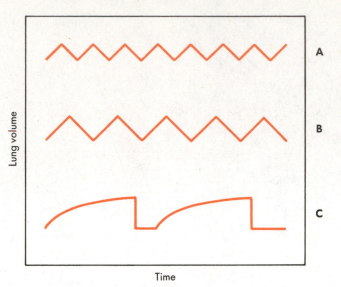

Fig. 40-2 ■ Patterns of breathing. **A,** Normal breathing. **B,** Section of the vagus nerves results in a prolongation in the duration of inspiration and expiration. **C,** Apneustic breathing characterized by deep and prolonged inspiration produced by lesions in the rostral pons.

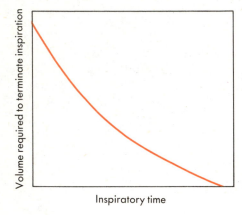

Fig. 40-3 ■ In anesthetized animals with intact vagi, lung inflation during the phase of inspiration can terminate inspiration by stimulating pulmonary stretch receptors. Early in inspiration a large volume inflation is required to terminate inspiration. Late in the inspiratory phase inspiration is terminated by small inflations.

by stimulating stretch receptors. The effect of lung inflation depends on the timing of the inflation and on the volume of gas introduced (Fig. 40-3). Early in inspiration large volume changes are needed to terminate inspiration, whereas late in inspiration only small volumes are necessary. After a vagotomy lung inflation has no effect on the duration of inspiration. Neurons whose firing is excited by lung inflation have been found near the NTS.

The regulation of the time in expiration (Te) is less well understood. In both spontaneously breathing animals and animals in which inspiratory time (Ti) is manipulated experimentally by lung inflation, a close direct correlation exists between Te and Ti, such that as one time becomes shorter, so does the other. Thus mechanisms that determine Ti probably also influence Te. This is not the whole story, however, because lung inflation during the first 75% of expiration prolongs Te without affecting the subsequent Ti. These reflex effects on respiratory timing may have important beneficial effects. Termination of inspiration with large lung inflations tends to prevent overinflation of the lung. It also facilitates the increases in respiratory frequency needed for adequate increases in ventilation when breathing is driven (for example, breathing CO_2-enriched gases or during exercise). When the resistance to inspiratory airflow is increased, the decreased rate of lung inflation prolongs inspiration. This allows more time for gas to enter the lung and thereby helps maintain a constant tidal volume. However, increases in lung volume or obstruction to airflow during expiration reflexly increases the time available for lung emptying by forestalling the onset of the next inspiratory effort.

In the vagotomized animal hypercapnia and hypoxia have no effect on breathing frequency, but they increase tidal volume and increase the rate of rise of activity in inspiratory nerves, like the phrenic nerve (Fig. 40-4).

These fundamental observations can be synthesized by means of an operational model of the brainstem mechanisms that produce the respiratory rhythm (Fig. 40-5). Signals

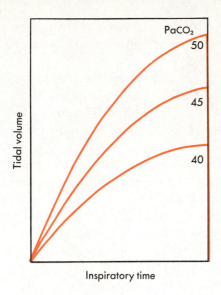

Fig. 40-4 ■ Effect of increases in Pa_{CO_2} on tidal volume and inspiratory time in vagotomized animals. Increasing the Pa_{CO_2} from 40 to 50 mm Hg results in a progressive increase in tidal volume. In vagotomized animals inspiratory time does not change with hypercapnia.

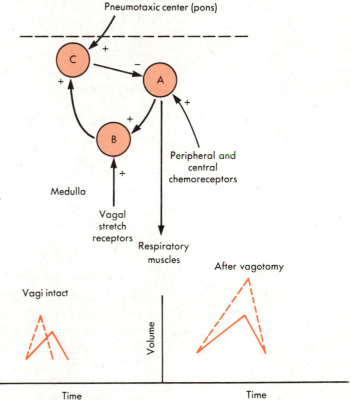

Fig. 40-5 ■ Brainstem mechanisms that regulate respiratory rhythm.

from the *central chemoreceptors* (in the medulla) and peripheral chemoreceptors (aortic and carotid bodies) impinge on a pool of inspiratory neurons, probably in the NTS *(pool A)*. These neurons project to the respiratory spinal motor neurons and evoke a ramplike increase in their activity *(central inspiratory activity)*. Increasing input from chemoreceptors (excited by hypoxia or hypercapnia) accentuates the steepness of the ramp. Central inspiratory activity, which is reflected by the rate at which phrenic motor activity rises, stimulates another pool of neurons *(pool B)*, probably also in the NTS. *Pool B* also receives signals from pulmonary stretch receptors. These stretch receptor signals, which increase with lung expansion, are summed with *pool A* input at *pool B*. *Pool B* in turn activates *pool C* of "inspiratory off-switch neurons," which inhibit *pool A* neurons. When a critical level of excitation occurs in *pool C*, *pool A* activity is extinguished, and expiration occurs. During this inhibitory period pulmonary stretch receptor excitation is able to excite further *pool C* neurons, thereby prolonging expiration.

An increase in chemoreceptor afferent input increases the "off-switch" threshold as well as the central inspiratory activity. Thus tidal volume can be increased even without any change in inspiratory duration. Ablation of the pneumotaxic center in the vagotomized animal increases the "off-switch" threshold without affecting central inspiratory activity, further lengthening inspiration.

This concept of the generation of respiratory rhythm suggests that the average rate of inspiratory airflow (that is, the ratio of tidal volume, V_T, to inspiratory time, T_i) is the mechanical analogue of central inspiratory activity. The ratio V_T/T_i seems to be better than the level of ventilation itself as an index of the controller response to chemoreceptor inputs.

Ventilation is conventionally expressed as the quotient of tidal volume and the duration of the breathing cycle (T_{Tot}). Alternatively ventilation can be determined from the product of V_T/T_i and the ratio of T_i to T_{Tot}, that is, $(V_T/T_i) \times (T_i/T_{Tot})$. As indicated previously, V_T/T_i is a measure of the level of central respiratory activity. The ratio T_i/T_{Tot}, on the other hand, reflects the activity of pulmonary stretch receptors. The representation of $(V_T/T_i) \times (T_i/T_{Tot})$ enables the separate determination of central respiratory and peripheral pulmonary stretch receptor effects on ventilation.

In recent studies in humans, the effects of hypoxia and hypercapnia on V_T/T_i, rather than on ventilation, have been used to assess chemosensitivity. In some cases depressed ventilatory responses probably were caused by altered mechanoreceptor function (which reduces the T_i/T_{Tot} ratio) rather than by depressed chemosensitivity.

Higher brain centers play an important role in modulating the function of the rudimentary controller. For example, inputs from stretch receptors have less influence in conscious than in anesthetized animals, and their effects can be abolished by visual or auditory stimulation. In contrast to their role in other species, in humans the stretch reflexes exert little or no influence on the regulation of resting breathing. Although recordings of stretch receptor activity in humans demonstrate that these receptors are active during tidal breathing, modulation of respiratory timing occurs only when tidal volumes exceed 1 L.

Anatomically separate neural pathways subserve the automatic (metabolic) and volitional (speaking, breath-holding, and voluntary hyperventilation) activities of the respiratory system. Efferent fibers from the NTS and NRA constitute the automatic pathways. They decussate at the level of the obex and travel contralaterally in the ventrolateral column to reach their respective spinal motor neurons. The fiber tracts to the expiratory and inspiratory muscles separate within the ventrolateral white matter of the spinal cord. The projections to expiratory muscles lie ventral and medial to the fibers to the inspiratory muscles.

Projections from the cerebral motor cortex constitute the volitional pathways. These fibers descend to the brainstem respiratory neurons via the corticobulbar tracts and to the spinal motor neurons via the corticospinal tracts, which are located in the dorsolateral columns of the cord. Selectively placed lesions in the spinal cord may eliminate rhythmic activity in the respiratory muscles, even though these muscles still can be contracted voluntarily.

■ Spinal integration

The descending respiratory drives that reach a given segment of the spinal cord are integrated with information arising both intrasegmentally and intersegmentally. Intracellular recordings from intercostal motor neurons in the ventral horn indicate the presence of descending excitatory and inhibitory inputs. Such influences alternately depolarize and hyperpolarize the motor cells, leading to a rhythmic increase and decrease of their excitability. These excitatory and inhibitory inputs are both present in the spinal neurons that innervate inspiratory muscles, but their activities are asynchronous.

Excitation and inhibition of inspiratory and expiratory muscles appear to involve segmental interneuronal networks, as well as descending influences. Inhibition of antagonist muscles takes place through interneurons that connect the motor neurons of inspiratory and expiratory muscles. Excitatory and inhibitory intersegmental reflexes that modulate intercostal and phrenic motor neuron output have been demonstrated in both anesthetized animals and conscious humans. Stretch of intercostal muscles or electrical stimulation of dorsal roots in the lowermost thoracic segments (T9 to T12) elicits a spinal intersegmental reflex that produces excitatory responses in intercostal and phrenic motor neurons. In contrast, similar stimuli applied to more rostral segments (T1 to T8) elicit a supraspinally mediated inhibition of phrenic motor neuron activity and a premature termination of inspiration.

The potential importance of spinal reflexes lies in their ability to stabilize the rib cage and prevent its collapse when intrathoracic pressure is markedly subatmospheric. They also serve to augment muscle force in the presence of increased respiratory resistance or decreased compliance. Inhibitory intercostal to phrenic reflexes that cut short inspiratory efforts may be beneficial in the newborn. The rib cage of the newborn is extremely compliant, and thus it is prone to substantial retraction when large subatmospheric pressures are produced in the thorax by forceful diaphragm contractions. The intercostal retraction reduces tidal volume; the inhibitory reflex cuts short ineffectual diaphragmatic contractions.

■ Chemical respiratory control
■ Effects of CO_2

When CO_2-enriched gas is inspired, ventilation increases. The greater the augmentation of ventilation caused by a change in inspired CO_2, the less the increase in arterial P_{CO_2}. In conscious persons *central chemoreceptors,* located within the medulla, account for 70% to 80% of the ventilation increases, and *peripheral chemoreceptors,* located in the carotid and aortic bodies, account for the remainder. Both central and peripheral chemoreceptors respond in proportion to the level of P_{CO_2}, but some studies suggest that the carotid body is also stimulated by increases in the rate of change of P_{CO_2}.

In humans and animals increases in P_{CO_2} over a wide range cause a virtually linear increase in ventilation (Fig. 40-6). At levels of arterial P_{CO_2} above 80 to 100 mm Hg the response to hypercapnia reaches a plateau and may even diminish. In anesthetized animals and humans, hyperventilation of room air progressively reduces P_{CO_2} and eventually produces apnea. However, in normal awake humans hyperventilation while breathing 100% O_2 rarely causes apnea. This has been attributed to a "wakefulness drive," which stems mainly from nonspecific environmental stimuli (such as noise, light, or touch) impinging on the cerebral cortex. However, in anesthetized or decerebrate animals in which hyperventilation is induced by massive excitatory inputs (for example, electrical stimulation of the carotid body nerves), phrenic nerve activity sometimes continues even at low levels of P_{CO_2}. Continued respiratory activity under these circumstances may be caused by reverberations in medullary respiratory neuron circuits. This effect may contribute to the wakefulness drive, and it helps stabilize breathing.

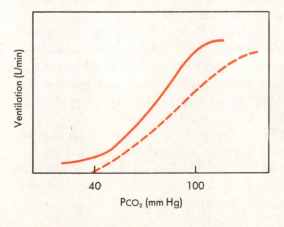

Fig. 40-6 ■ Ventilatory responses to increases in P_{CO_2}. Ventilation increases linearly with P_{CO_2} up to P_{CO_2} levels of 80 to 100 mm Hg. Thereafter ventilation levels off and may decrease as P_{CO_2} increases further. During wakefulness *(solid line)* ventilation persists even at low P_{CO_2} levels. In anesthetized animals and humans *(interrupted line)* apnea results as the P_{CO_2} is reduced below a threshold level.

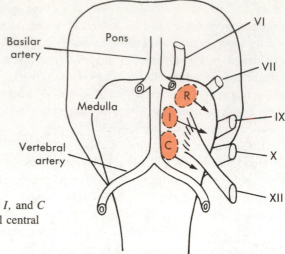

Fig. 40-7 ■ Chemoreceptors on the ventrolateral surface of the medulla. *R, I,* and *C* are the rostral, intermediate, and caudal central chemoreceptor areas.

The excitatory effects of CO_2 on breathing are mediated by specialized chemosensitive cells, the central chemoreceptors, in the medulla. These chemoreceptors project to the respiratory medullary neurons and are excitatory in nature (Fig. 40-7).

The exact location of the central chemoreceptors is still disputed, although there is substantial evidence that (1) they are distinct from the inspiratory motor neurons themselves, (2) they are not located in the dorsal and ventral groups of respiratory neurons described earlier, and (3) they respond to changes in the H^+ ion concentration of the brain interstitial fluid rather than to changes in arterial P_{CO_2} directly.

Studies with drugs and temperature probes suggest that the central chemoreceptors are located near the ventrolateral surface of the medulla. Striking ventilatory effects can be produced from the three superficial areas that are shown in Fig. 40-7 and that lie between the origins of the seventh and twelfth cranial nerves.

It is not yet possible to measure central chemoreceptor activity directly, even in animals. The function of the central chemoreceptors is assessed indirectly by enriching the content of the CO_2 in the inspired air and relating the increase in ventilation or phrenic nerve activity to the level of Pa_{CO_2}. Central chemoreceptor activity therefore is inferred from the effect of a CO_2 stimulus on the output of the respiratory neurons. This indirect estimation of central chemoreceptor activity is valid only under restricted circumstances and only if certain assumptions are made. After CO_2-enriched gas mixtures have been inspired for 10 to 20 minutes, changes in Pa_{CO_2} are assumed to reflect changes in the H^+ ion concentration of the brain.

Brain P_{CO_2}, and therefore the interstitial fluid pH, change much more slowly than does the blood P_{CO_2} when a CO_2-enriched gas mixture is breathed. Thus, although changes in Pa_{CO_2} may be complete within 1 to 2 minutes, ventilation continues to change for 10 to 20 minutes before a steady state is reached.

The assumption that changes in Pa_{CO_2} reflect a change in brain pH, even when a steady state has been reached, is not always valid. In metabolic acidosis or alkalosis neither arterial P_{CO_2} nor H^+ ion concentration reliably indicates the acid-base status of the interstitial fluid. HCO_3^- exchanges between blood and cerebrospinal fluid occur very slowly. As a consequence, changes in blood HCO_3^- levels will not be mirrored immediately in the brain interstitial fluid. The effects of chronic metabolic acidosis and alkalosis on the ventilatory response to CO_2 are shown in Fig. 40-8. It can be seen that at any given level of P_{CO_2} the ventilation is greater with metabolic acidosis and less with metabolic alkalosis, reflecting the altered level of HCO_3^- in the brain interstitial fluid. If the same ventilation results are plotted as a function of the H^+ concentration in the brain interstitial fluid, the response lines become identical.

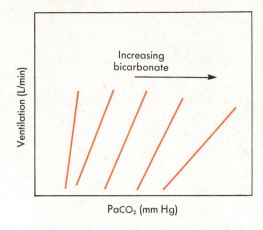

Fig. 40-8 ■ Effects of HCO_3^- ion concentrations on the ventilatory response to hypercapnia.

Acid injected directly into the blood actually will stimulate the peripheral chemoreceptors. Thus it will increase ventilation and lower the Pa_{CO_2}. Because the transfer of CO_2 between the brain interstitial fluid and the blood is faster than the transfer of H^+ or HCO_3^- ion, in the acute phase of metabolic acidosis the brain interstitial fluid may transiently become alkaline, thereby diminishing central chemoreceptor activity. With chronic acid-base disturbances H^+ changes in the cerebrospinal fluid usually occur in the same direction as those in the blood but are quantitatively less pronounced. Over the course of many hours therefore ventilation will steadily rise in metabolic acidosis as the brain interstitial pH decreases and approaches equilibrium with the blood. The converse is true when the blood pH is acutely increased during metabolic alkalosis.

The relationship between the arterial P_{CO_2} and the brain interstitial fluid P_{CO_2} also depends on the cerebral venous P_{CO_2} and consequently on the cerebral blood flow. The greater the cerebral blood flow, the less the difference between the arterial and interstitial fluid P_{CO_2}. Subnormal levels of cerebral blood flow produce a higher central P_{CO_2} for a given arterial value. Therefore reductions in cerebral blood flow increase the change in ventilation produced by a given rise in arterial P_{CO_2}. Because hypercapnia increases cerebral blood flow, the relationship between ventilation and Pa_{CO_2} may be influenced by the CO_2 responsivity of the cerebral vessels.

■ *Effects of hypoxia*

Hypoxia increases ventilation primarily by exciting sensors in the carotid body (innervated by the ninth cranial nerve) and to a lesser extent in the aortic body (innervated by the tenth cranial nerve). With decreases in Pa_{O_2} the consequent increase in carotid body activity stimulates ventilation. The relationship between Pa_{O_2} and ventilation is hyperbolic (Fig. 40-9, *A*). The respiratory response to hypoxia is therefore qualitatively different from the linear respiratory response to hypercapnia. If the carotid and aortic bodies were removed, hypoxia would depress breathing, presumably because the direct depressant effects of hypoxia on brain cells, including those in the medullary respiratory center, is unmasked. The carotid and aortic bodies also respond to a limited extent to change in P_{CO_2} and H^+ concentration. Such chemoreceptor responses are particularly important in the immediate increase in ventilation that occurs when blood pH is acutely reduced. The effect of hypercapnia and acidosis on peripheral chemoreceptor activity is accentuated by hypoxia (Fig. 40-9, *B*).

Conversely, changes in Pa_{CO_2} seem to enhance the ventilatory response to hypoxia, so that CO_2 and hypoxia interact multiplicatively. This interaction occurs both at the level of the carotid body itself and in the central nervous system, where the inputs from the central and peripheral chemoreceptors converge. It is important to remember that, when isocapnic conditions are not maintained and the Pa_{CO_2} is allowed to fall, the response to hypoxia is markedly attenuated.

Carotid body discharge varies during the breathing cycle. Arterial P_{O_2} and P_{CO_2}

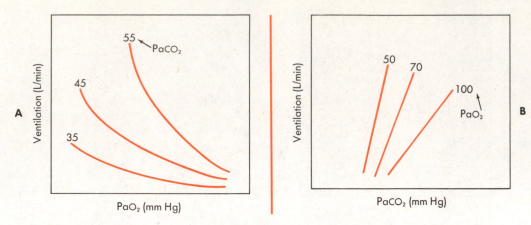

Fig. 40-9 ■ The effects of hypoxia on ventilation. **A,** At a given P_{CO_2}, ventilation increases in a hyperbolic fashion as the P_{O_2} is reduced. The ventilatory response to hypoxia is greater the higher the P_{CO_2}. **B,** The ventilatory response to hypercapnia is accentuated by hypoxia.

change with inspiration and expiration. These discharges can be enhanced by increasing the amplitude of the respiratory oscillations in arterial P_{CO_2}, even when the mean level of P_{CO_2} is unchanged. The ventilatory response to a given level of carotid body afferent activity is related to the respiratory phase during which the information is received by the central nervous system. Carotid body input is more effective in increasing ventilation when applied during inspiration than during expiration. Since the relationship between the phases of respiration and the cyclic variations in carotid body firing depends on the circulation time, changes in cardiac output may affect breathing. However, the magnitude of such effects is probably small.

■ *Carotid body structure and function*

In addition to nerve endings, there are at least two types of cells in the carotid body: type I cells, which contain large amounts of catecholamine, and type II cells, which do not. Either the type I cells or the afferent nerve endings are probably the actual hypoxia sensors.

Neural recordings from chemoreceptor afferent fibers show that carotid body chemoreceptors become more sensitive as hypoxia becomes more severe. The partial pressure of O_2 rather than the O_2 content of the arterial blood is the stimulus. The mechanism that allows the carotid body to respond to even relatively mild hypoxia has not been completely elucidated, but some details are known. Although the carotid body blood flow is unusually high, so is its metabolic rate. Vascular shunts through the carotid body as well as its high metabolic rate may produce local areas of hypoxia within the carotid body, even when the arterial blood is fully saturated with O_2. Measurements of carotid body P_{O_2} with microelectrodes have shown some regions with extremely low O_2 tensions, but the range of recorded tension is wide. Perfusion of the carotid body with hypoxemic blood causes acetylcholine accumulation, suggesting that acetylcholine may be the transmitter mediating the increased neural activity during hypoxia. Release of certain catecholamines, such as dopamine, also seems to occur with hypoxia. However, the bulk of the experimental evidence suggests that dopamine inhibits rather than augments the carotid body responses to hypoxia. Even mild hypoxia may limit ATP production in the carotid body. The biochemical mechanism underlying carotid body excitation may be a change in its phosphate potential, that is, the ratio of ATP to ADP + P_i. Within the carotid body, cytochrome enzymes that produce ATP may have an especially low affinity for O_2, thereby accounting for the sensitivity of the carotid body to even minimal changes in P_{O_2}.

Both the sensitivity of the peripheral chemoreceptors to alterations in P_{O_2} and the range of O_2 tensions over which the chemoreceptors are active may be modified by the central nervous system. For example, sympathetic nervous system regulation of carotid body blood flow affects the carotid body sensitivity to hypoxia. Decreases in carotid body blood flow increase its sensitivity to a given level of hypoxia. Also carotid body responses to hypoxia can be inhibited by direct efferent discharge from the central nervous system.

Changes in mechanical conditions of the thorax or changes in metabolic rate cause transient alterations in gas exchange and gas tensions. The change in blood gas tensions in such situations depends on the volume and arrangement of the body stores of O_2 and CO_2.

CO_2 exists in the body in large amounts as gas in the lungs and in solution in the body fluids but mainly in the form of HCO_3^- and $CO_3^=$ compounds in the blood and tissues. Oxygen, however, is stored in much smaller amounts in alveolar gas, in solution, and in combination with hemoglobin and myoglobin. Because of differences in storage capacity, acute changes in ventilation cause relatively small changes in Pa_{CO_2} and greater changes in Pa_{O_2}.

For example, during breath-holding Pa_{CO_2} will increase 3 to 8 mm Hg in the first minute, but Pa_{O_2} will fall nearly to the mixed venous level (a change of nearly 60 mm Hg). Rates of change of Pa_{O_2} and Pa_{CO_2} depend not only on the volume of gas stores but also on the rates of perfusion in the different body tissues, the tissue metabolic rate, and the ability of the tissues in each compartment to bind CO_2 and O_2.

The rate at which the peripheral and central chemoreceptors respond to changes in inspired CO_2 and O_2 will depend in part on the arrangement of the gas stores. The small size of the arterial compartment and the high rate of carotid body blood flow allow the peripheral chemoreceptors to respond in a matter of seconds to changes in both O_2 and CO_2. The larger CO_2 stores of the brain cause the central chemoreceptors to respond more slowly (many minutes) to changes in inspired CO_2. This difference in response time of the central and peripheral chemoreceptors has been used to distinguish the contribution of each set of receptors to the response to changes in inspired CO_2.

The lungs and airways contain sensory receptors whose natural stimuli, as in other hollow viscera, are irritation of the lining layers and changes in distending forces. The afferent fibers of these receptors are carried to the central nervous system in the tenth cranial nerve, the vagus.

There are basically three types of pulmonary receptors: *stretch receptors,* located within the smooth muscle layer of the extrapulmonary airways; *irritant receptors,* which ramify among airway epithelial cells and have a distribution similar to that of the stretch receptors; and *J (juxtacapillary) receptors,* situated in the lung interstitium near alveolar capillaries.

Stretch receptors are excited by an increase in bronchial transmural pressure and adapt slowly to a sustained stimulus. As the lung is inflated, they reflexly inhibit inspiration and promote expiration. They are responsible for the *Hering-Breuer reflex,* which produces expiratory apnea and augmentation of expiratory muscle contraction when the lung is inflated. The Hering-Breuer reflex appears to be weaker in humans than in animals. Newborn infants, however, may not display a Hering-Breuer reflex, presumably because descending inhibitory influences from the immature cortex have not yet developed.

Irritant receptors and J receptors are served by vagal fibers that are smaller than those which innervate the stretch receptors. Unlike the stretch receptors, these other receptors rapidly decrease their activity when subjected to a sustained stimulus.

Irritant receptors are stimulated mechanically by lung inflation, by increases in air-

■ *Body stores of O_2 and CO_2*

■ *Pulmonary mechanoreceptors*
■ *Receptors in the airways and lungs*

flow, and by changes in bronchial smooth muscle tone. They also are stimulated chemically by inhalation of noxious agents, such as particulate matter, sulfur dioxide, ammonia, nitrogen dioxide, or antigens. Irritant receptor stimulation reflexly augments inspiratory motor neuron activity. It leads to constriction of the airways, and it interacts with the stretch receptors to promote rapid, shallow breathing. This pattern of breathing, in combination with airway constriction, may limit penetration of dangerous agents into the lung and may prevent such agents from reacting with the gas-exchanging surfaces. These receptors also may stabilize lung compliance by initiating the periodic sighs that occur sporadically during normal breathing. This would serve to reexpand collapsed areas of the lung. The chemical mediators (such as histamine, slow-reacting substance of anaphylaxis, and bradykinin) released in the lung during allergic reactions stimulate irritant receptors. The augmentation of inspiratory activity and increases in breathing frequency produced by irritant receptor excitation may help maintain ventilation during asthmatic attacks, even when the work of breathing is severely increased.

J receptors are mainly stimulated by lung interstitial edema, but they also can be excited by certain chemicals, such as histamine, halothane, and phenyldiguanide. Activation of the J receptors causes laryngeal closure and apnea, followed by rapid, shallow breathing. When lung edema develops as a result of excessive exercise, J receptors seem to depress the activity of the exercising limbs by a somatic reflex involving the cingulate gyrus. J receptors, together with the irritant receptors, may be responsible for the tachypnea seen in patients with a pulmonary embolus, pulmonary edema, or pneumonia. Receptors that are similar to J receptors and which are innervated by unmyelinated fibers also can be found in the airways.

During breathing at rest the pattern of firing of irritant or J receptors is sporadic and unrelated to the phases of inspiration and expiration. Consequently it is believed that neither receptor plays a significant role in determining the pattern of resting breathing.

■ Chest wall proprioceptors

Three types of receptors in the chest wall are the *joint, tendon,* and *spindle receptors.* They signal to the respiratory neurons information about changes in the force exerted by the respiratory muscles and the movement of the chest wall. Specialized Ruffini receptors and pacinian and Golgi organs are present in the joints within the chest wall. Joint receptor activity varies with the degree and rate of change of rib movement.

Tendon organs in the intercostal muscles and the diaphragm monitor the force of muscle contraction and have an inhibitory effect on inspiration. It once was thought that tendon organ activity was provoked only by unusual levels of muscle activity, but it is now believed that tendon organs are stimulated by even small changes in force.

Muscle spindles, which are abundant in the intercostal muscles but scarce in the diaphragm, are involved in several kinds of intercostal respiratory reflexes. Spindles also help coordinate breathing during changes in posture and speech and to stabilize the rib cage when breathing is impeded by increases in airways resistance or decreases in lung compliance.

Fig. 40-10 shows the operation of the spindle and its neural connections. (See also Chapter 14.) Spindles are located on intrafusal muscle fibers, which are aligned in parallel with the extrafusal fibers that elevate the ribs. Motor innervation of the extrafusal fibers originates in α–motor neurons, whereas the intrafusal fibers receive motor innervation from γ(fusimotor)–motor neurons, which are under cerebellar control. Passive stretch of an intercostal spindle by lateral flexion of the trunk, for example, increases spindle afferent activity. Through a monosynaptic segmental reflex such spindle excitation causes contraction of the parent extrafusal fiber, which helps to restore the upright position. The spindles also can be stretched by efferent fusimotor discharge, which causes contraction of the intrafusal fiber itself. Some fusimotor fibers discharge phasically, so that their discharge rises during inspiration and falls during expiration;

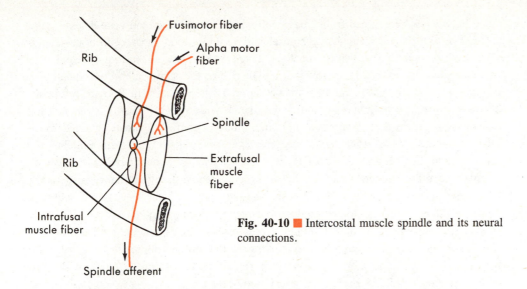

Fig. 40-10 ■ Intercostal muscle spindle and its neural connections.

other fusimotor fibers are tonically active. Without phasic fusimotor activity, spindle discharge would decrease when the extrafusal fibers in the external intercostal muscles contract during inspiration. Activation of fusimotor and α–motor neurons simultaneously causes the spindles to be under continuous stretch during inspiration, thereby enhancing the contribution made by the intercostal muscles to respiration. If inspiratory movements are impeded, afferent activity from a spindle innervated by a phasically active fusimotor fiber is heightened, thereby increasing inspiratory muscle force and helping to preserve tidal volume. Spindle afferents project all the way to the cerebral cortex. Such connections may provide the information that allows respiratory movements to be perceived consciously. It has been suggested therefore that the sensation of breathlessness results from an imbalance in the demand for muscle shortening and the degree of shortening that is actually achieved, as reflected in spindle afferent activity.

When lung function is normal, the sensitivity of the peripheral and central chemoreceptors to CO_2 can be evaluated by measuring the ventilatory response to inspired CO_2. In the conventional *steady-state test* the inspired CO_2 is increased in steps, and ventilation at each step is related to the change in arterial P_{CO_2}. Sensitivity to CO_2 is determined from the slope of the line relating ventilation to Pa_{CO_2}. Although CO_2 has easy access to the central chemoreceptors, the size of the cerebral CO_2 stores slows the rate at which central chemoreceptors equilibrate with inspired CO_2. Usually the inspired CO_2 concentrations at each step must be maintained constantly for 10 to 20 minutes to ensure equilibration. Relative rates of equilibration of P_{CO_2} in arterial and brain venous blood indicate that arterial P_{CO_2} reaches its steady-state level long before the venous P_{CO_2} (Fig. 40-11). It is apparent from the figure that, if ventilation is measured too soon, chemosensitivity will be underestimated.

■ *Tests of chemoreceptor sensitivity*

■ *Steady-state methods of hypercapnia*

Fig. 40-11 ■ Relative rates of equilibration of ventilation and arterial and cerebral venous P_{CO_2} following a step change in inspired P_{CO_2}.

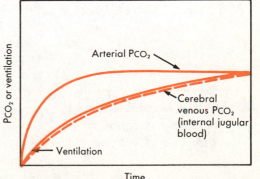

■ *Progressive hypercapnia*

When CO_2 is rebreathed from a bag continaing CO_2 at the mixed venous level (about 7% CO_2, with the rest being O_2), arterial and venous blood P_{CO_2} equilibrates rapidly (30 to 60 seconds), after which the P_{CO_2} levels in the arterial, cerebral venous, and mixed venous blood and in the expired air all rise at the same rate. Consequently the changes in P_{CO_2} in the expired air are the same as the changes in P_{CO_2} at the central chemoreceptors. Ventilatory measurements over a wide range of Pa_{CO_2} concentrations can be obtained in a few minutes, and such values agree with those obtained by the more prolonged steady-state technique. However, the rebreathing tests evaluate CO_2 sensitivity at much higher levels of Pa_{CO_2} than are usually encountered. Moreover differences have been noted between rebreathing and steady-state ventilatory responses to CO_2 when metabolic acidosis or alkalosis was present. With the steady-state technique moderate alkalosis and acidosis produce changes in the position of the ventilatory response line in the graph of ventilation versus arterial P_{CO_2} but only relatively small changes in its slope, whereas the reverse is true in the rebreathing tests. The explanation for the difference is obscure, but it may be related to the different levels of Pa_{CO_2} at which the steady-state and rebreathing tests are performed.

In normal individuals the average ventilatory response to inspired CO_2 is about 2.5 L/minute for each millimeter of mercury rise in Pa_{CO_2}. Is is somewhat less in women than in men, and it may decline with advancing age. There is considerable variability in the CO_2 response among individuals. Some of this variability is caused by differences in body size, genetic makeup, and even personality. Studies of identical twins indicate that genetic factors influence the tidal volume component of the CO_2 response, whereas environmental factors affect the frequency component. Some of the variability among individuals may be reduced when the CO_2 response is corrected for differences in vital capacity.

■ *Tests of peripheral chemoreceptors*

The peripheral chemoreceptors also respond to changes in arterial P_{CO_2}, and they contribute about 20% to 30% of the total ventilatory increase observed when CO_2 is inhaled. Since peripheral chemoreceptors react more rapidly to changes in inspired CO_2 than do central chemoreceptors, peripheral chemoreceptor responses to hypercapnia have been evaluated by measuring the immediate change in ventilation occurring in the first several breaths after addition or removal of a high-CO_2 gas mixture.

The response to hypoxia, like the response to hypercapnia, can be measured either by rebreathing or by steady-state techniques. Because of the prominent effects of CO_2 on breathing, it is important to keep arterial P_{CO_2} constant while the hypoxic response is being measured. Peripheral chemoreceptor responses to O_2 also can be evaluated by measuring the effect on ventilation of a few breaths of 100% N_2 or 100% O_2.

No matter how it is measured, the ventilatory response to hypoxia is curvilinearly related to the change in Pa_{O_2}, making quantification difficult. The response can be made linear, however, by relating ventilation to the reciprocal of Pa_{O_2} or to the arterial O_2 saturation.

The range of normal values for the ventilatory response to hypoxia is wide, greater than that to CO_2. In the same person the ventilatory response to hypoxia is influenced by the metabolic rate, the CO_2 response, and genetic and environmental factors.

Chronic hypoxemia, such as that which occurs at high altitudes, particularly if it begins at birth, is associated with depression of the ventilatory response to hypoxia, presumably because of adaptation at the receptor level.

■ *Effects of thoracic disease on ventilatory responses to hypercapnia and hypoxia*

Lung disease frequently depresses the ventilatory responses to chemical stimuli. The depressant effect seems to be greater for the response to hypercapnia than the response to hypoxia.

Studies in which subjects are required to breathe through an increased external resistance suggest that the inspiratory work of breathing at a given level of arterial P_{CO_2}

is fixed. Since the ratio of inspiratory muscle work to ventilation is increased by disease, ventilation is decreased at any level of Pa_{CO_2}. The hypercapnia observed in severe obstructive lung disease has been explained by the increase in work caused by chronic airway obstuction.

More recent studies indicate that small to moderate increases in airway resistance in normal subjects may have little effect on resting ventilation or on the CO_2 response, or the increases may even heighten ventilation. The mechanisms responsible for the preservation of ventilation in these circumstances could include (1) increased stimulation of the lung and chest wall mechanoreceptors, which augments ventilation, (2) improvement of the mechanical advantage and coordination of the respiratory muscles, (3) conscious efforts that increase breathing, and (4) the intrinsic properties of the muscles themselves, which allow them to increase their contractile force when the velocity of shortening is reduced.

Ventilation usually is used to assess the respiratory response to CO_2 and hypoxia. Because the performance of the chest bellows is impaired by lung disease that can limit ventilatory responses to chemical stimuli, other methods of assessing the response of respiratory motor neurons have been devised.

■ *Indices of respiratory efferent activity*

■ *Ventilation*

In the *occlusion pressure technique* the isometric force of contraction of the inspiratory muscles is measured. This test is based on the principle that under isometric conditions the force of contraction of skeletal muscle correlates closely with its electrical activity. This relationship has been confirmed experimentally for the respiratory muscles. To perform the test, the airway is momentarily blocked at the beginning of inspiration, and the negative pressure developed at the mouth during inspiration is measured. In conscious subjects the reproducibility of the response is greater when the occlusion of the airways is brief, about one tenth of a second. The occlusion pressure increases with hypercapnia and hypoxia and can be related to the change in Pa_{CO_2} and Pa_{O_2} to estimate chemosensitivity. Because occlusion pressure is measured under isometric conditions (that is, in the absence of airflow), it is not reduced by increases in airway resistance, whereas ventilation is impaired.

The tensions developed by the inspiratory muscles depend on their initial length. Thus occlusion pressure measurements in patients with lung disease could be affected by the alterations in inspiratory muscle length caused by changes in functional residual capacity. Increased functional residual capacity in animals reduces occlusion pressure responses. However, studies in conscious humans suggest that even fairly large changes in functional residual capacity (about 1 L) have little effect on occlusion pressure.

■ *Occlusion pressure*

Measurement of the electrical activity of the diaphragm is another way of evaluating central respiratory neuron output in humans. This can be accomplished by passsing electrodes down the esophagus and positioning them so that they straddle the diaphragm.

Various ways have been devised for quantitating the electrical activity of the diaphragm recorded in this way. In the current method the diaphragm activity is integrated over successive small intervals of time (100 to 200 msec), and the so-called moving average is computed. Electrical activity measured this way depends on the exact positions of the electrodes in relation to the diaphragm during breathing, so that it is difficult to compare one individual with another. However, it is possible to use this method to determine the effect of different therapeutic interventions in the same person. Unlike occlusion pressure, the diaphragm electromyogram evaluates the output of only one of the inspiratory muscles.

■ *Diaphragm electromyogram*

■ *Clinical implications of respiratory control abnormalities*

The most important cause of respiratory failure is a derangement of lung mechanics. However, only a fraction of patients with even severe impairment of pulmonary function develops respiratory failure. It has long been suspected that those patients who have the poorest chemosensitivity are the ones who are most likely to develop CO_2 retention when the performance of the chest bellows is reduced. Although direct evidence is lacking, there is considerable indirect evidence that supports this hypothesis. For example, offspring of patients in respiratory failure with obstructive lung disease have systematically lower responses to changes in inspired CO_2 than offspring of patients with normal values of Pa_{CO_2}. Children who have retained CO_2 because of upper airway obstruction caused by hypertrophy of the adenoids and tonsils have depressed CO_2 sensitivity, even after the tonsils and adenoids have been removed. Asthmatic patients who have retained CO_2 during an asthmatic attack also have persistently low ventilatory responses to CO_2 after recovery from the asthmatic episode.

There is also a small group of subjects who retain CO_2 even though lung function is normal. In some of these patients the cause of the depressed CO_2 sensitivity is not known. In others, however, it seems to be associated with certain metabolic abnormalities, such as alkalosis, or with the chronic administration of respiratory depressant drugs, like methadone.

■ *Abnormal breathing patterns*

Breathing is usually a smoothly recurring cycle of inspiration and expiration that occurs without interruption. However, in some diseases of the central nervous system recurrent episodes of respiratory arrest (apnea) occur.

In *Cheyne-Stokes breathing* there are repeated cycles in which tidal volume waxes and wanes, associated with recurrent periods of apnea (Fig. 40-12). This form of breathing occurs both in disorders of the central nervous system and in congestive heart failure. However, in some humans who have no obvious neurological defect, it appears on exposure to hypoxia, during sleep, and in the period immediately following voluntary hyperventilation. During these regular cycles blood gas concentrations of O_2 and CO_2 also cycle (see further).

Cheyne-Stokes breathing is a manifestation of instability in the ventilatory control system. Similar oscillatory changes in output occur in man-made feedback control systems in predictable circumstances. Delayed information transfer around the feedback system and excessive controller sensitivities are two important conditions that cause unstable control in a physical system.

In the respiratory system increased controller sensitivities can occur with hypoxia or with neurological diseases that increase the ventilatory responses to hypercapnia. Delays in information transfer occur when the circulation time is prolonged.

Fig. 40-12 ■ Cheyne-Stokes breathing. There is a waxing and waning of tidal volume. The Pa_{CO_2} is lowest and the Pa_{CO_2} highest at the point when ventilation is the greatest.

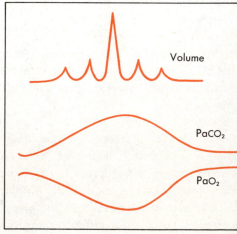

Volume

$PaCO_2$

PaO_2

Time

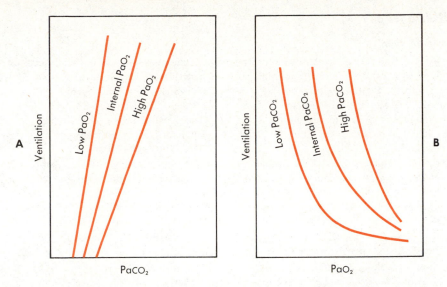

Fig. 40-13 ▪ Ventilatory response curves. **A,** Ventilatory responses to changes in Pa_{CO_2} at different levels of Pa_{O_2}. **B,** Ventilatory responses to changes in Pa_{O_2} at different levels of Pa_{CO_2}. The slope of each curve reflects the "sensitivity" of the respiratory controller. Note that the sensitivity does not vary with the level of the Pa_{CO_2}; each curve in **A** has a constant slope. However, the sensitivity does increase as the Pa_{O_2} decreases; the slope of each curve in **B** becomes more negative the lower the level of Pa_{O_2}.

The sensitivity of the respiratory controller is the slope of the line relating ventilation to Pa_{CO_2} and Pa_{O_2}. Over a broad range of values of Pa_{CO_2}, controller sensitivity is constant, since the ventilation changes are linear (Fig. 40-13, *A*). The changes in Pa_{CO_2} by themselves do not affect controller sensitivity.

However, ventilation increases nonlinearly with hypoxia. As hypoxia becomes more severe, controller sensitivity increases, that is, the slope increases as the Pa_{O_2} diminishes (Fig. 40-13, *B*). In the presence of hypercapnia, hypoxia causes even greater changes in controller sensitivity. Controller sensitivity can become so great that the controller overreacts and increases ventilation too much. For example, hyperventilation sufficient to drop Pa_{CO_2} below the threshold value will result in a subsequent period of apnea. During apnea CO_2 rises, and O_2 falls. When CO_2 tension rises to the threshold value, spontaneous ventilation begins, but at a very high level because of hypoxia. Although O_2 tension is quickly restored to normal values, the excessive ventilation again lowers the CO_2 tension below the threshold, apnea ensues, and the cycle continues.

Circulation delays are another factor contributing to Cheyne-Stokes breathing. Very simply stated, the respiratory controller needs to know the alveolar P_{CO_2} and P_{O_2} if it is to adjust ventilation correctly. When the time required for the blood to circulate from the lungs to the chemoreceptors is prolonged, the controller may react inappropriately. Congestive heart failure, which lengthens circulation time, is a cause of Cheyne-Stokes breathing.

■ *Bibliography*

Journal articles

Bradley, G.W., et al.: A model of the central and reflex inhibition of inspiration in the cat, Biol. Cybern. **19:**105, 1975.

Cherniack, N.S., and Longobardo, G.S.: Cheyne-Stokes breathing: an instability in physiological control, N. Engl. J. Med. **288:**952, 1973.

Cohen, M.I., and Feldman, J.L.: Models of respiratory phase switching, Fed. Proc. **36:**2367, 1977.

Cunningham, C.J.C.: The control system regulating breathing in man, Q. Rev. Biophys. **6:**433, 1974.

Kalia, M.: Central neural mechanisms of respiration, Fed. Proc. **36:**2365, 1977.

Paintal, A.S.: The nature and effects of sensory inputs into the respiratory centers, Fed. Proc. **36:**2428, 1977.

Rebuck, A.S., and Campbell, E.J.M.: A clinical method for assessing the ventilatory response to hypoxia, Am. Rev. Respir. Dis. **109:** 345, 1974.

Schlaefke, M.E., See, W.R., and Loeschcke, H.H.: Ventilatory response to alterations of H^+ ion concentration on small areas of the ventral medullary surface, Respir. Physiol. **10:**198, 1970.

von Euler, C., Herrero, F., and Wexler, I.: Control mechanisms determining rate and depth of respiratory movements, Respir. Physiol. **10:**93, 1970.

Weil, J.V., et al.: Hypoxic ventilatory drive in normal man, J. Clin. Invest. **49:**1061, 1970.

Books and monographs

Cherniack, N.S., and Altose, M.D.: Respiratory responses to ventilatory loading. In Hornbein, T.F., editor: Regulation of breathing, part II, New York, 1981, Marcel Dekker, Inc.

Hornbein, T.F., editor: Regulation of breathing, part I, New York, 1981, Marcel Dekker, Inc.

Widdicombe, J.G.: Reflex control of breathing. In Widdicombe, J.G., editor: Respiratory physiology, MTP International Review of Science Physiology Series One, vol. 2, London, 1974, Butterworths.

chapter 41

Environmental and developmental aspects of respiration

In the alert conscious human, stimuli from the environment, acting reflexly via higher brain centers, tend to exert a continuous excitatory effect on breathing and sustain breathing even at very low levels of chemical drive.

Fluctuations in the state of alertness occur regularly, even in awake subjects. However, they are most pronounced during sleep and affect respiration. The neural and biochemical mechanisms that produce sleep seem to involve an excitatory and inhibitory interplay among various areas of the brain. They probably involve changes in the balance of activity among noradrenergic, serotonergic, and cholinergic neurons. Sleep is not a homogeneous phenomenon, and it can be divided into a slow wave (SW) stage and a rapid eye movement (REM) stage, during which dreaming occurs. Each stage is associated with fairly characteristic changes in muscular, cardiovascular, and respiratory activity.

A reduction in the level of environmental stimulation and the withdrawal of higher brain center influences on respiration seem to account for the diminished ventilation and the consequent increase in arterial P_{CO_2} observed in SW sleep. Ventilatory responses to inspired CO_2 are somewhat attenuated in SW sleep, and the CO_2 response curve is shifted to the right (Fig. 41-1). Systemic blood pressure falls, and heart rate is reduced.

REM sleep has been divided into phasic and tonic stages. In tonic REM sleep breathing may maintain its regularity, but tidal volume may decrease. In addition, there may be a further reduction in the ventilatory responses to inspired CO_2. Studies in dogs and in humans suggest that the ventilatory responses to hypoxia may be better maintained in SW and REM sleep than are the responses to CO_2.

In phasic REM sleep external stimuli and changes in blood gas tensions are less effective in producing arousal than in SW sleep. The appearance of phasic REM sleep is associated with irregular breathing patterns. This could occur because the intrinsic activity of higher brain centers in this sleep stage, set into motion by impulses from pontine and mesencephalic neurons, dominates respiratory neuron activity and causes irregularities in breathing pattern.

Patients with depressed ventilatory responses to hypercapnia and hypoxia while awake seem to breathe less during sleep than do those with normal chemosensitivity. In some patients with lung or respiratory muscle disease the degree of hypoxemia developing during sleep seems disproportionate to the degree of hypoventilation. In these patients abnormalities in ventilation/perfusion ratios in some regions of the lung, produced by the recumbent position and discoordinated respiratory muscle movement, may contribute substantially to the arterial O_2 desaturation during sleep.

■ *Effects of sleep on respiration*

■ *Chemical regulation of breathing in sleep*

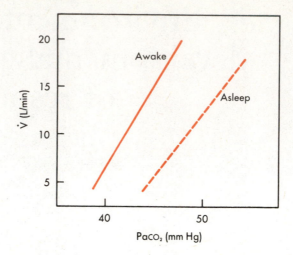

Fig. 41-1 ■ Ventilatory responses to hypercapnia during wakefulness *(solid line)* and during sleep *(interrupted line)*. With sleep the response line is shifted to the right, and the change in ventilation for a given change in Pa_{CO_2} is reduced.

■ Neural regulation of breathing during sleep

Mechanical stimulation of the airways in animals in either SW or REM sleep elicits reflex responses that may be different from those observed in the awake state. For example, laryngeal stimulation in awake dogs produces coughing, but it causes apnea when the dogs are in REM sleep. Pulmonary stretch reflexes seem to have a greater effect during REM sleep.

The phasic and tonic activity of skeletal muscle decreases during sleep, particularly in the REM stage. The activity of the diaphragm is less reduced than the activity of the intercostal and upper airway muscles. Because of the relative loss of tone in the upper airway and rib cage muscles, the negative pressure created by the diaphragm during inspiration may pull the rib cage inward and occlude the upper airway. The activity of the muscles of the upper airway, such as the posterior cricoarytenoid, which abducts the vocal cords, and the genioglossus, which protrudes the tongue, fluctuates with respiration. These fluctuations are affected by sleep, by hypoxia and hypercapnia, and by reflexes produced by mechanoreceptor stimulation. Brief periods of upper airway obstruction occur even in normal humans during sleep. Such obstruction may be caused by changes in the level of activity of these muscles, in their activation time, or in their response to chemical and mechanical stimuli.

■ Apneic pauses during sleep

Apneic periods occur during sleep in a third of normal individuals and are particularly frequent among older men. They may last for more than 10 seconds and are associated with falls in arterial O_2 saturation to 75% or less. These apneas appear during all sleep stages but are more common in the lighter stages of SW sleep and REM sleep than in the deeper stages of SW sleep.

Sleep apneas have been classified into two distinct categories: central and obstructive (Fig. 41-2). Central apnea is characterized by an absence of airflow past the nose and mouth because of a cessation of respiratory efforts. In obstructive apnea, despite persistent respiratory efforts, airflow ceases because of total upper airway obstruction. Snoring may be a manifestation of partial upper airway obstruction. Possible sites of upper airway obstruction during sleep include the larynx, the pharynx, or the oropharynx. Visualization of the upper airway during sleep has implicated all three of these locations. Arousal may be an important element, terminating sleep apnea. It may result from chemoreceptor excitation by hypoxia and hypercapnia.

Recently respiratory disturbances occurring during sleep have been recognized as the

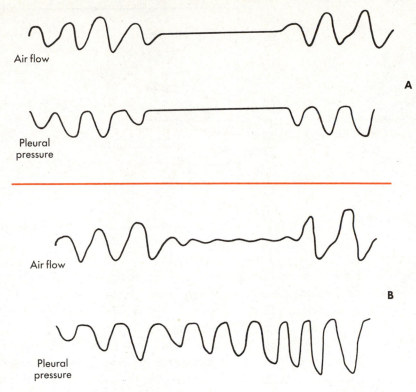

Fig. 41-2 ■ Sleep apneas. **A,** Central apneas are characterized by a cessation of airflow due to a temporary suspension of all respiratory efforts. The cessation of respiratory efforts is reflected in absence of any pleural pressure deflections. **B,** In obstructive apneas airflow ceases despite persistent respiratory efforts and pleural pressure swings.

primary factors in certain disease processes. In many of these patients prolonged and frequent obstructive apneas occur. The recurrent episodic periods of hypoxia and hypercapnia that occur with obstructive apneas may lead to polycythemia, right-sided heart failure, and pulmonary hypertension.

■ *Respiratory adjustments during exercise*

The ability to exercise depends on the capacity of the cardiovascular and respiratory systems, acting in coordination, to increase O_2 delivery to the tissues and to remove the excess CO_2. The circulatory adjustments are discussed in Chapter 36, and only the respiratory adaptations are considered here.

Cardiac output increases during exercise, for the reasons outlined in Chapter 36. Since virtually all the cardiac output passes through the lungs, pulmonary blood flow-increases by the same amount during exercise. Because of the low resistance and great distensibility of the pulmonary vascular bed, this increase in flow is accompanied by only a small rise in pulmonary vascular pressure. More capillaries are opened in the lung, and the area available for gas diffusion increases; the diffusing capacity for both O_2 and CO_2 rises. The alveolar arterial (A-a) PO_2 difference decreases slightly from the resting value in moderate exercise, reflecting the more even distribution of $\dot{V}/\dot{Q}$ ratios in the lung during exercise. However, as exercise intensifies, the A-a PO_2 difference widens, because mixed venous PO_2 is decreased and, in a few cases, because diffusion does become a limiting factor in O_2 exchange.

The airways distend slightly during exercise, increasing the anatomical dead space. However, the alveolar dead space is decreased, because of the improvement in ventilation/perfusion matching. Thus the absolute value of the physiological dead space is

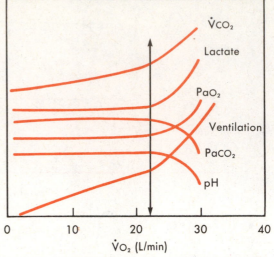

Fig. 41-3 ■ Changes in ventilation, CO_2 production ($\dot{V}CO_2$), $PaCO_2$, PaO_2, pH, and blood lactate during progressive exercise. The vertical line represents the anaerobic threshold.

nearly the same during exercise and rest. Because tidal volume is larger during exercise, the dead space/tidal volume ratio falls as exercise becomes more severe.

The ventilatory adjustments that take place are geared to the intensity of exercise and to its duration. In brief exercise, like sprinting, the breath frequently is held until the end of the exercise. With more prolonged exercise, however, ventilation exceeds the resting level, and it grows even greater as exercise becomes more strenuous.

Acid-base balance is normal during moderate exercise, when O_2 delivery to the mitochondria is adequate to meet energy requirements aerobically. However, with sufficiently severe exercise total energy requirements can be satisfied only by a combination of aerobic metabolism and anaerobic glycolysis. The lactic acid that is formed during glycolysis diffuses into the blood and increases the H^+ concentration. The chemical reaction of the acid with blood HCO_3^- forms CO_2, which adds to the CO_2 produced in the tissues. The highest level of work that can be performed without inducing a sustained metabolic acidosis is called the anaerobic threshold, which is higher in trained than in untrained subjects.

When the level of exercise is below the anaerobic threshold (Fig. 41-3), ventilation is linearly related to both CO_2 production and O_2 consumption. Arterial PO_2, PCO_2, and pH are virtually unchanged, although venous values may be altered. The decrease in arterial pH, which occurs when the level of exercise is above the anaerobic threshold, enhances the activity of the carotid body. Ventilation increases out of proportion to the rise in O_2 consumption, and PaO_2 may increase slightly. Since CO_2 formation is augmented, the relation between ventilation and $\dot{V}CO_2$ remains nearly constant at exercise levels above the anaerobic threshold, but $PaCO_2$ tends to decrease.

■ *Mechanism of exercise hyperpnea*

The increase in ventilation is the most obvious and important respiratory adjustment to exercise. It takes 4 to 6 minutes after the beginning of moderate exercise, reflected by a fivefold increase in $\dot{V}O_2$, before a steady rate of ventilation is reached; with severe exercise it may take even longer. The time course of the changes in ventilation that occur with exercise is shown in Fig. 41-4. Exercise increases ventilation in three fairly distinct phases: stage I, in which there is an abrupt ventilatory increase; stage II, in which ventilation increases more gradually; and stage III, in which ventilation is constant, because a steady state has been reached.

Both neural and chemical factors account for the changes in ventilation during exercise, just as they do during rest. Some of the stimuli that can increase ventilation during exercise are the following:

1. Cardiovascular receptors in systemic or pulmonary circulation
2. Thermal receptors

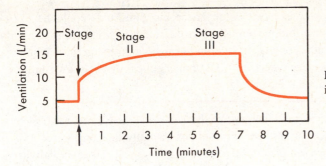

Fig. 41-4 ■ Time course of changes in ventilation during exercise.

Fig. 41-5 ■ At rest the level of ventilation and the Pa_{O_2} are designated by point *X*, at the intersection of the hyperbola describing the effect of ventilation on P_{CO_2}, and the line representing the ventilatory response to changes in P_{CO_2}. During exercise, $\dot{V}_{CO_2}$ increases and the hyperbola is shifted upward and to the right. If the ventilatory response to CO_2 were unchanged, the level of ventilation and the P_{CO_2} during exercise would be at point *Y*. In fact, the ventilatory response line is shifted to the left during exercise and intersects the exercise hyperbola at point *Z*. The adjustments in the ventilatory response to hypercapnia are such that the Pa_{CO_2} during exercise is virtually identical to that at rest.

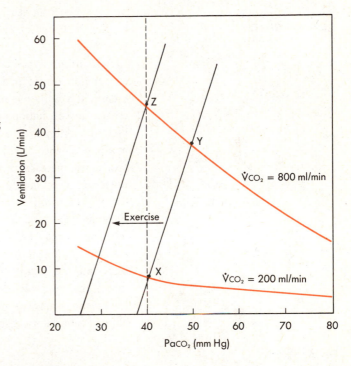

3. The central or peripheral chemoreceptors
4. Central nervous system excitations
5. Mechanoreceptors in muscle, for example, joint receptors or muscle spindles
6. Postulated receptors that monitor metabolism in the muscles or in the mixed venous blood

The size of the list indicates that there is no difficulty in explaining why ventilation increases during exercise. However, it is not at all obvious which of the signals are needed to permit the tight coupling of ventilation to metabolic rate.

Despite very large changes in CO_2 production and O_2 consumption during exercise, the arterial P_{O_2} and P_{CO_2} levels remain remarkably constant. This occurrence suggests that stimulation of chemoreceptors alone by changes in arterial blood gases is not totally responsible for exercise hyperpnea and that there must be some other stimulus to ventilation during exercise. This is explained in Fig. 41-5, where the response of the central controller to increasing Pa_{CO_2} is plotted, along with the two hyperbolas that show the effect of ventilation on P_{CO_2} at two different levels of metabolic rate (Chapter 40).

If the exercise-induced change in ventilation were caused by the ventilatory drive from the newly produced CO_2 accumulating in the bloodstream, the response line would intersect the curve for the higher $\dot{V}_{CO_2}$ at point Y, which defines the arterial P_{CO_2} that would occur during exercise if CO_2 sensitivity were the only factor responsible for the hyperpnea. In fact, during exercise P_{CO_2} intersects the $\dot{V}_{CO_2}$ curve at Z. The unchanging value of arterial P_{CO_2} during exercise (below the anaerobic threshold) could be explained by some factor sensitive to the metabolic rate that increases ventilation and so shifts the CO_2 response line. Studies of CO_2 inhalation during exercise suggest that such a shift in the CO_2 response line does take place. The nonchemical drives shown in the preceding list could be this additional factor.

■ Respiratory effects of high altitude and O_2 breathing
■ High altitude

Barometric pressure decreases with increasing altitude (Fig. 41-6); the relation is approximately exponential. Since the inspired air contains virtually the same percent of O_2 at high and low altitudes, inspired P_{O_2} and thus alveolar P_{O_2} must fall as an individual ascends from sea level. The resulting hypoxemia elicits a variety of compensatory responses. Some of these occur acutely, but others develop gradually during prolonged exposure.

The initial hyperventilation that occurs during exposure to high altitude is caused by stimulation of the peripheral chemoreceptors, particularly the carotid bodies. The increase in ventilation reduces the arterial P_{CO_2} and H^+ ion concentration, which decreases the excitation of central chemoreceptors and thus limits the degree of ventilatory increase. After 2 or 3 days at high altitude there is a further rise in ventilation. This increase is part of the process of acclimatization, which is poorly understood. This secondary increase in ventilation occurs in part by the following mechanisms: (1) a reduction in plasma HCO_3^- caused by the renal excretion of bicarbonate, which returns blood pH toward normal, (2) a direct excitatory effect of hypoxia on respiratory neurons, and (3) a reduction in HCO_3^- concentration in brain interstitial fluid. The last mechanism may be a result of some active process that expels HCO_3^- ions from the interstitial fluid into the body. There also may be passive transfer of HCO_3^- from the brain into the blood. Also anaerobic metabolism in the hypoxic brain allows lactic acid to accumulate, which lowers HCO_3^- levels.

Persons who live at high altitudes and who are chronically hypoxic seem to have a blunted ventilatory response to hypoxia. They breathe less at a given level of arterial

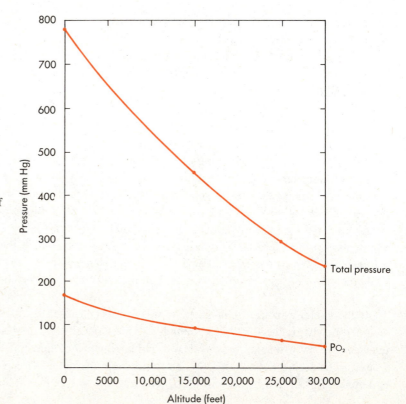

Fig. 41-6 ■ Changes in barometric pressure and the P_{O_2} of inspired air with altitude.

PO_2 than do those who live at sea level. The response to hypercapnia, on the other hand, is the same in people who live at high and low altitudes. The blunted response to hypoxia develops over years and is more likely to occur when chronic exposure to hypoxia begins in infancy. A similar blunted response to hypoxia occurs in patients who are hypoxemic because of congenital heart disease in which there are right to left shunts of blood in the heart. It is not clear whether the blunted ventilatory response is caused by some depressant effect of chronic hypoxia on the central nervous system or whether it originates at the peripheral chemoreceptors. At autopsy the carotid bodies of high-altitude natives are frequently enlarged and more vascular. It may be that at the same level of arterial PO_2 the PO_2 at the O_2 sensor is less in high-altitude natives than in those who live at sea level. The blunted response to hypoxia ultimately disappears when high-altitude natives stay at sea level for years or when the cause of cyanotic heart disease is corrected.

In addition to these depressant changes in the respiratory controller, other changes occur in the controlled system. Lung volume is the same in infants born at high altitude and at sea level. However, adult natives of high altitudes have disproportionately large lungs for their body size. The increase in size does not involve the airways, and it seems to occur mainly in the gas-exchanging units of the lung. This increase in lung size may be responsible for the increase in diffusing capacity observed in high-altitude natives.

High-altitude exposure also results in an increased hemoglobin concentration in the blood and consequently in an augmented O_2-carrying capacity. The increase in hemoglobin is caused by the release of *erythropoietin* from the kidney and possibly from other organs. This substance accelerates red cell production in the bone marrow.

During high-altitude exposure the concentration of 2,3-diphosphoglycerate also rises in the cell, which decreases the affinity of hemoglobin for O_2. The P_{50} of the blood is decreased. Although this change enhances the unloading of O_2 from hemoglobin in the tissues, it also interferes with the uptake of O_2 by hemoglobin in the lungs. Therefore its adaptative value is uncertain.

Circulatory changes also occur with exposure to high altitudes. Cardiac output, heart rate, and pulmonary artery pressure all increase. With prolonged exposure there may be an increase in the density of capillaries in many tissues. All these changes tend to improve O_2 transfer in the lung and O_2 delivery to tissue mitochondria.

Occasionally tolerance at high altitude disappears, and serious symptoms develop, which include ventilatory depression, polycythemia, and heart failure. This disease, called Monge's disease after its discoverer, is relieved by descent to a lower altitude or by administration of O_2.

■ Effects of breathing high O_2 mixtures

Breathing O_2-enriched gas mixtures for prolonged periods can produce injurious effects. Initially the pathological changes are exudative and are caused by damage to the capillary endothelium. Frank pulmonary edema can develop. If this stage is survived, the larger alveolar cells proliferate and form a continuous lining layer over the alveoli, replacing the membranous pneumocytes. In the newborn infant pure O_2 breathing may cause a necrotizing bronchiolitis. High O_2 breathing also can cause alveolar collapse. Alveoli that are served by airways which are obstructed tend to collapse, because the total pressure of the gases in the mixed venous blood is less than that in the alveoli, so that the alveolar gas is absorbed into the blood. The N_2 in the alveoli is more slowly absorbed than is the O_2. Inhalation of 100% O_2 eliminates the N_2 in the alveoli and accelerates collapse. In addition, O_2 breathing for even a few hours depresses ciliary function and mucus transport (p. 736).

In humans the pulmonary function abnormalities caused by breathing O_2-rich gas mixtures include a decreased vital capacity, a reduced lung compliance, a widened A-a PO_2 gradient, and a diminished diffusing capacity.

Animals can be made more resistant to O_2-induced injury by graded exposure to increasing concentrations of O_2 in the air. The tolerance developed seems to depend on certain enzymatic adaptations.

O_2 breathing also can injure nonpulmonary tissues. In infants 100% O_2 breathing may produce blindness by causing fibrous tissue to form behind the lens. The condition is known as *retrolental fibroplasia*.

Even brief exposures to hyperbaric O_2 at several atmospheres of pressure can produce convulsions. However, if care is used, hyperbaric O_2 therapy carried out in specially constructed chambers is sometimes useful in the treatment of gas gangrene and severe CO poisoning.

Divers breathe air at high pressure, which causes an increase in the partial pressure of O_2 and N_2 in the alveoli. The high partial pressure of N_2 forces the gas into solution in the tissues. If the diver is decompressed too rapidly, bubbles of N_2 may form in the tissues and blood, producing pain and neurological damage (the "bends") because the N_2 bubbles obstruct blood flow. Very high levels of N_2 also seem to depress the central nervous system and have a narcotic effect.

■ *Respiratory changes with age*
■ *The newborn*

One of the principal functions of the first few breaths is to transform the fluid-filled fetal lung to one containing air. High distending forces are needed in the first few breaths to move liquid out of the lung and to overcome surface forces that oppose the movement of air. With the first few breaths the end-expiratory lung volume (functional residual capacity) increases, and breathing becomes easier. The ability of the newborn to maintain air in the lungs depends on how well the chest wall can resist collapsing forces produced by contraction of the inspiratory muscles and how well surfactant is produced to reduce surface tension (Chapter 37). In the premature infant the chest is very pliable, and surfactant production may be poor, increasing the danger of lung collapse. Resistance to airflow is higher in the infant than in the adult. The recoil forces of the lung and chest wall are both low in the infant, but they increase with maturity. New alveoli are added until the child is about 8 years of age. Then the increase in lung size is achieved mainly by enlargement of existing alveoli.

Pulmonary vascular resistance is very high in the fetus, and the pulmonary blood vessels are very thick. After birth the vessel walls become thinner, so that by age 4 months the ratio of wall thickness to external diameter is about the same as in the adult.

Diffusing capacity increases during infancy as the alveolar surface area enlarges. By 6 years of age the diffusing capacity per square meter of body surface area is about the same as in the adult.

Breathing may be irregular at birth, particularly in the preterm infant. Preterm infants normally exhibit various patterns of breathing that range from regular respiration to frequent pauses or apneic episodes. In preterm infants apnea is predominantly central; that is, both respiratory movements and airflow at the mouth or nose are absent. In infants beyond the neonatal period, however, pure obstructive apnea has been documented frequently, and it may predispose the infant to the sudden infant death syndrome.

Both full-term and preterm infants increase minute ventilation in response to small increases in inspired CO_2 concentration. Responsiveness to CO_2 is less well developed in the more immature infant.

Preterm infants respond to a fall in inspired O_2 concentrations with a transient increase in ventilation over approximately 1 minute, followed by a sustained depression of ventilation. In full-term infants a similar response has been observed during the first week of life. This biphasic response to hypoxemia in the neonate is in marked contrast to the ventilatory response in adults, in whom a low Pa_{O_2} results in a sustained increase in ventilation. The characteristic response to low Pa_{O_2} in infants indicates an initial peripheral chemoreceptor stimulation followed by an overriding depression of the respiratory center as a direct result of the hypoxemia.

The effects of pulmonary stretch receptors in altering the timing of respiration at varying lung volumes are much more readily elicited in the newborn than in the adult. A small but sustained increase in lung volume causes a significant prolongation of the expiratory time and a concomitant decrease in the respiratory rate. The lung deflation reflex produces an increase in respiratory rate. The increase in breathing rate is primarily due to a shortening of the expiratory time, and it may assist the infant in maintaining an adequate functional residual capacity. Pulmonary irritant reflexes also have been elicited in the neonate. Direct stimulation of the tracheal wall with a fine catheter threaded through the endotracheal tube of intubated infants results in augmented respiratory efforts.

■ *Respiration in the elderly*

Pulmonary performance declines with advancing age. The changes proceed at a variable rate, depending both on the "aging process" and on the extent of exposure to noxious agents in the environment.

With advancing age the internal surface area of the lung decreases, leading to a decrease in diffusing capacity. The relative dead space volume also rises, since the volume of air in the alveoli relative to that in the airways is reduced.

Lung elasticity is diminished in the elderly, so that the transpulmonary pressure at any given lung volume is less in the aged than in the young adult (Fig. 41-7). The loss of lung elastic recoil leads to airway closure at higher lung volumes than in younger individuals and to the increase in residual volume noted in the elderly. Airway closure in dependent lung zones may occur at volumes exceeding functional residual capacity in men over 65 years old. This contributes to a greater unevenness of the distribution of ventilation in the aged.

There is an increasing spread of ventilation/perfusion ratios with advancing age. The widening of the alveolar-arterial O_2 gradient noted in older adults occurs because of a fall in Pa_{O_2} of about 0.3 mm Hg/year, which in turn is explained by a greater mismatching of ventilation and perfusion (Fig. 41-8). Pa_{CO_2} is the same in elderly and in young

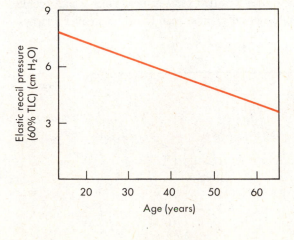

Fig. 41-7 ■ Effects of age on lung elastic recoil pressure.

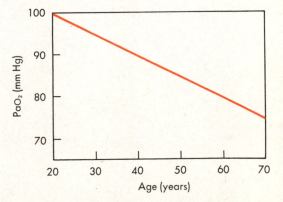

Fig. 41-8 ■ Effects of age on the partial pressure of O_2 in the arterial blood (Pa_{O_2}).

adults. Ventilatory responses to hypoxia and hypercapnia in elderly adults tend to be reduced in old age.

The strength of the inspiratory muscles tends to decrease in the elderly, and this may account for the reduction in vital capacity seen with advanced age. Expiratory muscle force and peak expiratory flow rates are also decreased. Chest wall compliance, unlike lung compliance, may be lower in the aged, and this may contribute to the changes in lung volumes in the elderly.

■ *Pulmonary defenses*

The lungs of a normal 70 kg man normally inspire approximately 6 L of air from the environment each minute, exposing the entire respiratory tree, down to the alveoli, to any toxic or pathogenic agents in the atmosphere. Maintenance of lung function depends on the presence of suitable defense mechanisms.

In humans defenses consist of (1) mechanical barriers (the nose, nasal hair, and turbinates), (2) the mucociliary escalator, (3) immunologic systems that operate both in the conducting airways and in the alveoli, and (4) phagocytic cells in the alveoli.

■ *Fate of inhaled particles*

The nostrils and turbinates effectively trap any particles with diameters greater than 10 μm. Thus particles in this range never reach the conducting airways. Particles between 2 and 10 μm become trapped on the mucus layers of the trachea, bronchi, and terminal bronchioles and are deposited by sedimentation onto the airway walls. This mucus layer normally moves the particles upward, toward the epiglottis and posterior pharynx. When the particles entrapped in the mucus layer reach the posterior pharynx, they are swallowed. The swallowed material either passes through the gastrointestinal system and is excreted or is metabolized in the stomach and small intestine.

■ *Mucociliary escalator*

Between 10 and 100 ml of mucous secretions are produced by tracheobronchial epithelium each day. These secretions spread out in a very thin (2 to 5 μm) layer (usually called the "gel" layer) over the ciliated surface and are propelled toward the glottis by the beating cilia that line the airways (Fig. 41-9). Beneath the gel layer there is an aqueous ("sol") layer (also called the "periciliary fluid") in which the cilia beat freely. The nature of the periciliary fluid and its source are unknown at the present time.

Tracheobronchial mucous secretions come from two sources: unicellular glands (goblet cells) in the surface epithelium and submucosal glands, mainly in the smaller airways.

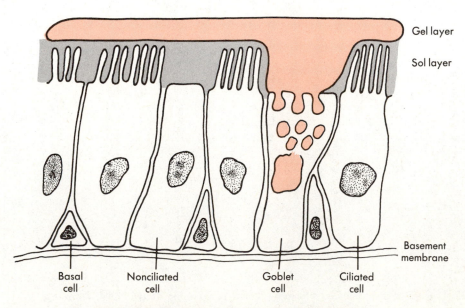

Fig. 41-9 ■ The airway epithelium and lining material.

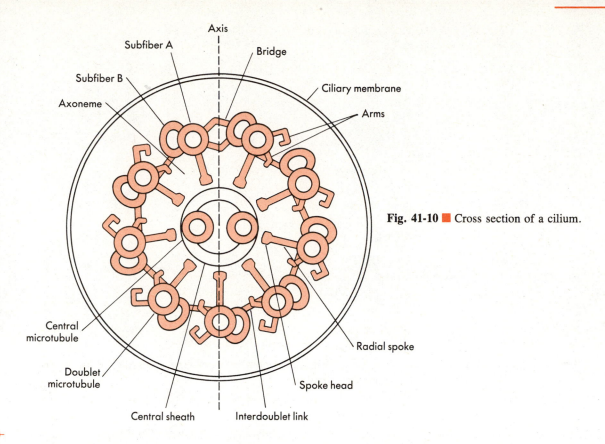

Fig. 41-10 ■ Cross section of a cilium.

These submucosal glands, which contain both mucus-secreting and serous cells, are distributed from the larynx to the smallest bronchi.

Vagal stimulation increases mucous secretion from the submucosal glands. The goblet cells respond mainly to local irritants, either physical or chemical, with an outpouring of mucus.

Most of the cells lining the tracheobronchial tract from the larynx to the terminal bronchioles are ciliated. Cilia beat from 12 to 20 times per second. Adjacent rows of cilia are in different phases of motion so that wavelike patterns are discernible on the surface of the epithelium.

Cilia are about 6 μm long and 0.3 μm in diameter and are anchored by basal bodies and rootlets in the apical cytoplasm. There are approximately 200 cilia per ciliated cell. Cross sections of these cilia demonstrate a series of microtubules, nine peripheral microtubule doublets arranged in a circle around two centrally positioned single tubules (Fig. 41-10). This microtubular system is called the axoneme. The microtubules contain a special ATPase called dynein (from "dynamic protein"). This ATPase catalyzes the release of energy from ATP, which diffuses into the cilia from the body of the cell. The peripheral microtubular doublets are loosely connected to a central tubule apparatus by radial spokes. When energy is supplied, each peripheral doublet can move longitudinally, relative to the central tubules. It is the differential movement of peripheral microtubules that results in bending of the cilia.

Both effective beating of the cilia and proper amounts of mucus with appropriate physical properties are needed for normal mucociliary transport. The mucous film is propelled at 0.5 to 1.5 cm/minute in the human trachea.

■ *Other defense mechanisms*

Lung secretions contain immunoglobulin, which inhibits viral infection, promotes clumping of some bacteria, and, in conjunction with lysozyme and complement, promotes phagocytosis.

Cellular elements in the lung include lymphocytes, which are important mediators

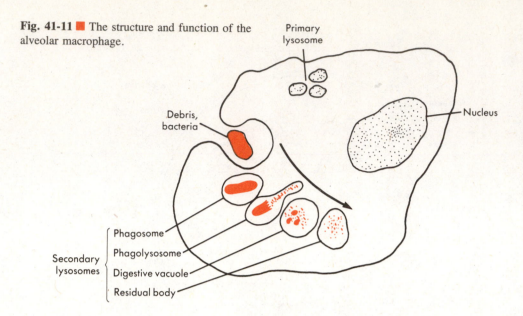

Fig. 41-11 ■ The structure and function of the alveolar macrophage.

of the immune reaction, and alveolar macrophages, which ingest and digest bacteria that reach the alveolar surfaces (Fig. 41-11).

The alveolar macrophage normally resides on the alveolar surface, but it is free to move about, migrating from alveolus to alveolus. Its activity can be random, phagocytosing whatever debris, particles, and organisms it encounters. The phagocytic activity of the macrophage probably also can become activated by chemical messengers released by the T(thymus derived)-lymphocytes.

In pulmonary inflammations white blood cells, alveolar macrophages, and destroyed bacteria release proteolytic enzymes, particularly elastase, into the lung. Elastase destroys the normal elastic fibers that support the lung's alveolar structure. This leads to the destruction of alveolar septa. Normally the level of α_1-antitrypsin in the plasma is sufficient to inactivate elastase. When a person suffers from a hereditary deficiency of α_1-antitrypsin, elastase is not effectively inactivated, alveolar septa are destroyed, and emphysema results. α_1-Antitrypsin deficiency is inherited as an autosomal recessive trait, and it occurs with a frequency of 1 per 1000.

■ *Respiratory sensation*

Dyspnea is one of the cardinal symptoms of diseases of the respiratory system. The term refers to the unpleasant or distressing sensation of labored or difficult breathing.

Dyspnea has been attributed to (1) a loss of respiratory reserve when the maximum breathing capacity is reduced to near the level of ventilation at rest, (2) increased work of breathing, (3) disordered action of the respiratory muscles, and (4) respiratory muscle fatigue. None of these theories adequately accounts for all the circumstances in which patients with respiratory disease complain of a distressing sensation of labored or difficult breathing.

■ *Breath-holding*

The sensation of dyspnea may in some instances be the same as that experienced during breath-holding. Hypoxia and hypercapnia contribute to the unpleasant sensation and eventual termination of breath-holding. However, alterations in blood gas tensions or H^+ ion concentration are not the sole factors limiting breath-holding time. If an individual at the breaking point of a breath-hold exhales and then inspires a mixture of gas with an even lower P_{O_2} and a higher P_{CO_2} than that of the alveolar gas just expired, a further period of breath-holding can be accomplished. Thus there seem to be a number of important nonchemical or mechanical factors that influence breath-holding time. Breath-holding time is longer at total lung capacity than at residual volume. This lung volume effect suggests that afferent impulses arising from the lung and traveling via the vagus nerves alter the sensation during breath-holding.

During breath-holding there is an initial period when the respiratory muscles are quiescent. Thereafter central respiratory activity sometimes produces diaphragmatic contractions, which may become more intense and frequent until the breath-hold breaks. The increasing tension in the isometrically contracting diaphragm and the resulting stimulation of mechanoreceptors in the muscle, presumably tendon organs, may well play an important role in the genesis of the sensation during breath-holding.

■ *Active breathing*

Respiratory sensations that occur during breathing have been used to investigate the factors that contribute to dyspnea. Studies of the ability of individuals to detect added ventilatory loads has led to the formulation of a theory that dyspnea is produced by an alteration in the relationship between the demand for ventilation and the effort required to achieve that level of ventilation. The responsible receptors are thought to be situated in the respiratory muscles or tendons and may include the muscle spindle.

The sensation of difficult or labored breathing seemingly is mediated in a manner similar to that of other kinesthetic sensations such as heaviness or force. Of critical importance is the sense of effort that stems from a motor command originating in the central nervous system. The level of respiratory efferent output thus appears to be a factor in the intensity of respiratory sensation.

■ *Behavioral effects*

Breathing serves both metabolic and behavioral functions, and the nervous system regulates these functions through different pathways. Behavioral effects constantly influence breathing. The behavioral system permits volitional acts such as talking, but it is probably also important in mediating respiratory adjustments to certain metabolic changes such as exercise, as well as to changes in ventilatory mechanical functions.

It is likely that behavioral responses are influenced by the sensations produced by breathing, and the intensity of the sensory input during breathing may be important in determining the magnitude of behavioral responses. When mechanical lung function is altered by disease, the intensity of sensation and hence behavioral responses may be particularly crucial in preserving ventilation and preventing the development of respiratory failure.

On the other hand, heightened levels of sensations leading to increases in respiratory efferent output may account for the extreme shortness of breath experienced by some patients who maintain adequate levels of alveolar ventilation.

■ *Bibliography*

Journal articles

Bakers, J.H.C.M., and Tenney, S.M.: The perception of some sensations associated with breathing, Respir. Physiol. **10**:85, 1970.

Barton, A.D., and Lourenco, R.V.: Bronchial secretions and mucociliary clearance, Arch. Intern. Med. **131**:140, 1973.

Bowden, D.H., and Adamson, I.Y.R.: The pulmonary interstitial cell as immediate precursor of the alveolar macrophage, Am. J. Pathol. **68**:521, 1972.

Campbell, E.J.M., and Howell, J.B.L.: The sensation of breathlessness, Br. Med. Bull. **19**:36, 1963.

Cohen, A.B., and Cline, M.J.: The human alveolar macrophage: isolation, cultivation in vitro and studies of morphologic and functional characteristics, J. Clin. Invest. **50**:1390, 1971.

Cotes, J.E.: Relationships of oxygen consumption ventilation and cardiac frequency to body weight during standardized submaximal exercise in normal subjects, Ergonomics **12**:415, 1969.

Daniel, R.P., et al.: Characterization of lymphocyte subpopulations in normal human lungs, Chest **67**(Suppl.):525, 1975.

Doershuk, G.F., and Matthews, L.W.: Airway resistance and lung volume in the newborn infant, Pediatr. Res. **3**:128, 1969.

Eliasson, R., et al.: The immotile cilia syndrome: a congenital ciliary abnormality as an etiologic factor in chronic airways infections and male sterility, N. Engl. J. Med. **297**:1, 1977.

Holland, J. et al.: Regional distribution of pulmonary ventilation and perfusion in elderly subjects, J. Clin. Invest. **47**:81, 1968.

Jones, N.L., et al.: Physiological dead space and alveolar-arterial gas pressure differences during exercise, Clin. Sci. **31**:19, 1966.

Knudson, R.J., et al.: The maximal expiratory flow volume curve; normal standards, variability and effects of age. Am. Rev. Respir. Dis. **113**:587, 1976.

Koch, G.: Alveolar ventilation, diffusing capacity and the A-aPo$_2$ difference in the newborn infant, Respir. Physiol. **4**:169, 1968.

Kronenberg, R.S., and Drage, C.W.: Attenuation of the ventilatory and heart rate responses to hypoxia and hypercapnia with aging in normal men, J. Clin. Invest. **52**:1812, 1973.

LaForce, F.M., Kelly, W.J., and Huber, G.L.: Inactivation of staphylococci by alveolar macrophages with preliminary observations on the importance of alveolar lining material, Am. Rev. Respir. Dis. **108**:784, 1973.

Lenfant, C., and Sullivan, K.: Adaptation to high altitiude, N. Engl. J. Med. **284**:1298, 1971.

Lourenco, R.V., Klimek, M.F., and Borowski, C.J.: Deposition and clearance of two micron particles in the tracheobronchial tree of normal subjects—smokers and non-smokers, J. Clin. Invest. **50**:1411, 1971.

Rhodin, J.A.G.: Ultrastructure and function of the human tracheal mucosa, Am. Rev. Respir. Dis. **93**(Suppl.):1, 1966.

Turner, J.M., Mead, J., and Wohl, M.E.: Elasticity of human lungs in relation to age, J. Appl. Physiol. **25**:664, 1968.

Wasserman, K.: Breathing during exericse, N. Engl. J. Med. **298**:780, 1978.

Wasserman, K., and Whipp, B.J.: Exercise physiology in health and disease, Am. Rev. Respir. Dis. **112**:219, 1975.

Whipp, B.J., and Wasserman, K.: Alveolar-arterial gas tension differences during graded exercise, J. Appl. Physiol. **27**:361, 1969.

Books and monographs

Asmussen, E.: Muscular exercise. In Fenn, W.O., and Rahn, H., editors: Handbook of physiology; Section 3: Respiration, vol. I, Washington, D.C., 1964, American Physiological Society.

Astrand, P.O., and Rodahl, K.: Textbook of work physiology, London, 1970, McGraw-Hill Book Co. (V.K.), Ltd.

Dejours, P.: Control of respiration in muscular exericse. In Fenn, W.O., and Rahn, H., editors: Handbook of physiology; Section 3: Respiration, vol I, Washington, D.C., 1964, American Physiological Society.

Muir, D.C.F.: Deposition and clearance of inhaler particles. In Muir, D.C.F., editor: Clinical aspects of inhaled particles, London, 1972, William Heinmann, Ltd.

Scarpelli, E.M.: Pulmonary physiology of the fetus, newborn and child, Philadelphia, 1975, Lea & Febiger.

THE GASTROINTESTINAL SYSTEM

Howard C. Kutchai

chapter 42

Gastrointestinal motility

The structure of the gastrointestinal tract varies greatly from region to region, but there are common features in the overall organization of the tissue. Fig. 42-1 depicts the general layered structure of the wall of the gastrointestinal tract.

The mucosa consists of an epithelium, the lamina propria, and the muscularis mucosae. The nature of the epithelium varies greatly from one part of the digestive tract to another. The *lamina propria* consists largely of loose connective tissue containing collagen and elastin fibrils. The lamina propria is rich in several types of glands and contains lymph nodules and capillaries. The *muscularis mucosae* is the innermost layer of intestinal smooth muscle and consists primarily of two thin layers of smooth muscle: an inner circular layer and an outer longitudinal layer. Contractions of the muscularis mucosae throw the mucosa into folds and ridges.

The submucosa consists largely of loose connective tissue with collagen and elastin fibrils. In some regions submucosal glands are present. The larger blood vessels of the

■ *Structure and innervation of the gastrointestinal tract*

■ *Structure of the wall of the gastrointestinal tract*

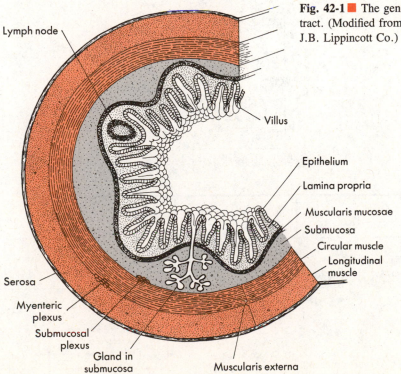

Fig. 42-1 ■ The general organization of the layers of the gastrointestinal tract. (Modified from Ham, A.W.: Histology, ed. 3, Philadelphia, 1957, J.B. Lippincott Co.)

Lymph node

Villus

Epithelium

Lamina propria

Muscularis mucosae

Submucosa

Circular muscle

Longitudinal muscle

Serosa

Myenteric plexus

Submucosal plexus

Gland in submucosa

Muscularis externa

intestinal wall travel in the submucosa. The submucosa contains the *submucosal plexus* (Meissner's plexus).

The *muscularis externa* characteristically consists of two substantial layers of smooth muscle cells: an inner circular layer and an outer longitudinal layer. In between the two layers lies the *myenteric plexus* (Auerbach's plexus), which plays a key role in coordinating the contractile activity of the muscularis externa. Contractions of the muscularis externa are responsible for mixing the contents in the lumen and for propelling them in a controlled fashion toward the anus.

The *serosa*, or adventitia, is the outermost layer and consists mainly of connective tissue covered with a layer of squamous mesothelial cells.

■ Innervation of the gastrointestinal tract

Sympathetic innervation. Sympathetic innervation of the gastrointestinal tract is primarily via postganglionic adrenergic fibers whose cell bodies are in prevertebral and paravertebral plexuses. The celiac, superior and inferior mesenteric, and hypogastric plexuses provide postganglionic sympathetic innervation to various segments of the gastrointestinal tract. Most of the sympathetic fibers terminate in the submucosal and myenteric plexus. Activation of the sympathetic nerves usually has an inhibitory effect on the ganglion cells of the plexuses. Some sympathetic fibers innervate blood vessels of the gastrointestinal tract, causing vasoconstriction. Other sympathetic fibers innervate glandular structures in the wall of the gut. Relatively few of the sympathetic fibers terminate in the muscularis externa. Stimulation of the sympathetic input to the gastrointestinal tract inhibits contraction of the muscularis externa but stimulates contraction of the muscularis mucosae. The inhibitory effect of the sympathetic nerves on the muscularis externa is not a direct action on the smooth muscle cells, since there are few sympathetic nerve endings in the muscularis externa. Rather the sympathetic nerves act to influence neurons in the intrinsic plexuses that provide input to the smooth muscle cells. This effect may be reinforced by the action of the sympathetic nerves in reducing blood flow to the muscularis externa. Other fibers that travel with the sympathetic nerves may

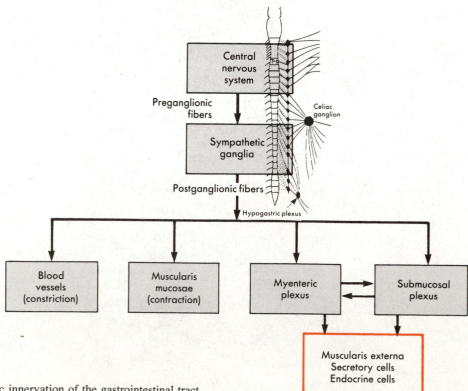

Fig. 42-2 ■ Major aspects of the sympathetic innervation of the gastrointestinal tract.

be cholinergic; still others release neurotransmitters that remain to be identified. Fig. 42-2 schematically summarizes the sympathetic innervation of the gastrointestinal tract.

Parasympathetic innervation. Parasympathetic innervation of the gastrointestinal tract down to the level of the transverse colon is provided by branches of the vagus nerve. The remainder of the colon receives parasympathetic fibers from the pelvic nerves by way of the hypogastric plexus. These parasympathetic fibers are preganglionic and predominantly cholinergic. Other fibers that travel in the vagus and its branches have other transmitters that have not been identified. Preganglionic fibers terminate predominantly on the ganglion cells in the intramural plexuses. The ganglion cells then function as postganglionic parasympathetic neurons that innervate the smooth muscle and secretory cells of the gastrointestinal tract. Parasympathetic input usually stimulates the motor and secretory activity of the gut. Fig. 42-3 schematically illustrates the parasympathetic innervation of the gastrointestinal tract.

Intramural plexuses. The myenteric and submucosal plexuses are the most well-defined plexuses in the wall of the gastrointestinal tract. Other plexuses also have been identified. The plexuses are networks of nerve fibers and ganglion cell bodies. Some of the incoming axons are preganglionic parasympathetic fibers, and others are postganglionic sympathetic fibers. Interneurons in the plexuses interconnect ganglion cells, with the result that the myenteric and submucosal plexuses can control a good deal of coordinated activity in the absence of extrinsic innervation to the gastrointestinal tract. Axons of plexus neurons innervate gland cells in the mucosa and submucosa, smooth muscle cells in the muscularis externa and muscularis mucosae, and intramural endo-

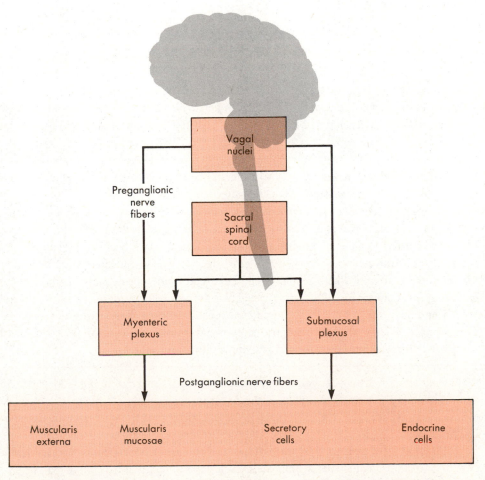

Fig. 42-3 ■ Major aspects of the parasympathetic innervation of the gastrointestinal tract.

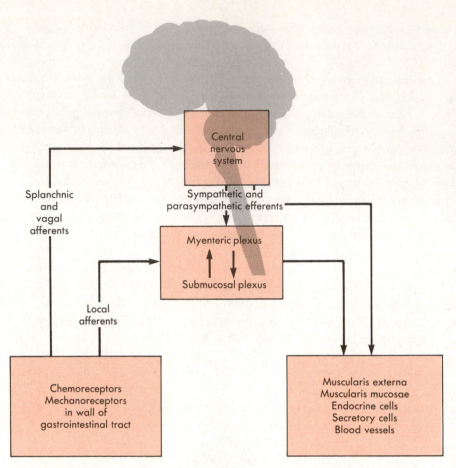

Fig. 42-4 ■ Local and central reflex pathways in the gastrointestinal system.

crine and exocrine cells. Afferent fibers from mechanoreceptors and chemoreceptors in the mucosa or deeper in the wall of the gastrointestinal tract synapse in the plexuses, so that local reflex activity is possible (Fig. 42-4). The functions of the intramural plexuses are discussed in more detail later in this chapter.

Afferent fibers. Afferent fibers in the gut provide the afferent limbs of reflex arcs that are both local and central (Fig. 42-4). Chemoreceptor and mechanoreceptor endings are present in the mucosa and in the muscularis externa. The cell bodies of some of these sensory receptors are located in the myenteric and submucosal plexuses. The axons of some of these receptor cells synapse with other cells in the plexuses to mediate local reflex activity. Others of these receptors send their axons back to the central nervous system. Still other sensory neurons have their cell bodies located more centrally. Those sensory afferent fibers in the vagus have cell bodies principally in the nodose ganglion, whereas sensory fibers that travel via the sympathetic nerves have their cell bodies in dorsal root ganglia. The number of sensory afferent fibers from the gastrointestinal tract is large, and its complex efferent innervation allows for fine control of secretory and motor activities by intrinsic and central pathways.

■ *Gastrointestinal smooth muscle*

■ *Smooth muscle cells of the muscularis externa*

Intestinal smooth muscle cells are long, slender cells that are about 200 μm long and less than 10 μm thick. Cells are separated from their neighbors by spaces of 50 to 80 nm. The extracellular spaces contain basement membrane material and an extracellular reticulum that is rich in collagen and elastin fibrils. The collagen network allows the contractile force generated by one smooth muscle cell to be transmitted to its neighbors and to connective tissues within the muscle layer. Neighboring cells have areas of close membrane contact that mediate electrical coupling of the adjacent cells.

The smooth muscle cells of the intestine are arranged in bundles that are separated and defined by connective tissue. The bundles branch and frequently anastomose with other bundles. The bundles are about 200 μm thick and contain several thousand cells in cross section. Because of the high degree of electrical coupling among the smooth muscle cells, the functional contractile unit is a bundle rather than a single cell.

■ *Electrophysiology of gastrointestinal smooth muscle cells*

The resting membrane potential. The resting membrane potential of gastrointestinal smooth muscle cells ranges from about -50 mV to around -80 mV. As discussed in Chapter 2, the relative membrane conductances of K^+, Na^+, and Cl^- are important in determining the resting membrane potential. Compared with skeletal muscle, gastrointestinal smooth muscle cells have higher Na^+ conductances. This contributes to the somewhat lower resting membrane potential of gastrointestinal smooth muscle.

If the relative conductances and the equilibrium potentials for K^+, Na^+, and Cl^- in guinea pig teniae coli (a well-studied preparation) are used in the chord conductance equation (equation 13, p. 34), a *predicted* resting membrane potential of -35 to -40 mV for the teniae coli is computed. When the resting membrane potential of teniae coli is measured with microelectrodes, a value of about -60 mV is obtained. The potential difference across the plasma membrane of the resting teniae coli is 20 to 25 mV larger than can be accounted for by diffusion of ions.

What is responsible for the extra 20 to 25 mV of polarization across the plasma membrane? The current view is that the Na^+, K^+ pump is *electrogenic,* and it contributes about this much to the resting membrane potential. Since 3 Na^+ are extruded for every 2 K^+ taken up, the pump produces a net outward flow of positive charge, which contributes to the membrane potential. (A process producing net current flow is called *electrogenic*.) There is experimental support for the idea that the electrogenic Na^+, K^+ pump is responsible for a significant part of the resting potential in gastrointestinal smooth muscle (Fig. 42-5). If the Na^+, K^+ pump of guinea pig teniae coli is inhibited with ouabain, the resting membrane potential changes from about -60 to near -40 mV. When ouabain is washed out, the resting potential returns to near -60 mV.

Variations in the membrane potential. In most other excitable tissues the resting membrane potential is rather constant in time. In gastrointestinal smooth muscle the resting membrane potential characteristically varies in time.

Slow waves. Slow waves are slow cyclical oscillations in the membrane potential that are characteristic of gastrointestinal smooth muscle. The frequency of the oscillations varies from about two per minute in the stomach to about 20 per minute in the

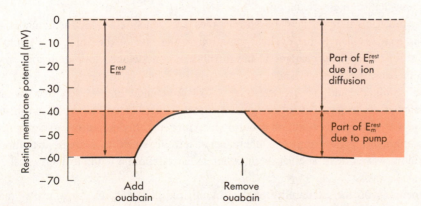

Fig. 42-5 ■ Idealized representation of an experiment suggesting that a significant part of the resting membrane potential of gastrointestinal smooth muscle is due to the electrogenic Na^+, K^+ pump.

Slow waves and resultant contractile tension

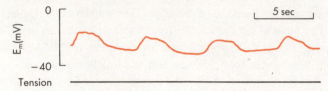

Slow waves and prepotentials that give rise to action potentials

Slow waves and prepotentials that fail to give rise to action potentials

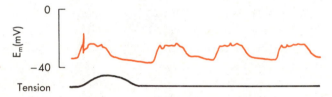

Fig. 42-6 ■ Electrical and mechanical activity of gastrointestinal smooth muscle. The upper tracing in each panel is the transmembrane electrical potential difference (millivolts), and the lower tracing shows contractile tension. The range of different electrical behaviors in a single tissue is noteworthy. The tissue contracts only in response to action potentials. (Modified from Bortoff, A.: Am. J. Physiol. **201**:203, 1961.)

duodenum. Fig. 42-6 shows almost sinusoidal slow waves in the rabbit jejunum. *Slow waves* also are referred to as the *basic electrical rhythm*.

In each segment of the gastrointestinal tract there is one or more groups of cells whose basic electrical rhythm is somewhat faster than that of the other cells. These cells serve as ''pacemakers'' for the slow waves in the region. Because the smooth muscle cells are well coupled electrically, the wave of depolarization (or repolarization) is spread throughout that segment of the gut.

The mechanism by which the slow waves are generated is not well understood. There is some evidence that cyclical variation in the rate of the Na^+, K^+ pump may play a role, but cyclical changes in the membrane conductance to certain ions may be the major cause of the slow waves.

The amplitude and, to a lesser extent, the frequency of the slow waves can be modulated in some segments of the gastrointestinal tract by the activity of intrinsic and extrinsic nerves and by circulating hormones. In general, circulating epinephrine and norepinephrine released from sympathetic nerve terminals tend to decrease the amplitude of the slow waves or to abolish them, whereas acetylcholine tends to increase the size of the slow waves.

Depending on the excitability of the smooth muscle cells, the slow waves may or may not cause action potentials to occur. If the peak of the slow wave is above threshold for the cells to fire action potentials, one or more action potentials may be triggered during the peak of the slow wave (Fig. 42-6).

The site of origin of the slow waves is a matter of controversy. It has been thought that the slow waves were generated by pacemaker cells in the longitudinal smooth muscle layer. However, some recent evidence suggests that smooth muscle cells in the circular layer just beneath the myenteric plexus may generate the slow waves. The generation of the slow waves does not require the participation of the intramural plexuses.

Prepotentials. Some gastrointestinal smooth muscle cells show spontaneous depolarizations that are much more rapid than the slow waves. These depolarizations resemble those in the pacemaker cells of the sinoatrial node of the heart and are known as *prepotentials* or *pacemaker potentials*. Prepotentials often, but not always, serve to depolarize the smooth muscle cell to threshold, whereupon an action potential is fired (Fig. 42-6).

The prepotential is believed to be caused by an increase in the membrane conductance to Na^+, Ca^{++}, or both. The slope of the prepotential may be altered by nervous and humoral influences. Sympathetic stimulation or circulating epinephrine may decrease or abolish prepotentials. Stimulation of parasympathetic nerves to the gut may increase the slope of the pacemaker potentials and increase their frequency. The degree of stretch of the tissue also can alter the frequency of slow waves and prepotentials.

Action potentials. Action potentials in gastrointestinal smooth muscle are more prolonged (10 to 20 msec duration) than those of skeletal muscle and have little or no overshoot. The rising phase of the action potential is caused by ion flow through channels that are relatively slow to open and probably conduct both Ca^{++} and Na^+. The Ca^{++} that enters the cell during the action potential plays a significant role in initiating contraction.

When the membrane potential of gastrointestinal smooth muscle reaches a threshold value, a train of action potentials (1 to 10 per second) occurs (Fig. 42-6). The extent of depolarization of the cells and the frequency of action potentials are enhanced by acetylcholine liberated from nerve endings. Circulating epinephrine and norepinephrine released from sympathetic nerve endings tend to hyperpolarize the smooth muscle cells and to abolish action potential spikes.

■ Electrical coupling between smooth muscle cells

Neighboring cells are said to be well coupled electrically if a perturbation of the membrane potential of one cell spreads rapidly, and with little decrement, to the other cell. The smooth muscle cells of the muscularis externa are well coupled.

The smooth muscle cells of the circular layer are somewhat better coupled than those of the longitudinal layer. The cells of the circular layer are joined by frequent gap junctions that are believed to provide low-resistance pathways that allow the spread of electrical current from one cell to another.

Since the electrical resistance of membranes is much higher than the resistance of the cytoplasm, the resistance along the long axis of smooth muscle cells is less than that in the transverse direction. Thus an electrical depolarization in the circular muscle is readily conducted circumferentially and quickly depolarizes a ring of circular muscle near the original site of depolarization. The ring of depolarization then spreads more slowly along the long axis of the gut.

In the intact gastrointestinal tract the longitudinal smooth muscle of the muscularis externa also behaves as though its cells were well coupled electrically. Because of the low electrical resistance along the length of the smooth muscle cells, an electrical disturbance is rapidly propagated along the long axis of the gut in the longitudinal layer. Thus the slow waves are rapidly conducted along the longitudinal layer, and they spread from the longitudinal layer inward to the circular layer of smooth muscle cells. Since contractions of the longitudinal smooth muscle are rather ineffectual in mixing and propelling the contents of the gastrointestinal tract, the major function of the longitudinal muscle may be to conduct electrical signals along the length of the gut. In this sense the function of the longitudinal layer may be analogous to that of the cardiac Purkinje fibers.

The high degree of electrical coupling in the longitudinal layer is somewhat puzzling. Gap junctions between cells are rare in the longitudinal layer, and the presence of other junctional complexes that mediate electrical coupling between cells has not been demonstrated in the longitudinal layer. Some investigators believe that electrical coupling is not intrinsic to the longitudinal muscle layer itself but requires the participation of the myenteric plexus and perhaps the circular layer as well.

Coupling between longitudinal and circular layers of smooth muscle. Typically the slow waves themselves do not elicit contractions of the smooth muscle layers. Contraction is evoked by the action potentials that are intermittently triggered near the peaks of the slow waves. The action potentials are also rapidly propagated along the long axis of the gut in the longitudinal smooth muscle layer and spread inward electrotonically to elicit contractions of the underlying circular layer. Neighboring cells of the circular and longitudinal layers may be joined by gap junctions, which may play a role in the high degree of electrical coupling between the two layers. As mentioned earlier, the site of the origin of slow waves is not certain, but regardless of which layer contains the pacemaker cells, the slow waves would be rapidly propagated to the other layer.

■ ***Neuromuscular interactions in the gastrointestinal tract***

Neuromuscular interactions in the gastrointestinal tract do not appear to involve true neuromuscular junctions with specialization of the postjunctional membrane, as occurs at the motor endplate. The neurons of the intramural plexuses send axons to the smooth muscle layers, and each axon branches extensively to innervate many smooth muscle cells.

The circular smooth muscle layer of the muscularis externa is heavily innervated by nerve terminals that form close associations with the plasma membranes of the smooth muscle cells. Neuromuscular gaps of about 20 nm are typical in the circular layer. The predominant innervation is by fibers that are *inhibitory* to the electrical and contractile activity of the smooth muscle cells.

The longitudinal smooth muscle cells are much less richly innervated by the neurons of the intrinsic plexuses, and the neuromuscular contacts are not so intimate. Gaps of about 80 nm separate nerve terminals from the plasma membrane of the smooth muscle cells they innervate. In contrast to the situation in the circular layer, the longitudinal layer appears to be totally devoid of inhibitory nerve endings. The sparse excitatory nerve fibers are believed to be predominantly cholinergic.

■ ***Excitation-contraction coupling***

As in skeletal and cardiac muscle, the level of intracellular Ca^{++} plays a central role in regulating the contraction of smooth muscle. In smooth muscle, contraction is initiated by the Ca^{++} that crosses the plasma membrane during the action potential, as well as by Ca^{++} released from the sarcoplasmic reticulum. The less well developed the sarcoplasmic reticulum, the more the smooth muscle is dependent on extracellular Ca^{++} for contraction.

As discussed in Chapter 22, regulation of contraction by Ca^{++} in smooth muscle occurs by a different mechanism than in skeletal muscle. Smooth muscle thin filaments lack troponin, and it appears that Ca^{++} regulation in smooth muscle centers on the myosin molecule itself. At present the best evidence supports the view that increased cytoplasmic Ca^{++} binds to calmodulin, which then activates a myosin light chain kinase, which phosphorylates one of the two myosin light chains. The phosphorylation allows the head of the myosin molecule to interact cyclically with actin of the thin filaments, causing muscle shortening.

The length/tension curve. The length/tension curve of gastrointestinal smooth muscle is similar in shape to that of skeletal muscle, but it has a much broader maximum. This gives gastrointestinal smooth muscle the ability to develop force effectively over a greater range of muscle length. This property may reflect the organization of muscle cells within the tissue and the organization of the contractile elements within the cells more than it reflects the intrinsic length/tension properties of the contractile elements themselves.

Relationship between membrane potential and tension. Gastrointestinal smooth muscle cells fire action potentials (Fig. 42-6), and the cells contract phasically in response to action potentials. The action potentials usually occur in bursts at the peak of the slow waves and cause phasic contractions that are superimposed on the baseline level of contraction. Because smooth muscle cells contract rather slowly (about 10 times slower than skeletal muscle), the individual contractions caused by each action potential in a burst are not visible as distinct twitches but rather sum temporally to produce a smoothly increasing level of tension. The increase in tension in response to a burst of action potentials is proportional to the number of action potentials in the burst.

Between bursts of action potentials the tension developed by gastrointestinal smooth muscle falls, but not to zero. This nonzero "resting," or baseline, tension developed by the smooth muscle is called *tone*. The tone of gastrointestinal smooth muscle may be altered by neurotransmitters, hormones, or drugs.

Responses to stretch. Gastrointestinal smooth muscle may respond to stretch or release of stretch. In many cases rapidly stretching gastrointestinal smooth muscle will lead to an immediate increase in the frequency of action potentials and an increase in contractile tension—a phenomenon known as *stress activation*. This may be followed by a decrease in tension back toward the original level—a phenomenon known as *stress relaxation*. Both these responses to stretch play roles in the motility of the various segments of the gastrointestinal tract.

■ *Contractility of intestinal smooth muscle*

Control of the contractile activities of gastrointestinal smooth muscle involves the central nervous system, the intrinsic plexuses of the gut, humoral factors, and electrical coupling among the smooth muscle cells. The motor behavior of particular segments of the gastrointestinal tract is discussed in the succeeding sections of this chapter.

This section outlines some of the general structural and functional properties of gastrointestinal smooth muscle and its innervation that underlie control and integration of contractile function. A major conclusion is that the gastrointestinal tract displays a great deal of intrinsic control. The intramural plexuses can mediate control and integration of most of the contractile behavior of the gut without assistance from or intervention by the central nervous system.

■ *Integration and control of gastrointestinal motor activities*

Circular layer. As mentioned earlier, the predominant innervation of the circular layer of gastrointestinal smooth muscle is inhibitory. In the absence of neural influences (as when the plexus neurons are inactivated by drugs) each slow wave elicits a burst of action potentials and a near-maximal contraction of the cells of the circular layer. Because of the phase relation among the slow waves in neighboring segments of the small intestine, the circumferential ring of contraction elicited by each slow wave travels along the small bowel toward the colon. This contrasts strongly with the behavior of the intact gut with a functional intrinsic innervation. In the functionally intact gut only every third or fourth slow wave may elicit action potentials and contractions, which are usually submaximal and may be quite localized. It appears that the normal activity of the intrinsic plexuses of the gut is largely inhibitory to the circular muscle layer, preventing the burst of action potentials and the maximal propagated wave of contraction that occur in response to each slow wave in the absence of the intrinsic neurons.

■ *Properties of gastrointestinal smooth muscle relevant to control of motility*

Longitudinal layer. In contrast to the circular layer, the longitudinal layer is believed to receive essentially no inhibitory innervation. The predominant innervation of the longitudinal layer is by cholinergic excitatory fibers from the intrinsic plexuses. For this reason action potentials and resultant contractions are less frequent in the functionally denervated longitudinal layer than in a more intact preparation.

Neural activity of the plexus neurons thus tends to inhibit the circular muscle but to stimulate the longitudinal muscle. This may be physiologically relevant, since the two layers are antagonistic in the sense that contraction of one layer opposes contraction of the other.

■ The intramural plexuses as an "enteric brain"

Recent studies of the myenteric and submucosal plexuses suggest that the plexuses resemble the central nervous system in structure and function and may be thought of as an "enteric brain" in some important respects.

Plexus neurons. Three different kinds of neurons have been distinguished in the plexuses on the basis of intracellular recordings: burst units, mechanosensitive units, and single-spike units.

Burst units can be further subdivided into "steady bursters" and "erratic bursters." The steady bursters discharge clusters of action potentials at rather regular intervals. The rhythm of the steady bursters is not dependent on their synaptic input. The steady bursters appear to play a central role in the activity patterns that are intrinsic to the gastrointestinal tract and do not depend on extrinsic innervation. The erratic bursters depend on synaptic input, perhaps from the steady bursters as well as from other neurons.

Mechanosensitive units, which also can be subdivided into groups based on their properties, are activated by mechanical distortion of the mechanosensitive plexus neurons themselves. The mechanosensitive neurons provide afferent input to other plexus neurons and also to the central nervous system. Perhaps this afferent information modulates the activities of the erratic burst neurons, both directly and via reflex pathways.

Single-spike units fire single action potentials at infrequent and inconsistent intervals. The pattern of discharge of the single-spike units may be modulated by synaptic input, but the discharge of these units is not absolutely dependent on synaptic input.

The three types of plexus neurons presumably interact in complex ways that allow for the high degree of local control of the motor and secretory activities of the gastrointestinal tract. Apparently the enteric brain contains pattern-generating neural circuits that are repeated along the length of the gastrointestinal tract and play a central role in the coordinated motor activities of the gut. The detailed neural mechanisms by which this control and integration occur remain to be elucidated.

■ Neural transmitters in gastrointestinal smooth muscle

Acetylcholine, acting predominantly on muscarinic receptors, is the primary neurotransmitter that mediates an excitatory effect directly on the smooth muscle cells of the gastrointestinal tract. Relaxing effects of acetylcholine at particular sites in the gastrointestinal tract have been proposed from time to time, but there is no convincing evidence that acetylcholine can have a relaxing effect.

Norepinephrine is released from sympathetic postganglionic nerve endings in the wall of the gut. Norepinephrine has predominantly inhibitory effects on electrical and contractile responses of the smooth muscle cells of the muscularis externa, but it stimulates contractions of the muscularis mucosae. There is little or no direct innervation of the smooth muscle cells of the tunica externa by adrenergic neurons, and the inhibitory effects primarily are exerted at the level of the intrinsic plexuses. Adrenergic neurons synapse on the somas or axons of plexus neurons that are excitatory to the smooth muscle. Norepinephrine released near the plexus neurons inhibits the activity of the excitatory neurons primarily via α-receptor-mediated events. Norepinephrine also may diffuse to smooth muscle cells from the plexuses or from enteric blood vessels that have adrenergic innervation. Norepinephrine then acts directly on the smooth muscle cells to inhibit their activity, primarily by β-receptor-mediated processes. Circulating epinephrine has effects similar to those of norepinephrine on plexus neurons and on smooth muscle cells of the tunica externa.

Serotonin (5-hydroxytryptamine) is a candidate neurotransmitter in the gastrointestinal tract. Serotonin is present in tissue from the stomach to the colon in quantities consistent with its having a physiological function. The key enzymes of serotonin biosynthesis have been found in the intestines of most mammals. Serotonin has been localized in enterochromaffin cells in the intestinal mucosa and in nerve cells in the intramural plexuses. The actions of serotonin are multiple and may vary in different gastrointestinal segments. It appears that serotonin has direct effects on gastrointestinal smooth muscle, but it also can affect smooth muscle behavior via effects on plexus neurons. In both cases the contractile behavior of the smooth muscle is enhanced. At the level of the plexuses serotonin facilitates contraction by acting on cholinergic neurons to increase their release of acetylcholine near smooth muscle cells. Acting via different receptors, serotonin directly stimulates smooth muscle cells to contract. It has been suggested that serotonin plays a role in the peristaltic response by sensitizing the reflex to elevated intramural pressure, but the mechanism of this effect remains to be determined.

Histamine is another candidate neurotransmitter that has both direct and indirect effects on the contraction of gastrointestinal smooth muscle. Histamine is present in the wall of the gastrointestinal tract from the stomach to the colon. In the stomach histamine appears to be located near the glands that are rich in parietal cells. There is much evidence that histamine plays an important role in stimulating gastric acid secretion. However, the details of histamine's role in acid secretion remain obscure. At the level of the intrinsic plexuses histamine acts to enhance acetylcholine release by cholinergic neurons that are excitatory to gastrointestinal smooth muscle. Direct action of histamine on smooth muscle contraction requires larger doses; thus it may be an unphysiological effect.

Substance P is a peptide of about 15 amino acids. It is present in the gastrointestinal tract and in certain parts of the central nervous system. Substance P has been postulated to be a neurotransmitter in the gastrointestinal tract. Substance P causes contraction of gastrointestinal smooth muscle and has been reported to stimulate the peristaltic reflex by a mechanism that is independent of serotonin's similar effect.

Other substances that may regulate gastrointestinal motility and may be released from nerve endings in the gastrointestinal tract are ATP, enkephalins, oxytocin, vasopressin, vasoactive intestinal peptide (VIP), and somatostatin. Much experimental work needs to be done to understand the role of these and other substances in the physiological regulation of the gastrointestinal tract.

■ *Chewing (mastication)*

Chewing can be carried out voluntarily, but it is more frequently a reflex behavior. Chewing serves to lubricate the food by mixing it with salivary mucus, to mix starch-containing food with the salivary α-amylase, and to subdivide the food so that it can be mixed more readily with the digestive secretions of the stomach and duodenum.

■ *Swallowing*

Swallowing can be initiated voluntarily, but thereafter it is almost entirely under reflex control. The swallowing reflex is a rigidly ordered sequence of events that results in the propulsion of food from the mouth to the stomach, at the same time inhibiting respiration and preventing the entrance of food into the trachea (Fig. 42-7). The afferent limb of the swallowing reflex begins with tactile receptors, most notably those near the opening of the pharynx. Sensory impulses from these receptors are transmitted to certain areas in the medulla. The central integrating areas for swallowing lie in the medulla and

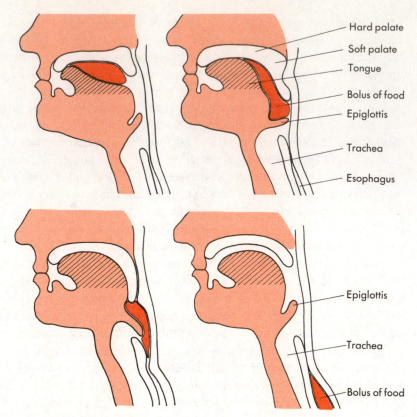

Fig. 42-7 ■ A simplified depiction of major events involved in the swallowing reflex. (Modified from Johnson, L.: Gastrointestinal physiology, ed. 2, St. Louis, 1981, The C.V. Mosby Co.)

lower pons; they are collectively called the "swallowing center." Motor impulses travel from the swallowing center to the musculature of the pharynx and upper esophagus via various cranial nerves.

The oral or voluntary phase of swallowing is initiated by separating a bolus of food from the mass in the mouth with the tip of the tongue. The bolus to be swallowed is moved upward and backward in the mouth by pressing first the tip of the tongue and later the more posterior portions of the tongue as well against the hard palate. This forces the bolus into the pharynx, where it stimulates the tactile receptors that initiate the swallowing reflex.

The pharyngeal stage of swallowing involves the following sequence of events, which occur in less than 1 second:

1. The soft palate is pulled upward, and the palatopharyngeal folds move inward toward one another. This prevents reflux of food into the nasopharynx and provides a narrow passage through which the food moves into the pharynx.
2. The vocal cords are pulled together, and the epiglottis covers the opening to the larynx. These actions prevent food from entering the trachea.
3. The upper esophageal sphincter relaxes to receive the bolus of food. Then the superior constrictor muscles of the pharynx contract strongly to force the bolus deeply into the pharynx.
4. A peristaltic wave, starting with the contraction of the superior constrictor muscles of the pharynx, is initiated and moves toward the esophagus. This forces the bolus of food through the relaxed upper esophageal sphincter.

During the pharyngeal stage of swallowing, respiration is reflexly inhibited.

The esophageal phase of swallowing also is partially controlled by the swallowing center. After the bolus of food passes the upper esophageal sphincter, the sphincter reflexly constricts. A peristaltic wave then begins just below the upper esophageal sphincter and traverses the entire esophagus in about 10 seconds. This initial wave of peristalsis, called *primary peristalsis,* is controlled by the swallowing center. Should the primary peristalsis be insufficient to clear the esophagus of food, the distension of the esophagus would initiate another peristaltic wave that begins at the site of distension and moves downward. This latter type of peristalsis, termed *secondary peristalsis,* is partially mediated locally, since it occurs (albeit more weakly) in the extrinsically denervated esophagus.

Before a peristaltic wave reaches the lower esophageal sphincter, the sphincter reflexly relaxes to allow the bolus to pass into the stomach. There is then a reflex relaxation of the stomach known as *receptive relaxation.*

■ *Esophageal function*

After food is swallowed, the esophagus functions as a conduit to move the food from the pharynx to the stomach. It is important to prevent air from entering at the upper end of the esophagus and to keep corrosive gastric contents from refluxing back into the esophagus at its lower end. This is particularly problematic because the pressure in the body of the resting esophagus closely approximates intrathoracic pressure. Thus it is less than atmospheric pressure and less than intraabdominal pressure. The pressure in the short abdominal section of the esophagus mirrors the intraabdominal pressure.

The structure of the esophagus follows the general scheme described earlier in this chapter, except that in the upper third of the esophagus both the inner circular and outer longitudinal muscle layers are striated. In the lower third of the esophagus the muscle layers are composed entirely of smooth muscle cells. In the middle third, skeletal and smooth muscles coexist, there being a gradient from all skeletal above to all smooth below.

The esophageal musculature, both striated and smooth, is primarily innervated by branches of the vagus nerve. Somatic motor fibers of the vagus nerve form motor end-plates on striated muscle fibers. Visceral motor nerves, primarily from the dorsal motor nucleus of the vagus nerve, are preganglionic parasympathetic fibers, and they synapse primarily on the nerve cells of the myenteric plexus. Neurons of the myenteric plexus are the postganglionic parasympathetic fibers. They innervate the smooth muscle cells of the esophagus and communicate with one another. The neural circuits that mediate control of the esophagus are schematized in Fig. 42-8.

The upper and lower ends of the esophagus function as sphincters to prevent the entry of air and gastric contents, respectively, into the esophagus. The sphincters are known as the *upper esophageal* (or pharyngeoesophageal) *sphincter* and the *lower esophageal* (or gastroesophageal) *sphincter*. The upper esophageal sphincter is formed by a thickening of the circular layer of striated muscle at the upper end of the esophagus; the sphincter also is known as the cricopharyngeus muscle. The lower esophageal sphincter is not identifiable anatomically, but the lower 1 to 2 cm of the esophagus function as a sphincter. In normal individuals the pressure at the lower esophageal sphincter is always greater than that in the stomach.

The function of the lower esophageal sphincter is of particular importance. As mentioned previously, it opens when a wave of esophageal peristalsis reaches it. The opening response is vagally mediated, and the transmitter involved is neither acetylcholine nor norepinephrine. In the absence of esophageal peristalsis the sphincter must remain tightly closed to prevent reflux of gastric contents, which would cause esophagitis and the sensation of heartburn. Acetylcholine liberated from nerve endings within the muscle increases the constriction of the sphincter. The hormone *gastrin,* released by the pyloric glandular mucosa, is a major physiological stimulus for gastric acid secretion. Gastrin also increases constriction of the lower esophageal sphincter. This *may* be an important control mechanism to increase constriction of the lower esophageal sphincter when secretion of gastric acid is elevated.

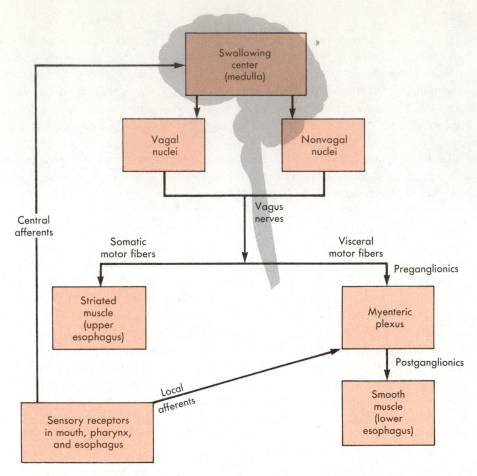

Fig. 42-8 ■ Local and central neural circuits involved in the control of esophageal motility.

In some individuals the lower esophageal sphincter fails to relax sufficiently during swallowing to allow food to enter the stomach, a condition known as *achalasia*. Therapy for achalasia may involve surgically weakening the lower esophageal sphincter. Individuals with *diffuse esophageal spasm* tend to have prolonged and painful contraction of the lower part of the esophagus after swallowing, instead of the normal esophageal peristaltic wave. Achalasia, incompetence of the lower sphincter, and diffuse esophageal spasm are the most common disorders of esophageal function.

■ *Gastric motility*

The motility of the stomach serves the following major functions: (1) to allow the stomach to serve as a reservoir for the large volume of food that may be ingested at a single meal, (2) to mix chyme with gastric secretions so that digestion can begin, and (3) to empty gastric contents into the duodenum. Fig. 42-9 shows the major anatomical divisions of the stomach.

The stomach has features that allow it to carry out each of these functions. The fundus and the body of the stomach can accommodate volume increases as large as 1.5 L without a marked increase in intragastric pressure. Contractions of the fundus and body are normally weak, so that much of the gastric contents remains relatively unmixed for long periods. In the antrum, however, contractions are vigorous and thoroughly mix antral chyme with gastric juice. The antral contractions serve to empty the gastric contents in small squirts into the duodenal bulb. The rate of gastric emptying is adjusted by a number of mechanisms so that chyme is not delivered to the duodenum more rapidly than it can be dealt with. The physiological mechanisms that underlie this behavior are discussed later.

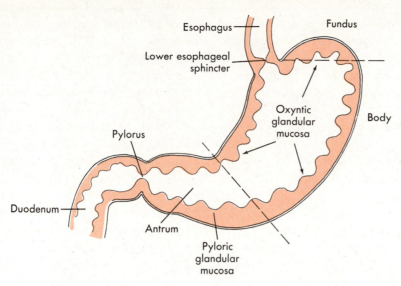

Fig. 42-9 ■ The major anatomical divisions of the stomach.

The basic structure of the gastric wall follows the scheme presented in Fig. 42-1. The circular muscle layer of muscularis externa is more prominent than the longitudinal layer. In addition to the external longitudinal muscle layer, there is an incomplete inner layer of obliquely oriented muscle cells present only on the anterior and posterior sides of the stomach. The muscularis externa of the fundus and body is relatively thin, but that of the antrum is considerably thicker, and it increases in thickness toward the pylorus.

The stomach is richly innervated by extrinsic nerves and by the neurons of the submucosal and myenteric plexuses. Axons from the cells of the intramural plexuses innervate smooth muscle cells and secretory cells of the stomach.

Extrinsic innervation comes via the vagus nerve and from the celiac plexus. In general, cholinergic nerves stimulate gastric smooth muscle motility and gastric secretions, whereas adrenergic fibers inhibit these functions. A number of important gastric transmitter substances remain to be identified.

Numerous afferent fibers leave the stomach in the vagus nerve, and some travel with sympathetic nerves. Other fibers are the afferent links of intrinsic reflex arcs via the intramural plexuses of the stomach. Some of these afferent fibers relay information to the central nervous system about intragastric pressure, gastric distension, or intragastric pH. Pain fibers are also present. The afferent fibers signaling gastric distension and intragastric pressure have been implicated in satiety. It is likely that gastric afferent fibers are significantly involved in other complex responses that are poorly understood at present.

When a wave of esophageal peristalsis reaches the lower esophageal sphincter, the sphincter reflexly relaxes. This is followed by a relaxation of the fundus and body of the stomach, known as *receptive relaxation*. The stomach also will relax if it is directly filled with gas or liquid. Both of these responses are greatly diminished if the vagus nerves are sectioned, so it is believed that the vagi are a major efferent pathway for reflex relaxation of the stomach. The vagal fibers that mediate this response are neither adrenergic nor cholinergic. When the stomach relaxes in response to filling, the sensory afferent fibers that report intragastric pressure and gastric distension are believed to provide the afferent limb of the response.

■ *Structure and innervation of the stomach*

■ *Responses to gastric filling*

■ Mixing and emptying of gastric contents

The muscle layers in the fundus and body are rather thin, and therefore weak contractions are characteristic of these parts of the stomach. As a result, the contents of the fundus and the body tend to form layers based on their density. The gastric contents may remain unmixed for as long as 1 hour after eating. Fats tend to form an oily layer on top of the other gastric contents. Fats thus tend to be emptied later than other gastric contents.

Gastric contractions of significant magnitude usually begin in the middle of the body of the stomach and travel toward the pylorus. The contractions increase in force and velocity as they approach the gastroduodenal junction. As a result, the major mixing activity occurs in the antrum, the contents of which are mixed rapidly and thoroughly with gastric secretions. Immediately after eating, the antral contractions are relatively weak, but as digestion proceeds, they become more forceful. Gastric contraction may be viewed as a peristaltic wave that begins in midbody and moves toward the pylorus. The contraction becomes stronger and more coordinated and travels faster as it nears the pylorus. Typically the frequency of gastric contraction is about three per minute.

Because of the acceleration of the peristaltic wave, the terminal end of the antrum and the pylorus contract almost simultaneously. This is known as *systolic contraction of the antrum*. The peristaltic wave pushes the antral contents ahead of the ring of contraction. Most waves are strong enough to cause a small fraction of antral contents to squirt into the duodenal bulb. The squirt is terminated by the abrupt closure of the pyloric sphincter. Because of its small diameter, the sphincter closes early in the systolic contraction of the antrum. The forceful contraction of the terminal end of the antrum rapidly forces antral contents back into the more proximal part of the antrum. The vigorous backward motion of antral content is called *retropulsion*. Retropulsion causes effective mixing of the antral contents with gastric secretions and serves to break up lumps in the antral contents.

Electrical activity that underlies gastric contractions. The gastric peristaltic waves follow the basic electrical rhythm (slow waves) of the stomach. A group of longitudinal smooth muscle cells located high on the greater curvature serves as a pacemaker for gastric slow waves. The slow waves are conducted over the rest of the stomach via the longitudinal smooth muscle and spread inward to the much thicker circular layer. Depending on the state of excitability of the circular smooth muscle cells, the slow waves will trigger more or fewer action potentials, which will cause more or less vigorous contractions of the circular smooth muscle. In humans the frequency of slow waves is about three per minute. In a given individual at rest the frequency of the slow waves is fairly invariant. Injections of gastrin can increase the frequency of the slow waves to almost four per minute. Secretin can slow the frequency of the slow waves or even abolish them. The physiological roles of the responses to gastrin and secretin have not been established.

As mentioned earlier, if the peak of the slow waves exceeds the threshold for action potential generation, bursts of action potential spikes will occur. The frequency with which the slow waves elicit action potentials and the number of action potentials per burst are determined by the excitability of the gastric smooth muscle. The excitability of gastric smooth muscle is enhanced by acetylcholine from cholinergic fibers and by gastrin. Excitability is inhibited by circulating epinephrine and secretin and by norepinephrine released from sympathetic nerve endings. Norepinephrine acts principally at the level of the myenteric plexus to inhibit neurons that are excitatory to the gastric smooth muscle. The amount of force generated by the gastric smooth muscle is directly proportional to the number of action potentials fired at the peak of the slow wave preceding contraction.

The pylorus separates the gastric antrum from the first part of the duodenum, the duodenal bulb (or cap). It is debatable whether the pylorus constitutes a true anatomical sphincter, but it functions physiologically as a sphincter in many respects. The circular smooth muscle of the pylorus is especially thick and is followed by a connective tissue ring that separates pylorus from duodenum. The mucosa, submucosa, and muscle layers of the pylorus and duodenal bulb are separate, with the exception of a few longitudinal muscle fibers that cross over the junction. There is, however, substantial continuity between the myenteric plexuses of the pylorus and duodenal bulb.

The duodenum has a basic electrical rhythm of 10 to 12 per minute, far faster than the 3 per minute of the stomach. The duodenal bulb appears to be influenced by the basic electrical rhythms of both the stomach and the postbulbar duodenum. It thus contracts somewhat irregularly.

The essential functions of the gastroduodenal junction are (1) to allow the carefully regulated emptying of gastric contents at a rate commensurate with the ability of the duodenum to process the chyme, and (2) to prevent regurgitation of duodenal contents back into the stomach. The gastric mucosa is highly resistant to acid, but it may be damaged by bile acids. The duodenal mucosa has the opposite properties. Thus too rapid gastric emptying may lead to duodenal ulcers, whereas regurgitation of duodenal contents often contributes to gastric ulcers.

■ *The gastroduodenal junction*

The emptying of gastric contents is regulated by both neural and humoral mechanisms. The duodenal mucosa has receptors that sense acidity, osmotic pressure, and fat content.

The presence of triglycerides, phospholipids, or fatty acids in the duodenum results in a dramatic decrease in the rate of gastric emptying. The rate of the gastric slow waves remains essentially unchanged, but the contractility of the antral smooth muscle is greatly diminished, so that the force of gastric contractions is much less.

The chyme that leaves the stomach is usually hypertonic. It becomes more so because of the digestive enzymes of the duodenum. Hypertonic solutions in the duodenum slow gastric emptying.

Duodenal contents with a pH less than about 3.5 result in slower gastric emptying. The presence of amino acids and peptides in the duodenum also may reduce the rate of gastric emptying.

As a result of these mechanisms, (1) fat is not emptied into the duodenum at a rate greater than that at which it can be emulsified by the bile acids and lecithin of the bile; (2) acid is not dumped into the duodenum more rapidly than it can be neutralized by pancreatic and duodenal secretions and by other mechanisms; and (3) in general, there is good correspondence between the rates at which the other components of chyme are presented to the small intestine and the rates at which the small intestine can process those components.

Mechanisms of regulation of gastric emptying. Both neural and hormonal mechanisms are involved in regulating the rate of gastric emptying. The duodenal receptors that initiate this response are discussed later. *Cholecystokinin* and *gastric inhibitory peptide* are released from the duodenal mucosa in response to components of the chyme. Both these hormones at physiological concentrations can decrease the rate of gastric emptying. They are believed to play physiological roles in the regulation of gastric emptying.

Both local and central circuitry are involved in the neural control of gastric emptying. The receptor afferent fibers either synapse with the neurons of the intramural plexuses and trigger intrinsic reflexes or leave the intestine via the vagus or sympathetic nerves to participate in extrinsic reflexes whose efferent pathways are in the vagus and

■ *Regulation of the rate of gastric emptying*

Fig. 42-10 ■ Neural and hormonal inhibition of gastric emptying.

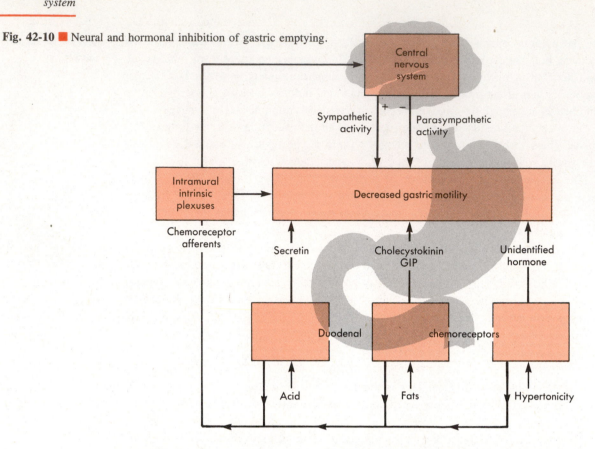

sympathetic nerves. Some of the vagal fibers that inhibit gastric emptying are neither adrenergic nor cholinergic. The neural and hormonal mechanisms that regulate gastric emptying are summarized in Fig. 42-10.

■ *Vomiting*

Vomiting is the expulsion of gastric (and sometimes duodenal) contents from the gastrointestinal tract via the mouth. Vomiting often is preceded by a feeling of nausea, a rapid or irregular heart beat, dizziness, sweating, pallor, and pupillary dilation. Vomiting usually is also preceded by *retching,* in which gastric contents are forced up into the esophagus but do not enter the pharynx. A series of retches of increasing strength often leads to vomiting.

Vomiting is a reflex behavior controlled and coordinated by certain medullary centers. Electrical stimulation of an area of the medulla known as the *vomiting center* elicits prompt vomiting without preliminary retching. Stimulation of another medullary locus leads to retching without subsequent vomiting. It appears that important interactions between these two areas in the medulla occur in typical vomiting. A large number of different areas in the body have receptors that provide afferent input to the vomiting center. Distension of the stomach and duodenum is a strong stimulus that elicits vomiting. Tickling the back of the throat, painful injury to the genitourinary system, dizziness, and certain other stimuli are among the diverse events that can bring about vomiting.

Certain chemicals, called *emetics,* can elicit vomiting. Some emetics do this by stimulating receptors in the stomach or more commonly in the duodenum. The widely used emetic ipecac works via duodenal receptors. Certain other emetics act at the level of the central nervous system on receptors in the floor of the fourth ventricle in an area known as the *chemoreceptor trigger zone*. Major aspects of the control of vomiting are schematically summarized in Fig. 42-11.

A *retch* is initiated by a forced inspiration against a closed glottis and strong contraction of the abdominal mucles. This results in decreased intrathoracic pressure but

Fig. 42-11 ■ Some aspects of control of vomiting.

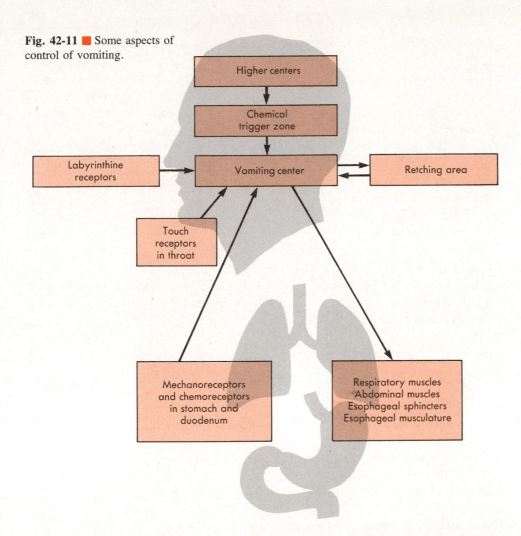

increased pressure in the abdomen. Thus it creates a large pressure difference between stomach and esophagus. The lower esophageal sphincter and the esophagus reflexly relax to receive gastric contents, but the upper esophageal sphincter remains closed, preventing vomiting. When the respiratory and abdominal muscles relax, the esophagus empties itself back into the stomach. Often a sequence of stronger and stronger retches precedes vomiting. A retch that will end in vomiting is terminated with a sharp increase in intraabdominal pressure caused by a sudden and strong contraction of the abdominal muscles that pushes the diaphragm up into the thorax to increase intrathoracic pressure. Then the larynx and hyoid bone are reflexly pulled forward to pull open the upper esophageal sphincter. This allows vomitus to be projected into the pharynx and mouth. This sequence of events may be repeated until the stomach is empty, or it may continue even after the stomach has been emptied. Retching and vomiting with an empty stomach can cause extreme abdominal pain.

The small intestine, particularly the duodenum and jejunum, is the site of most digestion and absorption. The functions of the movements of the small intestine are to mix chyme with digestive secretions, to bring fresh chyme into contact with the absorptive surface of the microvilli, and to propel chyme toward the colon.

The most frequent type of movement of the small intestine is termed *segmentation*. Segmentation is characterized by closely spaced contractions of the circular muscle layer that divides the small intestine into small neighboring segments. In rhythmic segmentation the sites of the circular contractions alternate, so that a given segment of gut

■ *Motility of the small intestine*

contracts and then relaxes. Segmentation is effective at mixing chyme with digestive secretions and bringing fresh chyme into contact with the mucosal surface. *Peristalsis* is the progressive contraction of successive sections of circular smooth muscle that moves along the gastrointestinal tract in an aboral direction. *Short peristaltic waves* do occur in the small intestine, but they usually involve only a short length of intestine. Peristaltic rushes that travel along much of the length of the small bowel are rare.

As in other parts of the digestive tract, the slow waves of the smooth muscle cells play an important role in determining the timing of intestinal contractions. The intramural plexuses can control segmentation and short peristaltic movements in the absence of extrinsic innervation. However, extrinsic nerves mediate certain long-range reflexes and modulate the excitability of small intestinal smooth muscle.

■ Electrical activity of small intestinal smooth muscle

Regular slow waves occur all along the small intestine (Fig. 42-12). The frequency is highest (11 to 13 per minute in humans) in the duodenum, and it declines along the length of the small bowel (to a minimum 8 or 9 per minute in humans in the terminal aspect of the ileum). As in the stomach, the slow waves may or may not be accompanied by bursts of action potential spikes during the depolarizing part of the slow waves. The action potentials, but not the slow waves, elicit the contractions of the circular smooth muscle that cause the major mixing and propulsive movements of the small intestine. Since the action potential bursts occur near the peak of the slow wave, the slow wave frequency determines the maximum possible frequency of intestinal movements. The time interval between intestinal contractions tends to be some multiple of the time interval between slow waves. Action potential bursts tend to be localized to short segments of the intestine. Thus they tend to elicit the highly localized contractions of the circular smooth muscle that cause segmentation.

From the duodenum toward the ileum, there tends to be a phase lag in the small intestinal slow waves. Thus it appears as though the slow waves were propagated in the orthograde direction. When peristaltic contractions do occur, they are propagated along the intestine at the apparent velocity of the slow waves.

The basic electrical rhythm of the small intestine is independent of extrinsic innervation; the extrinsically denervated intestine can carry out segmentation and short peristaltic movements. The frequency of the action potential spike bursts that elicit contractile behavior depends on the excitability of the smooth muscle cells of the small intestine. The excitability of the smooth muscle (and thus the frequency of spikes) is influenced by circulating hormones and by the autonomic nervous system. Excitability is enhanced by acetylcholine released from nerve endings in the smooth muscle layers and by preganglionic parasympathetic nerves acting to increase the activity of excitatory neurons in the intramural plexuses. The excitability is inhibited by sympathetic nerve activity acting primarily to inhibit excitatory neurons in the intramural plexuses. In these

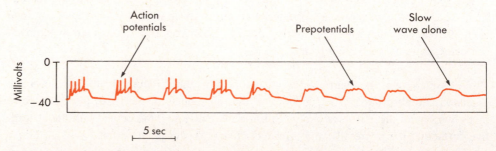

Fig. 42-12 ■ Intracellular microelectrode recordings of electrical activity in isolated longitudinal muscle of rabbit jejunum. Note that the depolarizing phase of the slow wave may elicit action potentials, prepotentials, or neither. (Redrawn from Bortoff, A.: Am. J. Physiol. **201:**203, 1961.)

ways, and probably by other mechanisms as well, the extrinsic innervation of the small intestine (parasympathetic via the vagus nerve and sympathetic nerves from the celiac and superior mesenteric plexuses) plays an important role in modulating contractile activity. The extrinsic neural circuits are essential for certain long-range intestinal reflexes discussed later.

Contractions of the duodenal bulb serve to mix chyme with pancreatic and biliary secretions, and they propel the chyme along the duodenum. Contractions of the duodenal bulb tend to follow antral systolic contractions. This helps prevent regurgitation of duodenal contents back into the stomach. The basic electrical rhythm of the stomach is about 3 per minute, and that in the duodenal bulb is about 11 per minute. Via the longitudinal muscle fibers that cross from stomach to duodenum, the gastric slow waves can influence the duodenal smooth muscle. Thus about every fifth duodenal slow wave is often somewhat increased in amplitude because of the propagated gastric slow wave. The extra depolarization increases the frequency of duodenal action potentials and thus increases the likelihood of a contraction of the duodenal bulb occurring.

Segmentation is the most frequent type of movement by the small intestine. Segmentation is characterized by localized contractions of rings of the circular smooth muscle, which tend to divide the small intestinal contents into oval segments (Fig. 42-13). When the contracted segments relax, neighboring segments may contract. Segmentation occurs at a frequency which is similar to that of the small intestinal slow waves: about 11 or 12 contractions per minute in the duodenum and 8 or 9 contractions per minute in the ileum. The frequency of segmental contractions is not as constant as that of the slow waves. A few sequential contractions tend to be followed by a 5- to 15-second

■ *Contractile behavior of the small intestine*

A

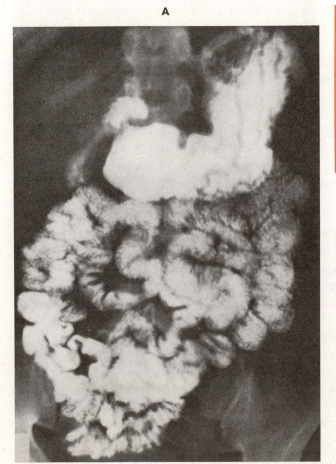

B

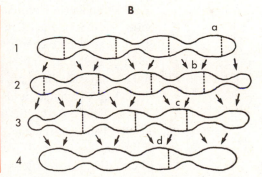

Fig. 42-13 ■ **A,** X-ray view of a normal individual showing the stomach and small intestine filled with barium contrast medium. Note that segmentation of the small intestine divides its contents into ovoid segments. **B,** The sequence of segmental contractions in a portion of a cat small intestine. Lines *1* through *4* indicate successive patterns in time (note the return in line *4* of the same pattern that existed in line *1*). The dotted lines indicate where contractions will occur next. The arrows show the direction of chyme movement. In this case segmentation occurred from 18 to 21 times per minute. (**A** from Gardner, E.M., et al.: Anatomy; a regional study of human structure, ed. 4, Philadelphia, 1975, W.B. Saunders Co.; **B** redrawn from Cannon, W.B.: Am. J. Physiol. **6:**251, 1902.)

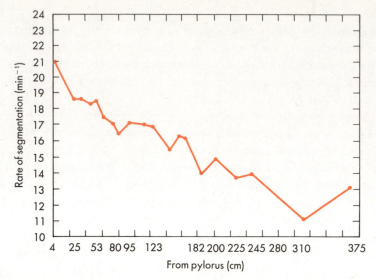

Fig. 42-14 ■ This graph shows the decreasing rate of segmentation along the length of the small intestine of the rabbit. The data were collected from nearly 30 rabbits. There is variability in segmentation rates among animals and with time in the same animal. (Redrawn from Alvarez, W.C.: Am. J. Physiol. **37**:267, 1915.)

period of rest. The decrease in segmental contraction frequency along the small intestine is illustrated in Fig. 42-14.

Segmentation causes a good deal of back-and-forth movement of intestinal contents. It efficiently mixes intestinal contents with digestive secretions and circulates the chyme across the surface of the mucosal epithelium. Because of the phase relationship among small intestinal slow waves, contractions tend to be followed by contraction of a neighboring segment further down the intestine. Such a sequence of contractions has the net effect of propelling chyme in an aboral direction. The decreasing frequency of segmentation as the chyme moves down the small intestine also contributes to net propulsion in the orthograde direction. Thus, whereas segmentation is most effective in mixing and circulating the chyme, it also has a net propulsive effect.

Short-range peristalsis also occurs in the small intestine, although much less frequently than segmentation. It is very rare for a peristaltic wave to traverse a large part of the small intestine. Typically a peristaltic contraction will die out after traveling about 10 cm. The relatively low rate of net propulsion of chyme in the small intestine allows time for digestion and absorption. Whereas intestinal motility is increased after eating, the net rate of chyme movement is actually decreased. This response depends on the extrinsic nerves to the gut.

■ *Intestinal reflexes*

Intestinal reflexes can occur along a considerable length of the gastrointestinal tract. These depend to some extent on the function of both intrinsic and extrinsic nerves.

When a bolus of material is placed in the small intestine, the response may be contraction oral to the bolus and relaxation aboral to the bolus. This may propel the bolus in an aboral direction. This response was first observed by Bayliss and Starling in pioneering investigations and is called the *law of the intestine*. Such a contraction may travel with the bolus like a peristaltic wave.

Overdistension of one segment of the intestine elicits relaxation of the smooth muscle in the rest of the intestine. This response is known as the *intestinointestinal reflex,* and it requires intact extrinsic innervation. There are reflex interactions between the stomach and the terminal aspect of the ileum. Distension of the ileum results in decreased gastric motility, a response called the *ileogastric reflex*.

Elevated secretory and motor functions of the stomach elicit increases in motility of the terminal part of the ileum and an enhanced rate of movement of material through the ileocecal sphincter. This response is called the *gastroileal reflex*. The hormone gastrin may play a role in this response, since gastrin, at physiological levels, increases ileal motility and relaxes the ileocecal sphincter.

Sections of the muscularis mucosae contract irregularly at a rate of about three contractions per minute. These contractions alter the pattern of ridges and folds of the mucosa, bring about local mixing of luminal contents, and bring different parts of the mucosal surface into contact with freshly mixed chyme. Especially in the proximal part of the small intestine, the villi themselves contract irregularly. This is believed to aid in emptying the central lacteals of the villi and thus enhance intestinal lymph flow.

■ *Contractile activity of the muscularis mucosae*

The *ileocecal sphincter* separates the terminal end of the ileum from the cecum, the first part of the colon. Normally the sphincter is closed, but short-range peristalsis in the terminal aspect of the ileum causes relaxation of the sphincter and allows a small amount of chyme to squirt into the cecum. Distension of the cecum causes the sphincter to contract and prevents additional emptying of the ileum. The gastroileal reflex, which enhances ileal emptying after eating, is described earlier. Under normal conditions the ileocecal sphincter allows ileal chyme to enter the colon at a slow enough rate so that the colon can absorb most of the salts and water of the chyme. The ileocecal sphincter is coordinated primarily by the neurons of the intramural plexuses.

■ *Emptying the ileum*

The colon receives about 1500 ml of chyme per day from the terminal part of the ileum. The regulation of transit through the ileocecal sphincter is discussed in the previous section. Most of the salts and water that enter the colon are absorbed; the feces normally contain only about 50 to 100 ml of water each day. Colonic contractions mix the chyme and circulate it across the mucosal surface of the colon. As the chyme becomes semisolid, this mixing resembles a kneading process. The progress of colonic contents is slow, about 5 to 10 cm/hour at most. One to three times daily a wave of contraction, called a mass movement, occurs. A *mass movement* resembles a peristaltic wave in which the contracted segments remain contracted for some time. Mass movements serve to push the contents of a significant length of colon in an aboral direction.

■ *Colonic motility*

As shown in Fig. 42-15, the major subdivisions of the large intestine are the cecum, the ascending colon, the transverse colon, the descending colon, the sigmoid colon, the rectum, and the anal canal. The structure of the wall of the large bowel follows the general plan presented earlier in this chapter, but the longitudinal muscle layer of the muscularis externa is concentrated into three bands called the *teniae coli*. In between the teniae coli the longitudinal layer is quite thin. The myenteric plexus is more dense under the teniae coli. The longitudinal muscle of the rectum and anal canal is substantial and continuous.

■ *Structure and innervation of the large intestine*

The extrinsic innervation of the large intestine is predominantly autonomic. Preganglionic parasympathetic innervation of the cecum and the ascending and transverse colon is via the vagus nerve; that of the descending and sigmoid colon, the rectum, and the anal canal is via the pelvic nerves from the sacral spinal cord. The preganglionic parasympathetic nerves end primarily on neurons of the intramural plexuses, some of which then function as postganglionic parasympathetic nerves. Sympathetic innervation to the large intestine is postganglionic and comes to the proximal part of the bowel via the superior mesenteric ganglion, to the distal part of the large intestine from the inferior mesenteric ganglion, and to the rectum and anal canal from the hypogastric plexus.

Stimulation of sympathetic nerves leads to cessation of colonic movements. Vagal stimulation causes segmental contractions of the proximal part of the colon. Stimulation

Fig. 42-15 ■ Major anatomical subdivisions of the colon.

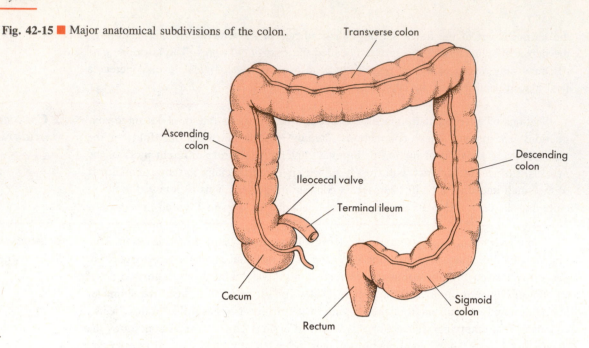

of the pelvic nerves brings about expulsive movements of the colon and sustained contraction of some segments.

The anal canal usually is kept closed by the internal and external anal sphincters. The internal anal sphincter is a thickening of the circular smooth muscle of the anal canal. The external anal sphincter is more distal, and it consists entirely of striated muscle. The external anal sphincter is innervated by somatic motor fibers via the pudendal nerves, which allow it to be controlled both reflexly and voluntarily.

At most times the tonic contraction of the puborectal muscle pulls the upper anal canal forward to cause a sharp angle between the rectum and the anal canal and prevent feces from entering the anal canal. The innervation of the purborectal muscle is the same as that of the internal anal sphincter.

■ *Motility of the cecum and proximal part of the colon*

As discussed in the previous section, the ileocecal sphincter allows intermittent entry of ileal chyme into the cecum. Most contractions of the cecum and proximal part of the large bowel are segmental, and they are more effective at mixing and circulating the contents than at propelling them. The mixing action facilitates absorption of salts and water by the mucosal epithelium.

Localized segmental contractions divide the colon into neighboring ovoid segments called *haustra* (Fig. 42-16). Hence segmentation in the colon is known as *haustration*. The most dramatic difference between haustration and the segmentation that occurs in the small intestine is the regularity of the segments (haustra) produced by haustration and the large length of bowel involved in haustration (Fig. 42-16). There are segmental thickenings of the circular smooth muscle that are probably an anatomical component of haustration. The pattern of haustra is not fixed, however. It fluctuates as segments of the large bowel relax and contract. This results in back-and-forth mixing of luminal contents, known as haustral shuttling. When a few neighboring haustra are emptied in a proximal to distal direction, net propulsion results; this is called *segmental propulsion*. Less frequently a few neighboring haustra will empty because of a concerted contraction of their smooth muscle; this is termed *systolic multihaustral propulsion*.

The net rate of chyme flow in the proximal part of the colon is only about 5 cm/hour in a fasting individual. Colonic propulsive motility increases after eating, and the net rate of flow increases to about 10 cm/hour. About one to three times daily a *mass movement* occurs that empties a large portion of the proximal part of the colon in an

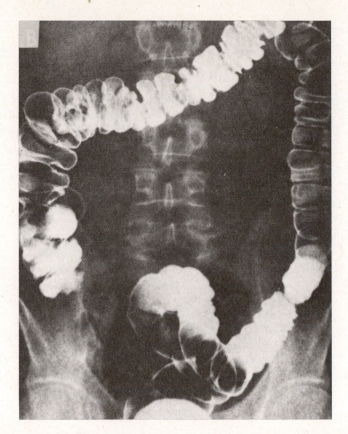

Fig. 42-16 ■ X-ray image of the colon of a normal individual showing a prominent haustral pattern. (Reproduced with permission from Keats, T.E.: An atlas of normal roentgen variants, ed. 2. Copyright © 1979 by Year Book Medical Publishers, Inc., Chicago.)

aboral direction. A mass movement resembles a peristaltic wave in which the contracted segments of the gut remain contracted for a longer period than in peristalsis.

In the normal course of events the distal part of the colon is filled with semisolid feces by a mass movement. Segmental contractions serve to knead the feces, facilitating absorption of remaining salts and water. About one to three times daily mass movements occur and sweep the feces toward the rectum.

■ *Motility of the distal part of the colon*

As in the other parts of the gastrointestinal tract, the motility of the colon is controlled by intrinsic and extrinsic nerves, hormones and neurotransmitters, and properties intrinsic to the smooth muscle of the gut.

Variations in resting membrane potential and in the frequency of action potential spikes do occur. The variations in membrane potential are not regular enough to constitute a basic electrical rhythm. As in the small intestine, much of the motility behavior can be controlled by the intramural plexuses: haustration and mass movements occur in the absence of extrinsic innervation. The effects of plexus neurons may be mostly inhibitory; in Hirschsprung's disease, in which the intramural plexuses are deficient, there is tonic contraction of the colon, which causes intestinal obstruction.

Hormones and neurotransmitters also influence colonic motility. Epinephrine and norepinephrine inhibit colonic motility. Gastrin in physiological doses enhances colonic motility, and it probably plays a role in the increased propulsive activity that follows a meal. Prostaglandins and other physiological substances affect colonic contractions, but their role in normal function remains to be determined.

■ *Control of colonic movements*

It is apparent that higher centers of the nervous system can influence colonic behavior. The efferent pathways for these influences are via the extrinsic nerves to the colon. The effects of emotional stress on colonic motility are well known.

■ The rectum and anal canal

The rectum is usually empty, or nearly so. The rectum is more active in segmental contractions than is the sigmoid colon. Thus rectal contents tend to move retrograde into the sigmoid colon. The anal canal is tightly closed by the anal sphincters. Prior to defecation the rectum is filled as a result of a mass movement in the sigmoid colon. The distension of the rectum when it fills brings about reflex relaxation of the internal anal sphincter and reflex constriction of the external anal sphincter (Fig. 42-17) and causes the urge to defecate. Persons who lack functional motor nerves to the external anal sphincter defecate involuntarily when the rectum is filled. The reflex reactions of the sphincters to rectal distension are transient. If defecation is postponed, the sphincters regain their normal tone, and the urge to defecate temporarily subsides.

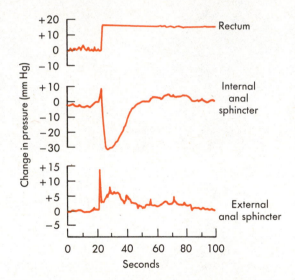

Fig. 42-17 ■ Responses of internal and external anal sphincters to a prolonged distension of the rectum. Note that the responses of the sphincters are transient. (Redrawn from Schuster, M.M., et al.: Bull. Johns Hopkins Hosp. **116:**79, 1965.)

■ Defecation

When an individual feels the circumstances are appropriate, he or she voluntarily relaxes the external anal sphincter to allow defecation to proceed. Defecation is a complex behavior involving both reflex and voluntary actions. The integrating center for the reflex actions is in the sacral spinal cord but is modulated by higher centers. The efferent pathways are cholinergic parasympathetic fibers in the pelvic nerves. The sympathetic nervous system does not play a significant role in normal defecation.

Before defecation the smooth muscle layers of the descending colon and the sigmoid colon contract in a mass movement to force feces toward the anus. The distension of the rectum signals the urge to defecate. The internal and external sphincters relax, the internal sphincter reflexly and the external sphincter voluntarily. The puborectal muscle reflexly relaxes and allows alignment of the rectum and the anal canal.

Voluntary actions are also important in defecation. The external anal sphincter is voluntarily held in the relaxed state. Intraabdominal pressure is elevated to aid in expulsion of feces. Evacuation is normally preceded by a deep breath, which moves the diaphragm downward. The glottis is then closed, and contractions of the respiratory muscles on full lungs elevates both the intrathoracic and the intraabdominal pressure. Contractions of the muscles of the abdominal wall further increase intraabdominal pressure, which may be as large as 200 cm H_2O and helps to force feces through the relaxed sphincters. The muscles of the pelvic floor are relaxed to allow the floor to drop. This helps to straighten out the rectum and prevent rectal prolapse.

■ *Bibliography*

Journal articles

Wood, J.D.: Intrinsic control of intestinal motility, Annu. Rev. Physiol. **43:**33, 1981.

Books and monographs

Atanassova, E., and Papasova, M.: Gastrointestinal motility. In Crane, R.K., editor: Gastrointestinal physiology II, Baltimore, 1977, University Park Press.

Davenport, H.W.: Physiology of the digestive tract, ed. 4, Chicago, 1977, Year Book Medical Publishers, Inc.

Duthie, H.L., and Wormsley, K.G., editors: Scientific basis of gastroenterology, Edinburgh, 1979, Churchill Livingstone.

Weisbrodt, N.W.: Gastrointestinal motility. In Guyton, A.C., Jacobson, E.D., and Shanbour, L.L., editors: Gastrointestinal physiology, London, 1974, Butterworths.

chapter 43

Gastrointestinal secretions

■ *Secretion of saliva*

Saliva lubricates food for greater ease of swallowing. It also facilitates speaking. Saliva contains ptyalin, an α-amylase, which begins the digestion of starch. In people lacking functional salivary glands, a condition called xerostomia (dry mouth), there is a prevalence of dental caries and infections of the buccal mucosa. Secretion of saliva is an active process. The ionic composition of the fluid produced by the acinar cells is similar to that of plasma. The epithelial cells that line the ducts of the salivary glands make considerable modifications in the ionic composition as the saliva flows by. The functions of the salivary glands are primarily controlled by the autonomic nervous system; both sympathetic and parasympathetic stimulation enhance the overall rate of salivary secretion. Physiologically the parasympathetic system plays by far the more important role.

■ *The major salivary glands and their structure*

In humans the parotid glands, the largest salivary glands, are entirely serous glands. Their watery secretion lacks mucins. The submaxillary and sublingual glands are mixed mucous and serous glands, and they secrete a more viscous saliva containing mucins.

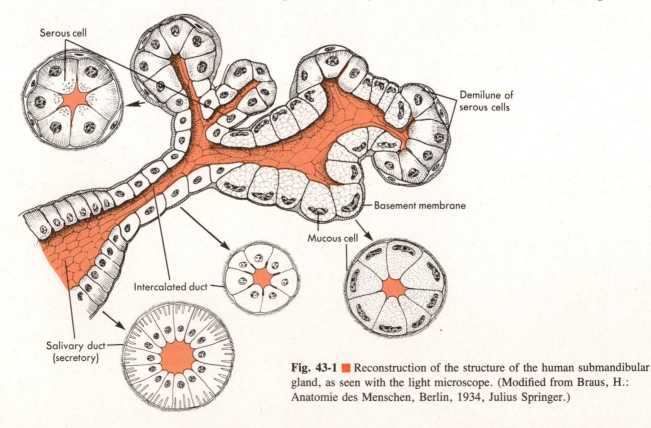

Serous cell

Demilune of serous cells

Basement membrane

Mucous cell

Intercalated duct

Salivary duct (secretory)

Fig. 43-1 ■ Reconstruction of the structure of the human submandibular gland, as seen with the light microscope. (Modified from Braus, H.: Anatomie des Menschen, Berlin, 1934, Julius Springer.)

Many smaller salivary glands are present in the oral cavity. The microscopic structure of mixed salivary glands is schematically depicted in Fig. 43-1. The salivary glands structurally resemble the exocrine pancreas in many respects. The serous acinar cells have zymogen granules that contain ptyalin and perhaps certain other salivary proteins as well. Mucous acinar cells secrete glycoprotein mucins into the saliva. The ducts that drain the acini are lined with columnar epithelial cells, which process the saliva to alter its ionic composition. The intercalated ducts drain the acini into somewhat larger ducts, which empty into still larger ducts, and so forth. A single large duct brings the secretions of each gland into the mouth.

The salivary glands have an extremely high blood flow. The capillaries that supply the ducts typically coalesce into venules, which in turn break up into another set of capillaries that supply the acini. The acini are thus partly supplied by a portal circulation.

■ *Functions of saliva*

The mucins (glycoproteins) produced by the submaxillary and sublingual glands lubricate food so that it may be more readily swallowed. The major digestive function of saliva results from the action of ptyalin on starch. Ptyalin is an enzyme that has the same specificity as the α-amylase of pancreatic juice. Ptyalin cleaves the internal α-1,4 glycosidic linkages in starch, but it cannot hydrolyze the terminal α-1,4 linkages or the α-1,6 linkages at the branch points. The major products of ptyalin action are maltose, maltotriose, and oligosaccharides containing an α-1,6 branch point (called α-limit dextrins). The pH optimum of ptyalin is about 7, but it has activity between pH levels of 4 and 11. Ptyalin action continues in the mass of food in the stomach, and it is terminated only when the contents of the antrum are mixed with enough gastric acid to lower the pH of the antral contents to below 4. More than half the starch in a well-chewed meal may be reduced to small oligosaccharides by ptyalin action. However, because of the large capacity of the pancreatic α-amylase to digest starch in the small intestine, the absence of ptyalin causes no malabsorption of starch.

■ *Metabolism and blood flow of salivary glands*

For their size the salivary glands produce a prodigious volume flow of saliva: the maximal rate in humans is about 1 ml/minute/g of gland. Salivary glands have a high rate of metabolism and a high blood flow, both being proportional to the rate of saliva formation. The blood flow to maximally secreting salivary glands is approximately 10 times that of an equal mass of actively contracting skeletal muscle. When parasympathetic nerves to salivary glands are stimulated, there is a prompt and massive increase in blood flow through the salivary glands that is caused by dilation of the vasculature of the glands. There is evidence that secreting glands release the protease *kallikrein*, which acts on the plasma protein kininogen to release the octapeptide vasodilator *bradykinin*. Bradykinin may contribute to vasodilation during the period of secretory activity.

■ *The ionic composition of saliva and secretion of water and electrolytes*

In humans saliva is always hypotonic to plasma. As shown in Fig. 43-2, the Na^+ and Cl^- concentrations are less than those of plasma, but K^+ and HCO_3^- concentrations are greater than those of plasma. There is considerable variation in the tonicity of saliva and in its ionic composition from species to species and from one salivary gland to another in the same species. The greater the rate of secretory flow, the higher the tonicity of the saliva; at maximal flow rates the tonicity of saliva in humans is about 70% of that of plasma. The pH of saliva from resting glands is slightly acidic. During active secretion, however, the saliva becomes basic, with pH approaching 8. The increase in pH with secretory flow rate is partly caused by the increase in salivary HCO_3^- concentration. The concentration of K^+ in saliva is almost independent of salivary flow rate over a wide range of flows.

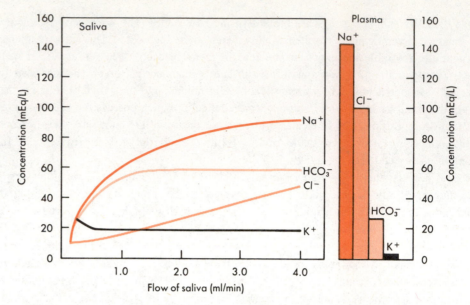

Fig. 43-2 ■ Average composition of the parotid saliva of three women as a function of the rate of salivary flow. (From Thaysen, J.H., et al.: Am. J. Physiol. **178:**155, 1954.)

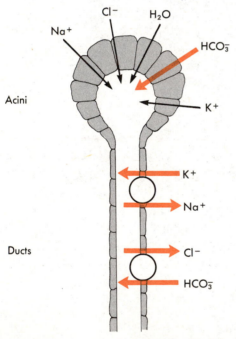

Fig. 43-3 ■ Some of the major ionic transport processes involved in the formation of saliva. Fluxes believed to be active are shown in color.

Studies in which small samples of fluid are taken from the intercalated ducts of rat salivary glands by micropuncture show that the intercalated ducts contain a fluid which resembles plasma in its concentration of Na^+, K^+, Cl^-, and HCO_3^-. Some investigators maintain that acinar fluid is hypertonic primarily because of levels of Na^+ and Cl^- which exceed those in plasma and that the hypertonicity induces water secretion into the lumens of the acini. This dilution and the equilibration of ions across the walls of the intercalated ducts apparently bring the ionic composition of the fluid in the intercalated ducts rapidly toward the ionic composition of plasma. At this point the saliva has most of its final water content. As the forming saliva flows down the intercalated ducts and down the larger ducts subsequent to them, ionic exchanges alter the ionic composition of saliva. In the ducts Na^+ is actively absorbed from the lumen in exhange for K^+, and HCO_3^- is actively secreted into the lumen partly in exchange for Cl^- (Fig. 43-3). Some investigators believe that some of the K^+ secreted is exchanged for H^+. The net effect is the absorption of more ions from the forming saliva than are secreted into it, making the saliva hypotonic. The entire duct system, including the larger ducts, appears to carry out these ionic exchanges. Aldosterone stimulates the absorption of Na^+ in exchange for K^+ in the ducts. The faster the flow rate of the saliva in the ducts, the less complete are the transport processes, and the closer to isotonicity is the saliva. In spite of the gradient for osmotic water flow out of the saliva, relatively little water is absorbed in the ducts, primarily because of the low permeability of the ductular epithelium to water.

■ *Secretion of ptyalin*

Serous acinar cells have zymogen granules that contain ptyalin in their apical cytoplasm. Formation of zymogen granules proceeds by the classic pathway elucidated by Palade, Siekevitz, and their co-workers. The enzyme is synthesized on the ribosomes of rough endoplasmic reticulum and enters the cisternae of the endoplasmic reticulum. Smooth vesicles containing the newly synthesized enzyme molecules move toward the Golgi apparatus, where the enzymes become encapsulated in membrane-bound vacuoles. The contents of the vacuoles are condensed, and the resulting vesicles take up residence in the apical cytoplasm of the cell, where they are recognizable as zymogen granules. When the gland is stimulated to secrete, the zymogen granules fuse with the plasma membrane. Their contents are released into the lumen of an acinus by exocytosis. Ptyalin also may be secreted by a pathway that is independent of the zymogen granules.

■ *Nervous control of
salivary gland function*

With the exception of the vasodilator action of bradykinin and the stimulation of ductal Na^+ and K^+ transport by aldosterone just mentioned, the physiological control of the salivary glands is solely effected by the autonomic nervous system. Stimulation of either sympathetic or parasympathetic nerves to the salivary glands stimulates salivary secretion, but the effects of the parasympathetic nerves is stronger and more long lasting. Interruption of the sympathetic nerves causes no defect in the function of the salivary glands, so the essential physiological control is by way of the parasympathetic nervous system. If the parasympathetic supply is interrupted, the salivary glands atrophy.

Postganglionic sympathetic fibers to the salivary gland come from the superior cervical ganglion. Preganglionic parasympathetic fibers come via branches of the facial and glossopharyngeal nerves (cranial nerves VII and IX, respectively), and they synapse with postganglionic neurons in or near the salivary glands. The acinar cells and ducts are supplied with parasympathetic nerve endings.

Parasympathetic stimulation increases the synthesis and secretion of ptyalin and mucins, enhances the transport activities of the ductular epithelium, greatly increases blood flow to the glands, and stimulates glandular metabolism and growth.

Sympathetic stimulation and circulating catecholamines also stimulate salivary secretion, primarily via β-adrenergic receptors. Sympathetic stimulation causes contraction of myoepithelial cells around the acini and ducts and constriction of blood vessels, with consequent reductions in salivary gland blood flow. The stimulation of salivary secretion that results from stimulation of sympathetic nerves is transient.

■ *Electrophysiological effects
of nerve stimulation and
stimulus-secretion coupling*

Acinar cells have a resting membrane potential that is typically between -20 and -35 mV (cytoplasm negative). The low resting membrane potential is apparently caused by a higher ratio of sodium conductance to potassium conductance than is typical of most electrically excitable cells. On stimulation of parasympathetic nerves or administration of cholinergic drugs there is a graded hyperpolarization of the acinar cells. This hyperpolarization is called the *secretory potential*. During the secretory potential the conductances of the acinar cell membrane to both Na^+ and K^+ are increased, with the K^+ conductance being increased more than the Na^+ conductance. The epithelial cells of the duct system have the opposite response to nerve stimulation; that is, they are depolarized, because the Na^+ conductance increases more than the K^+ conductance.

The linkage between the secretory potential in the acini and ducts and their transport of water and electrolytes is not well understood. Ca^{++} apparently is involved in the coupling of electrical and secretory events, since depletion of glandular Ca^{++} leads to cessation of fluid secretion with little or no change in the secretory potential elicited by nerve stimulation.

Stimulation of sympathetic nerves or administration of adrenergic agents (especially β-agonists) strongly potentiates the secretion of a saliva rich in K^+, HCO_3^-, and ptyalin. The stimulatory effect is only transient, however, in marked contrast to the prolonged effects of cholinergic stimulation.

β-Adrenergic agents and stimulation of sympathetic nerves cause an increase in the adenylate cyclase activity in acinar and duct cells and an increase in the intracellular concentration of cyclic AMP. Cholinergic stimulation, on the other hand, increases cyclic GMP levels in acinar and duct cells. Whether increases in cyclic AMP and cyclic GMP levels are responsible for the enhanced synthesis and secretion of proteins and transport of electrolytes is not known at present. Intracellular Ca^{++} is involved in the responses to both cholinergic and adrenergic stimulation. Perhaps Ca^{++}-calmodulin–regulated protein kinases play a role in stimulus-secretion coupling in salivary glands. The relationships among electrical events at the plasma membrane and processes influenced by cyclic nucleotides and Ca^{++} remain to be elucidated.

■ *Gastric secretion*

The stomach performs a number of functions, none of which are essential for normal digestion. The stomach serves as a reservoir, allowing the ingestion of large meals. It empties its contents into the duodenum at a rate consistent with the ability of the duodenum and small intestine to deal with them. The major secretions of the stomach are HCl, pepsinogens, intrinsic factor, and mucus. HC1 catalyzes the cleavage of pepsinogens to active pepsins and provides a low pH, at which *pepsins* can begin digestion of proteins. Pepsin is not required for normal digestion, since in its absence ingested protein is completely digested by pancreatic proteases and small-intestinal brush border peptidases. *Intrinsic factor* is a glycoprotein that binds vitamin B_{12} and allows it to be absorbed by the ileal epithelium. Intrinsic factor is the only gastric secretion required for normal health. The hormone gastrin is secreted by G cells in the antrum and plays an important role in regulating gastric acid secretion. *Mucous secretions* protect the gastric mucosa from mechanical and chemical destruction.

■ *Structure of the gastric mucosa*

The surface of the gastric mucosa is covered by columnar epithelial cells (which secrete mucus and an alkaline fluid) and is studded with gastric pits (Fig. 43-4) where gastric glands empty into the lumen. The pits are so numerous as to account for a significant fraction of the total surface area. Each pit contains the opening of a duct into which one or more gastric glands empty. Fig. 43-4, *B,* shows schematically the structure of a gastric gland. The surface epithelial cells extend a bit into the duct opening. In the narrow neck of the gland are the *mucous neck cells,* which secrete mucus. Still deeper in the gland, along its walls, are the *parietal* or *oxyntic cells,* which secrete HCl and intrinsic factor, and the *chief* or *peptic cells,* which secrete pepsinogen.

The mucosa may be divided into the oxyntic (acid-secreting) glandular region, above the notch, and the pyloric glandular area, below the notch. Parietal (oxyntic) cells are present primarily in the oxyntic glandular area, and they are especially numerous in the fundus. The glands in the pyloric glandular area have very few parietal cells; chief cells predominate in the glands there.

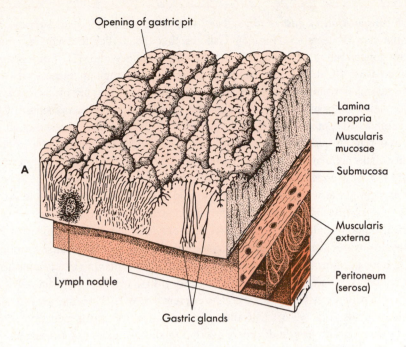

Opening of gastric pit

Lamina
propria

Muscularis
mucosae

Submucosa

A

Muscularis
externa

Peritoneum
(serosa)

Lymph nodule

Gastric glands

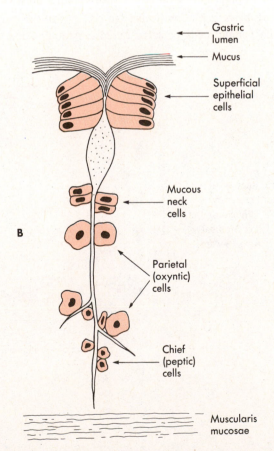

Gastric
lumen

Mucus

Superficial
epithelial
cells

Mucous
neck
cells

B

Parietal
(oxyntic)
cells

Chief
(peptic)
cells

Muscularis
mucosae

Fig. 43-4 ■ Structure of the gastric mucosa. **A,** Reconstruction of part of the gastric wall.
B, Schematic depiction of a gastric gland showing the different cell types. (**A** redrawn from
Braus, H.: Anatomie des Menschen, Berlin, 1934, Julius Springer; **B** redrawn from Johnson,
L.R., editor: Gastrointestinal physiology, 1977, St. Louis, The C.V. Mosby Co.)

■ *Gastric acid secretion*

Ionic composition of gastric juice. The ionic composition of gastric juice is a function of the rate of secretion. Fig. 43-5 shows that the higher the secretory rate, the higher the concentration of hydrogen ion. At lower secretory rates $[H^+]$ diminishes, and $[Na^+]$ increases. $[K^+]$ in gastric juice is always higher than in plasma, and consequently prolonged vomiting may lead to hypokalemia. At all rates of secretion Cl^- is the major anion of gastric juice. At high rates of secretion the composition of gastric juice resembles that of an isotonic solution of HCl. At low rates of secretion gastric juice is hypotonic to plasma. When secretion is stimulated, the gastric juice approaches isotonicity. The cellular mechanisms by which the ionic composition of gastric juice varies are not well understood. Gastric HCl serves to convert pepsinogen to pepsin, to provide an acid pH at which pepsin is active, and to kill most ingested bacteria.

Rate of secretion of gastric acid. The rate at which gastric acid is secreted varies considerably among individuals, partly because of variations in the number of parietal cells. Basal (unstimulated) rates of gastric acid production typically range from about 1 to 5 mEq/hour in humans. On maximal stimulation with histamine or pentagastrin, production rises to 6 to 40 mEq/hour. Patients with gastric ulcers secrete less HCl on the average, and patients with duodenal ulcers secrete more HCl on the average than do normal individuals (Fig. 43-6). Unfortunately the ranges of acid production in normal subjects and those with gastric disorders overlap to such an extent that measurements of gastric acid production often are not useful in diagnosis.

Morphological changes that accompany gastric acid secretion. Parietal cells have a distinctive ultrastructure. They have an elaborate system of branching *secretory canaliculi,* which course through the cytoplasm and are connected by a common outlet to the luminal surface of the cell. Microvilli line the surfaces of the canaliculi. In addition, the cytoplasm of the parietal cells contains extensive tubules and vesicles—the *tubovesicular system.*

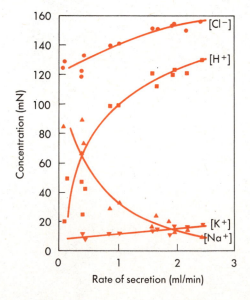

Fig. 43-5 ■ Concentrations of major ions in the gastric juice of a normal young man as a function of the rate of secretion. (Redrawn and reproduced with permission from Davenport, H.W.: Physiology of the digestive tract, ed. 5. Copyright © 1982 by Year Book Medical Publishers, Inc., Chicago; adapted from Nordgren, B.: Acta Physiol. Scand. **58**[suppl. 202]: 1, 1963.)

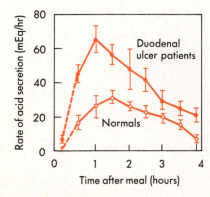

Fig. 43-6 ■ Rate of gastric acid secretion after a meal in six normal subjects and seven patients with duodenal ulcers. (Redrawn from Fordtran, J.S., and Walsh, J.H.: J. Clin. Invest. **52**:645, 1973 by copyright permission of The American Society for Clinical Investigation.)

When parietal cells are stimulated to secrete, a pronounced morphological change occurs. Tubules and vesicles of the tubovesicular system fuse with the plasma membrane of the secretory canaliculi, greatly diminishing the content of tubovesicles and greatly increasing the surface area of the secretory canaliculi. If, as some evidence suggests, the tubovesicles contain the HCl secretory apparatus, then the extensive membrane fusion that occurs on stimulation would greatly increase the number of HCl pumping sites available at the surface of the secretory canaliculi.

The cellular mechanism of gastric acid secretion. The mucosal surface of the stomach is always electrically negative with respect to the serosal surface. In the resting stomach the mucosa is -60 to -80 mV (negative with respect to the serosa). When acid secretion is stimulated, the potential difference falls to -30 to -50 mV. Thus Cl^-, the ultimate source of which is the plasma, is transported from the extracellular fluid into the lumen of the stomach against both electrical and concentration gradients. Hydrogen ion moves down an electrical gradient into the lumen of the stomach but against a much larger chemical concentration gradient. At maximal rates of secretion H^+ is pumped against a concentration gradient that is more than one million to one. Thus metabolic energy is required for transport of both H^+ and Cl^-.

One current model of the mechanism of gastric acid transport is shown in Fig. 43-7. The apical membrane of the parietal cell (the membrane facing the secretory canaliculi) contains a K^+-activated ATPase, which is capable of the active exchange of H^+ for K^+. This ATPase appears to be the primary H^+ pump. Both H^+ and K^+ are pumped against their electrochemical potential gradients.

When H^+ is pumped out of the parietal cell, this leaves behind an excess of HCO_3^-. Bicarbonate flows down its electrochemical potential difference across the basolateral plasma membrane. The protein that mediates HCO_3^- efflux transports another anion in the opposite direction. Since Cl^- is the major anion of the extracellular fluid, Cl^- exchanges for HCO_3^-. Chloride moves against its electrochemical potential gradient into the cell, the energy for the active transport of Cl^- coming from the downhill movement of HCO_3^-. The HCO_3^- that leaves the parietal cell is carried away in the blood. During active secretion the pH of venous blood leaving the stomach is elevated. This phenomenon is called the "alkaline tide."

Some of the entry of Cl^- into the parietal cell depends on the presence of Na^+ in the serosal fluid. Some investigators feel there is a Na^+/Cl^- cotransport system in the basolateral plasma membrane. Na^+ enters the cell down its electrochemical potential gradient and provides the energy to bring Cl^- into the cell against its electrochemical potential gradient. The existence of the Na^+ gradient depends on the action of the Na^+, K^+-ATPase in the basolateral plasma membrane.

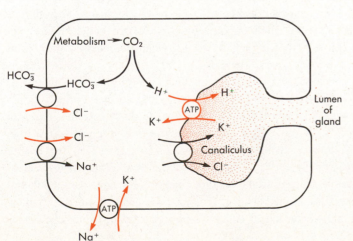

Fig. 43-7 ■ Hypothetical scheme for HCl secretion by the parietal cell. Ion fluxes against an electrochemical potential gradient are shown in color.

As a result of the action of the Cl^-/HCO_3^- exchange pump and the Na^+/Cl^- co-transport protein, Cl^- is concentrated in the cytoplasm of the parietal cell. It leaves the parietal cell at the apical membrane by facilitated transport. There is evidence that the efflux of Cl^- is coupled to the efflux of K^+ into the secretory canaliculus, with both ions moving down their electrochemical potential gradients. Fig. 43-7 shows a single protein mediating the coupled efflux of K^+ and Cl^-, but the existence of such a protein is controversial at present.

■ Pepsins

The pepsins are a group of proteases secreted primarily by the chief cells of the gastric glands and often collectively referred to as "pepsin." The pepsins fall into two electrophoretic classes: group I, secreted mostly in the oxyntic glandular mucosa, and group II, secreted throughout the stomach and by Brunner's glands of the duodenum.

Pepsins are secreted as inactive proenzymes known as *pepsinogens*. Cleavage of acid-labile linkages, because of gastric acidity, converts pepsinogens to pepsins; the lower the pH, the more rapid is the conversion. Pepsins also act proteolytically on pepsinogens to form more pepsins.

The pepsins have unusually low pH optima; they have their highest proteolytic activity at pH 3 and below. Pepsins may digest as much as 10% to 20% of the protein in a typical meal. When the duodenal contents are neutralized, pepsins are irreversibly inactivated.

Pepsinogens are contained in membrane-bound zymogen granules in the chief cells. The contents of the zymogen granules are released by exocytosis when the chief cells are stimulated to secrete. Chief cells without zymogen granules also can secrete pepsinogens.

■ Intrinsic factor

Intrinsic factor is a glycoprotein that has a molecular weight of about 55,000 and is secreted by the parietal cells of the stomach. Intrinsic factor is required for normal intestinal absorption of vitamin B_{12}. Vitamin B_{12} (in all its physiological forms) binds to intrinsic factor. The intrinsic factor–B_{12} complex is highly resistant to digestion. Receptors in the mucosa of the ileum bind the complex, and the B_{12} is taken up by the ileal mucosal epithelial cells. Intrinsic factor is released in response to the same stimuli that evoke secretion of gastric acid from the parietal cells.

■ Secretion of mucus

Secretions that contain glycoprotein mucins are viscous and sticky and are collectively termed mucus. Mucus adheres to the gastric mucosa and helps to protect the mucosal surface from abrasion by lumps of food and from chemical damage caused by the acid and digestive enzymes.

Mucous neck cells in the necks of gastric glands secrete a clear mucus sometimes called *soluble mucus*. Soluble mucus is not present in the resting stomach. Secretion of soluble mucus is stimulated by sham feeding and by some of the same stimuli that enhance acid and pepsinogen secretion, especially by acetylcholine released from parasympathetic nerve endings near the gastric glands.

The surface epithelial cells secrete a mucus containing different mucins. This mucus is cloudy in appearance and thus is termed *visible mucus*. The surface epithelial cells also secrete watery fluid with Na^+ and Cl^- concentrations similar to plasma, but with higher K^+ (four times) and HCO_3^- (two times) concentrations than in plasma. The high $[HCO_3^-]$ makes the visible mucus alkaline. Visible mucus is secreted by the resting mucosa and lines the stomach with a sticky, viscous, alkaline coat. When food is eaten, the rates of secretion of visible mucus and the alkaline secretion by the surface epithelial cells increase. Among the stimuli that enhance the rate of secretion are mechanical stimulation of the mucosa and stimulation of either sympathetic or parasympathetic nerves to the stomach.

The mucus, both soluble and visible, forms a coat on the mucosa that protects it from mechanical damage from chunks of food. The alkaline fluid it entraps protects against damage to the mucosa by HC1 and pepsin. The mucus and alkaline secretions are part of the *gastric mucosal barrier* that prevents damage to the mucosa by gastric contents.

■ *Control of acid and pepsin secretion*

There is usually a good correlation between the rates of secretion of gastric acid and pepsin. Acetylcholine, released by vagal fibers and by neurons of the intrinsic plexuses, is a strong stimulant of both acid and pepsinogen secretion. Acid itself, acting via local neural reflexes, enhances the release of pepsinogen from the chief cells. The antral hormone gastrin is also an important stimulus for acid secretion. Secretion of gastric acid is inhibited by neural and hormonal reflexes elicited by the presence of acid in the stomach and in the duodenum. To discuss in more detail the mechanisms that regulate gastric secretion, it is convenient to consider three phases: the *cephalic phase* (mechanisms elicited before food reaches the stomach), the *gastric phase* (mechanisms elicited while food is in the stomach), and the *intestinal phase* (mechanisms elicited when gastric contents are emptied into the duodenum).

The cephalic phase of gastric secretion is mediated almost entirely by impulses in the vagus nerves, the final common pathway for the response. Chemoreceptors triggered by the taste and smell of food and mechanoreceptors stimulated in chewing and swallowing send their afferent impulses to the brain and reflexly elicit vagal impulses to the stomach. The efferent fibers in the vagus are predominantly preganglionic and cholinergic. They synapse with neurons in the myenteric plexus and, to a lesser extent, in the submucosal plexus. These neurons serve as postganglionic parasympathetic nerves and are also predominantly cholinergic.

Release of acetylcholine near parietal and chief cells directly elicits the secretion of HC1 and pepsinogen, respectively. Acetylcholine released near the G cells in the antral mucosa elicits secretion of gastrin, which is released into the circulation and stimulates parietal cells to secrete acid. The response of the parietal cells to gastrin is enhanced in the presence of acetylcholine. In the absence of food in the stomach the pH of gastric contents rapidly falls, and acid juice in contact with the antral mucosa shuts off the release of gastrin, so that the total amount of gastrin released during the cephalic phase may be small.

The gastric phase of gastric secretion is brought about by the presence of food in the stomach, the principal stimuli being distension of the stomach and the presence of amino acids and peptides resulting from the actions of pepsins. Most of the acid secreted in response to a meal is secreted during the gastric phase.

When either the body or the antrum is distended, mechanoreceptors are stimulated. These mechanoreceptors serve as the afferent arms of both local and central reflexes that bring about secretory events. Both local and central responses are predominantly cholinergic. Central reflexes have both afferent and efferent pathways in the vagus nerve.

When the oxyntic glandular area is distended, local and central reflexes bring about release of acetylcholine near parietal and chief cells, which then release acid and pepsinogen, respectively. Distension of the body (oxyntic gland area) brings about gastrin release from the antral mucosa via a vagal reflex.

Distension of the pyloric glandular area (antrum) brings about enhanced gastrin release and, via a vagal reflex, increased acid secretion by the oxyntic glandular mucosa.

All the responses elicited by gastric distension can be blocked effectively by bathing the mucosal surface with an acid solution with pH 2 or less. Once the buffering capacity of gastric contents is saturated, gastric pH will fall rapidly and greatly inhibit further acid release. In this way the acidity of gastric contents regulates itself to some extent. In patients with duodenal ulcers, acid secretion is less inhibited by the presence of acid

TABLE 43-1

Major mechanisms for stimulation of gastric acid secretion

Phase	Stimulus	Pathway	Stimulus to parietal cell
Cephalic	Chewing, swallowing, etc.	Vagus nerve to 1. Parietal cells 2. G cells	Acetylcholine Gastrin
Gastric	Gastric distension	Local and vagovagal reflexes to 1. Parietal cells 2. G cells	Acetylcholine Gastrin
Intestinal	Protein digestion products in duodenum	1. Intestinal G cells 2. Intestinal endocrine cells	Gastrin Entero-oxyntin

Modified from Johnson, L.R., editor: Gastrointestinal physiology, ed. 2, St. Louis, 1981, The C. V. Mosby Co.; adapted from M.I. Grossman.

in the antrum. Amino acids and peptides, products of the actions of pepsins on ingested proteins, directly stimulate the parietal cells to secrete acid. The cellular mechanism of this response is not understood. Intact proteins do not have this effect. This effect is not cholinergically mediated, but it is blocked when gastric pH falls to 2 or below.

Pepsinogen secretion is tied to acid secretion in two ways. First, acetylcholine is the most potent stimulus for pepsinogen release. Most stimuli that elicit acetylcholine release do so at both parietal and chief cells. Second, acid in contact with the gastric mucosa elicits a local cholinergic reflex that enhances pepsinogen release. There is usually a high correlation between the rates of acid and pepsinogen release, which presumably ensures that pepsinogen is released when there is sufficient acid present to activate it. The hormone *secretin,* released by the duodenal mucosa in response to acid, stimulates pepsinogen release but inhibits gastric acid secretion.

In *the intestinal phase of gastric secretion* the presence of chyme in the duodenum brings about neural and endocrine responses that first stimulate and later inhibit secretion of acid and pepsinogen by the stomach. Early in gastric emptying, when gastric chyme is lower in acidity (pH 3 or above), the stimulatory influences predominate. Later, when the buffer capacity of gastric chyme is exhausted and the pH of chyme emptied into the duodenum falls below pH 2, inhibitory influences prevail. Tables 43-1 and 43-2 summarize major mechanisms that control gastric acid secretion.

Stimulation of gastric acid secretion is brought about by the presence of products of protein digestion (peptides and amino acids) in the duodenum via two endocrine mechanisms. The proximal duodenum is rich in G cells that release gastrin when stimulated by peptides and amino acids. The duodenal gastrin is carried in the blood to the parietal cells and stimulates them to secrete acid. There is also a second more recently discovered hormone that is released from the duodenum in response to the presence of chyme and that also acts directly on the parietal cells to stimulate acid secretion and to potentiate the effect of gastrin. This hormone has been named *entero-oxyntin,* and its physiological role in humans remains to be defined.

Inhibition of gastric acid secretion can be caused by at least five different mechanisms in the duodenum and jejunum (Table 43-2). The way in which these mechanisms interact physiologically remains to be determined. The stimuli for these mechanisms are the presence of acid, fat digestion products, and hypertonicity in the duodenum and proximal part of the jejunum.

TABLE 43-2

Major mechanisms for inhibition of gastric acid secretion

Region	*Stimulus*	*Mediator*	*Inhibit gastrin release*	*Inhibit acid secretion*
Antrum	Acid (pH<3.0)	None, direct	+, S*	
Duodenum	Acid	Secretin	+	+
		Nervous reflex		+
	Hyperosmotic solutions	Unidentified enterogastrone		+
Duodenum and jejunum	Fatty acids	Gastric inhibitory peptide	+	+, S
		Cholecystokinin		+
		Unidentified enterogastrone		+

From Johnson, L.R., editor: Gastrointestinal physiology, ed. 2, St. Louis, 1981, The C.V. Mosby Co.; adapted from M.I. Grossman.
*S, Physiologically significant mechanism. Others may prove significant.

Acid solutions in the duodenum cause the release of secretin into the bloodstream. Secretin inhibits gastric acid secretion in two ways: it inhibits gastrin release by G cells, and it inhibits acid secretion by a direct effect on the parietal cells. Acid in the duodenum also inhibits gastric acid secretion via a local nervous reflex.

The presence of *fatty acids* and certain other fat digestion products in the duodenum and proximal part of the jejunum causes the release of two hormones: *gastric inhibitory peptide* and *cholecystokinin*. Gastric inhibitory peptide is a potent inhibitor of acid secretion. It works both by suppressing gastrin release and by directly inhibiting secretion of acid from the parietal cells. Cholecystokinin inhibits the parietal cells, but this effect may not be physiologically important.

Hyperosmotic solutions in the duodenum elicit the release of another hormone, which remains to be identified, that causes inhibition of gastric acid secretion.

■ *Pancreatic secretion*

The human pancreas weighs less than 100 g, yet each day it elaborates 10 times its mass of pancreatic juice. The pancreas is unusual in having both endocrine and exocrine secretory functions. Its principal endocrine secretions are insulin and glucagon, whose functions are discussed in Chapter 50. The exocrine secretions of the pancreas play an important role in the digestive process. Pancreatic juice is composed of an aqueous component, rich in bicarbonate, that helps to neutralize duodenal contents, and an enzyme component that contains enzymes for digesting carbohydrates, proteins, and fats. Pancreatic exocrine secretion is controlled by both neural and hormonal signals, elicited primarily by the presence of acid and digestion products in the duodenum. Secretin plays a major role in eliciting secretion of the aqueous component, and cholecystokinin stimulates the secretion of pancreatic enzymes.

■ *Structure and innervation of the pancreas*

The structure of the exocrine pancreas resembles that of the salivary glands. Microscopic blind-ended tubules are surrounded by polygonal acinar cells whose primary function is to secrete the enzyme component of pancreatic juice. The acini are organized into lobules. The tiny ducts that drain the acini are called intercalated ducts (Fig. 43-8). The intercalated ducts empty into somewhat larger intralobular ducts. The intralobular ducts of a particular lobule drain into a single extralobular duct which empties that lobule into still larger ducts. The larger ducts converge into a still larger main collecting duct, which drains the pancreas and enters the duodenum along with the common bile duct.

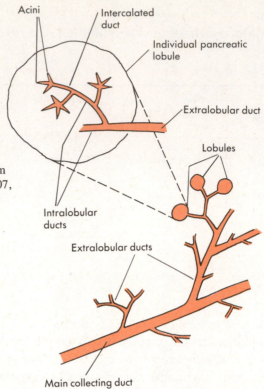

Fig. 43-8 ■ The duct system of the pancreas. (Redrawn from Swanson, C.H., and Solomon, A.K.: J. Gen. Physiol. **62**:407, 1973 by copyright permission of The Rockefeller University Press.)

The pancreas is supplied by branches of the celiac and superior mesenteric arteries. The portal vein is the exit pathway for pancreatic blood flow. The acini and islets are supplied by separate capillary nets, but some of those which supply the islets converge into venules, which then break up into second capillary networks around the acini.

The pancreas is supplied with preganglionic parasympathetic innervation by branches of the vagus. Vagal fibers synapse with postganglionic cholinergic neurons that are within the pancreas and that innervate both acinar and islet cells. Postganglionic parasympathetic nerves from the celiac and superior mesenteric plexuses innervate pancreatic blood vessels. In general, secretion of pancreatic juice is stimulated by parasympathetic activity and inhibited by sympathetic activity.

■ The aqueous component of pancreatic juice

The aqueous component of pancreatic juice is elaborated principally by the columnar epithelial cells that line the smaller ducts. It has Na^+ and K^+ concentrations which are similar to those in plasma. HCO_3^- and Cl^- are its major anions. The HCO_3^- concentration varies from about 30mM at low rates of secretion to over 100mM at high secretory rates (Fig. 43-9). HCO_3^- and Cl^- concentrations vary reciprocally, so that their sum equals the total cation concentration. As secreted by the duct cells, the aqueous component is hypertonic and has a high HCO_3^- concentration. As it flows down, the duct's water equilibrates across the duct walls to make the pancreatic juice isotonic, and some HCO_3^- exchanges for Cl^- (Fig. 43-10). The faster the flow rate, the less time there is for HCO_3^--Cl^- exchange, and the higher is the HCO_3^- concentration in the juice.

Micropuncture experiments suggest that under resting conditions the flow of the aqueous component is produced primarily by the intercalated and other intralobular ducts. When secretion is stimulated by secretin, however, the additional flow comes mostly from the extralobular ducts (Fig. 43-10). Secretin is the major physiological stimulus for secretion of the aqueous component. The secretin-stimulated juice secreted by the extralobular ducts is quite similar to the resting secretion produced by the intralobular ducts, but the extralobular secretion has a slightly higher HCO_3^- concentration.

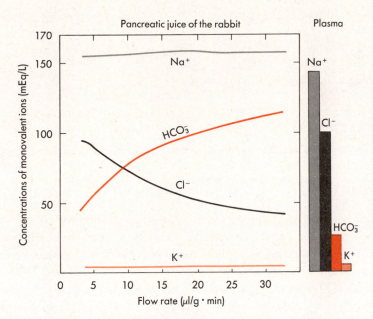

Fig. 43-9 ■ Concentrations of the major ions in pancreatic juice as a function of the secretory flow rate. The concentrations of these ions in plasma are shown for reference.

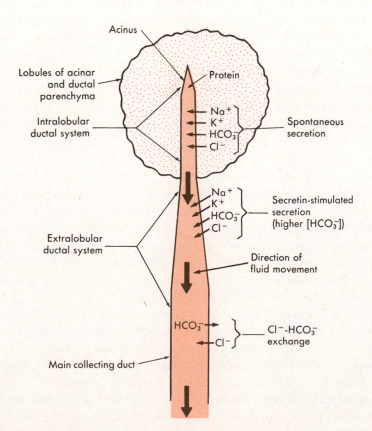

Fig. 43-10 ■ The locations of some of the transport processes involved in the elaboration of pancreatic juice. (Redrawn from Swanson, C.H., and Solomon, A.K.: J. Gen. Physiol. **62:**407, 1973 by copyright permission of The Rockefeller University Press.)

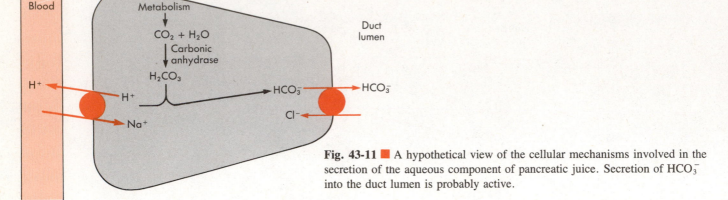

Fig. 43-11 ■ A hypothetical view of the cellular mechanisms involved in the secretion of the aqueous component of pancreatic juice. Secretion of HCO_3^- into the duct lumen is probably active.

The mechanism by which the duct cells secrete the aqueous component is not completely understood. One current view of the secretory mechanism is schematically depicted in Fig. 43-11. According to this scheme the duct cells operate somewhat like gastric parietal cells in reverse; they secrete H^+ into the blood and HCO_3^- into the lumen of the duct. Metabolically produced CO_2 is hydrated to form carbonic acid (H_2CO_3), a reaction catalyzed by carbonic anhydrase in the duct cells. H_2CO_3 dissociates partially into H^+ and HCO_3^-. In the model presented here H^+ is actively extruded from the basal surface of the cell in exchange for Na^+. The energy for the active transport of H^+ may be provided by the energy released by Na^+ flowing down its electrochemical potential gradient into the cell. At the surface of the cell that faces the duct lumen, HCO_3^- is actively pumped out of the cell into the lumen. It is generally accepted that HCO_3^- is actively transported into the duct lumen, but very little is known about the nature of this active transport process. Na^+ is actively secreted into the lumen, probably by a Na,K-ATPase in the luminal membrane. K^+ and Cl^- are believed to enter the lumen passively, moving down their electrochemical potential gradients.

■ *The enzyme component of pancreatic juice*

The enzyme component is secreted by the pancreatic acinar cells. This component contains enzymes important for the digestion of all the major classes of food stuffs. In the total absence of pancreatic enzymes, malabsorption of lipids, proteins, and carbohydrates occurs.

The proteases of pancreatic juice are secreted in inactive zymogen form. The major pancreatic proteases are *trypsin, chymotrypsin,* and *carboxypeptidase*. They are secreted as trypsinogen, chymotrypsinogen, and procarboxypeptidase, respectively. Pepsinogen is specifically activated by *enterokinase* (not a kinase but a protease) that is secreted by the duodenal mucosa. Trypsin then autocatalytically activates trypsinogen, chymotrypsinogen, and procarboxypeptidase. Trypsin and chymotrypsin cleave certain peptide bonds to reduce polypeptide chains to smaller peptides. Carboxypeptidase specifically removes amino acids from the C-terminal ends of peptide chains.

Pancreatic juice contains an α-amylase that is secreted in active form. Pancreatic amylase has the same substrate specificity as salivary amylase, and it cleaves only interior, α-1,4 links in starch to yield maltose, maltotriose, and various α-limit dextrins.

Pancreatic juice also contains a number of lipid-digesting enzymes, or *lipases*. Among the major pancreatic lipases are glycerol ester hydrolase, cholesterol ester hydrolase, and phospholipase A_2. Glycerol ester hydrolase (sometimes called pancreatic lipase) acts on triglycerides to cleave specifically the ester bonds at the 1 and 1′ positions to release two free fatty acids and leave a 2-monoglyceride. Cholesterol ester hydrolase acts on cholesterol esters to produce cholesterol and free fatty acids. Phospholipase A_2 specifically cleaves the fatty acyl ester bond at the 2 carbon of a phosphoglyceride to produce a free fatty acid and a 1-lysophosphatide.

Among the other enzymes contained in pancreatic juice are ribonuclease and deoxyribonuclease, which reduce RNA and DNA, respectively, to their constituent nucleotides. The nucleotides are absorbed to some extent and are also reduced to nucleosides and free purine and pyrimidine bases by brush border enzymes.

■ *Secretion of the enzyme component*

Pancreatic acinar cells contain numerous membrane-bound zymogen granules in their apical cytoplasm. These granules contain most of the enzymes of pancreatic juice. Typically the number of zymogen granules increases between meals and decreases after a meal when the zymogen granules release their contents by exocytosis into the duct lumen. Pancreatic enzymes and zymogens are synthesized on membrane-bound ribosomes, cross the membrane of the rough endoplasmic reticulum to enter the cisternae, and are concentrated somewhat in the cisternae of the rough endoplasmic reticulum. Smooth vesicles bud off from the rough endoplasmic reticulum. These vesicles carry the enzymes and zymogens to the Golgi apparatus, where they are packaged into zymogen granules (secretory vesicles). In response to stimulation of the acinar cell to secrete, the zymogen granules fuse with the plasma membrane to dump their contents into the lumen by exocytosis.

The zymogen granules are not a necessary route of secretion of pancreatic enzymes and zymogens. Acinar cells that have been completely depleted of zymogen granules by prolonged stimulation nonetheless can secrete pancreatic enzymes at high rates. We do not yet understand the way in which the zymogen granule secretory pathway and the granule-free pathway interact under physiological conditions.

■ *Control of pancreatic secretion*

The secretions of the aqueous and enzyme components of pancreatic juice are under separate control. As a result, the protein content of pancreatic juice can be less than 1% or as high as 10%. Both neural and hormonal influences regulate pancreatic secretion.

Secretion of the aqueous component is stimulated by the hormone secretin, which is contained in S cells in the duodenum and proximal part of the jejunum. Secretin is released in response to acid in the lumen of the duodenum and upper jejunum. Secretin is released when the luminal contents have a pH 4.5 or below. Below pH 4.5, the lower the pH, the greater is the release of secretin. Below pH 3 the determining factor becomes not pH but the amount of titratable acid present: the more titratable acid, the greater is the release of secretin.

There is an important interaction between the control of the secretion of the aqueous and the enzyme components of pancreatic juice. Cholecystokinin, which is the major hormonal stimulus to secretion of pancreatic enzymes, strongly potentiates the effect of secretin in evoking secretion of the aqueous component. By itself cholecystokinin does not stimulate secretion of the aqueous component.

Secretion of the enzyme component of pancreatic juice is under neural and hormonal control. Vagal fibers directly stimulate acinar cells to secrete enzymes and zymogens. This effect is mediated by acetylcholine. Vagotomy depresses the amounts of pancreatic enzymes and zymogens released in response to a meal. It is believed that a vagovagal reflex is responsible for the vagal component of pancreatic enzyme secretion, but the afferent stimuli which elicit this reflex have not been identified unequivocally. Mechanoreceptors in the stomach are candidates for the originators of this vagovagal reflex.

The hormone *cholecystokinin* (also known as *pancreozymin*) is the major physiological stimulus for secretion of pancreatic enzymes and zymogens. Gastrin, which has the same N-terminal amino acid sequence (for five residues) as cholecystokinin, will stimulate enzyme secretion when administered, but the physiological role of this effect of

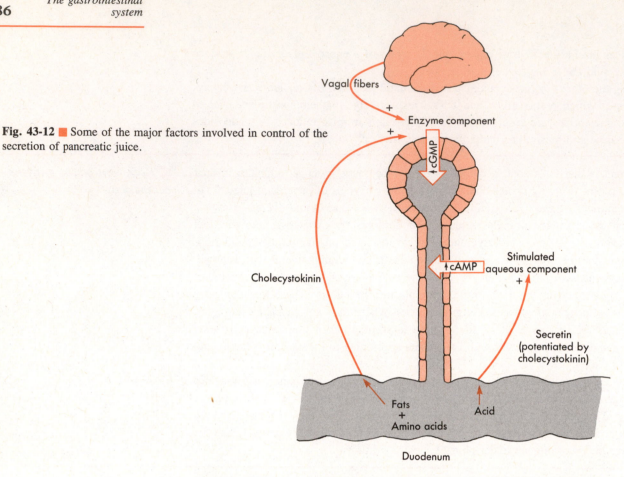

Fig. 43-12 ■ Some of the major factors involved in control of the secretion of pancreatic juice.

gastrin remains to be determined. Cholecystokinin is contained in mucosal cells of the duodenum and jejunum and is released in response to digestion products of proteins and fats in the duodenum and jejunum. Certain L-amino acids, particularly essential amino acids, and some peptides strongly stimulate release of cholecystokinin, as do long-chain fatty acids.

The presence of trypsin in the contents of the small bowel inhibits the release of cholecystokinin and thus exerts feedback control on secretion of pancreatic enzymes. The mechanism of this effect has not been worked out.

Intracellular mechanisms of pancreatic stimulation appear to involve cyclic nucleotides. When secretin stimulates duct cells to secrete the aqueous component, there is an increase in the activity of membrane-bound adenylate cyclase and a resultant increase in the intracellular level of cyclic AMP. The way in which the increased cyclic AMP stimulates secretion of the aqueous component is not known. Ca^{++} also may play a role in stimulus-secretion coupling in the duct epithelial cells.

Stimulation of acinar cells by acetylcholine or cholecystokinin (or gastrin) causes an increase in the activity of guanylate cyclase and a rise in intracellular cyclic GMP. Ca^{++} is released from intracellular stores to increase intracellular $[Ca^{++}]$. The relationships betwen cylic GMP and Ca^{++} are not understood at present. In ways that remain to be determined, cyclic GMP and Ca^{++} mediate electrical depolarization of the acinar cells and secretion of enzymes and zymogens both from zymogen granules and by the granule-independent pathway. The control of pancreatic secretion is schematically summarized in Fig. 43-12.

■ *Secretion of bile*

Bile is elaborated by the liver. It contains bile acids, cholesterol, lecithin, and bile pigments, which are all synthesized and secreted by hepatocytes into the bile canaliculi. This secretion is called the *bile acid–dependent fraction* of the bile. The bile canaliculi merge into ever larger ducts and finally into a single large bile duct. The epithelial cells

Fig. 43-13 ■ Overview of the enterohepatic circulation of bile.

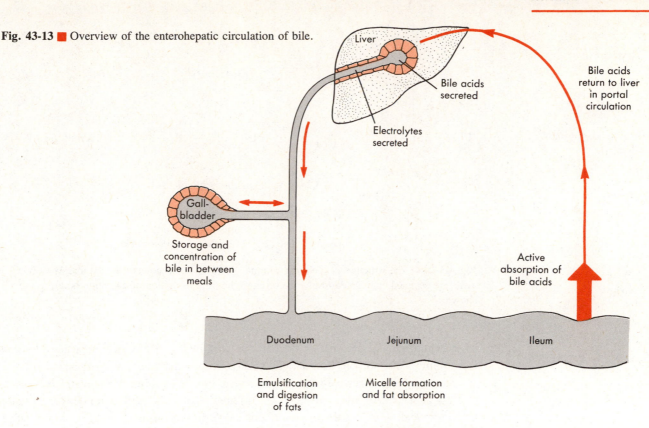

that line the bile canaliculi secrete a watery fluid that is rich in bicarbonate and contributes to the volume of bile that leaves the liver. The watery secretion is called the *bile acid–independent fraction* of bile.

In the periods between meals bile is diverted into the gallbladder. The gallbladder epithelium extracts salts and water from the stored bile, resulting in a five- to twentyfold concentration of bile acids. After an individual has eaten, the gallbladder contracts and empties its concentrated bile into the duodenum. The most potent stimulus for emptying of the gallbladder is the hormone *cholecystokinin,* which is released by the duodenal mucosa primarily in response to the presence of fats and their digestion products. From 250 to 1500 ml/day of bile enters the duodenum.

Bile acids emulsify lipids, thereby increasing the surface area available to lipolytic enzymes. Bile acids then form mixed micelles with the products of lipid digestion, which increases transport of lipid digestion products to the brush border surface and in this way enhances absorption of lipids by the epithelial cells. Bile acids are actively absorbed, chiefly in the terminal part of the ileum. A small fraction of bile acids escapes absorption and is excreted. The returning bile acids are avidly taken up by the liver and are rapidly resecreted during the course of digestion. The entire bile acid pool (approximately 2.5 g) is recirculated twice in response to a typical meal. The recirculation of the bile is known as the *enterohepatic circulation.* Fig. 43-13 summarizes some major aspects of the enterohepatic circulation.

■ *The bile acid–dependent fraction of bile*

The bile acids. Bile acids comprise about 50% of the dry weight of bile. Other important compounds secreted by the hepatocytes into the bile include lecithin, cholesterol, and bile pigments.

Bile acids have a steroid nucleus and are synthesized by the hepatocytes from cholesterol. The major bile acids synthesized by the liver are called *primary bile acids* (Fig. 43-14). These are cholic acid (3-hydroxyl groups) and chenodeoxycholic acid (2-hy-

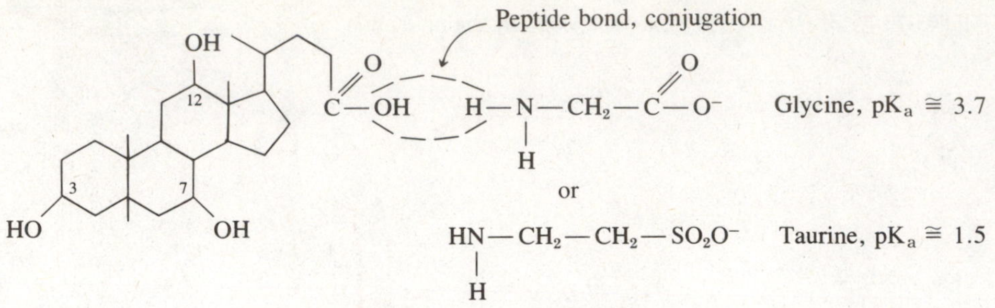

Cholic acid (primary bile acid)

No-OH on 12 = Chenodeoxycholic acid (primary)
No-OH on 7 = Deoxycholic acid (secondary)
No-OH on 7 and 12 = Lithocholic acid (secondary)

Fig. 43-14 ■ The structure of the most common primary and secondary bile acids. Most bile acids are secreted conjugated to glycine or taurine by the formation of a peptide bond, as shown.

droxyl groups). The presence of the carboxyl and hydroxyl groups makes the bile acids much more water soluble than the cholesterol from which they are synthesized.

Bacteria in the digestive tract dehydroxylate bile acids to form *secondary bile acids*. The major secondary bile acids (Fig. 43-14) are deoxycholic acid (from dehydroxylation of cholic acid) and lithocholic acid (from dehydroxylation of chenodeoxycholic acid). Bile contains both primary and secondary bile acids.

Bile acids normally are secreted conjugated with glycine or taurine (Fig. 43-14). *Conjugated bile acids* contain glycine or taurine linked by a peptide bond between the carboxyl group of an unconjugated bile acid and the amino group of glycine or taurine. The pK_a's of the carboxyl groups of unconjugated bile acids are near neutral pH, but the pK_a's of conjugated bile acids are considerably lower. Thus at the near-neutral pH of the gastrointestinal tract the conjugated bile acids are more completely ionized, and thus more water soluble, than the unconjugated bile acids. Conjugated bile acids therefore exist almost entirely as salts of various cations (mostly Na^+) and hence often are called *bile salts*.

The steroid nucleus of bile acids is roughly planar. In solution bile acids have their polar (hydrophilic) groups—the hydroxyl groups, the carboxyl moiety of glycine or taurine, and the peptide bond—all on one surface of the molecule. The other surface is quite hydrophobic (Fig. 43-15, *A*). This makes the bile acid molecule amphipathic, that is, having both hydrophilic and hydrophobic domains. Conjugated bile acids are more amphipathic than unconjugated ones. Because they are amphipathic, bile acids tend to form molecular aggregates, called micelles, by turning their hydrophobic faces inside and away from water and their hydrophilic surfaces toward the water (Fig. 43-15, *B*). Whenever bile acids are present above a certain concentration, called the *critical micelle concentration,* bile acid micelles will form. Above this concentration any additional bile acid will go into the micelles exclusively and not into molecular solution. In bile the bile acids are normally present at a concentration well above the critical micelle concentration.

Phospholipids in bile. Hepatocytes also secrete phospholipids into the bile, the most prominent class being lecithins. Cholesterol also is secreted into the bile, and this is the major route for cholesterol excretion. Being essentially insoluble in water, lecithin and cholesterol partition into (dissolve in) the bile acid micelles. Cholesterol, being apolar, partitions into the center of the micelle. Lecithin, because it is amphipathic, buries its fatty acyl chains in the micelle interior and leaves its polar head group near the micelle

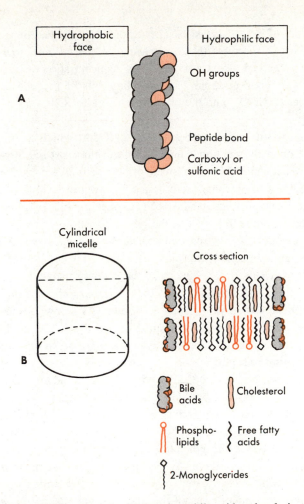

Fig. 43-15 ■ Structure of bile acids and micelles. **A,** A bile acid molecule in solution. The molecule is amphipathic in that it has a hyrophilic face and a hydrophobic face. The amphipathic structure is key in the ability of the bile acids to emulsify lipids and to form micelles. **B,** A model of the structure of a bile acid–lipid mixed micelle.

surface (Fig. 43-15, *B*). The presence of lecithin increases the amount of cholesterol that can be solubilized in the micelles. If more cholesterol is present in the bile than can be solubilized in the micelles, then crystals of cholesterol will form in the bile. These crystals are believed to play an important role in formation of cholesterol gallstones (the most common kind of gallstones) in the duct system of the liver or more commonly in the gallbladder.

Bile pigments. Bile pigments account for the greenish black color of the bile. The principal bile pigment is bilirubin, which is the major degradation product of heme porphyrin. Bilirubin is taken up by hepatocytes and secreted into the bile conjugated to glucuronic acid. Colonic bacteria convert bilirubin to other compounds that are responsible for the brown color of the stool.

Water and electrolytes are secreted by the hepatocytes along with the other components of bile. In addition, the epithelial cells that line the bile canaliculi and bile ducts contribute a watery secretion that contains only water and electrolytes and which may account for 30% or more of the volume of bile secreted. This secretion is called the *bile acid–independent fraction* of bile. Secretion of the bile acid–independent fraction is specifically stimulated by the hormone secretin, which increases bile volume without

■ *The bile acid–
independent fraction of bile*

increasing the flow of bile acids. At basal rates of bile flow the concentrations of Na^+, K^+, Cl^-, and HCO_3^- resemble those in plasma. When volume flow is stimulated by secretin, there is a marked increase in the HCO_3^- concentration of bile, with a concomitant fall in the Cl^- concentration. The epithelial cells of the bile canaliculi and bile ducts actively secrete HCO_3^- into the bile in exchange for Cl^-.

■ *Bile concentration and storage in the gallbladder*

In between meals the tone of the sphincter of Oddi, which guards the entrance of the common bile duct into the duodenum, is high. Thus most bile flow is diverted into the gallbladder. The gallbladder is a small organ, having a capacity of 15 to 60 ml (average about 35 ml) in humans. Many times this volume of bile may be secreted by the liver between meals. The gallbladder concentrates the bile by absorbing Na^+, Cl^-, HCO_3^-, and water from the bile, so that the bile acids can be concentrated from 5 to 20 times in the gallbladder. K^+ is concentrated in the bile when water is absorbed, and then K^+ is absorbed by simple diffusion.

The active transport of Na^+ seems to be the primary active process in the concentrating action of the gallbladder. Cl^- and HCO_3^- are absorbed to preserve electroneutrality. It has not been settled yet whether Cl^- and HCO_3^- flow because of the potential difference created by the electrogenic pumping of Na^+, or whether Cl^- and HCO_3^- are transported by a carrier that obligatorily couples the transport of one anion for each Na^+ pumped (electroneutral transport).

Because of its high rate of water absorption, the gallbladder serves as a model for water and electrolyte transport by tight-junctioned epithelia. The *standing gradient mechanism* for fluid absorption was first proposed for the gallbladder. A key initial observation was that, when fluid was being reabsorbed by the gallbladder, the lateral intercellular spaces between the epithelial cells were large and swollen. When fluid transport was blocked, for example, by poisoning the Na^+ pumps with ouabain, the intercellular spaces almost disappeared. These observations strongly suggested that the intercellular spaces are a major route of fluid flow during absorption. A current view is that Na^+ is actively pumped into the lateral intercellular spaces. The Na^+ pumps are believed to be especially dense near the mucosal (apical) end of the channel (Fig. 43-

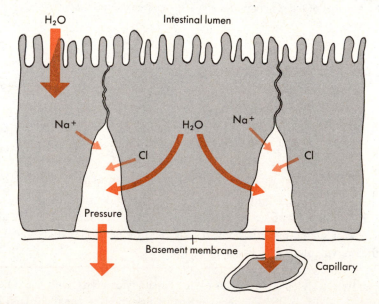

Fig. 43-16 ■ Water absorption from the gallbladder by the mechanism of the standing osmotic gradient. Na^+ is actively pumped into the lateral intercellular spaces; Cl^- follows. Water is drawn by osmosis to enter the intercellular spaces, elevating the hydrostatic pressure there. Water, Na^+, and Cl^- are filtered across the porous basement membrane and enter the capillaries.

16). Cl^- and HCO_3^- also are transported into the intercellular space, either because of an electrical potential created by Na^+ transport or because anion transport is obligatorily coupled to Na^+ transport. The high ion concentration near the apical end of the intercellular space causes the fluid there to be hypertonic. This produces an osmotic flow of water from the lumen via adjacent cells into the intercellular space. Water distends the intercellular channel because of increased hydrostatic pressure there. Because of water flow from adjacent cells, the fluid becomes less hypertonic as it flows down the intercellular channel, so that it is essentially isotonic when it reaches the serosal (basal) end of the channel. Ions and water flow across the basement membrane of the epithelium and are carried away by the capillaries.

■ *Emptying of the
gallbladder*

Emptying of the gallbladder contents into the duodenum begins within 30 minutes after a meal. Intermittent contractions of the gallbladder force bile through the partially relaxed sphincter of Oddi. During the cephalic and gastric phases of digestion, gallbladder contraction and relaxation of the sphincter are mediated by cholinergic fibers in branches of the vagus nerve. Stimulation of sympathetic nerves to the gallbadder and duodenum inhibits emptying of the gallbladder.

The highest rate of gallbladder emptying occurs during the intestinal phase of digestion, the strongest stimulus for the emptying being the hormone *cholecystokinin*. Cholecystokinin is released by the duodenal mucosa most strongly in response to the presence of fat digestion products in the duodenum. Cholecystokinin reaches the gallbladder via the circulation, and it causes strong contractions of the gallbladder and relaxation of the sphincter of Oddi. Substances that mimic the actions of cholecystokinin in promoting gallbladder emptying are called *cholecystogogues*.

Under normal circumstances the rate of gallbladder emptying is sufficient to keep the concentration of bile acids in the duodenum above the critical micelle concentration.

■ *Intestinal absorption
of bile acids and their
enterohepatic circulation*

The functions of bile acids in emulsifying dietary lipid and in forming mixed micelles with the products of lipid digestion are discussed in Chapter 44 in the section on digestion and absorption of lipids.

Normally, by the time chyme reaches the terminal part of the ileum, dietary fat is essentially completely absorbed. Bile acids are then absorbed. The epithelial cells of the distal part of the ileum actively take up bile acids against a large concentration gradient. The active transport system has a higher affinity for conjugated bile acids.

Bile acids also have a fair degree of lipid solubility. Thus they also can be taken up to a considerable extent by simple diffusion. Bacteria in the terminal part of the ileum and colon deconjugate bile acids and also dehydroxylate them to produce secondary bile acids. Both deconjugation and dehydroxylation lessen the polarity of bile acids, enhancing their lipid solubility and their absorption by simple diffusion.

Typically about 0.5 g of bile acids each day escape absorption and are excreted in the feces. This quantity is 15% to 35% of the total bile acid pool, and normally it is replenished by synthesis of new bile acids by the liver.

Bile acids, whether absorbed by active transport or simple diffusion, are transported away from the intestine in the portal blood, mostly bound to plasma proteins. In the liver, hepatocytes avidly extract the bile acids from the portal blood. In a single pass through the liver the portal blood is essentially cleared of bile acids. Bile acids in all forms, primary and secondary, both conjugated and deconjugated, are taken up by the hepatocytes. The hepatocytes reconjugate almost all the deconjugated bile acids and rehydroxylate some of the secondary bile acids. These bile acids are secreted into the bile along with newly synthesized bile acids. The circulation of bile acids from liver to intestine and back to the liver again is known as the *enterohepatic circulation* of the bile acids.

■ *Control of bile acid
synthesis and secretion*

The control of emptying of the gallbladder is discussed previously; this section discusses control of bile acid synthesis and secretion. The rate of return of bile acids to the liver is a major influence on the rate of synthesis and secretion of bile acids. Bile acids in the portal blood *stimulate the secretion* of bile acids by the hepatocytes but *inhibit the synthesis* of bile acids.

In the period long after a meal, when bile acids already have been returned to the liver, the level of bile acids in the portal blood is low. Thus there is little inhibition of synthesis of new bile acids, and synthesis proceeds at near maximal rates. The rate of secretion of bile acids is low because secretion is not being stimulated by bile acids in the portal blood, and the gallbladder fills rather slowly with bile.

When a meal is eaten, the presence of fat digestion products in the duodenum causes release of cholecystokinin, which brings about emptying of gallbladder contents into the duodenum and strongly stimulates secretion of bile acids by the hepatocytes. Secretin, released by acidic chyme in the duodenum, increases the rate of secretion of the bile acid–independent fraction. At this time the liver continues to synthesize bile acids at a high rate and to secrete them at a low rate. This continues until bile acids are absorbed from the terminal aspect of the ileum and return to the liver in the portal blood.

Bile acids returning to the liver inhibit the synthesis of new bile acids but stimulate high rates of secretion. Bile acids that are taken up are rapidly reconjugated (some secondary bile acids are rehydroxylated) and resecreted almost immediately. So powerful is the stimulus to resecrete returning bile acids that the entire pool of bile acids (1.5 to 3.5 g) recirculates twice in response to a typical meal. In response to a single meal with a very high fat content the bile acid pool may recirculate 5 or more times. Table 43-3 summarizes major aspects of control of gallbladder emptying and bile synthesis and secretion.

TABLE 43-3

The major factors affecting gallbladder emptying and bile synthesis and secretion

Phase of digestion	Stimulus	Mediating factor	Response
Cephalic	Taste and smell of food; food in mouth and pharynx	Impulses in branches of vagus nerve	Increased rate of gallbladder emptying
Gastric	Gastric distension	Impulses in branches of vagus nerve	Increased rate of gallbladder emptying
Intestinal	Fat digestion products in duodenum	Cholecystokinin	Increased rate of gallbladder emptying; increased rate of bile acid secretion
	Acid in duodenum	Secretin	Increased rate of secretion of the electrolytes and water of the bile acid–independent fraction of bile (this effect strongly potentiated by cholecystokinin)
	Absorption of bile acids in the distal part of ileum	High concentration of bile acids in portal blood	Stimulation of bile acid secretion; inhibition of bile acid synthesis
Interdigestive period	Low rate of release of bile to duodenum	Low concentration of bile acids in portal blood	Stimulation of bile acid synthesis; inhibition of bile acid secretion

The most common type of gallstones contain cholesterol as their major component. Cholesterol is essentially insoluble in water. When bile contains more cholesterol than can be solubilized in the bile acid–lecithin micelles, crystals of cholesterol form in the bile. Such bile is said to be supersaturated with cholesterol. The greater the concentration of bile acids and lecithin in bile, the greater is the amount of cholesterol that can be contained in the mixed micelles. Lecithin is important in this regard because lecithin-cholesterol mixed micelles can solubilize more cholesterol than can micelles of bile acids alone.

At night most normal individuals secrete bile that is supersaturated with cholesterol. This is because the rate of bile acid secretion is particularly low as a result of the absence of stimulation by bile acids returning in the portal blood. During the day, however, the average rate of bile acid secretion is higher, and bile is unsaturated with cholesterol in normal persons. Individuals with cholesterol gallstones tend to secrete bile that is supersaturated with cholesterol both night and day.

■ *Cholesterol gallstones*

■ *Intestinal secretions*

The mucosa of the intestine, from the duodenum through the rectum, elaborates secretions that contain mucus, electrolytes, and water. The total volume of intestinal secretions is about 1500 ml/day. The mucus in the secretions serves to protect the mucosa from mechanical damage. The nature of the secretions and the mechanisms that control secretion vary from one segment of the intestine to another.

■ *Duodenal secretions*

The duodenal submucosa contains branching glands that elaborate a secretion rich in mucus. The glands have ducts that empty into the crypts of Lieberkühn. The duodenal epithelial cells probably also contribute to duodenal secretions, but most of the secretions are produced by the glands. The duodenal secretion contains mucus and an aqueous component that does not differ greatly from plasma in its concentrations of the major ions. Gastrin, secretin, and cholecystokinin can stimulate duodenal secretion, but a physiological role for any of these hormones has not yet been established.

■ *Secretions of the small intestine*

Goblet cells, which lie among the columnar epithelial cells of the small intestine, secrete mucus that lubricates the mucosal surface and protects it against mechanical damage. During the course of normal digestion an aqueous secretion is elaborated by the epithelial cells at a rate only slightly less than the rate of fluid absorption by these cells. So the net absorption of fluid that normally occurs is the result of much larger unidirectional absorptive and secretory flows. As discussed in the section on absorption of water and electrolytes, cholera toxin greatly increases the rate of aqueous secretion by the small intestine, particularly by the jejunum.

■ *Secretions of the colon*

The secretions of the colon are smaller in volume, but richer in mucus, than small intestinal secretions. The mucus is produced by the numerous goblet cells of the colonic mucosa. The aqueous component of colonic secretions is alkaline because of active secretion of HCO_3^- in exchange for Cl^-, and the secretion contains high K^+ because of active secretion of K^+. The rate of production of colonic secretions is stimulated by mechanical irritation of the mucosa and by activation of cholinergic pathways to the colon. Stimulation of sympathetic nerves to the colon decreases the rate of colonic secretion.

■ *Bibliography*

Journal articles

Gardner, J.: Regulation of pancreatic exocrine function in vitro: initial steps in the actions of secretagogues, Annu. Rev. Physiol. **41:**55, 1979.

Jones, R.S., and Meyers, W.C.: Regulation of hepatic biliary secretion, Annu. Rev. Physiol. **41:**67, 1979.

Soll, A., and Walsh, J.H.: Regulation of gastric acid secretion, Annu. Rev. Physiol. **41:**35, 1979.

Books and monographs

Crane, R.K., editor: Gastrointestinal physiology II, Baltimore, 1977, University Park Press.

Davenport, H.W.: Physiology of the digestive tract, ed. 4, Chicago, 1977, Year Book Medical Publishers, Inc.

Duthie, H.L., and Wormsley, K.G., editors: Scientific basis of gastroenterology, Edinburgh, 1979, Churchill Livingstone.

chapter 44

Digestion and absorption

Plant starch, amylopectin, is the major source of carbohydrate in most human diets. There is no nutritional requirement for carbohydrate per se, but it is usually the principal source of calories. The amount of animal starch, glycogen, typically ingested varies widely among cultures and among individuals within a given culture. Sucrose and lactose are the principal dietary disaccharides, and glucose and fructose are the major monosaccharides. The capacity of a healthy digestive system to digest and absorb carbohydrates greatly exceeds the amount of carbohydrate presented to it under normal circumstances.

The structure of a branched starch molecule is schematically depicted in Fig. 44-1. It is a polymer of glucose consisting of chains of glucose units linked by α-1,4 glycosidic bonds. The α-1,4 chains have branch points formed by α-1,6 linkages, and the starch molecule is highly branched. The digestion of starch begins in the mouth with the action of the α-amylase ptyalin contained in salivary secretions. This enzyme catalyzes the hydrolysis of the internal α-1,4 links of starch but cannot hydrolyze the α-1,6 branching links. (The α-amylase secreted by the pancreas has the same specificity.) As shown in Fig. 44-1, the principal products of α-amylase digestion of starch are maltose, maltotriose, and branched oligosaccharides known as α-limit dextrins. The action of the salivary α-amylase continues until the food in the stomach is mixed with gastric acid. Considerable digestion of starch by the salivary α-amylase occurs normally, but this enzyme is not required for the complete digestion and absorption of the starch ingested. After the salivary α-amylase is inactivated by gastric acid, no further processing of carbohydrate occurs in the stomach.

The pancreatic secretions contain a highly active α-amylase. The products of starch digestion by this enzyme are the same as for the salivary α-amylase (Fig. 44-1), but the total activity of the pancreatic enzyme is considerably greater than the salivary amylase. The pancreatic α-amylase is most concentrated in the duodenum. Within about 10 minutes after entering the duodenum, starch is entirely converted to the following small oligosaccharides: maltose, maltotriose, α-1,4 linked malto-oligosaccharides (from four to nine glucose units long), and α-limit dextrins containing from five to nine glucose monomers.

The further digestion of these oligosaccharides is accomplished by enzymes that reside in the brush border membrane of the epithelium of the duodenum and jejunum (Fig. 44-2). The major brush border oligosaccharidases are *lactase,* which splits lactose into glucose and galactose, *sucrase,* which splits sucrose into fructose and glucose, α-dextrinase (also called isomaltase), which breaks the α-limit dextrins down to glucose monomers, and *glucoamylase,* which breaks malto-oligosaccharides down to glucose

■ *Digestion and absorption of carbohydrates*
■ *Carbohydrates in the diet*

■ *Digestion of carbohydrates*

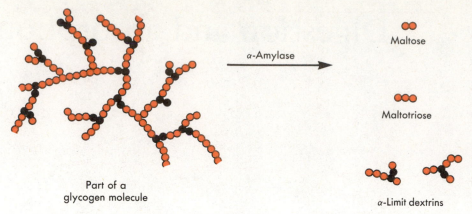

Fig. 44-1 ■ Structure of a branched starch molecule and the action of α-amylase. The circles represent glucose monomers. The dark circles show glucose units linked by α-1,6 linkages at the branch points. The α-1,6 linkages cannot be cleaved by α-amylase. Other sugars are linked by α-1,4 linkages.

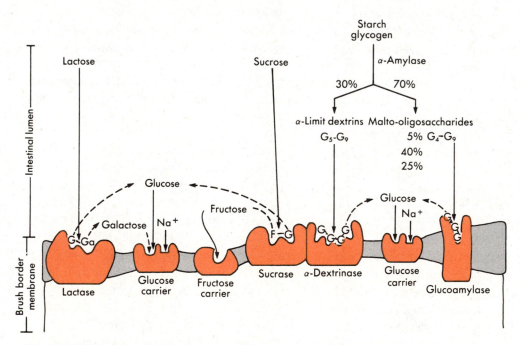

Fig. 44-2 ■ Functions of the major brush border oligosaccharidases. The glucose, galactose, and fructose molecules released by enzymatic hydrolysis are then transported into the epithelial cell by specific transport carrier proteins. Sucrase may also function as a special carrier for some of the glucose and fructose released from the sucrose it hydrolyzes. (From Gray, G.M: N. Engl. J. Med. **292:**1225, 1975. Reprinted by permission of The New England Journal of Medicine.)

units. The sucrase and isomaltase are noncovalently associated subunits of a single protein. Each enzyme is a single polypeptide chain. The two enzyme activities are not affected by their molecular interactions, so the function for this association remains obscure. The activities of these four enzymes is highest in the brush border of the upper jejunum, and they gradually decline through the rest of the small intestine.

■ *Absorption of
carbohydrates*

The duodenum and upper jejunum have the highest capacity to absorb sugars. The lower jejunum and ileum have capacities that are progressively less. The only dietary monosaccharides that are well absorbed are glucose, galactose, and fructose. Glucose and galactose are actively taken up by the brush border epithelial cells through a fairly

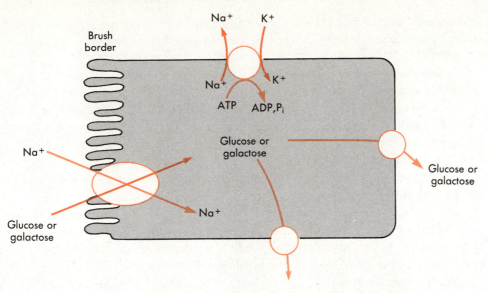

Fig. 44-3 ■ Major features of glucose and galactose absorption in the small intestine. Glucose and galactose enter the epithelial cell against a concentration gradient (Na^+ entry provides the energy). Glucose and galactose leave the cell at the basolateral membrane by facilitated transport.

well characterized transport system. Glucose and galactose compete for entry; other sugars are relatively ineffective competitors. Transport is inhibited by phlorhizin and by metabolic inhibitors.

The active entry of glucose and galactose into the intestinal epithelial cells is stimulated by the presence of Na^+ in the lumen. Conversely, the entry of Na^+ into the epithelial cell across the brush border membrane is stimulated by glucose or galactose in the lumen. The currently accepted interpretation is that Na^+ and glucose or galactose are transported into the cell by the same membrane protein, which has one Na^+-binding site and one sugar-binding site. Na^+ enters the cell down a large electrochemical potential gradient; both concentration and electrical forces drive it into the cell. It is believed that the energy released by the Na^+ flowing down its electrochemical potential is harnessed to force glucose or galactose into the cell *against* a concentration gradient for the sugar. The electrochemical potential gradient for Na^+ is due to Na^+, K^+ - ATPase molecules in the basal and lateral plasma membranes of the intestinal epithelial cells. Na^+ is said to be transported by *primary active transport* by the sodium-potassium pump. The sugars are said to be transported by *secondary active transport* because their active transport depends on the electrochemical potential gradient of another species (Na^+). Glucose and galactose leave the intestinal epithelial cell at the basal and lateral plasma membranes via a facilitated transport system and diffuse into the mucosal capillaries. Fig. 44-3 summarizes some major features of glucose and galactose absorption.

The absorption of fructose by the intestine is much less understood. Fructose does not compete well for the glucose-galactose system. Yet it is transported almost as rapidly as glucose and galactose and much more rapidly than other monosaccharides, so it is presumed that fructose uptake is mediated by a membrane protein. Active transport of fructose has been observed in rat intestine, but has not been demonstrated in the human intestine.

The absorption of sucrose has some unusual features. It appears that the sucrase is both a hydrolytic enzyme and a transport protein. When sucrase is isolated and reconstituted in an artificial lipid membrane, the protein binds sucrose at one face of the membrane and releases glucose and fructose on the other side of the membrane. It appears likely that sucrase also functions this way in vivo. The other brush border saccharidases are not believed to function as transport proteins, but rather to release glucose and galactose in the vicinity of the glucose-galactose active transport system.

■ *Carbohydrate malabsorption syndromes*

Lactose malabsorption syndrome. This is a common disorder of carbohydrate digestion and absorption. The syndrome is due to a deficiency of lactase in the brush border of the duodenum and jejunum. Undigested lactose cannot be absorbed. Thus it is passed on to the colonic bacteria, which avidly metabolize the lactose, producing gas and metabolic products that enhance colonic motility. Individuals with this disorder are said to be *lactose intolerant.* The symptoms of this disorder, and the other carbohydrate malabsorption syndromes as well, are gassiness, borborygmi, and diarrhea. Borborygmi are the gurgling noises made by the intestine as it mixes gas and liquid.

This condition seems to be largely genetically determined. In Oriental societies lactose intolerance among adults is almost universal. Northern European adults, on the other hand, are mostly lactose tolerant. Many black American adults are lactose intolerant. Many lactose-intolerant adults simply do not drink milk and certain milk products and thereby avoid the symptoms without being aware that they have the disorder. There is some evidence that the presence of lactose in the diet induces a higher level of intestinal lactase activity than would be present in the absence of dietary lactose.

Congenital lactose intolerance. This is rather rare. Infants who lack lactase have diarrhea when they are fed breast milk or formula containing lactose. The resulting dehydration and electrolyte imbalance is life threatening. Such infants do well when fed a formula containing sucrose or fructose.

Sucrase-isomaltase malabsorption syndrome. Rare individuals with deficient levels of sucrase-isomaltase are intolerant of ingested sucrose.

Glucose-galactose malabsorption syndrome. This is a rare hereditary disorder due to a defect in the brush border active transport system for glucose and galactose. Ingestion of glucose, galactose, or starch leads to flatulence and severe diarrhea. Fructose is well tolerated and can be fed to infants with this disorder. The level of the brush border saccharidases is normal in this disease.

Familial monosaccharide malabsorption syndrome. This is an extremely rare, inherited disorder. Patients with this disease do not tolerate glucose, galactose, *or* fructose. It is presumably due to coexisting defects in the glucose-galactose transport system and the transport system for fructose. Polymers of these monosaccharides are not tolerated either. Infants with this disorder must be fed a sugar-free diet, usually supplemented with intravenous sugar. Glycerol is sometimes used as a dietary carbohydrate for such patients.

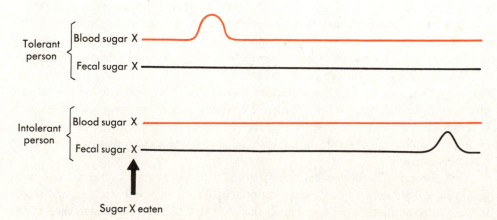

Fig. 44-4 ■ Results of oral tolerance tests. Sugar X is eaten at the *arrow* by a person who is tolerant of X and another who is intolerant of X. The resultant levels of sugar X in the blood and feces of both subjects are shown.

Some diagnostic procedures for carbohydrate intolerance. The most common diagnostic test is the *oral sugar tolerance test*. The patient is given an oral dose of the sugar in question, and the levels of that sugar in the patient's blood and feces are followed. If the patient is intolerant of the administered sugar, diarrhea will ensue and that sugar will fail to appear in the blood, but it will appear in the feces. Fig. 44-4 illustrates the possible outcomes of oral sugar tolerance tests. In suspected saccharidase deficiency the most definitive test for diagnosis is performed by taking a biopsy sample of the jejunal mucosa and assaying it for the deficient enzyme.

■ *Digestion and absorption of protein*

The amount of dietary protein varies greatly among cultures and among individuals within a culture. In some societies is is difficult for an adult to obtain the 0.5 to 0.7 g/day/kg body weight required to balance normal catabolism of proteins, and it is still more difficult for children to get the relatively greater amounts of protein required to sustain normal growth. In other societies, chiefly in industrially developed countries, a typical individual may ingest protein far in excess of the nutritional requirement.

In addition to ingested protein, the gastrointestinal tract is called on to deal with 10 to 30 g of protein per day contained in digestive secretions and a similar amount of protein in desquamated epithelial cells.

In normal humans essentially all ingested protein is digested and absorbed. Most of the protein in digestive secretions and desquamated cells is also digested and absorbed. The small amount of protein in the feces each day is derived principally from colonic bacteria, desquamated cells, and proteins in mucous secretions of the colon. In humans, ingested protein is essentially completely absorbed by the time the meal has traversed the jejunum.

■ *Digestion of proteins*

Digestion in the stomach. Pepsinogen is secreted by the chief cells of the stomach and is converted by hydrogen ions to the active enzyme *pepsin*. The extent to which pepsin hydrolyzes dietary protein is significant but highly variable. At most, about 15% of dietary protein may be reduced to amino acids and small peptides by pepsin. The duodenum and small intestine have such a high capacity to process protein that the total absence of pepsin does not impair the digestion and absorption of dietary protein.

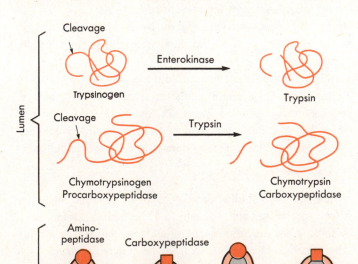

Fig. 44-5 ■ Major proteases in the small intestine, both in the lumen and in the brush border membrane.

Digestion in the duodenum and small intestine. Proteases secreted by the pancreas play a major role in protein digestion. The most important of these proteases are *trypsin, chymotrypsin,* and *carboxypeptidase.* The pancreatic juice contains these enzymes in inactive, proenzyme forms. The enzyme *enterokinase,* secreted by the mucosa of the duodenum and jejunum, converts trypsinogen to trypsin. Trypsin acts autocatalytically to activate trypsinogen and also converts chymotrypsinogen and procarboxypeptidase to the active enzymes. The pancreatic proteases are present at high activities in the duodenum and rapidly convert dietary protein to amino acids and small peptides. About 50% of the ingested protein is digested and absorbed in the duodenum.

The brush border of the duodenum and small intestine contains a number of proteases, including aminopeptidases and carboxypeptidases. These enzymes are richest in the proximal jejunum and function to reduce the oligopeptides resulting from the action of pancreatic proteases to single amino acids and small oligopeptides, which are then absorbed by the brush border transport systems. Fig. 44-5 schematically depicts the major proteases in the small intestine.

■ Absorption of the products of protein digestion

Intact proteins and large peptides. These are not absorbed by humans to an extent that is nutritionally significant, but amounts sufficient to trigger an immunological response can be absorbed. In ruminants and rodents, but not in humans, the neonatal intestine has a high capacity for the specific absorption of immune globulins present in colostrum. This plays a vital role in the development of normal immune competence in ruminants and rodents. Absorption takes place by a selective pinocytotic process that is currently being actively investigated.

Oligopeptides. It has only recently been appreciated that the human intestine has the capacity to absorb important quantities of small peptides, chiefly dipeptides, tripeptides, and tetrapeptides. Certain amino acids may be more rapidly absorbed from small peptides than as free amino acids. The peptides absorbed appear in the portal blood as free amino acids due to hydrolysis within the intestinal epithelial cells. Our knowledge of the transport systems that absorb small peptides is still rather fragmentary. The transport systems probably have fairly broad and overlapping specificities. The number of different peptide transport systems and the extent to which the peptide systems interact with the amino acid transport systems are unknown at the present time.

Absorption of amino acids. This process is carried on by at least three distinct transport systems present in the epithelial cells of the duodenum and small intestine. The three systems that have been identified are:

1. *Neutral system I,* which has affinity for all neutral amino acids
2. *Basic system,* which absorbs the basic amino acids (arginine, ornithine, and lysine) and handles cystine as well
3. *Neutral system II,* which has specificity for proline, hydroxyproline, sarcosine, and other *N*-methyl–substituted glycines. This system is also known as the *imino acid system.*

Neutral system I. This system has been studied more extensively than the other two transport systems. It has broad specificity, and virtually all the neutral amino acids compete for absorption. The more polar the amino acid side chain, the less its affinity for this transport system. The maximal rate of transport of neutral amino acids is at least 10 times greater than that for the basic amino acids. Absorption via neutral system I may be deficient in Hartnup's disease.

The mechanism and energetics of transport by neutral system I appear to have much in common with monosaccharide absorption. The presence of Na^+ in the intestinal lumen enhances the rate of absorption of neutral amino acids and vice versa. Na^+ enters the intestinal epithelial cell down a large electrochemical potential gradient. It is be-

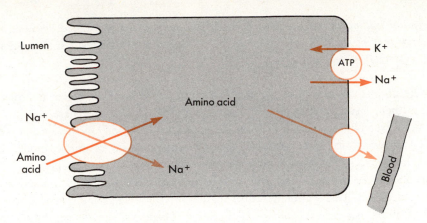

Fig. 44-6 ■ Intestinal absorption of neutral amino acids by neutral system I. Neutral amino acids are actively taken up at the mucosal surface of the epithelial cells by a carrier that uses the energy of the Na^+ gradient. Amino acids leave the cell at the basolateral surface via facilitated transport.

lieved that the energy released by a Na^+ ion moving down its electrochemical potential gradient is used to transport a neutral amino acid molecule from a lower concentration in the lumen to a higher concentration in the intestinal epithelial cell. Neutral system I is thus a *secondary active transport* system, since it depends on the electrochemical gradient of another actively transported species, Na^+. The electrochemical potential gradient of Na^+ is created by the Na, K pumps (Na, K-stimulated ATPase) in the lateral and basal plasma membranes of the epithelial cells. Neutral amino acids leave the epithelial cell via facilitated transport systems in the basolateral plasma membrane and diffuse into the intestinal capillaries. Fig. 44-6 illustrates present knowledge of absorption of neutral amino acids.

Basic system. This system is less understood than neutral system I. It has a high affinity for cystine, arginine, ornithine, and lysine, which mutually compete for transport by this system. This system has higher affinity for its substrates than does neutral system I but 10 to 20 times less maximal transport velocity. The basic amino acids can be concentrated in the intestinal epithelial cell. Whether this system depends on the Na^+ gradient has not yet been established.

Neutral system II (imino acid system). Neutral system II has high affinity for proline and hydroxyproline and other *N*-methyl–substituted glycines. Proline and hydroxyproline have much smaller affinities for neutral system I. The energetics and possible Na^+ dependence of imino acid transport remain to be determined. The other neutral amino acids have little affinity for this system and thus do not inhibit proline and hydroxyproline absorption by neutral system II.

Absorption of acidic amino acids. This process is relatively slow. There is evidence for a specific transport system for absorption of dicarboxylic amino acids (glutamate and aspartate) in rat intestine. Specific absorption of aspartate and glutamate has not yet been demonstrated in humans. Most of the glutamate and aspartate that enters the intestinal epithelial cells appears to undergo transamination, with pyruvate being the most common amino acceptor. The resulting α-ketoglutarate and oxaloacetate are believed to be metabolized primarily by the intestinal mucosa.

Cystinuria. In this disease there is defective renal reabsorption of cystine. In some cases there is also defective intestinal absorption of cystine and the basic amino acids.

Hartnup's disease. This rare hereditary disease involves defective renal transport of neutral amino acids. In some cases there is also decreased intestinal absorption via neutral system I.

■ *Defects of amino acid absorption*

Familial iminoglycinuria. This condition is rare and involves defective renal reabsorption of glycine, proline, and hydroxyproline. In some cases intestinal absorption via neutral system II is also defective.

■ *Intestinal absorption of salts and water*

Under normal circumstances humans absorb almost 99% of the water and ions presented to them in ingested food and in gastrointestinal secretions. There are thus normally net fluxes of water and ions from the lumen to the blood. In most cases the net fluxes of water and ions are the differences between much larger unidirectional fluxes from lumen to blood and from blood to lumen. The cellular and molecular mechanisms responsible for the flow of water and ions across the intestinal mucosa remain incompletely understood. This section describes the net fluxes of water and certain major ions that occur in the various segments of the intestine. There is also a discussion of current understanding of the cellular and molecular mechanisms that produce flows of water and ions.

■ *Absorption of water*

Typically, about 2 L of water are ingested each day and approximately 7 L/day are contained in intestinal secretions. Only about 50 to 100 ml of water per day are lost in the stool. Thus the gastrointestinal tract typically absorbs more than 8 L/day or more than 98% of the water presented to it.

Very little net absorption occurs in the duodenum, but the chyme is brought to isotonicity here. The chyme that is delivered from the stomach is often hypertonic. The

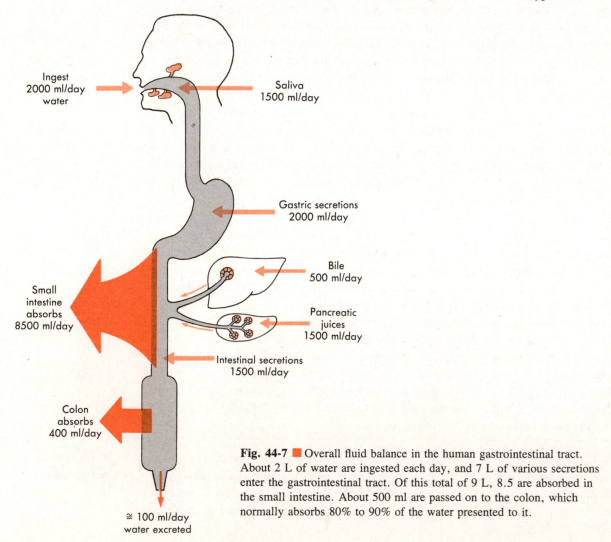

Fig. 44-7 ■ Overall fluid balance in the human gastrointestinal tract. About 2 L of water are ingested each day, and 7 L of various secretions enter the gastrointestinal tract. Of this total of 9 L, 8.5 are absorbed in the small intestine. About 500 ml are passed on to the colon, which normally absorbs 80% to 90% of the water presented to it.

action of digestive enzymes produces still more osmotic activity. The duodenum is highly water permeable, and very large fluxes of water occur from lumen to blood and from blood to lumen. Usually the net flux is from blood to lumen because of the hypertonicity of the chyme. Large net water absorption occurs in the *small intestine;* the *jejunum* is more active than the *ileum* in absorbing water. The net absorption that occurs in the *colon* is relatively small, about 400 ml/day. However, the colon can absorb water against a larger osmotic pressure difference than can the rest of the gastrointestinal tract. In all segments of the intestine the net water flux is the difference between much larger unidirectional fluxes from lumen to blood and from blood to lumen. Fig. 44-7 summarizes the handling of water by the gastrointestinal tract.

Na^+ is absorbed along the entire length of the intestine. As is the case with water, net absorption is the result of large, unidirectional fluxes from blood to lumen and from lumen to blood. The unidirectional fluxes are greater in the proximal gut than in the distal intestine, in keeping with greater brush border surface area per unit length in the jejunum, which diminishes toward the ileum and is still smaller in the colon. Na^+ crosses the brush border membrane down an electrochemical gradient and it is actively extruded from the epithelial cells by the Na, K-ATPase in the basal and lateral plasma membrane. Normally the contents of the small bowel are isotonic to plasma. Luminal contents have about the same Na^+ concentration as plasma, so that Na^+ absorption normally takes place in the absence of a significant concentration gradient. Na^+ absorption is active, however, and can occur against a small electrochemical potential difference for Na^+.

In the jejunum the net absorption of Na^+ is highest. Here the rate of Na^+ absorption is enhanced by the presence in the lumen of glucose, galactose, and neutral amino acids. It is believed that these substances and Na^+ cross the brush border membrane on the same transport proteins. Na^+ moves down its electrochemical potential gradient and provides the energy for moving the sugars (glucose and galactose) and neutral amino acids into the epithelial cells against a concentration gradient. Thus Na^+ enhances the absorption of sugars and amino acids and vice versa.

In the ileum the net rate of Na^+ absorption is smaller. Na^+ absorption is only slightly stimulated by sugars and amino acids because the sugar and amino acid transport proteins are less concentrated in the ileum. The ileum can absorb Na^+ against a larger electrochemical potential than can the jejunum.

In the colon Na^+ is normally absorbed against a large electrochemical potential difference. Sodium concentrations in the luminal contents can get as low as 25mM, compared with about 120mM in the plasma.

■ *Absorption of Na^+*

In the jejunum both Cl^- and HCO_3^- are absorbed in large amounts. By the end of the jejunum most of the HCO_3^- of the hepatic and pancreatic secretions has been absorbed. In the ileum Cl^- is absorbed, but HCO_3^- is normally secreted. If the HCO_3^- concentration in the lumen of the ileum exceeds about 45mM, then the flux from lumen to blood exceeds that from blood to lumen, and net absorption occurs. *In the colon* the situation is qualitatively similar to that in the ileum, in that Cl^- is absorbed and bicarbonate is usually secreted.

**■ *Absorption of Cl^-
and HCO_3^-***

As with the other ions, the net movement of potassium across the intestinal epithelium is the difference between large unidirectional fluxes from lumen to blood and from blood to lumen. *In the jejunum and in the ileum* the net flux is from lumen to blood. As the volume of intestinal contents is reduced due to the absorption of water, K^+ is concentrated, providing a driving force for the movement of K^+ across the intestinal mucosa and into the blood. There is no evidence for active transport of K^+ in the small

■ *Absorption of K^+*

TABLE 44-1

Transport of Na^+, K^+, Cl^-, and HCO_3^- in the large and small intestines

Segment of intestine	Na^+	K^+	Cl^-	HCO_3^-
Jejunum	Actively absorbed; absorption enhanced by sugars, neutral amino acids	Passively absorbed when concentration rises due to absorption of water	Absorbed	Absorbed
Ileum	Actively absorbed	Passively absorbed	Absorbed, some in exchange for HCO_3^-	Secreted, partly in exchange for Cl^-
Colon	Actively absorbed	Net secretion occurs when [K^+] concentration in lumen < 25mM	Absorbed, some in exchange for HCO_3^-	Secreted, partly in exchange for Cl^-

intestine. *In the colon* K^+ may be secreted or absorbed. Net secretion occurs when the luminal concentration is less than about 25mM; above 25mM, net absorption occurs. Under most circumstances there is net secretion of K^+ in the colon.

Because most absorption of K^+ is a consequence of its being concentrated in the lumen due to the absorption of water, there is likely to be significant K^+ loss in diarrhea. If diarrhea is prolonged, the K^+ level in extracellular fluid falls. Because of the importance of maintaining the normal K^+ level, especially for the heart and other muscles, life-threatening consequences such as cardiac arrhythmias may ensue. Table 44-1 summarizes the transport of Na^+, K^+, Cl^-, and HCO_3^- in the small and large intestines.

■ Mechanisms of salt and water absorption by the intestine

Structural considerations

The tight junctions. The epithelial cells that line the intestine are connected to their neighbors by tight junctions near their luminal surfaces. The tight junctions, however, are permeable to water and small ions. The tight junctions are leakiest in the duodenum, a bit tighter in the jejunum, still tighter in the ileum, and tightest in the colon.

Transcellular versus paracellular transport. Because the tight junctions are leaky, some fraction of the water and ions that traverse the intestinal epithelium passes between the epithelial cells, rather than passing through them. Transmucosal movement by passing through the tight junctions and the lateral intercellular spaces is called *paracellular transport.* Passage through the epithelial cells is termed *transcellular transport.*

Because the tight junctions in the duodenum are very leaky, much of the large unidirectional fluxes of water and ions that take place in the duodenum occurs via the paracellular pathway. The proportions of water or a particular ion that pass through the transcellular and paracellular routes are determined by the relative permeabilities of the two pathways for the substance in question. Even in the ileum, where the junctions are much tighter than in the duodenum, the paracellular pathway contributes more to the total conductance of the mucosa than does the transcellular pathway.

Villous versus crypt cells. As emphasized, the transmucosal fluxes of water and ions are usually the differences between much larger unidirectional fluxes from lumen to blood and from blood to lumen. Recent evidence suggests that the highly differentiated epithelial cells near the tips of the villi are specialized for absorption of water and ions, whereas the less differentiated cells in the crypts produce net secretion of water and ions.

Ion transport by intestinal epithelial cells. The movement of water across the intes-

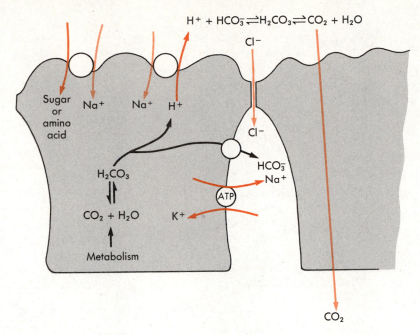

Fig. 44-8 ■ A summary of major ion transport processes that occur in the jejunum.

tinal epithelium is secondary to the movement of ions. Significant progress in characterizing the ionic transport processes that occur in intestinal epithelial cells has recently been made. Nevertheless, important issues remain unresolved. This section includes a current view of the ionic processes that occur in the different segments of the intestine. We wish to emphasize that this view is based on incomplete knowledge, and it is likely that certain aspects of the explanation will be altered by future investigations of intestinal ion transport.

Ion transport in the jejunum. The basolateral plasma membrane contains the Na^+, K^+-ATPase (Fig. 44-8). As a result of the active extrusion of Na^+ ions from the cytoplasm by the Na^+, K^+-ATPase, the electrochemical potential of Na^+ in the cytoplasm is much less than in the luminal fluid. Na^+ enters the epithelial cell by flowing across the brush border plasma membrane, moving down its large electrochemical potential gradient. Some of the Na^+ enters on the same transport proteins with sugars or amino acids, as described previously.

The luminal plasma membrane also contains a Na^+/H^+ exchange protein. Some of the energy released by Na^+ moving down its electrochemical potential gradient is harnessed to actively extrude H^+ into the lumen. Some of the H^+ in the lumen then reacts with HCO_3^- from bile and pancreatic juice to produce H_2CO_3, some of which then forms CO_2 and H_2O. The CO_2 diffuses readily across the epithelial cells and is carried away in the blood. Acidification of the luminal fluid appears to be the major mechanism for absorption of HCO_3^- in the jejunum.

The extrusion of H^+ from the cytoplasm creates a high cytoplasmic concentration of HCO_3^- in the epithelial cells. HCO_3^- is believed to leave the cell by crossing the basolateral plasma membrane by facilitated transport.

The electrogenic effect of the Na^+, $K^+-ATPase$ in the basolateral membrane and the electrogenic entry of Na^+ with sugars and amino acids at the luminal surface both tend to produce an electrical potential difference (lumen negative) across the epithelium. The electrical potential difference causes Cl^- to flow through the tight junctions into the lateral intercellular spaces. The tight junctions in the jejunum are quite leaky, so large fluxes of Cl^- occur by this route. The large paracellular flux of Cl^- almost short-circuits the electrogenic effects of Na^+ transport, and as a result the potential difference

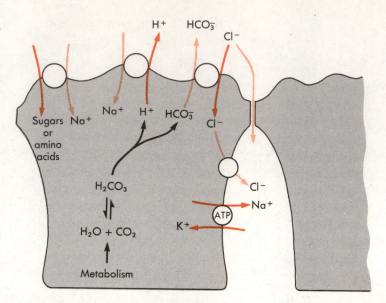

Fig. 44-9 ■ A summary of major ion transport processes that occur in the ileum.

across the jejunum is only a few millivolts (lumen negative). The flow of Cl^- via the tight junctions appears to be the major mechanism of Cl^- absorption in the jejunum.

Note that the only primary active transport is by the Na^+, K^+-ATPase. The resulting Na^+ gradient is responsible for the active extrusion of H^+. The extrusion of H^+ brings about absorption of HCO_3^-. The electrogenic effects of Na^+ transport create an electrical potential difference that powers Cl^- absorption in the jejunum.

Ion transport in the ileum. The ionic transport processes that occur in the ileum (Fig. 44-9) are somewhat similar to those in the jejunum. Basolateral Na^+, K^+-ATPase, electrogenic Na^+ entry with sugars and amino acids (to a lesser extent than in the jejunum), and luminal Na^+/H^+ exchange are all present in the ileum.

The brush border plasma membrane of the ileum contains a Cl^-/HCO_3^- exchange protein that is not present in the jejunum. As in the jejunum, the active extrusion of H^+ into the lumen is driven by Na^+ entry. The extrusion of H^+ elevates the intracellular concentration of HCO_3^-. The HCO_3^- then flows down its electrochemical potential gradient into the lumen, and part of the energy liberated is used to bring Cl^- into the epithelial cell against its electrochemical potential gradient. Cl^- then leaves the cell at the basolateral membrane by facilitated transport.

The net effect of Na^+/H^+ exchange and Cl^-/HCO_3^- exchange occurring together is the electroneutral entry of Na^+ and Cl^- into the epithelial cell. Since the extrusion of H^+ is driven by the downhill entry of Na^+, and the resulting downhill efflux of HCO_3^- drives the uphill entry of Cl^-, the energy for Cl^- entry comes, albeit indirectly, from the electrochemical potential gradient of Na^+. Thus the actual linkage of the ionic transport processes to metabolism in the ileum occurs via the Na^+, K^+-ATPase.

As in the jejunum, the electrogenic Na^+, K^+-ATPase and the electrogenic entry of Na^+ at the brush border cause a small electrical potential difference across the ileal mucosa (lumen negative). The electrical potential difference causes Cl^- to flow through the tight junctions into the intercellular spaces. The electrically driven flux of Cl^- in the ileum is smaller than it is in the jejunum because the ileal junctions are tighter than those in the jejunum.

Ion transport in the colon. Ion transport in the colon (Fig. 44-10) is characterized by luminal Na^+/H^+ and Cl^-/HCO_3^- exchange pumps and by basolateral Na^+, K^+-ATPase and facilitated Cl^- transport, as is the case in the ileum.

The Na^+/sugar and Na^+/amino acid cotransport systems are not present in the co-

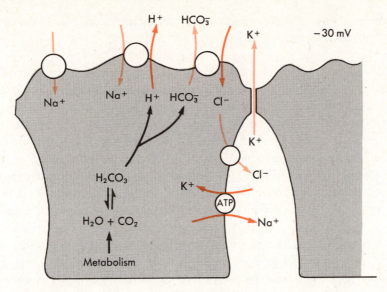

Fig. 44-10 ■ A summary of major ion transport processes that occur in the colon.

lon, but another process that mediates the electrogenic entry of Na^+ into the epithelial cells is present. The latter process is inhibited by amiloride. The electrogenic Na^+, K^+-ATPase and the luminal electrogenic entry of Na^+ cause an electrical potential difference (lumen negative) across the colonic mucosa. The transmucosal potential difference in the colon is about -30 mV, much larger than in the jejunum or the ileum. The greater electrical potential difference across the colonic mucosa is due partly to the presence of tighter junctions in the colon.

The tight junctions in all regions of the intestine are considerably more permeable to cations than to anions. The large electronegativity in the lumen of the colon causes K^+ to flow from the intercellular spaces to the lumen via the tight junctions. This may be the major mechanism for the net secretion of K^+ that usually occurs in the colon. Some investigators feel that there may also be facilitated transport of K^+ from cytoplasm to intestinal lumen across the luminal plasma membrane in the colon.

In the colon, as in the other segments of the intestine, the Na^+, K^+-ATPase is the only transport process that is directly linked to metabolic energy.

The absorption of water depends on the absorption of ions, principally Na^+ and Cl^-. Under normal circumstances, water absorption in the small intestine occurs in the absence of an osmotic pressure difference between the luminal contents and the blood in the intestinal capillaries. Water absorption by the colon typically proceeds against an osmotic pressure gradient. For a long time students of the gastrointestinal tract were puzzled by the absorption of water in the apparent absence of an osmotic pressure gradient to power the absorption.

Our current understanding is that water absorption occurs by a mechanism known as *standing gradient osmosis* (see Fig. 43-16). The major features of the standing gradient osmotic mechanism follow:

1. Active pumping of Na^+ into the lateral intercellular space by the Na^+, K^+-ATPase
2. Entry of Cl^- into the lateral intercellular space by flow from the lumen via the tight junctions or from the adjacent epithelial cells by facilitated transport
3. Presence of hypertonic fluid near the luminal ends of the lateral intercellular spaces
4. Entry of water by osmosis into the lateral intercellular spaces

■ *The mechanism of water absorption*

5. Hydrostatic flow of water and ions down the lateral intercellular space and across the epithelial basement membrane

Na^+, K^+-ATPase molecules are particularly concentrated in the basolateral plasma membrane that surrounds the luminal ends of the intercellular spaces. Because of the high rate of Na^+ pumping and the narrowness of the luminal ends of the intercellular spaces, the Na^+ concentration near the luminal ends of the intercellular spaces tends to reach high levels. Chloride enters the intercellular space from the lumen via the tight junctions (driven by the transmucosal electrical potential difference) and/or from the adjacent epithelial cells by facilitated transport across the basolateral membrane.

The NaCl concentration in the luminal ends of the lateral intercellular space is high enough that the fluid there is hypertonic to the luminal contents and to the cytoplasm of the adjacent epithelial cells. Because of the hypertonicity in the intercellular space, water flows by osmosis into the intercellular space from the adjacent epithelial cells and from the lumen via the tight junctions. The inflow of water causes an elevated hydrostatic pressure there that dilates the intercellular channels. Fluid flows down the intercellular space due to the hydrostatic pressure gradient, and water and ions flow across the basement membrane of the epithelium, which offers little resistance to their passage. As the hypertonic fluid flows down the intercellular channels, water continues to enter the channels by osmosis from the adjacent epithelial cells. By the time the fluid reaches the basement membrane, it is essentially isotonic to the cytoplasm of the epithelial cells. In this way isotonic fluid is taken up at the brush border and is discharged through the basement membrane, to be carried away by the intestinal capillaries. There is a gradient of tonicity in the lateral intercellular space. Fluid is hypertonic at the luminal end of the intercellular space and isotonic at the serosal end. This standing osmotic gradient gives the standing gradient mechanism of fluid absorption its name.

Since most water absorption takes place in the absence of a transmucosal osmotic pressure difference, the absorption of the end products of digestion, particularly sugars and amino acids, plays an important role in water absorption. The absorption of sugars and amino acids allows more water to be absorbed.

■ Pathophysiological alterations of salt and water absorption

The general causes of abnormalities in the absorption of salts and water include (1) deficiency of a normal ion transport system, (2) failure to absorb a nonelectrolyte normally with resultant osmotic diarrhea, (3) an enhanced rate of net secretion of water and electrolytes by the intestinal mucosa, and (4) hypermotility of the intestine leading to abnormally rapid flow of intestinal contents past the absorptive epithelium. Examples of each of these classes of abnormalities follow.

Failure to absorb an electrolyte normally. In congenital chloride diarrhea the Cl^-/HCO_3^- exchange pump that normally operates in the ileum and colon is absent or deficient. In this disease, ion transport and water absorption occur normally in the duodenum and jejunum, which normally lack the Cl^-/HCO_3^- exchanger, but the Cl^-/HCO_3^- exchange transport system in the brush border plasma membrane of the ileum and colon is missing or grossly deficient. As a result, chloride absorption is severely impaired. This leads to diarrhea in which the stools contain an unusually high chloride concentration. In this disease the concentration of Cl^- in the stool exceeds the sum of the concentrations of Na^+ and K^+. The Na^+/H^+ exchange pump continues to operate, so that H^+ is eliminated in the feces without HCO_3^- to neutralize it. The net loss of H^+, with retention of HCO_3^-, contributes to a metabolic alkalosis.

Failure to absorb a nutrient normally. In any of the carbohydrate malabsorption syndromes the sugar that is retained in the lumen of the small intestine contributes to additional osmotic pressure of the luminal contents. Water is retained as a result, and an increased volume of chyme is passed on to the colon. The increased volume flow may overwhelm the ability of the colon to absorb electrolytes and water, resulting in pronounced diarrhea. In addition, the high level of carbohydrates provide a medium that supports increased growth and metabolism of colonic bacteria. The increased production of CO_2 by colonic bacteria contributes to gassiness and borborygmi, and certain products of bacterial metabolism may inhibit absorption of electrolytes by the colonic epithelium.

Enhanced rate of secretion of electrolytes and water. Cholera is characterized by watery diarrhea that may be produced at rates of over 1 L/hour. In cholera the unidirectional fluxes of water and electrolytes from lumen to blood are relatively normal, but there are massive increases in the unidirectional fluxes of Na^+, Cl^-, HCO_3^-, and water from blood to lumen. The bacterium *Vibrio cholerae* produces a protein toxin that binds to receptors in the brush border plasma membrane of the small intestine. Binding of the toxin to its receptor results in activation of adenylate cyclase in the plasma membrane and leads to an elevated production of cyclic AMP. In ways that are not yet understood, the elevated levels of cyclic AMP in the intestinal epithelial cells lead to increased rates of secretion of water and electrolytes. Since absorption of nutrients and lumen-to-blood fluxes of electrolytes are not much changed, it may be that the cholera toxin acts specifically on the predominantly secretory epithelial cells in the crypts of Lieberkühn and leaves the absorptive epithelial cells near the tips of the villi relatively unaffected.

Hypermotility of the intestine. The causes of hypermotility of the intestine are not well understood. Hypermotility of the small intestine may result in electrolytes and water being delivered to the colon at faster rates than they can be absorbed by the colonic epithelial cells. Hypermotility of the colon may result in the elimination of feces before the maximum amount of salts and water can be extracted from them. Hypermotility may add to other factors that cause diarrhea. In cases of fat malabsorption, colonic bacteria metabolize lipids and produce certain waste products, such as hydroxylated fatty acids, that enhance the motility of the colon and inhibit salt and water absorption by the colonic epithelium.

■ *Absorption of calcium*

Calcium ions are actively absorbed by all segments of the intestine. The duodenum and jejunum are especially active and can concentrate Ca^{++} against a greater than tenfold concentration gradient. The rate of absorption of Ca^{++} is much greater than that of any other divalent ion, but still 50 times slower than Na^+ absorption.

The ability of the intestine to absorb Ca^{++} is regulated. Animals on a calcium-deficient diet increase their ability to absorb Ca^{++}. Animals on high-calcium diets have less ability to absorb Ca^{++}. Intestinal absorption of Ca^{++} is stimulated by vitamin D and slightly stimulated by parathyroid hormone. The mechanism by which parathyroid hormone exerts its effect is not yet understood.

Mechanism of absorption of Ca^{++}. A current view of the mechanism of the absorption of Ca^{++} by the small intestine is shown in Fig. 44-11. Ca^{++} crosses the brush border plasma membrane, moving down its electrochemical potential gradient into the cell. Calcium-binding proteins in the brush border membrane are apparently involved in transporting Ca^{++} into the cell, but the details of the transport mechanism remain unclear. In the cytoplasm Ca^{++} is bound to a cytoplasmic calcium-binding protein. This binding protein has a molecular weight of about 28,000 and binds 1 mole of Ca^{++} per mole of binding protein. The role of cytoplasmic calcium-binding protein in Ca^{++} absorption is not well understood. Nevertheless, the ability of the intestine to absorb Ca^{++} is highly correlated with the levels of the cytoplasmic calcium-binding protein in the intestinal mucosal epithelium. Calcium bound to the cytoplasmic calcium-binding protein may exchange with pools of calcium present in various organelles in the epithelial cell. At the basolateral surface of the intestinal epithelial cells, Ca^{++} is actively pumped out of the cell by two mechanisms. A basolateral Ca^{++}-stimulated ATPase,

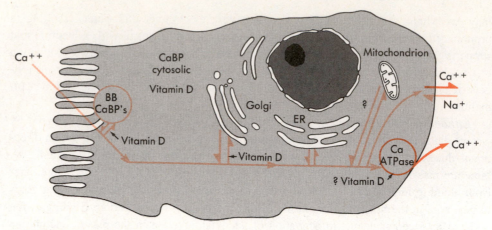

Fig. 44-11 ■ A current view of the major cellular processes involved in absorption of Ca^{++} by intestinal epithelial cells. At the brush border Ca^{++} enters the cell down its electrochemical potential gradient. Brush border calcium-binding proteins (BB CaBP's), some of which are vitamin D dependent, play a role in the transport of Ca^{++} across the brush border plasma membrane. In the cytoplasm Ca^{++} is bound to a cytoplasmic calcium-binding protein (CaBP). The synthesis of the cytoplasmic CaBP is specifically stimulated by vitamin D. At the basolateral plasma membrane Ca^{++} is actively pumped out of the cell by a Ca^{++}-stimulated ATPase and by a Na^+, Ca^{++} exchange pump. (Redrawn from Bronner, F., et al.: Fed. Proc. **41:**61, 1982.)

Fig. 44-12 ■ Effects of vitamin D_3 on absorption of Ca^{++} by chick duodenum. Control animals were fed a diet deficient in vitamin D *(lower curve)*. Upper curve, Ca^{++} absorption by animals fed the same diet, but administered vitamin D_3 24 hours before the experiment. (Redrawn from Wasserman, R.H.: J. Nutrition **77:**69, 1962. © J. Nutr. American Institute of Nutrition.)

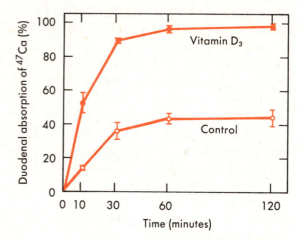

probably similar to the calcium pump of the sarcoplasmic reticulum, may be the principal mechanism for extruding Ca^{++} from the epithelial cells. In addition, there is a Na^+/Ca^{++} exchanger in the basolateral plasma membrane that uses the energy liberated when Na^+ enters the cell to pump Ca^{++} out. The Na^+/Ca^{++} exchanger seems to be quantitatively less important in extruding Ca^{++} from the cell than the Ca^{++}-stimulated ATPase.

Action of Vitamin D is essential for Ca^{++} absorption. In *rickets,* a disease due to vitamin D deficiency, there is a grossly low rate of Ca^{++} absorption. Fig. 44-12 shows the effects of administering vitamin D to chicks with rickets.

The mechanism by which vitamin D stimulates Ca^{++} absorption, and its other actions, is discussed in Chapter 51. Vitamin D stimulates the synthesis of the messenger RNA that codes for the cytoplasmic calcium-binding protein (Fig. 44-13). Some of the calcium-binding activity in the brush border membrane also is enhanced by the action of vitamin D.

One way in which the epithelial cells respond to the level of Ca^{++} in the diet is by

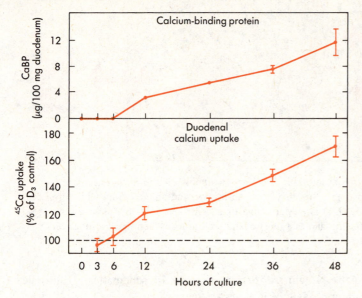

Fig. 44-13 ■ The correlation between the levels of calcium-binding protein (CaBP) and the rate of calcium uptake by chick duodenum in organ culture. At time zero, vitamin D_3 was added to the culture medium. (Redrawn from Corradino, R.A.: Endocrinology **94**:1607, 1974. © 1974, The Endocrine Society.)

altering the rate of synthesis of calcium-binding protein. Animals on a low-calcium diet have high levels of calcium-binding protein and vice versa.

The effect of parathyroid hormone on Ca^{++} absorption is not well understood. Parathyroid hormone weakly stimulates the pumping of Ca^{++} at the basolateral plasma membrane. This may be the basis for its effect on Ca^{++} absorption. Malabsorption of Ca^{++} does occur in hypoparathyroid individuals.

■ *Absorption of iron*

In the diet of a meat-eating American, heme is the major source of iron. A typical daily dietary intake is 15 to 20 mg of iron. Of this, 0.5 to 1 mg is absorbed by a typical adult male, and 1 to 1.5 mg is absorbed by a premenopausal woman.

About 30% to 50% of the iron in food is released in the stomach. The acid pH in the stomach prevents iron from forming insoluble complexes with phosphate and hydroxide. A glycoprotein secreted by the oxyntic cells binds iron. This protein, with a molecular weight of 350,000 is often called *gastroferrin*.

At the basic pH of the duodenum and the jejunum, which are the principal sites of iron absorption, there is a tendency for iron to form insoluble complexes with hydroxide, bicarbonate, and phosphate and thus to be made unavailable for absorption. Iron bound to gastroferrin and complexed by dietary ascorbate and fructose tends to escape this fate. Ascorbate also helps to maintain iron as Fe^{++}, which is much better absorbed than Fe^{+++}.

The mechanism of transport of iron into the cells of the intestinal epithelium is not yet well understood. Heme iron is taken up as heme presumably by a facilitated transport process. Fe^{++} is taken up by an active transport process at the brush border (Co^{++} and Mn^{++} compete for entry).

Inside the cell, iron is split from heme by a series of reactions involving xanthine oxidase. No intact heme is transported into the bloodstream. In the current view there are two pools of iron in the epithelial cells. One pool consists of iron hydroxide micelles bound to the protein *ferritin*. The other pool is soluble iron complexed by one or more intracellular carriers; certain amino acids may complex iron and serve as part of this soluble pool. The ferritin pool is not readily available for transport to the bloodstream, and most of the iron complexed with ferritin will be lost when the cell desquamates. It may be that the ferritin pool is available for absorption, but with a long course of time. The soluble pool is more rapidly transported actively out of the intestinal epithelial cells at their basal and lateral surfaces and is taken up by the mucosal capillaries. In the

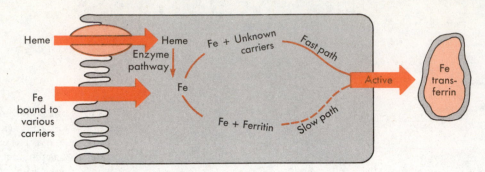

Fig. 44-14 ■ One view of iron absorption. There are two pools of iron in the intestinal epithelial cell, one that is available for fairly rapid absorption and another from which iron can be absorbed only very slowly (if at all). Many investigators believe the slow pathway involves iron bound to intracellular ferritin and that most of this iron is lost when the intestinal cell desquamates.

plasma, iron is bound to the protein *transferrin*. Fig. 44-14 schematically illustrates some of the features of a current view of iron absorption.

The rate of iron absorption is regulated in accordance with the body's need for iron. Iron absorption increases after a hemorrhage, with a time lag of 3 to 4 days. This is about the time required for epithelial cells to migrate from their sites of formation in the crypts of Lieberkühn to the tips of the villi, where they are most active in absorption. It is believed that the iron-absorbing capacity of the epithelial cells is programmed in the crypts in response to the plasma level of iron. Apoferritin synthesis is stimulated by iron at the translational level. The amount of apoferritin may determine how much iron can be trapped as ferritin in the epithelial cell and later be lost during desquamation. There is evidence that iron absorption by the more rapid pathway shown in Fig. 44-14 may also be regulated in some way by the level of iron in the body. There are limited opportunities for iron excretion. The menses is a major route for iron excretion in premenopausal women. By preventing absorption of more iron than is needed, the intestine plays a major role in homeostasis. *Siderosis* involves accumulation of excess iron in the body. This disease may sometimes result from a failure of the intestinal mechanisms that protect against absorption of too much iron.

■ *Absorption of other ions*

Magnesium. Magnesium is apparently absorbed along the entire length of the small intestine, with about half the normal dietary intake being absorbed. There is no evidence that Mg^{++} absorption is an active process, nor that the rate of Mg^{++} absorption is regulated to suit the body's need for it.

Phosphate. Phosphate is also absorbed all along the small intestine. Some phosphate may be absorbed by active transport. Little is known about the regulation of phosphate absorption.

Copper. Copper is absorbed in the jejunum, with approximately 50% of the ingested load being absorbed. Copper is secreted in the bile bound to certain bile acids, and this copper is lost in the feces. In individuals that fail to secrete sufficient amounts of copper in the bile, the body's copper pool grows and copper accumulates in certain tissues. The mechanisms by which absorption and excretion of copper are regulated are poorly characterized.

■ *Absorption of water-soluble vitamins*

Most water-soluble vitamins can be absorbed to a nutritionally significant extent by simple diffusion if they are taken in sufficiently high doses. Only relatively recently has it been recognized that specific transport mechanisms play important roles in the normal absorption of most water-soluble vitamins. Table 44-2 summarizes current knowledge of these transport mechanisms.

TABLE 44-2

Intestinal absorption of vitamins

Vitamin	Species	Site of absorption	Transport mechanism	Maximum absorption capacity in humans (per day)	Dietary requirements in humans (per day)
Ascorbic acid (C)	Humans, guinea pig	Ileum	Active	> 5000 mg	< 50 mg
Biotin	Hamster	Upper small intestine	Active	?	?
Choline	Guinea pig, hamster	Small intestine	Facilitated	?	?
Folic acid					
Pteroylglutamate	Rat	Jejunum	Facilitated	> 1000 μg/dose	100-200 ug
5-Methyltetrahydrofolate	Rat	Jejunum	Diffusion		
Nicotinic acid	Rat	Jejunum	Facilitated	?	10-20 mg
Pantothenic acid		Small intestine	?	?	(?)10 mg
Pyridoxine (B$_6$)	Rat, hamster	Small intestine	Diffusion	> 50 mg/dose	1-2 mg
Riboflavin (B$_2$)	Humans, rat	Jejunum	Facilitated	10-12 mg/dose	1-2 mg
Thiamin (B$_1$)	Rat	Jejunum	Active	8-14 mg	≈ 1 mg
Vitamin B$_{12}$	Humans, rat, hamster	Distal ileum	Active	6-9 μg	3-7 μg

Data from Matthews, D.M. In Smyth, D.H., editor: *Intestinal absorption,* vol. 4B (Biomembranes), London, 1974, Plenum; and Rose, R.C.: Annu. Rev. Physiol. **42:**157, 1980.

Ascorbic acid (vitamin C) Vitamin C is absorbed in the proximal ileum by active transport. The active transport of ascorbate depends on the presence of Na^+ in the lumen. It is believed that Na^+ and ascorbate are cotransported into the cell and that the energy for ascorbate transport comes from the energy in the electrochemical potential gradient of Na^+.

Biotin. Biotin appears to be actively taken up by the epithelial cells of the upper small intestine by a mechanism that depends on Na^+ in the lumen.

Folic acid (pteroylglutamic acid). Folic acid is absorbed by carrier-mediated transport (perhaps by active transport) in the jejunum. Pteroylpolyglutamates apparently share this transport pathway. Intracellularly the polyglutamates are cleaved to the monoglutamate derivative, which is transported into the blood. Another dietary source of folic acid, 5-methyltetrahydrofolate, is absorbed by simple diffusion.

Nicotinic acid. Nicotinic acid is absorbed by the jejunum partly by a Na^+-dependent, saturable mechanism. The details of nicotinic acid absorption remain to be elucidated.

Pyridoxine (vitamin B$_6$). Pyridoxine is one vitamin for which the evidence favors simple diffusion as the mechanism of absorption.

Riboflavin (vitamin B$_2$). Riboflavin is absorbed in the proximal small intestine, probably by facilitated transport (the evidence is not compelling). The presence of bile acids enhances the absorption of riboflavin by an unknown mechanism.

Thiamin (vitamin B$_1$). Thiamin is absorbed by a Na^+-dependent active transport mechanism in the jejunum. Some thiamin is phosphorylated in the jejunal epithelial cells, but the thiamin that appears in the blood is primarily free thiamin. The role of phosphorylation and dephosphorylation of thiamin in its absorption is not yet understood.

Vitamin B_{12}. A specific active transport process has also been implicated in the absorption of vitamin B_{12}. In the absence of vitamin B_{12} the maturation of red blood cells is retarded, and *pernicious anemia* ensues. Because of its medical importance, a good deal of attention has been paid to the absorption of vitamin B_{12}. As a result, we know much about the mechanism of its absorption. The dietary requirement for B_{12} is fairly close to the maximal absorption capacity for the vitamin (Table 44-2). There is no evidence that dietary restriction leads to an increase in the absorptive capacity. Enteric bacteria synthesize vitamin B_{12} and other B vitamins, but the colonic epithelium lacks specific mechanisms for their absorption.

Storage in liver. The liver contains a large store of vitamin B_{12} (2 to 5 mg). Vitamin B_{12} is normally present in the bile (0.5 to 5 µg daily), but about 70% of this is normally reabsorbed. Since only about 0.1% of the store is lost daily, even if absorption totally ceases the store will last for 3 to 6 years.

The role of intrinsic factor. The normal absorption of vitamin B_{12} depends on the presence of a glycoprotein with a molecular weight of about 55,000 that is known as *intrinsic factor* (IF). Human IF is secreted by the parietal (oxyntic) cells of the stomach. IF dimerizes in the presence of B_{12}, and the dimer binds two B_{12} molecules. IF binds all four physiologically important forms of B_{12} (cyanocobalamin, hydroxycobalamin, deoxyadenosylcobalamin, and methylcobalamin). The IF-B_{12} complex is highly resistant to digestion. Receptors on the mucosal membrane or in the glycocalyx of the epithelial cells that line the ileum bind the IF-B_{12} complex. The binding to the receptor is required for uptake of vitamin B_{12} into the cell. It is not known with certainty whether the IF-B_{12} complex or B_{12} alone enters the ileal epithelial cell. At the present time the most convincing evidence favors the hypothesis that B_{12} is split from IF extracellularly and that free B_{12} enters the cell by an active transport mechanism.

After the binding of the IF-B_{12} complex to the ileal receptors, it takes a long time for B_{12} to be transported through the epithelial cell and into the blood. Vitamin B_{12} does not appear in the blood until 4 hours after it is fed, and the peak B_{12} level in plasma occurs 6 to 8 hours after feeding. The reason for this delay is not well understood, but recent evidence suggests that for much of the lag period B_{12} is located predominantly in the mitochondria of the epithelial cells and that some conversion of cobalamin to deoxyadenosylcobalamin occurs there.

The exit of vitamin B_{12} from the cells of the ileal epithelium is even less well understood than its entry. Facilitated or active transport is presumably involved. Most of the B_{12} absorbed appears in the portal blood bound to transcobalamin II, a globulin. Transcobalamin II may be synthesized in the liver, but some evidence suggests that the ileal epithelium can also make this protein. The transcobalamin II–B_{12} complex is rapidly cleared from the portal blood by the liver.

Absorption in the absence of IF. In the complete absence of IF about 1% to 2% of an ingested load of B_{12} will be absorbed. If massive doses of B_{12} are taken (about 1 mg/day), enough can be absorbed to treat pernicious anemia. The IF-independent mechanism shows no maximal absorptive capacity, does not appear to be limited to the ileum, and shows a much shorter lag time (about 1 hour) than IF-dependent absorption.

Pernicious anemia and other diseases involving the malabsorption of vitamin B_{12}. In the absence of sufficient levels of B_{12} the maturation of red cells is retarded, and anemia results. Pernicious anemia is due to atrophy of the gastric mucosa, with almost complete inability to secrete HCl, pepsin, and IF. Most patients with pernicious anemia have serum antibodies against parietal cells, but it is not clear at this point whether the antibodies cause the disease or are the response to gastric damage from some other source.

Pernicious anemia in childhood is rare and has three forms: (1) an autoimmune type of pernicious anemia that has the characteristics just described, (2) congenital IF deficiency, in which pepsin and acid secretion are normal but IF secretion is deficient, and (3) congenital B_{12} malabsorption syndrome, in which there is normal gastric function and normal levels of IF are secreted, but B_{12} absorption is deficient due to a defect in the ileal IF-B_{12} receptors.

■ *Digestion and absorption of lipids*

The primary lipids of a normal diet are triglycerides. The diet contains smaller amounts of sterols, sterol esters, and phospholipids (Fig. 44-15). Because lipids are only slightly soluble in water, they pose special problems to the gastrointestinal tract at every stage of their processing. In the stomach, lipids tend to separate out into an oily phase. In the duodenum and small intestine, lipids are emulsified with the aid of bile acids. The large surface area of the emulsion droplets allows access of the water-soluble lipolytic enzymes to their substrates. The digestion products of lipids form small molecular aggregates, known as micelles, with the bile acids. The micelles are small enough to diffuse among the microvilli and allow absorption of the lipids from molecular solution at the intestinal brush border. The digestion and absorption of lipids are more complex than for any other class of nutrients and are thus more frequently subject to malfunction.

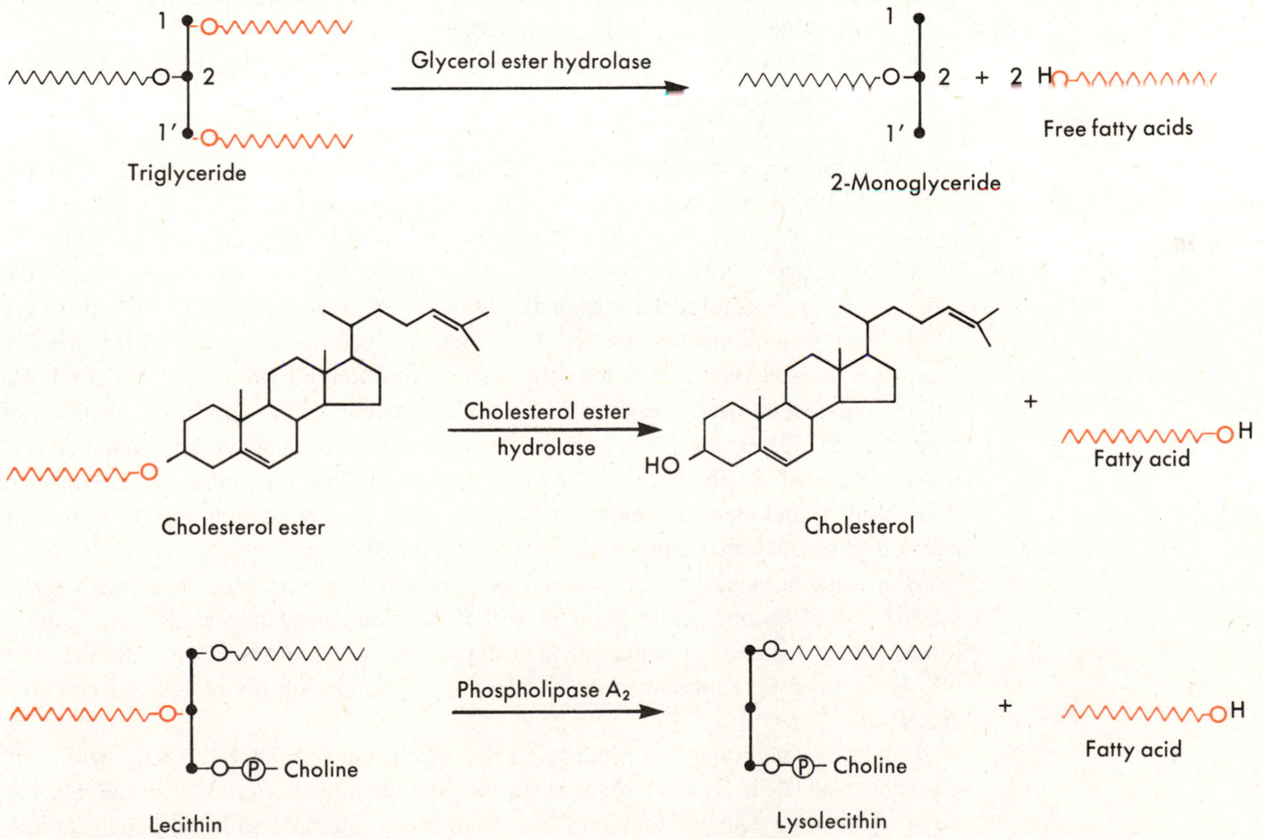

Fig. 44-15 ■ Action of major pancreatic lipases. The cleavage of lipids by glycerol ester hydrolase (pancreatic lipase), cholesterol ester hydrolase, and phospholipase A_2 is illustrated.

■ *In the stomach*

Because fats tend to separate out into an oily phase, they tend to be emptied from the stomach later than the other gastric contents. In spite of the presence of gastric lipase, little digestion of lipids occurs in the stomach. Any tendency to form emulsions with phospholipids or other natural emulsifying agents is inhibited by the high acidity. Fat in the duodenum is one of the most powerful stimuli to inhibit gastric emptying. This is important in ensuring that the fat is not emptied from the stomach more rapidly than it can be accommodated by the duodenal mechanisms that provide for emulsification and digestion.

■ *Digestion of lipids and micelle formation*

The lipolytic enzymes of the pancreatic juice are water-soluble molecules and thus have access to the lipids only at the surfaces of the fat droplets. The surface available for digestion is increased many thousand times by emulsification of the lipids. Bile acids themselves are rather poor emulsifying agents. However, with the aid of lecithin, which is present in high concentration in the bile, the bile acids produce an emulsion of dietary fats. The *emulsion droplets* are around 1 μm in diameter, and they have a large surface area on which the digestive enzymes can work.

The pancreatic secretions. Pancreatic juice contains the major lipolytic enzymes responsible for digestion of lipids (Fig. 44-15). The most important digestive enzymes follow:

1. *Glycerol ester hydrolase* (also called simply pancreatic lipase), which cleaves the 1 and 1′ fatty acids preferentially off a triglyceride to produce two free fatty acids and one 2-monoglyceride
2. *Cholesterol ester hydrolase,* which cleaves the ester bond in a cholesterol ester to give one fatty acid and free cholesterol
3. *Phospholipase A_2,* which cleaves the ester bond at the 2 position of a glycerophosphatide to yield, in case of lecithin, one fatty acid and one lysolecithin.

The formation of micelles of bile acids with the products of fat digestion, especially 2-monoglycerides, then occurs. The micelles are multimolecular aggregates (about 5 nm in diameter) containing about 20 to 30 molecules. 2-Monoglycerides and lysophosphatides tend to have their hydrophobic acyl chains in the interior of the micelle and their more polar portions facing the surrounding water. Bile acids are flat molecules that have a polar face and a nonpolar face (see Fig. 43-15). Much of the surface of the micelles is covered with bile acids, with the nonpolar face toward the lipid interior of the micelle and the polar face toward the outside. The exact structure of the micelles is a subject of current study. Extremely hydrophobic molecules, such as long-chain fatty acids, cholesterol, and certain fat-soluble vitamins, tend to partition into the interior of the micelle. Phospholipids and monoglycerides tend to have their more polar ends facing the outside aqueous phase. Micelles contain almost no intact triglyceride.

Bile acids must be present at a certain minimum concentration, called the *critical micelle concentration,* before micelles will form. Conjugated bile acids have a much lower critical micelle concentration than the unconjugated forms. In the normal state bile acids are always present in the duodenum at greater than the critical micelle concentration.

Lipids and lipid digestion products in the micelles are in rapid exchange with lipid digestion products in aqueous solution surrounding the micelle. In this way the micelles serve to keep the aqueous solution surrounding them saturated with 2-monoglycerides, various fatty acids, cholesterol, and lysophosphatides. These lipids are present in the aqueous solution at a low concentration because of their limited water solubility.

■ *Absorption of the products of lipid digestion*

Transport into the intestinal epithelial cell. The micelles play an important role in the absorption of the products of lipid digestion and most other fat-soluble molecules (such as the fat-soluble vitamins). The micelles diffuse among the microvilli that form the brush border. They keep the aqueous solution in contact with the brush border saturated with fatty acids, 2-monoglycerides, cholesterol, and other micellar contents.

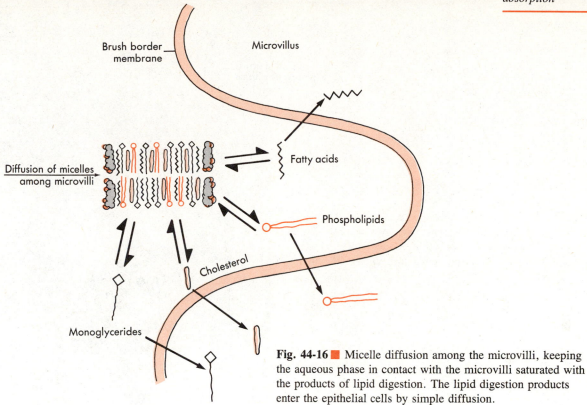

Fig. 44-16 ■ Micelle diffusion among the microvilli, keeping the aqueous phase in contact with the microvilli saturated with the products of lipid digestion. The lipid digestion products enter the epithelial cells by simple diffusion.

In this way the huge surface area of the brush border is made available for the absorption of the micellar contents (Fig. 44-16).

Because of their high lipid solubility, the fatty acids, 2-monoglycerides, cholesterol, and lysolecithin can readily diffuse across the brush border membrane. There are no known protein-mediated mechanisms for transporting the products of lipid digestion across the plasma membrane of the intestinal epithelial cells. Cholesterol is absorbed more slowly than most of the other constituents of the micelles, so as the micelles progress down the small intestine, they become more concentrated in cholesterol. The duodenum and jejunum are most active in fat absorption, and most ingested fat is absorbed by the midjejunum. The fat present in normal stools is not ingested fat (which is completely absorbed), but fat from colonic bacteria and from desquamated intestinal epithelial cells.

Inside the intestinal epithelial cell. Here the products of lipid digestion make their way to the smooth endoplasmic reticulum. A recently discovered cytoplasmic fatty acid–binding protein may play a role in transporting fatty acids to the smooth endoplasmic reticulum, or it may simply function to prevent fatty acids from forming fat droplets prematurely.

In the smooth endoplasmic reticulum, which is engorged with lipid after a meal, considerable chemical reprocessing goes on (Fig. 44-17). The 2-monoglycerides are re-esterified with fatty acids at the 1 and 1′ carbons to re-form triglycerides. Lysophospholipids are reconverted to phospholipids. Cholesterol is re-esterified to a considerable extent, although some free cholesterol remains. The processing of 2-monoglycerides and lysophospholipids is essentially complete, since the blood levels of these components are negligible. The intestinal epithelial cells are also capable of the de novo synthesis of lipids to some extent. Triglycerides can be made from fatty acids and glycerol. Limited processing of fatty acids occurs. Cholesterol can be synthesized from acetate; phospholipids can be made from fatty acids and glycerol.

Chylomicron formation and transport. The reprocessed lipids, along with those

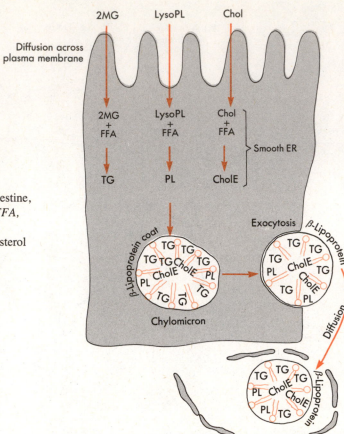

Fig. 44-17 ■ Lipid resynthesis in the epithelial cells of the small intestine, chylomicron formation, and subsequent transport of chylomicrons. *FFA,* Free fatty acid; *2MG,* 2-monoglyceride; *TG,* triglyceride; *lysoPL,* lysophospholipid; *PL,* phospholipid; *Chol,* cholesterol; *CholE,* cholesterol ester.

which are synthesized de novo, accumulate in the vesicles of the smooth endoplasmic reticulum. Phospholipids tend to cover the external surfaces of these lipid droplets, with their hydrophobic acyl chains in the fatty interior and their polar head groups toward the aqueous exterior. The lipid droplets, of the order of 1 nm in diameter at this point, are known as *chylomicrons*. About 10% of their surface is covered by β-lipoprotein, which is synthesized in the intestinal epithelial cells.

Chylomicrons are ejected from the cell by exocytosis (Fig. 44-17). It appears that the β-lipoprotein is important in this process; when it is absent, the intestinal epithelial cells become engorged with lipid, and lipid absorption is severely impaired. The chylomicrons leave the cells at the level of the nuclei and enter the lateral intercellular spaces. Chylomicrons are too large to pass through the basement membrane that invests the mucosal capillaries. However, they do enter the lacteals, which have sufficiently large fenestrations for the chylomicrons to pass through. The chylomicrons leave the intestine with the lymph, primarily via the thoracic duct, and are dumped into the venous circulation.

The absorption of bile acids. This process occurs chiefly in the terminal ileum by an active transport process. Bile acids are also absorbed to a significant extent by simple diffusion along the length of the intestine. Conjugated bile acids are the principal substrates for the active transport system; deconjugated and dehydroxylated (secondary) bile acids are not transported as well. The deconjugated and dehydroxylated bile acids are less polar, however, and are thus better absorbed by simple diffusion. Bile acids leave the gut in the portal blood and are returned to the liver, whose cells avidly take up the bile acids. The bile acids are then reprocessed and resecreted in the bile.

Malabsorption of lipids occurs more frequently than malabsorption of proteins or carbohydrates. Among the general causes of lipid malabsorption are bile deficiency, pancreatic insufficiency, and the intestinal mucosal atrophy that occurs in some disease states.

In cases of *bile deficiency* and *pancreatic insufficiency* the levels of bile acids and lipolytic enzymes, respectively, must be severely reduced before serious malabsorption occurs. In both cases the quantity of fecal fat is roughly proportional to the quantity ingested.

Even in the *complete absence of bile acids,* hydrolysis of triglyceride occurs to a significant extent. The rate of absorption of fatty acids from triglycerides may be 50% of normal. Cholesterol, cholesterol esters, and fat soluble vitamins are much less water-soluble than fatty acids, and their absorption is grossly deficient in the absence of bile acids.

In the *complete absence of pancreatic lipases* there is marked malabsorption of all lipid classes. This is probably due to the necessity of 2-monoglycerides and lysophosphatides (products of the action of pancreatic lipases) for the formation of mixed micelles with bile acids.

In *tropical sprue* and *gluten enteropathy* there is flattening of the intestinal epithelium and a decreased density of microvilli. Lipid malabsorption in these diseases is probably a consequence of the marked decrease in the surface area available for lipid absorption.

■ *Malabsorption of lipids*

■ *Absorption of fat-soluble vitamins*

Because of their solubility in nonpolar solvents, the fat-soluble vitamins (A, D, E, and K) partition into the mixed micelles formed by the bile acids and lipid digestion products. As is the case for other lipids, the fat-soluble vitamins enter the intestinal epithelial cell by diffusing across the brush border plasma membrane. There is no evidence for the mediation by membrane proteins of the passage of fat-soluble vitamins across the brush border plasma membrane. In general, the presence of bile acids and lipid digestion products enhances the absorption of fat-soluble vitamins. In the intestinal epithelial cell the fat-soluble vitamins enter the chylomicrons and leave the intestine in the lymph. In the absence of bile acids a significant fraction of the ingested load of a fat-soluble vitamin may be absorbed and leave the intestine in the portal blood.

Vitamin A. Vitamin A (retinol) is somewhat better absorbed than is provitamin A (β-carotene) and vitamin A aldehyde (retinal). In the epithelial cells of the intestine, β-carotene and retinal are converted to retinol. Vitamin A is present in thoracic duct lymph primarily as fatty acid esters of retinol. Vitamin A acid (retinoic acid) is absorbed independently of the bile acid–mixed micelles and leaves the intestine in the portal blood.

Vitamin D. Vitamin D is absorbed principally in the jejunum, primarily as the free vitamin. Most fatty acid esters of vitamin D are hydrolyzed in the intestinal lumen prior to absorption. With normal oral loads, 55% to 99% of the vitamin D ingested is absorbed.

Vitamin E. The absorption of vitamin E (α-tocopherol) requires the presence of bile acid–mixed micelles. Small oral doses of vitamin E are essentially completely absorbed. With larger doses, a significant fraction escapes absorption and appears in the feces. In the intestinal epithelial cell, vitamin E enters the chylomicrons and leaves the intestine in the lymph.

Vitamin K. Vitamins K_1 and K_2, which have hydrophobic side chains, partition into bile acid–mixed micelles, are absorbed from free solution, enter the chylomicrons, and leave the intestine in the lymph. Vitamin K_3 (2-methylnaphthoquinone) lacks a side chain, is absorbed independently of the mixed micelles, and leaves the intestine primarily in the portal blood.

■ *Bibliography*

Journal article

Rose, R.C.: Water-soluble vitamin absorption in intestine, Annu. Rev. Physiol. **42:**157, 1980.

Books and monographs

Crane, R.K., editor: Gastrointestinal physiology, ed. 2, Baltimore, 1977, University Park Press.

Czaky, T.Z., editor: Intestinal absorption and malabsorption, New York, 1975, Raven Press.

Davenport, H.W.: Physiology of the digestive tract, ed. 4, Chicago, 1977, Year Book Medical Publishers.

Krejs, G.J., and Fordtran, J.S.: Physiology and pathophysiology of ion and water movement in the human intestine. In Sleisenger, M.H., and Fordtran, J.S., editors: Gastrointestinal disease, ed. 2, Philadelphia, 1978, W.B. Saunders Co.

McColl, I., and Sladen, G.E., editors: Intestinal absorption in man, New York, 1975, Academic Press, Inc.

Rose, R.C.: Water-soluble vitamin absorption in intestine, Annu. Rev. Physiol. **42:**157, 1980.

Smyth, D.H., editor: Intestinal absorption, London, 1974, Plenum Press.

THE KIDNEY

Brian R. Duling

chapter 45

Components of renal function

The function of the kidney is, in simplest terms, to regulate the composition and the volume of the extracellular fluid, in spite of wide variations in an animal's environment and intake of food and water. Very simple organisms, living freely in aquatic surroundings, have a cellular environment that is stabilized by the relatively large fluid volumes around the cells. The excretory and regulatory needs of these organisms can be met by the simple expedient of direct exchange of waste products with the environment. More complex organisms must ingest foods, transform these into energy forms that can be used by various tissues, and then excrete the wastes generated in the process. Intake of food and the unavoidable loss of salts and water through processes such as respiration, perspiration, and defecation make it vital that some structure in the body be capable of adapting to varying exchanges between the organism and its environment. This adaptation to the needs of the organism is the primary role played by the kidney in homeostasis.

The kidney serves a variety of other needs, however, especially through its role in the production of hormones, most notably *angiotensin II, prostaglandins,* and the *kinins*. The kidney also is involved in the regulation of red cell synthesis through the formation of the hormone *erythropoietin*.

The formation of urine is usually thought of as a three-step process. The first step is *filtration* of a portion of the blood through a specialized ultrafilter, the *glomerulus*. The filtrate is free of both blood cells and large molecular weight materials in the plasma. The fluid formed by filtration then is modified as it flows through a series of very narrow tubes formed of epithelial cells, the *nephrons*. Some materials are *reabsorbed* from the filtrate and returned to the plasma. Other materials are removed from the plasma circulating through the kidney and added to the fluid that ultimately will be excreted as urine, that is, they are *secreted*. As a rule, the volume of the filtrate is reduced by about 99% before elimination from the body. The major fraction of filtered salts such as sodium also is reabsorbed. Materials such as ammonia may be secreted to produce urine concentrations hundreds of times greater than the plasma concentration.

Each of these processes is controlled so as to integrate renal function with the requirements for body homeostasis. Control is the result of both neural and humoral mechanisms.

■ Body fluid compartments

The primary function of the kidney is to stabilize the composition of the extracellular fluid and in so doing to regulate the composition of intracellular fluid. The kidney regulates the composition of the extracellular fluid indirectly, however, by adjusting the composition of plasma in the renal circulation. It is therefore essential to understand how the various fluid compartments within the body are interrelated to understand how a particular renal action may influence first plasma, then the extracellular fluid, and finally intracellular fluid.

Fig. 45-1 ■ Compartmentation of body water. Basic volumes are intracellular *(color)* and extracellular water *(gray)*. Extracellular volume is the sum of interstitial and plasma volumes (that is, 12% + 4.5% = 16.5%) and a slowly equilibrating component consisting of transcellular elements, cartilage, and bone (10.5%). Note the special position occupied by red cells in this scheme, that is, as part of the intracellular compartment that circulates in the blood. The membranes separating the various compartments (cell membrane and capillary endothelium) determine the type of tracer that can be used in the volume measurement (Table 45-1).

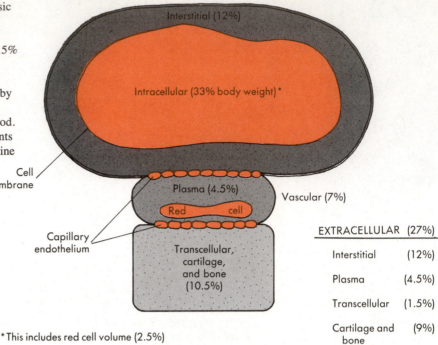

Fig. 45-1 illustrates the major body fluid compartments and the relations among these compartments. The figure emphasizes the relative volumes of the compartments and the fact that they are separated by membranes with different characteristics. A major concern of renal physiology is how body fluids behave, and therefore the figure reflects the distribution of the water in the body, not the total body mass. About 60% of the total body weight is water.

Body water is partitioned between *intracellular water*, which is 33% of body weight *(colored regions,* Fig. 45-1), and extracellular water, which is 27% of body weight *(gray regions*, Fig. 45-1). The cell membranes separate intracellular water from extracellular water, and the membrane properties and transport characteristic exert a major influence on the concentration differences between compartments.

Extracellular space is a combination of several volumes that are functionally different. *Interstitial fluid,* which represents about 12% of body weight, is that fluid in intimate contact with the cells, and together with the *plasma* it comprises a portion of the extracellular fluid that rapidly achieves equilibrium in various parts of the body. For this reason it is sometimes called the *fast extracellular fluid* (ECF,F). A significant fraction of the extracellular fluid is relatively inaccessible, being contained in closed cavities such as the joints or cerebral ventricles *(transcellular water)* or bound up in matrices of cartilage or bone. This amounts to about 10.5% of the body weight. The total extracellular volume is then the sum of interstitial volume, plasma volume, transcellular water, and bound water (27% of body weight). The compartment volumes are listed in Table 45-1 both as fractions of body weight and as fractions of total body water.

The volume of the compartments may change during adaptations to environmental stress or with disease, and therefore it has been necessary to develop a variety of methods for measuring body fluid volumes. A highly useful clinical method for measuring short-term changes in total body water is to simply measure the body weight. Over a period of a few days changes in body weight represent almost entirely gains or losses in body water. Thus careful records of weight changes can accurately reflect alterations in total body water.

Direct measurement of the various components of body water is more difficult and

TABLE 45-1

Measurement of the volumes of the various fluid compartments in the body

Compartment	Dilution indicator	Volume (percent of body weight)	Volume (percent of body water)
Vascular (V_B)	[P/(100 − hematocrit)]	7	11.7
Plasma (P)	Serum albumin	4.5	7.5
Interstitital (Int)	[ECF,F − plasma]	12	20
Extracellular, fast (ECF,F)	Inulin, thiosulfate	16.5	27.5
Extracellular, slow (ECF,S)	Thiocyanate	27	45
Intracellular (ICF)	[BW − ECF,S]	33	55
Total body H$_2$0 (BW)	DHO, antipyrine	60	100

From Koushanpour, E.: Renal physiology: principles and functions, Philadelphia, 1976, W. B. Saunders Co.

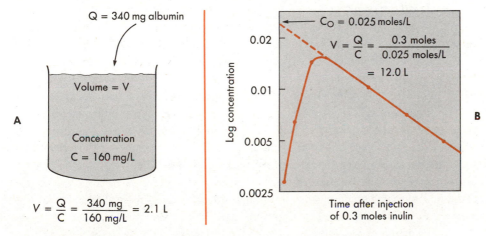

Fig. 45-2 ■ Use of the dilution principle in the determination of compartment volumes. **A,** Simple estimation of plasma volume uncorrected for loss. **B,** Correction for loss of indicator. Inulin is assumed to be lost in the urine during the time required for equilibration. Losses are corrected for by *back extrapolation* to zero time.

requires use of the *indicator dilution method*. The basic concept is quite simple and is illustrated in Fig. 45-2, *A*. It involves the injection of a known quantity (Q) of tracer into the body and the subsequent measurement of the resultant concentration (C) when the tracer comes into equilibrium with the fluid in the compartment. The volume of the compartment (V) then can be computed from the ratio of Q to C.

Two problems arise in the use of the indicator dilution method as shown. First, there are several fluid compartments in the body, and only an indicator that is confined to the appropriate compartment will be suitable. Second, loss of indicator by excretory processes during the measurement will lead to an overestimation of the compartment volume. The solution to the first problem has been accomplished largely through the selection of indicators with characteristics appropriately related to the behavior of the membranes that surround the compartments. Table 45-1 lists some of the indicators commonly used for the measurement of the various fluid compartments. Plasma volumes are most simply measured by using a labeled serum albumin molecule. Serum albumin crosses the capillary endothelium only very slowly and equilibrates within the plasma compartment within a few minutes.

The measurement of extracellular volume requires a molecule smaller than albumin, so it can readily enter interstitial space, but not so small or permeable that it can cross cell membranes. Because of the complex composition of the extracellular fluid, there is no single molecule that will completely define the extracellular fluid volume. There is a rapid and free exchange of material between the plasma and interstitial fluid, and molecules such as inulin and thiosulfate will readily trace these volumes, which comprise *fast extracellular fluid*. Smaller, more permeable, tracer molecules such as bromide and thiocyanate will equilibrate with the interstitial space and plasma also, but in addition, they will enter the cartilage as well as the transcellular compartments. The volume of these elements is referred to as the *slow extracellular volume* (ECF,S) because of the rate of equilibration.

To measure total body water, a compound is required that readily crosses cell membranes and is distributed uniformly in water. Deuterated or tritiated water and antipyrine are commonly used substances. The other body compartments must be computed by difference measurements (Table 45-1). Note that the total blood volume (V_B) is computed by dividing the measured plasma volume by 1 minus the measured red cell fraction, or 100 minus the hematocrit.

The second problem that arises in the use of the indicator dilution method is the loss of tracer. This is particularly difficult to deal with when using small molecules such as inulin, since these molecules are excreted rapidly by the kidney. Water leaves the body in urine, sweat, and expired air. To compensate for this loss, a commonly used technique is repeated measurement of the concentration of the indicator in the plasma following injection, and extrapolation of the measured concentrations to zero time. Fig. 45-2, *B,* illustrates this process for a typical estimation of extracellular volume. After injection of the tracer there is an initial mixing period, during which time the concentration of the tracer rises to a peak value. With continued passage of time renal losses of the tracer cause a progressive decrease in concentration. Typically this decrease occurs as an exponential function of time. As shown in Fig. 45-2, *B,* the concentrations are plotted on a log scale against time on a linear scale. The straight portion of the resulting curve is extrapolated to zero time, and the concentration at this point is used in the computation of the volume of the compartment.

■ Osmotic interactions between compartments

Through intake or loss body water can vary from day to day. It is important to understand how gains or losses in water may be distributed among the fluid compartments. This can become an extremely complex problem, but basically two relations govern water exchanges between compartments. First, the osmotic activity in all body compartments is very close to equilibrium. That is, in a steady state the osmotic activity of intracellular fluid, extracellular fluid, and plasma always will be very close to the same value. Thus removal of water from one compartment with a subsequent increase in osmolality will automatically result in the osmotic movement of water from the other compartments into the one with increased activity, quickly yielding the same osmolality in all compartments.

The second relation that permits one to understand fluid shifts among the body compartments is that shifts of body water will occur in proportion to the volumes of the compartments. For example, if a patient loses 1 L of pure water by respiratory losses and *insensible perspiration* (evaporation from skin without the appearance of fluid on the skin surface), osmotic activity can remain the same in all compartments only if this liter comes proportionally from both the intracellular and the extracellular spaces. Examination of Table 45-1 shows that 55% of the fluid must come from intracellular space and 45% from extracellular space if all the body fluid compartments are involved. However, in acute fluid shifts the water of the body cavities as well as that of bones and cartilage is not readily shifted, and therefore only interstitital water (20% of total water), plasma water (7.5% of total water) and intracellular water (55% of total body weight) are involved. Thus only about 82.5% (20% + 7.5% + 55%) of the total water exchanges on an acute basis. From this it can be calculated that about one third of the liter of water mentioned previously would come from extracellular fluid, and two thirds would come from intracellular fluid.

Ions are distributed in markedly nonuniform patterns among the various body compartments (Chapter 2). Table 45-2 summarizes the distributions of cations and anions between the intracellular and extracellular space. Note that concentrations are expressed both as mEq/L plasma and as mEq/L plasma water. This is necessary because plasma contains a high protein content (6% by weight), and thus only about 94% of the volume of plasma is actually water. Sodium is the major extracellular cation, and potassium is the major intracellular cation. Similarly the major intracellular anions are proteins and phosphate, whereas the major extracellular anions are chloride and bicarbonate. Chapter 2, on cell physiology, deals with the causes for these nonequilibrium distributions of the various anions and cations and relates these to the *Donnan effect* and the sodium-potassium pump.

The nonuniform distributions of ions have major implications for the ways in which changes in body fluids may alter the volumes of the compartments. For example, if an isotonic sodium chloride solution is infused into a subject, the fact that most sodium is localized in the extracellular space will cause most of the volume expansion induced by the sodium chloride solution to occur in the extracellular space.

Another important consequence of the relative sizes of the compartments and the distributions of ions is that the plasma composition is often a misleading index of the status of electrolyte balance within the body. The most striking example of this occurs in circumstances where potassium is depleted from the organism. Since potassium is largely an intracellular cation, and since intracellular fluid represents the majority of body water, the vast bulk of body potassium stores are found in the cells. Depletion of body potassium results in an exchange of potassium between the intracellular and extracellular stores. As a result, extracellular fluid potassium levels may fall little in spite of substantial losses in body potassium. Only by direct measurement of total body potassium or by a clear understanding of the source of the pathologic condition is it possible to gain insight into the changes in potassium levels in tissue cells.

■ *Ion distributions and body compartments*

TABLE 45-2

Distribution of ions in body compartments

	Intracellular (mEq/L)	*Interstitial (mEq/L)*	*Plasma (mEq/L)*	*Plasma water (mEq/L)*
Cations				
Na^+	10	147	142	153
K^+	140	4.0	5.0	5.4
Ca^{++}	5	2.5	5.0	5.4
Mg^{++}	27	2.0	3.0	3.2
Anions				
HCO_3^-	10	30	27	29
Cl^-	25	114	103	111
$PO_4^=$	80	2.0	2	2.2
$SO_4^=$	20	1.0	1	1.1
Organic acids	—	7.5	6	6.5
Proteins	47	1.0	16	17.2

From Koushanpour, E.: Renal physiology: principles and functions, Philadelphia, 1976, W. B. Saunders Co.

■ *Functional renal anatomy*

Form and function of the kidney are related in a most intricate fashion, and understanding of renal physiology is closely tied to understanding of renal anatomy and histology. The kidneys in the human are located bilaterally in the abdomen on either side of the aorta. The gross anatomical form is bean shaped. Each kidney is divided into seven to nine lobes, or *pyramids,* which are the largest functional units of the organ (Fig. 45-3, *A*). Each of these lobes is divided into a *cortical,* or outer, portion and a

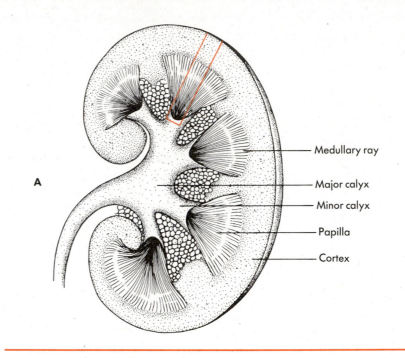

A

— Medullary ray

— Major calyx

— Minor calyx

— Papilla

— Cortex

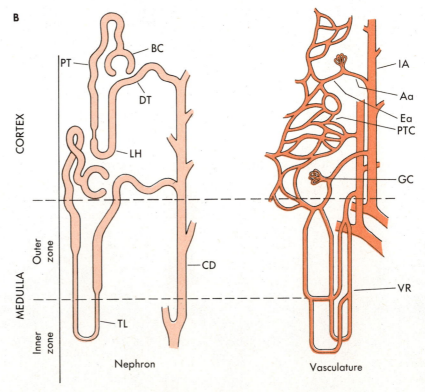

B

Fig. 45-3 ■ For legend see opposite page.

medullary, or inner, portion. The initial steps in urine formation occur in the cortex. Fluid flows subsequently through the medulla and finally exits the kidney through the *ducts of Bellini* on the surface of the papilla. The urine enters a minor calyx of the ureter and flows into the bladder.

The nephron, as shown in Fig. 45-3, *B*, is an extremely long, tubular structure that extends from the cortex into the medulla. Each nephron begins with a filtering unit, Bowman's capsule *(BC)*, and the *glomerulus (GC)* (more detail of this structure is shown in Fig. 45-4). Fluid filtered by the glomerulus initially flows through a highly convoluted section, the *proximal convoluted tubule (PT*, simplified in Fig. 45-3, *B)*. Fluid then enters into a straight section of the proximal tubule and flows into the medulla. After penetrating a variable distance into the medulla, the tubule turns back and reenters the cortex. The hairpin loop formed by this segment of the nephron is called the *loop of Henle (LH)* and is divided into a *descending limb* and an *ascending limb*.

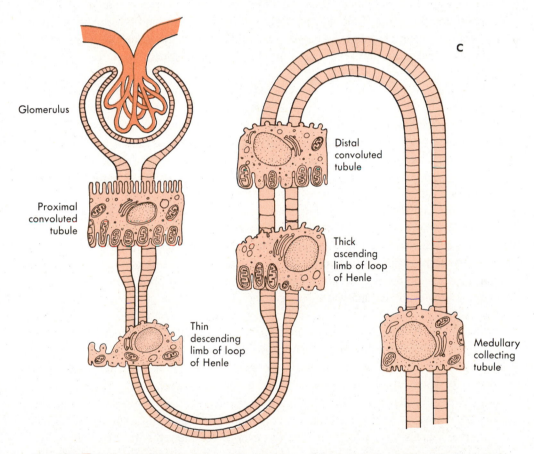

Fig. 45-3 ■ Anatomy of the nephrons and vasculature of the mammalian kidney. **A,** A sagittal section through the kidney showing gross anatomical relations. **B,** The segment from the renal pyramid shown by the colored lines in **A** is enlarged to show the relations among nephrons from cortical and juxtamedullary regions. The vascular and tubular anatomy are drawn to the same scale, and, if desired, a transparent copy of the tubular anatomy may be made to overlay the vascular anatomy. Note that only juxtamedullary nephrons possess long thin loops which penetrate into the interstitium of the medulla. Also it should be noted that the distal tubule of both the cortical and the juxtamedullary nephrons returns to close proximity with the afferent and efferent arteriole of the glomerulus from which it originated. (This association is called the juxtamedullary apparatus and is described in the section on renal blood flow.) Finally, note that the drainage from both cortical and juxtamedullary nephrons flows out through a common collecting duct. *PT,* Proximal tubule; *BC,* Bowman's capsule; *DT,* distal tubule; *LH,* loop of Henle; *TL,* thin limb; *CD,* collecting duct; *IA,* interlobular artery; *Aa,* afferent arteriole; *Ea,* efferent arteriole; *PTC,* peritubular capillary; *GC,* glomerular capillary; *VR,* vasa recta. **C,** The histology of the cells of the various segments of the tubule.

The length of Henle's loop varies with the site of origin of the nephron within the cortex. Nephrons originating in the superficial part of the cortex are called *cortical nephrons* and are short and thick over much of their length. Most loops of Henle penetrate only a short way into the medulla and then turn back into the cortex. In contrast, if the nephron originates more deeply within the cortex, it is referred to as a *juxtamedullary nephron*. The loops of these nephrons are much longer and have segments midway along their length that are very thin. The thin portions of Henle's loop penetrate the entire length of the medulla. The segment of the nephron following Henle's loop is called the *distal tubule (DT)*. It possesses both a straight and a convoluted section.

The distal tubule comes into close association with the glomerulus and the proximal convoluted tubule of the same nephron. The association between the distal tubule and the glomerulus forms a complex anatomical structure, the *juxtaglomerular apparatus*, which is described in detail later in relation to the control of renal blood flow. After passing through the distal tubule, urine from many nephrons enters a common *collecting duct (CD)* and flows out of the kidney. Note in Fig. 45-3, *B*, that both superficial and juxtamedullary nephrons may discharge their contents into the same collecting duct. Fluid in the final urine thus represents a mixture of urine formed by the two nephron populations.

The nephron is formed entirely of epithelial cells, but there are major differences in cell types along the tubules. The typical ultrastructure seen in the various tubular epithelial cells is shown in Fig. 45-3, *C*. Cells from the proximal convoluted tubule have large concentrations of mitochondria at the basal surface, marked basal membrane infoldings, and a well-developed brush border on the epithelium. The cells of the more distal straight portions of the proximal tubule and thick limbs of Henle's loop have fewer mitochondria and a more sparse brush border. In the thin segment of Henle's loop the cells have few mitochondria and only a very sparse brush border. Mitochondria and brush border reappear in the distal tubule, although not with the abundance observed in the proximal convoluted tubule. The collecting duct has an appearance that is intermediate between the distal convoluted tubule and the descending limb of Henle's loop. As explained later, the presence of mitochondria in abundance correlates with the transport of large quantities of salt and water, as does the presence of an abundant brush border on the epithelium.

Adjacent epithelial cells are joined by a variety of types of cell junctions that confer a wide range of permeability characteristics on the segments of the tubule; in general, the junctions of the proximal tubule are more open than those of the distal tubule, and therefore the epithelium is more permeable to solutes and water.

The pattern of the renal vasculature closely parallels the nephron structure, and the association between the two elements has important functional implications. Blood enters via the renal artery, which ramifies throughout the kidney with three successive right-angle branches. These branches finally yield a set of distributing vessels, the *interlobular arteries*, which pass radially from the corticomedullary border into the cortex and supply the glomerular circulation. The arterial blood supply to the glomerulus enters via the *afferent arteriole (Aa)*, passes through the *glomerular capillaries*, and exits through the *efferent arteriole (Ea*, Fig. 45-3, *B)*. In most of the cortex, blood leaving the glomerulus flows into a second capillary network, the *peritubular capillaries*, which is intertwined among the convolutions of the proximal and distal tubules.

The vascular pattern of the juxtamedullary nephrons is very similar to that of the superficial cortical nephrons, with the addition of a parallel circuit that enters the medulla. This network originates distal to the efferent arterioles of the glomeruli of the juxtamedullary nephrons, and it forms the *vasa recta*. The vasa recta are long thin capillary loops that match the pattern of the thin loops of the medullary Henle's loop.

Blood from juxtamedullary peritubular capillaries, superficial peritubular capillaries, and vasa recta capillaries combines and returns to the venous drainage that parallels the arterial supply.

Examination of Fig. 45-3, *B* and *C,* shows the reason for the different appearance of the three portions of the kidney: cortex, inner medulla, and outer medulla. The cortex is mainly glomeruli, convoluted tubules (both distal and proximal), and peritubular capillaries. The outer medulla is formed primarily of vasa recta and thick segments of Henle's loop. The inner medulla is formed of vasa recta, collecting ducts, and thin limbs of Henle's loop.

■ *Glomerular filtration*

It is possible to insert small glass pipettes into the beginning of the proximal tubule and withdraw fluid samples for chemical analysis. When this is done, it becomes apparent that the material entering the proximal tubule is an ultrafiltrate of plasma, that is, a fluid whose composition is essentially identical to that of plasma, with the exception that the large molecules, especially proteins, have been sieved out. These findings have been combined with the predictions of the Starling hypothesis for capillary exchange to yield quantitative predictions as to the nature of glomerular filtration.

■ *The Starling hypothesis*

The algebraic statement of the Starling hypothesis (Chapter 31) is:

$$\text{Glomerular filtration rate} = k[(P_c + \pi_t) - (P_t + \pi_c)]$$

where

P_c = glomerular capillary hydrostatic pressure
P_t = proximal tubular hydrostatic pressure
π_c = glomerular capillary oncotic pressure
π_t = proximal tubular oncotic pressure

The algebraic sum of the terms in brackets is called the *net filtration pressure.* k is a coefficient that describes how fast fluid moves across the glomerulus under a given net filtration pressure. k values can be estimated either for a single glomerulus or for the kidney as a whole. The combined effective k for both kidneys in the human is about 15 ml/minute/mm Hg. In other words, if the net driving pressure (the sum of the difference in hydrostatic and oncotic pressures across the glomerulus) is 1 mm Hg, a filtration rate of 15 ml/minute will be induced. This filtration rate reflects, of course, the sum of the individual filtration rates of all the approximately one million nephrons found in each kidney.

■ *Anatomical correlates of the filtration coefficient*

For capillaries in other vascular beds it appears that the anatomical correlate of the filtration coefficient is closely related to the space between endothelial cells. Fig. 45-4 illustrates the more complex anatomical basis of the filtration coefficient in the kidney. The glomerular membrane consists of a *fenestrated endothelium* overlying a *basement membrane,* which is a complex structure composed of collagen and glycoprotein (Fig. 45-4, *B*). Beneath the basement membrane lie the *foot processes* of the *podocytes.* The podocytes are specialized epithelial cells that are continuations of the epithelium of the proximal tubule, and the foot processes are extensions of these. The spaces between the epithelial cell foot processes are spanned by a diaphragm called the *filtration slit membrane.* In transmission electronmicrographs this diaphragm appears as a line connecting two podocytes with a dot midway across it. When viewed en face, that is, when rotated 90 degrees from the cross-sectional view shown in Fig. 45-4, *B,* the slit diaphragm is in fact a ladderlike structure with spaces between the rungs (Fig. 45-4, *C*).

Some understanding of the permeability characteristics of the glomerulus can be obtained by infusing molecules of varying sizes into the blood and analyzing plasma and proximal tubular fluid concentrations. Such an experiment yields data like those shown in Fig. 45-5. Note that molecules smaller than about 15 Å radius are found in

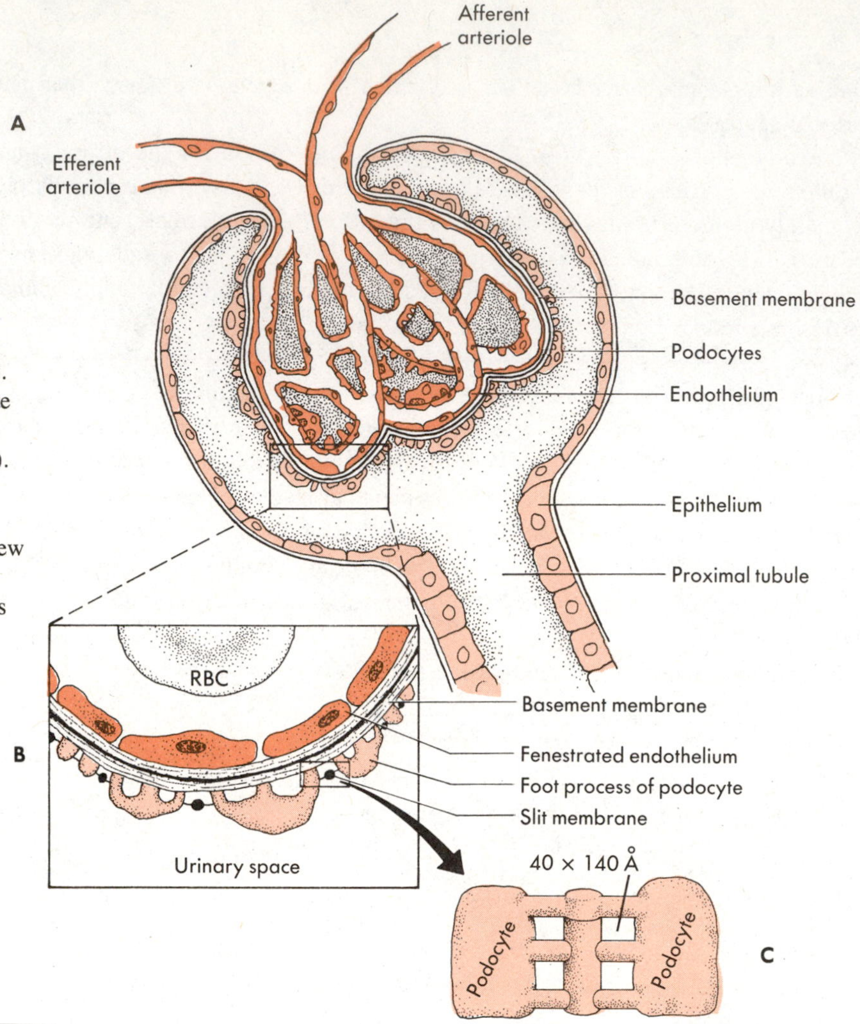

Fig. 45-4 ■ Anatomy of the mammalian glomerulus. **A,** A cross section through a single glomerulus. Note that the glomerular capillaries form an invagination into a capsule of epithelial cells (Bowman's capsule). The surface of Bowman's capsule covers the glomerular capillaries in the form of specialized epithelial cells called podocytes. **B,** An expanded view of the filtration membrane of the glomerulus. **C,** A view of one of the slit membranes rotated 90 degrees and viewed en face.

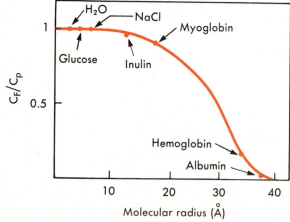

Fig. 45-5 ■ The relation between molecular size and passage of molecules through the glomerulus. C_F/C_P = concentration of a given molecule in the fluid filtered by the glomerulus related to the concentration that existed in the plasma at the time the measurement was made. Note that molecules much larger than albumin cannot cross the glomerular membrane and do not appear in the filtrate.

the filtrate in essentially the same concentration as in plasma. Above 15 Å the concentration of a molecule in filtrate declines rapidly, and at approximately 35 to 40 Å the concentration in the filtrate approaches zero. Comparison of Fig. 45-5 with Fig. 45-4, *C,* suggests that a large part of the size selectivity of the glomerulus resides in the dimensions of the holes in the slit membrane.

It is important to note that the structure of the basement membrane also contributes to the overall molecular selectivity of the glomerulus. The basement membrane is a glycosaminoglycan matrix and thus acts as a filter of large molecules, working in series with the foot processes of the podocytes. In addition, recent evidence suggests that the basement membrane also conveys a charge preference on the glomerulus. In experimental situations it is observed that anionic macromolecules enter the filtrate in lower concentrations than either uncharged molecules or cations. This charge selectivity probably reflects the presence of anionic sites in the basement membrane. Such sites would repell anions from the membrane, decrease their entry into the membrane, and thus reduce the rate of passage of anions into the filtrate.

The various elements in the filtration system all may be altered in relatively selective ways by pathological changes associated with a variety of clinical syndromes. For example, antigen-antibody reactions or certain kinds of systemic infections may produce thickening of the basement membrane and disruption of the foot processes, with resultant loss of molecular selectivity. Also one of the most common histological findings in diabetes is a thickening of the glomerular basement membrane.

■ *Filtration pressure*

Attention now can be focused on the magnitudes of the various pressure components as determinants of the rate of filtration. From Fig. 45-5 it should be apparent that tubular oncotic pressure must be close to zero, since few proteins will be present in the filtrate ($\pi_t = 0$). The proximal tubular hydrostatic pressure has been measured directly with micropipettes inserted into the tubular lumen, and it is found to be approximately 10 mm Hg ($P_t = 10$).

Recently great interest has been focused on the other two factors determining filtration pressure: the hydrostatic pressure within the glomerulus and the glomerular oncotic pressure. Normally the glomerulus lies deep within the cortex of the kidney and thus is inaccessible for puncture and direct pressure measurement within a glomerular capillary. However, in one strain of rats that possess superficial glomeruli, puncture of the capillaries has been possible, and the capillary hydrostatic pressure is about 45 mm Hg ($P_c = 45$). In contrast to the pressure in other capillaries, hydrostatic pressure in glomerular capillaries decreases little between afferent and efferent arterioles. This is the result of the short, wide geometrical configuration of the glomerular capillaries. Thus we can assume the hydrostatic pressure is close to 45 mm Hg along the entire length of the glomerular capillaries.

As is shown later, choosing a value for glomerular plasma oncotic pressure is difficult. This is because substantial quantities of fluid are filtered from the capillaries, thereby elevating protein concentration and oncotic pressure in the remaining plasma. However, a representative value for the oncotic pressure in glomerular plasma is about 27 mm Hg, that is, somewhat elevated above circulating plasma oncotic pressure.

Returning to the computation of the magnitude of glomerular filtration, we can see that the net pressure inducing filtration in this typical example would be 45 mm Hg (45 − 0), and the net pressure opposing filtration would be 37 mm Hg (10 + 27). Putting these numbers into the Starling equation, one obtains GFR = k(45 − 37) = 8k. Recall that k was 15 ml/minute/mm Hg, and thus GFR = 8 × 15 = 120 ml/minute.

As with other vascular beds, control of filtration is accomplished largely by changing hydrostatic pressure. For example, suppose that it is desirable to conserve fluid in a dehydrated subject. This could be accomplished by constricting the afferent arteriole, thereby lowering the glomerular hydrostatic pressure and reducing net filtration and ultimately flow rate. Assume that, after dehydration, constriction of the renal arteriole might reduce hydrostatic pressure in the glomerulus from 45 to 40 mm Hg, in which case:

$$\text{GFR} = \text{k}(40 + 0) - (10 + 27) = 3\text{k} = 45 \text{ ml/minute}$$

It is striking that almost a threefold reduction in glomerular filtration rate, from 120 to 45 ml/minute, was produced by a 5 mm Hg reduction in plasma hydrostatic pressure.

The finding mentioned earlier, namely that plasma oncotic pressure rises substantially along the length of the nephron, has generated significant uncertainty regarding the quantitative relations between renal blood flow, glomerular hydrostatic and oncotic pressure, and their influence on glomerular filtration rate. Fig. 45-6 shows the various

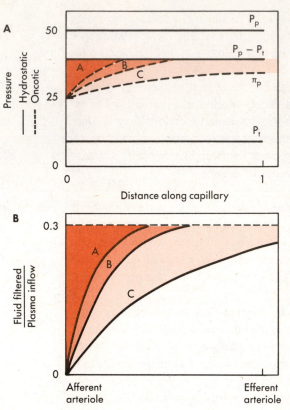

Fig. 45-6 ◼ **A,** Graphical analysis of the application of the Starling hypothesis to the glomerulus. P, Hydrostatic pressure; π, oncotic pressure; p, the value in plasma; t, the value in the tubule. $P_p - P_t$ is the net hydrostatic pressure across the glomerulus. The three lines, *A, B,* and *C,* represent observations made at progressively higher renal plasma flows. The area bounded by the net hydrostatic pressure line ($P_p - P_t$) and the oncotic pressure line (π_p) represents the net filtration pressure that drives glomerular filtration. **B,** Effect of elevated flow rate on filtration. Note that, as flow increases from *A* to *B,* the fraction filtered does not change. However, filtration of a constant fraction of the inflowing plasma represents an increase in the absolute rate of glomerular filtration when the flow rate is higher.

components of the filtration pressure and illustrates the nature of the uncertainty. The line $P_p - P_t$ represents the net hydrostatic pressure along the capillary. The oncotic pressure at three levels of flow is shown, and it is seen that the oncotic pressure rises as the water is filtered out of the plasma. Net filtration pressure is the difference between the hydrostatic pressures and the oncotic pressures, or the area bounded by the lines $P_p - P_t$ and π_p.

It has been found that in some cases so much water is filtered from the plasma that the oncotic pressure rises to equal the net hydrostatic pressure. This is refered to as *filtration equilibrium* (curves *A* and *B,* Fig. 45-6, *A*). Curves *A, B,* and *C* represent progressively higher flow rates of plasma through the glomerulus and illustrate the effects of flow on plasma oncotic pressure. As flow increases, the point along the capillary at which equilibrium is reached is displaced to the right, until at the highest flow *(C)* equilibrium is not reached at all. The significance of this derives from the fact that the areas proportional to net filtration pressure rise as flow increases, and therefore the net filtration rises. This is viewed in another way in Fig. 45-6, *B.* Here the fraction of the inflowing plasma that is filtered is plotted on the ordinate. For the two cases in which equilibrium occurs, the fraction filtered is the same, since equilibrium would be attained at the same oncotic pressure. However, since flow is higher, filtration of the same fraction must mean filtration of a larger absolute quantity of fluid. Thus glomerular filtration rate becomes dependent on flow as well as on hydrostatic pressure. It is known that the process of filtration equilibrium occurs in some species, but not in all. Data are inadequate to permit conclusions as to whether this happens in humans.

Traditionally renal physiologists have ascribed the control of glomerular filtration rate exclusively to changes in hydrostatic pressure in the glomerular capillaries, evoked by constriction of afferent and/or efferent arterioles. As just shown, however, flow itself may influence glomerular filtration rate. It also has been shown recently that the filtration coefficient may be under physiological control. Angiotensin II, norepinephrine, prostaglandins, and bradykinin all have been reported to decrease the filtration coefficient. The mechanism whereby these agents may act is not known, but it does not appear to depend on stimulation of the smooth muscle of the arterioles. It is likely to be the result of alteration in the number of capillaries in the glomerulus that are open. The precise significance of this in so far as physiological control is concerned remains to be established.

■ *Bibliography*

Journal articles

Chang, R.L.S., Deen, W.M., and Robertson, C.R.: Permeability of the glomerular capillary wall. III. Restricted transport of polyanions, Kidney Int. **8:**212, 1975.

Deen, W.M., Robertson, C.R., and Brenner, B.M.: Transcapillary fluid exchange in the renal cortex, Circ. Res. **33:**1, 1973.

Edelman, I.S., et al.: Body composition: studies in the human being by the dilution principle, Science **115:**447, 1952.

Farquhar, M.G.: The primary glomerular filtration barrier—basement membrane or epithelial slits? Kidney Int. **8:**197, 1975.

Kuntz, I.O., and Zipp, A.: Water in biological systems, N. Engl. J. Med. **297:**262, 1977.

Rodewald, R., and Karnovsky, M.J.: Porous substructure of the glomerular slit diaphragm in the rat and mouse, J. Cell Biol. **60:**423, 1974.

Tisher, C.C., Bulger, R.E., and Trump, B.F.: Human renal ultrastructure. III. The distal tubule in healthy individuals, Lab. Invest. **18:**655, 1968.

Books and monographs

Brenner, B.M., and Rector, F.C.: The kidney, vol. 1, Philadelphia, 1981, W.B. Saunders Co.

Koushanpour, E.: Renal physiology: principles and functions, Philadelphia, 1976, W.B. Saunders Co.

Orloff, J., and Berliner, R.W.: In Handbook of physiology; Renal physiology, Section 8: Washington, D.C., 1973, American Physiological Society.

Valtin, H.: Renal function: Mechanisms preserving fluid and solute balance in health, Boston, 1973, Little, Brown & Co.

Windhager, E.E.: Micropuncture techniques and nephron function, New York, 1968, Appleton-Century-Crofts.

chapter 46

Tubular mechanisms

The glomerular ultrafiltrate is modified by a variety of secretory and reabsorptive processes as fluid passes through the tubules. The task of understanding the movement of diverse materials such as sodium chloride, glucose, and protein across the tubular epithelium can be facilitated by viewing these processes within a framework of the basic concepts of transport and in the context of the general principles of the behavior of epithelia. Fig. 46-1 shows a boundary consisting of epithelial cells. Transport may occur across the boundary either from tubular lumen through peritubular space and into capillary blood, or vice versa. The processes and forces that determine how material will traverse the epithelium are fundamentally the same as those which determine cellular transport in other locations.

Lipid solubility, molecular size, charge, and concentration differences are the determinants of how easily a molecule may move across the tubular epithelium and its equilibrium condition. Very small, uncharged molecules, such as water, move more easily than do larger molecules, such as glucose, or charged molecules, such as sodium. A molecule the size of glucose, which has a low lipid solubility, can scarcely cross the tubular epithelium without some sort of specialized transport system.

The rate ($\dot{n}$) at which an uncharged molecule diffuses across the epithelium per unit time is equal to the product of the diffusion coefficient (D), the total area available for exchange (A), and the diffusion gradient, that is, the change in concentration over distance ($\Delta C/\Delta x$) (pp. 8-9). In biological systems the quotient of the diffusion coefficient and the diffusion distance are often considered together as the permeability (*P,* equation 1 in Fig. 46-1). In the case of charged molecules an additional term must be included to take into account the influence of electrical potential difference (ΔE) on the molecular motion.

If one wishes to understand the behavior of materials whose transepithelial movement is determined mainly by interaction with specific transport molecules, simply knowing that the molecules are transported by active or facilitated transport systems existing within the tubular epithelium suggests a number of generalizations. Following is a partial list of such generalizations:

Characteristics of active transport systems

1. Require energy (but not for facilitated transport)
2. Display saturation
3. Display competition for molecules of similar type
4. Sensitive to inhibitors

Typical classes of molecules transported

1. Monovalent ions: Na^+, Cl^-
2. Polyvalent ions: $PO_4^=$, Ca^{++}

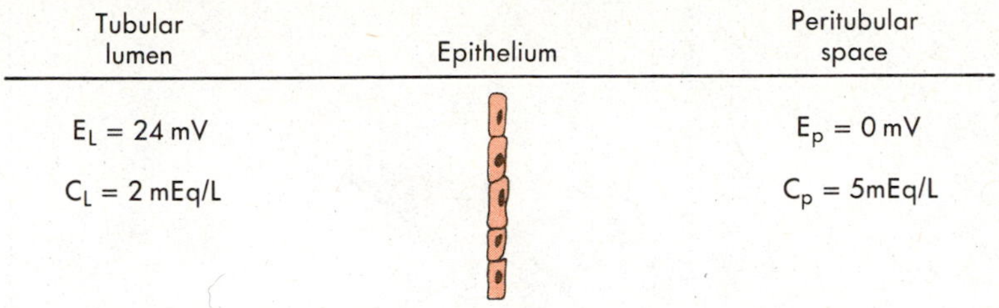

Tubular lumen	Epithelium	Peritubular space
$E_L = 24\,mV$		$E_p = 0\,mV$
$C_L = 2\,mEq/L$		$C_p = 5\,mEq/L$

(1) $\dot{n} = -DA \left[\dfrac{\Delta C}{\Delta x} + \dfrac{ZF}{RT} C \dfrac{\Delta E}{\Delta x}\right] = PA\,[\Delta C + 60C\,\Delta E]$

(2) $\Delta E = -\dfrac{60}{Z} \log \dfrac{C_L}{C_p}$

e.g., $\Delta E = -60 \log \dfrac{2}{5}$

$\Delta E = 23.9\,mV$

Fig. 46-1 ■ Quantitative relations among the variables determining passive movement of materials across epithelial cells. *E*, Potential difference; *C*, concentration; *Z, F, R,* and *T,* the valence, the Faraday constant, the universal gas constant, and the absolute temperatures, respectively; *ṅ,* the rate of movement of solute under the conditions described in equation 1; *L* and *P,* lumen and peritubular space, respectively. Data are typical for the potassium ion. Other variables are as described in the text.

Fig. 46-2 ■ Influence of substrate concentration on the rate of a biochemical reaction. In this case the rate is assumed to be the rate at which a transport carrier molecule might combine with and move a molecule across the epithelial membrane. V_{max} is the maximal rate of the reaction, and K_m is the substrate concentration that could drive the system at half the maximal rate.

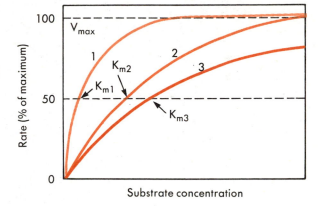

3. Aromatic acids: penicillin, para-aminohippuric acid (PAH)
4. Organic bases: quinine, choline, thiamine
5. Aliphatic acids: amino acids, pyruvate
6. Sugars: glucose, fructose

Understanding these characteristics of active transport systems allows one to generalize about various kinds of tubular transport processes.

Other generalizations can be made if one knows that tubular transport processes follow behavior patterns which are characteristic of a variety of biochemical interactions and enzyme systems. Fig. 46-2 shows an idealized version of the relation between the rate of a biochemical process (in this case the transport system) and the substrate concentration (in this case, the concentration of the molecule to be transported). As substrate concentration is increased from zero, the rate of the transport reaction rises and approaches a maximal rate (V_{max}), at which point transport is said to be saturated. The substrate concentration that will raise the rate of the reaction to half the maximal rate is called the K_m. Fig. 46-2 shows two processes with identical V_{max} (*1* and *2*) but with different K_m values. Curve *1* in Fig. 46-2 represents a system in which the K_m is low, that is, the affinity of the carrier molecule for the substrate is high, and we see that the reaction rate is accelerated to the V_{max} by a small increase in substrate concentration. Contrast this with line *2,* which will ultimately reach the same rate, but only if the substrate concentration is raised to much higher levels. One might expect to find transport processes in the kidney that become saturated at varying degrees of substrate availability and which have varying degrees of affinity between carrier and the material being transported. This probably explains in part the nature of the different kinds of transport systems observed in the kidney.

Two types of systems are commonly recognized by renal physiologists. These are

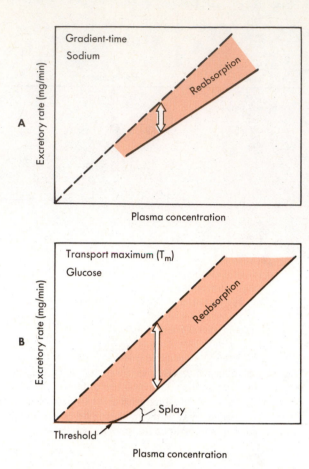

Fig. 46-3 ■ Comparison of gradient-time (**A**) and T_m transport (**B**) characteristics. The dashed line represents the rate at which a test molecule would be excreted if all filtered material passed through the tubule and was voided. The solid line represents excretion rates actually observed. The shaded areas and heavy arrows represent the rate of absorption of the solute (that is, the difference between filtration rate and excretion rate).

the *gradient-time systems* and the *transport maximum* (or T_m) systems. These designations have little significance in so far as tubular mechanisms are concerned but are commonly used terms to describe renal excretory patterns. When the plasma concentrations of various substances are changed and the resulting alterations in excretion are analyzed, the two broad categories of behavior shown in Fig. 46-3 are often apparent. The figure shows the excretory rate for a filtered solute plotted against the plasma concentration. For reference, the dashed lines show the amount of solute that is filtered per unit time as a function of plasma concentration. If there were no tubular modification of the filtrate, the excretory rate would be equal to the filtration rate. However, examination of the solid lines in the figure shows that at all plasma concentrations excretion for both sodium and glucose is much less than filtration. The colored area representing the difference between the two lines illustrates the rate of reabsorption of the substrate.

Note that a change in plasma concentration has a much different effect on glucose and sodium. Sodium typifies a group of materials whose transport is characterized by *gradient-time* transport. Glucose transport is said to be of the *transport maximum*, or T_m type. For the class of molecules that display T_m transport characteristics, progressive elevation of plasma concentration from zero initially produces no change in excretion. In the case of glucose, excretion is zero until the concentration reaches a plasma level of about 300 mg/dl of plasma, at which point the excretion begins to rise in a characteristic curvilinear fashion. The minimal plasma concentration at which excretion begins is termed the *plasma threshold*.

Contrast the behavior of glucose with a system displaying gradient-time characteristics, such as the sodium transport system. In the latter, excretion varies linearly with plasma concentration over a wide range, and there is no evidence of a threshold (Fig. 46-3, *A*). On comparing the solid and dashed lines in Fig. 46-3, we see that in the case of the T_m system (glucose), once the threshold is reached and a transition region (splay) is exceeded, the excretory rate and the filtration rate increase in parallel. In this region elevating the amount of material that is filtered leads to a matching increase in the

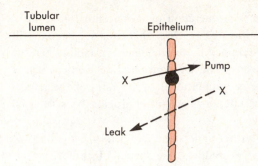

Fig. 46-4 ■ Pump-leak relations as determinants of net tubular transport. It is assumed that the molecule in question has a finite permeability and that pump activity will cause a concentration gradient that in turn will cause back leak into the lumen. The pump rate is assumed to be proportional to the concentration of the substrate [X]. ΔE is the potential difference across the epithelium.

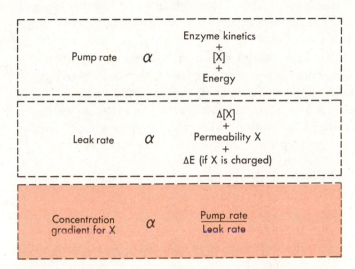

amount of the material that is excreted. In other words, the maximum transport rate for reabsorption has been reached, and this rate is the T_m. In contrast, in the gradient-time system (sodium) the reabsorption rate varies continuously with the plasma concentration. In the proximal tubule about 70% of filtered sodium is reabsorbed over a wide range of filtration rates.

The factors that generate the characteristics of the curves shown in Fig. 46-3 are complex and depend on many elements of renal function. However, a preliminary understanding of the processes can be developed by simply considering the enzyme kinetics likely to be associated with the transport process. The T_m system behaves as though the carrier molecules display kinetic behavior typified by a relatively low V_{max} and low K_m; that is, the carrier exhibits a high affinity for the substrate but is quickly saturated. In this case the carrier combines with the molecule to be transported at a low concentration and transports the molecule out of the tubule. A limiting transport rate is encountered when the V_{max} for the carrier is reached, and this corresponds to some degree to the threshold. In contrast, we can assume that the gradient-time system has a lower affinity and therefore does not become saturated until a higher concentration has been reached. The V_{max} for the gradient-time system is also greater, although this is not apparent in Fig. 46-2 because the rates are normalized to percent of maximum. Its kinetic behavior might be similar to that shown in Fig. 46-2, curve *3*. It should be noted that active intrarenal regulatory mechanisms also are involved in balancing the tubular reabsorption processes to the filtration rate; these are discussed later.

A second, basic tubular process that is likely to confer different behaviors on molecules of different types is "back leak" (Fig. 46-4). The concentration in peritubular interstitial fluid is held approximately equal to the concentration in plasma by diffusional exchange between the interstitium and the peritubular capillary blood. The tubule has a relatively small volume, and active transport of any significant quantity of material across the tubular epithelium will result in the generation of a concentration gradient between tubular fluid and peritubular interstitial space. Once this occurs, material will leak back across the epithelium at a rate determined by the tubular permeability and the chemical and electrical gradients. In the steady state the concentration gradient that is achieved across the tubular epithelium will be proportional to the ratio of the rate at which material is actively transported to the rate at which material leaks back (that is, the *pump-leak ratio*).

This pump-leak concept enables one to further understand the factors that contribute to the distinction between a T_m system and a gradient-time system. In a T_m system, such as that for glucose, the leak is small relative to the pump rate, since it is apparent that virtually all the material can be reabsorbed out of the tubule over a significant range of plasma concentrations (below threshold; Fig. 46-3, *B*). However, with a gradient-time system, such as that for sodium, the tubule has a relatively high permeability compared with the pump. Concentration gradients as great as those seen with T_m systems cannot be achieved because of back leak down the concentration gradient.

A second result of having a relatively low pump-leak ratio in a system in which the tubular volume is small, as in the kidney, is that the net transport in the system becomes flow dependent. When flow is relatively low, transport of material out of the tubule will quickly establish a limiting gradient (back diffusion equals pump rate). When flow is high, however, the tubular fluid spends a shorter time at each point along the tubule. Therefore the concentration gradient rises less rapidly, back leak is minimized, and net transport is favored.

These facts suggest the basis for the behavior of the *gradient-time system*. Under circumstances where affinity of the carrier is relatively low, the V_{max} is relatively high, and the tubular permeability is high compared with the transport rate, it is likely that a limiting gradient will be established before the end of the tubule is reached. The point at which this gradient is established will be a function of flow; that is, it will be a function of the time that the fluid remains in the tubule. Thus both *concentration gradient* and *time* become important determinants of reabsorptive processes.

■ Special transport systems

Renal physiology, as a rule, concerns itself most intensely with the transport of small molecules, such as sodium, chloride, and water. However, there are many molecules that are transported by specialized carrier systems located at various points along the nephron. This section presents general principles of the transport of these molecules and gives a few examples that have particular physiological relevance. The list on pp. 836-837 shows some types of molecules that are transported and lists some general characteristics of the transport systems. The importance of these transport systems can be appreciated by recalling that the glomerular filtration rate is approximately 180 L/day for an average man with 45 L of total body water. Many of the molecules in the list are small and would be filtered in a concentration equal to that in plasma (see Fig. 45-5). The tubular epithelium is not especially permeable to these molecules, and thus they would be excreted if no further processing by the kidney occurred. Continued excretion of essential materials such as phosphate or amino acids would be catastrophic. To prevent this, a primary activity of the renal epithelium is to reabsorb and return to the body those filtered materials which need to be conserved.

A second major function of the transport systems is to *secrete* materials that are not filtered in adequate quantities to permit elimination from the organism at the required rates. Hepatic metabolic end products frequently appear in the form of organic anions (for example, hippuric acid). As might be expected, there are appropriate transport systems in the tubular epithelium that permit the kidney to excrete these anions in the final urine, rather than allowing their accumulation in the body. (These transport sites for organic anions are frequently the means whereby drugs and other foreign substances, such as penicillin, are eliminated from the body.)

The amount of material with which the transport systems of the kidney must deal is usually quantified in terms of the *solute load*. The load is the quantity of solute per unit time that is presented to the tubule. Obviously the magnitude of this load ultimately will determine whether the tubule can meet the demand for transport. For a reabsorptive system the load is simply the quantity of material that is filtered per unit time. In other words, the glomerular filtration rate determines the volume of fluid per unit time that passes the transport system, and the concentration of the solute in the glomerular filtrate determines the solute quantity in each volume. Reabsorptive load is then simply the product of glomerular filtration rate and plasma concentration for small solutes that are freely filtered. Although not usually referred to by renal physiologists, a secretory load also might be determined. The secretory load is less easy to quantitate but conceptually simple; it is the peritubular capillary blood flow times the concentration of the solute in peritubular capillary blood.

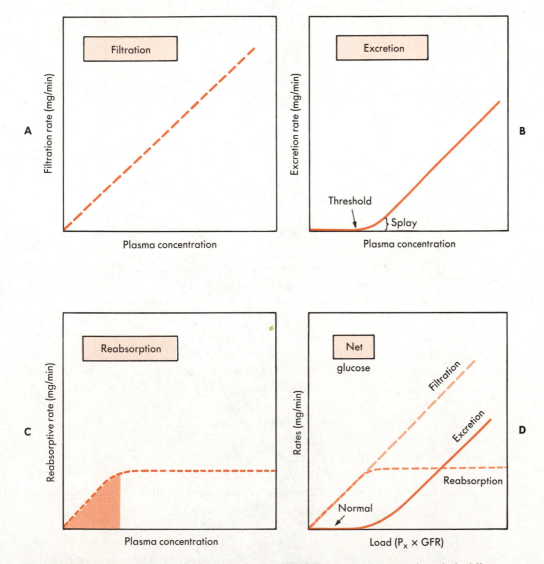

Fig. 46-5 ■ Graphical analysis of tubular transport. Solid lines represent excretion, dashed lines represent filtration, and dotted lines represent rate of tubular transport. **A,** The rate at which filtered solute enters the tubule as a function of plasma concentration. **B,** Excretion rate as a function of plasma concentration. **C,** Reabsorption, given filtration and excretion rates in **A** and **B.** The colored area represents the region in which filtration is exactly matched by reabsorption. When reabsorption cannot match filtration, excretion occurs, and the threshold is said to have been reached. **D,** The three curves plotted together.

A general equation for the tubular transport rate can be written as follows:

$$\text{Transport rate} = \text{filtration rate} - \text{excretion rate}$$

$$\text{TR} = (\text{GFR} \times P_x) - (U_x \times V) \qquad (1)$$

where

P_x = the plasma concentration of a solute
U_x = the corresponding urinary concentration
V = the *rate* of urine flow (Note that in renal physiology V rather than $\dot{V}$ is used to denote flow rate.)

If the transport rate has a negative value, then the excretion rate ($U_x \times V$) is in excess of the filtration rate ($\text{GFR} \times P_x$), and solute must have been added to the glomerular filtrate, that is, secretion took place. Conversely, if the transport rate has a positive value, then reabsorption must have taken place, since less material is in the final urine than had originally been filtered.

A graphical analysis is helpful in understanding the interrelations among filtration, reabsorption, and excretion (Fig. 46-5). This figure shows behavior of the glucose molecule. Fig. 46-5, *A*, shows the rate at which the filtered solute enters the tubule as a function of the plasma concentration (filtration). If glomerular filtration rate is constant, the plasma concentration will determine how much solute is filtered per unit time. Excretion rate as a function of plasma concentration is shown in Fig. 46-5, *B*. Fig. 46-5, *C*, shows what must happen to reabsorption, given the filtration and excretion rates shown in Fig. 46-5, *A* and *B*. Since the rate of filtration of glucose must increase linearly with increases in plasma glucose (Fig. 46-5, *A*) and there is no glucose excretion for plasma levels up to the threshold (Fig. 46-5, *B*), reabsorption must exactly match filtration over this range of plasma concentration (*lightly colored area,* Fig. 46-5, *C*). Once the threshold is reached and the range in which splay is observed is exceeded, a maximum rate of reabsorption will be achieved (T_m) at some predictable plasma concentration.

In Fig. 46-5, *D,* the three curves are plotted together on the same graph. The load ($\text{GFR} \times P_x$) replaces P_x as the abscissa for completeness.

▪ Phosphate reabsorption: physiological alteration in T_m

The behavior of the phosphate transport system typifies a system under humoral control. Fig. 46-6, *A*, shows the behavior of the phosphate transport system before and after the administration of *parathyroid hormone* (parathormone, or PTH). Parathormone is involved in the regulation of calcium and phosphate balance in the body, as described in Chapter 51. In general, these two ions are regulated in a reciprocal fashion. Parathormone causes an increase in the excretion of phosphate and a decrease in the excretion of calcium. This allows calcium levels in plasma to rise without danger of reaching a level that would cause precipitation of salts of calcium phosphate.

In contrast to the condition for glucose, the normal plasma concentration and filtered load for phosphate are above the threshold for renal excretion (for example, *A* in Fig. 46-6). Thus phosphate reabsorption is very much dependent on filtered load, and changes in both glomerular filtration rate and phosphate concentration can alter phosphate excretion.

Assuming that the normal position relating excretion and load is at point *A* on Fig. 46-6, *A,* after administration of parathormone the T_m for phosphate would be reduced. A reduction in the T_m results in a higher phosphate excretion at a given phosphate load (as shown by the transition *A-B* in Fig. 46-6, *A*). As excretion rises, the plasma phosphate concentration will fall *(B-C)*. The system will stabilize at some new, lower plasma phosphate concentration, since reduced plasma phosphate levels will reduce the filtration of phosphates. The new stable point will occur when plasma levels fall low enough to

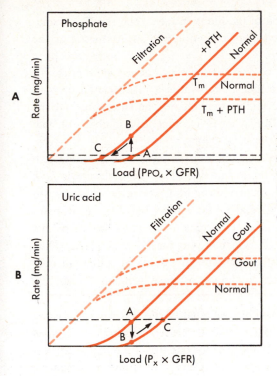

Fig. 46-6 ■ **A,** Phosphate reabsorption. *PTH,* Parathormone. *A* is the normal operating point for phosphate. Symbols are same as in 46-3. Curves are shown before (*normal*) and after (*+PTH*) administration of parathormone (PTH). **B,** Uric acid reabsorption. *A* is the normal operating point for the system. Gout elevates the T_m. Dashed black lines in **A** and **B** represent excretion required by normal dietary intake.

cause the excretion of phosphate to equal the daily intake of this ion. Note that in the steady state the phosphate excretion is the same before and after parathormone administration; the difference is that similar phosphate excretion rates are attained at lower plasma levels after the hormone has been given.

The effects of parathormone on renal function are extremely complex and also may reflect altered glomerular filtration rate, as well as altered tubular reabsorption of phosphate, sodium, and calcium. In general, the net result of parathormone addition is to decrease the reabsorption of phosphate and to increase the reabsorption of calcium.

Special transport systems also are subject to pathological changes. One of the more notable of these is the altered uric acid transport that occurs in gout. Fig. 46-6, *B,* shows the variations in uric acid reabsorption with load in a normal individual and in an individual with gout. The basic relations between filtration, excretion, and reabsorption are similar to those shown previously for glucose and phosphate. The normal operating point (*A*) for a human is such that uric acid is continuously excreted.

Gout is a disease in which uric acid accumulates in the plasma and interstitial fluid. The uric acid subsequently precipitates within the joints, thereby causing pain. The disease has a complex origin which includes changes in both the metabolism and the excretion of uric acid. Fig. 46-6, *B,* illustrates the change in tubular reabsorptive rate. In the gouty individual the reabsorptive T_m for uric acid is increased. This means that an elevated plasma concentration is required to raise the filtered load to a value such that the excretion of uric acid will equal its production. A typical pathway for pathophysiological changes in renal uric acid handling is shown in the path *A-B-C* in Fig. 46-6, *B.* Beginning with a stable situation at *A* the T_m increases pathologically, and excretion thus falls to *B.* Assuming there is a constant dietary intake of purines, reduced excretion will be followed by increasing plasma concentration and increasing excretion (*B-C*). A new, stable situation will be achieved when excretion reaches a level mandated by purine intake (*C*).

Uric acid crystals precipitate not only in the joints but also in the basement mem-

■ *Uric acid reabsorption: pathological alteration in T_m*

brane of the glomerulus, causing a reduction of the filtration rate. As a result, accumulation of uric acid is exacerbated by reduced glomerular filtration superimposed on the elevated T_m.

Treatment for gout consists of the use of antiinflammatory agents, appropriate dietary modifications, and therapy with drugs that diminish uric acid formation, as well as with drugs that enhance excretion of uric acid by the kidney. The latter drugs are the *uricosuric agents* such as *sulfinpyrazone*. This agent is a competitive inhibitor of both uric acid transport and the transport of other organic anions.

One of the peculiar characteristics of the uric acid transport system is that, although the net activity of tubular function is reabsorption of uric acid, the molecule is *both* secreted and reabsorbed during its passage through the nephron. The secretory and reabsorptive mechanisms tend to vary in importance along the tubule, but, in general, reabsorption is the predominant effect of the transport systems. As a consequence of this bidirectional transport, drugs that inhibit uric acid transport may decrease rather than increase the excretion of uric acid. This is the case with low doses of sulfinpyrazone. Obviously such an effect compromises the successful use of these agents in therapy.

◼ Penicillin secretion and the secretory system

There are a variety of secretory systems within the renal tubule that excrete organic anions, such as penicillin. It should be emphasized that the process of filtration removes from the blood a fluid with concentrations of most molecules that are almost identical to those in plasma. Thus, although plasma volume is reduced by filtration, the remaining plasma that flows through the peritubular capillaries and acts as a source for secreted penicillin has the same concentration as the systemic plasma. From the peritubular capillaries the solute can diffuse to epithelial cell transport sites, where secretion can occur.

A graphical analysis of the transport system for penicillin is shown in Fig. 46-7. As the load is increased, secretion rises and approaches a limit (T_m). In contrast to reabsorptive systems, the secreted molecules are added to the quantity of material filtered, and excretion exceeds filtration.

The organic anion secretory systems are extremely active, so much so that for some molecules, such as para-aminohippuric acid (PAH), virtually all the molecules that enter the peritubular capillary plasma are secreted and ultimately are excreted in the urine, along with the filtered components. As shown later, this becomes an important tool in the measurement of renal blood flow.

The avidity of the penicillin transport system is such that a large fraction of an administered dose of penicillin will be rapidly secreted and then excreted rather than circulating for therapeutic purposes. The dose therefore must be high, and administration must be frequent if therapeutic levels are to be maintained. When penicillin was

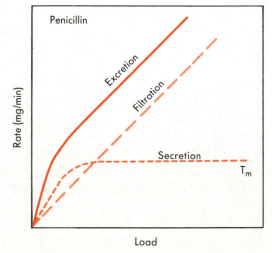

Fig. 46-7 ◼ Renal handling of penicillin. Note that excretion is the sum of filtration and secretion. Symbols are same as in 46-3.

originally introduced and was expensive, therapeutic levels were achieved more economically concomitantly by administering probenecid, a competitive inhibitor of organic anion transport, thereby blocking the secretory system for penicillin and diminishing its excretion.

In recent years it has become apparent that many of the transport processes alluded to in this section are dependent on cotransport with sodium. These transport systems appear to behave almost identically to those observed in the intestine (Chapter 44). For example, both a sodium ion and an organic solute must simultaneously occupy the carriers that transport diverse molecules such as glucose, amino acids, and PAH (Fig. 46-8, *B*). Even though this is important to an understanding of tubular function, it should be noted that sodium does not ordinarily limit transport of these compounds because its concentration is high at the carrier sites, unlike the situation in the intestine.

■ *Coupled transport and the tubular epithelium*

The transport of sodium in the proximal tubule can be characterized broadly as being a transport process that moves a large amount of material rather quickly through small electrochemical gradients. Fig. 46-8, *A*, summarizes the elementary salt and water transport in proximal tubular cells. The figure shows the tubular epithelial cell, with one side facing the tubular lumen and exposed to the filtrate and the other side facing the peritubular space and in close proximity to the peritubular capillaries. It should be noted that only a small distance separates epithelial cells and the peritubular capillaries and that these capillaries experience relatively high flow rates. As a result, concentrations of various small molecules within the renal interstitial space will approximate plasma levels, regardless of the reabsorptive processes.

Transport depends on the electrochemical potential difference that exists across the membrane to be traversed. Equation 1 in Fig. 46-1 describes the relation between the important variables and the movement of material in a purely passive system. For a monovalent cation this equation reduces to:

$$\dot{n} = pA \, (\Delta C + 60C \, \Delta E) \tag{2}$$

The equation indicates that net passive transport depends on both the electrical potential difference (ΔE) and the concentration difference. At equilibrium, movement of material is zero, and the relation between potential difference and the concentration gradient across the epithelium reduces to:

$$\Delta E = -60 \log C_L/C_P \tag{3}$$

C_L and C_P are concentrations in the lumen and plasma, respectively. The equation shows that a tenfold difference in concentration for a monovalent ion can be produced by a 60 mV potential difference across the membrane. From these two equations it is apparent that understanding tubular transport of small ions and water depends heavily on clearly defining concentration differences and potential differences across the tubular epithelium and across the various membranes of the tubular epithelium.

With the preceding concepts in mind, let us examine the reabsorption of filtered sodium (Fig. 46-8, *A*). Sodium in the lumen of the first portion of the proximal tubule has the same concentration as in plasma, since this fluid represents the glomerular filtrate. This concentration, approximately 140 mEq/L, is much greater than the 10 mEq/L concentration inside the cell. Thus a concentration gradient exists that acts to drive sodium from the lumen into the cell. A potential difference also favors cell entry for the sodium ion. The luminal potential over the length of the proximal tubule ranges from -2 mV in the first segments to $+2$ mV in the middle and end segments, and the intracellular potentials are of approximately -70 mV, yielding a net electrical gradient

■ *Proximal tubular salt and water transport*
■ *Sodium transport*

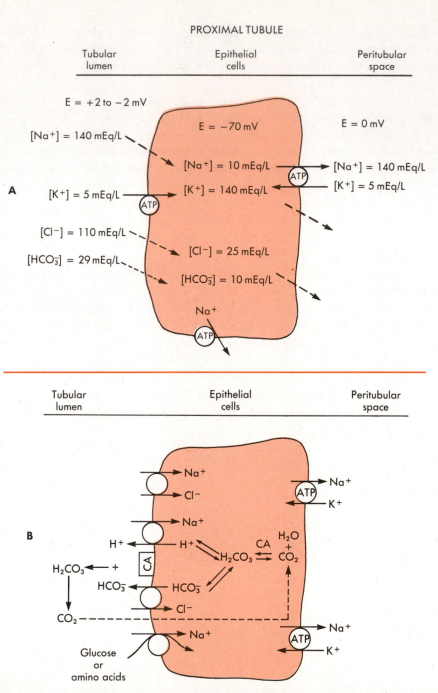

PROXIMAL TUBULE

Fig. 46-8 ■ A, Salt and water transport in a proximal tubular cell. Dashed arrows represent passive diffusion; solid arrows tangent to a circle represent active transport or facilitated diffusion. The former is indicated by *ATP* within the circle. The variations in luminal potential between +2 and −2 represent conditions found in different portions of the proximal tubule. **B,** Additional features of tubular function, especially newly recognized countertransport carriers for hydrogen and bicarbonate ions. *CA,* Carbonic anhydrase.

of 68 to 72 mV, which favors movement of sodium into the cell. It follows then, given these large electrical and chemical driving forces for sodium, that sodium entry into the epithelial cell is passive. There is evidence that sodium permeability at the luminal cell surface may be greater than that at the abluminal surface. Also, as discussed later, carriers that facilitate sodium entry are present at the luminal surface.

The exit of sodium at the abluminal side of the epithelial cell is quite different. Here the concentration gradient is from an intracellular level of 10 to 140 mEq/L in the peritubular space. The potential difference is between -70 mV in the cell and 0 mV in the peritubular space. Thus active sodium extrusion from the cell is required.

Sodium is actively transported out of the cells at the abluminal surface on the basal cell borders as well as into the space between adjacent epithelial cells. The bulk of the transport is accomplished by the coupled Na^+, K^+-ATPase on the basal cell border. This system transports two potassium ions into the cell for every three sodium ions removed.

Less is known of the transport system located on the lateral cell surfaces. This system requires energy, but coupling ratios and kinetics are unknown. What is known, however, is that transport results in net transfer of solute into the clefts between the cells; this transport system is extremely important in determining water reabsorption across the proximal tubular epithelium and is discussed later.

It must be emphasized that *net* reabsorption of sodium, or any other ion, will be determined by the total gradient across the tubular epithelium. If intraluminal concentration and potential are known and the concentration and potential in the interstitial space are known, then the Nernst equation can be applied to predict whether active transport and energy expenditure need be used. Whether molecular movement is active or passive at a particular cell surface need be considered only when trying to determine the location of transport processes. For the most part further discussion in this text is concerned with this net transport between lumen and renal interstitial space rather than a detailed examination of fluxes at particular cell borders. The intestine and the kidney show many characteristics of transport, and a more detailed treatment of epithelial transport processes is presented in Chapter 44.

One other feature that has important consequences for proximal tubular sodium and water transport is the fact that the gap junctions which connect adjacent epithelial cells are rather permeable, both to water and to small ions. Therefore only small concentration gradients can be generated across the tubular epithelium, since the high permeability means that back leaks will be large. This low pump-leak ratio means that concentrations of sodium in proximal tubular fluid are very close to those in plasma; only in the presence of osmotic diuretics are significant concentration gradients observed (as explained later).

In spite of the fact that only small concentration gradients are generated, large quantities of sodium are reabsorbed. As shown in Fig. 46-3, *A*, about 70% of the filtered sodium is reabsorbed, and the fraction is constant over a wide range of sodium filtration rates. This close coupling between sodium filtration and sodium reabsorption is due to many factors. Earlier it was pointed out that the kinetics of the transport system and the relatively high tubular permeability to sodium would both tend to contribute to the coupling. The coupling is also the result of a process in which filtration rate modifies tubular reabsorption, and vice versa. This process is called *glomerulotubular feedback* or *glomerulotubular balance.*

Glomerulotubular balance is physiologically significant as a means of maintaining body sodium levels constant in the face of changes in the glomerular filtration rate that may occur during normal daily activities, such as strenuous physical exertion or after a large food intake. The former causes a reduction in glomerular filtration rate, and the latter causes an increase in the rate. A link between sodium reabsorption and glomerular filtration tends to stabilize sodium excretion and minimize swings in plasma sodium levels.

Glomerulotubular balance also serves to balance filtration and reabsorption from nephrons of different lengths. There is no a priori reason why the glomerulus of a given nephron should filter at a rate exactly commensurate with the length of its proximal tubule. However, glomerulotubular balance ensures that, even in the face of anatomical disparities, filtration and reabsorption will be balanced. The phenomenon of splay in the glucose transport system (Fig. 46-5) is thought at least in part to be caused by uncorrected imbalances in filtration rate and transport rate for glucose.

■ *Potassium transport*

Continuing with an examination of proximal tubular function, we turn to a consideration of the reabsorption of filtered potassium. Proximal tubular potassium concentration remains relatively constant along the proximal tubule at about 5 mEq/L. (Recall that intracellular potassium concentration is in the vicinity of 140 mEq/L.) Thus potassium is absorbed into the cell against a concentration gradient but down an electrical gradient. It is known that some form of active transport for potassium exists at the luminal surface, but the details of this mechanism remain to be determined. Exit from the cell on the abluminal side is the result of passive diffusion down an electrochemical gradient.

The net transport of potassium involves both active and passive fluxes. In the latter part of the proximal tubule the positive intraluminal potential provides a passive driving force for potassium reabsorption. In the initial portions of the tubule, where the potential is negative, however, active reabsorption of potassium is required.

■ *Chloride and bicarbonate reabsorption*

The net reabsorptions of both chloride and bicarbonate in the proximal tubule are down electrochemical gradients and closely related to sodium movement and, to a lesser degree, to potassium reabsorption.

Cation reabsorption cannot occur without accompanying charge movement to maintain electroneutrality, and chloride and bicarbonate are mainly responsible for this in the proximal tubule. Recent evidence indicates that in the first part of the proximal tubule bicarbonate is the major anion accompanying sodium reabsorption. In contrast, in the middle and end of the proximal tubule chloride movement contributes more significantly to the maintenance of charge neutrality as sodium and potassium are transported.

■ *Coupled solute fluxes*

The elements of epithelial function shown in Fig. 46-8, *A,* are adequate for an overview of the determinants of the net flux of salt and water. However, important factors related to coupled solute fluxes and the membrane carriers that facilitate diffusion across the epithelial cell membrane are neglected. A number of carriers are present that do not use energy directly but rather are driven by energy derived indirectly from sodium concentration gradients. These gradients of course are generated at the expense of hydrolysis of ATP by the Na^+, K^+-ATPase.

Fig. 46-8, *B,* shows an epithelial cell that possesses all the features shown in Fig. 46-8, *A,* and an additional group of coupled carriers. Only the sodium-potassium pump is an active transport system.

Sodium ion efflux from the cell at the basolateral borders lowers intracellular sodium ion concentration and sets up a diffusion gradient for this ion. Sodium entry into the cell can drive the exchange carrier for hydrogen ion, thereby facilitating entry of luminal sodium into the cell and efflux of hydrogen ion. The hydrogen ion extruded may be derived either from carbon dioxide via formation of carbonic acid or from metabolically produced hydrogen ion.

The sodium gradient also accelerates the entry of sodium and glucose into the cell, as described earlier. This system appears to be electrogenic and responsible for much of the negative potential in the first portion of the proximal tubule.

Sodium transport also couples chloride reabsorption to sodium reabsorption by a rather complicated process. The Na^+, K^+-ATPase generates a sodium gradient that causes sodium influx. A portion of the sodium influx will occur via the sodium-hydrogen carrier, with the associated extrusion of cellular hydrogen ion into the lumen. Some of the hydrogen ion is presumed to be derived from the carbonic acid within the cells, that is, in equilibrium with cellular carbon dioxide. Removal of a hydrogen ion leaves an excess of bicarbonate ion in the cell, thereby generating a concentration gradient for bicarbonate ion between cell and lumen. Bicarbonate ion can then move down its concentration gradient in exchange for chloride ion, which moves into the cell. The abluminal surface of the epithelial cell is quite permeable to chloride, and it diffuses passively into the renal interstitial space.

Note the somewhat peculiar behavior of bicarbonate ion. Once in the lumen of the tubule, the hydrogen ion (from the sodium-hydrogen exchange carrier) and bicarbonate ion (from the chloride-bicarbonate exchange carrier), which originally were formed in the cell by dissociation of H_2CO_3, can recombine and ultimately form carbon dioxide in the lumen. Carbon dioxide, being highly permeable, can diffuse into the epithelial cell, thereby replacing the carbon dioxide presumed to have been used in the formation of the hydrogen and bicarbonate ions used in the original countertransport activities. Therefore the net result of the entire complex process is (1) sodium is transported across the abluminal border of the epithelium, (2) the resultant sodium gradient facilitates glucose transport, and (3) chloride ion moves with sodium ion across the cell. H^+ and HCO_3^- simply cycle to facilitate the transport processes.

The equilibrium between carbonic acid and carbon dioxide is attained quite slowly in saline solutions. To facilitate this and permit sufficiently rapid reabsorption of sodium chloride, the enzyme *carbonic anhydrase* (*CA*, Fig. 46-8, *B*) is present, both on the brush border and in the cytoplasm of the epithelial cells. This enzyme accelerates the reaction, and thus carbon dioxide can be quickly formed from the hydrogen ion and bicarbonate ion, in the tubular lumen. The high permeability of the carbon dioxide and water formed means that these molecules can rapidly traverse the epithelium.

These facts also explain why bicarbonate ion is reabsorbed more rapidly than chloride from the proximal tubule, in spite of the fact that the passive permeability of the chloride ion is greater than the permeability of bicarbonate ion. Bicarbonate is present in the glomerular filtrate at a concentration of about 25 mEq/L. Therefore a hydrogen ion that enters the lumen in exchange for a sodium ion is likely to combine with a filtered bicarbonate ion. When this happens, the resultant carbon dioxide can easily enter the cell and dissociate to form hydrogen and bicarbonate ions. The hydrogen ion essentially replaces the hydrogen ion exchanged for the sodium, and the bicarbonate ion is free to diffuse across the abluminal membrane in association with the sodium. The net result of the process is the reabsorption of $NaHCO_3$ by a very efficient process, rather than the somewhat slower absorption of sodium chloride.

Water is reabsorbed with solute in the proximal tubule (as described later), and, as a result, the concentration of remaining solutes rises, thereby generating diffusion gradients for these materials. Since the chloride ion is more permeable than the bicarbonate ion, its diffusion down its concentration gradient results in the generation of a slightly positive luminal potential in the end portions of the proximal tubule (a bi-ionic diffusion potential).

■ *Water reabsorption*

During the reabsorption of sodium, potassium, chloride, bicarbonate, and glucose in the proximal tubule the tubular fluid remains isotonic. The volume of this isotonic fluid is reduced by up to 70%; this must mean that solute and water reabsorption are in some way perfectly matched.

In trying to understand this process of *isotonic fluid reabsorption,* early workers attempted to induce water flux across the tubular epithelium by changing the osmotic gradient between tubular fluid and blood via the addition of solute to fluid in the tubular lumen. It was found that relatively large osmotic gradients were required to produce fluid movement of the magnitude observed during tubular water reabsorption. However, no such osmotic gradients could be measured experimentally when tubular fluid was sampled.

These facts have led to the hypothesis that water movement is somehow coupled to

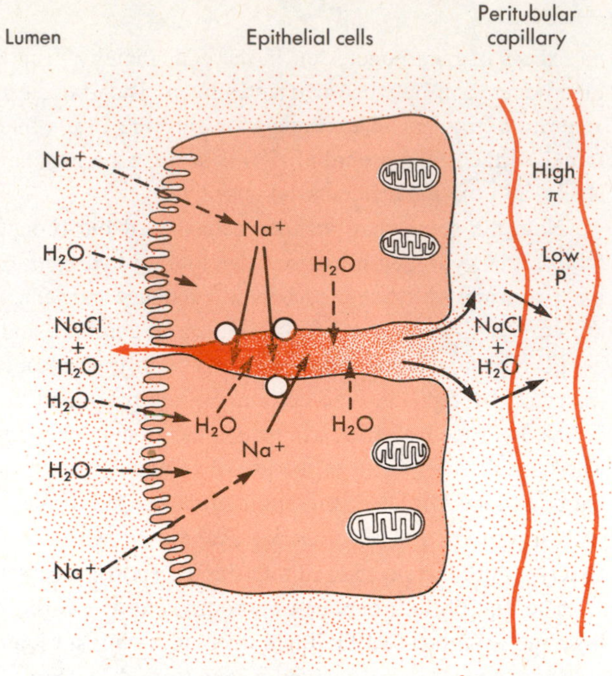

Fig. 46-9 ■ Elements of the standing gradient hypothesis. Note that sodium is confined to the restricted intercellular spaces, thus raising the osmolality (indicated by the density of shading).

sodium transport via a complex interplay between the anatomy of the epithelial cells and the localization of the transport processes. The process by which this coupling currently is thought to occur is now called the *standing gradient hypothesis for water transport*.

The operation of the standing gradient mechanism is shown in Fig. 46-9. The key element in the process is the location, previously referred to, of a sodium pump on the lateral cell membrane of the epithelial cell. In the tubular epithelium, especially in the proximal tubule, the spaces between epithelial cells are very long, narrow, and tortuous. The luminal end of these intercellular clefts is bounded by *gap junctions* of the epithelial cells. If sodium is transported into these intercellular spaces, the concentration of sodium can rise to relatively high levels because of the restricted nature of these spaces, and large osmotic gradients are generated. These osmotic gradients can pull water from the lumen, through the large surface area of the brush border of the epithelial cells, and into the intercellular spaces. The water and sodium then can flow out of the intercellular spaces, through the interstitial space, and into the peritubular capillary blood.

The water and sodium that enter the intercellular spaces must be removed by the peritubular capillaries. Entry of water and dissolved solutes into peritubular capillaries is ensured by two forces (Fig. 46-9). First, the peritubular capillary plasma oncotic pressure is high as a result of the prior removal of a protein-free fluid in the process of glomerular filtration—a process that concentrates the remaining proteins in the plasma. This concentrated plasma then circulates through the peritubular capillaries with a high oncotic pressure, encouraging fluid absorption. The second factor that facilitates water and salt reabsorption from the renal interstitium is the low hydrostatic pressure in the capillaries created by the pressure drop across the efferent arterioles.

The gap junctions between endothelial cells once were thought to be relatively tight structures. They now are recognized to be moderately permeable, both to low molecular weight solutes, such as sodium and chloride and to water. Since there is a significant permeability for water and salts at the epithelial gap junctions, some sodium and some water can leak back into the lumen rather than flow into the peritubular capillary blood. This fact, coupled with the dynamics occurring at the capillary, provides one possible explanation for glomerulotubular balance, the process coupling glomerular capillary function and tubular reabsorption.

Glomerulotubular balance may be produced by modifying fluid reabsorption into the peritubular capillaries. Assuming that filtration equilibrium is not attained (Fig. 45-6, *A*, line *C*), an increase in glomerular filtration will cause an increased protein oncotic pressure at the end of the glomerular capillary and therefore in peritubular capillary blood. This increase in plasma oncotic pressure then can be assumed to enhance reabsorption of fluid from the interstitial space surrounding the tubular epithelium. It therefore will reduce the hydrostatic pressure in this region and so diminish the amount of fluid and ions that leaks back through the junctions between epithelial cells. Thus the net effect of an increase in glomerular filtration is an increase in peritubular capillary oncotic pressure, with reduced back leak and a net increase in sodium and water reabsorption to balance the filtration.

■ *Osmotic diuretics*

The proximal tubular epithelium has a relatively high water permeability. This fact is in part responsible for the isotonic reabsorption of the filtrate. This high water permeability also means that the presence of nonreabsorbable solutes in the filtrate will severely restrict the amount of water that can be reabsorbed. As water is reabsorbed, the nonreabsorbable solute will be concentrated and exert an osmotic pressure that will tend to hold fluid in the tubule. In the normal process of sodium reabsorption water moves with the sodium. The isotonic fluid that is left behind contains sodium and the other normal constituents of the filtrate plus the nonreabsorbable solute. The water remaining behind balances the solute but in the process dilutes the sodium and generates a concentration gradient for this ion. As a result, proximal tubular sodium concentration may fall to 100 mEq/L when a nonreabsorbable solute is present in the lumen in significant concentrations. However, recall that the pump-leak ratio for sodium is low, so that at about 100 mEq/L of sodium in the lumen the back leak of sodium equals the transport rate, and further net transport can no longer occur. The fluid and solutes remaining in the tubule will then escape the proximal tubule.

Any substance that increases urine volume is called a diuretic. A nonreabsorbable solute that acts as a diuretic in the manner just described is called an *osmotic diuretic*. *Mannitol* is a low molecular weight, nonmetabolizable sugar-alcohol that often is used as an osmotic diuretic.

■ *Henle's loop ion and water transport*
■ *Sodium transport*

It is very difficult to study transport in Henle's loop, and thus detailed knowledge of the transport mechanisms is still scanty. It is known, however, that salt and water transport in this part of the nephron plays a vital role in determining the ability of the kidney to concentrate the urine; this process is discussed later.

In general, it appears that little net sodium transport occurs in the descending limb of Henle's loop, although there are large changes in composition induced by passive solute exchange with the fluid in the medullary interstitium. This portion of the nephron has a high permeability to both salts and water.

The thin ascending limb of Henle's loop has a lower permeability for salts and water, and there is substantial debate at present as to whether there is significant active transport of material. The epithelial cells in this portion of the nephron possess few mitochondria, and this is consistent with the idea that the thin ascending limb is largely a simple conduit for tubular fluid.

The major site of active transport in Henle's loop is in the thick ascending limb. Measurements of the luminal potential in isolated thick segments yield values ranging from approximately +9 mV in the medullary segments to +25 mV in the cortical segments. The sodium concentration over the medullary segments of the loop cannot be measured because the loop is buried in the parenchyma of the medulla. However, sodium concentration in the first part of the distal tubular fluids is found to be in the range of 65 to 75 mEq/L. Sodium exits the proximal tubule at 140 mEq/L, suggesting that substantial sodium reabsorption takes place in the thick ascending limb.

■ *Chloride transport*

Chloride concentration in the beginning distal tubular fluid is reduced approximately as much as sodium, and this, combined with the positive luminal potential, has led to the proposal that the chloride ion is the species which is actively transported. Current evidence, however, suggests that here, as well as in the proximal tubule, the chloride ion movement is secondary to activity of the Na^+, K^+-ATPase and primary sodium transport. The origin of the positive luminal potential is not yet established but may be a diffusion potential resulting from back leak of sodium into the lumen. This is consistent with the fact that sodium permeability is higher than chloride permeability in this segment of the nephron.

■ *Potassium transport*

About 30% of the filtered potassium leaves the proximal tubule and enters Henle's loop. There appears to be some entry of potassium into the descending limb of Henle's loop, but the sources of this potassium are poorly understood. In any case, when fluid samples from the beginning of the distal tubule are analyzed, it is found that only about 5% of the filtered potassium is present, thus showing that substantial net reabsorption of potassium occurs in the ascending limb of Henle's loop. A large part of this reabsorption is likely to be passive, resulting from the substantial positive intraluminal potential, and the energy is likely to be derived from sodium transport.

■ *Water transport*

The descending portion of Henle's loop appears to be permeable to water, whereas the ascending limb has a very low water permeability. As shown later, the interstitial fluid of the medulla has a very high osmotic activity (Fig. 46-11). These facts generate a peculiar pattern for water handling in Henle's loop.

During passage of fluid from the proximal tubule down the descending limb water is drawn osmotically out of the nephron into the medullary interstitium, thereby reducing the volume. The segment of the ascending limb of the loop where solute transport occurs is also the segment where water permeability is low. The result of reabsorption of sodium chloride out of the segment with low water permeability is to dilute the remaining fluid. Fluids entering the distal tubule have osmolalities less than 100 mOsm/L, as compared with the 300 mOsm/L seen in plasma and in the latter part of the proximal tubule. This characteristic of absorbing solute in excess of water seen in the ascending limb has led to its being called the *diluting segment* of the nephron.

■ *Distal tubular ion and water transport*
■ *Sodium transport*

Fig. 46-10, *A,* shows the general characteristics of the transport in a distal tubular cell. Many elements of distal tubular function are qualitatively similar to those seen in the proximal tubule, and many of the various carriers shown in Fig. 46-8, *B,* exist in the distal tubule as well. Perhaps the most significant element that distinguishes distal tubular function from proximal tubular function is a tighter epithelium with a more sparse brush border. These differences result in a lower permeability to water and to ions, thereby yielding a higher pump-leak ratio. Thus the luminal potential (-50 mV) that is found in the distal tubule is much larger than that found in the proximal tubule (± 2 mV). Also the sodium gradients can be much larger in the distal tubule, with tubular sodium concentrations falling to 40 mEq/L.

Fig. 46-10, *B,* shows how distal tubular sodium reabsorption varies as a function of sodium load in the distal tubule. Note that here sodium transport is more of the T_m type than the gradient-time type found in the proximal tubule. This has substantial functional implications. In the distal tubule the sodium load is, in fact, the amount of sodium entering from the proximal tubule and Henle's loop, not the sodium that was originally filtered. The fact that the distal sodium reabsorptive process is a T_m system means that, over a substantial range of sodium loads entering the distal tubule from the proximal tubule, distal tubular sodium reabsorption can match sodium entry, thus minimizing the escape of sodium from the tubule and loss from the body. However, when the T_m is

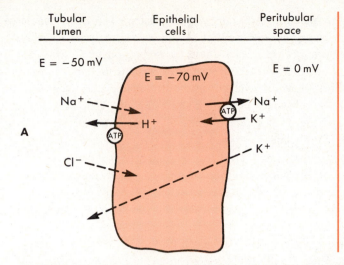

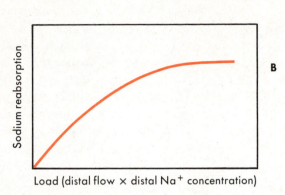

Fig. 46-10 ■ **A,** Distal tubular salt and water transport. Note the more negative luminal potential than in the proximal tubule and the fact that potassium is passively *secreted*. The hydrogen pump on the luminal side of the epithelial cell is electrogenic, and its activity can diminish the luminal potential. **B,** The relation between the sodium load entering the distal tubule from the proximal tubule and distal sodium reabsorption. Compare this with Fig. 46-3.

exceeded, further elevations in sodium load allow sodium to escape into the collecting duct.

The sodium gradient is dependent to some extent on the state of hydration and the associated changes in the antidiuretic hormone (ADH). During antidiuresis, that is, when water conservation is important, ADH level is high, and there are modest increases in the sodium pump rate and increases in the permeability of the distal tubule to water. The effects of ADH appear to be confined to the most distal segments of the distal tubule. As is explained later, the major effect of ADH with regard to water reabsorption has to do with its effect on tubular water permeability.

■ *Chloride reabsorption*

Charge neutrality in the distal tubule is maintained to a larger extent by chloride than by bicarbonate ion movement. This reflects the fact that a large fraction of the bicarbonate has been reabsorbed by the time the distal tubule is reached, and chloride now becomes the dominant anion accompanying sodium. There is some evidence for active distal tubular chloride reabsorption, but the quantitative significance of this remains to be shown.

■ *Potassium secretion*

Recall that most of the filtered potassium is reabsorbed in the proximal tubule and Henle's loop. Therefore urinary potassium largely reflects the quantity of potassium that is secreted during the passage of tubular fluid through the distal tubule and collecting duct. Potassium secretion in the distal tubule is largely passive. There is a relatively large potential difference across the tubular epithelium in the distal nephron (-50 mV), and, since most of the potassium is reabsorbed in the proximal tubule, fluid entering the distal tubule has a low potassium concentration; it follows that there is a substantial concentration difference driving potassium secretion. Thus the electrochemical potential that drives potassium secretion can be relatively high.

A number of factors can influence potassium secretion, especially the amount of potassium in the diet. Individuals maintaining a high potassium diet show much larger urinary secretion than normal. Conversely, when potassium is restricted in the diet, potassium actually may be reabsorbed rather than secreted over the course of the distal tubule.

Sodium and potassium transport processes are closely related. The distal tubular potential is heavily dependent on sodium reabsorption. Under conditions where sodium reabsorption is stimulated, for example, salt deprivation, the tubular potential difference will rise as sodium is more avidly reabsorbed. This increase in potential will enhance potassium secretion as a result of greater driving forces for transepithelial movement.

An important reciprocal relation between potassium and hydrogen ions is commonly observed. At one time it was believed that potassium and hydrogen competed for an exchange pump with sodium. This remains a useful device for remembering the interactions among these three ions, but it is now clear that this concept is not correct. The interaction between hydrogen ion and potassium ion is explained more accurately by the fact that the hydrogen pump in the distal tubule is electrogenic. This secretory pump is discussed in detail subsequently (p. 883), but for the present it can simply be noted that an increase in hydrogen secretion by virtue of the electrogenic nature of the pump will diminish the magnitude of the transepithelial potential difference in the distal tubule. This reduction in potential difference will reduce the electrochemical driving force for passive potassium secretion and thus reduce net potassium secretion.

■ *Water reabsorption*

An active solute pump and low sodium, chloride, and water permeability in the ascending limb of Henle's loop yield a hypotonic fluid in the first part of the distal tubule. This fluid may enter the distal tubule with an osmolality of about 100 mOsm/L. The distal portion of the distal tubule is sensitive to the presence of ADH, and the amount of water that is reabsorbed over the course of the distal tubule varies greatly depending on tubular water permeability. When ADH is absent, water permeability is low, and hypotonic fluid entering the distal tubule from the ascending limb remains hypotonic throughout the length of the distal tubule. However, when ADH is present, the hypotonic fluid in the lumen of the distal tubule and the isotonic fluid in the peritubular capillary blood around the distal tubule induce osmotic reabsorption of water and a progressive elevation of osmolality of distal tubular fluid toward isotonicity.

■ *The collecting duct*

Ion reabsorption. The collecting duct behaves qualitatively like the distal tubule. For the most part sodium, potassium, chloride, and hydrogen continue to be either reabsorbed or secreted in the collecting duct, as they were in the distal tubule. A major distinction between the two sites is that all of the collecting duct is sensitive to ADH. In fact it is now thought that the cells of the end of the distal tubule which respond to

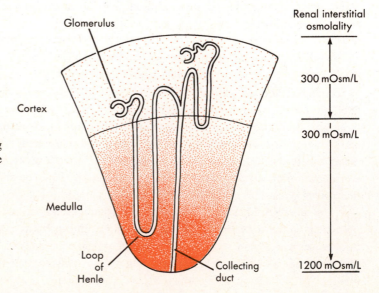

Fig. 46-11 ■ Osmolality of the renal interstitial fluid. This represents a cross section through a renal pyramid and shows cortical and juxtamedullary nephrons. The intensity of shading is proportional to solute concentration in the interstitium. Note the accumulation of solutes in the tip of the medulla.

ADH are actually the same cell type as those found in the collecting duct. It is in the collecting duct that large concentration gradients can be established and the final composition of the urine is determined.

Water reabsorption. In the collecting duct, water reabsorption is much the same as in the distal tubule with one notable exception (Fig. 46-11). By a process described in detail in a subsequent section, the interstitial fluid of the medulla is made extremely hypertonic; osmolalities as high as 1200 mOsm/L exist at the tip of the papilla. In the presence of ADH, when the latter part of distal tubular and collecting duct water permeability is high, osmotic withdrawal of fluid from the collecting duct can raise the final urine concentration as high as the osmolality of the interstitium. Thus the final urine concentration can range from something less than 100 mOsm/L (that is, the osmolality established by solute reabsorption without accompanying water in the ascending limb of Henle's loop) to 1200 mOsm/L, a value limited by the concentration of fluid in the medullary interstitium.

A variety of hormones influence the way in which salt and water are handled along the course of the nephron. One of the most important of these is *aldosterone,* a hormone that acts in the distal tubule to stimulate sodium reabsorption and potassium secretion (Chapter 54). Aldosterone is a steroid hormone released from the adrenal cortex under the influence of *angiotensin II* and *angiotensin III*. Also release is stimulated by low sodium or high potassium concentrations in the plasma. In general, it can be said that situations in which conservation of sodium is required or in which blood pressure falls stimulate the release of aldosterone.

■ *Hormones affecting the nephron*

The mechanism of action of aldosterone has been studied extensively, and several steps in the process are known. Aldosterone binds to a cytoplasmic receptor, and the combination enters the cell nucleus to stimulate DNA and thereby initiate the synthesis of membrane proteins (Chapter 54). These proteins act on the apical membrane of the distal tubular cell to increase sodium permeability and on the mitochondria to stimulate oxidative phosphorylation. It is thought that the increased cell membrane permeability allows sodium to enter the epithelial cell more readily and from there to be pumped into peritubular interstitial space.

Angiotensin II is an extremely important hormone in the control of renal function. It may have a modest direct action on the tubular epithelium, but its main activity is on the adrenal gland to stimulate aldosterone release, which subsequently acts on the epithelium. In addition, this hormone causes contraction of the smooth muscle of the renal arterioles.

Another hormone that may influence salt transport in the kidney is the *natriuretic hormone*. Natriuretic hormone is thought to be a small polypeptide that is released in response to expansion of extracellular fluid volume and to sodium loading. The site of formation of the hormone and the specific stimuli causing its release have yet to be determined. Its existence is deduced from the fact that elevated sodium excretion can be induced by sodium loading in animal preparations in which all other known influences on the sodium reabsorptive mechanism have been excluded. Furthermore, such a natriuretic factor has been recovered from the urine of laboratory animals and humans. It is thought that expansion of extracellular fluid volume causes release of this hormone, with a consequent inhibition of sodium reabsorption and resultant reduction in body sodium. This in turn reduces fluid volume of the extracellular space.

Two other hormones are known to inhibit sodium reabsorption, but their physiological importance remains to be established. *Parathormone* acts to inhibit salt and water reabsorption in the proximal tubule, and *prostaglandins* inhibit sodium reabsorption in the collecting duct.

ADH has been studied more intensively than almost any other renal hormone. Its site of action has been clearly defined, being primarily in the collecting duct and secondarily in the distal tubule. The effect of ADH is to increase water permeability, thereby allowing tubular fluid to equilibrate osmotically with interstitial fluid. The change in permeability is highly selective for water; other small molecules may be affected differently. For example, urea permeability is not increased in the end of the distal tubule and early collecting duct following application of ADH.

The mechanism of action of ADH depends on a hormone binding to the basolateral border of the renal tubular cell (Chapter 52). A membrane response is induced that causes the formation of cyclic AMP. This leads to an increase in myotubule formation in the epithelial cell. This latter response appears to be the specific change which generates an increase in water permeability of the luminal membrane, but the mechanism whereby myotubules increase epithelial cell water permeability is not known. It has been established, however, that associated with the increase in apical cell membrane permeability there is an increase in the aggregation of certain vesicular structures on the membrane surface.

■ *Passive transport mechanisms*

This section deals with the passive transport of two molecules, *urea* and *ammonia*. These molecules are presented as examples of the way various aspects of nephron function may indirectly influence the transport of molecules with finite tubular permeability. As seen previously, passive transport processes dissipate gradients that are formed as a result of the expenditure of energy used in actively transporting other materials. In the kidney the most common forces driving passive transport of a molecule are gradients generated secondary to sodium reabsorption and gradients generated by active secretion of hydrogen into the tubular lumen.

With previous emphasis directed so intently at active transport processes it is easy to lose sight of the fact that there is a finite permeability for small ions and small organic solutes in the nephron and that water reabsorption tends to increase the concentration of all these solutes, thereby generating a concentration gradient that favors reabsorption. Essentially all small solutes in the filtrate are influenced to some degree by the reabsorption of water and by the subsequent concentration gradients that are generated.

■ *Passive reabsorption secondary to sodium transport: urea reabsorption*

The reabsorptions of urea, water, chloride, and bicarbonate are highly dependent on movement of the sodium ion. Basically the motive force for the transport of each originates with the sodium transport processes described previously. Water, chloride, and bicarbonate reabsorption are considered elsewhere as special cases and are not dealt with here in detail, although the same principles described here do apply.

Transport of the urea molecule is easier to understand than the transport of some other molecules because of its lack of charge. Therefore it can be asserted that urea reabsorption will be mainly a function of the *permeability* of the tubule, the tubular *area* available for reabsorption, and the *urea concentration gradient*. The concentration gradient of course is induced by the reabsorption of water.

Two general statements may be made about the excretion of a molecule that is passively reabsorbed: (1) excretion will vary linearly with changes in plasma concentration, and (2) excretion will vary directly but curvilinearly with urine flow. It should be noted that these two characteristics are not exclusively the behavior of a molecule transported passively. For example, sodium excretion will vary linearly with plasma concentration under some circumstances in spite of the fact that transport is active.

Fig. 46-12, *A*, shows the relation between urea excretion and plasma urea concentration. The relation is shown at two different glomerular filtration rates. As plasma urea concentration is varied, the excretion varies linearly with the concentration; in other words, a constant fraction of the filtered urea is being reabsorbed. This relation between plasma concentration and urea excretion might be predicted when it is recalled that the

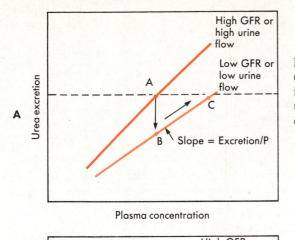

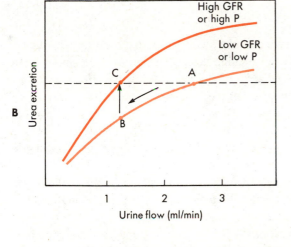

Fig. 46-12 ■ Effects of changes in plasma concentration (**A**) and urine flow rate (**B**) on urea excretion. Excretion is filtration minus reabsorption, and thus increasing excretion means decreasing reabsorption. The dashed line represents the urea excretion rate demanded by urea production. For a steady state to occur, urea excretion must equal the sum of the urea production and urea intake (*A* and *C*).

absence of a specific transport protein would preclude the system displaying saturation kinetics. Therefore variations in plasma concentration simply produce proportional variations in the quantity of urea filtered and excreted.

Note that the slope of the line relating excretion and plasma concentration is constant. The slope of this line is equal to the ratio of the urea excretion (U × V) to the plasma concentration (P), and this ratio (U × V/P) is known as the *clearance*. Fig. 46-12, *A*, also shows the effect of a change in GFR on urea excretion. A low GFR produces a decrease in the slope of the line relating urea excretion and plasma urea concentration, but the linear correspondence between urea excretion and plasma urea concentration is maintained.

Fig. 46-12, *B,* shows the effect of variations in urine flow on urea excretion; the form of this relationship is quite different from the relationship between urea excretion and plasma concentration. As urine flow is reduced, there is a disproportionate reduction in the amount of filtered urea that is excreted. In other words, the fraction reabsorbed increases at low flows.

If urea reabsorption is passive, then it follows that the nonlinear excretion must reflect some nonlinear change in tubular concentration as the urine flow rate changes. A hypothetical explanation for this nonlinear behavior can be derived by imagining that urea reabsorption is a two-part process: first, water is absorbed, thereby generating a urea concentration gradient; and second, a fixed time period is allowed for urea to diffuse out of the tubule down the concentration gradient. Fig. 46-13 shows what would happen if water were selectively extracted from 125 ml of solution (that is, the volume of water filtered in 1 minute) and the solute were left behind. As water is extracted, the concentration rises in a nonlinear fashion and increases sharply (approaching infinity) as

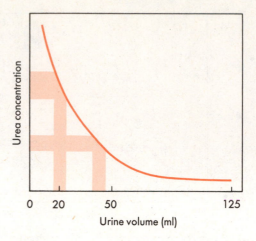

Fig. 46-13 ■ Hypothetical relation between urea concentration, which drives urea reabsorption, and the urine volume. Note the sharp rise in concentration at small urine volumes.

volume approaches zero. If this is imagined to be analogous to the pattern in the tubule, then the reason for the nonlinear behavior relating urine flow and urea excretion can be appreciated (Fig. 46-12, *B*). The colored areas of Fig. 46-13 show the effect on concentration of the removal of 5 ml of filtrate when the volume has been reduced to 48 ml and when the volume has been reduced to 22 ml. In other words, it represents the situation that might be expected when urine flow is reduced by 5 ml/minute, starting at an initial flow of 48 or at an initial flow of 22 ml/minute. It is apparent that there is twice as large a concentration increase when starting from the lower volume. Therefore it would be expected that equivalent reductions in urine flow starting from smaller initial urine flows might induce larger concentration gradients across the tubule and therefore induce larger fractional reabsorption of urea (smaller excretion).

The relations among plasma urea concentration, urine flow, and urea excretion have important consequences in terms of understanding renal function and its relation to plasma and urinary urea concentration. In a steady state *urea excretion must equal urea production*. Urea production is of course inextricably linked to intake and metabolism of proteins. Fig. 46-12, *A* and *B*, shows that glomerular filtration rate and urine flow will both alter the urea excretory rate. Furthermore it is well known that glomerular filtration rate and urine flow vary in relation to the state of body hydration. It follows that urea excretion must be related in some way to the state of hydration of the individual.

The net effect of an alteration in glomerular filtration rate can be easily ascertained from Fig. 46-12, *A*. Assume that a subject was operating normally, with a urea excretion rate indicated by the dashed line at the high glomerular filtration rate (point *A*). Assume also that, because of dehydration secondary to restricted fluid intake, the glomerular filtration rate was diminished. This would cause the urea excretion to diminish from *A* to *B* and thus become less than the rate of urea formation. However, if urea excretion were reduced, with continued intake of protein and formation of urea, the imbalance between excretion and production would result in a slow increase in plasma urea concentration and a return of urea excretion toward normal; it would follow the path shown by the line *B-C*. The system then would be stabilized at a lower glomerular filtration rate, a higher plasma urea concentration, and an excretory rate equal to the urea production rate.

Careful examination of Fig. 46-12, *A* and *B*, permits one to use similar, although more involved, logic to predict the effect of a reduction in urine flow on urea excretion and plasma urea concentrations. Suppose urine flow were reduced from 2.5 to 1.2 ml/minute (*A-B,* Fig. 46-12, *B*). This would lead to a reduction in urea excretion, and, in the face of continued constant protein intake, urea would begin to accumulate. Accumulation of urea would elevate plasma urea concentration, and this in turn would increase excretion. This is seen in Fig. 46-12, *A*, point *B*. Assume that the subject is operating in the low glomerular filtration rate range; the accumulating urea in plasma would elevate plasma concentration and then urea excretion from *B* to *C*. The situation in this case would stabilize at a point where urea excretion and urea production were equal, with a reduced urine flow and an elevated plasma concentration of urea (point *C*).

The importance of understanding the changes in urea excretion with urine flow and plasma concentration resides in the fact that urea concentration frequently is used as an index of renal function. Obviously plasma urea alone is an inadequate index of the adequacy of renal function in view of its dependence on physiological regulation of urine flow and glomerular filtration rate. Measurement of plasma urea levels alone does not permit a distinction between renal pathological conditions and changes in the state of hydration of the individual or a variety of other physiological adaptations.

Many secretory and reabsorptive processes in the kidney ultimately are related to the transport of hydrogen ion. The dependence of passive transport processes on hydrogen ion secretion is a reflection of the fact that the state of charge of weak acids and bases is dependent on pH and that the charge on small molecules has a major impact on their ability to traverse the renal epithelium. A charged molecule is much less permeable than an uncharged molecule.

A specific example of hydrogen ion–dependent transport is the secretion of ammonia (Fig. 46-14). Ammonia is produced by the tubular epithelial cells from a variety of amino acids, especially glutamine, by deamination and deamidation. The ammonia produced by these processes can diffuse into the tubular lumen, where it can combine with hydrogen ion secreted by the epithelial cells. This protonation of ammonia renders it less permeable and thus limits its exit from the lumen and its possible return to the epithelial cell. The result is that ammonium ion accumulates within the tubule in proportion to the luminal hydrogen ion concentration. Such a process is commonly referred to as *diffusion trapping*.

Secretion of ammonium is an important part of the maintenance of normal acid-base status of the organism because it provides a means whereby excess hydrogen ion can be

■ *Passive transport secondary to hydrogen ion secretion*

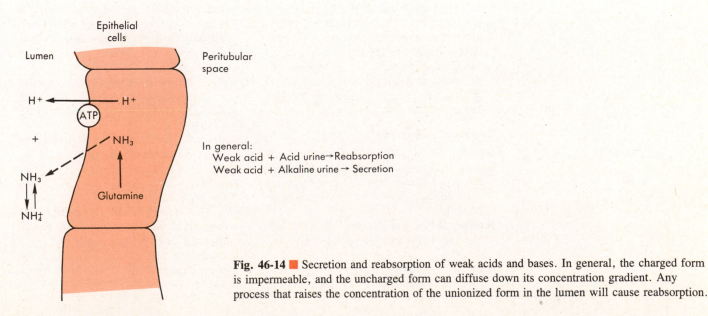

Fig. 46-14 ■ Secretion and reabsorption of weak acids and bases. In general, the charged form is impermeable, and the uncharged form can diffuse down its concentration gradient. Any process that raises the concentration of the unionized form in the lumen will cause reabsorption.

eliminated. As hydrogen ion secretion increases, the resultant fall in luminal pH automatically traps NH_4^+ within the lumen and ensures excretion of both ammonia and hydrogen ion.

The process of diffusion trapping by hydrogen ion is a general one that can apply to all weak acids and weak bases. It depends only on establishing a concentration difference for the uncharged form of a weak acid or base across the tubular epithelium. This concentration difference is induced by hydrogen ion secretion in a fashion predicted by the Henderson-Hasselbalch equation:

$$\text{Log} \frac{\text{Ionized form}}{\text{Unionized form}} = \text{pH} - \text{pK}' \tag{4}$$

This equation shows that a pH change of 1 unit will alter the ratio of ionized to unionized forms by a factor of 10. The essence of the process is that only the uncharged form passes the epithelial cell membrane. In the case of a simple weak acid, such as salicylic acid, reducing urine pH will increase the quantity of unionized acid in the tubular lumen and create a gradient leading to diffusion of the molecule from the lumen into the blood, that is, reabsorption. However, alkalinization of the urine, which may happen during hyperventilation or by ingesting a diet high in fruit juices, will result in a smaller concentration of the uncharged form of salicylic acid within the tubule. Therefore secretion will occur passively, and the excretion of salicylic acid will be increased. This fact can be used to assist in the elimination of salicylic acid from a patient who has received an overdose of aspirin or any other weak acid.

■ *Bibliography*

Journal articles

Bradley, S.E., editor: Hormones and the kidney, Kid. Intern., vol. 6, 1974.

DiBona, G.F.: Neural control of renal tubular sodium reabsorption in the dog, Fed. Proc. **37:**1214, 1978.

Giebisch, G., and Berliner, R.W.: Membrane transport in the kidney, Kid. Intern., vol. 9, 1976.

Giebisch, G., and Staton, B.: Potassium transport in the nephron, Annu. Rev. Physiol. **41:**241, 1979.

Schafer, J.A.: Salt and water absorption in the proximal tubule, Physiologist **25:**95, 1982.

Schmidt-Nielsen, B.: Urea excretion in mammals, Physiol. Rev. **38:**139, 1955.

Schnermann, J., Ploth, D.W., and Hermle, M.: Activation of tubuloglomerular feedback by chloride transport, Pfluegers Arch. **362:**229, 1976.

Books and monographs

Bijvoet, O.L.M.: Kidney function and calcium and phosphate metabolism. In Anioli, L.V., and Krane, S.M., editors: Metabolic bone disease, New York, 1977 Academic Press, Inc.

Brenner, B.M., and Rector, F.C.: The kidney, vol. 1, Philadelphia, 1981, W.B. Saunders Co.

Burg, M.B., and Bourdeau, J.E.: Function of the thick ascending limb of Henle's loop. In Vogel, H.G., and Ulrich, K.J., editors: New aspects of renal function, Amsterdam, 1978, Excerpta Medica.

Chonko, A.M., Irish, J.M., III, and Welling, D.J.: Microperfusion of isolated tubules. In Renal pharmacology, Methods in pharmacology series, vol. 4B, New York, 1978, Plenum Press.

Koushanpour, E.: Renal physiology: principles and functions, Philadelphia, 1976, W.B. Saunders Co.

Kriz, W., Barrett, J.M., and Peter, S.: The renal vasculature: anatomical-functional aspects. In Thurau, K., editor: Kidney and urinary tract physiology II, International review of physiology series, vol. 12, Baltimore, 1976, University Park Press.

Orloff, J., and Berliner, R.W.: Renal physiology. In Handbook of physiology, section 8, Washington, D.C., 1973, American Physiological Society.

Schultz, S.G.: Ion-coupled transport across biological membranes. In Andrioli, T.A., Hoffman, J.F., and Fanestil, D.P., editors: Physiology of membrane disorders, New York, 1978, Plenum Press.

Valtin, H.: Renal function: mechanisms preserving fluid and solute balance in health, Boston, 1973, Little, Brown & Co.

Windhager, E.E.: Micropuncture techniques and nephron function, New York, 1968, Appleton-Century-Crofts.

chapter 47

Integrated nephron function

The clearance calculation offers a means of comparing the excretion of molecules with differing excretory patterns under varying conditions. For example, suppose that a subject has a low urea excretory rate. How does one decide whether the rate is low because plasma urea concentration is low, the urine flow is low, or renal function is impaired? An important tool in deciding this kind of issue is the *clearance method,* which was previously referred to in the analysis of the behavior of urea. Fig. 46-12 shows the relation between urea excretion and plasma urea concentration. The slope of this line is constant; therefore the ratio of urea excretion to plasma urea concentration is constant. The constancy implies that this ratio may be a useful way of adjusting for variations in plasma concentration. Fig. 46-12, *A,* also shows a line obtained for a reduced urine flow. Recall that, when urine flow is reduced, urea excretion is also reduced, but it is reduced disproportionately (Fig. 46-12, *B*). Note, however, that at the lower urine flow, the urea excretion continues to vary linearly with plasma concentration, but the slope (U × V/P) of the line is less (Fig. 46-12, *A*). In each case, the slopes of the two lines in Fig. 46-12, *A,* represent the ratio of the excretion of a material (U × V) to its plasma concentration (P); this ratio is called the *clearance.*

As shown in Fig. 47-1, the excretion rate for any material is the quantity (milligrams or milliequivalents) of material that appears in the urine in a given unit of time; it is equal to the product of urine flow rate (V) and urine concentration (U). The plasma concentration is the quantity of material in a unit of plasma. The ratio of the excretory rate to the plasma concentration is then equal to the *minimum* volume of plasma at the prevailing plasma concentration that could have supplied the material appearing in the urine in the given time interval. In other words, the clearance is the volume of plasma per unit time that would have to be *cleared* entirely to provide the solute appearing in the urine.

Clearance appears to be a peculiar concept at first because it need not reflect the *real* plasma volume that is actually cleared of a material in a unit of time. The number calculated is a so-called *virtual volume* and gives only a lower limit on the behavior of the kidney. In other words, no less plasma than the computed volume could have supplied the excreted solute, but partial removal of the excreted substance from plasma would mean that more plasma was actually involved than indicated by the clearance computation. Obviously, if 20% of the solute were removed from each unit of plasma that passed through the kidney, the calculation of clearance would be a fivefold underestimate of the total volume of plasma that was acted upon.

Although the calculation of clearance does not yield a value equal to the true volume that has been stripped of a particular solute per unit of time, the calculation does provide an estimate of how rapidly concentration might change in plasma or interstitial space. For example, suppose sodium appears in the urine at a concentration of 150 mEq/L,

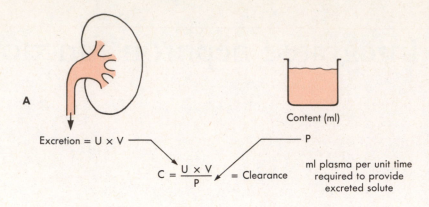

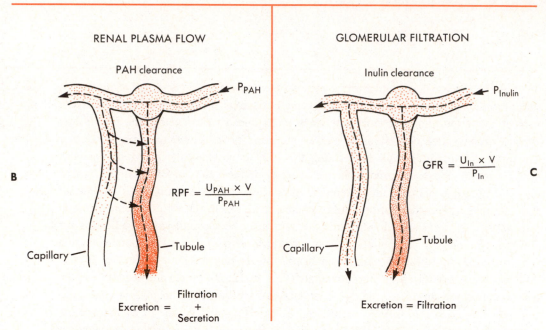

Fig. 47-1 ■ Clearance calculation and the estimation of flow. **A,** Basic concepts of the clearance calculation. **B,** Measurement of renal plasma flow. **C,** Measurement of glomerular filtration rate. The density of the color in the lower portions of the figure represents the concentration of either inulin, in the case of a glomerular filtration rate measurement, or PAH, in the case of a determination of renal blood flow. Note in **C** that inulin concentration rises solely because water is removed from the tubular fluid. PAH concentration rises both because of water removal and because of active secretion of PAH into the tubular lumen.

and the urine flow rate is 1 ml/minute; sodium excretion would equal $150 \times 0.001 = 0.150$ mEq/minute. The sodium clearance would then be equal to 0.15 mEq/minute divided by 150 mEq/L, or 0.001 L/minute or 1 ml/minute of plasma. This implies that sodium is being excreted at a rate that could be matched if all of the sodium were being removed from 1 ml of plasma per minute. If the total volume of the extracellular space is about 20 L and the excreted sodium is coming from the extracellular space, then the clearance computation suggests that sodium is being excreted at a rate approximately equal to $\frac{1}{20,000}$ or 0.005% of the total body sodium per minute. Thus we have some sense of how rapidly continued excretion of sodium by the kidney (without further replacement) might alter body composition.

■ *Renal blood flow and glomerular filtration rate measurements*

In spite of the fact that clearance, as a rule, represents a virtual volume per unit time, it is possible under two circumstances to use clearance to calculate a true volume rate of flow and in so doing to obtain estimates of *renal blood flow* or *glomerular filtration rate.* The measurement of renal blood flow is based on the idea that if there is a material that actually *is,* by some process, completely removed from the plasma by the kidney, then the rate at which plasma enters the kidney, that is, the renal plasma flow, would be equal to the clearance. In fact, a number of organic dyes are not only filtered but are also actively secreted by the kidney. Para-aminohippuric acid (PAH) is such a dye, and over 90% of this dye that enters the kidney via the renal artery will normally be excreted in the final urine as a result of filtration and secretion. Thus simply by measuring the PAH clearance, one can estimate, and be accurate to within 10%, the renal plasma flow (RPF) (Fig. 47-1, *B*).

$$RPF = \frac{U_{PAH} \times V}{P_{PAH}} \tag{1}$$

Extraction of PAH is not 100% because of (1) the lack of perfect transport systems in the proximal tubule and (2) the fact that a significant fraction of the blood flowing into the kidney does not come into sufficiently close contact with the proximal tubules where PAH secretion occurs. After passage through the efferent arterioles of the juxtamedullary nephrons, a fraction of the blood flows into the medulla via the vasa recta capillaries; no proximal tubules are in the medulla where the vasa recta capillaries are located. Therefore blood flowing through the vasa recta capillaries is never exposed to proximal tubules for secretion.

The second case in which clearance can be a true measure of a real volume rate of flow is the case of glomerular filtration (Fig. 47-1, *C*). After an extensive search, renal physiologists were able to find a marker that had no effect on renal blood flow or filtration rate, that was filtered in a concentration exactly equal to its concentration in plasma, and that was not metabolized, reabsorbed, or secreted by the tubules. Such a material can enter the tubule only by glomerular filtration, and whatever enters the tubule must be totally excreted in the urine. Thus, given the fact that the filtrate concentration is equal to the plasma concentration, the minimum volume of fluid that could have supplied the material in the final urine would be equal to the volume of fluid that was filtered with its contained solute. The substance that meets all of these criteria is *inulin,* a nonmetabolizable polysaccharide. Inulin clearance is equal to the glomerular filtration rate (GFR).

$$GFR = \frac{U_{In} \times V}{P_{In}} \tag{2}$$

Thus the clearance computation can be used to calculate two important values, glomerular filtration rate and renal plasma flow. In those two special cases a molecule was chosen that has very special characteristics that permit the clearance calculation to yield a real volume rate of flow, not a virtual volume.

In practice, a clearance measurement is made by infusing the appropriate substance at a constant rate until the plasma concentration is stable. Samples of urine and plasma are then taken, and the computation can be made.

It is now possible to insert a micropipette into the lumen of a single tubule and withdraw fluid for analysis. This process is called *micropuncture.* Within the kidney, solutes move into and out of the tubules, and water is reabsorbed. In order to understand the meaning of a micropuncture sample, fluid composition changes that result from water reabsorption must be distinguished from changes induced by solute transport. This is usually accomplished by using an indicator that is quantitatively retained within the tubular fluid and is filtered with a concentration equal to that of plasma; obviously, inulin is the appropriate indicator.

The fraction of the filtrate that remains at any point in the nephron is simply equal to the tubular fluid flow rate at that point divided by the glomerular filtration rate. In the case of the excreted urine, the fraction of filtered water excreted is simply equal to V/GFR. Recall that glomerular filtration rate is equal to the inulin clearance.

■ *Corrections for tubular water reabsorption—micropuncture analyses*

$$f = \text{fraction of filtered water excreted} = \frac{V}{\frac{(U_{In}V)}{P_{In}}} = \frac{P_{In}}{U_{In}} \tag{3}$$

$$\text{Fraction reabsorbed} = 1 - f = 1 - \left(\frac{P_{In}}{U_{In}}\right) \tag{4}$$

Thus simply by computing the plasma/urine ratio of inulin concentrations in a steady state, one can compute either the fraction of the filtered water that is excreted or, alternately, the fraction that has been reabsorbed.

The observed behavior of inulin can be used to correct the various solute concentrations for water reabsorption. The quantity of substance (x) filtered per unit time should be P_x times GFR, or P_x times C_{In}. The rate of excretion of solute in the urine is V times U_x. It follows that the fraction of the filtered material excreted must be

$$F = \frac{U_x \times V}{P_x \times C_{In}} = \frac{C_x}{C_{In}} \tag{5}$$

Because the urine flow term (V) is the same for substance x and inulin, equation (5) can be further simplified to:

$$\text{Fraction filtered solute excreted} = \frac{U_x/P_x}{U_{In}/P_{In}} \tag{6}$$

Thus the ratio of the clearance of an unknown substance (sodium, for example) to that of inulin gives the fraction of the filtered material that appears in the urine. If the clearance of a substance is greater than the clearance of inulin, this denotes secretion. If the clearance is less than the clearance of inulin, this implies reabsorption.

■ *Renal blood flow*

Fig. 47-2 is a schematic illustration of the anatomy of the renal circulation; it emphasizes the variety of pathways that can be taken through the vasculature of the kidney. There are basically three parallel networks: one supplying the cortical nephrons, one supplying the juxtamedullary nephrons, and one supplying the medulla. The networks

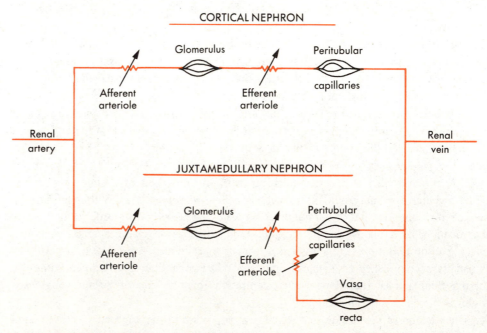

Fig. 47-2 ■ Network analysis of the renal circulation. Variable resistors represent arterioles. Black lines represent capillary networks.

within the cortical structures consist of two capillary beds in series, the glomerular capillaries and the peritubular capillaries. The glomerular capillaries are preceded and followed by arterioles, the afferent arteriole and efferent arteriole, respectively. Blood that perfuses the vasa recta represents blood that has passed through a juxtamedullary glomerulus, into the efferent arteriole, and, finally, into a vasa recta bundle.

The series arrangement of arterioles on both sides of the glomerulus confers stability on the glomerular filtration pressure. This is easily understood in terms of the simple resistive network shown in Fig. 47-2. The flow through the circuit is equal to the pressure difference between the input to the afferent arteriole and the output from the efferent arteriole, divided by the resistance. The total resistance is approximately equal to the sum of the resistances of the afferent and efferent arteriolar networks because capillary, that is, glomerular, resistance is low. Increasing the resistance of either arteriole will cause an increase in total resistance and, therefore, a decrease in flow, if arterial pressure is constant.

However, the two arterioles have directionally different effects on glomerular pressure. Constriction of the afferent arteriole will increase the pressure drop across this vessel and thus reduce pressure within the glomerulus. In constrast, constriction of the efferent arteriole will increase glomerular filtration pressure. This is the result of reduced flow through the afferent arteriole and the resultant smaller pressure drop prior to reaching the glomerulus. Because both arterioles will reduce flow when constricted, but the afferent and efferent arterioles have opposite effects on filtration pressure, renal blood flow and glomerular filtration can be controlled, to some extent, independently. Combined constriction of the afferent and efferent arterioles will produce minimal changes in capillary hydrostatic pressure and glomerular filtration rate but will reduce renal blood flow. Therefore changes in renal blood flow can be integrated into overall circulatory homeostasis without compromising glomerular filtration.

The presence of a high resistance (the efferent arterioles) downstream of the glomerulus but prior to the peritubular capillaries also ensures that hydrostatic pressure within the peritubular capillaries is reduced to relatively low levels (approximately 10 to 12 mm Hg). This low pressure within the peritubular capillaries facilitates reabsorption from proximal and distal tubules. The juxtamedullary nephrons have a similar vascular pattern in parallel with the superficial cortical glomeruli, although some differences exist in the distribution of the peritubular capillary networks.

Vasa recta blood flow originates distal to the efferent arteriole of the juxtamedullary nephrons. This point of origin becomes important in understanding the ability of the kidney to secrete various materials. Any blood that exits from the pathway through the juxtamedullary cortical circulation and flows into the vasa recta is not exposed to the environment surrounding proximal and distal tubules of these nephrons. Thus blood flowing through this pathway cannot contribute its solute to the secretory activities of the juxtamedullary nephrons. As mentioned, this is the major reason PAH is not fully cleared from the plasma circulating through the kidney.

The measurement of flow and filtration have been referred to previously; it was demonstrated that renal plasma flow could be approximated by measuring the extraction of PAH by the kidney. Renal blood flow is equal to renal plasma flow divided by 100 minus the hematocrit. A parameter that is frequently used to characterize renal function is the *filtration fraction* (FF), which is the fraction of the total renal plasma flow that is filtered:

$$FF = GFR/RPF = C_{In}/C_{PAH} \tag{7}$$

This fraction is proportional to the amount of protein-free water removed from plasma. Therefore, the greater this fraction, the greater will be the concentration of the plasma proteins in blood entering peritubular capillaries, and the larger will be the oncotic pressure available to ensure reabsorption of tubular fluid by peritubular capillaries.

The interplay mentioned previously between afferent and efferent arterioles allows a relatively independent control of glomerular filtration rate and renal plasma flow. This independent control implies that the filtration fraction may vary significantly during various conditions of renal blood flow.

■ Local control of blood flow

As with other circulations, the blood flow through the kidney is autoregulated. This is illustrated in Fig. 47-3, which shows that as perfusion pressure varies, there is only a small change in the renal blood flow rate. In the kidney this autoregulatory process is particularly effective; both renal blood flow and glomerular filtration rate are regulated. The fact that both are regulated implies that the afferent arteriole is involved in the regulatory process. Renal plasma flow might be regulated by the efferent arteriole alone, but simultaneous regulation of glomerular filtration rate would not be possible because constriction of the efferent arteriole has opposite effects on renal blood flow and on glomerular filtration rate, lowering the former and raising the latter. The afferent arteriole, in contrast, has similar effects on both flow and glomerular filtration rate. In fact, most recent evidence suggests that both the afferent and the efferent arterioles are involved in autoregulation.

The mechanism whereby autoregulation occurs is under debate, but an important component of the regulation appears to be the myogenic mechanism, as described in Chapter 32. As perfusion pressure is increased, the smooth muscle of the arterioles is stretched; this stretch induces a constriction of the smooth muscle and an increase in vascular resistance that offsets the increase in perfusion pressure. Another possible mechanism to explain autoregulation involves release of renin from the juxtaglomerular cells and formation of angiotensin, as described later.

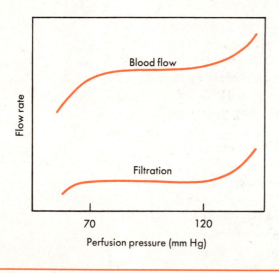

Fig. 47-3 ■ Renal autoregulation. Blood flow and glomerular filtration rate are stabilized in the face of changes in perfusion pressure.

■ Neural control

For purposes of simplification, the neural component in the regulation of renal blood flow can be viewed as subserving primarily the needs of the body as a whole, whereas local control mechanisms contribute to the stability of renal function. The renal circulation receives about 25% of the total cardiac output. Thus changes in blood flow can be used to regulate arterial pressure.

Neural control of renal blood flow is accomplished almost exclusively by release of norepinephrine from sympathetic adrenergic nerve terminals. Under normal circumstances, a small tonic discharge of nerve activity passes over the sympathetic nerves, resulting in partial contraction of the renal arterioles. Parenthetically, anesthesia results in a large increase in sympathetic drive and an exaggerated neural tone in the renal circulation.

The discharge of sympathetic nerves to the renal arterioles appears to be coordinated in such a way that a balanced constriction of the afferent and efferent arterioles is induced under conditions that require modest sympathetic activity. Such conditions would include changes in posture, modest emotional responses, and light physical activity. Consequently, glomerular filtration is reduced to a lesser extent than renal blood flow, and the filtration fraction is thereby increased. This is important in view of the fact that the reduction in renal blood flow can contribute to the maintenance of systemic blood pressure while allowing renal function to continue in a reasonably unaltered way because the glomerular filtration rate remains constant.

The stability of the filtration rate is not maintained at high rates of sympathetic discharge. In cases of hemorrhage or intense sympathetic activation, vasoconstriction exceeds the capacity of the kidney to dissociate renal blood flow and glomerular filtration rate, and both glomerular filtration rate and blood flow are reduced.

The sympathetic nervous system can modify the proportion of flow to the various parts of the kidney. Under conditions demanding water retention (e.g., dehydration or hemorrhage), a larger fraction of the renal blood flow passes through the juxtamedullary nephrons because of selective cortical nephron constriction.

There may also be a role for the sympathetic modulation of the blood flow through the vasa recta. A high vasa recta flow tends to wash out interstitial solutes and reduce urine-concentrating ability.

■ Renin-angiotensin system

There exists within the kidney a unique control system that is highly specialized, both with regard to its ability to influence renal blood flow and as a process that can alter systemic arterial blood pressure. This is the *renin-angiotensin system*. Renin is formed and stored in granules in specialized vascular smooth muscle cells of the afferent and efferent arterioles, the *juxtaglomerular cells* (Fig. 47-4, *A*). These cells are in contact with specialized epithelial cells of the nephron's distal tubule, the *macula densa*. The space delineated by the macula densa and the two arterioles is the *mesangial* region. In response to stimuli (described later), renin is released from the juxtaglomerular cells. Renin is an enzyme that can act upon an α_2-globulin, carried by the plasma, to cleave off the nonapeptide angiotensin I (Fig. 47-4, *B*). In the presence of *converting enzyme* and *chloride ion, angiotensin I* is then split to form *angiotensin II*. The *converting enzyme* is selectively localized on the surface of endothelial cells. In the adrenal cortex angiotensin II is converted to angiotensin III, which is active in stimulating secretion of aldosterone.

■ Control of renin release

Regulation of renin release and thereby angiotensin II formation is an extremely complex process that involves interactions among a number of hormones. There are several key elements in the process that are of major importance and can be clearly defined. First, low plasma sodium stimulates renin release. Second, reduced concentration of chloride ion in the tubular fluid at the macula densa will cause renin release. Third, reduced stretch of the juxtaglomerular cells, associated with reduced blood pressure in the afferent arterioles, will induce renin release. Fourth, stimulation of the renal sympathetic nerves can induce renin release via activation of β-receptors on the juxtaglomerular cells. In general, renin release is regulated in a manner that tends to stabilize blood pressure and extracellular fluid volume.

It is thought that the renin-angiotensin system may also contribute to the maintenance of *glomerulotubular balance*. Chloride concentration of the early distal tubular fluid at the *macula densa* is dependent on tubular flow rate. Recall that the fluid in the ascending limb of Henle's loop has a low chloride concentration. The osmolality of this fluid begins to rise and equilibrate with the surrounding interstitium as it exits the impermeable ascending limb of Henle's loop. The equilibration is a passive process, and

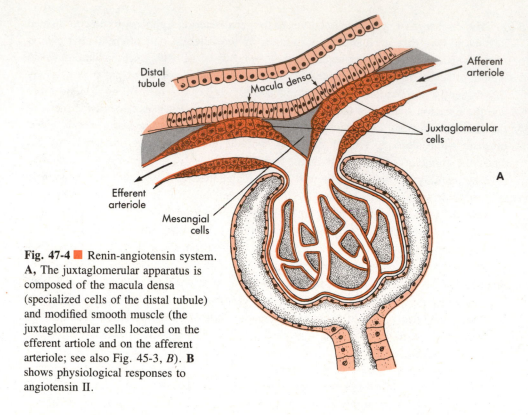

Fig. 47-4 ■ Renin-angiotensin system. **A,** The juxtaglomerular apparatus is composed of the macula densa (specialized cells of the distal tubule) and modified smooth muscle (the juxtaglomerular cells located on the efferent artiole and on the afferent arteriole; see also Fig. 45-3, *B*). **B** shows physiological responses to angiotensin II.

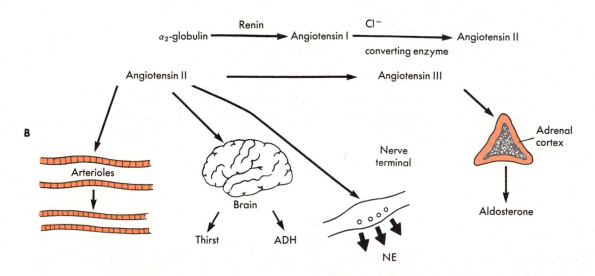

its degree is therefore expected to be flow dependent. Higher flow should reduce the time for water flux and thereby lower osmolality and chloride concentration at the macula densa. Low chloride in the macula densa stimulates renin release and angiotensin formation, and this in turn will cause constriction of the afferent arteriole. Thus if the glomerular filtration rate of a nephron is raised, macula densa chloride concentration falls, and angiotensin II formation is stimulated. This results in arteriolar constriction, reduced filtration, and stabilization of the relation between filtration and tubular reabsorption. The degree to which this mechanism is involved in autoregulation of renal blood flow still remains to be established.

Angiotensin II has wide-ranging and complex effects in a variety of other organs in the body. Five effects, however, appear to dominate the overall response observed in the presence of elevated angiotensin formation (Fig. 47-4, *B*). First, angiotensin II and angiotensin III, as mentioned previously, stimulate the release of *aldosterone* from the *adrenal cortex,* thus causing enhanced distal tubular sodium reabsorption. Second, angiotensin causes systemic vasoconstriction. Third, angiotensin acts on sympathetic nerve terminals to enhance neurotransmitter release. Fourth, angiotensin stimulates the release of antidiuretic hormone (ADH) from the posterior pituitary. Fifth, angiotensin acts within the brain to stimulate a sense of thirst, thereby increasing water intake. Note that each of these effects represents part of a coordinated pattern directed toward the expansion of extracellular fluid volume and elevation of blood pressure.

The effect of a pathological constriction of a renal artery illustrates the behavior of the renin-angiotensin system with respect to regulation of the circulation. Partial occlusion of the renal artery, as may occur with an atherosclerotic lesion, will lower intraluminal pressure in the renal circulation, and the stretch of the juxtaglomerular cells will be diminished. This causes renin release, angiotensin II formation, and ultimately stimulation of aldosterone secretion, systemic vasoconstriction, and enhanced sympathetic nerve discharge. These result in increased sodium reabsorption, increased peripheral resistance, and increased extracellular volume, all of which lead to the production of a hypertensive state. In other words, the kidney responds to renal artery stenosis in such a way as to raise arterial blood pressure high enough to achieve normal perfusion pressure distal to the stenosis.

It is thought that this process is a major factor in the etiology of hypertension in humans, especially in *renovascular hypertension,* but possibly also in other types of hypertension. However, it is clear that the process is complicated, since alleviation of a chronic renal artery constriction, both in humans and in experimental animals, in many cases fails to alleviate the hypertension. It would appear that other factors are capable of stabilizing a hypertensive state after it is initially established by the renin-angiotensin system.

A fair amount of confusion arises over the role of a high-salt diet in hypertensive states. In general, hypertension is not induced in a healthy person simply by the ingestion of large quantities of salt. However, if renal function is compromised, large salt intakes will result in an accumulation of salt within the individual. Secondarily, the retention of water within the extracellular space will expand the vascular volume. The expanded vascular volume will tend to increase cardiac output and thus increase systemic arterial blood pressure, but not by a mechanism secondary to angiotensin II–induced vasoconstriction.

■ *Intrarenal hormones*

The prostaglandins and kinins are hormones that are formed in the kidney and act on the kidney. The roles of these agents in determining renal function and their interaction at this time are not clear, but it appears that both glomerular filtration rate and sodium reabsorption may be under local hormonal controls. Also, it is likely that the prostaglandins may modify afferent neural sensitivity of the renal microvessels, that they modulate sensitivity of the vascular smooth muscle to angiotensin II, and that they inhibit the actions of ADH on collecting duct water permeability.

■ *Concentration and dilution of the urine*

Humans experience enormously wide variations in requirements for excretion of water. High dietary intake of fluids implies a requirement for excretion of urine with low solute content, that is, low osmolality. However, persons in an arid environment with little water to drink must reduce urinary losses as far as possible while continuing to excrete necessary waste solutes. The combined need to excrete solutes and retain water means that efficient mechanisms must be developed to produce a concentrated urine. The mammalian kidney that has evolved in the face of these needs is capable of producing urine osmolalities ranging from approximately 100 to 1200 mOsm/L. The renal mechanisms responsible for this ability are so efficient that some animals, such as the desert rat, can excrete urine so concentrated that no dietary water intake is necessary. They survive by acquiring necessary fluids from food and metabolic water. The process that has evolved to produce urine with such a wide range of concentrations is one of the more remarkable examples of form and function that has evolved in a biological system.

An underlying fact in the evolution of this complex process is that it is more efficient to move water by a passive osmotic process than by an active transport mechanism. Suppose that the osmotic difference between plasma and urine is 300 mOsm/L and that the difference is established by solute transport. The 300 mOsm/L difference represents 1.8×10^{23} solute molecules per liter of fluid transported (0.3 mole/L $\times$ 6×10^{23} molecules/mole). In contrast, primary water transport would require the movement of a larger number of water molecules because a 300 mOsm solution is approximately 55M with respect to water (1000/L $\div$ 18/mole = 55 moles/L), and 330×10^{23} molecules of water would have to be moved to produce a 300 mOsm osmotic gradient. In other words, approximately 180 times as many water as solute molecules would have to be moved to produce the observed osmotic gradient. Thus it is reasonable that the mechanism for concentration of the urine has evolved as one acting via salt transport rather than water transport.

The urinary concentrating mechanism is based on the existence of a concentrated fluid within the medullary interstitium, as shown in Fig. 46-11. Interstitial fluid osmolality in the medulla increases from 300 mOsm/L at the corticomedullary border to 1200 mOsm/L at the renal papilla. This concentrated fluid can be used to concentrate collecting duct fluid by osmotic withdrawal of water.

Fig. 47-5 shows the ratio of tubular fluid to plasma osmolality through the length of the nephron under conditions in which water is being conserved (*solid colored line*) and in which water excretion is required (*dashed colored line*), that is, in the presence and in the absence of ADH, respectively. It is striking that little difference is observed in the two cases until tubular fluid is well into the distal tubule. The difference between the formation of a concentrated urine and the formation of a dilute urine occurs within the distal tubule and especially within the collecting duct. Proximal tubular fluid is

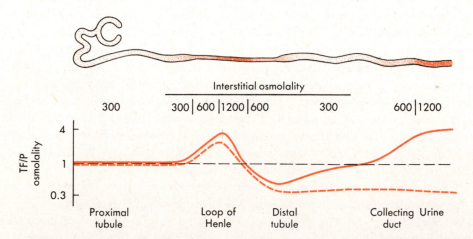

Fig. 47-5 ■ Tubular fluid osmolality. TF/P represents the ratio of the osmolality in the tubular fluid at any given location to the osmolality of plasma. Colored lines show osmolality in the presence (*solid*) or the absence (*dashed*) of ADH. The osmolality at various locations relative to the nephron is shown in the trace above.

isotonic, both in the presence and absence of ADH. There is a sharp rise in osmolality through Henle's loop (the origin of this will be considered later). The osmolality then falls and reaches *hypotonic* levels as the fluid passes through the ascending limb of Henle's loop.

In the absence of ADH (Fig. 47-5, *dashed line*), a dilute urine is excreted. The dilution is accomplished largely by transporting solute out of the tubules in a region where the epithelium is relatively impermeable to water. This is most prominent in the ascending limb of Henle's loop, but it continues to some extent throughout the distal tubule and collecting duct if no ADH is present.

Quite a different situation exists during the formation of concentrated urine (Fig. 47-5, *solid line*). The osmolality of the tubular fluid is similar to that observed during the formation of dilute urine until Henle's loop is reached, where there is a small increase in the osmolality in the ascending limb. The composition of the fluid entering the distal tubule is only slightly different from that observed during diuresis. Once fluid enters the distal tubule, however, the osmolality begins to rise, and it ultimately approaches the isotonic level observed in the proximal tubule. This rise in osmolality largely reflects the simple fact that ADH increases tubular permeability to water. Because the distal tubule is surrounded by capillaries that contain plasma at an osmolality of 300 mOsm/L, a high tubular water permeability will cause fluid to equilibrate osmotically between the lumen of the nephron and the lumen of the capillary.

In contrast to the environment surrounding the distal tubules, the fluid surrounding the collecting ducts in the medullary interstitium is strikingly hypertonic (up to 1200 mOsm/L). The formation of concentrated urine depends on osmotic equilibration between the fluid in the collecting ducts and the hypertonic medullary interstitium. Such osmotic equilibration occurs if ADH is present, and the urine can be concentrated to a level equal to that in the surrounding interstitium.

■ *Formation of a concentrated fluid in the medullary interstitium—countercurrent multiplication*

Much of the difficulty in understanding the urinary concentrating mechanism is involved with the question of how the high osmolalities in the medullary interstitium are generated. These concentrations are generated in Henle's loop by a process called *countercurrent multiplication*. This mechanism has evolved to meet the demands for the formation of a urine that is as much as 900 mOsm/L more concentrated than plasma and to accomplish this with a tubule that can sustain a gradient of only about 200 mOsm/L across its epithelium. Countercurrent multiplication allows the tubule to meet these two requirements by using highly selective localization of transport systems in the ascending limb of Henle's loop combined with the unique hairpin loop geometry and the great length of Henle's loop. Four facts should be assimilated before the mechanism is described:

1. Henle's loop reabsorbs sodium by an active process in the *ascending limb;* chloride accompanies sodium as the necessary anion.
2. The water permeability of the ascending limb is low, thus permitting the formation of a dilute fluid as chloride and sodium are removed.
3. Salt that is transported out of the ascending limb enters the medullary interstitium and causes the fluid both within the interstitium and within the descending limb to become more concentrated.
4. Descending limb permeability to water is high; the extraction of water combined with solute entry causes a rise in the osmotic activity of the fluid in the descending limb.

Fig. 47-6 illustrates how the geometry and transport characteristics of Henle's loop can produce a highly concentrated fluid in the medullary interstitium. The process is broken down into several steps: (1) a transport step by the tubular epithelial carriers, (2) a fluid equilibration step between the medullary interstitium and the descending limb of Henle's loop, and (3) a step in which new fluid is brought into Henle's loop from the proximal tubule. Initially the ascending limb, descending limb, and interstitium are at equilibrium, and all fluids have the same osmolality as that in the proximal tubule (Fig. 47-6, *a*). Next, sodium chloride is transported out of the ascending limb into the interstitium, and a gradient of 200 mOsm/L is established across the epithelium *(b)*. This is followed by equilibration between the interstitium and descending limb *(c)*. Water and solute diffusion into the descending limb of Henle's loop raise the solute concentration within the limb to 350 mOsm and simultaneously lower the medullary interstitial concentration to the same value. The process is continued by cycling new fluid in from the proximal tubule *(d)*. This forces fluid around the tip of the loop and reduces the gradient across the epithelium of the ascending limb *(asterisk)*. Pump activity can then establish a new gradient between the ascending limb fluid and the interstitial fluid *(e)*. This is followed by reequilibration of the fluid within the interstitium and the descending limb of Henle's loop *(f)*. With each cycle the concentration of fluid at the tip of Henle's loop rises. In the human kidney this process is reiterated until the salt concentration at the tip reaches approximately 600 mOsm/L (Fig. 47-6, *g*). This represents a steady-state condition in which the gradient across the epithelium at any level is no greater than 200 mOsm/L, but the longitudinal gradient from origin to tip is substantially larger. In some species the medullary interstitial salt concentration can rise to 15 times the plasma osmolality.

The osmolality generated by the sodium chloride acting in the countercurrent process is only about half the expected final osmolality of 1200 mOsm/L. The remainder of the

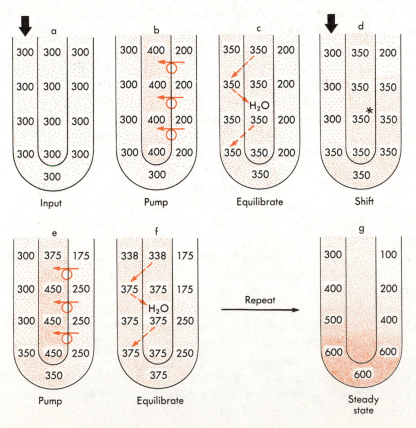

Fig. 47-6 ■ Countercurrent multiplication. Each figure represents a schematic of Henle's loop. Note that lateral active transport combined with flow through the parallel elements of the loop results in a longitudinal concentration gradient in the system. As the process is reiterated, the peak salt concentration in the loop reaches 600 mOsm/L.

osmotic pressure is generated by urea that is selectively deposited in the medulla by a process to be described later.

The countercurrent multiplication system will generate a concentrated medullary interstitium and produce the fluid that enters the distal tubule. The collecting duct permeability is then the ultimate determinant of the osmolality of the final urine. However, as described to this point, the system is not complete, and it would function only briefly. Water is osmotically withdrawn from the collecting duct and deposited in the medullary interstitium. Obviously this will result in a cumulative dilution of the concentrated salt solutions formed in the interstitial fluid. In order to prevent this, some means must be provided for removing the water that is reabsorbed from the collecting duct fluid. This must be accomplished, however, without washing out the solute that has accumulated in the medulla through the action of Henle's loop. These tasks are accomplished by the *vasa recta,* a highly specialized capillary network that complements Henle's loop. Fig. 47-7 shows schematically two hypothetical capillary networks passing through the medulla. One of these *(A)* is modeled after a typical capillary, such as might be found in skeletal muscle, that is, a linear capillary that enters and courses through the tissue and exits into a venule. The other capillary (Fig. 47-7, *B*) has an anatomy similar to that

■ *Stabilization of the medullary interstitial osmotic gradient—the vasa recta*

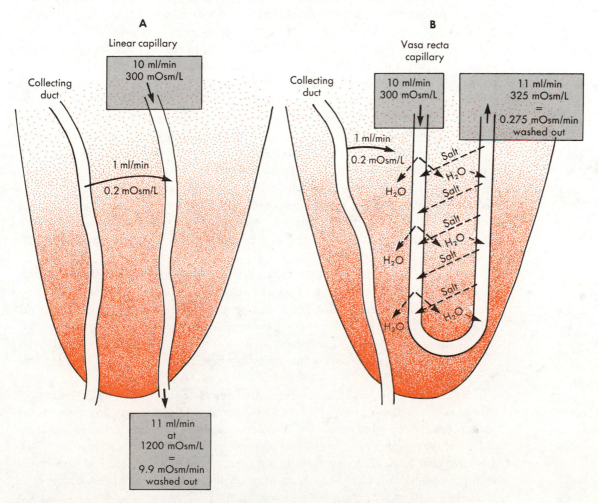

Fig. 47-7 ■ Countercurrent exchange. **A** and **B** represent two possible capillary organizations passing through the renal medulla. In both cases, 1 ml/minute of fluid is reabsorbed from the collecting duct and carried out in the effluent blood. Note the difference, however, in effluent solute concentration and rate of washout of solute with the flow-through linear capillary as compared to the countercurrent capillary of the vasa recta.

which would be found in the vasa recta, that is, a hairpin configuration. The diagram illustrates how the hairpin configuration of the vasa recta confers the ability to remove water and salts from the interstitium with minimal disruption of the osmotic concentration in the medullary interstitium.

It is assumed that the plasma in both capillaries equilibrates with interstitial fluid osmolalities by the capillary midpoint and that both capillaries have a flow of 10 ml/minute. In Fig. 47-7, *A*, equilibration at the midpoint would mean that the fluid within the capillary would exit at 1200 mOsm/L. In addition to solute, 1 ml of water, which was reabsorbed from the collecting duct, must be taken up so that 11 ml/minute leaves in the capillary blood. Given an exit concentration of 1200 mOsm/L (900 mOsm/L greater than entry), this would yield a washout of 9.9 mOsm/minute of solute.

Fig. 47-7, *B*, shows the configuration of the vasa recta capillary and a behavior that is quite different. In this capillary network, the concentration rises to 1200 mOsm/L at the midpoint of the capillary. However, because of the close association between the ascending and descending limbs, as blood turns and begins to flow out in the ascending limb, it comes into proximity with more dilute fluid entering from the input side of the vasa recta. Thus a concentration gradient exists across the intervening interstitium between the ascending limb and the descending limb. Solute and water movement can cause a progressive reduction of osmolality of the blood moving through the ascending limb of the vasa recta and out of the medulla. In this case, rather than exiting the organ at 1200 mOsm/L, as shown in Fig. 47-7, *A*, the fluid exits at 325 mOsm/L. In the latter case, the result is a washout of 0.275 mOsm/minute, rather than the 9.9 mOsm/L that would be expected if a more typical, linear capillary were involved (Fig. 47-7, *A*). Thus the hairpin loop configuration allows the solute and the water to be reabsorbed from the collecting duct and returned to the systemic circulation with much less dilution of the medullary interstitium than a more typical capillary system would induce.

■ *Role of urea in concentration of the urine*

Urea is a molecule with a multifaceted effect on urine osmolality. Being the end product of protein metabolism, it must be excreted each day in a quantity determined by the rate of protein intake. Thus increased dietary protein intake carries with it an obligatory requirement for increased urea excretion. The simplest effect of urea on urine osmolality derives from the fact that this solute must be excreted in whatever volume of urine is formed. Thus a large fraction of urinary solute is urea, and it will contribute to urine osmolality.

The kidney is specialized, however, in a way that ensures the ability to continue excreting large quantities of urea in spite of relatively low urine flow rates. This specialization derives from the differential urea permeabilities in various segments of the nephron; such characteristics cause the accumulation of urea within the medullary interstitium. In fact, as mentioned previously, with a normal protein intake, approximately half of the 1200 mOsm/L found in the medullary interstitium is contributed by the urea molecule.

The elements of tubular function that cause urea to accumulate in the interstitium of the medulla are shown in Fig. 47-8. Relative permeabilities for urea and water in the various tubular segments are shown. Urea permeabilities are high in the proximal tubule, and urea follows sodium chloride and water movement rather closely. In the descending limb of Henle's loop there is a modest permeability to urea and a relatively high permeability to water. Sodium chloride and urea diffuse into the descending limb of Henle's loop from the surrounding interstitium.

A key element in the sequence of events that leads to the accumulation of urea within the medullary interstitium is the fact that water permeability and urea permeability are quite different over the length of the distal tubule and collecting duct. In the distal tubule and early part of the collecting duct, urea permeability is low and is rela-

tively unaffected by ADH. Thus over this segment of the nephron, water from the dilute
solution created in the ascending limb of Henle's loop will be reabsorbed into peritubular capillaries. However, because the nephron is relatively impermeable to urea in
this portion, urea will fail to leave the tubule; therefore it becomes concentrated in
tubular fluid. In the medulla, the urea permeability of the collecting duct becomes high.
The tubular fluid urea concentration is relatively high as a result of the prior removal of
water in the distal tubule. Urea can therefore diffuse from the collecting duct into the
medullary interstitium. Once urea enters the medullary interstitium, it will tend to accumulate because the vasa recta system is rather inefficient for removing permeable
solutes. Accordingly, the interstitial concentration of urea will rise to a level that is high
enough to cause the rate of urea removal from the medullary interstitium to equal the
rate of entry from the collecting ducts.

It has been proposed that urea plays another, more central role in the formation of
concentrated urine. Fig. 47-8 illustrates the key elements of this mechanism in highly
simplified form. The mechanism was proposed because the epithelial cells in the thin
portions of Henle's loop have relatively few mitochondria and thus are probably not
involved in highly active transport processes (see Fig. 45-3, *C*). The hypothesis has
been advanced that, in this portion of the nephron, dilution of the ascending limb tubular
fluid is accomplished by using the urea accumulated in the medullary interstitium. A
proviso of the hypothesis is that the increasing concentration of tubular fluid that occurs
in the descending limb of Henle's loop is largely the result of water removal in response
to the osmotic gradient generated by the high concentrations of sodium, chloride, and
urea in the interstitium, not by solute entry. Thus the fluid that reaches the tip of Henle's
loop has high sodium, chloride, and urea concentrations. In the ascending thin limb, it
has been found that tubular permeability to sodium is significant, whereas permeabilities
to both water and urea are low. The sodium, which is concentrated by osmotic withdrawal of water from the descending limb of Henle's loop, can diffuse down a concentration gradient into the medullary interstitium. This dilutes the fluid in the ascending
limb of Henle's loop without requiring energy expenditure at this site. Once the fluid

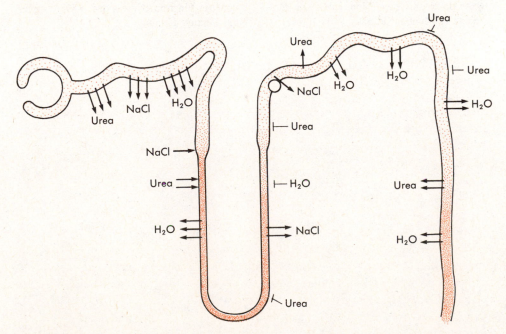

Fig. 47-8 ■ Permeability characteristics of the nephron. The number of arrows is proportional to
the magnitude of the permeability. A line with a crossbar indicates that the tubule is
impermeable. Direction of the arrows represents direction of movement of a given solute.

enters the thick segment of the ascending limb of Henle's loop, active sodium transport and passive chloride movement out of the tubule further dilute the fluid within the tubular lumen. The set of conditions needed for subsequent concentration of urea in tubular fluid are as described earlier.

In this model the active transport of salt that occurs in the thick portion of the ascending limb of Henle's loop is the source of energy for the system. This transport process results ultimately in an elevation of the urea concentration within the medullary interstitium. The urea acts with sodium chloride on the descending limb of Henle's loop to extract water, thereby generating the necessary concentration gradient to drive sodium reabsorption in the sodium-permeable segment of the thin ascending limb of Henle's loop. It is not known how important this process may be in urinary concentrating mechanisms relative to more classical views of the countercurrent system, in which all solute transport is assumed to be active.

■ Bibliography

Journal articles

Katz, A.I., and Lindheimer, M.D.: Actions of hormones on the kidney, Ann. Rev. Physiol. **39:**97, 1977.

Kokko, J.P.: Membrane characteristics governing salt and water transport in the loop of Henle, Fed. Proc. **33:**25, 1974.

Lewey, J.E., and Windhager, E.E.: Peritubular control of proximal tubular fluid reabsorption in the rat kidney, Am. J. Physiol. **214:**943, 1968.

Stephenson, J.L.: Central core model of the renal counterflow system, Kidney Int. **2:**85, 1972.

Books and monographs

Aukland, K.: Renal hemodynamics. In Proceedings of the Seventh International Congress on Nephrology, Montreal, 1978, University of Montreal Press.

Brenner, B.M., and Rector, F.C.: The kidney, vol. I, Philadelphia, 1981, W.B. Saunders Co.

Jamison, R.L.: Countercurrent systems. In Thurau, K., editor: Kidney and urinary tract physiology, Baltimore, 1974, University Park Press.

Koushanpour, E.: Renal physiology: principles and functions, Philadelphia, 1976, W.B. Saunders Co.

Orloff, J., and Berliner, R.W.: Renal physiology. In Handbook of physiology, section 8, Bethesda, Md., 1973, American Physiological Society.

Valtin, H.: Renal function: mechanisms preserving fluid and solute balance in health, Boston, 1973, Little, Brown & Co.

Windhager, E.E.: Micropuncture techniques and nephron function, New York, 1968, Appleton-Century-Crofts.

Regulation of the composition of extracellular fluid

The kidney must maintain the total volume of the extracellular fluid and its osmolality within very narrow limits. A host of regulatory processes act on the kidney and interact with each other to match tubular salt and water reabsorption to the requirements for maintenance of volume and osmolality. Few control systems within the body involve so many interacting features and such a complex network of neural, humoral, and local factors that act on a single common effector, in this case, the nephron of the kidney.

The *regulation of osmolality* is accomplished almost entirely by the action of antidiuretic hormone (ADH), or vasopressin, on the distal tubule and collecting duct (Fig. 47-8). The rate of formation and release of ADH is directly related to plasma osmolality. In contrast, the *regulation of extracellular fluid volume* is a more complex process involving interactions among ADH, aldosterone, natriuretic hormone, and neural and local mechanisms within the kidney. In each case a feedback system is involved that consists of (1) *receptors,* which sense the appropriate parameter, (2) *control elements:* hormones, nerves, or local events within the kidney, and (3) *effectors,* the tubules or renal arterioles. The following section describes the receptors for osmolality and for extracellular volume as well as the various control loops. Four forms of control are analyzed: (1) control of osmolality, (2) volume control via primary modification of water excretion (ADH), (3) volume control secondary to changes in solute excretion, and (4) volume control as a result of vascular adjustments.

Osmolality. Fig. 48-1 shows, in a highly schematic fashion, a sagittal section running through the optic chiasm and the anterior and posterior pituitary gland at the base of the brain (Chapter 52). The receptors that are sensitive to osmolality are found in the *supraoptic* and *paraventricular nuclei* in this region. Fibers that pass from this area to the posterior pituitary integrate sensing areas in the forebrain and secretory areas in the pituitary gland for the control of ADH secretion.

When hypertonic solutions are injected into the carotid artery or into the cerebrospinal fluid, a prompt increase in plasma ADH levels can be detected. The primary stimulus appears to be the result of osmotically induced water shifts in the brain. The hypertonic solution pulls water across the cerebral capillaries from the region of the paraventricular and supraoptic nuclei. This results in a reduction of interstitial and extracellular fluid volume in the vicinity of these nuclei. Simultaneously, the sodium concentration in the extracellular space surrounding the nuclei is changed. One or the other of these effects is the necessary and sufficient stimulus to evoke propagated action potentials along the fibers of the supraoptic-hypophyseal tract. The neural activity passes along this fiber tract, through the median eminence, and into the neural lobe of the posterior pituitary gland, resulting in the release of ADH, which is stored in this location. (See also Chapter 52 for details of synthesis and release.)

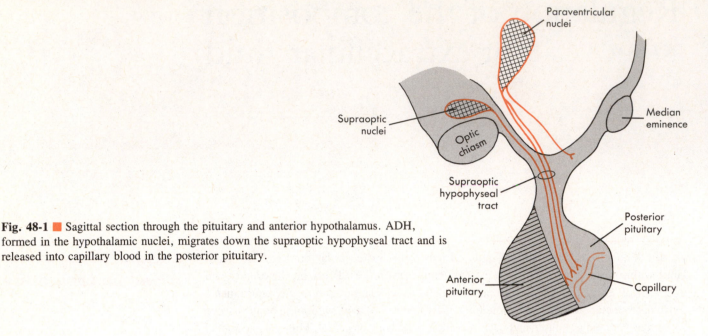

Fig. 48-1 ■ Sagittal section through the pituitary and anterior hypothalamus. ADH, formed in the hypothalamic nuclei, migrates down the supraoptic hypophyseal tract and is released into capillary blood in the posterior pituitary.

A variety of stimuli activate the system, leading to the release of ADH; fear, smoking, barbiturates, acetylcholine, and epinephrine all induce release of this material, whereas alcohol inhibits its release.

Extracellular volume. It appears that the receptors which serve to regulate extracellular volume in fact are not true receptors of extracellular volume but rather are detectors of the fullness of the vascular space. The receptors are basically stretch receptors, sensitive to distension of the atria, the great veins in the pulmonary circulation, and the aorta and carotid arteries. Distension of these receptors causes increased neural traffic over afferent neurons, which modify efferent neural outflow and the cerebral release of various hormones.

One must recognize that the control of extracellular volume originates from within the vascular compartment. This implies that the control of extracellular volume depends on the distribution and partition of extracellular fluid between the plasma and the interstitial space. Recall that this partition is a reflection of interactions among the factors in the Starling hypothesis for capillary exchange: the net hydrostatic and oncotic pressure difference across the capillaries. The consequences of this are simple. An elevation in hydrostatic pressure within the capillaries, all other things remaining equal, will result in filtration of fluid from vascular space into the interstitial space and fluid accumulation within this compartment. Thus a larger fraction of the total extracellular fluid will reside in the interstitial space. Since the sensors for extracellular volume reside within the vascular space, this will be interpreted as a reduction in total extracellular volume, and appropriate compensatory mechanisms will be initiated. In other words, any force that shifts fluid from the vasculature into the tissues, even without altering the total extracellular fluid volume, will be interpreted as a decrease in extracellular fluid volume by the receptors.

The fact that most of the receptors within the vascular compartment are in the cephalad portions of the body (thorax and neck) means that shifts of blood within the vascular compartment into and out of the thoracic circulation will have the same effects on volume receptors as would a total volume change. For example, immersion to the waist in water at neutral temperature will compress cutaneous veins in the legs and abdomen and shift fluid into the thoracic vasculature. This will be interpreted as an increase in extracellular fluid volume. Conversely, shifting from the recumbent to the upright po-

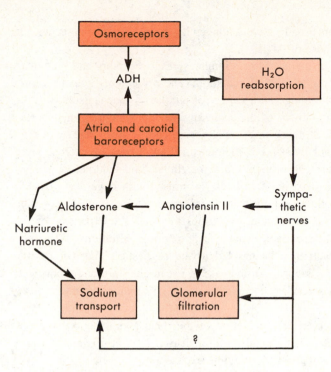

Fig. 48-2 ■ Regulation of volume and osmolality. Compare the single factor relating ADH and osmolality with the complex web of interactions concerned with volume control. Dark color shows receptors; light color shows altered function.

sition will cause a shift of blood out of the thorax and into dependent portions of the body. This will be interpreted as a reduction in extracellular volume.

In the following section control loops for fluid volume and osmolality are outlined. Little material that is found here is entirely new, since the control loops operate basically on mechanisms that have already been described. However, synthesis of the action of the control loops requires focusing on the coordinated activity of receptor, hormone or neural activity, and the nephron response (Fig. 48-2).

■ *Control of osmolality*

After an increase in osmolality that has been induced, for example, by an intravenous injection of a hypertonic fluid, the supraoptic nuclei are stimulated, thereby leading to an increase in ADH release from the pituitary gland. The actions of ADH on the distal tubules and collecting duct increase water reabsorption. ADH also may produce a small reduction in vasa recta blood flow, thus reducing the washout of solutes from the medulla. These effects would tend to reduce water excretion and restore osmolality toward normal. The opposite series of events would follow the infusion of hypotonic fluid.

■ *Control of volume*

Control of water excretion. Expansion of extracellular fluid volume with isotonic saline increases both the plasma and interstitial volumes. The former will increase the mean arterial pressure and the pulse pressure as a result of the increased filling of the cardiac ventricles and an increase in stroke volume. The increased arterial blood pressure will reduce the release of ADH via stimulation of the arterial baroreceptors, with a subsequent increase in water excretion.

A similar mechanism acting on the low-pressure side of the circulation would cause an increase in left atrial stretch following the expansion of volume. Left atrial distension would inhibit ADH release and thereby diminish water reabsorption and reduce the extracellular fluid volume toward normal.

Control of sodium excretion. As mentioned previously, volume control can be accomplished by affecting both sodium and water excretion. Effects on solute excretion tend to be slower and less precise than effects on water excretion, since the hormones that influence solute excretion usually have a longer lag time before release, and the effects are less prompt than those of ADH.

An increase in extracellular fluid volume stimulates atrial receptors and pulmonary receptors on the low-pressure side of the circulation. These receptors activate central nervous system pathways that impinge on a variety of regulatory mechanisms.

Control of aldosterone via activity of the renin-angiotensin system is an important element in volume regulation. Stimulation of renal nerves causes direct release of renin and subsequent formation of angiotensin II. Angiotensin II modulates aldosterone secretion and thereby influences salt reabsorption and extracellular fluid volume.

If the adrenal cortex is removed, or if animals are infused with large doses of aldosterone, expansion of extracellular volume is still followed by an increase in sodium excretion. This increase persists even under conditions in which the expansion of volume is accomplished without measurable change in glomerular filtration rate or other primary changes in renal function. This has led to the suggestion that a *natriuretic hormone* is involved in volume control. An increase in natriuretic hormone produces an increase in sodium excretion secondary to reduced proximal tubular sodium reabsorption.

Recently it has been observed that activation of the renal sympathetic nerves can directly inhibit proximal tubular sodium reabsorption. It may be that this mechanism is also involved in the natriuresis that follows volume expansion.

Renal vascular adjustments. Increased intravascular volume, acting via right atrial stretch receptors, can decrease sympathetic tone, increase glomerular filtration rate, and thus increase sodium and water excretion. The efficacy of a change in glomerular filtration rate probably depends on the operating level of the kidney. At low filtration rates a large fraction of the sodium load to Henle's loop, the distal tubule, and the collecting duct would be reabsorbed. Because the distal sodium transport mechanism is characterized by T_m kinetics (Fig. 46-10, *B*) changes in glomerular filtration rate would have little if any effect on sodium excretion. However, at high filtration rates a large sodium load is delivered to the distal tubule. Saturating quantities of sodium would be delivered to the tubular mechanisms, and thus the effects of a change in filtration rate would be substantial.

Another effect of activity in the renal sympathetic neurons is the tendency to redistribute flow within the kidney. A decrease in vascular volume elicits reflexes that tend to shift flow toward juxtamedullary nephrons and away from cortical nephrons. This brings a larger fraction of the glomerular filtrate through nephrons with long loops and higher concentrating ability.

■ Regulation of fluid and salt intake

All the mechanisms just outlined act to stabilize extracellular volume and osmolality via an effect on the salt and water excretory organ. However, intake of salt and water are also under the control of humoral mechanisms, presumably triggered by the receptors described. Thirst is clearly related to the volumes of vascular and extravascular space. Both hemorrhage and hypertonic saline will result in a profound sensation of thirst, which probably is intensified by the inhibition of the secretion by salivary and other oral mucosal glands.

Angiotensin II is a hormone of major importance in determining the sensation of thirst and the intake of water. When injected at appropriate sites within the brain, it results in a sensation of thirst and induces a prompt increase in water intake.

Behavioral modification probably also plays some role in the regulation of salt intake. When animals that have been deprived of water or salt are given free access to a variety of fluids that contain salt in various concentrations, they will select the fluids that contain a salt concentration appropriate to their dietary needs. Little is known about the feedback mechanisms that regulate this behavior.

As we have seen, water excretion can be controlled by relatively independent modification of renal tubular water reabsorption or by combined changes in solute and water reabsorption. It is important to be able to distinguish these two mechanisms if one wishes to determine whether urine flow at a given moment is concentrating or diluting the plasma and how fast the concentration or dilution may be occurring. Important tools in understanding these processes are the computation of the *osmolar clearance* (C_{osm}) and the *free water clearance* (C_{H_2O}).

The osmolar clearance is determined simply by applying the usual clearance equation to the total solute efflux in the urine; that is:

■ *Quantitation of the effects
of diuresis or antidiuresis
on animals*

$$C_{osm} = \frac{U_{osm} \times V}{P_{osm}} \tag{1}$$

Thus osmolar clearance reflects the rate at which solute particles, regardless of molecular species, are being excreted. For example, one might find during dehydration the excretion of a hypertonic urine of 1100 mOsm/L and a urine flow of 0.5 ml/minute. Plasma concentration might be slightly elevated, to about 308 mOsm/L. The osmolar clearance then would be equal to $1100 \times 0.5/308$, or 1.79 ml/minute. On the other hand, a very well-hydrated subject might excrete urine with an osmolality of 27 mOsm/L and have a urine flow of 20 ml/minute. This subject probably would have a slightly dilute plasma, 298 mOsm/L. The osmolar clearance in this case would be $27 \times 20/298$, or 1.79 ml/minute, which is the same rate as in the previous example.

The definition of clearance is *the minimum volume of plasma that could have provided all the solute excreted in a given time*. Thus in this case both the dehydrated subject and the overhydrated subject were removing solute from plasma at the same rate (that is, 1.79 ml/minute was being cleared). This, of course, would be expected if their dietary intake and their metabolic production rates were the same.

The clearance value of 1.79 ml/minute means that, if the solute excreted in the urine in 1 minute had been dissolved in 1.79 ml of fluid, the resulting urine would have been isotonic. The water and solute would have been excreted at exactly the same rate. If water is excreted in excess of solute, the excess water is called the *free water clearance*, or C_{H_2O}. The free water clearance can be computed simply as the difference between urine flow and osmolar clearance.

$$C_{H_2O} = V - C_{osm} \tag{2}$$

If the free water clearance is positive, then the flow is greater than that required to match the rate of solute excretion in the urine, and a dilute urine is being excreted. Excretion of a dilute urine implies that the plasma is being progressively concentrated. It follows that the magnitude of the free water clearance suggests how rapidly plasma concentration might be changing.

If the free water clearance is less than zero, that is, there is a *negative free water clearance* (usually designated as $T_{C_{H_2O}}$), this implies that a concentrated urine is being produced. Plasma is being diluted by the return of fluid to the circulation. In the case of the dehydrated subject the free water clearance equals 0.5 ml/minute − 1.79 ml/minute = −1.29 ml/minute. However, the overhydrated subject would have a free water clearance equal to $20 - 1.79 = 18.2$ ml/minute. Water is removed from the plasma of the overhydrated subject and restored to the plasma of the dehydrated subject by the action of the urinary concentrating process.

It should be noted that the extent to which the urine can be concentrated is basically limited by the excretory requirements for the solute that are generated by metabolism and dietary intake. These two factors determine the required excretory rate for solute, and the solute concentration in the medullary interstitium determines the minimum quantity of fluid that must accompany the excreted solute.

■ *Regulation of acid-base balance*

The concentration of hydrogen ion in plasma must be precisely regulated in spite of enormous variations in dietary intake and metabolic production of acids. The largest metabolic source of hydrogen ion arises from the metabolism of glucose, weak acids, and other carbohydrates, all of which on the average lead to a total production of approximately 13,000 mmole/day of carbon dioxide. No immediate requirement for renal excretion of hydrogen ion is engendered by this process, however, since the carbon dioxide formed is excreted in the lungs.

The metabolism of salts of weak acids, which are common components of fruits, typically results in the use rather than in the production of hydrogen ion. A typical diet might contain sufficient salts of weak acids to use approximately 35 mEq/day of hydrogen ion. Ingestion of proteins results in the ultimate formation of hydrogen ion and urea as the end products of amino acid metabolism; approximately 20 mEq/day of hydrogen ion are formed by this process. Oxidation of sulfhydryl groups of cysteine and methionine to form sulfuric acid contributes about 35 mEq/day, and incomplete oxidation of glucose to lactate generates about 35 mEq/day. Oxidation of fatty acid adds another 5 mEq/day to the total production of hydrogen ion. The net production of hydrogen ion is thus about 60 mEq/day.

The 60 mEq of hydrogen ion produced each day must be excreted in a urine volume of approximately 1.5 L. If hydrogen ion were excreted simply as the dissolved ion, a urinary concentration of about 40 mEq/L would result. A concentration of 40 mEq/L represents a pH of 1.4, which is not compatible with either comfort or life of mammalian cells. Therefore a complex series of processes have evolved that permit the mammalian kidney to excrete the necessary quantity of hydrogen ion without exceeding tubular hydrogen ion concentrations of about 1.5×10^{-5}M, that is, a pH lower than about 4.8. This is accomplished largely by forming complexes between hydrogen ion and other urinary constituents.

The elements of acid-base balance that are related directly to renal function consist of (1) extracellular buffering, (2) stabilization of extracellular bicarbonate, and (3) coordination of hydrogen ion secretion with other ions and with long term acid-base status.

■ *Buffering of metabolic acids*

Hydrogen ions are produced in a variety of tissues and are buffered in both intracellular and extracellular fluids. One of the most important extracellular buffers is the bicarbonate ion, which combines with hydrogen ion to form carbon dioxide; the carbon dioxide is ultimately excreted in the lungs. This, however, results in a net respiratory loss of the bicarbonate ion used to buffer the metabolic hydrogen ion. A pivotal element in renal function is the replenishment of these lost bicarbonate ions.

An important concept in understanding extracellular fluid buffering is the *isohydric principle,* which states that, if the ratio of one buffer pair in a mixture of several pairs is known, then the ratio of all buffer pairs also is defined. In general, one can write a dissociation constant K for the relation:

$$H^+ + A^- \rightleftharpoons HA \tag{3}$$

$$K = \frac{[H^+] [A^-]}{[HA]} \tag{4}$$

And for any three potential buffers in extracellular fluid:

$$H^+ = \frac{K_1[HA_1]}{[A_1^-]} = \frac{K_2[HA_2]}{[A_2^-]} = \frac{K_3[HA_3]}{[A_3^-]} \tag{5}$$

Examination of equation 4 shows that, if the dissociation constants are known, and if one of the ratios for the buffer pairs is known, the ratios of all other buffer pairs in the system are known as well. Thus considerations of acid-base status as a rule focus on the ratio of HCO_3^- to H_2CO_3 or more commonly to carbon dioxide. Knowing this ratio uniquely defines the pH as well as the ratios of all other buffer pairs in the system.

However, the isohydric principle does not mean that the other buffer pairs do not influence the way hydrogen ion behaves in the fluid. Other pairs within the system contribute to the total *buffering capacity* and therefore to the stability of the pH. A larger buffer capacity means that larger additions of acids or bases to the body can be tolerated, since smaller pH changes will be induced.

The relations between bicarbonate ion, carbon dioxide, and hydrogen ion are described by the well-known Henderson-Hasselbalch equation, which is simply the logarithmic form of the relations shown previously (Chapter 39).

$$pH = pK' + \log \frac{[HCO_3^-]}{0.03[PCO_2]} \tag{6}$$

In general, renal function is concerned with regulation of the concentration of bicarbonate, and the concentration of carbon dioxide is determined by respiratory function. The coordinated activity of the two organ systems determines the blood pH.

The fundamental cellular process involved in the renal-tubular handling of hydrogen ion is shown in Fig. 48-3. The diagram is very similar in overall form to that shown in Fig. 46-8. The interrelations among the various molecules and ions shown have been previously described. The mechanism provides a means whereby bicarbonate, a relatively large, charged anion, can be moved efficiently across the epithelial cell and be reabsorbed along with sodium or other cations.

Hydrogen ion is secreted into the lumen of the tubule, where it combines with filtered bicarbonate to form carbonic acid. For simplicity, countertransport of sodium ion by the hydrogen pump is omitted from this discussion (Fig. 46-8, *B*). The equilibration among carbonic acid, carbon dioxide, and water is relatively slow. An enzyme, *carbonic anhydrase,* which accelerates this process, is present both on the epithelial cell brush border and in the cytoplasm. Carbon dioxide is formed in the lumen from the filtered bicarbonate ion and the secreted hydrogen ion. The carbon dioxide can diffuse into the cell because the epithelial cell membrane is relatively permeable to it. Inside the epithelial cell the carbon dioxide combines with water to form carbonic acid; the process again is catalyzed by intracellular carbonic anhydrase. Carbonic acid can then dissociate into hydrogen ion and bicarbonate ion. The free bicarbonate ion now can

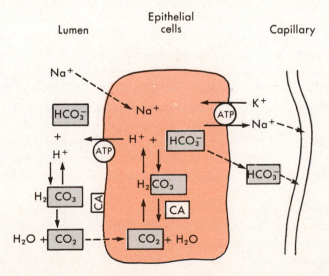

Fig. 48-3 ■ Reabsorption of bicarbonate. Note that one cycle uses secretion of the hydrogen ion to move bicarbonate into the cell, with no net change in the hydrogen ion that is available. The movement of the CO_2 moiety is indicated by the gray boxes. Solid arrows represent active transport; dashed arrows represent passive diffusion. *CA*, Carbonic anhydrase.

diffuse down a gradient and into the peritubular space to be returned to the circulation. The overall effect of the process is that a hydrogen ion secreted by the pump is used as a means of converting bicarbonate to a form (carbon dioxide) more readily permeable to the tubular epithelium, and finally the hydrogen ion is freed within the cell, where it can be used again by the pump. The net result is a facilitation of the movement of bicarbonate from lumen to plasma with no net change in hydrogen ion. This basic process forms a theme that is repeated in subsequent sections.

Secretion of hydrogen ion accomplishes two things. First, the bicarbonate that is continuously filtered in the glomerulus is returned to the circulation, and thus secretion of hydrogen ion is integral to *conservation of the filtered bicarbonate*. Hydrogen ion secretion also results ultimately in the *replenishment of bicarbonate stores,* referred to previously, that were lost in the process of buffering strong acids produced by the metabolic processes or by food intake.

■ Conservation of filtered bicarbonate

As mentioned previously, the mechanism shown in Fig. 48-3 results in the reabsorption of bicarbonate ion in parallel with the reabsorption of sodium. In the proximal tubule the carbonic anhydrase that is present on the brush border catalyzes the hydration of carbon dioxide within the tubular lumen and ensures the reabsorption of large quantities of bicarbonate against relatively small gradients. Although a similar process also takes place in the distal tubule, the bicarbonate reabsorption is less rapid. With this low flux rate of carbon dioxide, tubular formation of carbonic acid is adequate, and carbonic anhydrase need exist only within the cells. Little carbonic anhydrase is found on the brush border of the epithelium in the distal tubule.

Under normal conditions in excess of 99% of the filtered bicarbonate ion is reabsorbed by this process. In the distal tubule the sodium-hydrogen carrier also exists, and it is likely that an electrogenic hydrogen ion pump is active.

If the diet has a sufficiently high alkaline content or has the potential to produce alkali, bicarbonate is always in excess. Under this condition acid-base regulation can be accomplished relatively simply by varying the quantity of bicarbonate excreted. In fact it is possible to show that under some circumstances bicarbonate is being secreted rather than absorbed when high levels of bicarbonate are present in plasma. This situation exists in herbivores and in some human vegetarians. Normally, however, the diet generates acids, so the kidney must continuously replenish bicarbonate lost as carbon dioxide in the buffering process.

■ Restoration of depleted bicarbonate ion reserves

Titratable acids. Recall that the hydrogen ion which appears in the extracellular fluid, either through ingestion or metabolism, must be buffered and that buffering results in the respiratory loss of bicarbonate ion in the form of carbon dioxide (Fig. 48-4, *gray boxes*). This would cause progressive depletion of body stores of HCO_3^- were it not for renal replacement of lost bicarbonate *(colored arrows)*. Restoration of depleted bicarbonate is a process that is based on the hydrogen ion secretory mechanism described previously. In this case, however, the secreted hydrogen ion is buffered by combination with substances in the tubular lumen. Hydrogen ion is commonly combined with either a filtered buffer or with ammonia.

Fig. 48-4 outlines the total process. First, within extracellular fluid *(upper rectangle)* bicarbonate ion is used to buffer phosphates, presumed to be formed in the cell during metabolism *(upper left)*. The carbon dioxide so formed is excreted in the lungs with the loss of two bicarbonate ions for each mole of $HPO_4^=$ neutralized *(gray boxes)*.

The lower half of the figure shows reactions occurring in the kidney. If carbon dioxide is used as a source of bicarbonate and hydrogen ions in the tubule *(lower left)*, then, after active tubular secretion of hydrogen ion and luminal buffering of the H^+ by $HPO_4^=$, a bicarbonate ion, unpaired with a hydrogen ion, can be returned to extracellular

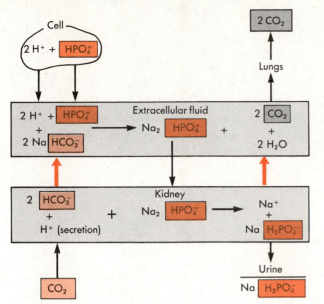

Fig. 48-4 ■ Buffering of strong acids by extracellular bicarbonate and replenishment of bicarbonate by the actions of the kidney. Extracellular events are shown in the upper half of the diagram and renal events in the lower half. Bicarbonate is used initially to buffer metabolically produced hydrogen ion in the extracellular fluid. The resultant carbon dioxide is exhaled from the lungs with the loss of two bicarbonate ions. Hydrogen ion secretion into the tubular lumen and subsequent combination of hydrogen ion with $HPO_4^=$ frees a bicarbonate derived from carbon dioxide to return to the extracellular fluid. The bicarbonate ion is reabsorbed with a sodium ion *(red arrows)*.

fluid *(left colored arrow)*. The hydrogen ion that entered the lumen of the tubule then can be excreted in the urine in association with the $HPO_4^=$ as $H_2PO_4^-$ *(lower right)*. The sodium thereby displaced from the phosphate can be returned to the plasma *(right colored arrow)*. To recapitulate, (1) extracellular buffering of metabolically produced strong acids leads to respiratory loss of bicarbonate, and (2) tubular buffering of secreted hydrogen ion frees a bicarbonate and a sodium ion for reabsorption to restore bicarbonate ion reserves.

The process shown in Fig. 48-4 is limited by the minimum urine pH that can be established, because the fractions of the phosphate ion that exist in the singly, doubly, and triply protonated form are determined by the ionization constant and the pH of the solution.

$$H_3PO_4 \xrightleftharpoons{K_1} H^+ + H_2PO_4^- \qquad pK_1 = 2.2$$

$$H_2PO_4^- \xrightleftharpoons{K_2} H^+ + HPO_4^= \qquad pK_2 = 6.8$$

$$HPO_4^= \xrightleftharpoons{K_3} H^+ + PO_4^{\equiv} \qquad pK_3 = 9.7$$

The Henderson-Hasselbalch equation predicts that the log of the ratio of the basic ($HPO_4^=$) to the acid ($H_2PO_4^-$) form of the phosphate, at a plasma pH of 7.4, equals the difference between the pH and the pK_2 (that is, $7.4 - 6.8 = 0.6$); the basic form is about 40% of the acidic form. Because pK_1 and pK_3 are so far from the plasma pH, essentially none of the phosphate exists either as H_3PO_4 or as $PO_4^{\equiv}$.

For the filtered phosphate to function effectively in the urine as a urinary buffer, the urine pH must be lower than plasma pH; this fact sets a limit on how much bicarbonate can be reclaimed by combining secreted hydrogen ion with phosphate. Urine pH cannot fall lower than about 4.8 because of the nature of the epithelium. At a pH of 4.8 about 99% of the phosphate will be in the form of $H_2PO_4^-$, and little will be in the form of H_3PO_4. Thus one hydrogen ion can be buffered by each $HPO_4^=$, and constantly one bicarbonate ion can be recovered. However, recall that two bicarbonate ions were lost in buffering the original H_3PO_4 formed (carbon dioxide loss, Fig. 48-4, *upper right*). Therefore titrating the filtered phosphate using secreted hydrogen ion can recover only half the lost bicarbonate ions. Further replenishment of bicarbonate stores requires the presence of other buffers in the urine or some other molecule to combine with the hydrogen ion. The secretion of ammonia by the mammalian kidney serves this function.

Ammonium secretion. Ammonia is produced by the metabolism of glutamine and

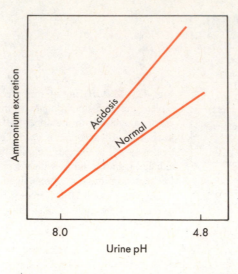

Fig. 48-5 ■ Relations between ammonium excretion and urine pH. Note that, in acidosis, ammonium excretion is raised at all urine pH's (see also Fig. 46-14).

other amino acids by the renal epithelial cells. Ammonia so formed diffuses into the tubular lumen because of its relatively high lipid solubility. It is protonated there by the hydrogen ion that is secreted by the tubular transport systems. The ammonia acts as a trap for the hydrogen ion, providing an additional means of freeing bicarbonate ions to be restored to plasma. Note also that the resultant formation of ammonia provides a cation that can take the place of either sodium or potassium, thus freeing these ions to be reabsorbed with bicarbonate or with chloride. The net result is the excretion of ammonium phosphate rather than sodium phosphate.

The formation of ammonium is affected by the state of acidosis of the subject. Fig. 48-5 shows the relation between ammonium excretion and urine pH. In a normal individual, as the pH rises, the ammonium excretion is increased, as would be predicted by the diffusion trapping process described earlier. During a period of prolonged acidosis, tubular epithelial cells are capable of adapting either by enzymatic changes or perhaps through changes in mitochondrial transport processes. The adaptation leads to an increase in formation of ammonia and therefore an increase in the rate of ammonium excretion at any given urine pH.

In summary, tubular mechanisms can be viewed as having conserved bicarbonate by three separate processes, all dependent on the activity of the hydrogen ion pump. The first of these is simply conservation of the filtered bicarbonate by reabsorption, secondary to hydrogen ion transport. The quantitative importance of this can be calculated to be simply the difference between the filtered bicarbonate and excreted bicarbonate. That is,

$$\text{Conservation of filtered bicarbonate} = (\text{GFR} \times [\text{HCO}_3^-]_P) - (V \times [\text{HCO}_3^-]_U)$$

The second process that stabilizes bicarbonate ion is buffering by filtered buffers. This is evaluated quantitatively by a measure called the *titratable acidity,* which is defined as the milliequivalents of strong base that must be added to the final urine to return the urine pH to that of plasma, typically 7.4. It is assumed that the hydrogen ion, being a strong acid, was added to the urine in a quantity sufficient to lower the pH from 7.4 to the observed urine pH. The quantity of strong alkali required to return the pH of the urine to that of plasma must equal the number of hydrogen ions added by the tubular transport process.

The third measure of urinary hydrogen ion excretion is the ammonium excretion, since this is a direct measure of the number of moles of hydrogen ion tied to the ammonium ion. Typical values of ammonium excretion and the formation of titratable acidity for a normal subject and a diabetic subject are shown in Table 48-1.

TABLE 48-1

Values of ammonium excretion and the formation of titratable acidity*

	Normal subject	*Diabetic subject*
Titratable acidity	5	100
Ammonium	25	300

*Values are in mEq of H^+/day excreted.

Note that ammonium excretion normally represents a larger fraction of excreted acid than does excretion of titratable acidity. In diabetic persons both mechanisms combat the acidosis induced by lipid metabolism. As mentioned, the augmented ammonium excretion is the result of stimulation of the metabolic production of ammonium by epithelial cells. The increased excretion of titratable acidity is related to the presence of ketoacids in the urine. Normally the low pK of the ketoacids limits their usefulness as buffers. In the diabetic subject, however, they may be excreted in such high quantities that they contribute significantly to buffering.

■ *Interactions between hydrogen ion and other ions*

Several molecules exert important effects on the processes described in the preceding section. An increase in plasma carbon dioxide tension, as might occur in respiratory acidosis, increases the fraction of the filtered bicarbonate that is reabsorbed (Fig. 48-6). This is caused by the elevation in carbon dioxide concentration inside the epithelial cell. This elevated carbon dioxide raises hydrogen ion concentration and stimulates hydrogen ion secretion, which in turn enhances bicarbonate ion reabsorption. This is obviously an important compensatory mechanism, because elevation of the fraction of bicarbonate reabsorbed will increase plasma bicarbonate levels in parallel with changes in carbon dioxide levels, thereby tending to stabilize plasma pH.

Potassium ion interacts in important ways with hydrogen ion in all cells of the body. An important general relation that should be borne in mind is the reciprocity which is commonly observed between the two ions, hydrogen and potassium. Elevation of hydrogen ion in the extracellular fluid, as occurs, for example, in metabolic acidosis, will result in movement of hydrogen into a cell, with a reciprocal exit of potassium from the cell and a diminution of cellular potassium concentration. The significance of this exchange is that the majority of buffers within the body are located in the intracellular compartment. Thus the hydrogen-potassium exchange allows the hydrogen to achieve access to protein buffer stores and minimizes the pH change in the body. However, in the process hyperkalemia, that is, elevated plasma potassium level, is induced by the potassium flux from cell to plasma, and this is often associated with metabolic acidosis.

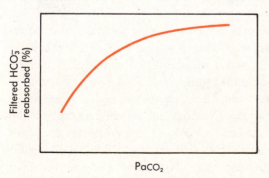

Fig. 48-6 ■ Rate of bicarbonate reabsorption by the renal tubules as a function of the plasma carbon dioxide.

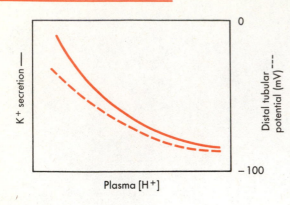

Fig. 48-7 ■ Relation between plasma hydrogen ion and the rate of potassium secretion. The hydrogen secretory pump is electrogenic, and therefore increased hydrogen ion pump activity causes more negative distal tubular potential.

Again note that only a small fraction of the large quantity of intracellular potassium need enter extracellular space to markedly alter extracellular concentration (see Tables 45-1 and 45-2). This reciprocity between hydrogen and potassium fluxes is reflected in the cellular responses to a variety of interventions and to a variety of physiological stimuli. It should be borne in mind in all considerations of both potassium balance and hydrogen balance.

The reciprocity between hydrogen and potassium ion fluxes occurs in the tubular epithelial cells as well, and elevation of plasma potassium ion level commonly results in exchange of hydrogen ion for potassium ion and a reduction in the amount of hydrogen available for transport by the epithelial secretory pump. The reduction in intracellular hydrogen causes a reduction in the tubular hydrogen pump secretory activity and a reduction in excretion. The reduced hydrogen secretion causes accumulation of hydrogen and systemic acidosis. On the other hand, acidosis causes reduced potassium secretion. Fig. 48-7 shows the inverse relation that is observed between hydrogen ion concentration and potassium secretion. This relation is dependent on the fact that the hydrogen ion secretory pump is electrogenic, and enhanced hydrogen ion secretion reduces the luminal potential *(dashed line)* and thus reduces the driving force for potassium secretion.

Sodium reabsorption is also importantly related to hydrogen ion balance of the body. The interrelations between sodium and bicarbonate excretion are largely tied to the fact that sodium reabsorption carries with it the requirement for reabsorption of anions; chloride and bicarbonate represent the major extracellular fluid anions. Thus an increase in sodium reabsorption will tend to increase bicarbonate reabsorption. As a rule, this process does not produce major effects on plasma pH. However, if tubular sodium reabsorption is stimulated markedly, as in prolonged sodium deprivation, increased bicarbonate reabsorption may induce an alkalosis in conjunction with the hyponatremia.

Chloride ion can affect acid-base status as well. Recall that chloride and bicarbonate movements are linked via an exchange carrier. An increase in chloride levels in plasma means that less bicarbonate need be reabsorbed with sodium ion. The reduced bicarbonate reabsorption leads to diminished plasma bicarbonate levels and acidosis associated with hyperchloremia.

Fig. 48-8 shows the integration of several elements of renal acid-base function that can be observed if an experimental subject is exposed to an acid load. In this experiment the subject ingested ammonium chloride, which was converted to urea and hydrogen ion in the liver. This process added an acid load to the body in the same way that the administration of hydrochloric acid would have. The required response to restore homeostasis is then renal elimination of the excess chloride and elimination of the hydrogen ion and urea. Elimination of urea, a neutral molecule, causes no particular problem as long as urine flow is adequate, which in this case was true. Chloride elimination poses no particular problem, since the anion can simply be excreted along with whatever urinary cations are present.

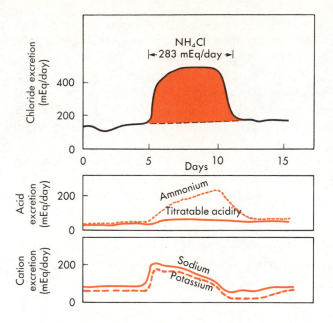

Fig. 48-8 ■ Effect of dietary acid on hydrogen ion balance. Ammonium is converted to urea and hydrogen ion. Early in the test period chloride ion is excreted with either potassium or sodium ion. As ammonia synthesis is stimulated, NH_4^+ replaces potassium and sodium ions, and the rate of potassium and sodium ion excretion falls.

In this experiment the chloride excretion exceeded the chloride ingestion. This loss of chloride represents the fact that the acidosis associated with NH_4^+ ingestion required an enhanced reabsorption of bicarbonate. Since the acid-base balance required an increased bicarbonate reabsorption, the maintenance of charge neutrality required less chloride reabsorption.

The most interesting responses to the administration of ammonium chloride are those displayed by the cations sodium and potassium. Immediately after the administration of the ammonium chloride, potassium and sodium excretion increased. This increase reflected the fact that cations had to be excreted in combination with the chloride ingested in the ammonium chloride. The cation portion of the original ammonium chloride was changed in form to a neutral compound, urea, and to hydrogen ion. Presumably the hydrogen ion would have been largely buffered and excreted as carbon dioxide via a pulmonary route, thus leaving a requirement for a cation to balance the chloride excreted. The cations would have been either sodium or potassium ion.

Both potassium and sodium excretion reached an early peak (Fig. 48-8, *lower segment*) and then diminished with time. This diminution was approximately the mirror image of a rise in ammonium excretion. The rise in ammonium excretion did not reflect excretion of the ammonium ion ingested, since most of this was metabolized to hydrogen ion and urea. Rather the ammonium excretion rose because of the acidosis induced and the enhanced ammonium secretion by the tubular epithelial cells (Fig. 48-5). This ammonium traps hydrogen ion, frees a bicarbonate to permit restoration of the bicarbonate reserves depleted in buffering the hydrogen ions, and finally permits conservation of cations. The latter occurs because the ammonium ion can now replace the sodium and the potassium ions being lost in the urine as the accompanying ions for chloride. After termination of the period of ingestion of ammonium chloride there is a prompt return of all variables to normal, with a slight undershooting in the excretion of potassium and sodium. This undershoot reflects the fact that some time is required for restoration of the ammonium production rates of the epithelial cells.

Titratable acidity also increases during the period of ingestion of ammonium chloride. The titratable acidity is dependent on the quantity of buffers that are present in the urine and on the urinary pH. Ingestion of ammonium chloride does not induce significant alteration in the quantity of filtered buffers. Thus the increase in titratable acidity reflects largely the fact that urine pH falls as a result of the acidosis.

Disturbances in acid-base status may originate either from altered respiratory function (respiratory acidosis or alkalosis) or altered metabolism or intake of food (metabolic acidosis or alkalosis). Compensation for a disturbance originating from altered respiratory function involves appropriate shifts in bicarbonate reabsorption by the kidney to stabilize the ratio of HCO_3^-/P_{CO_2}. Similarly, a metabolic disturbance can be compensated for by altering respiration rate and depth to adjust plasma P_{CO_2}. The interactions between respiratory function and renal function that control blood pH are shown in Fig. 39-13.

It should be noted that renal compensation for altered acid-base status is a relatively long-term process. More rapid adjustments can be made by alterations in respiratory function. Therefore respiratory compensation for metabolic acidosis or alkalosis can be quite prompt (within a few minutes), whereas renal compensation for a respiratory acidosis may take several days to become evident.

■ The urine-collecting system

■ Anatomy

Urine passes out of the body through a series of structures known as the *urine-collecting system*. This consists of the *calyces*, the *renal pelvis*, the *ureters*, the *bladder*, and the *posterior urethra*. The ureters are tubes 4 or 5 mm in diameter and 30 cm long and are connected to the renal pelvis. They enter the bladder bilaterally at a location toward its base and in close proximity to the origin of the urethra. The region comprising the two ureters and the urethra is known as the *trigone*. The bladder consists of two parts—the *fundus*, or body, and the *neck*. The neck is also known as the posterior urethra. In females the posterior urethra is the end of the urine-collecting system and the point of exit of urine from the body. In males a continuation of the urethra, the *anterior urethra*, extends through the penis. The anterior urethra, however, is embryologically part of the reproductive system rather than the urinary system.

The entire urine-collecting system is lined with transitional epithelium and surrounded by several layers of smooth muscle. The smooth muscle cells in much of the system are electrically coupled, fire action potentials and are capable of developing spontaneous tone and of responding to neurotransmitters. The orientation of the smooth muscle varies with location along the collecting system. In the renal calyces and pelvis the orientation is largely circular; in the ureter the orientation is more spiral. The ureters run obliquely for 2 or 3 cm within the wall of the bladder before opening into the bladder lumen. Where the ureters enter the bladder the smooth muscles of the ureters and the bladder are continuous. The smooth muscle of the bladder, known as the *detrusor muscle*, forms a basketlike meshwork that converges toward the urethra into fibers arranged in a more longitudinal pattern.

The smooth muscle and the epithelium, as well as the elastic lamina between the two, all are organized to permit substantial distention of the ureters and bladder. The epithelium is arranged in multicellular layers overlying the folded elastic lamina, and the smooth muscle possesses a substantial capacity to adapt its tension to altered wall stress. The inner elements of the urethra and bladder are thrown up into folds, or *rugae*. During expansion the rugae can be flattened, and the epithelial cells spread out into a single layer without disrupting the integrity of the epithelial lining. This pattern is exaggerated in the bladder, and as a result the volume of this structure can vary from 10 to 400 ml with a pressure change of only 5 to 10 cm H_2O.

■ Innervation

The innervation of the collecting system is a major element in determining function. The dominant neural input to the collecting system is the parasympathetic innervation of the bladder and the posterior urethra. The fibers arise in the second through the fourth sacral segments of the spinal cord and travel via the *pelvic nerve* to the bladder wall. Synapses located within the bladder give rise to postganglionic fibers that innervate the smooth muscle. Control of bladder function and the process of urination, or *micturition*, also greatly depends on voluntary activity, which in the collecting system is regulated by somatic motor nerves innervating the external sphincter of the urethra. These somatic motor nerves originate in the anterior horn cells of the spinal cord and travel via the *pudendal nerves* to the sphincter.

Sensory fibers detecting distention originate in the bladder and travel via sympathetic fiber tracts to the spinal cord. Parasympathetic fibers carry information about the presence of urine in the posterior urethra to the cord. Sympathetic and parasympathetic afferents synapse in the cord and participate in spinal reflexes as well as continuing to higher centers, including the pons and the sensory motor strip of the cortex.

■ *Passage of urine*

Urine flows passively out of the ducts of Bellini and into the calyces. The smooth muscle of the calyces and renal pelvis serves to drive urine into the ureter. With distention the circularly arranged smooth muscle surrounding these structures contracts and generates pressures of 6 to 10 mm Hg, which force urine into the ureter. Once urine enters the ureter, distention again serves as a trigger to induce contraction of the ureteral smooth muscle. This contraction is sufficiently vigorous to cause closure of the *ureteropelvic junction*. At the same time, a peristaltic wave is initiated, which moves down the ureter and thus induces movement of the urine toward the bladder. This process appears to be entirely autonomous and is related to the myogenic response of the smooth muscle. In other words, stretch per se initiates muscular contraction, which then propagates along the length of the ureter. Peristaltic waves are generated at frequencies of one to five per minute with four to six segmentations along the length of the ureter. Pressures vary with the presence of peristalsis and with the level of hydration of the subject, but they range from 20 to 100 mm Hg.

After voiding, the residual volume of the bladder is reduced to less than 10 mm Hg. As urine is formed and transported through the ureters into the bladder, it collects, and bladder volume expands. When the volume reaches 200 to 300 ml, stretch receptors in the bladder wall are activated, and a sense of fullness is perceived and an urge to urinate is felt. The maximum bladder capacity is generally on the order of two times the volume at which the urge to urinate is perceived.

■ *Neural control of micturition*

Whereas ureteral function is autonomous, that is, inherent within the smooth muscle surrounding the ureter, control of the bladder, posterior urethra, and external sphincter depends on complex neural mechanisms. Distention of the bladder leads to discharge over afferents of the pelvic nerve and the pudendal nerve to the spinal cord and thence to the pons and cerebral cortex. Afferent discharge results in activation of a spinal reflex involving the pelvic nerve. Pelvic nerve discharge induces contraction of the bladder and expulsion of urine. However, this reflex is easily inhibited by cortical efferents. Alternatively, when so desired, the spinal reflex mechanism that leads to bladder discharge can be facilitated by descending cortical impulses.

When the volume within the bladder initiates afferent nerve discharge and cortical inhibition is released, parasympathetic discharge to the detrusor muscle is initiated, the muscle contracts and bladder pressure is raised. Since the ureters course through the wall of the bladder, an increase in bladder pressure causes closure of the ureters and prevents reflux of urine into these structures. Contraction of the detrusor muscle also causes shortening and widening of the posterior urethra and entry of urine into this structure. Entry of urine into the posterior urethra triggers parasympathetic afferents; these in turn cause inhibition of neural traffic over the pudendal nerve, which normally maintains closure of the external sphincter. This permits relaxation of the external sphincter, and urine can then flow through the posterior urethra and exit the body. The smooth muscle of the bladder continues to contract, thereby maintaining high intrabladder pressure and forcing urine out of the body.

Once initiated, the process of micturition normally continues until the bladder is emptied. However, cortical influences can inhibit the process by triggering nerve activity over the pudendal nerve, which causes contraction of the external sphincter. This is then followed by relaxation of the detrusor muscle.

The process of micturition is assisted by contraction of a variety of voluntary muscles coordinated by cortical processes. Resistance to flow through the posterior urethra is reduced at the onset of micturition by contraction of the *levator ani* and *perineal* muscles. Coordinated with this process is the contraction of the abdominal muscles and a lowering of the diaphragm, both of which result in an elevation of intraabdominal pressure, which forces urine from the bladder.

■ *Incontinence*

Voluntary control of urination develops after birth, usually sometime between the second and third year of life. Until that time, voiding is entirely reflex and depends on bladder stretch. Interference with the reflex processes just described may severely handicap the ability of an individual to control the release of urine—a common clinical problem. If the pelvic nerves are damaged, urine is retained and the bladder becomes grossly distended. If damage is greater and pelvic as well as pudendal nerves are involved, the afferent input to the reflex arc is eliminated and the external sphincter no longer maintains tone. In this condition urine drips continuously from the urethra as a result of the combined high pressure in the distended bladder and the lack of control of the external sphincter. This condition is referred to as *incontinence*.

■ *Bibliography*

Journal articles

Bundy, H.F.: Review: carbonic anhydrase, Comp. Biochem. Physiol. **57**B:1, 1977.

Cheema-Dhadli, S., and Halperin, M.L.: Role of mitochondrial anion transporters in the regulation of ammoniagenesis in renal cortex mitochondria of rabbits, Eur. J. Biochem. **99:**483, 1979.

Schrier, R., editor: Water metabolism, Kid. Int. vol. 10, 1976.

Books and monographs

Balagura-Baruch, S.: Renal metabolism and transfer of ammonia. In Rouller, C, and Muller, A.F., editors: The kidney, New York, 1971, Academic Press, Inc.

Brenner, B.M., and Rector, F.C.: The kidney, vol. 1., Philadelphia, 1981, W.B. Saunders Co.

Davenport, H.W.: The ABC of acid-base chemistry, Chicago, 1958, University of Chicago Press.

Hills, A.G.: Acid-base balance, Baltimore, 1973 Williams & Wilkins Co.

Koushanpour, E.: Renal physiology: principles and functions, Philadelphia, 1976, W.B. Saunders Co.

Orloff, J., and Berliner, R.W.: Renal physiology. In Handbook of physiology, Washington, D.C., 1973, American Physiological Society.

Robinson, J.R.: Fundamentals of acid-base regulation, London, 1975, Blackwell Scientific Publishers.

Valtin, H.: Renal function: mechanisms preserving fluid and solute balance in health, Boston, 1973, Little, Brown & Co.

Windhager, E.E.: Micropuncture techniques and nephron function, New York, 1968, Appleton-Century-Crofts.

THE ENDOCRINE SYSTEM

Saul M. Genuth

chapter 49

General principles of endocrine physiology

Endocrinology was classically defined as a discipline concerned with the ''internal secretions of the body.'' The original concept—that a chemical substance called a hormone, liberated by one kind of cell, is carried by the bloodstream to act on a distant target cell—represented a major advance in physiological understanding. It suggested a basic mechanism for maintaining the stability of the internal milieu in the face of irregular nutrient, mineral, water, and thermal fluxes. A hormone was defined as a substance whose secretion was evoked by a specific alteration in the milieu; as a result of the hormone's action on its target cell(s), the alteration was counteracted, and the status quo was restored. However, this homeostatic notion has grown increasingly complex as the intricate operation of the endocrine system continues to be revealed.

The overall mission of the endocrine system currently is seen to include regulation of growth, maturation, reproduction, and behavior, as well as maintenance of chemical homeostasis. A diversity of endocrine cell types and of their organization and locations has been found; moreover the number of known hormones and their molecular variety has multiplied. Target cells of hormone action now include other hormone-producing cells, and some hormones now are known to require chemical modification at intermediate sites between the gland of origin and the target cell(s) before their mission can be accomplished. A complementary group of target cell substances—hormone receptors—has been found to play an essential role in mediating hormone action. Furthermore hormone action now is known to involve several different intracellular mechanisms. Much of the control of hormone secretion has been found to depend on the feedback principle described on p. 899. In addition, an intimate relationship between neural function and hormonal secretion has been established. Indeed the distinction between a hormone and a neurotransmitter has grown increasingly blurred. Because of this complexity and diversity, students are in danger of being overwhelmed unless they learn to relate the functional characteristics of each component of the endocrine system to a set of unifying principles. The major purpose of this chapter is to provide such a framework for the subsequent chapters to follow.

■ *Types of hormones*

Hormone molecules fall into three general chemical classes. The first to be discovered—the amines—includes thyroid hormone and catecholamines. Both originate from the amino acid tyrosine and retain the aliphatic α-amino group. Introduction of a second hydroxyl group in the ortho- position on the benzene ring is characteristic of the catecholamines, whereas iodination of the benzene ring distinguishes the thyroid hormones. The second group is composed of proteins and peptides. In some instances protein hormones with dissimilar missions nonetheless share identical structural sequences away from the biologically active core, suggesting a common branch point in evolution. In other instances a single progenitor protein gives rise to several hormone offspring of varying sizes, with some sharing amino acid sequences and some having overlapping action. The third group to be identified chemically is the steroids, which include adrenal cortical and reproductive gland hormones and the active metabolites of vitamin D. Cholesterol is the common precursor to this class. Modification of side chains, hydroxylation at various sites, and ring aromatization confer individual biological activities on the various steroid hormones.

■ *Hormone synthesis*

The protein and peptide hormones are synthesized on the rough endoplasmic reticulum in the same manner as other proteins. The appropriate amino acid sequence is dictated by specific messenger RNAs (mRNA) that originate in the nucleus. These have been isolated and characterized for a number of hormones, and the structures of the corresponding DNA molecules have been elucidated. By recombinant DNA technology, bacteria have been directed to synthesize a number of authentic mammalian protein hormones. In general, translation of the genetic message usually first results in the synthesis of large precursor proteins by the ribosomes. These "preprohormones" and "prohormones" contain extra peptide sequences either at one end of the hormone molecule (usually the N-terminal) or inserted as a temporary bridge between two peptide chains that will comprise the final molecule (Fig. 49-1). The function of an N-terminal leader sequence may be to direct the molecule across the membrane of the endoplasmic reticulum into the cisternal space for transport to the Golgi apparatus. The function of inserted sequences may be to ensure proper folding of the polypeptide chains to permit formation of other intramolecular linkages such as disulfide bonds.

The precursor protein migrates from the ribosomes (either before or after cleavage of accessory peptides) to the Golgi apparatus, where it is packaged for storage in secre-

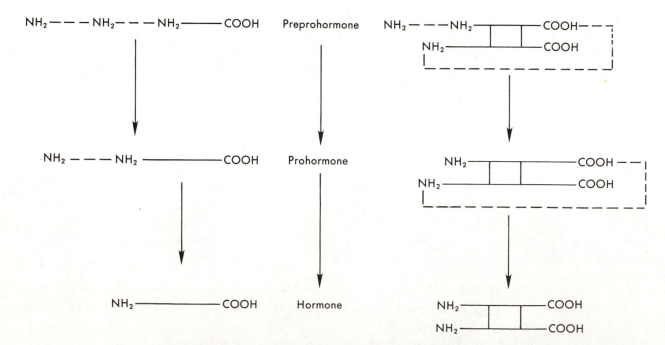

Fig. 49-1 ■ Peptide hormone synthesis begins in the endoplasmic reticulum with translation of an RNA message into a precursor protein within which the amino acid sequence of the hormone is encoded. This *preprohormone* undergoes synthesis and cleavage to a *prohormone* while still on the endoplasmic reticulum. The prohormone is further cleaved to the hormone itself in the Golgi apparatus and in the secretory granule. Two examples are shown.

tory granules. These granules also may contain proteolytic enzymes necessary for conversion of prohormone to hormone or elimination of accessory peptides. The theoretical possibility of genetic errors in protein and peptide hormone synthesis is appreciable. Single amino acid substitutions or deletions in active sites can alter the degree and/or specificity of biological action. If such errors occur in accessory peptides of preprohormones or prohormones, they may prevent normal processing to the active hormone. The difficulty in obtaining sufficient human hormone from any one individual for structural analysis undoubtedly has prevented an appreciation of the true frequency of such errors in humans, as compared, for example, with that of abnormal hemoglobins. However, a few such instances have been demonstrated.

The amine and steroid hormones are synthesized from tyrosine and cholesterol, respectively, through a sequential series of discrete enzymatic reactions. The intermediate products in the pathway may have hormonal properties of their own. This multiplicity of distinctive steps leads to the prediction that a variety of defects in hormone synthesis (and their clinical consequences) could arise from single gene-single enzyme deletions or from selective drug-induced enzyme inhibition. In fact numerous examples of congenital enzyme deficiencies in adrenal and gonadal steroid hormone and in thyroid hormone biosynthesis do occur. The resultant clinical states reflect both the deficiency of product hormone and the accumulation of hormone precursors.

■ *Hormone release*

Catecholamine and protein hormones share the property of being stored in secretory granules. Hormone release then is accomplished by the process of exocytosis. After migration through the cytoplasm and suitable positioning, the membrane of the secretory granule fuses with the immediately adjacent plasma cell membrane. The combined membrane material is lysed, and the contents of the granule are discharged into the extracellular space, leaving an empty core. The membrane material itself may be reprocessed. The movement of secretory granules to the cell membrane probably involves the participation of microtubular elements in the cytoplasm. Their activation often is mediated by prior generation of cyclic AMP and/or by uptake of calcium ions from extracellular fluid (Fig. 49-2). When secretory granules contain accessory peptides from the prohormone together with the enzymes that catalyze their cleavage, these materials are released simultaneously with the hormone. In the case of thyroid and steroid hormones storage does not take place in discrete granules, although the hormones may be compartmentalized in the cell. Once the hormones have appeared in free form within the cytoplasm, they apparently leave the cell by simple transfer through the plasma membrane.

These modes of hormone synthesis and release are essentially unicellular. However, more complicated patterns of hormone production also are encountered. Two adjacent cell types in a single gland may interact so that hormone A from cell A is modified in cell B to produce hormone B with an entirely different spectrum of biological effects. In this manner, for example, estrogens are produced from androgens in the gonads. In a further extension of this principle, peripheral tissues that are ordinarily considered to be nonendocrine may carry out similar conversions. A second mode of hormone production involves modification of a precursor molecule of low activity to one of higher activity by successive steps in several tissues. For example, a sterol synthesized in the skin requires actions by the liver and kidney to produce the most potent vitamin D hormone. Lastly, production of peptide hormones can even occur in the circulation itself from a protein substrate provided by one organ and then acted on by enzymes originating in other organs. A prototype for this pattern is the synthesis of angiotensin—a peptide hormone—from a globulin precursor released by the liver and acted on sequentially by enzymes from the kidney and the lung.

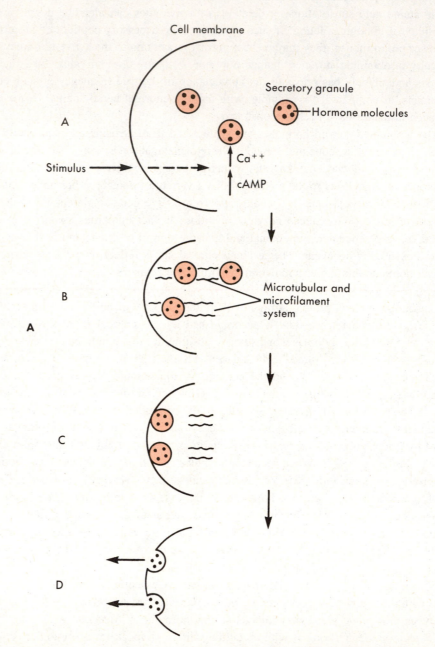

Cell membrane

Secretory granule

Hormone molecules

A

Stimulus

Ca⁺⁺

cAMP

B

Microtubular and microfilament system

A

C

D

Fig. 49-2 ■ **A,** Secretion of peptide hormones via exocytosis is initiated *(A)* by application of a stimulus that raises intracellular cyclic AMP levels and also usually results in a rise in cytosolic Ca^{++}. The secretory granules are lined up and translocated to the plasma membrane *(B)* via activation of a microtubular and microfilament system. The membrane of the secretory granule fuses with that of the cell *(C)*. The common membrane is lysed *(D)*, releasing the hormone into the interstitial space. **B,** Insulin secretory granules in a β-cell being stimulated by glucose. The arrow indicates a granule undergoing exocytosis. (From Lacy, P.E.: Beta cell secretion—from the standpoint of a pathobiologist, Diabetes **19**:895, 1970. Reproduced with permission from the American Diabetes Association, Inc.)

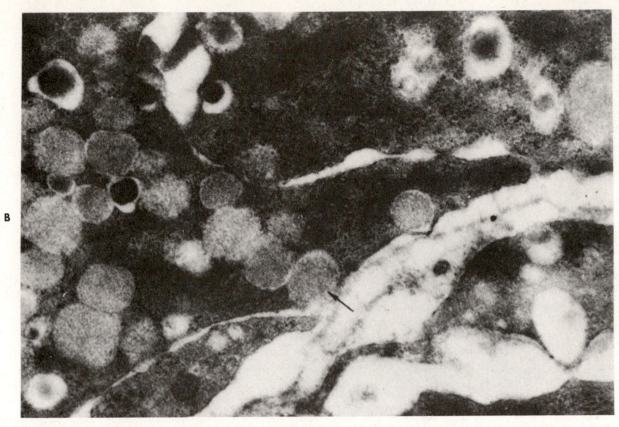

Fig. 49-2, cont'd ■ For legend see opposite page.

A number of general mechanisms that govern the secretion of hormones are the following:

Feedback control
    Hormone-hormone
    Substrate-hormone
    Mineral-hormone
Neural control
    Adrenergic
    Cholinergic
    Dopaminergic
    Serotoninergic

    Endorphinergic-enkephalinergic
    Gabergic
Chronotropic control
    Diurnal rhythm
    Sleep-wake cycle
    Menstrual rhythm
    Seasonal rhythm
    Developmental rhythm

The feedback principle is universally operative. Negative feedback is most common and acts to *limit* the excursions in output of each partner in the pair (Fig. 49-3). In the simplest instance hormone A, which stimulates secretion of hormone B, then in turn will be inhibited by an excess of hormone B. This straightforward mechanism dominates the relationship between hormones of the pituitary gland and its target glands. Secretion of a hormone that either accelerates the production or retards the utilization of a particular substrate, thus increasing its concentration in plasma, then will be inhibited by the resultant circulating excess of that substrate; conversely, secretion of the hormone will be stimulated by a circulating deficit of that substrate. On the other hand, the secretion of a hormone that impedes the production or accelerates the utilization of a particular substrate, thus decreasing its plasma concentration, will be stimulated by a circulating excess but inhibited by a circulating deficit of the substrate.

Positive feedback, which is less common, acts to *amplify* the initial biological effect

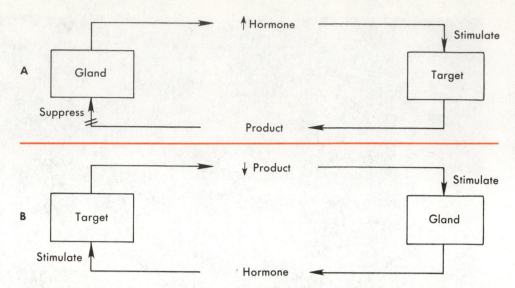

Fig. 49-3 ■ Negative feedback principle. **A,** A primary increase in hormone secretion stimulates a greater output of product from the target cell. The product then feeds back on the gland to suppress further hormone secretion. In this fashion hormone excess is limited or prevented. **B,** A primary decrease in output of product from the target cell stimulates the gland to secrete hormone. The hormone then stimulates a greater output of product from the target cell. In this fashion the product deficiency is limited or corrected.

of the hormone. Thus hormone A, which stimulates secretion of hormone B, in turn may be initially stimulated to greater secretion rates by hormone B, but only through a limited dose response range. Once sufficient biological momentum for secretion of hormone B has been obtained, other influences, including negative feedback, will reduce the response of hormone A to fit the final biological purpose.

Neural control acts to evoke or suppress hormone secretion in response to both external and internal stimuli. These may arise from visual, auditory, olfactory, gustatory, tactile, or pressure sensations and may be perceived consciously or unconsciously. Pain, emotion, sexual excitement, fright, injury, and stress all can modulate hormone secretion through neural mechanisms. Simple examples include the release of oxytocin, which fills the milk ducts in response to the stimulus of suckling, or the release of aldosterone, which augments the volume of the circulatory system in response to the upright posture. Certain patterns of hormone secretion appear to be dictated by rhythms that may be genetically encoded or acquired. Many hormones are secreted in distinct, independent diurnal rhythms. These may be tied to sleep-wake cycles or to light-dark cycles. Seasonal variation in hormone secretion also occurs, which may reflect the influence of temperature, sunlight, or tides. Some of these rhythms appear atavistic in humans, having probably served to fulfill biological needs of evolutionary ancestors. Perhaps the most intriguing of all are those patterns of hormone secretion which coincide with and are unique to developmental stages, such as the onset of puberty. A diverse and still growing number of neurotransmitter molecules carries these signals to the endocrine cells.

■ *Mechanisms of hormone action*

Although it is impossible to reconcile the great multiplicity of hormone effects with a single principle of action, several general mechanisms do exist. Because these relatively few mechanisms serve a large number of hormones, the specificity of any particular hormone–target cell interaction must rest on two factors: the uniqueness with which the hormone is recognized by the target cell and the uniqueness of the intracellular pathways with which a target cell can respond to hormonal stimulation (Fig. 49-4).

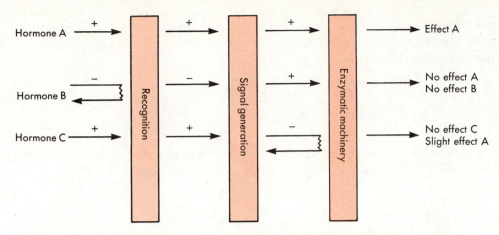

Fig 49-4 ■ Hormone-cell interaction. Hormone A is recognized by this cell through binding with its specific receptor. An intracellular signal is generated that stimulates the appropriate enzymatic machinery, and effect A is produced. Hormone B is not recognized because the cell lacks a receptor for it. Thus hormone B cannot produce effect A in this cell even though it might operate through an identical enzymatic machinery in its own target cell. Hormone C may be slightly recognized by this cell through individual overlap with the receptor to hormone A. Although a weak intracellular signal may be generated, no effect C results because the cell lacks the appropriate enzymatic machinery. To a minor extent, however, hormone C may produce effect A.

■ *Receptors*

Recognition of a hormone can take place in several sites. Receptors exist in the plasma cell membrane for catecholamine, peptide, and protein hormones, in the cytoplasm for steroid hormones, and largely in the nucleus for thyroid hormones. Receptors appear universally to be large protein molecules; plasma membrane receptors, in addition, contain carbohydrate and/or phospholipid moieties. There are generally 2000 to 100,000 receptor molecules per cell. In many instances this number is sufficient to guarantee that receptor availability will not be rate limiting for hormone action. It is not yet known whether the receptor molecule for a given hormone is identical in all its target cells or even whether in the same cell all the different actions of a hormone are initiated by binding to the same receptor molecule.

The association of a hormone with its receptor is a reversible reaction that appears to obey the following molecular chemical kinetics:

$$[H] + [R] \rightleftarrows [HR] \tag{1}$$

$$K_{assoc} = \frac{[HR]}{[H]\,[R]} = \text{Affinity constant} \tag{2}$$

$$\frac{[HR]}{[H]} = K_{assoc} \times [R] \tag{3}$$

where

$[H]$ = free hormone in solution
$[R]$ = unoccupied receptor
$[HR]$ = bound hormone = occupied receptor
R_o = initial receptor capacity = $[R] + [HR]$
K_{assoc} = affinity constant

If a fixed number of cells (equal to a constant amount of receptor) is incubated in vitro with increasing concentrations of hormone, the amount of bound hormone increases until receptor occupancy reaches 100%. At this point the number of bound hormone molecules equals the total number of originally available receptor molecules, that is, the receptor capacity (R_o). At the same time the ratio of bound hormone to free hormone has progressively decreased and approaches zero as the receptor occupancy approaches 100%. That is, as $[H] \rightarrow$ Infinity

$$[HR] \rightarrow R_o$$

$$\frac{[HR]}{[H]} \rightarrow 0$$

The data obtained by incubating cell receptors with increasing amounts of hormone can be plotted in a meaningful way by simple substitution in equation 3. Since

$$[R] = R_o - [HR]$$

$$\frac{[HR]}{[H]} = K_{assoc} \times (R_o - [HR]) \tag{4}$$

$$\frac{[HR]}{[H]} = -K_{assoc}\,[HR] + K_{assoc}\,R_o \tag{5}$$

$$\frac{\text{Bound hormone}}{\text{Free hormone}} = -K_{assoc} \times \text{Bound hormone} + K_{assoc} \times \text{Receptor capacity}$$

Plotting the ratio of bound hormone/free hormone as a function of bound hormone is called a Scatchard plot. This theoretically yields a straight line (Fig. 49-5, *A*). The slope of the line equals the negative of the association constant (K_{assoc}), and the x intercept equals R_o (the receptor capacity).

In practice many Scatchard plots of hormone binding to receptor yield exponential curves (Fig. 49-5, *B*). One interpretation of this phenomenon is that the cell possesses two classes of receptors, R_1 and R_2. R_1 has a higher affinity than R_2 (K_{assoc_1} is greater than K_{assoc_2}) but a lower capacity (R_{o_1} is less than R_{o_2}). Often the biological action of a hormone is quantitatively accounted for by interaction with the apparent high-affinity, low-capacity receptor, requiring occupancy of as little as 10% of the total receptor capacity of the cell. The function of the remaining ''spare receptors'' is not known for certain. However, they can act to increase the concentration of receptor-bound hormone (HR) when hormone concentration or receptor affinity is low. A second interpretation of nonlinear Scatchard plots is that the affinity of unoccupied receptor molecules for hormone is decreased by the presence of adjacent occupied receptor molecules. This phenomenon is termed *negative cooperativity*. It has the effect of moderating the increase in receptor-bound hormone and thus moderating the increase in hormone action should hormone concentration rise too precipitously.

Regulation of the hormone receptor provides another mechanism for regulating hormone action. Rearrangement of equation 5 to provide an expression for HR yields

$$[HR] = R_o \times \frac{K_{assoc}\,[H]}{K_{assoc}\,[H] + 1} \tag{6}$$

Thus it is seen that HR is directly proportional to R_o, the initial receptor number. An increase in receptor number (R_o) raises the maximum [HR] obtainable at saturating concentrations of hormone. This would raise the *maximum responsiveness* of the cell for those hormone effects in which receptor binding, rather than a later intracellular step in hormone action, is rate limiting. This may be seen where there are no spare receptors, as is generally the case for steroid and thyroid hormones. At submaximal hormone concentrations the increase in [HR] produced by an increase in R_o would enhance the *sensitivity* of the cell. This is commonly seen with peptide hormones.

It is interesting that receptor capacity often is regulated by its own hormone. In the usual instance the regulation is inverse, that is, a sustained excess of hormone decreases the number of its receptors per cell, which is *down regulation*. This acts to moderate

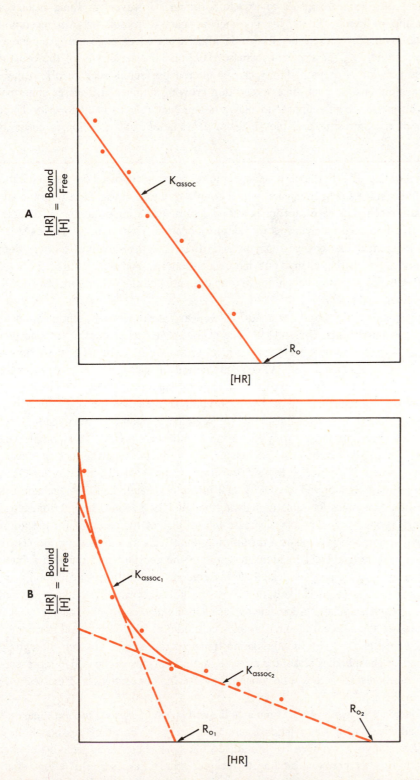

Fig. 49-5 ■ Scatchard plot. **A,** A linear plot results when the hormone reacts with a single receptor class and no cooperativity is present. **B,** An exponential plot results when the hormone reacts with multiple classes of receptors in the same cell or when the reaction with a single class of receptor is influenced by cooperativity. The association constant, K_{assoc}, equals the slope of the line. The receptor number, R_o, equals the intercept with the x axis.

the effect of chronic exposure to excess hormone. However, in some instances the relationship is direct, that is, the hormone appears to recruit its own receptors—*up regulation*—thereby amplifying its action after a period of exposure. An increase in receptor affinity (K_{assoc}) also will increase [HR] and the sensitivity of the cell to hormone stimulation. Receptor affinity can be altered by factors such as pH, osmolality, ion concentrations, and substrate levels. In a growing number of diseases abnormalities of the hormone receptor appear to play a significant role in pathogenesis. Therefore interest in receptor structure, biosynthesis, degradation, and function is currently intense.

■ Intracellular pathways of hormone action

The binding of a hormone to its specific receptor is followed by the generation of intracellular signals or messengers, which couple hormone recognition with the specific enzymatic machinery necessary for hormone action. For a large number of diverse hormones two universal mechanisms for coupling operate. One involves cyclic AMP as a second messenger that changes enzyme activities; the other involves transcriptional or translational effects that change enzyme concentrations. For other hormones, however, the link between receptor recognition and unique intracellular action remains speculative or completely unknown.

Many peptide or protein hormones, as well as catecholamines, stimulate an immediate rise in intracellular cyclic AMP. The cyclic AMP then acts as a ''second messenger'' that activates numerous enzymes critical to the various actions of these hormones. Cyclic AMP is generated from ATP by the action of adenylate cyclase, a membrane-bound enzyme. This enzyme consists of a catalytic and a regulatory subunit. As shown in Fig. 49-6, formation of the hormone receptor complex on the outer surface of the plasma membrane induces a change in the regulatory subunit of adenylate cyclase by causing the latter to complex with guanosine triphosphate (GTP). This complex in turn activates the catalytic subunit of adenylate cyclase on the inner surface of the plasma membrane where it can act on Mg-ATP to form cyclic AMP. In this process GTP may be simultaneously hydrolyzed to GDP (guanosine diphosphate); the GDP-regulatory subunit complex then becomes less able to activate the catalytic subunit. In addition, the GDP-regulatory subunit complex may decrease the affinity of the receptor for the hormone, providing still another mechanism for limiting the effects of excess hormone. The exact structural and geographical relationships among the receptor and the two subunits of adenylate cyclase are not yet clear, but it is evident that all three have mobility within the plasma membrane. It is likely that some but not all hormone receptors can activate a single adenylate cyclase. Adenylate cyclase activity also can be modulated by calcium, complexed to the calcium-binding protein, calmodulin. Therefore some hormones may amplify enzymatic activity by stimulating translocation of calcium into the cytoplasm. Inhibition of phosphodiesterase, the enzyme that hydrolyzes cyclic AMP, represents still another mechanism whereby a hormone may raise the concentration of the second messenger in the cell.

An increase in cyclic AMP stimulates the activation of various protein kinases; these in turn activate by phosphorylation a number of enzymes in numerous metabolic pathways. Alternatively, cyclic AMP–stimulated phosphorylation may deactivate other enzymes. Thus after hormone binding to receptor there is a cascade of effects that ultimately produces changes in the flux of metabolites in the cell. In the end either the storage or the release of an important metabolite may be facilitated. Activation or deactivation of reciprocal pathways by the same hormone can augment the result, for example, by simultaneously inhibiting the release pathway while stimulating the storage pathway. Adenylate cyclase and cyclic AMP are ubiquitously distributed. Therefore the specificity of enzyme responses to a hormone also may depend on the compartmentali-

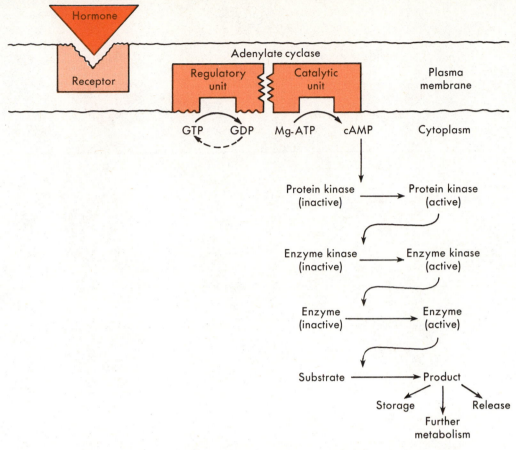

Fig. 49-6 ■ The schema for the mechanism of hormone action in which cyclic AMP acts as a second messenger. Although the example shown results in activation of a regulatory enzyme by phosphorylation, in other cases phosphorylation results in inactivation of a regulatory enzyme.

zation of cyclic AMP increases within the cell or on the proximity of the activated adenylate cyclase to target enzyme(s).

In contrast, steroid and thyroid hormones generate intracellular signals primarily by modulating transcription in specific areas of chromatin (Fig. 49-7). In this second major mechanism the steroid hormones combine with their receptors in the cytoplasm, and the complexes then enter the nucleus, whereas thyroid hormones enter the nucleus in the free state and then combine with their nuclear receptors. In both cases the hormone receptor complex interacts with target DNA molecules in the chromatin. Other proteins in the chromatin appear to facilitate acceptance by DNA, possibly by further altering the shape of the hormone receptor complex. As a result of these interactions, transcription generally is stimulated, and specific mRNA synthesis increases. The specific mRNAs in turn enter the cytoplasm where they direct the synthesis of specific proteins. These may be enzymes, structural proteins, receptor proteins, or proteins that are exported by the cell. In this manner enzymes are induced rather than activated, as they are by the cyclic AMP–mediated mechanism. However, it is also possible for this mode of hormone action to repress the synthesis of specific mRNAs and their complementary proteins. Therefore a metabolic pathway using a hormone-induced enzyme can either be accelerated or retarded. Likewise, other protein products may be either increased or decreased as a result of this type of hormone regulation. The rate and magnitude of hormone action via this mechanism may be limited (1) by specific factors such as hor-

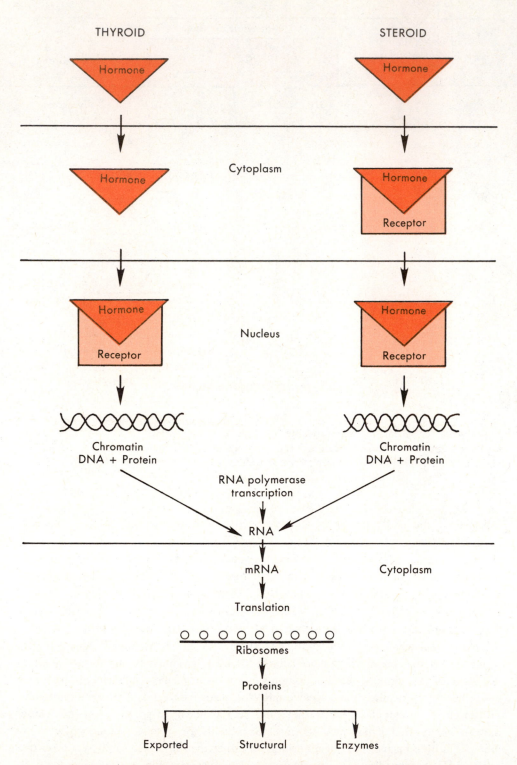

Fig. 49-7 ■ The schema for the mechanism of thyroid hormone *(left)* and steroid hormone *(right)* actions. The hormone receptor complexes undergo some form of activation or transformation prior to effectuating changes in DNA conformation and generation of mRNAs. Either induction or suppression of the synthesis of specific proteins can result.

mone receptor number and affinity or the number of chromatin acceptor sites in the nucleus or (2) by nonspecific factors, such as the concentrations of RNA polymerase, of the enzymes of protein synthesis, of transfer RNAs, and of amino acid substrates.

The cyclic AMP mechanism readily accounts for the frequently rapid (within minutes) effects produced by catecholamine, peptide, and protein hormones. The transcription mechanism logically explains the usually slower (hours to days) effects generally elicited by steroid and thyroid hormones. However, definite overlapping exists. Peptide hormones also cause effects that must invoke actions on protein synthesis at either a transcriptional or a translational level because they are blocked by selective inhibitors such as actinomycin and puromycin. Furthermore definite increases in enzyme mass, rather than simply in enzyme activity, have been documented after exposure to protein hormones such as insulin. There is now also considerable evidence for progressive internalization of the hormone–plasma membrane receptor complex, and receptors in the nucleus and mitochondrion for certain peptide hormones also have been described. Conversely, steroid and thyroid hormones can exhibit a few rapid effects—some on membranes—that cannot be readily accounted for by stimulation of transcriptional or translational events.

Hormone actions not readily explainable by either major mechanism certainly exist. These may involve activation of guanylate cyclase to alter cyclic GMP levels, phosphorylation of specific membrane proteins, or alterations in cellular and organelle permeability to ions, such as calcium. Finally in some instances prostaglandins appear to mediate the actions of a hormone. These may be related to or may be independent of effects of the prostaglandins on the adenylate cyclase system. The calcium-calmodulin complex may then act as an independent second messenger.

■ *Responsivity to hormones*

The final outcome of the interaction of a hormone with its target cell depends on a number of factors. These include hormone concentration, receptor number, duration of exposure, intervals between consecutive exposures, intracellular conditions such as concentrations of rate-limiting enzymes, cofactors or substrates, and the concurrent effects of antagonistic or synergistic hormones. Hormonal effects are not "all or none" phenomena. The dose-response curve for the action of a hormone is generally complex and often exhibits a sigmoidal shape (Fig. 49-8, *A*). An intrinsic basal level of activity may be observed independent of added hormone and long after any previous exposure. A certain minimum *threshold* concentration of hormone then is required to elicit a measurable response. The effect that is obtained at saturating doses of hormone defines the *maximum responsiveness* of the target cell. The concentration of hormone required to elicit a half-maximum response is an index of the *sensitivity* of the target cell. Alterations in the dose-response curve in vivo can take two general forms (Fig. 49-8, *B*). (1) A decrease in maximum responsiveness could be due to a decrease in the number of functional target cells, in the total number of receptors per cell, in the concentration of an enzyme being activated by the hormone, or in the concentration of a precursor essential to the final product of hormone action; it also could be due to an increase in the concentration of a noncompetitive inhibitor. (2) A decrease in hormone sensitivity could be caused by a decrease in the number or affinity of hormone receptors, alterations in the concentration of modulating cofactors, an increase in the rate of hormone degradation, or increases in antagonistic hormones.

The normal range of responsiveness to a hormone is usually rather broad and is in part a result of the physiological variability created by the aforementioned factors within and among normal individuals. However, the very exquisiteness with which hormonal effects can be modulated is an important component in achieving one major objective of hormonal regulation: metabolic stability.

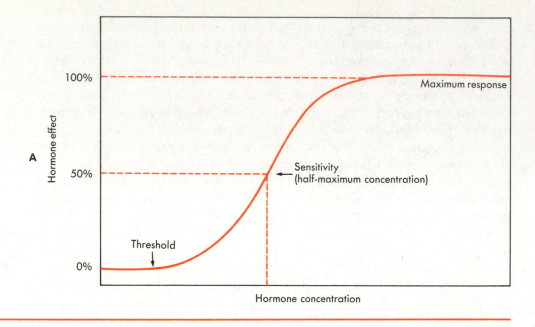

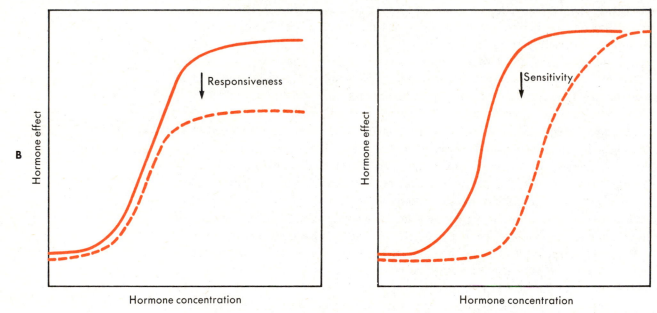

Fig. 49-8 ■ **A,** The general shape of a hormone dose-response curve. Sensitivity is most often expressed as the concentration of the hormone that produces a half-maximum response. **B,** Alterations in the dose-response curve can take the form of a change in maximum responsiveness *(left panel)* or of a change in sensitivity *(right panel)* or both.

■ *Hormone transport*

After secretion, hormones enter pools characterized by widely varying size, volume of distribution, degree of compartmentalization, and fractional turnover rates. Initially all hormones enter the plasma pool, where they may circulate either as free molecules or bound to specific carrier proteins. Ordinarily catecholamine, peptide, and protein hormones circulate unbound, although exceptions are noted. In contrast, steroid and thyroid hormones largely circulate bound to specific globulins that are synthesized in the liver. The extent of protein binding markedly influences the exit rates of hormones from plasma into interstitial fluid and thence into the intracellular space. As noted in examples presented in Table 49-1, the plasma half-life of a hormone is positively cor-

TABLE 49-1

Correlation of plasma half-life and metabolic clearance of hormones with their structure and degree of protein binding

Hormone	Protein binding (percent)	Plasma half-life (days)	Metabolic clearance (ml/minute)
Thyroid			
Thyroxine	99.95	6	0.7
Triiodothyronine	99.65	1	18
Steroids			
Cortisol	94	0.07	140
Testosterone	89	0.04	860
Aldosterone	15	0.016	1100
Proteins			
Thyrotropin (molecular weight 28,000)	Nil	0.034	50
Growth hormone (molecular weight 22,000)	Nil	0.017	220
Insulin (molecular weight 6000)	Nil	0.006	800

related with the percentage of protein binding. For example, thyroxine, a thyroid hormone, is 99.95% protein bound and has a plasma half-life of 6 days, whereas aldosterone, a steroid hormone, is only 15% bound and has a plasma half-life of 25 minutes. Larger and more complex protein hormones tend to have longer half-lives than do smaller proteins and peptides. Hormone exit from plasma does not have to be entirely irreversible. There is evidence in some instances that hormone molecules may return to plasma from other compartments, possibly after dissociation from cell membrane receptors. The return to plasma can occur by way of the lymphatic channels.

■ *Hormone disposal*

Irreversible removal of hormone is a result of target cell uptake, metabolic degradation, and urinary or biliary excretion. The sum of all removal processes is expressed in the term *metabolic clearance rate* (MCR). In a steady state this is defined as volume of plasma cleared/unit time, which equals mass removed/unit time divided by circulating mass/unit volume, that is,

$$\text{MCR} = \frac{\text{mg/minute removed}}{\text{mg/ml of plasma}} = \frac{\text{ml cleared}}{\text{minute}} \tag{7}$$

MCR is one expression of the efficiency with which a hormone is removed from plasma. It is most conveniently determined by tracer techniques. A minute quantity of a valid radioactive tracer of the hormone is infused at a constant rate $\left(\dfrac{\text{CPM}}{\text{minute}}\right)$ until equilibrium is reached, that is, until the plasma concentration of tracer $\left(\dfrac{\text{CPM}}{\text{ml}}\right)$ becomes constant. At this point the outflow of tracer from the plasma must equal the inflow of tracer. Thus

$$\text{MCR} = \frac{\text{CPM/minute infused}}{\text{CPM/ml of plasma}} \tag{8}$$

The ratio of MCR to the volume of distribution of a hormone is a measure of its fractional turnover rate (K). The plasma half-life, which is inversely related to K, is a cruder but more conveniently determined index of hormone disappearance. As shown in Table 49-1, MCR is negatively correlated with plasma half-life and usually also negatively correlated with the percentage of protein binding.

The kidney and liver are usually the major sites of hormone extraction and degradation. Renal clearance of hormone is reduced markedly by protein binding to globulins in the plasma. For example, less than 1% of secreted cortisol appears unchanged in the urine because only the small, free fraction of plasma cortisol is filtered by the glomerulus. On the other hand, about 30% of cortisol metabolites are excreted in the urine, since these are generally unbound or only loosely bound to protein. Many peptide and smaller protein hormones are filtered to some degree by the glomerulus. Usually, however, they subsequently undergo tubular reabsorption and degradation within the kidney, so that only a minute amount appears in the final urine.

Metabolic degradation occurs by enzymatic processes that include proteolysis, oxidation, reduction, hydroxylation, decarboxylation, and methylation. Virtually all hormones are extracted from the plasma and degraded to some extent by the liver, which often is quantitatively the most important site. In addition, glucuronidation and sulfation of hormones or their metabolites may be carried out, with the conjugates subsequently excreted in the bile or the urine. Some hormonal degradation appears to take place during interaction with target tissues. As noted previously, internalization of a portion of the hormone–plasma membrane receptor complex does occur, and hormone plus receptor may be degraded together by lysosomes within the cell. It is difficult to tell at present to what extent hormone action and hormone degradation are quantitatively linked in these situations.

■ Hormone measurement

In large part the history of endocrinology is the history of methodology for hormone measurement. Initially hormones all were measured by biological assay, that is, by measuring an effect that they produced. When whole animals served as test subjects, sensitivity was necessarily low, and precision was greatly influenced by many variables that were difficult to control. As more knowledge of hormone action developed, biological assays could be carried out on organs in vivo and later on tissues in vitro. The sensitivity improved, less replication was needed as precision was enhanced, and the amount of sample required decreased. Of equal importance, specificity was increased. For example, a biological effect such as elevation of plasma glucose concentration could be produced by two very dissimilar hormones via different mechanisms. A whole animal assay with plasma glucose as an end point might not distinguish between the two. However, if one hormone worked on the liver and the other on muscle, selective tissue assays could discriminate between the two. Nevertheless tissue assays still put practical limitations on sensitivity and on the number of samples that could be handled and thus on the amount of information that could be obtained. Physicochemical methods, such as spectrophotometry and fluorometry, were developed that offered greater ease. Their sensitivity permitted their use only for catecholamine, thyroid and steroid hormones, or their metabolites. Even then often only the relatively higher concentrations in tissue, urine, or effluent blood from the gland of origin could be accurately determined. Furthermore laborious chromatographic procedures sometimes were required to separate hormones with similar chemical groupings.

■ Radioimmunoassay

A dramatic advance in the acquisition of information and understanding in endocrinology followed the development of radioimmunoassay and related methods in 1957. In one stroke, sensitivity was brought to the level of normal plasma concentrations for virtually all hormones, precision to the level where changes of 25% could be reliably determined, and ease, cost, and sample volume to the point where as many as 50 values could be obtained from an individual in a single day. Specificity also was vastly enhanced initially; for example, two hormones, such as glucagon and epinephrine, which both raise blood glucose (in vivo bioassay) and tissue cyclic AMP levels (in vitro bioassay), have absolutely no overlapping in a radioimmunoassay. It subsequently has be-

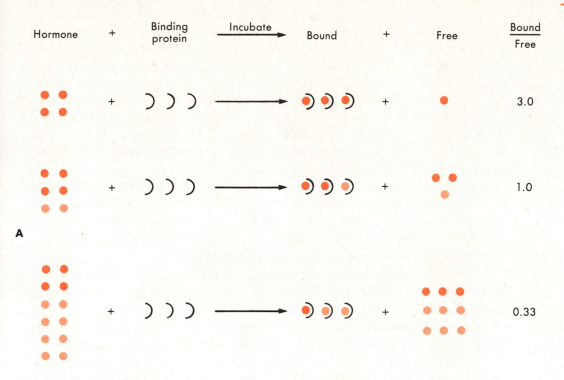

Hormone + Binding protein →Incubate→ Bound + Free | Bound/Free

A

- Tracer hormone
- Hormone in sample or standard
-) Binding protein

Fig. 49-9 ■ **A,** The principle of radioimmunoassay. Note that concentrations of tracer hormone and binding protein are fixed. Addition of increasing amounts of unlabeled hormone in sample or standard in effect dilutes the tracer hormone molecules and displaces them from the binding protein. This results in a decreasing ratio of bound/free tracer hormone. **B,** A typical radioimmunoassay standard curve is exponential. The curve, however, often can be made linear by reciprocal or logarithmic transformation of the data.

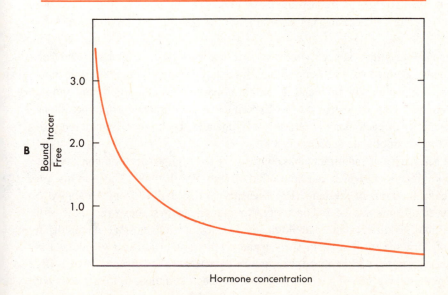

B

come possible to design radioimmunoassays that can reliably discriminate between a hormone and its precursor (for example, insulin and proinsulin) or a hormone and its slightly modified metabolites (for example, thyroxine, triiodothyronine, and reverse triiodothyronine).

The principle of radioimmunoassay is shown in Fig. 49-9. The sample, which may be plasma, urine, cerebrospinal fluid, or a tissue extract, is incubated with a fixed amount of radioactively labeled hormone (the tracer) and a fixed amount of a specific hormone-binding protein, most often an antibody. The nonradioactive hormone molecules in the sample *compete* with the tracer hormone molecules for the binding sites on the protein. The concentration of binding sites is fixed and limiting. Therefore progressive increases in the number of nonradioactive hormone molecules in the sample will displace more and more of the tracer hormone molecules from the binding sites. At the end of the incubation the bound tracer hormone molecules are separated from those which are free. The radioactivity in the individual bound and free tracer fractions is counted. If the hormone concentration in the sample is relatively high, the percentage of radioactivity remaining in the bound fraction will be relatively low, and vice versa (Fig. 49-9, *A*). The absolute amount of hormone in the sample is calculated by comparison with the standard curve generated by incubating varying amounts of authentic hormone with identical amounts of tracer and binding proteins (Fig. 49-9, *B*).

The specificity of radioimmunoassay rests on the reaction of a hormone with a binding site that uniquely recognizes the hormone molecule. Most commonly the binding protein is an antibody produced by immunizing an animal. Protein hormones frequently possess sufficient antigenic power by themselves. Peptides, catecholamines, steroids, and thyroid hormones must be coupled, as haptens, to a larger protein, such as albumin, to render them sufficiently antigenic. In addition to antibodies, naturally occurring binding proteins also may serve in radioassays. These include proteins obtained from plasma and receptor proteins obtained from cells. The latter offers the special advantage that radioassay specificity almost always will parallel the specificity of biological action. Alterations in structure that render a hormone devoid of biological activity may still permit its easy recognition by an induced radioimmunoassay antibody whose determinants happen to be removed from the biologically active sites. Therefore elevated levels of the hormone may be measured by a radioimmunoassay employing such an antibody even though clinical signs of hormone deficiency exist. This is less likely to occur if a normal tissue receptor replaces the induced antibody as the binding protein in the radioassay.

Radioimmunoassays may not always distinguish completely or sufficiently between similar hormones secreted by the same gland (for example, two adrenal steroid hormones), between a peptide hormone and its prohormone, or between a hormone and its metabolic products. This may give rise to experimental or clinical misinterpretations unless the specificity of each assay is carefully documented. Preliminary separation procedures sometimes can obviate such problems.

The great sensitivity of radioimmunoassay rests on the high association constants of antigen-antibody reactions, which can be around 10^{10} to 10^{12} L/mole. This permits use of very low concentrations of reactants. High sensitivity also derives from the ability to measure very small amounts of radioactivity. In clinical assays, however, it is still sometimes difficult to differentiate pathologically low levels from normal unstimulated basal levels of hormone in plasma. In such instances the conclusion that glandular hypofunction exists may have to be made from the inference that the hormone level is not elevated, despite the presence of a strong negative feedback stimulus, or that it does not increase to the normal extent with pharmacological stimulation.

The precision of radioimmunoassay stems from the simplicity of handling and its suitability for automation. When samples are analyzed in replicate in a single assay, the coefficient of variation, an index of precision, is 5% to 10%. When the same sample is measured repeatedly in separate assays, a slightly higher coefficient of variation is observed. Samples with high concentration of hormone are diluted to bring them into the most precise portion of the standard curve.

Absolute quantitation of the output of a single hormone by an individual gland can only be accomplished in vivo by catheterization of the blood supply to the gland. The arterial (A) and venous (V) concentrations of hormone, in mass per unit volume, and the blood flow (BF) across the gland, in volume per unit time, must be measured. The rate of secretion in mass per unit time is then $(V - A) \times BF$. This method for assessing secretion rate is suitable for animal studies but rarely can be performed in clinical circumstances and certainly not often enough to obtain daily secretion rates. However, sampling of veins that drain glands is a useful clinical procedure for localizing the source of excess hormone production. This is true for multiple glands, such as the parathyroid glands, or for hormones, such as steroids, that may be secreted from different glands, such as the adrenal glands and the ovaries.

■ *Estimates of hormone secretion*
■ *Direct measurement*

A less direct but frequently satisfactory method to estimate hormone secretion is the measurement of the blood production rate (BPR) in mass per unit time. This is the total amount of the hormone entering the peripheral circulation each day; in a steady state it will be equal to the total amount of hormone leaving the circulation. Therefore blood production rate is determined by measuring plasma concentration (P) and the MCR (p. 909):

$$BPR = P \times MCR \qquad (9)$$

Although this measurement primarily is carried out under research conditions in animals and humans, it can be used in select circumstances for clinical diagnosis.

■ *Blood production rate*

A simple plasma concentration provides a valid index of blood production rate when the metabolic clearance of the hormone is within normal limits *and can be taken as a constant;* since $BPR = P \times MCR$, BPR is proportional to P if MCR is a constant. This is the theoretical basis for employing plasma hormone measurements alone as an index of activity of the gland of origin. However, the release of many hormones is characterized by diurnal variation as well as by episodic spurts. It therefore may be hazardous to conclude too much from a single plasma value. Multiple daily measurements or measurements taken at different times of the day may be needed. To reduce the number of laboratory analyses, multiple samples of equal volume can be pooled, and a single careful measurement of this pool then yields the average plasma concentration over the interval in which the samples were collected.

■ *Plasma levels*

Measurement of urinary hormone excretion is cumbersome because it requires accurately timed collections. However, it offers the advantage of, in effect, averaging plasma fluctuations over the collection period. Furthermore in some instances the quantity of a hormone metabolite in the urine far exceeds the plasma or urinary level of the hormone itself and is therefore more easily measurable. Urinary excretion (UV) is equal to the urinary concentration (U) times the urine volume (V). Urinary excretion of the hormone is related to blood production rate and the renal clearance (C) of the hormone (Chapter 47), as follows:

$$C = \frac{UV}{P}; \; P = \frac{UV}{C} \qquad (10)$$

By substitution into equation 9:

$$BPR = UV \times \frac{MCR}{C} \qquad (11)$$

Hence BPR is proportional to UV when MCR/C can be taken as a constant. Thus urinary excretion of a hormone is a valid index of blood production rate and hence

■ *Urinary excretion*

secretion rate when the following two conditions are fulfilled: (1) the kidney must be contributing its usual fraction of the total metabolic clearance; and (2) within the kidney itself the usual proportioning of hormone between intrarenal degradation and urinary excretion must be maintained. The chief sources of error in employing urinary excretion as a reflection of hormone secretion are general impairment in renal function, a change in the pattern of degradation to a metabolic product excreted differently by the kidney, and incomplete collections of urine. When there is little diurnal variation in hormone secretion and degradation, incomplete collections can be partially compensated for by indexing hormone excretion to simultaneously determined creatinine excretion.

It is important that these principles of hormone measurement and of estimating hormone secretion rates be kept in mind when interpreting the results of physiological experiments or of clinical diagnostic studies.

■ *Bibliography*

Journal articles

Alford, F.P., et al.: Temporal patterns of circulating hormones as assessed by continuous blood sampling, J. Clin. Endocrinol. Metab. **36:**108, 1973.

Baxter, J.D., et al.: Hormone receptors, N. Engl. J. Med. **301:**1149, 1979.

Gordon, P., et al.: Internalization of polypeptide hormones; mechanism, intracellular localization and significance, Diabetologia **18:**263, 1980.

Lacy, P.E.: Beta cell secretion—from the standpoint of a pathobiologist, Diabetes **19:**895, 1970.

Sherwin, R.S., et al.: A model of the kinetics of insulin in man, J. Clin. Invest. **53:**1481, 1974.

Tait, J.F.: The use of isotopic steroids for the measurement of production rates *in vivo,* J. Clin. Endocrinol. **23:**1285, 1963.

Books and monographs

Rasmussen, H.: Organization and control of endocrine systems. In Williams, R.H., editor: Textbook of endocrinology, Philadelphia, 1974, W.B. Saunders Co.

Yalow, R.S., et al.: Introduction and general considerations. In Odell, W.D., and Daughaday, W.H., editors: Principles of competitive protein-binding assays, Philadelphia, 1971, J.B. Lippincott Co.

chapter 50

Whole body metabolism and the hormones of the pancreatic islets

The hormones of the pancreatic islets, especially *insulin* and *glucagon,* are among the most important regulators of fuel metabolism. Therefore before a detailed consideration of the endocrine function of the islets some fundamental aspects of whole body metabolism and substrate flux are reviewed.

A healthy adult human being ingests and expends an average of 2500 kcal/day* in the form of carbohydrate, protein, and fat and maintains a rather constant body weight. This energy turnover can vary from 1500 to 5000 kcal, depending on the individual's size, habitus, and physical activity. For adults, total daily energy expenditure averages 39 kcal/kg in men and 34 kcal/kg in women. Of this, approximately 20 kcal/kg are expended in *basal metabolism,* that is, the energy consumed in all the chemical reactions and mechanical work (such as breathing and heart action) when the individual is at complete rest. Basal metabolism may be increased up to 15% by eating, the so-called *specific dynamic action* of food, which is mostly attributable to chemical reactions (deamination and ureagenesis) initiated by absorption of amino acids. The remainder of energy expenditure results from voluntary or involuntary activity. If daily activity is light, the values for total energy expenditure are 10% to 15% less than just noted. If activity is moderately heavy, these values are 15% to 20% greater; and if very heavy, 25% to 35% greater. Energy expenditures for a number of common activities are listed in Table 50-1.

The approximate composition of a normal 70 kg adult human, viewed from the standpoint of energy content, is shown in Fig. 50-1. It is clear that fat forms the major energy storage depot, both because of its greater mass and because of its high caloric value (Table 50-2). Certain features of the relationship between the individual components of energy stores and their turnover are noteworthy. The average daily turnover of carbohydrate is approximately 250 g/day (1000 kcal/day). This is of the same magnitude as the total carbohydrate stores of 450 g (1800 kcal) and exceeds the rapidly mobilizable stores of 100 g contained in the extracellular fluid and liver glycogen. Thus a daily intake of at least 150 g of carbohydrate is desirable; if this ceases, a prompt mechanism for synthesis of glucose from other sources is required. The major source of endogenous glucose production is protein, with a smaller fraction available from the glycerol contained in fat.

In humans the average daily turnover of body protein is 2 g/kg, or 150 g/day (600 kcal). In contrast to carbohydrate, this represents only 1.5% of total body protein (10 kg) and only 2.5% of the metabolically mobilizable portion (6 kg). Since daily release of amino acids from the main endogenous source, the muscle mass, has been estimated

---

* 1 Kilocalorie (kcal) = 1 Calorie ("C") = 1000 calories ("c"); 1 kcal = 4184 joules.

TABLE 50-1

Estimates of energy expenditure in adults

Activity	Calorie expenditure (kcal/minute)
Basal	1.1
Sitting	1.8
Walking, 2.5 miles/hour	4.3
Walking, 4.0 miles/hour	8.2
Climbing stairs	9.0
Swimming	10.9
Bicycling, 13 miles/hour	11.1
Household domestic work	2.0 to 4.5
Factory work	2.0 to 6.0
Farming	4.0 to 6.0
Building trades	4.0 to 9.0

Data from Kottke, F.J.: Animal energy exchange. In Altman, P.L., editor: Metabolism, Bethesda, Md., 1968, Federation of American Society for Experimental Biology.

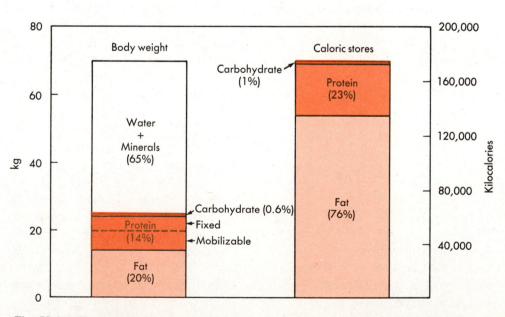

Fig. 50-1 ■ The composition of an average 70 kg human is shown in terms of weight *(left)* and caloric stores *(right)*. Note the trivial proportion of carbohydrate stores relative to fat stores.

TABLE 50-2

Energy equivalents of foodstuffs

	Kilocalories produced per gram	*O_2 used (L/g)*	*Kilocalories produced per liter of O_2*	*Respiratory quotient*
Carbohydrate	4.2	0.84	5.0	1.00
Fat	9.4	2.00	4.7	0.70
Protein	4.3*	0.96*	4.5	0.80
Typical fuel mix	—	—	4.8	0.90

*Each gram of protein oxidized yields 0.16 g of urinary nitrogen. Thus these values should be multiplied by 6.25 to express them per gram of urinary nitrogen.

to yield about 50 g/day a dietary protein intake of 100 g/day is ample. In fact the minimum daily requirement is much less, only 0.5g/kg, or a total of 35 g. This is because of efficient reuse for protein synthesis of those amino acids derived from proteolysis but retained within the cells of tissues such as muscle and liver. Although amino acids can be used as a source of oxidative energy, it is obviously advantageous to conserve them for the renewal of functional and structural protein molecules. The daily turnover of fat is around 100 g/day (900 kcal), representing less than 1% of body stores. Whereas carbohydrate and protein stores are rather narrowly fixed by the limitations of hepatic size and function and by the adult body habitus, respectively, fat stores are subject to great flexibility. The percentage of body weight accounted for by fat can vary from as little as 5% to as much as 80% in pathologically obese humans.

In the resting postabsorptive state the average 70 kg human being is expending approximately 1.0 to 1.2 kcal/minute in the process of basal metabolism. This requires the use of approximately 200 to 250 ml of oxygen per minute (Table 50-2). This basal metabolic rate is linearly related to body surface area. In the process of oxidizing substrates to meet basal energy needs the proportion of carbon dioxide produced ($\dot{V}CO_2$) to oxygen used ($\dot{V}O_2$) varies according to the fuel mix. The proportion of $\dot{V}CO_2$ to $\dot{V}O_2$ is known as the *respiratory quotient* (RQ) (Chapter 39). As indicated by the following equations, RQ equals 1.0 for oxidation of carbohydrate (for example, glucose), whereas RQ equals 0.70 for oxidation of fat (for example, palmitic acid).

For carbohydrates:

$$C_6H_{12}O_6 + 6\ O_2 \rightarrow 6\ CO_2 + 6\ H_2O$$
$$\textbf{Glucose}$$

$$RQ = \frac{6\ CO_2}{6\ O_2} = 1.0$$

For fats:

$$C_{15}H_{31}COOH + 23\ O_2 \rightarrow 16\ CO_2 + 16\ H_2O$$
$$\textbf{Palmitic acid}$$

$$RQ = \frac{16\ CO_2}{23\ O_2} = 0.70$$

The RQ for protein reflects that of the individual RQs of the amino acids and averages 0.80. Under ordinary circumstances protein is a minor energy source. Its small contribution can be corrected for by measuring the urinary excretion of the nitrogen that results from the metabolism of amino acids. The amount of protein oxidized equals 6.25 × grams of nitrogen. The amount of O_2 consumed by oxidation of 1.0 g protein equals 0.96 L (Table 50-2); the amount of CO_2 produced equals 0.80 × the amount of O_2

consumed. Thus, for each gram of urinary nitrogen excreted, 5.9 L of O_2 are consumed, and 4.8 L of CO_2 are produced.

By determining the nonprotein RQ, one can then calculate the proportion of carbohydrate and fat in the fuel mix being oxidized. For example, when the O_2 consumed = 200 ml/minute, and CO_2 produced = 180 ml/minute:

$$RQ = \frac{180}{200} = 0.90$$

C is the proportion of carbohydrate, and F is the proportion of fat in the fuel mix; then

$$1.0 \times C + 0.70 \times F = 0.90$$

Since

$$C + F = 1.0$$

$$1.0 \times C + 0.70 \times (1.0 - C) = 0.90$$

$$C = 0.67$$

$$F = 0.33$$

With a knowledge of the actual rate of energy expenditure in kilocalories per minute and the standard values of 4 kcal/g of carbohydrate and 9 kcal/g of fat, one can calculate the actual rates of oxidation of carbohydrate and fat, respectively, in grams per minute.

$$200 \text{ ml } O_2 \text{ consumed/minute} = 1.0 \text{ kcal/minute}$$

$$4.0 \times g \text{ Carbohydrate} + 9.0 \times g \text{ Fat} = 1.0$$

Since

$$\frac{g \text{ Fat}}{g \text{ Carbohydrate}} = \frac{0.33}{0.67} = 0.5$$

then

$$4.0 \times g \text{ Carbohydrate} + 9.0 \times (0.5 \times g \text{ Carbohydrate}) = 1.0$$

$$g \text{ Carbohydrate} = 0.118$$

$$g \text{ Fat} = 0.059$$

In the resting adult the rate of glucose oxidation is approximately 110 mg/minute (160 g/day), whereas that of fat oxidation is approximately 50 mg/minute (90 g/day). Thus in caloric equivalents glucose supplies about 45% of the energy required for basal metabolism. The central nervous system is an obligate glucose consumer and uses the major portion of this fuel. Estimates of cerebral glucose utilization are about 125 g/day. In contrast, the large muscle mass oxidizes primarily fatty acids in resting individuals. However, like all tissues, the muscle mass uses at least some glucose, thereby maintaining sufficient concentrations of Krebs cycle intermediates for efficient disposal of acetyl CoA and completion of fatty acid oxidation.

The overall rate of glucose utilization in adults averages 2 to 2.5 mg/kg/minute, or 200 to 250 g/day. The excess of glucose used over the glucose oxidized to carbon dioxide and water is largely a result of those glucose molecules which only undergo glycolysis to lactate. The latter then returns to the liver for resynthesis into glucose (Cori cycle). In the postabsorptive state the only significant supplier of glucose for peripheral utilization is the liver. Therefore the normal basal rate of hepatic glucose production equals that of glucose utilization (2 to 2.5 mg/kg/minute). About 70% to 80% of this postabsorptive glucose production results from glycogenolysis and only 20% to 30% from gluconeogenesis. Hepatic uptake and utilization of circulating lactate account for about half the glucose production from gluconeogensis. Most of the remainder is attributable to uptake and utilization of amino acids, especially alanine. Circulat-

*Whole body metabolism
and the hormones of the
pancreatic islets* **919**

ing glycerol and pyruvate ordinarily make only minimal contributions. The supply of lactate largely comes from glycolysis in red blood cells, white blood cells, and muscle. The amino acid precursors come from muscle proteolysis. The rate of release of alanine from muscle far exceeds that of all other amino acids. However, in this respect alanine is functioning largely as a cyclic carrier of amino groups transferred to pyruvate from other amino acids by transamination rather than as the major provider of new carbon atoms from protein for hepatic glucose synthesis.

The turnover of free fatty acids in the basal state largely represents molecules released from adipose tissue by lipolysis and taken up by muscle, liver, and other parenchyma. Only about half the turnover is accounted for by terminal oxidation. The other half represents re-esterification to triglycerides within muscle and liver. The hepatic triglycerides are then recirculated as very low-density lipoprotein, from which the free fatty acids may be extracted and stored again in adipose tissue.

Changes in caloric expenditure, largely because of muscle activity, are compensated for by increases or decreases in caloric intake so as to prevent undue weight loss or gain. If normal humans are fed liquid diets whose caloric density (kilocalories per milliliter) is systematically raised or lowered without their knowledge, they will quickly adjust the volume they drink to maintain an isocaloric intake. The mechanisms that regulate appetite and stabilize body weight appear to be multifactorial and still are not fully understood. The crucial role of specific areas of the hypothalamus in the brain appears well established. Current concepts include a *hunger center,* which stimulates food seeking and probably is located in the ventrolateral hypothalamus, and a *satiety center,* which tonically inhibits the hunger center and probably is located in the ventromedial area of the hypothalamus.

Evidence exists in support of various pathways regulating caloric intake: (1) A decrease in plasma glucose concentration, in body glucose stores, or in intracellular glucose metabolism in the hypothalamus reduces the firing rate of the satiety center. This in turn releases the hunger center from its tonic inhibition and initiates food seeking. This would key total energy and nutrient input to the need for one critical substrate: glucose. (2) A decrease in total adipose tissue mass, sensed in the hypothalamus via neural or chemical signals, decreases satiety center activity. This would key caloric input to the level of the major energy stores. (3) Increased thermogenesis following the oxidation of fuels is sensed by the hypothalamus, which then suppresses eating. This would key caloric intake to energy expenditure and to the need to maintain body temperature.

Other factors less obviously relevant to total body metabolism or energy considerations also may influence appetite significantly. Increased adrenergic input to the hypothalamus may stimulate eating, relating it to "stress." Increases in endorphins—polypeptides associated with pain relief— may increase eating, possibly relating it to "pleasure." Increases in brain concentrations of gastrointestinal hormones, such as cholecystokinin, may decrease eating, possibly relating it to feedback from the process of digestion and absorption of food.

It is also evident that primary or physiologically inappropriate increases or decreases in caloric intake can be compensated for, to some extent, by responsive increases or decreases in caloric expenditure. Long-term feeding of a caloric excess to volunteers causes weight gains almost entirely accounted for by fat storage. However, the rate of weight gain is less than that expected for the caloric excess that is administered. Furthermore maintenance of the new increased body weight requires 50% more calories per day than did maintenance of the original weight. These data cannot be explained by any observable increase in physical activity or in cost of work. Therefore an adaptive increase in thermogenesis, in either the basal or postprandial state, has been suggested as a normal physiological response to overexpansion of adipose tissue mass. The dissipa-

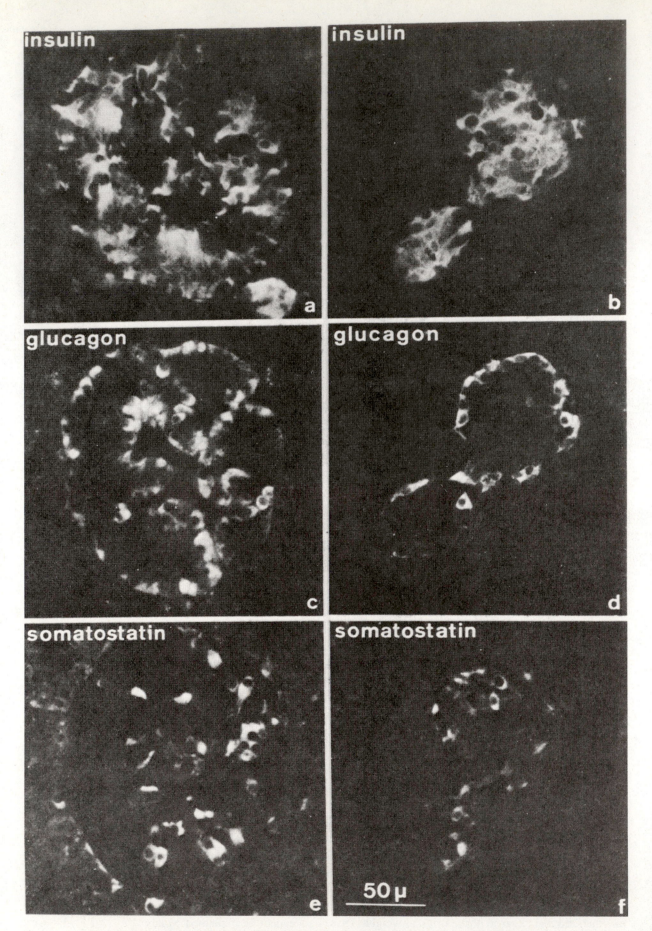

Fig. 50-2 ■ A human islet stained by immunohistochemical methods shows the central distribution of β-cells, the peripheral distribution of α- and δ-cells, and the close proximity of the latter two cell types (**a, c, e**). The lobular infolding of another islet, bringing α- and δ-cells into closer contact with the β-cells, is shown (**b, d, f**). (From Unger, R.H., et al.: Reproduced with permission from the Annual Review of Physiology, volume 40. Copyright © 1978 by Annual Reviews, Inc.)

*Whole body metabolism
and the hormones of the
pancreatic islets*   **921**

tion of extra ingested energy as heat could come about in part from hypertrophy of "futile metabolic cycles," which waste high-energy ATP, for example, glucose → glucose-6-phosphate → glucose, or fatty acids → triglyceride → fatty acids. The ability of islet hormones to regulate the key enzymes in such cycles (as discussed later) may be relevant to mechanisms that relate fuel turnover to fuel storage.

Insulin, glucagon, somatostatin, and *pancreatic polypeptide* are the four hormones currently known to be released by the *islets of Langerhans* in the pancreas. Each is synthesized, stored, and secreted by a different cell type. The islets are discrete bodies that are scattered throughout the pancreas and comprise 1% to 2% of its weight. They are more numerous in the tail than in the body or head of the organ. The adult human pancreas contains approximately 900,000 islets. The β-cells, which are the source of insulin, make up the bulk of the islet mass (60%) and in humans are located in the center of the islet. α-Cells, the source of glucagon, make up 25%, and δ-cells, the source of somatostatin, comprise 10% of the islet mass. These latter two cell types are located in the periphery of the islet. However, penetration of α- and δ-cells into the central core appears to occur in lobules. This arrangement allows for more zones in which β-, α-, and δ-cells may form a functional syncytium (Fig. 50-2). The cells responsible for pancreatic polypeptide release are concentrated in islets located in the head of the pancreas. The islets are heavily vascularized, and the capillary endothelial cells are fenestrated, which permits rapid exchange across them. Innervation to the islets is by both adrenergic and cholinergic fibers of the autonomic nervous system.

The endocrine cells of the islets appear to develop from pancreatic ducts, which are endodermal in origin. Although they show some characteristics similar to those of neuroectodermal cells of the neural crest, there are no conclusive data that islet cells first migrated from the neural crest to the foregut. In the course of evolution of the intestinal mucosa the β-cells were the first to agglomerate into discrete masses resembling islets in cyclostomes (for example, hagfish). Shortly thereafter the δ-cells began to shift into the islets. Not until the phylogenetic stage of cartilaginous fish did the α-cells migrate from their diffuse intestinal loci into the islets. Thus, the close anatomical association of these cells serves a relatively late evolutionary function. In the human fetus the islets are identified by 4 weeks of gestation. They are capable of synthesis and secretion of islet hormones by 10 weeks. However, responsiveness to stimulation is markedly blunted compared with that of adult islets, probably as a result of the stable supply of substrates from the mother.

The islet cells are arranged in cords along capillary channels. Their plasma membranes contain various specialized areas. Tight junctions are linear lines of fusion between plasma membranes of adjacent cells. These may serve to form intercellular compartments protected from the general interstitial fluid of the islets but permitting the secreted hormone of one cell to act on the membrane of its neighbor (Fig. 50-3). Gap junctions are bridging channels through adjacent plasma membranes that may permit direct passage of low molecular weight substances from the cytosol of one islet cell to the cytosol of its neighbor.

By electron microscopy islet cells are seen to contain secretory granules with smooth membranes within which hormone is stored. A system of microtubules is present, often lying in parallel bundles that separate linear rows of secretory granules. In addition, microfilaments containing myosin and actin form a web adjacent to the plasma membrane and in association with the microtubules. Agents that destroy microtubules or prevent their function (such as colchicine and deuterium oxide) inhibit hormone release, whereas agents that cause hypercontraction of microfilaments (such as cytochalasin B) enhance hormone release. Therefore it has been proposed that, when an islet cell is stimulated, secretory granules are actively transported to the plasma membrane, guided

■ *Anatomy of pancreatic islets*

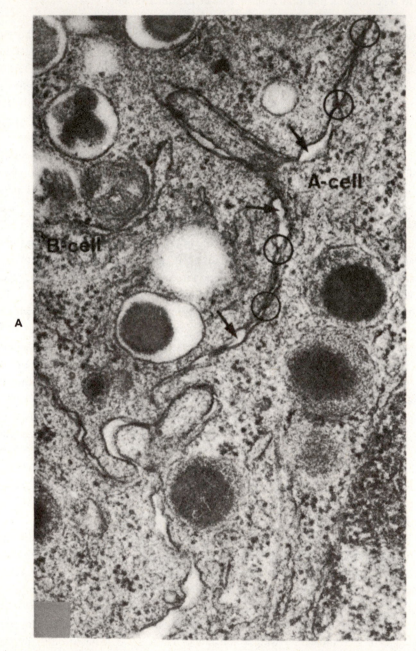

Fig. 50-3 ■ **A,** Electron microscopic photograph of adjacent β- and α-cells, showing tight junctions *(arrows)* and gap junctions *(circled areas)*. **B,** Further magnification of **A** showing the discrete structural characteristics of the tight junctions *(TJ)* and gap junctions *(GJ)*. (From Orci, L., et al.: Cell contacts in human islets of Langerhans, J. Clin. Endocrinol. Metab. **41:**841, 1975. Copyright 1975. Reproduced by permission.)

B

Fig. 50-3, cont'd ■ For legend see opposite page.

by the microtubules and pulled by the microfilaments. This has been most extensively studied by cinemicroscopy, which has shown movement of insulin-containing granules through the cytoplasm of β-cells at a rate of 1.5 μm/second.

Among the peptide hormones, insulin has been of preeminent historical, physiological, and clinical importance. It was the first, in 1921, to be isolated from animal sources in pure enough form to be administered therapeutically—and with dramatic life-saving effects. It was the first to have its amino acid sequence and tertiary structure elucidated. Insulin was the first peptide hormone for which a mechanism of action on the cell membrane was demonstrated. It was also the first to be measured by radioimmunoassay; indeed the entire concept of this landmark analytical method developed from studies of the metabolism of radiolabeled insulin in humans. The biosynthesis of peptide hormones from larger precursor molecules was observed initially in insulin. Finally, it was the first mammalian peptide hormone whose biosynthesis in bacteria by recombinant DNA technology was shown, in 1980, to result in a fully active product in humans. When one adds to this record the central role insulin plays in the rapid modulation of all major fuel fluxes, its importance to endocrinology can hardly be overstated.

■ *Insulin*
■ *Structure and synthesis*

Among the peptide hormones, insulin has been of preeminent historical, physiological, and clinical importance. It was the first, in 1921, to be isolated from animal sources in pure enough form to be administered therapeutically—and with dramatic life-saving effects. It was the first to have its amino acid sequence and tertiary structure elucidated. Insulin was the first peptide hormone for which a mechanism of action on the cell membrane was demonstrated. It was also the first to be measured by radioimmunoassay; indeed the entire concept of this landmark analytical method developed from studies of the metabolism of radiolabeled insulin in humans. The biosynthesis of peptide hormones from larger precursor molecules was observed initially in insulin. Finally, it was the first mammalian peptide hormone whose biosynthesis in bacteria by recombinant DNA technology was shown, in 1980, to result in a fully active product in humans. When one adds to this record the central role insulin plays in the rapid modulation of all major fuel fluxes, its importance to endocrinology can hardly be overstated.

Insulin is a peptide consisting of two straight chains. Its molecular weight is 6000. The A chain, containing 21 amino acids, and the B chain, containing 30 amino acids, are linked by two disulfide bridges. In addition, the A chain contains an intrachain

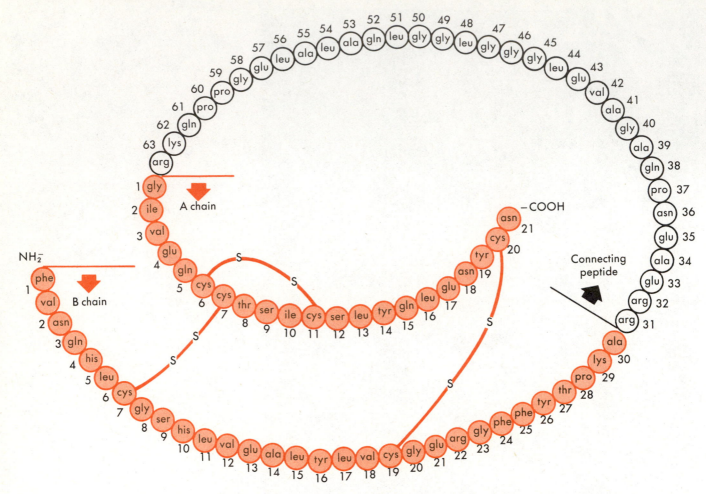

Fig. 50-4 ■ The structure of porcine proinsulin. The solid area is the insulin molecule released by cleavage of the connecting peptide. (Redrawn from Shaw, W.N., and Chance, R.E.: Effect of porcine proinsulin in vitro on adipose tissue and diaphragm of the normal rate, Diabetes **18:**737, 1968. Reproduced with permission from the American Diabetes Association, Inc.)

disulfide ring (Fig. 50-4). The multiple amino acid differences among insulins from various species change the biological activity little; mammalian and most fish insulins are virtually equipotent in humans. The features most conserved in vertebrate evolution are the positions of the three disulfide bonds, the N-terminal and C-terminal amino acids of the A chain, and the hydrophobic character of the amino acids at the C-terminal of the B chain. These areas determine secondary and tertiary structure, which are critical to biological activity. Insulin monomers (molecular weight 6000) readily form dimers. In the presence of zinc three of the latter in turn form a crystalline hexameric unit with a threefold axis passing through two zinc atoms. Crystalline zinc insulin is the basic pharmaceutical preparation of greatest importance in therapy.

Synthesis of insulin occurs as shown in Fig. 50-5. A preproinsulin of molecular weight 11,500 is first synthesized on the ribosomes at the direction of specific messenger RNA (mRNA). Preproinsulin may be thought of as consisting of four sequential peptides: an N-terminal prepeptide, the B chain of insulin, a connecting peptide, and the A chain of insulin (Fig. 50-5). The N-terminal 23 amino acid residue is rapidly cleaved at the site of synthesis to yield the proinsulin chain. The function of the evanescent prepeptide may be to bring the ribosomes into association with the membrane of the endoplasmic reticulum and permit guided passage of proinsulin to the Golgi apparatus.

*Whole body metabolism
and the hormones of the
pancreatic islets*

925

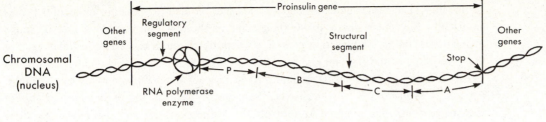

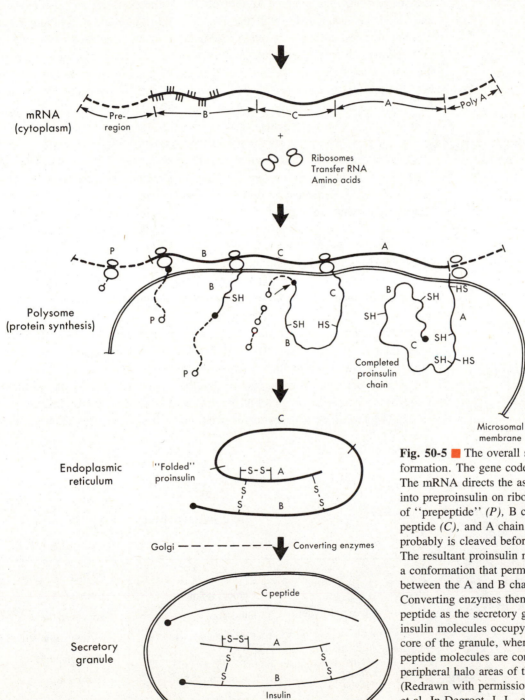

Fig. 50-5 ■ The overall sequence of insulin formation. The gene codes for preproinsulin. The mRNA directs the assembly of amino acids into preproinsulin on ribosomes in the sequence of "prepeptide" *(P),* B chain, connecting peptide *(C),* and A chain. The prepeptide probably is cleaved before the A chain is added. The resultant proinsulin molecule is folded into a conformation that permits the disulfide linkage between the A and B chains of insulin to form. Converting enzymes then cleave the connecting peptide as the secretory granule is formed. The insulin molecules occupy the electron dense core of the granule, whereas the connecting peptide molecules are concentrated in the peripheral halo areas of the granule (Fig. 50-3). (Redrawn with permission from Steiner, D.F., et al. In Degroot, L.J., et al.: Endocrinology, vol. 2, New York, 1979, Grune & Stratton, Inc.)

Establishment of the disulfide linkages yields the ''folded'' proinsulin molecule. Proinsulin has a molecular weight of 9000. It contains the disulfide bonded A and B chains of insulin linked to a connecting peptide (C peptide) through two basic residues each at the C terminal of the B chain and the N-terminal of the A chain (Fig. 50-4). During its transport from the endoplasmic reticulum and its packaging into granules by the Golgi apparatus, proinsulin is slowly cleaved. Separate trypsinlike and carboxypeptidase-like enzyme activities are required to split off the arg-arg and lys-arg residues (Figs. 50-4 and 50-5). The generated insulin and C-peptide molecules are each retained in the granules and released during secretion in equimolar amounts. The association of insulin with zinc takes place as the secretory granules mature. The zinc insulin crystals then form the dense central core of the granule, whereas C peptide is present in the clear space between the membrane and the core. The overall process of insulin synthesis is stimulated by glucose and feeding, which increase mRNA for preproinsulin. Release of insulin occurs by exocytosis (Chapter 49).

■ *Regulation of secretion*

A large number of factors modulate insulin secretion (Table 50-3). The major principle, however, is that *insulin secretion is stimulated under circumstances of fuel excess and is inhibited under circumstances of fuel deficiency*. Glucose is the stimulant of greatest importance in the human being. Since insulin in turn stimulates the utilization of glucose, this substrate-hormone pair forms a feedback system for close regulation of plasma glucose levels. The relationship between plasma insulin and plasma glucose is sigmoidal (Fig. 50-6). Virtually no insulin is secreted below a plasma glucose threshold of about 50 mg/dl. A half-maximum insulin secretory response occurs at a plasma glucose level of about 150 mg/dl and a maximum insulin response at levels of 300 to 500 mg/dl.

Both in vitro and in vivo insulin secretion exhibits a biphasic response to a continuous glucose stimulus (Fig. 50-7). Within seconds of exposure of the β-cells to an increased glucose concentration there is an immediate pulse of insulin release, peaking at 1 minute and then returning toward baseline. After 10 minutes of continuous stimulation a second phase of secretion begins with a slower rise of insulin to a second plateau, which can be maintained for many hours in normal individuals. The genesis of

TABLE 50-3

Insulin secretion

Stimulators		Inhibitors
D-Glucose	Secretin	Fasting
Galactose		
Mannose	Cholecystokinin	Exercise
Glyceraldehyde	Glucagon	Somatostatin
Protein	Enteroglucagon	α-Adrenergic stimuli
Arginine		
Lysine	Calcium	Prostaglandins
Leucine	Potassium	Diazoxide
Alanine		
Free fatty acids	Vagal activity	Phenytoin
Ketoacids	β-Adrenergic stimuli	
Gastric inhibitory	Acetylcholine	
polypeptide	Sulfonylurea drugs	
Gastrin		

*Whole body metabolism
and the hormones of the
pancreatic islets* **927**

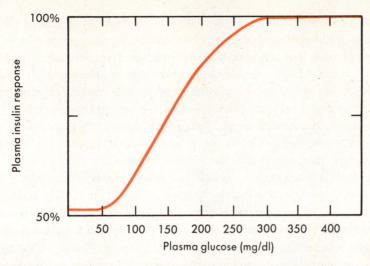

Fig. 50-6 ■ Approximate in vivo relationship between plasma glucose and insulin secretion, the latter being assessed by the plasma insulin response to stepwise infusion of glucose in humans. No insulin is secreted below a plasma glucose level of 50 mg/dl. Half maximum secretion occurs at 125 to 150 mg/dl. (Redrawn from Karam, J.H., et al.: "Staircase" glucose stimulation of insulin secretion in obesity; measurement of beta cell sensitivity and capacity, Diabetes **23:**763, 1974. Reproduced with permission from the American Diabetes Association, Inc.)

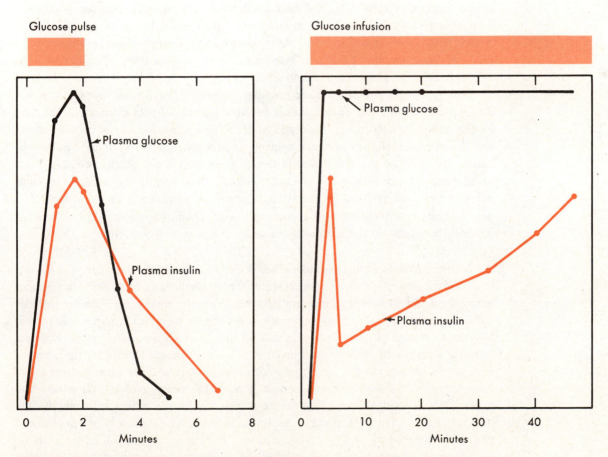

Fig. 50-7 ■ In vitro insulin response to a glucose pulse *(left panel)* shows no β-cell memory. If glucose infusion is continued *(right panel),* the initial burst of insulin is not sustained, but a second wave of secretion develops soon after.

this biphasic response remains controversial. Although initially described for insulin, it also is seen with stimulation of other peptide hormones by appropriate agents. A number of explanations have been suggested: (1) two storage compartments exist—labile and stable—that contain granules with different sensitivities to glucose; (2) the second phase of insulin secretion represents newly synthesized insulin resulting from glucose stimulation of the synthetic process; (3) a feedback inhibitor of insulin release is rapidly generated after glucose stimulation and then slowly removed. Whatever the mechanism, it is important to note that loss of the initial rapid phase of insulin secretion is the most characteristic feature of insulin-deficient diabetes mellitus.

When glucose is given orally, a greater insulin response is elicited than when plasma glucose is comparably elevated by intravenous administration. This is accounted for by release of a number of gastrointestinal hormones that normally accompany the digestive process and that are capable of potentiating glucose-stimulated insulin secretion. Gastric inhibitory polypeptide (GIP) is probably the most important of these insulinogogues, but in high concentrations gastrin, secretin, cholecystokinin, pancreatic glucagon, and enteroglucagon all share this property. This prompt gastrointestinal mechanism of insulinogenesis moderates the early rise in plasma glucose that follows the ingestion and absorption of a carbohydrate meal.

Insulin secretion also is stimulated by oral protein; this is mediated by the amino acids resulting from digestion of the protein. The basic amino acids, arginine and lysine, are the most potent stimulants; leucine, alanine, and others contribute modestly to this effect. Glucose and amino acids are synergistic in their actions, so that the rise in insulin which follows a meal represents more than the additive effect of its carbohydrate and protein content. Triglycerides and fatty acids have little, if any, stimulatory effect in humans but may contribute more to sustaining insulin secretion in other species. Ketoacids at concentrations that prevail during prolonged fasting stimulate insulin secretion modestly; however, this effect may help sustain a critical low level of insulin when β-cell stimulation by glucose and amino acids is reduced.

Both potassium and calcium are essential for normal insulin responses to glucose. Thus relative insulin deficiency occurs in subjects depleted of potassium or calcium, and insulin excess is seen in hypercalcemic subjects. Sympathetic nervous system activity and the adrenal medullary hormone epinephrine stimulate insulin secretion via their β-adrenergic receptors but inhibit insulin secretion via their α-adrenergic receptors. Parasympathetic activity also increases insulin release, as shown by stimulating the vagus nerve or by exposing β-cells to acetylcholine in vitro. The neural routes of regulation may be important in mediating a cephalic phase of insulin secretion, which in animals has been demonstrated to immediately precede the entrance of food into the gastrointestinal tract.

A large number of other hormones indirectly produce hyperplasia of the β-cells and lead to chronic increases in insulin secretion. These hormones largely do so by antagonizing the action of insulin or increasing its need by peripheral tissues. The list includes cortisol, growth hormone, estrogen-progesterone, human placental lactogen, and thyroid hormones. A class of drugs known as sulfonylureas, which are useful in the treatment of some patients with diabetes mellitus as well as in diagnostic studies for the presence of insulin-secreting tumors, acutely stimulates insulin release. Of great interest is the presence within the islets of somatostatin, a powerful inhibitor of insulin release that may exert an important local effect on the β-cell. Insulin secretion also is inhibited by prostaglandins. A feedback effect of insulin on its own secretion has been demonstrated that is independent of its hypoglycemic action. During an insulin infusion plasma C-peptide levels (representing β-cell activity) will decrease even though plasma glucose is held constant by concomitant glucose administration.

The net result of these many physiological influences is to maintain in normal hu-

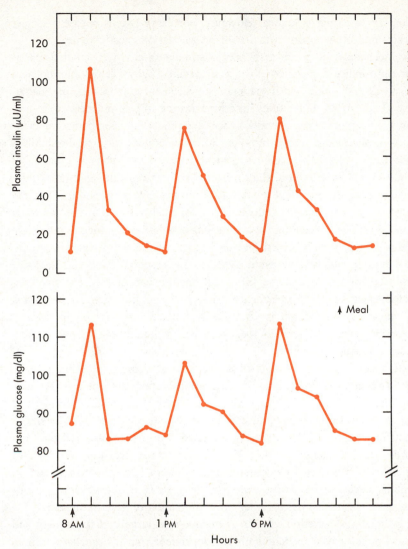

mans an average basal peripheral plasma insulin level of 10 μU/ml (7×10^{-11}M). Under conditions of total fasting this value declines 50%, largely because of a decrease in plasma glucose. A similar decline also occurs during prolonged exercise. Plasma insulin increases three- to tenfold after a typical meal, usually peaking 30 to 60 minutes after eating is initiated (Fig. 50-8). A larger prandial pulse of insulin secretion is seen with breakfast than with subsequent meals of identical carbohydrate and calorie content. Because the pancreatic vein drains into the portal vein, the liver is regularly exposed to insulin concentrations two to three times higher than those of other organs in the basal state and five to ten times higher after acute β-cell stimulation.

Although C peptide is secreted in amounts that are equimolar to insulin, its basal peripheral plasma levels are approximately fivefold higher, averaging 1 ng/ml (3×10^{-10}M). This difference is caused by a lower rate of metabolic clearance for C peptide. In contrast to insulin, C peptide undergoes no significant hepatic extraction. Despite its lack of biological activity, plasma C-peptide measurements provide useful information about β-cell function. This is particularly so when a small increase in insulin secretion may not be detected by a change in peripheral plasma levels, because of its hepatic extraction, or when measurement of plasma insulin is rendered unsatisfactory by prior insulin treatment. Proinsulin also is released by the β-cell and forms up to 20% of total insulin immunoreactivity in the basal state. During normal β-cell stimulation absolute proinsulin levels increase more slowly and to a lesser degree than does insulin. No biological function for proinsulin is known. However, neoplastic β-cells release a greater proportion of proinsulin, a phenomenon that is of occasional diagnostic value.

The intimate mechanism of β-cell response to stimulants has been studied extensively. Thus far only indirect evidence exists for specific D-glucose or L-amino acid receptors on the β-cell membrane. Other evidence suggests that glucose must be metabolized within the β-cell to an unidentified intermediate for full stimulation of insulin release to occur. For example, D-glyceraldehyde—a three-carbon compound metabolized like glucose—is an equally potent stimulant. Mannoheptulose, a sugar that inhibits glucose phosphorylation, and 2-deoxyglucose, an analogue that is phosphorylated but blocks glucose utilization, both inhibit glucose-stimulated insulin release. Intracellular cyclic AMP levels increase during glucose-stimulated insulin secretion, and this nucleotide also potentiates the response to glucose. The electrical activity of the β-cell is altered by glucose, which in stimulating concentrations depolarizes the membrane. Most important, calcium is rapidly translocated into the cytosol of the β-cell during stimulation; this translocation is enhanced by cyclic AMP and is essential for β-cell response to occur. It has been postulated that the influx of calcium causes contraction of the microtubular-microfilament system. The latter then moves secretory granules to the plasma cell membrane, where they fuse with it and extrude insulin and C peptide by exocytosis. The fused membrane material returns to the cytoplasm as microvesicles and can be either recycled or degraded.

Once secreted, insulin normally circulates unbound to any carrier protein. Its half-life in plasma is 5 to 8 minutes, its volume of distribution is about 20% of body weight, and its metabolic clearance rate is 800 ml/minute (1200 L/day). In humans estimates of basal insulin delivery rate to the peripheral circulation are about 0.5 to 1.0 units/hour (20 to 40 μg/hour).* During meals delivery rate increases up to tenfold, and the total daily peripheral delivery of insulin is about 30 units. Since the liver removes 50% of portal vein insulin on the first pass, the actual β-cell secretory rate is approximately 60 units/day. The initial hepatic extraction is decreased by glucose administration, permitting a greater escape of insulin to the periphery for stimulation of glucose uptake.

Once in the peripheral circulation, insulin is metabolized largely in the kidney and liver by a specific protease and by a glutathione-dependent transhydrogenase, which splits the disulfide bonds, producing separate A and B chains. Very little insulin is excreted unchanged in the urine. Some degradation of insulin also occurs in association with its plasma membrane receptor in target cells. In patients who are treated chronically with animal insulins, circulating antibodies to insulin are produced. A majority of the plasma insulin content is bound to these antibodies, which can influence the rate of availability of the hormone to target cells and degradative processes.

■ Hormone actions: intracellular mechanisms

Our present understanding of insulin's many actions is based in part on observations following in vivo administration of the hormone, in part on changes that occur after selective withdrawal of the hormone by inhibition or destruction of the β-cells, and, to a lesser degree, on direct in vitro demonstrations of hormone effects. Therefore it should be emphasized that certain well-defined consequences of insulin action may be, in the final analysis, only indirectly attributable to the hormone. For example, changes in substrate or cofactor levels in one pathway that is directly regulated by insulin in turn may affect critically the activities of enzymes and the resultant rate of flux through another pathway.

The action of insulin is initiated by reversible association with its specific plasma cell membrane receptor. The latter appears to be a glycoprotein with a molecular weight of around 350,000, composed of symmetrical α- and β-subunits linked by sulfhydryls

---

*1 mg crystalline zinc insulin equals 25 units of biological activity.

*Whole body metabolism
and the hormones of the
pancreatic islets* **931**

in a structure reminiscent of immunoglobulins. Binding to receptor occurs at insulin concentrations well within the physiological range of 10 to 150 μU/ml, with a half-maximum level observed at the upper end of that range. For certain actions of the hormone full biological activity is expressed with only 5% to 10% of the receptor sites occupied; therefore spare receptors exist. Insulin receptors appear to be grouped, rather than randomly distributed, on the cell surface, and they also exhibit mobility within the plane of the plasma membrane. The number and affinity of insulin receptors are capable of modulation. Insulin itself down regulates the number of receptors. For example, obese humans and animals, who have elevated plasma insulin levels, have reduced numbers of receptors. When their plasma insulin levels are lowered by fasting, the number of insulin receptors increases. This effect on its own receptors is a direct action of insulin which can be shown in vitro in tissue culture.

Some intracellular actions of insulin are demonstrable within seconds or minutes. Examples of such rapid effects include stimulation of a membrane glucose transport system and the activation or deactivation by covalent modification of certain enzymes involved in glucose metabolism. One concept for which there is growing experimental support is that insulin stimulates the dephosphorylation of a number of enzymes. This action may be mediated by a plasma membrane product that has been shown to be generated by the interaction of insulin with its receptor. This product of currently unknown structure is then thought to act as a second messenger which modulates the actions of specific phosphoprotein phosphatases. The latter in turn could modulate the activities of regulatory enzymes in various pathways by dephosphorylating them. To date, the best studied example of this proposed mechanism of rapid insulin action is the activation of pyruvate dehydrogenase. Other key enzymes that may be activated by insulin in such a manner include glycogen synthase, pyruvate kinase, and acetyl CoA carboxylase. Conversely, the enzyme phosphorylase may be rapidly inactivated by insulin in this manner. These acute effects of insulin are of particular regulatory importance when eating causes abrupt changes in nutrient supply, especially that of carbohydrate.

Other effects of insulin develop more slowly—over hours to days—and are associated with increases or decreases in the actual tissue concentrations of enzymes and other macromolecules such as RNA. It is now well documented that some of the insulin–plasma membrane receptor complex becomes internalized after its formation. Furthermore insulin binding to the Golgi apparatus, to the endoplasmic reticulum, and to nuclei has been reported. These observations suggest routes whereby insulin—directly or indirectly—may cause changes in translation or transcription. For example, mRNA levels for the enzyme phosphoenolpyruvate carboxykinase are decreased by insulin, whereas mRNA levels for albumin are increased by the hormone. Thus insulin also may have tonic effects on metabolism that wax and wane gradually in adjustment to slowly developing changes in the milieu, as occurs with prolonged fasting. However, in many instances the rapid and slow effects of insulin reinforce each other in that they produce the same directional change in flow of an important metabolite, such as glucose.

■ *Actions on flow of fuels*

Insulin is the hormone of abundance. When the influx of nutrients exceeds concurrent energy needs and rates of anabolism, the insulin secreted in response permits efficient storage of the excess while suppressing mobilization of endogenous substrates. The stored nutrients then can be made available during subsequent fasting periods to maintain glucose delivery to the central nervous system and free fatty acid delivery to the muscle mass and viscera. The major targets for insulin action are the liver, the adipose tissue, and the muscle mass. Fig. 50-9 displays the overall flow of substrates produced by insulin, and Fig. 50-10 the important metabolic control points where in-

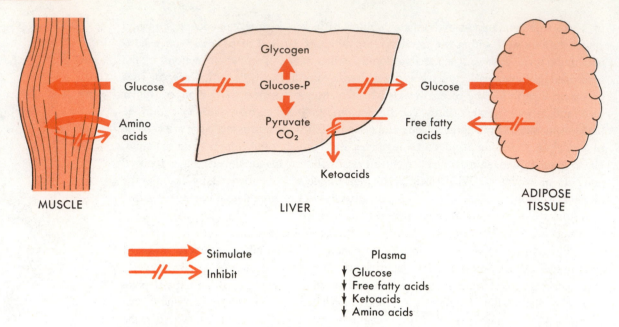

Fig. 50-9 ■ Effect of insulin on the overall flow of fuels results in tissue uptake and sequestration of glucose, fatty acids, and amino acids with a resultant decrease in their plasma levels.

sulin is currently thought to act. The numbered loci in Fig. 50-10 are referred to in the text that follows.

Carbohydrate metabolism. Insulin stimulates glucose utilization and storage and simultaneously inhibits glucose production. Therefore insulin either lowers the basal circulating glucose concentration or limits the rise in plasma glucose that results from a dietary carbohydrate load. This is accomplished by a number of insulin effects:

1. In the *liver*, insulin increases glucose uptake and its storage in the form of glycogen. Insulin facilitates uptake of glucose by inducing hepatic glucokinase *(4)* and promotes storage of glucose by rapidly activating the glycogen synthase enzyme complex *(2)*. Insulin reduces glucose output, immediately by inhibiting glycogenolysis through decreasing glycogen phosphorylase activity *(1)* and gradually by decreasing glucose-6-phosphatase levels *(3)*. Insulin also inhibits gluconeogenesis. This is accomplished by decreasing the availability and the hepatic uptake of amino acids *(15, 16, 17)* and by decreasing the levels and/or activities of pyruvate carboxylase *(9)*, phosphoenolpyruvate carboxykinase *(10)*, and fructose-1,6-diphosphatase *(6)*. Furthermore insulin stimulates glycolysis through increasing the activities of phosphofructokinase *(5)*, pyruvate kinase *(7)*, and pyruvate dehydrogenase *(8)*. As a result of the latter two effects, the supply of phosphoenolpyruvate for gluconeogenesis is diminished.

Many of these hepatic effects of insulin require concurrent administration of glucose and may be reinforced by generation of products of glucose metabolism. Some also may be mediated or at least augmented by other consequences of insulin action. For example, since insulin decreases free fatty acid delivery to the liver *(21)*, β-oxidation and generation of intramitochondrial acetyl CoA are decreased; by increasing the conversion of extramitochondrial acetyl CoA to malonyl CoA through activation of acetyl CoA carboxylase *(12)*, insulin further decreases hepatic levels of acetyl CoA. Since acetyl CoA is an important allosteric activator of pyruvate carboxylase, the activity of this gluconeogenic enzyme is thus indirectly lowered by insulin. Finally, as noted later, virtually all these hepatic actions of insulin are directly countered by glucagon. Since insulin also inhibits the secretion of this antagonist, the effect of increasing insulin levels on glucose

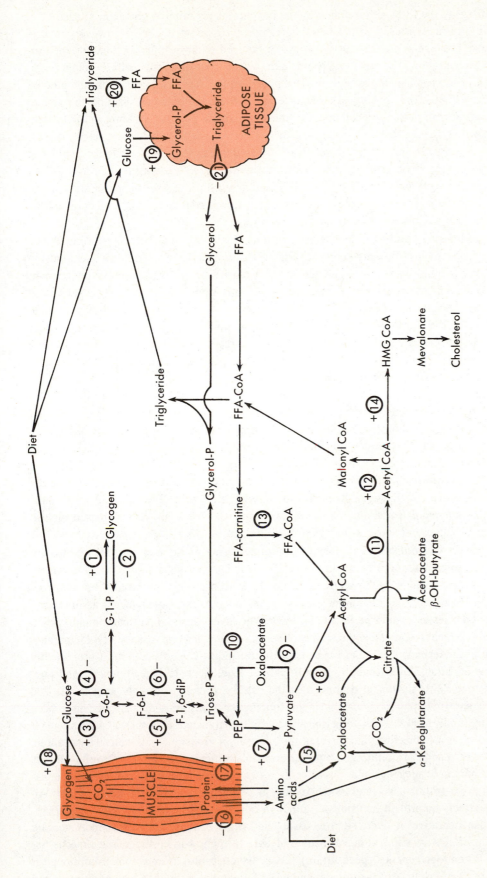

LIVER

Fig. 50-10 ■ A more detailed overview of metabolic reactions that are stimulated (+) or inhibited (−) by insulin. In some instances the mechanism involves activation or deactivation of enzymes, in others induction or repression. The central focus of the figure is insulin action in the liver.

+ Increases
− Decreases

handling by the liver are amplified by the concurrent restraint of glucagon release.

If insulin is infused in the basal state, plasma glucose declines; however, hepatic glucose production recovers from its insulin-induced nadir despite continuation of insulin administration. This recovery begins after a short time and reflects three counterregulatory processes: (1) an autoregulatory response of the liver to a lowered plasma glucose concentration, (2) a sensing of hypoglycemia in the hypothalamus, leading to stimulation of glucose output via activation of the sympathetic nervous system, and (3) a sensing of hypoglycemia by the α-cells of the pancreatic islets, which overcomes the inhibition by insulin and causes secretion of its antagonist, glucagon. Thus the actions of insulin on the liver—so important for assimilation of an excess of glucose—are not permitted to go unchecked for very long should insulin increase inappropriately in the absence of exogenous glucose. These mechanisms are essential to protection of the central nervous system from glucose deprivation.

2. In *muscle* insulin stimulates the transport of glucose into the cell by a membrane carrier system in a process known as *facilitated diffusion (18)*. The exact relationship between the insulin receptor and the glucose carrier in the membrane is not known. Some animal evidence suggests that insulin not only rapidly activates the transport system but also may gradually increase the number of carrier units. Even though a small portion of the translocated glucose subsequently undergoes glycolysis and oxidation, insulin specifically directs a major fraction into storage as muscle glycogen by activating glycogen synthase. The increased stores of muscle glycogen then are available for oxidative support of short-term exercise.

3. In *adipose tissue* insulin also stimulates the transport of glucose into the adipose cells in a manner analogous to its effect on muscle *(19)*. This glucose is then metabolized to α-glycerophosphate, which is used in the esterification of fatty acids, permitting their storage as triglycerides. On subsequent lipolysis the glycerol that is released can then serve as a gluconeogenic precursor in the liver.

The overall results of insulin deprivation on carbohydrate metabolism therefore are depleted stores of glycogen in liver and muscle, high basal levels of glucose in the circulation, retarded assimilation of an exogenous glucose load, and increased mobilization of glucose precursors from peripheral tissues.

Protein metabolism. Insulin actions on protein and amino acid metabolism also are directed toward sequestration, that is, anabolism. During the assimilation of a protein meal the increase in insulin secretion limits the rise in plasma amino acid levels, especially of the branched chain amino acids valine, leucine, and isoleucine. If administered in the basal state, insulin preferentially lowers the levels of these particular amino acids. In *muscle* insulin stimulates the uptake of certain amino acids across the cell membrane, an action independent of its effect on glucose transport. This effect is most prominently on the essential amino acids, especially valine, leucine, isoleucine, tyrosine, and phenylalanine *(17)*. Insulin also stimulates the rate of protein synthesis in general. The mechanism involves modification of the function of ribosomes that already have been charged with mRNA, so as to permit more efficient translation of the RNA message. It is not clear whether insulin increases the number of ribosomes as well. The anabolic effects of insulin are reinforced by the hormone's anticatabolic effects. Insulin inhibits proteolysis and suppresses the release of the essential branched chain and aromatic amino acids *(16)*. Insulin also inhibits the oxidation of the branched chain amino acids. All these effects on protein metabolism, although quantitatively most important for muscle, are also operative in liver, adipose tissue, and other organs. Of great importance are similar effects on stimulation of macromolecule formation in cartilage and osseous tissue. Taken together, this spectrum firmly establishes insulin as an essential hormone for growth. The insulin-deprived young animal or human has a reduced lean body mass and is retarded in height and maturation.

It should be noted that certain other peptides and proteins, variously called insulin-like growth factors (IGF-1 and IGF-2), somatomedins, multiplication-stimulating activity, nerve growth factor, and epidermal growth factor, are derived from diverse animal and human tissues and circulate in plasma. These display some of the aforementioned anabolic actions of insulin, have certain similar amino acid sequences, and cross-react modestly with insulin receptors. However, they have distinct plasma membrane receptors of their own. In general, these peptides have less effect on carbohydrate and fat metabolism. Their endocrine status and exact role in human physiology are under active investigation.

Fat metabolism. The metabolism of both endogenous and exogenous fat is profoundly influenced by insulin. The net overall effect is to enhance storage and to block mobilization and oxidation of fatty acids. Thus insulin lowers the circulating levels of free fatty acids and ketoacids within minutes of administration and subsequently reduces the level of triglycerides.

In *adipose tissue* the deposition of fat is stimulated by insulin in several ways. Exogenous triglycerides from a meal and endogenous triglycerides produced by the liver circulate as lipoprotein complexes, which cannot directly enter fat cells. They first must be split within the capillary endothelium by the enzyme lipoprotein lipase released from adjacent adipose cells. The resultant free fatty acids associated with albumin then can enter the adipose tissue cells. Insulin induces this key lipoprotein lipase *(20)*. Once in the adipose cell the free fatty acids are converted to triglycerides by esterification with α-glycerophosphate. The latter must be generated from glucose transported into the adipose cell under insulin stimulation *(19)*, since this cell lacks the glycerol kinase necessary to convert free glycerol to its phosphate ester. In addition, glucose itself is converted to fatty acids by insulin activation of glycolysis, pyruvate dehydrogenase, and acetyl CoA carboxylase. Although this is a very minor quantitative pathway in the adipose tissue of humans, it is of considerable significance for storage of carbohydrate as fat in rodents.

Most important, insulin profoundly inhibits hormone-sensitive adipose tissue lipase activity *(21)*. It may do this by decreasing the levels of cyclic AMP and thereby inhibiting cyclic AMP–dependent protein kinase. By suppressing lipolysis of stored triglycerides and the release of free fatty acids, insulin lowers their rate of delivery to the liver and to peripheral tissues. A major consequence of decreasing free fatty acid supply to the liver is a marked reduction in the generation of ketoacids. In addition, insulin stimulates the utilization of ketoacids by the peripheral tissues. Thus insulin is the major and perhaps the sole *antiketogenic* hormone.

In the *liver* insulin is also antiketogenic and lipogenic in its effects. Free fatty acids entering from the circulation are shunted away from β-oxidation and from ketogenesis. They are instead reesterified with glycerophosphate, derived either from insulin-stimulated glycolysis or from glycerol via the enzyme glycerophosphate kinase. The rate-limiting step in the de novo synthesis of free fatty acids from acetyl CoA appears to be its conversion to malonyl CoA by acetyl CoA carboxylase *(12)*. As indicated previously, the activity of this cytosolic enzyme is increased by insulin. Therefore, when glucose is available and glycolysis is stimulated by insulin, acetyl CoA generated in the mitochondria from pyruvate is incorporated into fatty acids under the influence of insulin after transfer to the cytoplasm. Furthermore insulin increases the activity of the hexose monophosphate shunt. This generates the supply of reduced triphosphopyridine nucleotide (TPNH) needed for fatty acid synthesis. The fact that insulin increases the formation of malonyl CoA from acetyl CoA may mediate in part insulin's intrahepatic antiketogenic action. Malonyl CoA inhibits carnitine acyltransferase *(13)*, the enzyme system responsible for transferring free fatty acids from the cytoplasm into the mitochondria for oxidation and conversion to ketoacids. Finally, insulin also favors hepatic sequestration of cholesterol by activating hydroxymethylglutaryl CoA reductase *(14)*, the rate-limiting step in cholesterol synthesis. Thus the net effect of insulin on lipid metabolism in the liver is to decrease free fatty acid oxidation and ketogenesis, to increase the synthesis of free fatty acids and cholesterol, and to increase the storage and decrease the release of triglycerides and cholesterol.

In summary, the insulin-deprived animal or human has high circulating levels of free fatty acids, ketoacids, cholesterol, and triglyceride and has diminished adipose tissue lipid stores. Whether hepatic lipid content is increased or decreased depends on the relative rates of augmentation of free fatty acid delivery versus free fatty acid oxidation. In the previously well-nourished subject the former may overwhelm the latter and lead to excess hepatic accumulation of triglyceride.

Other actions. Both the storage of glucose as glycogen and protein anabolism require concomitant cellular uptake of potassium, phosphate, and magnesium. Insulin stimulates translocation of all three of these minerals into muscle cells and of at least potassium and phosphate into the liver. The facilitation of muscle potassium uptake is mediated by an insulin action on membrane polarization that is independent of the hormone's effect on glucose transport. Insulin secreted in response to a carbohydrate load causes significant declines in serum potassium, phosphate, and magnesium, whereas inhibition of basal insulin secretion (by somatostatin) causes a rise in serum potassium. Thus insulin is considered to be one of the normal regulators of potassium balance. Another effect of insulin on electrolyte balance is to increase reabsorption of potassium, phosphate, and sodium by the tubules of the kidney. These renal effects contribute to anabolism by conserving vital intracellular electrolytes, as well as sodium, which is necessary for formation of the additional extracellular fluid required when lean body mass is being expanded.

The overall consumption of glucose by the central nervous system is independent of insulin. However, there is growing evidence that selected areas of the brain, in particular in the hypothalamus, are insulin responsive. Radioactively labeled insulin binds to hypothalamic cells or to the adjacent capillary endothelium, and insulin receptors can be demonstrated in hypothalamic plasma membranes. The glucose analogue, gold thioglucose, which selectively destroys the ventromedial nucleus (the satiety center), requires insulin to act. And, interestingly, insulin administration initially increases the firing rate of single neurons from the same area of the hypothalamus before the suppressive effect of insulin-induced hypoglycemia appears. All these data suggest that insulin, secreted in response to carbohydrate, also may facilitate a hyperglycemic feedback signal to the hypothalamus, which then shuts off hunger by stimulating the activity of the satiety center.

The insulin sensitivity of tissues relates well to prevailing plasma insulin levels in various physiological states. Fuel mobilization, that is, lipolysis, ketogenesis, and proteolysis, is significantly but only partially inhibited at basal peripheral plasma insulin concentrations of 10 μU/ml. This permits a regulated increase in substrate flow during the daily nocturnal fasting period. Shortly after eating, as concentrations rise to 20 to 30 μU/ml, hepatic glucose output becomes totally suppressed. With a further increase to peak postprandial levels of 50 to 100 μU/ml, glucose and amino acid uptake by peripheral tissues is strongly stimulated, facilitating their storage at a time of abundance. Under maximum insulin stimulation glucose utilization increases from 2 to 12 mg/kg/minute. Thus the β-cell responds to the physiological need of the moment with insulin delivery rates that provide appropriate hormone concentrations for regulating substrate fluxes.

■ *Glucagon*
■ *Structure and synthesis*

Glucagon is a single straight-chain peptide hormone of 29 amino acids and molecular weight 3500. The amino acid composition and sequence are identical in mammalian glucagons thus far studied, including the human, porcine, and bovine hormones. Even in lower species, amino acid substitutions are infrequent, and both the structure and

*Whole body metabolism
and the hormones of the
pancreatic islets* **937**

function of glucagon appear highly conserved in evolution. The N-terminal residues 1 to 6 are essential for receptor binding and biological activity.

Glucagon is synthesized by α-cells from a preprohormone with a molecular weight of 18,000. A prohormone of 100 amino acids and molecular weight of 12,000, known as glycentin, contains both N-terminal and C-terminal extensions from the glucagon sequence. Glycentin is localized to the peripheral halo of the secretory granule, whereas glucagon is localized to its dense core. Further details of the cleavage process are lacking, but release of the hormone from its secretory granule is accomplished by exocytosis. The biosynthesis of glucagon is stimulated when glucose concentrations are low and is progressively diminished as glucose concentrations are raised. In humans glucagon is found exclusively in the pancreatic islets. However, in other species, such as the dog, glucagon also is produced by α-like cells scattered in the walls of the stomach. In addition, glucagon-like peptides, including glycentin, are synthesized and secreted by small intestinal cells in humans and other species.

The most important principle governing glucagon secretion appears to be the maintenance of normoglycemia. In contrast to insulin, glucagon is secreted in response to glucose lack and acts to increase circulating glucose levels. Basal plasma glucagon concentration in humans averages 100 pg/ml (3×10^{-11}M). Hypoglycemia causes a two- to fourfold increase in these levels, whereas hyperglycemia lowers them approximately 50%. That glucose directly regulates α-cell secretion is shown by in vitro studies. However, the effect of glucose appears to be modulated by insulin. Glucagon secretion is stimulated more by low glucose levels if insulin is absent. Conversely, the presence of insulin potentiates the suppressive effect of high glucose levels on the α-cell. Neither the exact mechanism through which glucose is directly sensed by the α-cell nor how it is modulated by insulin is known. In general, the specificity of α-cell responses to glucose analogues, metabolites, and blockers of intracellular glucose metabolism is similar to that exhibited by β-cells.

Glucagon secretion also is stimulated by a protein meal and, most powerfully, by amino acids such as arginine and alanine. However, the α-cell response to protein is greatly dampened if glucose is administered concurrently. This interaction is partly via insulin; glucagon responses to amino acids are restrained by insulin excess and augmented by insulin deficiency. In species other than humans a triglyceride meal also can be shown to stimulate glucagon release modestly. However, in humans and animals free fatty acids—like glucose—exert a suppressive effect, whereas glucagon secretion increases following a rapid decline in plasma free fatty acids. There is also evidence that glucagon responses to orally ingested nutrients (as opposed to their intravenous delivery) may be reinforced by the release of gastrointestinal glucagon secretagogues. The sum of all these individual influences is that the ingestion of ordinary meals in humans produces much less variation in plasma glucagon than in plasma insulin levels. This is at least partly explained by the offsetting effects of the carbohydrate and protein portions of the meal on the α-cell, as compared with the synergistic effects of these nutrients on the β-cell. It is also in keeping with the secondary role of glucagon versus the primary role of insulin in the disposal of exogenous nutrients.

Fasting for 3 days increases plasma glucagon twofold; however, the mechanism may involve a reduction in the metabolic clearance of glucagon, as well as a possible increase in secretion. Exercise of sufficient intensity and duration also increases plasma glucagon. Neural mechanisms may mediate some of these α-cell responses. In particular vagal stimulation and acetylcholine acutely increase glucagon secretion. Of importance to the physician is that a variety of stresses, including infection, toxemia, burns, tissue infarction, and major surgery, all increase glucagon secretion promptly. This phenomenon is probably mediated by the adrenergic nervous system via sympathetic outflow

■ *Regulation of secretion*

from the hypothalamus to α-adrenergic receptors in the β-cells. The excess of glucagon often leads to clinically significant hyperglycemia. The neurohormone somatostatin inhibits the secretion of glucagon, as it does that of insulin. Finally, various other protein hormones (growth hormone, human placental lactogen) and steroid hormones (cortisol, estrogen-progesterone) affect glucagon secretion indirectly.

Monomeric glucagon circulates essentially unbound in plasma. It has a half-life of 6 minutes and a metabolic clearance rate of about 600 ml/minute. The daily secretion rate of glucagon is estimated to be 100 to 150 μg. A portion of glucagon immunoreactivity is caused by higher molecular weight substances, one of which resembles a proglucagon; the physiological significance of these fractions is unclear. The ratio of portal vein to peripheral vein glucagon concentrations is about 1.5 in the basal state. In contrast to insulin, only 25% of glucagon is extracted by the liver on a single passage. The kidney and liver are the major loci of glucagon degradation. Less than 1% of the glucagon filtered by the glomerulus is excreted in the urine.

■ *Hormone actions*

In almost all respects the actions of glucagon are exactly opposite to those of insulin. Glucagon promotes mobilization rather than storage of fuels, especially glucose (Fig. 50-11). Both hormones act at numerous similar control points in the liver. Therefore reference is made to the numbered loci of Fig. 50-10 in the following discussion.

Glucagon binds to plasma membrane receptors in the liver. The glucagon-receptor complex causes a rapid increase in intracellular cyclic AMP (Chapter 49). A specific enzymatic cascade ensues. Protein-phosphokinase catalytic activity increases, converting inactive phosphorylase kinase to active phosphorylase kinase. The latter then converts inactive phosphorylase to active phosphorylase. A number of other enzymes important to glucose and fat metabolism, whose functional state is dependent on interconversion of phosphorylated and dephosphorylated forms, also are regulated by glucagon. These include phosphofructokinase *(5),* pyruvate kinase *(7),* acetyl CoA carboxylase *(12),* and hydroxymethlyglutaryl CoA reductase *(14).* In these instances, however, the catalytic activity of the enzyme is decreased by the phosphorylation that results from glucagon action.

Unequivocally the most important mission of glucagon is to promote and sustain hepatic glucose output. Thus the dominant effect of glucagon is in the liver. Its actions on adipose tissue and muscle appear to be of minor significance unless insulin is vir-

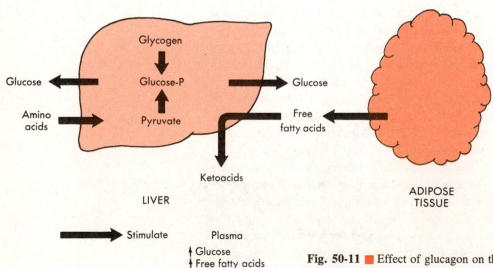

Fig. 50-11 ■ Effect of glucagon on the overall flow of fuels results in tissue release of glucose, fatty acids, and ketoacids into the circulation and hepatic uptake of amino acids for gluconeogenesis.

tually absent. Glucagon exerts an immediate and profound glycogenolytic effect through activation of glycogen phosphorylase *(1)*. The glucose-1-phosphate released is prevented from undergoing resynthesis to glycogen by a simultaneous inhibition of glycogen synthase *(2)*. Glucagon also stimulates gluconeogenesis by at least several mechanisms. The hepatic extraction of amino acids, especially alanine, is increased. The activities of the gluconeogenic enzymes, pyruvate carboxylase *(9)*, phosphoenolpyruvate carboxykinase *(10)*, and fructose-1,6-diphosphatase *(6)* are increased, whereas that of the glycolytic enzyme pyruvate kinase *(7)* is decreased.

The crucial importance of glucagon to the maintenance of basal hepatic glucose output is shown by the marked decline of 75% that follows the selective inhibition of glucagon secretion with somatostatin. The powerful glycogenolytic and hyperglycemic action of glucagon is exhibited at plasma hormone concentrations of 150 to 500 pg/ml and occurs even in the presence of insulin levels somewhat above basal levels (20 to 30 μU/ml). However, this action is transient; the acute initial stimulation of hepatic glucose output, which occurs during glucagon administration, wanes after about 30 minutes, probably because of counterregulatory influences. Glucagon has little or no influence on glucose utilization by peripheral tissues. Thus hyperglucagonemia has no effect on the plasma glucose levels that are generated by an exogenous glucose load as long as the insulin response is normal.

Another intrahepatic action of glucagon is to direct incoming free fatty acids away from triglyceride synthesis and toward β-oxidation. Thus glucagon is a ketogenic as well as a hyperglycemic hormone. Recent evidence suggests that glucagon inactivates acetyl CoA carboxylase *(12)*, the rate-limiting step in free fatty acid synthesis from cytoplasmic acetyl CoA. This results in lower levels of malonyl CoA, an allosteric inhibitor of carnitine acyltransferase *(13)*. In turn, this would allow a faster rate of influx of fatty acyl CoA into the mitochondrion for conversion to ketoacids. Simultaneously the greater rate of β-oxidation of free fatty acids increases the level of intramitochondrial acetyl CoA, which activates pyruvate carboxylase *(9)*, helping to increase gluconeogenesis. Glucagon also is capable of activating adipose tissue lipase *(21)*, thereby increasing lipolysis, the delivery of free fatty acids from adipose tissue to the liver, and ketogenesis. However, glucagon's ketogenic actions often have been demonstrated with doses above the usual physiological range and are easily nullified by rather small amounts of insulin, particularly at the adipose tissue locus. Finally, hepatic hydroxymethylglutaryl CoA reductase activity *(14)* is inhibited by glucagon, thereby decreasing hepatic cholesterol synthesis.

Other actions of glucagon include inhibition of renal tubular sodium resorption, causing natriuresis, and activation of myocardial adenylate cyclase, causing a moderate increase in cardiac output. The latter effect has been of rare therapeutic value in refractory heart failure. The possibility that glucagon—like insulin—may act as a local central nervous system hormone in the regulation of appetite also has been suggested.

From the foregoing it is apparent that fuel fluxes are finely regulated by coordinated secretion of the β-cells and α-cells in a manner appropriate to the physiological situation. Indeed, as discussed previously, many common enzyme and substrate control points for glycogenolysis, gluconeogenesis, and ketogenesis, as well as for glycogen and fat synthesis, are affected by both hormones but in exactly the opposite fashion. Therefore it may be the ratio of insulin to glucagon (I/G ratio), rather than the absolute level of each, that is of critical importance in metabolic regulation.

■ *The insulin/ glucagon ratio and metabolism*

When fasting is prolonged beyond the usual overnight period, there is a rapid fall in the molar ratio of I/G from 2.0 to 0.5, or less. This stimulates glycogenolysis, which is the initial source of support for the plasma glucose level. Since glycogen stores in the

■ *Fasting*

liver are only about 75 g and quickly depleted, a rapid enhancement of gluconeogenesis also is necessary to maintain delivery of glucose for the metabolism of the central nervous system and blood cells. About 75 to 100 g of muscle protein are then catabolized daily to provide the bulk of glucose precursors. Enhanced hepatic extraction of amino acids, especially of alanine, facilitates the increasing gluconeogenesis from protein. Another 15 to 20 g of glucose precursors are provided by the glycerol that is released with free fatty acids as lipolysis is dramatically increased by the lowered I/G ratio. The increased rate of fatty acid oxidation boosts the production of the ketoacids acetoacetate and β-hydroxybutyrate from basal levels of 10 g/day, equivalent to 60 kcal, to as much as 80 g/day, yielding 500 kcal. These ketoacids are then used by peripheral tissues, partly replacing glucose. Conservation of glucose for central nervous system use is further aided by a reduced rate of glucose utilization by the muscle mass and viscera; therefore RQ decreases. The net result of this rapid lowering of I/G is reflected in changing levels of plasma substrates and in rates of urinary nitrogen excretion (Fig. 50-12). Plasma glucose declines 15 to 30 mg/dl but then stabilizes. Plasma free fatty acids, glycerol, ketoacids, and branched chain amino acids all increase. Plasma alanine decreases as hepatic extraction of this amino acid exceeds its release from muscle. Urinary nitrogen excretion rises consequent to the increased catabolism of endogenous protein.

If fasting continues beyond a few days, further important adaptive changes occur, although the I/G ratio remains at the same low level. The basal rate of metabolism diminishes 15% to 20%, conserving body stores of energy. This adaptation may be due to diminished sympathetic nervous system activity and/or to a reduction in effective thyroid hormone activity. Glucose utilization by the central nervous system declines by two thirds, to 45 g/day, and ketoacids replace glucose as the major oxidative fuel of the

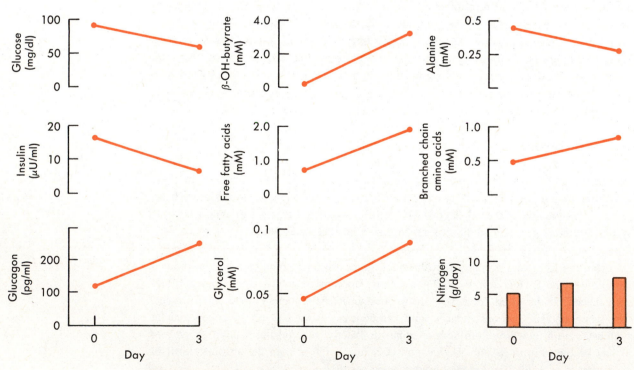

Fig. 50-12 ■ Changes in plasma hormone and substrate levels and urine nitrogen during 3 days of fasting in humans. Note that as the ratio of insulin to glucagon declines parallel to plasma glucose, there is a rise in lipid-derived fuels and an increase in protein catabolism. (Redrawn in part from Felig, P., et al.: J. Clin. Invest. **48**:584, 1969. Reproduced by copyright permission of the American Society for Clinical Investigation.)

*Whole body metabolism
and the hormones of the
pancreatic islets*   **941**

brain. This key change permits a large reduction in gluconeogenesis, which declines to match the lowered rate of glucose utilization. In turn, this greatly spares muscle protein, the catabolism of which now only needs to supply 20 to 25 g of glucose precursors per day. This is reflected in a much lower rate of urinary nitrogen excretion (3 to 4 g/day). Plasma substrate concentrations of glucose, free fatty acids, and glycerol remain essentially the same as at 3 days of fasting. Ketoacid levels, however, increase still further. Alanine and other amino acid levels are sharply reduced. With these additional accommodations and an ample supply of fluid a normal-weight individual can survive up to 2 months of total fasting, whereas very obese individuals have been kept without food for up to 1 year for therapeutic purposes.

■ *Exercise*

Exercise represents a special example of rapid fuel mobilization, geared to supplying an excess of substrate for muscle oxidation while preserving a steady delivery of glucose to the central nervous system. Although carbohydrate utilization by resting muscle is generally at a low rate, the situation changes dramatically with exercise. The initial response is a shift to carbohydrate oxidation. Breakdown of muscle glycogen (present at concentrations of 9 to 16 g/kg wet weight) provides an immediate source of glucose-6-phosphate for glycolysis and oxidation. However, this must soon be augmented. Within 10 minutes glucose uptake from blood may increase seven- to fifteenfold, and by 60 minutes, twenty- to thirtyfold, depending on the muscle group and the intensity of exercise. It is important to emphasize that this increased uptake of glucose by exercising muscle does not require insulin. Oxidation of fatty acids also increases with time, sustained by increased uptake of free fatty acids from plasma. During steady, moderate leg exercise about one third of the energy is supplied by oxidation of glucose and two thirds by oxidation of free fatty acids.

It is to supply these needed fuels that glucagon secretion increases, insulin secretion decreases, and the I/G ratio falls during exercise (Fig. 50-13). This promotes an increase in hepatic glucose output of up to 500%, depending on the intensity and duration of the exercise. In mild exercise of short duration glycogenolysis is the dominant source of the extra glucose, with gluconeogenesis contributing only 20% to 30%. If exercise is more

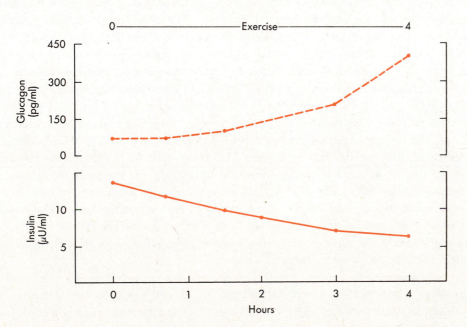

Fig. 50-13 ■ Effect of prolonged exercise on plasma insulin and glucagon in humans. (Redrawn from Ahlborg, G., et al.: J. Clin. Invest. **53:**1080, 1974. Reproduced by copyright permission of the American Society for Clinical Investigation.)

strenuous and glucose demand greater, absolute rates of glycogenolysis and gluconeogenesis both increase initially, but the fraction contributed by gluconeogenesis declines to 10%. If exercise is more moderate but is prolonged for hours, liver glycogen stores become depleted, and gluconeogenesis then becomes the major source of hepatic glucose production. Thus there is increased muscle proteolysis, and the amino acids so generated are taken up by the liver at an accelerated rate. Alanine is again the dominant amino acid that turns over in this process.

The free fatty acids for muscle oxidation come from the increased rate of lipolysis generated by the low I/G ratio. Ketoacid production also increases slightly but contributes little to the energy needs of exercising muscle. Plasma substrate levels during exercise depend on the relative rates of production and utilization and on the coordinated changes in the I/G ratio. Most important, plasma glucose is maintained constant for a long time, although it may eventually decline a modest 10 to 15 mg/dl. Plasma alanine initially increases but eventually also declines below basal level. The same is true for branched chain amino acids if exercise is strenuous. Plasma free fatty acids increase strikingly to levels up to 400% over the basal level. Plasma lactate, pyruvate, glycerol, and ketoacids all increase moderately.

■ The newborn period

The I/G ratio declines within hours of birth in the human, and it remains low for several days. This period marks a transition from dependence on maternal fuels to dependence on the endogenous stores of the neonate. A rapidly falling plasma glucose level provides the signal for the appropriate set of islet hormones. Initially glycogenolysis is stimulated, but once this limited store is gone, the low I/G ratio promotes development of the immature gluconeogenic capacity of the newborn. Until this occurs, the enhanced release of free fatty acids and their ketoacid products temporarily takes up the fuel slack, the ketoacids being of special importance to the central nervous system, as in prolonged fasting. Once regular feeding becomes established, the I/G ratio slowly returns to that characterizing the postabsorptive adult.

■ Carbohydrate intake

When a pure load of carbohydrate is ingested, the I/G ratio rises strikingly from 2 to 30. This ensures efficient retention of the glucose. Of this, 60% to 80% is initially taken up by the liver and rapidly phosphorylated. Part of it is converted to glycogen, thus becoming part of the hepatic glucose stores that will sustain plasma glucose levels many hours later. Part is metabolized to acetyl CoA, which in turn serves as the substrate for de novo free fatty acid synthesis. In this way extra carbohydrate calories can be more efficiently stored as triglyceride. The high I/G ratio also immediately suppresses endogenous glucose output from the liver by inhibiting glycogenolysis and gluconeogenesis. Muscle proteolysis is inhibited as the need to provide amino acid precursors for glucose is diminished, and lipolysis also is suppressed. About 30% of the exogenous glucose escapes hepatic capture; 20% to 25% is oxidized by the brain, whereas only 5% to 10% is oxidized by muscle or taken up by adipose tissue for conversion to glycerol in triglyceride. The RQ rises, reflecting the increased oxidation of glucose and decreased oxidation of fat. Were it not for the high I/G ratio in the portal vein, hepatic storage of the carbohydrate load would be less efficient, plasma glucose would rise unduly, more of the glucose would be dissipated by utilization in muscle, and some might even be lost by excretion in the urine.

■ Protein intake

When a protein load is ingested, plasma insulin and glucagon both increase; the resultant I/G ratio increases slightly. The liver preferentially removes only a small fraction of the absorbed amino acids for use in intrahepatic protein synthesis and to sustain the continuing need for gluconeogenesis. The bulk of amino acids escapes hepatic sequestration, particularly the essential branched chain amino acids (valine, leucine, and

*Whole body metabolism
and the hormones of the
pancreatic islets*   **943**

isoleucine). They subsequently enter muscle, where they are synthesized into structural and contractile proteins, as well as into a labile protein pool from which glucose precursors can come during a later fasting period. Other organs also extract amino acids required for specific functions or for protein renewal. Muscle release of gluconeogenic precursors such as alanine—now not needed by the liver—is greatly reduced after a protein load. The dual islet hormone response to protein smoothly coordinates these processes. The increase in insulin promotes the anabolic disposal of the amino acids, largely into muscle. The increase in glucagon prevents a fall in hepatic glucose output and the hypoglycemia that would ensue if insulin action were unopposed.

■ *Fat intake*

A pure load of fat is seldom ingested by a human, so that the necessary hormonal influences on fat disposition are generated by the accompanying carbohydrate and protein in the meal. Mixed meals increase the I/G ratio because of the sharp rise in insulin release (Fig. 50-8). Most of the ingested fat enters the peripheral circulation as chylomicrons via lymphatics and the thoracic duct (Chapter 44). Clearance of the triglycerides in the chylomicrons from plasma is dependent on hydrolysis by the enzyme lipoprotein lipase, located on the surface of capillary endothelial cells. The fate of the released free fatty acids depends on the site of hydrolysis. In adipose tissue they are promptly re-esterified with glycerophosphate and stored as triglyceride. In muscle they are largely oxidized. In the liver they may be either oxidized or re-esterified into triglyceride and stored for later release as very low-density lipoprotein. The increase in insulin secretion facilitates these processes, both by increasing the key enzyme lipoprotein lipase and by stimulating uptake of glucose to supply the glycerophosphate needed for intracellular triglyceride synthesis.

■ *Somatostatin
secretion and action*

The neurohormone somatostatin was originally discovered in hypothalamic extracts as an inhibitor of growth hormone secretion (Chapter 52). However, subsequent studies showed that somatostatin is a profound inhibitor of insulin and glucagon release as well. This led to the finding that somatostatin also is secreted by the δ-cells of the pancreatic islets. The hormone is a single-chain peptide containing 14 amino acids, synthesized via a prosomatostatin, and released by exocytosis. There is evidence that cyclic AMP may mediate this process in δ-cells. Preliminary reports suggest that somatostatin circulates in low concentration in peripheral plasma with a half-life of approximately 2 minutes. The secretion of δ-cell somatostatin is stimulated by glucose, amino acids, and free fatty acids, by gastrointestinal hormones, such as secretin and cholecystokinin, and by glucagon. It is inhibited by catecholamines, by acetylcholine, and probably by insulin. In humans peripheral plasma somatostatin levels increase modestly following ingestion of a mixed meal.

The administration of somatostatin to experimental animals or humans decreases the rate of assimilation of all nutrients from the gastrointestinal tract. This is accomplished by inhibitory actions at all levels of gastrointestinal function. Somatostatin inhibits gastric, duodenal, and gallbladder motility; it reduces the secretion of hydrochloric acid, pepsin, gastrin, secretin, and intestinal juices, as well as pancreatic exocrine function; it also inhibits the absorption of glucose, xylose, and triglycerides across the mucosal membrane. It seems reasonable to believe that somatostatin of δ-cell origin participates in a feedback arrangement, whereby entrance of food into the gut stimulates the release of the hormone so as to prevent rapid nutrient overload. The anatomical relationships between α-, β-, and δ-cells and the existence of tight junctions and gap junctions between them, as described previously, have further stimulated the hypothesis that all three islet hormones may be influencing each other's secretion by local or *paracrine* effects (Fig. 50-14). This could improve the coordination between the bulk movement, digestion, and absorption of nutrients with the insulin and glucagon responses necessary

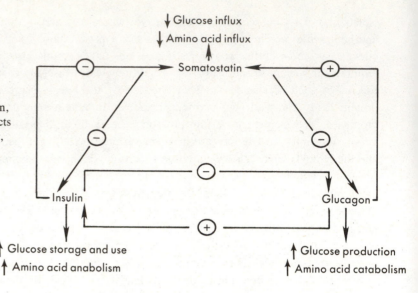

Fig. 50-14 ■ A schema interrelating the effects of somatostatin, insulin, and glucagon on each other's secretion with their effects on glucose and amino acid metabolism. (Redrawn from Unger, R.H., et al. Reproduced with permission from the Annual Review of Physiology, volume 40. Copyright © 1978 by Annual Reviews, Inc.)

for their proper disposition in the liver and other organs. In addition, by coordinating its inhibiting effects on exogenous nutrient influx with an inhibitory paracrine effect on insulin secretion, somatostatin could reduce fat deposition and act as an antiobesity hormone.

■ *Pancreatic polypeptide*

Pancreatic polypeptide, an islet hormone, was isolated from pancreatic digests, its structure was analyzed, a radioimmunoassay was developed for it, and it was localized to a specific islet cell type—all without any clear idea of its function or purpose. Nonetheless it is present in plasma, and its concentration increases markedly with ingestion of food. Moreover it is suppressed by glucose infusion and stimulated by hypoglycemia. Insulin-deficient animals and humans show hyperplasia of its cell of origin and generally exhibit augmented secretion of the hormone in response to stimuli. Although the metabolic function of pancreatic polypeptide in normal humans currently is not known, patients with pancreatic islet cell tumors often show elevated plasma levels of the hormone. It therefore may serve as a useful means of diagnosing and monitoring the therapy of such neoplasms.

■ *Clinical syndromes of islet cell dysfunction*
■ *Insulin*

Tumors of the β-cells produce a primary excess of insulin. The cardinal manifestation is a low plasma glucose level (< 50 mg/dl) in the fasting state. Because the sympathetic nervous system is actuated quickly by abrupt hypoglycemia, bursts of insulin hypersecretion produce episodes of rapid heart rate, nervousness, sweating, and hunger. With chronic insulin excess and persistent hypoglycemia central nervous system function is disturbed, leading to bizarre behavior, defects in cerebration, loss of consciousness, convulsions, and permanent brain damage if not treated soon enough. The need to ingest large amounts of carbohydrate to offset the autonomous action of insulin combined with the lipogenic effect of the hormone produces weight gain. The restraining effect of insulin on glucagon secretion prevents an adequate compensatory response of its physiological antagonist, aggravating the tendency to hypoglycemia. The diagnosis is established by demonstrating plasma insulin levels inappropriately high for the prevailing glucose levels during a short-term fast. Removal of the tumor cures the condition. Failing that, drugs that inhibit insulin secretion palliate the hypoglycemia.

Primary deficiency of insulin is the consequence of β-cell destruction. This disorder, known as *diabetes mellitus,* results from a genetically conferred vulnerability to an insult by a currently unspecified environmental agent(s). The pathophysiological picture in many respects resembles starvation (fasting) in an accelerated and aggravated form. The exceedingly low I/G ratio is responsible for the markedly distorted metabolic state. Glucose production is augmented, and peripheral glucose utilization is reduced, until a new equilibrium between these processes is reached at a very high plasma glucose level (300 to 2000 mg/dl). An increased rate of gluconeogenesis is supported by increased proteolysis. Muscle anabolism is correspondingly reduced; thus nitrogen balance becomes negative, and lean body mass decreases. Excess mobilization of adipose tissue triglycerides elevates plasma free fatty acids. Hepatic lipid content and plasma triglycerides increase. Ketogenesis is markedly stimulated, whereas peripheral ketoacid use is inhibited, thereby greatly elevating plasma ketoacid levels (10 to 20 mM/L). These carboxylic acids raise the hydrogen ion concentration of the blood. As they are neutralized by sodium bicarbonate, the primary buffer, carbonic acid is formed, which dissociates to carbon dioxide and water. Pulmonary hyperventilation is stimulated, which removes some of the excess carbon dioxide. Despite this compensatory reduction in carbon dioxide content, blood pH may finally fall to lethal levels (less than 6.8), and death in diabetic ketoacidosis coma ensues.

Before this terminal point a classic constellation of symptoms is observed. Because of the high plasma glucose levels, the filtered load of glucose exceeds the renal tubular capacity for reabsorption. Glucose therefore is excreted in the urine in large quantities (up to 400 g/day), causing, by its osmotic effect, increased excretion of water and salts. The patient notes increased frequency of urination. Thirst is stimulated by the hyperosmolality of the plasma and the tendency to hypovolemia. The loss of glucose is also a caloric drain, which the patient attempts to recoup by eating more. This causes further exaggerated rises in plasma glucose, as the liver is unable to store the influx of carbohydrate from the gut efficiently in the presence of the very low I/G ratio. Despite the increase in appetite and caloric intake, loss of lean body mass, adipose tissue, and body fluids ensues. Deficits of nitrogen, potassium, phosphate, magnesium, and other intracellular components develop as these are excreted in the urine in amounts that exceed their intake. Fatigue is a common symptom, and muscle performance may be impaired during exercise because of the reduction in both muscle and liver glycogen stores.

Osmotic fluid shifts secondary to the high plasma glucose, and its conversion to other hexoses, such as sorbitol, adversely affect other tissues. For example, the lens of the eye swells, causing blurred vision, and even opacification. Other cellular malfunctions whose pathogenesis is less clear include initially increased filtration through the renal glomeruli, slowed transmission of nerve impulses, decreased resistance to infectious agents, and long-term complications such as retinal disease, kidney failure, and accelerated atherosclerosis.

The biochemical abnormalities and symptoms of this form of diabetes mellitus are rapidly reversed by insulin replacement. The typical pattern of biochemical response is shown in Fig. 50-15. The I/G ratio increases not only because plasma insulin rises with treatment but also because plasma glucagon decreases as the α-cell hyperactivity is inhibited by insulin. Sequentially, plasma glucose and urine glucose excretion declines, lipolysis and ketosis diminish, and protein catabolism is replaced by anabolism, as indicated by the decline in urinary nitrogen excretion. Hydration is normalized, body weight regained, and vigor restored. It should be pointed out that a modest improvement in the I/G ratio in decompensated diabetic subjects also can be effected solely by decreasing glucagon secretion, for example, by administration of somatostatin. This does reduce plasma and urine glucose significantly in undertreated patients and can moderate the otherwise rapid development of diabetic ketoacidosis if insulin is withdrawn completely.

Another even more common form of diabetes mellitus often is associated with obesity. In this situation there is resistance to the action of insulin in its major target tissues,

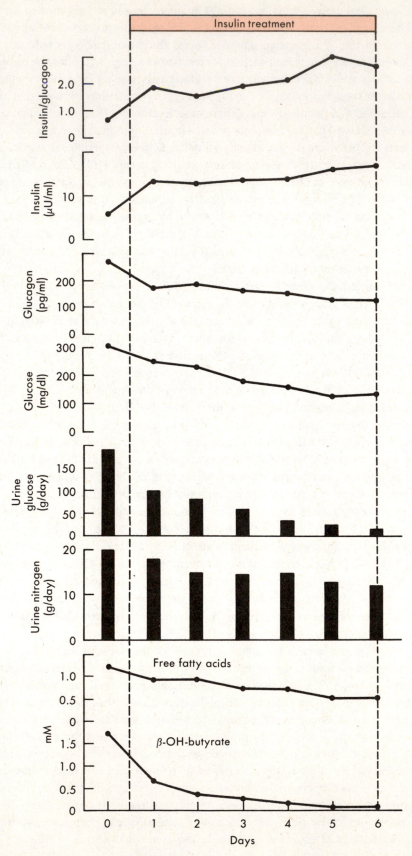

Fig. 50-15 ■ The effects of insulin replacement to insulin-deficient diabetic humans on plasma hormone and substrate levels and urine losses of glucose and nitrogen.

in part because of a decrease in the number of insulin receptors. In addition, there is a derangement in β-cell recognition of glucose as a stimulus, so that insulin secretion is delayed and inadequate. The effective I/G ratio is less markedly reduced in these patients, so that the major biochemical manifestations are those of elevated plasma and urine glucose, particularly following exogenous carbohydrate intake. Accelerated mobilization of endogenous lipids with ketogenesis is not seen in the absence of intercurrent illness, and nitrogen balance is generally maintained. Treatment of this form of diabetes does not ordinarily require insulin administration. Instead caloric regulation, weight reduction if obesity is present, and occasionally use of sulfonylurea drugs simultaneously improve tissue responsiveness to endogenous insulin and β-cell responsiveness to glucose. Hyperglycemia and glycosuria then abate.

■ *Obesity*

In a very small number of cases human obesity clearly results from destruction or infiltration of the hypothalamic areas regulating appetite. In another small fraction obesity results from primary disturbances in cortisol, thyroid hormone, or growth hormone secretion. However, in the vast majority of cases no primary cause for morbid obesity can be identified. Nonetheless a considerable body of physiological and pathophysiological data does suggest a role for insulin either in the genesis or the maintenance of obesity. To begin with, obese animals and humans are generally resistant to the actions of insulin on carbohydrate metabolism. Second, they almost invariably secrete excess amounts of insulin in the basal state and in response to nutrients. Their β-cells are more numerous, and their insulin stores are enlarged. In a model strain of mice with genetic obesity, hyperinsulinism can be detected at virtually the same time as body lipid stores begin to enlarge compared with those of normal littermates. Even though caloric restriction can prevent the excess gain in weight, the percentage of body weight accounted for by fat remains abnormally high; however, chemical destruction of the β-cells of such obese mice does diminish their adiposity. Since the action of insulin is clearly to facilitate storage of calories as triglycerides, it seems reasonable to postulate that their hyperinsulinism is involved in the genesis of their obesity. Interestingly, in all models of obesity studied, as well as in human obesity, the excessive secretion of insulin is decreased by caloric deprivation and weight reduction as insulin resistance also diminishes. Thus hyperinsulinism must be deemed a phenomenon secondary to the ingestion of excess calories, but one that clearly acts to prevent dissipation of the calories and to promote their storage.

■ *Glucagon*

Primary glucagon excess is produced by α-cell tumors. The catabolic action of the hormone is exhibited clinically by loss of weight and by a peculiar destructive skin lesion. Plasma levels of glucose and of ketoacids are elevated. The marked increase in gluconeogenesis causes a generalized reduction of plasma amino acids and an increase in urinary nitrogen. Severe diabetes mellitus is uncommon because of the compensatory increase in insulin secretion. The predictable consequence of primary glucagon deficiency would be fasting hypoglycemia. In fact, however, such cases only rarely have been reported with adequate documentation. It is likely that other counterregulatory hormones, such as epinephrine, take up the slack when glucagon is absent.

■ *Somatostatin*

A primary excess of somatostatin is seen rarely with certain islet cell tumors. The patients exhibit weight loss caused by inhibition of nutrient absorption. Plasma glucagon and insulin levels are low, as anticipated. Because of the latter deficiency, hyperglycemia is seen after carbohydrate loading.

An isolated deficiency of δ-cell somatostatin is as yet unknown. In a few cases of hyperinsulinism in infancy, hyperplasia of β-cells has been accompanied by an absence of δ-cells. It has been suggested that loss of a local inhibitory effect of somatostatin might be responsible for the unrestrained secretion of insulin.

■ *Bibliography*

Ahlborg, G., et al.: Substrate turnover during prolonged exercise in man; splanchnic and leg metabolism of glucose, free fatty acids, and amino acids, J. Clin. Invest. **53:**1080, 1974.

Björntorp, P., et al.: Carbohydrate storage in man: speculations and some quantitative considerations, Metabolism **27:**1853, 1978.

Bray, G.A., et al.: Metabolic factors in the control of energy stores, Metabolism **24:**99, 1975.

Cahill, G.F.: Starvation in man, N. Engl. J. Med. **282:**668, 1970.

Chiasson, J.L., et al.: Gluconeogenesis from alanine in normal postabsorptive man; intrahepatic stimulatory effect of glucagon, Diabetes **24:**574, 1975.

Curry, D.L., et al.: Dynamics of insulin secretion by the perfused rat pancreas, Endocrinology **83:**572, 1968.

Czech, M.P.: Insulin action, Am. J. Med. **70:**142, 1981.

Exton, J.H.: Gluconeogensis, Metabolism **21:**945, 1972.

Felig, P.: The glucose-alanine cycle, Metabolism **22:**179, 1973.

Felig, P., et al.: Amino acid metabolism during prolonged starvations, J. Clin. Invest. **48:**584, 1969.

Felig, P., et al.: Concentrations of glucagon and the insulin: glucagon ratio in the portal and peripheral circulation, Proc. Soc. Exp. Biol. Med. **147:**88, 1974.

Felig, P., et al.: Fuel homeostasis in exercise, N. Engl. J. Med. **293:**1078, 1975.

Floyd, J.C., et al.: Insulin secretion in response to protein ingestion, J. Clin. Invest. **45:**1479, 1966.

Frank, H.J.L., et al.: A direct in vitro demonstration of insulin binding to isolated brain microvessels, Diabetes **30:**757, 1981.

Garber, A.J., et al.: Hepatic ketogenesis and gluconeogenesis in humans, J. Clin. Invest. **54:**981, 1974.

Genuth, S.M.: Plasma insulin and glucose profiles in normal, obese, and diabetic persons. Ann. Intern. Med. **79:**812, 1973.

Gerich, J.E., et al.: Characterization of the glucagon response to hypoglycemia in man, J. Clin. Endocrinol. Metab. **38:**77, 1974.

Gerich, J.E., et al.: Effects of physiologic levels of glucagon and growth hormone on human carbohydrate and lipid metabolism, J. Clin. Invest. **57:**875, 1976.

Guillemin, R., et al.: Growth hormone-releasing factor from a human pancreatic tumor that caused acromegaly, Science **218:**585, 1982.

Hornnes, P.J., et al.: Simultaneous recording of the gastro-entero-pancreatic hormonal peptide response to food in man, Metabolism **29:**777, 1980.

Itoh, M., et al.: Antisomatostatin gamma globulin augments secretion of both insulin and glucagon in vitro; evidence for a physiologic role for endogenous somatostatin in the regulation of pancreatic α- and β-cell function, Diabetes **29:**693, 1980.

Jefferson, L.S.: Role of insulin the regulation of protein synthesis, Diabetes **29:**487, 1980.

Karam, J.H., et al.: "Staircase" glucose stimulation of insulin secretion in obesity; measure of beta-cell sensitivity and capacity, Diabetes **23:**763, 1974.

Liljenquist, J.E., et al.: Evidence for an important role of glucagon in the regulation of hepatic glucose production in normal man, J. Clin. Invest. **59:**369, 1977.

Olefsky, J.M.: Insulin resistance and insulin action: an in vitro and in vivo perspective, Diabetes **30:**148, 1981.

Orci, L., et al.: Cell contacts in human islets of Langerhans, J. Clin. Endocrinol. Metab. **41:**841, 1975.

Owen, O.E., et al.: Liver and kidney metabolism during prolonged starvation, J. Clin. Invest. **48:**574, 1969.

Owen, O.E., et al.: Human forearm metabolism during progressive starvation, J. Clin. Invest. **50:**1536, 1971.

Passmore, R., et al.: Human energy expenditure, Physiol. Rev. **35:**801, 1955.

Perley, M.J., et al.: Plasma insulin responses to oral and intravenous glucose: studies in normal and diabetic subjects, J. Clin. Invest. **46:**1954, 1967.

Reichard, G.A., Jr., et al.: Ketone-body production and oxidation in fasting obese humans, J. Clin. Invest. **53:**508, 1974.

Runcie, J., et al.: Energy provision, tissue utilization, and weight loss in prolonged starvation, Br. Med. J. **2:**352, 1974.

Shaw, W.N., et al.: Effect of porcine proinsulin in vitro on adipose tissue and diaphragm of the normal rat, Diabetes **17:**737, 1968.

Stricker, E.M.: Hyperphagia, N. Engl. J. Med. **298:**1010, 1978.

Unger, R.H., et al.: Studies of pancreatic alpha cell function in normal and diabetic subjects, J. Clin. Invest. **49:**837, 1970.

Unger, R.H., et al.: Insulin, glucagon, and somatostatin secretion in the regulation of metabolism, Annu. Rev. Physiol. **40:**307, 1978.

Unger, R.H., et al.: Glucagon and the A cell. I. Physiology and pathophysiology, N. Engl. J. Med. **304:**1518, 1981.

Verspohl, E.J., et al.: Evidence for presence of insulin receptors in rat islets of Langerhans, J. Clin. Invest. **65:**1230, 1980.

chapter 51

Endocrine regulation of calcium and phosphate metabolism

The maintenance of calcium and phosphate homeostasis is dependent on major contributions from three organ systems: the intestinal tract, the skeleton, and the kidneys, with minor but essential contributions from the skin and the liver. Before detailing the hormonal mechanisms of regulation it is necessary to present an overall view of calcium and phosphate metabolism with special emphasis on the role played by the bone mass. Without such an initial overview it is difficult to tie together the individual endocrine components of the system.

The calcium ion is of fundamental importance to all biological systems. Calcium participates in numerous enzymatic reactions. It is a vital component in the mechanism of hormone secretion and a mediator of hormonal effects. Calcium is intimately involved in neurotransmission, in muscle contraction, and in blood clotting. It is the major cation in the crystalline structure of bone and teeth. For these and other reasons it is vital that cells be bathed with fluid in which the calcium concentration is kept within the narrow limits of physiological tolerance.

The normal range of calcium in the plasma is 8.6 to 10.6 mg/dl (4.3 to 5.3 mEq/L). For any single individual, however, the variation from day to day is generally well under 10%. Approximately 50% of plasma calcium is in ionized, biologically active form; 10% is complexed in nonionic but ultrafilterable forms, such as calcium bicarbonate; and 40% is bound to proteins, mainly albumin. The equilibrium between ionized and protein-bound calcium is shifted toward increased binding as the blood pH increases.

Fig. 51-1 details in simplified fashion the normal turnover of calcium in the body. Daily dietary calcium intake can range from as little as 200 to as much as 2000 mg. The percentage of dietary calcium that is absorbed from the gut is inversely related to intake in a curvilinear manner. Thus an adaptive increase in fractional absorption is one important mechanism for maintaining normal body calcium stores in the face of dietary deprivation, whereas an adaptive decrease prevents overload in the face of dietary surfeit. At an average daily intake of 1000 mg about 35% is absorbed. The same amount of calcium, 350 mg, ultimately must be excreted to maintain balance. About 150 mg of calcium is secreted back into the intestine and excreted in the stools, along with the unabsorbed fraction from the diet. The remaining 200 mg is excreted in the urine. The kidney filters about 10,000 mg of calcium per day (non-protein-bound calcium concentration $\times$ glomerular filtration rate = 60 mg/L $\times$ 170 L/day). However, approximately 98% is reabsorbed in the tubules. Alterations in the small fraction of filtered calcium that is finally excreted provide another sensitive means of maintaining overall calcium balance.

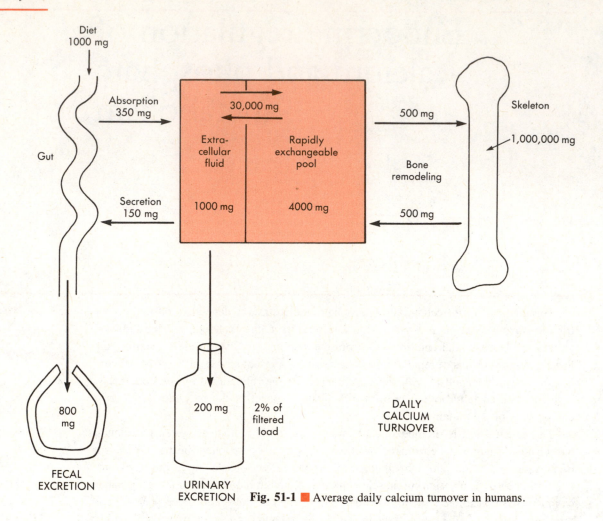

Fig. 51-1 ■ Average daily calcium turnover in humans.

Calcium enters and exits from an extracellular pool of 1000 mg. This in turn is in equilibrium with a rapidly exchanging pool of several times that size, which probably represents the surface of recently or partially mineralized bone. Although it has been estimated that 30,000 to 40,000 mg of calcium may move back and forth in these internal pools daily, only 500 mg is "irreversibly" removed from the extracellular space by bone formation, and 500 mg is returned to it by bone resorption in the process that constitutes normal bone remodeling. The tremendous store of skeletal calcium (approximately 1 kg) makes it quantitatively the most important depot in maintaining the stability of plasma calcium.

The phosphate ion is also of critical importance to all biological systems. Phosphate is an integral component of all glycolytic compounds from glucose-1-phosphate to phosphoenolpyruvate. It is part of the structure of high-energy transfer compounds, such as ATP and creatine phosphate, of cofactors such as NAD, NADP, and thiamin pyrophosphate, and of lipids such as phosphatidylcholine. It functions as a covalent modifier of numerous enzymes. And of course phosphate is a major anion in the crystalline structure of bone.

The normal concentration of phosphate in the plasma is 2.5 to 4.5 mg/dl (0.81 to 1.45 mmole/L). Because the valence of phosphate changes with pH, it is less meaningful to express phosphate concentration in milliequivalents per liter. The turnover of phosphate is shown in Fig. 51-2. In contrast to calcium, the percentage of phosphate absorbed from the diet is relatively constant. Thus net absorption of phosphate from the gut is linearly related to intake over a wide range, and adaptive regulation at this site is

*Endocrine regulation of
calcium and phosphate
metabolism*   **951**

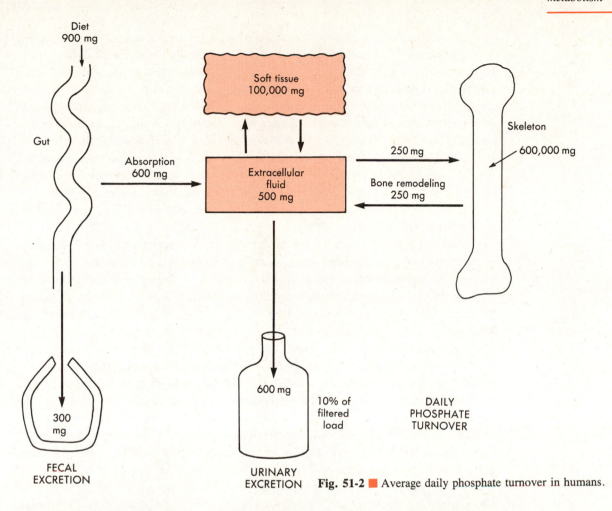

Fig. 51-2 ■ Average daily phosphate turnover in humans.

of minor importance. Therefore urinary excretion provides the major mechanism for preserving phosphate balance. Of the daily filtered load of approximately 6000 mg (plasma concentration × glomerular filtration rate = 36 mg/L × 170 L/day), renal tubular reabsorption can vary from 70% to 100%, with an average of 90%. This provides the needed flexibility to compensate for large swings in dietary intake. Soft tissue stores of phosphate, such as in the muscle mass, are of major magnitude; transfer between them and the extracellular fluid is an important factor in minute-to-minute regulation of plasma phosphate. About 250 mg, or half the total extracellular fluid pool of 500 mg, enters and leaves the bone mass daily in the process of remodeling.

The divalent cation magnesium (Mg^{++}) is related in some metabolic respects to calcium and phosphate. Magnesium has an essential role in neuromuscular transmission and serves as a cofactor in numerous enzyme reactions, most notably those involving energy transfers via ATP and those concerned with ribosomal protein synthesis. The normal range of magnesium in plasma is 1.8 to 2.4 mg/dl (1.5 to 2.0 mEq/L). One third of plasma magnesium is bound to protein. The body content is about 25,000 mg, of which 50% is present in the skeleton. Virtually all the rest is present in the intracellular fluid, where magnesium functions with potassium as a major cation. The average daily intake is about 300 mg. Forty percent of this is absorbed, and in a steady state the same amount, 120 mg, is excreted in the urine.

A detailed account of the development of the skeleton, its structural properties, and its mechanical function is beyond the purview of endocrine physiology. However, a brief review of the organization of bone in relation to its function as a mineral reservoir is essential to understanding hormonal regulation of calcium and phosphate metabolism.

■ *Bone dynamics*

Throughout life the bone mass is continuously turning over by a well-regulated coupling of the processes of formation and resorption. This coupling is achieved in all types of bone within individual microscopic units, called osteons. The chemical or mechanical signals that coordinate local rates of formation and resorption remain mysterious. During the growth years formation exceeds resorption, and skeletal mass increases. Linear growth occurs at the ends of long bones by replacement of cartilage in specialized areas known as epiphyseal plates. These close off at the end of puberty when adult height is reached. Increase in bone width occurs by apposition to the outer surfaces under the connective tissue covering, the periosteum. A peak in bone mass is reached between ages 20 and 30 years. Thereafter equal rates of formation and resorption stabilize the bone mass until age 40 to 50 years, at which time resorption begins to exceed formation, and the total mass slowly decreases. This process of bone turnover in the adult is known as *remodeling;* it is one of the major mechanisms for maintaining calcium homeostasis. As much as 15% of the total bone mass normally turns over each year in the remodeling process. Endocrine diseases that disrupt the coupling of formation and resorption are more florid in their manifestations when they are superimposed on the natural disequilibrium that exists during the early phase of growth and the late phase of senescence. Women are severely affected by such diseases more often than men because they have a smaller peak bone mass and a more rapid physiological rate of senescent loss.

Three major cell types are recognized in histological sections of bone: *osteocytes, osteoblasts,* and *osteoclasts* (Fig. 51-3). The first two arise from osteoprogenitor cells within the connective tissue of the mesenchyme. The osteoclasts may arise from the same stem line or from circulating macrophages. Bone formation is carried out by active osteoblasts, which synthesize and extrude collagen into the adjacent extracellular space. The collagen fibrils line up in regular arrays, producing an organic matrix known as osteoid within which calcium is then deposited as amorphous masses of calcium phosphate. The mineralization process requires normal plasma concentrations of calcium and phosphate and is dependent on vitamin D. The enzyme alkaline phosphatase and possibly other macromolecules from the osteoblast also participate in this process. Gradually hydroxide and bicarbonate ions are added to the mineral phase, and mature hydroxyapatite crystals are formed. These have a calcium/phosphate ratio of 2.2 by weight and 1.7 by moles. As this completely mineralized bone accumulates and surrounds the osteoblast, that cell decreases its synthetic activity and becomes an interior osteocyte (Fig. 51-3). Osteoblastic activity therefore is observed only along the surfaces of bone, whether they are the concentric lamellae of the cortex or the linear lamellae of the interior bridging trabeculae (Fig. 51-4). Along these surfaces other osteocytes, in a resting state, appear to be potential osteoblasts.

Within each bone unit, minute fluid-containing channels, called canaliculi, traverse the mineralized bone; within these channels the interior osteocytes remain connected with surface osteocytes and osteoblasts via syncytial cell processes (Fig. 51-3). This arrangement provides an enormous surface area for transfer of calcium from the interior to the exterior of the bone units and thence into the extracellular fluid. This transfer process, which is carried out by the osteocytes, is known as osteocytic osteolysis. It probably does not actually decrease bone mass but simply removes calcium from the most recently formed crystals.

In contrast, the process of resorption of bone does not merely extract calcium, but it also destroys the entire matrix, thereby diminishing the bone mass. The cell responsible for this is the osteoclast, which is a giant multinucleated cell formed by fusion of several osteoprogenitor cells (Fig. 51-3). The osteoclast contains large numbers of mitochondria and lysosomes. It attaches to the surface of the bone unit. At the point of attachment a ruffled border is created by infolding of the plasma membrane, and in this zone the process of bone dissolution is carried out by collagenase, lysosomal enzymes,

*Endocrine regulation of
calcium and phosphate
metabolism*   **953**

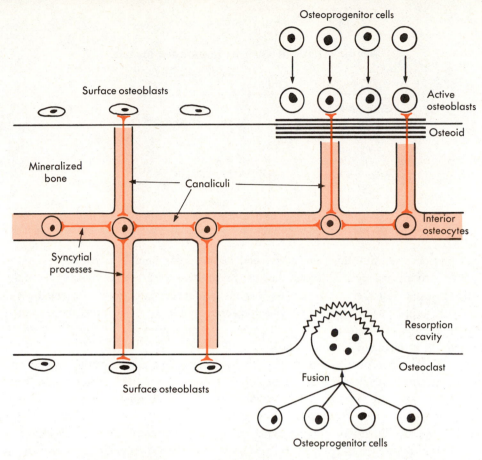

Fig. 51-3 ■ The relationships between bone cells and bone surfaces. The canaliculi provide a huge interface between the interior surfaces of mineralized bone and intercellular fluid. This permits efficient osteolysis with transfer of calcium and phosphate to the exterior via syncytial processes connecting interior and surface osteocytes. (Redrawn from Avioli, L.V., et al.: Bone metabolism and disease. In Bondy, P.K., and Rosenberg, L.E.: Metabolic control and disease, Philadelphia, 1980, W.B. Saunders Co.)

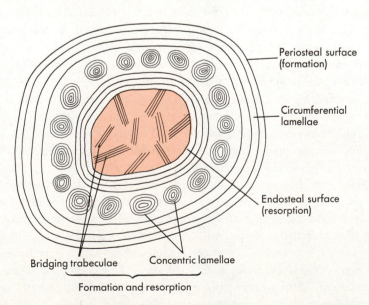

Fig. 51-4 ■ Cross section of a long bone. The outer cortex consists of circumferential lamellae within which other concentric lamellae (haversian canals) are present. These contain nutrient vessels. The inner portion consists of bridging trabeculae or spicules of bone in lamellar arrays, between which bone marrow elements and connective tissue cells are found.

TABLE 51-1

Major effects of various hormones on bone

Bone formation	Bone resorption
Stimulated by	*Stimulated by*
Growth hormone	Parathyroid hormone (PTH)
Insulin	Vitamin D
Androgens	Thyroid hormone
Vitamin D (?)	Cortisol
Inhibited by	Prostaglandins
Cortisol	*Inhibited by*
	Calcitonin
	Estrogens

and phosphatase. Calcium, phosphate, and amino acids unique to collagen, such as hydroxyproline and hydroxylysine, thus are released into the extracellular fluid. In this way the osteoclast literally tunnels its way into the mineralized bone. As noted previously, resorption and formation are closely coordinated locally. Therefore the resorption cavity created by the osteoclast is often the site of subsequent osteoblastic activity, which fills in the cavity with new bone.

The conversion of osteoprogenitor cells to either osteoblasts or osteoclasts, as well as the activity of each cell type, is greatly influenced by an array of hormones (Table 51-1). As a general principle, whether the primary effect of a hormone is on formation or resorption of bone, the phenomenon of coupling will secondarily alter the other process in the same direction. For example, whereas the primary effect of an excess of parathyroid hormone (PTH) is to increase bone resorption, a compensatory increase in bone formation follows. The net effect of any endocrine abnormality will depend on the degree to which the total bone mass is defended by the process of coupling.

■ *The parathyroid glands*

The parathyroid glands are major regulators of plasma calcium concentration and calcium flux in the body. The paramount effect of PTH is to increase or sustain the plasma calcium level. This is accomplished by causing entry of calcium into the plasma from the gastrointestinal tract, the bone mass, and the tubular urine. The four parathyroid glands develop at 5 to 14 weeks of gestation from the third and fourth branchial pouches. They descend to lie posterior to the thyroid gland. The lower pair actually is derived from the third branchial pouch and traverses a longer embryological descent. Ectopic locations in the neck and mediastinum are not rare, a fact that surgeons must appreciate in searching for abnormal parathyroid glands. The total weight of adult parathyroid tissue is about 130 mg, and that of any one gland is 30 to 50 mg. Their blood supply is from the thyroid arteries, and venous drainage is into the jugular and inferior thyroid veins. Samples drawn from veins on the left and the right can localize the site of unilateral parathyroid gland hyperfunction.

The histological appearance of the parathyroid glands changes with age. The predominant cell, known as the *chief cell,* is present throughout life and is the normal source of PTH. The appearance of individual chief cells on electron microscopy varies with their state of activity. Resting cells have abundant glycogen, an involuted Golgi apparatus, and a few clusters of secretory granules. Active cells have little glycogen, a large convoluted Golgi apparatus with vacuoles and vesicles, and a granular endoplasmic reticulum. During active hormone secretion numerous granules may be seen undergoing exocytosis. A second cell type—the *oxyphil cell*—is distinguished by an eosinophilic cytoplasm. This cell first appears at puberty. Its normal function is not known, but its ultrastructural appearance suggests energy production and storage rather than hormone synthesis. Nonetheless under some circumstances it does appear to be

*Endocrine regulation of
calcium and phosphate
metabolism*   **955**

capable of hormone release. Either of these cell types may undergo hyperplasia or form neoplasms, giving rise to hypersecretion of PTH.

■ *Synthesis and release
of PTH*

PTH is a single-chain protein of 9000 molecular weight, containing 84 amino acids. Although there is species variation in structure, PTH from bovine, porcine, and human glands shows a great deal of immunological and biological cross-reactivity. The biological activity of PTH resides in the N-terminal portion of the molecule within amino acids 1 to 34. This fragment has 77% of the activity of native hormone. Alteration, elimination, or backward extension of the N-terminal amino acid markedly decreases biological activity. The function of the remainder of the molecule with its carboxy terminus is not known.

The synthesis of PTH begins with prepro-PTH, a 115–amino acid protein that contains all the structural information encoded in the gene for PTH. This conclusion has been strongly supported by reversed synthesis of the DNA transcript from messenger RNA (mRNA) for PTH. As the peptide chain of prepro-PTH grows to its complete length on the ribosomes, first 2 and then 23 amino acids are enzymatically removed from the N-terminal, leaving pro-PTH. This 90–amino acid molecule is then transported to the Golgi apparatus, within which another 6 amino acids are removed by a tryptic enzyme. Some of the resulting PTH is packaged for storage in mature secretory granules, and it will be released later by the process of exocytosis. There is also evidence that on acute stimulation some newly synthesized PTH may be transported, still in Golgi vesicles, directly through the cell for immediate release. Degradation of PTH also occurs within the parathyroid glands, so that not all of the synthesized molecules reach the circulation.

The dominant regulator of parathyroid gland activity is the plasma calcium level. The hormone and the ion form a negative feedback pair. Secretion of PTH is inversely related to the plasma calcium concentration in a sigmoidal fashion (Fig. 51-5). Maximum secretory rates are achieved below a calcium concentration of 7 mg/dl (3.5 mEq/ L). As calcium concentration increases to 11 mg/dl (5.5 mEq/L), PTH secretion is progressively diminished. However, above 11 mg/dl there is a persistent basal rate of PTH secretion that is not suppressible by further elevation of plasma calcium. It is actually the ionized fraction of plasma calcium that regulates PTH secretion. Alterations in PTH secretion in response to changes in plasma calcium occur within minutes. In vitro, after exposure to hypocalcemia, exocytosis of secretory granules is mediated by activation of adenylate cyclase and a resultant rise in intracellular cyclic AMP levels. Exocytosis also is promoted by inhibition of phosphodiesterase, the enzyme that degrades cyclic AMP.

In addition to rapidly regulating secretion, calcium also modulates PTH synthesis and degradation within the glands. Prolonged exposure to a high ambient calcium con-

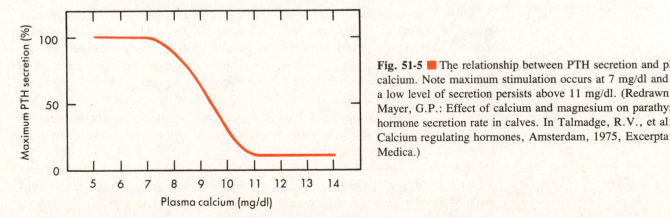

Fig. 51-5 ■ The relationship between PTH secretion and plasma calcium. Note maximum stimulation occurs at 7 mg/dl and that a low level of secretion persists above 11 mg/dl. (Redrawn from Mayer, G.P.: Effect of calcium and magnesium on parathyroid hormone secretion rate in calves. In Talmadge, R.V., et al.: Calcium regulating hormones, Amsterdam, 1975, Excerpta Medica.)

centration eventually depresses the rate of PTH synthesis; the exact control point is not known. In contrast, intraglandular degradation of PTH is enhanced by elevating the calcium concentration. Thus the net effect of increasing ambient calcium is to decrease both the glandular stores and the release rates of PTH. Conversely, hypocalcemia increases PTH stores and secretory rates and ultimately stimulates hyperplasia of the glands as well.

The divalent cation Mg^{++} modulates PTH secretion in a manner analogous to that of Ca^{++}, although it is only half as effective on a molar basis. Therefore Mg^{++} is of much less importance in its normal physiological range (1.5 to 2.5 mEq/L). However, chronic hypomagnesemia strongly inhibits PTH synthesis. In severely magnesium-depleted individuals this phenomenon often predominates, leading to a reduced rate of PTH release.

Because of its close physiological relation to calcium, much work has been done to determine the influence of phosphate on PTH secretion. No direct effects of phosphate on the parathyroid glands have been demonstrated. However, by physicochemical mechanisms a rise in plasma phosphate causes an immediate fall in ionized calcium levels, which in turn does stimulate PTH secretion. As is elaborated later, this resultant increase in PTH helps moderate hyperphosphatemia by increasing renal phosphate excretion.

Certain vitamin D metabolites may exert an inhibitory influence on the parathyroid glands. This mechanism needs further clarification, although some feedback of vitamin D on PTH secretion is a teleologically attractive hypothesis. Epinephrine—a catecholamine hormone—stimulates PTH secretion through its β-adrenergic receptor, probably by activating adenylate cyclase. Histamine also stimulates PTH release via specific histamine (H_2) receptors. Finally, some circadian variation in PTH secretion occurs, probably independent of plasma calcium concentration, with the highest levels of PTH occurring in the evening.

Under normal circumstances neither prepro-PTH nor pro-PTH is secreted from the glands. Furthermore neither of these molecules appears to have intrinsic biological activity. One or more products of the intraglandular degradation of PTH, however, are released. In addition, PTH undergoes rapid metabolism in the peripheral tissues, with the kidney and liver being the major sites. As a result, several fragments of PTH circulate, and these cross-react in radioimmunoassays that employ antibodies raised against native PTH sequence 1 to 84. So studies of plasma concentration and turnover of PTH are hampered by the fact that often less than half the total measured immunoreactivity is caused by the intact hormone. In bovine PTH equivalents the plasma concentration of total immunoreactive PTH in humans is around 0.5 ng/ml (10^{-10}M). The major circulating species is a 6000 molecular weight carboxy terminal fragment that is considered biologically inactive. Whether amino terminal fragments generated during cleavage of PTH also circulate is currently controversial, as is any putative biological activity they might have in various target tissues.

There is no evidence for carrier protein binding of circulating PTH or its metabolites. The plasma half-life of the intact hormone is 20 to 30 minutes. The carboxy terminal fragment has a much longer plasma half-life than does the native hormone. For this reason radioimmunoassays that are specifically directed toward either the carboxy terminus or another portion of the 6000 molecular weight species have been very useful for determining chronic states of hyperparathyroidism and hypoparathyroidism. Assays directed at the N-terminal of the molecule have been especially useful for dynamic studies of parathyroid gland function, since they measure only the biologically active hormone.

■ *PTH actions*

PTH action is initiated by binding to plasma cell membrane receptors; the determinants of binding lie in a restricted zone of the biologically active 1 to 34 sequence. In all target cells activation of adenylate cyclase follows with augmentation of intracellular

Endocrine regulation of
calcium and phosphate
metabolism   **957**

cyclic AMP levels. Most of the effects of PTH in vivo can be mimicked by infusions of cyclic AMP or a suitable analogue that can enter cells. The subsequent intracellular events mediated by cyclic AMP are not known. At present it can only be speculated that cyclic AMP may trigger a protein kinase cascade (Fig. 49-6) which ultimately leads to phosphorylation of a protein necessary for enhanced transport of calcium and other ions.

Independent of its action on cyclic AMP levels, PTH also stimulates the uptake of calcium by the cytosol from the bathing bone fluid and from mitochondrial stores. Whether this calcium is itself another intracellular second messenger or whether it only acts to modulate the adenylate cyclase response by counterregulatory inhibition remains moot. The initial uptake of calcium is reflected in a slight transient hypocalcemia, which follows PTH administration and precedes the classic hypercalcemic response. Of great significance is that the presence of vitamin D is required for the exhibition of the full spectrum of PTH actions. It remains to be determined whether vitamin D is an intracellular cofactor, whether it stimulates synthesis of an essential intracellular protein mediator, or whether it only augments the pool of extracellular calcium from which the rapid early uptake occurs. A sufficient intracellular concentration of magnesium is also necessary for maximum PTH responsiveness.

The overall effect of PTH is to increase plasma calcium and decrease plasma phosphate by acting on three major target organs: bone, kidney, and gastrointestinal tract. All three actions ultimately increase calcium influx into the plasma, raising the concentration. In contrast, the actions on bone and gut, which increase phosphate influx, are overwhelmed by the action on the kidney, which increases phosphate efflux, so that plasma phosphate concentration falls (Fig. 51-6).

Bone. PTH accelerates removal of calcium from bone by at least two processes. Its initial effect is to stimulate osteolysis by surface osteocytes, causing a transfer of calcium from the bone canalicular fluid into the osteocyte and thence out the opposite side

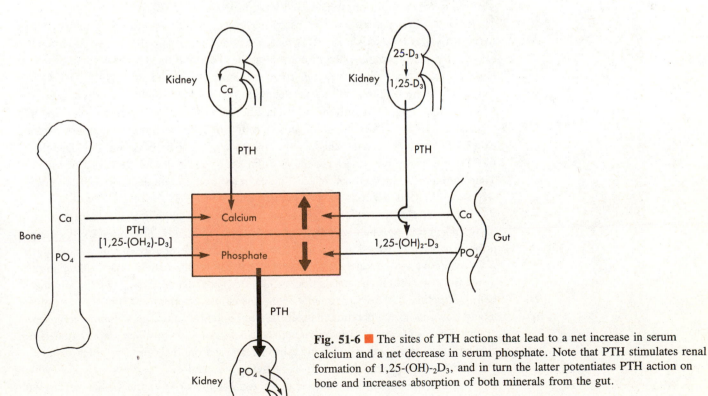

Fig. 51-6 ■ The sites of PTH actions that lead to a net increase in serum calcium and a net decrease in serum phosphate. Note that PTH stimulates renal formation of $1,25\text{-}(OH)\text{-}_2D_3$, and in turn the latter potentiates PTH action on bone and increases absorption of both minerals from the gut.

of the cell into the extracellular fluid. Replenishment of calcium in the canalicular fluid probably then occurs from the surface of partially mineralized bone. Phosphate does not appear to be mobilized with calcium in this process. A second, more slowly developing effect of PTH is to stimulate the osteoclasts to resorb completely mineralized bone. In this process both calcium and phosphate are released for transfer into the extracellular fluid, and the organic bone matrix is hydrolyzed by increased activity of collagenase and of lysosomal enzymes. A PTH-stimulated increase in lactic acid production, with a resultant lowered ambient pH, also may contribute to the resorptive process. As a result of this destructive action of PTH, hydroxyproline is released into the extracellular fluid and from there excreted in the urine. PTH not only activates osteoclasts but also stimulates their proliferation. The increased numbers of giant osteoclasts then create large resorption cavities in both cortical and trabecular bone. In addition to these effects on osteoclasts, PTH also inhibits the synthesis of collagen by osteoblasts. These catabolic effects of PTH on bone are achieved by the elevated concentrations of hormone that result from stimulation of the parathyroid glands by hypocalcemia; thus they are appropriate for rapidly restoring the plasma calcium level to normal. These actions also require the concomitant presence of a specific metabolite of vitamin D, $1,25\text{-}(OH)_2\text{-}D_3$.

There is also evidence that at lower concentrations PTH has an anabolic action on bone. This occurs under circumstances in which plasma calcium is normal or even increased, but compensatory hypersecretion of PTH has existed for a long time. In bone cultures low doses of PTH increase the number of osteoblasts and collagen synthesis. In bone biopsies there may be seen an increase in woven (as opposed to lamellar) bone and a proliferation of fibroblasts adjacent to the areas of increased resorptive activity. The plasma level of alkaline phosphatase, an osteoblastic enzyme whose activity parallels bone formation, is often increased by PTH. Thus the net effect of sustained increases in PTH may be either a decrease or an increase in total skeletal mass. This undoubtedly depends on concomitant factors that affect bone remodeling, such as the availability of calcium, phosphate, and vitamin D.

Kidney. PTH increases the reabsorption of calcium from the distal tubule of the kidney. By this mechanism the PTH that is secreted in response to hypocalcemia helps raise a depressed plasma calcium concentration. The relationship between urinary calcium excretion and plasma calcium concentration is shifted to the right by PTH (Fig. 51-7). Therefore suppression of PTH secretion by a sudden calcium load will help prevent hypercalcemia by permitting a greater fraction of the filtered calcium to escape into the urine. The net effect of prolonged alteration in PTH secretion on urinary calcium excretion, however, is dominated by the influence of PTH on bone and gut. Deficiency of PTH eventually will lower plasma calcium and, with it, the filtered load of calcium. Therefore the absolute amount of calcium excreted in the urine decreases. The converse sequence of events occurs with a prolonged excess of PTH.

The most dramatic effect of PTH on the kidney is to inhibit the reabsorption of phosphate in the proximal tubule. Within minutes of its administration PTH increases urinary phosphate excretion. As befits its second messenger role in this action, cyclic AMP excretion into the urine also increases just before that of phosphate. The phosphaturic effect of PTH allows disposition of the extra phosphate released by PTH-stimulated bone resorption. Otherwise simultaneous plasma elevation of calcium and phosphate would occur, with the potential danger of precipitating calcium-phosphate complexes in critical tissues. In contrast, under circumstances of primary phosphate deprivation serum calcium tends to rise, thereby suppressing PTH secretion. This in turn increases tubular phosphate reabsorption, conserving this essential mineral.

PTH also stimulates the synthesis in the renal tubules of vitamin D metabolites with bone-resorbing action. The decrease in plasma and renal phosphate content caused by

Endocrine regulation of
calcium and phosphate
metabolism   **959**

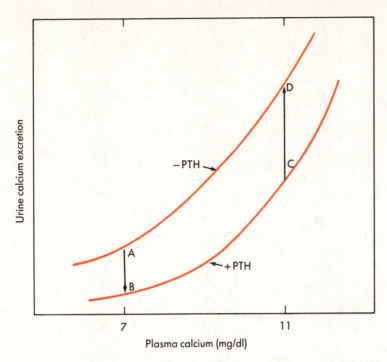

Fig. 51-7 ■ Effect of PTH on the relationship between urine calcium excretion and plasma calcium. At a low plasma calcium, PTH secretion is stimulated, and the hormone shifts urine calcium from point *A* to point *B,* thus conserving calcium. At a high plasma calcium, PTH secretion is suppressed, and urine calcium is shifted from point *C* to point *D,* thus disposing of excess calcium. (Redrawn from Nordin, B.E.C., et al.: Lancet **2:**1280, 1969.)

PTH fortifies its direct action on renal vitamin D metabolism (as explained later); this contributes significantly to the major function of PTH: an increase in plasma calcium. PTH also inhibits the resorption of bicarbonate in the proximal tubule in a manner parallel to that of phosphate. This action may prevent the occurrence of metabolic alkalosis, which could result from the release of bicarbonate during the dissolution of hydroxyapatite crystals in bone.

Intestine. The absorption of calcium as well as phosphate from the gastrointestinal tract is enhanced by PTH. This action, however, is indirect. It comes about because PTH stimulates the synthesis of $1,25\text{-}(OH)_2\text{-}D_3$, a potent vitamin D metabolite that directly increases active transport of calcium across the intestinal mucosa.

Miscellaneous. Alterations in the function of the central nervous system, of peripheral nerves, of muscles, and of other endocrine glands are seen in states of PTH excess or deficiency. Most of these effects can be attributed to the concomitant changes in plasma calcium. However, the possibility exists that PTH or one of its metabolites may have direct action on these tissues, either by stimulating calcium transfer across their cell membranes or by some calcium-independent mechanism still to be discovered.

Overall action of PTH. The summated biochemical effects of PTH are illustrated in Fig. 51-8, which demonstrates the results of administering the hormone to a hypoparathyroid patient. There is a prompt increase in plasma calcium and a decrease in plasma phosphate. The renal tubular reabsorption of phosphate falls. Urinary excretion of calcium initially declines as its tubular reabsorption increases. However, as plasma calcium continues to increase, the filtered load goes up, and urinary calcium excretion also rises. Urinary excretion of hydroxyproline increases as a result of PTH-stimulated bone resorption.

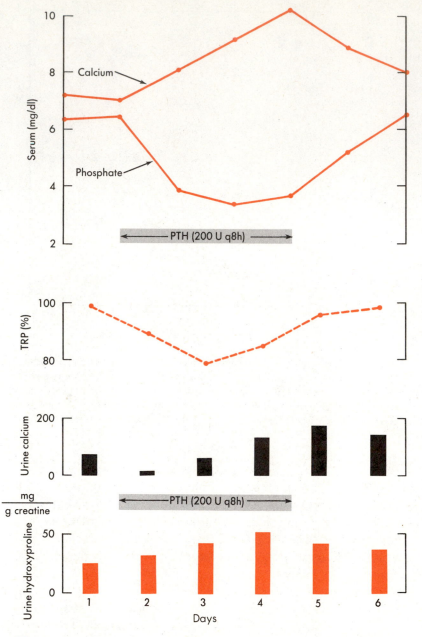

Fig. 51-8 ■ Effect of administration of PTH to a PTH-deficient human. *TRP,* Renal tubular resorption of phosphate. Note initial decrease in urine calcium followed by an increase as plasma calcium rises.

■ *Calcitonin*

■ *Synthesis and release*

The parafollicular, or C, cells of the thyroid gland secrete another protein hormone, *calcitonin,* which influences calcium metabolism. Whereas PTH acts to increase plasma calcium, calcitonin acts to lower it. Although uncertainty exists as to the significance of its physiological role in humans, calcitonin is likely to be an important regulator of plasma calcium in lower animals that live in an aquatic environment high in calcium. The parafollicular cells are of neural crest origin. In lower animals they form a discrete ultimobranchial gland from the sixth branchial pouch. In humans they are concentrated in the central portion of the lateral lobes of the thyroid gland. They are relatively large cells with a pale cytoplasm that contains small secretory granules enclosed in membranes. Parafollicular cells give rise to thyroid gland neoplasms, which also contain

*Endocrine regulation of
calcium and phosphate
metabolism*   **961**

large amounts of a nondescript protein called amyloid. Pathological secretion of calcitonin also is seen with other tumors that are of neural crest origin and that secrete amine or peptide hormones.

Calcitonin is a straight-chain peptide composed of 32 amino acids. It has a molecular weight of 3400. The hormone contains a seven-membered disulfide ring at the N-terminal and prolineamide at the C-terminal. There is considerable species variation in structure, but both fish and animal calcitonins are active in humans. The biologically active core of the molecule probably resides in its central region, although some evidence suggests that the entire peptide sequence is required. As in the case of PTH, the synthesis of calcitonin proceeds from a preprohormone through a prohormone to the monomeric form, which is packaged in granules.

The major stimulus to secretion of calcitonin is a rise in plasma calcium. However, the degree of response seen in various species is related to their need to prevent hypercalcemia. Vertebrates that originated in fresh water (of low calcium concentration) but migrated into the sea (with a calcium concentration of 40 mg/dl) were the first to require and to develop a calcium-lowering hormone. When vertebrates moved to land, the emphasis in calcium economy shifted toward defense against hypocalcemia rather than against hypercalcemia. PTH was developed, and the importance of calcitonin in plasma calcium regulation probably declined. Nevertheless calcitonin circulates in humans at concentrations of 10 to 100 pg/ml (10^{-11}M) and increases two- to tenfold after an acute increase in serum calcium of as little as 1 mg/dl. Much larger responses of the hormone to calcium infusion are elicited in patients with calcitonin-secreting tumors. Likewise, in such patients a sharp reduction in ionized calcium lowers plasma calcitonin. The stimulating effect of calcium on calcitonin secretion has been verified in vitro. It appears to involve an increase in intracellular cyclic AMP. Ingestion of food stimulates calcitonin secretion without elevating plasma calcium. This is mediated by several gastrointestinal hormones, of which gastrin is the most potent. The physiological significance of this phenomenon is not yet clear, but gastrin responsiveness provides a useful diagnostic test for states of calcitonin hypersecretion.

Circulating calcitonin is heterogeneous; immunoreactive molecules both larger and smaller than the native hormone have been found in the plasma of patients with tumors. Whether precursors are secreted normally is not known. Calcitonin appears to be largely cleared and degraded by the kidney. Its plasma half-life is less than 1 hour.

■ *Actions*

Binding of calcitonin to plasma membrane receptors is followed by an elevation of intracellular cyclic AMP. This second messenger initiates at least a portion of calcitonin action in all target cells. The subsequent intracellular events are obscure. Sequestration of calcium in the mitochondria, thus lowering cytosol calcium concentration, has been advanced as a mechanism for reducing the efflux of calcium from bone cells into the extracellular fluid. The major effect of calcitonin administration is a rapid fall in plasma calcium. The magnitude of this decrease is directly proportional to the baseline rate of bone turnover. Thus young growing animals are most affected, whereas in adults, who have more stable skeletons, only a minimal response is seen. Inhibition of osteolysis by osteocytes and inhibition of bone resorption by osteoclasts, particularly when these are stimulated by PTH, are the immediate mechanisms of the hypocalcemic action. An escape phenomenon often is noted after 12 hours, possibly caused by down regulation of calcitonin receptors. However, continued provision of calcitonin eventually decreases the number of osteoclasts and alters their morphology, as well as their activity. More dense bone with fewer resorption cavities eventually results. Calcitonin has minor effects, the reverse of those of PTH, on calcium handling by the kidney and gut.

Calcitonin is clearly a physiological antagonist to PTH with respect to calcium. However, with respect to phosphate, it has the same net effect as PTH; that is, it causes a decrease in plasma phosphate. The fate of the phosphate leaving the extracellular fluid under the action of calcitonin is still not clear, but most likely it enters bone. There is also a small, somewhat inconsistent increase in urinary phosphate excretion after calcitonin administration. The hypophosphatemic effect has been shown to be independent of the hypocalcemic effect. Nonetheless it has been suggested that enhanced entrance of phosphate into bone may mediate the hypocalcemic action of calcitonin, at least in part, by complexing calcium with phosphate within the bone cells.

The importance of calcitonin to normal human calcium economy is unclear. Under ordinary circumstances the absorption of dietary calcium loads produces little, if any, elevation of plasma calcium. Whether the increase in calcitonin provoked by eating helps prevent the development of postprandial hypercalcemia is not clear. Calcitonin deficiency resulting from complete removal of the thyroid gland does not lead to hypercalcemia, although there is some evidence that disposition of an acute calcium load is retarded. A chronic excess of calcitonin, generated either by tumor secretion or by exogenous administration, does not produce hypocalcemia. At present it may be most reasonable to conclude that any effects of calcitonin deficiency or excess are easily offset by appropriate adjustment of PTH and vitamin D levels.

On the other hand, the possibility that calcitonin participates significantly in the regulation of bone remodeling cannot be easily excluded. The fact that plasma calcitonin is lower in women than in men and also declines with aging may imply a functional role for the hormone in the common development of accelerated bone loss after the menopause. In lower vertebrates and mammals a role for calcitonin appears likely in diverse calcium-related processes, such as lactation, production of egg shells, and protection of the skeleton from the calcium drain of pregnancy. Calcitonin has found clinical uses in the acute treatment of hypercalcemia and in certain bone diseases, where a sustained reduction in osteoclastic resorption is therapeutically beneficial. Finally, the discovery of calcitonin in a number of locations throughout the body—within cells of neural crest origin that contain amine neurotransmitters—has raised the possibility that calcitonin also may have a neurotransmitter function.

■ *Vitamin D and its metabolism*

There are two sources of vitamin D in humans: that produced in the skin by ultraviolet irradiation (D_3) and that ingested in the diet (D_2). In this sense vitamin D is not a classic hormone of endocrine gland origin. But its intimate and hormonelike relationship with calcium and phosphate metabolism, its pathway of molecular modification to yield active metabolites, and its mechanism of action similar to that of bona fide steroid hormones all justify classifying it as a hormone. Certainly no discussion of the endocrine regulation of calcium and phosphate metabolism would be possible without including the role of vitamin D.

The structures of vitamin D_3, its precursor, and its metabolites are shown in Fig. 51-9. Vitamin D_2, which differs only in having an additional double bond at the 21 to 22 position, is derived from the plant sterol, ergosterol, by ultraviolet radiation and is a major dietary source in many countries. It has identical biological actions to vitamin D_3, and henceforth the term *vitamin D* is used to indicate both forms. Under the influence of sunlight 7-dehydrocholesterol is photoconverted to a previtamin D_3 of unknown structure, which spontaneously transforms to vitamin D_3. The process occurs in the epidermis, but a specific cell of origin has not been identified. In summer endogenous vitamin D is plentiful, and exogenous sources are unimportant. In winter or in sunlight-poor climates dietary vitamin D is essential for health. The most important sources are fish, liver, and irradiated milk. The minimum daily dietary requirement is approximately 2.5 μg (100 units), and the recommended daily intake is 10 μg. Because of its fat solubility, absorption of vitamin D from the gut is mediated by bile salts and occurs via the lymphatic glands.

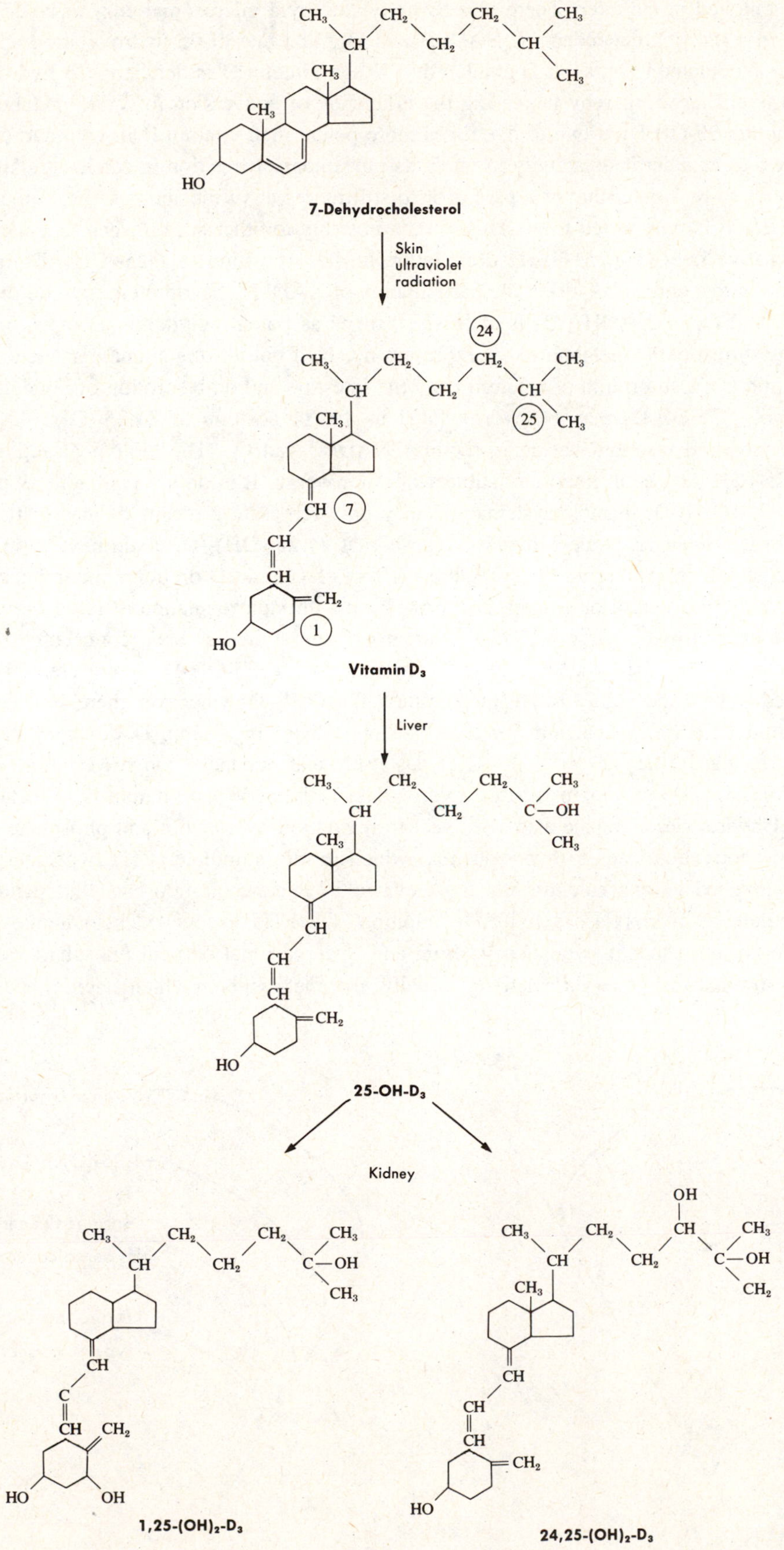

*Endocrine regulation of
calcium and phosphate
metabolism*   **963**

Fig. 51-9 ■ Structures of vitamin D₃, its precursor, and its metabolites.

Once either exogenous or endogenous vitamin D has entered the circulation, it is concentrated in the liver. There it is hydroxylated by a microsomal enzyme to 25-OH-D_3, in a reaction dependent on NADPH and O_2. The rate of 25-hydroxylation appears to be modulated by vitamin D availability. When vitamin D is deficient, 25-hydroxylation is enhanced, thereby increasing the efficiency of conversion to active metabolites. Although 25-OH-D_3 is two to five times more potent than vitamin D in vivo and can be shown to have intrinsic activity in vitro, its physiological function in calcium regulation appears to be largely that of a precursor to still more active metabolites. From the liver 25-OH-D_3 is transported to the kidney (and possibly to other sites), where it undergoes alternative fates (Fig. 51-9). Hydroxylation in the 1 position to 1,25-$(OH)_2$-D_3 occurs in the mitochondria, requiring the participation of NADPH, flavoprotein, cytochrome P-450, and O_2. 1,25-$(OH)_2$-D_3 is at least 10 times as potent as vitamin D in vivo. It is unquestionably the metabolite that expresses much, if not all, the stimulatory activity of vitamin D on absorption of calcium from the intestine and on resorption of bone. Alternatively, 25-OH-D_3 may be hydroxylated in the 24 position to 24,25-$(OH)_2$-D_3 in a mitochondrial reaction that again requires NADPH and O_2. The biological activity of 24,25-$(OH)_2$-D_3 is an important subject of controversy. It is only a twentieth as potent as 1,25-$(OH)_2$-D_3 in most systems and may only represent a means of inactivating vitamin D. However, some evidence suggests that 24,25-$(OH)_2$-D_3 could have a separate specific role in expressing the stimulatory effect of vitamin D on bone mineralization.

Endocrine control of vitamin D actions occurs through regulation of renal 1-hydroxylase and 24-hydroxylase activities. Numerous animal studies have contributed to the development of the concepts illustrated in Fig. 51-10. 25-OH-D_3 is preferentially directed toward the highly active metabolite 1,25-$(OH)_2$-D_3 whenever there is a lack of vitamin D, a lack of calcium, or a lack of phosphate. In vitamin D–deficient states it may be partly the lack of 1,25-$(OH)_2$-D_3 itself that enhances 1-hydroxylation, since 1,25-$(OH)_2$-D_3 is a suppressor of 1-hydroxylase activity. When vitamin D is sufficient, the 1-hydroxylase enzyme also is subject to regulation by calcium and phosphate. Calcium deprivation leads to hypocalcemia, which in turn stimulates PTH hypersecretion. The lowered plasma calcium and the elevated PTH concentration each independently stimulates 1-hydroxylase activity. In addition, the PTH excess leads to a phosphate diuresis; the resultant hypophosphatemia and lowered renal cortical phosphate content also stimulate 1-hydroxylase activity. Finally, in experiments in which calcium and PTH

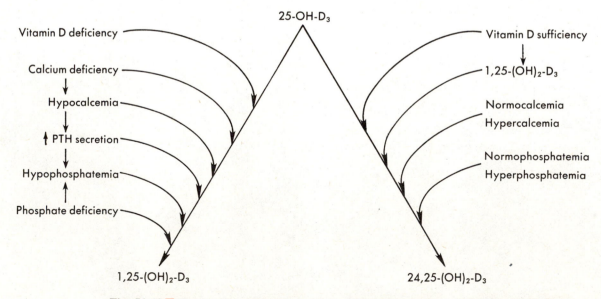

Fig. 51-10 ■ Factors that regulate conversion of 25-OH-D_3 to either 1,25-$(OH)_2$-D_3 or 24,25-$(OH)_2$-D_3. The former is increased by either calcium or phosphate lack.

*Endocrine regulation of
calcium and phosphate
metabolism*   **965**

are held constant, the imposition of phosphate deprivation leads to hypophosphatemia and directly to an increase in 1-hydroxylase activity.

The regulation of 24-hydroxylase activity is essentially the mirror image (Fig. 51-10). A sufficiency of vitamin D, via 1,25-$(OH)_2$-D_3, and normal or supernormal calcium and phosphate concentrations all stimulate the synthesis of 24,25-$(OH)_2$-D_3. Thus the supply of active 1,25-$(OH)_2$-D_3 is augmented whenever the mobilization of calcium or phosphate from intestine and bone into the extracellular fluid is needed. Conversely, augmentation of the supply of relatively inactive 24,25-$(OH)_2$-D_3 is favored when calcium and phosphate are plentiful and bone accretion can be sustained. The exact intracellular mechanism whereby these hydroxylations are regulated has not been defined. However, in all these instances many hours are required for the changes to be effected, implying that enzyme synthesis probably is involved.

Vitamin D, 25-OH-D_3, 1,25-$(OH)_2$-D_3, and 24,25-$(OH)_2$-D_3 all circulate bound to an α-globulin of about 50,000 molecular weight. The concentrations, approximate half-lives, and estimated daily production rates for the key metabolites in humans are shown in Table 51-2. It is evident that 1,25-$(OH)_2$-D_3 has by far the lowest concentration and the shortest half-life of the three. Certain relationships among the plasma concentrations are of physiological significance. 1,25-$(OH)_2$-D_3 concentration is ordinarily independent of 25-OH-D_3 concentration. Except in vitamin D–deficiency states, the regulatory factors previously outlined appear to maintain the appropriate concentration of this active metabolite irrespective of the supply of precursor. In contrast, 24,25-$(OH)_2$-D_3 concentration is ordinarily directly proportional to 25-OH-D_3 concentration. This suggests that 24-hydroxylation is a key sluice for disposing of the excess precursor that reaches the kidney. Both 1,25-$(OH)_2$-D_3 and 24,25-$(OH)_2$-D_3 can be cross-hydroxylated in the kidney to 1,24,25-$(OH)_3$-D_3, a compound of lesser activity. Further inactivation of vitamin D metabolites occurs by hydroxylation at the 26 and 27 positions.

1,25-$(OH)_2$-D_3 acts through the general mechanism outlined for steroid hormones (Fig. 49-7). After combination with a cytosol receptor the hormone receptor complex enters the nucleus, where it stimulates transcription of mRNA for at least one identified product. This is a calcium-binding protein that can be isolated from the intestinal mucosa and from kidney cells. A 12-hour lapse between administration of vitamin D and its earliest measurable effects on the intestine is consistent with the transcriptional mechanism. Absorption of calcium from the intestinal lumen against a concentration gradient is the major action of 1,25-$(OH)_2$-D_3. In some manner the increased quantity of calcium-binding protein facilitates transfer of calcium across the brush border into the cytoplasm. Calcium subsequently is extruded on the opposite side of the cell and enters the bathing capillary blood. This action of 1,25-$(OH)_2$-D_3 is responsible for the adaptation, described previously, by which intestinal calcium absorption adjusts to alterations in dietary calcium. By an independent, still unknown mechanism active absorption of phosphate across the intestinal cell membrane also is stimulated by vitamin D.

1,25-$(OH)_2$-D_3, but not 25-OH-D_3, stimulates bone resorption in vitro and in vivo. A cytosolic receptor for 1,25-$(OH)_2$-D_3 has been found in bone cells, and the resorptive effect can be blocked by actinomycin, a DNA transcription inhibitor. In this action 1,25-

TABLE 51-2

Vitamin D metabolism in humans

	Plasma concentration ($\mu g/L$)	Plasma half-life (days)	Estimated production rate ($\mu g/day$)
1,25-$(OH)_2$-D_3	0.03	1 to 3	1
24,25-$(OH)_2$-D_3	2	15 to 40	1
25-OH-D_3	20	5 to 20	10

$(OH)_2$-D_3 synergizes critically with PTH by an unknown mechanism. There is additional evidence that exceedingly low concentrations of $1,25$-$(OH)_2$-D_3 also can stimulate osteolysis and mobilizaton of bone calcium, independent of PTH.

The normal mineralization of newly formed osteoid in regular fashion along a calcification front is also critically dependent on vitamin D. In the absence of the vitamin, excess osteoid accumulates from osteoblastic activity, and the bone so formed is weakened. It remains enigmatic whether this action of vitamin D is entirely accounted for by augmentation of the supply of calcium and phosphate in the fluid bathing the osteoblast, or whether $1,25$-$(OH)_2$-D_3 or another metabolite, perhaps $24,25$-$(OH)_2$-D_3, acts in a direct and specific manner on the bone cells themselves to hasten mineralization. The presence of high-affinity receptors for $1,25$-$(OH)_2$-D_3 in osteoblasts supports a direct action. In any case the first observable effect of vitamin D replacement on the bone from vitamin D–deficient animals and humans is the reappearance of a normal mineralized front.

Despite much contradictory evidence, it is likely that $1,25$-$(OH)_2$-D_3 has direct renal effects similar to those of PTH. However, these probably are far outweighed by the action of PTH and the effects of ambient calcium and phosphate concentrations themselves on renal tubular function. $1,25$-$(OH)_2$-D_3 also accumulates in the parathyroid glands, and a vitamin D–dependent calcium-binding protein can be found there. It is likely, but not well established, that direct feedback inhibition of PTH secretion by either $1,25$-$(OH)_2$-D_3 or another vitamin D metabolite occurs.

A major storage site for vitamin D is in muscle tissue. Profound muscle weakness is a prominent result of vitamin D deficiency. Although no direct effect of vitamin D on muscle cells has yet been discovered, there is some evidence that 25-OH-D_3 may play a role in the maintenance of muscle metabolism and function.

■ Integrated hormonal regulation of calcium and phosphate

From all the foregoing it should be clear that a complex interplay of several hormones acting on a number of tissues is responsible for maintenance of normal concentrations of calcium and phosphate in body fluids. This integrated system is best visualized by tracing the compensatory responses to deprivation of calcium and of phosphate (Figs. 51-11 and 51-12).

Calcium deprivation with hypocalcemia as the signal, primarily stimulates PTH secretion. PTH increases urinary phosphate excretion, thereby decreasing serum phosphate and renal cortical phosphate content. All three factors—excess PTH, hypocalcemia, and hypophosphatemia—act to stimulate the production of $1,25$-$(OH)_2$-D_3. The latter raises plasma calcium toward normal by increasing absorption of calcium from the gastrointestinal tract and, in concert with PTH, by increasing osteocytic and osteoclastic bone resorption. Only PTH further acts to return calcium to the plasma by increasing its reabsorption from the renal tubular urine. Thus this beautifully integrated response to calcium deprivation increases the flux of calcium into the extracellular fluid. Simultaneously the extra phosphate that enters with the calcium from the bone and gut is disposed of by excretion in the urine. The recovery of plasma calcium to normal will shut off PTH hypersecretion by negative feedback. $1,25$-$(OH)_2$-D_3 synthesis will then decline, $24,25$-$(OH)_2$-D_3 synthesis will increase, and the whole sequence will diminish. As a further safety valve, should the compensatory rise in plasma calcium exceed normal levels, stimulation of calcitonin secretion would act to moderate it as well.

In a contrasting sequence phosphate deprivation via hypophosphatemia will directly stimulate $1,25$-$(OH)_2$-D_3 production. The latter increases the flux of phosphate into the extracellular fluid by stimulating its absorption from the gut and by stimulating bone resorption. The extra calcium that simultaneously enters the extracellular fluid raises plasma calcium. Hypophosphatemia, by retarding bone formation, also directly helps raise plasma calcium. The increase in plasma calcium suppresses PTH secretion. The absence of PTH causes tubular reabsorption of phosphate to increase, conserving urinary

*Endocrine regulation of
calcium and phosphate
metabolism*   **967**

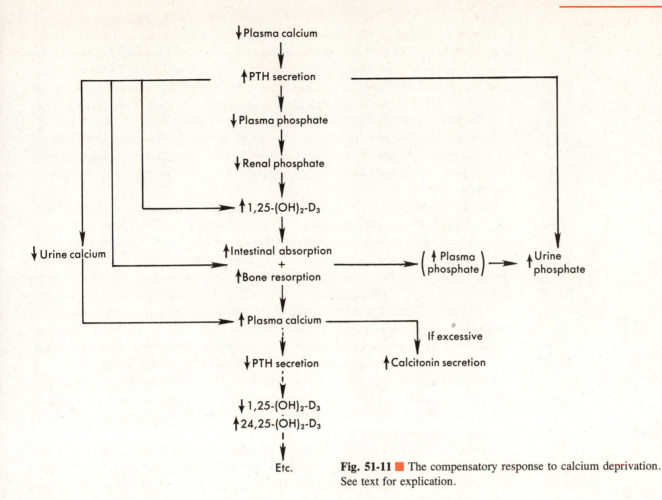

Fig. 51-11 ■ The compensatory response to calcium deprivation.
See text for explication.

Fig. 51-12 ■ The compensatory response to phosphate deprivation. See text for explication.

phosphate and aiding in the restoration of plasma phosphate levels to normal. At the same time the lack of PTH permits easier disposal of the extra calcium that was mobilized by diminishing its renal tubular reabsorption and increasing its excretion in the urine. As plasma phosphate returns to normal, $24,25\text{-}(OH)_2\text{-}D_3$ production will be favored, $1,25\text{-}(OH)_2\text{-}D_3$ levels will decline, and the whole process will be reversed.

Thus this combined arrangement of dual hormone regulation and dual hormone action permits selective defense of either plasma calcium or plasma phosphate, without creating a circulatory excess of the other. Obviously the same principles apply in reverse to imposition of excess calcium or excess phosphate loads of either endogenous or exogenous sources.

Certain characteristics of the homeostatic systems for adjustment of body calcium and phosphate deserve final emphasis. The renal responses to PTH provide the most rapid (within minutes) defense against perturbations of both calcium and phosphate. As PTH excretion ranges from very high to very low levels, the rate of urinary calcium excretion can rise 25-fold from approximately 0.05 to 1.2 mg/minute, and that of phosphate can fall from 2 to 0 mg/minute. A sudden 2 to 3 mg/dl increase in either calcium or phosphate concentration in the extracellular fluid can be corrected within 24 hours by the kidney, acting under the appropriate alteration in PTH levels. In the face of complete deprivation renal conservation of phosphate is complete, whereas that of calcium is not quite so. The gastrointestinal component of this homeostatic system is both slower and narrower in range. As a result of variations in $1,25\text{-}(OH)_2\text{-}D_3$, the absorption of dietary calcium increases from 20% to 70% as calcium intake decreases from 2000 to 200 mg/day. Thus absorbed calcium can effectively range from 140 to 400 mg/day. Hormonal effects on phosphate absorption are even less striking, since the latter is virtually a linear function of dietary intake. Bone responses to regulatory fluctuations in both PTH and $1,25\text{-}(OH)_2\text{-}D_3$ are rapid when produced by osteocytic osteolysis and relatively slow when caused by osteoclastic resorption. However, the capacity for compensatory calcium and phosphate uptake and release is enormous. For example, in humans tenfold variations in calcium turnover have been observed. Finally, an important major difference between the renal and gastrointestinal mechanisms on the one hand and the bone mechanisms on the other hand must be borne in mind. The compensatory responses of the kidney and the gut have the virtue of defending total body *and bone* stores of calcium and phosphate against erosion or inundation. In contrast, the skeletal mechanisms of defense against perturbations of plasma calcium and phosphate have the disadvantage that their long-term employment eventually sacrifices the chemical and structural integrity of the bone mass.

■ Clinical syndromes of hormone dysfunction
■ PTH

Hyperparathyroidism most commonly results from a single parathyroid adenoma and less often from hyperplasia of all four glands. The clinical picture is largely dominated by the effects of hypercalcemia, which depresses neuromuscular excitability. This produces symptoms of dulled mentation, lethargy, anorexia, constipation, and muscle weakness. If urinary excretion of calcium is excessive, renal stones may form. Peptic ulcers can result from stimulation of gastric acid secretion by calcium. In long-standing cases with marked hypersecretion of PTH, massive irregular bone resorption, demonstrable on x-ray films, causes weakening of the bones, pain, fractures, and deformities. Persistent hypercalcemia also can lead to reduced renal function and to irreversible renal failure, secondary to deposition of calcium in the kidney cells. Hyperparathyroidism must be distinguished from numerous other causes of hypercalcemia. The diagnosis is established by demonstrating an elevated plasma total calcium or ionized calcium, a decreased plasma phosphate level, and an elevated PTH level. A mild hyperchloremic acidosis is also frequently present. Treatment is surgical removal of the hyperfunctioning tissue.

*Endocrine regulation of
calcium and phosphate
metabolism*   **969**

Hypoparathyroidism most often is caused by autoimmune idiopathic atrophy or inadvertent surgical removal of the glands. In rare cases the problem is end organ resistance to PTH actions. Calcium absorption from the gut is diminished, because of defective synthesis of $1,25\text{-}(OH)_2\text{-}D_3$, and there is also decreased bone resorption. The resultant hypocalcemia increases neuromuscular excitability. This produces hyperactive reflexes, spontaneous muscle contractions, convulsions, and laryngeal spasm with airway obstruction. Formation of ocular cataracts is common but not well explained. Low plasma total calcium or ionized calcium and elevated plasma phosphate levels are essential to the diagnosis. Plasma PTH levels are low when hypoparathyroidism is caused by destruction of the glands. Plasma PTH is elevated when target tissue unresponsiveness is the problem. In the former case administration of exogenous PTH for diagnostic purposes causes prompt increases in urinary excretion of cyclic AMP and phosphate; in the latter case it does not. Although treatment with PTH would be logical, this is impractical because of the development of serum antibodies to the bovine hormone preparations currently available. Instead replacement with $1,25\text{-}(OH)_2\text{-}D_3$ and calcium supplements is used, with acceptable clinical effectiveness.

■ *Vitamin D*

Toxicity occurs from excessive administration of vitamin D or, in a few diseases, from abnormal sensitivity to ordinary amounts of the vitamin. Absorption of calcium from the gut and resorption of bone are enhanced. This results in elevated plasma and urinary calcium, with the same clinical consequences noted previously for hyperparathyroidism. In contradistinction to the latter, however, vitamin D excess raises, rather than lowers, serum phosphate because the hypercalcemia shuts off PTH secretion. Tubular reabsorption of phosphate therefore is increased, and phosphate is retained. The duration of toxicity is greatest from exogenous vitamin D itself because of its large storage capacity and low turnover rate; it is least for $1,25\text{-}(OH)_2\text{-}D_3$, which is rapidly removed from the body. Treatment consists of blocking the effects of vitamin D by administration of calcitonin or cortisol while waiting for the excess vitamin D to be cleared.

Deficiency of vitamin D action can result from inadequate sunlight, lack of dietary intake, diminished absorption, or defective hydroxylation in the liver or kidney. This leads to decreased gastrointestinal absorption of calcium and phosphate. The resultant fall in plasma calcium is buffered by stimulation of PTH secretion. This in turn strongly accentuates the fall in plasma phosphate by decreasing its tubular reabsorption in the kidney and increasing its excretion in the urine. Urinary calcium excretion, on the other hand, is reduced by the PTH excess. The strongly negative phosphate balance and more modestly negative calcium balance together cause a decrease in the rate of bone mineralization. This is further aggravated by the lack of the putative direct action of a vitamin D metabolite on calcification of osteoid. An excess of osteoid accumulates on bone surfaces, and no regular line of calcification can be seen.

Two clinical pictures result from vitamin D deficiency. In children the centers of endochondral ossification at the epiphyseal plates are most critically affected, producing growth failure and the characteristic deformities and x-ray pictures of *rickets*. In adults softening and bending of long bones, with fracture lines along nutrient arteries, creates the condition known as *osteomalacia*. The typical diagnostic pattern is slightly reduced plasma calcium, greatly reduced plasma phosphate, and elevated plasma PTH. The level of alkaline phosphatase is also high in the plasma because of overactivity of the osteoblasts. In adults bone biopsy may be needed to clarify subtle cases. Treatment of simple deficiency consists of replacement doses of vitamin D itself, with supplemental calcium. In cases caused by biosynthetic defects (for example, hepatic or renal disease) or target tissue unresponsiveness large doses of vitamin D, $25\text{-}OH\text{-}D_3$, or $1,25\text{-}(OH)_2\text{-}D_3$ may be needed.

■ *Bibliography*

Journal articles

Austin, L.A., et al.: Calcitonin: physiology and pathophysiology, N. Engl. J. Med. **304:**269, 1981.

Bordier, P., et al.: Vitamin D metabolites and bone mineralization in man, J. Clin. Endocrinol. Metab. **46:**284, 1978.

Cheung, W.Y.: Calmodulin plays a pivotal role in cellular regulation, Science **207:**19, 1979.

DeLuca, H.F.: Recent advances in our understanding of the vitamin D endocrine system, J. Lab. Clin. Med. **87:**7, 1976.

DeLuca, H.F.: Vitamin D: revisited 1980, Clin. Endocrinol. Metab. **1:**3, 1980.

Haddad, J.G., et al.: Circulating 25-hydroxyvitamin D in man, Am. J. Med. **57:**57, 1974.

Harrison, H.E.: Parathyroid hormone and vitamin D, Yale J. Biol. Med. **38:**393, 1966.

Kaminsky, N.I., et al.: Effects of parathyroid hormone on plasma and urinary adenosine 3′,5′-monophosphate in man, J. Clin. Invest. **49:**2387, 1970.

Mawer, E.B.: Clinical implications of measurements of circulating vitamin D metabolites, Clin. Endocrinol. Metab. **9:**63, 1980.

Nordin, B.E.C., et al.: Role of kidney in regulation of plasma-calcium, Lancet **2:**1280, 1969.

Parfitt, A.M.: The actions of parathyroid hormone on bone: relation to bone remodeling and turnover, calcium homeostasis, and metabolic bone disease. I. Mechanisms of calcium transfer between blood and bone and their cellular basis: morphologic and kinetic approaches to bone turnover, Metabolism **25:**809, 1976.

Parfitt, A.M.: The actions of parathyroid hormone on bone: relation to bone remodeling and turnover, calcium homeostasis, and metabolic bone disease. II. PTH and bone cells: bone turnover and plasma calcium regulation, Metabolism **25:**909, 1976.

Parthemore, J.G., et al.: Calcitonin secretion in normal human subjects, J. Clin. Endocrinol. Metab. **47:**184, 1978.

Raisz, L.G.: Direct effects of vitamin D and its metabolites on skeletal tissue, Clin. Endocrinol. Metab. **9:**27, 1980.

Rasmussen, H.: Cell communication, calcium ion, and cyclic adenosine monophosphate, Science **170:**404, 1970.

Reiss, E., et al.: The role of phosphate in the secretion of parathyroid hormone in man, J. Clin. Invest. **49:**2146, 1970.

Russell, R.G.G., et al.: Physiological and pharmacological aspects of 24,25-dihydroxycholecalciferol in man, Adv. Exp. Med. Biol. **103:**487, 1978.

Sherwood, L., et al.: Evaluation by radioimmunoassay of factors controlling the secretion of parathyroid hormone; intravenous infusions of calcium and ethylene diamine tetraacetic acid in the cow and goat, Nature **209:**52, 1966.

Books and monographs

Avioli, L.V., et al.: Bone metabolism and disease. In Bondy, P.K., and Rosenberg, L.E.: Metabolic control and disease, Philadelphia, 1980, W.B. Saunders Co.

Mayer, G.P.: Effect of calcium and magnesium on parathyroid hormone secretion rate in calves. In Talmadge, R.V., et al.: Calcium regulating hormones, Amsterdam, 1975, Excerpta Medica.

Neer, R.M.: Calcium and inorganic phosphate homeostasis. In Degroot, L.J., editor: Endocrinology, New York, 1979, Grune & Stratton, Inc.

chapter 52

The hypothalamus and the pituitary gland

The hypothalamus-pituitary consortium forms the most complex and dominant portion of the entire endocrine system. Its internal anatomical and functional relationships are elaborate and subtle. The output of the hypothalamus-pituitary unit regulates the function of the thyroid, adrenal, and reproductive glands and also shares in the control of somatic growth, lactation, milk secretion, and water metabolism. Two hormones, *antidiuretic hormone (ADH* or *vasopressin)* and *oxytocin,* are synthesized in the hypothalamus but are stored and secreted by the *posterior pituitary gland,* or *neurohypophysis.* A group of *tropic hormones,* adrenocorticotropic hormone (ACTH), thyroid-stimulating hormone (TSH), luteinizing hormone (LH), follicle-stimulating hormone (FSH), growth hormone, and prolactin, are synthesized, stored, and secreted by the *anterior pituitary gland,* or *adenohypophysis*. However, a set of releasing and inhibiting hormones that are synthesized in the hypothalamus and travel to the adenohypophysis regulates the synthesis and secretion of these tropic hormones. When one considers that all of this emanates from a mass of only 500 mg of pituitary in association with the adjacent brain tissue, the performance is truly awesome.

■ Anatomy

A knowledge of the embryological development of the pituitary gland is crucial to understanding its anatomy and function. The fully developed gland is really an amalgam of hormone-producing glandular cells (the adenohypophysis) and neural cells with secretory function (the neurohypophysis). The anterior endocrine portion of the pituitary develops from an upward outpouching of ectodermal cells from the roof of the oral cavity (Rathke's pouch). This pouch eventually pinches off and becomes separated from the oral cavity by the sphenoid bone of the skull. The lumen of the pouch is reduced to a small cleft. The posterior neural portion of the pituitary develops from a downward outpouching of ectoderm from the brain in the floor of the third ventricle. The lumen of this pouch is obliterated inferiorly as the sides fuse into the infundibular process. Superiorly, the lumen remains contiguous with and forms a recess in the adult third ventricle. The upper portion of this neural stalk expands to invest the lowest portion of the hypothalamus and is called the *median eminence.* The cleftlike remnant of Rathke's pouch demarcates the interwoven anterior and posterior portions of the pituitary. In some animals, but not in humans, cells in the area of Rathke's pouch and adjacent to the neurohypophysis form a distinct intermediate lobe of variable size. The entire pituitary gland sits in a socket of sphenoid bone called the *sella turcica*. A reflection of the dura mater, called the *diaphragm,* extends across the top of the sella turcica and separates the bulk of the pituitary gland from the brain. However, the neural stalk penetrates the diaphragm, maintaining its continuity with the brain. These anatomical relationships are shown in Fig. 52-1, *B*. The recent ability to visualize the human pituitary gland by computerized axial tomography (CAT scanning) has greatly facilitated detection of disease in this area (Fig. 52-1, *A*).

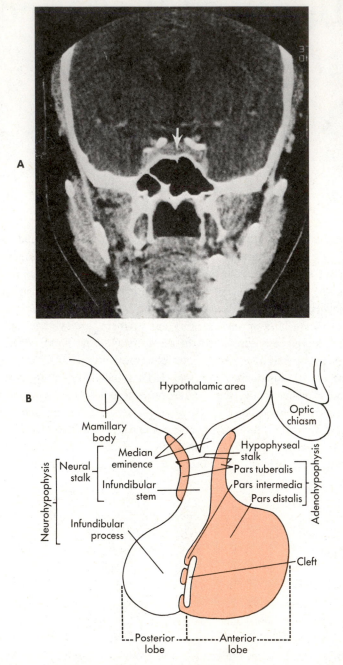

Fig. 52-1 ■ **A,** CAT scan of the skull showing the location of the pituitary gland (arrow) in the sella turcica above the sphenoid sinus. **B,** Diagram of the pituitary gland showing its division into the adenohypophysis and neurohypophysis. (**B** adapted from an original painting by Frank H. Netter, M.D., from The CIBA Collection of Medical Illustrations copyright by CIBA Pharmaceutical Co., Division of CIBA-Geigy Corporation.)

The blood supply to this complex of neural and endocrine tissue is equally important. In the posterior pituitary, the neural tissue of the infundibular process derives its blood mostly from the inferior hypophyseal artery, and the capillary plexus thereof drains into the dural sinus. The neural tissue of the upper stalk and of the median eminence is supplied largely by the superior hypophyseal artery. After investing the axons in these areas, the capillary plexus emanating from this artery forms a set of long portal veins that carry the blood downward into the anterior pituitary. There these portal

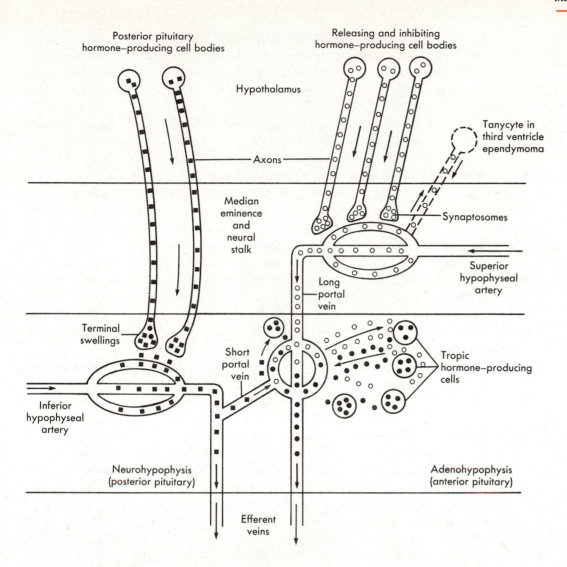

Posterior pituitary hormone-producing cell bodies

Releasing and inhibiting hormone-producing cell bodies

Hypothalamus

Axons

Tanycyte in third ventricle ependymoma

Median eminence and neural stalk

Synaptosomes

Superior hypophyseal artery

Long portal vein

Terminal swellings

Short portal vein

Tropic hormone-producing cells

Inferior hypophyseal artery

Neurohypophysis (posterior pituitary)

Adenohypophysis (anterior pituitary)

Efferent veins

■ Posterior pituitary hormones
● Anterior pituitary tropic hormones
○ Releasing or inhibiting hormones

Fig. 52-2 ■ Anatomical and functional relationships between the hypothalamus, the pituitary gland, and its blood supply. Arrows indicate direction of movement of hormone molecules. Note that the adenohypophysis has no direct arterial supply but receives blood from the median eminence, which contains hypothalamic releasing and inhibiting hormones.

veins give rise to a second capillary plexus that supplies the endocrine cells with the majority of their blood, which is then drained off into the dural sinus. The anterior pituitary receives its remaining blood via a set of short portal veins originating in the capillary plexus of the inferior hypophyseal artery within the neural stalk. Thus there is little or no direct arterial blood supply to the adenohypophyseal cells. Furthermore, it should be noted that the pituitary gland lies outside the blood-brain barrier.

The reasons for this involved anatomical arrangement become apparent when the functional relationships are examined. These are illustrated in block diagram form in Fig. 52-2. The entire neurohypophysis represents a collection of axons whose cell bodies lie in the hypothalamus. Peptide hormones synthesized in the cell bodies of these hypothalamic neurons travel down their axons in neurosecretory granules to be stored in the nerve terminals lying in the posterior pituitary gland. After neural stimulation of the cell bodies, the granules are released from the terminals by exocytosis; the peptide hormones then enter the peripheral circulation via the capillary plexuses of the inferior hypophyseal artery. Thus a single cell performs the entire process of hormone synthesis, storage, and release.

In contrast, the adenohypophysis is a collection of glandular cells that are regulated by bloodborne stimuli originating in neural tissue. Cell bodies of hypothalamic neurons synthesize *releasing hormones* and *inhibiting hormones,* which travel in packets down their axons only as far as the median eminence. Here they are stored as neurosecretory granules in the nerve terminals (synaptosomes). After stimulation of these hypothalamic cells, the releasing or inhibiting hormones are discharged into the median eminence and enter the capillary plexus of the superior hypophyseal artery. Having now left neural tissue, they are transported down the long portal veins and exit from a second capillary plexus to reach their specific glandular target cells in the anterior pituitary. These cells then react to the releasing hormones or inhibiting hormones by increasing or decreasing their output of tropic hormones. The latter enter the same second capillary plexus through which they ultimately reach the peripheral circulation. Thus two cells, one neural and one glandular, participate in the processes leading to synthesis and release of the anterior pituitary tropic hormones.

The preceding description implies an entirely unidirectional arrangement. However, recent studies have suggested that this may not be entirely the case. It has been shown that not all the venous drainage from the anterior pituitary necessarily empties directly into the dural sinus and systemic circulation. The short portal veins may act as conduits for reverse flow of blood from the anterior pituitary cells through the neurohypophyseal capillary plexus back up to the neurons in either the median eminence or the hypothalamus itself. This direction of flow would permit anterior pituitary tropic hormones to bathe the neurons in high concentration without impedance from the blood-brain barrier. It is also possible that two-way traffic between the cerebrospinal fluid and both the neurohypophysis and adenohypophysis may exist. Specialized ependymal cells in the inferior recess of the third ventricle send long processes down into the portal veins of the median eminence and pituitary gland (Fig. 52-2). These cells, known as *pituitary tanycytes,* could facilitate transfer of regulatory substances from the cerebrospinal fluid to the pituitary. They could also allow posterior pituitary peptide hormones, hypothalamic releasing or inhibiting hormones, or even anterior pituitary tropic hormones to have access to the brain via the cerebrospinal fluid. Further definition of the physiological significance of these possible relationships is awaited.

■ *Hypothalamic function*

The hypothalamus clearly plays a key role in regulating pituitary function. It can be considered a central relay station for collecting and integrating signals from diverse sources and funneling them to the pituitary. The hypothalamus receives afferent nerve tracts from the thalamus, the reticular activating substance, the limbic system (amygdala, hippocampus, and habenula), the eyes, and remotely from the neocortex. Some of the connections are multisynaptic. Through this input, pituitary function can be influenced by pain, sleep or wakefulness, emotion, fright, rage, olfactory sensations, light, and possibly even thought. It can be coordinated with patterned behavior and mating responses. The proximity of other hypothalamic nuclei that govern autonomic nervous system function also allows coordination between the output of pituitary hormones and either augmentation or reduction of sympathetic or parasympathetic activities. Hypothalamo-hypothalamic tracts exist that may help to integrate multiple simultaneous pituitary responses with each other, as well as to regulate pituitary function in accordance with change in temperature, caloric status, or fluid balance. The neurotransmitters involved in afferent impulses to the hypothalamus are largely norepinephrine and serotonin. Dopamine (along with acetylcholine or γ-aminobutyric acid) appears to act as the chief neurotransmitter for efferent nerve tracts to the median eminence. These impulses regulate the discharge of releasing hormones or inhibiting hormones from the nerve terminals in the median eminence (Fig. 52-2). In addition, neurotransmitters from the hypothalamus may reach the portal vein blood and directly influence anterior pituitary tropic hormone output. The presence of neurotransmitter receptors in adenohypophyseal cells supports this possible regulatory route.

The hypothalamus-pituitary axis is under the influence of blood-borne substances and neural input. Virtually all of the tropic hormones from the adenohypophysis cause changes either in the concentrations of peripheral gland hormones (thyroid, adrenal, gonadal) or of substrates, such as glucose or free fatty acids. Conditions exist for at least three levels of *humoral feedback,* as illustrated in Fig. 52-3. Peripheral gland hormones or substrates arising from tissue metabolism can exert feedback control on both the hypothalamus and the anterior pituitary gland. This is known as *long-loop feedback* and is usually negative, although it can occasionally be positive. Negative feedback can also be exerted by the tropic hormones themselves on the synthesis or discharge of hypothalamic releasing or inhibiting hormones. This is known as *short-loop feedback.* Since tropic hormones do not ordinarily cross the blood-brain barrier,

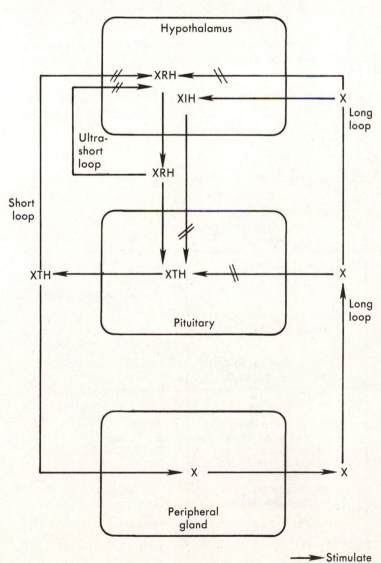

Fig. 52-3 ● Negative feedback loops regulating hormone secretion in a typical hypothalamus–pituitary–peripheral gland axis. *X,* Peripheral gland hormone; *XTH,* pituitary tropic hormone; *XRH,* hypothalamic releasing hormone; *XIH,* hypothalamic inhibiting hormone.

short-loop feedback may occur either by specialized transport across fenestrated endothelial cells of the capillaries that bathe hypothalamic neurons or by retrograde flow through the short portal veins, as just described. Finally, there is some evidence that hypothalamic releasing hormones may even inhibit their own synthesis and discharge. This is called *ultrashort loop feedback*. It could occur under unusual circumstances in which hypothalamic releasing hormones might gain access to the peripheral circulation in high concentrations, or it could be mediated by transport of the releasing hormone to the cerebrospinal fluid via the pituitary tanycytes. Examples of each of these feedback loops will be noted as the individual pituitary hormones are discussed.

The large neurons responsible for posterior pituitary peptide hormone synthesis are localized to two well-defined areas of the anterior hypothalamus, namely, the *supraoptic* and *paraventricular nuclei*. Their axons form definable tracts to the posterior pituitary gland. The smaller neurons responsible for releasing and inhibiting hormone synthesis are both clustered in defined areas and scattered throughout the hypothalamus. Varying degrees of localization of specific anterior pituitary releasing or inhibiting hormones to particular areas of the hypothalamus have been observed, although considerable overlap also exists. Furthermore, the arrangement of hormone-specific axons within the tracts to the median eminence remains to be worked out. Current mapping data also leave open the question as to whether a single hypothalamic cell type is responsible for production of each individual hormone.

Table 52-1 lists the currently known and strongly suspected hypothalamic releasing or inhibiting hormones. All appear to be peptide in nature except for dopamine. Studies thus far suggest their synthesis is carried out by ribosomes in classic fashion. In at least one instance, that of somatostatin, a prohormone is likely to be involved. Further studies are needed to elucidate the mechanisms for packaging these hormones into secretory

TABLE 52-1

Hypothalamic hormones and factors

		Structure	*Target tropic hormones*
Thyrotropin-releasing hormone	TRH	pyroglu-his-pro-NH$_2$	Thyrotropin Prolactin Growth hormone (pathological)
Luteinizing hormone–releasing hormone	LHRH	pyroglu-his-trp-ser-tyr-gly-leu-arg-pro-gly-NH$_2$	LH FSH Growth hormone (pathological)
Growth hormone–inhibiting hormone	Somatostatin	ala-gly-cys-lys-asn-phe-phe-trp-lys-thr-phe-thr-ser-cys	Growth hormone Prolactin Thyrotropin ACTH (pathological)
Corticotropin-releasing factor	CRF	ser-gln-glu-pro-pro-ile-ser-leu-asp-leu-thr-phe-his-leu-leu-arg-glu-val-leu-glu-met-thr-lys-ala-asp-gln-leu-ala-gln-gln-ala-his-ser-asn-arg-lys-leu-leu-asp-ile-ala-NH$_2$	ACTH β-Lipotropin Endorphins
Prolactin-inhibiting factor	PIF	Dopamine (?)	Prolactin
Prolactin-releasing factor	PRF	Not established	Prolactin
Growth hormone–releasing factor	GHRF	Recently proposed	Growth hormone

granules and transporting them to the median eminence, for relating their synthesis to that of possible carrier proteins, and for coordinating their synthesis with that of neurotransmitters within the same cell of origin. All the currently identified hypothalamic releasing and inhibiting hormones have been named on the basis of the anterior pituitary hormone whose secretion they were originally discovered to influence. Although it was initially presumed that each tropic hormone might be under the control of a unique hypothalamic releasing or inhibiting hormone and that each hypothalamic hormone would have only one target anterior pituitary cell, the actual physiology has proven to be much more complex. The tripeptide conventionally known as *thyrotropin-releasing hormone (TRH)* can also stimulate secretion of prolactin from normal cells and secretion of growth hormone from neoplastic pituitary cells. Somatostatin, discovered as a growth hormone–inhibiting factor, can also inhibit the secretion of thyrotropin and prolactin. Undoubtedly other examples of overlap will be discovered.

The anterior pituitary contains a collection of at least five endocrine cell types. These cannot be completely distinguished by conventional histological staining or by localization to specific areas. However, in recent years, the development of immunohistochemical techniques employing hormone-specific antisera has permitted each type to be identified specifically in normal pituitary tissue and in functioning pituitary tumors. Fig. 52-4 demonstrates the results of immunohistochemical staining of serial sections of a

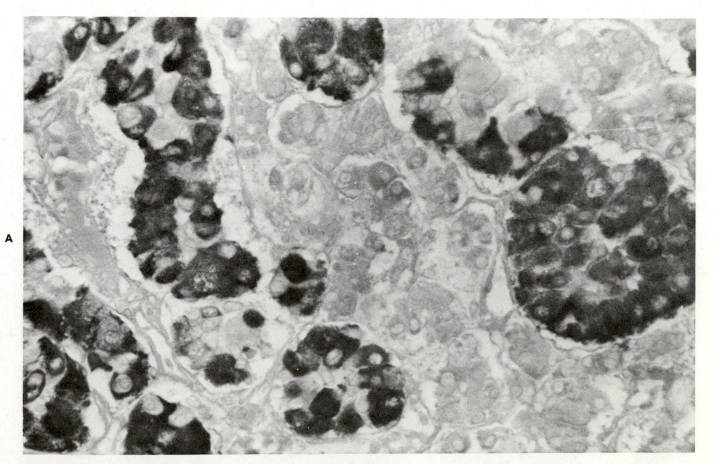

A

Fig. 52-4 ■ Adjacent sections of a human anterior pituitary gland stained by immunohistochemical techniques for ACTH **(A)**, HGH **(B)**, and prolactin **(C)**. Note the different patterns of distribution of the three hormones. (Courtesy Dr. Manuel Velasco, Case Western Reserve University, School of Medicine, Cleveland, Ohio.)   *Continued.*

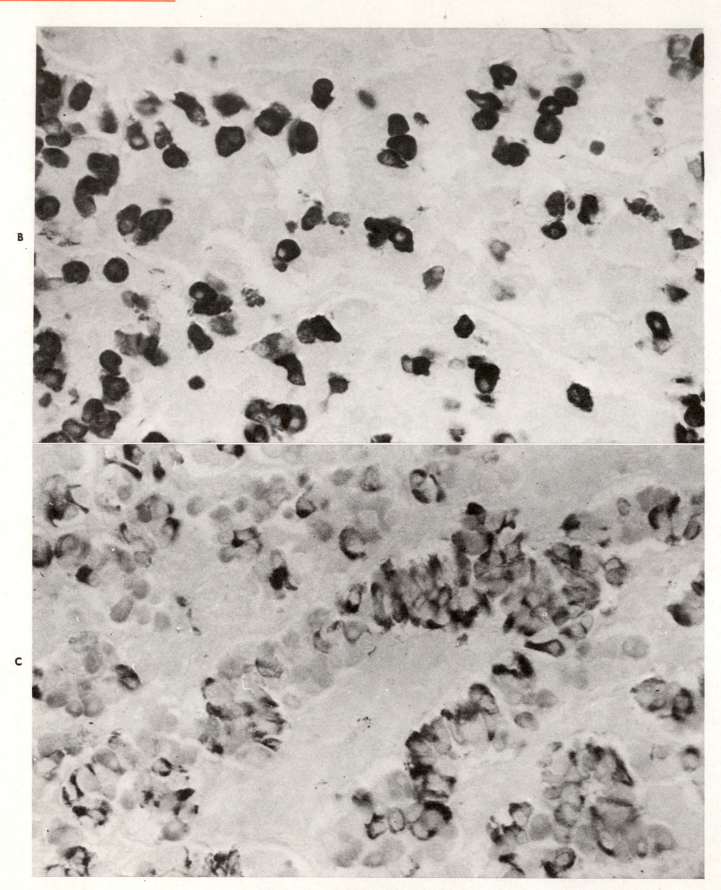

Fig. 52-4, cont'd ■ For legend see p. 977.

TABLE 52-2

Anterior pituitary cells and hormones

Cell	Pituitary population (%)	Products, molecular weight	Targets
Corticotroph	20	ACTH, 4500 β-Lipotropin, 11,000	Adrenal cortex, adipose tissue, melanocytes
Thyrotroph	3 to 5	TSH, 28,000	Thyroid gland
Gonadotroph	5	LH, 28,000 FSH, 33,000	Gonads
Somatotroph	30 to 40	Somatotropin (growth hormone, HGH), 22,000	Multiple tissues
Mammotroph	3 to 5	Prolactin, 23,000	Breasts, gonads

normal anterior pituitary gland and gives some idea of the cell distributions. Table 52-2 summarizes the basic features of each cell.

TSH is a glycoprotein hormone whose function is to regulate the growth and metabolism of the thyroid gland and the secretion of its hormones, thyroxine (T_4) and triiodothyronine (T_3). The TSH-producing cells normally form 3% to 5% of the adult human anterior pituitary population, and they are found predominantly in the anteromedial area of the gland. These cells develop at about 13 weeks of gestation at the same time that the fetal thyroid gland is beginning to secrete thyroid hormone in response to TSH. In the adult pituitary, TSH is stored in small secretory granules of 125 to 200 nm in diameter.

TSH has a molecular weight of 28,000 and contains 15% carbohydrate bound covalently to the peptide chains. The hormone is made of two subunits associated by noncovalent forces. The α-subunit of 96 amino acids is nonspecific, being a component also of two other anterior pituitary hormones (FSH and LH), as well as of a placental hormone (human chorionic gonadotropin). The β-subunit of 110 amino acids confers the specific biological activity on the molecule; however, by itself it is essentially inactive. Species variation among β-subunits exists, but biological activity overlaps considerably. Bovine TSH, for example, is very active in humans. The synthesis of TSH proceeds through sequential steps. The α- and β-subunits are individually synthesized by prohormones on separate ribosomes and subsequently combined. Addition of the carbohydrate moieties is then enzymatically mediated. TSH synthesis is stimulated by the hypothalamic TRH and is suppressed by thyroid hormone.

The secretion of TSH is predominantly regulated by two factors. TRH increases the rate of secretion, whereas thyroid hormone decreases it in negative feedback fashion (Fig. 52-5). As a result of the balance between TRH stimulation and thyroid hormone inhibition, TSH is secreted in a rather steady fashion. This is in contrast to certain other hormones like growth hormone or ACTH, whose secretion fluctuates greatly. This befits the role of TSH, which is to stimulate a target gland whose own output is meant to be steady because the actions of its hormones (thyroxine and triiodothyronine) wax and wane slowly. In other words, although the hypothalamic-pituitary-thyroid gland axis is set with great precision, it moves to higher or lower levels in hours or days rather than in minutes.

TRH reaches the thyrotropic cells via the median eminence and long portal veins. It binds to a specific plasma membrane receptor, causing cyclic AMP levels to increase and TSH to be released by exocytosis. In humans, within minutes of intravenous ad-

■ *Anterior pituitary
hormones*
■ *Thyrotropic hormone*

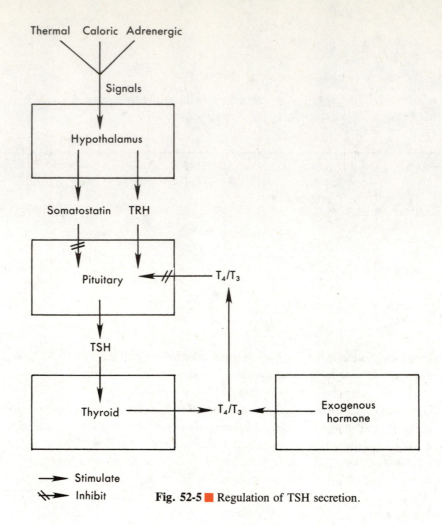

Fig. 52-5 ■ Regulation of TSH secretion.

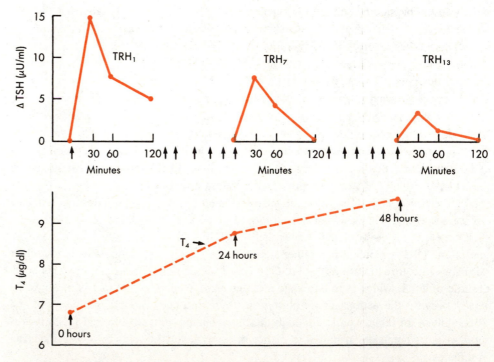

Fig. 52-6 ■ Pituitary and thyroid gland responses to repetitive injections of TRH every 4 hours for 48 hours in humans. Note that as plasma thyroxine (T_4) increases as a result of stimulation by TSH, the TSH responses to TRH are progressively blunted. (Redrawn from Snyder, P.J.: J. Clin. Invest. **52:**2305, 1973. Reproduced by copyright permission of The American Society for Clinical Investigation.)

ministration of TRH, plasma TSH levels rise as much as tenfold and return toward baseline levels by 60 minutes (Fig. 52-6). With repeated TRH injections, the TSH response diminishes over time because the secondarily stimulated thyroid gland increases its output of T_4 and T_3 (Fig. 52-6). This demonstrates vividly the negative feedback regulation of TSH secretion. These small increments of thyroid hormone concentration suppress TSH secretion by blocking the stimulatory action of TRH; conversely, small decrements of thyroid hormone augment TSH responsivity to TRH. Significant modulation of TSH secretion is associated with variations in plasma thyroid hormone concentrations of only 10% to 30% above or below the individual's baseline level.

The intracellular mediator of thyroid hormone's effect on TSH is probably T_3. Furthermore, there is evidence that T_3 generated within the pituitary cell from T_4 may be more effective and important in this regard than is T_3, which enters from the circulation. The acute suppressive effect of thyroid hormone on TSH release has a half-life of days, and it is prevented if protein synthesis by the pituitary thyrotroph is inhibited. This suggests that T_3 induces the synthesis of a protein with TSH-suppressing properties. Although thyroid hormone clearly inhibits TSH secretion and synthesis at the pituitary level, there is a little evidence that it also acts at the hypothalamic level to affect the synthesis or release of TRH. Because of negative feedback, individuals with thyroid diseases that result in chronic deficiency of thyroid hormone *(hypothyroidism)* have very high plasma TSH levels, whereas the opposite is true of patients with *hyperthyroidism*. Hyperplasia of the thyrotrophs sufficient to produce enlargement of the adenohypophysis can also result from hypothyroidism.

Physiological modulation of TSH secretion (and consequently of thyroid hormone output) occurs in at least two circumstances. TSH responsiveness to TRH and possibly TRH release itself are diminished during total fasting. This coincides with a decrease in metabolic rate and appears teleologically useful. In animals, TSH secretion is augmented by exposure to cold, but this has been demonstrated only infrequently in humans. Since TSH will increase thermogenesis via stimulation of the thyroid gland, this is a logical response. Other hormonal and neural influences have been noted. A slight diurnal variation in TSH secretion has been observed, with the highest levels occurring at night. A tonic inhibitory effect on TSH secretion, of physiological significance, is exerted by the hypothalamic peptide somatostatin. Cortisol (an adrenocortical hormone) decreases both TRH and TSH secretion. Growth hormone also reduces TSH secretion. Dopaminergic pathways are inhibitory as well. The importance of these influences on TSH secretion remains to be elucidated.

TSH normally circulates in plasma at a concentration of 1 to 10 μU/ml. Because of discrepancies and variations in bioassays, an exact equivalent molar concentration cannot be given, but it is of the order of 10^{-11}M, similar to other protein hormones. Daily TSH production is about 165,000 μU, which is equivalent to the entire content of one normal pituitary gland. (The metabolic clearance rate of TSH is 50 L/day.) In normal individuals, the α-subunit is also individually secreted and circulates at low levels. When TSH secretion is chronically hyperstimulated in response to deficient function of the thyroid gland, both β- and α-subunits circulate individually in elevated amounts.

The only TSH actions of importance are those exerted on the thyroid gland. TSH, as its name implies, is tropic; that is, it promotes growth of the gland and stimulates all aspects of its function. The glandular uptake of iodide, its organification, the completion of thyroid hormone synthesis, and the subsequent release of thyroid gland products are all stimulated by TSH. These effects are described in detail in Chapter 53. TSH binds to a plasma membrane receptor, and cylic AMP is the intracellular mediator of many of the hormone's effects by activating a protein kinase in the thyroid gland. The substrates for this enzyme and their subsequent connection with all the steps in thyroid hormone synthesis and release are not yet known. In tissue culture preparations, cyclic AMP may also mediate the TSH-induced differentiation of thyroid cells into follicles.

The sole pathological effects of an excess or deficiency of TSH are those of increased or decreased thyroid gland function, as described in Chapter 53.

■ *Adrenocorticotropic hormone*

ACTH is an anterior pituitary polypeptide hormone whose function is to regulate the growth and secretion of the adrenal cortex. Its most important target gland hormone is cortisol. The ACTH-producing cells form 20% of the anterior pituitary population. Although these are largely localized to the pars distalis of the anterior lobe (Fig 52-1), there is some evidence in animals for existence of ACTH-producing cells in the pars intermedia as well. The hormone is stored in secretory granules that are about 375 to 550 nm in diameter and have a clear space or halo between the contents and the membrane. The ACTH-producing cells are distinguished ultrastructurally by the presence of large numbers of microfilaments. In humans, ACTH synthesis and secretion begin at about 10 to 12 weeks of gestation, just before the subsequent rapid enlargement and development of the fetal adrenal cortex.

ACTH is a straight-chain peptide with 39 amino acids and a molecular weight of 4500. The N-terminal 1 to 24 sequence contains full biological activity; the remaining C-terminal portion probably only prolongs the hormone's action by protecting it against enzymatic degradation. Sequence 5 to 10 of ACTH is critical for stimulating the adrenal cortex.

Synthesis of ACTH. Our developing knowledge of the biosynthesis of ACTH forms a fascinating chapter of modern endocrinology, in which may lie clues to the basic meaning of the close connection between neural and hormonal events. The current concept of biosynthesis of ACTH and related peptides is shown in Fig. 52-7. A single gene appears to control the transcription of RNA, which directs the synthesis of proopiomelanocortin, an ACTH precursor of 31,000 molecular weight. Proopiomelanocortin also contains the sequences of a number of other biologically active peptides, including β-lipotropin, γ-lipotropin, β-endorphin, β-melanocyte-stimulating hormone (β-MSH), and α-melanocyte-stimulating hormone (α-MSH). The processing of proopiomelanocortin varies in different cell types and animal species. In humans, the adenohypophyseal cells of the pars distalis cleave proopiomelanocortin primarily to ACTH and β-lipotropin and, to a lesser degree, to γ-lipotropin and β-endorphin (Fig. 52-7). There is strong immu-

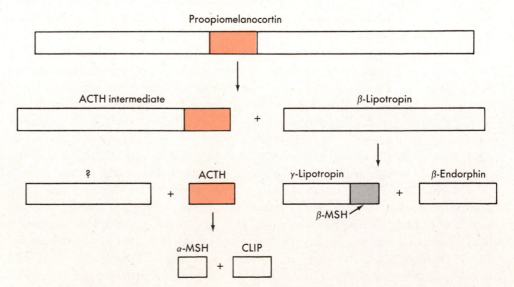

Fig. 52-7 ■ The processing of proopiomelanocortin. In the anterior lobe of the human pituitary, ACTH is the end product. In some species, ACTH is further cleaved to α-MSH and CLIP in the intermediate lobe.

nohistological evidence that these peptides are present in the same cell and they are frequently, if not always, released together. In other animal species, but not in humans, the cells of the pars intermedia further cleave ACTH to α-MSH and a second product known as *corticotropin-like intermediate peptide (CLIP)*. The factors that regulate the rate of synthesis of proopiomelanocortin and its subsequent cleavage by pituitary cells remain to be elucidated.

Certain structural relationships between ACTH and melanocyte-stimulating peptide are noteworthy. The heptapeptide sequences 4 to 10 in ACTH and 47 to 53 in the β-MSH portion of α-lipotropin are identical, and α-MSH is equivalent to ACTH 1 to 13. Because there is little or no evidence for the independent presence of either α-MSH or β-MSH in the pituitary tissue or plasma of humans, they may depend on these sequences in ACTH (or possibly in the lipotropins) for their melanocyte-stimulating activity.

In various species, ACTH, α-MSH, and β-endorphin have also been found in the brain, or in neuroendocrine cells of the pancreatic islets and gastrointestinal tract where they are independently synthesized. Furthermore, the N-terminal pentapeptide of β-endorphin is identical to met-enkephalin, with which it shares the analgesic and mood-modifying effects of opiates. The enkephalins, however, are synthesized via a different route and do not arise by cleavage of β-endorphin. Nonetheless, the widespread distribution of and the structural similarities among the ACTH-related peptides have helped to focus attention on the intimate relationship between endocrine and neural function.

Secretion of ACTH. The regulation of ACTH secretion is among the most complex of all the pituitary hormones (Fig. 52-8). The hormone exhibits circadian rhythms, cyclic

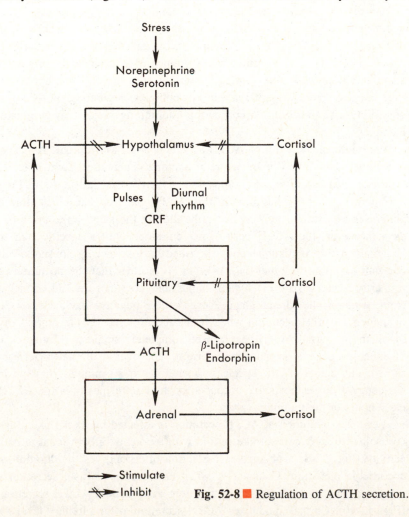

Fig. 52-8 ■ Regulation of ACTH secretion.

TABLE 52-3

Regulation of ACTH secretion

Stimulation	*Inhibition*
Cortisol decrease	Cortisol Increase
Adrenalectomy	
Metyrapone	Enkephalins
	Opiates
Sleep-wake transition	
	ACTH
Stress	
Hypoglycemia	
Anesthesia	
Surgery	
Trauma	
Infection	
Pyrogens	
Psychiatric disturbance	
Anxiety	
Depression	
α-Adrenergic agonists	
β-Adrenergic antagonists	
Serotonin	
ADH	
γ-Aminobutyric acid	

bursts, feedback control, and responses to a wide variety of stimuli (Table 52-3). Although the mechanisms for each form of control are not yet clear, a hypothalamic corticotropin-releasing factor (CRF) is the important mediator. Recent evidence indicates that CRF is a peptide that has 41 amino acids and that stimulates the release of ACTH and β-endorphin. Certain other neuropeptides of hypothalamic origin, such as ADH, also exhibit corticotropin-releasing activity in vitro and in vivo, but for various reasons they are less likely to participate in physiological regulation of ACTH secretion. The possibility also exists that a corticotropin-inhibiting factor may originate in the hypothalamus.

One dominant rhythm of ACTH secretion is clearly diurnal. As shown in Fig. 52-9, a peak occurs 2 to 4 hours before awakening. Thereafter, the average level decreases to a nadir just before or after falling asleep. A rise and fall in the major adrenocortical hormone, cortisol, is entrained in this ACTH pattern. The clock time of the diurnal pattern can be shifted by systematically altering the sleep-wake cycle for a number of days; however, the ACTH peak is not entrained with a specific stage of sleep. The circadian rhythm is diminished or abolished by loss of consciousness, blindness, or by constant exposure to either dark or light. It is clear that the nocturnal ACTH peak is primarily generated by the hypothalamus and does not depend on negative feedback from its target, the adrenal gland. Nevertheless, negative feedback is itself more potent in inducing ACTH secretion when it is superimposed on the diurnal rhythm; furthermore, the nocturnal ACTH peak can be completely suppressed by an excess of cortisol of either exogenous or endogenous origin. The overall diurnal pattern is composed of short cyclic bursts of ACTH release lasting 10 to 20 minutes. It is an increase in the frequency of these bursts that accounts for the rise of mean plasma ACTH levels in the early hours of the morning.

Feedback inhibition of ACTH secretion is effected by its peripheral target hormone, cortisol, or by any cortisol-like steroid with a potency proportional to its glucocorticoid activity (Fig. 52-8). The suppressive action often outlives the duration of plasma glucocorticoid elevation. Conversely, when endogenous cortisol secretion is reduced by disease or is acutely decreased by the drug metyrapone, ACTH secretion is stimulated (Fig. 52-10). Cortisol modulates ACTH secretion at the pituitary level by blocking the

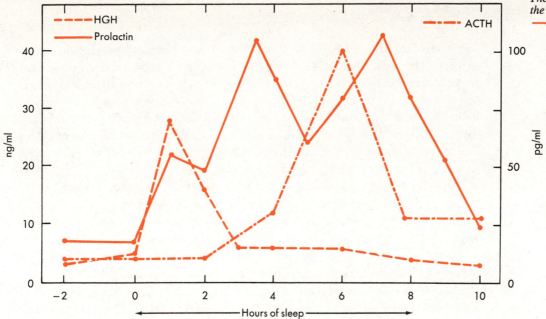

Fig. 52-9 ■ Nocturnal release of ACTH, HGH, and prolactin. Note the distinctive pattern for each hormone. (Redrawn from Takahashi, Y., et al.: J. Clin. Invest. **47:**2079, 1968; Berson, S.A., et al.: J. Clin. Invest. **47:**2725, 1968; reproduced by copyright permission of The American Society for Clinical Investigation; and Sassin, J.F., et al.: Science **177:**1205, 1972. Copyright 1972 by the American Association for the Advancement of Science.)

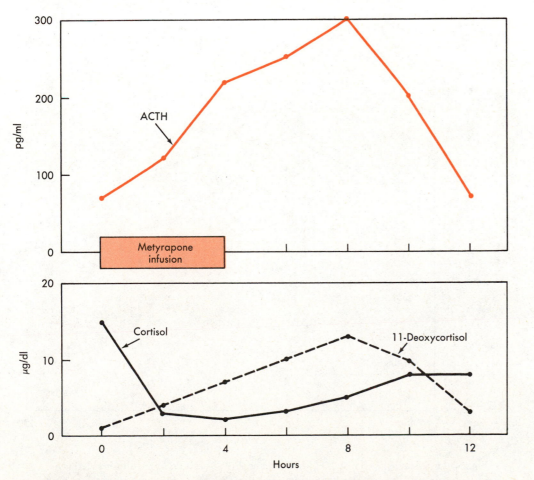

Fig. 52-10 ■ Negative feedback stimulation of *ACTH* release by metyrapone, a drug that blocks the conversion of 11-deoxycortisol to cortisol. (Redrawn from Jubiz, W., et al.: Arch. Intern. Med. **125:**468, 1970. Copyright 1970, American Medical Association.)

stimulatory action of CRF. In addition, cortisol may decrease the secretion of CRF. Evidence also exists that ACTH may inhibit its own secretion by decreasing CRF release, an example of short-loop feedback. Chronic deficiency of cortisol leads to persistent elevation of plasma ACTH, but the diurnal and pulsatile patterns are preserved, indicating their basic nonfeedback origin. Chronic autonomous hypersecretion of cortisol or long-term therapeutic administration of cortisol analogues for various diseases leads to functional atrophy of the CRF-ACTH axis. Several months may be required for recovery of this axis after the suppressive influence has been removed.

ACTH secretion responds most strikingly to stressful stimuli, a response that is critical to survival. Numerous factors that elicit the stress reaction are noted in Table 52-3. All have been demonstrated in humans. The response to insulin-induced hypoglycemia is illustrated in Fig. 52-11. In some instances, such as major abdominal surgery or severe psychiatric disturbance, the stress-induced hypersecretion of ACTH completely overrides negative feedback, and it cannot be suppressed by even the maximum level of cortisol secretion of which the adrenal cortex is capable. Stress also often obliterates the regular diurnal variation of ACTH, although secretory spurts may still be observed. The pathways by which each stress is signaled, sensed, and then stimulates CRF secretion vary. Among the monoamine neurotransmitters, both norepinephrine (via α-adrenergic receptors) and serotonin have been implicated in modulating the ACTH

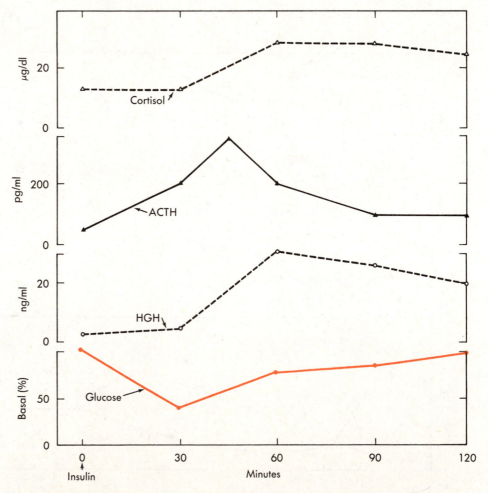

Fig. 52-11 ■ Stimulation of *ACTH*, cortisol, and *HGH* secretion by insulin-induced hypoglycemia in humans. (Redrawn from Ichikawa, Y., et al.: Plasma corticotropin, cortisol and growth hormone responses to hypoglycemia in the morning and evening, J. Clin. Endocrinol. Metab. **34:**895, 1972. Copyright 1972. Reproduced by permission.)

response to insulin-induced hypoglycemia. Other stress responses may be modulated by acetylcholine, since they can be blocked by atropine.

Once secreted, ACTH circulates in plasma unbound to protein. Basal concentrations at 6 AM range from 20 to 100 pg/ml (average 50 pg/ml or 10^{-11}M). Accurate measurements of daily production rate are not available. An estimate of 100 to 300 μg/day has been derived from the integrated diurnal plasma curve and the plasma half-life of the hormone (about 10 minutes), with an assumed volume of distribution equal to 40% of body weight. This estimate compares to an average adult human pituitary content of 250 to 500 μg. However, the bulk of the ACTH is secreted in a limited period of each day, which allows time for recovery of stores.

Action of ACTH. ACTH stimulates the growth of those specific zones of the adrenal cortex concerned with secretion of cortisol and androgenic steroids. In this respect, the effect is more to increase the size rather than the number of adrenal cells. In the absence of ACTH, profound atrophy of the relevant adrenal zones occurs. ACTH action follows binding to a specific plasma membrane receptor and activation of adenylate cyclase. A number of steps in the synthesis of adrenal steroids are thereby stimulated by ACTH, and these are detailed in Chapter 54. Because of the rapidity of synthesis, ACTH promptly causes secretion of adrenocortical hormones. Adrenal responsiveness to ACTH is attenuated and delayed by prior chronic underexposure to the tropic hormone; conversely, responsiveness is accentuated by prior chronic overexposure.

In animals, a number of extraadrenal actions of intact ACTH or of fully active fragments such as the 1 to 24 peptide, have been described. These require much larger doses than those needed for adrenal stimulation and are difficult to demonstrate in humans. Stimulation of lipolysis by ACTH may be due to the previously noted structural overlap with β-lipotropin. Exogenous ACTH also stimulates insulin secretion with consequent hypoglycemia. This could be an experimental indicator of some regulatory influence of endogenous ACTH generated within the pancreatic islets.

ACTH, because of its structural overlap with MSH, increases skin pigmentation. In amphibians, MSH acts on melanocytes, causing the dispersal of melanin pigment granules within these cells and their dendrites. This action, which is probably mediated by cyclic AMP, results in darkening of the skin. In humans, it is more likely that peptides with MSH activity cause hyperpigmentation by stimulating melanin synthesis and the transfer of melanin from the melanocytes to epidermal cells.

Secretion and actions of β-lipotropin and related peptides. β-Lipotropin is secreted by the anterior pituitary and is present in human plasma. A fourfold increase in the plasma level of β-lipotropin occurs after stimulation of the corticotroph by insulin hypoglycemia or by the administration of ADH. This parallels the concurrent rise in plasma ACTH. β-Lipotropin is suppressed by exogenous cortisol or other glucocorticoids and is stimulated by acute or chronic cortisol deficiency, indicating a pattern of negative feedback control identical to that of ACTH. Although the synthetic sequence outlined predicts equimolar secretion of the two peptides, the molar ratio of β-lipotropin to ACTH in plasma is only 0.3. This suggests that β-lipotropin has a higher metabolic clearance rate. β-Lipotropin was originally discovered as a distinct lipolytic factor in the pituitary gland. Its physiological role in fatty acid mobilization from human adipose tissue requires further exploration. γ-Lipotropin also circulates in plasma, but its actions are unknown.

β-Endorphin is found in very low concentrations in human plasma. However, the plasma level clearly increases in a parallel manner with ACTH when the corticotroph is stimulated by either stress or a lack of cortisol. The function of circulating β-endorphin is unclear. Since it penetrates the blood-brain barrier poorly, it is unlikely that β-endorphin secreted by the pituitary contributes to the peptide that is present in the brain and responsible for analgesic activity. When β-endorphin is administered systemically in pharmacological amounts, it causes stimulation of prolactin, insulin, and glucagon se-cretion and a small rise in plasma glucose levels. These observations may reflect para-crine actions of endogenous β-endorphin that is generated within the brain or pancreatic islets. They may also explain similar previously described effects of morphine. Abnor-malities in β-endorphin production have been proposed as possible causes of obesity and of noninsulin-dependent diabetes.

Clinical syndromes of ACTH dysfunction. The pathological effects of a primary excess or deficiency of ACTH are essentially those due to increased or decreased secre-tion of adrenocortical hormones. These are described in Chapter 54. Hyperpigmentation of the skin characterizes those diseases in which large increases in ACTH secretion occur.

■ *Gonadotropic hormones*

LH and FSH are glycoproteins whose function is to regulate the growth, pubertal maturation, reproductive cycles, and sex steroid secretion of the gonads of either sex. Currently, both hormones are believed to be secreted by a single cell type, the gonado-troph, which forms about 5% of the anterior pituitary population and is scattered throughout the gland. With immunocytologic examination, the same cell in the human can be shown to stain for both FSH and LH, although with electron microscopy, two somewhat dis-tinct types of granules have been noted. Granule diameters are 275 to 375 nm. Both hormones are present by 10 to 12 weeks of fetal life; however, neither appears to be required for intrauterine gonadal development or for sexual differentiation.

LH, with a molecular weight of 28,000, and FSH, with a molecular weight of 33,000, have similar structures. Each is composed of the common pituitary hormone α-subunit (molecular weight, 14,000; 96 amino acids) and a unique β-subunit, which differentiates the two hormones from each other, as well as from TSH and HCG (Fig. 52-12). The α- and β-subunits are held together by noncovalent forces. The carbohy-drate moieties are about 15% by weight and contain oligosaccharides composed of man-nose, galactose, fucose, galactosamine, and sialic acid. The carbohydrate groups seem to function in receptor attachment, whereas the sialic acid residues protect the hormone from rapid degradation in the circulation. Neither the β-subunit of LH nor that of FSH is biologically active by itself.

Intimate details of LH and FSH biosynthesis are still lacking. Ribosomal assembly of the two peptide chains is followed by addition of the carbohydrate moieties in the endoplasmic reticulum and Golgi apparatus. In women, the pituitary stores of both LH and FSH fluctuate throughout the menstrual cycle, being highest just before ovulation (Chapter 55). Only a single releasable pool of FSH appears to exist; in contrast, the responses of LH to its releasing hormone strongly suggest the existence of a rapidly releasable pool and a slowly releasable pool. Whether these reflect two types of granules differing in quality, size, or proximity to the cell membrane is not apparent. Hormone secretion occurs by exocytosis.

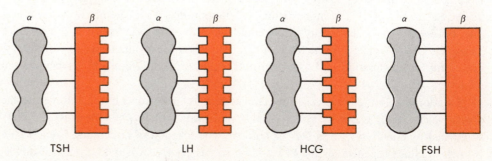

Fig. 52-12 ■ Structural similarities among *TSH, LH, HCG,* and *FSH* are depicted schematically. Note all share the same α-subunit.

Secretion of LH and FSH. The regulation of LH and FSH secretion is highly complex, embodying pulsatile, periodic, diurnal, and cyclic elements. It is also different in women and men. The main factors controlling gonadotropin secretion will be discussed in this chapter, and their reproductive function in both sexes and their relationship to the menstrual cycle will be reiterated and amplified in Chapter 55. The secretion of LH and FSH are both stimulated primarily by a single hypothalamic hormone, *luteinizing hormone–releasing hormone (LHRH)*. As its name implies, it causes a much greater increase in LH than FSH secretion, but because of the dual response, the term *gonadotropin-releasing hormone (GnRH)* is also sometimes used. Whether a separate hypothalamic releasing hormone with greater specificity for FSH exists remains uncertain. Human LHRH is a decapeptide (Table 52-1) that has been synthesized chemically and is available for clinical testing purposes. The cells of origin of LHRH are found predominantly in the arcuate nucleus and the preoptic area of the hypothalamus. Whether these two clusters have different functional roles is not yet known. After transport to the median eminence, LHRH is stored in small granules.

Detailed chemical and anatomical studies have shown that LHRH neurons are under the influence of dopaminergic, serotoninergic, and noradrenergic input to the hypothalamus. In addition, LHRH neurons and dopamine neurons are in close association within the arcuate nucleus of the hypothalamus. In rats, dopamine is probably stimulatory and serotonin inhibitory, whereas the effects of norepinephrine are contradictory. However, extrapolation of such data to the situation in humans is, at present, unclear. For example, dopamine infusion depresses human LH secretion; however, this might be due to a direct effect on the anterior pituitary gonadotroph rather than on the hypothalamus or median eminence. Much clinical experience, nonetheless, attests to the reality of neurotransmitter regulation of human gonadotropin secretion. Neural input from the retina to the hypothalamus probably accounts for the influence of light/dark cycles on gonadotropin secretion. Connections from the olfactory bulb probably transfer reproductive signals received from another individual by way of *pheromones,* which are airborne or waterborne chemical exciters or inhibitors. Loss of menstrual function in women is also a common event during prolonged physical or psychic stress.

LHRH binds to specific plasma membrane receptors in the gonadotroph and operates, at least in part, through cyclic AMP as a second messenger. In humans, an infusion of LHRH causes a biphasic response in plasma LH; the initial peak is reached at 30 minutes, followed by a secondary rise beginning at 90 minutes and continuing for hours thereafter (Fig. 52-13). Since LHRH also stimulates synthesis of LH, this action may account for the second phase of secretion. In contrast, LHRH infusion causes only a uniphasic progressive rise in FSH (Fig. 52-13). In normal women, LH is spontaneously secreted in pulses. These are characterized by a 15-minute upsurge and a falloff with a half-life of 60 minutes. Thus peaks of the plasma LH level are produced with periodicity, varying from 1 to 7 hours, depending on the phase of the menstrual cycle. The amplitude of the pulses can be equivalent to 100% changes in the plasma LH level, except at the time of ovulation, when it may be significantly greater. Normal men also exhibit eight to ten secretory bursts of LH per day (Fig. 52-14).

Much evidence indicates that the pulsatile secretion of LH is due primarily to pulsatile secretion of LHRH into the portal veins of the anterior pituitary rather than to a rapidly fluctuating sensitivity of the gonadotrophs to the releasing hormone. Moreover, pulsatility does not depend on the presence of sex steroid hormones from target glands, since agonadal individuals and postmenopausal women exhibit, if anything, even sharper spikes of the plasma LH level. Pulsatile secretion of LH is not found in young children but it makes its appearance just before puberty, at first occurring only at night. During the initial stages of puberty, this produces a nocturnal peak of LH. Although this diurnal pattern lasts only 1 or 2 years, disappearing as puberty is completed, the pulsatility of

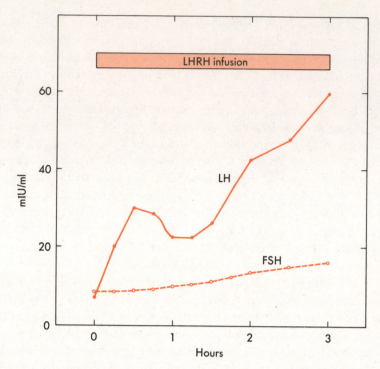

Fig. 52-13 ■ Stimulation of gonadotropin release by LHRH. Note biphasic response of LH and uniphasic response of FSH. (Redrawn from Wang, C.F., et al.: The functional changes of the pituitary gonadotrophs during the menstrual cycle, J. Clin. Endocrinol. Metab. **42:**718, 1976. Copyright 1976. Reproduced by permission.)

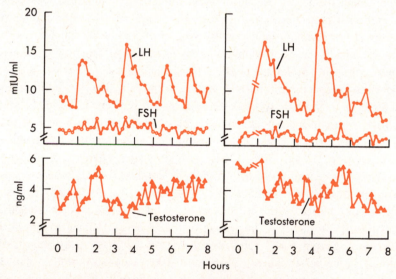

Fig. 52-14 ■ Pulsatile fluctuations in plasma LH levels and its target hormone, testosterone, in men. (Redrawn from Naftolin, F., et al.: Pulsatile patterns of gonadotropins and testosterone in man: the effects of clomiphene with and without testosterone, J. Clin. Endocrinol. Metab. **36:**285, 1973. Copyright 1973. Reproduced by permission.)

LH secretion becomes fixed. The most striking feature of LH secretion in women, as opposed to men, is its monthly cyclicity. This probably results from a complex interaction between the hypothalamus and sequential changes in ovarian steroid secretion.

FSH secretion also exhibits a pulsatile pattern, usually synchronized with LH, but of lesser magnitude (Fig. 52-14). The possibility of a separate hypothalamic releasing hormone for FSH has been suggested by the fact that the ratio of FSH to LH levels in

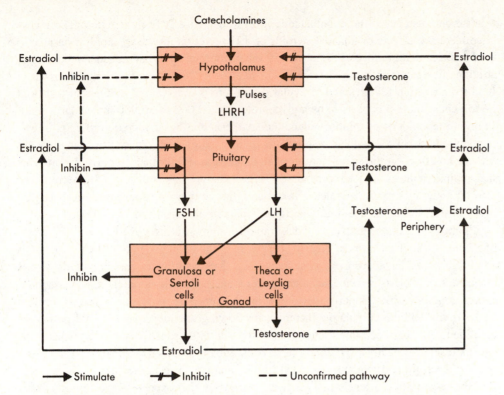

Fig. 52-15 ■ Negative feedback loops regulating hormone secretion in the hypothalamu-pituitary-gonadal axis. Estradiol is the primary gonadal product that inhibits LH in women; testosterone is the primary gonadal product that inhibits LH in men. Inhibin regulates FSH in both sexes.

plasma fluctuates. It is much less than one during the ovulatory gonadotropin spike, but it is greater than one in other phases of the menstrual cycle and in functionally agonadal individuals of either sex. These changes have not been consistently or completely reproduced simply by sex steroid modulation of pituitary responsiveness to LHRH.

Feedback regulation of gonadotropins. The secretion of both LH and FSH are clearly regulated by gonadal products. However, the patterns and mechanisms are more complex than those which have thus far been described for TSH and ACTH. It is best to consider first the basic framework, which is of the classic negative feedback type. The existence of negative feedback is proven by the straightforward observation that plasma levels of FSH and LH are both significantly elevated by surgical or functional removal of the gonads in either humans or experimental animals. FSH, however, is usually increased proportionally more than LH. A number of gonadal products from at least two gonadal cell types normally act to restrain the secretion of each gonadotropin. The basic schema is depicted in Fig. 52-15.

The major androgen, testosterone, from the interstitial cells of the ovary and Leydig cells of the testis, inhibits the release of LH. It does so more by decreasing the frequency of the LH pulses than their amplitude. The major estrogen, estradiol, which arises from the follicular cells of the ovary and the Sertoli cells of the testis as well as by conversion from testosterone in peripheral tissues, also inhibits the release of LH. Estradiol, however, decreases more the amplitude of the LH pulses than their frequency. Both testosterone and estradiol administration blunt the response of the gonadotroph to acute administration of LHRH. Conversely, in estradiol-deficient women and in testosterone-deficient men, LH responses to LHRH are exaggerated. Thus both sex steroids must act at the pituitary level. It is likely, but not absolutely proven, that estradiol and testosterone also act at the hypothalamic level to decrease LHRH secretion. These negative feedback effects of gonadal steroids on LH secretion are fairly rapid in onset.

Feedback inhibition of FSH secretion is carried out by other gonadal products. A specific substance, now called *inhibin,* has been isolated and partially characterized in fluid from ovarian follicles and testicular seminiferous tubules and cultures of testicular Sertoli cells. Inhibin appears to be a glycoprotein of 10,000 to 30,000 molecular weight. Purified preparations of inhibin reduce the synthesis and basal secretion of FSH in cultured pituitary cells and their response to LHRH stimulation. In contrast, inhibin has little or no effect on the same features of LH secretion. Preliminary reports also indicate that inhibin reduces LHRH secretion by hypothalamic cell cultures. FSH secretion is also inhibited by estradiol, which acts to block the pituitary response to LHRH. Higher doses of the steroid are required than for comparable inhibition of LH release.

Engrafted on this negative feedback framework are striking *positive feedback* effects of sex steroids. This is most clearly seen with estradiol in women. When estradiol is administered in an appropriate dose range and for a sufficient number of days, LH response to LHRH is *augmented* rather than reduced. Furthermore, if LHRH is administered repetitively to properly estradiol-primed women, both the response to the first LHRH pulse and the cumulative increments following the subsequent LHRH pulses are amplified. This has been interpreted to mean that both the sensitivity of the gonadotroph and its reserve capacity (LH stores) have been enhanced by estradiol treatment. Moreover, in such primed women, a sudden further increase in estradiol itself causes a significant rise in plasma levels of LH. It is noteworthy that aspects of positive feedback and negative feedback can be observed simultaneously. This is seen when estradiol-deficient agonadal women are given initial estradiol replacement therapy. After 7 days of treatment, the originally elevated basal levels of LH (and FSH) decline (negative feedback) yet the capacity to respond to repetitive doses of LHRH actually increases (positive feedback).

Progesterone, another major steroid product of the ovary, also participates in modulating LH release. When progesterone is administered alone in repository form, it induces a rise in the plasma level of LH 24 to 48 hours later. However, progesterone can either blunt or enhance the positive feedback effects of estradiol on LHRH responsiveness, depending on the timing with which the various hormones are administered. All of these complex interactions between gonadal steroids and gonadotropin secretion help to shape the typical patterns observed throughout female menstrual cycles.

Two other inhibitory influences on LH secretion are noteworthy. Prolactin, a mammotropic hormone from the anterior pituitary, also appears to inhibit LHRH release and to lower basal secretion of LH and FSH. Although the physiological significance of the effect is not well understood, it is of importance in explaining many cases of infertility and pathological loss of menses. Finally, animal experiments suggest that LH is capable of inhibiting secretion of its own releasing hormone, LHRH, via short-loop negative feedback. The route of access of LH to the hypothalamic LHRH neuron may be retrograde flow in the pituitary portal veins.

LH and FSH both circulate unbound to plasma proteins. The concentrations of both as measured by radioimmunoassay, are in the range of 5 to 25 mIU/ml in men and in reproductive-age women. In the latter, the levels of both hormones are higher in the first half of the menstrual cycle than in the second half; in addition, both hormones show sharp, single-day peaks at the time of ovulation. Currently, most laboratories still report their results in terms of the bioassay values rather than in their mass units. Basal plasma concentrations of each hormone are, however, of the order of 10^{-11}M. The metabolic clearance rates of LH and FSH, respectively, are 36 and 20 L/day, and their half-lives in plasma are approximately 1 and 3 hours. Estimates of daily secretory rates are 1100 IU for LH and 200 IU for FSH except on the day of ovulation. Metabolism of gonadotropins largely occurs in the liver and kidney. In contrast to the trivial excretion of other peptide hormones, 10% of the daily production of LH appears in the urine. This permits employment of urinary LH measurements as a reflection of integrated plasma concentrations. Such measurements are particularly useful when plasma levels are low, as in children. The α-subunit common to LH and FSH is secreted separately and circulates in most normal individuals. In contrast, the specific β-subunits of LH and FSH are secreted only by hyperstimulated gonadotrophs or by neoplastic cells.

Actions of gonadotropins. Both LH and FSH bind to plasma membrane receptors. At least part of the gonadotropins' actions are mediated by activation of adenylate cyclase with cyclic AMP as the second messenger. The latter activates a protein kinase that is important in one or more steps of steroid hormone synthesis. Each gonadotropin has specific primary target cells in the gonads of women and men. Granulosa cells in the ovary and Sertoli cells in the testis are stimulated by FSH. Interstitial cells in the ovary and Leydig cells in the testis are stimulated by LH. In addition, in the ovary, LH also acts on FSH-primed granulosa cells and luteal cells. The major effect of FSH on the testis is to stimulate spermatogenesis. Its major effect on the ovary is to stimulate follicle development and estradiol synthesis and to induce responsiveness to LH. The major effect of LH in the testis is to stimulate testosterone (and estradiol) synthesis. In the ovary, LH stimulates estradiol, progesterone, and testosterone synthesis and ovulation. The relationship between the specificity and timing of each gonadotropin's actions on ovarian and testicular function is complex and will be discussed in the sections on the gonads.

Abnormalities in secretion. Disorders (usually neoplasms) of the hypothalamus or pituitary gland may produce deficiency of one or both gonadotropins. With rare exceptions, this leads to a loss of reproductive capacity in adults and can cause some regression of already established secondary sexual characteristics. If gonadotropin deficiency occurs before the onset of puberty, the development of secondary sexual characteristics, the expected rapid growth spurt, and skeletal maturation are all prevented. Slow growth may continue for a long time, producing the eunuchoidal habitus, that is, a tall, juvenile-appearing adult. A primary excess of FSH or LH secretion is exceedingly rare, and a unique clinical picture cannot be described. In some women, for unknown reasons, a persistently high ratio of LH to FSH secretion exists. This is associated with noncycling and elevated estradiol levels, increased androgen secretion, loss of ovulation, infrequent or absent menses, and multiple cysts of the ovary. This condition, known as the *polycystic ovary syndrome,* is a common cause of infertility.

■ *Growth hormone (somatotropin)*

Growth hormone is a protein hormone that stimulates postnatal somatic growth and development. In addition, it has numerous actions on protein, carbohydrate, and fat metabolism. The hormone originates in anterior pituitary cells that make up 30% to 40% of the cell population of the gland in adult humans. It is stored in large dense granules, 350 to 500 nm in diameter. Typically, somatotrophs stain with acidophilic dyes, such as eosin. In humans, these cells may form tumors that hypersecrete growth hormone and produce a highly distinctive disease called *acromegaly*.

Human growth hormone (HGH) is a single-chain polypeptide with a molecular weight of 22,000. It contains 191 amino acids and 2 disulfide bridges. Neither the three-dimensional structure nor the exact sites or sequences essential for biological activity are known, but considerable activity is retained by products of tryptic digestion. HGH is active in many animals, but only primate growth hormones are active in humans. Until recently, this required that HGH–deficient individuals be treated with hormone extracted from human pituitaries, thereby limiting the available therapeutic supply. The development of recombinant DNA technology has now made it feasible to produce the HGH in bacteria.

The synthesis of HGH involves a prohormone of approximately 28,000 molecular weight. The regulation of HGH synthesis has not been well-characterized; in tissue-cultured pituitaries, synthesis is induced by cooperative effects of thyroid hormone and cortisol. Stored HGH is released by exocytosis, before which a microtubular system appears to be involved in bringing the secretion granules to the plasma cell membrane. Activation of adenylate cyclase and elevation of intracellular cyclic AMP levels mediate the response to stimulators of HGH secretion.

Secretion of HGH. As seen in Table 52-4, HGH secretion is under many different influences. An acute fall in plasma levels of either of the major energy-yielding substrates, glucose or free fatty acids, produces an increase in the plasma level of HGH. For example, when insulin is administered intravenously, the plasma level of HGH rises twofold to tenfold 30 to 60 minutes after the plasma glucose level has declined to below 50 mg/dl (Fig. 52-11). Conversely, a carbohydrate-rich meal or a pure glucose load causes a prompt decrease in the plasma HGH level of at least 50%. Responses to alterations in free fatty acid levels are generally slower and smaller. A high protein meal or the infusion of a mixture of amino acids produces a rise in the plasma HGH level; arginine is the most consistent amino acid stimulator. However, prolonged protein calorie deprivation or total fasting also stimulate HGH secretion by as yet undetermined mechanisms. Exercise, such as running and cycling, and stresses of varying severity, including blood drawing, anesthesia, fever, trauma, and major surgery, all produce rapid increases in HGH secretion. In addition to spikes in the plasma levels of HGH produced by all of these factors, a regular nocturnal peak occurs 1 or 2 hours after the onset of deep sleep (Fig. 52-9). This correlates with stage 3 or stage 4 sleep, as indicated by the electroencephalogram.

It is difficult to implicate a single final common pathway for all these somatotropic stimuli. There is clear evidence for the existence of a hypothalamic HGH-releasing factor, and its amino acid sequence has been recently established from a human tumor. The neurotransmitters dopamine, norepinephrine, and serotonin all increase HGH secretion. They probably act at the hypothalamic level either by stimulating the hypothalamic

TABLE 52-4

Regulation of HGH secretion

Stimulation	*Inhibition*
Glucose decrease	Glucose increase
Free fatty acid decrease	Free fatty acid increase
Amino acid increase (arginine)	Cortisol
Fasting	Obesity
Prolonged caloric deprivation	Pregnancy
Stage 4 sleep	Somatostatin
Exercise	HGH
Stress (Table 52-3)	
Estrogens	
Dopamine	
Serotonin	
α-Adrenergic agonists	
γ-Aminobutyric acid	
Enkephalins	

releasing factor or by suppressing somatostatin, the HGH-inhibiting peptide. The HGH responses to exercise, stress, hypoglycemia, and arginine are reduced by α-adrenergic blockade and are augmented by β-adrenergic blockade. This suggests that these responses are facilitated by α-adrenergic receptors and inhibited by β-adrenergic receptors in hypothalamic neurons. In contrast, the sleep-induced rise in HGH is unaffected by the adrenergic system and is more likely stimulated by serotoninergic pathways from the brainstem. The latter may also modulate the response to hypoglycemia. The physiological role of dopamine pathways is unclear. HGH responsiveness is greater in women than in men and is greatest just before ovulation. This is explained by a well-documented augmenting effect of estrogens on HGH secretion. Daily HGH secretion is somewhat greater in childhood than in adult life. However, the role of HGH in promoting the growth spurt of puberty remains to be worked out. Stimulation of HGH secretion by pharmacological doses of glucagon and vasopressin appears to have little physiological relevance.

HGH secretion is also under a number of negative influences. It is specifically inhibited by somatostatin, a 14–amino acid single-chain peptide synthesized and released in the hypothalamus. Recently, a 28–amino acid somatostatin molecule has been described that may be a prosomatostatin but that has equal biological activity. When somatostatin is administered intravenously, it blocks growth hormone responses to many of the stimuli listed in Table 52-4. It is presumed, although not proven, that somatostatin is the hypothalamic mediator of such negative influences on HGH secretion as hyperglycemia. It is even possible that HGH is under tonic inhibition by somatostatin and that many stimuli merely lift the inhibition by suppressing somatostatin release.

HGH also acts to inhibit its own secretion. This is shown by the fact that administration of exogenous HGH dampens subsequent endogenous HGH responsiveness to a number of stimuli, such as hypoglycemia. The mechanism may involve a short-loop feedback, inasmuch as HGH appears to stimulate the synthesis and release of somatostatin in vitro. However, somatostatin synthesis and release may also be similarly modulated by somatomedin, a peripheral peptide mediator of HGH action. Some of these relationships are illustrated in Fig. 52-16. Cortisol decreases HGH secretion, an action that may contribute modestly to its negative influence on growth. An unexplained de-

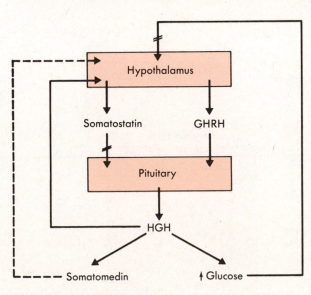

Fig. 52-16 ■ Regulation of HGH secretion. Note that *negative* feedback from the periphery can be effected by *stimulation* of somatostatin.

cline in HGH secretion occurs during the latter part of pregnancy, despite the presence of high estrogen levels. Finally, obese animals and humans exhibit dampened HGH responses to all stimuli; these are reversed by return to a normal weight.

The normal basal plasma HGH concentration is 1 to 5 ng/ml (about 10^{-10}M). This may increase as much as tenfold to fiftyfold under various stimuli. The plasma half-life of HGH is 20 minutes and the metabolic clearance rate is 350 L/day. Daily secretion in normal adults is approximately 500 μg, or 5% of the average pituitary content. A minor portion of the circulating immunoreactive HGH consists of larger aggregated forms with little biological activity.

HGH actions. HGH is a hormone with profound anabolic action. In its absence, experimental animals show stunted growth. When it is administered to hypophysecto-mized animals, it causes nitrogen retention, hypoaminoacidemia, and decreased urea production because the amino acids are diverted from oxidation to protein synthesis as growth ensues.

The multiplicity of HGH targets and effects is indicated in Fig. 52-17. The most striking and specific effect is the stimulation of linear growth that results from HGH action on the epiphyseal cartilage plates of long bones. All aspects of the metabolism of the cartilage-forming cells, the chondrocytes, are stimulated. This includes the incorporation of proline into collagen and its conversion to hydroxyproline and the incorporation of sulfate into the proteoglycan chondroitin, which, together with collagen, forms the resilient extracellular matrix of cartilage. In addition, HGH stimulates the general

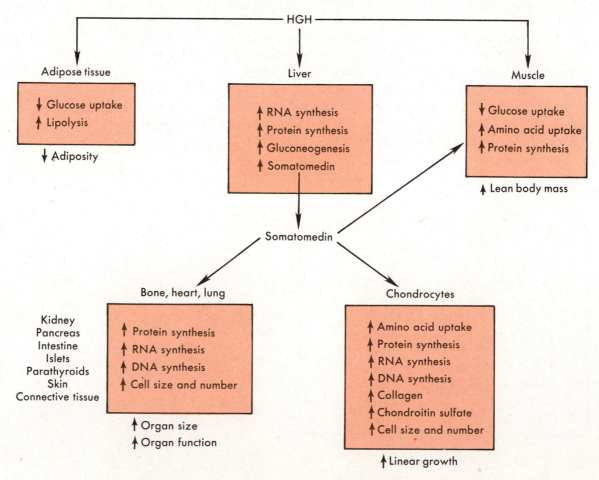

Fig. 52-17 ■ Biological actions of HGH. The effects on linear growth, organ size, and lean body mass are mediated by somatomedin produced in the liver.

synthesis of proteins, RNA and DNA in chondrocytes, as well as their proliferation. In support of the accelerated rate of protein synthesis, HGH also stimulates cellular uptake of amino acids. Other tissues share in the anabolic response to HGH. The width of bones increases as a result of enhanced growth at the periosteal surface. Visceral organs (liver, kidney, pancreas, intestines), endocrine glands (adrenals, parathyroids, pancreatic islets), skeletal muscle, heart, skin, and connective tissue all undergo hypertrophy and hyperplasia in response to HGH. In most instances, this is reflected in an enhanced functional capacity of the enlarged organ. For example, glomerular filtration, cardiac output, and hepatic clearance of test substances are all increased by HGH.

The actions of HGH on carbohydrate and lipid metabolism are bimodal. As noted previously, HGH has a tropic effect on the pancreatic islets; in its absence, insulin secretion in response to β-cell stimulation declines. However, the predominant effect of a prolonged growth hormone *excess* is to increase plasma glucose levels, despite a compensatory increase in insulin secretion. This occurs because HGH induces resistance to the action of insulin and inhibits glucose uptake by muscle and adipose cells. HGH also enhances lipolysis (in the presence of cortisol) and antagonizes insulin-stimulated lipogenesis. These actions lead to increased plasma free fatty acid levels and to a generalized decrease in adipose tissue. If insulin secretion is deficient, HGH can even cause ketosis. Thus on balance, HGH is a classic diabetogenic hormone in animals and humans.

Certain additional effects of HGH remain to be better elucidated. HGH increases intestinal calcium absorption, urinary calcium excretion, and urinary hydroxyproline excretion. Sodium retention by the kidney is increased, possibly by augmenting the secretion of aldosterone, a sodium-retaining adrenal steroid. This last action leads to a general increase in interstitial fluid volume in states of HGH excess. HGH is involved in spermatogenesis in some manner, and it has sufficient structural overlap with prolactin to give it definite mammotropic activity.

Mechanisms of HGH action. The mechanism of HGH action on cells is complex. Specific plasma membrane HGH receptors can be demonstrated, but thus far no intracellular second messenger has been clearly identified. However, a number of earlier observations suggested that the anabolic, growth-promoting effects of HGH required an intermediary. For example, HGH had little or no direct effect on cartilage in vitro. When the hormone was administered in vivo, a time lapse of 12 hours was necessary before its action could be demonstrated. However, at that time, the plasma harvested from the HGH-treated animal did stimulate cartilage metabolism in vitro.

It has now been shown that many, but not all, of the activities of HGH require the prior generation of one or more of a family of peptides known as *somatomedins*. These peptides have a molecular weight of about 7000 and bear considerable structural resemblance to proinsulin. In fact, the term *insulin growth factors (IGF)* has been applied to them. They originate primarily, if not solely, in the liver. Several hours after HGH administration to the whole animal, there is a significant release of somatomedins by the perfused liver. In contrast to the fluctuation of plasma levels of HGH, somatomedin concentrations are relatively stable. For example, they do not change significantly when the plasma glucose level is varied, whereas HGH does. The plasma half-life of somatomedins is also longer, being of the order of 3 to 5 hours. This may be due, in part, to the fact that they circulate bound to a large carrier protein. This protein is also synthesized in the liver and appears to be under HGH regulation.

At present, the growth-promoting effects of HGH can be accounted for by the somatomedins, which have been shown to stimulate typical HGH responses in cartilage, muscle, adipose tissue, fibroblasts, and tumor cells in vitro. Somatomedins bind to specific plasma membrane receptors in responsive tissues. These receptors can also bind insulin and proinsulin, although with much lower affinities. Conversely, bona fide insulin receptors also bind somatomedins weakly. These cross-reactivities may assume biological importance when very high concentrations of either somatomedins or of insulin exist. For example, some patients with tumors that secrete somatomedins develop spontaneous hypoglycemia, presumably because of insulin receptor activation by the somatomedins. Other patients with insulin receptor deficiency who have in compensation extremely high plasma insulin levels develop excessive soft tissue growth, presumably because of somatomedin receptor activation by insulin. As is the case with insulin, the roles of cyclic AMP and cyclic GMP as intracellular mediators of somatomedin action are still obscured by contradictory data.

Plasma somatomedins are increased by administration of HGH, and they disappear from HGH-deficient animals or humans. Somatomedin levels increase during adolescence and correlate with the progression of pubertal growth. Although HGH itself is not necessary for fetal growth, one or more of the somatomedins may be. Presumably, their production in utero is stimulated by non-HGH factors. Somatomedin production is reduced by factors that can override HGH. Fasting, protein deprivation, and insulin deficiency all lead to diminished liver production of somatomedins and to a decrease in their plasma levels, despite increases in growth hormone secretion. Indeed, in these pathophysiological states, the lack of somatomedin may be the cause of the elevated HGH levels through negative feedback. Estrogens also decrease somatomedin production; this may account for their antagonism to HGH action despite their stimulation of growth hormone secretion. Cortisol also diminishes somatomedin levels.

Overall role of HGH in substrate flow. It is useful to review the interactions between HGH and insulin in common physiological circumstances, as presented in Fig. 52-18. When protein (and energy) intake is ample, it is appropriate to utilize the absorbed amino acids for protein synthesis and to stimulate growth. Hence, both HGH and insulin secretion are stimulated by amino acids and together they augment the production of somatomedins. The latter, in turn, stimulate accretion of cartilage and bone. (These actions are probably directly enhanced by insulin as well.) The insulin antagonistic effect of the HGH molecule itself on carbohydrate metabolism is also useful at this time; it helps to prevent hypoglycemia, which might result from insulin stimulation in the absence of carbohydrate. On the other hand, when a carbohydrate load is ingested and insulin secretion is correspondingly increased, HGH secretion is suppressed. In this circumstance, accelerated generation of somatomedins is not needed because protein anabolism is not advantageous in the absence of amino acid inflow. Neither is insulin antagonism necessary; on the contrary, unrestrained expression of insulin action permits efficient storage of the excess carbohydrate calories in the liver, muscle, and adipose tissue. Finally, when an individual is totally fasting, insulin secretion falls, partly because of a fall in plasma glucose levels. Although this increases HGH secretion, the significant deficiency of insulin predominates in the liver, and somatomedin production falls. Again, this is appropriate in a situation where an increase in protein anabolism is disadvantageous and protein catabolism is essential. However, the increase in HGH may still be beneficial during fasting, since it enhances lipolysis, decreases peripheral glucose utilization, and increases gluconeogenesis. Thus, like glucagon, HGH may help to provide glucose for central nervous system needs.

Clinical syndromes of HGH dysfunction. Deficiency of HGH in children can result from hypothalamic dysfunction, pituitary tumors, a biologically incompetent HGH molecule, a failure to generate somatomedins normally, or receptor deficiency. Short stature and correspondingly delayed bone maturation are the consequences. Mild obesity is common and puberty is usually delayed. In the adult, no physical signs are evident. In both children and adults, the lack of insulin antagonism from HGH may lead to rare episodes of hypoglycemia. The diagnosis of absolute HGH deficiency is established by demonstrating low plasma HGH levels, which fail to rise after stimulation with insulin

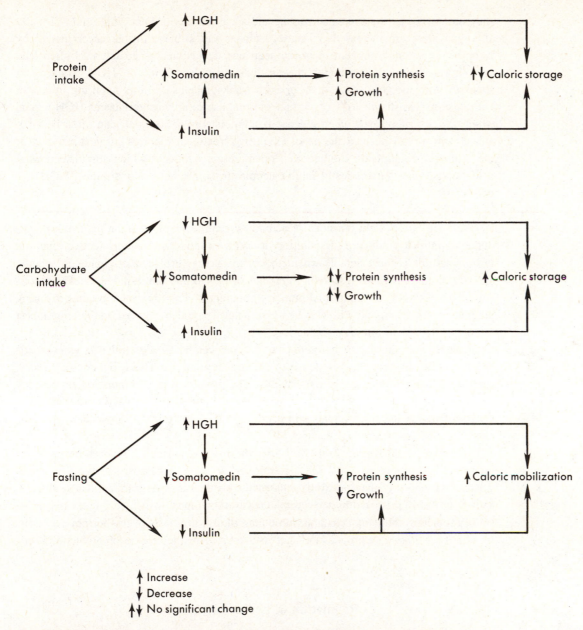

Fig. 52-18 ■ Complementary regulation of HGH and insulin secretion coordinate nutrient availability with anabolism and caloric flux. Note that both hormones are increased by protein, and both stimulate protein synthesis.

hypoglycemia, arginine, or exercise or at night. Replacement treatment with HGH causes nitrogen retention and an increase in growth velocity. Pubescence occurs and fertility is established.

Sustained hypersecretion of HGH results from pituitary tumors and produces a unique syndrome called *acromegaly*. If it begins before puberty has been completed and before the bony epiphyses are closed, the individual grows very tall and has long arms and legs. This condition has been referred to as *giantism*. In adults, only periosteal bone growth can be increased by HGH, leading to enlarged fingers, toes, hands, and feet, prominent bony ridges above the eyes, and a prominent lower jaw. Facial features are coarsened by accumulation of excess soft tissue. The nose is bulbous, the tongue enlarged, and the skin is thick, whereas subcutaneous fat is sparse. The overall appearance is usually so characteristic as to allow diagnosis at a glance. Nerves may be entrapped and compressed by the excess soft tissue and accumulation of interstitial fluid. Virtually all organ sizes are increased; in some, such as the kidney, this is accompanied by increased function. Enlargement of the heart and accelerated atherosclerosis often lead to a shortened life span. Finally, the insulin antagonistic effect of HGH produces an abnormal tolerance to carbohydrate or even frank diabetes mellitus requiring treatment with insulin. The diagnosis is confirmed by demonstrating elevated plasma HGH levels, which are not suppressed when glucose is administered. The tumor can often be visualized with a CAT scan of the head. Definitive treatment requires surgical removal of the tumor or ablation by irradiation. Dopaminergic agonists, which stimulate normal somatrophs, may paradoxically inhibit neoplastic somatrophs and diminish HGH hypersecretion.

■ Prolactin

Prolactin is a protein hormone principally concerned with stimulating breast development and milk production. In addition, it exerts an influence on reproductive function; in humans this is most notable when present in excess. Prolactin originates in specific anterior pituitary cells with secretory granules of 275 to 350 nm in diameter. These cells comprise 3% to 4% of the normal pituitary population. They increase in number during pregnancy, lactation, and with estrogen treatment. They also give rise to the commonest tumor of the human pituitary gland.

Prolactin is a single-chain protein of molecular weight 23,000, with 198 amino acids and 3 disulfide bridges. It has extensive structural similarity to HGH, but does not cross-react with it in specific radioimmunoassays. The biological and immunological potency of various mammalian prolactins is rather similar. Synthesis of prolactin proceeds via a prohormone that has a 29–amino acid peptide on the N-terminal. This leader peptide is cleaved before secretion.

Secretion of prolactin. Table 52-5 lists the most important influences on prolactin secretion. Consistent with its essential role in lactation, prolactin secretion increases steadily during pregnancy, leading to plasma levels that are twentyfold elevated at term. This is probably mediated by the large increase in estrogen, which stimulates hyperplasia of prolactin-producing cells and, more than likely, synthesis of the hormone as well. In addition, although estrogen does not itself stimulate the release of prolactin, it en-

TABLE 52-5

Regulation of prolactin secretion

Stimulation	Inhibition
Pregnancy	Dopamine
Estrogen	Dopaminergic agonists
Nursing (breast manipulation)	Bromergocriptine Apomorphine
Sleep	L-Dopa
Stress (Table 52-3)	Prolactin
TRH	
Dopaminergic antagonists Phenothiazines Metoclopramide	
Histamine antagonists (H₂) Cimetidine	
Adrenergic antagonists Reserpine α-Methyldopa	

hances responsiveness to other stimuli. If a new mother fails to nurse her child, the plasma level of prolactin declines 3 to 6 weeks after delivery to the normal range of nonpregnant women. Nursing, however, maintains elevated levels of prolactin secretion, especially for the first 8 to 12 weeks (Fig. 52-19). This effect may be mimicked by the use of a breast pump or by other nipple manipulation.

Like other tropic hormones, prolactin secretion rises at night, possibly because of entrainment with sleep (Fig. 52-9). The first peak appears 60 to 90 minutes after the onset of slow-wave sleep, and subsequent peaks occur later, after cycles of REM (rapid eye movement) sleep. The function of this association with sleep is unknown. Stress of various sorts, including anesthesia, surgery, insulin-induced hypoglycemia, fear, and mental tension, all cause prolactin release. Whether this represents a "spillover" phenomenon or whether the prolactin participates in accommodating to stresses is unclear.

The pathways for regulating prolactin release in each physiological circumstance remain to be worked out. However, uniquely among the pituitary hormones, prolactin secretion ordinarily appears to be under tonic *inhibition* by the hypothalamus. Disruption of the hypothalamic-pituitary connection produces prompt and enduring increases in plasma prolactin, often with biological consequences. Dopamine has many characteristics that qualify it for the role of primary *prolactin-inhibiting factor (PIF),* although it is not a hypothalamic peptide. This catecholamine is a potent inhibitor of prolactin release, either when generated within the brain in vivo, or when applied to pituitary tissue in vitro. Furthermore, a dopaminergic tract runs from the hypothalamus to the median eminence, so that dopamine has access to the anterior pituitary by way of the portal veins. However, the possibility of a separate peptide PIF, whose synthesis and release might be under dopaminergic control, has not been excluded. In any case, the profound prolactin-inhibiting effect of dopamine is exploited therapeutically by the use of agonists, such as bromergocriptine, in pathological states of prolactin excess. Conversely, hyperprolactinemia and sometimes milk secretion may result from the use of dopaminergic antagonists, such as phenothiazine compounds, in the treatment of psy-

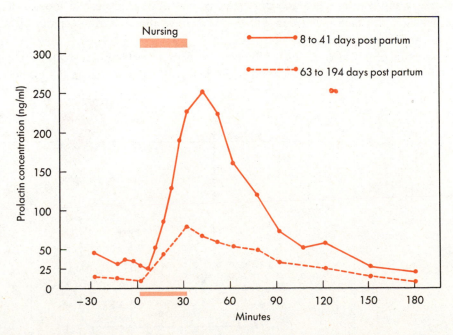

Fig. 52-19 ■ Stimulation of prolactin secretion by nursing. Note decreased responses with increasing interval of time from delivery. (Redrawn from Noel, G.L., et al.: Prolactin release during nursing and breast stimulation in postpartum and nonpostpartum subjects, J. Clin. Endocrinol. Metab. **38**:413, 1974. Copyright 1974. Reproduced by permission.)

chiatric disorders. The latter are also employed diagnostically to test the integrity of prolaction secretion.

The hypothalamus clearly has positive and negative effects on prolactin. TRH is a powerful stimulator of prolactin release. Within minutes of intravenous administration, plasma levels of prolactin increase twofold to fivefold. When TSH secretion is chronically increased because of loss of negative feedback from the thyroid gland, prolactin also tends to increase modestly, probably because of increased endogenous TRH. It is doubtful, however, that TRH is the sole or perhaps even the most important positive regulator of prolactin. During nursing, for example, an acute rise in plasma levels of TSH does not accompany that of prolactin, as would be expected if TRH were the mediator. In addition, hypothalamic extracts have been prepared that have little TRH content but considerable prolactin-releasing activity. The physiological significance of the positive effects of serotonin and the negative effects of norepinephrine on prolactin release remains to be established.

Normal basal plasma concentrations of prolactin are about 10 ng/ml (5×10^{-10}M). In addition to the natural hormone, which accounts for 70% to 90% of immunoreactivity in plasma, larger molecular weight species, termed "big prolactin," also circulate. Whether this fraction represents a secreted precursor molecule remains unsettled. There is no evidence for protein binding of plasma prolactin. The half-life of the hormone is 20 minutes. Its metabolic clearance rate is 110 L/day, and the daily production is estimated at 350 μg. Its metabolic fate is not worked out, but the kidney is one likely organ of degradation, since patients with renal failure often have high plasma prolactin levels. It is interesting that relative to plasma levels, concentrations of prolactin are high in amniotic fluid and milk and low in cerebrospinal fluid.

Biological effects of prolactin. Prolactin participates in stimulating the original development of breast tissue and its further hyperplasia during pregnancy. It is the principal hormone responsible for lactogenesis. During prepubertal and postpubertal life, prolactin, together with estrogens, progesterone, cortisol, and growth hormone, stimulates the proliferation and branching of ducts in the female breast. During pregnancy, prolactin, along with estrogen and progesterone, causes development of lobules of alveoli within which milk production will occur. Finally, after parturition, prolactin with cortisol stimulates milk synthesis and secretion.

The action of prolactin begins by combination with a specific plasma cell membrane receptor (Fig. 52-20). Very recent evidence suggests that this results in the generation

Fig. 52-20 ■ Schema of prolactin effect on lactogenesis. A second messenger coupling the prolactin–plasma membrane receptor interaction with the nuclear effect has been proposed. Its structure has not been identified.

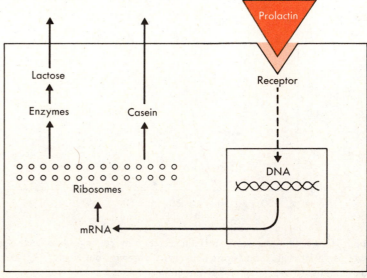

of a second messenger from the plasma membrane. This intermediary, which is not cyclic AMP, causes increased transcription of specific messenger RNA for the milk protein casein. This is quickly followed by increased synthesis of casein itself. Galactosyltransferase and *N*-acetyllactosamine synthetase, enzymes necessary for lactose synthesis, are concurrently induced. Prolactin augments the number of prolactin receptors, this being an example of up regulation. Estrogen also increases the number of prolactin receptors; nonetheless, it directly antagonizes the stimulatory effect of prolactin on milk synthesis (Fig. 52-20).

The second major effect of prolactin is on the reproductive axis. In women, prolactin appears to block the synthesis and release of LHRH, causing a loss of normal spurts of LH and preventing ovulation. In female rodents and in women, low concentrations of prolactin help to sustain ovarian progesterone secretion; but at higher concentrations, inhibiting effects on gonadal steroidogenesis appear to predominate in women and men. Certain behavioral effects of prolactin have been described, such as inhibition of libido in humans and stimulation of parental protective behavior toward the newborn in animals. Minor degrees of anabolic and osmoregulatory activity have been reported, but they appear to be of no importance in mammals.

Clinical syndromes of prolactin dysfunction. In women prolactin deficiency produces the inability to lactate. No other clinical consequences are known for certain. Prolactin excess results from hypothalamic dysfunction or from pituitary tumors. In women, this causes loss of menses, anovulation, and infertility. Less often, lactation unassociated with pregnancy (galactorrhea) occurs. In men, decreased testosterone secretion and sperm production result from prolactin excess. Stimulation of breast development is uncommon and lactation is rare. In both sexes, decreased libido is noted. The diagnosis is established by demonstrating a high plasma prolactin level that often fails to increase further after stimulation with TRH or a dopamine antagonist. Therapy may consist of surgical ablation of tumor tissue. However, in many instances, treatment with dopaminergic drugs reduces prolactin secretion to normal, reversing all adverse effects on reproduction and galactorrhea.

■ *Posterior pituitary hormones*

Two nonapeptides of homologous structure (Fig. 52-21), ADH and *oxytocin,* are secreted from the posterior pituitary gland. The primary role of ADH is to conserve body water and regulate the tonicity of body fluids. The primary role of oxytocin is to eject milk from the lactating mammary gland. Although their functions are different, the synthesis, storage, and mode of secretion of the two hormones are similar and will be discussed together. The general schema is illustrated in Fig. 52-22. Both hormones are synthesized in the cell bodies of hypothalamic neurons. ADH largely originates in the supraoptic nucleus and oxytocin largely in the paraventricular nucleus of the hypo-

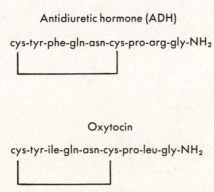

Antidiuretic hormone (ADH)

cys-tyr-phe-gln-asn-cys-pro-arg-gly-NH₂

Oxytocin

cys-tyr-ile-gln-asn-cys-pro-leu-gly-NH₂

Fig. 52-21 ■ Structures of posterior pituitary peptides. The alternate term for ADH is arginine vasopressin (AVP).

thalamus, although small amounts of each hormone are synthesized in the alternate site. Each nonapeptide is assembled on ribosomes as part of a large precursor protein of 20,000 molecular weight. This inactive precursor is translocated to the Golgi apparatus where the biologically active nonapeptide is split off and packaged into neurosecretory granules. At the same time, for each nonapeptide, an individual carrier protein of 10,000 molecular weight called *neurophysin* is cleaved from the same precursor. This very close relationship between the synthesis of the hormone and its neurophysin is shown by the fact that a strain of rats with congenital absence of ADH also lacks its neurophysin. Neurophysin-1 for oxytocin and neurophysin-2 for ADH serve to transport their respective hormones down the axon of the cell where they are stored within separate neurosecretory vesicles in both nonterminal and terminal swellings. The latter in the posterior pituitary are called *Herring bodies*. The rate of transport of the neurosecretory granules from the cell body to the Herring body is 2 or 3 mm/day.

Secretion occurs when a nerve impulse is transmitted from the cell body in the

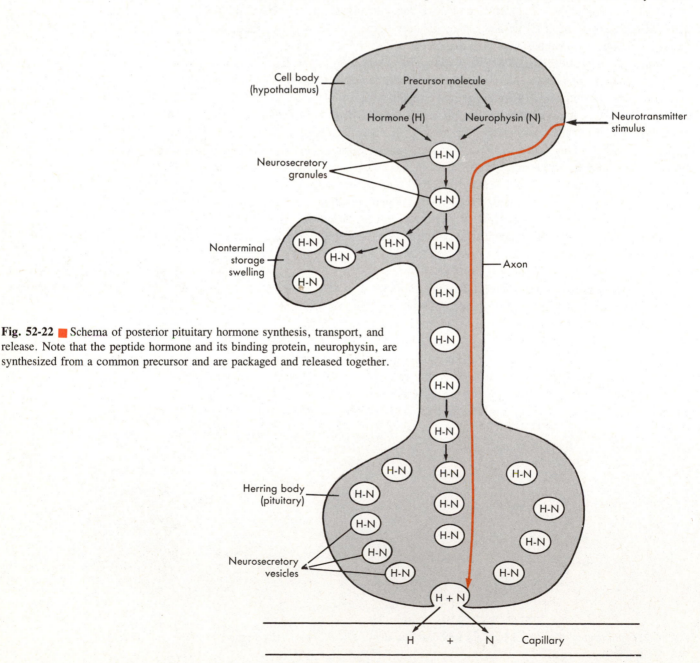

Fig. 52-22 ■ Schema of posterior pituitary hormone synthesis, transport, and release. Note that the peptide hormone and its binding protein, neurophysin, are synthesized from a common precursor and are packaged and released together.

hypothalamus down the axon, where it depolarizes the neurosecretory vesicles within the terminal Herring body. An influx of calcium into the neurosecretory vesicle then results in hormone secretion by exocytosis. During this process, the nonapeptide hormone dissociates from its neurophysin, and separately, each enters the closely adjacent capillary. Subsequent passage of the hormone into the bloodstream is by endocytosis into the endothelial cell and by diffusion through pores in the fenestrated capillary endothelium.

Consonant with its role in water metabolism, secretion of ADH is primarily regulated by osmotic and volume stimuli (Table 52-6). Water deprivation produces an increase in the osmolality of plasma and hence of the fluids bathing the brain. This causes the loss of intracellular water from osmoreceptor neurons in the hypothalamus. These may be identical with the neurons that synthesize and secrete ADH, or they may be a distinct population with connections to the latter. In either case, the shrinkage of cell volume causes ADH to be released immediately in the manner just described. Conversely, water ingestion causes a decrease in plasma osmolality, which suppresses osmoreceptor firing and consequently shuts off ADH release. If plasma osmolality is directly increased by administration of solutes, only those which do not freely or rapidly penetrate cell membranes, such as sodium, cause ADH release. Substances, such as urea, that enter cells rapidly do not stimulate ADH secretion, because they do not produce osmotic disequilibrium between extracellular and intracellular fluids. An increase in sodium concentration of the cerebrospinal fluid also increases ADH secretion, possibly through a specific sodium-sensing membrane. The hypothalamic osmoreceptors are extraordinarily sensitive, being responsive to changes in osmolality of only 1%. If water deprivation is prolonged, ADH synthesis and secretion are increased.

ADH release is also stimulated by a decrease of 5% to 10% in total circulating blood volume, central blood volume, or cardiac output. Hemorrhage is a potent stimulus to ADH release. Quiet standing and positive pressure breathing, both of which reduce cardiac output and central blood volume, also increase ADH secretion. Conversely, administration of blood or isotonic saline, which increases total circulating blood vol-

■ *Secretion of ADH*

TABLE 52-6

Regulation of ADH secretion

Stimulation	*Inhibition*
Extracellular fluid osmolality increase	Temperature decrease
Volume decrease	α-Adrenergic agonists
Pressure decrease	Ethanol
Cerebrospinal fluid sodium increase	Cortisol (?)
Pain	Thyroid hormone (?)
Stress (Table 52-3)	
Temperature increase	
β-Adrenergic agonists	
Drugs	
Nicotine	
Opiates	
Barbiturates	
Sulfonylureas	
Antineoplastic agents	

ume, or immersion to the neck in water, which increases central blood volume, all suppress ADH release. Hypovolemia is perceived by a number of pressure (rather than volume) sensors. These include carotid and aortic baroreceptors, stretch receptors in the walls of the left atrium and pulmonary veins, and possibly the juxtaglomerular apparatus of the kidney. The afferent impulses of this neurohumoral arc are carried by the ninth and tenth cranial nerves to their respective nuclei in the medulla and then by way of the midbrain to the supraoptic nuclei of the hypothalamus. Since the pressure receptors normally maintain tonic inhibition of ADH release, a decrease in circulating blood volume actually decreases the flow of impulses from the baroreceptors to the hypothalamus and in this manner increases ADH secretion. Hypovolemia also stimulates the generation of renin and angiotensin directly within the brain; angiotensin, in addition to stimulating thirst, may also mediate the release of ADH.

The separate effects of changing osmolality and changing blood volume and the interaction between these two stimuli are shown in Fig. 52-23. The effect of increasing or decreasing volume is to reinforce the osmolar responses by raising or lowering, respectively, the threshold for osmotic release of ADH. The slope relating plasma levels of ADH to plasma osmolality, that is, the sensitivity of the response, is not affected by volume alteration in humans (although it is in other animals). An increase in plasma osmolality of 3 mOsm/kg produces an increase in plasma levels of ADH of about 1 pg/ml.

Secretion of ADH is also influenced by a number of other conditions (Table 52-6). Pain, emotional stress, heat, and a variety of drugs are stimulators. Ethanol is a commonly encountered inhibitor; as little as 30 to 90 ml of whiskey is sufficient to suppress secretion completely. Cortisol and thyroid hormones appear to restrain ADH release in a permissive manner; in their absence, ADH may be secreted even though plasma osmolality is low.

ADH circulates at concentrations of about 1 pg/ml (10^{-12}M). The plasma half-life, determined by radioimmunoassay, is 6 to 10 minutes, although the half-life of biological

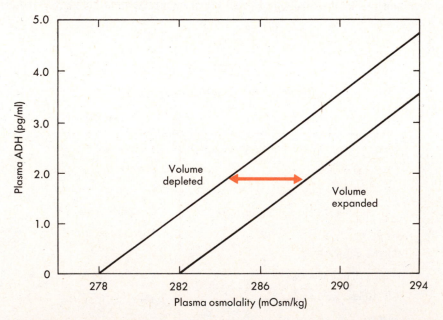

Fig. 52-23 ■ Regulation of ADH secretion by the interaction between plasma osmolality and plasma volume in humans. Increases or decreases in plasma volume respectively increase or decrease the threshold for ADH release in response to osmolality. (Redrawn from Robertson, G.L., et al.: J. Clin. Endocrinol. Metab. **42:**613, 1976. Copyright 1976, The Endocrine Society.)

effect may be up to 20 minutes. Degradation occurs in the kidney and liver. Estimates of daily secretion under ordinary circumstances are of the order of 1 μg, or about 4% of the posterior pituitary content. During water deprivation for 48 hours, secretion increases threefold to fivefold. Transient increases of fiftyfold can occur with hemorrhage (Chapter 36) or severe pain. Neurophysin-2 also circulates in plasma and its levels rise and fall parallel with ADH. No functional role for extrapituitary neurophysin has been identified.

■ *Actions of ADH*

The major action of ADH is on renal cells that are responsible for reabsorbing free (i.e., osmotically unencumbered) water from the glomerular filtrate. These ADH-responsive cells line the distal convoluted tubules and collecting ducts of the renal medulla. ADH binds to a specific plasma membrane receptor on the capillary side of the cell where it activates adenylate cyclase. The increase in intracellular cyclic AMP activates a protein kinase on the luminal side of the cell. This phosphorylates a currently unidentified membrane protein, following which the permeability of the cell membrane to water is enhanced. The participation of microtubular and microfilamentous elements of the cell in this process has also been suggested. The increase in membrane permeability permits back diffusion of water along an osmotic gradient from the hypotonic tubular urine that emerges from the loop of Henle to the hypertonic interstitial fluid of the renal medulla. The mechanisms for establishing this gradient are independent of ADH and are discussed in Chapter 46. The net result of ADH action is to increase the osmolality of urine to a maximum which is fourfold greater than that of the glomerular filtrate or plasma. In other words, ADH significantly reduces free water clearance by the kidney.

Water deprivation stimulates ADH secretion, resulting in decreased free water clearance and enhanced conservation of water. Thus ADH and water form a negative feedback loop. A water load decreases ADH secretion, resulting in increased free water clearance and more efficient excretion of the load. The dose-response relationship between plasma levels of ADH and urine osmolality is shown in Fig. 52-24. The relationship is sigmoidal, the majority of biological effect being observed between 2 and 5 pg/ml of ADH. In this range, urine osmolality correlates directly with plasma ADH con-

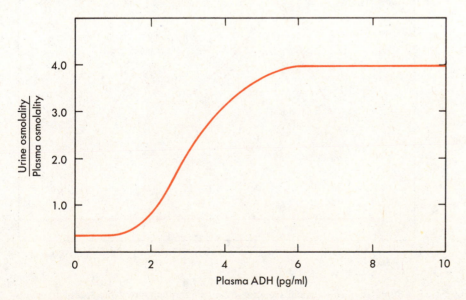

Fig. 52-24 ■ Dose-response curve for the effect of ADH to increase renal tubular reabsorption of free water, expressed as the ratio of urine to plasma osmolality. (Data from Moore, W.W.: Fed. Proc. **30**:1387, 1971.)

centrations. The generation of much higher plasma levels of ADH in humans may be related to other functions.

A number of factors blunt the action of ADH on the tubular cell: solute diuresis, chronic water loading (which reduces medullary hypertonicity), prostaglandin E (which interferes with ADH activation of adenylate cyclase), cortisol, potassium deficiency, calcium excess, and lithium. Certain sulfonylureas, used in the treatment of diabetes mellitus, are prominent among a number of agents that potentiate ADH action.

ADH may subserve other functions in addition to its primary role in water metabolism. Axonal projections of ADH neurons onto the median eminence provide access of the hormone to the anterior pituitary via the portal veins. By this route, ADH may function as a CRF. Other axons transfer ADH into the third ventricle, where it may facilitate long-term memory. It probably plays a contributory role in the cardiovascular adjustments to hemorrhage (Chapter 36). When administered systemically in pharmacological doses, ADH causes elevation of the blood pressure, coronary vasoconstriction, and intense splanchnic vasoconstriction. The last effect has been exploited therapeutically in controlling persistent, serious gastrointestinal bleeding.

■ *Oxytocin secretion and actions*

Suckling is the major stimulus for oxytocin release. Afferent impulses are carried from sensory receptors in the nipple via nerves to the spinal cord, where they ascend in the spinal thalamic tract. From relays in the brain stem and midbrain, they reach the paraventricular nuclei of the hypothalamus, and, via a cholinergic synapse, they trigger oxytocin release within seconds from the neurosecretory vesicles in the posterior pituitary. As suckling is continued, oxytocin synthesis and transfer down the hypothalamic axon are also stimulated. As shown in Fig. 52-25, the stimulus of suckling is specific for oxytocin, since no release of ADH is noted. Correspondingly, the various stimuli for ADH secretion do not appear to cause appreciable oxytocin release in humans. Oxytocin circulates unbound and exhibits a plasma half-life of 3 to 5 minutes. It is degraded by the kidneys and liver.

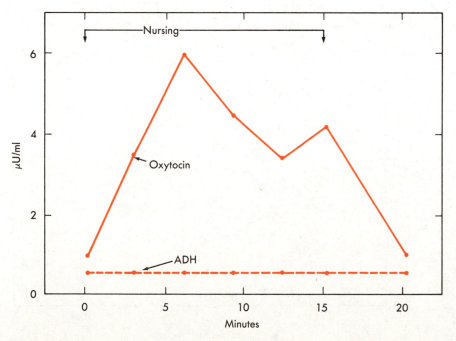

Fig. 52-25 ■ Stimulation of oxytocin secretion by nursing in humans. Note specificity of response as no release of ADH is observed. (Redrawn from Weitzman, R.E., et al.: The effect of nursing on neurohypophyseal hormone and prolactin secretion in human subjects, J. Clin. Endocrinol. Metab. **41:**836, 1980. Copyright 1980. Reproduced by permission.)

The unique effect of oxytocin is to cause contraction of the myoepithelial cells of the alveoli of the mammary glands. This forces milk from the alveoli into the ducts, from where it is evacuated by the infant. No other hormone has this action. Oxytocin combines with plasma membrane receptors in the breast, and receptor binding is increased by estrogen. Catecholamines, on the other hand, can block the action of oxytocin. At present, no analogous action of oxytocin has been identified in men, although their plasma levels of the hormone are similar to those in women.

Oxytocin also has a powerful contracting action on the uterus. Rhythmic contractions of the myometrium are stimulated by very small doses, which act by slightly lowering the threshold for membrane depolarization in the smooth muscle. Large doses lower the threshold still further, prevent repolarization and spiking discharges, and induce a sustained tetanic contraction. Despite the ability to stimulate rhythmic uterine contractions, there is little evidence that oxytocin is, in fact, an essential hormone of labor. It may, however, play a role in the sustained postpartum contractions that help to maintain hemostasis after evacuation of the placenta. Since oxytocin has only 0.5% to 1% the antidiuretic activity of ADH, it has no role in the therapy of ADH deficiency. Oxytocin is used clinically to induce labor in women who are physiologically ready, and it is also used therapeutically to decrease immediate postpartum bleeding.

■ *Clinical syndromes of*
posterior pituitary hormone
dysfunction

ADH deficiency is caused by destruction or dysfunction of the supraoptic and paraventricular nuclei of the hypothalamus. The posterior pituitary gland alone can be removed without seriously affecting the availability of the hormone because the sectioned axons acquire the ability to secrete ADH. The inability to produce a concentrated urine is the hallmark of ADH deficiency, a condition called *diabetes insipidus*. Whereas in normal humans, water deprivation can be compensated for by an increase in urine osmolality to 1000 to 1400 mOsm/kg, individuals who totally lack ADH cannot achieve osmolalities higher than that of plasma (290 mOsm/kg) and often not higher than 50 to 200 mOsm/kg. Since a typical diet generates 600 to 900 mOsm of solute per day for obligatory excretion by the kidney, urine volumes as high as 10 to 12 L may be required in diabetes insipidus, in contrast to the usual 1 to 3 L. The patient therefore urinates frequently both day and night and must also drink fluids constantly to replace the loss of water. Laboratory studies generally show a chronic elevation of serum osmolality (greater than 290 mOsm/kg) and of serum sodium (greater than 145 mEq/L). If, for any reason, the patient loses access to fluids, hypertonicity, severe dehydration, circulatory collapse, and death may ensue. Replacement of ADH (available as arginine vasopressin or analogues) prevents this sequence and relieves the symptoms of thirst and frequent urination.

ADH excess, inappropriate to either the tonicity or volume of the body fluids, can result from a variety of conditions: (1) increased secretion because of central nervous system disease, trauma, or psychosis; (2) ectopic production of the hormone by tumors of the lung and pancreas; or (3) potentiation of hormone action by drugs. The reduction in free water clearance caused by ADH, combined with voluntary or involuntary fluid intake, leads to water retention. Plasma sodium concentration and osmolality are significantly lowered, whereas urine osmolality is increased. Characteristically, sodium excretion in the urine is also increased, despite the hyponatremia. Both intracellular and extracellular body fluids are expanded. The swelling of brain cells and hypoosmolality cause headache, nausea, lethargy, somnolence, convulsions, and coma. These symptoms are not usually seen until plasma osmolality has declined to below 250 mOsm/kg, and plasma sodium to below 120 to 125 mEq/L. Water restriction is a logical and effective acute treatment. It may have to be supplemented in very sick patients with agents that induce a water diuresis or with hypertonic sodium chloride solutions to raise osmolality more rapidly.

Oxytocin deficiency leads to difficulty in nursing because of poor milk ejection. Oxytocin excess is unknown as a clinical syndrome.

■ *Hormones as neurotransmitters*

The preceding description of the hypothalamic-pituitary axis is part of a growing body of evidence from diverse sources that reveals an intimate relationship between neural and endocrine function. Similar concepts are emerging from studies of gastrointestinal tract secretions (Chapter 43). Thus it is appropriate to review briefly at this point the possible role of peptide hormones as neurotransmitters.

The classic concept of a hormone is a substance that is secreted into the bloodstream and that travels long distances to act on a distant target cell. The classic concept of a neurotransmitter is a substance that is liberated from a nerve terminal and travels a very short distance across a synapse to act on another membrane or effector cell. This distinction became blurred as the functioning of the hypothalamic-pituitary axis was worked out. The hypothalamic peptides represent molecules synthesized and stored in neurons in a manner analogous to neurotransmitters, but secreted into the bloodstream to act on distant targets, as do hormones. In a sense, such peptides are hybrids and may be called *neurohormones* or *neuropeptides*.

Further studies of the distribution of neurohormones have revealed that they are not limited to the hypothalamus. For example, 80% of the total TRH in the rat brain is found outside the hypothalamus. There appear to be TRH-containing fibers in many areas of the nervous system, including the motor nuclei of cranial nerves and the spinal cord. TRH is also present in cells of the pancreatic islets. Somatostatin-containing cells have been found in the cerebral cortex, the caudate nucleus, the amygdala, and sensory neurons. Furthermore, this peptide is also synthesized in neurons of the intestinal tract and in other cells of neural crest origin, as in δ-cells of the pancreatic islets and calcitonin-producing C-cells of the thyroid gland. LHRH-containing fibers originating in the hypothalamus project to other areas of the brain, such as the mammillary bodies. Fibers containing neurophysins (and therefore likely also to contain ADH and oxytocin) project not only to the posterior pituitary, but to autonomic centers in the brain stem and spinal cord.

In addition, peptides, originally established to be hormone products of classic endocrine cells, have now been found in various and often widespread distributions in the brain. These include ACTH, MSH, β-endorphin, LH, FSH, and HGH from the adenohypophysis, insulin and glucagon from the pancreatic islets, and calcitonin from the thyroid C-cells. Secretory products, such as vasoactive intestinal peptide (VIP) and cholecystokinin-gastrin, initially discovered and characterized as gastrointestinal hormones, have also subsequently been found in the brain. Such peptides as neurotensin and substance P, originally identified as neurotransmitter candidates, have turned out to have hormone-like or hormone-stimulating properties. In many of these instances, complementary receptors for these peptides have also been found in the brain. Finally, histochemical and immunohistological observations have suggested that the same cell may even produce both a classic peptide hormone and a classic neurotransmitter, for example, ACTH and serotonin.

The functional significance of these widespread distributions is under very active investigation in animals and humans. Many effects of neurohormones, beyond their classic ones, have already been found. Examples include mood enhancement and behavior excitation produced by TRH, increased sexual behavior produced by LHRH, increased memory retention produced by ADH, and increased satiety produced by cholecystokinin-gastrin.

From these studies, the concept of a diffuse neuroendocrine system of related cells is emerging. Employing a variety of amine and peptide molecules as messengers, the cells of this system transmit signals to other cells. According to the route by which it is transmitted, the same messenger molecule may function as a neurotransmitter (conveyance by axon), an endocrine hormone (conveyance by the bloodstream), or a paracrine hormone (conveyance by local channels between adjacent cells). The effect produced by the signal then depends on the target cell and the intracellular mechanisms that are activated. Thus a hypothalamic cell may secrete somatostatin into pituitary portal blood to decrease HGH secretion by pituitary somatotrophs, one brain cell may transmit somatostatin to a synapse with a second brain cell so as to alter behavior, and a pancreatic islet δ-cell may release somatostatin into the fluid bathing adjacent β-cells to inhibit insulin secretion.

This concept may help to explain heretofore obscure physiological phenomena or the pathogenesis of certain diseases. An important example of the latter is that of endocrine neoplasms or hyperplasia occurring in multiple sites in the same patient. In one syndrome, adenomas of the adrenal medulla, of the C-cells of the thyroid gland, and of the parathyroid glands occur in association with neuromas. In another syndrome, adenomas of the pituitary gland, pancreatic islets, and parathyroid glands are grouped. Further understanding of this neuroendocrine system undoubtedly will lead to the development of analogues for replacement, substitution, or selective inhibition of neurohormones in other neuroendocrine disorders. It is exciting to contemplate the fulfillment of some of these possibilities by the time these words are being read.

■ Bibliography

Journal articles

Bala, R.M., et al.: Serum immunoreactive somatomedin levels in normal adults, pregnant women at term, children at various ages, and children with constitutionally delayed growth, J. Clin. Endocrinol. Metab. **52**:508, 1981.

Bergland, R.M., et al.: Can the pituitary secrete directly to the brain? (affirmative anatomical evidence), Endocrinology **102**:1325, 1978.

Berson, S.A., et al.: Radioimmunoassay of ACTH in plasma, J. Clin. Invest. **47**:2725, 1968.

Clemmons, D.R., et al.: Reduction of plasma immunoreactive somatomedin D during fasting in humans, J. Clin. Endocrinol. Metab.: **53**:1247, 1981.

Gallagher, T.F., et al.: ACTH and cortisol secretory patterns in man, J. Clin. Endocrinol. Metab. **36**:1058, 1973.

Greenwood, F.C., et al.: The plasma sugar, free fatty acid, cortisol, and growth hormone response to insulin. I. In control subjects, J. Clin. Invest. **45**:429, 1966.

Grossman, A., et al.: New hypothalamic hormone, corticotropin-releasing factor, specifically stimulates the release of adrenocorticotropic hormone and cortisol in man, Lancet **1**:921, 1982.

Hökfelt, T., et al.: Peptidergic neurons, Nature **284**:515, 1980.

Ichikawa, Y., et al.: Plasma corticotropin, cortisol and growth hormone responses to hypoglycemia in the morning and evening, J. Clin. Endocrinol. Metab. **34**:895, 1972.

Jubiz, W., et al.: Plasma metyrapone, adrenocorticotropic hormone, cortisol, and deoxycortisol levels, Arch. Intern. Med. **125**:468, 1970.

Keye, W.R., Jr., et al.: Modulation of pituitary gonadotropin response to gonadotropin-releasing hormone by estradiol, J. Clin. Endocrinol. Metab. **38**:805, 1974.

Knopf, R.F., et al.: The normal endocrine response to ingestion of protein and infusions of amino acids: sequential secretion of insulin and growth hormone, Trans. Assoc. Am. Physicians **79**:312, 1966.

Krieger, D.T., et al.: Human plasma immunoreactive lipotropin and adrenocorticotropin in normal subjects and in patients with pituitary-adrenal disease, J. Clin. Endocrinol. Metab. **48**:566, 1979.

Krieger, D.T., et al.: Brain peptides. I. N. Engl. J. Med. **304**:876, 1981.

Krieger, D.T., et al.: Brain peptides. II. N. Engl. J. Med. **304**:944, 1981.

Lasley, B.L., et al.: The effects of estrogen and progesterone on the functional capacity of the gonadotrophs, J. Clin. Endocrinol. Metab. **41**:820, 1975.

Merimee, T.J., et al.: Growth hormone secretion in starvation: a reassessment, J. Clin. Endocrinol. Metab. **39**:385, 1974.

Moore, W.W.: Antidiuretic hormone levels in normal subjects, Fed. Proc. **30**:1387, 1971.

Mortimer, C.H., et al.: Effects of growth-hormone release-inhibiting hormone on circulating glucagon, insulin, and growth hormone in normal, diabetic, acromegalic, and hypopituitary patients, Lancet **1:**697, 1974.

Naftolin, F., et al.: Pulsatile patterns of gonadotropins and testosterone in man: the effects of clomiphene with and without testosterone, J. Clin. Endocrinol. Metab. **36:**285, 1973.

Noel, G.L., et al.: Prolactin release during nursing and breast stimulation in postpartum and nonpostpartum subjects, J. Clin. Endocrinol. Metab. **38:**413, 1974.

Pelletier, G., et al.: Identification of human anterior pituitary cells by immunoelectron microscopy, J. Clin. Endocrinol. Metab. **46:**534, 1978.

Phillips, L.S., et al.: Somatomedins. I. N. Engl. J. Med. **302:**371, 1980.

Phillips, L.S., et al.: Somatomedins. II. N. Engl. J. Med. **302:**438, 1980.

Rapoport, B., et al.: Suppression of serum thyrotropin (TSH) by L-dopa in chronic hypothyroidism: interrelationships in the regulation of TSH and prolactin secretion, J. Clin. Endocrinol. Metab. **36:**256, 1973.

Robertson, G.L., et al.: The interaction of blood osmolality and blood volume in regulating plasma vasopressin in man, J. Clin. Endocrinol. Metab. **42:**613, 1976.

Sassin, J.F., et al.: Human prolactin: 24-hour pattern with increased release during sleep, Science **177:**1205, 1972.

Segar, W.E., et al.: The regulation of antidiuretic hormone release in man, J. Clin. Invest. **47:**2143, 1968.

Silva, J.E., et al.: Contributions of plasma triiodothyronine and local thyroxine monodeiodination to triiodothyronine to nuclear triiodothyronine receptor saturation in the pituitary, liver, and kidney of hypothyroid rats: further evidence relating saturation of pituitary nuclear triiodothyronine receptors and the acute inhibition of thyroid-stimulating hormone release, J. Clin. Invest. **61:**1247, 1978.

Snyder, P.J., et al.: Inhibition of thyrotropin response to thyrotropin-releasing hormone by small quantities of thyroid hormones, J. Clin. Invest. **51:**2077, 1972.

Snyder, P.J., et al.: Repetitive administration of thyrotropin-releasing hormone results in small elevations of serum thyroid hormones and in marked inhibition of thyrotropin response, J. Clin. Invest. **52:**2305, 1973.

Snyder, S.H.: Brain peptides as neurotransmitters, Science **209:**976, 1980.

Stewart-Bentley, M., et al.: The feedback control of luteinizing hormone in normal adult men, J. Clin. Endocrinol. Metab. **38:**545, 1974.

Takahashi, Y., et al.: Growth hormone during sleep, J. Clin. Invest. **47:**2079, 1968.

Thompson, R.G., et al.: Integrated concentrations of growth hormone correlated with plasma testosterone and bone age in preadolescent and adolescent males, J. Clin. Endocrinol. Metab. **35:**334, 1972.

Vale, W., et al.: Characterization of a 41-residue ovine hypothalamic peptide that stimulates secretion of a corticotropin and β-endorphin, Science **213:**1394, 1981.

Weitzman, R.E., et al.: The effect of nursing on neurohypophyseal hormone and prolactin secretion in human subjects, J. Clin. Endocrinol. Metab. **41:**836, 1980.

Yen, S.S.C., et al.: Modulation of pituitary responsiveness to LRF by estrogen, J. Clin. Endocrinol. Metab. **39:**170, 1974.

Books and monographs

Frohman, L.A., et al.: The physiological and pharmacological control of anterior pituitary hormone secretion. In Dunn, A., and Nemeroff, C., editors: Behavioral neuroendocrinology, New York, 1983, Spectrum Publications, Inc.

Guillemin, R.: Neuroendocrine interrelations. In Bondy, P., and Rosenberg, L.E., editors: Metabolic control and disease, Philadelphia, 1980, W.B. Saunders Co.

Harris, G.W.: The pituitary gland, London, 1955, Edward Arnold, Ltd.

Savoy-Moore, R.T., et al.: Differential control of FSH and LH secretion. In Greep, R.O., editor: Reproductive physiology, III, Baltimore, 1980, University Park Press.

Scharrer, E., et al.: Secretory cells within the hypothalamus. In Fulton, J.H., editor: The hypothalamus, vol. 20, New York, 1940, Hafner Press.

chapter 53

The thyroid gland

The thyroid gland is of historic importance in endocrinology. The symptom complex associated with its hyperfunction was recognized as early as 1825, in part because the location of the gland provided visible and palpable appreciation of its enlargement. Shortly thereafter the appearance of a distinctive symptom complex of opposite character, which developed with the gland's surgical absence or spontaneous atrophy, was noted. The early speculation that this was caused by the deficiency of an internal secretion was subsequently borne out when crude extracts of the thyroid gland subsequently became the first example of hormonal replacement therapy in 1891. Thyroid hormone secretion is unusually steady, and its actions are very general. Its most important mission is to regulate the overall rate of body metabolism; in addition, it is critical for normal growth and development. Because diseases of the thyroid gland are common and are almost always curable, a sound understanding of thyroid physiology is vital to every practicing physician.

The thyroid gland develops from endoderm associated with the pharyngeal gut. It descends to the anterior part of the neck, where it lies on either side of the trachea. Abnormalities in its developmental descent may lead to final locations anywhere from the base of the tongue to the anterior mediastinum. By 11 to 12 weeks of gestational age the gland is capable of synthesizing and secreting its own thyroid hormones under the stimulus of fetal thyroid-stimulating hormone (TSH). Both are absolutely required for subsequent normal intrauterine development of the central nervous system and skeleton (although not for body growth), because neither maternal TSH nor maternal thyroid hormone can reach the fetus to any significant extent.

The two lobes of the adult thyroid gland together weigh approximately 20 g. They receive a rich blood supply from the thyrocervical arteries and innervation from the autonomic nervous system. The histological structure is shown in Fig. 53-1, *A*. The hormone-producing, cuboidal epithelial cells form circular follicles 200 to 300 μm in diameter. Within their lumens newly synthesized hormone is stored in the form of a colloid material. The base of each cell is covered by a basement membrane, and tight junctions connect adjacent cells at both their basal and apical (luminal) portions. When the gland is under intensive stimulation, the endocrine cells enlarge and assume a more columnar shape with their nuclei at the base (Fig. 53-1, *B*). The lumens of the follicles then appear scalloped because of active proteolysis of the hormone-containing colloid (Fig. 53-1, *B*). Scattered within the gland in close association with the epithelial cells is a separate line of parafollicular cells, called *C-cells*. These are the source of the polypeptide hormone, calcitonin, which is discussed in the section dealing with calcium metabolism in Chapter 51.

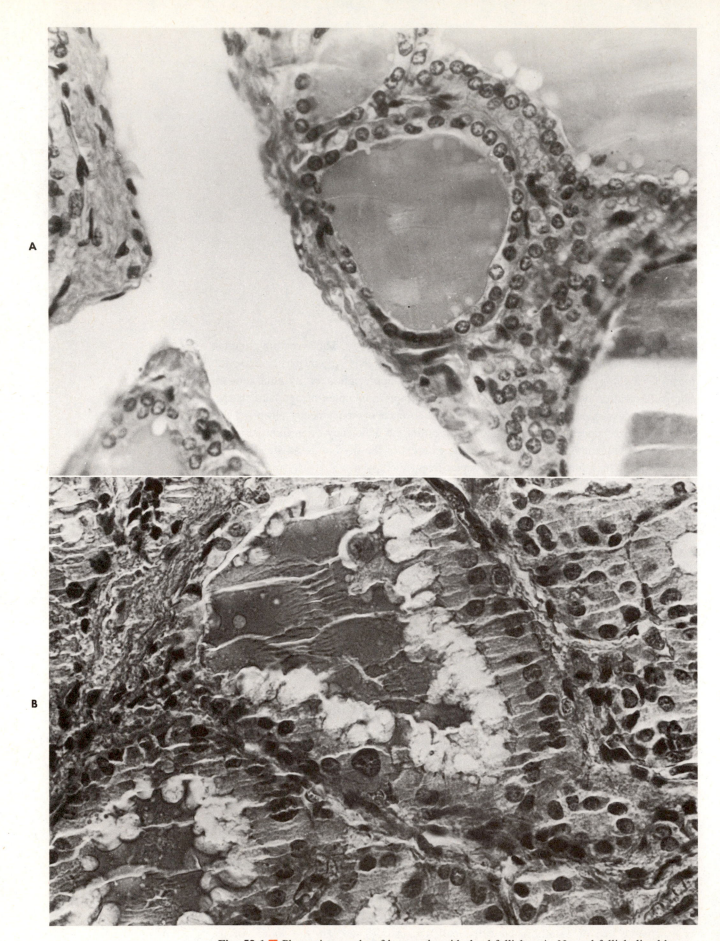

Fig. 53-1 ■ Photomicrographs of human thyroid gland follicles. **A,** Normal follicle lined by cuboidal cells and filled with colloid. **B,** TSH-stimulated follicle lined by hypertrophied columnar cells with scalloping of the colloid caused by active proteolysis. (Courtesy Dr. William Hawk, The Cleveland Clinic.)

The secretory products of the thyroid gland are *iodothyronines,* a series of compounds resulting from the coupling of two iodinated tyrosine molecules. Eighty to ninety percent of the output is 3,5,3′,5′-tetraiodothyronine (*thyroxine,* or T_4), 10% is 3,5,3′-triiodothyronine (T_3), and less than 1% is 3,3′,5′-triiodothyronine (reverse T_3, or rT_3). Normally these three compounds are secreted in the same proportions as they are stored in the gland.

Because of the unique role of *iodide* in thyroid physiology, a description of thyroid hormone synthesis properly begins with a consideration of iodide (Fig. 53-2). An average of 400 µg of iodide per person is ingested daily in the United States. In a steady state virtually the same amount is excreted in the urine. Iodide is actively concentrated in the thyroid gland, the salivary glands, and gastric glands. About 70 to 80 µg of iodide is taken up daily by the thyroid gland. This represents 10% to 35% of the circulating pool of iodide, which ranges from 250 to 750 µg. If this extrathyroidal iodide pool is labeled with a small dose of radioactive iodine (^{123}I or ^{131}I), the percentage uptake of this tracer in 24 hours (10% to 35%) gives a clinically useful dynamic index of thyroid gland activity. The total iodide content of the thyroid gland averages 7500 µg, virtually all of which is in the form of iodothyronines. In a steady-state condition 70 to 80 µg of iodide, or about 1% of the total, also is released from the gland daily. Of this amount, 75% is secreted as thyroid hormone, and the remainder is free iodide. The very large ratio (100:1) of iodide stored in the form of hormone to the amount turned over daily initially protects the individual from the effects of iodide deficiency

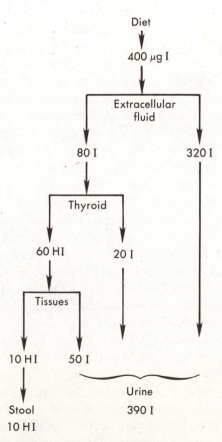

Fig. 53-2 ■ Average daily iodide turnover in humans (United States). Note that 20% of the intake is taken up by the thyroid gland, and 15% turns over in hormone synthesis and disposal. The unneeded excess is excreted in the urine. *I,* Iodide; *HI,* hormonal iodide.

for at least 2 months. Further conservation of iodide then is effected by a marked reduction in its renal excretion as the circulating concentration and filtered load fall.

Iodide is actively transported into the gland against a normal thyroid/plasma free iodide ratio of 30. This so-called iodide trap requires energy generation via oxidative phosphorylation, but its chemical nature remains obscure. Some evidence links it to a Na, K-ATPase. The iodide trap is markedly stimulated by TSH. There is evidence that this is mediated through cyclic AMP and also that it involves stimulation of the synthesis of a specific protein, possibly the iodide carrier itself. The trap displays saturation kinetics. However, the K_m of this transport system is above the estimated normal plasma concentration of iodide ($2 \times 10^{-8}M$). Thus small to moderate increases in iodide intake lead to increases in thyroidal iodide, and the gland/plasma ratio of 30 is still maintained. Initially this acts to increase the rate of thyroid hormone synthesis to above normal. However, as the dosage of iodide exceeds 2 mg/day, an intraglandular concentration of iodide or of some organic iodide product is reached that is inhibitory to the iodide trap and to the biosynthetic mechanism, so that hormone production declines back to normal. This autoregulatory phenomenon is known as the Wolff-Chaikoff effect. In unusual instances the inhibition of hormone synthesis by iodide can be great enough to induce thyroid hormone deficiency. A primary reduction in dietary iodide intake causes depletion of the iodide pool and greatly enhanced activity of the thyroid iodide trap. Under these circumstances the percentage uptake of a radioactive iodide tracer can increase to 80% to 90%. If the lack of iodide is severe enough, thyroid hormone deficiency results.

A number of anions, such as thiocyanate (CNS^-) and perchlorate ($HClO_4^-$), act as competitive inhibitors of active iodide transport. If for any reason iodide cannot be rapidly incorporated into tyrosine after its uptake by the cell, then administration of one of these competitive anions will, by blocking further uptake, cause a rapid discharge of the iodide from the gland in accordance with the high thyroid/plasma concentration gradient. This discharge can be demonstrated by monitoring the thyroid gland in vivo after labeling the pool with radioactive iodide, a maneuver that may assist in the diagnosis of biosynthetic defects. Another competitive inhibitor, the pertechnetate ion, also is taken up like iodide, but it cannot be significantly incorporated into organic molecules. In its radioactive form as $^{99m}TcO_4$, it is a useful substitute for radioactive iodide in the measurement of the trapping function and in the visualization of thyroid gland anatomy by external isotope scanning with a photon detector.

Once within the gland, iodide rapidly moves to the apical surface of the cell and into the lumen. Iodide (I^-) is immediately oxidized to iodine (I^0) as a postulated intermediate and is virtually simultaneously incorporated into tyrosine molecules (Fig. 53-3). The latter are not free in solution, but they exist in peptide linkages as components of *thyroglobulin*. This is a 19S glycoprotein with a molecular weight of 670,000. Thyroglobulin is synthesized on the rough endoplasmic reticulum as peptide units that polymerize and then have carbohydrate units added in transit to the Golgi apparatus. The completed protein, incorporated in small vesicles, moves to the plasma membrane and then into the adjacent lumen of the follicle.

Just within the follicle, iodide is incorporated into the thyroglobulin. As a result of iodination, both monoiodotyrosine (MIT) and diiodotyrosine (DIT) are formed (Fig. 53-3). Thereafter two DIT molecules are coupled to form T_4, or one MIT and one DIT molecule are coupled to form T_3. Very little rT_3 is synthesized. This entire sequence of reactions is catalyzed by *thyroid peroxidase,* an enzyme complex adjacent to or within the cell membrane bordering the follicular lumen. The immediate oxidant (electron acceptor) for the reaction iodide → iodine is hydrogen peroxide. The mechanism by which hydrogen peroxide is itself generated in the thyroid gland remains debatable, but some evidence suggests reduction of oxygen by NADPH or NADH via cytochrome reduc-

$$2I^- + H_2O_2 \longrightarrow I_2$$

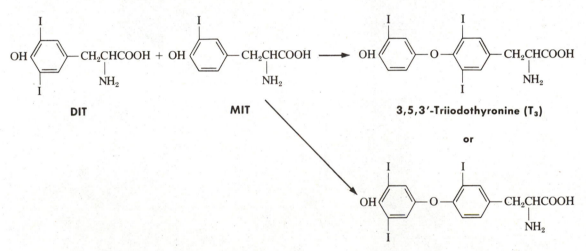

Fig. 53-3 ■ Overall pathway of thyroid hormone synthesis.

tases. The intimate chemistry of tyrosine iodination and iodotyrosine coupling likewise still is unsettled. One suggested mechanism of coupling involves the direct transfer of one iodinated phenolic ring to a second, leaving a serine residue behind in peptide linkage in the thyroglobulin. The special structural characteristic of the thyroglobulin molecule could facilitate this. A second mechanism would require a transamination of MIT or DIT to the respective iodophenylpyruvic acid, the latter then serving as the donor of the iodinated phenol to another DIT molecule.

The usual distribution of iodoaminoacids, as residues per molecule of thyroglobulin, is MIT, 6.5; DIT, 4.8; T_4, 2.3; and T_3, 0.2. Approximately one third of the iodine in thyroglobulin is in the form of calorigenic hormone (T_4 and T_3). Certain factors regulate the proportion of T_3 to T_4 that is synthesized. When iodide availability is restricted, the formation of T_3 is favored. Because T_3 is three times as potent as T_4, this response provides more active hormone per molecule of organified iodide. The proportion of T_3 also is increased when the gland is hyperstimulated by TSH or other activators. Finally, certain autonomous thyroid adenomas preferentially secrete T_3 for unknown reasons.

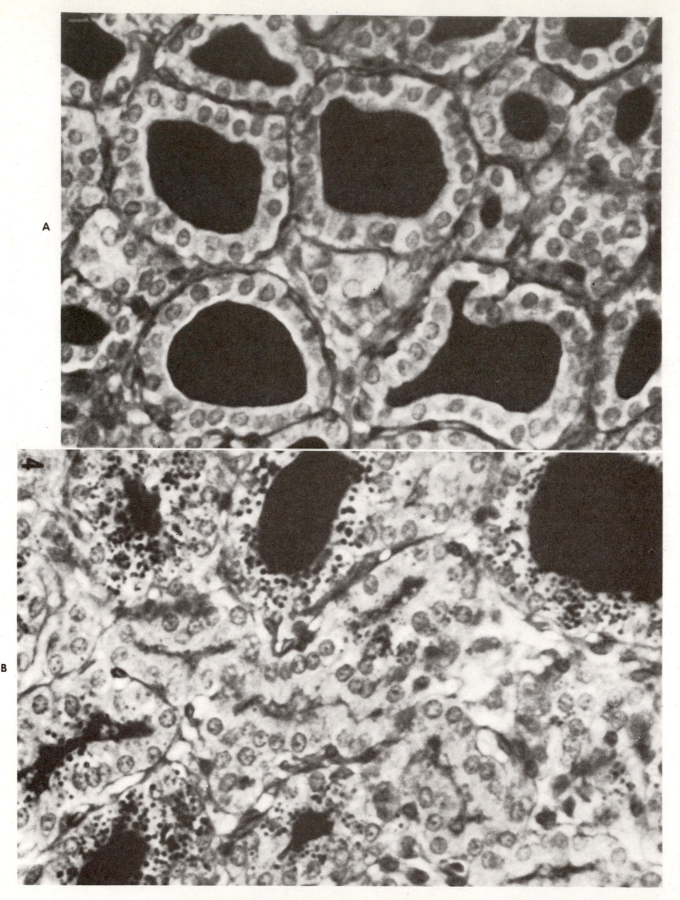

Fig. 53-4 ■ Histological demonstration of the process of resorption of colloid. **A,** Unstimulated follicles. **B,** Within minutes of TSH administration colloid droplets are seen inside the follicular cells. **C,** Close up showing individual colloid droplets *(D)* undergoing endocytosis. Light gray area is the cytoplasm of the cell. (From Wollman, S.H., et al.: J. Cell Biol. **21:**191, 1964. Reproduced by copyright permission of The Rockefeller University Press.)

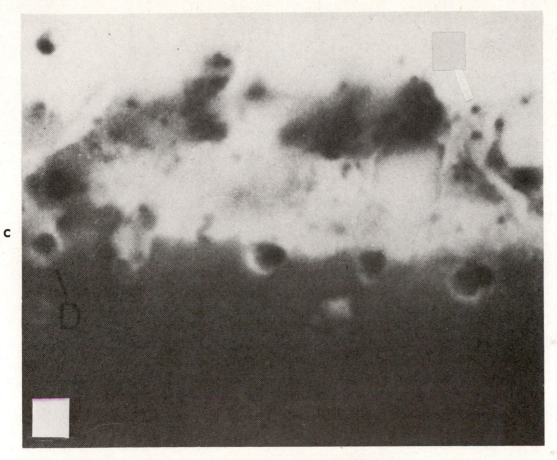

c

Fig. 53-4, cont'd ■ For legend see opposite page.

Once thyroglobulin has been iodinated, it is translocated into the lumen of the follicle, where it is stored as colloid. Release of the peptide-linked T_4 and T_3 into the bloodstream requires proteolysis of the thyroglobulin. Histochemical and radiographic studies have demonstrated that the colloid is retrieved from the lumen of the follicle by the epithelial cell through the process of endocytosis (Fig. 53-4). The plasma cell membrane forms pseudopods that engulf a pocket of colloid. After this portion of the luminal content has been pinched off by the plasma cell membrane, it appears as a colloid droplet within the cytoplasm. The droplet moves through the cytoplasm in a basal direction, probably as a result of microtubule and microfilament function. At the same time lysosomes move from the base toward the apex of the cell and fuse with the colloid droplets that are moving to meet them. The action of the lysosomal proteases then releases free T_4 and T_3, which leave the cell through the plasma membrane at the basal end and enter the bloodstream via the adjacent rich capillary plexus.

The MIT and DIT molecules, which also are released on proteolysis of thyroglobulin, are rapidly deiodinated within the follicular cell by the enzyme deiodinase. Since these compounds are metabolically useless and would be lost in the urine if secreted, their deiodination retrieves the iodide for recycling into T_4 and T_3 synthesis. Only minor amounts of intact thyroglobulin leave the follicular cell under normal circumstances. However, in pathological processes such as neoplasms, which disrupt the normal integrity of thyroid cells, and in hyperthyroid states there is a greater rate of thyroglobulin escape and much higher circulating concentrations of the storage protein. Each of the preceding steps in the sequence of thyroid hormone synthesis and release is discrete, so

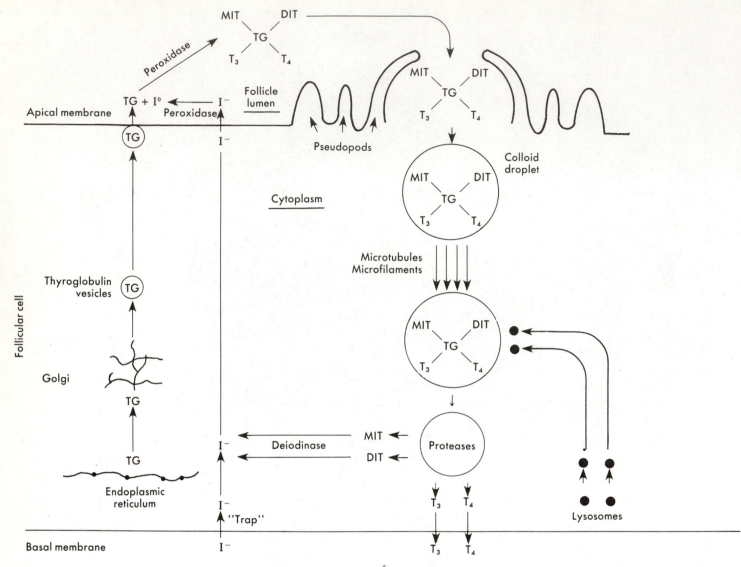

Fig. 53-5 ■ Overall schema of thyroid hormone synthesis and release. Note that iodide is incorporated into tyrosine molecules, which are in peptide linkage in thyroglobulin *(TG)*. The stored T_4 and T_3 are retrieved by endocytosis of colloid from the follicular lumen followed by protoeolysis with lysosomal enzymes. Iodide is recovered from MIT and DIT by the action of deiodinase.

that specific congenital biosynthetic defects occur which result in thyroid hormone deficiency. The entire process of secretion is summarized in Fig. 53-5.

■ *Regulation of thyroid gland activity*

The most important regulator of thyroid gland function and growth is TSH. Because TSH secretion shows only minimal diurnal or day-to-day variation, thyroid hormone secretion and plasma concentrations are also relatively constant. Small nocturnal increases in secretion of TSH and release of T_4 have been described; they are of unknown physiological significance. TSH stimulates the process of iodide trapping and of each step in T_4 and T_3 synthesis, as well as the endocytosis of colloid, the proteolysis of thyroglobulin, and the release of T_4 and T_3. Sustained TSH stimulation leads to hypertrophy and hyperplasia of the follicular cells (Fig. 53-1, *B*). The enlarged cells show an increased volume of endoplasmic reticulum, increased numbers of ribosomes, and a larger and more complex Golgi apparatus. An increase in DNA synthesis is shown by increased incorporation of tritiated thymidine. Proliferation of capillaries also is observed, and thyroid blood flow increases. In the absence of TSH marked atrophy of the gland occurs, although a low basal level of thyroid hormone production and release usually continues in humans, seemingly independent of TSH.

The effects of TSH are exerted through a multiplicity of actions. The initial step in TSH action is binding to a specific plasma membrane receptor. This is followed by activation of membrane adenylate cyclase and a subsequent increase in intracellular cyclic AMP levels. Within minutes colloid droplets appear in the cytoplasm as thyroglobulin stored within the follicular lumen undergoes endocytosis. Shortly thereafter an increase in iodide uptake and in peroxidase activity can be demonstrated. Coincident with these actions on hormone synthesis and release, TSH also stimulates glucose oxidation, especially via the hexose monophosphate shunt. This may be the means for generating the NADPH needed for the peroxidase reaction. Further effects of TSH on the thyroid gland are seen after a delay of hours to days. Nucleic acid and protein synthesis is increased, and effects on both transcription and translation have been described. Phospholipid synthesis also is stimulated. These actions seem less likely to be mediated by the rise in intracellular cyclic AMP, and their mechanism remains to be further elucidated. They probably relate to the effects of TSH on promoting growth of the gland.

The regulation of thyroid hormone secretion by TSH is under exquisite feedback control (Figs. 52-5 and 52-6; Chapter 52). Circulating T_4 and T_3 each produce feedback to the pituitary to decrease TSH secretion; as their levels fall, TSH secretion increases. It is free T_4 and T_3, not the protein-bound portions, that regulate pituitary TSH output. Since the pituitary gland is capable of deiodinating T_4 to T_3, the latter may be the final effector molecule in turning off TSH. In situations in which the diseased thyroid gland produces thyroid hormones autonomously or in which exogenous thyroid hormone is administered chronically, plasma TSH will be low.

Another important regulator of thyroid gland function is iodide itself, which has a biphasic action. At relatively low levels of iodide, thyroid hormone synthesis is positively correlated with iodide availability. However, beyond the levels needed for maximum synthesis an important negative effect of iodide on hormone secretion occurs. A large excess of iodide strongly inhibits the synthesis of stored T_4 and T_3. In addition, probably by an independent mechanism, an excess of iodide eventually also inhibits its own binding by the follicular cells. Additional modes of autoregulation may play a role in preventing excessive responses to TSH stimulation. Thyroglobulin has been shown to inhibit binding of TSH to its receptors, as well as the response of adenylate cyclase to the tropic hormone. In addition, T_4 and T_3 also exhibit direct inhibitory effects on the thyroid gland in vitro. The close proximity of sympathetic nerve fibers to the capillary plexuses in the gland and to the basal membrane of the epithelial cells suggests a regulatory influence. Thyroid hormone secretion is somewhat stimulated by epinephrine, via β-adrenergic receptors. However, the physiological significance of this, as well as an inhibitory influence of acetylcholine, is unknown. Similar clarification is awaited for the role of prostaglandins, which are present in the thyroid and can mimic some effects of TSH.

Thyroid hormones increase energy expenditure and heat production. Therefore it would be logical to expect that thyroid hormone availability would respond to the body's changing caloric and thermal status. In fact ingestion of an excess of calories, particularly in the form of carbohydrate, produces increases in T_3 production and plasma concentration and in the metabolic rate, whereas prolonged fasting leads to corresponding decreases. However, similar fluctuations in T_4 do not occur. Therefore, since most T_3 arises from circulating T_4 (Table 53-1), peripheral mechanisms are more important in mediating these changes than are alterations in thyroid gland secretion. In animals exposure to cold increases thyroid gland activity. In humans this is clearly seen only in

TABLE 53-1

Thyroid hormone turnover

	T_4	T_3	rT_3
Daily production (μg)	90	35	35
From thyroid (percent)	100	25	5
From T_4 (percent)	—	75	95
Extracellular pool (μg)	850	40	40
Plasma concentration			
Total (μg/dl)	8.0	0.12	0.04
Free (ng/dl)	2.0	0.28	0.20
Half-life (days)	7	1	0.8
Metabolic clearance (L/day)	1	26	77
Fractional turnover per day (percent)	10	75	90

the neonatal period when the infant suddenly becomes responsible for maintaining his or her own body temperature. At birth an acute rise in TSH secretion is followed by a rise in plasma T_4 to levels well above those of adults. Over the ensuing weeks or months plasma T_4 then subsides to a range that remains stable throughout life.

Pharmacological inhibition of thyroid gland activity is of considerable therapeutic importance. A class of drugs known as *thiouracils* (for example, propylthiouracil and methimazole) blocks the synthesis of T_4 and T_3 by acting on the peroxidase complex. After continuous administration for weeks the stores of thyroid hormone (and of iodide) become depleted. Because of the inhibition of organification, iodide taken up by activity of the trap is rapidly discharged again, as shown by studies with radioactive iodine. These drugs are very effective in the treatment of hyperthyroidism. Lithium salts, in the same doses used to treat manic depressive illness, inhibit the release of thyroid hormone and probably, secondarily, their synthesis. Lithium may act by blocking adenylate cyclase and cyclic AMP accumulation. Finally, a large excess of iodide, in addition to the effects previously noted, also can significantly and very promptly inhibit thyroid hormone release. Although this is a transient action, the administration of iodide to severely thyrotoxic individuals may produce an important degree of early benefit.

■ *Metabolism of thyroid hormones*

Table 53-1 shows the daily production rates, pool sizes, plasma concentrations, half-lives, metabolic clearances, and fractional turnovers of T_4, T_3, and rT_3. T_4 is clearly the dominant secreted and circulating form of thyroid hormone. It is important to note that the major portion of T_3 and virtually all of rT_3 come secondarily from circulating T_4, rather than primarily from thyroid gland secretion. Thus T_4 serves as a prohormone for T_3, in addition to providing some intrinsic intracellular action of its own. This "storage" function of plasma T_4 also is reflected in its much lower metabolic clearance and fractional turnover rates, compared with those of T_3 or rT_3. The concentration of intact thyroglobulin in plasma, averaging about 5 ng/ml, is much lower than that of the thyroid hormones.

Secreted T_4 and T_3 circulate in the bloodstream almost entirely bound to proteins. Normally only 0.03% of total plasma T_4 and 0.3% of total plasma T_3 are in the free state (Table 53-1). However, these are the critical fractions that are *biologically active,* not only in exerting thyroid hormone effects on peripheral tissues but in pituitary feedback as well. The major binding protein is *thyroxine-binding globulin* (TBG). This is a glycoprotein α-globulin with a molecular weight of 63,000 which is synthesized in the liver. Some 70% to 75% of circulating T_4 and T_3 is bound to TBG; the remainder is bound to albumin and to a thyroid-binding prealbumin (TBPA). Compared with TBG,

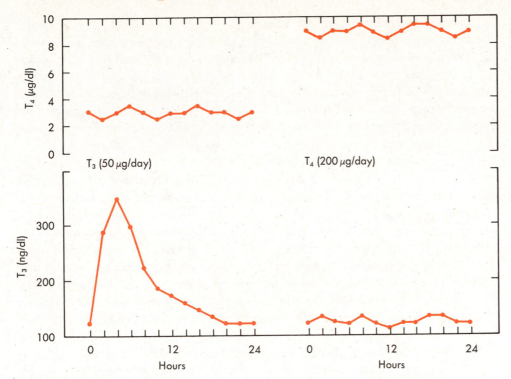

Fig. 53-6 ■ Effect of administering a single day's supply of calorigenic hormone by mouth to hypothyroid individuals. Note that 50 μg of T_3 causes a marked elevation in plasma T_3 levels for many hours, whereas 200 μg of T_4 causes no significant change in either plasma T_4 or T_3. This is because of the much larger pool size and tighter protein binding of T_4 than of T_3. (Redrawn from Saberi, M., et al.: Serum thyroid hormone and thyrotropin concentrations during thyroxine and triiodothyronine therapy, J. Clin. Endocrinol. Metab. **39**:923, 1974. Copyright 1974. Reproduced by permission.)

these additional binding proteins have much lower affinities but much higher capacities for T_4 and T_3. Ordinarily, however, only alterations in TBG concentration significantly alter total plasma T_4 and T_3 levels.

Two biological functions have been ascribed to TBG. First, it maintains a large circulating reservoir of T_4, which buffers against acute changes in thyroid gland function. Even the addition to the plasma of the calorigenic hormone needed for an entire day would cause only a 10% increase in the total T_4 concentration (Fig. 53-6). Conversely, after removal of the thyroid gland it would take 1 week for the plasma T_4 concentration to fall as much as 50%. Second, the binding of plasma T_4 and T_3 to large proteins prevents the loss of these relatively small hormone molecules into the urine and thereby helps conserve iodide.

The reservoir function of TBG is best understood by examining the chemical equilibrium between T_4 and TBG that governs the distribution of the hormone between the free (T_4) and bound ($T_4 \cdot$TBG) forms.

$$T_4 + TBG \rightleftharpoons T_4 + TBG \tag{1}$$

$$Keq = \frac{[T_4 \cdot TBG]}{[T_4]\,[TBG]} \tag{2}$$

$$\frac{[T_4]}{[T_4 \cdot TBG]} = \frac{Free\ T_4}{Bound\ T_4} = \frac{1}{Keq\,[TBG]} \tag{3}$$

$$[T_4] = [T_4 \cdot TBG] \times \frac{1}{Keq\,[TBG]} \tag{4}$$

A temporary decrease in free T_4, caused by a drop in thyroid gland output or accelerated uptake by target cells, can be rapidly compensated for by a dissociation of bound T_4 ($T_4 \cdot TBG$), until the new ratio of $T_4/T_4 \cdot TBG$ returns to that required by Keq (equation 3). A temporary increase in free T_4, caused by endogenous secretion or exogenous administration, can be rapidly compensated for by association of the excess with TBG, because normally only 30% of the available T_4 binding sites on TBG are occupied. Of course sustained decreases or increases in T_4 supply, resulting from thyroid disease, must eventually lead to sustained decreases or increases in free T_4, because the latter is directly proportional to $T_4 \cdot TBG$ (equation 4).

It is important to note that a primary change in TBG concentration will also disturb the ratio of free to bound T_4 (equation 3). In this circumstance it is the normal thyroid gland that must increase or decrease its rate of hormone secretion appropriately, until the new equilibrium state restores the free T_4 level to normal. TBG concentration can decrease because of reduced hepatic synthesis (liver disease) or excessive loss in the urine (kidney disease). Free T_4 then will increase temporarily. In compensation, pituitary TSH secretion will be suppressed by negative feedback. T_4 output by the thyroid gland then will decrease until the new, lower steady-state level of $T_4 \cdot TBG$ yields a normal level of free T_4 (equation 4). Hepatic synthesis of TBG also can be stimulated (for example, by estrogen administration or pregnancy), leading to an increase in TBG concentration. Free T_4 will decrease temporarily; this will stimulate pituitary secretion of TSH, and consequently T_4 output by the thyroid gland will increase. This will continue until the elevated level of $T_4 \cdot TBG$ is sufficient to restore the free T_4 level to normal in a new steady-state condition.

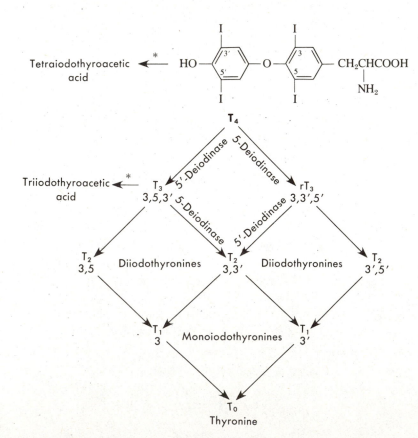

Fig. 53-7 ■ Peripheral metabolism of thyroxine (T_4) by successive deiodinations. A key regulatory step is the proportion of T_4 undergoing the initial deiodination to metabolically active T_3 versus metabolically inactive rT_3. *, Oxidative deamination and decarboxylation.

Identical qualitative considerations govern the circulating levels of free and bound T_3. However, the buffering action of TBG is less effective, because the Keq for T_3 is an order of magnitude lower than that for T_4 (2×10^9 versus 2×10^{10}, respectively), and because the total extrathyroidal pool of T_3 is only a twentieth that of T_4 (Table 53-1). Thus rapid addition of T_3 equivalent to the calorigenic hormone needed for an entire day produces greater swings in the concentrations of total and free T_3 (Fig. 53-6). Although alterations in TBG are not usually caused by thyroid gland disease, they must be taken into account when plasma thyroid hormone levels are measured for diagnostic purposes.

The peripheral metabolism of circulating thyroid hormones is outlined in Fig. 53-7. Degradation by deiodination accounts for the majority of the 80 to 90 μg of T_4 that is released by the gland daily. However, approximately 15% is irreversibly excreted in the bile as the various iodothyronines in glucuronide or sulfate conjugates. The quantitative importance of tetraiodoacetic acid and triiodoacetic acid as metabolites remains to be defined. The liver, kidney, and skeletal muscle are the major sites of degradation. The overall rate of disposal of T_4 is directly related to the free T_4 concentration in the plasma and to the intracellular T_4 content, particularly of the liver. Thus T_4 increases its own metabolism. The entire cascade of products—T_3, T_2, and T_1—of the sequential deiodination steps is regularly increased in the plasma in hyperthyroidism and is usually decreased in hypothyroidism.

■ *Relationship between hormone metabolism and hormone action*

The initial step in T_4 metabolism, the intracellular conversion of T_4 to either T_3 or rT_3, is of critical importance to thyroid hormone action. T_3 is the hormone of greatest biological activity, whereas rT_3 has almost no apparent calorigenic action. Therefore factors that regulate the relative rates of inner ring versus outer ring monodeiodination (Fig. 53-4) also determine the final biological effect of secreted T_4. In humans the normal distribution of T_4 products is 45% T_3 and 55% rT_3. An increase in T_4 concentration leads to a decrease in its conversion to T_3. Thus the biological effects of T_4 excess or deficiency are automatically mitigated to a slight extent by accelerated or retarded metabolic inactivation, respectively. Other states and factors are associated with reduced conversion of T_4 to T_3 and usually with a reciprocally enhanced conversion of T_4 to rT_3. These include the gestational period, fasting, stressful states, catabolic diseases, hepatic disease, renal failure, glucocorticoids, thiouracil drugs, and β-adrenergic blockade. A single common denominator responsible for switching T_4 away from T_3 and toward rT_3 has not been identified with certainty in all these conditions. However, the inhibition of 5′-monodeiodinase is an attractive possibility. This would simultaneously decrease production of T_3 from T_4 and decrease degradation of rT_3 to $3,3′$-T_2 (Fig. 53-7).

The biological effects of T_4 are largely a result of its intracellular conversion to T_3. When administered exogenously, T_3 is three to four times more potent than T_4 in humans. However it is still debatable how much intrinsic biological activity of its own is possessed by T_4. Evidence favoring activity of T_4 is found in situations in which a state of clinical thyroid deficiency exists with normal plasma T_3 but low plasma T_4 concentrations and in which a clinically normal state exists with low plasma T_3 but normal plasma T_4 concentrations. Furthermore, in the absence of endogenous thyroid gland function the maintenance of a euthyroid state requires doses of T_3 that sustain supranormal plasma T_3 levels, whereas only doses of T_4 that sustain normal plasma T_4 levels are required. The intrinsic biological activity of T_4 could lie in the greater efficiency of T_3 that has been generated from T_4 intracellularly as compared with the efficiency of T_3 reaching its intracellular sites of action from the circulation; rT_3 is almost without calorigenic activity. Moreover, by competing for the enzyme 5′-monodeiodinase, it inhibits the conversion of T_4 to T_3 and the calorigenic activities of both. The physiological significance of this action in normal humans remains to be determined.

■ *Intracellular actions of thyroid hormone*

T_4 and T_3 appear to enter cells freely by passive diffusion. Within the cell most if not all the T_4 undergoes 5′-monodeiodination to T_3 (or rT_3). Unlike steroid hormones, T_3 and T_4 require no cytoplasmic receptor for subsequent transport into the nucleus (Fig. 49-7). Instead they are bound directly to a specific nuclear receptor protein that is associated with chromatin. T_3 binds with much greater affinity than T_4 to this receptor. Histones may participate in enhancing this affinity. The T_3 receptor complex interacts with DNA to stimulate transcription of messenger RNAs (mRNA). The latter then directs increased synthesis of specific proteins. This sequence of events has been well documented in several in vitro systems. A recently well-studied example is the synthesis of growth hormone by cultured pituitary cells. T_3 stimulates the production of specific mRNA for growth hormone, followed by an increase in the protein hormone itself.

The responsiveness of tissues to T_3 correlates well with their nuclear receptor capacity and with the degree of receptor saturation. In the euthyroid state about half the available T_3 receptor sites are occupied. In humans several indices of T_3 action have been shown to correlate linearly with the percentage occupancy of nuclear receptors; occupancy was calculated from serum T_3 concentration and the known affinities of human liver and kidney T_3 receptors. Such observations strongly support the physiological importance of nuclear receptor binding for T_3 action.

Thyroid hormone appears to act largely through influencing transcription, and many of its effects can be blocked by inhibitors of protein synthesis, such as puromycin. This mechanism accounts for the 12- to 48-hour delay before most of the hormone's effects become evident in vivo. Indeed several weeks of T_4 replacement are required before all the consequences of the hypothyroid state are eliminated.

A voluminous amount of experimental data has been obtained in the attempt to explain the multitude of thyroid hormone actions on an intracellular basis. A large catalogue of thyroid-induced enzyme and substrate changes can be listed. To date, however, no single biochemical mechanism or final common pathway can be offered incontrovertibly as a unifying theory of hormone action. More likely, actions exist at multiple loci, which may vary in different tissues. Effects in the nucleus include stimulation of RNA polymerase and phosphoprotein kinases and the synthesis of other nuclear proteins. These nuclear effects are followed or paralleled by an increase in the size and number of mitochondria with a corresponding increase in their rate of respiration. Direct effects of T_4 and T_3 on mitochondrial RNA and protein synthesis also have been noted. Key respiratory enzyme activities, such as NADPH cytochrome C reductase and cytochrome oxidase, are increased. α-Glycerophosphate dehydrogenase, malic enzyme, and pyridine nucleotide transhydrogenases, important in regulating the levels of pyridine nucleotide cofactors, are likewise increased. The activities of a host of other enzymes concerned with glucose oxidation and gluconeogenesis also are augmented by thyroid hormone.

The clinical observation that thyroid hormone excess appears to increase the rate of oxygen use without increasing useful work output suggests that T_4 and T_3 might decrease the efficiency with which high-energy phosphate bonds are formed during aerobic respiration. Early in vitro evidence appeared to support the attractive hypothesis that thyroid hormone acts like dinitrophenol to uncouple oxidative phosphorylation. However, supraphysiological doses of thyroid hormone often were required for such an effect, and it also could be obtained with thyroxine analogues that had no thyroid hormone activity in vivo. Furthermore normal P/O ratios of approximately 3 subsequently were observed in muscle from hyperthyroid humans—an observation weakening the support for this theory. Thyroid hormone also increases the activity and amount of plasma membrane Na⁺, K⁺-ATPase. Ouabain, an inhibitor of this enzyme, also blocks the action of thyroid hormone on respiration. It therefore has been suggested that the increased oxygen use produced by thyroid hormone may be a response to accelerated dissipation of high-energy phosphate bonds. The extra ADP generated from the increased Na⁺, K⁺-ATPase activity might then stimulate oxygen use in the mitochondria.

In those tissues, such as the brain, in which oxygen consumption is not stimulated, the action of thyroid hormones to increase the synthesis of specific structural or functional proteins via transcriptional or translational effects must be invoked. In brain and other tissues thyroid hormone stimulates the transport of amino acids across the cell membrane, thereby facilitating protein synthesis. On the other hand, proteolytic and lysosomal enzyme activities also are increased by thyroid hormone, especially in muscle.

The most obvious in vivo effect of thyroid hormone, already appreciated in 1895, is to increase the rate of oxygen consumption and heat production (Fig. 53-8). This action is demonstrable in all tissues except the brain, gonads, and spleen. Oxygen use at rest in humans ranges from about 150 ml/minute in the hypothyroid state to about

■ *Whole body actions of thyroid hormone on metabolism*

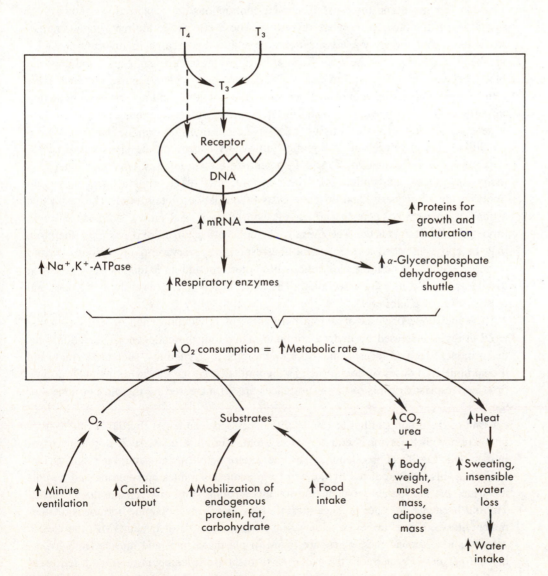

Fig. 53-8 ■ Overall schema of thyroid hormone effects. The upper portion represents intracellular actions, the lower portion whole body actions.

400 ml/minute in the hyperthyroid state, around a euthyroid mean of about 250 ml/minute. When converted to its caloric equivalent and standardized to body surface area, the basal metabolic rate ranges from -40% to $+80\%$ of normal at the clinical extremes of thyroid function. Thyroid hormone also augments the increase in oxygen consumption that accompanies exercise. Of necessity, thermogenesis increases concomitantly with oxygen use. Thus increases and decreases in body temperature parallel fluctuations in thyroid hormone availability. The potential increase in body temperature, however, is moderated by a compensatory increase in heat loss through appropriate thyroid hormone–mediated increases in blood flow, sweating, and ventilation.

Thyroid hormone could not regulate oxygen utilization for long without also influencing oxygen supply to the tissues. Thus T_4 and T_3 increase the resting rate of ventilation sufficiently to maintain a normal arterial P_{O_2} in the face of increased oxygen utilization and a normal P_{CO_2} in the face of increased carbon dioxide production. Additionally, a small increase in red blood cell mass is effected—enhancing the oxygen-carrying capacity—probably through stimulation of erythropoietin production.

Most important, thyroid hormone increases cardiac output, ensuring sufficient oxygen delivery to the tissues. The resting heart rate and the stroke volume are both increased. The speed and force of myocardial contractions are enhanced, as shown by a shortening of the preejection phase of systole. These effects are indirect, via adrenergic stimulation, and direct, via thyroid hormone–induced increases in myocardial Ca^{++} uptake and adenylate cyclase activity. Systolic blood pressure is augmented and diastolic blood pressure is decreased, so that the net result is a widened pulse pressure. This reflects the combined effects of the increased stroke volume and a substantial reduction in peripheral vascular resistance resulting from blood vessel dilation. The latter in turn is secondary to the increase in tissue metabolism that thyroid hormone induces.

Stimulation of oxygen utilization must ultimately depend on the provision of necessary substrates for oxidation. T_4 and T_3 potentiate the stimulatory effects of other hormones on glucose absorption from the gastrointestinal tract, on gluconeogenesis, on lipolysis, on ketogenesis, and on proteolysis of the labile protein pool. The respective actions of epinephrine, norepinephrine, glucagon, cortisol, and growth hormone on these processes are all enhanced. The overall metabolic effect of thyroid hormone therefore may be aptly described as accelerating the response to starvation. In addition, thyroid hormone stimulates both the biosynthesis of cholesterol and its oxidation, its conversion to bile acids, and its biliary secretion. The net effect is to decrease the body pool and plasma level of cholesterol.

The metabolic disposal of steroid hormones, of B vitamins, and of many administered drugs is increased by thyroid hormone. Therefore the endogenous secretion rates of hormones, such as cortisol, the dietary requirement of vitamins, such as thiamin, and the administered doses of drugs, such as digoxin, that are necessary to maintain normal or effective plasma levels of these substances are all increased by thyroid hormone.

■ *Thyroid hormone and sympathetic nervous system activity*

One of the prominent but incompletely understood features of thyroid hormone is its interaction with the sympathetic nervous system. Certain effects, such as the increases in metabolic rate, heat production, heart rate, motor activity, and central nervous system excitation, also are produced by the adrenergic catecholamines epinephrine and norepinephrine. An indisputable explanation for this striking similarity remains to be found. Thyroid hormone does not produce increased levels of catecholamine hormones or their metabolites in blood, urine, or tissues. But increased levels of cyclic AMP, the classic catecholamine second messenger, are found in plasma, urine, and muscle from hyperthyroid persons. Furthermore the cyclic AMP response to epinephrine in cultured myocardial cells is augmented by T_3. This suggests that T_4 and T_3 may modulate the sensitivity to catecholamine hormones. Earlier studies appeared to show augmented cardiovascular responses to epinephrine in the presence of excess thyroid hormone. More recent detailed dose-response studies have failed to confirm thyroid hormone enhancement of cardiac catecholamine effects in humans. One mechanism for interaction between the two, however, is suggested by the observation that T_3 increases the number of β-adrenergic receptors in heart muscle. Synergism between catecholamines and thyroid hormones also may be required for maximum thermogenesis, lipolysis, glycogenolyis, and gluconeogenesis to occur. The importance of this physiological issue, which is still to be resolved, is underscored by the fact that adrenergic blockade, particularly of the β-receptors, is very effective in decreasing some of the cardiovascular and central nervous system manifestations of hyperthyroidism.

■ *Thyroid hormone effects on growth and development*

Another major effect of thyroid hormone in vivo is on growth and maturation. Perhaps the most spectacular example is seen in the stimulation of the metamorphosis of tadpoles. Addition of thyroid hormone to the fluid bathing tadpoles accelerates all aspects of their metamorphosis to adults, including limb growth, tail resorption, shortening of the gastrointestinal tract, and induction of hepatic ureagenesis. In humans and other mammals thyroid hormone stimulates endochondral ossification, linear growth, and the maturation of the epiphyseal growth centers in bone. The effects of T_4 and T_3 on growth may be caused in part by facilitation of the secretion of growth hormone, as noted previously. The regular progression of tooth development and eruption is dependent on thyroid hormone, as is the normal cycle of growth and maturation of the epidermis and its hair follicles. Because the normal degradative processes in these structural and integumentary tissues also are stimulated by thyroid hormone, excess exposure to T_4 and T_3 also can cause resorption of bone, rapid desquamation of skin, and hair loss.

Normal development of the central nervous system is critically dependent on thyroid hormone. Growth of the cerebral and cerebellar cortex, branching of axons and dendrites, and the process of myelinization all are decreased when thyroid hormone is deficient in utero. This may result in irreversible damage if not recognized promptly after birth. Thyroid hormone also enhances wakefulness, alertness, responsiveness to various stimuli, awareness of hunger, memory, and learning capacity. Normal emotional tone also is dependent on proper thyroid hormone availability. The speed and amplitude of peripheral nerve reflexes are increased by thyroid hormone, as is motility throughout the gastrointestinal tract.

In both women and men thyroid hormone plays at least a permissive role, although an important one, in the regulation of reproductive function. The normal ovarian cycle of follicular development, maturation, and ovulation, the homologous testicular process of spermatogenesis, and the maintenance of the healthy pregnant state all are disrupted by significant deviations of thyroid hormone from the normal range. In part these may be caused by alterations in the metabolism or interconversion of steroid hormones.

Normal function of skeletal muscles also requires normal amounts of thyroid hormone. This may well be related to the regulation of energy production and storage in this tissue. Concentrations of creatine phosphate are reduced by an excess of T_4 and T_3; the inability of muscle to take up and phosphorylate creatine leads to its increased urinary excretion.

Kidney size, renal plasma flow, glomerular filtration rate, and tubular transport maximums for a number of substances also are increased by thyroid hormone.

■ *Clinical syndromes of thyroid dysfunction*

Hyperthyroidism results from a number of causes. Most commonly the entire gland undergoes hyperplasia and increases its secretion of T_4 and/or T_3 *(Graves' disease)*. This is because of development of an autoantibody, thyroid-stimulating immunoglobulin, which binds to the normal TSH receptor and initiates all the biological activities of the natural tropic hormone. The production of this "antibody to self" is thought to reflect dysfunction of thymus-derived lymphocytes, which normally suppress the development of such autoantibodies. Next most commonly, one or more areas of the thyroid form benign neoplasms or autonomous nodules, which escape from the normal hypothalamic-pituitary regulation. Least common causes are inflammation of the thyroid, excessive pituitary secretion of TSH, or ingestion of exogenous T_4 or T_3.

The patient suffering from an excess of thyroid hormone presents one of the most striking pictures in clinical medicine. The large increase in metabolic rate causes the highly characterisic combination of weight loss concomitant with an increased intake of food. The increased heat that is generated in the hypermetabolic state causes discomfort in warm environments, fever when the condition is severe, excessive sweating, and a greater intake of water. The increase in adrenergic activity is manifested by a rapid heart rate, atrial arrhythmias, hyperkinesis, tremor, nervousness, and a wide-eyed stare. Weakness is due to loss of muscle mass as well as a specific thyrotoxic myopathy. Other symptoms include diarrhea, emotional lability, and breathlessness during exercise. The patient also may have difficulty swallowing or breathing because of compression of the esophagus or trachea by the enlarged thyroid gland *(goiter)*.

The diagnosis is established by demonstrating an elevated serum T_4 and/or T_3 level (appropriately corrected for any abnormalities in TBG concentrations). In most instances the thyroid uptake of iodine (labeled with ^{131}I or ^{123}I) is excessive. Serum TSH levels are low and cannot be increased by administration of thyroid-releasing hormone (TRH), since the pituitary is inhibited by the high levels of T_4 and T_3. Various modalities of treatment are available. β-Adrenergic antagonists, such as propranolol, provide relief from tachycardia, tremor, and nervousness. Iodide inhibits release of thyroid hormones, but it is only useful temporarily. Thiouracil drugs block hormone synthesis and generally are given for at least 1 year. The most definitive treatment is ablation of thyroid tissue, either by radiation effects of ^{131}I or by surgery.

Hypothyroidism most often results from idiopathic atrophy of the gland—often thought to be preceded by a chronic inflammatory reaction. This form of *lymphocytic thyroiditis* is also an autoimmune condition. In this case the antibodies are produced to various thyroid cell antigens, including thyroglobulin, and some may have cytotoxic properties. Other causes of hypothyroidism include radiation damage, surgical removal, nodular goiters, hypothalamic or pituitary destruction, and, in certain areas of the world, iodide deficiency. Rarest of all is resistance to the action of thyroid hormones, possibly because of deficiency of receptors.

The clinical picture is, in many respects, the exact opposite of that seen in hyperthyroidism. The lower than normal metabolic rate leads to weight gain without appreciable increase in caloric intake. The decreased thermogenesis lowers body temperature and causes intolerance to cold, decreased sweating, and dry skin. There is decreased adrenergic activity, with bradycardia, a generalized slowing of movement, speech, and thought, lethargy, sleepiness, and a lowering of the upper eyelids (ptosis). An accumulation of mucopolysaccharides—ground substance—in the tissues causes an accumulation of fluid, as well. This *myxedema* produces puffy features, an enlarged tongue, hoarseness, joint stiffness, effusions in the pleural, pericardial, and peritoneal spaces, and entrapment of peripheral and cranial nerves with consequent dysfunction. Constipation, loss of hair, menstrual dysfunction, and anemia are other signs. Notably hypothyroidism in infancy or childhood causes marked retardation of growth and even greater slowing in the maturation of the epiphyseal growth centers of the bones. If hypothyroidism is present at birth and remains untreated, the central nervous system will not undergo its normal maturation process in the first year of life. Developmental milestones, such as sitting, standing, and walking, will be late, and severe irreversible mental retardation can result unless treatment is begun within 2 to 4 weeks after birth.

TABLE 53-2

Some congenital defects in thyroid hormone synthesis

Defect	Diagnostic pattern
Iodide trap	Decreased uptake of radioactive iodine; decreased salivary/blood ratio of radioactive iodine
Peroxidase	Increased early uptake of radioactive iodine*; rapid discharge by perchlorate
Deiodinase	Increased uptake of radioactive iodine*; increased MIT and DIT in urine
Coupling	Increased uptake of radioactive iodine*; increased MIT and DIT and decreased T_4 and T_3 in thyroid tissue

*Radioactive iodine uptake is increased because of increased TSH secretion, which stimulates the iodide trap.

The diagnosis is made by finding a low serum T_4. Serum TSH will be elevated because of negative feedback unless the hypothyroidism is caused by hypothalamic or pituitary disease. If the pituitary is at fault, TSH levels will be low and will not respond to administration of TRH. Other common laboratory abnormalities include elevation of serum cholesterol, because of slowed degradation, and of muscle enzymes such as creatine phosphokinase. Replacement therapy with T_4 is curative. T_3 is not needed, since it will be generated intracellulary from the administered T_4.

The stepwise nature of thyroid hormone synthesis offers multiple possibilities for congenital hypothyroidism caused by specific enzyme deficiencies. Clinically they are characterized by the symptoms of childhood hypothyroidism noted previously, plus thyroid gland enlargement (congenital goiter) resulting from persistent hypersecretion of TSH. Table 53-2 lists the best understood of these syndromes, with the biochemical findings that point to the lesion. Inability to carry out active transport of iodide into the gland is deduced from the finding of a very low uptake of radioactive iodine. Adequate synthesis of T_4 can be achieved by administration of large amounts of iodine, which is effective treatment. In peroxidase deficiency iodide is concentrated by the gland but not incorporated into tyrosine. Therefore, after radioactive iodine is rapidly taken up, it is gradually discharged again as radioactive iodide levels in the blood fall. This discharge is markedly accelerated by administration of perchlorate ion, which competes for iodide in the trap, thus preventing replenishment of the thyroidal free iodide level after it begins to fall. Treatment of peroxidase deficiency requires T_4 replacement.

A deficiency of thyroid deiodinase acts to produce a state of intraglandular iodine deficiency. The MIT and DIT molecules released during proteolysis of thyroglobulin leak out of the gland, carrying with them iodine; they then are irretrievably lost in the urine. Thus this defect also is corrected by administration of excess iodine. Radioactive iodine uptake is high, as it would be in iodide-deficient states. The diagnosis is established by finding large amounts of MIT and DIT in the urine. The diagnosis of a coupling defect is suspected when radioactive iodine uptake by the gland is high but there is no evidence for peroxidase or deiodinase deficiency. The diagnosis can be established only by demonstrating the presence of large amounts of MIT and DIT but very little T_4 and T_3 in a specimen of thyroid tissue. Treatment with T_4 is required.

■ *Bibliography*

Journal articles

Amidi, M., et al.: Effect of the thyroid state on myocardial contractility and ventricular ejection rate in man, Circulation **38**:229, 1968.

Asano, Y., Liberman, U.A., and Edelman, J.S.: Thyroid thermogenesis; relationships between Na^+-dependent respiration and Na^+ + K^+-adenosine triphosphatase activity in rat skeletal muscle, J. Clin. Invest. **57**:368, 1976.

Bantle, J.P., et al.: Common clinical indices of thyroid hormone action: relationships to serum free 3,5,3'-triiodothyronine concentration and estimated nuclear occupancy, J. Clin. Endocrinol. Metab. **50**:286, 1980.

Carter, W.J., et al.: Effect of thyroid hormone on metabolic adaptation to fasting, Metabolism **24**:1177, 1975.

Chopra, I.J.: An assessment of daily production and significance of thyroidal secretion of 3,3',5'-triiodothyronine (reverse T3) in man, J. Clin. Invest. **58**:32, 1976.

Cook, P.B., et al.: The effects of thyrotoxicosis upon the metabolism of calcium, phosphorus, and nitrogen, Q. J. Med. **23**:505, 1959.

Crowley, W.F., Jr., et al.: Noninvasive evaluation of cardiac function in hypothyroidism; response to gradual thyroxine replacement, N,. Engl. J. Med. **296**:1, 1977.

Dumont, J.E.: The action of thyrotropin on thyroid metabolism, Vitam. Horm. **29**:287, 1971.

Karlberg, B.E., et al.: Cyclic adenosine 3',5'-monophosphate concentration in plasma, adipose tissue and skeletal muscle in normal subjects and in patients with hyper- and hypothyroidism, J. Clin. Endocrinol. Metab. **39**:96, 1974.

Larsen, P.R., et al.: Inhibition of intrapituitary thyroxine to 3,5,3'-triiodothyronine conversion prevents the acute suppression of thyrotropin release by thyroxine in hypothyroid rats, J. Clin. Invest. **64**:117, 1979.

Martial, J.A., et al.: Regulation of growth hormone gene expression; synergistic effects of thyroid and glucocorticoid hormones, Proc. Natl. Acad. Sci. U.S.A. **74**:4293, 1977.

O'Conner, J.F., et al.: The 24-hour plasma thyroxine profile in normal man, J. Clin. Endocrinol. Metab.**39**:765, 1974.

Onaya, T., et al.: New *in vitro* tests to detect the thyroid stimulator in sera from hyperthyroid patients by measuring colloid droplet formation and cyclic AMP in human thyroid slices, J. Clin. Endocrinol. Metab. **36**:859, 1973.

Oppenheimer, J.H., et al.: Stimulation of hepatic mitochondrial α-glycerophosphate dehydrogenase and malic enzyme by L-triiodothyronine; characteristics of the response with specific nuclear thyroid hormone binding sites fully saturated, J. Clin. Invest.**59**:517, 1977.

Ridgway, E.C., et al.: Acute metabolic responses in myxedema to large doses of intravenous L-thyroxine, Ann. Intern. Med. **77**:549, 1972.

Saberi, M., et al.: Serum thyroid hormone and thyrotropin concentrations during thyroxine and triiodothyronine therapy, J. Clin. Endocrinol. Metab. **39**:923, 1974.

Schimmel, M., et al.: Thyroidal and peripheral production of thyroid hormones, Ann. Intern. Med. **87**:760, 1977.

Shapiro, L.E., et al.: Thyroid and glucocorticoid hormones synergistically control growth hormone mRNA in cultured GH_1 cells, Proc. Natl. Acad. Sci. U.S.A. **75**:45, 1978.

Stocker, W.W., et al.: Coupled oxidative phosphorylation in muscle of thyrotoxic patients, Am. J. Med. **44**:900, 1968.·

Vagenakis, A.G., et al.: Control of thyroid hormone secretion in normal subjects receiving iodides, J. Clin. Invest. **52**:528 1973.

Vagenakis, A.G., et al.: Effect of starvation on the production and metabolism of thyroxine and triiodothyronine in euthyroid obese patients, J. Clin. Endocrinol. Metab. **45**:1305, 1977.

Wolff, J.: Iodide goiter and the pharmacologic effects of excess iodide, Am. J. Med. **47**:101, 1969.

Wolff, J., et al.: The role of microtubules and microfilaments in thyroid secretion, Recent Prog. Horm. Res. **29**:229, 1973.

Wollman, S.H., et al.: Localization of esterase and acid phosphatase in granules and colloid droplets in rat thyroid epithelium, J. Cell Biol. **21**:191, 1964.

Books and monographs

Greer, M.A., et al.: Thyroid secretion. In Handbook of physiology; Section 7, Endocrinology, vol. III, Thyroid, Baltimore, 1974, American Physiological Society.

chapter 54

The adrenal glands

The adrenal glands are complex, multifunctional endocrine organs, the essential nature of which has been known since the mid-1800s. At that time, both clinical observations and animal experiments demonstrated that severe illness resulted from their atrophy, and death followed their complete removal. The catecholamine *epinephrine* was discovered in 1901; the androgenic steroids and the glucocorticoids *(cortisol* and *corticosterone)* were identified in the period between 1930 and 1950, and the mineralocorticoid *aldosterone* was identified in 1952. The firm association of specific clinical syndromes resulting from excesses and deficiencies of each type of adrenal hormone has gone hand in hand with their isolation and with knowledge of their biosynthetic pathways.

The adrenal glands are located in the retroperitoneum just above each kidney and within that organ's investing fat. Their total weight is 6 to 10 g. The adrenal gland is really a combination of two separate functional entities (Fig. 54-1). The outer zone, or *cortex,* which comprises 80% to 90% of the gland, is derived from mesodermal tissue and is the source of steroid hormones. The inner zone, or *medulla,* comprising the other 10% to 20%, is derived from neuroectodermal cells of the sympathetic ganglia and is the source of catecholamine hormones.

The adrenals are exceedingly well vascularized, receiving arterial blood from branches of the aorta, the renal arteries, and the phrenic arteries; they have one of the body's

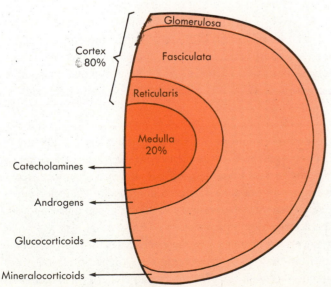

Fig. 54-1 ■ Schematic representation of the zones of the adrenal gland and their main secretory products.

highest rates of flow per gram of tissue. Arterial blood enters the cortex and breaks up into sinusoidal capillaries; the blood then drains down into medullary venules, an arrangement that exposes the medulla to relatively high concentrations of corticosteroids from the cortex. The right adrenal vein drains directly into the inferior vena cava, whereas the left drains into the renal vein on that side. Catheterization of each adrenal vein for individual sampling or radiographic visualization is possible, but it is more difficult on the right. The human adrenal glands can now also be well visualized by computerized axial tomography (CAT scanning).

■ *The adrenal cortex*

The major hormones of the cortex are (1) the *glucocorticoids,* cortisol and corticosterone, which are critical to life by virtue of their effects on carbohydrate and protein metabolism, (2) a *mineralocorticoid,* aldosterone, which is vital to maintaining sodium and potassium balance, and (3) the *sex steroids, androgens* and *estrogens* and their precursors, which contribute to establishing and maintaining secondary sexual characteristics. Particular interest in cortisol and in glucocorticoids in general was heightened greatly by the discovery of their potent antiinflammatory effects. In supraphysiological doses, the glucocorticoids have been used to treat a wider variety of disease than has any other single group of drugs. This has produced a far greater incidence of iatrogenic disorders of glucocorticoid excess than that resulting from spontaneous adrenal diseases.

The cortical portion of the adrenal gland differentiates by 8 weeks of intrauterine life and is initially much larger than the adjacent kidney. At this time the cortex contains two zones. The *peripheral neocortex,* comprising 15% of the cortex, is undifferentiated and relatively inactive. The inner 85%, known as the *fetal cortex,* is highly active and is responsible for fetal adrenal steroid production throughout almost all of intrauterine life. Shortly after birth, the fetal cortex begins to involute; hemorrhage and necrosis occur, and this zone disappears completely in 3 to 12 months. Concurrently, the thin outer zone enlarges, differentiates, and becomes the permanent three-layered adrenal cortex of the normal human (Fig. 54-1).

The histological appearance of the three adult cortical zones differs. The outermost *zona glomerulosa* is the narrowest zone and consists of small cells with numerous elongated mitochondria that possess lamellar cristae. The middle *zona fasciculata* is the widest zone and consists of columnar cells that form long cords. The cytoplasm is highly vacuolated and contains lipid droplets. The mitochondria are distinguished by their large size and numerous vesicular cristae within their membranes or matrices. The innermost *zona reticularis* contains networks of interconnecting cells with fewer lipid droplets. Their mitochondria are similar to those of fasciculata cells. Under stimulation by adrenocorticotropic hormone (ACTH), the size and the number of cells in the fasciculata and reticularis increase. The mitochondria of these cells become more numerous and larger and develop central ribosomes, vesicular cristae, and polylamellar membranes that extend to nearby cholesterol-containing vacuoles. In addition, the endoplasmic reticulum increases. These changes relate to ACTH effects on steroid hormone synthesis as detailed later.

■ *Synthesis of adrenocortical hormones*

All hormones of the adrenal cortex represent chemical modifications of the steroid nucleus shown in Fig. 54-2. Potent glucocorticoids require the presence of a ketone at the 3 position and hydroxyl groups at the 11 and 21 positions. Potent mineralocorticoids require an oxygenated carbon at the 18 position. Potent androgens are characterized by the elimination of the C_{20-21} side chain and the presence of an oxygenated carbon at the 17 position. Estrogens are characterized by aromatization of the A ring.

The precursor for all adrenocortical hormones is *cholesterol.* This 27-carbon steroid molecule is actively taken up from the plasma by adrenal cells. Specific plasma membrane receptors bind circulating low-density lipoproteins rich in cholesterol. After trans-

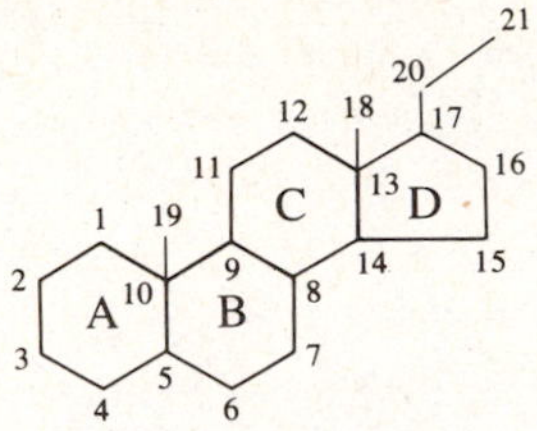

Fig. 54-2 ■ The adrenocorticosteroid nucleus.

fer into the cell by endocytosis, the cholesterol is largely esterified and stored in cytoplasmic vacuoles. Cholesterol is also synthesized in adrenal cells from acetylcoenzyme A (acetyl CoA) by the usual pathway. Under basal circumstances, free cholesterol from plasma is the most direct and major source used for hormone synthesis. When production of corticosteroids is stimulated, however, stored cholesterol becomes an increasingly important precursor. Cholesterol synthesized de novo is always a minor component of the pool.

Most of the synthetic reactions from cholesterol to active hormones involve cytochrome P-450 enzymes, which are mixed oxygenases that catalyze steroid hydroxylations. These hydrophobic hemoproteins are localized in the lipophilic membranes of the endoplasmic reticulum and mitochondrial cristae. Molecular oxygen is split so that one oxygen atom interposes between the carbon and hydrogen of the steroid site; the other oxygen atom is reduced by a hydrogen to H_2O. NADPH and, to some extent, NADH, which are generated by oxidation of a variety of substrates, are the ultimate donors of the hydrogen. A flavoprotein, adrenoxin reductase, and an iron-containing protein, adrenoxin, are intermediates in the transfer of hydrogen from NADPH to the P-450 enzymes.

Glucocorticoids. The synthesis of glucocorticoids occurs largely in the zona fasciculata with a smaller contribution from the zona reticularis. The sequence of reactions for synthesis of cortisol and corticosterone is shown in Fig. 54-3. The intracellular localization of the various steps is illustrated in Fig. 54-4. Cortisol is the glucocorticoid for humans and most mammals, whereas corticosterone serves this function in some rodents. However, if cortisol synthesis is blocked but the pathway to corticosterone is open, increased synthesis of the latter can provide the necessary glucocorticoid activity in humans. The initial reaction converting cholesterol to Δ^5-pregnenolone is catalyzed by a P-450 enzyme complex known as *desmolase*. This intramitochondrial complex, which carries out successive hydroxylations followed by cleavage of the cholesterol side chain, is rate limiting for adrenal steroidogenesis in general. The Δ^5-pregnenolone is then converted to 11-deoxycortisol by successive steps within the endoplasmic reticulum. The latter is hydroxylated in the 11 position after transfer back to the mitochondria. The end product, cortisol, rapidly diffuses out of the organelle and then out of the cell. The last and critical step for glucocorticoid synthesis, 11-hydroxylation, is very efficient in humans; 95% of the 11-deoxycortisol formed is converted to cortisol.

The order of hydroxylations from Δ^5-pregnenolone to 11-deoxycortisol is not invariant. Even the 3β-ol-dehydrogenase and $\Delta^{4,5}$-isomerase reactions can occur after all of the hydroxylations, rather than before. However, the order of hydroxylation presented in Fig. 54-3 *(17-21-11)* is compatible with the usual pattern of precursor accumulation that occurs when the various hydroxylases are either chemically blocked or congenitally deficient. It is important to note that the critical hormone, cortisol, and its precursors are not stored in the adrenocortical cell to any significant extent. Hence, an acute need for increased amounts of circulating cortisol requires rapid activation of the entire synthetic sequence and most particularly that of the rate-limiting desmolase reaction.

Androgens and estrogens. The synthesis of sex steroids occurs largely in the zona reticularis, although the zona fasciculata possesses the same capabilities. The 17-hydroxylated derivatives of Δ^5-pregnenolone and progesterone are the starting points for

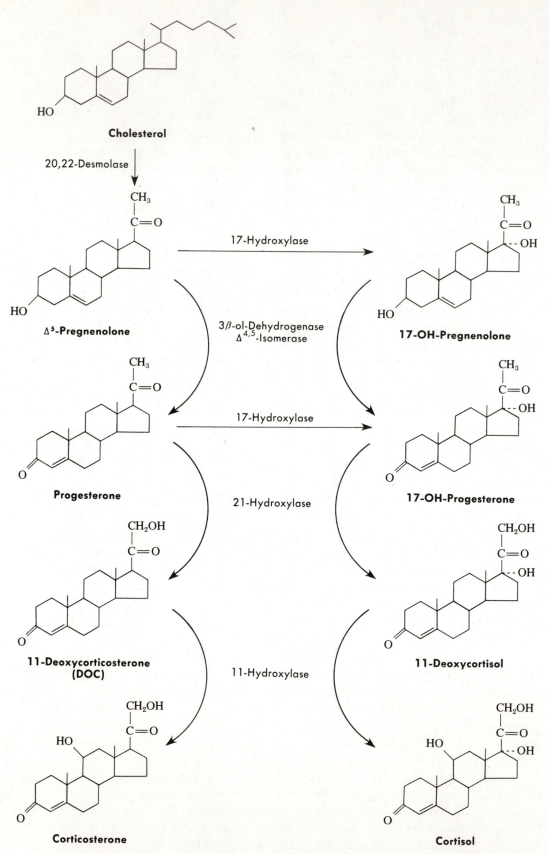

Fig. 54-3 ■ Synthesis of glucocorticoids in the zona fasciculata. Cortisol is the major product in humans.

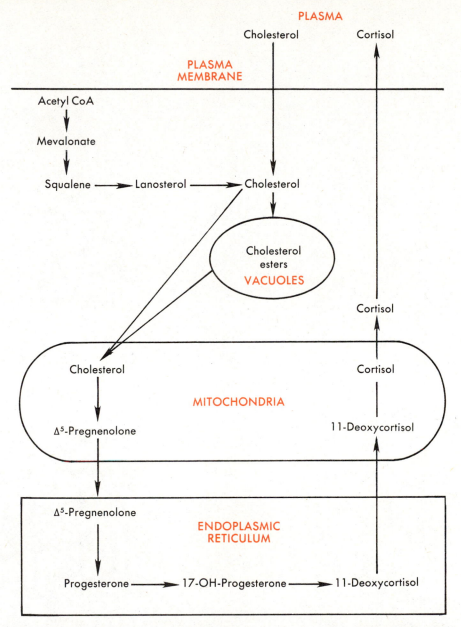

Fig. 54-4 ■ Intracellular localization of the reactions involved in cortisol biosynthesis.

androgen and estrogen synthesis. Removal of the C_{20-21} side chain by a microsomal desmolase-like reaction is the key step yielding dehydroepiandrosterone (DHEA) and androstenedione, respectively, as shown in Fig. 54-5. DHEA is sulfated by a specific enzyme; the sulfate donor is 3'-phosphoadenosine 5'-phosphosulfate. Dehydroepiandrosterone sulfate (DHEA-S) and DHEA are the major androgenic products of the adrenal. Although rather weak androgens themselves, they are converted to the potent androgen *testosterone* in peripheral tissues. Smaller amounts of androstenedione, an androgen of limited potency, and tiny amounts of testosterone itself are synthesized and secreted by the zona reticularis.

In women the adrenals supply 50% to 60% of the androgenic hormone requirements. Adrenal androgens are of little biological importance to men because the testes produce a large quantity of testosterone. The further conversion within the adrenal cortex of testosterone and androstenedione to estradiol and estrone, respectively, is of little quan-

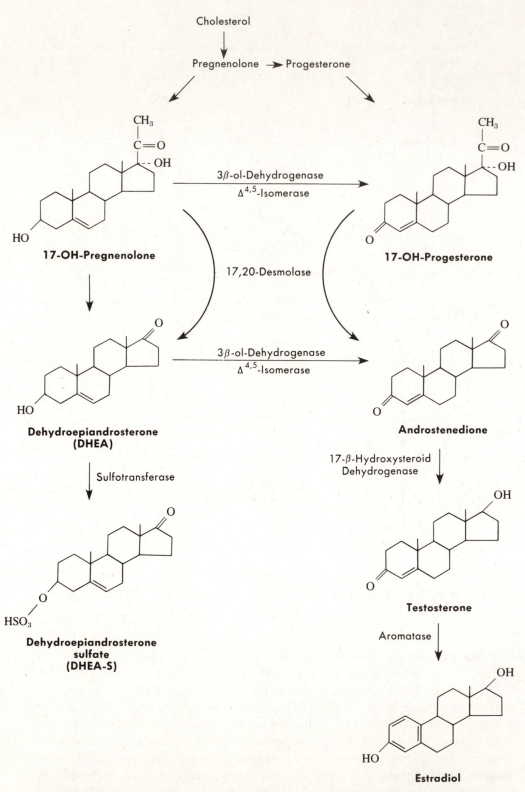

Fig. 54-5 ■ Synthesis of androgens and estrogens in the zona reticularis. DHEA-S is the major product. Normally only trivial amounts of testosterone and estradiol are produced.

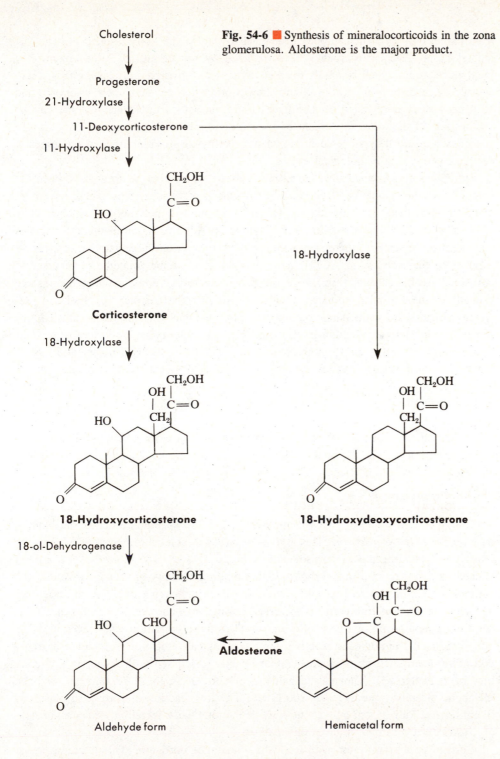

Fig. 54-6 ■ Synthesis of mineralocorticoids in the zona glomerulosa. Aldosterone is the major product.

titative significance in women until the ovaries are removed or cease to function after menopause. Then estrogens secreted directly from the adrenal or arising in the periphery from adrenal precursors become the only source for this biological activity. It is very important to note that 17-hydroxylation is the last reaction common to the synthesis of cortisol and the adrenal androgens (compare Fig. 54-3 with Fig. 54-5). For this reason, when cortisol synthesis is impaired at any point beyond this step, the accumulation of 17-hydroxypregnenolone and 17-hydroxyprogesterone leads to greatly increased androgen synthesis.

Mineralocorticoids. The synthesis and secretion of aldosterone, the major mineralocorticoid, is carried out exclusively by the zona glomerulosa (Fig. 54-6). The sequence from cholesterol to corticosterone is identical to that in the zona fasciculata. The C_{18} methyl group of corticosterone is then oxidized by mitochondrial P-450 enzymes to yield aldosterone, which is rapidly released. Whether 18-hydroxycorticosterone is a necessary intermediate is a point of controversy. Corticosterone, deoxycorticosterone, and 18-hydroxydeoxycorticosterone also have some mineralocorticoid activity and can be synthesized in the zona fasciculata under ACTH stimulation. However, only in rare circumstances are they secreted in sufficient quantity to provide either physiologically adequate action or a pathological excess.

Inhibitors of adrenocortical hormone synthesis. A number of drugs have been developed that are capable of blocking steroid synthesis at various steps. These have both diagnostic and therapeutic usefulness. Of particular note is metyrapone, which inhibits 11-hydroxylation, the last step in cortisol synthesis. Administration of this drug creates acute cortisol deficiency, thereby stimulating ACTH secretion via negative feedback. As a result of the increase in ACTH, adrenal production of the immediate precursor to cortisol, 11-deoxycortisol (Fig. 54-3), increases markedly, demonstrating the reserve capacity of the normal hypothalamic–pituitary ACTH axis (Chapter 52). Failure to respond indicates a hypothalamic-pituitary disease that has diminished or ablated this function. Aldosterone secretion is also decreased by metyrapone by virtue of its additional inhibitory effect on 18-oxygenation. Aminoglutethimide is a potent inhibitor of the desmolase reaction, thereby decreasing all adrenal steroid synthesis and causing accumulation of cholesterol in the gland. The aromatase reaction is also blocked by aminoglutethimide. This drug has been useful in treating patients with excessive cortisol and androgen output, as well as in diminishing estrogen production in women with breast cancer. Another drug, 2,2-*bis*(2-chlorophenyl, 4-chlorophenyl)-1-1-dichloroethane, known as *o,p'*-DDD, predominantly destroys the zona fasciculata and reticularis, resulting in cortisol deficiency. This drug is effective in therapy of adrenocortical cancer.

■ Metabolism of adrenocorticosteroids

Cortisol circulates in plasma 75% to 80% bound to a specific corticosteroid-binding α_2-globulin called *transcortin*. This glycoprotein has a molecular weight of 52,000 and binds a single molecule of cortisol with a K_{assoc} of 3×10^7 moles/L. The normal concentration of transcortin is 3 mg/dl, with a binding capacity of 20 µg cortisol/dl. An additional 15% of plasma cortisol is bound to albumin, leaving only 5% to 10% free. The concentration of transcortin is increased during pregnancy and by estrogen administration; thus total plasma cortisol is increased in these conditions. However, because bound cortisol is biologically inactive, the physiological effects of an increase in transcortin are determined by principles similar to those discussed with regard to thyroxine binding (Chapter 53). The plasma half-life of cortisol is about 70 minutes, and the metabolic clearance rate averages 200 L/day. The free fraction of cortisol in plasma is filtered by the kidney, but only about 0.3% of the total daily secretion, or approximately 50 µg, is excreted in the urine in this manner.

Cortisol is in equilibrium with its 11-keto analogue cortisone. This interconversion is catalyzed by 11β-ol-dehydrogenase, which is present in many tissues. The glucocorticoid activity of cortisone is dependent on its conversion to cortisol. The vast majority of cortisol and cortisone is metabolized in the liver; the metabolites then are conjugated and excreted in the urine as glucuronides. Each undergoes a parallel series of reactions that reduces ring A to tetrahydro derivatives and the 20-keto group to a hydroxyl group. About half of these excretory products are normally derived from cortisol and half from cortisone. The sequence is illustrated in Fig. 54-7.

The measurement of urinary metabolites provides a reliable index of cortisol secretion as long as hepatic and renal functions are intact. Of particular use is the 17,21-dihydroxy-20-ketone configuration of tetrahydrocortisol and tetrahydrocortisone, which

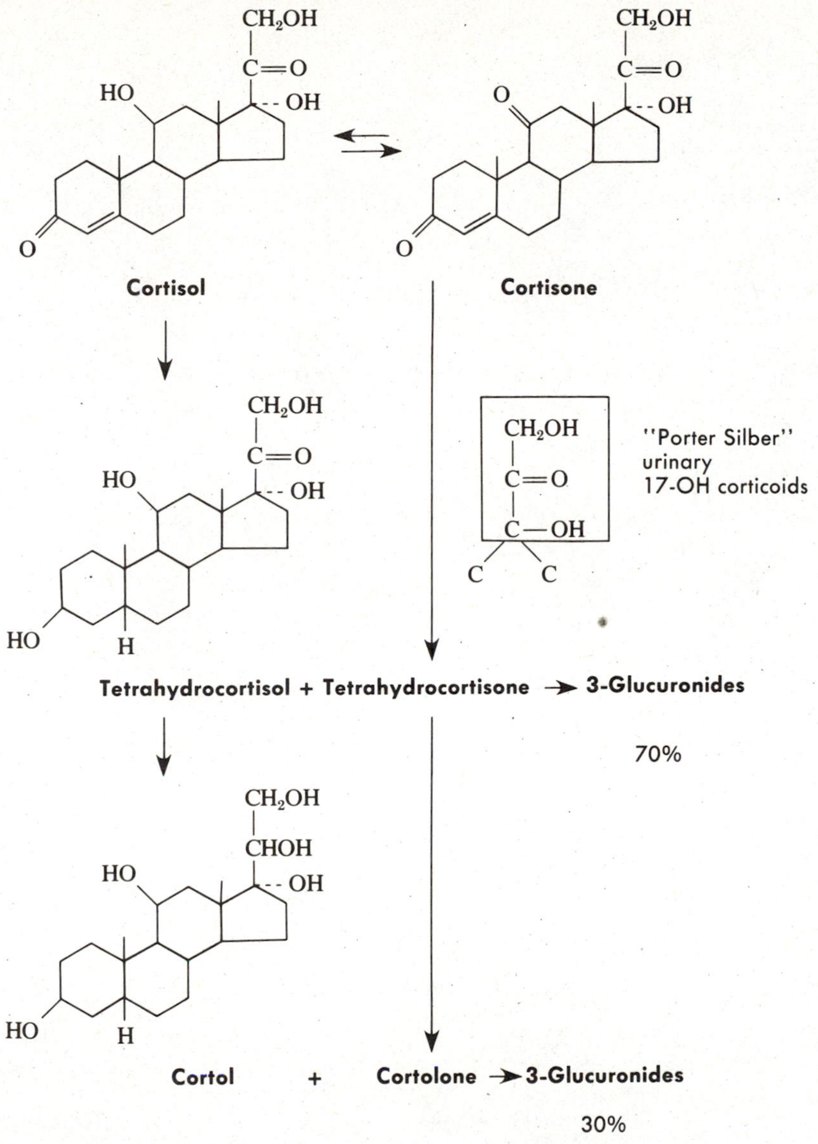

Fig. 54-7 ■ Major pathways of cortisol metabolism. The analogous compounds are formed from cortisone. The ratio of cortisol to cortisone metabolites is normally about 1:1.

forms the basis for a long-employed convenient procedure (Fig. 54-7). The Porter-Silber method measures the urinary "17-hydroxycorticoids," which represent up to 50% of the total daily cortisol secretion. Normal values of 17-hydroxycorticoids range from 2 to 12 mg/day. The values are slightly higher in men than in women, as are cortisol secretion rates. Daily urinary 17-hydroxycorticoid excretion may be corrected for differences in lean body mass by expressing the result as a ratio to the urinary creatinine excretion. Measurement of urinary 17-hydroxycorticoids (or of urinary free cortisol) is used to assess adrenocortical responsiveness to ACTH stimulation or its suppressibility by exogenous synthetic glucocorticoid. These are valuable tests for adrenocortical insufficiency or for autonomous adrenocortical hyperfunction, respectively. In addition, because 11-deoxycortisol also possesses the Porter-Silber reaction grouping and is excreted as a tetrahydrometabolite, the 17-hydroxycorticoid fraction will normally increase in the urine following administration of the 11-hydroxylase inhibitor metyrapone, as previously described. Hence, the integrity of the negative feedback response of the corticotropin-releasing factor (CRF)–ACTH axis may also be assessed in this manner.

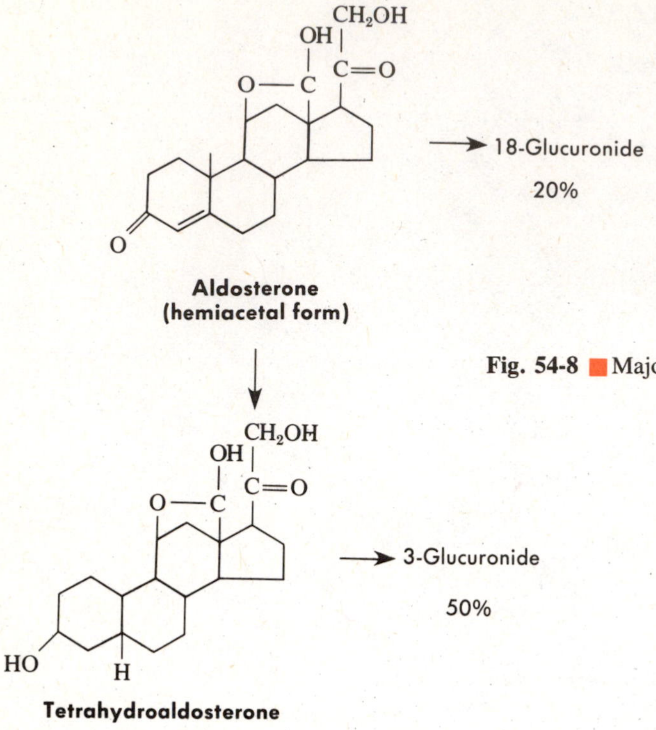

CH2OH
OH
C=O
O—C
18-Glucuronide
20%
O
Aldosterone
(hemiacetal form)
CH2OH
OH
C=O
O—C
3-Glucuronide
50%
HO
H
Tetrahydroaldosterone

Fig. 54-8 ■ Major pathway of aldosterone metabolism.

O
O
Androstenedione
OH
O
Testosterone
+
O
O
HO
H
Androsterone
HO
H
Etiocholanolone
17-Ketosteroids
+
11-β-OH-Androsterone and 11-β-OH-Etiocholanolone
+
Dehydroepiandrosterone sulfate (DHEA-S)

Fig. 54-9 ■ Major pathways of androgen metabolism. Together all of the urinary metabolites of testosterone make up the 17-ketosteroid fraction. Note, however, that the potent androgen testosterone is not a 17-ketosteroid.

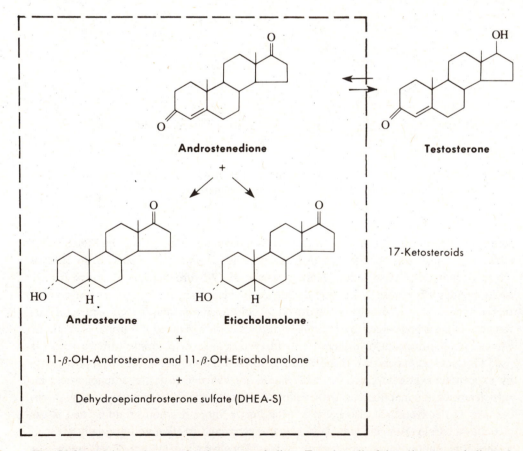

The cortisol precursors, progesterone and 17-hydroxyprogesterone, are metabolized to cortols analogous to those of cortisol. The excretory products are known as pregnanediol and pregnanetriol, respectively. In adult females, measurement of these urinary metabolites reflects both adrenal and ovarian secretion. In prepubertal children, however, elevation of urinary pregnanetriol specifically indicates increased secretion of adrenal 17-hydroxyprogesterone and is, therefore, a valuable marker for congenital blocks in cortisol secretion.

Aldosterone circulates in plasma bound to a specific aldosterone-binding globulin, to transcortin, and to albumin. Overall binding is weaker than for cortisol. Hence, the plasma half-life is only 20 minutes, and the metabolic clearance rate is 1600 L/day. Ninety percent of aldosterone is cleared by the liver in a single passage. There, aldosterone is reduced to tetrahydroaldosterone, the major metabolite that is excreted in the urine as the 3-glucuronide (Fig. 54-8). A smaller portion of aldosterone is excreted simply as the 18-glucuronide. The latter, however, is the urinary form that is commonly measured. The values in subjects with a normal sodium diet range from 5 to 20 μg/day.

The metabolism of androgens, in general, involves reduction of the 3-ketone group and the A ring in the liver. The two isomers formed, androsterone and etiocholanolone (Fig. 54-9), are then excreted in the urine. These metabolites, however, are not specific for the adrenal because they arise from gonadal androgens as well. DHEA-S is entirely excreted directly in the urine. Together these make up the major part of a urinary fraction called 17-ketosteroids. They are measured by a single colorimetric procedure known as the Zimmerman reaction. Normal values range from 5 to 14 mg/day in women and 8 to 20 mg/day in men. Two thirds of this fraction is normally derived from adrenal and one third from gonadal androgen secretion. However, when urinary 17-ketosteroid excretion is greatly elevated, this almost always indicates an adrenal abnormality. This is especially useful diagnostically in women or children who show evidence of virilization. Similar information can be obtained by measurement of plasma or urinary DHEA-S.

The secretion of cortisol by the zona fasciculata is, for all practical purposes, under the exclusive physiological control of pituitary ACTH. The secretion of adrenal androgens is likewise influenced by ACTH, but some evidence suggests the possible existence of a separate tropic hormone for the zona reticularis. In the complete absence of its tropic hormone, human adrenocortical secretion virtually ceases. ACTH initiates its action by binding to an adrenal plasma membrane receptor. The composition of the receptor includes protein, carbohydrate, sialic acid, and phospholipid. It is of interest that smaller analogues of ACTH have been synthesized that react well with the receptor but have virtually no biological activity. This suggests that sites on the ACTH molecule other than those involved in receptor binding may be required to couple the occupied receptor to subsequent steps in hormone action. ACTH activates adenylate cyclase in a Ca^{++}-dependent manner, and cyclic AMP (cAMP) levels rise. It is not clear, however, that all the actions of ACTH are mediated by cAMP. There is evidence suggesting that one subsequent event is the phosphorylation of ribosomal protein or proteins; the latter, in turn, may control the translation from messenger RNA of another labile regulatory protein capable of activating a key cytochrome P-450 enzyme. The nature of the regulatory protein is unknown. ACTH also increases adrenal phospholipids, which may mediate the stimulation of steroid synthesis.

The most important action of ACTH is to stimulate the rate-limiting desmolase reaction (cholesterol $\rightarrow$ Δ^5-pregnenolone). In addition, 11-hydroxylase activity is increased by ACTH. The tropic hormone also increases availability of cholesterol by stimulating its active uptake from plasma, its de novo synthesis from acetyl CoA, and the esterase activity that liberates cholesterol from its stored esterified form. Finally, it

■ Regulation of zona fasciculata and zona reticularis functions

has been suggested but not proven that ACTH also directs mitochondrial uptake of cholesterol and its binding to the cytochrome P-450 desmolase system.

As seen in Fig. 54-10, plasma levels of cortisol, adrenal androgens, and the precursors of both rise within minutes of intravenous ACTH administration to humans. A sustained increase in cortisol secretion of two- to fivefold is obtained when 250 μg of the fully active 1 to 24 sequence of ACTH is infused over 8 hours. Plasma ACTH concentrations of about 300 pg/ml, or 6 times the basal level, appear to be maximally effective in the short term. However, when the adrenal gland is chronically hyperstimulated by endogenous or exogenous ACTH, it undergoes hyperplasia. With the increase in adrenal mass, the capacity for cortisol secretion rises up to twentyfold.

As detailed in Chapter 52, all of the factors that increase or decrease ACTH secretion likewise affect cortisol secretion; plasma levels of the latter generally follow those of the former by 15 to 30 minutes. Thus cortisol secretion, like that of ACTH, exhibits distinct diurnal variation with a peak just before awakening in the morning and a nadir at or near zero just after falling asleep (Fig. 54-11). However, when plasma cortisol concentrations are measured every 20 minutes instead of every hour and individual normal subjects are scrutinized (Fig. 54-12), the deceptively smooth diurnal curve is seen to consist of 7 to 13 pulses or episodes of secretion per day. Within the major nocturnal pulse before awakening, 50% of the day's total cortisol is secreted. The reason for the daytime bursts of cortisol is not known. Animal studies and at least one human observation suggest cortisol secretion may be entrained with feeding patterns, although not necessarily in response to the resultant plasma substrate changes. Calculations based on cortisol half-life and volume of distribution have suggested that plasma cortisol peaks are determined by the frequency and duration of secretory bursts rather than by changes in the absolute rate of cortisol secretion in milligrams per minute. This interpretation implies that the basal unstimulated rate of secretion must be near zero and that ACTH produces an all-or-none adrenal response. Other adrenal steroids, such as DHEA, show profiles parallel to that of cortisol when their plasma levels are measured frequently; any disparities reflect differences in metabolic clearances.

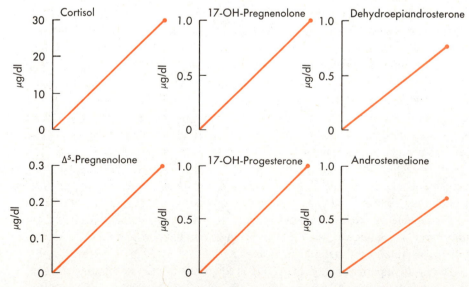

Fig. 54-10 ■ Plasma adrenocortical hormone responses to a 1-hour infusion of ACTH in humans. Note increase of precursors as well as hormonally active products indicating activation of the entire biosynthetic sequence by ACTH. (Redrawn from Lachelin, G.C.L., et al.: Adrenal function in normal women and women with the polycystic ovary syndrome, J. Clin. Endocrinol. Metab. **49:**892, 1979. Copyright 1979. Reproduced by permission.)

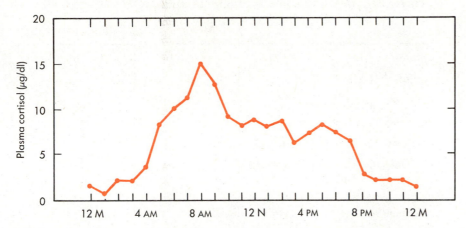

Fig. 54-11 ■ Diurnal variation in plasma cortisol as assessed from mean values of hourly sampling. Note the early morning peak and compare with the plasma ACTH profile in Fig. 52-9. (Redrawn from Weitzman, E.D., et al.: Twenty-four hour pattern of the episodic secretion of cortisol in normal subjects, J. Clin. Endocrinol. Metab. **33:**14, 1971. Copyright 1971. Reproduced by permission.)

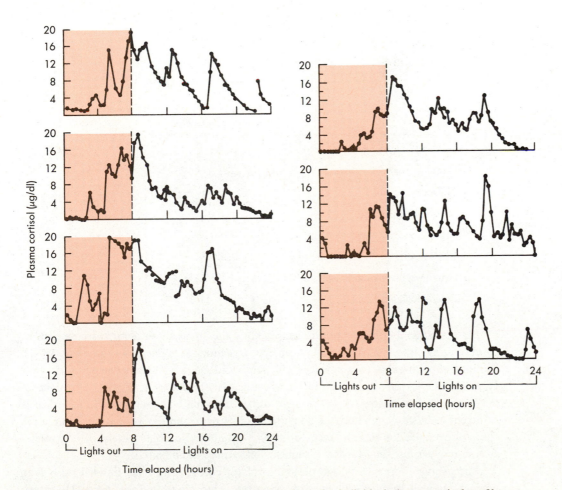

Fig. 54-12 ■ Pulsatile nature of cortisol secretion is shown by individual plasma cortisol profiles obtained with sampling every 20 minutes. (Redrawn from Weitzman, E.D., et al.: Twenty-four hour pattern of the episodic secretion of cortisol in normal subjects, J. Clin. Endocrinol. Metab. **33:**14, 1971. Copyright 1971. Reproduced by permission.)

TABLE 54-1

Relative glucocorticoid and mineralocorticoid potency of natural corticosteroids and some synthetic analogues in clinical use*

	Glucocorticoid	Mineralocorticoid
Cortisol	1.0	1.0
Cortisone (11-keto)	0.8	0.8
Corticosterone	0.5	1.5
Prednisone (1,2 double bond)	4	<0.1
6α-Methylprednisone (Medrol)	5	<0.1
9α-Fluoro-16α-hydroxyprednisolone (triamcinolone)	5	<0.1
9α-Fluoro-16α-methylprednisolone (dexamethasone)	30	<0.1
Aldosterone	0.25	500
Deoxycorticosterone	0.01	30
9α-Fluorocortisol	10	500

*All values are relative to the glucocorticoid and mineralocorticoid potencies of cortisol, which have each been set at 1.0 arbitrarily. Cortisol actually has only 1/500 the potency of the natural mineralocorticoid aldosterone.

TABLE 54-2

Average 8:00 AM plasma concentration and secretion rates of adrenocortical steroids in adult humans

	Plasma concentration (μg/dl)	Secretion rate (mg/day)
Cortisol	13	15
Corticosterone	1	3
11-Deoxycortisol	0.16	0.40
Deoxycorticosterone	0.07	0.20
Aldosterone	0.009	0.15
18-OH Corticosterone	0.009	0.10
Dehydroepiandrosterone sulfate	115	15
Dehydroepiandrosterone	0.5	15

Plasma cortisol concentration is increased by the stress of surgery, infection, fever, psychosis, electroconvulsive therapy, or hypoglycemia. It is decreased promptly by the administration of exogenous synthetic glucocorticoids such as dexamethasone (Table 54-1), which suppresses ACTH secretion via negative feedback. A single dose of dexamethasone that is biologically equivalent to twice the daily secretion rate of cortisol is sufficient to block completely the nocturnal ACTH peak and the morning rise in plasma cortisol level. Under circumstances of stress, the normal diurnal pattern of cortisol secretion may be lost, and feedback suppressibility may be impaired.

The average normal 8:00 AM plasma levels of cortisol and other adrenal steroids in humans, as well as their estimated secretion rates, are given in Table 54-2. The dominance of cortisol over corticosterone as a glucocorticoid and of DHEA-S as a source of androgenic activity is evident. Under severe stress, the maximum rate of cortisol secretion is 300 to 400 mg/day. Therefore this amount is usually provided to patients who lack adrenal function and are either acutely ill or must undergo surgical stress. There is little variation of plasma cortisol concentration with age, but secretion rates are correlated with lean body mass. The age-dependent changes in adrenal androgens are detailed later.

■ *Actions of glucocorticoids*

Cortisol is one of the few hormones essential for life. Although provision of carbohydrate and of a pure mineralocorticoid or sodium chloride can postpone death, human beings cannot survive total adrenalectomy for long without glucocorticoid replacement. The exact reasons for this—the most critical site of life-preserving action—are still difficult to explain, despite knowledge of many important effects of cortisol. The hormone is involved in maintaining glucose production from protein, facilitating fat metabolism, supporting vascular responsiveness, and modulating central nervous system function. In addition, the hormone affects skeletal turnover, hematopoiesis, muscle function, immune responses, and renal function. The sum total of its metabolic actions is catabolic or antianabolic. The term *permissive* has been used to describe many of cortisol's actions, implying that the hormone may not directly *initiate* so much as *allow* certain processes to occur. This concept may help to explain the need for the hormone's critical yet unfocused presence. As Ingle enunciated*:

> The general role of the adrenocortical hormones in body economy is to act as general tissue hormones which support the capacity of all tissues to adapt. . . . [They] do not have obligatory actions upon the metabolic processes which they affect but rather they support the capacity of the tissues to attain peak rates of metabolic processes when such are required to sustain homeostasis.

This role is manifested in at least three ways. First, cortisol may amplify the effect of another hormone on a process that it does not affect by itself, for example, the stimulation of glycogenolysis by glucagon. Second, cortisol may synergize with another hormone in a regulatory step that it does itself affect; for example, cortisol and glucagon individually increase the activity of the gluconeogenetic enzyme, phosphoenolpyruvate carboxykinase, but their combined effect is more than additive. Third, cortisol may act to facilitate a process that is independent of other hormonal influences and that would eventually occur even without the glucocorticoid, for example, the maturation of the fetal lung.

In vitro and in vivo, the effects of cortisol may be evident within minutes (inhibition of ACTH release) or hours (increase in plasma glucose) or may require days to be expressed (induction of glucose 6-phosphatase). Cortisol enters target cells freely and is then bound to a specific receptor in the cytoplasm. This receptor has been demonstrated in virtually every tissue, which is consistent with the hormone's widespread actions. The cortisol receptor complex is translocated to the nucleus where it interacts with the DNA and/or the protein portion of the nuclear chromatin (as illustrated in Fig. 49-7), thereby stimulating transcription of messenger RNAs. The specificity of hormone response is not dependent on interaction with the cytosolic receptor but rather on the subsequent nuclear events and the particular messenger RNAs that have been augmented or diminished. For example, progesterone and androgenic steroids bind to the cytosolic receptor for cortisol but do not initiate most glucocorticoid effects; indeed, they can inhibit certain ones. Phosphorylation and acetylation of chromatin protein and increases in template RNA, transfer RNA, and ribosomal RNA have all been observed in one system or another after exposure to cortisol. It is difficult at present to pinpoint which of these effects is the critical one for a particular action of the hormone. Increased synthesis of a wide variety of enzymes follows, and these at least partly explain the

*From Ingle, D.J.: J. Endocrinol. **8**:23, 1952.

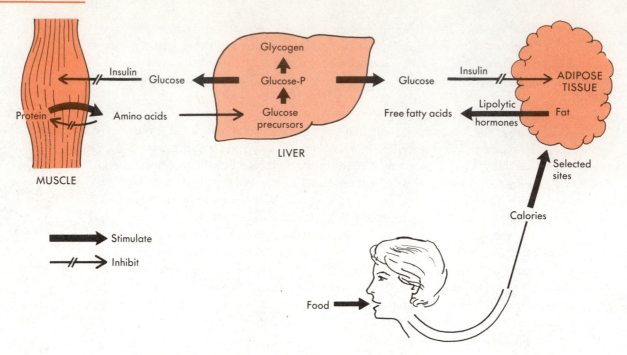

Fig. 54-13 ■ Effect of cortisol on the overall flow of fuels facilitates release of amino acids as well as both storage and release of glucose and fatty acids.

effects of cortisol on amino acid turnover and glycogen accumulation. Other intracellular mechanisms of cortisol action also probably exist. Cortisol does not generally alter intracellular cAMP levels, but it does appear to synergize with the nucleotide in a number of situations. Cortisol might also exert its permissive effects by altering the phospholipid component of various intracellular membranes, particularly those to which enzymes are bound. The fact that even in a single cell type the various enzyme changes produced by cortisol are not necessarily linked bespeaks multiple mechanisms of action.

Effects on metabolism. The most important overall action of cortisol is to facilitate the conversion of protein to glycogen (Fig. 54-13). Cortisol enhances the mobilization of muscle protein for gluconeogenesis by accelerating protein degradation and inhibiting protein synthesis (Fig. 54-14). Although this combined catabolic and antianabolic action of cortisol in normal amounts is physiologically beneficial, an excess of endogenous cortisol or any exogenous glucocorticoid produces a continuous drain on body protein stores, most notably in muscle, bone, connective tissue, and skin. This drain cannot be compensated for by dietary protein because of the inhibition of protein synthesis. Cortisol further stimulates the transformation of the proteolytically derived amino acids into glucose precursors (Fig. 54-13). Table 54-3 lists a number of liver enzymes induced by cortisol that are involved in this process.

Glucocorticoids are critical for the survival of a fasting animal or human. Without them there is no increase in proteolysis during fasting, as evidenced by lack of increase in urinary nitrogen excretion. Therefore when liver glycogen stores are used up, gluconeogenesis from protein is deficient, and death from hypoglycemia may ensue. The secretion of cortisol is not increased to any great degree, if at all, by fasting. It is the *previous* exposure to normal levels of cortisol and the maintenance of those levels during fasting that "permit" augmented amino acid mobilization to occur.

Cortisol plays a similar role in the defense against hypoglycemia as may be evoked

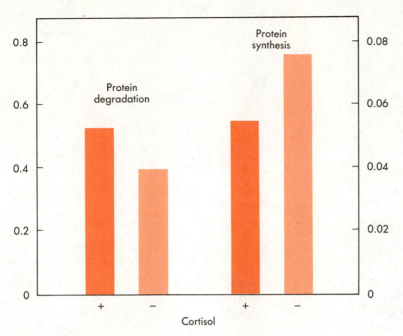

Fig. 54-14 ■ The effect of cortisol replacement on muscle protein turnover in adrenalectomized fasted rats. Cortisol increases protein degradation and decreases protein synthesis. +, With cortisol; −, without cortisol. (Data from Goldberg, A.L., et al.: Fed. Proc. **39**:31, 1980.)

TABLE 54-3

Enzymes whose activities are increased by cortisol*

Provide carbon precursors	Convert pyruvate to glycogen	Release glucose	Dispose of ammonia liberated from amino acids in urea cycle
Alanine transaminase	Pyruvate carboxylase	Glucose 6-phosphatase	Arginine synthetase
Tyrosine transaminase	Phosphoenolpyruvate carboxykinase		Arginosuccinase
Tryptophan pyrrolase	Phosphoglyceraldehyde dehydrogenase		Arginase
Threonine dehydrase	Aldolase		
Serine dehydrase	Fructose 1,6-diphosphatase		
	Phosphohexoisomerase		
	Glycogen synthetase		

*Increase varies from 130% for glycogen synthetase to 1000% for alanine transaminase. Induction time varies from 3 hours for tryptophan pyrrolase to 4 days for aldolase.

acutely by insulin. Although it is the acutely released glycogenolytic hormones, glucagon and epinephrine, that are primarily responsible for the rapid recovery of plasma glucose, the *prior* action of cortisol has permitted sufficient glycogen stores to be built up. Furthermore, cortisol amplifies the glycogenolytic responses to glucagon and epinephrine, as well as their gluconeogenic actions. Although the major impact of cortisol is on liver glycogen, an excess of the hormone will eventually increase plasma glucose levels hours after administration. This is because cortisol also antagonizes the actions of insulin on glucose metabolism by inhibiting insulin-stimulated glucose uptake in adipose tissue and muscle and by reversing insulin suppression of hepatic glucose production.

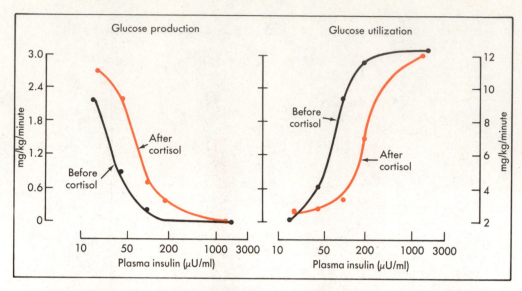

Fig. 54-15 ■ The effect of cortisol on glucose turnover in response to increasing levels of insulin in a 24-year-old man. Cortisol decreases the sensitivity to insulin (the dose response curve is shifted to the right) both with regard to insulin inhibition of glucose production and insulin stimulation of glucose use. (Redrawn from Rizza, R.A., et al.: Am. J. Med. **70:**169, 1981.)

As shown in Fig. 54-15, cortisol decreases tissue sensitivity but not maximal responsiveness to insulin. Some studies suggest cortisol may do this, in part, by decreasing the number of insulin receptors.

In part, cortisol plays an analogous role in fat metabolism (Fig. 54-13). Although only weakly lipolytic itself, the presence of cortisol is necessary for maximum stimulation of fat mobilization by epinephrine, growth hormone, and other lipolytic peptides. Thus during fasting, cortisol permits accelerated release of energy stores as well as protein stores. Along with the free fatty acids, the glycerol released from adipose tissue provides yet another gluconeogenic substrate. Total fat stores are increased in adrenalectomized animals, and fasting ketogenesis is reduced. However, cortisol actions on body fat are quite complex. The hormone increases appetite and caloric intake and can stimulate lipogenesis in certain areas of the adipose tissue mass, especially those in which brown fat or neonatal fat is usually found. Therefore an excess of cortisol finally results in obesity with a peculiar redistribution of fat that favors the trunk and face but spares the extremities.

Overall, cortisol is an important diabetogenic, antiinsulin hormone. Its hyperglycemic, lipolytic, and ketogenic actions are usually exhibited only when its secretion is stimulated by stress. Then cortisol potentiates and extends the duration of the hyperglycemia evoked by glucagon, epinephrine, and growth hormone, while further accentuating loss of body protein. These diabetogenic and catabolic actions are markedly amplified when insulin secretion is deficient and are commonly encountered in that clinical setting.

Other effects. Cortisol has actions on the structure and function of numerous organs. Many of these have been deduced from observations in patients with disorders of cortisol secretion or in experimental animals that have been rendered either cortisol-deficient or have been treated with excess hormone.

Skeletal muscle. Cortisol exerts a dual action on muscle function. In the absence of the hormone, the contractility and work performance of skeletal and cardiac muscle decline, implying a facilitative, permissive action. Yet an excess of cortisol causes increased catabolism, protein wastage, a consequent reduction of muscle mass, and muscle weakness.

Bone. Cortisol acts to inhibit bone formation by several mechanisms. First, the synthesis of collagen, the fundamental component of bone matrix, is reduced by cortisol. Second, the rate of differentiation of osteoprogenitor cells to active osteoblasts is decreased (Chapter 51). Third, the absorption of calcium from the intestinal tract is decreased by an antagonism of cortisol to the action of $1,25\text{-}(OH)_2\text{-}D_3$ and possibly also by a reduction in the synthesis of this active vitamin D metabolite. In addition, the excretion of calcium by the kidney is enhanced by cortisol. The net result of the latter effects is to reduce the availability of calcium for bone mineralization. Finally, cortisol also increases the rate of bone resorption. Thus one major consequence of a cortisol excess is an overall reduction in bone mass.

Connective tissue. Inhibition of collagen synthesis by cortisol produces thinning of the skin and of the walls of capillaries. The consequent fragility of the capillaries leads to their easy rupture and intracutaneous hemorrhage.

Vascular system. Cortisol is required for the maintenance of blood pressure. The hormone permits normal responsiveness of arterioles to the constrictive action of catecholamines. Cortisol also helps to sustain blood volume by decreasing the permeability of the vascular endothelium and by a weak mineralocorticoid action that enhances sodium conservation. The mass of red blood cells is also increased by cortisol.

Kidney. The rate of glomerular filtration is increased by cortisol. The hormone is also essential for the rapid excretion of a water load. In the absence of cortisol, the secretion of antidiuretic hormone and its action on the renal tubule are enhanced. Therefore free water clearance is diminished, and maximum dilution of the urine cannot occur.

Central nervous system. Receptors for cortisol are present in the brain, especially in the limbic system. In an unknown manner cortisol modulates perceptual and emotional functioning. Auditory, olfactory, and gustatory acuity are accentuated by cortisol deficiency; this suggests that the hormone normally has a damping effect. That may be important in preventing sensory overload and disorganized responses. However, an excess of cortisol interferes with normal sleep and can either elevate the mood strikingly or depress it. In addition, the threshhold for seizure activity may be lowered.

Fetus. Cortisol has important permissive effects which facilitate in utero maturation of the central nervous system, retina, skin, gastrointestinal tract, and lungs. The latter two have been best studied. The digestive enzyme capacity of the intestinal mucosa changes from a fetal pattern to a mature adult pattern under the influence of cortisol. This permits the newborn to use disaccharides present in milk. Timely preparation of the fetal lung in order to permit satisfactory breathing immediately after birth is facilitated by cortisol. The rate of development of the alveoli, flattening of the lining cells, and thinning of the lung septa are increased by the hormone. Most importantly, during the last weeks of gestation, the synthesis of surfactant, a phospholipid vital for maintaining alveolar surface tension, is increased. The effect is mediated by increasing the activity of key enzymes in the surfactant biosynthetic pathway, including phosphatidyl acid phosphatase and choline phosphotransferase.

Inflammatory and immune responses. Cortisol has a very special and important influence on the set of reactions evoked by tissue trauma, chemical irritants, foreign proteins, and infection. The immediate local reaction to injury consists of dilation of capillaries and changes in the endothelial cell membranes that enhance the trapping of circulating leukocytes at the site of injury. These reactions, mediated by prostaglandins and thromboxanes, are profoundly inhibited by all glucocorticoids, possibly through depression of the synthesis and release of the mediators. In addition, glucocorticoids stabilize lysosomes, thereby reducing the local release of proteolytic enzymes and hyaluronidase. The recruitment of circulating neutrophils to the site of trauma or infection is inhibited by cortisol. The hormone decreases margination of leukocytes from blood vessels and their adherence to capillary endothelium. Although conflicting data have been reported, cortisol probably also acts to decrease the phagocytic and bactericidal activity of leukocytes. Because cortisol increases the release of neutrophils from bone marrow, their circulating number actually increases, although their effectiveness decreases. Cortisol also decreases the proliferation of fibroblasts, the synthesis of collagen, and the deposition of fibrils, which form the basis for the more chronic response to injury. The net result of these actions is to impede the ability to deal locally in an effective manner with irritants or organisms and to prevent the walling off of infection.

Cortisol also influences immune responses to foreign substances. The hormone decreases the number of circulating thymus-derived lymphocytes, their transport to the site of antigenic stimulation, and their function. This cell-mediated immunity, as manifested by the classical delayed hypersensitivity response to the products of bacteria that cause tuberculosis or by the rejection of transplanted tissue, is markedly inhibited by the hormone. On the other hand, neither the production of antibodies by bursa-derived lymphocytes nor the reaction of antibodies with their specific antigens (the humoral response to invasion) is directly affected by cortisol. The helper function of thymus-derived lymphocytes, however, may be decreased, and secondarily, this may reduce the function of bursa-derived lymphocytes.

The actions of cortisol on inflammatory and immune responses generally require supraphysiological doses. They represent a two-edged therapeutic sword in clinical medicine. When the symptoms of tissue injury resulting from disease are severe, functionally disabling, or life threatening or when the rejection of transplanted organs or tissues must be prevented, then the use of glucocorticoids is dramatically beneficial. This is exemplified by their employment in severe asthma, in virulent forms of dermatitis, and in the preservation of kidney transplants. However, if glucocorticoids are administered therapeutically for very long, they may increase the susceptibility to infections or allow their dissemination, and they may prevent normal wound healing after injury. These serious adverse effects enjoin physicians to use glucocorticoids as pharmaceutical agents cautiously and only when no safer form of treatment can succeed.

■ Action of adrenal androgens

The adrenal steroids DHEA-S, DHEA, and androstenedione are relatively weak androgens. Their physiological function is largely expressed by their peripheral conversion to the potent androgen testosterone. In females, testosterone of ultimate adrenal origin sustains normal pubic and axillary hair. It may possibly also contribute to red blood cell production. When present in excess, adrenal androgens are of considerable importance in females (p. 1058). In males, testosterone of testicular origin far exceeds that of adrenal origin, rendering the latter physiologically unimportant.

■ Regulation of zona glomerulosa function

The principal function of *aldosterone,* the major product of the zona glomerulosa, is to sustain extracellular fluid volume by conserving body sodium. Hence, aldosterone is largely secreted in response to signals arising from the kidney when a reduction in circulating fluid volume is sensed. As shown in Fig. 54-16, when sodium depletion is produced by dietary restriction, the fall in extracellular fluid and plasma volume causes a decrease in renal arterial blood flow and pressure. The juxtaglomerular cells of the kidney respond to this change by secreting the enzyme *renin* into the peripheral circulation. As detailed in Chapter 47, renin acts on its substrate, *angiotensinogen* (an α_2-globulin of hepatic origin), to form the decapeptide *angiotensin I*. The latter is then further cleaved by a converting enzyme of pulmonary origin to the octapeptide *angiotensin II*. This extremely potent vasoconstrictor binds to specific receptors in the zona glomerulosa and directly stimulates the synthesis and release of aldosterone (Fig. 54-17).

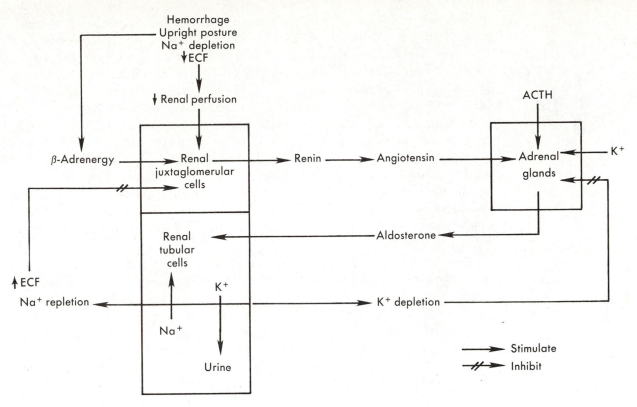

Fig. 54-16 ■ Renin-angiotensin-aldosterone axis and the regulation of aldosterone secretion. Sodium deprivation and hypovolemia stimulate secretion of aldosterone (via renin and angiotensin), which, in turn, causes renal sodium retention and restoration of circulating volume. Potassium directly stimulates secretion of aldosterone, which, in turn, causes renal potassium excretion. The feedback loop is closed when either sodium repletion or potassium depletion then inhibits aldosterone secretion.

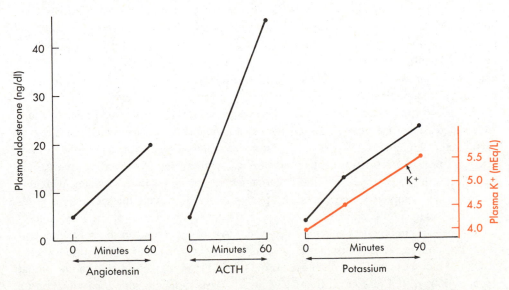

Fig. 54-17 ■ Plasma aldosterone responses to stimulators of zona glomerulosa function in humans. Note that an increase in plasma potassium in the physiological range stimulates aldosterone secretion. (Redrawn from Dluhy, R.G., et al.: J. Clin. Invest. **51**:1950, 1972; Horton, R.: J. Clin. Invest. **48**:1930, 1969. Reproduced by copyright permission of The American Society for Clinical Investigation.)

Both early and late steps in the biosynthesis of aldosterone are stimulated by angiotensin. After 5 days of only 10 mEq sodium intake, aldosterone secretion rates increase four- to eightfold. Conversely, when excess sodium is ingested and extracellular fluid volume expands, renin release, angiotensin II generation, and aldosterone secretion are all suppressed. The renin and aldosterone response to hypovolemia can also be evoked rapidly by hemorrhage, by assumption of the upright posture for several hours, or by an acute diuresis. Such maneuvers increase plasma aldosterone two- to fourfold. Thus the juxtaglomerular cells and the zona glomerulosa form a physiological feedback system. Sodium deprivation induces aldosterone hypersecretion via renin and angiotensin. When the additional aldosterone has caused sufficient sodium retention to restore extracellular fluid and plasma volume to normal, renal release is dampened, and aldosterone hypersecretion ceases. In this manner, daily aldosterone secretion ranges from 50 μg with a dietary sodium intake of 150 mEq to 250 μg with a dietary sodium intake of 10 mEq.

The release of renin from the juxtaglomerular cells is additionally enhanced by increased sympathetic neural activity, which is induced by hypovolemia. Norepinephrine, released at the nerve endings, reacts with β-adrenergic receptors in the kidney to stimulate renin release. Hence, β-adrenergic antagonists (such as propranolol) depress renin and aldosterone responses to sodium depletion. The release of renin also appears to be mediated by local prostaglandins; therefore prostaglandin synthesis inhibitors, such as indomethacin, also reduce aldosterone responses. Short-loop feedback inhibition of renin release is exerted by angiotensin II, but there is no direct feedback on the juxtaglomerular cells by aldosterone.

Aldosterone also participates in a vital physiological feedback relationship with potassium (Fig. 54-16). Because another major function of aldosterone is to facilitate the clearance of potassium from the extracellular fluid, potassium acts as an important stimulator of aldosterone secretion. In humans an acute infusion of potassium that raises plasma levels only 0.5 mEq/L immediately increases plasma aldosterone threefold (Fig. 54-17). An increase in dietary potassium from 40 to 200 mEq/day increases plasma aldosterone sixfold in supine resting subjects. Conversely, potassium depletion lowers aldosterone secretion (Fig. 54-16). The effect of potassium on the zona glomerulosa is a direct one, being demonstrable in vitro, and several steps in aldosterone biosynthesis are enhanced. However, potassium also directly inhibits the release of renin. Thus, to some extent, the stimulation of aldosterone secretion by potassium loading will be limited by a simultaneous reduction in renin (provided sodium intake and volume are held constant). Conversely, the diminution in aldosterone secretion caused by potassium depletion will be moderated by an increase in renin release.

Zona glomerulosa function is also stimulated by ACTH. In doses similar to those which increase cortisol release from the zona fasciculata, ACTH causes an acute rise in plasma aldosterone (Fig. 54-17). ACTH stimulates aldosterone synthesis at the desmolase step, probably acting through an increase in cAMP levels. However, in vivo ACTH stimulation of aldosterone secretion wanes after several days. The mechanism of this escape is not completely understood; it appears likely, however, that as sodium is retained and extracellular fluid volume rises as a result of the action of the mineralocorticoid, there is suppression of renin release and loss of its action on the zona glomerulosa. In addition, ACTH also tends to inhibit the conversion of deoxycorticosterone (DOC) to aldosterone. The physiological role of ACTH in regulating aldosterone output appears limited to a tonic one; that is, when ACTH is deficient, the zona glomerulosa is less able to respond to the primary stimulus of sodium depletion. This debility is seldom critical in patients with hypopituitarism. ACTH also stimulates the secretion of DOC from the zona fasciculata, as well as from the zona glomerulosa, and the secretion of 18-hydroxy DOC (18-OH-DOC) from the latter (Fig. 54-6). Under rare pathological circumstances, these steroids can generate clinical syndromes of mineralocorticoid excess.

Stimulation of aldosterone secretion by the three major factors noted earlier is interrelated. A low sodium intake potentiates aldosterone responsiveness to angiotensin, potassium, and ACTH. The increased sensitivity to angiotensin that results from sodium depletion is explained partly by increased binding of the peptide hormone to its receptors in the zona glomerulosa and partly by enhanced activity of the biosynthetic pathway. Conversely, if adrenal cell potassium content is depleted, the responses to angiotensin, ACTH, and cAMP are diminished. There is evidence for still other regulatory factors of physiological significance. The existence of a pituitary tropic hormone other than ACTH that is specific for the zona glomerulosa has long been suspected. Recent studies suggest that lipotropin or a related peptide, which is secreted in concert with ACTH, may increase aldosterone secretion. In addition, dopamine appears to exert tonic inhibition of aldosterone secretion by a direct action on the zona glomerulosa. This was revealed by the observation that aldosterone responses to other stimuli are enhanced by dopamine antagonists.

In humans the plasma aldosterone level shows a definite diurnal fluctuation, with the highest concentration occurring at 8:00 AM and the lowest at 11:00 PM. Although this profile correlates with similar directional changes in plasma renin and plasma cortisol levels, the diurnal pattern of aldosterone appears to be independent of variation in sodium intake, posture, ACTH suppression by exogenous glucocorticoids, or serum potassium.

■ *Actions of aldosterone and other mineralocorticoids*

The kidney is the major site of mineralocorticoid activity. Aldosterone binds to a cytosolic receptor in renal tubular cells and is transferred with the receptor to the nucleus, where the steroid effects transcriptional changes. A protein of still undetermined structure or locus of action is induced, which apparently mediates the hormone's effects. A lag of 1 to 2 hours is therefore required between exposure to aldosterone and its onset of action. Aldosterone stimulates the active reabsorption of sodium from the distal tubular urine; the sodium is transported through the tubular cell and back into the capillary blood. Thus net urinary sodium excretion is diminished, and the vital extracellular cation is conserved (Fig. 54-18). Because water is passively reabsorbed with the sodium, there is little increase in serum sodium concentration, and extracellular fluid volume expands in an isotonic fashion. Although only 3% of total sodium reabsorption is regulated by aldosterone, deficiency of this hormone does produce a negative sodium balance. The mechanism of aldosterone action is discussed extensively in Chapter 48. In brief, aldosterone is thought to act at one of three loci in the distal renal tubular cell: (1) at the apical (luminal) surface to activate a permease that facilitates sodium entry into the cell along an electrochemical gradient, (2) at the basal (capillary) surface of the cell to activate Na^+, K^+-ATPase, which pumps the sodium out, or (3) in the mitochondria, stimulating several reactions important in generating the needed energy for extrusion of sodium into the interstitial fluid and capillary blood.

Concurrently, aldosterone stimulates the active secretion of potassium out of the tubular cell and into the urine (Fig. 54-18). This does not constitute a stoichiometric *exchange* of potassium for sodium. Nonetheless, the active reabsorption of sodium is thought to create an electronegative condition in the tubular lumen, which facilitates the passive transfer of potassium into the tubular urine. Therefore the extent of kaluresis is greatly dependent on the delivery of sodium to the distal tubule. Aldosterone does not increase potassium excretion in a sodium-depleted subject because virtually all of the sodium will be reabsorbed proximally. Conversely, a high sodium intake will greatly exacerbate urinary potassium losses caused by aldosterone. Ordinarily most of the potassium that is excreted daily does result from distal tubular secretion. Hence, the pres-

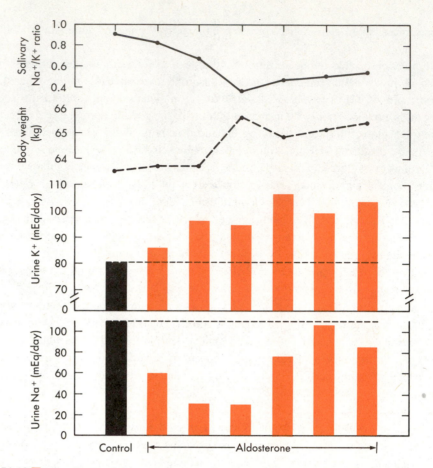

Fig. 54-18 ■ The effects of aldosterone administration in a normal human. Note eventual escape from sodium retention with stabilization of body weight yet continuing loss of potassium in the urine. Dashed lines represent levels of sodium and potassium intake. (Redrawn from August, J.T., et al.: J. Clin. Invest. **37**:1549, 1958. Reproduced by copyright permission of The American Society for Clinical Investigation.)

ence of aldosterone or some other mineralocorticoid is critical for disposal of the daily dietary potassium load. Potassium flux, unlike that of sodium, does not entrain the movement of water; therefore, in the absence of aldosterone, potassium retention can result in a dangerous rise in serum potassium levels, and an excess of the hormone causes a significant decrease in serum potassium levels. Continued administration of aldosterone in the face of a normal sodium intake produces sodium retention, weight gain (Fig. 54-18), and an increase in blood pressure as a result of the expanded extracellular fluid volume. However, after several days and an accumulation of 200 to 300 mEq of sodium, retention ceases, balance is achieved, and body weight stabilizes. This escape is thought to be caused by a depression in proximal sodium reabsorption secondary to expansion of the extracellular fluid and possibly mediated by a natriuretic hormone. Nonetheless, the mineralocorticoid-induced potassium loss continues because sodium delivery to the distal tubule is maintained.

In addition to its effects on potassium secretion, aldosterone enhances tubular secretion of hydrogen ion as sodium is reabsorbed. Therefore aldosterone excess leads to the development of a mild systemic metabolic alkalosis, which can be further aggravated by the depletion of potassium. Ammonium excretion is also increased; however, the final urine pH is usually alkaline because the expansion of extracellular fluid volume inhibits bicarbonate reabsorption. In contrast, a deficiency of aldosterone produces a metabolic acidosis. Finally, aldosterone also stimulates the excretion of magnesium.

Aldosterone additionally affects mineral transport in other organs. The hormone stimulates sodium reabsorption from the gastrointestinal tract while enhancing potassium excretion in the feces. Similarly, the hormone decreases the ratio of sodium to potassium in perspiration and in saliva (Fig. 54-18). These actions, however, have little importance in overall cation balance. Whether aldosterone significantly affects sodium and potassium exchange between the extracellular fluid and the intracellular fluid of the muscle mass is controversial.

An effect of clinical importance is the increased blood pressure that results from an excess of aldosterone. This appears to be an indirect consequence of the retention of sodium, expansion of the extracellular fluid volume, and a slight increase in cardiac output. In addition, the sodium and water content of the arteriolar cells may increase; the resultant swelling narrows the arteriolar lumen and increases peripheral resistance. In contrast, blood pressure falls below normal in aldosterone-deficiency states as a result of hypovolemia, although there is usually a compensatory renin-stimulated increase in the vasoconstrictor angiotensin II.

Aldosterone actions in the kidney are blocked by progesterone and 17-hydroxyprogesterone, which therefore have natriuretic activity. An important inhibitor in clinical usage as a diuretic and antihypertensive agent is spironolactone. The mechanism of inhibition by these steroids appears to involve competition with aldosterone for binding to its renal tubular receptor.

■ Clinical syndromes of adrenocortical dysfunction

Hypofunction. Destruction of the adrenal cortex, *Addison's disease,* is ultimately incompatible with life. Except in cases of surgical removal or sudden infarction of the gland, Addison's disease usually progresses slowly, permitting the gradual development of glucocorticoid, adrenal androgen, and mineralocorticoid deficiencies. A lack of cortisol leads to loss of appetite with weight loss, malaise, fatigue, lethargy, muscle weakness, nausea, vomiting and abdominal pain, fever, poor tolerance of minor medical or surgical stress, fasting hypoglycemia, and an increase in circulating lymphocytes and eosinophils with a reduction in neutrophils. A loss of adrenal androgens may contribute to anemia and, in females, a reduction in pubic and axillary hair. Because of negative feedback, the secretion of ACTH and β-lipotropin increases as the cortisol levels decline. The melanocyte-stimulating activity inherent in ACTH and possibly other peptides produces striking hyperpigmentation of the skin. The diagnosis of destruction of the zona fasciculata and zona reticularis is established by demonstrating low plasma cortisol levels and decreased urinary excretion of 17-hydroxycorticoids and 17-ketosteroids. Plasma ACTH levels are elevated, and if exogenous ACTH is administered, cortisol and its metabolites fail to increase in response. Deficiency of aldosterone is marked by polyuria (which is caused by natriuresis), dehydration, hypotension, hyperkalemia, hyponatremia, and metabolic acidosis. Plasma and urinary aldosterone levels are low. Plasma renin and angiotensin levels are elevated consequent to the stimulus of sodium depletion. In most cases of adrenal disease, the entire adrenal cortex is involved. However, primary selective loss of either glucocorticoid or mineralocorticoid function can occur.

Adrenal insufficiency may be secondary to ACTH deficiency, resulting from disease of the hypothalamus or pituitary. The clinical picture, then, is that described earlier for loss of cortisol and adrenal androgen. Hyperpigmentation does not occur because plasma ACTH levels are low. When ACTH is replaced exogenously, plasma cortisol and urinary 17-hydroxycorticoid levels gradually increase.

Treatment of acute life-threatening adrenal insufficiency (adrenal crisis) consists of large doses of intravenous cortisol, glucose, and sufficient isotonic sodium chloride infusion to restore normal extracellular fluid volume and lower serum potassium levels. For lifetime maintenance, patients require oral cortisol, a generous salt intake, and usually a synthetic mineralocorticoid such as 9α-fluorocortisol (Table 54-1).

Hyperfunction. The most common cause of endogenous adrenocortical hormone excess is bilateral hyperplasia of the adrenal cortex secondary to hypersecretion of ACTH. Tumors giving rise to autonomous hormone secretion also occur in the various adrenocortical zones. The major manifestations of increased glucocorticoids include (1) obesity with a peculiar distribution of fat involving the cheeks (moon facies), the supraclavicular areas, the posterior cervicothoracic junction (buffalo hump), the trunk, and the abdomen—the extremities being spared; (2) a loss of bone mass (osteoporosis), leading to back pain, vertebral fractures, and necrosis of the hips; (3) a loss of connective tissue integrity, leading to fragile capillaries, easy bruisability, and thin skin through which the underlying blood vessels may be seen (purple striae); (4) increased protein catabolism, resulting in atrophy and weakness of the muscles of the trunk and extremities, poor wound healing, and stunted growth; (5) abnormal carbohydrate metabolism, sometimes sufficient to produce frank diabetes; (6) impaired response to infections, particularly those produced by staphylococci, *Mycobacterium tuberculosis,* and fungi; and (7) headache, insomnia, and psychosis. All these pathological consequences can be produced iatrogenically, as well, by the therapeutic administration of synthetic glucocorticoids.

Manifestations of adrenal androgen hypersecretion in adult females are varying degrees of masculinization. This includes loss of regular menses, regression of breast tissue, hirsutism, acne, deepening of the voice, enlargement of the clitoris, increased muscularity, and heightened libido. There are virtually no clinically detectable changes in men.

The diagnosis of endogenous glucocorticoid excess depends on demonstrating elevated plasma levels of cortisol and loss of its normal diurnal variation. In parallel, there is increased urinary excretion of cortisol and its 17-hydroxycorticoid metabolites. The concurrent or independent existence of adrenal androgen excess is marked by elevated urinary excretion of 17-ketosteroids and elevated plasma levels of DHEA, DHEA-S, androstenedione, and, in women, testosterone. Negative feedback suppression of plasma gonadotropins may be observed.

Once it has been determined that endogenous glucocorticoid and/or adrenal androgen excess exists, the next step is to determine the cause. If plasma ACTH is normal or only modestly elevated, then a pituitary (or hypothalamic) origin is indicated, usually a small ACTH-producing neoplasm. The dependence of the adrenocortical excess on pituitary ACTH is shown by demonstrating that a large dose of exogenous glucocorticoids (8 mg dexamethasone) suppresses endogenous cortisol secretion, whereas a smaller dose (2 mg dexamethasone) does not; however, the latter dose is sufficient to suppress cortisol secretion in normal persons. If plasma ACTH levels are low and adrenocortical secretion cannot be suppressed by dexamethasone, then an autonomous adrenal neoplasm is present. If plasma ACTH levels are very high but the adrenal gland cannot be suppressed, then an extraadrenal, extrapituitary neoplasm secreting the ACTH is implicated.

A primary excess of aldosterone produces a clinical syndrome characterized by hypertension, an expanded extracellular fluid volume, minimal edema, headache, hypokalemia with metabolic alkalosis, and slight hypernatremia. The diagnosis is established by demonstrating that plasma and urinary aldosterone (or rarely, DOC or 18-OH-DOC) are elevated when the patient has a high sodium intake (200 to 300 mEq). In the majority of cases, the lesion is a neoplasm arising in the zona glomerulosa; a lesser number have bilateral hyperplasia. Hence, the secretion of aldosterone is essentially autonomous, and plasma renin levels are low, being suppressed by the expanded extracellular fluid. It is important to distinguish primary from secondary aldosteronism, where the hypersecretion of the mineralocorticoid is secondary to an excess of renin. This may be generated by obstructive lesions of the renal arteries, which reduce perfusion pressure, or by secretion from a tumor of the juxtaglomerular cells.

Treatment of adrenocortical excess syndromes resulting from adrenal tumors is best accomplished by removal of the neoplasm. If the basic lesion is within the pituitary gland, modern microsurgical techniques permit removal of the small adenomas, usually without disturbing the remaining function of the pituitary gland. Ectopic production of ACTH by malignancies requires treatment of the primary lesion, but palliative treatment may be afforded by chemical blockade of adrenocortical secretion with drugs such as aminoglutethimide. Definitive treatment of hyperaldosteronism secondary to renal artery stenosis requires restoration of renal perfusion by vascular grafting or removal of the offending kidney. Symptomatic relief from any form of hyperaldosteronism is obtained by treatment with the aldosterone antagonist spironolactone.

Biosynthetic defects. A number of congenital enzyme deficiencies in the pathways of adrenocortical hormone synthesis occur. These include deficiencies of desmolase, 3β-ol-dehydrogenase, 21-hydroxylase, 11-hydroxylase, 17-hydroxylase, and 18-hydroxylase. *Adrenogenital syndrome* is the term given to this group of disorders. The consequences of each biosynthetic defect can be predicted (Fig. 54-3); the product of the reaction will be deficient, and the precursor or precursors will be increased enormously as a result of negative feedback to the pituitary gland or, occasionally, to the juxtaglomerular cells as well. Detailed descriptions of each of these syndromes is beyond the scope of this discussion. However, a single example will be presented in order to emphasize the importance of understanding the detailed pathways of adrenal hormone biosynthesis.

The most common biochemical lesion encountered clinically is that of 21-hydroxylase deficiency. Plasma levels of cortisol and 11-deoxycortisol tend to be low, whereas plasma levels of the immediate precursor, 17-hydroxyprogesterone, will be greatly elevated (Fig. 54-3). The corresponding urinary metabolites show the same pattern; that is, 17-hydroxycorticoid levels are low while pregnanetriol is increased. The block in cortisol synthesis leads to increased ACTH secretion, which stimulates all open pathways and causes immense hyperplasia of the zona fasciculata and zona reticularis. As seen in Fig. 54-5, 17-hydroxyprogesterone serves as the major precursor for adrenal androgens. Therefore in 21-hydroxylase–deficient patients, androgens will be secreted in great excess as a result of hyperstimulation by ACTH. Plasma androgen levels (DHEA, DHEA-S, androstenedione, testosterone) will be high, as will the excretion of their urinary metabolites (17-ketosteroids).

The clinical consequences of this block and its overflow are dramatic. The high adrenal androgen levels in female fetuses cause a masculinized pattern of development of the external genitalia. Thus they have penilelike clitorides and scrotallike labia. The ambiguous genitalia can lead to incorrect gender assignment at birth. If not treated promptly, the androgen excess will cause early acceleration of linear growth and early appearance of pubic and axillary hair but suppression of gonadal function and normal puberty in females and males. Eventually the patient will be short as a result of premature closure of the growth centers in the bones. If the 21-hydroxylase block is severe, cortisol and aldosterone secretion may be so impaired as to cause episodes of adrenal crisis (p. 1057). This entire sequence can be reversed by supplying appropriate amounts of cortisol and an aldosterone substitute.

■ *The adrenal medulla*

The adrenal medulla is the source of the circulating catecholamine hormone *epinephrine*. It also secretes small amounts of *norepinephrine*—nominally a neurotransmitter—which in select circumstances may also function as a hormone. These compounds have diverse effects on metabolism as well as on virtually all organ systems in the body. The adrenal medulla represents essentially an enlarged and specialized sympathetic ganglion. However, the neuronal cell bodies of the medulla do not have axons; instead they discharge their catecholamine hormones directly into the bloodstream, thus functioning as endocrine rather than nerve cells. The adrenal medulla is formed in parallel with the peripheral sympathetic nervous system. At about 7 weeks of gestation, neuroectodermal cells from the neural crest invade the anlage of the primitive adrenal cortex. There they develop into the medulla, which by birth is completely functional.

Adrenal medullary tissue in the adult weighs about 1 g and consists of chromaffin cells (so named for their affinity for chromium stains). These are organized in cords and clumps in intimate relationship with venules that drain the adrenal cortex and with nerve endings from cholinergic preganglionic fibers of the sympathetic nervous system. Within the chromaffin cells are numerous granules of 100 to 300 nm diameter, similar to those found in postganglionic sympathetic nerve terminals. These granules consist of the catecholamine hormones epinephrine and norepinephrine (20% by weight), adenosine triphosphate and other nucleotides (15%), protein (35%), and lipid (20%). They also contain enkephalins. About 85% of the chromaffin granules store epinephrine and 15% norepinephrine. Small clumps of similar chromaffin cells can also be found outside the adrenal medulla along the aorta and the chain of sympathetic ganglia. Although such cells are physiologically inconsequential, they can give rise to functioning chromaffin tumors.

The adrenal medulla is usually activated in association with the rest of the sympathetic nervous system and acts in concert with it. Many of the actions of the neurotransmitter norepinephrine, which is released locally at the effector site of the postganglionic sympathetic nerve ending, are duplicated and amplified by the hormone epinephrine, which reaches similar sites via the circulation. However, epinephrine has unique effects of its own, some of which modulate those of norepinephrine. Furthermore, under certain circumstances, for example, during hypoglycemia, the adrenal medulla is probably activated selectively. A complete description of the sympathetic nervous system and its function is presented in Chapter 20. This section will focus on adrenal medullary function.

■ Synthesis and storage

The catecholamine hormones are synthesized within the chromaffin cell by a series of reactions shown in Fig. 54-19. The first, catalyzed by the enzyme tyrosine hydroxylase, is the rate-limiting step in the sequence and occurs in the cytoplasm. The conversion of tyrosine to dihydroxyphenylalanine (dopa) requires molecular oxygen, a tetrahydropteridine, and NADPH. Norepinephrine inhibits this reaction by negative feedback. The conversion of dopa to dopamine is catalyzed by a nonspecific aromatic L-amino acid decarboxylase that employs pyridoxal phosphate as a cofactor. The dopamine thus formed in the cytoplasm must be taken up by the chromaffin granule before it can be acted on further. The next enzyme in the sequence, dopamine β-hydroxylase, is present exclusively within the granule, both in membrane-bound and soluble form. In the presence of molecular oxygen and a hydrogen donor, it catalyzes the formation of norepinephrine from dopamine. In approximately 15% of the granules, the sequence ends here, and the norepinephrine is stored. In 85% of the granules, norepinephrine diffuses back into the cytoplasm. There it is *N*-methylated by phenylethanolamine *N*-methyltransferase using *S*-adenosylmethionine as the methyl donor. The resultant epinephrine is then taken back up into the chromaffin granule where it is stored as the predominant adrenal medullary hormone. The uptake of dopamine, norepinephrine and epinephrine by the secretory granules is an active process that requires adenosine triphosphate (ATP) and magnesium. The storage of the catecholamine hormones at such high intragranular concentration also requires energy in the form of ATP. One mole of the nucleotide is present in a complex with 4 moles of catecholamine and a specific protein known as chromogranin.

The synthesis of epinephrine and norepinephrine is regulated by several factors. Acute stimulation of the sympathetic innervation to the medulla activates tyrosine hydroxylase, possibly by decreasing cytoplasmic catecholamine levels and relieving the

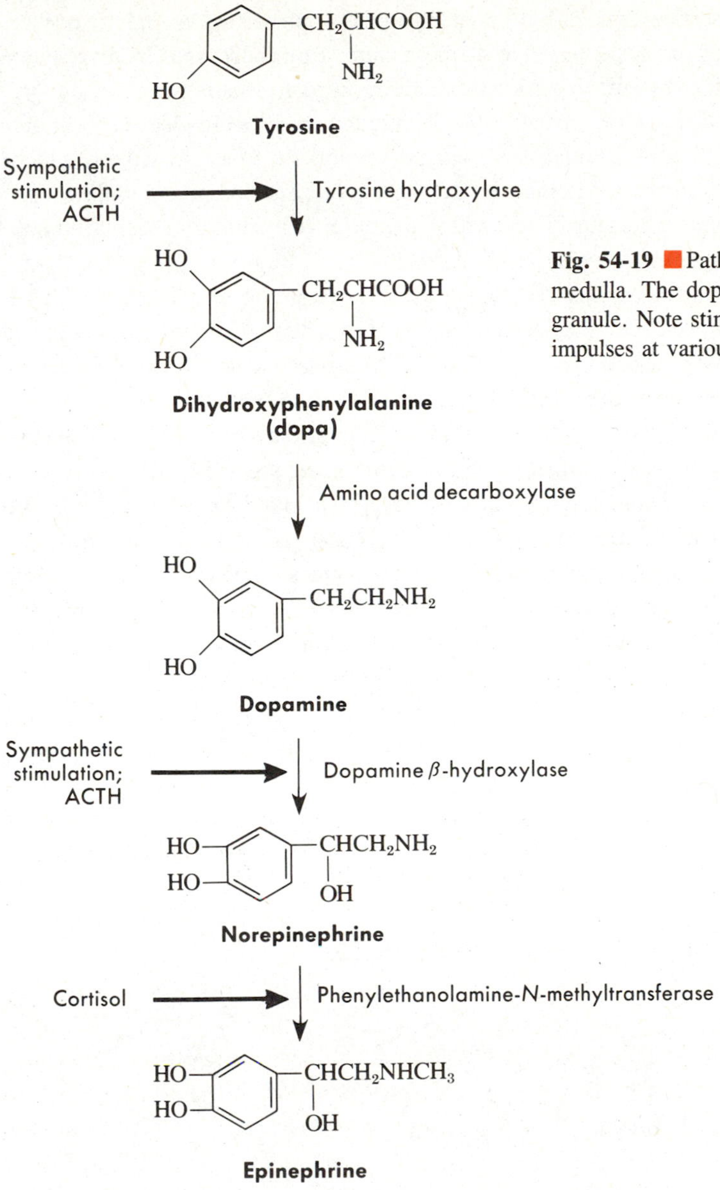

CH₂CHCOOH structure (Tyrosine)

Tyrosine

Sympathetic stimulation; ACTH → Tyrosine hydroxylase

Fig. 54-19 ■ Pathway of catecholamine hormone synthesis in the adrenal medulla. The dopamine β-hydroxylase reaction occurs within the secretory granule. Note stimulatory effects of ACTH, cortisol, and sympathetic nerve impulses at various points.

Dihydroxyphenylalanine (dopa)

Amino acid decarboxylase

Dopamine

Sympathetic stimulation; ACTH → Dopamine β-hydroxylase

Norepinephrine

Cortisol → Phenylethanolamine-*N*-methyltransferase

Epinephrine

feedback inhibition. Chronic stimulation of the preganglionic fibers induces increased concentrations of both tyrosine hydroxylase and dopamine β-hydroxylase, thus helping to ensure maintenance of catecholamine output in the face of continuous demand. ACTH, acting directly, helps to sustain the levels of the same two enzymes under stressful conditions. In contrast, cortisol specifically induces the *N*-methyltransferase and therefore selectively stimulates epinephrine synthesis. The anatomical relationship between the medulla and the cortex subserves this action, since blood from the cortex with a high concentration of cortisol directly perfuses the chromaffin cells.

Essentially all of the circulating epinephrine is derived from adrenal medullary secretion. In contrast, most of the circulating norepinephrine is derived from sympathetic nerve terminals and from the brain, having escaped immediate local re-uptake from synaptic clefts. However, the metabolic fate of epinephrine and norepinephrine merges into one or two major excretory products.

Epinephrine and norepinephrine have extremely short life spans in the circulation, allowing rapid turnoff of their dramatic effects. Half-lives are in the range of 1 to 3

■ *Metabolism of catecholamines*

minutes. The metabolic clearance rate of epinephrine is reported to be 3.5 to 6.0 L/minute, and that of norepinephrine is 2.0 to 4.0 L/minute. Both hormones increase their clearance rates still further by activating β-adrenergic receptors, another mechanism that helps to limit their actions. Only 2% to 3% of catecholamines are disposed of unchanged in the urine, giving a normal total daily excretion of about 50 μg, of which 20% is epinephrine and 80% is norepinephrine. The vast majority of circulating epinephrine and norepinephrine is metabolized within the adrenal medullary chromaffin cell itself when synthesis exceeds the capacity of storage.

The catecholamine hormones are metabolized by the reaction sequences shown in Fig. 54-20. The key enzymes are catecholamine *O*-methyltransferase and the combination of monamine oxidase and aldehyde oxidase. *O*-Methylation and oxidative deamination can be carried out in either order, giving rise to several products that are then excreted in the urine. *O*-Methylation alone yields an average daily excretion of metanephrine (from epinephrine) plus normetanephrine (from norepinephrine) of 300 μg. In contrast, the excretion of the common deaminated products, vanillylmandelic acid (VMA) and methoxyhydroxyphenylglycol (MOPG), averages 4000 μg and 2000 μg, respectively. It is important to note that under normal circumstances epinephrine accounts for only a very minor proportion of urinary VMA and MOPG. Because the majority is derived from norepinephrine, urinary VMA and MOPG levels largely reflect activity of

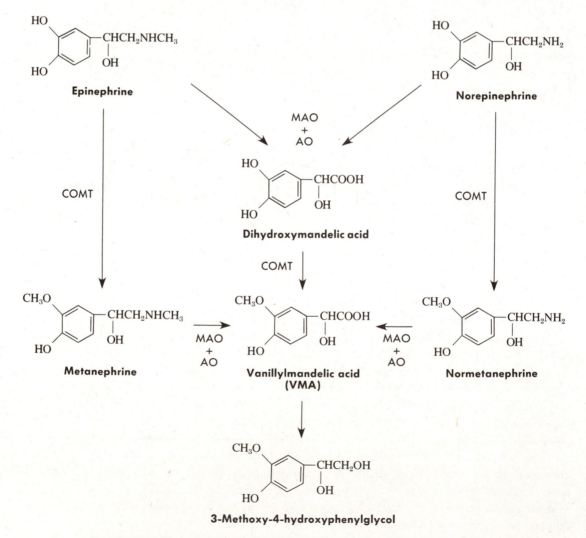

Fig. 54-20 ■ Metabolism of catecholamine hormones. VMA is quantitatively the main product. *MAO*, Monamine oxidase; *AO*, aldehyde oxidase; *COMT*, catecholamine O-methyltransferase.

the sympathetic nervous system rather than that of the adrenal medulla. The latter can only be assessed specifically by measurement of urinary free epinephrine or plasma epinephrine levels. However, measurement of urinary VMA and total metanephrines is of importance in the detection of hyperfunctioning adrenal medullary neoplasms that secrete either epinephrine or norepinephrine.

Secretion from the adrenal medulla forms an integral part of the "fight-or-flight" reaction evoked by stimulation of the sympathetic nervous system. Thus perception or even anticipation of danger or harm (anxiety), trauma, pain, hypovolemia from hemorrhage or fluid loss, hypotension, anoxia, extremes of temperature, hypoglycemia, and severe exercise cause rapid secretion of epinephrine (and probably norepinephrine) from the adrenal medulla. These stimuli are sensed at various higher levels in the sympathetic nervous system. However, the final common effector pathway activating the adrenal medulla is cholinergic preganglionic fibers in the greater splanchnic nerve. Upon stimulation, acetylcholine is released from the nerve terminals. The neurotransmitter depolarizes the chromaffin cell membrane and induces an influx of calcium ion. The latter probably produces microfilament contraction, which draws the chromaffin granules to the cell membrane. There they fuse with it and discharge their contents via exocytosis. Epinephrine, norepinephrine, ATP, the enzyme dopamine β-hydroxylase, and soluble granule proteins are all simultaneously released into the circulation. The membranous material, however, is retained and probably recycled. Exocytosis can also be induced by histamine and glucagon in chromaffin cell tumors.

Basal plasma epinephrine levels are 25 to 50 pg/ml (6×10^{-10}M). A daily basal delivery rate of 150 μg can be estimated. It was formerly believed that the concentration of epinephrine required to produce physiological effects was seldom, if ever, achieved in humans. Recent data, shown in Table 54-4, clearly demonstrate otherwise. For example, under the stimulation of a modest fall in plasma glucose from 90 to 60 mg/dl, adrenal medullary secretion causes a rise of endogenous epinephrine to 230 pg/ml. If epinephrine is infused exogenously at a rate sufficient to maintain its plasma level at 150 to 200 pg/ml, a rise in plasma glucose results. Hence, the adrenal medulla is quite capable of secreting physiologically meaningful amounts of epinephrine to contribute to glucose homeostasis. The same is true of cardiovascular responses to the hormone. An

■ *Regulation of adrenal medullary secretion*

TABLE 54-4

Comparison of circulating concentrations of catecholamine hormones with biologically effective concentrations

Physiological state	Relevant biological action	Plasma epinephrine (pg/ml)		Plasma norepinephrine (pg/ml)	
		Observed	Effective range for relevant biological action	Observed	Effective range for relevant biological action
Basal	—	34	—	228	—
Upright position	↑ Heart rate and blood pressure	73	50 to 125	526	1800
↓ Plasma glucose	↑ Plasma glucose	230	150 to 200	262	1800
Severe hypoglycemia	—	1500	—	770	—
Diabetic ketoacidosis	↑ Lipolysis and ketosis ↓ Insulin	510	100 to 400	1270	1800

Data from Clutter, W.E., et al.: J. Clin. Invest. **66**:94, 1980; Silverberg, A.B., et al.: Am. J. Physiol. **234**:E252, 1978; and Christensen, N.J.: Diabetes **23**:1, Jan. 1974.

increase in heart rate and systolic blood pressure can be produced by the concentrations of epinephrine that are generated endogenously by assumption of the upright position (Table 54-4). In addition, the high concentrations of epinephrine that occur in such illnesses as diabetic ketoacidosis are quite capable of contributing to the pathological state by stimulation of lipolysis and ketosis. Thus epinephrine functions as a true hormone in all these situations. In contrast, circulating norepinephrine levels do not generally increase with stimulation to levels sufficient to produce relevant biological actions (Table 54-4). Therefore, under normal circumstances, norepinephrine does not appear to function as a true hormone, although it may do so in severe, stressful illnesses such as myocardial infarction. Instead, norepinephrine's effects on similar metabolic processes, such as glucose production or lipolysis, result from its role as a neurotransmitter, wherein the necessary high concentrations are generated locally at the effector site.

■ *Actions of catecholamines*

Catecholamines exert their effects via plasma membrane receptors. The adrenergic receptors can be grouped into certain categories on the basis of functional similarity as well as reactivity with various agonists and antagonists. Epinephrine reacts well with β_1- and β_2-receptors but exhibits diminished potency when reacting with α-receptors. Norepinephrine reacts predominantly with α-receptors, less strongly with β_1-receptors, and virtually not at all with β_2-receptors. Numerous therapeutically useful catechol-

TABLE 54-5
Some actions of catecholamines

β-Receptor mediated (Epinephrine > norepinephrine)	α-Receptor mediated (Norepinephrine > epinephrine)
↑ Glycogenolysis	↑ Gluconeogenesis
↑ Lipolysis and ketosis	
↑ Calorigenesis	
↓ Glucose utilization	
↑ Insulin secretion	↓ Insulin secretion
↑ Glucagon secretion	
↑ Muscle K^+ uptake	
↑ Arteriolar dilation (muscle)	↑ Arteriolar vasoconstriction (splanchnic, renal, cutaneous, genital)
↑ Cardiac contractility (β_1)	
↑ Heart rate (β_1)	
↑ Cardiac conduction velocity (β_1)	
↑ Muscle relaxation (gastrointestinal, urinary, bronchial [β_2])	↑ Sphincter contraction (gastrointestinal, urinary) Sweating ("adrenergic") Dilation of pupils
↑ Renin secretion	↑ Growth hormone secretion
↑ Parathyroid hormone secretion	
↑ Thyroid hormone secretion	
Antagonists	
Propranolol	Phentolamine
Practolol (β_1)	Phenoxybenzamine
Butoxamine (β_2)	Prazosin

amine agonists and antagonists with enhanced receptor selectivity have been synthesized and find a wide range of use in clinical medicine.

Catecholamine interaction with β-receptors activates adenylate cyclase, and cAMP is the second messenger for β-actions. A cAMP-dependent protein kinase then initiates a cascade of enzyme activations or deactivations (Fig. 49-6). In contrast, catecholamine α-receptor interactions, if anything, decrease cAMP levels. Some evidence supports a role for calcium or cyclic guanosine monophosphate (cGMP) as a second messenger for α-receptor activities. Continuous stimulation of catecholamine release or exposure to synthetic catecholamine agonists, such as isoproterenol, down-regulates receptor number and induces partial refractoriness to hormone action. Conversely, sympathectomy produces an increase in receptor number and enhanced sensitivity to catecholamines.

A listing of important and endocrinologically relevant actions of epinephrine and norepinephrine is presented in Table 54-5. Via β-adrenergic receptors, epinephrine increases glucose production. The catecholamine stimulates glycogenolysis in the liver by activating phosphorylase through the same cAMP-initiated cascade as is produced by glucagon. Glycogen synthase activity is concurrently restrained. The epinephrine response to hypoglycemia is not needed as long as glucagon secretion is intact. However, if the secretion of glucagon is blocked, for example, by somatostatin, then the catecholamine action becomes essential for recovery from hypoglycemia. Epinephrine also stimulates gluconeogenesis, in part by a direct hepatic effect and in part by stimulating muscle glycogenolysis, which increases plasma lactate levels and provides additional substrate to the liver. Simultaneously, epinephrine inhibits insulin-mediated glucose uptake by muscle and adipose tissue. Epinephrine also stimulates glucagon secretion while inhibiting insulin secretion. All these actions help restore plasma glucose and its delivery to the central nervous system. At the same time, epinephrine activates adipose tissue lipase, thereby increasing plasma free fatty acids, their β-oxidation in muscle and liver, and ketogenesis. These actions in humans are demonstrated in Fig. 54-21.

When the catecholamine hormones are secreted during exercise, they promote (1) use of muscle glycogen stores by stimulating phosphorylase, (2) efficient hepatic reutil-

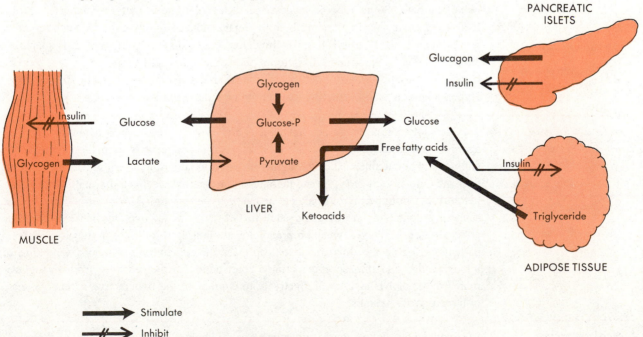

Fig. 54-21 ■ Metabolic action of epinephrine. The hormone stimulates glucose production and inhibits glucose use. It also stimulates lipolysis and ketogenesis. Insulin secretion is inhibited. The net effect is a rise in plasma glucose, free fatty acids, and ketoacids.

ization of lactate released by the exercising muscle, and (3) provision of free fatty acids as alternate fuels. When epinephrine secretion is stimulated by ''stress,'' such as during illness or surgery, its actions contribute significantly to induction of hyperglycemia and ketosis; that is, it is a diabetogenic hormone. These diabetogenic effects are generally dependent on the presence of cortisol through its permissive actions. Epinephrine also increases the basal metabolic rate. This action increases nonshivering thermogenesis and is an important part of the response to cold exposure. The mechanism may involve enhancement of Na^+, K^+-ATPase activity, as well as the increased mobilization of fuels. In all of these metabolic effects, epinephrine is at least 10 times as potent as norepinephrine (Table 54-4). The latter may, nonetheless, contribute to the metabolic actions via the simultaneous activity of the sympathetic nervous system.

The cardiovascular effects of secreted epinephrine reinforce its metabolic actions. Cardiac output rises as a result of an increase in heart rate and contractile force; on the other hand, arteriolar constriction is selectively produced in the renal, splanchnic, and cutaneous beds, whereas the muscle arterioles are dilated. Systolic blood pressure increases, whereas diastolic blood pressure remains unchanged or decreases slightly. The net effect of these changes is to shunt blood toward exercising muscles while maintaining coronary and cerebral blood flow. This guarantees delivery of substrate for energy production to the critical organs in the fight-or-flight situation. During exposure to cold, constriction of cutaneous vessels helps to conserve heat, reinforcing epinephrine's thermogenic action. The inhibition of gastrointestinal and genitourinary motor activity, the relaxation of bronchioles to prevent expiratory airway obstruction, and the dilation of pupils to permit better distant vision are of obvious benefit to the endangered person.

The diverse circumstances that evoke marked epinephrine and norepinephrine responses are characterized by increased total energy requirements. The role of basal or altered adrenal medullary secretion in ordinary nutrient flux is less clear. There is no evidence that during fasting, when fuel mobilization is enhanced, either epinephrine secretion or sympathetic nervous system activity is increased. If anything, animal studies indicate that the latter is reduced, probably contributing to the reduction in basal metabolic rate. Some reports suggest that catecholamine activity is increased during feeding, a factor that might contribute to dietary-induced thermogenesis. Such an effect might be mediated by epinephrine's stimulatory action on the thyroid gland. Epinephrine enhances renin (and hence aldosterone) secretion via β-adrenergic receptors, thereby helping to prevent sodium depletion and dangerous potassium accumulation during stress. Hyperkalemia is also averted by stimulation of muscle potassium uptake.

■ Therapeutic usage and endocrine implications

Catecholamine agonists and antagonists are in widespread use in medicine. Various agonists, called *amphetamines,* are used as nasal decongestants, appetite suppressants, and general stimulants. All of these may be prescribed or sold over the counter. Rarely, they may increase plasma glucose in diabetics, exacerbate tachycardia, palpitations, and nervousness in hyperthyroid patients or cause hypertension. β-Adrenergic antagonists, such as propranolol, are of great value in the treatment of hypertension and coronary artery disease. However, when propranolol is administered for these purposes to diabetic patients taking insulin, this antagonist can mask the epinephrine-generated symptoms of hypoglycemia and impede epinephrine's contribution to the recovery from hypoglycemia. Propranolol is also used effectively to counteract the hyperactive adrenergic state in hyperthyroid patients.

■ Pathological secretion of catecholamines

Spontaneous deficiency of epinephrine is unknown as an adult disease, and adrenalectomized patients do not require epinephrine replacement. There is evidence, however, that in some young children idiopathic hypoglycemia may result from epinephrine deficiency, and it may respond to treatment with longer-acting preparations of the hormone.

Hypersecretion of epinephrine and norepinephrine from tumors of the chromaffin cells *(pheochromocytomas)* results in well-defined syndromes. In two thirds of the patients, sustained hypertension is produced; in one third blood pressure elevation is sporadic and accompanies dramatic clinical episodes produced by sudden spurts of excess catecholamines. These bursts of secretion can result from any stress, such as anesthesia, or from a rapid change in posture. Typically, the patient notes sudden severe headache, palpitations, chest pain, extreme anxiety with a sense of impending death, cold perspiration, pallor of the skin caused by vasoconstriction, and blurred vision. When examined during an attack, blood pressure may be extremely high, for example, 250/150. If primarily epinephrine is being secreted, the heart rate will be increased; if norepinephrine is the predominant hormone, the heart rate will be decreased reflexly in response to the marked hypertension. Attacks may end spontaneously in minutes or may last hours. In addition to these episodes, chronic epinephrine excess may produce weight loss as a result of an increased metabolic rate and hyperglycemia caused by inhibition of insulin secretion. The diagnosis is established by detecting high levels of plasma epinephrine or norepinephrine at rest in the recumbent position. In addition, urinary excretion of catecholamines, metanephrines, and VMA will usually be increased. Definitive treatment requires removal of the adrenal medullary tumor. Symptomatic treatment during attacks is provided by α-adrenergic antagonists such as phentolamine or prazosin, which rapidly lower the elevated blood pressure. A β-adrenergic antagonist, such as propranolol, is used to reduce tachycardia, but it may exacerbate hypertension unless simultaneous α-adrenergic blockade is provided.

■ *Bibliography*

Journal articles

August, J.T., et al.: Response of normal subjects to large amounts of aldosterone, J. Clin. Invest. **37**:1549, 1958.

Axelrod, J., et al.: Catecholamines, N. Engl. J. Med. **287**:237, 1972.

Clutter, W., et al.: Epinephrine plasma metabolic clearance rates and physiologic thresholds for metabolic and hemodynamic actions in man, J. Clin. Invest. **66**:94, 1980.

Cryer, P.E.: Physiology and pathophysiology of the human sympathoadrenal neuroendocrine system, N. Engl. J. Med. **303**:436, 1980.

Dluhy, R.G., et al.: Studies of the control of plasma aldosterone concentration in normal man. II. Effect of dietary potassium and acute potassium infusion, J. Clin. Invest. **51**:1950, 1972.

Edelman, I.S.: Mechanism of action of aldosterone: energetic and permeability factor, J. Endocrinol. **81**:49P, 1979.

Eigler, N., et al.: Synergistic interactions of physiologic increments of glucagon, epinephrine, and cortisol in the dog. A model for stress-induced hyperglycemia, J. Clin. Invest. **63**:114, 1979.

Fauci, A.S., et al.: Glucocorticosteroid therapy: mechanisms of action and clinical considerations, Ann. Intern. Med. **84**:304, 1976.

Goldberg, A.L., et al.: Hormonal regulation of protein degradation and synthesis in skeletal muscle, Fed. Proc. **39**:31, 1980.

Hamburg, S., et al.: Influence of small increments of epinephrine on glucose tolerance in normal humans, Ann. Intern. Med. **93**:566, 1980.

Horton, R.: Stimulation and suppression of aldosterone in plasma of normal man and in primary aldosteronism, J. Clin. Invest. **48**:1230, 1969.

Horton, R.: Aldosterone: review of its physiology and diagnostic aspects of primary aldosteronism, Metabolism **22**:1525, 1973.

Katz, F.H., et al.: Diurnal variation of plasma aldosterone, cortisol, and renin activity in supine man, J. Clin. Endocrinol. Metab. **40**:125, 1975.

Lachelin, G.C.L., et al.: Adrenal function in normal women and women with the polycystic ovary syndrome, J. Clin. Endocrinol. Metab. **49**:892, 1979.

Melby, J.C.: Systemic corticosteroid therapy: pharmacology and endocrinologic considerations, Ann. Intern. Med. **81**:505, 1974.

Merry, B.J.: Mitochondrial structure in the rat adrenal cortex, J. Anat. **119**:611, 1975.

Parker, L.N., et al.: Evidence for existence of cortical androgen-stimulating hormone, Am. J. Physiol. **236**:E616, 1979.

Rizza, R.A., et al.: Cortisol-induced insulin resistance in man: impaired suppression of glucose production and stimulation of glucose utilization due to a postreceptor defect of insulin action, J. Clin. Endocrinol. Metab. **54**:131, 1982.

Santiago, J.V., et al.: Epinephrine, norepinephrine, glucagon, and growth hormone release in association with physiological decrements in the plasma glucose concentration in normal and diabetic man, J. Clin. Endocrinol. Metab. **51:**877, 1980.

Silverberg, A., et al.: Norepinephrine: hormone and neurotransmitter in man, Am. J. Physiol. **234:**E252, 1978.

Thompson, E.B., et al.: Unlinked control of multiple glucocorticoid-induced processes in HTC cells, Mol. Cell Endocrinol. **15:**135, 1979.

Weitzman, E.D., et al.: Twenty-four hour pattern of the episodic secretion of cortisol in normal subjects, J. Clin. Endocrinol. Metab. **33:**14, 1971.

Williams, G.H., et al.: Studies of the control of plasma aldosterone concentration in normal man. I. Response to posture, acute and chronic volume depletion, and sodium loading, J. Clin. Invest. **51:**1731, 1972.

Books and monographs

Baxter, J.D., et al.: Glucocorticoid hormone action, New York, 1979, Springer-Verlag New York, Inc.

Nelson, D.H.: The secretion of the adrenal cortex and steroid biosynthesis. In Smith, L.H., Jr., editor: The adrenal cortex: physiological function and disease, vol. 18, Major problems in internal medicine, Philadelphia, 1980, W.B. Saunders Co.

chapter 55

The reproductive glands

The endocrine glands that have been discussed thus far are essential to the maintenance of the life and well-being of the individual. In contrast, the endocrine function of the gonads is primarily concerned with maintaining the life and well-being of the species. The evolution of sexual reproduction has required the development of highly complex patterns of gonadal function. These are concerned with the nurturance and maturation of the individual male and female germ cells, their successful union, and the subsequent growth and development of the newly created individual within the body of the mother. There are many obvious differences between male and female gonadal function, but there are also important basic conceptual similarities and operational homologies. Therefore the subject of human gonadal endocrinology is approached as a single unit in the following sequence: (1) sexual differentiation, (2) common aspects of gonadal structure and function, (3) testicular function, (4) ovarian function, and (5) endocrine aspects of pregnancy.

■ *Sexual differentiation*

Any discussion of reproductive endocrine physiology should begin with a consideration of the process of sexual differentiation, that is, the pattern of development of the gonads, genital ducts, and external genitalia. It is convenient and logical to divide sexual differentiation into three components: genetic sex, gonadal sex, and genital, or phenotypic, sex.

■ *Genetic sex*

The normal female chromosome complement is 44 autosomes and two sex chromosomes, XX. Both of the X chromosomes are active in germ cells and are essential for the genesis of a normal ovary (Fig. 55-1). In addition, evidence suggests that there is participation of autosomes in ovarian development, since rare individuals with a normal complement of XX sex chromosomes inherit defective gonads as an autosomal recessive trait. In contrast, differentiation of the genital ducts and external genitalia along normal female lines requires that only a single X chromosome be active in directing transcription within the cell. The second X chromosome of a normal XX female is genetically inactive in all extragonadal tissues. Therefore individuals with an XO sex chromosome complement have very abnormal gonads but still develop as normal phenotypic females.

The presence of a Y chromosome is the single most consistent determinant of maleness. The normal male has a chromosome complement of 44 autosomes and two sex chromosomes, XY (Fig. 55-2). The presence of additional X chromosomes does not alter the fundamental maleness dictated by the Y chromosome, even though the gonads usually are rendered dysfunctional. With rare exceptions, the absence of the Y chromosome precludes normal testicular development, as well as masculinization of the

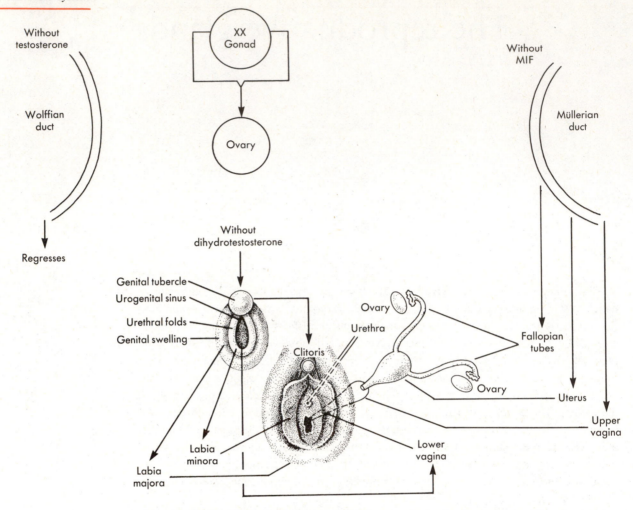

Fig. 55-1 ■ Development of the human female reproductive organs and tract. Note the independence from hormonal products of the gonad. In the absence of any gonads the female format results. *MIF,* Müllerian inhibiting factor.

genital ducts and external genitalia. However, even though the Y chromosome is virtually essential, it is not in itself sufficient for maleness. There is evidence that genetic material located on the X chromosome participates in directing the organization of the gonad into a testis and in sensitizing the genital ducts and external genitalia to the masculinizing effects of androgenic hormones (Fig. 55-2).

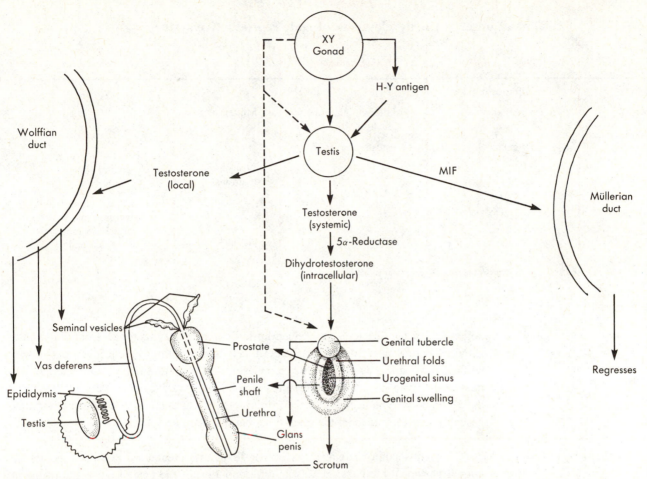

Fig. 55-2 ■ Development of the human male reproductive organs and tract. Note the dependence on hormonal products of the gonad. *MIF*, Müllerian inhibiting factor.

During the initial 5 weeks of fetal life the gonads develop along sexually indistinguishable lines. The primordial germ cells migrate to the genital ridge and associate themselves with mesonephric tissue, an assembly that consists of an outer cortex and inner medulla. In a normal boy at 6 weeks the seminiferous tubules begin to form, followed by the Sertoli cells at 7 weeks and the Leydig, or interstitial, cells at 8 to 9 weeks (Table 55-1). At 9 weeks a recognizable testis is present, and testosterone secretion is established. The medulla of the early testis encloses the germ cells while the cortex regresses. The organization of the gonadal anlage into the characteristic seminiferous or spermatogenic tubules of the male is directed by a specific substance, known as the H-Y antigen (Fig. 55-2). This surface glycoprotein is found in all male cells, and its presence correlates absolutely with the presence of a testis in all species. Synthesis of the H-Y antigen appears to be normally under the control of the Y chromosome. There are rare instances of XX individuals with testes. In these cases there is either sex chromosome mosaicism (XX/XY), or the H-Y antigen may have been translocated to an X chromosome or even to an autosome.

In the normal female, differentiation of the indifferent gonad into an ovary does not start until 9 weeks of age. At this time *both X chromosomes* within the germ cells become activated (Fig. 55-1). The germ cells begin to undergo mitosis, giving rise to oogonia, which continue to proliferate. Shortly thereafter meiosis is initiated in some

■ *Gonadal sex*

TABLE 55-1

Time of onset of the development of male reproductive system

Time (weeks)	Testicular cells	Hormone produced	Ducts	External genitalia masculinizes
6	Seminiferous tubules			
6 to 7	Sertoli cells	Müllerian inhibiting factor (MIF)	Müllerian regresses	
8 to 9	Leydig cells	Testosterone		
9 to 10			Wolffian grows	Urogenital sinus*
10 to 11				Urogenital tubercle,* swelling, and folds*; genitalia enlarge;
12 to 39				testes descend

*Dihydrotestosterone synthetic capacity is present in these tissues at 6 to 10 weeks.

oogonia, and they become surrounded by follicular cells and stroma, from which interstitial cells subsequently appear. The germ cells are now known as primary oocytes and remain in the first stage of meiosis (prophase) until ovulation many years later. In contrast to the male gonad, the cortex predominates in the developed ovary while the medulla regresses. The capacity of the primitive ovary to synthesize estrogenic hormones probably develops at about the same time that testosterone synthesis begins in the testis.

■ *Genital (phenotypic) sex*

Up to this point in fetal development sexual differentiation does not require any known hormonal products. However, differentiation of the genital ducts and of the external genitalia definitely do. The guiding principle is that positive hormonal influences from the gonad are required to produce the masculine format. In the absence of any gonadal hormonal input the feminine format will result.

During the sexually indifferent stage, from 3 to 7 weeks, two genital ducts develop on each side. In the male at about 9 to 10 weeks, the wolffian, or mesonephric, ducts begin to grow and eventually give rise to the epididymis, the vas deferens, the seminal vesicles, and the ejaculatory duct (Fig. 55-2). This constitutes the system for delivering sperm from the testis to the penis. The differentiation of the wolffian ducts is preceded by the appearance of Leydig cells in the testis and the production of testosterone by those cells (Table 55-1). It is the steroid hormone *testosterone* that stimulates the growth and differentiation of the wolffian ducts in the male. Furthermore the testosterone produced by each testis acts unilaterally on its own wolffian duct (Fig. 55-2), as shown by gonadal transplantation experiments or by testosterone implantations. Testosterone does not have to be converted to its hormonally active product, dihydrotestosterone, to act within the wolffian duct cells, as it does in some other tissues (described later). Indeed these cells do not develop the 5α-reductase activity necessary for this conversion until after they have fully differentiated. In the female the wolffian ducts, lacking testosterone stimulation, begin to regress at 10 to 11 weeks.

The müllerian ducts arise parallel to and in part from the wolffian ducts on each side. In the male these ducts begin to regress at 7 to 8 weeks, about the same time that the Sertoli cells of the testis appear. These cells produce a large glycoprotein called *müllerian-inhibiting factor* (MIF), which causes the atrophy of the müllerian ducts. Although MIF still can be found in the postnatal testis, it has no known role after birth. There is preliminary evidence that the secretion of MIF may be inhibited by the appearance of fetal follicle-stimulating hormone (FSH). In the female, lacking MIF, the müllerian ducts grow and differentiate into fallopian tubes at the upper ends, whereas at the lower ends they join to form the uterus, cervix, and upper vagina (Fig. 55-1). This process is not completed until 18 to 20 weeks. It does not require any ovarian hormone.

The external genitalia of both sexes begin to differentiate at 9 to 10 weeks. They are derived from the same anlage: the genital tubercle, the genital swelling, the urethral, or genital, folds, and the urogenital sinus. In the normal female or in the absence of any gonads these develop without apparent hormonal influences into the clitoris, labia majora, labia minora, and lower vagina, respectively (Fig. 55-1). The possibility that steroid hormones derived from the placenta might be involved in imprinting the normal female phenotype cannot be excluded at present. If the normal female fetus is exposed to an excess of testosterone or other potent androgens during differentiation of the external genitalia, a male pattern can result. However, once the female pattern of differentiation has been achieved, exposure to testosterone cannot change the external genitalia to the male pattern, although it can cause enlargement of the clitoris.

In the male, testosterone must be secreted into the circulation of the fetus and subsequently must be converted to *dihydrotestosterone* within the cells of the anlage tissues for normal differentiation of the external genitalia to occur. As a result of dihydrotestosterone stimulation, the genital tubercle grows into the glans penis, the genital swellings fold and fuse into the scrotum, the urethral folds enlarge and enclose the penile urethra and corpora spongiosa, and the urogenital sinus gives rise to the prostate gland. Although the intracellular mechanism of action of dihydrotestosterone is discussed later in more detail, it is pertinent to note here that this steroid hormone requires a cytoplasmic receptor for its action.

The hormone production necessary for sexual differentiation does not seem dependent on fetal pituitary gonadotropins. Chorionic gonadotropin from the placenta undoubtedly stimulates testosterone production by the Leydig cells. It also has been suggested that placental steroid hormone precursors, such as pregnenolone, could serve as a source of fetal gonadal androgens or estrogens. This would obviate the necessity for gonadotropin stimulation of the reactions from cholesterol to pregnenolone (see later discussions).

Other aspects of phenotypic sexual differentiation are not evident until long after birth. These include the constant pattern of gonadotropin secretion in the male versus the cyclic pattern in the female, the different degree of breast development, and the psychological identification with a unique gender. It is difficult at present to be certain what factors imprint or regulate these traits in humans. There is evidence from rodent studies that androgens induce the fetal hypothalamus to set a constant pattern of gonadotropin secretion in the postpubertal male. In the absence of androgens the cyclic pattern of the female results. This would constitute another instance in which the female pattern was the "neutral pattern," whereas the male pattern required an action ultimately derived from the Y chromosome.

Mammary gland development in the rodent embryo also is clearly under androgen regulation. In its absence a normal female breast develops; in its presence the ductal system is suppressed. However, in the human, male-female differences in breast development are not apparent before puberty. At that time the hormonal milieu in the female induces growth and differentiation of breast tissue, whereas that in the male suppresses it. A large body of clinical evidence suggests that psychological gender identification is independent of hormonal regulation or even of the phenotype of the genitalia; instead it appears to depend on rearing cues. However, exceptions to this are noted in certain cases of male pseudohermaphrodites raised as girls. In such individuals significant growth of the penis under pubertal testosterone stimulation seems to cause a reversal of psychosocial gender from female to male.

■ *Abnormalities of sexual differentiation*

Based on the preceding brief review, it is instructive to predict the anatomical aberrations to be expected from certain of the better understood genetic errors. These are listed in Table 55-2. Sexual differentiation can be distorted by abnormalities in either sex chromosomes or autosomes.

The XO chromosomal karyotype produces individuals with only a vestigial gonadal streak, because they do not have either the ovarian organizational input of two active X chromosomes or the testicular organizational input of the Y chromosome. The absence of testicular function in turn leads to müllerian duct development, female external genitalia, and wolffian duct regression.

XY individuals who completely lack the capacity to respond to androgenic hormones because of receptor deficiency (the X-linked testicular feminization syndrome) will develop a testis because of the presence of the Y chromosome. They will demonstrate müllerian duct regression, caused by the presence of MIF. However, they will show no growth or development of the wolffian ducts nor masculinization of their external genitalia, since without receptors there is no effective testosterone or dihydrotestosterone action.

XY individuals who have one of five known defects in testosterone biosynthesis will develop a testis, because of the presence of the Y chromosome, and will show müllerian duct regression, because of the presence of MIF. However, depending on the severity of the testosterone lack, the wolffian duct structures will show variable degrees of underdevelopment, whereas the external genitalia will vary from a completely female pattern to simple failure of the urethral folds to fuse completely.

XY individuals who are deficient only in the conversion of testosterone to dihydrotestosterone will have a normal testis because of the presence of the Y chromosome. They will demonstrate müllerian duct regression because of the presence of MIF. The development of the epididymis, vas deferens, and seminal vesicles will be normal because of the presence of testosterone. However, they will have external genitalia that vary from a slight to a complete female pattern because of the lack of dihydrotestosterone.

XX individuals with a deficiency of adrenal 21- or 11-hydroxylase enzymes will overproduce androgens in utero (Chapter 54). They will have ovaries because of the

TABLE 55-2

Examples of abnormal development of the reproductive system

Genetic state	Gonad	Müllerian duct	Wolffian duct	External genitalia
XY, normal ♂	Testis	Regressed	Developed	♂
XX, normal ♀	Ovary	Developed	Regressed	♀
XO, Turner's syndrome	Streak*	Developed	Regressed	♀
XY, loss of X-linked gene for androgen receptor	Testis	Regressed	Regressed	♀
XY, deficient testosterone synthesis	Testis	Regressed	Regressed to variably developed	♀/♂
XY, deficient 5α-reductase	Testis	Regressed	Developed	♀/♂
XXY, Klinefelter's syndrome	Dysgenetic testis	Regressed	Developed	♂
XX, adrenal 21- or 11-hydroxylase deficiency	Ovary	Developed	Regressed	♀/♂

*A fibrous streak essentially devoid of germ cells.

presence of two X chromosomes and normal müllerian duct development because of the absence of MIF. Their wolffian structures regress because of the absence of local gonadal testosterone and the relatively late exposure to adrenal androgen excess. However, depending on the severity of the enzymatic block and the degree of resultant androgen hypersecretion, the external genitalia of such XX individuals show variable degrees of the male pattern. This ranges from mild enlargement of the clitoris to complete scrotal fusion of the labia and a persistent urogenital sinus. Similar although less severe masculinization of otherwise normal XX fetuses has resulted from transplacental passage of excess androgenic hormones from the mother.

Individuals with supernumerary X chromosomes develop testes if a Y chromosome is also present and ovaries if it is not. Their genital ducts and external genitalia develop normally. However, spermatogenesis and seminiferous tubule development are markedly deficient in individuals (males) with XXY chromosomes *(Klinefelter's syndrome)*. Individuals (females) with XXX chromosomes may have shortened reproductive lives. The mechanisms whereby extra X chromosomes damage germ cell function are unknown.

■ Common aspects of gonadal structure and function

It is helpful to review certain homologous aspects of gonadal structure and function in the two sexes before discussing the numerous differences that exist. As seen in Fig. 55-3, the primordial germ cells generate the oogonia and spermatogonia that undergo eventual reductional division and maturation into large numers of ova and sperm, respectively. Only a few of each will eventually unite with each other to reproduce the species, in a manner guaranteeing an almost infinite variety of individual characteristics. One cell line of the indifferent gonad becomes the *granulosa cells* of the ovarian follicle and the *Sertoli cells* of the seminiferous tubules. The function of these cells is homologous: to sustain or "nurse" the germ cells, foster their maturation, and guide their movement into the genital duct system. This cell line is probably the main source of estrogenic hormones in both sexes. Another cell line, the *interstitial cells,* gives rise to *theca cells* in the ovary and *Leydig cells* in the testis. The primary function of this cell line is to secrete androgenic hormones. These are essential for sperm production in the testis and as precursors for estrogen synthesis in the ovary.

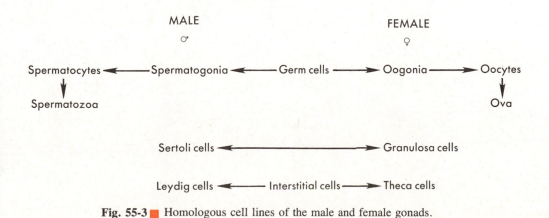

Fig. 55-3 ■ Homologous cell lines of the male and female gonads.

■ Pathway of gonadal steroid synthesis

Both sexes use a common pathway of steroid hormone biosynthesis in gonadal tissue. It is essentially identical to that of the adrenal cortex. The enzyme characteristics and cofactor requirements are also those previously described for the adrenal glands in Chapter 54. As shown in Fig. 55-4, cholesterol, either generated by de novo synthesis from acetyl CoA or taken up from the circulating plasma pool, is the starting compound. In the gonads, in situ cholesterol synthesis may be quantitatively more important than in the adrenal glands. Cholesterol side chain cleavage (the 20, 22-desmolase step) is localized to the mitochondria, and it appears to be rate limiting for synthesis of progesterone, of the androgens testosterone, dihydrotestosterone, and androstanediol, and of the estrogens estradiol and estrone. The 17-hydroxylase, 17, 20-desmolase, 3-β-ol-dehydrogenase, and $\Delta^{4,5}$-isomerase activities are all located in the microsomes of the smooth endoplasmic reticulum. The first two and the second two form closely associated pairs

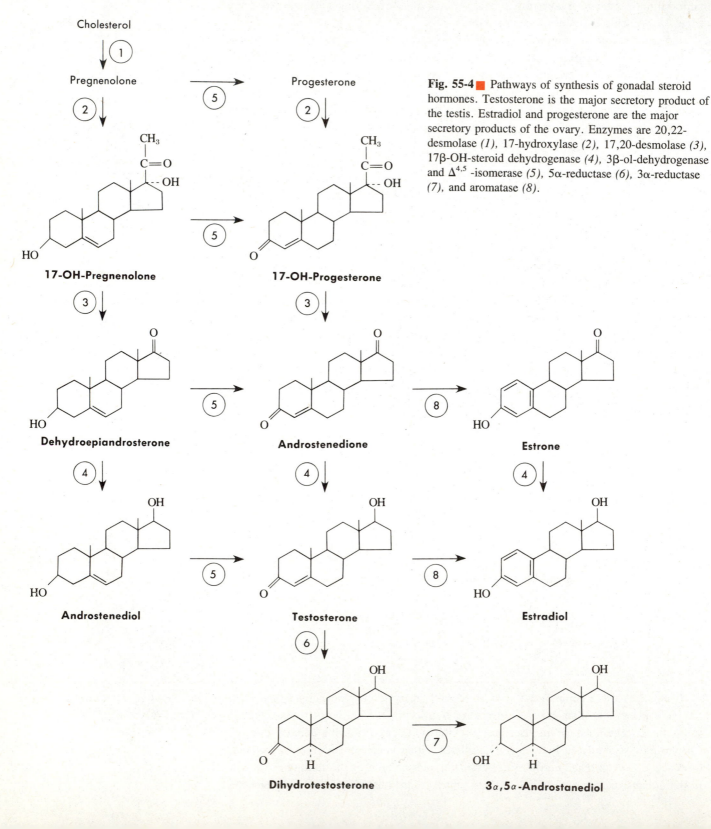

Fig. 55-4 ▪ Pathways of synthesis of gonadal steroid hormones. Testosterone is the major secretory product of the testis. Estradiol and progesterone are the major secretory products of the ovary. Enzymes are 20,22-desmolase *(1)*, 17-hydroxylase *(2)*, 17,20-desmolase *(3)*, 17β-OH-steroid dehydrogenase *(4)*, 3β-ol-dehydrogenase and $\Delta^{4,5}$-isomerase *(5)*, 5α-reductase *(6)*, 3α-reductase *(7)*, and aromatase *(8)*.

within the microsomal membranes. Two parallel pathways to testosterone are evident (Fig. 55-4). Oxidation of the A ring by the 3β-ol-dehydrogenase–isomerase complex can take place at any level from pregnenolone to androstane. The factors that determine which route will be taken are not known. A small quantity of testosterone undergoes 5α-reduction to dihydrotestosterone and a further α-reduction of the 3-ketone position to 5α-androstanediol (Fig. 55-4).

The conversion of androgens to estrogens occurs by aromatization of the A ring. The aromatase complex is a cytochrome P-450 enzyme that sequentially hydroxylates the 19-methyl group, oxidizes it to the carboxylic acid, and decarboxylates it. The 1 to 2 position then is reduced by removal of two hydrogens, creating the characteristic benzene ring. This series of steps occurs in the endoplasmic reticulum. Estradiol and estrone result from testosterone and androstenedione, respectively. The two estrogens also may be interconverted via 17-hydroxysteroid dehydrogenase.

<hr>

■ *Mechanisms of gonadotropin action*

Luteinizing hormone (LH) stimulates the interstitial cell line of male and female gonads to secrete androgens and possibly estrogens. The mechanism of action of LH begins with binding to a plasma membrane receptor; the complex activates adenylate cyclase and raises intracellular cyclic AMP levels. The interaction of LH with its receptors is exquisitely sensitive. As little as 1% receptor occupancy may be sufficient for stimulation, and 5% to 10% occupancy may produce maximum cellular responsiveness to the hormone. Cyclic AMP formation and LH receptor occupancy usually correlate well. Continued stimulation of gonadal cells by LH leads to down regulation of its receptors and reduced responsivity to the hormone. Prostaglandins have been implicated as additional intermediaries in LH action, possibly potentiating cyclic AMP effects.

As a result of protein kinase activation by cyclic AMP, one or more specific regulatory proteins concerned with steroidogenesis are activated. The major locus of acute LH action is on the mitochondrial conversion of cholesterol to pregnenolone. This may involve facilitating the movement of cholesterol from its esterified storage form into the mitochondrial matrix and the activation of the 20,22-desmolase reaction. Longer term stimulation with LH may increase the microsomal steroidogenic enzymes, as well.

FSH acts on ovarian granulosa cells and testicular Sertoli cells also via plasma membrane receptors. The subsequent increase in cyclic AMP is followed by an increase in steroidogenesis and in the rate of the aromatase reaction. The mechanism of these FSH actions is less well understood. Another important effect of FSH is to increase the number of LH receptors in target cells, thereby amplifying their sensitivity to LH.

In addition to their actions on steroidogenesis, LH and FSH produce diverse metabolic effects on their target gonadal cells. Glucose oxidation and lactic acid production are increased, and this may lead to local vasodilation. The tropic effects of the two hormones appear to depend on stimulation of amino acid transport, RNA synthesis, and general protein synthesis.

Both testosterone and estradiol have negative feedback effects on LH and FSH secretion in both sexes (Chapter 52). It is pertinent to recall here that plasma LH and FSH levels will be increased in individuals whose gonads are functionless or surgically removed. In addition, the natural cessation of ovarian function in menopausal women, as well as the slower decline of testicular function in men in the seventh and eighth decades of life, is associated with increases in both plasma gonadotropins. In all these instances loss of inhibin secretion plays a specific role in the elevation of plasma FSH (Chapter 52).

<hr>

■ **The testes**
■ *Anatomy*

The human testes are normally situated in the scrotum, where they are maintained at a temperature somewhat below that of the body core temperature. Each testis weighs about 40 g and has a long diameter of 4.5 cm. The testes receive blood from the spermatic arteries, which arise directly from the aorta. The right spermatic vein drains

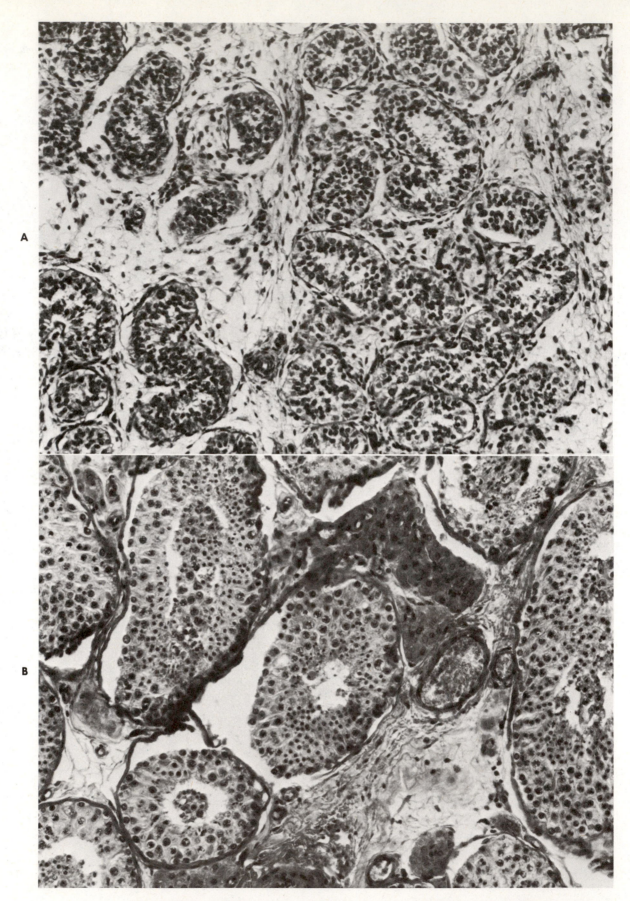

Fig. 55-5 ■ Histological sections of the testis at low (**A** and **B**) and high (**C** and **D**) magnification. **A** and **C** are from a prepubertal testis. Note absence of Leydig cells and absence of active spermatogenesis, with only spermatogonia being evident in the tubules. **B** and **D** are from a postpubertal testis. Note clumps of Leydig cells adjacent to tubules and the presence of active spermatogenesis, as evidenced by the progression of cell types between the basement membrane and lumen of the tubules and the many mitotic figures. (Courtesy Dr. Howard Levin, The Cleveland Clinic.)

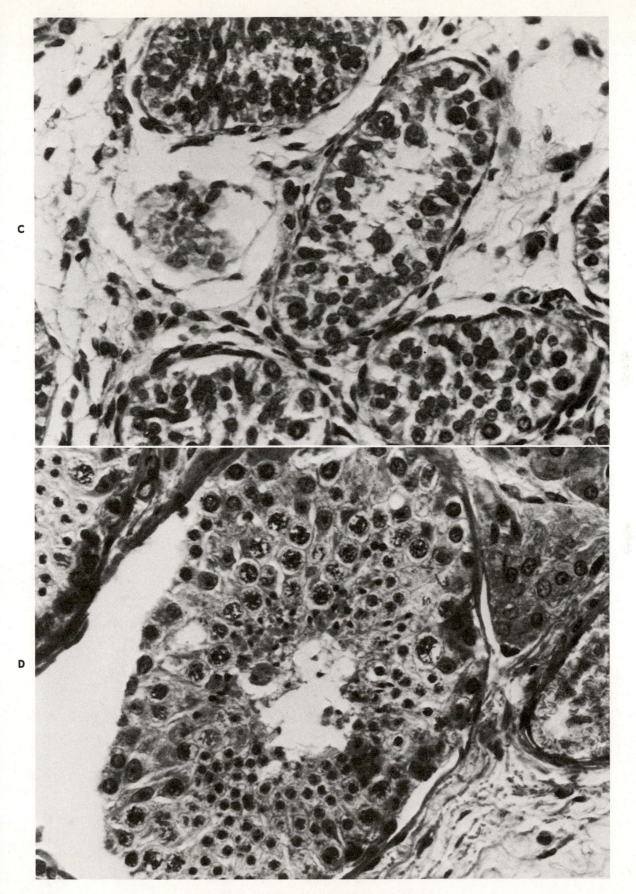

Fig. 55-5, cont'd ■ For legend see opposite page.

into the inferior vena cava, whereas the left drains into the ipsilateral renal vein. Eighty percent of the adult testis is made up of the seminiferous tubules; the remaining 20% is composed of supportive connective tissue, throughout which the Leydig cells are scattered (Fig. 55-5). The seminiferous tubules are a coiled mass of loops; each loop begins and ends in a single duct, the *tubulus rectus.* The tubuli recti, in turn, anastomose in the *rete testis* and eventually drain via the *ductuli efferentes* into the *epididymis.* The latter constitutes a storage and maturation depot for spermatozoa. From the epididymis the spermatozoa are carried via the *vas deferens* and *ejaculatory duct* into the penis, to be emitted during copulation.

The structure of the adult seminiferous tubule is shown schematically in Fig. 55-6. Each seminiferous tubule is bounded by a basement membrane, separating it from the Leydig cells and the surrounding connective tissue. Immediately next to the basement membrane are spermatogonia and Sertoli cells. As the spermatogonia divide and develop into spermatocytes and spermatids, a column of cells is formed that reaches from the basement membrane to the lumen of the tubule and culminates in the spermatozoa. In contrast, the cytoplasm of each Sertoli cell extends all the way from the basement membrane to the lumen. This cytoplasm invests the spermatogonia and its germ cell line successors (Fig. 55-6). Special processes of the Sertoli cell cytoplasm fuse into *tight junctions,* which create two compartments of intercellular space between the basement membrane and the lumen of the tubule. The spermatogonia lie within the *basal compartment,* whereas the spermatocytes and subsequent stages in spermatozoon development lie in the *adluminal compartment.* This compartmentalization accomplishes two things. In effect the Sertoli cell cytoplasm forms a blood-testis barrier that excludes a variety of circulating substances from the fluid bathing the maturing germ cells and from the seminiferous tubular fluid. Conversely, products from the later stages of spermatogenesis are prevented from diffusing back into the bloodstream and, if recognized as foreign, are prevented from producing antibodies.

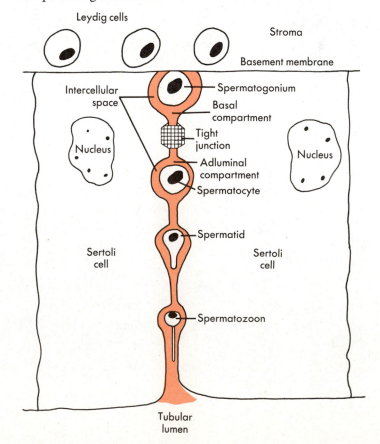

Fig. 55-6 ■ Geographical relationships between Leydig cells, Sertoli cells, and germ cells in the testis. Note that the tight junction between the two Sertoli cells seals off the intercellular space surrounding the spermatocytes and spermatids from contact with the general intercellular space and thus with plasma fluid contents. (Redrawn from Fawcett, D.W.: Ultrastructure and function of the Sertoli cell. In Hamilton, D.W., and Greep, R.O., editors: Handbook of physiology; Section 7, vol. 5, Bethesda, Md., 1975, The American Physiological Society.)

Correlating with its anatomy, the testis may be considered to consist of two functional elements. The Leydig cells are pure steroid-secreting cells whose major product, testosterone, has both vital local effects on germ cell replication and actions on distant target cells. The seminiferous tubules carry out the process of spermatogenesis while bathed in locally generated testosterone and estradiol. It seems best to describe first the process of spermatogenesis and its hormonal regulation and, second, the effects of testicular hormone secretions throughout the rest of the body.

The production of sperm is an ongoing process throughout the reproductive life of the male. Approximately 2 million sperm are produced daily. It is of crucial importance in generating this large number that the spermatogonia also renew themselves by cell division. This situation differs fundamentally from that in the female, who at birth has a fixed number of oocytes, which can then only decrease throughout her life.

The extraordinary metamorphosis from spermatogonium to spermatozoon, progressing by stages of histological appearance from the basement membrane to the lumen of the tubule, is depicted in Fig. 55-7. The first two mitotic divisions of a spermatogonium

■ *The biology of spermatogenesis*

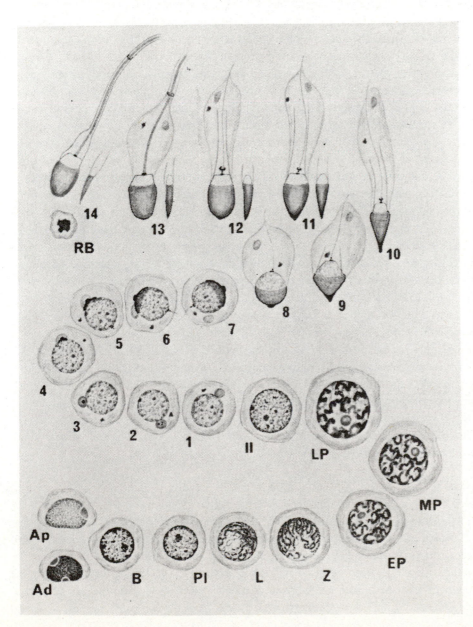

Fig. 55-7 ■ Development of spermatozoa from spermatogonia in the subhuman primate. *Ad,* Dark spermatogonium; *Ap,* pale spermatogonium; *B,* type B spermatogonium; *PI,* preleptotene primary spermatocyte; *L,* Leptotene spermatocyte; *Z,* Zygotene spermatocyte; *EP, MP, LP,* early, middle, and late pachytene spermatocytes; *II,* secondary spermatocyte; *1-7,* early spermatids; *8-13,* late spermatids with progressive formation of flagellum; *14,* spermatozoon; *RB,* residual body. (Redrawn from Clermont, Y.: Physiol. Rev. **52:**198. 1972.)

give rise to four cells: a single resting cell *(Ad)* that will eventually serve as the ancestor of a later generation of sperm and three active cells *(Ap)*. The latter divide by further mitoses to yield type B spermatogonia, which then give rise to a number of primary spermatocytes. These cells enter the prophase of meiosis, the first reduction division, in which they remain for about 20 days. After completion of meiosis their daughter cells, the secondary spermatocytes, immediately divide again. The products, called spermatids, now each contain 22 autosomes and either an X or a Y sex chromosome. The spermatids lie near the lumen of the seminiferous tubule, attached to the abutting Sertoli cells and connected with each other through intercellular bridges. Here they eventually develop into flagellated spermatozoa, which are finally extruded into the tubular lumen in a process called *spermiation*. During the process of spermiation much of the cytoplasm of the spermatid is discarded, leaving a linear structure, the spermatozoon. There are several components to the spermatozoon. The head contains the nucleus and an acrosomal cap in which are concentrated hydrolytic and proteolytic enzymes that facilitate penetration of the ovum and possibly also the mucous plug of the female cervix. The middle piece, or body, contains mitochondria, which generate the motile energy of the spermatozoon. The chief piece of the tail contains stored ATP and pairs of contractile microtubules down its entire length, one pair in the center and nine pairs around the circumference. An ATPase releases the stored energy in a way that allows the microtubules to impart flagellar motion to the spermatozoon.

In man approximately 70 days are required for the entire sequence of development from spermatogonia to spermatozoa. However, individual resting spermatogonia do not start into the process of spermatogenesis randomly. Careful studies have shown that cycles of spermatogenesis exist with distinct cycle times. Groups of adjacent resting spermatogonia initiate a new cycle about every 16 days, thus constituting one "generation." At about the same time that the primary spermatocytes of one cycle enter prophase, a second cycle of spermatogonia is activated. A third cycle begins approximately synchronously with the appearance of the spermatids from the first cycle. By the time these spermatids have completed their transformation into spermatozoa, a fourth cycle of spermatogonia has been started. Around the circumference of any individual seminiferous tubule several spermatogenic cycles may be in process simultaneously. Since within each cycle approximately six stages of cellular development can be identified histologically, this gives rise to a picture of several specific cellular constellations existing side by side (Fig. 55-5, *B*). In some mammals, but not in man, spermatogenic cycles are repeated in a defined topographical relationship to each other along the length of each seminiferous tubule. This has been termed the *wave of spermatogenesis*.

There may not be total separation of the individual germ cells that constitute the successive descendents of type B spermatogonia and which lie within the adluminal compartment of the tubule. Continuity of cytoplasm and possibly cell-to-cell intercommunication may exist. Because of these possibilities and because of the regular topographical association of particular stages of spermatogenesis in neighboring cycles, it appears likely that products of germ cells in one stage of spermatogenesis may initiate or regulate events in earlier or later stages.

■ *Sexual functioning*

After spermiation the spermatozoa reach the epididymis, which they traverse over a number of days. During this time they undergo further maturation. It is still unclear to what extent this is preprogrammed within the spermatozoon and to what extent it depends on specific secretions of the epididymal cells. By the time they reach the vas deferens, the spermatozoa have acquired motility but not the ability to fertilize an ovum. They then may be stored viably in the vas deferens for several months. Delivery of spermatozoa into the female genital tract occurs by ejaculation from the vas deferens. To the contents of the vas deferens are added fluid from the accessory organs of reproduction. Fluid from the seminal vesicles contains fructose, which serves as an important oxidative substrate, and also prostaglandins, which may stimulate contractions of the uterus and fallopian tubes that help propel the spermatozoa toward the ovum. Fluid from the prostate gland contains citrate, calcium, 5'-nucleotidase, and acid phosphatase. The alkalinity of prostatic fluid helps neutralize the acid pH of the vaginal and cervical secretions.

During the process of erection the venous sinuses of the penis fill with blood, converting it into a firm organ for penetration. This is caused by arteriolar dilation coupled with venous constriction and is under parasympathetic control. Ejaculation occurs as a result of sympathetic nervous system impulses. A typical emission contains 200 million to 400 million spermatozoa in a volume of 3 to 4 ml. Once within the vagina, the spermatozoa's rate of flagellated movement is up to 4 mm/minute. The life span of spermatozoa in the female genital tract is approximately 2 days. As a result of contact with the chemical milieu of the vagina, the spermatozoa gain the ability to fertilize an ovum. This process is termed *capacitation*.

■ *Hormonal control of spermatogenesis*

Testosterone and estradiol, along with LH, FSH, and possibly growth hormone and prolactin, all participate in the regulation of spermatogenesis. Present knowledge still does not permit pinpointing the exact role of each in the spermatogenic cycle; therefore only a tentative description can be offered.

The transformation of the primordial germ cells to the primitive type A spermatogonia may be stimulated by testosterone during ontogeny of the fetal testis at 8 to 20 weeks. From then until puberty these spermatogonia remain in a resting stage. The initial division into type B spermatogonia and possibly the subsequent divisions into primary spermatocytes currently are thought to be independent of gonadotropins or gonadal steroid hormones but possibly to be dependent on growth hormone. The completion of the long prophase of the primary spermatocytes does appear to be dependent on the high concentration of testosterone that is produced locally by LH action on the Leydig cells. In men lacking Leydig cells administration of testosterone in amounts sufficient to restore its other biological actions is insufficient to promote spermatogenesis.

The terminal steps of the differentiation of spermatids into spermatozoa in man require the continuous presence of FSH. In other species, such as rodents, apparently only an initial period of exposure to FSH early in life is required. Currently it is conjectural whether the effects of FSH are mediated through stimulation of estradiol production by the Sertoli cells, with the steroid then conditioning the spermatids. The process of spermiation appears to be stimulated by LH acting either directly or indirectly. The testis possesses receptors for prolactin, and this hormone in physiological concentrations appears to synergize with LH in some of its actions. Finally, the role of inhibin, which is produced by the Sertoli cells, remains to be clarified. Since selective destruction of the spermatogenic cells causes a rise in FSH secretion, it is presumed that some product from the germ cell line normally directs the Sertoli cells to secrete inhibin. This protein hormone in turn may help coordinate the availability of FSH with the needs of a particular stage of the spermatogenic cycle.

Despite the cyclic nature of spermatogenesis locally, the testis as a whole is continuously releasing spermatozoa throughout its entire length. Furthermore, even though gonadotropin release is pulsatile, the mean daily plasma levels of FSH and LH are essentially constant in adult men. It is speculative at present whether local differences in gonadotropin sensitivity or regional differences in gonadotropin distribution contribute to the cyclic and topographical nature of the spermatogenic process. The situation is clearly different from that of the ovary, in which a phasic pattern of gonadotropin secretion, highlighted by a single distinct burst, produces the single monthly release of an ovum.

■ *The role of the Sertoli cell*

In early fetal life the Sertoli cells secrete MIF. Their subsequent function until puberty is not known for certain, although presumably they are secreting inhibin, thereby helping keep pituitary FSH secretion low. After puberty the Sertoli cells do not undergo any further cell divisions. In association with the cycle of spermatogenesis they do undergo regular changes in the activity and shape of the nucleus, in the size, shape, and branching of cytoplasmic processes, in concentrations of lipid and glycogen, in mitochondrial function, and in enzyme content. In some way these changes relate to the processing of the germ cells, but whether this is in a controlling, a facilitative, or a reactive manner still remains to be determined. The cytoplasmic processes of the Sertoli cells, extending from the basement membrane to the lumen of the seminiferous tubule, act as conduits, between which the various stages of germ cells move in their passage to the lumen (Fig. 55-6). In some regular fashion the tight junctions must open to permit the maturing primary spermatocytes to pass, and then the junctions must close again behind them. The Sertoli cell cytoplasm also acts as a filter, permitting only certain substances, presumably those advantageous to spermatogenesis, to reach the spermatocytes.

Sertoli cells probably synthesize estradiol and estrone in response to FSH stimulation. Testosterone and androstenedione, which have diffused in from the Leydig cells, serve as precursors (Fig. 55-4). A unique, FSH-dependent function of the Sertoli cells is the secretion of androgen-binding protein. This is a protein with a molecular weight of 90,000 that has very similar properties to those of plasma sex steroid–binding globulin (SSBG) (as discussed later). The androgen-binding protein complexes testosterone, dihydrotestosterone, and estradiol with high affinity. It is thought to have several possible functions. It effectively concentrates testosterone in the Sertoli cell, creating a storage form of the hormone for controlled release by this cell during appropriate stages of spermatogenesis. Androgen-binding protein also is secreted into the fluid of the seminiferous tubule, where it may serve to prevent reabsorption of testosterone from this fluid. This would ensure the availability of testosterone and possibly of estradiol, as well, to the spermatozoa during their maturational sojourn in the epididymis. It also has been suggested that androgen-binding protein, by complexing estradiol, prevents this steroid from diffusing back into the Leydig cell and inhibiting testosterone synthesis.

Either by mechanical or chemical means the Sertoli cell also is clearly responsible for the ejection of the spermatozoa into the lumen. In this process the nucleus of the spermatozoon is oriented toward the base of the tubule. The bulk of the cytoplasm then is squeezed out past the nucleus and shed as the residual body, while the spermatozoon is cast free. The residual body and other fragments are then phagocytosed by the Sertoli cells and subsequently degraded. Finally, the Sertoli cells generate the fluid drive that sweeps the spermatozoa through the rete testis into the epididymis.

■ *Androgenic hormones*

Secretion. Testosterone, the major androgenic hormone, is synthesized as described previously (Fig. 55-4). Its synthesis and release by the Leydig cells are regulated by LH. In adult men LH is secreted in a pulsatile pattern produced by luteinizing hormone–releasing hormone (LHRH); correspondingly, plasma testosterone levels also show small pulses throughout the day (Fig. 52-14). There is, in addition, a superimposed diurnal trend, such that plasma testosterone is 20% to 25% lower at 8:00 PM than at 8:00 AM. The response to prolonged stimulation with exogenous LH is a rapid rise in plasma testosterone, followed by a brief decline and then a second plateau (Fig. 55-8). The temporary decrease may be due to down regulation of LH receptors by the peak concentrations of gonadotropin.

Testosterone also gives rise to two other potent androgens: dihydrotestosterone and 5α-androstanediol (Fig. 55-4). Although both of these may be secreted by the testis in small amounts, the major fraction of circulating dihydrotestosterone and androstanediol

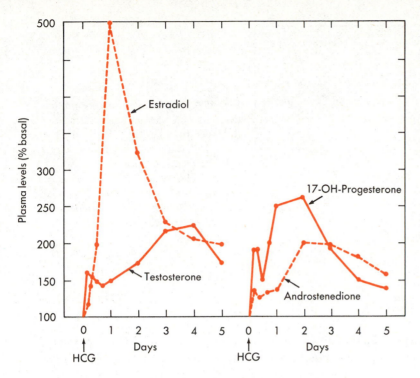

Fig. 55-8 ■ Plasma hormone responses to stimulation of testes in normal men by human chorionic gonadotropin (HCG), with LH-like activity. Note biphasic response of testosterone. Note also the increase in precursor levels on right, indicating activation of an early step in the synthetic pathway. (Redrawn from Forest, M.G., et al.: Kinetics of human chorionic gonadotropin–induced steroidogenic response of the human testis, J. Clin. Endocrinol. Metab. **49**:284, 1979. Copyright 1979. Reproduced by permission.)

TABLE 55-3

Turnover of gonadal steroids in adult men

Steroid	Plasma concentration (ng/dl)	Blood production rate (μg/day)	Metabolic clearance rate (L/day)
Testosterone	650	7000	1100
Dihydrotestosterone	45	300*	600
5α-Androstanediol	12	200*	1800
Androstenedione	120	2400	2000
Estradiol	3.0	50†	1700
Estrone	2.5	60†	2500

*About 60% to 80% produced peripherally from testosterone.
†About 80% to 90% produced peripherally from testosterone and androstenedione, respectively.

is derived from the reduction of testosterone in peripheral tissues. The plasma levels, blood production rates, and metabolic clearances of these androgens are shown in Table 55-3. The testosterone precursor, androstenedione, also is secreted in major amounts by the Leydig cells (Fig. 55-8), but it contributes little per se to androgen action. The two estrogens—estradiol and estrone—are produced in significant amounts in men. However, only 10% to 20% of the daily production is by direct testicular secretion in response to LH (Fig. 55-8). The majority is derived from circulating testosterone and androstenedione by aromatization in various peripheral sites, most notably in the adipose tissue and liver.

Leydig cell function varies distinctively during the life span of the individual. As shown in Fig. 55-9, plasma testosterone rises to levels of 400 ng/dl in the fetus at the time that the external genitalia are undergoing differentiation to the masculine pattern. By birth, however, these levels have declined to less than 50 ng/dl. Very soon thereafter plasma testosterone begins to rise, reaching a peak of 150 to 200 ng/dl at 4 to 8 weeks

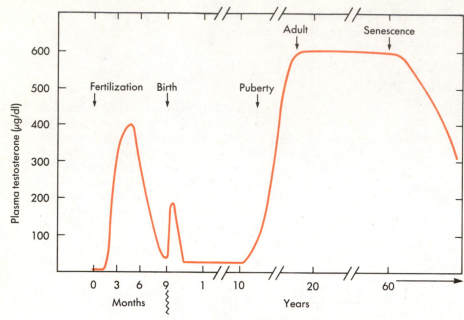

Fig. 55-9 ■ Plasma testosterone profile during the life span of a normal male. (Redrawn from Griffin, J.E., et al.: The testis. In Bondy, P.K., and Rosenberg, L.E.: Metabolic control and disease, Philadelphia, 1980, W.B. Saunders Co.; and Winter, J.S.D., et al.: Pituitary-gonadal relations in infancy, J. Clin Endocrinol. Metab. **42**:679, 1976. Copyright 1976. Reproduced by permission.)

of age. This peak is stimulated by a corresponding rise in plasma LH. Its physiological significance is not known. After age 2 months plasma testosterone and LH again fall to low levels, and thereafter, throughout childhood, Leydig cells cannot be identified in the testes (Fig. 55-5, *C* and *D*). At about age 11 plasma testosterone begins a steep rise, reaching an adult plateau at about age 17 (Fig. 55-9). This is sustained for about 50 years. The pubertal increase and adult plateau of testosterone secretion correspond with the function of the hormone in helping to initiate and then to maintain spermatogenesis. During the seventh and eighth decades of life plasma testosterone gradually declines about 50% because of loss of Leydig cell responsiveness to stimulation. Because of negative feedback, plasma LH levels slowly rise at this time.

Metabolism. Testosterone circulates largely bound to a sex steroid–binding globulin (SSBG), also known as testosterone-estradiol-binding globulin. This is a β-globulin glycoprotein with a molecular weight of 94,000. Fifty to sixty percent of circulating testosterone is complexed to SSBG, and most of the remainder is bound to albumin and other proteins. SSBG also binds dihydrotestosterone and 5α-androstanediol. Only the free and the loosely bound albumin fractions of testosterone and the other androgens are biologically active. The SSBG-bound fractions serve as circulating reservoirs, similar to those of thyroid hormone and cortisol. The concentration of SSBG is increased by estrogens and is decreased by androgens. Reciprocally, then, estrogen reduces the percentage of free testosterone, whereas androgen increases it. About 1% of the daily production of testosterone (70 μg) is excreted daily in the urine as glucuronide. Most of the remainder is metabolized to products that are excreted in the urine as part of the 17-ketosteroid fraction (Chapter 54). In men only 30% of the total urinary 17-ketosteroids derives from testosterone; most arises from adrenal sources. Therefore measurement of plasma testosterone (and occasionally urine testosterone) is the mainstay for assessing Leydig cell function.

Actions. The extratesticular effects of testosterone and related androgens can be divided into two major categories: those which pertain specifically to reproductive function and secondary sexual characteristics and those which pertain more generally to stimulation of tissue growth and maturation. At present the evidence suggests that similar intracellular mechanisms are involved in both categories. In general, the model for steroid hormone effects is applicable (Fig. 49-7).

Testosterone diffuses freely into cells. In many but not all target cells it rapidly undergoes reduction to dihydrotestosterone and, in some, to 5α-androstanediol (Fig. 55-4). The relevant hydroxysteroid dehydrogenases are microsomal in location and employ

NADPH as the reductant. A single cytoplasmic protein receptor binds all three steroids, with affinities in the order dihydrotestosterone > testosterone >> 5α-androstanediol. The absolute requirement for the androgen receptor is best illustrated by the syndrome of *testicular feminization* in humans and mice. Lacking the gene for androgen receptor synthesis, XY individuals with testes show complete failure to masculinize their genital ducts or external genitalia.

The steroid receptor complex moves into the nucleus, where it interacts with chromosomal DNA and nuclear proteins. The result is a significant stimulation of RNA polymerase, of various messenger RNAs, and of the synthesis of proteins. In addition, enzymes concerned with DNA synthesis, such as thymidine kinase and DNA polymerase, are increased. Virtually all actions of androgens are blocked by inhibitors of RNA or protein synthesis. Therefore they require induction of new enzyme molecules, as opposed to allosteric or covalent activation of existing enzyme molecules.

In target tissues, such as the prostate gland and seminal vesicles, polyamine (for example, spermine and putrescine) synthesis is stimulated by androgens, and these compounds in turn enhance RNA synthesis. Androgens also stimulate remarkable growth of these accessory organs of reproduction, characterized by hypertrophy and hyperplasia of the epithelial cells, stromal components, and blood vessels.

By far the major circulating androgen is testosterone (Table 55-3). Testosterone in part can be considered a prohormone for dihydrotestosterone and 5α-androstanediol, much as thyroxine is a prohormone for triiodothyronine. However, unlike thyroxine, testosterone definitely can be shown to have intrinsic hormonal activity of its own in tissues that lack the enzyme 5α-reductase. An interesting and major source of such information comes from studies of a group of individuals who are genetic and gonadal males but who have congenital 5α-reductase deficiency. These individuals have feminized external genitalia at birth, but during puberty they undergo selective masculinization in response to the rising testosterone secretion.

A presumptive classification of androgen effects, according to the probable actual effector hormone, is shown in Table 55-4. Dihydrotestosterone is specifically required in the fetus for the differentiation of the genital tubercle, genital swellings, genital folds, and urogenital sinus into the penis, scrotum, penile urethra, and prostate, respectively. It is required again during puberty for growth of the scrotum and prostate and in adult life for the stimulation of prostatic secretions. Dihydrotestosterone or 5α-androstanediol mediates stimulation of the hair follicles, producing the typical male pattern. This consists of beard growth, a diamond-shaped pubic escutcheon, relatively large amounts of body hair, and the recession of the temporal hairline, which in some men culminates in baldness. Increased production of sebum by the sebaceous glands also is brought about by dihydrotestosterone or 5α-androstanediol.

TABLE 55-4

Major actions of androgenic hormones

Testosterone	*Dihydrotestosterone*
Fetal development of	Fetal development of
Epididymis	Penis
Vas deferens	Penile urethra
Seminal vesicles	Scrotum
	Prostate
Pubertal growth of	
Penis	Pubertal growth of
Seminal vesicles	Scrotum
Musculature	Prostate
Skeleton	Sexual hair
Larynx	Sebaceous glands
Spermatogenesis	Prostatic secretion

Testosterone, on the other hand, specifically stimulates the differentiation of the wolffian ducts into the epididymis, vas deferens, and seminal vesicles. During puberty testosterone causes enlargement of the penis and of the seminal vesicles to adult size. It also causes enlargement of the larynx and thickening of the vocal cords, resulting in a deeper voice. Most important, testosterone is the local hormone required for initiation and maintenance of spermatogenesis, as demonstrated by the fact that spermatozoa production is normal in postpubertal individuals lacking dihydrotestosterone because of 5α-reductase deficiency. Testosterone itself also first stimulates the pubertal growth spurt and then causes cessation of linear growth by closure of the epiphyseal growth centers. Testosterone is the anabolic hormone that enlarges the muscle mass in boys during puberty. In subsequent adult life administration of testosterone causes nitrogen retention in both sexes. It is noteworthy that the hypothalamus lacks significant 5α-reductase activity. Thus suppression of LH secretion by negative feedback is largely a direct function of testosterone, with a possible small additional effect from circulating dihydrotestosterone.

Certain other diverse androgenic actions can be ascribed to testosterone. These include (1) initiation of sexual drive (libido) and the ability to achieve a physiologically complete erection (potency), (2) suppression of mammary gland growth, (3) stimulation of hematopoiesis and maintenance of a normal red blood cell mass, (4) stimulation of renal sodium reabsorption, (5) stimulation of aggressive behavior, and (6) suppression of hepatic synthesis of SSBG, cortisol-binding globulin, and thyroxine-binding globulin.

■ *Male puberty*

From the early newborn period to the onset of puberty the testis is endocrinologically dormant (Fig. 55-5, *C* and *D*). Beginning at an average age of 10 to 11 and ending at an average age of 15 to 17, males develop full reproductive function, Leydig cell pro-

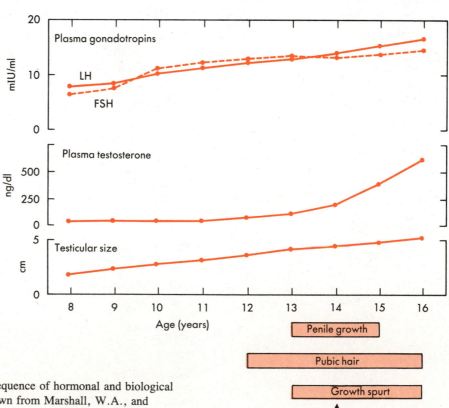

Fig. 55-10 ■ Average chronological sequence of hormonal and biological events in normal male puberty. (Redrawn from Marshall, W.A., and Tanner, J.M.: Arch. Dis. Child. **45**:13, 1970; and Winter, J.S.D., et al.: Pediatr. Res. **6**:126, 1972.)

liferation, and adult levels of androgenic hormones (Fig. 55-5, *A* and *B*). Secondary to activation of the testis, males acquire adult size and function of the accessory organs of reproduction, complete secondary sexual characteristics, and adult musculature. They undergo a linear growth spurt, and the epiphyses close when they attain adult height. A composite picture of the measurable and visible portions of this sequence is shown in Fig. 55-10. It must be stressed that this process can start as early as age 8 and as late as age 20, without any evidence of disease.

The events of puberty begin with the secretion of increasing amounts of gonadotropin. Prior to age 10 plasma gonadotropin levels are low, despite the very low plasma concentrations of testicular steroids. Therefore either the negative feedback system is inoperative or the hypothalamus and pituitary gland are exquisitely sensitive to testosterone or estradiol. Thus one explanation offered for the onset of puberty is the gradual maturing of a hypothalamic gonadostat, leading to increased synthesis and release of LHRH. The rate of this maturational process may well be genetically preprogrammed. However, a simple resetting of the negative feedback axis does not explain all the observed phenomena. For example, during early and middle puberty, and at no other time of life, a nocturnal peak in LH secretion is observed (Fig. 55-11). At around the same time, a pulsatile pattern of LH secretion develops, which will then remain for life. These events suggest some other central nervous system influences on the initiation of puberty. Various animal and clinical studies implicate input from the amygdala, the pineal gland, or the olfactory bulb as triggers of pubertal hypothalamic functioning.

The gonad itself is not necessary for activation of gonadotropin secretion, since castrated or functionally agonadal individuals demonstrate a rise in gonadotropin concentrations and the appearance of pulsation at the usual age of puberty. The fact that

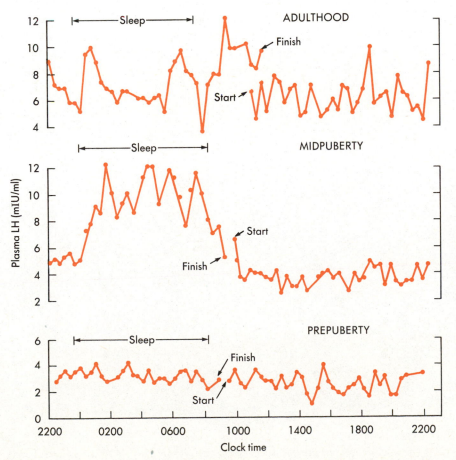

Fig. 55-11 ■ The development of pulsatile secretion of LH and the transient appearance of a nocturnal peak during puberty. The same sequence occurs in males and females. (Redrawn from Boyar, R.M., et al.: Reprinted, by permission of the New England Journal of Medicine **287**:582, 1972.)

dehydroepiandrosterone sulfate (DHEA-S) levels increase, coincident with the onset of puberty, has suggested that this adrenal steroid may play an important role in the maturing of the hypothalamus. However, this seems unlikely, because boys with adrenal cortical insufficiency do undergo normal pubescence (with proper cortisol replacement), whereas boys with high adrenal androgen secretion as a result of congenital adrenal hyperplasia or adrenal tumors generally do not.

The responsiveness of the pituitary gland also changes. In childhood the response of LH secretion to acute LHRH stimulation is less than that of FSH secretion, but it gradually increases during puberty. This may be explained by an increasing synthesis and storage of LH in response to activation of LHRH secretion. Responsiveness of the Leydig cells to LH stimulation, although present in childhood, also is augmented during puberty. Thus the pubescent period can be viewed as a cascade of increasing responsiveness from the hypothalamic level to the pituitary gland to the testis.

Enlargement of the testis is the first and most important clinical sign of puberty. This represents principally an increase in the volume of the seminiferous tubules, and it is preceded by small increases in plasma FSH. Leydig cells appear, and testosterone secretion is stimulated, as plasma LH increases. Plasma testosterone then climbs rapidly over a 2-year period, during which time pubic hair appears, the penis enlarges, and peak velocity in linear growth is achieved (Fig. 55-10). An additional clinical feature seen in about one third of boys is an early transient stimulation of breast growth and tenderness. This probably reflects increased production of estradiol secondary to LH stimulation. As testosterone levels continue to climb, the breast tissue regresses. One to two years after adult testosterone levels are reached, closure of the epiphyseal growth centers ends puberty.

■ *The ovaries*

The ovaries, fallopian tubes, and uterus comprise the internal reproductive organs of the female and are situated in the pelvis. Each adult ovary weighs approximately 15 g and is attached to the lateral pelvic wall and to the uterus by ligaments, through which run the ipsilateral ovarian artery, vein, lymphatic vessels, and nerve supply. The ovary consists of three distinct zones (Fig. 55-12). The dominant zone is the cortex, which is

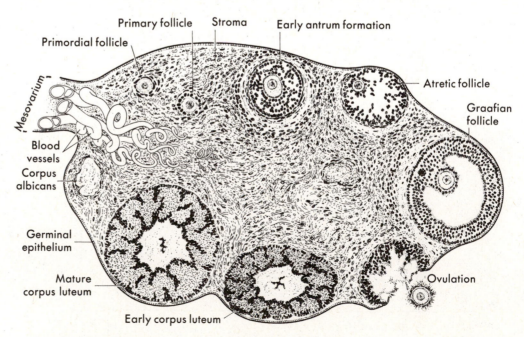

Fig. 55-12 ■ The microscopic anatomy of the human ovary. (From Ham, A.W., and Leeson, T.S.: Histology, ed. 4, Philadelphia, 1961, J.B. Lippincott Co.)

lined by germinal epithelium and contains all the oocytes, each enclosed within a follicle. Follicles in various stages of development and regression can be seen through the cortex during the reproductive years (Fig. 55-12). Interposed between the follicles is the stroma, composed of supporting connective tissue elements and interstitial cells. The other two zones of the ovary are the medulla, consisting of a heterogeneous group of cells, and the hilum, at which the blood vessels enter. These zones contain scattered steroid-producing cells, whose normal function is unknown. Neoplasms of the ovary can arise from any one of its three zones and their individual cell lines. During physical examination the ovaries can be palpated manually through the abdominal wall and also can be well visualized by ultrasonography and computerized axial tomography (CAT scanning).

As a hormone-secreting organ, the ovary functions in two ways. First, the ovarian sex steroids function locally to modulate the complex events in the development and extrusion of the ova. Second, these steroids are secreted into the circulation and act on diverse target organs, including the uterus, vagina, breasts, hypothalamus, pituitary gland, adipose tissue, bones, and liver. Many but not all of the distant effects are closely involved in the reproductive sequence.

■ *Oogenesis*

The primordial germ cells migrate from the yolk sac of the embryo to the genital ridge at 5 to 6 weeks of gestation. There, in the developing ovary, they produce oogonia by mitotic division until 20 to 24 weeks, when the total number of oogonia has reached a maximum of 7 million. Beginning at 8 to 9 weeks some oogonia start into the prophase of meiosis, becoming primary oocytes. This process continues until 6 months after birth, when all oogonia have been converted to oocytes. Almost from the start, however, there is also a process of attrition, so that by birth only 2 million primary oocytes remain, and by the onset of puberty the number falls to 400,000. Thus, in contrast to the man, who is continuously producing spermatogonia and primary spermatocytes, the woman cannot manufacture new oogonia and must function with a continuously declining number of primary oocytes from which ova can mature. At or soon after menopause there are few, if any, oocytes left, and reproductive capacity ends.

At the time meiosis begins oocytes are 10 to 25 μm in diameter. They grow to 50 to 120 μm at maturity, the size of the nucleus and cytoplasm having increased proportionately. The first meiotic division is not completed until the time of ovulation; thus primary oocytes have life spans up to 50 years. The lengthy suspension of the oocyte in prophase apparently is dependent on the hormonal milieu provided by its surrounding sustaining cells.

First stage. The first stage of follicular development occurs very slowly, over a period usually not less than 13 years and to as long as 50 years. As an oocyte enters meiosis, it induces a single layer of spindle cells from the stroma to surround it completely. These cells are the precursors of the female follicular cells, the *granulosa cells.* Cytoplasmic processes from these cells attach to the plasma membrane of the oocyte. In addition, a membrane called the *basal lamina* forms outside the spindle cells, delimiting the complex from the surrounding stroma. This constitutes the *primordial follicle,* which is about 25 μm in diameter (Fig. 55-13). Beginning at 5 to 6 months of gestation some of these follicles enter the next phase of development. The spindle-shaped cells become cuboidal and begin to divide, creating several layers of granulosa cells around the oocyte. This complex is called the *primary follicle.* The granulosa cells secrete mucopolysaccharides, which form a protective halo, the *zona pellucida,* around the oocyte (Fig. 55-13). The cytoplasmic processes of the granulosa cells, however, continue to penetrate the zona pellucida, evidently providing nutrients and chemical signals to the maturing primary oocyte within. Thus cytoplasm of the granulosa cells, like that of the male Sertoli cells, forms a filter through which plasma substances must pass before reaching the germ cell.

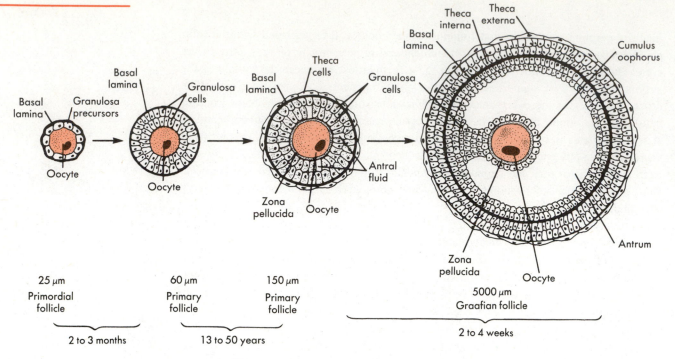

Fig. 55-13 ■ The development of an ovarian follicle (not to scale). The graafian follicle has more layers of granulosa and theca cells than shown here.

The primary follicle continues to grow, reaching a diameter of 150 μm. At this point the oocyte has reached its maximum size, averaging 80 μm in diameter. There are two other concurrent developments. Another layer of spindle interstitial cells is recruited outside the basal lamina and forms the *theca interna*, while the granulosa cells begin to extrude small collections of fluid between themselves. This completes the first stage of follicular development. It is also the maximum degree of development ordinarily found in the prepubertal ovary.

Second stage. In contrast to the first stage, the second stage of follicular development is much more rapid, requiring 2 to 4 weeks for completion. This stage takes place after menarche, that is, after the onset of menses. During each menstrual cycle approximately six to twelve primary follicles enter the next sequence. The small collections of follicular fluid coalesce into a single central area called the *antrum* (Fig. 55-13). The fluid in the antrum contains mucopolysaccharides, plasma proteins, electrolytes, gonadal steroid hormones, inhibin, and other factors. The steroid hormones reach the antrum by direct secretion from granulosa cells and by diffusion from the theca cells outside the basal lamina. A nonsteroidal substance that is capable of inhibiting oocyte meiosis probably also is secreted into the antral fluid and is believed to be of granulosa cell origin.

The granulosa cells continue to proliferate and displace the oocyte into an eccentric position on a stalk, where it is surrounded by a distinctive layer, two to three cells thick, called the *cumulus oophorus*. The theca cells also proliferate, and those nearest the basal lamina are transformed into cuboidal steroid-secreting cells—the *theca interna*. Additional peripheral layers of spindle cells from the stroma form around the theca interna and together with vascular spaces comprise the *theca externa*. By the end of this stage the entire complex, called a *graafian follicle* (Fig. 55-13), has reached an average diameter of 5000 μm (5 mm). Although graafian follicles may be found rarely in premenarchal ovaries, they never achieve this size or degree of development.

Third stage. The third and final stage of follicular development is the most rapid, being completed within 48 hours. It occurs only in the postmenarchal reproductive ovary. Several days before ovulation a single graafian follicle from that cycle achieves dominance. In this particular graafian follicle the granulosa cells significantly increase the production of antral fluid. The colloid osmotic pressure of the fluid also increases because of depolymerization of the mucopolysaccharides. The granulosa cells spread apart, and the cumulus oophorus loosens. At the same time the vascularity of the theca increases greatly. The total size of this follicle reaches 10 to 20 mm within 48 hours. The portion of the basal lamina adjacent to the surface of the ovary then is subjected to proteolysis. The follicle gently ruptures, releasing the oocyte with its adherent cumulus oophorus into the peritoneal cavity. At this time the initial meiotic division is completed. The resultant secondary oocyte is drawn into the closely approximated fallopian tube, and the first polar body is discarded. In the fallopian tube fertilization causes completion of the second meiotic division, resulting in the haploid (23 chromosome) ovum and the second polar body.

Corpus luteum formation. The residual elements of the ruptured follicle next form a new endocrine structure, the *corpus luteum* (Fig. 55-12). This unit will provide the necessary balance of gonadal steroids that optimizes conditions for implantation of the ovum, should fertilization occur, and for subsequent maintenance of the zygote until the placenta can assume this function. The corpus luteum is made up of granulosa cells, theca cells, thecal capillaries, and fibroblasts. The granulosa cells comprise 80%. They hypertrophy to a diameter of 30 μm, become arranged in rows, and undergo striking changes. The mitochondria develop dense matrices with tubular cristae, numerous lipid droplets form within the cytoplasm, and the smooth endoplasmic reticulum proliferates. This process, called *luteinization,* is precipitated by the exit of the oocyte from the follicle.

The remaining 20% of the corpus luteum consists of theca cells arranged in folds along its outer surface. The theca cells exhibit similar although less dramatic changes of luteinization. The basal lamina between the granulosa cells and theca cells disappears. The antrum may become temporarily engorged with blood from hemorrhaging thecal vessels, but a clot quickly forms and is subsequently lysed. If fertilization and pregnancy do not ensue, the corpus luteum begins to regress after a 14-day life span. In this process, known as *luteolysis,* the endocrine cells undergo necrosis, and the structure is invaded by leukocytes, macrophages, and fibroblasts. Gradually the former corpus luteum is replaced by an avascular scar known as the *corpus albicans* (Fig. 55-12).

Atresia of follicles. During the reproductive life span of the average woman only 400 to 500 oocytes (one per month) will undergo the complete sequence of events culminating in ovulation. The remaining millions disappear in a process called *atresia,* which begins almost as soon as the first primordial follicles appear in the fetal ovary. In first-stage follicles atresia is a relatively simple process. The oocyte becomes necrotic, its nucleus becomes pyknotic, and the granulosa cells degenerate. This accounts for the vast majority of oocytes. In more advanced follicles atresia is a more complex process. In some follicles the granulosa cells furthest from the oocyte first undergo necrotic changes. Loss of their function may actually precipitate a resumption of meiosis in the oocyte to the point of extrusion of the first polar body. Eventually the granulosa cells in the cumulus oophorus also die, the protective zona pellucida disappears, and the oocyte degenerates. In other follicles degeneration of the oocyte may initiate the process. Eventually fibroblasts invade the follicle, and everything inside the basal lamina collapses into an avascular scar, called a *corpus albicans.* Outside the basal lamina the theca cells dedifferentiate and return to the pool of interstitial cells from which they came.

■ **Hormonal patterns during the menstrual cycle**

The menstrual cycle is divided physiologically into three sequential phases. The *follicular phase* begins with the onset of menstrual bleeding and is of variable length. The *ovulatory phase* is of 1 to 3 days' duration and culminates in ovulation. The *luteal phase* has a constant length of 13 to 14 days and ends with the onset of menstrual

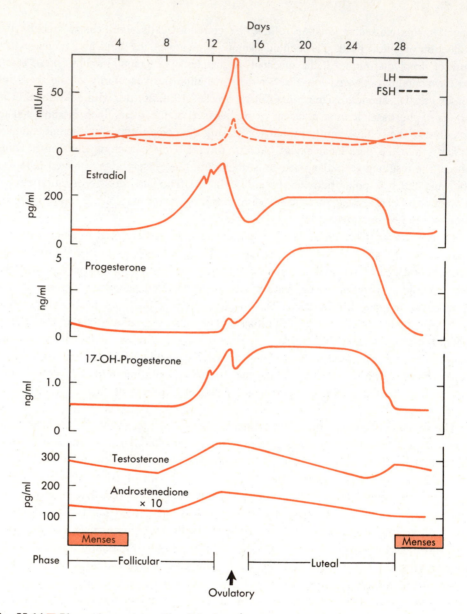

Fig. 55-14 ■ Plasma hormone profile of the human menstrual cycle. Note that increase in estradiol precedes ovulatory LH peak. (Redrawn from Yen, S.S.C., and Jaffe, R.B.: Reproductive endocrinology, Philadelphia, 1978, W.B. Saunders Co.)

bleeding. The overall duration of a normal menstrual cycle can vary from 21 to 35 days, depending mostly on the length of the follicular phase.

A series of cyclic changes in gonadal steroid hormone production characterizes adult ovarian function (Fig. 55-14; Table 55-5). This monthly steroid hormone profile results from cyclic changes in pituitary gonadotropins (Fig. 55-14). However, the pattern of gonadotropin secretion in turn is critically regulated by both negative and positive feedback from gonadal steroids (Chapter 52). In a sense the menstrual cycle might be viewed as a stone rolling up and down a series of hills; the momentum derived from each phase of the cycle powers the next phase, and so on into the next cycle.

Toward the end of the luteal phase and beginning of the follicular phase, plasma FSH and LH are at their lowest levels (Fig. 55-14). The LH/FSH ratio is slightly greater than 1. During the first half of the follicular phase FSH levels rise, followed somewhat later by LH levels. The estrogens (estradiol and estrone) increase almost imperceptibly, whereas progesterone and 17-hydroxyprogesterone levels remain constant and low dur-

TABLE 55-5

Turnover of gonadal steroids in adult women

Steroids	Plasma concentration (ng/dl)	Production rate (µg/day)	Metabolic clearance rate (L/day)
Estradiol			
Early follicular	6	80	1400
Late follicular	50	700	
Middle luteal	20	300	
Estrone			
Early follicular	5	100	2200
Late follicular	20	500	
Middle luteal	10	250	
17-Hydroxyprogesterone			
Early follicular	30	600	2000
Late follicular	200	4000	
Middle luteal	200	4000	
Progesterone			
Follicular	100	2000	2200
Luteal	1000	25000	
Testosterone	40	250	700
Dihydrotestosterone	20	50	400
Androstenedione	150	3000	2000
DHEA	500	8000	1600

Modified from Lipsett, M.B.: Steroid hormones. In Yeh, S.S.C., and Jaffe, R.B., editors: Reproductive endocrinology, Philadelphia, 1978, W.B. Saunders Co.

ing this critical 6 to 8 days. The same is true for the androgens androstenedione and testosterone. During the second half of the follicular phase FSH levels fall modestly, whereas LH levels continue to rise very slowly. The LH/FSH ratio therefore increases to about 2. Concurrently, estradiol and estrone production and plasma levels rise sharply, reaching peaks fivefold to ninefold higher just prior to the ovulatory phase. The estradiol in the plasma is secreted directly by the ovaries, whereas the estrone largely arises from peripheral conversion of estradiol and androstenedione. Plasma progesterone and 17-hydroxyprogesterone remain low until just before the ovulatory phase, when progesterone begins to increase due to ovarian secretion. Androstenedione and testosterone also rise modestly in parallel with 17-hydroxyprogesterone. About half the plasma androgens are derived from ovarian androstenedione secretion and half from adrenal androstenedione secretion.

The large increment in estradiol secretion during the latter half of the follicular phase arises from the granulosa cells of the dominant follicle, which previously were stimulated by FSH and then by LH. The rising plasma estradiol, augmented by follicular inhibin, suppresses FSH secretion by negative feedback. The rise in ovarian androgen secretion toward the end of the follicular phase results from LH stimulation of the theca cells.

The succeeding ovulatory phase is characterized by a very sharp, transient spike in plasma gonadotropin levels. LH increases much more than FSH (Fig. 55-14), so that the LH/FSH ratio rises to about 5. Plasma estradiol levels plummet from their peak at the same time that LH and FSH are on their ovulatory upswing. Estrone, 17-hydroxyprogesterone, androstenedione, and testosterone also now decrease but much more gradually than estradiol. In contrast, a small rise in progesterone begins during the ovulatory phase.

After ovulation LH and FSH both continue to decline during the luteal phase, reaching their lowest points in the cycle toward its end, prior to the onset of menses. The most distinctive and important feature of the luteal phase is a tenfold increase in progesterone, which emanates from the corpus luteum. Estradiol, estrone, and 17-hydroxyprogesterone, partly of corpus luteum origin, also increase, providing broad second peaks of these steroids through the middle of the luteal phase. Androstenedione and testosterone, however, continue to decline during the luteal phase. If pregnancy does not occur, the cycle ends as the gonadal steroids decrease dramatically to their lowest levels and menstrual bleeding starts.

■ *Hormonal regulation of oogenesis*

The preceding sections have provided separate descriptions of the processes of follicular development and atresia and of the changing hormonal milieu. It remains a challenge to reproductive endocrinologists to integrate these areas into a physiological whole. The following discussion presents a current consensus view.

The entire first stage, from the primordial follicle to the primary follicle, *can* proceed in the absence of the pituitary gland. However, studies of the ovaries of individuals with congenital gonadotropin deficiencies suggest that the midgestation surge of fetal FSH and LH, as well as the low levels of gonadotropins secreted during childhood, may increase the rate of growth or the number of follicles undergoing development. Nonetheless first-stage growth appears to be largely a local phenomenon in which one or more factors from the oocyte stimulate granulosa cell development. In turn the granulosa cells probably initiate thecal development and arrest the size and maturation of the oocyte once it has reached 80 μm in diameter. Estradiol and an oocyte maturation–inhibiting factor are the probable granulosa cell hormones involved in these functions.

There is little doubt that second-stage follicular development is initiated by FSH. Two crops of follicles probably begin antral formation during each menstrual cycle. The first group follows the FSH surge of the ovulatory phase, reaches its peak size of 3 to 5 mm 8 to 9 days later, possibly contributes to ovarian estrogen production in the luteal phase, and then undergoes atresia by the end of that phase. The second group of follicles is initiated by the rise in FSH that begins just before the start of the follicular phase. This group, plus a few survivors of the first group, reaches a peak size of 5 mm at 10 to 14 days after the onset of menstrual bleeding. It is from this group that the single dominant follicle, from only one of the two ovaries, ordinarily emerges to undergo ovulation.

The initial action of FSH on the primary follicles is to stimulate growth of the granulosa cells (Fig. 55-15). Additionally, aromatase activity is increased, so that estrogen synthesis from androgen precursors is enhanced. The increasing *local estradiol* causes proliferation of its own receptors and reinforces FSH actions, both by increasing FSH receptors and by synergizing with the gonadotropin in stimulating further granulosa cell hyperplasia and hypertrophy (Fig. 55-15). This in turn further boosts estradiol production. Thus the initiation of second-stage follicular development may be viewed as a self-propelling mechanism that involves fine coordination between the pituitary gland and ovary and that yields *exponential* rates of follicular growth and estradiol production. Two other important actions, which develop somewhat later, contribute to this autocatalytic process. (1) FSH, along with estradiol, induces LH receptors on the granulosa cells. (2) The slowly rising plasma estradiol levels appear to condition the hypothalamic gonadotropin axis so as to maintain or slightly increase plasma LH while plasma FSH is decreasing. Furthermore pituitary LH stores are enhanced by estradiol. This is reflected in the fact that administration of exogenous pulses of LHRH produces greater LH responses in the second half of the follicular phase than in the first half (Fig. 55-16).

Stimulation of the theca cells by LH causes them to produce increasing amounts of

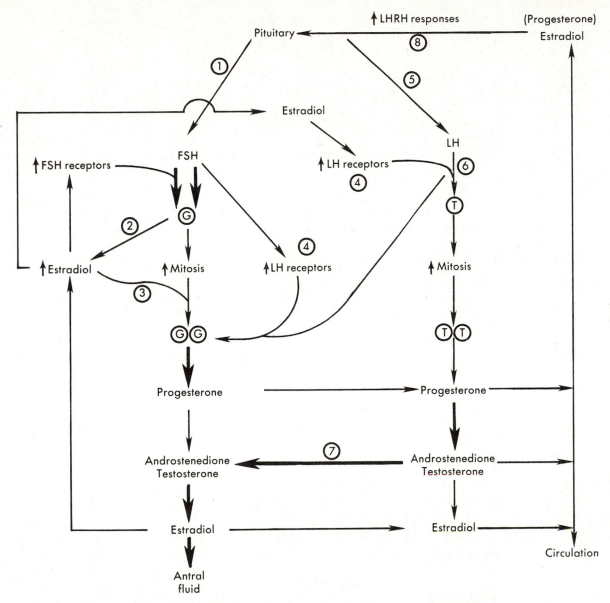

Fig. 55-15 ■ Hormonal regulation of follicular development in the human. Note self-reinforcing effects of FSH and estradiol on the granulosa cells ② and ③, two-way steroid hormone traffic between the ganulosa and theca cells, facilitating estradiol production by the former ⑦, and eventual positive feedback effect of estradiol on the pituitary gland, leading to augmented LHRH responsiveness prior to ovulation ⑧. *G,* Granulosa cell; *T,* theca cell.

androstenedione and testosterone. These steroids diffuse across the basal lamina, where they serve as substrates for granulosa cell aromatase and sustain the augmented estradiol production (Fig. 55-15). This is of considerable importance because isolated granulosa cells are relatively inefficient producers of estradiol. In addition, LH stimulates the granulosa cells to produce progesterone, some of which diffuses back into the theca cells to serve as a substrate for androgen synthesis (Fig. 55-15). Thus, although individually granulosa cells and theca cells possibly can synthesize both androgens and estrogens to some extent, their proximity and the two-way traffic of steroids between them appear to greatly increase the overall efficiency of the follicle.

During the latter half of the follicular phase a single follicle outstrips the others. Exactly how this follicle is selected or selects itself remains unknown. The key may be

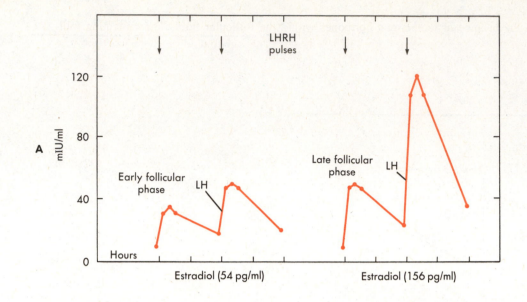

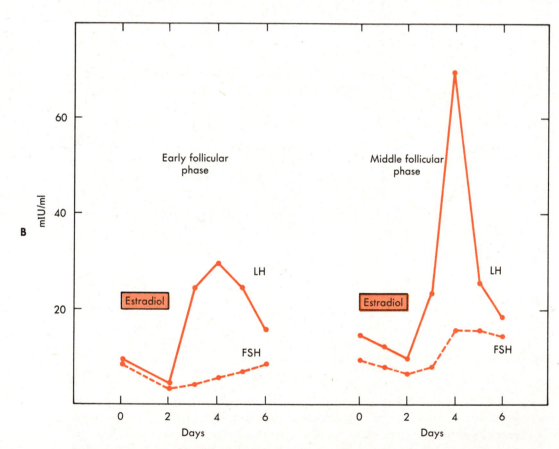

Fig. 55-16 ■ **A,** Increased responsiveness of pituitary gonadotrophs to LHRH in the late follicular phase of the menstrual cycle when endogenous estradiol levels are increased **B,** Plasma LH and FSH after exogenous estradiol administration. After the initial decrease caused by negative feedback, plasma LH rebounds well above baseline when estradiol is discontinued. This positive effect of estradiol also is accentuated as the follicular phase of the menstrual cycle progresses. (**A** redrawn from Wang, C.F., et al.: The functional changes of the pituitary gonadotrophs during the menstrual cycle, J. Clin. Endocrinol. Metab. **42:**718, 1976. Copyright 1976. Reproduced by permission. **B** redrawn from Yen, S.S.C., et al.: Causal relationship between the hormonal variables. In Ferin, M., et al., editors: Biorhythms and human reproduction, New York, 1974, John Wiley & Sons, Inc.)

in the balance of factors and hormones within the follicular fluid. The remaining follicles of the group undergo atresia. The most consistent finding in fluid sampled from smaller or frankly atretic follicles is a relatively high concentration of androgens and relatively low concentrations of estradiol and FSH. An atretic fate may be determined by insufficient FSH receptors or by a relative inability of the granulosa cells to aromatize the supply of androgens from its own theca cell partners (and possibly from the theca cells of the more rapidly growing dominant follicle).

In any event the point is reached after 8 to 12 days when estrogen secretion by the dominant follicle is rapidly augmented. Plasma estradiol rises abruptly in a sawtooth fashion (Fig. 55-14). This surge of estradiol triggers an acute release of LH and FSH (Fig. 55-15). A much smaller preovulatory rise in progesterone may synergize with estradiol in this action. The locus of this striking positive feedback effect is in both the pituitary gland and the hypothalamus. The surge of LH then triggers the process of ovulation previously described. The exact mechanism is not known, but one relevant phenomenon that follows the LH surge is a rapid loss of LH receptors because of down regulation. This desensitizes the granulosa and theca cells to LH, and a rapid fall in estradiol and androgen production ensues. The essential causative role of LH in ovulation has been well confirmed. Either cyclic administration of LHRH or administration of an LH equivalent (human chorionic gonadotropin, or HCG) induces ovulation in women who have been properly primed with FSH. The LH surge also may contribute to the beginning luteinization of the granulosa cells. However, evidence suggests that luteinization occurs primarily as a result of physical separation of the granulosa cells from the oocyte, which frees them from a luteinization-inhibiting factor found in preovulatory follicular fluid.

The dispatched oocyte is not known to be under any immediate further hormonal influence. However, the organization and growth of the corpus luteum and its secretory pattern probably are under hormonal control. Current evidence is conflicting, and a great deal of species variation exists. In the human LH is essential for normal corpus luteum function. If the declining LH levels of the late luteal phase are not replaced by the equivalent placental hormone, HCG, the corpus luteum regresses, and its secretion of progesterone and estradiol ceases completely by 14 days. Prolactin may contribute to sustaining progesterone output by increasing LH receptors. However, there is little systematic variation in plasma prolactin throughout the menstrual cycle, and corpus luteum function has been observed in prolactin-deficient women. Estradiol also may play a role in maintaining the corpus luteum.

The steadily increasing progesterone and estradiol output of the corpus luteum exerts negative feedback on the pituitary gland. The resultant gradual decline in both LH and FSH to their low luteal phase levels withdraws support from the postovulatory group of follicles that had entered second-stage development. This probably causes their atresia. After the eighth postovulatory day the corpus luteum in the nonpregnant female begins to regress. This process is marked by increasing corpus luteum concentrations of cholesterol as progesterone synthesis declines. Luteolysis is mediated by prostaglandins made within the corpus luteum. By the twelfth postovulatory day progesterone and estradiol levels have fallen low enough to release the pituitary gland from negative feedback inhibition, and the FSH rise of the next cycle begins.

The hormonal regulation of the female reproductive cycle, as just described, has left open an important question: what determines the monthly cyclicity of the LH/FSH surge and the resultant ovulation? Although the concept of a primary central nervous system clock is inherently attractive, considerable evidence now suggests that in the human it is the ovary which determines the basic rhythm. Five observations support this view. (1) No cyclic release of LH/FSH is observed in women whose ovaries never functioned, in women whose ovaries were removed in the midst of their reproductive years, or in postmenopausal women after follicular development had ceased. (2) The follicular phase of the cycle has a variable length. The determinant of this interval seems to be the rate of maturation of the follicle through the second stage of development; that is, the ovulatory gonadotropin pulse does not occur until the dominant follicle has reached the appropriate stage of development. (3) In some otherwise normal women who are consistently not ovulating, brief administration of clomiphene citrate—an antiestrogen drug— stimulates a small rise in FSH and LH by negative feedback. Some 10 to 14 days later, after sufficient time has elapsed for follicular development, a spontaneous LH/FSH surge and ovulation ensue. This suggests that the pattern of gonadal steroid secretion which characterizes a normal follicular phase conditions the hypothalamus and the pituitary gland each month to respond at the appropriate time with an acute LHRH and gonadotropin discharge. The ovarian steroids may specifically recruit the function of a cyclic center in the hypothalamus, in contradistinction to a center for tonic gonadotropin control. (4) Administration of estradiol and progesterone in a particular format to castrated or postmenopausal women has been shown to induce a rise in plasma LH that resembles somewhat the normal preovulatory surge. (5) In a monkey whose pituitary gland has been completely severed from its hypothalamus, any central nervous system regulation of gonadotropin secretion is prevented. However, if LH and FSH secretion is sustained by an *unchanging* program of exogenous LHRH infusion, cyclic ovarian function and gonadotropin release are observed.

It must be borne in mind, however, that ovarian signals can be either overridden or reinforced by other influences on the hypothalamus. Loss of cyclic gonadotropin secretion can occur in a number of situations which suggest that the hypothalamus is responding to a caloric, thermal, photic, olfactory, or emotional signal. This is seen in women who are calorically deprived and thus lose considerable amounts of adipose tissue and lean body mass, so that they would be incapable of sustaining a fetus if they became pregnant. It is also seen in women who undergo physical translocation, climatic change, or emotional deprivation. Conversely, it has been observed that women living in close physical proximity can adopt a common cycle, possibly because of pheromones. These are chemical signals emitted by one individual which produce effects in another. In the human there is no evidence that ovulation is stimulated by sexual behavior. However, some, but not all, reports suggest that sexual activity may increase in women around the time of ovulation.

■ *Extraovarian actions of gonadal steroids*

The cyclic changes in estradiol and progesterone secretion produce effects on the uterus, fallopian tubes, vagina, and breasts. These coordinate precisely with the expectation of conception and the institution of a pregnancy.

Uterus. The function of the uterus is to house and nurture the developing fetus and then evacuate him or her safely at the appropriate time. It is a muscular organ, enclosing a cavity that is lined with a special mucous membrane called the *endometrium*. At the beginning of the follicular phase in each menstrual cycle the uterus is in the process of shedding its lining and is therefore incapable of receiving a conceptus. The endometrium is thin, and its glands are sparse, straight, and narrow lumened and exhibit few mitoses (Fig. 55-17). After the menstrual slough has ceased, the increase in estradiol secretion during the follicular phase produces a threefold to fivefold increase in endometrial thickness. Mitoses appear in the glands and stroma, the glands become somewhat tortuous, and elongation occurs in the spiral arteries that supply the endometrium. This is termed the *proliferative phase* of the endometrium. The mucus elaborated by the cervix also changes dramatically during this phase from a scant, thick, viscous material to a copious, more watery, but more elastic substance that can be stretched into a long fine thread. It also produces a characteristic fernlike pattern when dried on a glass slide. In this estrogen-stimulated condition the cervical mucus creates a myriad of channels in

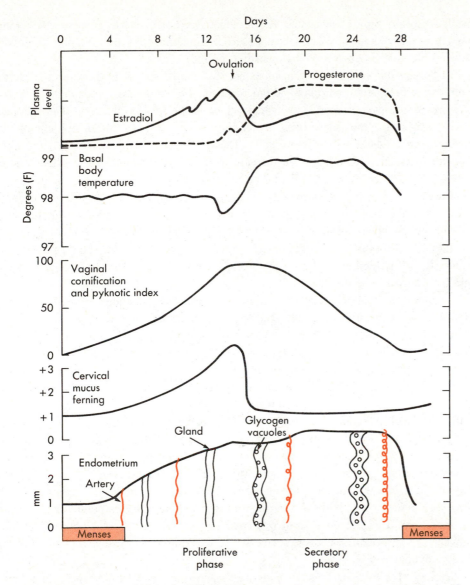

Fig. 55-17 ■ Correlation of biological changes throughout the menstrual cycle with the profiles of plasma estradiol and progesterone. (Redrawn by permission from Odell, W.D.: The reproductive system in women. In Degroot, L.J., et al., editors: Endocrinology, vol. 3, New York, 1979, Grune & Stratton, Inc.)

the opening of the cervix that facilitate the entrance of the sperm and direct their motion forward into the uterine cavity.

Shortly after ovulation the rise in progesterone produces marked alterations (Fig. 55-17). Rapid proliferation of the endometrium is slowed, and mitotic activity is reduced. The uterine glands become much more tortuous, and they begin to accumulate glycogen in large vacuoles at the base of each cell. As the luteal phase of the cycle progresses, the vacuoles move toward the lumen, and the glands greatly increase their secretions. The stroma of the endometrium becomes edematous, the originally straight spiral arteries elongate further and become coiled. This is termed the *secretory phase* of the endometrium. These changes all enhance its ability to support an implanted conceptus. At the same time progesterone decreases the quantity of the cervical mucus, causes it to return to its original thick, nonelastic state, and inhibits the ferning pattern.

If pregnancy does not occur and the corpus luteum regresses, the abrupt loss of estradiol and progesterone causes spasmic contractions of the spiral arteries, probably mediated locally by prostaglandins. The resultant ischemia produces necrosis, the stroma condenses and degenerates, and the superficial endometrial cells are sloughed along with sludged blood. This comprises the menstrual period.

Fallopian tubes. The fallopian tubes are the normal site of fertilization. These bilateral structures are 10 cm long and emerge from the uterus. Each tube ends in finger-like projections called *fimbriae,* which lie close to the ipsilateral ovary. The fallopian tube consists of a muscular layer surrounding a mucosa lined by an epithelium that contains both ciliated and secretory cells. The cilia beat toward the uterus. During the follicular phase estradiol causes an increase in the number of cilia and in their rate of beating, as well as in the number of actively secreting epithelial cells. In addition, the fimbriae become more vascularized, and, as ovulation approaches, they begin to undulate to draw the shed ovum into the tube. During the luteal phase progesterone probably maximizes the ciliary beating. Progesterone also increases the secretion of materials nutritious to the ovum, to any incoming sperm, and to the zygote, should fertilization occur. A well-studied example of such a progesterone effect is the secretion of ovalbumin by the oviduct of the hen.

Vagina. The vaginal canal is lined with a stratified squamous epithelium that is highly sensitive to estradiol. In the absence of the hormone there is only a thin layer of basal and parabasal cells. For the first few days of the follicular phase of the menstrual cycle the epithelium is relatively thin, and smears taken from the surface show cells with vesicular nuclei from the intermediate layer. As the cycle progesses to the ovulatory phase, more layers of epithelium are added, and the maturing cells accumulate glycogen. Vaginal smears at this point show many large, eosinophilic staining, cornified cells with small pyknotic or absent nuclei. The percentage of these cells on a vaginal smear provides a sensitive index of estrogenic activity (Fig. 55-17). Progesterone, on the other hand, reduces the percentage of cornified cells. Vaginal secretions are increased by estradiol, and they form an essential element in the events leading to fertilization. In a manner not currently understood vaginal, as well as uterine and tubal, secretions enhance the ability of spermatozoa to penetrate the zona pellucida and fertilize the ovum. This function, known as *capacitation,* requires 4 to 5 hours of contact between the spermatozoa and the secretions. Capacitation is probably maximum at the time of ovulation.

Breasts. The mammary glands consist of a large series of lobular ducts lined by an epithelium that is capable of secreting milk. These ducts empty into larger milk-conveying ducts that converge at the nipple. These glandular structures are imbedded in supporting adipose tissue, and the breasts are separated into lobules by connective tissue. The development of adult-sized mammary glands is absolutely dependent on estrogens. Before puberty the breasts grow only in proportion to the rest of the body. After the onset of increased estrogen secretion that occurs with pubescence, the growth of the lobular ducts is accelerated, and the area around the nipple (the areola) enlarges. Estrogens also cause a selective increase in the adipose tissue of the breast, giving it its distinctive female shape. The lobular ducts are capable of outpouching to form numerous secretory alveoli. This process is stimulated by progesterone. During the menstrual cycle proliferation of the lobules occurs primarily in parallel with estradiol levels but possibly is further fortified by progesterone. This causes swelling of the breasts; however, by the end of the luteal phase breast size and tenderness diminish.

Other tissues. During puberty estradiol is to the female what testosterone is to the male. Estradiol causes almost all the changes that result in the normal adult female phenotype. In addition to stimulating growth of the internal reproductive organs and breasts, estrogens cause pubertal enlargement of the labia majora and labia minora. Linear growth is accelerated by estradiol. However, because the epiphyseal growth centers are more sensitive to estradiol than to testosterone, they close sooner. For this reason the average height of women is less than that of men. The hips enlarge, and the pelvic inlet widens, facilitating future pregnancy. The predominance of estradiol over testosterone as a gonadal steroid in women is responsible for the fact that their total body adipose mass is twice as large as that of men, whereas their muscle and bone mass is only two thirds that of men. The specific deposition of fat about the hips is another effect of estradiol.

The adult skeleton, the kidney, and the liver are also target tissues of estrogens. Estrogen inhibits bone resorption; the loss of this hormone action after menopause contributes to a declining bone mass (osteoporosis) and a resultant increased frequency of fractures. Reabsorption of sodium from the renal tubules is stimulated by estradiol, and this may contribute to the cyclic fluid retention noted by some women. The hepatic synthesis of a number of circulating proteins is increased, including thyroxine-binding globulin, cortisol-binding globulin, sex steroid–binding globulin, the renin substrate, angiotensinogen, and very low-density lipoproteins. Estrogens also elevate plasma glucose in susceptible patients.

Only a few systemic actions of progesterone are known. Body temperature is increased by this steroid, accounting for the 0.5° C rise that occurs shortly after ovulation (Fig. 55-17). Central nervous system actions include an increase in appetite, a tendency to somnolence, and a heightened sensitivity of the respiratory center to stimulation by carbon dioxide. Because progesterone is an aldosterone antagonist, it can induce natriuresis.

■ *Mechanism of action of gonadal steroids*

Estradiol, estrone, other estrogens, and progesterone all enter cells freely and bind to cytoplasmic receptors of about 200,000 molecular weight. No further intracellular metabolism of estrogens or progesterone is required for their actions. After a conformational change that is essential for activity, the hormone-receptor complex enters the nucleus. There it initiates transcriptional changes that ultimately lead to increased or decreased synthesis of specific proteins. Spare receptors generally are not present, so the sensitivity and the maximum responsiveness of various tissues to gonadal steroids both are proportional to receptor concentrations and binding affinities. An important clinical exception is found in a group of synthetic steroids that have a high affinity for estrogen receptors but which act as antiestrogen agents by blocking the access of estradiol and estrone to their receptors. Although the antiestrogen-receptor complex is translocated to the nucleus, its conformation must be abnormal because it maintains prolonged association with the nuclear acceptor site without initiating any biological action. Two important examples are tamoxifen, a compound widely used in the treatment of estrogen-sensitive breast cancer, and clomiphene, an agent used to stimulate LH and FSH release. The latter results from negative feedback because the clomiphene blockade of the estrogen receptor mimics deficiency of the hormone.

Receptors are also a site for estrogen and progesterone interactions. Estrogen increases the number of progesterone receptors, thus priming target tissues (for example, the uterus) for sequential estrogen and progesterone effects. In contrast, progesterone can decrease the number of estrogen receptors, accounting in part for its antiestrogen actions.

■ *Metabolism of gonadal steroids*

Estradiol and estrone bind to SSBG, but their affinities are much lower than that of testosterone. Therefore the estrogens circulate largely bound loosely to albumin, and they have relatively high metabolic clearance rates (Table 55-5). In menstruating women most of the circulating estradiol is derived from ovarian secretion; a minor fraction is formed from testosterone in adipose tissue, liver, and other sites. Most of the circulating estrone is derived from estradiol by peripheral 17-hydroxysteroid dehydrogenases. In postmenopausal women estrone is the dominant circulating estrogen, and it is formed from adrenal and theca cell androgens. Estrone is further metabolized to estriol, a compound with some estrogenic activity (Fig. 55-18).

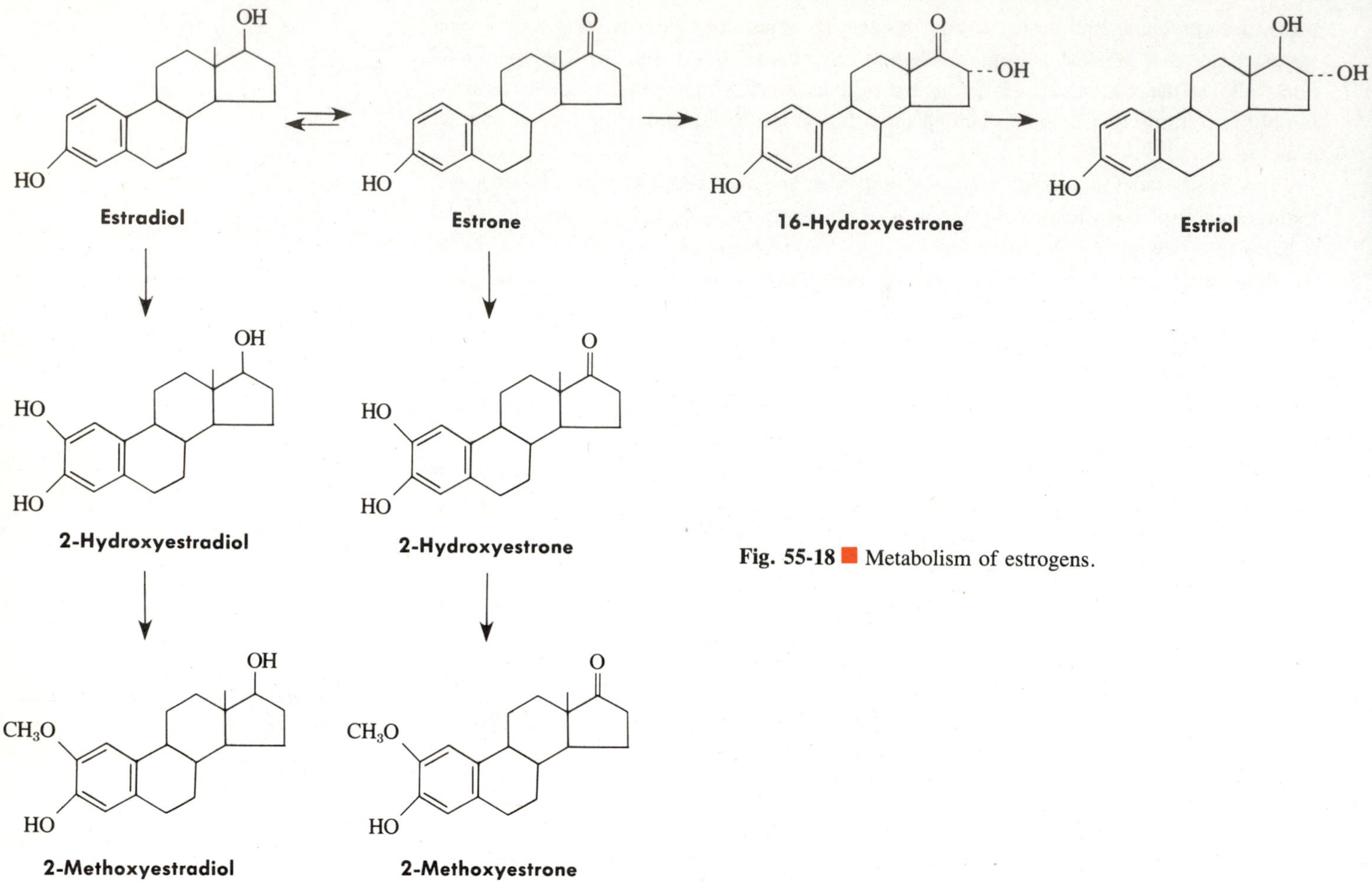

Fig. 55-18 ■ Metabolism of estrogens.

Sulfated and glucuronidated derivatives of all three estrogens are excreted in the urine. Total daily urinary estrogen excretion reflects ovarian function. Values range from 20 μg during the early follicular phase to 65 μg at the preovulatory peak. An additional pathway of estrogen metabolism involves 2-hydroxylation and produces the so-called catechol estrogens (Fig. 55-18). These compounds resemble the catecholamine neurotransmitters norepinephrine and dopamine in their hydroxylated benzene rings. Because 2-hydroxylase activity is present in the hypothalamus, it has been suggested that catechol estrogens generated within the brain might modulate estradiol effects on LHRH release. Two separate hypotheses have been proposed. The first is that catechol estrogens might competitively inhibit catecholamine methyltransferase, an enzyme that inactivates norepinephrine by methylation (Chapter 54). Such inhibition could raise norepinephrine to stimulating levels in critical LHRH hypothalamic neurons. A second hypothesis is that catechol estrogens, which bind to estradiol receptors but do not have estradiol actions, in effect are natural antiestrogen agents that (like clomiphene) increase LHRH by negative feedback.

Progesterone can bind to cortisol-binding globulin, but this is largely prevented by the much higher plasma cortisol concentration. Therefore progesterone circulates loosely bound to albumin. Its metabolism to the urinary metabolite, pregnanediol, is described in Chapter 54. During the follicular phase of the cycle about half the circulating progesterone is secreted by the ovary and half by the adrenal glands. During the luteal phase, however, the vast majority originates in the ovary.

In women 70% to 80% of circulating testosterone is derived from peripheral conversion of DHEA and androstenedione. About half the daily production stems from adrenal activity and half from ovarian precursors. Therefore normally relatively little testosterone or dihydrotestosterone is secreted by the ovary. In pathological situations, however, ovarian cells can secrete amounts of these potent androgens that approach those of normal men and cause virilization.

The possible initiating factors of puberty, the development of pulsatile gonadotropin secretion, and the transient phenomenon of nocturnal gonadotropin peaks are all similar in females to those already discussed in males. Reproductive function begins after an increase in gonadotropin secretion from the low levels of childhood (Fig. 55-19). Females differ from males somewhat in more clearly demonstrating an earlier rise in FSH than in LH (compare Figs. 55-19 and 55-10). Budding of the breasts is the first observable physical sign of puberty and coincides with the first detectable increase in plasma estradiol, as ovarian secretion commences. The onset of menses occurs approximately 2 years later, after LH levels have risen more sharply. Irregularity of menstrual cycles is not uncommon initially. This is because ovulation usually does not occur in the first few cycles. The menstrual bleeding is initially induced by the withdrawal of estrogen as graafian follicles undergo atresia. The growth spurt and the peak velocity of growth are characteristically earlier in girls than in boys. Further increase in height usually ceases 1 to 2 years after the onset of menses. The development of pubic hair precedes menses, and it correlates best with rising levels of adrenal androgens, especially DHEA-S. All stages of female puberty have wide ranges that may be influenced by factors of race, climate, and individual heredity.

■ *Female puberty*

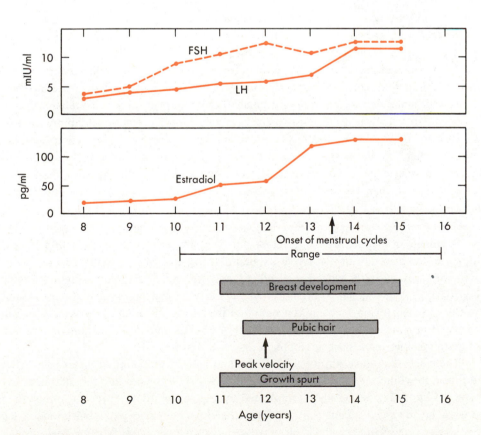

Fig. 55-19 ■ Average chronological sequence of hormonal and biological events in normal female puberty. (Redrawn from Lee, P.A., et al.: Puberty in girls, J. Clin. Endocrinol. Metab. **43**:775, 1976; copyright 1976; reproduced by permission; and Marshall, W.A., and Tanner, J.M.: Arch. Dis. Child. **45**:13, 1970.)

■ *Sexual functioning*

The desire for sexual activity is increased by androgens, but there is little evidence that the normal variation in plasma testosterone or dihydrotestosterone is of physiological significance in this regard. During sexual intercourse vascular erectile tissue beneath the clitoris is activated by parasympathetic impulses. This causes the introitus to be tightened around the penis. Simultaneously these impulses stimulate copious secretion of mucus by glands located beneath the labia minora and in the vagina. The secretions lubricate the vagina and help it produce a massaging effect on the penis. Female orgasm results from spinal cord reflexes similar to those involved in male ejaculation. Involuntary contractions of the skeletal muscle of the perineum, of the musculature of the vagina, uterus, and tubes, and of the rectal sphincter occur. The clitoris retracts against the symphysis pubis. After orgasm the cervix remains widely patent for 20 to 30 minutes, permitting entrance of sperm into the uterus. The first wave of sperm may reach an ovum in the fallopian tube within 10 minutes. However, orgasm is not required for this rapid internal propulsion.

■ *Menopause*

The reproductive capacity of women begins to wane in the fifth decade of life, and menses terminate at an average age of 50. For several years before menopause the frequency of ovulation decreases. This is followed by irregular menses at variable intervals and with decreased flow. With the disappearance of virtually all follicles, ovarian estrogen secretion declines to low levels, eventually to cease completely. From then on maintenance of the lowered plasma estradiol concentrations characteristic of menopause is dependent on peripheral conversion of precursors secreted by the adrenal glands.

During the last few years of reproductive life follicular sensitivity to gonadotropin stimulation diminishes, and plasma FSH and LH gradually increase in compensation. Once menopause occurs, gonadotropin levels average 4 to 10 times those of the normal follicular phase, and the LH/FSH ratio falls to less than 1. Although the cyclicity of gonadotropin secretion is lost, pulsatile secretion persists.

As a consequence of decreased estradiol availability in the postmenopausal years, there is thinning of the vaginal epithelium and loss of its secretions, a decrease in breast mass, and an accelerated resorption of bone. More controversial in their relationship to estrogen deficiency are phenomena such as vascular flushing, emotional lability, and an increase in the incidence of coronary vascular disease. Hormone production does not stop completely in the postmenopausal ovary. The stromal interstitial cells continue to secrete androstenedione and testosterone. In some women proliferation of these cells leads to enough increase in plasma androgens that mild stimulation of hair growth in a masculine pattern can occur.

■ *Pregnancy*

Pregnancy is marked by the development of a new endocrine organ with a limited life span: the placenta. In addition to its vital nutritive role, the placenta functions as an extraordinarily versatile endocrine gland, capable of synthesizing and secreting a wide variety of protein and steroid hormones that affect both maternal and fetal metabolism. These hormones exhibit regular concentration profiles in maternal plasma and also can be found in fetal plasma or amniotic fluid. Fig. 55-20 summarizes the temporal pattern of gestational hormone concentrations in maternal plasma.

The placenta originates from an early stage in the differentiation of the zygote. After fertilization within the fallopian tube the zygote slowly advances into the uterine cavity over a 3- to 5-day period and is implanted by the seventh day. From the initial solid mass a layer of cells, known as *trophoblasts,* separates. The trophoblasts attach the blastocyst to the secretory endometrium and initiate the complex fetal-maternal circulatory arrangements that will support the conceptus. By the ninth day the trophoblasts also have begun to secrete the first key pregnancy hormone.

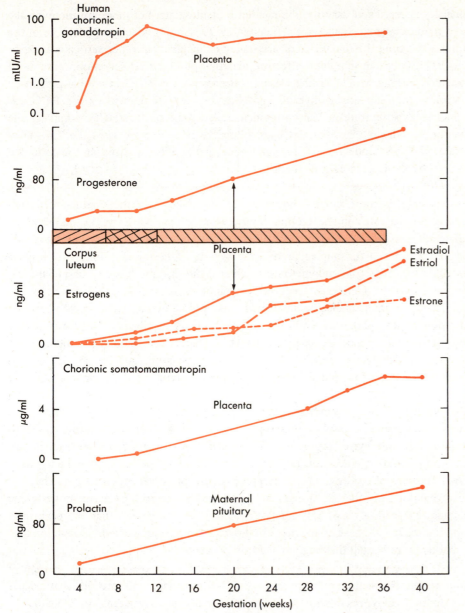

Fig. 55-20 ■ Profile of plasma hormone changes during normal human pregnancy. Note logarithmic scale for HCG. Also note shift from corpus luteum to placenta as the source of estrogens and progesterone between 6 and 12 weeks of gestation. (Redrawn from Goldstein, D.P., et al.: Am. J. Obstet. Gynecol. **102**:110, 1968; Rigg, L.A., et al.: Am. J. Obstet. Gynecol. **129**:454, 1977; Selenkow, H.A., et al.: Measurement and pathophysiologic significance of human placental lactogen. In Pecile, A., and Finzi, C.: The foeto-placental unit, Amsterdam, 1969, Excerpta Medica; and Tulchinsky, D., et al.: Am. J. Obstet. Gynecol. **112**:1095, 1972.)

■ *Human chorionic gonadotropin*

Human chorionic gonadotropin (HCG) is a glycoprotein of 39,000 molecular weight with two subunits. The α-subunit is identical to that of thyroid-stimulating hormone (TSH), LH, and FSH, whereas the β-subunit strongly resembles that of LH. HCG can be detected in maternal plasma and urine 9 days after conception by radioimmunoassays developed specifically to measure the β-subunit. This permits discrimination of HCG from LH. Detection of HCG is the most commonly employed and specific test for pregnancy. Maternal plasma levels of HCG increase at an exponential rate, reach a peak at 9 to 12 weeks, and then decline to a stable plateau for the remainder of the pregnancy (Fig. 55-20). After delivery HCG disappears from maternal plasma with a half-life of 12 to 24 hours.

HCG acts to maintain the function of the corpus luteum beyond its usual life span of 14 days. The placental gonadotropin stimulates ovarian secretion of progesterone and estrogens by mechanisms essentially identical to those previously described for LH.

When the placenta itself assumes the synthesis of these steroids, thereby relieving the fetus of its dependency on the corpus luteum, HCG secretion declines. The role of the gonadotropin during the remainder of pregnancy is unclear, but other actions are attributed to it. HCG has an inhibitory effect on maternal pituitary LH secretion. Because of its structural overlap with TSH, the plasma concentrations of HCG in normal pregnancy may stimulate an increase in thyroid gland activity. In pathological excess HCG can even induce hyperthyroidism. Recent evidence also suggests that HCG stimulates the production of relaxin (as discussed later). Finally, HCG that reaches the fetus may stimulate DHEA-S production by the fetal zone of the adrenal gland and testosterone production by the Leydig cells of the testis.

■ *Progesterone*

Progesterone is the hormone most directly responsible for the establishment and sustenance of the fetus in the uterine cavity. During the first 2 weeks progesterone stimulates the tubal and endometrial glands to secrete nutrients on which the free-floating zygote depends. After implantation of the zygote, progesterone, along with trophoblastic activity, causes the endometrium to change into a thick layer of tissue, the *decidua,* on which the conceptus feeds directly until the placental circulation is established. Thereafter maintenance of the decidua may provide an effective barrier between the invading fetal trophoblastic cells and the uterine layer of muscle. Two other actions of progesterone are important during pregnancy. The steroid inhibits uterine contractions, preventing premature expulsion of the fetus. It also stimulates the development of the alveolar pouches of the mammary glands, greatly magnifying their eventual capacity to secrete milk.

The placenta begins to synthesize progesterone at about 6 weeks, and by 12 weeks it is producing enough to replace the corpus luteum for this purpose. During this transition period the otherwise progressive rise in plasma progesterone reaches a temporary plateau (Fig. 55-20). Cholesterol, present in low-density lipoproteins, is extracted from maternal plasma and serves as the major precursor for placental progesterone. The synthetic pathway is like that of the adrenal gland and the ovary. By term progesterone production reaches a level of 250 mg/day, which is tenfold greater than the peak rates seen during the luteal phase of the menstrual cycle. Urinary pregnanediol excretion also rises markedly, reflecting this huge increase in production.

■ *Estrogens*

Augmented production of estrogens (estradiol, estrone, and estriol) also occurs throughout pregnancy, resulting in several important actions. Estrogens stimulate (1) the continuous growth of the uterine myometrium, preparing it for its role in labor, (2) the further growth of the ductal system of the breast, out of which the alveoli will develop, and (3) enlargement of the external genitalia. In addition, estrogen, along with relaxin, causes relaxation and softening of the pelvic ligaments and the symphysis pubis of the pelvic bones, allowing better accommodation of the expanding uterus.

Like progesterone, estrogens are initially produced by the corpus luteum under stimulation by HCG. The placenta then assumes this function. Lacking the enzyme 17-hydroxylase, the placenta uses DHEA-S, derived from the maternal and fetal adrenal glands, as the major precursor for estrogen. The sulfate is removed, and DHEA is converted to testosterone and androstenedione, which are then aromatized (Fig. 55-4). About 50% of estradiol and estrone production is derived from maternal DHEA-S and 50% from fetal DHEA-S in this way. In contrast, the placental synthesis of estriol is 90% dependent on fetal precursors. The fetal adrenal gland synthesizes DHEA-S from cholesterol. DHEA-S is subsequently hydroxylated in the 16 position by the fetal liver. 16α-Hydroxy-DHEA-S is then desulfated and aromatized in the placenta to estriol.

One third of the unconjugated estrogens circulating in maternal plasma at term is accounted for by estriol (Fig. 55-20). In the form of sulfate and glucuronic acid conjugates, estriol represents 90% of the total estrogen excreted in maternal urine. Daily urinary estriol levels correlate quite well with placental weight. Because estriol is derived almost entirely from the fetal placental unit, measurement of this estrogen in maternal plasma or urine provides a valuable clinical index to the state of the fetus. A rapid decrease in urinary estriol suggests fetal distress or placental insufficiency. In conjunction with the results of other monitoring procedures, low urinary estriol levels may dictate active intervention and prompt delivery of a fetus at risk.

■ *Human chorionic somatomammotropin*

Another protein hormone, unique to pregnancy, is *human chorionic somatomammotropin* (HCS), also called human placental lactogen (HPL). The synthesis of HCS by the placental trophoblasts can be detected at about 4 weeks of gestation. The maternal plasma concentration rises steadily to a peak of 6 µg/ml at term (Fig. 55-20). The HCS production rate of 1 to 2 g/day far exceeds that of any other human protein hormone. HCS concentrations correlate quite well with placental weight. Therefore measurement of maternal plasma HCS provides another indicator of placental function that has been used to monitor threatened pregnancies. After delivery the hormone rapidly disappears from maternal plasma with a half-life of 20 minutes.

HCS has a very close structural similarity to human growth hormone (HGH) and prolactin. In several in vitro and in vivo test systems HCS exhibits anabolic nitrogen-retaining activity like that of HGH, although at only one hundredth the potency. HCS also has lactogenic activity similar to that of prolactin and some luteotropic activity, as well. Most important, HCS stimulates lipolysis and, like HGH, exerts antagonism to insulin actions on carbohydrate metabolism; this tends to raise plasma glucose. Plasma HCS is slightly increased by fasting or by insulin-induced hypoglycemia, suggesting that placental secretion may be under some degree of feedback regulation by glucose. As is detailed later, the primary function of HCS appears to be that of directing maternal metabolism to maintain a continuous flow of substrates, especially glucose, to the fetus. HCS levels in fetal plasma are far below those in maternal plasma, and there is no known direct role for HCS in the fetus.

■ *Prolactin*

Another hormone secreted in excess during normal pregnancy is maternal pituitary prolactin. Plasma levels rise linearly, by term reaching values eightfold to tenfold higher than those of nonpregnant women (Fig. 55-20). Prolactin is essential for expression of the mammotropic effects of estrogen and progesterone. More specifically, prolactin stimulates the lactogenic apparatus (Chapter 52). Limited formation of milk begins at about 5 months of gestation, but significant lactation is inhibited by the great excess of estrogen and progesterone. Lactation is initiated after delivery by the precipitous drop in steroid hormones, and it is maintained thereafter in the nursing mother by prolactin through the stimulus of suckling. Although basal prolactin concentrations gradually decline over the next 4 to 8 weeks, they are acutely elevated during each period of suckling (Chapter 52), helping to sustain milk secretion. Prolactin also suppresses reproductive function in the nursing mother. During the first 7 to 10 days post partum, plasma FSH and LH levels remain low. FSH then rises to above normal follicular phase levels, but LH still remains low. The responsiveness of the ovaries to FSH probably is reduced by prolactin, and LH secretion by the pituitary probably is inhibited by prolactin. A decrease in circulating prolactin, because of either cessation of nursing or administration of a dopaminergic agonist, will trigger LH release and initiate cycling.

■ *Relaxin*

In addition to gonadal steroids, the corpus luteum of pregnancy secretes a polypeptide hormone called *relaxin*. Interestingly, its structure is insulin-like. Plasma levels of relaxin rise early, peak in the first trimester, and then decline somewhat. The production of relaxin by the corpus luteum appears to be stimulated by HCG. There is also evidence that relaxin may be produced by uterine tissue and by the placenta. In addition to its relaxing effect on pelvic bones and ligaments, relaxin inhibits myometrial contractions and softens the cervix. Thus it may function to ensure uterine quiescence and prevent early abortion of the pregnancy.

■ Other hormones

Other hormones have been detected in placental extracts, including TRH, LHRH, prolactin, an adrenocorticotropic hormone (ACTH)-like material, and a thyrotropin (TSH)-like substance. There is evidence of varying degrees of certainty that these hormones actually are synthesized by trophoblastic cells. It also remains to be established whether they are secreted into maternal or fetal plasma and whether they function as local hormones within the placenta itself or act on distant maternal or fetal target cells.

■ Other maternal hormonal changes

The pregnant state induces a characteristic series of changes in a number of other maternal hormones. One of the most significant alterations is in pancreatic islet β-cell function. Insulin secretion, in response to glucose challenge or to meals, increases after the third month of pregnancy. This hypersecretion reaches its peak during the last trimester, coinciding with the peaks of placental weight and of plasma HCS. During this same period maternal sensitivity to insulin is greatly diminished. Therefore, to some extent, insulin hypersecretion may be considered compensatory. In contrast, basal glucagon levels and responses to stimulation do not change significantly, although suppressibility of the α-cells by glucose is somewhat enhanced.

Aldosterone secretion increases significantly throughout pregnancy, reaching a six-fold to eightfold elevation by term. This is because of increased function of the renin angiotensin system; plasma renin and the renin substrate, angiotensinogen, both are augmented by pregnancy. Various reasons have been advanced for the hyperaldosteronism of pregnancy. It may be stimulated by a reduction in *effective* circulating blood volume that results from the large placental blood pool. It also could represent a response to the antialdosterone action of progesterone on the renal tubules. Hyperaldosteronism contributes to the positive sodium balance that is needed to maintain a high total maternal plasma volume and to build the extracellular fluid of the fetus.

The plasma total cortisol level is elevated because of the estrogen-induced increase in cortisol-binding globulin. However, plasma free cortisol also rises modestly. The enhanced glucocorticoid activity may contribute to maternal adipose tissue gain and to mammary gland development. It also may be responsible for the plethoric face, thin skin, and susceptibility to bruising of the pregnant woman.

Parathyroid hormone secretion is increased in response to the continuous drain on maternal calcium created by the growing fetal skeleton. The hyperparathyroidism augments plasma levels of the active vitamin D metabolite, $1,25\text{-}(OH)_2\text{-}D_3$, which in turn increases dietary calcium absorption. The enhanced supply of calcium is especially important during the third trimester.

Total plasma thyroid hormones, thyroxine (T_4) and triiodothyronine (T_3), are elevated because of estrogen-induced increases in thyroid-binding globulin. Plasma free T_4 remains within the normal range. Nonetheless there are increases during pregnancy in maternal thyroid gland size, radioactive iodine uptake, basal metabolic rate, and resting pulse rate, all of which are compatible with some augmentation of thyroid gland activity. This may be due to the thyrotropic activity of HCG or to secretion of another placental thyrotropin.

Growth hormone secretion is decreased in response to various stimuli during pregnancy, possibly because the anabolic functions of maternal pituitary HGH are carried out by placental HCS. Maternal LH and FSH secretion also is suppressed by the high levels of estrogen and progesterone.

During normal pregnancy the average gain in maternal weight is 11 kg. About half of this is attributable to changes in maternal tissues and half to the conceptus. The typical distribution of the excess weight is shown in Fig. 55-21. Approximately 75,000 extra calories (250 to 300 kcal/day) must be ingested to support this weight gain; 65,000 kcal support fetal metabolism and growth, and 10,000 kcal are stored in maternal fat. An extra protein intake of 30 g/day ensures adequate supplies for maternal needs and for the accumulation of fetal protoplasm. At birth the protein content of the fetus has reached 400 to 500 g.

From the metabolic standpoint, pregnancy can be divided into two phases. For approximately the first half the mother herself is in an anabolic phase, whereas the conceptus represents an insignificant nutritional drain. During the second half, and especially the final third, fetal and placental weight increases at an accelerated rate. These demands cause the mother to shift into a state aptly described as "accelerated starvation."

The maternal anabolic phase is characterized by normal or even increased sensitivity to insulin. Some evidence suggests that progesterone and estrogen may facilitate insulin actions at this stage. Maternal plasma levels of glucose, amino acids, free fatty acids, and glycerol are normal or slightly reduced. Carbohydrate and amino acid loads are readily assimilated. Lipogenesis is favored and lipolysis braked in maternal adipose tissue, glycogen stores are increased in liver and muscle, and protein synthesis is enhanced. The net effect is to stimulate growth of the breasts, uterus, and essential musculature in the mother, while preparing her to withstand the metabolic demands of later fetal growth.

During the second phase the metabolism of the mother shifts into a mode that effectively accommodates the accelerating needs of the fetus. Insulin sensitivity is replaced by insulin resistance. The assimilation of dietary carbohydrate, protein, and fat by maternal tissues is slowed, so that postprandial plasma levels of glucose and amino acids are elevated. This increases the rate of glucose diffusion and of facilitated amino acid transport across the placenta into the fetus. Glucose is the major fuel of the fetus, and the amino acids are required for fetal protein synthesis. By term, the fetus, who is using glucose at a rate of 5 mg/kg/minute compared with the maternal rate of 2.5 mg/kg/

■ *Maternal-fetal metabolism*

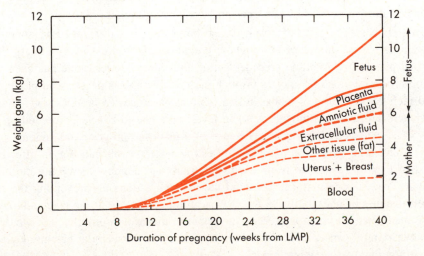

Fig. 55-21 ■ Pattern and components of maternal weight gain during normal pregnancy. (Redrawn from Pitkin, R.M.: Obstetrics and gynecology. In Schneider, H.A., Anderson, C.E., and Coursin, D.B.: Nutritional support of medical practice, New York, 1977, Harper & Row Publishers, Inc.)

minute, must be supplied with up to 25 g/day. During fasting intervals maternal plasma glucose and amino acid levels fall more rapidly than in nonpregnant women, because of continued fetal siphoning of these substances. Conversely, lipolysis is accelerated, and maternal plasma free fatty acids, glycerol, and ketoacids rise more rapidly than in nonpregnant women. This ensures alternate oxidative fuels for the mother; in addition, ketoacids and, to a lesser extent, free fatty acids cross to the fetus, where they may be used instead of some glucose. Placental HCS is probably the key hormone responsible for insulin resistance and for facilitating lipid mobilization during fasting in the latter stage of pregnancy. The increase in plasma free cortisol also may contribute to these actions.

Along with the other changes in maternal metabolism, plasma cholesterol and triglyceride levels rise throughout pregnancy. The cholesterol is partly used for estrogen and progesterone synthesis. The increased circulating triglycerides are largely the result of an increased hepatic synthesis of very low-density lipoprotein, which is stimulated by estrogens. Some of the triglycerides are stored in the breasts in preparation for milk production. They are shifted away from less specific storage elsewhere by a marked reduction in adipose tissue lipoprotein lipase levels.

■ Parturition

Just as the maintenance of the pregnant state is dependent on a unique hormonal milieu, its termination probably also depends on specific hormonal changes. Roles for glucocorticoids, estrogen, progesterone, relaxin, oxytocin, prostaglandins, and catecholamines in the initiation and final achievement of uterine evacuation have been suggested by various lines of evidence. Because much species variation exists, it is difficult to extrapolate the results of the numerous studies in subprimates to humans. Thus a completely proven sequence of events cannot be presented.

Once the conceptus has reached a critical size, distension of the uterus itself and stretching of the muscle fibers increase their contractility. In the human, uncoordinated uterine contractions begin at least 1 month before the end of gestation. The inherent contractility of the uterus probably in itself would cause eventual evacuation of the conceptus. There is also considerable evidence that some signal from the fetus initiates active labor, and in sheep a fetal adrenal product, probably cortisol, has been strongly implicated. This fits with the fact that fetal cortisol production rises sharply during the last few weeks of gestation. It has been postulated that cortisol may be acting to direct placental output more toward estrogens and less toward progesterone.

The relative availability of estrogens and progesterone is thought to be important in maintaining uterine quiescence throughout pregnancy. In some species a sharp drop in maternal plasma progesterone precedes parturition. Since progesterone has been shown to inhibit transmission of impulses through the myometrium, a decrease in progesterone could facilitate the onset of labor. In humans plasma progesterone does not fall significantly before delivery, although estrogen levels continue to rise near the end of gestation. Some evidence suggests the existence of a placental progesterone-binding protein whose concentration may be increased by estrogen near term, causing an effective removal of local progesterone from the myometrium and thereby decreasing uterine quiescence. Progesterone may act to inhibit uterine contractions by preventing the release from lysosomes of phospholipase A_2, the rate-limiting enzyme in prostaglandin synthesis. With a decrease in progesterone the concentrations of prostaglandins would rise. The latter in turn, causes an increase in myometrial free intracellular calcium concentrations, triggering uterine contractions.

Relaxin probably synergizes with progesterone in suppressing uterine contractility throughout gestation. However, relaxin also may assist in parturition in two ways. First, it acts to soften the cervix, which is critical for permitting eventual passage of the fetus. Second, relaxin causes an increase in oxytocin receptors, which may contribute to the greatly increased sensitivity of the uterus to stimulation by oxytocin at term. Once labor is initiated, both maternal and fetal oxytocin may help sustain uterine activity and increase the strength of uterine contractions. However, labor can proceed in the absence of this peptide. Lastly, both α- and β-adrenergic receptors are present in the myometrium; α-stimuli cause contraction, and β-stimuli cause relaxation. Therefore circulating catecholamines also may contribute to the final hormonal cascade of parturition.

Once labor has begun, it proceeds in three clinically recognized stages. In the first stage, lasting a variable number of hours, the uterine contractions, which originate at the fundus and sweep downward, force the head of the fetus against the cervix. This progressively widens and thins the opening to the vaginal canal. In the second stage, lasting less than 1 hour, the fetus is forced out of the uterine cavity and through the cervix and is delivered from the vagina. In the third stage, lasting 10 minutes or less, the placenta is separated from the decidual tissue of the uterus and forcefully evacuated. Myometrial contractions at this point act to constrict the uterine vessels and prevent excessive bleeding. Once the placenta has been removed, all its hormonal products disappear from maternal plasma according to their characteristic half-lives. In general, by 48 to 72 hours the steroid and protein hormone concentrations have reached nonpregnant levels.

■ *Endocrine state of the fetus*

No maternal protein or peptide hormone effectively traverses the placenta, with the exception of TRH and possibly antidiuretic hormone. The same is essentially true for maternal thyroid hormones. Steroid hormones, in contrast, can move readily from the mother to the fetus, and vice versa. Catecholamines also cross to the fetus.

The ontogeny of each of the endocrine glands is individually described in previous sections. Fetuses are dependent on their own insulin for anabolism and for deposition of adipose tissue. Their own thyroid hormones are required for normal central nervous system and skeletal maturation. Linear growth of fetuses does not appear to depend on their own growth hormone, even though fetal plasma HGH levels are very high. Somatomedins and other insulin-like growth factors also are present in fetal plasma, but their origin and function are obscure. Fetal ACTH probably is not essential for the first 12 to 20 weeks, although later it definitely stimulates production of steroids by the fetal zone of the adrenal cortex. The newborn mounts an immediate stress response; if endogenous ACTH and cortisol cannot be secreted at this time, death will ensue unless replacement therapy is provided. The roles of fetal FSH and LH, the plasma levels of which peak in midgestation, remain to be worked out. Prolactin concentrations are high in fetal plasma and in amniotic fluid. It has been suggested that prolactin may contribute to fetal growth, and there is some evidence that the hormone also may stimulate the adrenal cortex. The free diffusion of calcium across the placenta negates the need for fetal parathyroid hormone secretion, but fetal calcitonin levels are relatively elevated. A role for the latter hormone in bone formation has been postulated. Finally, current studies hint at the existence of other placental protein hormones with specific functions in fetal development.

■ *Bibliography*

Journal articles

Abraham, G.E.: Ovarian and adrenal contribution to peripheral androgens during the menstrual cycle, J. Clin. Endocrinol. Metab. **39**:340, 1974.

Boyar, R.M., et al.: Synchronization of augmented luteinizing hormone secretion with sleep during puberty, N. Engl. J. Med. **287**:582, 1972.

Boyar, R.M., et al.: Human puberty: simultaneous augmented secretion of luteinizing hormone and testosterone during sleep, J. Clin. Invest. **54**:609, 1974.

Bryant-Greenwood, G.D.: Relaxin as a new hormone, Endocr. Rev. **3**:62, 1982.

Clermont, Y.: Kinetics of spermatogenesis in mammals: seminiferous epithelium cycle and spermatogonial renewal, Physiol. Rev. **52**:198, 1972.

Forest, M.G., et al.: Kinetics of human chorionic gonadotropin–induced steroidogenic response of the human testis. II. Plasma 17 α-hydroxyprogesterone, Δ⁴-androstenedione, estrone, and 17 β-estradiol: evidence for the action of human chorionic gonadotropin on intermediate enzymes implicated in steroid biosynthesis, J. Clin. Endocrinol. Metab. **49**:284, 1979.

Goebelsmann, U.: Protein and steroid hormones in pregnancy, J. Reprod. Med. **23**:166, 1979.

Goldstein, D.P., et al.: Radioimmunoassay of serum chorionic gonadotropin activity in normal pregnancy, Am. J. Obstet. Gynecol. **102**:110, 1968.

Harman, S.M., et al.: Reproductive hormones in aging men. I. Measurement of sex steroids, basal luteinizing hormone, and Leydig cell response to human chorionic gonadotropin, J. Clin. Endocrinol. Metab. **51**:35, 1980.

Imperato-McGinley, J., et al.: Male pseudohermaphroditism: the complexities of male phenotypic development, Am. J. Med. **61**:251, 1976.

Judd, H.L., et al.: Endocrine function of the postmenopausal ovary: concentration of androgens and estrogens in ovarian and peripheral vein blood, J. Clin. Endocrinol. Metab. **39**:1030, 1974.

Kalkhoff, R.K., et al.: Carbohydrate and lipid metabolism during normal pregnancy: relationship to gestational hormone action, Semin. Perinatol. **2**:291, 1978.

Korenman, S.G., et al.: Further studies of gonadotropin and estradiol secretion during the preovulatory phase of the human menstrual cycle, J. Clin. Endocrinol. Metab. **36**:1205, 1973.

Lee, P.A., et al.: Puberty in girls: correlation of serum levels of gonadotropins, prolactin, androgens, estrogens, and progestins with physical changes, J. Clin. Endocrinol. Metab. **43**:775, 1976.

Lipsett, M.B.: Physiology and pathology of the Leydig cell, N. Engl. J. Med. **303**:682, 1980.

McNatty, K.P., et al.: The production of progesterone, androgens, and estrogens by granulosa cells, thecal tissue, and stromal tissue from human ovaries in vitro, J. Clin. Endocrinol. Metab. **49**:687, 1979.

McNatty, K.P., et al.: The microenvironment of the human antral follicle: interrelationships among the steroid levels in antral fluid, the population of granulosa cells, and the status of the oocyte in vivo and in vitro, J. Clin. Endocrinol. Metab. **49**:851, 1979.

Marshall, W.A., and Tanner, J.M.: Variations in the pattern of pubertal changes in boys, Arch. Dis. Child. **45**:13, 1970.

Marshall, W.A., et al.: Variations in the pattern of pubertal changes in girls, Arch. Dis. Child. **44**:291, 1969.

Nathanielsz, P.W.: Endocrine mechanisms of parturition, Annu. Rev. Physiol. **40**:411, 1978.

Ohno, S.: The role of H-Y antigen in primary sex determination, J.A.M.A. **239**:217, 1978.

Quagliarello, J., et al.: Induction of relaxin secretion in nonpregnant women by human chorionic gonadotropin, J. Clin. Endocrinol. Metab. **51**:74, 1980.

Rich, B.H., et al.: Adrenarche: changing adrenal response to adrenocortiocotropin, J. Clin. Endocrinol. Metab. **52**:1129, 1981.

Rigg, L.A., et al.: Pattern of increase in circulating prolactin levels during human gestation, Am. J. Obstet. Gynecol. **129**:454, 1977.

Thoburn, G.D., et al.: Endocrine control of parturition, Physiol. Rev. **59**:863, 1979.

Tsang, B.K., et al.: Androgen biosynthesis in human ovarian follicles: cellular source, gonadotropic control, and adenosine 3′,5′-monophosphate mediation, J. Clin. Endocrinol. Metab. **48**:153, 1979.

Tulchinsky, D., et al.: Plasma estrone, estradiol, estriol, progesterone, and 17-OH progesterone in human pregnancy, Am. J. Obstet. Gynecol. **112**:1095, 1972.

Tulchinsky, D., et al.: Plasma human chorionic gonadotropin, estrone, estradiol, estriol, progesterone, and 17-OH progesterone in human pregnancy, Am. J. Obstet. Gynecol. **117**:884, 1973.

Wang, C.F., et al.: The functional changes of the pituitary gonadotrophs during the menstrual cycle, J. Clin. Endocrinol. Metab. **42**:718, 1976.

Wilson, J.D.: Sexual differentiation, Annu. Rev. Physiol. **40**:279, 1978.

Winter, J.S.D., et al.: Pituitary-gonadal relations in male children and adolescents, Pediatr. Res. **6:**126, 1972.

Winter, J.S.D., et al.: Pituitary-gonadal relations in infancy. II. Patterns of serum gonadal steroid concentrations in man from birth to 2 years of age, J. Clin. Endocrinol. Metab. **42:**679, 1976.

Yen, S.S.C., et al.: Acute gonadotropin release induced by exogenous estradiol during the mid-follicular phase of the menstrual cycle, J. Clin. Endocrinol. Metab. **34:**298, 1972.

Books and monographs

Bardin, C.W.: Pituitary-testicular axis. In Yen, S.S.C., and Jaffe, F.B.: Reproductive endocrinology, Philadelphia, 1978, W.B. Saunders Co.

Fawcett, D.W.: Ultrastructure and function of the Sertoli cell. In Hamilton, D.W., and Greep, R.O., editors: Handbook of physiology; Section 7, vol. 5, Bethesda, Md., 1975, The American Physiological Society.

Fisher, D.A.: Fetal endocrinology: endocrine disease and pregnancy. In Degroot, L.J., et al., editors: Endocrinology, vol. 3, New York, 1979, Grune & Stratton, Inc.

Griffin, J.E., et al.: The testis. In Bondy, P.K., and Rosenberg, L.E.: Metabolic control and disease, Philadelphia, 1980, W.B. Saunders Co.

Odell, W.D.: The reproductive system in women. In Degroot, L.J., et al., editors: Endocrinology, vol. 3, New York, 1979, Grune & Stratton, Inc.

Pitkin, R.M.: Obstetrics and gynecology. In Schneider, H.A., Anderson, C.E., and Coursin, D.B.: Nutritional support of medical practice, New York, 1977, Harper & Row Publishers, Inc.

Ross, G.T., et al.: The ovary. In Yen, S.S.C. and Jaffe, R.B.; Reproductive endocrinology, Philadelphia, 1978, W.B. Saunders Co.

Selenkow, H.A., et al.: Measurement and pathophysiologic significance of human placental lactogen. In Pecile, A. and Finzi, C.: The foeto-placental unit, Amsterdam, 1969, Excerpta Medica.

Yen, S.S.C.: The human menstrual cycle (integrative function of the hypothalamic-pituitary-ovarian-endometrial axis). In Yen, S.S.C., and Jaffe, R.B.: Reproductive endocrinology, Philadelphia, 1978, W.B. Saunders Co.

Index

Page numbers in *italics* indicate illustrations.
Page numbers followed by *t* indicate tables.